The Agricultural Notebook

Originally compiled by Primrose McConnell in 1883 this standard work has been completely re-written and enlarged by the contributors listed on p. vii, under the editorship of R. J. Halley.

Primrose McConnell's

The Agricultural Notebook

17th edition

Edited by

R. J. Halley
Seale-Hayne College

Butterworth Scientific

London Boston
Sydney Wellington Durban Toronto

First published, 1883
17th edition, 1982

© Butterworths & Co. (Publishers) Ltd., 1982

British Library Cataloguing in Publication Data
McConnell, Primrose
Primrose McConnell's the agricultural notebook.
—17th ed.
1. Agriculture—Great Britain—Handbooks, manuals, etc.
I. Title   II. Halley, R.J.
630′.941   S513
ISBN 0–408–10701–4

Typeset by CCC,
printed and bound in Great Britain
by William Clowes (Beccles) Limited,
Beccles and London

Preface

The first edition of *The Agricultural Notebook* appeared in 1883. It was compiled by Primrose McConnell, a tenant farmer of Ongar Park Hall, Ongar, Essex, who found, when he was an agricultural student under Professor Wilson at Edinburgh University, the great want of a book containing all the data associated with the business of farming. This is the seventeenth edition and all but coincides with the centenary of the publication. The Agricultural Industry has changed so much since the last edition it was decided to abandon the 'pocket book' size and use the new, easy-to-read, format.

Clearly no one person is capable of writing on all the aspects of agriculture necessary for achieving the objectives of *The Agricultural Notebook*. The Editor is most grateful to the team that have contributed to the new edition. Contributors have not been constrained to a common editorial mould but rather, following discussion with the Editor, have been encouraged to let their own informed aspirations guide the content and layout of their particular sections. Contributors have collaborated and, where appropriate, adequate cross-referencing is given. Compared with the previous edition, some new topic areas have been added and some eliminated to reflect development within the industry and to contain the total volume within an acceptable limit.

The book is aimed at meeting the needs of students of agriculture, be they full-time students in colleges and universities, or practising farmers and advisers. It is hoped the reader will perceive that the essentials of modern science, technology, management and business studies have been carefully blended to give the information which experience suggests is necessary for successful farming today.

Contributors

Peter Beale, BSc (Hons), PhD, MIBiol, MBIM.

Present appointment: Principal Lecturer in Landscape Planning and Applied Ecology and Head of Environmental Studies Section, Seale-Hayne College, Newton Abbot, Devon.

Held appointments as a river authority fisheries officer in Cornwall, then Devon between 1964 and 1969. This was followed by research at Exeter University, leading to a PhD. Appointed assistant county conservation officer and ecologist to the planning department of the Devon County Council. A lectureship at Seale-Hayne College followed. Now head of the environmental studies section and director of the HND course in Rural Resources and their Management. Keenly interested in landscape and wildlife conservation in relation to the effects of land uses.

John F. Birtles, FRICS, FIAgS, MBIM, MRAC.

Present appointment: Director of The Agricultural Computer Centre Ltd (ACC); Director of Multiple Accounting Services Ltd (MAS), Bank Chambers, Market Place, Chipping Norton, Oxon.

Land agent; trained in management accounting on the Stock Exchange. Managing director of MAS Ltd, and of The Agricultural Computer Centre Ltd. MAS through its subsidiary ACC provides an independent accounting and recording bureau service for farms and estates throughout the UK.

P. H. Bomford, BSc (Hons), MSc, CEng, MIAgrE, MBIM.

Present appointment: Senior Lecturer, Engineering Section, Seale-Hayne College, Newton Abbot, Devon.

Educated at Reading University and the University of Newcastle upon Tyne. Previous appointments have included assistant lecturer in Farm Mechanisation, Wye College (University of London), lecturer in farm mechanisation, Essex Institute of Agriculture and head of Engineering Department, Ridgetown College of Agricultural Technology (Canada). Directed and carried out research into field drainage, environmental control, vegetable crops mechanisation, farm machine performance, reversible tractors, maize harvesting, mulch laying. Current professional interests are cereal and forage harvesting and storage, tractors and power units and alternative sources of energy.

Paul Brassley, BSc (Hons), BLitt.

Present appointment: Senior Lecturer in Agricultural Economics, Seale-Hayne College, Newton Abbot, Devon.

After reading agriculture and agricultural economics at the University of Newcastle upon Tyne went on to research at Oxford. Following this he farmed in Devon until appointed to his present post at Seale-Hayne College. Current research interests are in agricultural policy and agricultural history.

J. S. Brockman, BSc, PhD.

Present appointment: Principal Lecturer in Agriculture, Seale-Hayne College, Newton Abbot, Devon.

After obtaining a PhD for a residual study of the classic Treefield experiment at Cockle Park, he helped to found Fisons North Wyke Grassland Research Centre in Devon. Has published many papers on the nutrition and utilisation of grassland. Since 1978 has been at Seale-Hayne. Is an associate editor of *Grass and Forage Science* and President of the British Grassland Society, 1982–83.

Peter Brooks, BSc, PhD.

Present appointment: Head of the Agriculture Department, Seale-Hayne College, Newton Abbot, Devon.

Educated at Nottingham University and obtained PhD for work on pig nutrition and reproduction. Spent a further two years as a postdoctoral fellow at Nottingham before moving to Seale-Hayne College as lecturer in animal production. Appointed head of department in 1974. Researched many aspects of pig production but particular research interest in the breeding female, her nutrition and management.

J. L. Carpenter, BSc, MSc, CEng, NDAgrE, FIAgrE.

Present appointment: Senior Lecturer in Agricultural Structures, Seale-Hayne College, Newton Abbot, Devon.

Educated at Leeds and Newcastle upon Tyne Universities. Taught horticultural mechanisation and farm buildings at Writtle Agricultural College. Now teaches courses on farm structures and environment. Leads a group investigating the energy used by milking parlour equipment. Past holder of Top Fruit Cultivation Scholarship from Worshipful Company of Fruiterers.

John Collins, NCA, NDA, CDFM.

Present appointment: Now farming but formerly tutor for the one year course in Gamekeeping and Waterkeeping, Hampshire College of Agriculture, Sparsholt, Winchester, Hants.

Studied agriculture and farm management at Seale-Hayne College and Hampshire College of Agriculture. After several years in practical farm management, obtained recent appointment. Until recently involved in the training of students in game conservation, riverkeeping, practical gamekeeping and allied subjects, including game and agriculture, wildlife management and fishery management.

G. R. Cooke, BSc (Hons), MIBiol.

Present appointment: Freelance writer on agricultural matters, based at Holsworthy, Devon.

Educated at King's College, Newcastle upon Tyne (Durham University). Has lectured in crop production at both Seale-Hayne and Writtle Agricultural Colleges. At Writtle he was also Farm Director. Appointments previously held include head of technical advisory service with a fertiliser manufacturer. Main interests are soils and fertilisers, and grassland and forage crop production.

Alan Cooper, CDA (Hons), NDA, MSc.

Present appointment: Senior Lecturer in Animal Production, Seale-Hayne College, Newton Abbot, Devon.

Previous appointments have included lectureships at Shropshire Farm Institute and the University of Malawi. Research interests cover growth promoters and reproduction and animal behaviour.

P. J. Davies, MSc.

Present appointment: Head of Department of Agricultural and Horticultural Science, Writtle Agricultural College, Writtle, Essex.

Is a graduate of Bangor University where he also pursued postgraduate work in the Department of Biochemistry and Soil Science on metabolic disorders of cattle and sheep. Formerly a member of staff at Seale-Hayne College he has been at Writtle since 1963. Is a member of the British Society of Animal Production and of the Nutrition Society and has contributed papers on sheep nutrition to meetings of both societies.

Ivan H. Fincham, MRCVS.

Present appointment: Divisional Veterinary Officer (Veterinary Investigation), ADAS, Starcross, Devon.

Diagnostic and consultative interests developed at the Veterinary Investigation Centre at Wye College, Kent. After secondment to Jamaica for three years, returned to the MAFF Central Veterinary Laboratory, Weybridge, Surrey, thence to the VI Centre at Starcross, then serving Devon and Cornwall, becoming director, as Divisional Veterinary Officer, in 1955. Special interests are diseases common to man and animals, also the promotion of sound animal husbandry to control diseases, involving talks, discussion groups and radio programmes.

G. W. Furness, MSc, Dip Farm Management

Present appointments: Chief Agricultural Economist, Department of Agriculture for Northern Ireland, Belfast and Professor of Agricultural Economics, The Queen's University, Belfast.

Spent nine years as an agricultural economist at the University of Manchester, and was lecturer, then senior lecturer and Head of the Department of Farm Business Management at Seale-Hayne College. In 1967, joined the Department of Agriculture for Northern Ireland as Deputy Chief Agricultural Economist and the Queen's University, Belfast as lecturer, later senior lecturer, in agricultural economics. Appointed to present positions in 1977, he is responsible for the assembly of official agricultural statistics and farm income data in Northern Ireland, and for research and teaching in agricultural economics, farm management and marketing, with particular interest in the role of agriculture in the regional economy.

J. S. Goddard, BSc, MSc, PhD.

Present appointment: Head of Department of Fishery Management, Hampshire College of Agriculture, Sparsholt, Winchester, Hampshire.

Educated at the University of Nottingham and the University College of North Wales, Bangor. Obtained a PhD for research in the fields of fish physiology and nutrition. Took up present post following a four year period of research and development in fish farming based in Yorkshire. Current professional interests involve fishery management training, trout and salmon culture, and development work with tropical species of food-fish.

R. John Halley, BSc, MSc, MA.

Present appointment: Vice-Principal, Seale-Hayne College, Newton Abbot, Devon.

After gaining first degree was awarded a Ministry of Agriculture postgraduate scholarship to the University of Reading and studied at the National Institute for Research in Dairying. First appointment was at Seale-Hayne College where teaching duties were combined with research on grassland utilisation. Appointed as university demonstrator, School of Agriculture, Cambridge and also acted as assistant to farm director. Managed a large farming estate in Hertfordshire before returning to Seale-Hayne as senior lecturer in agriculture and farm manager. Later appointed vice-principal at Seale-Hayne. Mr Halley co-edited the sixteenth edition of *The Agricultural Notebook* with the late Dr Ian Moore, and is the editor of this new edition.

Richard P. Heath, NDA, NOAgrE, TEng (CEI), MIAgrE.

Present appointment: Senior Lecturer, Seale-Hayne College, Newton Abbot, Devon.

Trained at Shuttleworth and Writtle Colleges. Joined the engineering section staff at Seale-Hayne in 1970, teaching at TEC diploma, HND and degree levels. Interests are mechanisation aspects of crop production to harvest and dairying. Undergraduate study and materials/structures study extends his teaching interests. A member of the Institution of Agricultural Engineers he has held branch and national officer positions.

A. D. Hughes, BSc (Hons), PhD.

Present appointment: Soil Scientist, ADAS, Coley Park, Reading.

Trained in soil science at the University of Newcastle upon Tyne and subsequently researched aspects of phosphate nutrition at Leeds University. From 1970 to 1981 he was a soil scientist with ADAS at the Starcross Sub-Centre, advising farmers throughout Devon and Cornwall. Has recently moved to the ADAS Regional Centre at Reading. Whilst working in the south west of England was closely involved with the problems of arable cropping on heavy soils under high rainfall conditions.

David Iley, BSc (Hons).

Present appointment: Lecturer in Zoology, Seale-Hayne College, Newton Abbot, Devon.

Trained at King's College, Newcastle upon Tyne and Imperial College Field Station, Sunninghill. Professional interests include development of integrated systems for control of crop pests, apiculture and landscape and wildlife conservation.

J. A. Kirk, BSc, PhD.

Present appointment: Senior Lecturer in Animal Production, Seale-Hayne College, Newton Abbot, Devon.

Spent nine years farming before reading agriculture with agricultural economics at the University College of North Wales, Bangor. Postgraduate research on the growth and development of Welsh Mountain lambs led to a PhD. Current research interests are the growth and development of lambs and bull beef.

G. Moule, BSc (Hons), MSc.

Present appointment: Senior Lecturer in Botany and Crop Pathology, Seale-Hayne College, Newton Abbot, Devon.

Has held present appointment since 1971. Main interests include crop pathology and weed biology, with particular interest in the epidemiology of crop fungal and viral pathogens.

R. M. Orr, BSc, PhD, MIBiol.

Present appointment: Senior Lecturer, Seale-Hayne College, Newton Abbot, Devon.

Trained at the University of Aberdeen, then undertook studies on voluntary feed intake in ruminants at Edinburgh University which led to a PhD. Currently teaches animal nutrition to both degree and diploma students.

M. F. Ponting, BVSc, MRCVS, NDA.

Present appointment: General Practitioner, Senior Lecturer in Veterinary Science, Royal Agricultural College, Cirencester.

Studied at the Royal Agricultural College, Cirencester and the Veterinary School, University of Bristol. Was in general practice in Somerset and Devon before obtaining an appointment as senior lecturer in veterinary medicine at Seale-Hayne College. Currently is a senior lecturer at the Royal Agricultural College as well as being the principal of a mixed veterinary practice in Gloucestershire with specific interest in equine veterinary medicine and surgery.

John I. Portsmouth, NCP, NDP, DipPoult, NDR, FPH.

Present appointment: Executive Director, Peter Hand (GB) Ltd, Stanmore, Middlesex.

Had been employed extensively in the poultry feed industry before present appointment. Is author of several books and many articles on poultry and rabbit feeding and management. Presented original research on calcium metabolism in pre-lay hens at World Poultry Science Association meetings. Spent last fifteen years studying the micronutrient requirements of farm livestock, particularly poultry.

Charles L. Pugh, BSc, PhD.

Present appointment: Farmer and freelance computer consultant, Falmouth, Cornwall.

Obtained first degree from Reading University, then studied for PhD by research on computer simulation techniques. After a period in Canada working on the CANFARM Data System, returned to lectureship in farm management at Wye College. Continued research into computers, specialising in the microcomputer. In 1978 left Wye to set up dairy farm in Cornwall. Since then has also been acting as an independent consultant, on farm computers, to the agricultural industry.

John Sobey, MIPM, MBIM, MRIPHH.

Present appointment: Academic Registrar and Senior Lecturer in Management, Seale-Hayne College, Newton Abbot, Devon.

Had been involved in staff training schemes in large companies before joining Seale-Hayne College. Main professional studies are now in incentive payment systems in agriculture and related industries.

G. H. T. Spring, MA.

Present appointment: Senior Lecturer in Law, Department of Business Studies, Plymouth Polytechnic, Drake Circus, Plymouth.

After reading law at Cambridge, spent several years in East Africa in government and business employment. Has taught for the last twelve years in the Business Studies Department of Plymouth Polytechnic and also at Seale-Hayne College. Is an active member of the Agricultural Law Association.

R. D. Toosey, BSc, MSc, NDA.

Present appointment: Principal Lecturer in Crop Production, Seale-Hayne College, Newton Abbot, Devon.

Graduated from Reading University. Did postgraduate research on sprouting of potatoes and subsequent work on brassica fodder crops. Has spent nine years in practical agriculture and managed Seale-Hayne College Farm for six years. Formerly lectured at Rodbaston and Writtle Colleges of Agriculture before taking up present appointment. Is author of several publications.

John C. Weeks, CBE, FIAgrE.

Present appointment: Retired. Formerly Director of Health and Safety in Agriculture and HM Chief Agricultural Inspector for Great Britain.

Until his retirement in 1980 he was HM Chief Agricultural Inspector and Director of Agricultural Policy Branch in the Health and Safety Executive. Is Chairman of the Committee of the Association of the British Society for Research in Agricultural Engineering, the Ergonomics Study Group of the National Institution of Agricultural Engineering, and a member of the Engineering Advisory Committee of the Governing Body of the NIAE and of the Court of the Cranfield Institute of Technology. Was awarded the CBE in 1981, is a fellow of the Institution of Agricultural Engineers and was President from 1978–1980.

David Wright, BSc, PhD.

Present appointment: Lecturer in Crop Production, University College of North Wales, Bangor, Gwynedd.

After completing a PhD was appointed as a lecturer in crop production at Seale-Hayne College, before taking up present appointment. Current research interests include developmental physiology of winter and spring barley, factors affecting yield and utilisation of fodder root crops, tuber initiation and growth in potatoes.

Contents

Part 4   Farm management   573

Part 1

Crop production

1

Soils

A. D. Hughes

WHAT IS A SOIL?

A soil is an ordered combination of minerals, organic matter, air, water and living organisms.

Simple chemical analysis tells us the gross composition of soil but takes no account of the ordered nature of its constituents. Such analysis therefore obscures more useful knowledge than it reveals.

The relative proportions of mineral material, organic matter and 'pore spaces' (which may contain either air or water) vary within very wide limits. For example, peat soils are almost devoid of mineral material, whilst some sands have only very low levels of organic matter. Subsoils usually have lower organic matter levels and fewer pore spaces than topsoils. A useful figure to remember is that in fertile agricultural soils, pore spaces account for about 50% of the total volume. The subdivision of the other 50% between mineral and organic fractions is complicated by the difficulty of knowing the volume occupied by organic matter in soil. By weight (dry soil basis), organic matter levels are typically in the range 2–10% for mineral soils. *Figure 1.1* shows the proportions of the various soil constituents on a percentage volume basis.

Formation of soils

Soils originate from the rocks of the earth's crust. Rocks are classified as follows:

(1) *igneous* formed by cooling of 'molten magma', e.g. granite, diorite and gabbro;
(2) *sedimentary* formed by consolidation of sediments from some previous weathering cycle, e.g. sandstone, shale, dolomite and limestone;
(3) *metamorphic* formed by action of heat and pressure on sedimentary rock, e.g. slate, marble and quartzite.

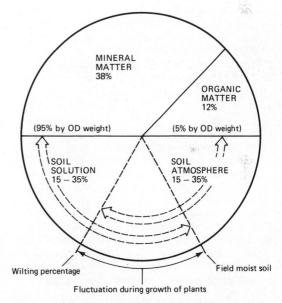

Figure 1.1 Percentage composition of soil on a volume basis (after Jackson, 1964) (From *Chemistry of the Soil*, p. 72, ASC Monograph No. 160, 2nd edn. Ed. by F. E. Bear. New York: Reinhold Publishing Co., reproduced by courtesy of the author and publishers)

Note that igneous rocks are formed *de novo* from the earth's molten core whereas material in the other (secondary) types must at some time have existed as igneous rock.

Weathering

Weathering of exposed rock surfaces is due to both physical and chemical agencies.

3

Physical weathering

Forces of expansion and contraction induced by diurnal temperature variation cause rock shattering, whilst freezing and thawing of any entrapped water greatly accelerates the process. In glaciated areas the major physical agency is crushing and movement by ice and water.

Chemical weathering

Rock minerals are slightly soluble; dissolution of cations and their removal over thousands of years leads to a gradual weakening of the rock's fabric. Chemical weathering is intensified by the acidity resulting from dissolution of atmospheric carbon dioxide, and from organic acids produced during the decay of vegetation. Lichens (which are a symbiotic association of an alga and a fungus) have the ability to fix atmospheric nitrogen as well as being able to fix carbon by photosynthesis. During their growth they accelerate weathering by acidifying the 'soil solution', and rapidly removing the products of weathering as inorganic nutrients. Following their death, carbon is returned to the embrionic soil and provides its first addition of organic matter.

The processes described above can be observed at any time by careful study of the stonework of many ancient buildings and monuments.

Peat soils

A special kind of soil development occurs when very wet conditions prevent the normal breakdown of organic matter by aerobic decomposition. Instead, anaerobic decomposition occurs and the end product is peat. There are two distinct peat formation mechanisms.

Blanket bogs occur in upland areas of the UK where temperatures are low and rainfall exceeds evapotranspiration for at least a part of the year. Heathers and sphagnum moss decay slowly and the organic residues are acid. Any bases (nutrient cations) which may be present are continually leached away by the high rainfall.

Basin peats form in areas where drainage water collects. Native plants are typically sedges and grasses. The incoming drainage water provides ample nutrients so the peat is fertile, having a high base status. Such fen peats are ideal for agricultural cropping when drained, but the act of drainage initiates 'peat wastage', since normal oxidative decay becomes possible. Raised bogs occur when the level of a basin peat surface exceeds the water table height and rainfall is sufficiently great for the blanket bog type vegetation to take over, leading eventually to a blanket bog over the top of fen (sedge) peat.

The soil profile

The vertical section of soil seen in the sides of a pit is the soil profile. Individual layers composing it are called horizons.

Two set of processes are involved in soil formation:

(1) physical and chemical weathering giving rise to 'parent material';
(2) profile development from parent material.

Usually the former precedes the latter but they may proceed simultaneously. Soils where development has proceeded without disturbance exhibit distinctive profiles, their characters being utilised for soil classification and survey.

The upper layer called the A horizon (or 'plough' layer) generally contains appreciable amounts of organic matter which produces a dark colour. This horizon merges into a layer markedly weathered, but comparatively free from organic matter, known as the B horizon. At its base the B horizon, or subsoil, merges into the C horizon or parent material. The upper part of the C horizon is often considerably weathered and merges gradually into unaltered rock.

THE MINERAL COMPONENTS OF SOIL

The mineral matter in soil is derived by weathering of pre-existing rock—the soil parent material. The mineral fraction of the soil can be subdivided in two principal ways:

(1) according to particle size; into sand, silt or clay;
(2) according to mineral composition.

The two are closely related, e.g. fine particles (clay) are composed predominantly of clay minerals, whilst the coarse particles (sand) are normally composed of primary minerals, derived from rock without chemical modification. For practical agricultural purposes in the UK, particle size is the more important attribute.

Sand is largely inert and a very sandy soil is usually well drained, drought prone and lacking in nutrient reserves. Sand grains are usually composed of unaltered primary minerals inherited from the soil parent material; quartz (SiO_2) and feldspars ($MAlSi_3O_8$—aluminium silicates of basic cations M^+) are the most abundant of these. The so-called accessory minerals also occur in smaller amounts in the sand fraction. They have a high density and are therefore easily separated from the rest of the sand fraction, and can be used as 'fingerprints' to help identify soil parent material. Weathering of some accessory minerals releases nutrients of importance to agricultural crops. Pyroxenes and amphiboles provide a number of minor nutrients (trace elements) whilst phosphate is released by the weathering of apatite.

Clay is composed of very fine particles with an enormous surface area and great chemical activity.

Clay soils hold large amounts of water, much of which may not, however, be available to plants. They are frequently poorly drained, but have good reserves of nutrients.

Silt is insufficiently fine to have significant surface chemical activity, but also lacks the agriculturally attractive properties of sand. Hence silty soils have many of the undesirable properties of both sands and clays and the desirable ones of neither.

Real soils are always a mixture of sand, silt and clay, with the relative proportions governing soil properties. An ideal soil would have balanced proportions of sand, silt and clay, giving adequate water holding capacity, free drainage, stable structure and a plentiful supply of nutrients.

Soil texture and particle size distribution

In the UK a distinction is drawn between particle size distribution (psd—an objective laboratory determination) and a subjective assessment of 'soil texture' made by assessing the feel of a soil when worked moist between thumb and fingers.

Soil texture determined by hand texturing is defined in terms of the sensation experienced by the person doing the 'texturing'. A key to the hand texturing system used by ADAS is given in *Table 1.1*.

Table 1.1 The assessment of soil texture based on the feel of a ball of moist soil

Soils		*Symbol*
Sands		
Gritty, no cohesion	Sands	
Kinds of sand:	coarse sand	CS
sea shore sand	(medium) sand	S
dune sand	fine sand	FS
grains only just visible to eye	very fine sand	VFS

The sands in the other sandy soils below are classified as above.

Gritty, very slight cohesion, 'ball' of soil falls apart		
readily	loamy sands	LS
Loams		
deform readily and are gritty	sandy loams	SL
deforms readily and is silky, smooth	silty loam	ZyL
binds together strongly, silky, some polish	silt loam	ZL
binds together strongly, takes polish	clay loam	CL
binds together strongly, takes polish, gritty	sandy clay loam	SCL
binds together strongly, takes polish, silky	silty clay loam	ZyCL
binds together strongly, not gritty, not silky, will		
not polish	loam	L
Clays		
takes very marked polish	clay	C
silky as well	silty clay	ZyC
gritty as well	sandy clay	SC
Influence of organic matter		
increases cohesion of light soils		
reduces cohesion of heavy soils		
Influence of chalk		
may give silky feeling or grittiness but the fact that a soil is known to be chalky		
should not influence the texturing. However it may be qualified, e.g. calc. ZL.		

Other soil textures		
Peat	contains over 45% organic matter	Pt
Loamy peat	contains 35–45% organic matter	Lmy Pt
Peaty loam	contains 25–34% organic matter	Pty L
Organic 'mineral' soils	contain 8–24% organic matter	Org.

The prefix 'organic' may be added to the full range of mineral soil textures.

The 'feel' is influenced by the wetness of the soil so a standard wetness is essential. The soil should be worked down thoroughly to eliminate the effects of structure at a consistency just wet enough to mould.

Binding and cohesion		
none or very slight		sands
moulds readily into a cohesive ball		loams
difficult to deform		clays

From *Soil Field Handbook*, Agricultural Development and Advisory Service. London: MAFF with permission.

Since hand texturing is a subjective assessment it is less precise (reproducible) than a laboratory determination of psd; however it is quicker and therefore cheaper and most professional advisers agree that it is better, since it is possible to feel characteristics of great agricultural significance. These are not necessarily reflected in psd data.

Particle size distribution (psd)

The process of determining the psd of a soil is referred to as 'mechanical analysis'. The principle is that soils are dispersed in an aqueous solution and separated into size fractions by a combination of sieving (down to about 0.05 mm) and timed sedimentation.

Stokes' equation states

$$V = \frac{g(\sigma - \rho)d^2}{18\eta}$$

where V is the settling velocity in cm/s,
  g is the acceleration due to gravity (about 980 cm/s^2),
  σ is the density of the particle,
  ρ is the density of the water,
  η is the viscosity of water (about 0.010 at 20 °C),
  d is the diameter of the particle (cm).

If an aqueous suspension of soil is thoroughly agitated, then left to stand, the various components will settle at different velocities in accord with Stokes' equation. By carefully sampling from a given depth in the suspension at successive intervals of time, it is possible to calculate the proportion of particles with a given 'equivalent spherical diameter' (esd). (To make this possible particle density is assumed, by convention, to be 2.6 g/cm^3, so that the term $(\sigma - \rho)$ is numerically 1.60.)

Various standard procedures differ in detail, for example:

(1) pre-treatment—drying and sieving techniques may differ,
(2) dispersion—use of ultrasonic equipment, treatment with deflocculating chemicals or exchange resins, destruction of organic matter or dissolution of carbonates,
(3) sedimentation and sampling techniques are fairly standard but an alternative procedure (the Bouyoucos method) uses a hydrometer to measure the density of the suspension at a given depth.

Procedures in common use are described by Russell (1973) and the method used for routine purposes by The Soil Survey of England and Wales and by ADAS is described in MAFF (1973) and Avery and Bascomb (1974). There is no universal agreement on the boundaries of the particle size classes. Systems in common use are illustrated in *Figure 1.2*.

Since numerical percentage composition data are not easily appreciated in speech, it is customary to use a triangular diagram to relate psd data to named textural classes. Again, standardisation is lacking and a number of systems have been used. The current Soil Survey of England and Wales triangular diagram is reproduced as *Figure 1.3*.

The use of a triangular diagram to assign textural

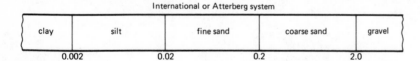

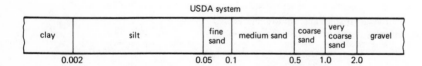

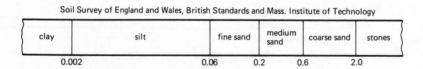

Diameter (mm) (log scale)

Figure 1.2 Particle size class systems in common use (White, R. E., 1979 *Introduction to the Principles and Practice of Soil Science*. Oxford: Blackwell Scientific, reproduced by courtesy of the author and publishers)

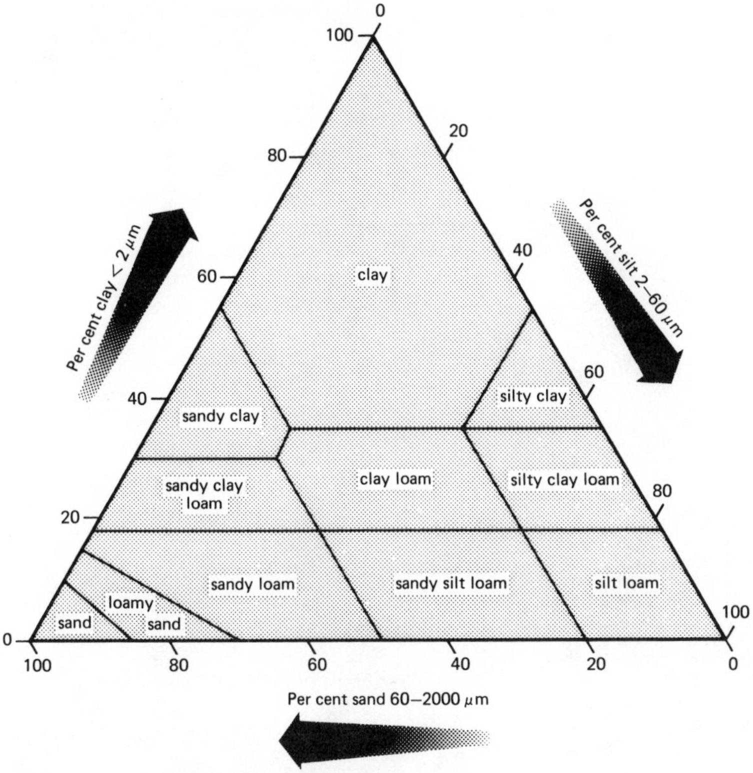

**Figure 1.3 Triangular diagram of soil textural classes
(Hodgson, J. M., 1974 *Soil Survey Field Handbook*, p. 23.
Soil Survey of England and Wales, Rothamsted
Experimental Station, Harpenden, reproduced by courtesy of
the author and publishers)**

class names can lead to confusion—whatever the
name used (e.g. silty loam) it is important to know
whether it was arrived at by 'hand texturing' or
derived from a laboratory determination of psd. These
are determinations of two separate properties, not
always giving the same result. To add to the
possibilities for confusion, skilled soil surveyors
frequently use the hand texturing method to assess
psd, avoiding the need for laboratory analysis. This is
not hand texturing in the sense described above, and
is as used by advisory officers in the UK.

CLAY MINERALS

The minerals of the clay fraction may be primary,
inherited from the parent material or secondary,
formed as a result of chemical processes in the soil.
The term 'clay mineral' is conventionally reserved for
clays formed as a result of chemical changes in the
soil.

The study of clay minerals is a very complex subject
and much of the information contained in specialist
publications on the subject is of limited practical use

to agriculturalists. Agriculturally important properties
of soil clays are:

(1) cation exchange capacity and ability to absorb
    anions;
(2) surface area available for chemical reactions;
(3) ability to shrink and swell in response to changes
    in moisture content.

Structure of clay minerals

The commonly occurring clay minerals are layer
silicates and may be thought of as being built up from
planes of oxygen or hydroxyl units. There are two
types of oxygen/hydroxyl planes.

(1) *Complete planes* in which each oxygen touches six
    others (*see Figure 1.4*) and some or all the oxygen
    position may be occupied by hydroxyls.
(2) *Hexagonal planes* which differ from the 'complete'
    arrangement (*see Figure 1.5*) in that (a) there are
    no hydroxyls, (b) every fourth oxygen position is
    vacant, (c) the inter-oxygen separation is reduced
    in the ratio $\sqrt{3}:2$, so that the two types of plane

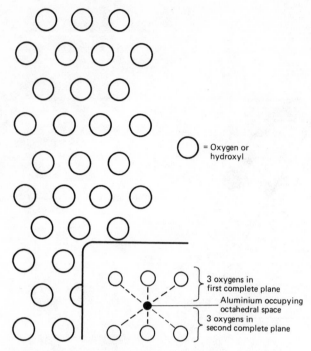

Figure 1.4 Complete plane of oxygens or hydroxyls. Inset—aluminium occupying the octahedral space between oxygens of two complete planes

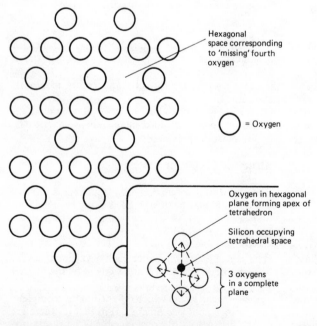

Figure 1.5 Hexagonal plane of oxygens. Inset—silicon occupying the tetrahedral space between oxygens of a complete plane and a hexagonal plane

contain the same number of oxygens : hydroxyls per unit area.

Tetrahedral sheets

If a single oxygen is placed on top of a complete plane it will fit comfortably in a position above a space in the plane and beneath it there will be what is called a tetrahedral space (so called because it is between four spheres whose centres may be thought of as the apices of a tetrahedron). If, instead of one single oxygen, a hexagonal plane is placed above a complete one, a network of tetrahedral holes will occur. In clay minerals these holes contain either silicon (usually) or aluminium cations (occasionally). The two planes and their associated cations are said to form a tetrahedral sheet.

Octahedral sheets

If two complete planes are brought together, as well as tetrahedral spaces, a network of octahedral spaces will be formed between three oxygens (or hydroxyls) of one plane and three of the other. It is a feature of clay minerals that none of the tetrahedral spaces are filled in these cases, but a proportion (ranging from 66 to 100%) of the octahedral ones are occupied by either aluminium or magnesium cations. The two planes and associated cations are said to form an octahedral sheet. If magnesium is the predominant cation all the octahedral spaces are filled and the sheet is said to be tri-octahedral (sometimes called a brucite sheet—brucite is $Mg(OH)_2$).

If the predominant cation is aluminium, two-thirds of the spaces are filled and the sheet is said to be di-octahedral (sometimes called a gibbsite sheet—gibbsite is $Al(OH)_3$).

Isomorphous replacement

In the sheet structure just described the arrangement of oxygens, hydroxyls and cations is such that electrical neutrality is maintained. In 'real' clay minerals some cations are replaced by others of similar size but different charge. This replacement does not alter the fundamental structure (form) of the layer but does leave the layer with a net charge.

Thus, aluminium can replace a proportion (up to about 5%) of silicon in tetrahedral sheets. The larger octahedral spaces can be occupied by a wide range of cations. Isomorphous replacement normally leaves the sheet with a negative charge since every silicon replaced by an aluminium is conceptually the removal of one unit of positive charge, equivalent to a gain of one unit of negative charge. In the same way every octahedral aluminium replaced by a divalent cation contributes one unit of negative charge. Note that isomorphous replacement is a feature of a sheet incorporated at the time of its formation. 'Replace-ment' implies a departure from an idealised model *not* a dynamic soil process.

Combinations of sheets—layer lattice minerals

The clay minerals may now be visualised as combinations of tetrahedral and octahedral sheets, some of which may be negatively charged due to isomorphous replacement.

Nomenclature

Note that from following recent discussions (*see* Bailey *et al.*, 1971):

(1) atoms forming a continuous array in two dimensions constitute a *plane* (e.g. oxygen—hydroxyl planes, above);
(2) combinations of planes form a *sheet* (e.g. tetrahedral and octahedral sheets, above);
(3) combinations of sheets form a *layer* and successive layers may be separated by *interlayer materials*.

Interlayer bonding

The bonding between adjacent layers may be only oxygen–hydroxyl linkages between uncharged layers which are close together (*see* Kaolinite group *below*), or by means of cations in the interlayer space neutralising negative charges on the sheets. This type of bonding is readily subdivided into three categories on the basis of the magnitude of the negative charge.

High charge—mica type bonding

The outer sheets of adjacent layers are close together and arranged so that potassium ions can be located where 'surface depressions' in adjacent sheets are opposite one another. This gives a strong bond since inter-ionic distances are small and the basal spacing is 1 nm.

Medium charge—vermiculite type bonding

The adjacent layers are further apart than in the mica type and the interlayer space is occupied by two molecular layers of water containing magnesium or aluminium ions. The basal spacing is 1.4–1.5 nm and the larger inter-ionic distance gives a weaker bond, although the interlayer water contributes some hydrogen bonding.

Low charge montmorillonite or smectite type

The structure is similar to vermiculite, but the bond is weaker and the interlayer cation can be exchanged with solution cations.

Typical layer lattice clay minerals (Figure 1.6)

Kaolinite is known as a 1:1 clay mineral composed of one tetrahedral and one octahedral sheet. There is no

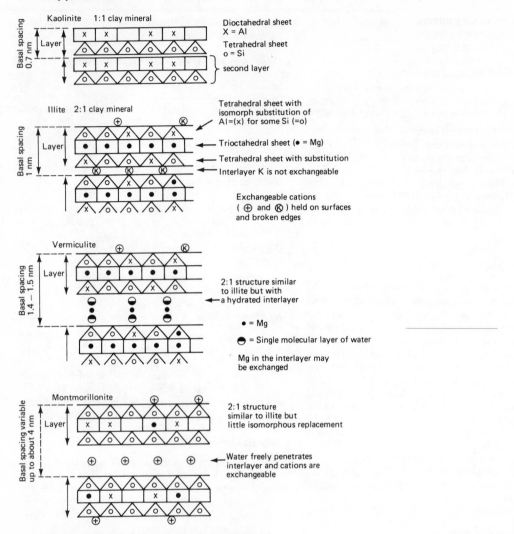

Figure 1.6 Typical layer lattice clay minerals (after Mengel and Kirkby, 1978 *Principles of Plant Nutrition*, p. 38. Bern: International Potash Institute, reproduced with permission of the authors and publishers)

interlayer and successive layers are held by oxygen–hydroxyl linkages. Water is not able to move between the units and only exposed surfaces and broken edges contribute to the cation exchange capacity.

Illite is a 2:1 mineral (two tetrahedral sheets sandwiching one octahedral sheet). Adjacent layers are bonded by potassium which forms an integral part of the structure and is not exchangeable.

Vermiculite is similar to illite, but interlayer bonding is due to magnesium (Mg) in association with two monolayers of water. The Mg^{2+} is exchangeable, so intergrades occur and leaching with KCl solution leads to replacement of Mg^{2+} by K^+ and adoption of the illite structure (i.e. 1 nm basal spacing).

Montmorillonite is the commonest of the 'smectite' or 2:1 swelling clays. Lattice structure is again a variation of that of illite; the charge on the layers is low (little isomorphous replacement) and bonding is sufficiently weak that interlayer cations are readily exchangeable, and water may enter the interlayer so that there is no unique basal spacing.

Hydrous oxides

Iron (Fe) and aluminium (Al) form hydrous oxides with no recognisable structure (amorphous) and a gross composition represented by $Fe_2O_3 \cdot H_2O$ and $Al_2O_3 \cdot H_2O$. These hydrous oxides occur principally as surface films on clay minerals in soils of temperate lands, and contribute to the cation exchange capacity (CEC) and phosphate fixing capacity. In tropical soils the hydrous oxides may dominate the colloid fraction; phosphate fixation capacity is then extremely large.

Clays in soils

The clay particles present in soils differ from the pure examples already described. In particular, soil clay minerals are formed *in situ* during natural weathering processes, so in all but the oldest and most mature soils intermediate stages of clay mineral formation, as well as end products, are found. In general, soil clays have a less well ordered structure than the pure varieties (usually of geological origin) and both the layers and interlayer ions may differ within a single clay particle. Similarly adjacent clay particles are not expected to have the same composition, and the picture is further complicated by the ubiquitous occurrence of hydrous oxide films.

Because of this large range of variability, soil clays should be thought of as smectites (montmorillonite type), vermiculite type, illite type or kaolinite type. When successive layers (and interlayers) have different forms and compositions the mineral is said to be interstratified.

Figure 1.7 shows the important features of a soil clay complex formed by weathering of mica.

Clays and soil shrinkage

Laboratory observations show that minerals of the smectite group expand as they are wetted and contract on drying. Casual examination of field soil surfaces or soil profile pits during dry weather reveals the presence of cracks which are entirely absent under wet conditions. Since the cracking observed in the field is most extensive on clay soils, it is tempting to assume that it is due to the same mechanism as the expansion and contraction seen under laboratory conditions. This may be the case in hot climates with a long 'dry season', but recent studies, particularly the work of Reeve *et al.* (1980), have shown that the proportion of smectite in the clay fraction is not related to the field cracking behaviour of UK soils. The explanation for this is that the drying experienced in the field is not sufficiently severe to remove the interlayer water removed by laboratory drying. The observed cracking in the field is largely due to 'bulk shrinkage' caused by reduction in the size of soil pores as the soil dries out.

SOIL ORGANIC MATTER AND THE CARBON CYCLE

Soil organic matter (OM) is an extremely emotive topic, associated in many people's minds with 'the goodness of the soil' and man's presumed ill treatment of it. It is also a difficult subject for study. This is because OM is not a single entity—it is a complex system whose individual components are not all capable of precise characterisation. Furthermore, the soil's organic fraction is never static—transformation from one form to another is one of its fundamental characteristics. At any one time, some 2% of the soil's organic matter is accounted for by living organisms.

The formation of organic matter

Living organisms (plant and animal, including micro-organisms) are the sole precursors of the soil organic fraction. Visible plant remains are not usually considered as 'soil organic matter' but micro-organisms are

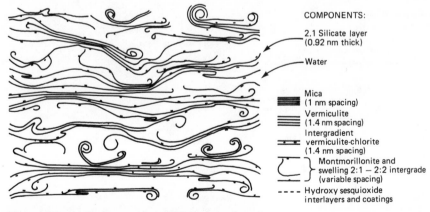

COMPONENTS:

2.1 Silicate layer
(0.92 nm thick)

Water

Mica
(1 nm spacing)
Vermiculite
(1.4 nm spacing)
Intergradient
vermiculite-chlorite
(1.4 nm spacing)
Montmorillonite and
swelling 2:1 – 2:2 intergrade
(variable spacing)
Hydroxy sesquioxide
interlayers and coatings

Figure 1.7 Important features of a soil clay complex formed by weathering of mica (from Jackson, M.L., 1964 *Chemistry of the Soil*, p. 93, ACS Monograph No. 160. Ed. by F. E. Bear. New York: Reinhold Publishing Co, reproduced with permission of the author and publishers)

included, if only for want of any means of isolating them from the soil matrix.

Chemical analysis of organic matter

In a previous section it was noted that simple chemical analysis of soil obscures more useful knowledge than it reveals. This is particularly true for OM.

Traditional proximate analysis of a number of soil organic matter fractions gave the results shown in *Table 1.2*.

Note that the groups into which the OM was separated are somewhat ill defined, and for example, the fractions labelled hemicellulose and cellulose also include many other compounds, some containing nitrogen. Most of the OM is accounted for by the residue which is resistant to the methods of the analytical organic chemist, and therefore only readily characterised by its total elemental analysis.

Table 1.2 Proximate chemical composition of soil organic matter from different soils (from Waksman and Stevens, 1930)

Ether-soluble material (fats, waxes) (%)	Alcohol-soluble material (resins) (%)	Hydrolysis with 2% HCl (hemicelluloses) (%)	Hydrolysis with 80% H_2SO_4 (cellulose) (%)	Residue (lignin-humus complex) (%)	N in residue by Kjeldahl (organic nitrogenous complexes) (%)	Sum of the constituents accounted for (%)
3.56	0.58	5.44	3.55	43.37	33.78	90.28
4.71	1.53	8.60	5.22	40.81	34.74	95.61
1.94	3.10	12.59	5.36	35.18	35.77	93.94
0.80	0.82	5.53	4.12	41.87	37.35	90.49
1.02	0.88	6.96	3.50	42.05	33.25	87.66
0.46	0.84	8.54	2.83	42.83	33.36	88.86
0.52	0.63	10.66	3.38	46.50	33.13	94.82
0.62	0.61	8.61	3.64	49.29	30.38	92.75

From *Soil Science* (1930) **80**, 97 by courtesy of authors and the publisher Williams and Wilkins: Baltimore, Maryland

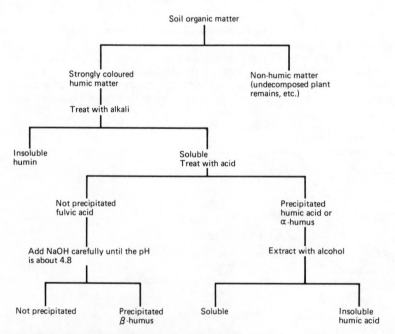

Figure 1.8 Fractionation of soil organic matter (from Moretensen and Himes, 1964, in *Chemistry of the Soil*, p. 216. Ed. by F. E. Bear. New York: Reinhold Publishing Co., reproduced with permission of the author and publishers)

Extracts of organic matter

If soil is extracted with sodium hydroxide (NaOH), a proportion of its OM is rendered soluble. This fraction can be further subdivided by subsequent chemical treatments and one such scheme of fractionation is shown in *Figure 1.8*.

The fraction not removed by NaOH is known as humin. Such fractionations are subject to three serious limitations.

(1) NaOH is quite alien to the normal soil environment and its use almost certainly causes major chemical changes in the materials being extracted. An alternative procedure using pyrophosphate extraction is less likely to produce such artefacts, but removes a smaller proportion of the total organic matter.
(2) The fractions named in *Figure 1.8* do not correspond to any particular chemical composition. Both humic and fulvic fractions are characterised by a wide range of molecular weights; for fulvic acids typically less than 10000. Humic acid molecular weights range from 5000 to several million.
(3) The fraction not extracted by NaOH is christened 'insoluble humin' and thereafter ignored, even though it accounts for about 30% of total carbon.

Properties and functions of organic matter in the soil

The 'chemical' classifications of OM described in the previous two sections will be seen to be of limited relevance to agricultural problems. The serious agriculturist should beware of being misled by the great volume of specialist literature relating to them.

Organic matter and soil structure stability

Naturally occurring gums and particularly polysaccharides can be shown to have a very important role in stabilising the structure of some arable soils. More complex 'humic colloids' fulfil the same function in established grassland and under deciduous forest. The compounds responsible for these effects can be visualised as 'rope', or 'net' structures having the mechanical effect of binding together clay particles. The linkage between the organic colloid and the clay can be either due to Van der Waal forces (neutral colloid), electrostatic attraction at broken clay edges or 'bridging', whereby an exchangeable divalent cation, e.g. calcium, uses one of its valency bonds to link with the clay and the other to link to a carboxylate group on the colloid.

Recent work by Hamblin and Greenland (1977) has cast some doubts on the universal importance of polysaccharide in such stabilisation and they draw attention to the importance of organic compounds

capable of forming strong bonds to clay particles through association with iron and aluminium.

Exchange capacity

Many of the chemical properties of soils are dependent upon cation exchange capacity (CEC—sometimes called base exchange capacity). Organic substances contain carboxyl groups which can dissociate to provide negatively charged sites. The charge is typically 2–4 mEq/g compared with 0–2 mEq/g for clay minerals. Thus even though the OM content of most soils accounts for a small proportion of the total soil mass, it may make a significant contribution to the CEC.

Mineral nutrition of crops

Since nitrogen, phosphorus and sulphur are important constituents of OM, eventual breakdown by microorganisms releases quantities of these elements to the soil in a plant available state. (Nitrogen, phosphorus and sulphur cycles are discussed more fully on pp. 25–30.) Some organic substances form complexes with copper, and hence its availability to crops is also influenced by OM formation and breakdown.

Organic matter and agricultural practice

Whilst there may be argument about the chemical nature of OM and the mechanisms whereby it affects soil properties, there is no doubt that high OM levels are generally associated with ease of management of soil (*see* for example Low, 1972).

Soil organic matter level

In the laboratory, soil organic matter is normally determined by wet oxidation with a dichromate/sulphuric–phosphoric acid solution. The dichromate remaining after oxidation is determined by titration with ferrous sulphate solution (*see* MAFF, 1973). The method gives a value for percentage organic carbon: by convention this is multiplied by 1.72 to give 'organic matter percentage'.

The oxidation reaction is invariably incomplete since a proportion of the OM is protected by intimate association with clay, but the figures obtained are well established in the literature and serve adequately for most comparative purposes.

The OM level in a soil at any given time is the result of a balance between additions and losses over its recent history. The factors affecting the balance are as follows.

Climate

In higher rainfall areas or where drainage is poor, the rate of decomposition is lower and OM levels are

Table 1.3 Quantities of organic material returned by roots of various crops (after Davies, Eagle and Finney, 1972)

Crop	kg/ha of dry roots in top 20 cm	% increase in total organic matter in 20 cm of soil before decomposition
1-year grass ley	4500–5500	0.2–0.3
3-year grass ley	6750–9500	0.3–0.5
Winter cereals	2500	0.1
Spring cereals	1450	less than 0.1
Sugar beet	550	less than 0.1
Potatoes	300	less than 0.1
Red clover	2200	0.1
25 tonnes farmyard manure	4500	0.2

From *Soil Management*, Table 27, p. 194 by courtesy of authors and the publishers Farming Press: Ipswich.

higher. Thus for any given soil texture and cropping sequence the OM level will be higher in the west of the UK, and at higher altitudes.

Soil type

Because of the intimate association between humus and clay particles, the OM in clay soils is better protected against attack and degradation than in sandy soils. Also clay soils are in general more likely to remain moist for longer periods of the year. Both these factors lead to clay soils having higher OM levels than lighter soils under otherwise similar conditions.

Cultivation

Frequent cultivation is thought to increase the rate of OM loss. This is not now believed to be a simple effect of better aeration leading to faster oxidative breakdown of organic materials. Powlson (1980) has suggested that the disruptive forces of cultivation result in exposure to micro-organisms, of organic matter which had previously been inaccessible to them. Also some micro-organisms are killed when soil is disturbed and their premature decay contributes to the loss of OM.

Less soil disruption and consequent slower OM breakdown in the surface horizon may be a beneficial effect attributable to 'non-plough' cultural methods.

Cropping

There has been interest in the effects of crop remains returned to the soil since chemists first started to look at agriculture. Typical quantities involved are shown in *Table 1.3*.

The main point illustrated by this table is that the amounts of organic material returned to the soil from crop roots or manure dressings are too small to affect soil OM levels appreciably, even if no adjustment is made in recognition of the fact that the majority of organic residues returned to the soil are lost by respiration of micro-organisms, and only a small proportion is eventually incorporated into humus.

Note also that contrary to common belief, the root remains from a good winter cereal crop make a sizable contribution to the soil's OM economy (equivalent to about 14 tonnes/ha of farmyard manure).

Effects of leys on soil levels have been studied by a series of experiments—at the ADAS (formerly NAAS) experimental husbandry farms. Results are summarised in *Table 1.4* and show that the increases due to a three-year ley are small and those following nine years of ley are only modest; note also that the gains of nine years' grass were virtually wiped out by a mere three years of arable cropping.

These results are consistent with the view that the structure stabilising powers of leys are due to gums produced by actively growing soil bacteria, the clay binding powers of bacteria themselves, and the mechanical effects of both grass roots and fungal hyphae. All of these agents are ephemeral and their decline following return to arable cropping would be reflected by a fall in total OM level.

Permanent pasture usually has higher OM levels than leys under similar conditions, and at least a proportion of it is of a more durable nature, so that it is lost only slowly following a change to arable cropping.

The carbon nitrogen (C:N) ratio of organic matter of

Table 1.4 Effect of leys on level of organic matter in soils from ley fertility experiments (from Davies, Eagle and Finney, 1972)

Experimental husbandry farm	Location	Soil	% organic matter in topsoils			
			at start of trial	after 3 year ley	after 9 year ley	after 3 year arable following 9 year ley
Boxworth	Cambs	Clay loam	3.1	+0.1	+0.5	+0.0
Bridgets	Hants	Chalk loam	4.4	+0.1	+0.6	+0.1
Gleadthorpe	Notts	Sand	1.6	+0.1	+0.36	+0.3
Rosemaund	Hereford	Silt loam	3.5	+0.2	+1.0	+0.0
High Mowthorpe	Yorks	Chalk loam	3.8	+0.4	+1.0	+0.5

From *Soil Management* Table 28, p. 196 by courtesy of the authors and the publisher Farming Press: Ipswich.

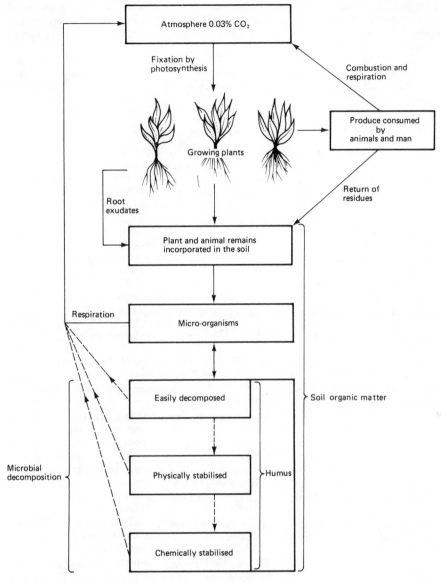

Figure 1.9 The carbon cycle

cultivated soils in temperate regions is fairly constant, having a value between 10:1 and 12:1. When manures with wider ratios than this are added to soil, either a disproportionate amount of carbon must be lost by microbial respiration, or additional N taken from some other source to compensate. (This is dealt with in more detail on p. 26).

The carbon cycle

The transformations of carbon from gaseous carbon dioxide in the atmosphere to plant, animal, and thence soil OM and finally back to the atmosphere, form a cycle depicted in *Figure 1.9*.

Carbon dioxide (CO_2) in the atmosphere is the source of carbon for plant growth, and the 0.03% CO_2 in the world's atmosphere represents a total quantity of about 7×10^{11} tonnes of carbon. This is a relatively small amount by comparison with the estimated 9.6×10^{13} tonnes of carbon in fossil fuels (coal and oil) and 12×12^{15} tonnes in sedimentary rock (Ehrlich *et al.*, 1977).

The rate of fixation by photosynthesis depends on climatic conditions, soil and crop. Maximum rates (world-wide) are about 50 g of dry weight/(m² d) (200 kg/(ha d) of carbon). Under UK conditions the maximum attainable winter wheat yield is thought to be about 12 tonnes/ha of grain (Austin, 1978). This crop makes the majority of its growth in the period April–July (120 d) and the average rate of increase in grain dry weight over this period is 0.1 tonnes/(ha d) or 10 g/(m² d). Making allowance for leaf and root growth, 20 g/(m² d) is a reasonable figure for dry matter (DM) production in the UK. This corresponds to a carbon fixation rate of 80 kg/(ha d).

Plant and animal remains (including excreta) are the source of carbon from which soil organic matter is formed and quantities returned by roots of various crops have already been quoted (*Table 1.3*). Additionally, root systems contribute to soil organic matter by production of root exudates. These comprise both soluble (sugars, amino acids, etc.) and insoluble compounds (polysaccharides, sloughed-off cells, etc.). They make a direct contribution to the nutrition of micro-organisms in the rhizosphere and are thought to be responsible for the very high level of microbial activity in that zone.

Because of the close association of soil, micro-organisms, roots and their exudates, measuring the quantity exuded has proved extremely difficult. Estimates vary, but 10% of the total above ground dry weight over a season seems a likely order of magnitude.

This input of organic matter to the soil may be visualised as a flux of carbon being 'processed' through a volume of soil according to the equation

$$\frac{dC}{dt} = A - kC$$

This assumes that decomposition is what is known as a 'first order reaction' and k is its 'rate constant'; A is the annual addition of residues and dC/dt is the rate of change of the soil's organic carbon level (by convention organic matter (% OM) is derived from a laboratory determination of % C by multiplication by 1.72). For an equilibrium state $dC/dt=0$ and the 'turnover time' is given by

$$\frac{C}{A} = \frac{1}{k}(\text{years})$$

Annual inputs of carbon for old arable soils at Rothamsted are typically 5% of soil organic carbon, so turnover time is calculated to be 20 years.

Radioactive carbon-14 dating techniques have allowed the mean age of carbon in existing soil OM to be measured, and results are in the region of 2000 years. Closer study has shown that different components of soil OM have differing turnover times.

Easily decomposed humus has a turnover time of about three years.

For physically stabilised humus—that which is protected from easy microbial attack by its location in fine pore spaces—the average figure is about 100 years.

Chemically stabilised 'ligno humus complexes', which may be protected by adsorption onto clay mineral surfaces, have mean turnover times as large as 4000 years.

The population of micro-organisms at any one time accounts for 2–3% of the soil's total organic carbon and this fraction has a turnover time of two to three years.

The 'priming action' of fresh organic matter additions

At one time it was thought that the enhanced microbial activity occurring when fresh OM was added to the soil could lead to accelerated breakdown of the old and more stable humic substances. By use of carbon-14 tracer techniques, Jenkinson (1966) was able to show that this 'priming effect' generally had a negligible effect on the overall rate of decomposition, any losses being compensated for by the biomass of the enlarged population of micro-organisms.

CHEMICALLY ACTIVE SURFACES AND THE CHEMISTRY OF SOIL COLLOIDS

The seat of chemical activity in the soil is the colloid fraction. This comprises the clay minerals and organic matter.

The soil's power to hold (adsorb) cations

Cations, represented as M^+ (monovalent, e.g. Na^+, K^+) or M^{2+} (divalent, Ca^{2+}, Mg^{2+}) are held by electrostatic attraction to negatively charged sites in the soil colloid fraction.

The negative charge on soil colloids is due to:

(1) *isomorphous replacement* of trivalent by divalent cations within clay mineral lattices, leading to a deficit of charge;
(2) *unsatisfied charges* at broken edges of lattices;
(3) *dissociation* of protons
   (a) from organic acid groups

   $-COOH \rightarrow COO^- + H^+$

   (b) from hydrous oxide surfaces

   $M-OH \rightarrow M-O^- + H^+$

(M is Si (tin), Mn (manganese), Fe (iron) or Al (aluminium).

Note that such dissociations are pH dependent, so that the negative charge (i.e. the CEC) is greater at high pH, and some protonation may occur at low pH to give sites of positive charge capable of interacting with anions (*see* the section 'The soil's power to hold anions', *below*).

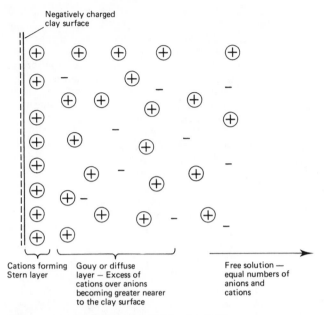

Negatively charged
clay surface

Cations forming Stern layer

Gouy or diffuse layer — Excess of cations over anions becoming greater nearer to the clay surface

Free solution — equal numbers of anions and cations

Figure 1.10 The double layer system associated with a negatively charged clay surface

Exchange of cations

Ions held by the electrostatic forces at charged surfaces are also subject to the kinetic forces of thermal motion, and the strength of their retention is the result of the balance between the two. At equilibrium a characteristic distribution pattern described by the Gouy–Chapman model is observed.

The Gouy–Chapman model says, in effect, that distribution of charged species within a solution is distorted near to a negatively charged colloid surface, there being a disproportionate concentration of cations near to the surface. An inner conceptual layer, next to the charged surface, is composed solely of cations (Stern layer) with a gradual transition (the Gouy or diffuse layer), to free solution. As shown in *Figure 1.10*, these two layers have an excess positive charge, balanced by the negative charge on the surface and cations in the 'double layer' are in equilibrium with those in the solution. Increasing the concentration of the external solution compresses the Gouy layer, so that the change from the all-cation Stern layer to free solution occurs over a smaller distance. Changing the composition of the solution leads to corresponding changes in the double layer system. Thus, if a clay is saturated with Ca^{2+} and in equilibrium with a $CaSO_4$ solution addition of a K^+ salt will lead to replacement of some Ca^{2+} by K^+ ($2K^+$ replacing each Ca^{2+}), i.e. cation exchange. Some cations exchange (or form Stern layers) more readily than others. Small, highly charged ions are held preferentially to large ones with low charge, but small, strongly hydrated ions behave similarly to larger ones.

Exchange reactions are reversible and the thermodynamically less favourable ones are able to occur, providing the concentration in the external solution is great enough. Thus strongly bound Ca^{2+} will be replaced by less strongly bound K^+, provided that the K^+ concentration in the external solution is high enough (i.e. a mass action effect).

The Gapon Equation is a physical relationship that governs the equilibrium between exchangeable and free cations of two species. For a monovalent cation C^+ and a divalent cation C^{2+} the equation states:

$$\frac{C^+ \text{ads}}{C^{2+}\text{ads}} = k \frac{a_{c+}}{\sqrt{a_{c2+}}}$$

C^+ads = adsorbed monovalent cation
C^{2+}ads = adsorbed divalent cation
a_{c+} = activity of monovalent cation in solution
a_{c2+} = activity of divalent cation in solution

$\dfrac{a_{c+}}{\sqrt{a_{c2+}}}$ is called the activity ratio, AR.

Hence

$$\frac{C^+ \text{ads}}{C^{2+}\text{ads}} = k \cdot AR$$

k is primarily dependent upon the strengths of adsorption of the two species.

Adsorption of water

Water is a dipolar molecule with an asymmetrical distribution of charge:

$$\delta^+ \quad \underset{H}{\overset{H}{\diagdown}} O \; \delta^-$$

Thus a layer of water molecules may be attracted to a negatively charged colloid surface and they will all be orientated with their δ^+ ends towards the surface. They will thus expose a new surface of δ^- charges which may in turn attract a second layer, and so on. The proximity to the negative colloid surface will increase the effective charge on the innermost layers of water by induction, but this effect decreases with distance from the colloid and so the strength of bonding also decreases towards an outer surface at which electrostatic attraction is balanced by the forces of thermal motion.

The soil's power to hold anions

Positively charged colloid surfaces

Although soil colloids are predominantly negatively charged, localised positive charged sites also occur, particularly on clay mineral edges and on hydrous oxide surfaces.

Of the nutritionally important anions in the soil NO_3^- is not adsorbed to any appreciable extent and sulphate is adsorbed only slightly. Although phosphate is fixed and tightly held by soils, it is incorrect and misleading to think of phosphate being exchanged by negative surfaces (*see below*, 'Ligand exchange'), so overall the positive charge sites on soil colloids are relatively unimportant.

Ligand exchange

Phosphate, silicate, fluoride and molybdate are bound by ligand exchange. This reaction is very important as the mechanism for 'phosphate fixation'. On iron or aluminium hydroxide surfaces, some oxygen atoms are less strongly bound to iron or aluminium than those in the body of the material. They may be replaced by negatively charged species such as phosphate anions. This reaction is characterised by an increase in the negative charge on the surface. It is not reversible and the exchanged species can only be displaced by another anion capable of further increasing the negative charge.

Quantity intensity relationships

Although roots take up nutrients from the soil solution, as already explained, the solution is in equilibrium with a solid phase capable of replenishing losses. Thus nutrient availability to plants depends not only upon concentration in the solution, but also on the ability of the solid phase to sustain that concentration against depletion. The term Quantity (Q) refers to the amount of a nutrient, and Intensity (I) reflects the strength of the retention whereby it is held. Concentration in the equilibrium solution is a convenient measure of intensity, since if a species is weakly adsorbed it will support a high concentration in the equilibrium solution. *Figure 1.11* shows two contrasting Q/I relationships, or isotherms. Going from R to S along A shows that the soil is able to sustain the intensity of X as uptake proceeds and the quantity (Q) in the reserve is reduced. From T to V along B a small uptake (reduction in Quantity) leads to a large fall in intensity and the soil is said to be 'weakly buffered with respect to X'.

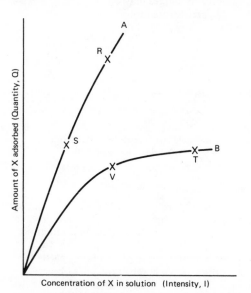

Figure 1.11 Quantity/Intensity relationships for adsorption of 'X' by two contrasting soils

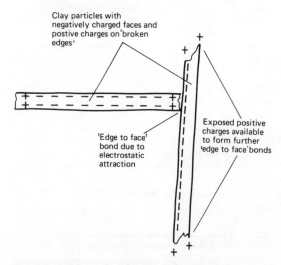

Figure 1.12 'Edge to face' bonding in flocculated clay

Flocculation and deflocculation of clays

Clays have properties that enable their particles to either repel one another, or to attract one another and form stable groupings called 'flocs'. Flocculation of clay is vitally important in agricultural soils and fortunately deflocculation rarely occurs in the clay of UK soils.

The mechanism of flocculation

A clay mineral may be visualised as a lattice structure carrying negative charges. It has already been pointed out that positive charges may occur on broken edges. If two such particles are brought together edge to face they will attract one another and many such would flocculate into larger units (*see Figure 1.12*). This mechanism is responsible for the structural stability of clay soils.

If anything happens to suppress the positive charges, flocculation is prevented. Consider, for example, a clay saturated with Na^+ ions in a strong NaCl solution (e.g. flooded with sea water). A Gouy–Chapman double layer will form where there are negative surface charges, and it will be compact due to the concentrated solution (*see Figure 1.13a*).

If the external solution becomes less concentrated (e.g. rainfall, or irrigation after flooding) the double layer expands and the clay's positive charge sites virtually become part of it and are quite definitely 'blocked' by it as shown in *Figure 1.13b*. In this condition individual clay particles repel one another

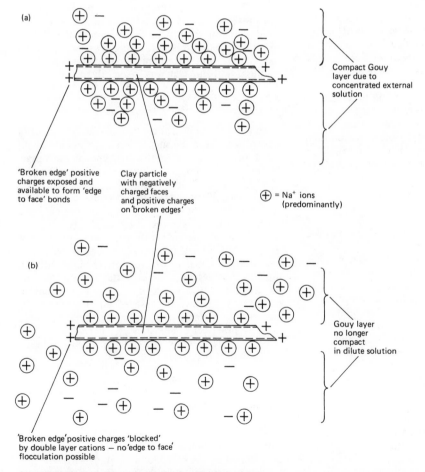

(a)

Compact Gouy layer due to concentrated external solution

'Broken edge' positive charges exposed and available to form 'edge to face' bonds

Clay particle with negatively charged faces and positive charges on 'broken edges'

$\bigoplus$ = Na^+ ions (predominantly)

(b)

Gouy layer no longer compact in dilute solution

'Broken edge' positive charges 'blocked' by double layer cations — no 'edge to face' flocculation possible

Figure 1.13 (a) 'Broken edge' charges available for bond formation when the external solution has a high Na^+ concentration. (b) 'Broken edge' charges 'blocked' by the cations of the double layer with a weak Na^+ external solution

and the clay's contribution to soil structural stability is lost.

Calcium on an exchange complex is tightly bound, Ca^{2+} ions are to be found in the Stern layer rather than in the double layer. Consequently Ca^{2+} does not prevent flocculation in the way that sodium does (Ca^{2+} forms a compact double layer system, even with a dilute external solution). Thus the clay in a well-limed soil can be relied upon to remain flocculated and therefore stable.

Prevention of deflocculation after salt water flooding

Deflocculation in salt water flooded soils is prevented by applying gypsum ($CaSO_4$) to ensure that the Na^+ is replaced by Ca^{2+} *before* the electrolyte concentration in the soil solution is allowed to fall.

Soil acidity and pH

The soil's acidity, measured on the pH scale, is of fundamental importance to farmers. When pH is too high, the availability of trace elements to crops is reduced, whilst at low pH (acid conditions) crop yields are reduced drastically, due to excessive availability of aluminium and manganese.

In formal terms, the pH of a solution is defined as the negative logarithm of the activity of H^+ ions. Soil pH values range from about 4.0 (very acid) to around 8.0. The chemist's concept of pH 7.0 being neutrality is not appropriate to soils, and pH 6.5 is the value at which a soil normally exhibits neither acidic nor alkaline properties.

Measurement of soil pH

A sample of soil is shaken with 2.5 times its own volume of demineralised water and the pH of the suspension determined with an electronic meter which relies on the effect of H^+ ions on a thin glass membrane. Alternatively a coloured dye solution which changes colour according to solution pH may be used. There are a number of difficulties relating to measurement and interpretation of soil pH.

pH is particularly variable within fields and it is important to have a representative sample (*see* p. 35).

Duplicate pH measurements at the same point may vary over a time scale of days or weeks due to changes in moisture content, soil solution concentration or carbon dioxide level in the soil atmosphere.

The presence of a Gouy–Chapman double layer means that H^+ ion concentration is not uniform throughout the whole suspension.

Thus interpretation of a pH reading is imprecise—no one can know whether a particular root environment corresponds to the external solution or something near to the composition of the Stern layer.

Large variations in the soil:water ratio affect the result obtained. On average, the pH sensitive surface of a glass electrode is nearer to the inner (Stern) layer in a more concentrated suspension and so records a lower value (most H^+ ions) than in a more dilute suspension.

An electrical junction effect due to clay particles may occur in the reference electrode system of an electronic pH meter, giving a slightly erroneous reading. This effect is most pronounced when measurements are made in thick suspensions or pastes.

Hydrogen ions are cations and therefore subject to the Gapon equation (*see* p. 17) so that their activity is controlled in part by the activity of other ions in the system. If, for example, K^+ ions are added to the solution, H^+ ions will transfer from the exchange sites into the solution, increasing its acidity (i.e. lowering the pH). When pH is measured by the water suspension method described, the cations in the external solution are derived entirely from the sample itself. The cation concentration will be governed by the previous history of the soil and the soil:water ratio. This uncertainty and its consequent effect on pH may be reduced by measuring pH in a salt extract having a concentration similar to that of a typical soil solution; 0.01 mol $CaCl_2$ is frequently used.

pH measurement using 0.01 mol CaCl₂

Results obtained by using 0.01 mol $CaCl_2$ in place of water for making the suspension are generally about 0.5–1 pH unit lower at the acid end of the range. Differences are smaller at high pHs. The Soil Survey of England and Wales routinely uses both methods and an idea of the relationship between them can be obtained from the analytical data on typical soils (*Table 1.20*). Despite the theoretical advantage of $CaCl_2$ as an extractant for pH measurement, it is not used for agricultural advisory purposes in the UK— most people working with crops are familiar with the numerical values given by the 'water method', and the advantages of $CaCl_2$ are not thought sufficient to justify any change. Because of all the sources of possible error in measurement or interpretation, there are few practical circumstances where it would be advantageous to know a pH value to an accuracy better than ± 0.2 pH units.

The effects of soil properties on activity and degradation of pesticides

Adsorption

The concentration of a pesticide in soil solution is determined by the extent and strength with which it is adsorbed by soil colloids. This in turn determines its effectiveness or toxicity to weeds or insects, its rate of degradation, leachability and uptake.

Organic matter (OM) content is the soil property

with the greatest influence on pesticide adsorption, so % OM must be taken into account when deciding the dose necessary to give the required activity for pest or weed control without causing crop damage. At one extreme, on soils of high OM level, the dose necessary to give the required effect may be prohibitively expensive. For many pesticides adsorption by peat soils may be virtually complete, so the pesticide is entirely without effect. By contrast, on some sandy soils with low OM levels, certain soil-acting herbicides cannot be used because of the risk of damage to the crop.

The bipyridyl herbicides, paraquat and diquat, are strongly and irreversibly adsorbed by clay minerals and are thereby rendered inactive on contact with the soil. However adsorption onto organic surfaces is reversible, so that bipyridyl herbicides adsorbed by peats or humose layers can subsequently be taken up, and may damage crops.

Soil texture is closely related to both the soil's colloid surface activity and its ability to retain moisture. Consequently, texture is important in determining suitability and application rates for pesticides. For these purposes soil textural classes are combined into 'groups' as shown in *Table 1.5*.

Pesticide degradation

Most pesticides in common use are degraded by microbial action. Bacteria, actinomycetes and fungi are able to utilise even chemically stable pesticides as

Table 1.5 Soil textural groups for pesticide recommendation purposes

Textural class	Symbol	Textural group
Coarse sand	CS	
Sand	S	
Fine sand	FS	Sands
Very fine sand	VFS	
Loamy coarse sand	LCS	
Loamy sand	LS	
Loamy fine sand	LFS	Very light soils
Coarse sandy loam	CSL	
Loamy very fine sand	LVFS	
Sandy loam	SL	Light soils
Fine sandy loam	FSL	
Very fine sandy loam	VFSL	
Silty loam	ZyL	Medium soils
Loam	L	
Sandy clay loam	SCL	
Clay loam	CL	
Silt loam	ZL	Heavy soils
Silty clay loam	ZyCL	
Sandy clay	SC	
Clay	C	Very heavy soils
Silty clay	ZyC	

From MAFF, Agricultural Chemicals Approval Scheme 1980 with permission.

carbon sources. Typically, such microbial degradation is preceded by a 'lag phase' and this is assumed to correspond to the time taken for a population capable of performing the particular degradation to build up in the soil.

Some pesticides may be protected from microbial attack by adsorption onto soil organic colloids. Thus residue levels are frequently much greater in peaty soil than in mineral soils. Because these residues are adsorbed, they do not normally support high concentrations in the soil solutions, so are not taken up by roots.

SOIL ORGANISMS

Chemical processes in the soil can be imagined as a number of cycles. Carbon is cycled between atmospheric carbon dioxide and plant tissue, and nitrogen is fixed from its gaseous form, incorporated in living tissue then released again to the atmosphere. Similarly, other elements are taken up by plants, become part of their structure and are eventually released to be utilised over and over again. All such cyclic processes in soil are dependent on the processes of decay and degradation brought about by living organisms.

Recently dead plant (or animal) remains are either consumed by animals, worms, slugs, etc. (the fauna) or decomposed *in situ* by the smaller micro-organisms. Fungi generally mount the initial attack on resistant organic remains (which include the excreta of the faunal population), whilst bacteria and actinomycetes require chemically less complex nutrients such as simple sugars. Algae, at least when found on or near the soil surface, carry out photosynthesis in the same way as green plants, and some are able to fix nitrogen. The smallest of the fauna, the protozoa, feed largely by capture of other micro-organisms, particularly bacteria, whilst the larger macro-fauna (e.g. earthworms) feed on plant and animal remains. They also have a mechanical effect on the soil, improving aeration and drainage as well as mixing soil constituents.

Note that the various soil organisms are not to be regarded in isolation. All compete for food; some are predatory on others. Whilst some exist together in symbiotic relationships, others exude antibiotics which may confer some competitive advantage. The guts of saprophytic invertebrates commonly support a protozoal population to facilitate cellulose decomposition, and there are numerous examples of symbiosis between micro-organisms and the plant population.

Nutrition of soil micro-organisms

Historically micro-organisms have been thought of as either autotrophs, needing no organic substances, or heterotrophs needing organic compounds of various degrees of complexity. This classification leads to confusion, for example, when dealing with bacteria *capable* of functioning without any organic substances, but only showing maximum growth when obtaining energy from oxidation of carbon compounds. Such difficulties are avoided when it is understood that nutrients have three separate functions:

(1) provide the building blocks from which the organism is formed;
(2) provide a source of energy;
(3) accept electrons released during the oxidation reaction which provides energy.

Autotrophic organisms are either chemoautotrophs, obtaining energy from inorganic oxidation reactions, or photoautotrophs carrying out a photosynthetic reaction similar to that of higher plants. As well as the complexity of their carbon source requirements, heterotrophs have differing electron acceptor requirements. In aerobes the electron acceptor is oxygen, whilst anaerobes utilise some organic products of their own metabolism or an inorganic substance.

Classification of soil micro-organisms

Classification may be according to taxonomy, morphology (form as revealed by observation) or physiology (function). In the case of soil bacteria the activity of the various species is well documented. Hence, from the viewpoint of soil studies, it is profitable to adopt a classification based on physiological criteria. When dealing with actinomycetes, fungi, algae and protozoa it is not always possible to ascribe clearly defined functions to the various species, and only taxonomic or morphological classifications are used.

Bacteria

Due to difficulties in reliably distinguishing bacteria from the rest of the soil, estimates of their population differ widely. The commonly accepted range is in the region of 10^7–10^9 organisms/g of soil, equivalent to a live weight of about 3 tonnes/ha (i.e. less than 1% of the soil organic matter).

Morphological classification distinguishes between cocci (diameter about 0.5μm × 3μm length) and rods of diverse shape, but is of little use since many bacteria are able to change from one form to another and there is no correlation between morphology and function. A more useful type of classification for heterotrophs is that proposed by Lockhead and Chase (1943), based on nutritional requirements and summarised in *Table 1.6*.

It is observed that fertile soils contain a greater proportion of groups 5, 6 and 7 (having complex

Table 1.6 Lockhead and Chase's classification of heterotrophic bacteria according to complexity of nutrient requirement

Group	Types included
1	Those growing in a simple glucose–nitrate medium containing mineral salts.
2	Those that can only grow when ten amino acids are added to this medium.
3	Those that can only grow when cysteine and seven growth factors (aneurin, biotin, etc.) are added.
4	Those needing both the amino acids and the growth factors.
5	Those needing yeast extract.
6	Those needing soil extract.
7	Those needing both yeast and soil extract.

requirements), whereas the immediate root zone of plants is characterised by the presence of groups 2, 3 and 4. Large dressings of farmyard manure have been shown to increase populations of bacteria having the more complex requirements, whereas inorganic fertilisers are without any detectable effect.

The chemoautotrophic bacteria may be similarly classified on the basis of the oxidation reaction used for energy supply.

(1) Nitrogen compounds oxidised
   (a) Ammonium oxidised to nitrite   *Nitrosomonas*
   (b) Nitrite oxidised to nitrate   *Nitrobacter*
(2) Inorganic sulphur compounds converted to sulphate   *Thiobacillus*
(3) Ferrous iron converted to ferric   *Thiobacillus ferroxidans*
(4) H_2 oxidised   Several genera

The microbial oxidation of nitrogen compounds, in particular, is vital for crop nutrition *see* p. 25.

Bacterial fixation of nitrogen

Free living bacteria notably *Azotobacter* and *Beijerinckia* are capable of fixing atmospheric nitrogen, but quantities fixed are small. The process is inefficient in terms of nitrogen fixed per unit of carbohydrate oxidised and the rate of fixation appears to be limited by the carbohydrate supply. It is not therefore surprising that when such bacteria establish themselves in close association with the roots of plants which exude carbohydrate, they perform rather better. On maize, rice and some tropical crops, such associations can give fixation rates in excess 1 kg/(ha d) of nitrogen (N). Quantities of N fixed in this way under UK conditions are minimal.

The notion, once popular in Russian literature, that N fixation could be increased by inoculation with *Azotobacter* has been shown to be fallacious—it is almost universally true that if the rate of a process is limited *only* by the size of a microbial population, that population will increase until some other factor, such as substrate availability, becomes limiting.

Symbiotic nitrogen fixation is a characteristic of legumes and a small number of non-leguminous plants. Agriculturally, interest is confined to the legumes and in the UK the crops concerned are clovers, lucerne, peas and beans. The fixation occurs in root nodules formed from plant tissue, in response to invasion by *Rhizobium* bacteria. Up to 200 kg/ha of N may be made available to a UK clover crop in this way and rather lesser amounts to peas and beans. In grass–clover swards, grass is able to benefit from the 'clover nitrogen', although the mechanism involved has not been clearly established. In such mixed swards the *Rhizobium* can only start to fix N in the spring time once clover growth has become sufficiently vigorous to provide a carbohydrate supply. The grass in the sward will thus respond to an early dressing of nitrogen. At other times (or on any 'legume only' crop) application of fertiliser nitrogen will depress *Rhizobium* activity.

It should be noted that not all strains of *Rhizobium* are able to infect particular legumes and in some cases crop yields may suffer due to the absence of a suitable strain. Use of inoculated seed is a means of guarding against this possibility.

Genetic engineering, attempting to introduce the capacity to fix nitrogen into existing agricultural crops, is frequently talked about at the present time. This may be possible in the future, although not for some years. It must be remembered that it will only be possible for N fixation to occur at the expense of photosynthesate, i.e. crop yield. At *present* crop and fertiliser prices it seems unlikely that such fixed nitrogen could compete commercially with fertiliser N.

Bacteria and soil structure

Polysaccharide gums are thought to be of great importance in stabilising soil structure (*see* p. 13). Most of these are thought to be formed as products of bacterial metabolism.

Actinomycetes

In any strict sense actinomycetes are classified within the bacteria, but on the basis of morphology and physiology they represent a transition between true bacteria and the fungi. A large number of genera of actinomycetes have been recognised in the soil. Many of these are of interest because of their ability to synthesise antibiotics and the soil has been the primary source of organisms for commercial production of these agents in the pharmaceutical industry.

In agricultural soils, abundance is in the order 10^5–10^8/g but only *Streptomyces* (70–90% of the population) and *Nocardia* (10–30%) are present in sufficient numbers to be of importance in soil studies. The nocardias are sufficiently similar to bacteria that they are frequently (and perhaps inadvertently) included in estimates of bacterial populations. Some of their number have the ability to modify their metabolism quite rapidly, to decompose synthetic chemicals.

Streptomyces are responsible for the characteristic smell of freshly turned earth. The role of actinomycetes in soil processes is still not clearly defined, but it appears that they have a lesser biochemical importance than bacteria and fungi. They have been observed to attack resistant components of plant and animal tissues some time after addition of fresh remains, presumably when competition from organisms with more fastidious nutritional requirements has declined. Their metabolites are of interest since they are very similar to the complex molecules described as 'humus' when occurring in the soil. Actinomycetes are favoured by abundance of organic materials, relatively dry soil conditions and high soil pH. Under waterlogged conditions or at pHs less than 5.0 they are virtually absent and some authorities claim that it is possible to estimate the pH of arable soil on the basis of the smell associated with *Streptomyces*.

Thermoactinomyces and some species of *Streptomyces* have an important degradative function at elevated temperatures (50–65 °C) in compost heaps.

Fungi

Fungi occur in all soils and they are commonly associated with the initial breakdown stages of organic debris. Many are saprophytes, tolerant of acidic conditions and capable of flourishing in the litter layer of forest soils. They are highly efficient at converting organic carbon into fungal tissue, even to the extent of taking nitrogen from the soil solution in order to metabolise organic matter with a large C:N ratio. Since fungal tissue is quite resistant to breakdown, N locked up in this manner may be unavailable for some time. Like the actinomycetes, fungi can incorporate N into humic-like substances which normally have a very long half-life in the soil.

Fungi are heterotrophs and require oxidisable carbon compounds as energy sources. The requirements of various species range from simple carbohydrates, sugars, organic acids and starch through cellulose and lignin (a substance resistant to attack by most other organisms) to those requiring complex chemical growth factors, or only capable of growing as parasites on living plant tissue.

The soil is a highly competitive environment and there are many organisms capable of metabolising simple sugars. Those of the fungi which have sugar as substrate maintain a competitive advantage by growing very rapidly, so consuming the sugar before it is lost to other organisms. The cellulose decomposers, Ascomycetes, Fungi Imperfecti and Basidiomycetes face less competition and have slower growth rates.

Since there is virtually no competition for lignin as a substrate, the higher Basidiomycetes which function as lignin decomposers are able to grow very slowly and appear to build up a food reserve prior to attacking any new lignin source. Apparently the initiation of decomposition of lignified material is an energy-consuming process.

Common parasitic fungi cause several well known diseases of crop plants (*see* p. 251). Some other fungi are normally saprophytic, but may occasionally invade living tissue and produce disease symptoms. The mechanisms controlling this 'occasional parasitism' are not understood, but adequately fertilised, vigorously growing crops enjoying good cultural conditions rarely seem to be affected.

Symbiotic fungi—mycorrhiza

For many years it has been appreciated that a symbiotic relationship exists between some fungi and forest trees. The commercially important relationships such as occur between Scots pine and fungi such as *Boletus, Amanita* and *Lactarius* have been carefully studied. Generally the fungus is one requiring sugar as substrate and the most abundant supplies of sugars are found in trees whose protein synthesis is limited by nitrogen or phosphate deficiency. Thus the fungus invades the root tissue of the deficient tree and then 'repays' the tree by increasing the efficiency of uptake of N and P. The relationship is in delicate balance: if the fungus is too successful supplying $N+P$ to the tree its sugar supply will be reduced, whilst if the growth of the tree is checked for any reason, the fungus may become parasitic.

At the time of writing there is considerable interest in the role of vesicular-arbuscular mycorrhiza in the nutrition of some crop plants. The commercial importance of these in UK agriculture is not immediately clear since they are only normally active in soils of low chemical fertility.

Fungi and soil structure

Although fungi lack any significant capacity to produce 'gummy' substances, some have considerable power to stabilise soil structure by the mechanical effect of their mycelia binding soil particles together.

Algae

Algae are photoautotrophs using light to provide energy and absorbing atmospheric carbon dioxide. Thus their preferred habitat is the soil surface. Strictly, many soil algae are facultative photoautotrophs having the ability to metabolise some carbohydrates, and significant numbers are found in sub-surface habitats. Since algae only have any recognisable competitive advantage when exposed to light at the soil surface, the significance of these sub-surface populations is not understood. Algae form a relatively small part of the soil biomass, typically 10–300 kg/ha (but up to 1500 kg/ha where a visible surface bloom occurs). When growing at the surface, multiplication of algae is only limited by lack of light, moisture or inorganic nutrients and competition for other factors does not affect them. Thus an algal bloom on soil is indicative primarily of moist surface conditions and a lack of crop competition for light. Algae have no significance in UK agriculture, but the blue-green algae in particular, are important as nitrogen fixers in rice paddies. Algae and lichens (symbiotic alga + fungus combinations) are important as primary colonisers of bare rock and they provide the sole source of organic matter in embryonic soils.

Protozoa

The Protozoa are the smallest of the soil fauna, distinguishable from all the organisms previously described by having means of locomotion. They are very small, 10–80 μm in length, to enable them to move in soil water films. Nutrition may be photoautotrophic (e.g. *Euglena*) and some may exist as saprophytes. However, the major mode of nutrition is by predation on bacteria, small algae, yeasts, etc. The estimation of protozoal populations is extremely difficult, but estimates of low precision are acceptable since their sole claim to agricultural significance is their role in reducing the numbers of the organisms that constitute their prey. 'Order of magnitude' calculations indicate that the weight of the protozoal population is about 1% of the weight of bacteria in the soil.

Other animal life

The fauna is capable of near infinite sub-division. For the present purpose it merely needs to be pointed out that anything that eats, be it elephant or earthworm, is participating in the chain of events whereby vegetation is transformed to yield soil organic matter and plant nutrients. Those animals which have their being within the soil mass, rather than above it, fulfil the additional function of physically mixing the soil constituents. In this context, earthworms are worthy of particular mention. They ingest organic remains and excrete an intimate mixture of soil and fresh 'humus', whilst leaving behind them channels which provide easy access for plant root growth, and improve both aeration and drainage. Where earthworms thrive, they dominate the faunal population, exceeding the sum of all other fauna by up to five-fold, on a weight basis. The three million or so worms which may be found under one hectare of deciduous forest weigh about two tonnes. Where earthworms do not dominate the fauna, total faunal populations are normally much smaller and the soil appears 'lacking in life' when examined in a profile section.

THE MAJOR NUTRIENT ELEMENTS

These are nitrogen (N), phosphorus (P), potassium (K), magnesium (Mg) and sulphur (S). They are present in soils and are taken up by plants in relatively large quantities. Calcium (Ca) is also a major element, but is dealt with separately (p. 30) because of its importance in relation to liming.

The major elements have a role in building the structure of plants, whereas the micronutrients (*see* p. 33) are important in enzyme systems and contribute to the plant's function rather than its structure.

Nutrient offtake and nutrient cycles

The quantities of the major nutrients removed by normal yields of crops are given in *Table 1.7*. (These figures should not be used to calculate fertiliser applications, *see* p. 98.)

According to the type of farming, a part, or nearly all of the crop's major element uptake may eventually be returned to the soil in plant remains or animal excrement.

Nitrogen

Nitrogen exists in the soil either in the inorganic form, as ammonium (NH_4^+) and nitrate (NO_3^-) ions, or as organic compounds.

Inorganic nitrogen (also called mineral nitrogen)

This is the form of nitrogen used by crops. Nitrate (NO_3^-) is the main form taken up by plants, and ammonium (NH_4^+) is quickly converted in the soil to nitrate by the action of autotrophic bacteria:

$$NH_4^+ \xrightarrow{Nitrosomonas} NO_2^- \text{ (Nitrite)}$$

$$NO_2^- \xrightarrow{Nitrobacter} NO_3^- \text{ (Nitrate)}$$

These reactions can only occur under aerobic conditions. Since the second is potentially able to proceed faster than the first, NO_2^- does not normally accumulate in soils.

Nitrate ions are not held by soil colloids, so if not taken up by roots, nitrate is 'leached' by rainfall and lost in drainage water. Ammonium ions, however, are adsorbed by negatively charged exchange sites and so are held against leaching. Adsorption does not provide protection against oxidation to nitrate, so adsorbed ammonium is readily lost from the soil when conditions (temperature and aeration) are suitable for nitrification. Thus, with the exception of the driest parts of eastern England, it may be reliably assumed that between one autumn and the following spring, virtually all the inorganic nitrogen in a soil will be leached from the rooting zone unless taken up by a crop.

Levels of total mineral N ($NH_4^+ + NO_3^-$) in soils fluctuate widely over short time periods, so although it is a relatively simple matter to determine total mineral N in the laboratory, the result is of no use for determining fertiliser application rates. Levels normally fall in the range 5–50 mg/ℓ (mg of N/ℓ of soil), corresponding to about 10–100 kg/ha of N. The highest levels are frequently found under heavily fertilised grassland at the end of the growing season, before the onset of winter rains.

Organic nitrogen

Soil reserves of organic nitrogen are large. A soil with an OM level of 5% would have an organic N content of about 0–3% to 6 tonnes/ha of N. This is some 60–600 times larger than the total mineral N.

The N in organic compounds is not available to plants until it has been released by the decomposition of organic matter. The fraction of the soil OM most actively involved in decomposition reactions contains most of its N in the form of proteins, and many microorganisms are able to break these down to amino acids and then to ammonium ions, a process known as ammonification. In more general terms the conversion of nitrogen from organic to inorganic forms is referred to as mineralisation.

Table 1.7 Major nutrient removals by normal yields of common crops

	Wheat		Barley		Turnip roots	Potatoes tubers	Meadow hay
	Grain	Straw	Grain	Straw			
Yield (kg/ha, harvested)	7000	4000	5500	3300	30000	30000	5000
Total offtake (kg/ha) of							
Nitrogen (N)	145	23	100	22	40	63	86
Phosphorus (P)	34	3	20	4	9	15	9
Potassium (K)	36	34	25	40	82	94	73
Calcium (Ca)	3	7	3	12	16	3	36
Magnesium (Mg)	18	4	5	3	4	6	16
Sulphur (S)	8	6	8	7	12	5	14

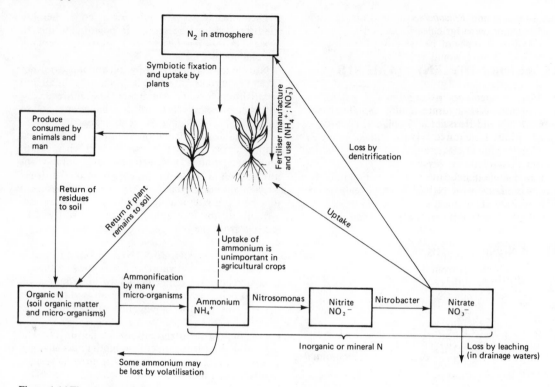

Figure 1.14 The nitrogen cycle

Since the small amount of mineral N is derived from the large heterogeneous reserves of organic N by biological degradative processes, it is not surprising that mineral N levels are highly variable. In general, soils with a high total N content are likely to need less N fertiliser to achieve maximum yields than those with low N levels, but the relationship is not good enough for predictive purposes. A 'total N' determination measures the reservoir of nitrogen potentially available to be mineralised, but tells nothing of the rate at which the micro-organisms are likely to carry out the mineralisation.

Nitrogen in crop residues and organic manures is not available to crops until it has been converted to ammonium. Subsequent conversion to nitrate is normally rapid and build-up of ammonium N only occurs under anaerobic (waterlogged) conditions. The ratio of C:N in the organic matter of agricultural soils in temperate lands is remarkably constant at between 10:1 and 12:1. If a manure or crop residue has a ratio less than this (i.e. more N relative to C) microbial action will release N quite rapidly. Materials with a large C:N ratio, such as cereal straw, require an additional source of N for their decomposition. In the absence of any other source, micro-organisms decomposing high C:N ratio materials will remove mineral N from the soil, frequently in competition with growing crops. To prevent such adverse effects it is

good practice to apply about 10 kg of N for every tonne of cereal straw incorporated into the soil. Note that N utilised by micro-organisms in decomposing organic residues of large C:N ratio is not lost—it is released again when the carbon is used up and the microbial biomass decays.

The nitrogen cycle

A diagrammatic representation of the N cycle appears in *Figure 1.14*.

The total nitrogen in the earth's atmosphere is estimated at about 4×10^{15} tonnes. It exists as molecular N_2 (di-nitrogen—the gas that makes up about four-fifths of the atmosphere). This is only available to plants after *fixation*, a process which involves large inputs of energy.

The Haber–Bosch process is used industrially. N_2 and H_2 are reacted together at high temperature and pressure to form ammonia, which is the precursor of modern nitrogen fertilisers.

Seventy MJ of energy is needed for each kg of fertiliser N produced (1 kg oil = about 45 MJ).

Total world fertiliser N use is of the order of 40 million tonnes of N (1974 figure). The corresponding UK figure was 520 000 tonnes. UK consumption in 1978/79 was 640 000 tonnes.

Microbial N fixation involving the symbiosis between a legume and *Rhizobium* bacteria can make a useful contribution to agricultural production.

Peas and most types of beans are grown without any fertiliser N and would not respond if it were given. A grass/clover ley without N yields about as much dry matter as an all grass sward given 120 kg/ha of N. This nitrogen is not 'free' to the farmer since the *Rhizobium* uses carbohydrate that would otherwise contribute to yield. Russell (1973) states that 15–20 g of carbohydrate is required for the fixation of 1 g of N. Thus 100 kg/ha of clover N is obtained at the cost of about 1500–2000 kg of dry matter (20 000 MJ energy equivalent if made into silage). The production of the same amount of N by the fertiliser industry would use 7000 MJ of energy.

Soil OM breakdown typically provides about 100 kg/(ha annum) of N to agricultural crops. The quantity is greater on soils with higher OM levels and where the OM is of recent origin. On cultivated fen peat soils the N supply from the soil organic matter is great enough that little or no fertiliser N may be needed for cereal crops.

Nitrogen is lost from the soil by leaching and denitrification. Leaching losses may be determined experimentally by collecting drainage water from large confined cylinders of soil called lysimeters. Over winter most N (including any that is mineralised during the winter) is lost by leaching. In the growing season, leaching is greatest in areas of high rainfall. Only small quantities of N are lost from grass crops but shallower and less vigorously rooting arable crops allow more N to be leached. Losses are greatest under fallow and are particularly large when land used for heavily manured early crops is fallowed through the summer, as is common practice in the Scilly Isles and Channel Islands.

Denitrification occurs when soils are waterlogged whilst biological activity is high. Most of the soil micro-organisms are aerobes utilising oxygen as final electron acceptor. Under anaerobic conditions many are able to utilise NO_3^- reducing it to nitrous oxide (N_2O) or di-nitrogen (N_2). Both of these are gases and are therefore immediately lost.

Waterlogging in winter does not cause anaerobic conditions to develop rapidly, since biological activity is low, so winter waterlogging does not lead to denitrification.

The denitrification process is difficult to study experimentally but it is usually assumed that N losses are large during short periods of waterlogging in the growing season.

Phosphorus (P)

Phosphorus is unusual amongst the major nutrient elements in that concentrations in the soil solution are frequently extremely low, in the order of 1 μg/ml, but it is taken up by crops in large amounts. This is explained by the replenishment of soil solution phosphate from solid phase reserves.

The soil's reserves of phosphate

In addition to phosphate in solution it is possible to recognise by experiment two fairly distinct pools of phosphate.

Labile phosphate

This is held on colloid surfaces but remains in rapid equilibrium with soil solution phosphate. ^{32}P labelled phosphate added to the soil solution is found to exchange readily with the non-labelled P in this fraction.

Insoluble phosphate

This part of the soil's reserve of P is not readily available to crops, but it can release P slowly into the 'labile pool' and conversely labile pool P may be 'fixed'.

Phosphate fixation and fertiliser placement

Phosphate fixation is the process whereby soil solution/labile pool P is rendered insoluble by reactions occurring at colloid surfaces. Under suitable conditions phosphate reacts with calcium surfaces, to form layers having properties similar to the almost insoluble calcium phosphate minerals. Under more acidic conditions, iron and aluminium surfaces predominate, and ligand exchange reactions occur (*see* p. 18). It appears that phosphate films on iron surfaces are stable, but phosphate adsorbed into aluminium surfaces is able to move into the solid phase and become part of a crystal structure, leaving a fresh aluminium surface for further adsorption of phosphate.

Older texts on soils and fertilisers place great emphasis on fixation of applied phosphate fertiliser. Thus, it was said that phosphate in cereal seedbeds should be placed (i.e. combine drilled) near to the seed, to prevent fixation throughout the whole bulk of the soil. Similarly, a granule, which gave rise to a discrete volume of high P availability, into which roots could grow, was preferable to a powder. Further, it was argued that since water soluble P quickly became part of the insoluble phosphate reserves if not taken up immediately by plants, the same effect could be achieved at a lesser cost by use of insoluble phosphate fertilisers.

In advanced agricultural countries such as the UK, there is now a sufficient history of phosphate fertiliser use and phosphate import in animal feeds (much of which eventually finds its way into the soil in animal manures) that the most powerful sites of phosphate fixation have already been satisfied, and newly applied P is fixed less strongly than was the case in years gone by. Placement of phosphate thus assumes a lesser importance and should only be regarded as desirable where soil phosphate reserves are known to be low.

Quantity/intensity isotherms for phosphate (*see* p. 18) have a characteristic shape such that at low solution concentrations it is difficult to increase the concentration because any P added is rapidly adsorbed. At higher concentrations this adsorption is less pronounced and thus less phosphate needs to be added to the system to bring about a further increase. This effect is illustrated by curve B, *Figure 1.11.*

The interpretation of phosphate isotherms in terms of adsorption mechanisms is the subject of active debate in the soil science literature (*see*, for example, Posner and Bowden, 1980 and Ryden, McLaughlin and Syers, 1977). It is not clear whether the changes of slope of isotherms represent transitions from one type of adsorption site to another, or whether the whole curve can be interpreted in terms of a single model.

Phosphate in soil organic matter

Organic matter contains phosphate and so its decay eventually liberates phosphate into the soil. This phosphate contributes to the labile pool and is, of course, liable to fixation. Where soil phosphate levels are very low, release of P from OM assumes great importance; the growing plant may be regarded as competing with unsatisfied fixation capacity for any available phosphate, and organic matter decomposition favours the plant by releasing P during periods of high biological activity, when the plant's needs are likely to be greatest. The organic fraction thus protects phosphate from fixation in these circumstances.

A schematic representation of soil phosphate relationships is given in *Figure 1.15*. No crops grown in the UK normally remove more than 70 kg/ha of P_2O_5 per annum and most take considerably less. Thus fertilisers used at the rates recommended by ADAS (*see* p. 98) are more than sufficient to replace offtake.

Phosphate in the soil solution, even in fertile soils, amounts to only 1–5 kg/ha of P_2O_5. Since a rapidly growing crop can take up about 3 kg/(ha d), the solution phosphate must be capable of rapid replenishment, more than once per day in some cases. The size of the labile pool is largely governed by previous fertiliser history, but for most UK soils it is in the order of several hundreds of kg/ha.

Soil OM has a P:C ratio of 1:100–1:200, so that the organic P in a soil of 5% OM represents a reserve of up to 1700 kg/ha of P_2O_5 and a small proportion of this is made available each year as OM is mineralised.

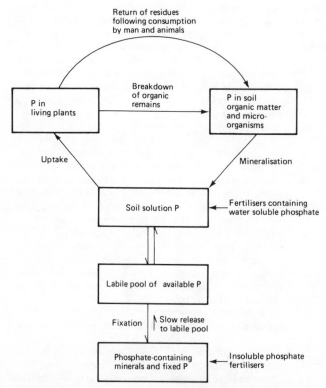

Figure 1.15 Schematic representation of soil phosphate relationships

Potassium (K)

In the absence of applied fertilisers, K in the soil is derived from weathering of minerals in which potassium is either an integral part of the mineral structure, as in the feldspars, or forms a tightly held interlayer as in illite (*see* p. 10). Potassium thus released becomes part of the soil's reserve of exchangeable K.

Exchangeable K

The K^+ cation may in theory occupy any exchange site, satisfying the negative charge on a soil colloid and maintaining an equilibrium with other soil solution cations as required by the Gapon equation (*see* p. 17). Experiments have shown that two different Gapon coefficients must be used to account for the behaviour of exchangeable K.

Weakly held K

Gapon coefficient 2.2 (mmol ℓ)/2 relative to Mg^{2+}. This is the K adsorbed onto external layer silicate surfaces and other sites which are not specific for K.

Strongly held K

Gapon coefficient 102 (mmol ℓ)/2. This is the K adsorbed at the edges of exposed interlayers in illite-type structures and these sites are specific for K^+, since larger ions could only be accommodated by a separation of the layer structure.

Potash fixation

To a degree, the weathering processes may be considered as slowly reversible, and K^+ in the soil solution may diffuse into the interlayer of vermiculite-type 2:1 minerals. Eventually, if sufficient K^+ enters the interlayer (or more correctly a part of an interlayer), it may revert to an illite-type structure with 1 nm basal spacing, and the K is then 'fixed'. This is sometimes represented as involving a third type of exchange site (with infinite Gapon coefficient), but there is little justification for use of the term exchange since, in the absence of any weathering, the reaction is irreversible.

Schematic representation of soil K relationships

Potassium is peculiar amongst crop nutrients in that although it is retained in the soil in an available form (*see Figure 1.16*), annual crop uptake may be quite large in relation to soil reserves. Hence the soil 'K index' may change rapidly and this is particularly so under intensive grassland. For example, grass DM may contain up to 3% K (3.6% K_2O) so that a conserved crop of 10 tonnes/ha dry matter could remove 360 kg/ha of K. Two or three years of removal at this rate would easily take a sandy or silty soil from

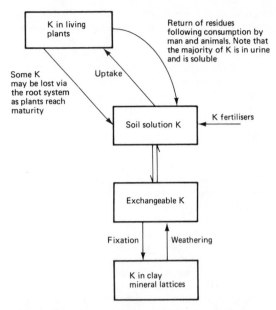

Figure 1.16 Schematic representations of soil potassium relationships

ADAS K index 2, down to index 0. Clay soils are able to release K from clay minerals so their reserves are not so easily reduced, and for crops which are not responsive to high levels of K it is good practice to keep 'available K' at only a moderate level. Maximum benefit can then be obtained from clay mineral derived K.

Sodium in soils

Sodium may contribute to the nutrition of some crop plants, and is occasionally used as a fertiliser. In the quantities used for this purpose the behaviour of Na^+ in the soil may be considered analogous to K^+. At the higher concentrations which may be achieved in laboratory experiments, or occur in soils which have been flooded by sea water, sodium in the exchange complex brings about very profound changes (*see* p. 19).

If a fertiliser dressing of salt (sodium chloride) remains in the surface few cm of an unstable soil it can produce surface capping and reduced germination. Autumn application of salt, and subsequent cultivation, both reduce this risk.

Magnesium (Mg)

Like potassium, magnesium is a constituent of the minerals of the clay fraction and is released by weathering. Water soluble and exchangeable forms are available to plants, but Mg^{2+} is relatively easily lost from the soil by leaching. In clayey soils, weathering can usually be relied on to keep pace with magnesium, but sandy soils are frequently deficient. Where a potential deficiency is recognised, it is normal practice to apply Mg to the soil by use of liming materials containing magnesium, and rates of addition can be quite large. Mg^{2+} typically accounts for 10–20% of the soil's cation exchange capacity, and contributes to the control of soil pH in the same way as calcium.

Sulphur (S)

Although crop offtakes of sulphur are of a similar order of magnitude to phosphate offtakes, there has been little interest in soil sulphur until quite recently. Sulphur has been described as the forgotten element. For many years it has been applied unwittingly as an impurity in commercial fertiliser, particularly superphosphate (12%S) and ammonium sulphate (24%S). These two materials are not constituents of modern compound fertilisers and no appreciable quantities of S now come from inorganic fertilisers.

It is thought that a large proportion of present day crop S need is being provided by S in rainfall (i.e. atmospheric pollution from industrial processes and combustion of fossil fuel). Typically, inputs from this source in industrialised countries are in the range 10–20 kg/ha, greater near to industrial centres, and possibly considerably less in more remote areas. Sea spray carried by wind is another source of S, and in the UK at present it appears that either proximity to industry or to the sea ensures adequate supply of S everywhere, but this may not always be the case.

In Southern Ireland, levels of atmospheric sulphur are low (probably the lowest in Europe) and there are recent reports (Hanley and Murphy, 1970; Murphy, 1975) of yield responses of grass and clover to added S at inland sites.

Sulphur in the soil occurs in both organic and inorganic forms, but organic S provides the major reserve. Soil C to S ratios are about 100:1.

Approximate calculations assuming a soil OM level of 5% and a mean turnover time for OM of 20 years, show that annual mineralisation from OM provides about 30 kg/ha of S.

Hydrogen sulphide (H_2S) is produced during microbial breakdown of organic matter. Under aerobic conditions this is oxidised (autoxidation) to $SO_4{}^{2-}$, but in anaerobic soils it is oxidised to elemental S by the bacteria *Beggiatoa* and *Thiothrix*. These bacteria and *Thiobacillus*, in aerobic soil, can also oxidise S to H_2SO_4, the mechanism whereby additions of sulphur can be utilised to lower soil pH.

CALCIUM AND LIMING

On the basis of its role in plant nutrition, calcium (Ca) should be included amongst the list of major nutrient elements, but for practical agricultural purposes it is hardly ever considered as a nutrient at all. This is because calcium deficiency is unknown to agriculture (it occurs occasionally in horticultural crops grown in small containers of peat compost). Calcium has two distinct functions:

(1) an essential nutrient for plants;
(2) the dominant ion in determining soil pH.

The second of these is the critical one, and depletion of soil calcium is always evidenced by acidity damage to the crop, long before the level is low enough for Ca deficiency to occur.

Calcium in the soil

Calcium exists in water soluble form in soils, in equilibrium with exchangeable Ca^{2+} held on soil colloids. It also occurs in solid form as $CaCO_3$ and as a constituent of clay minerals.

Levels of calcium in the soil solution are governed by chemical equilibria between solution, exchangeable and solid forms. Weathering of primary Ca containing minerals and dissolution of added lime ($CaCO_3$), is greatly assisted by 'carbonic acid'—CO_2 (produced by respiration of organisms), dissolved in water.

$$CO_2 + H_2O \rightarrow H_2CO_3$$

$CaCO_3$ is only very slightly soluble (about 0.3 mmol) but in the presence of carbonic acid:

$$CaCO_3 + H_2CO_3 \rightarrow Ca(HCO_3) \text{ (calcium bicarbonate)}$$

This bicarbonate is much more soluble than $CaCO_3$.

Exchangeable calcium is the dominant ion on soil colloid surfaces and its function of replacing H^+ ions from exchange sites governs the H^+ concentration in the soil solution. It therefore determines soil pH.

Particles of calcium carbonate ($CaCO_3$) in soil may be of natural origin, derived from weathering of limestone, chalk or chalky boulder clay, or they may arise from use of agricultural lime.

Feldspars, amphiboles and calcium phosphates of the clay fraction release Ca on weathering, and constitute the main source of Ca in soils other than those formed from chalk and limestone parent materials.

Lime in the soil

Wherever rainfall exceeds evapotranspiration, a steady downward movement of water through the soil removes (leaches) cations from the soil exchange complex. In its simplest form this reaction may be represented as follows.

$$\underset{\text{clay surface}}{Ca^{2+}} + 2H_2CO_3 \rightarrow 2Ca(HCO_3)_2 + \underset{\text{clay surface}}{H^+ H^+}$$

The calcium remains in solution as Ca^{2+} ions and is lost in drainage water. Eventually Ca (and other cations) are lost, and H^+ takes their place on exchange sites, i.e. the soil becomes acid.

This is a natural tendency and, under UK weather conditions, all soils whose parent materials are incapable of releasing cations from the clay fraction at an adequate rate, eventually become acid unless

(1) lime is applied,
(2) the soil is subject to flushing by water of a high base status,
(3) bases are re-cycled by deep rooting trees (as leaf fall) or added in the excrement of animals.

pH and liming

The acidity of soils is expressed on the pH scale. *Table 4.7* shows the pH requirements of various crops grown on mineral soils (slightly lower values are acceptable on peats).

Determination of pH may be by use of either an electronic meter (*see* p. 20) or a coloured dye solution called indicator. Whichever method is chosen, taking the sample from the field is more likely to lead to significant error than is any shortcoming in 'technique'. Careful attention should be paid to points mentioned on p. 35. When using indicator on field moist soil, mixing and sub-sampling is not generally feasible and the correct procedure is to examine *each core* with indicator and take an average value as the pH of the field.

The lime requirement is the amount of lime (expressed as $CaCO_3$) needed to restore the pH to 6.5 for arable crops, or 6.0 for grassland. Although the lime requirement is expressed in tonnes/ha of ground limestone ($CaCO_3$), equivalent quantities of any other lime may be used and the magnesium in magnesium limestones is also able to replace H^+ ions. It therefore has a liming effect similar to that of Ca^{2+}. Lime requirement is not directly related to pH and in the laboratory it is determined as a separate item. In general, for any given pH, clay soils and soils with a high organic matter will have higher lime requirements than sandy or low OM soils. In the absence of a laboratory lime requirement determination, lime requirements may be estimated by use of information in *Table 1.8*. This method is most likely to be adequate where a farmer knows his land by previous experience and pH levels are not very low. Where pH levels are below about 5.5 and previous liming history is unknown, it is unwise to attempt to estimate lime requirement.

Overliming can lead to serious financial losses. At high pH values trace element availability is drastically reduced. Deficiency of boron in brassicas is frequently induced by over-zealous use of lime. For the dairy farmer, too much lime can lead to low herbage manganese levels and this may contribute to problems

Table 1.8 Lime requirement related to soil texture (after Batey, unpublished)

pH Value	I	II	III	IV	V
6.0	2.5	3.1	3.8	5.0	6.2
5.5	5.0	6.2	8.2	10.0	12.5
5.0	7.5	9.4	12.0	15.0	18.0
4.5	10.0	12.0	16.3	20.0	24.0
4.0	12.5	15.0	20.0	25.0	30.0
3.5	15.0	18.0	24.0	30.0	36.0

Lime requirement in tonnes/ha of $CaCO_3$.
  I. All loamy sands.
 II. All sandy loams.
III. Loams, silty loam, all organic sandy loams.
IV. All clays loams, all organic loams and silt loams, peaty loams.
 V. All clays, organic clay loams, peat.

of infertility in the herd. Most livestock are susceptible to deficiency of copper and this condition also may be aggravated by overliming.

In cereal crops, deficiencies of manganese and copper which remain untreated may lead to serious yield loss. Overliming is the most frequent cause of these deficiencies. Iron deficiency is also induced by overliming, but no major agricultural crops are affected. Note that lime-induced deficiencies are common where the pH has been increased by liming at high application rates; problems are less widespread on soils where pH is naturally high.

Loss of lime from the soil

Calcium is removed from the soil whenever crops are harvested, since it is a constituent of all plant tissue. Amounts removed by individual crops are given in *Table 1.7. Table 1.9* shows lime removals associated with four different cropping systems in terms of kg/ha $CaCO_3$ removed. Note that these amounts are small by comparison to normal rates of liming and crop removal does not make a major contribution to lime loss.

Loss in drainage water occurs all the time due to the effects of slightly acidic rainwater (containing small quantities of dissolved CO_2, and SO_2). The amount lost depends on rainfall, soil texture and the amount of calcium in the soil. Losses are estimated to range from 100 to 1000 kg/ha of calcium carbonate annually.

Since lime moves downwards in drainage water, it is quite possible for the surface soil to become acid, irrespective of the nature of the subsoil. The movement of drainage water tends to be uneven and some areas of soil lose lime more rapidly than others; this may lead to variations in acidity and hence the first observable effects of acidity in a crop are often patchy.

Use of fertilisers affects loss of lime from the soil. Nitrogen fertilisers have an acidifying effect whenever ammonium nitrogen is converted in the soil to nitrate nitrogen. Urea breaks down to ammonium salt in the soil and thereafter behaves similarly. Aqueous ammonia and anhydrous ammonia increase acidity in the same way. Experimentally determined values for

Table 1.9 Lime removals associated with four different farming systems

Farming system		Calcium carbonate removed/annum (kg/ha)
Arable	70 ha cereals, 20 ha sugar beet, 10 ha potatoes ⎰ all straw retained	14
	⎱ all straw sold	24
Mixed	30 ha cereals, all straw retained, 10 ha cash roots, 60 ha temporary grass, milk, calves and culled stock sold from 110 milking cows, plus followers	44
Dairy	All grass; milk, calves and culled stock from herd of 160 cows plus followers	56
Grass drying	All grass, very high nitrogen fertiliser usage; all herbage sold	220

From MAFF Reference book No. 35 'Lime and Liming' with permission.

the amount of lime ($CaCO_3$) needed to counter the effect of 1 kg of nitrogen range from 1.4 to 2.9 kg.

Some ammonium nitrate fertiliser is sold mixed with lime. Research in the Netherlands has shown that 100 kg of an ammonium nitrate–lime compound containing 20.5% N has a liming effect equivalent to 8 kg $CaCO_3$. A product of this type containing 23% N has only a slight effect on soil pH whilst the 26% N product commonly sold in the UK may be expected to have a slight acidifying effect.

Potassium fertilisers have an acidifying effect on the soil, but it is much less than that due to nitrogen, and in practice can be ignored.

Phosphate fertilisers do not cause any lime loss from the soil. Basic slag contains lime, and any phosphate fertiliser containing basic slag will act as a liming material.

Normal dressings of farmyard manure and slurries have very little, if any, effect on soil acidity.

Soil texture and organic matter level influence the retention of lime. Light, sandy soils have the lowest capacity for holding calcium, and, as they are usually freely drained, it is rapidly lost. On these soils troubles from acidity are most common and most acute, but easily remedied. Since comparatively small amounts of lime will raise the exchangeable calcium content towards saturation point, small dressings generally suffice to restore very acid sandy soils to neutrality. By contrast, clay soils or those with high organic matter levels hold very much larger quantities of calcium. This calcium is tightly held and many years may be required for its removal. When liming becomes necessary, heavier dressings are needed.

Lime and soil structure

It is widely held by farmers that liming improves the structure of heavy soils, reduces stickiness, lightens cultivations and makes it easier to break down clods and obtain a satisfactory tilth. In the days of horse cultivations, it was often said that after very heavy applications of burnt lime, 'four-horse land' became 'three-horse land' or even better. In experiments it has been observed that soil structure can be improved on clay soils by heavy applications of lime, to give a pH of over 7. It is also well known that naturally calcareous clays are generally much less intractable and better drained than naturally acid clays. However, it is usually inadvisable to purposely overlime, because nutrient deficiencies may be induced.

Other beneficial effects of maintaining the optimum pH are indirect. The addition of lime to an acid soil increases the activity of the carbon cycle bacteria and enhanced decomposition of organic matter assists soil crumb formation and increases soil structural stability.

TRACE ELEMENTS AND MICRONUTRIENTS

Elements that occur in very small amounts in soils are called trace elements. A sub-set of these, the micronutrients, are essential for plant growth.

Toxicity

Most trace elements (including micronutrients) are toxic to plants at high concentrations; they are also toxic to animals, but poisoning of farm livestock is usually attributable to causes other than the consumption of home-grown feedingstuffs having high tissue levels of trace elements.

Occurrence

Table 1.10 gives median trace element contents of soils in England and Wales (*see* p. 36 for explanation of the term 'extractable').

Table 1.10 Median trace element content of soils in England and Wales (total values expressed as mg/kg; extractable as mg/ℓ air dry soil) (from Archer, 1980)

Element	Median	Range	Number farms	Number samples
TOTAL				
Boron	33	7–71	72	227
Cadmium	<1.0	0.08–10	204	689
Cobalt	8	<1.0–40	125	421
Copper	17	1.8–195	226	751
Nickel	26	4.4–228	226	752
Lead	42	5–1200	226	752
Zinc	77	5–816	225	748
Selenium	0.6	0.2–1.8	34	114
Mercury	0.04	0.008–0.19	17	53
ACETIC ACID EXTRACTABLE				
Nickel	1.0	0.12–22.7	198	647
Zinc	6.6	0.4–97.6	204	664
EDTA EXTRACTABLE				
Copper	4.4	0.5–74.0	201	662
Selenium	0.04	0.01–0.59	32	112
HOT WATER EXTRACTABLE				
Boron	1.0	0.1–4.7	153	493

From F. C. Archer. In *Inorganic pollution and agriculture* Table 1, p. 184–190 MAFF Reference Book No 326. London: HMSO with permission.

Since trace elements are largely bound in mineral lattices and released by weathering, the parent material very largely determines the abundance of individual elements in soil. Copper, manganese, zinc, cobalt, nickel and lead are associated with the ferromagnesian minerals of basic rocks. Boron is derived from the weathering of the mineral tourmaline. Soils derived from unmineralised granite and coarse grained sandstones tend to have low levels of most trace elements; levels are high on serpentine derived soils such as those of the Cornish Lizard Peninsula.

Trace elements not essential for plant growth

Iodine and selenium (which are essential to animals), arsenic (the chemistry of which has similarities to phosphorus in the soil) and a host of 'heavy metals' such as cadmium, chromium, nickel, tin and lead are not required by plants. Interest in the heavy metals stems largely from their occurrence in sewage sludges originating from urban areas. Because of the possibility of heavy metal contamination, professional advice should be sought before applying sewage sludge to farm land (*see* DOE, 1977).

The essential micronutrients

Copper, zinc, manganese, boron, molybdenum and iron are essential for plants. Deficiencies in agricultural crops may be due to:

(1) an absolute lack of the element (e.g. soils derived from granite). This type of deficiency is common on peats where the current soils surface is too far removed from weathering minerals for replenishment by weathering to be effective;
(2) inadequate availability, frequently due to overliming or adverse soil physical conditions.

Information on individual elements is summarised in *Table 1.11*.

Table 1.11 Summary of information on micronutrients

Element	Forms in the soil	Remarks	Threshold values[1] for susceptible crops (mg/ℓ)
Copper (Cu)	Clay mineral lattices Organic compounds Exchangeable Cation Cu^{2+} In soil solution	Soil solution concentrations are low. Cu occurs in clay mineral lattices but this form is not available. Cu^{2+} is tightly held by inorganic colloids and this form is not easily available. Complexes with soil organic matter largely control availability to plants. This is why the chelating agent EDTA is frequently used to determine 'available' copper. Availability is reduced by excessive liming.	0–1.6 deficiency likely 1.6–2.4 slight deficiency likely 2.5–4.0 deficiency unlikely >4.0 adequate For chalk soils deficiency levels are related to organic matter content OM>6% deficiency possible when Cu>2.5 OM 0–5% deficiency possible when Cu>1.0
Zinc (Zn)	Substituted for Fe^{2+} and Mg^{2+} (isomorphous replacement) in ferromagnesian minerals. Exchangeable Zn^{2+} Organic complexes	Zinc deficiency is virtually unknown in the UK.	—

[1] These values relate to analysis by standard 'available' methods.

Table 1.11—*cont.*

Element	Forms in the soil	Remarks	Threshold values[1] for susceptible crops (mg/ℓ)
Manganese (Mn)	Clay mineral lattices Nodules and manganese containing iron pans Exchangeable Mn^{2+} and oxides of trivalent and tetravalent Mn	Mn^{2+} is taken up by plants and this is the form stable under reducing conditions. Mn^{2+} stability decreases 100-fold for each one unit increase of pH, and formation of Mn-organic complexes unavailable to plants is greatest at high pH. Thus Mn availability is governed by soil reaction (pH) and aeration, more than by the presence of 'adequate quantities of Mn'. Deficiency is frequently related to weather conditions, being common during warm periods following cold spells. Lack of consolidation of the seedbed predisposes a crop to Mn deficiency. Chemical determination of manganese is useless as a predictor for deficiency.	—
Boron (B)	Contained in tourmaline and may substitute for Al or Si in clay minerals. Adsorbed onto soil colloid surfaces.	Adsorption of B by soil colloids decreases as pH is lowered. Availability to plants is minimal at high pH values, hence B deficiency can be induced by liming. The range between deficiency and toxicity is very small.	< 0.5 severe deficiency likely 0.6–1.0 severe deficiency possible 1.1–2.0 satisfactory 2.1–4.0 high > 4.1 toxicity possible
Molybdenum (Mo)	Adsorbed on clay mineral surfaces. As calcium molybdate. May also occur in organic forms and as hydrated Mo oxides	Occurring in anionic form Mo behaves differently to other micronutrients. Availability is high at high pHs and in soils derived from high Mo sediments (Teart soils). Toxicity in livestock (Mo induced copper deficiency) is consequent upon use of lime. Deficiency does not normally occur unless pH is below 6.0, and is rectified by liming.	In relation to possible deficiency in crops Low 0.1 Satisfactory 0.2 In relation to possible Mo induced copper deficiency in stock Normal 4 Induced deficiency possible 4–8 Induced deficiency likely > 8
Iron (Fe)	Ubiquitous in soils—an essential constituent of ferromagnesian minerals	Soluble Fe levels are very low compared with the total Fe present. Except where microbiological activity under anaerobic conditions leads to reduction to Fe^{2+}, iron availability in the soils is governed by the solubilities of Fe^{3+} hydrous oxides. Oxide precipitation in strongly pH dependent and Fe^{3+} activity in solution decreases 1000-fold for every unit increase in pH. Thus deficiency occurs on soils of high pH and where lime use has been excessive. Iron deficiency does not occur in agricultural crops in UK but some fruit trees are seriously affected.	—

[1] These values relate to analysis by standard 'available' methods.

SOIL SAMPLING, SOIL ANALYSIS AND SOIL NUTRIENT INDICES

The methods of calculating fertiliser requirements for common UK crops are described on p. 98. Knowledge of the soil's 'index' of P, K and Mg is basic to accurate and reliable calculations. Soil pH and lime requirements must also be known if liming policy is to have a rational basis.

Soil sampling

Where any chemical analysis is employed as a basis for reaching a decision, the decision can only be as good as the sample submitted to the analyst.

The essence of a good soil sample is that it is representative of the whole situation. In order to achieve this, account must be taken of soil variability.

Variability and sampling error

Variability is a fundamental property of biological materials and thus constitutes a potential major source of error when sampling.

Variability in soils has been studied extensively. If a large number of soil cores were taken from a 'uniform' field and analysed individually, a spread of results would be obtained. More useful information may be obtained by mixing the soil cores (sub-samples) and then doing a single analysis. Such a 'composite sample' may be described as representative if there is a high mathematical probability that another similar sample obtained in the same way would give similar results. Experiments have shown that 25 cores or sub-samples will constitute a representative sample provided that the following instructions are followed carefully.

Method of taking a soil sample

The area to be sampled should be examined carefully and, if necessary, sub-divisions made to accommodate areas which are not uniform in crop growth, soil drainage, soil texture or colour, or areas which have been cropped, limed or fertilised differently, and areas of marked variations in acidity and alkalinity (which can be delineated by use of soil indicator).

Soil should be taken from at least 25 spots in the field. To ensure even distribution of the sample spots, follow a 'W' shaped path as illustrated in *Figure 1.17*.

The sampling technique must be appropriate to the particular crops being grown.

Arable soils and short-term leys (including permanent grass about to be re-seeded) should be sampled to a depth of 15 cm using a screw auger (approx. 25 mm diameter), tubular corer, or a cheese-type corer for very dry powdery soils. The 25 cores obtained may be put straight into the sample bag to give an acceptable size of sample.

Permanent grass and long leys (unless they are soon to be ploughed up) should be sampled to a depth of 7.5 cm, preferably with a tubular corer, a cheese-type corer or pot corer. The core should be of uniform diameter and must include the surface mat or turf. This requirement cannot be met using a screw auger. The 25 cores obtained may be put straight into the sample bag to give an acceptable size of sample.

Direct drilled crops require special attention. When sampling either grassland or the stubble of a previously direct drilled crop, prior to direct drilling, it is a wise precaution to make a separate examination of the surface 2–3 cm of soil. This avoids the possibility of undetected acidity in the immediate surface, adversely affecting crop establishment. It may best be done using soil indicator, but where doubt still remains a separate sample to 2–3 cm should be taken. Be especially careful to detect possible patches of acidity in such cases.

Avoidance of contamination

Soil samples for chemical analysis should not be put

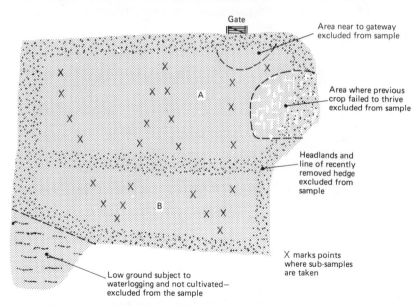

Note — If the soils of parts A and B are known to be different, or the sampler believes them to have been farmed differently in the recent past, two separate samples should be obtained.

Figure 1.17 Path to be followed when soil sampling a field

into old fertiliser bags. Even bags which have been washed will usually contain sufficient phosphate to ensure that any analytical result is completely erroneous.

Samples for trace element analysis must be carefully protected from possibility of contamination by sealing in new polythene bags. Such samples should not come into contact with copper, brass or other metals which might interfere with the analysis.

Samples for special purposes

The foregoing sections refer specifically to samples to be used as a basis for making fertiliser recommendations. For other purposes (e.g. diagnosis of a crop disorder or determination of a herbicide residue), requirements may be different. Instructions for taking samples in most circumstances likely to be met by farmers have been given by Hughes (1979).

Soil analysis

On receipt at the laboratory, it is normal practice to air-dry soil samples, crush them mechanically to separate soil aggregates without breaking stones, and then retain as 'fine earth' all the soil which will pass a 2 mm sieve. This may be sub-sampled and analysed to ascertain pH, lime requirement and a variety of nutrient levels. Elements may be determined either as 'total' or 'available'.

Determination of 'total' element levels

Because of the occurrence of the major nutrient elements as constituents of mineral lattices, their total levels in soil have no bearing on potential for crop growth. Total levels of major nutrients only need to be known by those studying soil formation processes.

In the case of trace elements, knowledge of total levels can be of use in predicting possible deficiency situations, and where use of different techniques for 'available' analysis would otherwise invalidate comparisons.

Determination of 'available' element levels

For most elements, the amount available to plants comprises that which is water soluble, together with some exchangeable and probably also some insoluble form (which may come into solution or enter the exchange complex during the crop's growth). To date it has not proved possible to draw any clear distinction between such 'available nutrients' and the non-available remainder. Because of this, developments in soil analysis have relied on the fact that even if chemists don't know what is available to plants, the plants themselves do. Over the years, a number of

empirical procedures for determining available nutrients have been tested by comparison with crop performance in field experiments.

Those which correlate best with crop performance have been retained and are considered to be standard techniques within the UK. Different techniques are frequently used in other countries or for special purposes in UK. Whilst various techniques have their protagonists, it should always be remembered that, owing to the empirical nature of 'available nutrient' analysis, the procedure for which the largest and most reliable range of interpretive information is available is likely to be of greatest use. Brief details of standard ADAS procedures are given in *Table 1.12*.

The methods are published in full by MAFF (1973).

Determination of nitrogen in soil

There is no satisfactory chemical means of determining the nitrogen likely to become available to a crop during its growth and so chemical analysis is not normally of use for determining rates of application of nitrogen.

The nutrient index system

The nitrogen index

ADAS advice on nitrogen fertiliser use is based on the Nitrogen Index. This is calculated from knowledge of previous cropping and automatically makes allowances for the likely contributions of crop residues to the supply of available nitrogen under normal circumstances.

How to determine the nitrogen index of a soil

(1) If lucerne, long leys (three or more years) or permanent pasture have not featured in the last five years cropping, use *Table 1.13* only.
(2) If the last crop grown was lucerne, long ley or permanent pasture, use *Table 1.14* only (use the 'Years since ploughing out' column 0 only).
(3) If one of the above crops has been grown in the last five years, but was not the last crop, look up the values from both *Table 1.13* and *1.14*. The higher of the two values found is the Nitrogen Index.

Phosphorus, potassium and magnesium indices

Indices for these elements are derived from analytical data as shown in *Table 1.15*.

Interpretation of results of soil micronutrient analyses

Soil analysis may be used to predict likely deficiency of boron, molybdenum and copper in crops. The data in *Table 1.11* is helpful in interpreting these results. Soil analysis may also be used to assist in diagnosis of

Table 1.12 Summary of standard ADAS laboratory methods for analysis of soils

Determinand	Extraction procedure	Analytical finish
Available phosphorus	Shake for 30 min with 0.05 mol sodium bicarbonate solution at pH 8.5, containing polyacrylamide.	Spectrophotometric (880 nm). Phosphomolybdate formed when acid ammonium molybdate reacts with phosphate, is reduced by ascorbic acid to a blue coloured complex.
Available potassium	Shake for 30 min with 1 mol ammonium nitrate.	Flame photometry on filtrate.
Available magnesium	As above	Atomic absorption spectrophotometry at 285 nm using a strontium chloride releasing agent.
Available copper	Shake for 1 h with 0.05 mol EDTA diammonium salt at pH 4.0.	Atomic absorption spectrophotometry at 325 nm using calcium nitrate compensation in standard solutions.
Water soluble boron	Boil for 5 min with water (detailed instructions must be followed).	Spectrophotometric, as a methylene blue complex with fluroborate (in 1, 2-dichloroethane extract) after removal of nitrates and nitrites.
Available molybdenum	Shake for 16 h with ammonium oxalate/oxalic acid solution.	Spectrophotometric, as a complex formed by Mo, iron and thiocyanate in the presence of a reducing agent.
pH	pH measurement is described on p. 20.	
Lime requirement (LR)	Uses the same suspension as pH determination. LR is calculated from the new equilibrium pH attained when *para*-nitrophenol buffer is added to the extract.	

Table 1.13 Nitrogen Index—based on last crop grown (from MAFF, 1979)

Last crop	Nitrogen index
Any crop receiving farmyard manure or slurry	1
Any crop receiving large frequent dressings of farmyard manure or slurry	2
Beans	1
Cereals	0
Forage crops, removed	0
Forage crops, grazed	1
Leys (1–2 year), cut	0
Leys (1–2 year), grazed, low N[1]	0
Leys (1–2 year), grazed, high N[2]	1
Maize	0
Oil seed rape	1
Peas	1
Potatoes	1
Sugar beet, tops ploughed in	1
Sugar beet, tops removed	0
Vegetables receiving less than 200 kg/ha N	0
Vegetables (intensive) receiving more than 200 kg/ha N	1

From Lime and Fertiliser recommendations—booklet 2191. Pinner, Middlesex: MAFF/ADAS with permission.
[1] Low N = less than 250 kg/ha N per year or low clover content.
[2] High N = more than 250 kg/ha N per year or high clover content.

Table 1.14 Nitrogen Index—based on past cropping with lucerne, long leys and permanent pasture (from MAFF, 1979)

Crop	Years since 'ploughing out'				
	0	1	2	3	4
	Nitrogen index				
Lucerne	2	2	1	0	0
Long leys low N[1]	1	1	0	0	0
Long leys high N[2]	2	2	1	0	0
Permanent pasture—poor quality					
Permanent pasture—matted	0	0	0	0	0
Permanent pasture—average	2	2	1	1	0
Permanent pasture—high N[2]	2	2	2	1	1

From Lime and Fertiliser recommendations—booklet 2191. Pinner, Middlesex: MAFF/ADAS with permission.
[1] Low N = less than 250 kg/ha N per year or low clover content.
[2] High N = more than 250 kg/ha N per year or high clover content.

Table 1.15 Soil indices for phosphorus, potassium and magnesium (from MAFF, 1979)

Index	Phosphorus (mg/ℓ)	Potassium (mg/ℓ)	Magnesium (mg/ℓ)
0	0– 9	0– 60	0– 25
1	10– 15	61– 120	26– 50
2	16– 25	121– 240	51– 100
3	26– 45	245– 400	101– 175
4	46– 70	405– 600	176– 250
5	71–100	605– 900	255– 350
6	101–140	905–1500	355– 600
7	141–200	1510–2400	610–1000
8	205–280	2410–3600	1010–1500
9	over 280	over 3600	over 1500

From *Fertiliser Recommendations* publication GF1 MAFF 2nd Edition (1979). London: HMSO with permission.

deficiencies of copper and cobalt in grazing livestock, but the interpretation of the results obtained is not straightforward.

Levels of manganese (Mn) found by analysis of soil do not correlate with Mn uptake by crops. Soil analysis for Mn is of no use to farmers, since results obtained are incapable of interpretation.

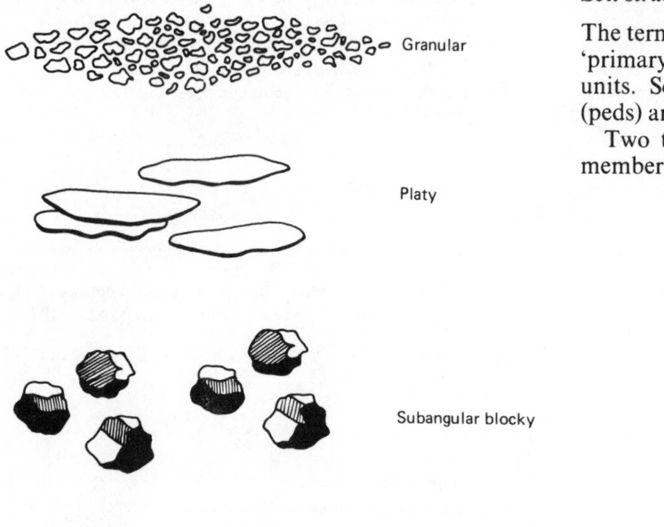

PHYSICAL PROPERTIES OF THE SOIL

Soil physical properties affect crops as follows:

(1) directly—soil structure, porosity, aeration, moisture holding capacity are part of the root environment;
(2) indirectly—since soil physical properties determine suitability and timing of cultivations.

Soil structure, air and water

Soil structure

The term 'soil structure' relates to the arrangement of 'primary particles' of sand, silt and clay into ordered units. Some examples of types of structural units (peds) are illustrated in *Figure 1.18*.

Two types of structure may be regarded as end members of a continuum.

Granular

Platy

Subangular blocky

Angular blocky

Prismatic
(similar, but with rounded upper
surfaces are described as 'columnar')

Figure 1.18 Some examples of common types of structural units (after *Modern Farming and the Soil*, 1970, Agricultural Advisory Council, MAFF. London: HMSO, reproduced with permission)

Single grain structure (of sandy soils) is exemplified by dune sand and the bulk of the soil is not organised in any perceptible manner. Roots may grow through the intergranular spaces and these spaces afford adequate aeration. Water holding capacity is very low, since the only zones where water can be retained against gravity are adjacent to points of contact of the individual grains.

Massive structure is the opposite extreme. In this case the soil constituents are tightly packed together to form a continuous mass with no visible planes of weakness. Soils of widely differing textural composition are capable of assuming a massive structure, usually in response to ill-timed trafficking or cultivation.

Soils with massive structure do not allow development of root systems, and although they may contain considerable water within their small pore spaces it is very tightly held by capillary forces and not available to plants (*see* p. 41).

Granular, subangular blocky, angular blocky and prismatic structures represent the middle range of the continuum and have features conducive to biological activity and crop growth. These structures are further subdivided on the basis of size (*see Table 1.16*).

The stability of structure is an important agricultural property. The ideal soil for crop production has a highly stable structure not easily destroyed by cultivation, water movement or treading by livestock. Structural stability is favoured by high organic matter levels, loamy or clayey soil texture, and high levels of calcium. Silty soils with low levels of organic matter are particularly unstable.

Adequate aeration with rapid drainage through the large spaces between the 'peds', (the name for the individual structural unit) and water retention in the smaller voids, is provided by a fine granular structure. Since roots are readily able to explore the spaces between peds they extract both water and nutrients from the whole volume of the soil.

Aeration, drainage and ability to provide water and nutrients to plants generally decrease as structural units become larger, and the regular-faced angular blocky and prismatic units are less favourable to root activity than the irregular surfaces of granular peds.

Structure of subsoils

Most crops in the UK remove water from the soil to a depth of at least 1 m, and the deeper-rooted ones noted for drought resistance (e.g. lucerne), to 2 m or more. It is thus important that the subsoil provides a satisfactory rooting environment. Adequate cracks and fissures must be present for roots to grow through, and to allow free movement of water. Many subsoils in their natural states have prismatic structures which are perfectly adequate, provided that the spaces between prisms are large enough.

Usually, in subsoils, the faces of one ped more or less fit those of its neighbours. When the fit is good, changes in the width of fissures between peds, as a result of shrinking and swelling, become important.

Narrow fissures may close completely on swelling, causing impeded drainage. This is evidenced by polished ped surfaces indicating contact between adjacent peds, and the mottled colours associated with drainage impedence.

Subangular and columnar structures are frequently of significance, since the absence of the more pronounced angular features of prismatic and blocky peds may indicate a degree of instability, with the possibility of displaced material flowing between the peds and causing serious drainage impedence. A transition from angular blocky or prismatic to this more rounded type of ped is usually indicative of transient waterlogging and the cause should be sought and rectified, since the effect is a self-perpetuating one.

Platy structures are particularly detrimental, since roots can only pass through them by very tortuous paths, and they restrict water movement. In arable soils a platy layer can quickly develop into a serious

Table 1.16 Size classification of soil structural units (after SSEW, 1974)

	Platy[1]	Prismatic[2]	Angular blocky[3]	Subangular blocky[3]	Granular[3]
Fine	Fine platy; < 2 mm	Fine prismatic; < 20 mm	Fine angular blocky; < 10 mm	Fine subangular blocky; < 10 mm	Fine granular; < 2 mm
Medium	Medium platy; 2 to 5 mm	Medium prismatic; 20 to 50 mm	Medium angular blocky; 10 to 20 mm	Medium subangular blocky; 10 to 20 mm	Medium granular; 2 to 5 mm
Coarse	Coarse platy; 5 to 10 mm	Coarse prismatic; 50 to 100 mm	Coarse angular blocky; 20 to 50 mm	Coarse subangular blocky; 20 to 50 mm	Coarse granular; 5 to 10 mm
Very coarse	Very coarse platy; > 10 mm	Very coarse prismatic; > 100 mm	Very coarse angular blocky; > 50 mm	Very coarse subangular blocky; > 50 mm	Very coarse granular; > 10 mm

From Soil Survey Technical Monograph No 5 *Soil Survey Field Handbook* (by Hodgson J. M.) Harpenden 1974 (Table 3, p. 31) by courtesy of the Soil Survey of England and Wales, Rothamsted Experimental Station, Harpenden
[1] Thickness of plate.
[2] Diameter in the horizontal plane.
[3] Diameter.

pan, since soil above the platy layer will be subject to transient waterlogging and is therefore likely to 'slake' into its constituent fine particles. These move downwards with drainage water and progressively build up a massive impermeable layer.

Platy structures are rare in unfarmed soils as they are formed by loading with machinery, or the smearing action of cultivation equipment.

Timely and correctly executed cultivations should always lead to an overall improvement of the structural state of a soil profile.

Recognition of poor soil structure

Soil conditions can only be studied by examination of a soil profile. A drainage trench or a pit purposely excavated by machine provide ideal opportunities for soil profile inspection. Much additional information is often obtained by using a spade to dig a rectangular pit to a depth of 45 cm, or deeper if interest centres on possible 'natural' subsoil conditions, as well as those produced by agricultural practices. In the writer's experience this is best done during the months of April and May when it is possible to follow the distribution of roots of actively growing crops.

The following points should be observed.

Poor or weedy patches in the crop, often accompanied by a hard and rutted or very cloddy surface layer, indicate structural problems.

Overgrown, blocked or infilled ditches indicate that the problem may be primarily due to lack of maintenance of an artificial drainage system.

The plough layer should be friable and of uniform colour. On arable land, rusty staining on root channels indicates waterlogging (some rusty staining is normal under grass on heavy soils) and any mottling (colour variation) on structure faces indicates transient waterlogging. Man- (or machine-) made clods are very dense, not able to be explored by roots and they do not subdivide along natural planes of weakness. Their occurrence is clear evidence of inappropriate or ill-timed cultivation.

The base of the plough layer should merge gradually into subsoil. Any abrupt change is easily identified with the spade and a platy structured layer, or plough pan, is recognised as 'difficult digging'. Such layers are best examined by gently disturbing the soil in an exposed profile, using a penknife. The stronger platy layer will then stand out from the rest, and be clearly visible.

The subsoil should have adequate large pores and fissures. In heavy textured soils, ped faces should be examined for evidence of polishing due to adjacent units rubbing together. Mottled colours on structural surfaces indicate transient waterlogging. When a ped is broken open by hand, the exposed internal surface should be of a similar colour to the structure face and either living roots, or the channels of dead roots and worms should be visible. A uniform grey internal colour and absence of visible channels is indicative of waterlogged conditions. Beware of relying on colour criteria when examining soils developed on strongly coloured parent materials such as Kauper marl.

Symptoms of waterlogging may be due to either a soil structural problem or the need for artificial drainage, and it is often difficult to identify the primary cause. If symptoms are found within 50 cm of the surface, it is always worth examining the subsoil to a greater depth; if no symptoms are found in any deeper horizon the problem is almost certainly one of soil structure.

Soil air

That part of the total soil porosity which is not occupied by water contains the soil atmosphere. This differs from 'air' in two main ways.

(1) Its carbon dioxide level is many times higher.
(2) Its composition varies.

The magnitude of the variation of composition of soil air is shown in *Figure 1.19* which gives data obtained from a dunged wheat plot on the classic Broadbalk field at Rothamsted in 1913 and 1914.

The importance of oxygen in the soil

Most of the soil micro-organisms which are beneficial to agricultural crops (e.g. those involved in organic matter production, and the bacteria of the nitrogen cycle) require aerobic conditions. When conditions are anaerobic (lacking oxygen) the following occur:

(1) normal oxidative decomposition of carbohydrate gives way to fermentation and putrefaction. Organic acids and ethylene may accumulate and seriously damage crop roots.
(2) denitrification occurs. Nitrate (NO_3) is reduced microbially to N_2O or N_2 which are lost to the atmosphere (*see* p. 27),
(3) other inorganic species are reduced (*see* p. 22).

Aeration and soil structure

The main mechanisms whereby oxygen is supplied to sites of biological activity in the soil, and carbon dioxide removed, is diffusion through soil pores. Diffusion is restricted when:

(1) pores are few in number,
(2) pores are small,
(3) pores are very tortuous or not continuous,
(4) pores are filled with water.

In arable soils particularly, such limitations on diffusion often ensure that the internal zones of soil peds remain anaerobic. Since such zones are not

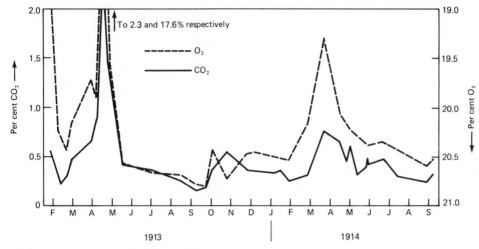

Figure 1.19 The oxygen and carbon dioxide content of the soil air in the dunged plot of Broadbalk under wheat (after Russell, E. W., 1973, *Soil Conditions and Plant Growth*, p. 411. London: Longmans, reproduced with permission of the author and publishers)

normally explored by roots this may be of little consequence, apart from being a waste of rooting volume. However, any nitrate in such anaerobic zones is likely to be lost by denitrification.

Cultivations are often undertaken 'to improve soil aeration'. Whilst any movement of soil clearly increases the flow of air to the surface of structural units that are disturbed, cultivations frequently do nothing to improve air movement *within* structural units, and problems may be exacerbated by the production of dense impermeable clods.

The major means that a farmer has at his disposal for improving soil aeration is the improvement of drainage.

The use of water by crops, soil water reserves and irrigation

Water and crop growth

In order to 'fix' atmospheric carbon by photosynthesis, plants need to maintain their stomata in the open position so that carbon dioxide gas can be taken up. As a consequence, water vapour is lost by diffusion in the opposite direction. Since water is vital to a plant's well-being, in times of stress it protects itself from further water loss by closing stomata. As well as preventing further (potentially catastrophic) loss of water, this stops photosynthesis. Inhibition of photosynthesis through drought normally occurs during bright sunny weather when the potential for photosynthesis is at its greatest, so crop loss may be considerable.

The efficiency with which crops utilise water depends on many factors and for grass it ranges from

1 kg of dry matter produced per 1000 kg of water transpired (low N input) to 1 kg per 400 kg of water with high nitrogen use. Efficiencies quoted for other UK crops generally fall within this range.

Water in the soil

Unless a soil is so wet that water will drain freely from it by gravity, water may only be removed from the soil by the expenditure of energy. Work must be done to remove water, and the drier the soil the more work must be done to remove a given quantity of water. Conversely, when water is added to dry soil it is distributed within the soil so as to maximise its attraction to the soil; the system achieves a maximum reduction of free energy, and this may be measured as the soil water potential, ψ.

The soil water potential (ψ)

At any point in the soil this may be defined as the energy required to transfer to that point, 1 mole of water from a free liquid state at the soil surface, and free from dissolved solutes. Since such a transfer would release energy, ψ is always negative.

Units of measurement

The most appropriate units for ψ, on grounds of theoretical chemistry, would be joules per mole of water (J/mol). In practice the normal unit is J/m^3 since values expressed in this unit are numerically equal to pressures expressed in Newtons per square metre (N/m^2). Any pressure or suction may be expressed in terms of the height of a water column, usually measured in cm. For soil suctions it is

Table 1.17 Water potential, hydraulic head and pF of a soil of varying wetness (after White, 1979)

Soil moisture condition	Water potential (ψ) (bars)	Head (m)	pF	Equivalent radius of largest pores that would just hold water (μm)
Moisture held after free drainage (field capacity)	−0.05	0.51	1.7	30
Approximately the moisture content at which plants wilt	−15	153	4.2	0.1
Soil at equilibrium with a relative vapour pressure of 0.85 (approaching air-dryness)	−220	2244	5.4	0.007

From White, R. E. *Introduction to the principles and practice of Soil Science* (1969) p. 68, by courtesy of Blackwell Scientific Publications, Oxford

traditional to avoid large numbers by use of the term pF when $pF = \log_{10} h$ (h measure in cm). A convenient unit for measurement is the bar (10^5 N/m^2) which approximates to the pressure exerted by a column of water 10 m high. The equivalence of the various units is indicated in *Table 1.17*.

When water is progressively added to dry soil, initial increments are adsorbed onto soil colloid surfaces (*see* p. 17). Subsequent additions are held in the soil pores by forces of attraction. The finest pores are filled first and coarser ones at later stages. The equivalent radius of largest pores that would just hold water are given in *Table 1.17*.

Field capacity is the moisture state at which free drainage just ceases following thorough wetting of the soil. The concept is strictly applicable only to freely draining soils.

Plant available water has a potential between −0.05 bars and −15 bars (field capacity—permanent wilting point).

Available water capacity (*AWC*) is a soil's capacity to store plant available water.

It may be expressed as a percentage of soil volume (Av%) or, more usefully when it is intended to perform computations involving both available water and rainfall—mm/m depth of soil (mm/30 cm is also used since 30 cm represents a reasonable depth of plough layer). To determine the AWC of a soil profile it is necessary to sum the individual values for identified soil horizons to the maximum depth of rooting. *Table 1.18* provides all the necessary information. Typical values for common soil series are given in *Table 1.19*.

Water uptake by crops

Water is not removed from soils in a uniform manner. During spring growth uptake is normally from the surface horizon, and appreciable amounts of water start to be removed by roots at greater depth as the work done in extracting the water from surface soil increases. Where a water table exists within the depth of a profile, root systems may use this water, provided that they can get near enough to it. A water table is of no use to crops if its depth varies during the growing season in such a way as to cause transient waterlogging within the rooting zone.

The ease of uptake of water is governed by the water potential, ψ. Water in coarse pores in the topsoil is more readily available than water held tightly by fine pores deep in the subsoil.

Potential transpiration (PT), is the amount of water lost from a complete canopy of a uniform green crop adequately supplied with water. It is not easily measured but may be calculated from readily available meteorological data. The method for doing this is described in MAFF (1967). Monthly values for England and Wales are tabulated in MAFF (1976) and range from 0 (January in NE England) to 100 mm+ (June and July in SE England).

Current values are calculated daily by the UK Met. Office and published as weekly summaries.

Actual transpiration is less easily determined, since a real crop may transpire either at the potential rate, or at some lesser rate due to:

(1) incomplete crop cover,
(2) crop water use reduced by water stress.

Calculation of irrigation need

In order to plan an irrigation programme and calculate how much water to apply and when to apply it, a water balance must be constructed. The following information is needed.

The amount of water stored in the soil at the outset. For simplicity this is recorded as a soil moisture deficit, SMD (i.e. the amount already lost). Initially SMD may be taken as zero (i.e. soil at field capacity), or some other value calculated from historic data. Note that undisturbed bare soils are unable to transpire more than 25–30 nm of water, irrespective of weather conditions.

Table 1.18 Average available water (Av) per cent for mineral soils in relation to horizon, particle size, class and packing density (after SSEW, 1977)

| Particle-size class[3] | Horizon | $Av(\%)$[1] for different packing density classes[2] | | |
		Low (<1.40 g/cm³)	Medium (1.40–1.75 g/cm³)	High (>1.75 g/cm³)
Clay	A	22	18	(19)
	E,B,C	(19)	15	13
Sandy clay	A	—	—	—
	E,B,C	—	—	14
Silty clay	A	23	17	—
	E,B,C	—	16	12
Sandy clay loam	A	(25)	17	—
	E,B,C	—	17	16
Clay loam	A	25	20	(17)
	E,B,C	19	15	12
Silty clay loam	A	27	20	—
	E,B,C	21	17	12
Silt loam	A	—	(25)	—
	E,B,C	—	(22)	—
Sandy silt loam	A	23	21	—
	E,B,C	20	18[4]	—
Sandy loam	A	20	16[4]	(19)
	E,B,C	(20)	15[4]	11[4]
Loamy sand	A	13	14	—
	E,B,C	(16)	12[4]	(8)
Sand	A	—	—	—
	E,B,C	—	9[4]	(4)

From Soil Survey Technical Monograph No 9 *Water retention, porosity and density of field soils*, by D. G. M. Hall, M. J. Reeve, A. J. Thomasson and V. F. Wright, by courtesy of The Soil Survey of England and Wales, Rothamsted Experimental Station, Harpenden

[1] 10 mm/m = 1%.
[2] Packing density is derived from bulk density as follows: Packing density = Bulk density $+0.009$ (%₀ clay). Packing density is considered easier to estimate in the field than bulk density.
[3] Particle-size classes are determined by use of the triangular diagram of *Figure 1.3*.
[4] If mainly fine sand add 2%: if the majority of the fine sand is of 60–100 μm grade Av will be 20%₀ except for sandy silt loam (about 30%).
Brackets ( ) indicate limited data.
A dash indicates insufficient information.

A record of rainfall and irrigation inputs of water. Rainfall *must* be measured 'on site'—figures obtained as near as 1 km away can lead to major errors. Similarly, adequate metering of irrigation water is essential, since errors are cumulative throughout the season.

A figure for potential evaporation is needed in order to know how soon an SMD will reach some critical value if irrigation is not applied. In practice two figures are needed:

(1) An average daily rate of potential transpiration. This is taken from MAFF (1976). Since potential transpiration values change only gradually from place to place and do not vary excessively between successive dry days, this figure may be used for daily calculations.

(2) At the end of a 7-d period a weekly potential transpiration figure is provided by ADAS and the UK Meteorological Office, based on daily climatological measurements. This figure is used to apply a correction to the previous 7-d calculations.

A strategy is needed to take account of departures of actual transpiration from the potential rate, due, for example, to incomplete crop cover. Also threshold SMD values at which irrigation should be applied must be selected on the basis of soil and crop.

Detailed recommendation for various crops will be found in ADAS (1979). Differing programmes are given for soils of low, medium and high AWC (less than 60 mm, 60–100 mm and greater than 100 mm within 500 mm depth of soil, respectively).

A soil water balance is conveniently kept by daily updating of a simple graph. A sheet of squared paper is prepared as in *Figure 1.20* by ruling lines to represent the long-term average PT for each month. Each dry day the line is extended parallel to the pre-ruled line, and on wet days rainfall is entered as a decrease in SMD. Until the crop cover reaches 15% water loss from bare ground sets an upper limit of 25 mm SMD. Between 15 and 75% crop cover record keeping is exactly as described above (irrigation is counted as rainfall). After 75% ground cover is attained the long-term average PT figures cease to be sufficiently reliable and each week the calculations

Table 1.19 Mean values of available water for some common soil series (after SSEW, 1977)

Soil series	Depth of integration (to 100 cm or to rock if shallower) (cm)	Total available water (0.05–15 bar) (mm)	Easily available water (0.05–2 bar) (mm)
Aberford	60	110	70
Ardington	100	150	100
Bardsey	100	130	80
Bridgnorth	80	85–105[1]	65–90[1]
Bromsgrove	80	105	75
Clifton	100	150	100
Crewe	100	140	75
Denchworth	100	160	82
Eardiston	60	105	70
Elmton	30	65	40
Flint	100	120	70
Gresham	100	140	80
Marcham	35	45	30
Newport	100	75–150[1]	65–130[1]
Oak	100	130	65
Ragdale	100	130	75
Ross	100	155	110
Salop	100	135	80
Sherborne	35	55	30
Spetchley	100	120	65
Tickenham	100	165	120
Urchfont	100	195	160
Wick-Arrow-Quorndon	100	100–200[1]	70–160[1]
Wilcocks	100	215	135
Worcester	100	115	60

From Soil Survey Technical Monograph No 9 *Water retention, porosity and density of field soils*: Appendix II (p. 75) by D. G. M. Hall, M. J. Reeve, A. J. Thomasson and V. F. Wright by courtesy of The Soil Survey of England and Wales, Rothamsted Experimental Station, Harpenden
[1] Varies greatly according to content of 60–100 μm sand and surface organic matter levels.

should, if necessary, be corrected to allow for any discrepancy between the *previous weeks* ADAS/Met. Office PT figure and the long-term average. (Note that this correction must be done retrospectively since a very large SMD could build up before the Met. office figure becomes available.) Note that in these calculations it is assumed that irrigation will be applied when needed. If a large SMD is allowed to build up transpiration rates need to be adjusted to allow for lower transpiration by crops under stress.

Figure 1.20 gives an example of a water balance record using fictitious data designed to illustrate the various steps involved. The crop was potatoes (maincrop) grown on a Class B (medium AWC) soil in South Devon (Agroclimatic Area 43S). A 15% crop cover was attained on 2 June and 75% on 24 June. The irrigation plan was 25 mm applied at a 35 mm SMD. The following notes are numbered to correspond with *Figure 1.20*:

(1) Although the first 20 d in April were dry the SMD did not exceed 25 mm since this figure represents the maximum loss under bare ground conditions.
(2) SMD returned to zero on 28 April. When this

occurs it can only depart from zero following a day in which PT exceeds rainfall.
(3) An irrigation requirement arose in the period 3–9 June, but no water was applied since the crop was not thought to have reached a responsive stage.
(4) Irrigation applied 27 June. Note that 10 mm SMD remained unsatisfied. If the entire irrigation need had been applied, the rainfall on 29 June would have been wasted.
(5) On 13 July information was received that the actual PT for the previous week was 5 mm less than average. An adjustment was made on 13 July to compensate. A similar situation occurred on 16 August.
(6) The planned deficit was reached during the last week in August, but irrigation was delayed since the crop was near to harvest and rainfall following irrigation could have led to difficult harvest conditions.

Soil and cultivations

Traditionally there have been many reasons for cultivating soils, with weed control and burial of trash amongst the most important. Now that herbicides can take care of much of the weed problem, the reasons

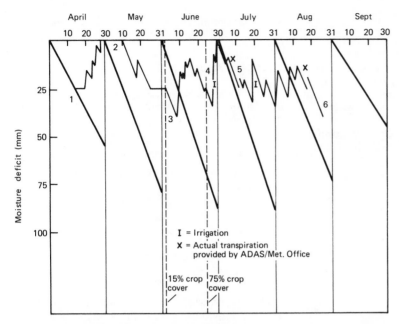

Figure 1.20 Soil water balance for maincrop potatoes grown in a class B soil in South Devon (Agroclimatic area 43 S)

for continuing to cultivate are to do with soil structure and rooting conditions. Cultivations should normally:

(1) reduce the bulk density (mass per unit volume) of soil, thereby increasing the proportion of pore spaces (particularly large pores) able to contribute to water movement and aeration;

(2) break up any soil pans allowing greater root penetration;

(3) break up clods to form a fine tilth. This increases the proportion of the soil volume which can be explored by roots.

The success or otherwise of cultivation operations is strongly influenced by soil moisture content.

The lower plastic limit is the minimum moisture content at which puddling is possible and the maximum moisture at which the soil is friable. It is usually defined as the moisture content at which the soil can be rolled into a 'worm' about 3 mm diameter without breaking. The optimal working range for most implements is below the lower plastic limit. Smearing and soil compaction (loss of structure) will occur if cultivations are carried out when the soil is too wet.

It is important to estimate the working properties of soils at the depth at which an implement will work. Surface soils dry very rapidly in windy weather, but below plough depth the subsoil is likely to remain too moist to support the weight of a tractor without compacting.

Primary cultivations

Ploughing inverts the soil and buries trash. In so doing the surface structure is disrupted and under good conditions the ploughed soil may need little additional disturbance to form a seedbed.

When soil is too wet, the furrow slice is smeared by the mouldboard and clods are formed. The tractor wheel running in the furrow bottom causes smearing and compaction. On heavy soils the use of crawler tractors running on unploughed ground avoids this problem.

Ploughing year after year at the same depth will eventually cause soil compaction in most soils.

Use of special shallow ploughs working to only 12–20 cm depth can achieve burial of trash without such a great risk of causing soil structural damage under poor conditions.

Powered digging machines at present under development show promise of being able to work satisfactorily under moist conditions where ploughing would not be desirable.

Chisel ploughs and tine cultivators disturb the upper layer of soil without inversion and burial of trash. The action of tines has been studied by Stafford (1979). As with other cultivation equipment, the action in any particular soil depends on moisture content. Working at excessive depths in soils which are too wet produces clods which are difficult to break down by subsequent secondary cultivations.

Secondary cultivations

Following the primary cultivation, some secondary operation is usually needed to produce as seedbed. Such secondary cultivations should be kept to the bare minimum since even when no compactive forces are produced by the implement, frequent tracking by a tractor quickly reduces the beneficial effects of primary cultivations.

Harrows and tine cultivators used at shallow depths break down the coarse structural units left by primary cultivations.

Disc harrows and some rollers are used for the same purpose but involve more drastic disruptive forces. These implements have a compactive effect on soil and should not be used unnecessarily.

Direct drilling

Where the soil structural conditions following harvest of an above ground crop (i.e. not potatoes, swedes, sugar beet, etc.) remain adequate to support a further crop, no cultivation is necessary and the seed may be sown by 'direct drilling'. The technique has several advantages.

Work rates are high, so there is no temptation to work under adverse conditions and crops can be established rapidly.

The risks of causing soil structural damage due to cultivation when soils are too wet are removed.

Fuel use is reduced.

Winter cereal crops are particularly suited to establishment by direct drilling. Not all soils give consistently reliable results when direct drilled. Cannell *et al.* (1978) have classified soils according to their suitability for sequential direct drilling.

Disturbance of the subsoil

Subsoiling is undertaken to disrupt the subsoil to a greater depth than achieved during normal cultivations. Compacted layers or pans may be shattered and water movement and root penetration greatly improved. Spoor and Godwin (1978) have studied the working of subsoilers and shown that in a particular soil any implement has a critical working depth. Working at a shallower depth, subsoiling is achieved, whereas below the critical depth upward movement of soil is prevented and compaction occurs along with the formation of a channel, sometimes called a 'square mole'. The correct subsoiling action can be achieved when working at depth, by loosening the surface soil beforehand (this may be by means of tines immediately preceding the subsoiler). Such loosening has the effect of increasing the critical depth. In general terms subsoilers work most satisfactorily in dry soils (critical depth is influenced by soil plasticity).

Mole drainage is really a drainage treatment but is frequently confused with subsoiling. When mole draining, the mole plough should be operated at a depth exceeding the critical depth for subsoiling, so that a channel is formed without appreciable upward heave.

Soil temperature

With the important exception of the effect of drainage, soil temperature is beyond the control of the farmer for practical purposes.

The effect of land drainage on soil temperature

There are two reasons why wet land warms up slowly in spring:

(1) heat that would raise the temperature of a dry soil is used (as latent heat) in evaporating water from the surface of a wet soil;
(2) the specific heat of water is high and additional heat is required to raise the temperature of the water, over and above that which would be needed to warm a similar dry soil.

Heavy land which has been artificially drained is frequently up to two weeks 'earlier' in spring than its undrained counterpart.

Variations in soil temperature

Through the year mean temperatures of surface soils follow a cyclical pattern roughly in phase with increasing solar radiation. At depth the magnitude of the cycle is reduced and lag occurs in the timing of maxima and minima.

The nature of the surface affects the reflection of incoming radiation and hence temperatures of the surface soil. Temperatures are lower under vegetation than bare soils. In experiments, darkened soil surfaces (spread with soot) warm up more rapidly than where lime is used to whiten the surface. This effect is illustrated in *Figure 1.21*.

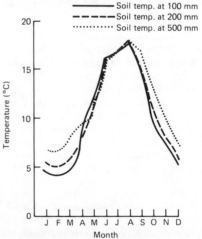

Figure 1.21 Average underground temperatures at 100, 200 and 500 mm (after Dougall, B. M.)

SOIL SURVEY AND LAND CLASSIFICATION

Soil survey

It is common experience that soils differ from one another. It is thus to be expected that our understanding of them will be enhanced by systematic classification.

Soil series

All classifications of soil are founded on the soil surveyor's perception of the 'unit of soil', referred to as the 'pedon' in the United States Department of Agriculture (USDA) nomenclature. This is the soil as observed in a single pit, usually 1 m square and up to 1.5 m depth (shallower if parent material is encountered at a lesser depth). A second 'pedon' would certainly be different from the first, but it is possible to define a 'soil series' such that the pedons comprising it have sufficient in common to be thought of as a homogeneous entity.

Producing a soil map

Depending on the local complexity of soils in the landscape and the scale of mapping, the soil surveyor may use the series as 'mapping unit'; alternatively he may be unable to distinguish individual series at the particular scale of mapping, and so use a 'complex mapping unit'. In either case, the boundaries of the mapping units must be located—these are what actually form the map. Two methods may be used.

(1) *Free survey* involves the soil surveyor using his knowledge of the relationship of soil with landscape features to guide him to areas where boundaries are expected. He then employs his efforts in making auger borings around such areas to obtain clearly defined boundaries. Areas that the surveyor considers to be uniform are given relatively less attention. Where work is undertaken by skilled surveyors, this approach gives quite accurate results and work rates are high. There is always the danger of a change in soil type not related to visible features passing undetected, and very little information is obtained about the degree of purity of the mapping units. (High purity implies a uniform mapping unit with few variations or inclusions of different soils.)

(2) *Grid survey* as the name implies, involves collecting data (usually soil descriptions based on auger borings) at systematically chosen points (grid line intersections) throughout the area to be mapped. Mapping units are subsequently identified, and boundaries inserted on the map so as to group like with like and delineate differences. This type of survey does not demand the employment of highly trained staff on field work, but speed of mapping (per person employed) is less than using free survey. It has the considerable advantage that the field records may be retained and consulted at a later date, or entered into a computer store to facilitate use of numerical classification techniques (*see* for example Webster, 1978).

Soil survey publications

The usefulness of any soil survey is greatly enhanced if it is accompanied by text giving background information. The Soil Survey of England and Wales (SSEW) publish 'Memoirs' and 'Records'. These provide a description of the methods used by the surveyor, major features of the landscape and much else relevant to the particular locality. All contain a number of specimen soil profile descriptions and analytical data. SSEW data relating to topsoils of a variety of soil series is reproduced in *Table 1.20*.

Table 1.20 SSEW published data on a selection of topsoils

Soil series	Wye Downs	Malling	Batcombe	Denchworth	Highweek
Topsoil texture	SL to SCL	L	ZL or ZyCL	ZyCL to C	CL
Classification	Brown calcareous	Brown earth	Gleyed brown	Surface-water,	Brown earth
Analyses (%)	soil		earth	gley soil	
Sand { 500 µm–2 mm	1.2	2.5	3.3	} 6.4	9
Sand { 200–500 µm	8.3	16.2	8.7		8
Sand { 100–200 µm	51.3	28.5	4.8	} 5.2	10
Sand { 50–100 µm	9.3	4.4	5.9		
Silt 2–50 µm	12.3	28.5	52.8	51.6	46
Clay >2 µm	17.7	20.5	24.5	36.2	27
Loss on ignition	8.0	4.3	9.1	10.2	—
CaCO$_3$ equivalent %	0.6	Nil	Nil	—	—
Organic carbon	—	1.3	—	3.1	3.1
pH in water	7.2	6.1	6.1	5.3	7.0
pH in CaCl$_2$	7.0	5.6	6.1	4.6	6.9
CEC	—	15.3	—	22.6	31.6
Depth of 'A' horizon (cm)	7	23	20	19	—
Available water in 'A' horizon (cm)	1.3	4.1	4.6	4.6	—

— = Data not available.

Soil classification

As an aid to understanding differences between soils, or to facilitate mapping at small scales (e.g. 1 : 1 000 000 soil map of England and Wales) a system of soil classification is needed. This may be based on:

(1) inferred processes of soil formation (genetic classification),
(2) observable or easily detectable attributes (morphological classification),
(3) suitability for one or more specific purposes (suitability classification).

The soil series is normally the lowest unit of classification and the higher units, groups of series, have different names in the various classification systems. In England and Wales the classification is into subgroups, groups, and major groups.

Genetic classification was used almost exclusively in Europe and the UK until the 1960s. Such systems have the considerable advantage of focusing attention on the natural processes of soil formation.

Morphological systems have become more popular in recent years, because the need for rapid surveying of soils throughout the world has demanded systems which are precise, and do not rely on the surveyor's judgement to interpret his observations in terms of soil formation processes. There has been a change of emphasis to use 'observable or easily detectable attributes'—the philosophy basic to the USDA 'Comprehensive System' (Soil Survey Staff, 1975). There is general agreement amongst soil scientists that morphological systems are most likely to be of use if the diagnostic criteria are chosen so as to be related to processes of soil formation. Such choices of criteria will of course provide some links with genetic systems of classification. In the UK, the change to a morphological type of system has been made in a way that retains a large measure of consistency with previous surveys and at the same time facilitates easy reconciliation with similar systems in use elsewhere (Avery, 1973).

Suitability classification has frequently been a major part of soil survey work. Some of the earliest soil surveys in England were concerned with suitability for fruit cropping. This type of work is nowadays regarded as a part of land capability classification.

The United States Department of Agriculture (USDA) soil taxonomy (Soil Survey Staff, 1975)

This is probably the most completely developed system in the world, with some 10 500 known series grouped into families, subgroups, great groups, suborders and orders. It was developed by a process of publication, distribution to working soil scientists and modification in the light of comments received, so giving rise to successive 'approximations'. It was the first major morphological classification with soils assigned to the various series, and series assigned to the higher units of classification, on the basis of the occurrence of diagnostic horizons which may be recognised by the surveyor, using only techniques likely to be readily available to him in the field. This system has the virtue that, properly applied by any surveyor, it will provide the same unique classification of any soil.

The world soil classification

This classification was produced as an adjunct to the FAO–Unesco Soil Map of the World, published in 1974. Classification is on the basis of 'diagnostic horizons' as in the USDA 'Comprehensive System'. Higher order names are traditional ones with unambiguous, internationally accepted meanings, or newly coined ones which can be relied upon to translate without change of meaning, so the system is highly suitable for communicating soil information across national boundaries.

Soil classification in England and Wales

The classification used in England and Wales (by The Soil Survey of England and Wales) has been evolved over many years through successive refinements, based on the concepts of Robinson (1943). Since the early 1960s, further modifications have been made to change the basis of classification from genetic to morphological, without loss of continuity. The system used at present is described by Avery (1973).

In Scotland, soil survey work is undertaken by the Scottish Soil Survey and soil information in Northern Ireland is the responsibility of the provinces' Department of Agriculture.

The concept of good land

It is a common misconception that the quality of land can be assessed by a chemist working on a soil sample at his laboratory bench. Nothing could be further from the truth. Most (but not all) chemical properties of soil can be modified by agricultural practice, usually at minimal expense. Physical properties are of prime importance to the farmer and cannot be changed readily.

The following are the attributes of good land:

A deep soil to facilitate adequate depth of rooting and storage of moisture. The best soils have 1 m or more of depth favourable to roots.

A satisfactory soil texture so that moisture retention is adequate, but cultivations are possible throughout much of the year.

Adequate drainage so that the soil may dry quickly and root exploration will not be restricted by anaerobic conditions.

Artificial drainage may be installed to alleviate problems but

(1) it is expensive,
(2) maintenance and eventual replacement is necessary,
(3) even the best artificial drainage is seldom an equal to natural good drainage.

A stable soil structure to facilitate arable cropping.
Freedom from risk of erosion.
A climate suitable for crop growth.
Level or with uniform gentle slope and relatively stone free, to enable machinery to be operated safely and efficiently.
Freedom from any concentration of noxious chemical.

Land classification

The aim of land classification is to provide an interpretation of the results of soil survey work, combined with knowledge of physical factors that affect land use. The aim is to locate 'good land' within an area, usually using some numerical scale that relates to 'degree of goodness'. Different classifications have different objectives. They are either general purpose or special purpose.

General purpose classifications use some generalised concept of 'good land'. The ADAS/Soil Survey of England and Wales system provides an example of a general purpose classification.

Special purpose classifications relate to specific soil properties (frequently electrical conductivity in arid lands) or some proposed land use.

Land classification in England and Wales

Two systems of land classification are in use in England and Wales at the present time (again the situation is different in Scotland and Northern Ireland).

The Agricultural Land Classification of England and Wales divides agricultural land into five grades on the basis of the type of physical limitations to land use already discussed. Existing maps at a scale of 1 : 63 360 (1 inch to 1 mile) cover all agricultural land.

The ADAS/Soil Survey of England and Wales Land Capability System classifies land according to similar principles but uses seven capability classes. These classes are further subdivided into subclasses on the basis of the limitations which exclude the land from higher classes.

For example, much of inland Devon and Cornwall (excluding the high moors and areas affected by poor drainage) is classified as 3c since only the climate (high rainfall and exposure, in this case) prevents classification in class 2. A further subdivision into capability units is made, such that all land in any one capability unit responds in a similar way to management and improved agricultural practices. It should be noted that this classification deals with *potential* land use, so relies on skilled judgement of likely responses to different levels of input. For example, an overgrown waterlogged area of alluvial soil may be quite worthless under present land use; its potential would be assessed assuming it had been drained if this were judged practicable, and class 2 could well be appropriate. In judging the level of input to assume (e.g. drainage in the above example) the appropriate yardstick is 'under a moderately high level of agricultural management'. The system has been described in detail by Bibby and Mackney (1969).

Future work on land classification in the UK is likely to involve a combination of the two systems described, and a major publication on the topic is expected in the near future (MAFF, in preparation).

References

ADAS (1979). *Irrigation Guide*, Booklet 2067. Pinner, Middlesex: MAFF

AUSTIN, R. B. (1978). *ADAS Quarterly Review* **29**, 76–87

AVERY, B. W. (1973). *Journal of Soil Science* **24**, 324–338

AVERY, B. W. and BASCOMBE, C. L. (1974). Soil Survey Laboratory Methods. *Soil Survey Technical Monograph No 6*. Harpenden

BAILEY, S. W., BRINDLEY, G. W., JOHNS, W.D., MARTIN, R. T. and ROSS, M. (1971). *Clays and Clay Mineralogy* **19**, 129–132

BIBBY, J. S. and MACKNEY, D. (1969). Land Use Capability Classification. *Soil Survey Technical Monograph No 1*. Harpenden

CANNELL, R. Q., DAVIES, D. B., MACKNEY, D. and PIDGEON, J. D. (1978). *Outlook on Agriculture* **9**, 306–316

DOE (1977). *Report of the Working Party on the Disposal of Sewage Sludge to Land*, London: National Water Council

EHRLICH, P. R., ERHLICH, A. H. and HOLDEN, J. P. (1977). *Ecoscience*, San Francisco: W. H. Freeman & Co.

HAMBLIN, A. P and GREENLAND, D. J. (1977). *Journal of Soil Science* **28**, 410–416

HANLEY, P. K. and MURPHY, M. D. (1970). Crop Responses to Sulphur in Ireland. In *Sulphur in Agriculture*. Dublin: An Foras Talúntais

HUGHES, A. D. (1979). *Sampling of Soils, Soilless Growing Media, Crop Plants and Miscellaneous Substances for Chemical Analysis*. Booklet 2082. Pinner, Middlesex: MAFF

JENKINSON, D. S. (1966). *Journal of Soil Science* **17**, 280–302

LOCKHEAD, A. G. and CHASE, F. E. (1943). *Soil Science* **55**, 185

LOW, A. J. (1972). *Journal of Soil Science* **23,** 363–380

MAFF (1976). *The Agricultural Climate of England and Wales.* Technical Bulletin No. 35. London: HMSO

MAFF (1973). *The Analysis of Agricultural Materials.* Technical Bulletin No. 27. London: HMSO

MAFF (1967) *Potential Transpiration.* Technical Bulletin No 16. London: HMSO

MURPHY, M. D. (1975). Sulphur on Grassland. *Soil Research Report,* 26–28. Dublin: An Foras Talúntais.

POSNER, A. M. and BOWDEN, J. W. (1980). *Journal of Soil Science* **31,** 1–10

POWLSON, D. S. (1980). *Journal of Soil Science* **31,** 77–85

REEVE, M. J., HALL, D. G. M. and BULLOCK, P. (1980). *Journal of Soil Science* **31,** 429–442

ROBINSON, G. W. (1943). *Discovery* **4,** 118–121

RUSSELL, E. W. (1973). *Soil Conditions and Plant Growth,* 10th edition. London: Longmans

RYDEN, J. C., MCLAUGHLIN, J. R. and SYERS, J. K. (1977). *Journal of Soil Science* **28,** 79–92

SOIL SURVEY STAFF (USDA) (1975). *Soil Taxonomy Agricultural Handbook.* Washington: US Department of Agriculture No 436

SPOOR, G. and GODWIN, R. J. (1978). *Journal of Agricultural Engineering Research* **23,** 243–258

STAFFORD J. V. (1979). *Journal of Agricultural Engineering Research* **24,** 41–56

WEBSTER, R. (1978). *Journal of Soil Science* **29,** 389–402

Further reading and sources of information

There are a number of modern texts which have achieved the status of 'standard works' in soil science. Probably the most comprehensive is *Soil Conditions and Plant Growth* by E. W. Russell (10th Edition, Longmans, London 1973).

Chemical aspects are covered by D. J. Greenland and M. H. B. Hayes in *The Chemistry of Soil Constituents* (Wiley–Interscience, Chichester 1978), K. Mengel and E. A. Kirkby in *Principles of Plant Nutrition* (International Potash Institute, Berne 1978) and F. E. Bear in *Chemistry of the Soil* (American Chemical Society Monograph No 160, Reinhold Publishing Corp, New York 1964). The latter publi-

cation includes a description of methods for chemical analysis of soils. C. A. Black in *Methods of Soil Analysis, Agronomy No 9* (American Society of Agronomy, Madison USA 1965) deals with physical, chemical and microbiological analysis of soil. The 'standard' chemical analyses used for soil science advisory work in UK are described in *The Analysis of Agricultural Materials* (MAFF Technical Bulletin 27, HMSO, London 1973).

A wide range of Government publications on agricultural subjects including many concerning soils are listed in the catolugue of MAFF publications issued periodically and available from MAFF (Tolcarne Drive, Pinner, Middlesex HA5 2DT).

Much of interest to those concerned with soil structure and the general well-being of soil is to be found in *Modern Farming and the Soil—Report of the Agricultural Advisory Council on Soil Structure and Soil Fertility* (HMSO, London 1970). *Soil Management* by D. B. Davies, D. J. Eagle and J. B. Finney (Farming Press Ltd, Ipswich 1972) is a guide to the maintenance of overall soil fertility by farming practices.

Microbiological aspects of soils are dealt with by M. Alexander in *Introduction to Soil Microbiology* (John Wiley, London 1977).

A wide range of information on the properties and distribution of individual soils is provided by Soil Survey of England and Wales publications. Sixty-four existing 'records' cover selected 10 km-sided grid squares with a further 35 or so in preparation, and a number of 'memoirs' describe the soils of more extensive areas. The present publication programme aims to produce soil maps and accompanying texts on a county basis. These are expected to be available by 1984.

A large number of student texts are available. A recent addition is R. E. White's *Introduction to the Principles and Practice of Soil Science* (Blackwell Scientific, London 1979). W. N. Townsend's *An Introduction to the Scientific Study of the Soil* (5th Edition, Edward Arnold, London 1973) contains a chapter devoted to 'Soil Science Literature' and those requiring a more comprehensive guide to the subject are referred thereto.

To various degrees, all the above have been consulted during the preparation of this section of 'The Notebook' and acknowledgement is due to the authors concerned.

2

Drainage

G. Cooke*

Land drainage work can be divided into two main categories: arterial and field drainage.

Arterial drainage is the concern of the Water Authorities (*see* p. 67) and comprises work on major rivers and watercourses to prevent or control flooding over relatively large areas.

Field drainage refers to work on individual holdings or limited areas of land in order to improve its condition for farming or other activities such as forestry and amenity use.

This section is concerned primarily with field drainage.

FUNDAMENTAL ASPECTS

The need for drainage

Land which is excessively wet will impose limits on choice of farming systems and operations and will prevent optimum crop growth.

Inadequate soil drainage may result in a number of problems:

(1) reduction in soil temperature leading to later and reduced crop growth; water increases the specific heat of soil thereby increasing the amount of heat energy needed to raise its temperature (*see Table 2.1*);

Table 2.1 Effect of water on specific heat of soil

Soil water status	J/kg (°C)
Dry soil	840
Soil + 20% water	1386
Soil + 30% water	1596
Soil saturated with water	2940

* The machinery aspects contained within this chapter were written by Mr P. H. Bomford.

(2) interference with desirable micro-organic life in the soil; this may lead to reduction in the rate of decomposition of organic matter, to denitrification and often to a reduction in pH;

(3) lack of aeration may restrict crop root development thus limiting nutrient uptake and leading to drought susceptibility at a later stage; under some conditions the accumulation of ethylene in waterlogged soils may cause direct root damage;

(4) predisposition to certain crop and livestock troubles; examples include liverfluke, husk and foot problems in livestock, slug damage and certain diseases in crops, e.g. cereal foot rots, club root in brassicas and soft rots in potatoes;

(5) many weeds and low-value plants are favoured by wet soil conditions, e.g. rushes and sedges, tussock grass and horsetails;

(6) the liability to damage to soil structure through cultivations, the use of heavy machinery and poaching by livestock is increased; compaction and smearing caused by working or travelling over soils when wet can accentuate existing surface drainage problems;

(7) restriction of farming operations—the time available for cultivations, root harvesting, stocking grassland and other farming activities is reduced; in extreme circumstances this factor may determine the types of crop and even the systems of farming which are practicable.

Field drainage aims at alleviating or removing these problems and is essential for modern, intensive, mechanised farming on soils not naturally well-drained. The economic justification for field drainage work is considered in a later section (p. 65).

Factors determining drainage requirements

An understanding of the climatic, geomorphic and soil factors affecting the water regime of a site is an important prerequisite to the diagnosis of any drainage problem and the designing of an effective scheme for its alleviation.

Climate

Climate is clearly the basic factor governing drainage requirements since, for a given set of soil and topographical conditions, climate and particularly those climatic factors involving water, i.e. rainfall, evaporation, will determine drainage needs.

The main climatic factors involved in drainage planning and design are the relationship between rainfall and evaporation and the seasonal patterns of the two. This determines the duration of the drainage 'winter' period when the soil is above field capacity. More detailed considerations such as the duration of suitable conditions for drainage operations, e.g. moling and subsoiling, can also be based on climatic data.

Overall drainage need

This depends on the degree to which rainfall exceeds evaporation (as expressed by potential transpiration, PT). In all parts of the UK the total annual rainfall exceeds total PT mainly because of relatively high rainfall and low PT figures during the winter months. The difference in the driest part of the country, the south east, is very small (less than 50 mm), whereas in certain areas of the west and north the excess of rainfall over evaporation can exceed 1000 mm, particularly in upland districts. A significant proportion of precipitation escapes as surface run-off or into underground aquifers so calculation of the annual excess of rainfall over evaporation is an over-simplification of the situation in drainage terms. It does however give an indication of the overall drainage requirement.

Duration of 'winter'

The period when the soil is above field capacity is when the benefits of effective drainage will be felt. More specifically the ratio termed the 'drainage climatic index' provides a useful measure of drainage need:

$$\text{draining climatic index (mm/d)} = \frac{\text{excess 'winter' rain in mm}}{\text{duration of 'winter' (d)}}$$

It represents the amount of rainfall per day to be removed during 'winter'.

Climatic data and drainage work planning

Much greater emphasis has been given in recent years to carrying out drainage work when soil conditions are suitable. Requirements for moling and subsoiling (*see* p. 63) are fairly specific in terms of soil moisture status at 0.5–0.6 m (approximately, soil moisture deficits > 50 mm for moling and > 100 mm for subsoiling). Although the soil moisture status is not so critical for pipe draining, better results are obtained on many soils under drier conditions. The 'date of return to field capacity' in the autumn provides a useful guide to the end of the drainage 'summer' and can be predicted from existing records. Clearly, appreciable variation occurs from year to year so that knowledge of the range of variation recorded in the above factors provides an additional aid to the drainage planner.

Full areal data for all the above factors will be found in MAFF (1976). Climatic data may also be used for more detailed design planning. Here the important factor is the short-term pattern, and in particular, intensities of rainfall. Drainage designs need to take account of the probability and frequency of predicted rainfall intensities. Designs may be based on 5-d or 1-d periods; the former is based on the facts that the upper layers of the soil provide some short-term buffer capacity and that short-term waterlogging is not serious during winter. Thus for drainage design purposes the total rainfall over any 5-d period is assumed to arrive at the drains uniformly over the whole 5 d. So, for example, if it is assumed that 60 mm is expected in 5 d, the design would be based on 12 mm/d or 0.5 mm/h. Areal data on this factor are also provided by MAFF (1976) as well as figures relating to varying 'return periods' or frequency of occurrence. For certain situations, e.g. mole schemes, 1-d rainfall figures are more appropriate.

Topography

Since water moves mainly by gravity, low lying situations are inevitably more vulnerable to drainage problems, particularly if the other factors referred to in this section are unfavourable. Areas of relatively flat land, river valleys and plateaux will tend to lie wet. Large river basins and coastal marshes at, or even below, sea level—such as the Fens and Romney Marsh—give rise to more acute drainage difficulties.

Serious drainage problems can occur also at higher elevations where a combination of high rainfall, impervious rock and unfavourable topography can result in wet, boggy conditions as found in many upland and moorland areas in Great Britain.

Geomorphology

Combinations of topography and geology may result in land being more or less uniformly wet over relatively large areas. In other situations excess wetness only occurs in patches; this may be due to variation in the depth below the surface of less-permeable strata. The

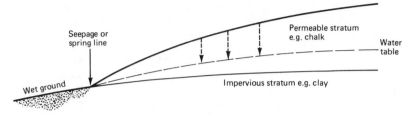

Figure 2.1 Simplified cross-section illustrating occurrence of a seepage or spring line

water table—the level below which the soil or rock is saturated—may therefore approach or reach the surface at certain points only. This phenomenon is commonly found on sloping ground where a permeable stratum overlies a less-permeable layer. Water tends to seep to the surface at the junction of the two, leading to a seepage or spring line (*see Figure 2.1*).

Springs may occur also where there is a fault in a water-bearing stratum below the soil surface. In practice a great variety of combinations of sub-surface conditions occurs leading to a range of drainage problems of varying complexity.

Table 2.2 Hydraulic conductivity values of soils (K)

Texture	Drainage rate	K value (m/d)
↑ Increasing clay % / Decreasing number of large pores	Very slow	<0.03
	Slow	0.03–0.1
	Moderately slow	0.1–0.5
	Moderate	0.5–1.5
	Moderately fast	1.5–3.0
	Fast	3.0–6.0
	Very fast	>6.0

Nature of soils and subsoils

The readiness with which soils and rocks allow water to percolate through will clearly have a major effect on drainage considerations.

Hydraulic conductivity

The rate of water movement through a soil or subsoil is known as its *hydraulic conductivity*. This factor depends partly on textural properties as determined by the derivation of the soil. Highly permeable rocks and soils derived therefrom include sandstones and sands, chalk and soft limestone and gravels. No artificial drainage system is likely to be required on such areas other than in exceptional situations. At the other end of the scale, relatively impervious rocks include clays, shales, igneous and metamorphic rocks such as granite. Between these extremes lies a range of rocks and subsoils of moderate permeability such as hard limestones, heavy silts and lighter clays.

The hydraulic conductivity (in terms of metres of water movement per day, known by the symbol K) of a soil in practice depends largely on its clay content; this situation and the range of values found are summarised in *Table 2.2*.

The hydraulic conductivity and drainage needs of soils will depend also on their structure and in particular, on the crack properties. Soils with naturally low K values will depend on measures to improve their structure to raise hydraulic conductivity.

Drainable porosity

Another soil factor which may affect drainage design considerations is drainable porosity. This is the difference between the water content of the soil when saturated and that at field capacity (i.e. when drainage ceases and no further water can be removed by gravity). It governs the degree to which the water table will rise or fall for a given amount of rain, e.g., 25 mm of effective rain will raise the water table by 500 mm if the drainable porosity is 5% but by only 250 mm if it is 10%. Clays can have values as low as 2%.

Permeability of the topsoil can be influenced by day-to-day management through such factors as the timing and nature of cultivations, tractor and machinery traffic, use of organic materials and exposure to the weather.

Structure and permeability in the subsoil depends more on natural factors but can be improved by draining and by deep cultivations such as subsoiling and mole ploughing (*see* p. 64).

FIELD DRAINAGE PRACTICE

The four stages involved in field drainage work are:

(1) site investigation and diagnosing the problem;
(2) designing and budgeting the scheme;
(3) carrying out the necessary work;
(4) maintenance.

Farmers and land owners should seek specialist assistance with the first three stages at least. Advice is available from the ADAS Land and Water Service staff, drainage contractors and specialist private consultants.

Site investigation and problem diagnosis

The initial aim is to identify the *cause(s)* of the wetness problem.

The physical environment of the site as outlined above should first be considered. Two broad categories of problem occur—*surface or top water* and *ground water*. A surface water problem arises from low hydraulic conductivity of the subsoil and topsoil so that the surface remains wet. This situation arises mainly on clay-type soils. With a ground water problem the water table rises from below to cause waterlogging near, or at the surface (ponding). This phenomenon may occur even on freely permeable soil and subsoils in certain topographical situations, e.g. flood plains of rivers.

A *perched water table* may occur where an impervious layer, e.g. impermeable subsoil or artificial pan, lies relatively close to the surface.

Site investigations in the field need to include a check on any previously installed drainage system. Nearly all lowland farmland on soils requiring drainage will almost certainly have been drained at some time in the past. Current problems may therefore arise through the failure of part at least of such an earlier system.

Field investigations

These should start with an inspection of ditches to determine whether or not they are fully functional. In particular, outfalls from an existing underdrainage scheme should be checked since, if they become blocked through silting up of the ditch or any other reason, water will back up the drains and cause wet areas in the field.

Wet patches in the field may be due to springs or to blocked or broken drains in an old system. A spade is essential in checking drainage problems—not only for investigating faults in old systems but for determining the general subsoil conditions.

In order to expose the subsoil a pit of about 1 m depth is required. This is termed a '*profile*' *pit* and it should have at least one clean vertical face. The following features should be noted:

(1) soil colour—uniform brown or brownish shades throughout the profile indicate that no significant water problem exists; dark grey, blue or blackish colours suggest that the soil is waterlogged for long periods; rust-coloured mottling indicates waterlogging at certain times of the year;

(2) subsoil structure—the profile should be examined for any compacted or structureless layers, e.g. cultivation pan, rust-coloured iron pan, surface compaction.

(3) the soil series present should, if possible, be identified.

Other key points in an existing drainage system such as culverts, inspection pits, and inlets should also receive early attention. Wherever available, relevant maps and plans are an asset in investigating drainage problems. Old plans of previous drainage systems (local MAFF offices may be able to assist with these if previous work was grant-aided) and Soil Survey maps can provide valuable evidence. The area of the country mapped by The Soil Survey is now extensive and since the drainage characteristics of many soil series are well understood, knowledge of the soil series on a site make the designing of a drainage scheme more straightforward. Summing up, new drainage problems are generally found to fall into one of four main categories:

(1) permeable soils (hydraulic conductivity 6 m + /d) with a high water table;
(2) clay soils with a surface water problem due to low permeability;
(3) soils with natural or cultivation pans;
(4) spring and seepage lines.

In many situations this grouping represents an oversimplification of the problem and combinations of two or more of the above categories may well occur. The experience and expertise of the specialist adviser is required to interpret the available evidence as a basis for formulating an appropriate remedial scheme.

Designing the scheme

Principles

Considerable strides have been made in recent years in the scientific approach to drainage design as a replacement for old rule-of-thumb methods; the problem still exists, however, of finding practical and cost-effective ways of applying the knowledge now available.

The steps in drainage design are to firstly decide what alternatives are technically possible in the particular circumstances and, secondly, decide which offers the best cost/benefit ratio. Appreciable problems exist in practice in making these decisions partly because obtaining the required physical data on the soils on a site is difficult and expensive and partly because there is limited experimental evidence available.

Modern practices in drainage design therefore have to accept some compromise and are based on three main approaches.

(1) Identification of the soils on the site and applying known data on their physical characteristics and drainage requirements. Wherever possible the basis would be to use the soil series classification of The Soil Survey of England and Wales. Generalised drainage designs for a large number of soil series have been worked out and can be applied where appropriate; if such information is not available on soils on a drainage site then an examination of the texture and structure of the soils via profile pits may enable them to be related to similar soils with working drainage designs.

(2) The use of inferred values for hydraulic conductivity. This parameter is the main physical property of the soil which will influence drainage design. Unfortunately it is difficult and costly to measure K values at individual drainage sites. Estimated values can be applied when the soils have been identified as a basis for drain specifications such as distance apart and depth. Trafford (1977 and 1978) has suggested a rough-and-ready basis for relating drainage design to hydraulic conductivity, shown in *Table 2.3*.

Table 2.3 Approximate basis for relating drainage design to hydraulic conductivity (Trafford 1977 and 1978)

Hydraulic conductivity value (m/d)	Type of drainage likely to be effective and likely range of drain spacing
>10	Very wide spacing; probably ditches only
1.0–10.0	Normal pipe drainage, spacing more than 10 m probably 20 m or more
0.1–1.0	Pipe drainage, practicable in deep soils, otherwise rather closely spaced drains needed
0.01–0.1	Mole drainage likely to be necessary
<0.01	No subsurface drainage possible without a physical change in the subsoil

Many of our claylands have K values of less than 0.1 and on some the theoretical drain spacing is as low as 2 m. Mole draining is the only economic way of achieving such close spacing.

(3) The degree of protection required. Most drainage designs are a compromise between what is technically desirable and what is economically viable. The latter will depend in part on the future planned use of the land and on the degree of protection against waterlogging that is appropriate. Reference has been made on p. 52 to designing drainage schemes on the basis of climatic data. Full protection, designed to drain rainfall intensities which occur very infrequently—say over 20 years—are unlikely to be justified economically and partial protection is the normal aim, i.e. ability to drain the land with rainfall intensities which occur relatively frequently. This factor relates to the intended future use of the land which will, in turn, affect the return on the investment in drainage. A more intensive scheme will be justified where the intention is to change the cropping towards a more intensive system than where the objectives are more limited. As a guide, it would be reasonable to work on rainfall frequencies of once in ten years for expensive horticultural or other intensive cropping, two years for other arable land and more intensive grassland, whilst for less-intensively used grassland a design rainfall rate which was exceeded, on average once a year could be acceptable.

Once the general design plan has been formulated the detailed layout must be prepared by surveying the area to establish falls and to determine the appropriate location of the drain lines. In order to facilitate the operation of modern drainage machinery, the usual aim is to achieve simplicity with the minimum of short 'runs' and junctions. Specialist advice should again be sought at this stage.

Drainage systems and specifications

The main types of field drainage are

(1) open—ditches and surface furrows, and
(2) underdrainage—tiling, plastic drains, mole draining and combined systems.

Open drainage

Ditches

These form an integral part of most drainage systems; they may act as drains in themselves, intercepting and collecting water from surrounding ground or they may serve as receivers and carriers of water from an underdrainage system. As drains they have the advantage of being adaptable, whilst their condition can readily be observed. The main drawbacks of ditches are the loss of land and interruption of cultivations involved, extra fencing costs in livestock areas and the liability to obstruction by vegetation, weathering banks and the activities of vermin.

Ditch specifications

Appropriate dimensions depend mainly on the volume of water the ditch is expected to carry at peak flow times. This will depend on catchment area, soil type and gradient. Although formulae have been devised for calculating the required cross-sectional area of a ditch, most are constructed on a rule-of-thumb basis.

Depth will mainly be determined by the necessity to ensure that the floor is at least 250 mm below any drain outlets. In certain situations with a ground water problem, deeper channels or dykes are needed in order to hold the ground water level well below the surface.

Bank slope must be sufficient to ensure stability and is mainly determined by soil type. The usual rule-of-thumb guide is to make the top width equal to the sum of the bottom width and the depth. On heavy clay land sides can be steeper than this but on lighter, sandy soils the batter should be about 30 degrees from horizontal.

In practice the type of machinery used is a factor determining bank slope. *Table 2.4* shows recommended dimensions for typical ditches.

Table 2.4 Summary of ditch dimensions

	Bottom width (mm)	Top width (m)	Depth (m)
Small field ditch	450	1.5	1.0
Large field ditch	500	2.0	1.5
Large collecting ditch	650	2.5	2.0

Gradients should be as uniform as possible and not too steep (leading to scouring) or too shallow (resulting in shoaling and silting up). A minimum of 0.25% is required but 0.5% is more satisfactory where a larger volume of water is carried. Topography will, however, determine the gradient limits that are possible.

Ditches should be cut as straight as possible and resulting spoil spread well away from the edges (at least 600 mm).

Guard fencing will be required in fields to be used by livestock. The fence should be erected at least 450 mm back from the ditch; posts should be no more than 4.6 m apart. Wire (plain or barbed) should be 650–850 mm from the ground depending on the type of livestock involved. Two strands are preferable for containing young stock.

Culverts

Culverts will be needed at farm crossings. Pipes used (usually concrete) must be sufficiently large to ensure minimum impedance of water flow; size will depend ultimately on the area of land draining into the ditch; 230 mm is the acceptable minimum size suitable for up to 12 ha and approximate specifications for larger catchments are as follows:

300 mm diameter, up to 24 ha
375 mm diameter, up to 40 ha
450 mm diameter, up to 65 ha.

Pipes should be firmly bedded and packed tightly with stone-free soil; a minimum cover of 600 mm is needed. The culvert should be well buttressed with headwalls built well into the bank on either side.

A typical culvert design is shown in *Figure 2.2*.

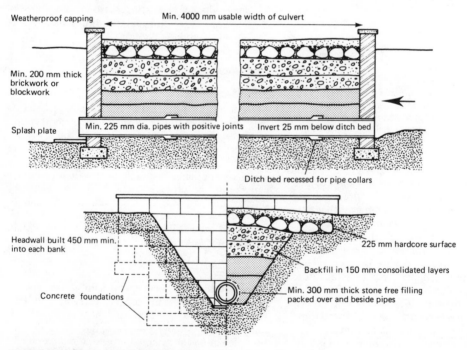

Figure 2.2 A typical culvert design. (By courtesy of the Ministry of Agriculture, Fisheries and Food. From *Technical Notes on Workmanship and Materials for Field Drainage Schemes*)

The length of the culvert must be sufficient to allow room for large modern equipment to turn into and out of the crossing without having to travel too near the headwall. A minimum length of 5 m is advisable but, if large machinery is to use the crossing, 9 m is preferable.

Hill drains

Open furrows or 'grips' are used for draining difficult open hill and moorland sites. A hill draining or gripping plough is used to cut tapered V-shaped channels across the slope. These channels are about 450–500 mm deep, 650–750 mm wide at the top and with a bottom width of 200–250 mm. The grips discharge into main outfall channels or natural watercourses; their spacing depends on the topography and subsequent use of the land. They are normally cut at 15–20 m intervals (approximately 500–650 m/ha). The plough moves the spoil some 600–800 mm away from the channel on the lower side.

Ditch elimination

With modern trends towards the use of larger tractors and machinery and with the general aim of economy of scale, many smaller fields are being amalgamated, with intervening ditches being piped in. This practice also has the merit of increasing the area of land for cultivation and eliminating the labour cost of hedge and ditch maintenance. It is, though, a practice to be approached carefully and with the needs of conservation in mind. Not all ditches are suitable for piping in any case; those which act as interceptors of surface run-off, major carriers taking a large volume of water and storage ditches should not be piped. A careful appraisal of the situation needs to be made by a drainage technician taking into account the functions and design of all local drains and watercourses, before a decision is made to pipe in a ditch.

Inlets A well constructed inlet from a remaining ditch may be required to prevent silt, trash and vermin entering the pipe. The main requirements for this are a silt trap with a hard floor 300 mm below the bottom of the pipe, head and sidewalls of durable material, a hinged grating with bars not more than 25 mm apart over the end of the pipe and, on large watercourses or those carrying a lot of debris, a trash grid further away from the pipe with bars about 50 mm apart (*see Figure 2.3*).

Size of pipe will depend on the catchment area of the pipe, the soil type and gradient. The minimum will be 150 mm but many sites will require larger diameter 230 or 300 mm pipes.

Installation Pipes (usually concrete) must be laid on a firm bed to an even gradient and as straight as possible. To this end it may be preferable to site the piped drain in firm ground to the side of the existing ditch, particularly if the latter is irregular in direction or gradient.

Other key points are to ensure that all existing outfalls into the ditch are efficiently connected to the new pipe and that there is adequate cover over the pipe (minimum 275 mm) so that disturbance does not occur when levelling the site nor in subsequent field operations. An inspection chamber is to be recommended at significant changes of gradient or direction and at junctions. A properly constructed outlet is necessary. These and the inspection chamber are similar to those described for underdrainage (p. 58), but, as main water carriers are involved, a high standard of materials and construction is important.

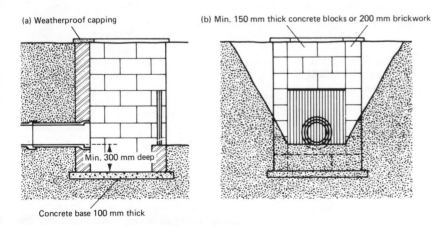

(a) Weatherproof capping

(b) Min. 150 mm thick concrete blocks or 200 mm brickwork

Min. 300 mm deep

Concrete base 100 mm thick

Figure 2.3 Drain inlet design (a) with silt trap, (b) with grating. (By courtesy of the Ministry of Agriculture, Fisheries and Food. From *Technical Notes on Workmanship and Materials for Field Drainage Schemes*)

Underdrainage

Underdrainage involves the installation of a system of artificial underground channels for removing excess water.

There are two types of underdrainage channel,

(1) walled—either with tiles (mainly clay or concrete) or plastic materials;
(2) unwalled—formed by a mole plough.

Pipe drainage

This is the traditional method of underdrainage practised for some 200 years and it is still the predominant system in practice.

Clay tiles   Clayware tiles are normally cylindrical and 250 mm long. The outer surface may be plain, corrugated or octagonal/hexagonal, this last-mentioned type being particularly suitable for pallet stacking and mechanised handling. A range of pipe diameters is available, the most widely used being 75 mm, 100 mm and 150 mm. The lower sizes are mainly used for lateral or minor drains and the larger size for mains; where diameters in excess of 150 mm are required, concrete pipes are more commonly used.

Good quality tiles are important since one faulty or broken tile can render a whole drain line ineffective; they should therefore be checked for uniformity and regularity in shape and they should be free from cracks and flaws, be well-baked and have clean, square ends.

A range of special purpose tiles is available, particularly for use at junctions, including both Y-shaped, acute-angled junctions and T-pieces.

Plastic materials   In recent years plastics—polyethylene and PVC—have been used on an increasing scale to provide walling for underdrains as an alternative to clayware. The advantages of plastics for this purpose are:

(1) easier and lighter to handle than clayware (a 200 m coil weighs approximately only 36 kg compared with 1350 kg for an equivalent length of tiles);
(2) more straightforward to lay mechanically;
(3) fewer joins and less easily displaced—they are particularly valuable on sites too unstable to hold tiles;
(4) a narrower trench is possible when laying plastics, which can mean lower costs and less ground disturbance;
(5) controlled slot size for water entry and smooth surfaces result in more efficient water removal.

On the other hand plastics have still to stand the test of time and there is still some conservatism in their use in certain areas. Nevertheless, nationally, about one-third of new drainage work is carried out using plastics; in some areas the proportion is as high as 50%.

Plastic drainage pipes are of two main types:

(1) 6 m lengths, the ends of which are either tapered or enlarged so that the lengths slot into each other to form a continuous line;
(2) coils of about 200 m in length which are unwound and laid automatically by trenchless machines.

They are made either with smooth or corrugated walls and contain a number of slots of specific dimensions cut either cross- or lengthwise in the pipe to permit entry of water. The number, size and arrangement of slots in the pipe varies according to the manufacturer's design requirements.

A range of pipe diameters is available from 50 mm upwards; the larger diameters have, to date, tended to be particularly expensive and not competitive with tiles although this situation is tending to change.

Most plastic-pipe manufacturers also provide a range of accessory items such as jointing sleeves, reducers, various types of junction pieces and outfall pipes. Filter wrapped plastic pipes are also available; coconut or other fibre is wrapped around the pipe and wired to it. The object is to prevent pipe siltation on sites susceptible to it, e.g. some silts. They are unnecessary and too costly for general use.

Installation

Most pipe installation is now carried out mechanically using trenchers or trenchless machines; about 15% of new work involves the latter (*see* p. 62).

Depth   The depth at which pipes are to be laid must be carefully pre-determined and will depend on the nature of the water problem and hence the purpose of the drains as well as on the subsoil characteristics. On clay soils, with low hydraulic conductivity, where the problem is surface water, drains need to be relatively shallow (750–900 mm). For ground water problems drains will normally require to be laid deeper; actual depth will depend on the particular circumstances of the site but will normally be in the range of 900–1500 mm. For cost reasons drains should not be placed deeper than necessary but 600 mm should be regarded as the minimum owing to the risk of displacement by machinery and deep cultivations.

Distance apart   Depth and distance apart of drain lines are usually interrelated, shallower drains being placed closer than deeper drains.

This is a reflection of the effect of drains on the water table in more permeable soils and subsoils and on the need for more intensive draining on clay-type soils of low hydraulic conductivity. A diagrammatic representation of the effect of drain lines on the water table level is shown in *Figure 2.4.*

Figure 2.4 Effect of drains on the water table level

Traditional tile systems were laid 6–8 m apart on heavy soils and 10–14 m apart under more permeable conditions. Such intensive piping would be too expensive today in most farming systems so some compromise has to be effected between technical efficiency and economy; the use of cheaper secondary treatments such as moling and subsoiling is used often therefore to complement pipe draining.

The minimum economic spacing is likely to be about 10 m and this would only apply to limited-scale work on high-value land. The appropriate spacing is 20 m for less permeable conditions but the costs of such a scheme would also be high; the commercial validity of such an intensive scheme would need thorough budgeting. The modern trend, wherever feasible, is to lay pipes wider apart (about 40–60 m) as mains and mole drain or subsoil over (*see* Mole draining, p. 63). Such an approach is supported by experimental results in recent years which have tended to show that, on clay soils, the spacing of drains is not absolutely critical *provided secondary treatments are carried out*. Crop yield improvement has been similar over a range of drain spacings.

Given the necessary input data from field investigations, theoretical distances apart and depth of drains can be calculated mathematically. Unfortunately on clay soils in the UK the theoretical spacing of drains required is often so close (2–3 m) as to be impossible in practice other than with mole drains.

Falls   These must be carefully pre-determined by surveying the site, particularly on land with poor natural gradients. Silting up of pipe systems is more likely with shallow falls while, if too steep, some scouring and tile displacement may occur. Falls will depend partly on site features—natural slope, length of drain run and ditch size, and, within limits, are not critical. The minimum fall for laterals is 0.2% but the minimum workable fall is about 0.25%. Fall in main drain lines should be at least 0.35% while the maximum fall to be recommended for most purposes is 2%.

Length of run   This is not critical but it is generally accepted that excessive lengths should be avoided owing to the tendency for the effects of any faults to accumulate; 250 m can be considered the maximum for lateral drains but slope, pipe diameter and other factors will determine this figure.

Size of pipes   The diameter of pipes to be used will depend on a number of factors and will relate to other design specifications. The volume of water to be carried is the main determinant of pipe diameter; this will in turn depend on such factors as rate of drainage required, soil type, catchment area and gradient. Using clay tiles 75 mm and 100 mm diameters are most commonly used for lateral or minor drain lines. A larger number of pipe diameters is available in plastic; the nature and design of plastic pipes (the smoothness of the walls minimising friction, the longer lengths of unit used and the regular arrangement of the slots) means that, at a given diameter, plastic pipes have better water flow characteristics than clay tiles. Slightly narrower diameters can be used therefore in designs based on plastic pipes; 60 and 80 mm are commonly used for independent plastic drain lines.

When pipes act as mains and receive water from a number of minor main drains, larger diameters will be required, normally. The total catchment area of the main will determine largely the pipe diameter. As a broad guide the following relationships between mains catchment area and the pipe diameter apply:

up 2.5 ha—80 mm plastic
          100 mm tile
over 2.5 ha up to 7.0 ha —120 mm plastic
          150 mm tile
over 7.0 ha up to 20.0 ha—170 mm plastic
          225 mm tile.

Porous or permeable backfill   Permeable materials are recommended for use over and around pipes in certain circumstances. Materials available for this purpose include washed gravel, stone, clinker and lightweight aggregate; these materials should be size graded between 5 and 50 mm. Miscellaneous on-farm materials such as straw and brushwood are occasionally used for this purpose.

The object in using permeable backfill is mainly to assist water percolation to the drain; it will, however, add significantly to the cost of the scheme, often by as much as 100%. Its use is to be recommended in the following circumstances:

(1) any situation where drains lie in relatively impermeable subsoil;
(2) where mole draining or subsoiling is to be used in conjunction with pipes;
(3) in some interceptor drain situations;
(4) as a connector with old drainage systems;

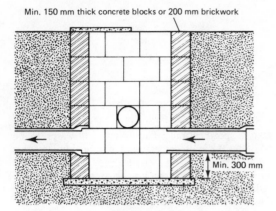

Min. 150 mm thick concrete blocks or 200 mm brickwork

Min. 300 mm

Min. 50 mm thick factory made precast reinforced concrete floor slab or 100 mm thick concrete if cast *in situ*

Figure 2.5 Cross-section of typical inspection chamber. (By courtesy of the Ministry of Agriculture, Fisheries and Food. From *Technical Notes on Workmanship and Materials for Field Drainage Schemes*)

(5) at junctions and other vulnerable points in the system.

The quantity required depends on trench width and depth and how near to the surface the fill is brought (usually ± 375 mm). For trenched drains about 1.5 m^3/20 m length of drain is required but the trenchless system requires less.

Owing to the high cost of permeable backfill material experiments have recently been carried out testing plastic polymers to produce more stable aggregates around the drains as an alternative; results to date, however, have not been promising.

Junctions It is frequently necessary to join drain lines as, e.g. when lateral spurs feed into a main. The most satisfactory junction is a purpose-built Y- or T-pipe. If these are not available satisfactory junctions can be made by chipping pipes to make a good connection and by covering with permeable fill. Lateral drains should preferably enter the main at an acute angle in order not to impede the water flow and cause silting up and/or displacement. At sites where a new drainage system traverses an existing operational system, pipes must be connected either directly or, if at different levels, by using a 'bridge' of permeable filling. Where main junctions occur an inspection chamber and silt trap are advisable. The bottom of this should be at least 300 mm below the outlet pipe; the walls should be impervious up to outlet level but may be pervious or honeycomb above. The top must be at ground level or above and covered with a removable slab. The site should be well marked to avoid traffic damage and regular inspection and clearance of silt is essential. *Figure 2.5* shows a recommended inspection chamber.

Outfalls The outlet at which a drain discharges into a ditch is a key point in any system. If this were to become blocked water would back-up and render the drain line ineffective. Special care is needed therefore in its construction. For stability, the last 1.5 m of pipe should be continuous, durable and frost resistant (e.g. plastic, pitch fibre or metal). It should discharge some 250 mm above the ditch floor, clear of the bank. The pipe should be supported by a durable headwall, e.g. concrete slab, and covered with a grating to prevent vermin entering the drain. A typical outfall is shown in *Figure 2.6*.

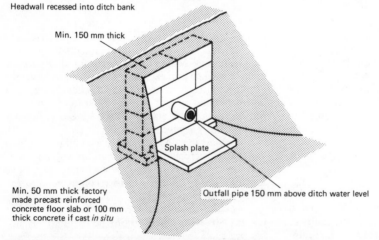

Headwall recessed into ditch bank

Min. 150 mm thick

Splash plate

Min. 50 mm thick factory made precast reinforced concrete floor slab or 100 mm thick concrete if cast *in situ*

Outfall pipe 150 mm above ditch water level

Figure 2.6 A typical outfall design. (By courtesy of the Ministry of Agriculture, Fisheries and Food. From *Technical Notes on Workmanship and Materials for Field Drainage Schemes*)

Care must be taken when cleaning out the ditch sides not to disturb the outlet pipe. Retractable outfalls are available for avoiding this problem.

Layout of system

The positioning of the drains and their combination into a pattern to achieve efficient drainage of an area is a matter for tailoring to site requirements. For this purpose it is prudent to seek specialist advice so that the best layout design to meet a particular problem can be determined particularly now that drainage work is so costly.

Overall drainage of an area is usually achieved by running a series of parallel drains at the pre-determined distance apart into a collecting main or direct into a ditch. Such a layout lends itself readily to mechanical installation. At some sites, it may not be necessary to drain the whole area and 'spot' drainage using a small number of strategically-placed drains may suffice. The old herringbone layout using a central main drain 'spine' with fishbone-arranged laterals discharging into it, is less popular today for overall schemes as it is not so amenable to mechanised pipelaying; it may still be useful in draining wet patches on a 'spot' basis.

Ground water problems on sloping ground (spring and seepage lines) can sometimes be dealt with by carefully siting single drains across the slope to intercept the downward movement of water and carry it away to a ditch (*see Figure 2.7*).

Laying the pipes

Pipes need to be bedded carefully and firmly with clay tiles being pushed as tightly together as possible. Pipe-laying machinery can achieve good results provided it is operated carefully under the right conditions. Better and more lasting results are achieved if drain laying can be carried out under relatively stable, dry conditions. If operations are carried out when the soil and subsoil is very wet, compaction, smearing and slurrying can occur and the drains may never function properly. Much pipe installation work is now carried out during the spring and summer which suits the contractor who must keep his expensive machinery in work for as long as possible during the year. Work can be carried out in growing crops since the long-term advantages can easily outweigh the loss of crop involved. This will be small if the work is done early in the growing season. To minimise damage when draining in growing crops, the work needs to be planned systematically to reduce the traverse of machinery across the site. Tracklaying trenchless machines are usually employed for this purpose so that soil disturbance is minimised.

The two main mechanical systems for laying pipes are the self-grading trencher/tile layer and the trenchless pipe-laying machine. The former is still the most widely used (some 75% of current new work); it excavates the trench and lays the pipes along it.

The trenching machine excavates by means of a power-driven vertical endless chain with blades attached to its links. It can produce a trench 150–450 mm wide and up to 1.5–2.5 m deep. Rates of work can be over 30 m/min for the most powerful machines. Spoil is brought up by the chain and pushed out to either side of the trench by horizontal augers.

Directly behind the digging chain is a hollow box as wide as the trench. This prevents loose soil from falling back. A V-shaped shoe shapes the trench bottom. Clay or plastic pipes may be slid down the inside of the box, to emerge at the bottom of the trench as the machine moves forward. Where plastic is being installed, a weighted wheel holds the springy pipe down into the bottom of the trench while a pair of angled discs turn some loose soil back on top of the pipe to retain it in position.

The complete digging and pipelaying mechanism can be raised or lowered hydraulically. Self-propelled machines are generally fitted with a storage rack for carrying containers of clay pipes, which may be moved by a hydraulic crane built onto the trencher. Individual pipes are fed by hand into the curved V-shaped track inside the hollow box.

Smaller trenchers can be fitted to tractors of 40–80 kW, which must also be equipped with a 10:1 reduction gear and preferably a four-wheel drive. Self-propelled machines run on long crawler tracks and are powered by engines of 60–250 kW. The trench is subsequently filled, usually with a tractor-mounted blade. If permeable backfill is used a separate hopper/dispenser running behind the trencher pours

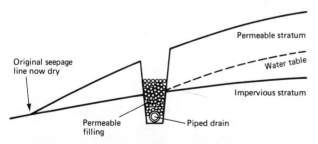

Figure 2.7 Diagram showing principle of interceptor drain

the backfill over the pipe via a lateral conveyor before the trench is covered in.

Trenchless machines Whilst a trenching machine can be used for installing clay and rigid or flexible plastic tubing, in sizes of 50–300 mm, the more recently developed trenchless machines can deal only with materials up to 150 mm internal diameter (100 mm on smaller machines). The original concept was to install plastic tubing in continuous lengths but most machines can now be adapted to handle clay pipes in 300 mm units.

The trenchless machine is based on the mole plough, which it resembles on a much larger scale. The width of the leg is greater, to allow the pipe to pass down its centre, and this is followed by a hopper which can discharge porous fill into the slot above the pipe. The plastic tube is carried in a reel on the forward part of the machine, and feeds down through the blade as the machine moves along. It is important that the pipe unrolls and feeds freely through all parts of the machine since any tension will stretch and thereby weaken the corrugated tubing. The machine may be mounted on a tracked chassis identical to those used for trenchers (in some models the two digging attachments are interchangeable), or may be mounted on a wheeled chassis and drawn by a winch. Winching minimises soil damage under wet conditions and allows a smaller power unit to be used. The leg can be raised or lowered hydraulically and is mounted on a linkage to isolate it from any pitching movements of the chassis.

Rates of work of 1.5–4 km/d are possible with plastic tubing; with clay pipes this is reduced due to the extra handling involved in transporting and feeding the pipes into the machine. Power requirements range from 75 kW for winched machines to 250 kW for the largest self-propelled machines. Depths to 2 m are possible. If porous fill is to be used, a gravel trailer must travel alongside the machine at all times in order to keep the small hopper on the machine topped up. This operation demands considerable skill from the operator as well as good organisation to avoid delays while the trailer is being refilled.

The leg of the trenchless machine displaces a great deal of soil and may leave a ridge up to 1.5 m wide and 0.5 m high. In pastures the ridge may be flattened by running the track of the machine over it or it may be dispersed on arable land by subsequent cultivations. Flat fronted, forward curved blades, which have the lowest draught, tend to bring up chunks of subsoil which may be a nuisance on grassland; a knife-fronted blade is unlikely to do this.

Apart from the great advantage of working with plastic tubing in terms of reduced material handling and ease of connecting, a further benefit of the trenchless system is that the work is completed rapidly, in a single operation. There can be a saving of 50% or more in the amount of porous fill material needed because of the narrow trench. This represents a considerable financial saving and there is no loss of performance since a band of porous fill as narrow as 50 mm is still adequate.

One drawback of the trenchless system is that, where large diameter pipes are required to be installed as mains a trenching machine may be required on the site as well. Since there is no open trench, junctions between mains and laterals must be formed in excavations made by a backacter, a further complication. For this reason the trenchless system is more suitable for larger, overall schemes laid out on a parallel pattern; it is not so flexible as tiling for smaller 'spot' schemes.

Backacter trenchers About 10% of drainage systems are laid in trenches excavated by hydraulic backacters. While this is a slower method than those described above and it is not easy to have precise control of gradient or to produce such neat work, these machines offer several advantages:

(1) they can be used without damage in very rocky ground such as reclaimed mining sites and in shallow soil overlying rock;
(2) they are cheaper and therefore more widely available than specialised drainage machines;
(3) they can be used for other excavation work, including the installation of junctions and outfalls and for ditch maintenance;
(4) many farmers already own light duty tractor-mounted backacters which can be adapted for small scale trenching work for the modest cost of a narrow digging bucket.

Maintenance of gradient in drains

A correct and even gradient is essential for the satisfactory performance and long life of underground drains. Any flattening of the slope, however short, will lead to silt deposition in the channel with consequent loss of performance and possibly complete blockage. By regulating the working depth of the machine as it moves forward, the operator can ensure that the pipes are laid at a constant gradient. Apart from the need to maintain a minimum soil cover of 600 mm above the pipes, the position of the drain is independent of the soil surface.

Traditionally, 'levelling' is carried out in two stages. First, a row of surveyor's poles is set up along the line of the proposed drain, and an adjustable crossbar on each pole is moved to a height above the required drain depth which corresponds to the 'machine constant'. The operator of the trencher then adjusts the working depth of his machine so that an optical sight on the machine is aligned with the row of crossbars. The bottom of the trench is then at the correct depth and can be kept so by small adjustments as the machine moves along. This method can produce very satisfactory results, but the setting-up of the poles is time consuming and requires at least two people. At the high operating speeds of which the larger machines are capable the machine operator's task demands great skill and concentration.

The use of laser levelling can cut the time taken to set up reference levels and relieve the machine operator of the task of keeping the machine at the correct depth.

A very narrow laser 'plane' is established above the area to be drained. By means of a spinning mirror, the laser beam is rotated round a central 'command post', rather like a whirling weighted string. This 'plane' is tilted to the required gradient and direction of the drains. On an undulating site, several different 'planes' may be required as the work proceeds.

A mast carrying three detectors is attached to the working end of the drainage machine. The height of the mast is adjusted so that the central detector is at the height of the laser plane with the pipe at the correct depth at one end of the run. From this point, electrohydraulic controls maintain working depth as the machine is driven forward. If the depth varies outside quite narrow limits, the laser will strike either the upper or the lower sensor on the mast and the automatic control system will detect this and will make a correction to return the machine to the correct working level.

Mole drainage

Mole draining involves forming a channel through the soil with no artificial walls by using a mole plough. It has the advantage of being a cheap and fast method of draining and has the valuable subsidiary effect of breaking-up and fissuring the soil, so aiding water percolation. Its drawbacks and limitations are that it is only effective on certain stone-free soils of reasonable slope and it has a relatively short life.

The mole plough The main frame of a contractor's mole plough is a heavy steel beam up to 3.5 m in length. At right angles to this beam projects a knife-edged 'leg' up to 35 mm by 200 mm in section. In work, the beam is drawn along the ground by a crawler or four-wheel-drive tractor of 60–90 kW. The leg cuts into the soil below the beam by an amount which can be adjusted to give the desired working depth, normally between 400 and 750 mm.

At the bottom of the leg is attached a cylindrical 'bullet', pointed at the front, 75 mm in diameter by 300 mm long. Behind the bullet, connected by a few links of heavy chain, is the 90 mm diameter 'expander', which forms the final shape of the channel.

A wheeled frame with hydraulic or cable lifting gear carries the beam when not in work. Attachments are available for drawing-in water pipes or electric cables behind the machine.

Most mole ploughs in current use are based on models designed for use with steam traction engines. Attempts to reduce draught have resulted in the following variants:

(1) the use of a split beam similar to the runners of a sleigh, with the leg carried on a central bridge structure; this allows the soil to lift freely ahead of the leg;
(2) the use of a flat-fronted leg, narrowing towards the rear, which reduces soil friction acting on the sides of the leg;
(3) angling the leg forward at 45 degrees, which lifts and loosens the soil.

Mole draining is only suitable for heavy soils with a high clay-particle content (over 35%) and low sand fraction (<20%) since the plastic properties of clay are essential for holding open a stable channel. Subsoils must also be free from stones and from pockets of sand and gravel which would cause the mole channels to collapse. Mole drains are usually spaced 2.5–3.0 m apart so that most of the soil area is fissured or disturbed. Depth of moling is normally between 500 and 750 mm and the diameter of the channels 75–100 mm. Mole draining should preferably be carried out when the surface is relatively dry but the subsoil at moling depth is moist (14–16% moisture content) so that the clay is sufficiently plastic. The mole plough should ideally be drawn up the slope so that the fissures formed drain the water down the slope into the channel. It is important to use a tractor of adequate power to enable the channels to be drawn at adequate depth and careful setting of the plough is required if the channels formed are to be smooth-walled and long lasting.

Layout Mole channels are normally drawn on a systematic parallel layout over the whole area to be drained. Although it is possible to connect them from ditch to ditch, mole channels usually discharge over a previously-laid piped main using porous fill as a connection (*see Figure 2.8*).

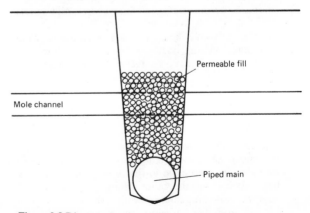

Figure 2.8 Diagram showing moling over piped main

The mole system should be designed to avoid, if possible, lengths of run exceeding 200 m and to achieve falls lying between 0.5 and 1.75%. Falls outside these limits will result in a reduced length of life for the system.

An efficiently installed system will last for upwards of five years on arable land and longer on grassland but as it is such a cheap operation once main drains are installed, more frequent moling is to be recommended.

Gravel tunnel drainage   A system termed *gravel tunnel drainage* has been developed in Ireland for use on sites where the stability of a straight mole drain would be suspect. A special mole plough is used to cut a large mole-type channel which is filled with 12–20 mm gravel or stone chips from a dispenser on the plough as it proceeds. The system is relatively expensive and has not been taken up to any extent in Great Britain.

Pipe-mole combination

The usual way in which mole ploughing is incorporated in a drainage system is in combination with pipes. Such a combined system is now virtually standard on clay type soils; it is a compromise system necessitated by the very high cost of intensive piping. It consists of a skeletal system of pipe drains which act as permanent mains over which mole drains are drawn.

The precise layout adopted must take account of the site conditions but a common arrangement is for 100 mm pipe drains to be laid on a parallel basis about 40–60 m apart and at a depth of 750–900 mm. Porous fill is used over the tiles to within 350 mm of the surface. Mole channels are then drawn across at approximately right angles at a depth of 500–600 mm and 2.5–3.0 m apart; they therefore pass through the porous fill connections for discharge of water into the mains. The pipes are usually laid across the general fall of the land, the lowest being at least 10 m in from the ditch or field boundary to allow the mole plough to reach full moling depth before passing over the main. Experimental work has shown that mole draining is likely to be the most cost-effective system for clay soils and should be used in all appropriate situations. Moling over pipes has been shown to increase crop yields compared with pipes alone. It is the only economic method of installing closely spaced drains.

Subsoiling

The use of the subsoiler can form a valuable drainage treatment when carried out over piped mains. It provides an alternative to the mole plough on soils less suitable for the latter. Being non-specialised equipment it can be operated cheaply by the farmer himself and can therefore be repeated regularly.

Details of subsoils and subsoiling will be found in Chapter 24. Subsoiling is not generally so effective as moling in drainage terms nor in effects on the yields of following crops.

Pumped drainage

The installation of land drainage pumps may be an advantage. The main place for pumps is in low-lying marshland situations where drainage to outfalls by gravity is inadequate owing to unsatisfactory gradients. The use of pumps should be avoided, wherever possible, owing to the substantial extra costs of installation and maintenance.

Land drainage pumps have to lift relatively large quantities of water through relatively small heights. Electrically-driven impeller pumps are normally used but in larger schemes, Archimedean screw pumps are sometimes installed. A pumped scheme requires care in design and installation and professional services should be used for this purpose. In particular, decisions will require to be made on such aspects as the selection of pump (type and capacity), siting of pump and associated works, weed and trash removal, protection of equipment against flooding, frost protection, the need for sluices and many other factors. Above all, the cost effectiveness of such a scheme needs to be explored carefully before detailed planning and work start.

Maintenance of drainage systems

All too often 'out of sight' means 'out of mind' and little attention is paid to aftercare of drains until faults occur and wet areas or spots appear. However in view of the heavy investment involved in drainage work it is important that the system is kept effective for as long as possible. Ditches are the first essential since, if these become ineffective, the rest of the system will also become defunct. They should be inspected regularly and particular attention paid to the following features:

(1) development of vegetation which will slow water flow and cause silting up;
(2) rubbish such as plastic bags, paper, old oil drums, etc;
(3) bank slips and erosion;
(4) shoals forming in the ditch bottom;
(5) condition of the guard fencing, where present;
(6) the condition and functioning of piped outfalls.

Appropriate clearance and maintenance should be carried out on a regular basis. A reasonable interlude between ditch bottom cleaning is five years, with annual attention to the vegetation on the banks.

The ditch bottom can best be cleaned by drawing a V-shaped scoop or bucket along it. This gives a smooth finish, and ensures that the bottom of the ditch remains narrow, an important consideration if summer flows are low. A flat-bottomed ditch carrying little water is an ideal site for summer weed growth, and this makes a fine trap for silt later on. Spoil from ditch cleaning should be spread well clear of the ditch bank.

Few machines which work from the side of the ditch can be operated in the way described, an exception being the dragline excavator. The backacter is normally used for ditch cleaning, with its boom working at right angles to the ditch. The use of a wide digging bucket makes it easier to keep the ditch straight, but a narrower bucket is necessitated if much depth of digging is required. The greatest skill and care must be applied if any semblance of a V-shaped ditch bottom is to be maintained.

The hydraulic backacter or backhoe is available in many sizes, from those which may be attached to the rear of a farm tractor up to large self-propelled machines which can move many tons of soil at a bite. The machine consists of two horizontally hinged arms, the end of one arm being mounted on a swivelling pivot on the tractor, while the outer end of the second arm carries one of a variety of digging buckets. Hydraulic rams lift or extend the boom, slew it from side to side through 180 degrees or more and change the angle of the bucket. Retractable hydraulic legs steady the machine while it is at work. In very muddy conditions the boom may be used to drag the tractor along if the wheels have lost traction.

A typical farmers' machine has a reach of 3.5–4.5 m, and can dig to a depth of 2.5–3 m with various buckets 140 mm–1.5 m in width. Some machines have their own hydraulic system powered by a pump connected to the tractor's power take-off shaft, while others rely on the tractor's hydraulic system; 30–50 kW is required. In several machines the bucket can be replaced by a hedgecutting attachment, which makes a very versatile unit.

Hedgecutting machines are very suitable for controlling the vegetation on accessible ditch banks. Piped drain outfalls must be clearly marked in order to avoid damage during this operation.

Catch pits and ditch inlets should also be checked regularly and cleaned out as necessary; if they are not the whole point of installing them is lost. It is advisable also to check the functioning of individual drain runs on a systematic basis; faults will soon show up as wet areas on the surface. Dyes such as potassium permanganate can be used to trace water movement and identify trouble spots. Particular attention should be paid to drains near trees and hedges since roots can invade and block drains very quickly.

It may be possible to clear blocked drains for some distance from the outfall by rodding, but this must be done carefully to avoid displacement of tiles. Water-jet equipment for cleaning field drains is also available; it is more applicable to plastic drains owing to the danger of displacement of tiles.

ADMINISTRATIVE AND ECONOMIC ASPECTS OF LAND DRAINAGE

Economics

As the costs of draining farm land escalate it becomes more pertinent to consider the justification for the investment involved. Costs can be budgeted quite accurately; it is, however, very difficult to estimate and quantify the increased returns which will accrue. In the past drainage has tended to take the form of remedial 'fire brigade' action; all too infrequently has its part in a constructive farm plan been estimated. With current grant-aid schemes, a practical budgetary approach is more common and is to be encouraged despite the difficulties imposed by the intangibles involved, e.g. the increase in crop yield which can be expected; the length of time the scheme will remain effective. In the economic context two main benefits accrue from field drainage.

An increase in the yields of subsequent crops

There is unfortunately relatively little information on this point deriving directly from experimental work. This is mainly owing to the inherent difficulties in obtaining statistically valid data from drainage experiments. Among the problems in this respect are the need for large plot sizes making for difficulty in achieving uniformity, finding true, undrained control plots and the need for long-term monitoring owing to marked seasonal variations. Few drainage experiments have therefore been completely satisfactory from the statistical point of view. Evidence in this country has been available in recent years from the work of the Field Drainage Experimental Unit based at Cambridge. At Drayton Experimental Husbandry Farm and other clayland sites yield data have been obtained from cereals and other crops in drainage experiments. The gain in yield of winter wheat, although variable from year to year has been in the order of 0.6–1.0 tonne/ha. These results confirm previous figures obtained mainly from farm survey work suggesting gains in the yield of wheat in the order of 0.75–1.00 tonne/ha after drainage has been carried out. Higher responses to nitrogen fertilisers have been obtained following drainage. Few experiments have attempted to quantify gains in output from draining grassland; one such Ministry experiment in Devon showed increased liveweight gains of the order of 135 kg/ha but the financial implications were not clear cut. Improved botanical composition in grassland swards following drainage has been recorded so that, given satisfactory management in other respects, e.g. use of fertilisers, gains in grass yields should follow.

Increases in yield of arable crops following drainage work occur not only through improved crop growth but also through greater timeliness in field operations such as cultivations and drilling. Studies have shown that over 20 extra working days in both autumn and spring are possible after draining heavy land. Likewise the utilised output from grassland can increase due to the greater length of the grazing season and ability of the land to withstand higher stocking rates.

From a negative point of view it can be argued that loss of crop yield due to flooding or waterlogging can be prevented. Many crops are surprising susceptible to damage by waterlogging. The following data (*Table 2.5*) arising from work in Europe indicate the approximate scale of losses expected from this cause.

Although these data do not necessarily apply directly to UK conditions they do provide a useful guide to the problem. Any calculation of the cost effectiveness of drainage in this respect must clearly estimate the frequency of expectation of such losses occurring at a given site.

Greater flexibility in the farming system

Field drainage may permit a wider range of crops to be grown and provide greater flexibility for the farm plan as a whole. There are many examples in practice of a change from permanent grassland to rotational ley/arable cropping following drainage with consequent benefit to the farm budget as a whole. Ministry surveys and farm case studies have provided numerous examples of significant increases in farm gross margins following drainage work.

Grants

Grant aid has been paid to owners and occupiers of agricultural land towards the cost of field drainage work for some 40 years. Current schemes fall within the EEC pattern and grants for drainage, including

Table 2.5 Percentage reduction in crop yield due to waterlogging (Greyon, 1970)

Month	Length of water-logging (d)	Grass	Sugar beet	Potatoes	Winter wheat
January	3	–	–	–	–
	7	–	–	–	5
	15	–	–	–	15
February	3	–	–	–	–
	7	–	–	–	5
	15	–	–	–	20
March	3	–	10	30	5
	7	–	50	80	15
	15	10	100	100	50
April	3	–	10	30	10
	7	10	50	80	25
	15	30	100	100	70
May	3	–	10	40	20
	7	15	50	90	40
	15	50	100	100	100
June	3	–	10	50	20
	7	20	40	100	50
	15	50	100	100	100
July	3	–	10	50	–
	7	20	40	100	–
	15	50	100	100	20
August	3	–	10	50	–
	7	10	40	100	–
	15	30	100	100	–
September	3	–	10	20	–
	7	–	40	40	–
	15	10	100	80	–
October	3	–	–	–	–
	7	–	10	–	–
	15	–	50	–	20
November	3	–	–	–	–
	7	–	–	–	–
	15	–	–	–	20
December	3	–	–	–	–
	7	–	–	–	5
	15	–	–	–	20

both underdrainage and ditching, may be claimed under one of two schemes.

The Agriculture and Horticulture Grant Scheme (AHGS)

Under this scheme the Ministry offers grants on individual items of capital investment. These are known as Investment Grants.

The Agriculture and Horticulture Development Scheme (AHDS)

This offers grants under approved development plans. These are known as Development Grants. A farmer may apply for either one or the other but not for both. One major change from previous schemes is that prior approval is no longer required for an Investment Grant.

The standard rate of grant under the AHGS scheme is currently (Autumn, 1980) 37.5%. A final plan of drainage work must be submitted when claiming grants.

Under the AHDS scheme a grant of 50% is currently payable for field drainage work which must be part of an approved farm development plan. Individual items within the plan do not require specific prior approval.

Under both schemes the current rate of grant for field drainage in officially-designated 'less favoured areas' is 70%.

Full details of grant aid schemes involving drainage may be found in leaflets available from local MAFF Divisional Offices.

Organisations concerned with land drainage

The Ministry of Agriculture, Fisheries and Food (MAFF)

The Ministry has a major interest, both financial and technical, in land drainage. Control is centred in London at the headquarters Drainage Division in the Ministry. Its work is carried out in the country by the Land and Water Service of the Agricultural Development and Advisory Service (ADAS). Its staff are based in Divisional Offices and are responsible for advisory, development and promotional work in the field as well as work in connection with grant schemes.

The MAFF Publications Unit publishes a series of technical and advisory leaflets on land drainage. These are available locally from Divisional Offices.

The Ministry is responsible also for the Field Drainage Experimental Unit based at Trumpington, Cambridge. This unit is responsible for carrying out experimental and developmental surveys and studies in the sphere of field drainage both in the laboratory and in the field. The systematic scientific study of drainage matters is comparatively recent in this country. Its work is providing a more scientific and rational basis for drainage design and practice in future.

The Soil and Water Management Association (SAWMA)

This organisation is based at the National Agricultural Centre at Stoneleigh, Warwickshire and concerns itself with soil management matters including drainage. It publishes a journal, *Soil and Water*, twice a year.

Farmer's Weekly

For a number of years *Farmers Weekly* has taken a particular interest in land drainage by sponsoring an annual National Drainage Demonstration.

This is held usually in May at a farm site which varies from year to year across the country. At this demonstration drainage techniques and equipment can be seen in working situations.

Water Authorities

Under the Water Act, 1973, ten Regional Water Authorities were set up. Membership of these authorities comprises nominees of the Department of the Environment, the MAFF and appropriate County and District Councils. They have responsibilities in connection with land drainage (arterial), water supply and pollution, fisheries, water-borne recreation and sewage. They are financed by grant from the MAFF and by precept on constituent County Councils.

Water Authority powers with specific reference to land drainage are vested in the Regional Land Drainage Committee, members of which are appointed by the MAFF, Water Authority and County Councils. Local Land Drainage Committees have also been set up within each Authority area.

The Water Authorities have direct control of drainage and other works involving designated 'main rivers'. The responsibility for 'non-main rivers' falls primarily on District Councils.

Sea defences in coastal areas also normally come under the aegis of the Water Authority. Areas with special drainage problems may be designated Internal Drainage Districts administered by an Internal Drainage Board comprising landowners and occupiers of land and property within the Board's area. These Boards are largely autonomous and specify the water courses for which they are prepared to accept responsibility; a drainage rate is levied on all occupiers of land within a Drainage District. Water Authorities have powers to make and implement relevant bye-laws; they are also required to undertake periodic surveys of land drainage requirements within their area.

The powers and functions of Water Authorities and other related bodies in respect of land drainage together with relevant legislation was consolidated in the Land Drainage Act, 1976.

The national drainage situation

The extent of field drainage requirements nationally has been reviewed from time to time by means of surveys. It is difficult to establish a precise picture since much agricultural land is still relying on old drainage systems laid mainly between 1840 and 1880. Such systems are liable to failure, so increasing the area of land requiring new drainage. Estimates have suggested that the rate of failure in these old systems is probably of the order of 50 000 ha/annum. If this is so then the current position (1980) in England and Wales is approximately as summarised in *Table 2.6*.

The position is therefore not really satisfactory in

Table 2.6 Summary of the national drainage situation (England and Wales)

	10^6 ha
Total area of agricultural land	11.00
Land not requiring artificial drainage (good natural drainage, e.g. chalk)	4.25
Land not likely to justify drainage on economic grounds	1.05
Net area requiring drainage	5.70
Old systems still working	1.55
Area drained with grant aid since 1940	1.75
Net area requiring new drainage	2.40

that there is a substantial backlog of drainage work required nationally.

The annual area drained in recent years has been around 100 000 ha; if the assumption regarding the annual failure rate in old systems is right it means that the net area with working drainage systems is increasing by only about 50 000 ha/annum. When this is set against the current national target, the problem can be seen to be substantial, with about 50 years being required at the present rate of progress to clear the backlog.

There is considerable regional variation in drainage work; broadly speaking the arable areas in the eastern half of England are in a much better position than the predominantly grassland areas of the west. About 70% of drainage work in recent years has been on the eastern side of the country, mainly owing to the greater economic advantages accruing on arable land.

A Grassland Research Institute–ADAS survey of permanent grassland has suggested that more than 40% of permanent grassland in England and Wales is inadequately drained for intensive production.

References

GUYON, G. (1970). Bulletin Technique de Génie Rural No 102, Paris: Republique Francaise Ministere de l'Agriculture

MAFF (1976). *Climate and Drainage*. Technical Bulletin No 34

TRAFFORD, B. D. (1977 and 1978). Recent Progress in Field Drainage, Parts 1 and 2. *Journal of the Royal Agricultural Society of England* **138** and **139**

3

Crop physiology

D. Wright

INTRODUCTION

Growing a crop demands certain operations (e.g. cultivation, weed control) and inputs (e.g. seed, fertiliser) and the objective of crop husbandry research is to define the optimum combination of these that will maximise yield. Although this approach provides data which can be translated into field recommendations for farmers, it does not *explain why* particular treatments did or did not affect yield.

The yield obtained from a crop depends upon the *genotype* (species and variety) of the plants and the *environment* (soil and aerial, modified by weather and husbandry) in which they are growing. In order to explain differences in harvested yield which are observed between varieties, in different seasons and between different husbandry treatments, it is necessary to examine the response of the crop plants throughout their growth. Results of such investigations can then be used by the agronomist to explain differences in yield observed in experiments, and by the plant breeder to identify characteristics associated with yield that could be used as selection criteria in a breeding programme.

The aim of crop physiology is to understand the ways in which the various physiological processes occurring in crop plants are integrated, how they are affected by the genotype and the soil and aerial environment, and how they themselves affect growth, development and yield.

This chapter outlines some of the general principles and applications of crop physiology. Examples, where necessary, are drawn from the major arable crops. The reader is also referred to the textbooks of Milthorpe and Moorby (1974) and Evans (1975).

SEED PHYSIOLOGY AND CROP ESTABLISHMENT

Seed factors which influence field performance

Viability

Purchased crop seed has a guaranteed minimum germination percentage specified by the UK Seed Certification Scheme. Failure of seed to germinate will not only result in the establishment of a suboptimal plant population, but there may also be an associated deterioration in the performance of the remaining plants (Roberts, 1972a). Seed viability is considerably affected by the conditions of seed storage.

Seed size

Seeds of any one crop show a wide variation in size due to the effects of weather, variety, crop husbandry and soil type on the growth of the plant from which they were derived. 'Larger' seeds may have larger embryos and/or greater amount of endosperm. When sown therefore, larger seeds may emerge quicker and produce heavier seedlings with larger cotyledons than small seeds. These initial differences in plant size may or may not persist until harvest (Wood, Longden and Scott, 1977).

Dormancy

Dormancy is the term used to describe the condition in which a viable seed does not germinate, even though it is under environmental conditions normally considered suitable for germination. Several different types of dormancy are recognised, all are of considerable agricultural significance (Roberts, 1972b).

Innate dormancy prevents the embryo of the new seed germinating when the seed is still attached to the parent plant. Under *enforced dormancy* the seed fails to germinate because some environmental factor is unfavourable. Once the limiting factor is removed the seed germinates. Weed seeds buried deep in the soil may therefore fail to germinate because of lack of oxygen, but will do so once they are disturbed by cultivation. Where the seed remains dormant for some time after the limiting factor has been removed, it is said to be under *induced dormancy*. Seeds of certain species (including cereals and grasses) require to undergo a period of dry storage after harvest ('after ripening') before they will germinate.

Stage of maturity at harvest

The developing and growing seed on the parent plant undergoes a complex sequence of biochemical and physiological changes as well as increasing in size and dry weight. The time at which harvesting is carried out can therefore influence the maturity and size of the harvested seed. The problem is greatest in indeterminate crops (with no terminal inflorescence) such as dried peas in which flowering takes place over a prolonged period, and where there may be differences in degree of ripening between seeds at different positions on the plant. Variations in degree of seed maturity within the crop create problems in deciding time of harvest and subsequently drying difficulties. In addition, the field performance of 'immature' seed may be inferior to that of 'mature' seed.

Seed-borne micro-organisms

Seed-borne micro-organisms may affect the performance of the seed (e.g. covered smut of cereals) or the crop derived from infected seed (e.g. virus diseases of potatoes).

Seed vigour

Different seed lots of the same crop variety and germination percentage may show different field emergence percentages (i.e. number of plants established per 100 seeds sown) when sown under identical field conditions. These differences are often attributed to differences in *seed vigour*. Most definitions of 'vigour' refer to the ability of vigorous seeds to germinate and produce a plant under adverse environmental conditions. Differences in seed vigour may arise as a result of any one of the aforementioned seed factors influencing field performance (Heydecker, 1972).

Pattern of germination and emergence

In addition to the *number* of seeds germinating, the *pattern* of germination and emergence is also import-

ant. It is desirable that all seeds emerge within a short time interval so that the component plants of the crop are as uniform as possible. This is particularly important with crops such as carrots and potatoes for canning, where final plant size and economic yield is dependent upon inter-plant competition for light, nutrients and water restricting individual plant size. Unevenness of plant size may also create difficulties in deciding when to carry out husbandry operations which have to be carried out at particular growth stages, e.g. herbicide applications. Root crops which consist of plants of widely differing size may present mechanical difficulties at harvest.

Effect of environmental factors on seed germination and emergence

The environmental factors which affect *germination* are moisture, oxygen and temperature. For satisfactory *emergence* (visible appearance of the shoot above the ground surface) the seed must also withstand the attack of soil pathogens and have sufficient strength to overcome the physical resistance to growth imposed by the soil. Differences in number of plants established can occur as a result of variations in germination percentage and/or variation in emergence percentage.

Temperature can affect the time of onset of germination and the rate of germination. The concept of cardinal temperatures involves minimum and maximum temperatures at which there is no germination, and an optimum temperature at which the maximum number of seeds germinate. 'Optimum' temperatures for germination are usually higher than those encountered in the soil at the normal time of sowing. Temperature can also affect the rate of germination (number of seeds germinating per unit time), independently of an effect on number of seeds germinating.

A supply of water is necessary for the mobilisation of seed reserves and extension growth of the embryo. Excess water supply restricts the availability of oxygen to the seed. Therefore soil physical properties can have a marked effect on germination and emergence.

The germination of seeds of many species is stimulated by light (e.g. lettuce, tobacco, tomato), although light sensitivity is often confined to seeds germinated at relatively high temperatures.

Improving the performance of seed by seed treatment

Various types of seed treatment can be applied to seed bulks after harvest with the intention of improving the performance of either the seed or the resulting plants. These include:

(1) size grading,
(2) application of pesticides,

(3) dormancy breaking treatments,
(4) treatment with hormones or nutrients,
(5) pelleting,
(6) seed priming (including pre-sprouting of potatoes and pre-germination prior to fluid drilling).

These treatments are reviewed in papers by Graham-Bryce (1973), Heydecker and Coolbear (1977) and Wurr (1978).

In addition, standard germination tests and various vigour tests assess the suitability of the seed for sowing. Seedlots with performance which is lower than a specified threshold can be rejected as being unsuitable for seed sowing.

CROP GROWTH

Crop *growth* involves an increase in size or weight of individual plants. It is most reliably measured as an increase in dry weight as crops contain variable amounts of water. Crop *development* describes the series of anatomical and morphological changes undergone by the individual plants between the sowing of one 'seed' and the harvesting of the economically important part of the crop, which may itself be a seed.

The carbon economy of the plant

The bulk of the dry matter of crops consists of carbon compounds derived from *photosynthesis*. Metabolic activity within the plant requires energy. This energy is provided in the form of a high energy phosphate compound (adenosine triphosphate, ATP) by *respiration*. During respiration sugars produced by photosynthesis are oxidised, and carbon dioxide and water are released. Crop yield (net production of carbohydrate) therefore depends upon the net balance between photosynthesis and respiration.

Photosynthesis

The first stage involved in photosynthesis is the diffusion of carbon dioxide from the atmosphere through the stomata to the internal tissues of the plant. Light energy intercepted by the chloroplasts is then used to split the water molecule to produce molecular oxygen, energy (in the form of ATP) and reducing power (nicotinamide adenine dinucleotide phosphate, NADPH). This process occurs only in the light. The NADPH and ATP produced are then used to reduce carbon dioxide to carbohydrate. This reaction is not light dependent. The overall process of photosynthesis can be represented by the equation:

$$CO_2 + 2H_2O \xrightarrow[\text{Chlorophyll}]{\text{Light}} (CH_2O) + H_2O + O_2$$

Further details of the biochemistry of photosynthesis are given in many recent textbooks of plant physiology. In most temperate species the carbon dioxide initially combines with ribulose diphosphate to produce two molecules of phosphoglyceric acid, which has three carbon atoms in its molecular structure. This pathway is called the Calvin cycle or C_3 pathway. In many grasses of tropical or arid areas a different pathway of carbon dioxide fixation is involved. In these species the carbon dioxide combines with phosphoenol pyruvate to form oxaloacetate and other compounds with four carbon atoms in their molecular structure. This is called the Hatch–Slack or C_4 pathway. In C_3 species between 20 and 50% of the carbon fixed by photosynthesis is immediately respired. This is termed *photorespiration*. C_4 species do not exhibit photorespiration and are therefore capable of producing higher crop yields, because photorespiration reduces net photosynthesis (rate of uptake of carbon dioxide—rate of evolution of carbon dioxide from respiration) in C_3 plants.

Environmental factors affecting crop photosynthetic rate

The rate of crop photosynthesis depends upon light intensity, carbon dioxide concentration, temperature and the availability of water. When one of these factors is present in limiting quantities, increasing its supply increases the rate of photosynthesis up to a point at which the rate becomes independent of that factor, i.e. the availability of some other factor is now limiting.

Gross photosynthetic rate of single leaves is relatively insensitive to temperature over the range of 20–30 °C. It is strongly dependent upon light intensity, and approximately proportional to the concentration of carbon dioxide in the air surrounding the leaves (Milthorpe and Moorby, 1974). Although the concentration of carbon dioxide in the atmosphere (0.03%) ultimately limits photosynthesis, it cannot be artificially increased except in the glasshouse.

The carbon dioxide concentration within the canopy is rarely depleted to such an extent that it has a large effect on gross photosynthesis. Water stress reduces the rate of net photosynthesis, mainly by increasing stomatal resistance to entry of carbon dioxide as the guard cells lose turgor.

Rate of crop photosynthesis shows a seasonal trend which reflects variation in the amount of incoming solar radiation. There are also diurnal trends in rate of photosynthesis. These reflect changes in light intensity associated with the elevation of the sun, and changes in stomatal aperture associated with the build up of internal plant water deficits around noon (Milthorpe and Moorby, 1974) (*see* p. 79).

Studying crop growth: growth analysis

Measurements of final yield alone provide no information about the variation in *growth rate* between sowing and harvest which may cause yield differences. The techniques of growth analysis are widely used in agronomic and crop physiological research. For a fuller treatment the reader is referred to Hunt (1978).

In growth analysis, small known areas of the crop or each plot of the experiment are harvested at regular intervals. The plant material is then separated into suitable components (e.g. leaves, stems and ears of cereals; roots, crown and tops of sugar beet; tubers, roots, stolons, stems and leaves of potatoes) and the numbers of appropriate parts determined. Because photosynthesis is essential to crop growth, growth analysis frequently involves estimating the total amount of photosynthesising tissue present. Usually this is measured as plan green area. The harvested material is then dried to constant weight. Assuming that a plant or crop increases from a dry weight of W_1 g/m^2 ground surface and green plan leaf area L_1 m^2 at time t_1 to a dry weight of W_2 g/m^2 and leaf area L_2 m^2 at time t_2, then:

(1) *Crop growth rate* (CGR) $= \dfrac{W_2 - W_1}{t_2 - t_1}$

(mean crop growth rate over time interval $t_1 \rightarrow t_2$)
Units: g/m^2 per unit time.

(2) *Leaf area index* (LAI) $= L_1$ at time t_1 and L_2 at time t_2

Units: leaf area index is calculated as the number of m^2 of leaf (actively photosynthesising tissue)/m^2 of ground, and hence it is usually expressed as a number without any units.

(3) *Relative growth rate* (RGR)
$= \dfrac{2.303 \, (\log_{10} W_2 - \log_{10} W_1)}{t_2 - t_1}$

The mean net assimilation rate over the time period $t_1 \rightarrow t_2$ describes the efficiency of the leaf area in producing dry matter (increase in dry weight per unit leaf area per unit time).
Units: g/g^2 per unit time.

(4) *Net assimilation rate* (NAR or E)
$= \dfrac{2.303 \, (\log_{10} L_2 - \log_{10} L_1)}{L_2 - L_1} \times \dfrac{W_2 - W_1}{t_2 - t_1}$

The mean net assimilation rate over the time period $t_1 \rightarrow t_2$ describes the efficiency of the leaf area in producing dry matter (increase in dry weight per unit dry weight per unit time)
Units: g/m^2 per unit time.

Physiological factors which limit crop growth and yield

Leaf area and light interception

During the early stages of crop growth a large amount of incoming solar radiation is wasted because it hits bare ground. The production of the leaf canopy depends upon:

(1) the rate of seed germination and crop emergence,
(2) the rate of initiation and expansion of leaves.

The main environmental factor affecting these processes is temperature (Terry, 1968). Higher yields could be achieved by selecting varieties which have faster rates of germination and leaf production at the temperatures normally encountered in the field at the time of sowing. Sugar beet, sown in the UK at the end of March, does not reach a LAI of 1 until mid-June. By transplanting seedlings raised in a glasshouse early leaf growth and sugar yield can be increased (Scott and Bremner, 1966).

Bolting resistance is an important varietal characteristic associated with suitability for early sowing. The bulk of the UK sugar beet crop is now drilled before mid April, compared with late April 25 years ago, when varieties were more susceptible to bolting. Other attempts to promote early establishment of sugar beet have involved sowing 'large' seeds (Scott *et al.*, 1974), fluid drilling (Longden *et al.*, 1979) and autumn sowing (Wood and Scott, 1975). Radiation interception during early crop growth can be increased by increasing sowing density (Goodman, 1966), but this results in an unfavourable partition of dry matter between shoots and roots. As leaf area index increases, the proportion of incoming radiation which is intercepted increases, resulting in a higher crop growth rate. The value of leaf area index at which crop growth rate is at its maximum value is often referred to as the 'critical leaf area index'. At this time the crop is intercepting approximately 95% of the incoming solar radiation. Maximum crop growth rates have been observed with leaf area indices of 9–10 in wheat and barley (Watson *et al.*, 1963) and of 3–5 in potatoes and sugar beet (Bremner and Taha, 1966; Scott and Bremner, 1966).

With further increases in leaf area index, a situation may ultimately be reached at which photosynthesis is light saturated in leaves in the upper part of the crop canopy and light limited in leaves in the lower part of the crop canopy. These leaves are potentially a drain on overall resources, since they will continue to respire. However, these leaves have a lower respiration rate, so that crop net photosynthesis tends to plateau out at high leaf area indices (Yoshida, 1972). Differences in canopy structure can also affect yield, higher yields often being achieved with more erect leaves (Trenbath and Angus, 1975). This is because in canopies with erect leaves, the distribution of light is more uniform, and photosynthesis in the lower leaves is less light limited. In addition, for canopies with erect leaves, it may also be possible to increase yield by being able to sow more plants per unit land area without mutual shading. At low leaf area indices greater rates of canopy photosynthesis are achieved with horizontal leaves. Maximum dry matter production might therefore be achieved by having horizontal leaves at early stages of development and erect leaves at later stages of development.

Because light availability is a major factor determining the photosynthetic production of a mature crop canopy, the potential production of crop dry matter depends upon how seasonal charges in leaf area index coincide with seasonal changes in the amount of radiation received from the sun. Wheat and barley reach a peak leaf area index (approx. 10) during June and July when temperature and light intensity are at their maximum values, i.e. when the environment is potentially most suitable for photosynthesis. Conversely, potatoes and sugar beet reach peak leaf area index (approx. 3–5) values much later, when the environment is becoming progressively less favourable for photosynthesis (Watson, 1947a). Relative growth rate is not constant throughout the life of a crop, but changes according to plant development. Superimposed on this will be any seasonal weather influence. The concept of relative growth rate assumes that all of the dry weight is equally as effective in adding new dry weight. This is not so. As plants develop the proportion of structural and non-photosynthetic material increases. The concept of net gain in dry weight per unit leaf area (net assimilation rate) is potentially more biologically meaningful. Net assimilation rate varies with the seasonal trend in climatic factors, highest values being found in the long, hot, bright days of mid-summer (Watson, 1952). However, the effects of fertilisers, season, species and variety on crop growth rate are more closely correlated with differences in leaf area index, rather than with the much smaller differences observed in net assimilation rate (Watson, 1947a and b). As leaf area index increases net assimilation rate decreases because of increased mutual shading. As crops mature crop growth rate declines rapidly. This is because:

(1) older leaves are less photosynthetically efficient,
(2) the proportion of respiring plus non-photosynthetic tissue increases,
(3) leaf area index decreases rapidly because of senescence.

Yields could be increased if leaves survived longer, assuming that the associated delay in maturity did not interfere with the normal harvesting procedure of the crop. When crop senescence is induced prematurely, e.g. by drought, frost or disease, yields are often reduced.

The integral of the leaf area index over the whole growth period is termed the *leaf area duration*. It is measured in units of time (because leaf area index has no units) and takes into account the amount and duration of photosynthetic tissue. Not all of the total leaf area present during the growing season contributes directly to yield. Cereal grains are only formed at anthesis, and hence there is often a good correlation between leaf area duration after ear emergence and grain yield (Thorne, 1974). Accumulation of dry matter by roots of sugar beet and tubers of potatoes starts when leaf area index is small, and continues throughout most of the growth period. With potatoes, total tuber yields are correlated with leaf area duration throughout the whole growth period (Gunasena and Harris, 1969).

More recently, Monteith (1977) has analysed crop yields in terms of the amount of incoming solar radiation which the crop intercepts and the amount of dry matter produced per unit of radiation intercepted. The proportion of radiation intercepted depends upon the leaf area index and the spatial arrangement and orientation of the photosynthetic organs. The total amount of radiation intercepted depends upon seasonal changes in crop canopy size and structure, and the total amount of radiation incident on the crop.

Linear relationships between crop growth rate and amount of intercepted radiation have been reported for potatoes (Allen and Scott, 1980), barley, wheat, maize and soya beans (Biscoe and Gallagher, 1977). For crops which received adequate amounts of nutrients and water, Monteith (1977) observed a very close relationship between total annual dry matter yields and the amount of intercepted radiation. The efficiencies of conversion of incident radiation to dry matter of apples, barley, potatoes and sugar beet were very similar (1.4 g dry matter/MJ intercepted solar radiation).

Assimilate distribution

Photosynthetic assimilates are transported in the conducting tissues of the plant from a *source* (site of synthesis or storage) to a *sink* (site of metabolic activity). Individual plant organs (e.g. seeds) may be a 'source' during one phase of development (germination and seedling growth) and a sink during another (grain filling). The pattern of distribution of assimilates within the plant is not static, but changes in phase with the requirements of crop growth and development.

The use of radioactive carbon dioxide ($^{14}CO_2$) greatly facilitates the study of assimilate distribution in field crops. The technique involves exposing a plant organ or whole canopy to a gas mixture (usually air) containing $^{14}CO_2$ for a certain length of time. The amount of radioactivity in different plant organs (e.g. ears, stems, leaves, tillers, roots of a cereal plant), including the original organ that was fed (e.g. flag leaf), is determined at different times after feeding. This can be done by preparing autoradiographs by exposing the tissues directly to X-ray film (e.g. Quinlan and Sagar, 1972) or by using a counting procedure (e.g. Makunga *et al.*, 1978).

The majority of such studies of assimilate distribution have been carried out on graminaceous crops including wheat (Rawson and Hofstra, 1969), barley (Porter, Pal and Martin, 1950) and several grasses (Williams, 1964; Ryle, 1970; Ryle and Powell, 1972). More recently studies have been extended to examine the effect of environment and husbandry techniques such as nitrogen nutrition (Makunga *et al.*, 1978) and defoliation (Gifford and Marshall, 1973) on assimilate distribution.

Pattern of assimilate distribution in grasses and cereals

During vegetative growth young (expanding) leaves retain all the (^{14}C-labelled) assimilates they produce and import from older (expanded) leaves. Fully expanded leaves export photosynthetic assimilates to the stem apex, tiller buds, small tillers and roots, as well as to expanding leaves.

At the onset of reproductive growth the pattern of assimilate distribution changes, more assimilates being diverted to the stem in order to satisfy the demands of internode elongation. The overall pattern of assimilate distribution from a plant organ is determined by its photosynthetic activity and by the strength, size and proximity of the various sinks. Thus in wheat, as successive leaves emerge, the function of supplying assimilates to these leaves passes from one expanded leaf to the next younger one in order on the stem. In general, upper leaves export assimilates to the stem and ear, lower leaves to the tillers and roots.

The bulk of the dry matter in the grain of temperate cereals is derived from the products of photosynthesis after anthesis of green tissue above the flag leaf node (Thorne, 1974; Evans *et al.*, 1975). Shortly after anthesis when the grains are still small there may be temporary storage of material in the stem, which is later retranslocated to the ear. Under normal conditions the contribution of assimilates fixed before anthesis to grain filling is small, but it may be increased when the plant is stressed (Gallagher *et al.*, 1975; Makunga *et al.*, 1978).

The point at which tillers become independent of their parent shoot is not yet clear. In both wheat (Rawson and Hofstra, 1969) and ryegrass (Ryle and Powell, 1972) some movement of labelled assimilates from the main shoots to the tiller ears has been observed. When grass plants are defoliated, assimilates are diverted to the re-growing tillers (Gifford and Marshall, 1973). Provided that vascular connections are maintained therefore, established independent tillers may again become dependent on the parent shoot when placed under stress. A tiller will not become completely independent until it has a root system sufficient for uptake of moisture and nutrients, as well as sufficient leaf area to produce enough assimilates to satisfy the demands of respiration and growth.

Source–sink relationships and crop yields

Because storage sites (sinks) are often the economi-cally useful part of the crop (tubers of potato, grains of cereals, roots of sugar beet) many workers have attempted to identify whether source activity or sink capacity limits yield. Yield may also be limited by the capacity of the translocation system to transport assimilates from source to sink. The good correlation observed between grain yield of wheat and leaf area duration after anthesis suggests that photosynthetic capacity may limit grain yield (Thorne, 1974). Results of other experiments where grain yield was reduced in response to shading or defoliation also lend support to this hypothesis. Artificial reduction in grain number per ear results in a compensatory increase in grain size (Bingham, 1967), suggesting that grain size may be limited by assimilate supply. Sink capacity is dependent upon the number and size of sinks; in cereals, on the number and size of grains per ear. Bingham (1966) fertilised half of the spikelets of an F_1 hybrid between the wheat varieties Holdfast and Cappelle with Holdfast pollen and half with Capelle pollen, thereby creating sinks of different genetic constitution on the same plant. The resulting two types of grain had different average weights, and therefore sink capacity can limit yield. Semi-dwarf and other modern varieties of winter wheat produce more grains per unit ground area than older, lower yielding varieties (Pearman *et al.*, 1978).

Because the mechanism of assimilate translocation is as yet unknown, it is difficult to identify whether the capacity of the transport mechanism limits yield. Grain growth of cereals and tuber bulking of potatoes are often observed to be linear with respect to time, despite variations in green leaf area (declining due to senescence), weather (over the period of organ growth) and sink size (increase in size during growth). Whether source or sink limits yield is likely to depend upon the environmental conditions during development. In cereals the size of the source (flag leaf and ear) and the potential size of the sink (number of florets per unit ground area) are determined before anthesis, but grain filling is dependent upon photosynthesis and environmental conditions after anthesis. There is also evidence that the photosynthetic activity of the source can be affected by the capacity or activity of the sink. Cooling or removal of potato tubers results in a reduction in net assimilation rate or rate of photosynthesis of tops (Milthorpe and Moorby, 1974). In reciprocal grafts between tops and roots of sugar beet (large storage root) and spinach beet (small storage root) the net assimilation rate of the tops was always greatest when grafted onto sugar beet roots (Thorne and Evans, 1964), independent of the source of the scion leaf. The proximity of the sink to the source of supply may affect its 'strength', relative to other sinks on the same plant. Also, the time of initiation of a sink relative to other sinks with which it is competing may affect its strength. In the ear of wheat the largest grains are found in the basal florets of central spikelets, which are the first to reach anthesis (Rawson and

Evans, 1970). Development of grain in more advanced (basal) florets can inhibit grain set in distal florets (Evans, Bingham and Roskams, 1972), which are less favourably placed with respect to assimilate supply.

Biological yield and harvest index

The total amount of material produced by a crop is termed *biomass*. The economically useful part of the crop is only a fraction of the total biomass. The ratio: economic yield/biomass, is termed the *harvest index*, e.g. for cereals: wt of grain/(wt of grain + straw + roots). In the case of a grass crop the bulk of the above ground material is harvested (grazed or cut) and therefore the harvest index (excluding roots) is approximately 1. However, as grass is grown for feeding to livestock, one could calculate a harvest index for grass as

$$\frac{\text{wt of grass contributing to animal output (meat and milk)}}{\text{biomass}}$$

This ratio will depend upon the nutritive value of the herbage and the proportion of the grass digested. Modern wheat varieties outyield older ones because they have a higher harvest index, not because of increased total dry matter production (Austin *et al.*, 1980).

A physiological approach to increased yield

To summarise three possible ways of increasing yield are:

(1) to increase the size (leaf area), activity (rate of net photosynthesis) or duration of the source,
(2) to increase the number and capacity of the sinks,
(3) to increase the proportion of assimilates translocated from the source to the sink, at the expense of that going to other parts of the plant.

CROP DEVELOPMENT

Crop *development* describes the series of anatomical and morphological changes undergone by the individual plants between the sowing of one 'seed' and the harvesting of the economic part of the crop, which may itself be a seed. In many crops it is useful to identify certain periods of development (incorrectly referred to as 'growth' stages) at which particular husbandry operations must be carried out (e.g. application of herbicides) and hence various keys (e.g. Large, 1954; Tottman and Makepeace, 1979) for recording crop 'growth stages' have been devised. Because the environment, particularly climate, has such a large effect on the rate of crop development, use of calendar dates gives only an approximate guide

to the timing of husbandry operations. Use of crop 'growth stages' offers a much more reliable and repeatable approach.

The pattern of crop development varies between different crop species, and for further information the reader is referred to standard texts (e.g. Gill and Vear, 1980).

Crop development and yield components

In order to gain greater insight into the causes of yield variation agronomists and crop physiologists divide crop yields (per unit area or per plant) into separate *yield components*. The nature of the individual yield components varies according to the botany of the crop concerned, but in each case total yield is equal to the product of the separate yield components:

Crop	Yield	Yield components
Sugar beet	Sugar yield/ha	Root no. × average root size × sugar%
Cereals	wt grain/ha	Ear no. × grains/ear × mean wt/grain
Potatoes	Tuber yield/ha	Tuber no. × average tuber size

The magnitude of the individual yield components is determined by different aspects of the development of the crop, e.g. with cereals:

Yield component	Developmental process
Ear no./unit area	Tiller production and survival
Grain no./ear	Ear development and grain set
Average wt/grain	Grain growth

Variations in yield can thus be attributed to effects on specific developmental processes and yield components. Very often the developmental processes determining the individual yield components occur only over a limited period at some time in the life cycle of the crop. Husbandry or climatic factors at that time may have large effects on the magnitude of the particular yield component being determined. The application of fungicides to winter wheat after ear emergence is unlikely to affect ear number, but, by preventing pathogen induced senescence, may have large effects on grain growth and final grain size. Crops which branch have a remarkable capacity for yield component compensation, a deficiency in one yield component being compensated for by increases in other yield components.

Seasonal aspects of development

In temperate climates plant development and hence cropping is adapted to seasonal weather changes. This is most pronounced in annuals, in which reproductive development occurs during the summer months (favourable environmental conditions), so that a resting stage ('seed' which is frost hardy) is achieved before the onset of adverse winter weather conditions. Those annuals which are frost hardy can be sown in the previous autumn, and are often higher yielding than their spring sown counterparts. These types intercept more radiation in spring, and reach peak leaf area index earlier. Winter-sown types often pass through comparable developmental stages earlier than spring-sown types when environmental conditions are more favourable. They may also benefit from a more extensive and deeper rooting system, enabling them to extract water and nutrients from the lower regions of the soil profile as the surface horizons dry out during the summer months.

Flowering

All annuals, weeds and crops, need to flower and set seeds in order to survive. The term 'flowering' is used here to describe the stage of development at which the shoot apex undergoes a transition from the vegetative to the reproductive state. Flowering is an important factor affecting crop yields. In cereals, peas, beans, oilseed rape and other crops in which the seed or grain is the economic product, not only is flowering essential for yield, but the efficiency of flowering will have a direct effect on yield. In root and fodder crops flowering (bolting) is detrimental to yield or quality, but is still necessary from the point of view of seed multiplication. In grass the bulk of a silage crop consists of stem material and hence flowering is essential, yet at advanced stages of reproductive development the nutritive quality of the crop is markedly reduced.

Vernalisation

Many plants require to experience a period of cold before they will flower. This is known as vernalisation. Certain species have an absolute cold requirement, e.g. winter wheat and winter oilseed rape, and are unsuitable for late spring sowing. Other species have a quantitative requirement for vernalisation (flowering hastened by chilling). The cold stimulus is perceived by the shoot apex. The length of the cold period necessary depends upon the temperatures involved, a short period at a low temperature can be as effective as a longer period at a relatively higher temperature. Certain species (e.g. wheat) can be vernalised by subjecting inbibed seeds to low temperatures, others need to have attained a certain plant size. Vernalised plants can be devernalised if subsequently grown at higher temperatures.

Photoperiodism

Many species need to experience a certain critical length of exposure to daylight (photoperiod) before they will flower. The stimulus is perceived by the leaves and transferred to the stem apex although the chemical nature of the 'message' is unknown. Many plants which have a vernalisation requirement also have a photoperiodic requirement. Plant species can be divided into three groups on the basis of their photoperiodic requirement:

(1) long-day plants which require to experience a photoperiod longer than the critical period in order to flower,
(2) short-day plants which require to experience a photoperiod shorter than the critical period in order to flower,
(3) day-neutral plants which are insensitive to photoperiod.

Some species need to achieve a certain size before they are competent to respond to inductive photoperiods, i.e. they pass through a 'juvenile' phase during which they are insensitive to photoperiod.

With many plant species only one inductive photoperiod of the appropriate length is required in order to induce flowering, others may require several inductive cycles. In general, as more inductive cycles are given, so flowers are formed earlier and the percentage of individuals flowering increases.

For chrysanthemums which are to be sold as cut blooms it is necessary to have a certain amount of vegetative growth before flowering occurs. Chrysanthemum is an obligate short-day plant and requires a light period shorter than the critical period in order to flower, i.e. a night period longer than the critical period. When grown under short days chrysanthemum readily flowers. However, by interrupting the night period with a short period of illumination (a 'night-break') it is possible to delay flowering, so that plants with the necessary vegetative growth can be produced all year round.

Regulation of growth and development

Growth and development in plants involves an orderly sequence of phases of cell division, extension and differentiation. The integration and coordination of growth and developmental processes is controlled by *plant growth hormones*. These are chemical substances, produced within the plant, that can be transported through the plant and which influence the growth and differentiation of the tissues with which they come into contact. Plant growth hormones are conveniently divided up into several groups; auxins, gibberellins, cytokinins, ethylene and abscisic acid (Wareing and Philips, 1978). Knowledge of the methods and sites of synthesis, transport mechanisms and modes of action of each of the plant growth hormones is extremely fragmentary. Much of this knowledge has been obtained by assaying endogenous hormone levels or by studying the effects of exogenously applied hormones; correlating the concentrations involved with the plants response. Plant growth hormones have been shown to affect seed germination; root, stem and leaf growth; seed and fruit development (Wareing and Philips, 1978). These observations have necessarily been made at the level of the individual plant.

Chemical manipulation of crop growth and development

More recently, man has attempted to manipulate the growth or development of his crop plants by application of synthetic hormones, or *plant growth regulators*. Having identified a specific 'target' (e.g. rate of leaf photosynthesis, stem extension), a suitable chemical is sprayed on to the crop, at a suitable time and concentration, in order to produce the desired change in the plant's normal metabolism. Instances of such techniques being used are becoming more common. Many of the herbicides used for weed control are synthetic auxins (e.g. 2, 4-D, 2,4,5-T and MCPA). The advantage of these materials is that they are selective; they can be used to kill weeds without damaging surrounding crop plants. Selectivity is based on differing sensitivity of the cells of the crop and the weed to the synthetic auxin.

During germination the seed reserves of the endosperm are converted to sugar by the enzyme alpha amylase. Exogenously applied gibberellins stimulate alpha-amylase synthesis. Gibberellins can therefore be used to speed up germination of barley seed during malting.

A wide range of 'growth retardants' is available. These chemicals restrict stem extension; in cereals with the aim of minimising the risk of lodging and in ornamental plants to produce more compact plants which are more desirable in appearance. One of the most widely used growth retardants is CCC (Chlormequat). This acts by inhibiting gibberellin biosynthesis, thereby restricting cell division and extension in internodes.

Senescence, at least in detached leaves, is associated with a decline in leaf protein content because of a reduced capacity for protein synthesis. This can be arrested by the application of synthetic cytokinins. In future such materials may be used to prolong leaf area duration, or to prevent senescence in cut flowers and vegetables.

Developing fruits evolve ethylene. Ripening of the fruits of many species, particularly citrus, can be accelerated by exposure to ethylene. The attractiveness of plant growth regulators is that by their use it may be possible, in a single step, to overcome a 'limitation' that would take many years of breeding and selection to remove. The majority of plant growth regulators are the product of a screening programme, i.e. of testing large numbers of potential compounds on a wide range of species. In the future, when more is known of the hormonal regulation of plant growth, it may be possible to synthesise, chemically, compounds which have specific effects on particular processes.

COMPETITION

Plants grown in close proximity to one another will compete with each other for any environmental resource which is available in limited amounts. The components of the environment for which adjacent plants compete are light, nutrients and water. In a field crop competition may occur

(1) between the component plants of the crop,
(2) between the crop plants and weeds.

Competition between the aerial parts of the plants (shoot competition) will be for light; roots compete for nutrients and water.

Husbandry practices affect the degree and type of competition occurring in the crop. Techniques of weed control aim to eliminate competition between crops and weeds. Grassland swards often consist of a number of species, the relative proportions of which can be changed by different systems of defoliation and fertiliser practice. Since seed rate determines the number of plants established per unit area it has a direct effect on the time of onset and degree of competition occurring within the crop. At high seed rates, not only are there more individuals competing, but also the individuals will start to compete with each other at an earlier stage of development. Interplant competition is often used as a means of restricting individual plant size, such as in the production of potatoes and carrots for canning.

The supply of nutrients and water to the individual plant can be increased by judicious use of fertilisers and irrigation. Use of supplementary lighting may only be practical and profitable in the glasshouse.

The topic of weed control is covered elsewhere in this volume, and therefore emphasis here is placed upon the relationships between plant population and crop yield.

Relationships between plant population and crop yield

A study of the relationships between plant population and crop yield is of fundamental importance. The farmer wishes to know:

(1) the plant population associated with maximum yield,
(2) the plant population associated with maximum profitability,
(3) whether the 'optimum' plant population is affected by change in variety, sowing date, intended usage of the crop and other husbandry practices.

Plant *population* is described by two components:

(1) plant *density*, the number of individual plants per unit area;

(2) spatial *arrangement*, the arrangement of these plants on the ground.

The farmer can influence plant density by changing seed rate; spatial arrangement depends upon seed rate and row width.

Both 'yield' and 'density' may require further qualification. 'Yield' may be the total amount of dry matter produced by the crop (biomass), the economic yield, or the yield within a given size grading. 'Density' depends upon the unit of plant population. This is usually the number of individual plants, but may be the number of independent plant units (tillers of cereals, stems of potatoes). Choice of plant unit will depend upon the degree of physiological interdependence between these units.

Effect of plant density on crop yield

As plant density increases, yield per plant decreases because of inter-plant competition. Where yield per unit area approaches a maximum value at high density, but does not reach it nor decline, the relationship is asymptotic. An asymptotic relationship between yield and density is often observed when 'yield' is the total amount of dry matter produced by the crop, or where it consists of the vegetative part of the crop (e.g. potato tubers).

Where yield per unit area declines at high densities the relationship is *parabolic*. Parabolic relationships between yield and density are often observed when 'yield' comprises the reproductive portion of the crop (grains and seeds), or when the produce of the crop is graded.

The relationships between plant population and yield of arable crops have been reviewed by Holliday (1960).

With cereals, both asymptotic and parabolic relationships between grain yield and seed rate have been observed. Decline in yield at high densities is often attributable to increased lodging or disease incidence, but not always. Plant density affects the individual yield components. As density increases ear no./m^2 increases, and ear no./plant, no. of grains/ear and average grain weight all decrease (Kirby, 1967; Darwinkel, 1978).

The ability of cereal crops to compensate for reductions in plant density is due to their capacity to tiller. The seed gives rise to a single shoot, the main stem. The buds in the axils of the leaves on the main stem may develop into new shoots (tillers). Many more tillers are produced than actually survive to bear an ear at harvest. Increased sowing density is associated with greater tiller production (increased maximum no. of tillers per unit area) but a greater proportion of the tillers which are produced die (Kirby, 1967; Darwinkel, 1978).

Recent developments in cereal varieties and changes in husbandry techniques may necessitate a re-evaluation of the relationship between seed rate and grain yield:

(1) the introduction of short, stiff strawed varieties which are less prone to lodging,
(2) the availability of preventative fungicides for disease control,
(3) the use of greater amounts and split-dressings of nitrogen fertiliser,
(4) the introduction of varieties with more erect leaves, and varieties with a reduced tillering habit (e.g. Armada winter wheat).

The effects of plant density on the yield of potatoes are particularly important. In this crop seed costs amount to approximately 30–50% of the total cost of growing the crop. Plant density also affects tuber size and hence size grading, which may affect the suitability of the crop for different outlets. The weight of potato seed planted per unit area depends upon the number of tubers planted and their average size. The relationship between total tuber yield and number and size of tubers planted is asymptotic. When grading is imposed on total tuber yield, the relationship becomes parabolic, the optimum tuber size or tuber number for maximum 'ware' yield decreasing as the riddle size increases. The effects of tuber size, within row spacing and seed rate on the yield of potatoes are now interpreted via their effects on the number of above ground stems (Allen, 1978). The individual eyes on the potato tuber possess one main bud, and possibly other lateral buds. The buds develop into sprouts and later into stems. As stem density increases, the average number of tubers produced per stem decreases. This reduction in number of tubers per stem is offset by the increase in stem number per unit area, so that an increase in number of above ground stems results in a linear increase in number of tubers produced per unit area. Because total tuber yield is found to be fairly constant over a wide range of stem densities, manipulation of stem number and hence tuber number permits manipulation of tuber size and hence suitability for different outlets. The optimum stem populations for canning (700 000/ha) are higher than those where the crop is destined for the domestic market (300 000/ha). There are varietal differences in response to stem density. The variety Majestic tends to produce oversize tubers if planted at too low a population; the variety King Edward tends to produce large numbers of undersize tubers if planted at too high a density.

When the length of the growing season is reduced, in order to produce the maximum number of saleable tubers inter-tuber competition and hence stem density must be reduced.

The optimum plant population for maximum yield of sugar from sugar beet is 75 000 plants/ha (30 000 plants/ac) (Hull and Jaggard, 1971). The relationship between sugar yield, root yield and sugar percentage is asymptotic; yields obtained from populations of 100 000 plants/ha being similar to those obtained from populations of 75 000 plants/ha.

Effects of spatial arrangement on crop yield

Any given plant density can be achieved by sowing plants at widely different patterns of arrangement. For a crop drilled in rows spatial arrangement is measured as *plant rectangularity*, which is the ratio of the distance between plants within the row to the distance between the rows. A low rectangularity is associated with a more even distribution. The effect of plant rectangularity on yield depends upon the plasticity of the individual plant. In general, as rectangularity increases, either by increasing seed rate or row width, yield per unit area gradually declines (Holliday, 1960, 1963). With potatoes and sugar beet, husbandry requirements demand that the crop is grown in fairly wide rows. However, as new mechanisation techniques are introduced, then so seed rates and row widths may be adjusted so as to increase yields.

Uniformity of distribution can also affect yield. Ideally all plants should be equally spaced within the rows. An uneven plant distribution may result in an unevenness of competition and result in variations in plant size at harvest. This problem is of particular importance in the sugar beet crop with the advent of genetic monogerm seed, use of drilling-to-a-stand and elimination of hand-hoeing. At high densities an uneven distribution may result in large numbers of small roots being produced which are unharvestable by machine. At low densities an uneven distribution results in lower yields because more of the incident radiation falls on to bare ground.

Interaction between plant population and other factors

At high densities yield is limited by inter-plant competition for light, nutrients and water, and therefore water and nutrient availability may well affect the response of the crop to increasing plant population. In particular, it may be possible to increase yields at high populations by the use of extra fertiliser and irrigation water.

CROP–WATER RELATIONS

Over three-quarters of the fresh weight of actively growing crops consists of water; and hence an adequate supply of water is essential for plant growth. Water forms a major constituent of physiologically active tissues and maintains cell turgor. It is a reagent in photosynthesis and many other biochemical reactions. It also provides the medium in which materials are transported throughout the plant.

Effects of water deficits on crop growth

Water stress results in a reduction in the size of plant organs. This is mainly due to an effect on cell extension rather than on cell division. Reductions in leaf area as a result of water stress will result in a reduced crop growth rate and hence lower yield. The ratio of root dry weight : shoot dry weight increases under water stress, extra assimilates being diverted towards the roots, which forage deeper in the soil profile for water. Water stress results in stomatal closure. This reduces the rate of carbon dioxide uptake and hence the rate of photosynthesis, which may also result in reduced yield. Rate of respiration is less sensitive to water stress. Water stress also affects floral development. It reduces the rate of initiation of primordia at the growing point, and hence may result in fewer grain sites per ear. Water stress can also affect meiosis and reduce fertilisation (either by dehydration of pollen grains, poor germination of pollen, or impaired pollen tube growth) and hence reduce seed set. Plants under water stress often produce smaller seeds. This is due to both a reduced rate of photosynthesis and a shorter duration of seed filling, as water stress tends to hasten senescence.

Moisture sensitive stages of crop growth

There is abundant evidence to show that crop yields are particularly sensitive to water stress during certain periods of growth or development. Early work on this subject has been comprehensively reviewed by Salter and Goode (1967). In general, an individual plant organ is most sensitive to water stress during its period of most rapid growth.

Interaction between water stress and nutrient supply

Water stress results in a depression in nutrient uptake, particularly of nitrogen and phosphorus. This may contribute towards reduced yield. In addition, although the crop may have roots penetrating the deeper and wetter parts of the soil profile, nutrients concentrated in the dry surface soil may be unavailable. This may also limit growth and yield.

Crop and soil water status

Soil water deficit (the amount of water that needs to be added in order to return the soil to field capacity) and absolute soil water content (volume of water per unit volume of soil) do not provide a reliable estimate of the availability of water to plants or degree of stress being experienced. Early experiments on soil–plant–water relations suffered from the limitation that they were based on irrigation cycles or *soil* water deficits and not *plant* water deficits, which depend upon complex interactions between soil, plant and atmospheric factors.

Evaporation

Evaporation of water is a physical phenomenon. For evaporation to occur three conditions must be satisfied:

(1) there must be a supply of water,
(2) there must be sufficient energy (latent heat of vaporisation of water),
(3) there must be a means of removing the water vapour, otherwise a layer of saturated vapour accumulates over the evaporating surface and evaporation stops.

The *rate* of evaporation depends upon meteorological factors (temperature, wind speed, hours of bright sunshine, humidity, radiation). By the use of weather records one can calculate the total amount of evaporation that has occurred potentially over a given time period (Smith, 1967). *Potential evaporation* (ET) so calculated depends upon the three conditions specified above being satisfied, it is a purely meteorological quantity based on atmospheric demand. Similarly *potential soil water deficit* does not indicate the availability of the moisture in soil to the growing crop. It indicates the extent to which atmospheric demand exceeds water supply (rainfall plus irrigation).

Actual evaporation vs potential evaporation

Providing that the soil surface remains moist evaporation continues. Once the rate of water loss from the surface becomes greater than the rate of upward movement of water from the lower soil horizons a dry crust begins to form on the surface, acting as a barrier to further evaporation. This is the basis for the one year fallow/one year cropping type of farming carried out in many of the drier parts of the world. When ground cover is incomplete actual evaporation will again be less than potential evaporation, evaporation taking place from both crop and soil. Provided that corrections are made for the reduction in potential evaporation when the soil is bare or when ground cover is incomplete, then there is a close relationship between potential evaporation and actual evaporation (French and Legg, 1979). Under continued drought, a point may be reached at which the roots can no longer supply moisture to the leaves in order to maintain evaporation at the potential rate. Ultimately actual evaporation may cease, but potential evaporation and potential soil moisture deficit (based purely on meteorological measurements) continue to increase. Penman (1970) has defined a *limiting potential soil water deficit* (D1) as a threshold of soil dryness at which there is a complete check to growth. If the potential soil water deficit is less than D1 there is no

check to growth. If the potential soil water deficit is greater than D1 there is no growth. The maximum observed potential soil moisture deficit (Dm) gives an indication of the total amount of stress experienced by the crop during the growing season. When water is freely available there is a linear relationship between accumulated growth (crop dry matter) and accumulated potential evaporation (Penman, 1970). The loss of crop yield, due to water stress, is therefore proportional to (Dm − D1). Values of the limiting potential soil water deficit have been reported for a series of crops grown at Rothamsted (silty clay loam) and Woburn (sandy loam) by French and Legg (1979) and Penman (1970). The limiting potential soil water deficits for crops grown at Rothamsted were higher than the deficits for the same crops grown at Woburn. This is because the soil at Rothamsted has a much greater available water capacity.

Development of internal plant water deficits

Water moves through the soil–plant–atmosphere system along a gradient of decreasing *water potential* (Slayter, 1967). The total water potential (ψ) of water in soils or plants is made up of separate components due to the presence of dissolved solutes, hydrostatic or turgor pressure and interfaces. Water moves from high to low water potential until equilibrium is reached, at which point the net flow is zero.

Evaporation of water from the mesophyll cells of the leaves via the stomata lowers the water potential in these cells, and sets up a gradient of water potential between these and adjacent cells. Further evaporation extends the gradient into the conducting tissues of the plant and down to the roots, so that there is a net flow of water from soil to atmosphere. Because plants can only extract water from the soil when leaf water potential is lower than soil water potential, actively transpiring plants are always experiencing a certain degree of water stress. As the evaporative demand of the atmosphere increases during the day, the rate of water loss from the leaves becomes greater than the rate of water supply from the roots and an internal plant water deficit develops. Later in the day, when evaporative demand has fallen and transpiration becomes less than water uptake, tissues recover from water deficit. By the following morning, leaf water potential again equals soil water potential, but this equilibrium value is lower than on the previous day because of the net water loss from the system.

Soil water potential is approximately zero when the soil is wet. At this point only a small difference between leaf water potential (becomes negative) and soil water potential is required to sustain flow. However with continued evaporation the soil dries out (ψ soil becomes more negative) and a progressively larger difference between leaf water potential and soil water potential is required to maintain flow.

Stomata remain fully open until a certain critical value of leaf water potential is achieved, and then close rapidly. During a drying cycle therefore, there is

(1) a diurnal fluctuation in plant water status, highest internal plant water deficits being achieved at the time of day when evaporative demand is greatest,
(2) a gradual decrease in soil water potential which sets the limit of recovery possible by the plant tissues during the following evening.

As leaf water potential falls, leaf cell turgor declines, and the leaves begin to wilt. Once the soil water potential is such that the leaves cannot recover turgor, then this soil water content is the permanent wilting percentage. The physiological significance of crop water deficits has been recently reviewed by Begg and Turner (1976).

CROP NUTRIENT RELATIONS

Of the total amount of dry matter produced by a crop, only a small percentage of this consists of mineral material derived from the soil. However, nutrient availability is a critical factor affecting crop yields and is readily manipulated by correct fertiliser application.

The supply of nutrients to crop tissues depends upon

(1) the release of nutrients from a solid 'reserve' to a soluble form, in the soil solution,
(2) the transfer of nutrients in the soil solution to the crop root surfaces,
(3) nutrient uptake by roots,
(4) transport of nutrients from the roots to the other tissues of the plant.

A reduction in the rate of any one of these processes could result in a deficiency being experienced by the plant. The nature of the solid 'reserve' and the processes involved in the release of plant nutrients vary according to the nutrient concerned. This subject is comprehensively reviewed in standard soil science textbooks. One of the fundamental problems of soil chemistry is to devise simple chemical or biological tests that will indicate the amounts of 'available' nutrients in soils.

Two processes are involved in the transfer of nutrients in the soil solution to the root surfaces. The flow of water towards the roots, set up by the transpiration stream, carries nutrients with it. The supply of nutrients by this *mass flow* process depends upon the factors which affect the movement of water in soils (pore size distribution, suction gradient). If the rate of nutrient uptake becomes greater than the rate of supply of nutrient from the body of the soil, a concentration gradient is set up, and nutrients may be supplied to roots by *diffusion*.

Both mass flow and diffusion take place in the liquid phase, and hence nutrient supply to crops is reduced in dry conditions. Crops usually receive most of their nitrate by mass flow. Diffusion of phosphorus and potassium is an important source of supply of these nutrients.

The method of nutrient uptake by plants is not yet completely understood. Roots do show some selectivity in uptake. This is apparent because the ionic composition of plants is not the same as the ionic composition of the soil solution. Root uptake may involve active processes. This may be important for ions which occur in the soil solution at very low concentrations. The total amount of nutrient in the crop is equal to the product of the nutrient concentration (percentage of the dry matter) and the total amount of crop dry matter. In general, the nutrient concentration of the major plant nutrients increases after sowing, reaching a maximum during early growth and then declining. The total amount of nutrient increases throughout the growing season. Greatest rates of nutrient uptake are observed when crops are growing most rapidly. The total amount of nutrient in the plant may decrease at later stages of development. This may be attributable to either loss of plant material (tillers and dead leaves), leaf leaching, or root efflux. In cereals and grasses, large amounts of potassium taken up during growth, may be lost prior to harvest.

During senescence, much of the mineral material in plant tissues may be retranslocated to other parts of the plant (Wareing and Philips, 1978).

The relationships between nutrient supply and crop yields are discussed elsewhere in this volume.

References

ALLEN, E. J. (1978). Plant density. In *The Potato Crop*, pp. 279–326. Ed. by P. M. Harris. London: Chapman and Hall Ltd

ALLEN, E. J. and SCOTT, R. K. (1980). *Journal of Agricultural Science, Cambridge* **94**, 583–606

AUSTIN, R. B., BINGHAM, J., BLACKWELL, R. D., EVANS, L. T., FORD, M. A., MORGAN, C. L. and TAYLOR, M. (1980). *Journal of Agricultural Science, Cambridge* **94**, 675–689

BEGG, J. E. and TURNER, N. C. (1976). *Advances in Agronomy* **28**, 161–218

BINGHAM, J. (1966). *Nature* **209**, 940

BINGHAM, J. (1967). *Journal of Agricultural Science, Cambridge* **68**, 411–422

BISCOE, P. V. and GALLAGHER, J. N. (1977). Weather, dry matter production and yield. In *Environmental Effects on Crop Physiology*, pp. 75–100. Ed. by C. V. Cutting and J. J. Landsberg. Proceedings of the 5th Long Ashton Symposium, 1975. London: Academic Press

BREMNER, P. M. and TAHA, M. A. (1966). *Journal of Agricultural Science, Cambridge* **66**, 241–252

DARWINKEL, A. (1978). *Netherlands Journal of Agricultural Science* **26**, 383–398

EVANS, L. T. (1975). *Crop physiology—some case histories.* Cambridge: Cambridge University Press

EVANS, L. T., BINGHAM, J. and ROSKAMS, M. A. (1972). *Australian Journal of Biological Sciences* **25**, 1–8

EVANS, L. T., WARDLAW, I. F. and FISCHER, R. A. (1975). Wheat. In *Crop physiology—some case histories*, pp. 101–150. Ed. by L. T. Evans. Cambridge: Cambridge University Press

FRENCH, B. K. and LEGG, B. J. (1979). *Journal of Agricultural Science, Cambridge* **92**, 15–37

GALLAGHER, J. N., BISCOE, P. V. and SCOTT, R. K. (1975). *Journal of Applied Ecology* **12**, 319–336

GIFFORD, R. M. and MARSHALL, C. (1973). *Australian Journal of Biological Sciences* **26**, 517–26

GILL, N. T. and VEAR, K. C. (1980). *Agricultural Botany; Volume 1, Dicotyledonous crops; Volume 2, Monocotyledonous crops.* London: Duckworth

GOODMAN, P. J. (1966). *Agricultural Progress* **41** 89–107

GRAHAM-BRYCE, I. J. (1973). *Proceedings of the 7th British Insecticide and Fungicide Conference* **3**, 921–932

GUNASENA, H. P. M. and HARRIS, P. M. (1969). *Journal of Agricultural Science, Cambridge* **73**, 245–259

HEYDECKER, W. (1972). Vigour. In *Viability of seeds*, pp. 209–252. Ed. by E. H. Roberts. London: Chapman and Hall Ltd.

HEYDECKER, W. and COOLBEAR, P. (1977). *Seed Science & Technology* **5**, 353–425

HOLLIDAY, R. (1960). *Field Crop Abstracts* **13**, 159–167; 247–254

HOLLIDAY, R. (1963). *Field Crop Abstracts* **16**, 71–81

HULL, R. and JAGGARD, K. W. (1971). *Field Crop Abstracts* **24**, 381–390

HUNT, R. (1978). *Plant Growth Analysis. The Institute of Biology's Studies in Biology No. 96.* London: Edward Arnold

KIRBY, E. J. M. (1967). *Journal of Agricultural Science, Cambridge* **68**, 317–324

LARGE, E. C. (1954). *Plant Pathology* **3**, 128–129

LONGDEN, P. C., JOHNSON, M. G., DARBY, R. J. and SALTER, P. J. (1979). *Journal of Agricultural Science, Cambridge* **93**, 541–552

MAKUNGA, O. H. D., PEARMAN, I., THOMAS, S. M. and THORNE, G. N. (1978). *Annals of Applied Biology* **88**, 429–437

MILTHORPE, F. L. and MOORBY, J. (1974). *An Introduction to Crop Physiology.* Cambridge: Cambridge University Press

MONTEITH, J. L. (1977). *Philosophical Transactions of the Royal Society of London, Series B* **281**, 277–294

PEARMAN, I., THOMAS, S. M. and THORNE, G. N. (1978). *Journal of Agricultural Science, Cambridge* **91**, 31–45

PENMAN, H. L. (1970). *Journal of Agricultural Science, Cambridge* **75**, 69–73; 75–88; 89–102

PORTER, H. K., PAL, N. and MARTIN, R. V. (1950). *Annals of Botany* **14**, 55–67

QUINLAN, J. D. and SAGAR, G. R. (1962). *Weed Research* **2**, 264–273

RAWSON, H. M. and EVANS, L. T. (1970). *Australian Journal of Biological Sciences* **23**, 753–764

RAWSON, H. M. and HOFSTRA, G. (1969). *Australian Journal of Biological Sciences* **22**, 321–331

ROBERTS, E. H. (1972a). Loss of viability and crop yields. In *Viability of seeds*, pp. 307–320. Ed. by E. H. Roberts. London: Chapman and Hall Ltd

ROBERTS, E. H. (1972b). Dormancy, a factor affecting seed survival in the soil. In *Viability of Seeds*, pp. 321–359. Ed. by E. H. Roberts. London: Chapman and Hall Ltd.

RYLE, G. J. A. (1970). *Annals of Applied Biology* **66**, 155–167

RYLE, G. J. A. and POWELL, C. E. (1972). *Annals of Botany* **36**, 363–375

SALTER, P. J. and GOODE, J. E. (1967). *Crop Responses to Water at Different Stages of Growth. Research Review No. 2.* Farnham Royal: Commonwealth Agricultural Bureaux

SCOTT, R. K. and BREMNER, P. M. (1966). *Journal of Agricultural Science, Cambridge* **66**, 379–388

SCOTT, R. K., HARPER, F., WOOD, D. W. and JAGGARD, K. W. (1974). *Journal of Agricultural Science, Cambridge* **84**, 517–530

SLATYER, R. O. (1967). *Plant–water relationships.* London: Academic Press

SMITH, L. P. (1967). *Potential Transpiration.* Ministry of Agriculture, Fisheries and Food, Technical Bulletin, No. 16. London: HMSO

TERRY, N. (1968). *Journal of Experimental Botany* **19**, 795–811

THORNE, G. N. (1974). *Rothamsted Experimental Station. Report for 1972*, Part 2, 5–25

THORNE, G. N. and EVANS, A. F. (1964). *Annals of Botany* **28**, 499–508

TOTTMAN, D. R. and MAKEPEACE, R. J. (1979). *Annals of Applied Biology* **93**, 221–234

TRENBATH, B. R. and ANGUS, J. F. (1975). *Field Crop Abstracts* **28**, 231–244

WAREING, P. F. and PHILIPS, I. D. J. (1978). *The Control of Growth and Differentiation in Plants.* Oxford: Pergamon Press

WATSON, D. J. (1947a). *Annals of Botany* **11**, 41–47

WATSON, D. J. (1947b). *Annals of Botany* **11**, 375–407

WATSON, D. J. (1952). *Advances in Agronomy* **4**, 101–145

WATSON, D. J., THORNE, G. N. and FRENCH, S. A. W. (1963). *Annals of Botany* **27**, 1–22

WILLIAMS, R. D. (1964). *Annals of Botany* **28**, 419–426

WOOD, D. W., LONGDEN, P. C. and SCOTT, R. K. (1977). *Seed Science & Technology* **5**, 337–352

WOOD, D. W. and SCOTT, R. K. (1975). *Journal of Agricultural Science, Cambridge* **84**, 97–108

WURR, D. C. E. (1978). 'Seed' tuber production and management. In *The Potato Crop*, pp. 327–354. Ed. by P. M. Harris. London: Chapman and Hall Ltd

YOSHIDA, S. (1972). *Annual Review of Plant Physiology* **23**, 437–464

4

Crop nutrition

J. S. Brockman

All the essential nutrients required by green plants can be taken up by the plant in inorganic form, unlike animals where some organic compounds are also needed. There are 16–19 elements known to be essential for plant growth and these are listed in *Table 4.1*, together with a summary of their main functions in the plant and the form in which they are taken up by the plant. Much further information on the uptake and role of plant nutrients is contained in Mengel and Kirkby (1978).

An essential nutrient is defined as one that is required for the normal growth of the plant, is directly involved in the nutrition of the plant and cannot be substituted by another nutrient. Not all plants have the same essential requirements, and sodium (Na), silicon (Si) and cobalt (Co) are needed by some plant species only. Sometimes essential plant elements are divided into the macronutrients carbon (C), hydrogen (H), oxygen (O), nitrogen (N), sulphur (S), phosphorus (P), potassium (K), calcium (Ca), magnesium (Mg), sodium (Na), silicon (Si) and micronutrients boron (B), manganese (Mn), copper (Cu), zinc (Zn), molybdenum (Mo), iron (Fe), chlorine (Cl), cobalt (Co). This division is based on the probable content of each element in the plant, but plants differ greatly in their nutrient composition so such a distinction is not always meaningful. Also, the importance of a nutrient is not related to the quantity contained in plant tissues, so that the economic significance of the nutrient and the probability of deficiency symptoms cannot be related to the level of plant requirement.

SUMMARY OF NUTRIENT FUNCTIONS OF SOIL-SUPPLIED ELEMENTS AND DEFICIENCY SYMPTOMS

Nitrogen (N)

Nitrogen is an essential constituent of all proteins and so is vital to any living plant. Broadly, there are three stages in the utilisation of N by the plant:

(1) *stage 1* uptake of NO_3^- and NH_4^+ ions, with N present in the plant in inorganic form;
(2) *stage 2* N present in low molecular weight organic compounds such as amino acids, amides and amines;
(3) *stage 3* N in high molecular weight organic compounds such as proteins and nucleic acids.

If there is plenty of available N in the soil, the rate of N uptake (stage 1) can exceed the rate of protein synthesis (stage 2–stage 3). Whilst this is not harmful over a short period because protein production can catch up with a short-term flush of N uptake, if the imbalance continues for a long period there can be:

(1) excessive elongation of vegetative cells, leading to soft weak growth, e.g. lodging in cereals;
(2) decrease in carbohydrate production; this can lead to unpalatability in grazed grass, fermentation problems in grass cut for silage, and poor quality in plants that store carbohydrate in roots or tubers, such as sugar beet and potatoes.

Deficiency symptoms

These show first in older leaves, as young tissues get priority for any limited N supply available to the plant: older leaves become senile prematurely, show a general yellow/brown colour and probably become detached from the stem. Nitrogen deficient plants become spindly, lack vigour and are dwarfed (i.e. low yielding); also they have a pale colour and a low leaf:stem ratio. Often N deficient plants ripen earlier than those receiving adequate N nutrition. Nitrogen deficiency in young plants can be corrected rapidly by

Table 4.1 Summary of essential nutrients in crop nutrition, form of uptake and function in plant

Element	Form of nutrient uptake	Main functions in plant
C	CO_2 from atmosphere (HCO_3^- from soil solution)	major constituent of organic material, can account for 40% of dry weight of plant
H	H_2O from soil (and from atmosphere if humid)	linked with C in all organic compounds
O	CO_2 and H_2O, O_2 during respiration	with C essential in carboxylic groups, with H in oxidation–reduction processes
N	NO_3^- and NH_4^+ from soil solution (N_2 from air if fixed by micro-organisms)	essential constituent of proteins
S	SO_4^{2-} from soil solution. SO_2 absorbed from atmosphere	constituent of some essential amino acids, e.g. cysteine, cystine and methionine
P	HPO_4^{2-} and $H_2PO_4^-$ from soil solution	vital constituent of living cells, associated with storage and transfer of energy within the plant
K	K^+ from soil solution	essential for efficient water control within the plant and for formation and translocation of carbohydrate
Na	Na^+ from soil solution	partially interchangeable with K^+ in most plants, essential in some plants only, e.g. those of marine origin
Ca	Ca^{2+} from soil solution	essential role in biological membranes
Mg	Mg^{2+} from soil solution	essential constituent of chlorophyll, some enzymes and some organic acids
Si	probably $Si(OH)_4$ from soil solution	used in cellulose framework and interacts with P in the plant (mechanism uncertain)
B	H_3BO_3 or BO_3^{3-} from soil solution	assists in carbohydrate synthesis, uptake of Ca^{2+} and absorption of NO_3^-. Big species variation in need for B
Mn	Mn^{2+} from soil solution (availability to plant heavily dependent on soil pH)	associated with chlorophyll formation and some enzyme systems
Cu	Cu^{2+} or copper chelates[1] in soil solution	small quantities needed for enzyme systems converting NO_3^- to protein
Zn	Zn^{2+} from soil solution	assists with starch formation and some enzyme systems
Mo	MoO_4^{2-} from soil solution (or as chelates[1])	small quantity essential for enzymes controlling N nutrition (also for N-fixing bacteria)
Fe	Fe^{2+} or Fe^{3+} or Fe chelates[1]	essential in chlorophyll and enzyme activities, associated with enzyme activities in photosynthesis
Cl	Cl^- from soil solution	involved in evolution of O_2 during photosynthesis (excess of Cl more common problem than deficiency)
Co	Co^{2+} or Co chelates[1] from soil solution	probably not essential for most plants, it is essential for N fixation by bacteria and could be needed in N nutrition by some plants

[1] In a chelate the metal atom concerned is bound to an organic compound, forming a highly stable and water-soluble compound in which the nutrient availability to the plant is largely unaffected by soil pH.

application of nitrate-containing fertilisers, but over-application of these fertilisers should be avoided as this will lead to the imbalance problems listed above.

Phosphorus (P)

Phosphorus is associated with the transfer and storage of energy in the plant, as such it is vital for living processes within the cell. A supply of P is essential to the plant during the early stages of growth and also during seed formation—the latter because P is found in phytin seed reserves that can supply P to the seedling in the next generation. Most P is taken up from the soil in inorganic form, but can be converted to organic forms in a matter of minutes: hence most of the P found in young tissue is in organic compounds. Phosphorus in older tissue is mainly in inorganic form.

Deficiency symptoms

These are not easily detected from appearance. Early stages of deficiency cause a reduced growth rate and limitation in root development, often with leaves looking a rather healthy dark green-bluish green colour: also there will be poor seed formation. The plant will appear stunted only when the deficiency is severe, and at this stage the stem can be reddish in colour. Phosphorus deficiency is hard to correct in existing plants once visual symptoms are found, because these symptoms appear at a late stage in the diagnosis. Where plants are to be sown in P-deficient soil, it is essential some water-soluble P is placed near the seed at the time of sowing.

Potassium (K) and sodium (Na)

Potassium is essential for efficient water relationships in the plant, formation and translocation of carbohydrate and activation of some enzymes. Plant requirements for K vary from high to very high, depending on the species. As the size of the plant increases, so does the need for K, so that $N \times K$ interactions can occur, where the full effect of N in increasing plant size may be restricted by insufficient available K.

Under some conditions Na can replace K, particularly in plants of marine origin, where some Na is essential, e.g. sugar beet. However, Na cannot replace K entirely as Na does not enter into enzymatic activity in place of K.

Deficiency symptoms

For K these are seen in a loss of plant turgidity as water control mechanisms become impaired. Also growth rate slows and older leaves show yellow/brown spots on leaves and /or whole leaves become brown at the edge.

In potatoes and brassicas initial K deficiency is shown by a blue-green colour, leading to bronzing and marginal necrosis of the leaves. In barley the necrosis of leaves can be spotted as well as on the leaf margins and can be whitish-brown or purplish-brown. In legumes the first symptoms are often whitish spots on leaves followed by marginal browning or bronzing. Generally application of fertilisers containing muriate or sulphate of potash will correct the deficiency provided the deficiency has not reached an advanced stage.

Sulphur (S)

Sulphur is required for the production of the essential amino acids cysteine, cystine and methionine. Because of this, lack of S in the plant may be linked with ineffective use of N, i.e. restriction of protein synthesis and a high inorganic N content in the tissues.

Deficiency symptoms

These are rarely seen in agricultural crops in UK at the present time because the theoretical deficit in supply from fertilisers to meet plant demand is overcome from atmospheric pollution depositing S arising from the burning of fossil fuels. However brassicas and grass receiving high N rates are sensitive to S deficiency and can be regarded as indicator crops. Also most legumes have a high S-requirement as S is essential for N fixation by *rhizobial* bacteria in root nodules.

Calcium (Ca)

Calcium is essential for plant growth, being used in cell walls and biological membranes, and plants need Ca in substantial quantities. However, Ca is the major element controlling soil pH, and so low levels of Ca in the soil will inhibit plant growth via increased soil acidity long before Ca becomes limiting as a nutrient in its own right. Whilst vegetative growth of plants is not restricted by lack of Ca, in some plants poor translocation of Ca within the plant can cause *calcium deficiency* in fruit, e.g. bitter pit in apples and blossom-end rot in tomatoes. Prevention of Ca deficiency in fruit is by spraying calcium nitrate on the leaves when the fruit is setting.

Strontium

Strontium is chemically related to calcium and can be taken up and deposited in cell walls in much the same way as Ca. This fact is important if strontium-90 from nuclear fall-out is present, as it will accumulate in the plant.

Magnesium (Mg)

Some 20% of the Mg in the plant is found in chlorophyll and much of the rest is in enzyme systems or associated with the movement of anions (mainly phosphates). Magnesium uptake is rapid in the early stages of growth. Where the plant has access to abundant levels of K^+ and NH_4^+ ions, the uptake of Mg can be reduced, leading to induced Mg deficiency, particularly in grass, potatoes, sugar beet and tomatoes.

Deficiency symptoms

These are always seen first in older parts of the plant, as young tissues get priority. Typical deficiency symptoms are intervenal chlorosis, with the leaf veins standing out as dark green. In sugar beet the chlorosis first shows on the margins of older leaves and can be followed by black necrotic areas. In potatoes, mottling on the leaf can be followed by reddish/purple tinting and loss of older leaves.

Low levels of Mg can cause foliar symptoms without reduction in yield, although yield loss has been found in severe deficiency in sugar beet and potatoes. In grass, Mg deficiency does not affect the yield of grass, but it can lead to Mg deficiency in stock eating the grass (hypomagnesaemia).

Where Mg deficiency is caused by a genuine lack of available Mg in the soil, the application of 250 kg/ha of magnesite (45% MgO) or 400 kg/ha of kieserite (27% MgO) will help maintain Mg supplies. For rapid action, foliar application of 25–35 kg/ha of Epsom salts (16% MgO) in 400 ℓ water is recommended. If Mg deficiency is *induced* by K^+ or NH_4^+, then foliar application is essential for plant deficiency to be corrected, although with grass it is better and more certain to feed an Mg supplement to the stock at risk.

Boron (B)

Boron acts in a similar way to P in hydrolysis reactions and it facilitates sugar translocation. It is an important minor element and its availability is much reduced at high soil pH.

Deficiency symptoms

They have not been found in UK in cereals or grass, but they occur commonly in root crops (particularly swedes and sugar beet), legumes and leafy brassicas. Boron deficiency is characterised by the death of the apical growing point of the main stem and failure of lateral buds to develop shoots. Leaves may become thickened and sometimes they curl. In sugar beet, fodder beet and mangolds mis-shaped young leaves are the first sign, followed by death of primary leaves and a 'rosette' of secondary leaves covering a scarred root crown. Hollowing may extend from the crown into the root and fungal decomposition can occur. In swedes and turnips plant growth may appear normal, but at harvest the roots may have brown hollow areas known as 'crown rot'. In lucerne terminal leaves become yellow or red, plants are stunted and adopt a rosette form with death of terminal buds.

Treatment is by use of a boronated fertiliser or borax at a rate equivalent to 20 kg/ha of borax. Alternatively, foliar application can be made using 'solubor' at 5–10 kg/ha in 250–500 ℓ/ha of water using a wetting agent. Care must be taken *not* to use boronated fertilisers on cereals as they are sensitive to boron *toxicity*, which results in browning of leaf tips and rapid necrosis of the whole leaf.

Manganese (Mn)

Manganese is involved in chlorophyll formation and enzymatic control of oxidation-reduction processes. Most UK soils contain adequate available Mn, but some organic soils of high pH are deficient and so are some heavily leached podzols. Manganese deficiency can also be found in badly drained soils. It can be induced by over-liming or deep ploughing of calcareous soils.

Deficiency symptoms

They are chlorosis of leaves. In oats this deficiency is called 'grey speck'; in barley the leaves have brown spots and streaks; in wheat there are intervenal white streaks. In all three of these cereals maturity is delayed and ear emergence reduced with high incidence of blind ears. In sugar beet the chlorosis is called 'speckled yellows' and is most severe in early stages of growth, accompanied by an upright leaf habit with curling edges. Brassicas show chlorotic marbling; potatoes have stunted leaves with small terminal leaves rolled forward; peas and beans often show

brown lesions on the inner surface of cotyledons, the term 'marsh spot' referring both to the condition on the plant and its occurrence on badly drained organic soils. Although Mn deficiency may be controlled by soil application of 125–250 kg/ha of manganese sulphate, this treatment has no long-lasting effect as the Mn is oxidised rapidly. Recommended treatment is foliar application of 6–10 kg/ha of manganese sulphate in 225–1000 ℓ/ha of water using a wetting agent.

Copper (Cu)

The function of Cu in the plant is uncertain, but it acts as a catalyst in enzymes associated with the metabolic pathways converting N to protein and the redox processes in cells.

Deficiency symptoms

They can be found in fruit and some cereal crops, particularly wheat growing on some peats in well-defined areas, on leached acid sandy soils such as heaths and on some shallow chalks with high organic matter content. In cereals, deficiency symptoms are not seen until the plants are well-developed, at which time the symptoms can change rapidly from yellowing of the tips of youngest leaves to spiralling of leaves: ears have difficulty in emerging, have white tips and are devoid of grain. Wheat grown on copper-deficient chalk soils specifically can show blackening of ears and straw.

Treatment can be by application of 60 kg/ha of copper sulphate to the soil—a treatment that can last two to three years on peaty and sandy soils but not on the deficient chalks: otherwise foliar application of 2–3 kg/ha of copper oxychloride or cuprous oxide can be effective when applied at a fairly late stage of growth (and this is when symptoms may first show).

Iron (Fe)

Iron is essential for the proper functioning of chlorophyll and related photosynthetic activity. Iron uptake is strongly related to soil pH and some species such as sugar beet, brassicas and beans can show lime-induced Fe deficiency.

Deficiency symptoms

These are chlorotic markings on younger leaves (in contrast to Mn deficiency which can affect all leaves irrespective of age). Iron deficiency is found more widely in horticultural crops than agricultural ones, being found particularly in fruit and calcifuge plants such as azalea and rhododendron. The only effective treatment for Fe deficiency is foliar application of an iron chelate (*see Table 4.1*).

Molybdenum (Mo)

Molybdenum is essential for the enzymes nitrogenase and nitrate reductose. It is also important for N fixation by *rhizobial* bacteria. Molybdenum deficiency can occur on some acid soils and on soils of high pH. Excessive availability of Mo can induce low Cu uptake in grass, causing 'teart pastures' with Cu deficiency in grazing stock.

Deficiency symptoms

They are found only to a marked extent in cauliflowers, where hearts fail to form in plants nearing maturity, leading to a condition known as 'whiptail'. Sometimes liming will alleviate whiptail, otherwise plants should be sprayed early in their growth with 0.25–0.5 kg/ha of sodium molybdate solution.

Zinc (Zn)

Plants need Zn in very small quantities and in the UK only fruit trees show deficiency symptoms; deficiency can be induced by overliming. Excess Zn can be toxic to plants, causing induced Fe deficiency. Some sewage sludges contain appreciable quantities of Zn and regular use of such materials in the same area of the farm can lead to an undesirable build-up of Zn in the soil. Sensitive crops are beans, clover, lucerne and cereals. Grass is not sensitive, but care should be taken if grass with a high Zn concentration is eaten by dairy cows as high levels of Zn can occur in milk.

Silicon (Si)

Silicon is the second most abundant element in the lithosphere after oxygen, and so it is not likely to limit plant growth. Silicon is used by plants in the cellulose framework of cells and plants grown deliberately in the absence of Si have a very limp habit. It has been shown that application of 450 kg/ha of sodium silicate to some soils can increase availability of P.

Cobalt (Co)

Cobalt is essential for micro-organisms that fix N but it is probably not essential for higher plants. Excess Co can induce deficiency of both Fe and Mn.

Chlorine (Cl)

Chlorine is essential to plants in a small quantity only, where it is used in processes connected with evolution of O_2 during photosynthesis.

The effect of excess Cl is relatively common, particularly on soils affected by salt. Chlorine toxicity is seen as burning of leaf tips, bronzing and premature yellowing of leaves. Sugar beet, barley, maize and tomatoes are tolerant of high available Cl, but potatoes, lettuce and many legumes are sensitive. Where there is a risk of a sensitive crop being affected, K should not be applied as muriate of potash (KCl) but as sulphate of potash.

ORGANIC MANURES

Livestock manures

On stock farms large quantities of faeces and excreta are accumulated from housed animals. Quite apart from the environmental and health aspects of the storage and disposal of these products, all animal manures have some value as plant nutrients. The application of these manures to farm land can represent both a relatively safe and a money-saving method of disposal. Before outlining the probable nutrient value of animal manures (including slurry) it must be stressed that all such materials can be extremely variable in composition for three main reasons:

Variation in nutrient content

Main factors affecting the nutrient content of the manure as collected for field distribution are:

(1) the quantity of excreta produced, based on the size and type of animal involved (*see Table 4.2*);
(2) the composition of the excreta is influenced by the animals' diet;
(3) the method of collection and storage of excreta, involving the degree of dilution by rainwater and/or washing-down water or straw.

Table 4.2 Approximate quantity of excreta from livestock

Type of stock	Mean body weight (kg)	Faeces+ urine (ℓ/d)	Moisture content (%)
Dairy cow	500	41	87
Two-year bullock	400	27	88
One-year bullock	220	15	88
Pig (dry meal feed)	50	4	90
Pig (2.5:1 water:meal)	50	4	90
Pig (4:1 water:meal in feed)	50	7	94
Pig (whey fed)	50	14	98
1000 laying hens	2000	114	75

From MAFF (1979a) with permission

Variation in losses during storage

Main factors are:

(1) gaseous loss of N as ammonia; the quantity lost will vary with the conditions of storage and temperature. Loss of 10% N is average but loose-

stacked farmyard manure (FYM) that is turned prior to spreading can lose up to 40%. Where slurry is stored for long periods, 10–20% of N can be lost and agitation before removal will aggravate the loss.

(2) leaching losses from FYM stored in the open are in the range 10–20% of N, 5–8% of P_2O_5 and 25–35% of K_2O: the flatter the heap the greater the likely loss.

(3) seepage losses from slurry can result in 20–25% loss of N, a little loss of P_2O_5 and 20–30% loss of K_2O.

The above sources of loss can be additive, so that material that has been stored outdoors for a long period will have a greatly reduced manurial value due to gaseous, leaching and seepage losses.

Proportion of nutrients available to plant roots

Not all the nutrients contained in manure are available to plants, as much of the material is in complex organic compounds that will form part of the soil organic cycle—perhaps having an eventual nutrient value, depending on the activity of the soil biomass. *Table 4.3* sets out the standard ADAS guidelines on the composition of livestock manures: these values represent the 'agreed' figures and as such are used by ADAS and others in preparing recommendations.

The nutrient values in *Table 4.3* are based on assumed spring application, as this is normally the most effective time for nutrients contained in manures—particularly those liable to leaching over winter. If manures are applied at other times in the year, then the N value should be reduced to the following comparative effectiveness:

if spring application = 100, then
autumn = 0–20
early winter = 30–50
late winter = 60–90

Phosphate and potash values are not affected by leaching and should not be reduced.

Whenever livestock manures are applied for a crop, its nutrient value should be recognised and an adjustment made in the quantity of inorganic fertilisers used. Standard fertiliser recommendations normally allow for this, for example:

fertiliser recommendation for maincrop potatoes

	kg/ha to apply		
	N	P_2O_5	K_2O
fertiliser nutrients without FYM	220	300	300
available from 50 tonnes/ha of FYM	75	100	200
fertiliser nutrients with FYM	145	200	100

Risks from application of livestock manures

Hypomagnesaemia in grazing stock

Cattle FYM and slurry are rich in potash and where high rates of such manures are applied in late winter and spring to grassland, there will be an increase in the potash content of the grass. Spring herbage is naturally low in Mg, and as a high plant content of K depresses Mg content, then there is an enhanced risk of hypomagnesaemia in animals grazing grass that has received cattle manures. Where such manures have to be applied to grass that is allocated to spring grazing, then it should be applied well before Christmas.

Animal disease

The main hazard that exists is from the infection of pasture with bacteria of the *Salmonella* group—normally as a result of applying infected cattle or poultry manures and slurries. The following advice is offered to minimise the risk:

Table 4.3 Composition of manures and nutrients available to crops from spring application[1]

Type	Composition				Available nutrients		
	approx. DM%	% of fresh weight			N	P_2O_5	K_2O
		N	P_2O_5	K_2O			
FYM					(kg/tonne of material)		
cows and cattle	25	0.6	0.3	0.7	1.5	2.0	4.0
pig	25	0.6	0.6	0.4	1.5	4.0	2.5
poultry (deep litter)	70	1.7	1.8	1.3	10.0	11.0	10.0
poultry (broiler litter)	70	2.4	2.2	1.4	14.5	13.0	10.5
poultry (dried house droppings)	70	4.2	2.8	1.9	25.0	17.0	14.0
Slurry (undiluted)					(kg/m³, or kg/1000 ℓ)		
cows and cattle	10	0.5	0.2	0.5	2.5	1.0	4.5
pig (dry meal fed)	10	0.6	0.4	0.3	4.0	2.0	2.7
pig (pipeline fed)	6–10	0.5	0.2	0.2	2.0–3.5	1.0–2.0	1.5–2.7
pig (whey fed)	2–4	0.3	0.2	0.2	0.8–1.6	0.4–0.8	0.8–1.5
poultry	25	1.4	1.1	0.6	9.1	5.5	5.4

From MAFF (1979a) with permission
[1] For available nutrients from autumn and winter application, *see* text.

(1) as most problems arise where fresh material is applied to grass, store the manure for two to three weeks before application;
(2) allow rain to wash the manure from the herbage before grazing;
(3) ensure water courses are kept free from contamination.

Toxic elements

The main risk is from manure obtained from fattening pigs, as this can contain high levels of Cu and Zn, both derived from feed additives. Copper and Zn build up slowly in the soil and frequent application of pig manure to the same field can lead to potential toxicity problems. Particular risk arises on grazed grassland, where evidence suggests that stock suffer more from slurry contamination on the herbage than high Cu and Zn levels in the herbage itself: so physical contamination should be allowed to disappear prior to grazing.

Specifically, sheep should *never* be allowed to graze grass contaminated with Cu as extensive liver damage can occur.

Pollution

All animal manures have a high moisture content and under some storage conditions this can lead to loss of effluent from the store. To minimise the risk of effluent it is vital that additional water (rain water, washing-down water) is kept out of manure and slurry stores.

There are strict legal requirements concerning the discharge of effluent into water courses, quite apart from the loss of potential plant nutrients that results. The legal aspects of effluent also apply to fields where heavy dressings of manure lead to effluent being lost into field drains or from surface run-off. Consequently it is unwise to:

(1) apply very heavy dressings of manures during autumn and winter;
(2) apply any manure at all in winter to sloping fields near ditches and water courses.

Management aspects of use of livestock manures

All-grass farms

These produce more manure per total farm hectare than mixed grass/arable farms and have fewer suitable crop situations for its use. Suggested areas during the year are:

(1) *November/December* apply to next year's spring grazing area,
(2) *January–March* apply to areas for spring conservation cuts,
(3) *March–September* plough-in prior to reseeding,

(4) *May/June* apply thinly to areas cut for conservation in spring.

Other points to note are:

(1) High rates applied to grass can cause physical shading of the grass and loss of stand, particularly if the grass is not growing rapidly at the time.
(2) Cattle manures are high in potash and its use in spring can predispose hypomagnesaemia.
(3) Most manures cause taint and subsequent refusal problems by grazing stock: some farmers find sheep graze behind cattle slurry better than cattle.
(4) There is a *Salmonella* risk where fresh cattle slurry is applied to grass regrazed by cattle.
(5) Pig slurry is often high in Cu and Zn: grass should be free from slurry contamination before grazing, particularly with sheep.

Pig and poultry units

These are always 'exporters' of manure and this manure requires disposal at regular intervals throughout the year as the stock are nearly always housed. A mixed farm can more easily accommodate and efficiently utilise pig and poultry manure than either an all-grass or all-arable farm. Poultry manure has a higher N content than other manures and is valuable for application to grassland in the March–September period.

Straw contamination

Where manure contains much fresh, unrotted straw, it is possible that the soil bacteria breaking down the straw need more N than is contained in the manure: thus the material can *deplete* N status in the short term. To correct this, additional fertiliser N must be used at the rate of 10 kg of N for every one tonne of fresh straw used: note that this N should be applied to the next crop and *not* to the manure or the soil at the time of manure application.

Other organic manures

Straw and composts

Straw can be ploughed in, preferably after it has been chopped, and this will provide some phosphate and potash but as noted above, it may cause a deficiency of nitrogen that will need correction.

Where straw is burnt a small quantity of phosphate and potash residues are left.

Composts can be made from straw alone, straw mixed with another crop waste, or from other wastes alone. As a general rule all such materials suffer from lack of sufficient nitrogen for proper bacterial decomposition and it is recommended that ammonium nitrate is added at a rate of 30 kg of fertiliser/tonne of composting material. Subsequent rotting down may well release N as ammonia unless the mass is kept under anaerobic conditions, and so the resulting compost will still be deficient in N. For this reason good compost is made from very short-chopped material that is well-consolidated and kept in a compact heap. Even so, composts are not highly rated for their nutrient value and it is often easier to apply waste materials to suitable fields at the earliest opportunity as a disposal exercise, plough them in and let decomposition take place within the soil: in this way any available nutrients find their way into the soil/plant complex at minimum cost.

The use of straw and composts should be regarded as a 'fertility builder' and no deduction from standard fertiliser recommendations should be made. Also it follows that it is sound practice to apply such composts to poorer fields, including those where the physical aspects of strawy materials may help alleviate a structure problem.

Seaweed

Seaweed is rich in potash (and sodium) as shown in *Table 4.4,* and a dressing of 25 tonnes/ha would provide some 300 kg/ha of K_2O. Sometimes seaweed is used to make composts, but like so many organic manures, it is bulky material and expensive to transport. Thus seaweed is used only in close proximity to the coast and is valuable for some horticultural crops and potatoes. Also it is used to provide humus on some light soils.

Table 4.4 Analysis of some organic materials

	DM%	% of fresh weight N	P_2O_5	K_2O
Fresh seaweed (typical)	20	0.2	0.1	1.2
Straw compost	25–30	0.4	0.2	0.3
Dried sewage sludge	60–90	0.5–2.0	0.5–3.0	trace–0.5
Town refuse compost	50–70	0.8	0.5	0.3
Night soil	5–20	0.8	0.4	0.2

Note: As with all organic manures, the composition is highly variable: the above analyses are typical of those that can be expected but they should not be regarded as average. Ideally each specific source of supply should be analysed.

ORGANIC FERTILISERS

Organic fertilisers are derived from either plant or animal materials: compared with inorganic sources of nutrients, organic sources have the following features:

(1) not immediately soluble in water and so not readily leached;
(2) because they have to break down to become partially soluble, they can act as a slow-release source of nutrient to the plant;
(3) they can be applied at heavy rates without risk of injury to roots or germinating seeds as they have little ionic activity;

(4) they can stimulate microbial activity in the soil;
(5) are much more costly per unit of plant food;
(6) the recovery of nutrients contained in the materials by plant roots is low.

Because of the above features, organic fertilisers are little used in agriculture but they do have a place in market gardening and horticulture, where the slow release characteristics have application for some of the high-value crops that are grown.

Organic nitrogen fertilisers

Hoof and horn meals, hoof meal, horn meal

These contain 12–14% N and can be obtained either coarsely or finely ground: fine materials release N more quickly. They should be worked into the soil before planting or sowing.

Dried blood

This contains 12–13% N, is very expensive but of great value in glasshouse crops where it is quick-acting.

Shoddy

Analysis varies from 3–12% N, depending on the proportion of wool contained amongst other wastes. Shoddy is a very slow release material, should be worked into the soil before planting and should be analysed before purchase.

Organic phosphate fertilisers

Bone meals

These contain 20–24% P_2O_5 (insoluble in water) together with 3–4% N. The phosphate acts very slowly and is of most value on acid soils.

Steamed bone meals and flours

These contain 26–29% P_2O_5 (insoluble in water) together with about 1% N. These materials are made from bones that are steamed to obtain glue-making substances and a good deal of the N is removed in the process. The bones are generally ground after extraction and the phosphate acts more quickly than in ordinary bone meal.

Meat and bone meal (Also *meat guano* or *tankage*)

Contains 9–16% P_2O_5 (insoluble in water) together with 3–7% N. These are made from meat and bone wastes and analysis varies; the phosphate has slow availability.

Fish and meals and manures (also *fish guano*)

These contain 9–16% P_2O_5 (insoluble in water) together with 7–14% N. They are waste products from fish processing.

Organic potash fertilisers

Potassium does not occur in organically chemical form in fertiliser materials but it does occur in some organic products such as *bird guano*, which is the collected excreta of birds. As such it is very similar in nature to poultry manure mentioned in the previous section. The potash contained in these materials is very readily available to plants once it is placed on the soil.

INORGANIC FERTILISERS

Inorganic fertilisers form an important part of the basis of modern crop production for very few soils are able to supply sufficient quantities of all the nutrients necessary for high yields and quality in crops. However, profitable use of inorganic fertilisers depends on careful assessment of the total level of nutrients required by the crop on the one hand and the extent of supply from the soil and organic sources on the other.

This section lists the commonly available inorganic fertilisers and concludes by summarising guidelines for their profitable use.

The three nutrients needed from inorganic fertilisers in greatest quantity are nitrogen, phosphorus and potassium. By law every bag of inorganic fertiliser that contains one or more of these three nutrients must state the content of that nutrient on the bag—in terms of percentage composition of N, P_2O_5 or K_2O. There are strict EEC regulations that specify the narrow permitted tolerance between the stated composition on the bag and the actual analysis of the contents, thus it can be assumed that there is great consistency in the composition of different batches of a similar fertiliser.

It is important to realise that fertilisers do not contain N, P_2O_5 or K_2O; these are merely a convenient way of giving a common basis for expressing the composition of all inorganic fertilisers. Nitrogen is usually in the form of nitrates or ammonium compounds, P_2O_5 in the form of phosphates and K_2O in the form of potassium chloride or sulphate. In the case of phosphates, some are water-soluble (and rapidly available in the soil, at least for a short period) and some are insoluble in water. If these latter are to have some agronomic value, they must become available in the soil and sometimes the solubility of these materials in 2% citric acid or neutral ammonium citrate is given as a guide to probable value as a fertiliser.

In some specific crop and soil situations other nutrients are needed, such as Mg, B or S. Other nutrients are either applied on their own or mixed with one of the normal fertilisers, in either case the composition must be stated.

Nitrogen fertilisers

The major N fertilisers are as follows.

	Formula	%N
ammonium nitrate	NH_4NO_3	34.5
ammonium sulphate	$(NH_4)_2SO_4$	21
calcium ammonium nitrate	$NH_4NO_3 + CaCO_3$	21–26
ammonium nitrate sulphate	$NH_4NO_3 \cdot (NH_4)_2SO_4$	26
potassium nitrate	KNO_3	14
urea	$CO(NH_2)_2$	46
anhydrous ammonia	NH_3	82
calcium cyanamide	$CaCN_2$	21
aqueous ammonia	$NH_3 + H_2O$	25–40

Ammonium nitrate

This is an important fertiliser material and is sold both as a straight fertiliser or as a component in compound fertilisers. Half the N is in ammonium form and half in nitrate form; in this way it has only half the acidifying effect of ammonium sulphate for any given amount of N applied. Practically pure ammonium nitrate is sold as a straight fertiliser containing 33.5–34.5% N: this material is very soluble in water and also hygroscopic, so it must be stored in a dry place in sealed bags. It is a very powerful oxidising agent and if subject to heat or flame it can explode: thus straight ammonium nitrate and compounds containing a high proportion of ammonium nitrate should not be stored in barns with hay or straw and *never* stored in bulk, as the risk of explosion is thus enhanced.

Being soluble in water, ammonium nitrate is often used as a major N source in liquid fertilisers.

Ammonium sulphate

Ammonium sulphate was once the most important source of inorganic N; it is soluble in water and although all the N is in NH_4^+ form it is quick-acting under field conditions. However, because of its high ammonium ion content it has an acidifying action in the soil and also its N concentration is limited to about 21%. Ammonium sulphate has been replaced by ammonium nitrate in UK and some other temperate countries and by urea in tropical and subtropical countries and some high-rainfall temperate areas.

Calcium ammonium nitrate (CAN)

This is a mixture of ammonium nitrate with calcium carbonate and is sometimes called nitrochalk. It has a higher N content than ammonium sulphate and has little acidifying action in the soil. Also the hygroscopic and explosive properties of ammonium nitrate are nullified by the presence of chalk, making it easy and safe to store, even in bulk. However, CAN tends to be more expensive per unit of N purchased than ammonium nitrate.

Urea

Urea is the most concentrated solid source of N that is currently available on any wide scale, it is very soluble in water and hygroscopic. When urea is applied to many soils it breaks down rapidly to form ammonia and this ammonia can cause problems in a number of circumstances:

(1) If the ammonia is lost to the atmosphere then the fertiliser is less effective as a source of N for crop nutrition; this loss of ammonia is most likely when the soil is alkaline and if the urea is applied to the soil surface in dry weather.
(2) In seedbeds the high concentration of ammonia near germinating seeds can cause toxicity and serious loss of plant stand.
(3) On poorly structured soils liable to surface capping the released ammonia can be trapped in the soil and not only kill germinating seedlings but also severely check the growth of roots of established plants.

It follows that urea is best used on acid-to-neutral soils, surface-applied to established crops during periods of frequent rainfall. Thus in the British Isles urea has a place on grassland.

Compared with the cost per unit of N in ammonium nitrate, the price of urea varies widely throughout the world: in some countries urea is much the cheaper and is well worth using despite its limitations, whereas in other countries (including UK) urea and ammonium nitrate are very similar in price and as a result very little urea is used in solid fertilisers.

Urea is used widely as a major N source in liquid fertilisers, as liquid application overcomes some of the physical problems of urea noted above and when mixed with phosphoric acid to make a compound fertiliser the acid nature of the material further reduces the risk of ammonia loss.

Anhydrous ammonia

Anhydrous ammonia is the most concentrated source of N available, being pure ammonia. At normal temperature and pressure this material is a gas, but when stored at high pressure it is a liquid containing the equivalent of 82% N. At this high concentration it is very economical to transport on a weight-to-value basis, but the whole process of its storage, transport and application needs special equipment, both to maintain up to the point of application the high pressure needed and to prevent hazards from uncontrolled loss of ammonia. Anhydrous ammonia is

applied below the surface of the soil using special injection equipment and because of the high cost of this operation, it is economical to apply only fairly high rates of N (at least 100 kg/ha of N) at each application. Also it is essential that loss of ammonia to the atmosphere is kept to a minimum during application by careful sealing of the slits. For these reasons the use of anhydrous ammonia is best restricted to row crops that need high individual N dressings. Although widely used in some countries where suitable crops are grown in stone-free, friable soils, anhydrous ammonia has not established itself in UK; mainly because it was not found suitable for the potentially major market on grassland due to sward damage caused by injection equipment and loss of ammonia from the slits in the soil.

The very high concentration of ammonia released in the soil in the vicinity of the slits causes partial sterilisation of the soil and has the effect of retarding the action of the bacteria that convert NH_4^+ to NO_3, so that anhydrous ammonia can act as a slow release fertiliser.

Aqueous ammonia

Aqueous ammonia is a solution of ammonia in water, a solution of normal pressure containing about 25% N. Aqueous ammonia has most of the advantages of the anhydrous form except high N concentration, but to offset that it does not need such specialised equipment. In crop situations where a liquid straight N fertiliser is required, then aqueous ammonia has an important place. Its concentration can be increased by:

(1) partial pressurisation, where storage and application under only a modest pressure can enable concentration to increase to 40% N;
(2) mixture with urea and/or ammonium nitrate to give 'no pressure' solutions of about 30% N.

To prevent gaseous loss of ammonia all these materials should be either injected into the soil or worked in immediately after application: they should *not* be used on calcareous soils.

Potassium nitrate and sodium nitrate

These are both very quick-acting and highly soluble forms of N that are expensive but valuable in some horticultural crops and in particular for use in foliar feeds.

Calcium cyanamide

This is not important in the UK but is a fertiliser containing N in both amide and cyanide forms. It is soluble in water and in the soil it is converted to urea: during this conversion some toxic products are released that can kill germinating weeds and slow the rate of nitrification. Breakdown requires water and so the product is not effective in dry conditions.

Slow release fertilisers

Most of the commonly used and cheaper forms of N are rapidly available to the growing plant and yet many crops would gain greatest benefit from applied N if it were available over a period covering most of the plant's vegetative growth. There are three ways in which this objective can be met:

(1) Apply frequent small dressings of conventional N fertilisers during vegetative growth, for example as with spring N applications on some winter wheat crops and on grassland.
(2) Use a conventional fertiliser material that has received a coating to reduce its rate of breakdown into plant-available forms, for example sulphur-coated urea.
(3) Application of complex compounds of N that require considerable chemical change in the soil before they are available to the plant.

There is much research to find suitable materials in the last of the above categories. To be successful such a compound must be economic in price and supply available N at the rate required by the crop. The materials on the market tend to be expensive, make available to the crop a relatively low proportion of the total N they contain and have rates of release controlled by soil temperature and moisture so that during humid weather most of the N can be released too rapidly for full crop benefit.

The two most common slow-release fertilisers at present are *urea formaldehyde* ('ureaform') and *isobutylidene diurea* ('IBDU'). Both these compounds hydrolyse to give urea and a significant part of the initial breakdown is controlled by microbial processes, which in turn are affected by soil moisture and temperature. Ureaform and IBDU are used in commercial glasshouse production, because not only is the slow rate of N release of particular value to some of the crops grown, but also the rate of breakdown can be controlled to a large extent by adjustment to the management of the house.

Phosphate fertilisers

The major P fertilisers are:

	Formula	% P_2O_5
Water soluble		
superphosphate	$Ca(H_2PO_4)_2 + CaSO_4$	18–22
triple superphosphate	$Ca(H_2PO_4)_2$	45–47
mono-ammonium phosphate	$NH_4H_2PO_4$	48–50
di-ammonium phosphate	$(NH_4)_2HPO_4$	54
Water insoluble		
basic (or Thomas) slag	$Ca_3P_2O_8 \cdot CaO + CaO \cdot SiO_2$	10–20
Rhenania (sinter) phosphate	$CaNaPO_4 \cdot Ca_2SiO_4$	25–30
ground rock phosphate	$CaPO_4$ (apatite)	30
Senegal rock phosphate		30

In addition to the above, which can all be used in solid forms, there are a number of liquid products based on *polyphosphates*. The basic component of these solutions is superphosphoric acid, made from orthophosphoric acid and one of a number of polyphosphoric acids. An ammonium compound is used to neutralise the appropriate acid and NP-solutions of ratios in the order 11:37:0 can be obtained. These products are expensive but do provide a highly-concentrated source of available P in liquid fertilisers. There can be problems when potassium salts are mixed with polyphosphates, as they can cause crystallisation and precipitation.

Superphosphate

Superphosphate is made by treating ground rock phosphate with sulphuric acid and producing water-soluble monocalcium phosphate and calcium sulphate. Usually some rock phosphate remains in the product as the quantity of sulphuric acid is restricted to prevent the final product containing free acid. Superphospate was once the most widely used form of water-soluble phosphate and formed the basis of most compound fertilisers. Because of its relatively low P_2O_5 content it limited the concentration of phosphate in compounds, but because of the $CaSO_4$ present, it did apply some sulphur.

Triple superphosphate

This is made by treating rock phosphate with phosphoric acid, this produces mainly water-soluble monocalcium phosphate with no calcium sulphate. Thus triplesupers contains nearly two-and-a-half times as much phosphate as supers, but no sulphur. Because of its higher concentration, triplesupers is used widely in the production of high-analysis compound fertilisers.

Ammonium phosphate

This is made by adding ammonia to phosphoric acid, producing both mono- and di-ammonium phosphates. These compounds are very soluble in water, quick-acting in the soil and supply both N and P in a highly available chemical combination. They are not used as straights to any extent because of the few crop situations that require a low N:high P fertiliser, but they do form an increasingly important base for compound fertilisers.

Basic slags

Basic slags are by-products of the steel industry and contain phosphates, lime and trace elements. In the smelting process, P contained in iron ore is held in the furnace bound to CaO, becoming a solid 'slag' as the material cools. The phosphate is in complex chemical combination with calcium and has to be ground before it can be used as a fertiliser. Although the phosphate in slag is not water soluble, it has been found that it can release phosphate slowly over a period of years on soils of pH range 4–7, although the rate of release is greater on acid soils. The trace element content of basic slag is also important, particularly Mg, Fe, Zn, Si and Cu.

Recent changes in the steel-making process have resulted in a drastic reduction in the quantity of slag available with a reasonable phosphate content. Whereas some half-million tonnes of basic slag were sold in UK each year for many years, the supply is now very limited. Farmers have thus lost access to a very cheap and agronomically acceptable source of phosphate and several alternative products have come on the market, however none of these can meet all the useful characteristics of basic slag, which were:

(1) steady release of available P over several years on both acid, neutral and alkaline soils;
(2) supply of valuable trace elements;
(3) some liming value;
(4) cost per unit of P about half the cost of water-soluble P.

Rock phosphates

These are now being offered to farmers as a source of phosphate where rapid release of P is not essential but soil P status needs to be maintained. Rock phosphates will only release P under acid conditions, although some experiments have shown that soil pH is not necessarily a true indication of pH that might exist around root hairs, so that at times rock phosphate has shown some value on neutral soils. Rock phosphates vary considerably in their composition and potential agronomic value, and in all cases they must be ground finely to have any value at all, current legislation being that 90% of the material should pass a 100 mesh sieve.

Some of the hard crystalline apatites have very little fertiliser value even when crushed finely and used under acid conditions. On the other hand, soft rock phosphates such as Gafsa do have value and are regarded as the best of untreated rock phosphates.

Rhenania phosphate is produced by disintegrating rock phosphate with sodium carbonate and silica in a rotary kiln: this 'sintered' product contains much of the P in the form of calcium sodium phosphate and as such the P is rendered a little more available than in the original rock.

Senegal rock phosphate contains aluminium phosphate and when this is heated in a kiln and then allowed to cool in a humid atmosphere, an expanded type of rough granule is produced that greatly enhances the surface area of the material in contact with the soil after application. This calcined Senegal rock is claimed to be an effective P source, even on alkaline soils.

No rock phosphate has a rapid release of available P, so that these materials should not be used in situations where the crops' need for P is rapid, for example at crop establishment or on crops that have a big demand for P such as potatoes. Also the crop recovery of P from rock phosphates is less than the recovery from water-soluble sources, so that the amount of *useful* P in rock phosphate is less than in the water-soluble types: this fact should be remembered when the costs of rival products are considered.

Potash fertilisers

The major K fertilisers are:

	Formula	*% K_2O*
muriate of potash	KCl	60
sulphate of potash	K_2SO_4	50

other potassium containing fertilisers sometimes used are:

potassium nitrate	KNO_3 (13%N)	44
potassium metaphosphate	KPO_3 (27%P_2O_5)	40
potassium magnesium sulphate	K_2SO_4, $MgSO_4$ (18% MgO)	22
magnesium kainite	$MgSO_4$ + KCl + NaCl (6% MgO; 18% Na)	12

Muriate of potash (potassium chloride)

This is mined and is sold in either a powdered form or a fragmented (granular) form. It is the source of K in nearly all compound fertilisers and its use as a straight fertiliser in UK is very limited. Where crops require substantial amounts of K it should be remembered that equal quantities of Cl are also applied if muriate of potash is used. In a crop like potatoes it is sometimes advisable to use potassium sulphate instead of muriate.

Sulphate of potash

This is made by treating muriate with sulphuric acid and so is a more expensive source of K. However for Cl-sensitive crops the extra cost of the sulphate salt is recommended. Some manufacturers make a series of compound fertilisers containing potassium sulphate and these are used widely by horticulturists, not only because of some sensitive crops but because of the risk of a build-up of Cl^- ions in the soil where regular heavy manuring is carried out.

Potassium nitrate (saltpetre)

This is expensive and its main fertiliser use is in foliar applications on fruit trees and some horticultural crops.

Potassium metaphosphate

This compound is a water-insoluble source of K (unlike all the other sources mentioned which are soluble in water), and it can be used where it may be necessary to keep ionic concentrations at a low level in the root vicinity.

Potassium magnesium sulphate and magnesium kainite

These are sometimes used where some Mg is required in addition to K; also kainite is often used for sugar beet because of the Na content.

Agricultural salt (mainly NaCl)

This can replace KCl in some situations, for example where sugar beet or mangolds are grown. Sodium chloride should not be applied to poorly structured heavy soils as it may predispose deflocculation, and it should not be applied less than three weeeks before sowing lest it affect germination: in fact it is often applied in autumn and ploughed in. Application of Na reduces the need for K.

FORMS OF FERTILISER AVAILABLE
Solids

This is the most common form in which fertilisers are available and normally manufacturers go to great lengths to ensure that material stores without becoming compacted and spreads evenly and freely at time of application. Water-insoluble materials, such as some types of phosphate, have to be in solid form, but most other materials could be sold in liquid form if necessary. Main advantages of solid types are:

(1) high concentration of nutrients in the material, reducing transport and storage costs and leading to high rate of work at spreading;
(2) when packed in 50 kg plastic bags, very cheap storage is possible and the material will keep in good condition for many months;
(3) when sold in bulk, relatively simple modifications to existing buildings will enable successful storage and make possible spreading systems that have low labour requirements;
(4) the farmer can have a range of fertiliser types available, can use only that quantity which is needed at a given time, and can judge the application rate easily by counting the number of bags used per hectare.

Where water-soluble ingredients are used in solid fertilisers, it normally takes little moisture to render these ingredients available in the soil. Similarly, where the same ingredients are applied in liquid form to a dry soil, the quantity of water applied with the fertiliser is so minute that it has no irrigation effect. So it can be assumed that '*once brought into contact with the soil, liquids behave in the same way as comparative solid fertilisers, and generally no differences are observed in relation to growth and crop yields*'. (Mengel and Kirkby, 1978).

Solid materials are sold as powders, granules, prills or fragments.

Powders

These can vary in degree of fineness and for some materials (e.g. water-insoluble phosphates) powders are essential to ensure activity of the material in the soil: powders can be difficult to spread evenly and this is critical if relatively low rates/ha are needed; under these conditions full-width spreaders are advised.

Granules

They are made during manufacture by passing the fertiliser powder through a rotary drum, often with an inert granulating and coating agent; sometimes several fertiliser materials are mixed at the same time, so that each granule contains a mixture of materials. All granules made in this way are designed to break down quickly in the soil and release their nutrients, but sometimes the inert carrier remains visible in the soil for some days. The size and texture of the granules will have a marked effect on the sowing rate and spreading width of fertiliser distributors, and both these granule characteristics are specific to the type of material being granulated and the conditions under which the granulator is working: it follows that granules will vary in spreader performance and each consignment should be checked when being used.

Prills

Prills are made by the rapid cooling of a hot fertiliser liquid, are homogeneous in character and usually have a very smooth surface. Like granules, they can vary in size and great care is necessary at spreading time as prills usually flow more freely than granules and spreader-setting is critical to get correct rates and even spread. Prills normally consist of one fertiliser material only but this may contain more than one nutrient, for example a prilled ammonium phosphate would contain both N and P.

Fragments

Fragments are made by very coarse grinding of a solid material, usually followed by a screening process to produce material of a certain size range, for example some samples of muriate of potash are made in this way. Fragments behave in a similar way to very uneven granules (but note that granules are always spherical in shape because of the way they are made), but fragments and granules should not be mixed and spread together as marked segregation will occur both in the spreader and during the spread itself. Some fragmented material has a very abrasive action on working parts in the spreader and can cause rapid wear.

Most manufacturers take great trouble to ensure compatibility within each fertiliser type, but mixing of different fertiliser types on the farm prior to or during application can lead to poor spreading performance.

Liquids

Some materials must be used in liquid form, either because they do not exist as solids at normal temperature and pressure (such as polyphosphates) or because their intended use precludes solid form (e.g. foliar nutrient application).

The main features of liquid fertilisers are:

(1) easy handling on the farm, as they can be pumped in and out of store and spread by sprayer fitted with non-corrosive working parts and nozzles;
(2) relatively low concentration of nutrients, as some chemicals will crystallise out during storage and at low temperatures if concentration is raised;
(3) storage tanks required, which can be expensive; this can mean that storage capacity is restricted so that an annual requirement for fertilisers cannot be purchased at the lowest price;
(4) the possibility that herbicides can be mixed with fertiliser to further reduce spreading costs.

In UK liquid fertilisers hold a fairly small share of the total fertiliser market, unlike the situation in some other countries. There is no simple guide as to whether a farmer should use liquids or solids as the decision rests on several aspects of general farm management, such as fertiliser purchasing policy, availability of existing stores for solid fertiliser storage, contacts for contra-trading with merchants selling either solid or liquid fertilisers, types of crop grown and labour and/or machine availability for spreading.

It is likely that the use of liquid fertilisers will continue to increase, both as a result in changing farm management practices, and as a consequence of developments in the production of liquid fertilisers. For example, there has been much research in USA on the use of fertiliser *suspensions*: these are not true liquids as not all the material is completely dissolved, but they do offer the possibility of much more concentrated fluid fertilisers than at present. However, these materials will need very carefully controlled transport, storage and spreading, so that they will be based on a contractor service rather than a farm operation.

Gases

As mentioned earlier, anhydrous ammonia can be used as a source of N (but because of the high pressure needed for storage and the special injection equipment needed, this is a contractor-based service) and is valuable on relatively few crops and soils in UK.

COMPOUND FERTILISERS

When fertilisers are sold with a single main chemical ingredient they are known as *straights*: examples are ammonium nitrate, superphosphate and muriate of potash. Sometimes straights can supply more than one nutrient, for example ammonium phosphate supplies N and P and potassium nitrate supplies N and K. Occasionally straights may be comprised of a mixture of ingredients, but if they still supply only one main nutrient they are still known as straights, for example some phosphatic fertilisers contain a mixture of water-soluble and water-insoluble phosphate, but still supply only P.

Many crops need more than one of the three major nutrients, N, P and K, and in most of these situations it is convenient to apply all the nutrients at the same time. For this purpose, manufacturers have produced a wide range of *compound* fertilisers, that is fertilisers that contain more than one ingredient and supply more than one nutrient. Compound fertilisers are more expensive than straights, but many farmers find the extra cost is justified because if several straights were used the cost of application would be increased and it is difficult to mix and effectively spread straights on the farm.

Because exact NPK ratios required vary with the crop, soil and farming system, there are many compound fertilisers on the market. The number is increased because manufacturers compete in the market by attempting to offer unique or unusual ratios to meet a specific type of crop situation.

On any individual farm it should be possible to rationalise the number of fertiliser materials needed, with the aim of storing no more than about four fertilisers unless some fairly unusual crops are grown. Not only does this simplify storage and use, but it also enables sizable quantities to be purchased leading to the chance of keen prices being quoted.

Table 4.5 gives examples of the main compound fertiliser types and gives some suggestions on how a selection could be made for the major crops. The details given are for *example only*, but show how similar plans can be made on the farm. Most manufacturers make a range of specific fertilisers that fit into the general descriptions given in *Table 4.5*.

FERTILISER PLACEMENT

Most fertilisers are spread evenly over the surface of the soil or crop, and are used at the period when the crop is most in need of the nutrients contained. For example, fertilisers are worked into a seedbed prior to sowing and N fertilisers are top-dressed on to some growing crops, particularly winter cereals in the spring and to grassland.

In some row-crops, placing fertiliser in a band near to the zone of the soil in which the majority of the

Table 4.5 Examples of compound fertilisers and their use

Code	Type	Approx. ratio			Typical analysis			Code	Type	Approx. ratio			Typical analysis		
		N	P_2O_5	K_2O	N	P_2O_5	K_2O			N	P_2O_5	K_2O	N	P_2O_5	K_2O
A	extra high N	5	1	1	29	5	5	B	high N	2	1	1	20	10	10
					27	6	6						22	11	11
C	NPK	1½	1	1	20	14	14	D	universal	1	1	1	15	15	15
					20	15	15						17	17	17
E	low N	⅓	1	1	9	24	24	F	NP	2	1	0	30	13	0
					8	20	16								
G	NK	2	0	1	25	0	16	H	PK (i)	0	1	1	0	24	24
					26	0	13		(ii)	0	1	2	0	14	28
I	high K	1	1	1½	15	15	21	J	high PK	1	1½	1½	15	21	21
					13	13	20								
K	low P	4	1	3	20	8	14								
					24	4	15								
					17	8	24								

Crop	Compounds for normal conditions	Adaptation for low P soils	Adaptation for low K soils
Grassland	A or B, plus G or K after cutting (also straight N for top-dressing and D or E for reseeding)	use of water-insoluble P every 3–4 years as well as normal compounds	*regular* use of ferts with K, particularly G (except in spring)
Winter cereals	E or H in seedbed (N in spring)	higher rate of H	use H (ii) specifically
Spring cereals	*either* D or E in seedbed (plus N later) *or* B in seedbed	higher rate of D or E C instead of B	higher rate of D or E C instead of B
Potatoes (maincrop)	I	J	H (ii) in seedbed plus I at planting
Sugar beet	*without kainit* K	I	use kainit
	with kainit B	B or C	B or K
Forage legumes	H (i)	higher rate of H (i)	H (ii)

Notes: on soils high in P, use much K and G (instead of B and D); on soils high in K, use much F (instead of B); on soils high in P and K, use much straight N (instead of A and B)

plant roots will feed can both reduce the total quantity of fertiliser needed and check weed growth outside the crop rows. Fertiliser placement in this way tends to reduce the speed of fertiliser application and if carried out at sowing time by *combine-drilling* will also slow down the rate at which drilling can be carried out. Also, if the fertiliser rate is too high or too close to the seed or plant root, the material can cause scorch: scorch can reduce plant survival and check growth in established plants. Band placement of fertiliser can be advantageous in the following circumstances:

(1) where soil nutrient status is very low: for example experiments have shown that on poor soils, responses to placed fertiliser are greater than to very high rates of fertiliser evenly spread over the field;
(2) on crops that need very high rates of nutrient application and are grown in discreet rows set well apart: placement not only reduces the overall amount of fertiliser needed but also ensures the roots come into contact with sufficient material. A good example is the main-crop potato grown on normal mineral soils;
(3) where the applied nutrient has limited mobility in the soil: this applies particularly to P which

scarcely moves in the soil even when applied in water-soluble form.

From the above it can be concluded that band replacement should be considered for row crops that need P and are grown on soils with a low P availability: under these conditions placement is advised for cereals and potatoes.

Some soils are reasonably well supplied with nutrients but further fertiliser application is recommended to maintain the adequate level: this situation applies particularly to P and K. It is thus suggested that it may not be necessary to use P and K fertilisers for *every* crop but adopt a *rotational* fertiliser programme where PK fertilisers are used periodically and used immediately before a PK sensitive crop is grown. This approach can save money but needs careful monitoring to ensure that soil nutrient levels do not decline: also it should be noted that PK fertilisers will be needed on some fields on the farm each year as the rotation dictates.

In dry weather the growth of crops slows down, mainly due to lack of water for growth and transpiration but partly because the roots cannot absorb the soil nutrients which are located near the surface. Experiments have shown that *deep-placement* of fertilisers can maintain a faster growth rate in crops in dry weather than where the same nutrients are placed on the surface in a conventional way. It should be noted that for this deep-placement to work, the nutrients should be at least 150 mm below the surface and this is deeper than normal injection of anhydrous or aqueous ammonia.

Currently, experiments are in progress to study the possibility of combining *rotational* and *deep-placement* use of fertiliser by periodically ploughing-in substantial quantities of P and K. However, this approach will not work with N as this nutrient is readily leached from the soil (and deep placement in high rainfall areas could accelerate leaching).

FERTILISER RECOMMENDATIONS

Figure 4.1 illustrates the principles used to link knowledge of soil nutrient availability with understanding of crop nutrient need, to reach a fertiliser recommendation.

For all nutrients other than N, soil analysis will give a good indication of the quantity of readily available nutrients in the soil: assuming a crop needs the same total quantity of nutrient for optimum growth, *Figure 4.1* also shows how the need for inorganic fertiliser supplementation will vary with soil index and the use of organic manures. The same principle is used in deciding N recommendations, except that the soil index is determined by the previous cropping history (*see* Chapter 1).

Soils differ in their ability to release available nutrients and crops vary in their nutrient requirements. ADAS and major fertiliser manufacturers publish crop fertiliser recommendations that take into account not only soil differences but also differences in the management of the crop: for example whether grass is cut or grazed, whether cereal yields are

consistently above or below average, whether straw is burnt, removed or ploughed in. It should be remembered that fertiliser recommendations are not based only on chemical aspects of plant nutrition but also on the results of many field trials and farmer experience: thus recommendations are not clinically correct but best approximations that are updated from time to time as new information is obtained.

For example, the ADAS recommendations for wheat, given in *Table 4.6* to illustrate the way in which recommendations should be used, are taken from their booklet 2191 published in 1979, and the recommendations differ in several important respects from those published in 1978 and 1976.

Calculating product use from recommendations

By law every bag of fertiliser must contain on it a statement of the plant food value of its contents, and under EEC regulations the phosphate value must be as P_2O_5 and the potash value as K_2O. Thus a compound fertiliser might be called 20:10:10 and state on the bags that its contents were:

20% N   10% P_2O_5 10% K_2O.

Most bags weigh 50 kg and if so, then dividing the percentage value on the bag by two gives the number of kg of the nutrients in that bag. So the bag mentioned above contains the equivalent of 10 kg of N, 5 kg of P_2O_5 and 5 kg of K_2O: if a farmer were to apply 5 bags/ha of this compound then the nutrient application would be 50 kg/ha of N, 25 kg/ha of P_2O_5 and 25 kg/ha of K_2O.

The following two examples are based on *Table 4.6* and are designed to help those who wish to check their ability to convert recommendations into actual product use.

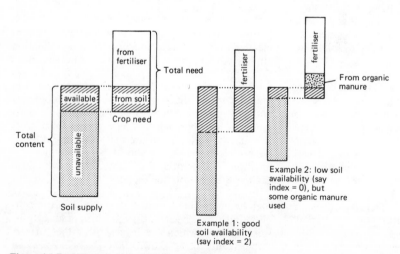

Figure 4.1 Basis for calculating fertiliser recommendations

Table 4.6 ADAS fertiliser recommendations for wheat

Phosphorus and potassium index	kg/ha of P_2O_5 or K_2O to apply				
	Straw ploughed or burnt		Straw removed		
	P_2O_5	K_2O		P_2O_5	K_2O
Farm yields—average or less[1]					
0	75	75		75	75
1	40	40		40	60
2	40	40		40	60
3	40	nil		40	nil
over 3	nil	nil		nil	nil
Farm yields—consistently above average[1]					
0	75	75		75	100
1	60	60		60	80
2	60	60		60	80
3	60	nil		60	nil
over 3	nil	nil		nil	nil

Nitrogen index	kg/ha of N to apply				
	Sandy soils, shallow soils over chalk and limestone	Med. Fen and heavy silts and clays	All other mineral soils	Light fen peats loamy peats	All other peats and organic soils
Winter wheat					
0	175(a)	150	150	50	90
1	125	75(b)	100	nil	45
2	75	nil	40	nil	nil
Spring wheat					
0	150(a)	125	125	40	70
1	100	50(b)	75	nil	35
2	50	nil	30	nil	nil

From MAFF (1979b) with permission
[1] Normal average can be taken as 5 tonnes/ha of grain.
Notes: (a) in dry summers water stress will limit response to N to a lower level. (b) increase by 25 kg/ha in poor soil conditions, e.g. after sugar beet: on deep silty soils such as Fen silts and brickearths winter wheat will need only 50 kg/ha.

Example 1

Winter wheat is to be grown on a normal mineral soil where soil index for both P and K = 1, normally yields are above average, straw is burnt and N index is also 1. The farmer has available a 0:24:24 compound and a 34.5% N straight; calculate the quantities he should apply.

From the recommendations, N usage should be 100 kg/ha (and this should be applied in the spring) and 60 kg/ha each of P_2O_5 and K_2O. For phosphate and potash application, each bag of the compound available contains 12 kg of nutrient, so the rate to apply is

$$\frac{60}{12} = 5 \text{ bags/ha of fertiliser (sometimes expressed as 250 kg/ha of fertiliser).}$$

For nitrogen, each bag of the N fertiliser contains 17.25 kg of N, so the spring topdressing should be at the rate of

$$\frac{100}{17.25} = 5.8 \text{ bags/ha of fertiliser (or 290 kg/ha of fertiliser).}$$

Example 2

Spring wheat is to be grown on a farm where yields are average and in the field in question soil N index = 0 and P and K index = 2: straw is burnt and it is a mineral soil. The farmer intends to use all the necessary phosphate and potash in the seedbed, with some N, and topdress the remainder of the N later: fertilisers available are 15:15:15 and 34.5% N.

From the recommendations, N usage should be 125 kg/ha and P_2O_5 and K_2O usage should be 40 kg/ha.

The 15:15:15 fertiliser will be used in the seedbed, and as each bag contains 7.5 kg of each nutrient, then the rate to apply will be

$$\frac{40}{7.5} = 5.3 \text{ bags/ha}$$

this fertiliser will apply all the P_2O_5 and K_2O needed and also $5.3 \times 7.5 = 40$ kg/ha of N. So the amount of N needed in the topdressing is $125 - 40 = 85$ kg/ha. As each bag of N fertiliser contains 17.25 kg N, then the rate to use is

$$\frac{85}{17.25} = 4.9 \text{ bags/ha.}$$

Note that in the examples given, the fertilisers applied have met the recommendation exactly. In practice this will not always happen: the essential nutrient to get nearly right is N, followed by either phosphate *or* potash depending on which has a lower soil index and to which the crop grown may have a greater sensitivity.

Some textbooks give phosphate and potash recommendations in terms of P and K and not P_2O_5 and K_2O. Great care should be taken when reading literature to ensure confusion does not occur and the following conversions can be used:

(1) to convert P to P_2O_5 multiply by 2.29 (P_2O_5 to P multiply by 0.44);
(2) to convert K to K_2O multiply by 1.20 (K_2O to K multiply by 0.83).

Optimum fertiliser rates

Fertiliser recommendations are based on consideration of both physical response and financial return.

Physical response

This is obtained from field experiments, where the shape of the crop response curve to increasing levels of nutrient input are studied. *Figure 4.2* gives a typical crop response curve: although the exact shape of this curve will vary with crop, soil type and climate, the following principles hold good:

(1) Under field conditions, some yield is obtained where no fertiliser nutrient is used, the yield A depending on the level of available nutrient in the soil. Fertiliser recommendations take this into account by allowing for soil index and making specific recommendations for each crop.
(2) There is a maximum yield (B) which is obtained from a certain level of nutrient input (M): using more nutrient than M will not give a greater yield—it may give the same yield as M or less depending on the nutrient and the crop.
(3) The shape of the response curve is such that the yield return per each extra kg of nutrient declines: for example it can be seen from *Figure 4.2* that the first half of the nutrient dressing M gives a much greater yield increase than the second half of the rate concerned. Thus maximum yield is reached at a point where yield return per unit of extra nutrient is so low that it would not pay for the extra nutrient used.

Financial return

Financial return is considered in establishing fertiliser recommendations by giving monetary value to the crop yield and nutrient cost. *Figure 4.3* is based on the

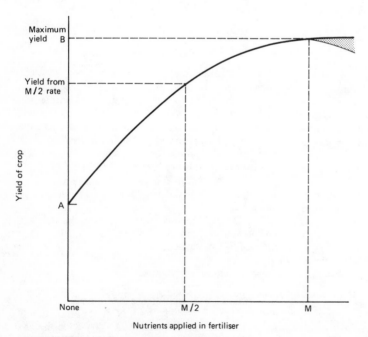

Figure 4.2 Response of crop yield to increasing rate of fertiliser nutrient

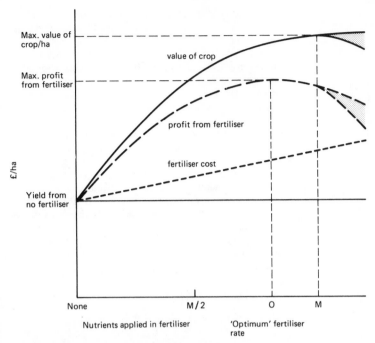

Figure 4.3 Calculation of optimum fertiliser rate

data shown in *Figure 4.2* and shows a curve for 'profit from fertiliser' which is taken as (value of crop— nutrient cost). The point O is very important as it is the highest rate of nutrient application that will show an increasing financial return. It is called the *optimum rate* and is below the maximum rate (M). Optimum rate depends on the ratio of crop value: nutrient cost, as well as on the shape of the agronomic response curve.

For most crops grown in UK there is a large bank of experimental data that gives reliable information on nutrient responses over a wide range of soil and climatic conditions. Thus ADAS and manufacturers' recommendations are sound guides to probable optimum rates of application. Optimum rates are not very sensitive to small changes in nutrient costs and the periodic update of recommendations takes steady trends into account.

Where a new crop situation develops, either by the extension of an existing crop into a new soil type, climate area or by the introduction of a new crop into the country, then soil scientists and agronomists soon conduct experiments to establish agronomic criteria for economic optima to be calculated. Good examples in recent years have been the work on forage maize and oilseed rape, where recommendations were updated regularly for several years and now the fertiliser recommendations for these crops in UK are different from those suggested in other countries where soil, climate and cropping systems are not the same as in UK.

LIMING

Adequate lime status is essential for the following reasons:

(1) it reduces soil acidity, and most plants will not thrive under acid conditions (*Table 4.7* gives pH guidelines for major crops);
(2) it increases the availability of certain plant nutrients in the soil;
(3) adequate soil pH encourages soil biological activity, thus enhancing the organic matter cycle in the soil, often increasing available N;
(4) it improves the physical condition of the soil by increasing the stability of crumb structures.

As explained in Chapter 1, there is a constant loss of lime from the soil, partly by crop removal and partly through leaching. Thus it is necessary to have a regular liming policy on all soils other than those with a natural Ca content. The frequency of liming will depend on the soil type, rate of leaching (soil texture, rainfall, fertiliser N use), and sensitivity of crops grown. Soil analysis gives a good indication of soil pH and each field should be sampled on a regular basis—say every four years.

On grassland farms soil pH should not be allowed to fall below 5.5 and it should be limed to bring the pH up to 6.0.

On arable farms soil pH must be kept above the level required by the most sensitive crop grown and

Table 4.7 Soil pH and crop tolerance

pH	Agricultural crops	pH	Horticultural crops
6.2	lucerne	6.6	mint
	sanfoin	6.3	celery
6.1	trefoil	6.1	lettuce
6.0	beans	5.9	asparagus
5.9	barley		beetroot
	peas		peas
	sugar beet	5.8	spinach
	vetches	5.7	brussels sprouts
5.8	mangolds		carrot
5.7	alsike clover		onion
5.6	rape	5.6	cauliflower
	white clover	5.5	cucumber
	maize		sweet corn
	wheat	5.4	cabbage
5.4	kale		mustard
	linseed		parsnip
	mustard		rhubarb
	swede		swede
	turnip		turnip
5.3	cocksfoot	5.1	chicory
	oats		parsley
	timothy		tomato
4.9	potato	4.1	hydrangea
	rye		
4.7	wild white clover		
	fescue grasses		
	ryegrass		

At a pH below the indicated level the growth of the crop may be restricted

general advice is to apply lime when pH falls to 6.0 and bring the pH up to 6.5.

Where grass is grown in a rotation, it is essential that soil pH is kept up in preparation for succeeding arable crops, as it is difficult to restore rapidly a low pH. If acid grassland is ploughed for arable crops, not only must some lime be worked into the soil, but sensitive crops such as barley should not be grown for several years.

Liming policy can affect the two diseases 'club root' of brassicas (*Plasmodiophera brassicae*) and 'common scab' of potato (*Streptomyces scabies*): club root is less severe on well-limed soils whereas common scab is more prevalent where lime has been applied recently. Thus in a rotation where brassicas and potatoes are grown, lime that may be required should be applied before the brassica crop, leaving at least two years before potatoes are grown.

Purchasing lime

Under the Agriculture Act 1970 the purchaser of liming materials must be given a warranty by the vendor: under present regulations there are three main considerations:

(1) fineness of grinding,
(2) neutralising value,
(3) permissible variation (normally $\pm 5\%$).

Fineness of grinding affects the speed of action of the lime and the total quantity of the lime applied that may become useful: the finer the material the more rapid its action and some very coarse materials may never break down sufficiently to be of value. Not only is fineness important, but also the hardness of the material; it is more essential for hard material such as limestone to be ground fine than a softer form such as chalk.

Neutralising value (NV) is the standard basis for comparing liming materials. NV can be determined by laboratory analysis and is expressed as a percentage of the effect that would be obtained if pure calcium oxide (CaO) had been used. For example if a sample of ground limestone has NV = 55, then 100 kg of this material would have the same neutralising value as 55 kg of CaO.

Lime recommendations, based on soil pH and the buffering capacity of the soil, are given in terms of the weight of CaO needed.

Example

A field has a lime requirement of 1.0 tonne/ha of CaO and two materials are available, ground limestone (NV 55) at £15/tonne delivered and hydrated lime (NV 70) at £28/tonne delivered. The two materials can be compared by calculating the cost of NV:

$$\text{ground limestone} = \frac{1500}{55}$$
$$= 27.3 \text{ p/\textit{l} CaO equivalent}$$
$$\text{hydrated lime} = \frac{2800}{70}$$
$$= 40.0 \text{ p/\textit{l} CaO equivalent.}$$

Thus the ground limestone is the cheaper buy, except that the hydrated lime may be easier to apply (if not being applied by contractor) and is quicker-acting, factors that may over-ride price differences in some circumstances.

Some liming materials contain magnesium as well as calcium; where the addition of Mg is considered beneficial, then the use of such material may be justified, even though the cost of each unit of NV is greater than a straight Ca lime.

Liming materials

Liming materials occur naturally over a wide area of UK and they are mostly sedimentary rocks laid down as a result of:

(1) accumulation of shells and skeletal remains of animals and plants;
(2) deposition by marine calcareous algae;
(3) precipitation of Ca (or Mg) carbonates from water;

(4) aggregation of fragments from pre-existing limestones.

Ground limestone

This is a sedimentary rock consisting largely of calcium carbonate and containing not more than 15% of MgO equivalent. One hundred per cent of material must pass a 5 mm sieve, 95% must pass a 3.35 mm sieve and 40% through a 150 µm sieve. NV is about 50 (maximum for pure $CaCO_3$ is 56). Legal declaration is NV and % through 150 µm sieve.

Screened limestone (limestone dust)

This is a by-product of grading limestone for non-agricultural purposes such as road metal and railway ballast. One hundred per cent must pass 5 mm sieve, 95% must pass 3.35 mm sieve and 30% through a 150 µm sieve. NV is about 48 and legal declaration is NV and % through 150 µm sieve.

Coarse screened limestone (coarse limestone dust)

This is similar to the above but coarser in that 100% must pass 5 mm sieve, 90% pass 3.35 mm sieve and 15% through a 150 µm sieve. Legal declaration is as above and expected NV is about 48.

Ground magnesian limestone

This is similar to the above materials, and subject to the same legal requirements, except that it must contain not less than 15% magnesium expressed as MgO. These materials are useful where there is a long-term need to maintain soil Mg level, however they do not provide rapid treatment for an existing Mg deficiency.

Ground chalk

This is natural chalk (cretaceous limestone) ground so that at least 98% will pass through a 6.3 mm sieve. The legal declaration is for NV only, which is usually about 50. Some chalks are quite wet when quarried and have to be dried before grinding: the NV value should refer to the material as supplied (i.e. partially dried) and not to a completely dried sample.

Screened chalk

This is produced from chalk outcrops by scarifying the surface with tractor-drawn implements and passing the collected loose material through a screen. This material is produced widely in south and east England. Ninety-eight per cent must pass a 45 mm sieve and the legal declaration is for NV only (usual value is about 45). However producers often guarantee fineness to pass 25.4, 12.7 or 6.3 mm sieves.

Calcareous sand (shell sand)

This is found on some beaches, particularly in Cornwall and Scotland, and arises because of the large proportion of shell fragments in the sand. In such localities, this material is cheap and effective for arable and grass crops. All the material must pass 6.3 mm sieve and the legal declaration is for NV only (variable in the range 25–40).

Calcareous marine algae (lithothanmion calcareum)

These exist in some coastal waters and this material, of coral-like consistency, is dredged, screened to remove shells and either sold as graded material or dried and sold as a milled powder. A number of properties have been ascribed to these products from time to time, but they are very effective liming materials, with an NV of 40–50. They tend to be much more expensive than other liming materials.

By-product liming materials

A number of industrial by-products can be used as liming materials; their availability will depend on production in the locality as transport costs can soon cancel out any price/tonne advantage of these materials.

Sugar beet factory sludge

This contains a large quantity of calcium carbonate, but is very wet with about half its weight as water. Nevertheless, this wet material has an NV of 16–22 and contains some N, P and Mg as well: it can be spread by normal farm muck spreaders.

Sometimes this sludge is dried; it is then more expensive and has an NV of 50.

Water works waste

This waste occurs in some areas and consists of a very wet material that contains both calcium carbonate precipitated from water and some calcium hydroxide. Some works partially dry the waste to a sludge containing 25–30% water.

Whiting sand

Whiting sand is a by-product of the manufacture of whitening. It is a pure form of calcium carbonate arising from the process where chalk is ground under water, but it normally has a high water content.

Other materials

They include waste from soap works, paper works and bleach works. Many of these materials are wet, but can be spread by contractor and they will dry out to form a fine powder. Some have a slight caustic quality and should be applied to the soil rather than to a crop.

Historically other liming materials have been used. For example when fuel was cheap, it was common practice to burn limestone or chalk in kilns to produce burnt lime (calcium oxide). With an NV of 100, this material was the cheapest form of lime to transport, but it had to be kept free from contact with water. When water is added to burnt lime it forms hydrated (or slaked) lime (calcium hydroxide) and the chemical reaction can evolve considerable heat if allowed to proceed without control. Hydrated lime has an NV of about 70, and although more expensive than the liming materials mentioned earlier, it can be used in cases of emergency as it is extremely fine and quick acting.

Blast furnace slag and basic slag

These materials can have an NV of 35–45 and were widely available until changes in the steel-making processes rendered scarce slags with good NV.

Overliming

Overliming can cause problems and arises when either a regular liming policy is carried on without regard to soil pH or when the soil is so acid that substantial quantities of lime are needed. In the latter case, it is essential that some lime is worked into the soil, so that extreme acidity can be alleviated and no part of the soil is overlimed. A good rule is never to apply more than 3 tonnes/ha of CaO equivalent at one dressing.

Overliming can cause the following deficiencies: Mn—causing grey speck in oats, speckled yellows in sugar beet and marsh spot in peas; B—causing heart-rot in sugar beet and crown rot in swedes; Cu—causing severe stunting, particularly in cereals; Fe—causing chlorosis in many crops, particularly fruit-bearing species. Overliming also encourages scab in potatoes.

References

MAFF (1979a). *Profitable utilisation of livestock manures.* ADAS Booklet 2081. London: HMSO

MAFF (1979b). *Lime and fertiliser recommendations.* ADAS Booklet 2191. London: HMSO

MENGEL, K. and KIRKBY, E. A. (1978). *Principles of Plant Nutrition*, 1st edn. Berne: International Potash Institute

5

Arable crops

R. D. Toosey

CROP ROTATION

Rotational practice is valuable for

(1) weed control;
(2) control of certain soil- and residue-borne pests and diseases;
(3) maintenance of soil structure and organic matter, and recycling of plant nutrients.

Many soil- and residue-borne diseases are specific to a crop or group of crops and related weeds, which must be present for pest multiplication. Growing one crop too frequently allows pests to build up to infestation levels, so that subsequent crops of the same type fail completely or produce uneconomic yields. The pests must then be starved out by growing crops not susceptible to the pest or disease in question. Unfortunately some pests and diseases survive in the ground without a host plant for many years and starving out becomes a long-term project of dubious practicability; growing field-resistant or immune varieties (if available) is of value in certain cases. Hence it is cheaper and safer to adopt a sound rotation, with other measures to prevent disease building up to serious proportions. Other measures include:

(1) sound general sanitation, using clean seed, avoiding infected tubers or transplants and carrying infected material from one field to another, e.g. feeding clubroot (*Plasmodiophora brassicae*)-infected swedes on leys);
(2) providing suitable conditions for vigorous crop growth and where possible making the environment unsuitable for development of pests, e.g. by good drainage, enough lime, ample nitrogen;
(3) destroying all alternative sources of carry over,

e.g. trash, self-sown progeny or ground keepers and susceptible weeds.

Rotating crops around the whole farm ensures all fields benefit from restorative crops like leys, clover, lucerne, applications of farmyard manure and deeper cultivations. The three to four year ley improves structure and though it scarcely lives up to the results originally claimed, farmers who have leys in rotations rarely suffer soil structure problems. It is possible to manure many crops exclusively with purchased artificial fertilisers but cost is so high that maximal conservation and re-cycling of nutrients is of prime economic importance. Residues of leguminous crops, heavily stocked leys, folded roots and green crops provide substantial quantities of nitrogen and cut fertiliser bills considerably. Slurry and farmyard manure, properly conserved and applied, provide substantial quantities of all nutrients.

CHOICE OF CROP AND SEQUENCE

When deciding whether to introduce a new crop check:

(1) climatic requirements—temperature and water needs, tolerance of exposure and wind;
(2) suitable soil types, depth and subsoil requirements; interaction of soil and rainfall on particular site; crop tolerance of acidity;
(3) total and seasonal labour requirements, availability of suitable casual labour, e.g. vegetable production; likelihood of labour peaks coinciding with other crops. Feasibility of mechanising production and harvesting, storage and capital requirements. Availability of contractors;

(4) market demands and transport, marketing costs. It is pointless to produce crops which cannot be sold at a profit possibly because of excessive transport and marketing costs. Some crops are subject to area quotas, e.g. potatoes and sugar beet; others can only be grown safely on contract, e.g. herbage seeds, oilseed rape, vegetables for freezing;

(5) suitability of fodder crops to type of livestock, farming system, method of utilisation, length of growing period and time of year feed is provided;

(6) soil- or residue-borne pests and diseases already on farm, e.g. clubroot (*Plasmodiophora brassicae*) may preclude production of brassica vegetables—brussels sprouts, cabbage, cauliflower. Some pests may attack a range of unrelated crops, e.g. stem and bulb nematode (*Ditylenchus* spp.). Various races may attack onions, peas, beans, vetches and oats; sugar beet cyst nematode (*Heterodera schachtii*) attacks *all* members of beet family, e.g. mangels, fodder beet and red beet and is harboured by *all* brassica crops; frequency of cropping of beet and brassica is controlled by the British Sugar Corporation (BSC).

CROP DURATION AND IMPORTANCE
Maincrops

These occupy land for the majority if not the whole of the growing season, are the most important financially and take precedence over all others. Maincrops include all major crops, e.g. cereals for grain, sugar beet, maincrop potatoes, brussels sprouts. Always stick to maincrops when it is difficult to establish a second crop, especially in dry areas (e.g. East Anglia and on heavy soils).

Doublecrops

Double cropping refers to growing two crops in a year when both are equally important. The system applies to intensive situations (e.g. quick growing vegetable crops, Italian ryegrass for silage followed by kale). Second crops are highly dependent on adequate moisture and good tilth for satisfactory germination or establishment and the system is only suitable where there is ample rainfall or irrigation and a free working soil. Timely removal of first crop is essential.

Catch crops

These are of secondary importance and are snatched between two maincrops when the land would be otherwise unoccupied. They are usually grown to provide short-term fodder: e.g. *Year 1* winter barley harvested in July, then sow continental turnips to graze in autumn. *Year 2* spring barley. Only fast growing crops are suitable as catch crop.

Summer catch crops

These are sown in summer, make all their growth during summer and autumn and are usually consumed by Christmas. Use forage brassicas, selected according to date of clearance of first crop, e.g. after early potatoes there is a wide choice; after winter barley in southern England the choice is restricted to continental or grazing turnips or forage rape; rape kale for early spring keep, sown early in August is also suitable.

Winter catch crops

These are sown in late summer or early autumn, must be frost hardy to stand the winter and produce a large bulk the following spring. Mainly forage rye and Italian ryegrass.

Effect on maincrops

Catch crops must not interfere with maincrops, particularly:

(1) they must not harbour pests or diseases of any maincrop in rotation: e.g. rape is highly susceptible to clubroot (*Plasmodiophora brassicae*) which affects all maincrop brassicas; rye or Italian ryegrass should be preferred as catch crops in rotations including swedes, cauliflowers or oilseed rape. Avoid brassica catch crops in rotation containing sugar beet;

(2) catch crops must not delay sowing of subsequent maincrop by occupying land too long; or

(3) dry out soil.

Resources to grow catch crops

Plan well ahead for early harvested preceding maincrop to ensure timely sowing of catch crop, give adequate fertiliser to catch crop and only catch crop clean land. Catch cropping is unsuitable for dry areas where there is no irrigation, on heavy soils or where the growing season is very restricted, e.g. much of the North and cold uplands. If the crop is to be grazed or folded, use only well drained land with good access.

Brassica crops

The group includes:

(1) brassicas grown primarily for market as vegetables: brussels sprouts, cabbages, cauliflowers;
(2) brassicas grown primarily for fodder: kales, rapes, fodder radish, white mustard, turnips and swedes (including 'market' swedes);
(3) brassicas grown for producing seeds for manufacture or processing: oilseed rape and mustard for condiment manufacture.

GENERAL

Rotation

Do not grow brassicas more frequently than once in five years at most, preferably longer, to avoid build up of persistent diseases in soil, notably clubroot and dry rot or canker (*Phoma lingam*). Infected land must be rested much longer.

Pests and diseases

Clubroot (Plasmodiophora brassicae)

This is worst in wet and/or acid soils, i.e. high rainfall areas and poor drainage.

Prevention

Good sanitation; feed infected root crops only *in situ*. Ensure clean transplant beds. Improve drainage and ensure soil pH of at least 6.5 for all brassica crops by liming if needed.

Susceptible varieties

All horticultural brassicas, oilseed and forage rapes (variety Nevin is tolerant) and many varieties of swedes and turnip are highly susceptible and should not be grown on infected land.

Degree of resistance

This depends on races of clubroot present and level of infestation and is not absolute. Marrow stem and thousand head kales, fodder radish and some varieties of continental and hardy yellow turnips are highly resistant; Marian swede and some other sorts moderately resistant.

Dry rot or canker (Phoma lingam)

This attacks many brassicas, causing concave necrotic areas followed by rotting and collapse of plants. Infection can be seed borne, carried in soil or may arise from brassica waste or farmyard manure. Best seedsmen build up stocks from seed given hot water treatment or fungicidal soak. Sow clean or treated seed. Sanitation as for clubroot.

Beet cyst eelworm (Heterodera schachtii)

Brassicas act as host, but though rarely damaged, are treated as 'excluded crops' in rotations containing sugar beet (*see* p. 167 Sugar beet, rotations).

Brassica cyst eelworm (Heterodera cruciferae)

Adopt long rotation between brassica crops.

Other general pests and diseases

Numerous pests and diseases attack brassicas; the following are most important.

Flea beetle (Phyllotreta spp.)

Dress seed of *all* brassica crops with combined insecticide–fungicide and inspect regularly until well into rough leaf stage; apply insecticide as soon as symptoms of attack appear.

Cabbage aphid (Brevicoryne brassicae)

Feeds on leaves during summer and then infests developing brussels sprouts; even with small infestations crop becomes unmarketable. Routine inspection and treatment of sprout and other crops with approved aphicides are essential. Destroy overwintering hosts before May, where practicable.

Cabbage root fly (Erioischia brassicae)

Larvae feed on main roots, leading to wilt and death. Causes serious damage in horticultural brassicas, i.e. seedbeds, transplanted and direct drilled crops and early drilled swedes or cattle cabbage. Later generations may seriously disfigure market swedes and brussels sprouts at base of stem. Protect seedbeds of horticultural brassicas, transplants and direct drilled crops with approved granular insecticide, e.g. carbofuran, chlorfenvinphos and brussels sprout bottoms with trichlorphon if needed. Applicators are necessary.

Caterpillars (several spp.)

Damage occurs mostly from July to September. Inspect crops regularly and apply approved insecticide when required. Failure to do so may result in serious damage.

Wood pigeons (*Columba palumbus*)

Wood pigeons can strip young crops of brassicas and severely damage and soil older crops with droppings. Avoid isolated fields surrounded by woods. Regular shooting useful.

Slugs (*Agriolimax reticulatus*)

Slugs can strip young seedlings, seriously disfigure brussels sprout crops and spoil cabbage. Broadcast slug pellets as soon as damage is noticed.

BRASSICAS GROWN PRIMARILY AS VEGETABLES

Crops of brussels sprouts, cabbages and cauliflowers may be:

(1) raised in a plant bed or individual peat blocks or pots and then transplanted, or
(2) the seed is drilled directly into its final cropping position in the field (known as 'direct drilling' to horticulturalists).

This process should not be confused with 'direct drilling' agricultural crops into undisturbed or shallow cultivated ground after applying non-persistent total herbicide (e.g. paraquat).

Raising plants under glass or polythene protection

Essential for earliest spring sowings of summer cabbage and brussels sprouts and overwintered summer cauliflower. May be undertaken in frames or covered beds or borders, requiring specialised techniques and knowledge. Pots or peat blocks, made with block-making machines from horticultural peat and peat balancer avoid root disturbance and result in quick 'take' and rapid growth of plants. Horticultural peat is acid and deficient in plant nutrients; use of balancer, which must be *thoroughly* mixed with peat, is essential. Proprietary brands are available or details of suitable mixes may be obtained from ADAS. Plants raised in peat blocks require special transplanters; irrigation facilities are essential.

Raising plants outdoors

Choose a *fresh* clean site annually for the plant bed, which must be *known* to be free from persistent diseases of brassicas, NB clubroot (*P. brassicae*); (arrange for ADAS test if uncertain); soil pH should be *at least* 6.5, preferably 7.0. Ensure that no infective material is carried onto plantbed or future plantbed areas. Land previously in grass for several years is

useful. Plough in previous autumn for well settled bed with frost mould; work to produce fine firm level seedbed. Sturdy well developed and rooted plants are required. 'Bed' system gives most uniform results, drilled in rows 150–330 mm apart according to type of plant required, 20 mm deep and cut into soil moisture. Dress seed with combined fungicide/insecticidal dressing and precision drill if possible; if not sow *thinly* with brush or similar drill. Watch for flea beetle (*Phyllotreta* spp.) after emergence and treat again if necessary. For precise details *see* individual crops.

Direct drilling

Widely used for spring greens, spring cabbages, close spaced brussels sprouts and autumn cabbage. Except for closest spaced crops most economical method is to drill groups of three seeds at required spacings, e.g. 500 mm apart, individual seeds in group separated by 25–35 mm in row or 25 mm abreast. Ignore any field factor*. Unwanted plants are chopped out; use dressed graded seed and protect against cabbage root fly (*Erioschia brassicae*). Superfine firm level seedbed essential; work requires highest degree of precision.

Brussels sprouts

Brussels sprouts are grown for marketing fresh or freezing. Crops are planted specifically for picking over, which caters mainly for the fresh market or for single harvest, which includes processing and prepacking.

Picking over Sprouts are picked as they mature at intervals of three to five weeks, beginning at bottom of stem. Season starts with earliest varieties, raised under protection, picked in late August or early September and continued until late March. Period of availability of a given variety is much longer than with a single harvest.

Single harvest All sprouts are taken from stem at same time; stems cut in field and stripped or removed elsewhere.

Varieties

Sprouts produced commercially now come almost entirely from F_1 or other hybrids, whose sprouts are much denser and of better quality than from older open pollinated varieties. Seed is very expensive. Development of new varieties is rapid, resulting in

---

* *Field factor* is an allowance for differences between the number of viable seeds sown and the number of seedlings which emerge, which depends on seedbed conditions (e.g. soil, temperature, moisture, structure, pest damage). It is used calculating seedrate required, 0.5 = poor field conditions to 0.9 good field conditions. For details *see* different crops.

previous varieties being outdated quickly. Current varieties are listed in: Vegetable Growers Leaflet No. 3, *List of Brussels Sprouts* (NIAB).

Varieties overlap considerably in maturity, forming a continuous series to provide a season-long succession. For picking over: *earlies*: late August or early September to end November or early December. *Early midseason*: mid September or early October to late December or early January. *Midseason*: mid October or early November to late January or early February. *Late*: November to end of March. Shorter season for single pick, from about early October to late January. Many varieties are suited to both single pick harvest and picking over. When selecting varieties, attention should be paid to height and susceptibility to lodging, ease of picking or deleafing, quality and disease resistance.

Climate and soil

Commercial sprout production can be successfully undertaken in many areas in England and Wales on land below 275 m altitude but only in south west England or west Wales if varieties resistant to ringspot (*Mycosphaerella brassicicola*) are grown. Soil must be well drained, with pH 6.5 or above. Deep moisture retentive mineral soils are best, notably well structured calcareous clays and silts. Avoid peaty soils and shallow or sandy soils liable to dry out.

Cultural methods for crops raised in seedbed and subsequently transplanted

Prepare fine level evenly consolidated seedbed immediately before sowing as previously outlined. Work in 80–120 kg/ha or 35–70 g/m² 18% P_2O_5 superphosphate. *Sowing* and *transplanting* dates depend on variety and harvest date.

Variety	Early	Early midseason
Sow	Mid–end Feb (under protection)	First week March, in open (under protection in cold areas)
Transplant	Beginning–mid May	Mid May
Single pick harvest	Late Aug–end Sept	Late Sept–end Oct

Variety	Midseason	Late
Sow	First week March, in open (under protection in cold areas)–early April	End March–mid April
Transplant	Mid May–first week June	First week June
Single pick harvest	End Oct–end Jan	Late Dec–early March

Transplanting

Plough land in early winter to ensure stale furrow and well weathered, settled soil. Provide fine firm moist uniform tilth, not compacted underneath, soft enough on surface to allow penetration of transplanting machine. Plants pulled from seedbed should be placed in shallow trays and protected immediately from sun; plant at once. Insecticides for cabbage root fly (*Erioschia brassicae*) control should be applied at transplanting. Replace any failure within 7–14 d of transplanting. Irrigation immediately after transplanting ensures rapid root development; essential with plants raised in peat blocks. Apply 25 mm of irrigation at 25 mm moisture deficit. Glasshouse raised plants are brittle and need very careful handling.

Seedrate for seedbed drilling in rows

Seedrate required is determined jointly by seed count or size (i.e. number of seeds per 10 g in thousands), laboratory germination percentage and field factor, which depends on soil, moisture and temperature and seedbed conditions. Field factor:

	Drilling period		
	March	April (1–14)	(15–30)
Tilth in seedbed in open:			
good	0.6	0.7	0.8
cloddy	0.5	0.6	0.7
uneven compaction	0.5	0.6	0.6

Seedrate (kg/ha) of seedbed
$$= 10 \times \frac{\text{Required plant population/m}^2}{\text{No. seed/10 g } (10^3) \times \text{field factor} \times \text{germination \%}}$$

Aim for population of 39 plants/m run of row.

At 39 plants/m of row, the following row widths will give plant populations/m² as follows:

Row width (mm)	Population/m²
150	260
200	195
250	156
300	130

Cultural methods for direct drilled crops

Widely used for close spaced crops, occasionally at wider spacings. Plough at least 250 mm deep before end of December, having previously subsoiled if necessary, when soil was dry enough to shatter readily.

Table 5.1. Number of plants required/100 m² (1/100 ha) according to spacing

Planting distance cm (i.e. mm/10)	10	15	20	25	30	40	50	60	70	80	90	100	125	150
10	10 000	6666	5000	4000	3333	2500	2000	1666	1428	1250	1111	1000	800	666
15	6666	4444	3333	2666	2222	1666	1333	1111	942	833	740	666	533	444
20	5000	3333	2500	2000	1666	1250	1000	833	714	625	555	500	400	333
25	4000	2666	2000	1600	1333	1000	800	666	571	500	444	400	319	266
30	3333	2222	1666	1333	1111	833	666	555	476	416	370	333	266	222
35	2857	1905	1428	1143	952	714	571	476	408	357	317	286	228	190
40	2500	1666	1250	1000	833	625	500	416	357	312	277	250	200	166
45	2222	1481	1111	889	740	556	444	370	317	278	247	222	178	148
50	2000	1333	1000	800	666	500	400	333	285	250	222	200	160	133
55	1818	1212	909	727	606	454	363	303	259	227	202	182	145	121
60	1666	1111	833	666	555	416	333	277	238	208	185	166	133	111
65	1538	1026	769	615	513	385	308	256	220	192	171	154	124	103
70	1428	942	714	571	476	357	285	238	204	178	158	142	114	95
75	1333	889	667	548	444	333	267	222	190	167	148	133	107	89
80	1250	833	625	500	416	312	250	208	178	156	138	125	100	83
90	1111	740	555	444	370	277	222	185	158	138	123	111	88	74
100	1000	666	500	400	333	250	200	166	142	125	111	100	80	66
120	833	556	417	333	278	208	167	139	119	104	92	83	67	56
125	800	533	400	319	266	200	160	133	114	100	88	80	64	53
150	666	444	333	266	222	166	133	111	95	83	74	66	53	44

In spring prepare a fine firm level seedbed with minimal working and compaction for accurate drilling. Precision drills are most satisfactory. Sow groups of three spaced seeds centred at required final spacing 25–35 mm within row or abreast 25 mm apart 20 mm deep, removing unwanted plants later.

Drilling date

This depends on required date of harvest and variety. Very high populations, 44 000 plants/ha and over mature about a month later than those between 30 000–36 000 plants/ha.

Type of variety	Drill	Single pick harvest
Early	1st week April	Late August–end September
Early mid-season	1st week April	Late September–end October
Mid-season	Throughout April	End October–end January
Late	End April–mid May	Late December–mid March

Plant population

Increasing plant population reduces sprout size, improves uniformity of sprout development and delays maturity; length of stem is increased and sprouts are formed further apart on the stem. Level of plant population is more important than precise arrangement. Choice of population depends on market and size of sprout required, method of harvest and variety. Populations may vary from 12 000 to 50 000 plants/ha although populations over 36 000 plants/ha do not necessarily give greater yields of sprouts. Some non hybrid types are unsuitable for close spacing, while many hybrids are unsuitable for traditional wide spacings of 12 000 plants/ha (900 × 900 mm spacing). As an approximate guide aim for 42 000–

Table 5.2 Manuring of brussels sprouts for fresh picking or single harvest (kg/ha)

N, P or K index	N	P₂O₅	K₂O
0	300	200	300
1	250	125	250
2	200	60	150
3	—	60	50
Over 3	—	Nil	Nil

ADAS figures.

45 000 plants/ha (450 × 500 mm) for small sprouts for freezing and 26 000–28 000 plants/ha (600 × 600 mm) for picking over, according to variety. For single pick prepack sprouts populations range from 36 000 plants/ha for early shorter varieties to 28 500 for later varieties. For spacing and plant population *see Table 5.1.*

Manuring (Table 5.2)

Fertilisers are applied during preparation of direct drilled seedbeds or while soil is being worked prior to transplanting. Where N exceeds 150 kg/ha, excess should be top dressed within one month of crop emergence or transplanting on light soils. Top dressing of further N may also be necessary when rainfall substantially exceeds transpiration rate within two months of base application. Excessive applications of N may result in loose sprouts.

Stopping

Stopping is done to improve uniformity of sprouts in crops to be taken for single harvest up to end of November, but not for later crops. Apical bud is given a sharp downward blow with soft rubber headed hammer or a piece 30 mm diameter, including apical bud, is removed from four to ten weeks before harvest, when 50% of basal buds reach 12 mm diameter. Stopping too early may cause 'blowing' of upper sprouts.

Market and saleable yield

Sprouts for *fresh market* must be clean, sound, tight and fresh in appearance, and free from aphids or pest damage, graded into EEC classes I, II or III and packed into nets, holding 10 kg or boxes holding 6–14 kg or in small prepacks of 0.5–1 kg.

Yield

Poor	12.5 tonnes/ha
Average	20–22.5 tonnes/ha
Good	30 tonnes/ha or more

Sprouts for freezing must be clean, tight and blemish free and are closely graded into sizes: small, standard and large. Usually packed in 250 kg returnable bulk bins for delivery to freezing plant. Being smaller than the fresh pick market, yield is much lower.

Poor	5 tonnes/ha
Average	7.5–10 tonnes/ha
Good	12.50 tonnes/ha

Sample grading of 11 tonne crop:

Small: 13–29 mm	10%
Standard: 29–35 mm	85%
Large: 35+ mm	5%

Cabbages

Cabbages are produced for both animal and human consumption. Traditional cattle cabbages, once widely grown, are now of little importance. Modern hybrid cabbages bred for human consumption, are frequently grown for livestock as they produce high yields without rosettes. Cabbages for human consumption or market must be regarded as a group of crops rather than a single crop. Crop residues and unsaleable crops make useful stockfeed or green manure.

Spring cabbage

Crop may be grown as spring greens or collards, which are unhearted and as semi-hearted or fully hearted cabbage. Peak marketing period occurs between March and May, although with favourable weather in milder areas it continues throughout winter. Crop is speculative, as it may be unsaleable if there is a surplus of other vegetables or a glut in mild winter. Hearted cabbage may be cut selectively as they heart but it is often wise to cut some crops before maturity if prices are very high or growth is sappy and liable to frost damage.

Varieties

All varieties have pointed hearts.

Spring greens Preference for large leafy cabbage. Large leafy pale leaved varieties, e.g. Offenham–Scarisbrick Special or smaller darker leaved varieties, e.g. Durham Early.

Hearted cabbage Early small headed varieties, e.g. April, medium maturity and head, e.g. Offenham compacta; late maturity, medium to large head, e.g. Avon Crest. Varieties are listed in Vegetable Growers Leaflet No. 2, *List of Cabbages* (NIAB).

Soil, cultivations and establishment

Choose a free draining soil of light to medium texture which warms up early in spring but retains adequate moisture. Minimum pH 6.0 but preferably 6.5–7.0. Prepare fine firm seedbed with good tilth in adequate time for sowing between late July and September, taking care to conserve moisture. If irrigation is needed apply after completion of cultivations and allow soil to become dry enough to drill. To avoid compaction and improve uniformity many growers use the bed system. Nowadays spring cabbages and greens are almost entirely direct drilled (i.e. seed is drilled directly into cropping position); raising in plantbeds and transplanting only used on very small areas.

For spring greens aim for 28–40 plants/m^2 precision drilled, 20 mm deep using seed dressed with an approved fungicide/insecticide, in 350 mm rows to produce plants 70–100 mm apart; hearted spring cabbage requires 10 plants/m^2 in rows 400 mm apart drilled to produce plants 250 mm apart.

Manuring

Seedbed/base applications, *see Table 5.3*. Crop requires a steady supply of N throughout growing period but excess ruins winter hardiness. For spring harvested crops base dressing should not exceed 75 kg/ha N but for greens harvested before Christmas increase N to 125 kg/ha. Rate of spring N top dressing

Table 5.3 Manuring of cabbages for human consumption (kg/ha)

N, P or K index	N					P_2O_5	K_2O
	Spring, hearted and greens	Summer and autumn	Winter white for storage	Winter and savoy		All cabbage types	
				Pre-Christmas cutting	Post-Christmas cutting		
0	75	300	150	300	150	200	300
1	50	250	125	250	125	125	250
2	25	200	100	200	100	60	150
3	—	—	—	—	—	60	50
Over 3	—	—	—	—	—	Nil	Nil
N top dressing	200–400				0–75		

ADAS figures.

depends on expected size of crop; fully mature cabbage harvested in late spring can use up to 380 kg/ha whereas lighter crops of greens need less than half this quantity. Dressings of 100–200 kg/ha N at any one time are normally adequate. Timing depends on weather conditions and potential marketing period. To avoid scorch and subsequent infection by grey mould fungus (*Botrytis cinerea*) top dress when crop is dry or very wet to minimise adhesion to foliage.

Cutting date

Determined by sowing date and variety.
   Spring greens: December to April,
   Hearted cabbage: April to early June.

Yield

Extremely variable and difficult to quantify, income/ha main consideration. Crops may be cut completely for unhearted greens, cut over partly for greens and residue left to heart or cut over hearted, according to market. Average: 12.5 tonnes/ha hearted cabbage.

Summer and autumn cabbage

Grown to produce a succession of hearted cabbage to market between late May and late October, for which

sequential sowings and a range of varieties are necessary. Varieties are listed in Vegetable Growers Leaflet No. 2, *List of Cabbages* (NIAB). Sowing in February and March for late May, June and July harvesting are made under protection with heat; April and May sowings can be made outdoors. Growers often prefer to continue sowing under protection for harvests from July to October to ensure continuity of supply. Varietal types, and periods of sowing, transplanting and cutting are given in *Table 5.4*. Sowings made in April and May can also be drilled direct into final cropping position, where suitable.

Soil, cultivations and establishment

Soil must have stable structure to permit extensive root development with at least 600 mm effective rooting depth; good drainage essential, especially autumn cabbage. Earliest summer crops come from sandy loams, loams or peats. For later crops water retentive medium to heavy soils and silts are suitable.
   Crops respond well to irrigation; on loamy sands it is usually essential. Soil pH should be 6.5 on mineral soils, 6.0 on peats.
   Prepare a fine level tilth but not overworked or compacted. Subsoil if required and plough during previous autumn. Spring ploughing is unsatisfactory on heavier soils; allow ample time for working and

Table 5.4 Summer and autumn cabbage—types and times of sowing, transplanting and cutting

Type of cabbage	Sow	Method of raising	Plant out[1]	Shape of head	Cutting period	Direct drilled
Summer–early	late February	P	mid April	Pointed, e.g. Hispi Fi[2] or round, e.g. Golden Acre	late May–late June	—
Summer–midseason	late March to mid April	P	late May	Round[2]	July–August	—
Summer–late	April	O or P	[1]	Round[2]	September	} May to mid June
Autumn	late May	O or P	mid July	Round[2]	September–late October	

[1] For spring and early summer sowings six week old protected plants and eight week old open ground plants give better yield and quality than older plants.
[2] Fi hybrid varieties usually produce more uniform crops than open pollinated varieties or selections but seed is dearer.
P = raised under protection; O = raised in open.

weathering before planting. Direct drilled cabbage require rather more working than transplanted crops; with the latter a single pass with a spring tine cultivator should be adequate. For direct drilling use only the best soils, ploughed in autumn.

Plant density and spacing

This depends on maturity and size of cabbage required. Summer cabbages for market allow 9 plants/m^2, rows 380 mm apart and 300 mm between plants; for autumn cabbages allow 3–4 plants/m^2 rows 600 mm and plants 400–600 mm apart, according to size required.

Manuring

Fertiliser is thoroughly worked into soil while preparing tilth, except where N topdressing is appropriate, (*Table 5.3*). Where heavy dressings of P and K are required (e.g. index 0) part should be applied before ploughing.

Yield

Variable. Good growers achieve 20–30 tonnes/ha.

Winter cabbage and savoys

The group includes hearted cabbage grown for marketing from October until hearted spring cabbage are ready.

Types and varieties

Winter white cabbages   These are round, dense, hard cabbages, white inside ready for cutting up to end of November or early December. Varieties are suitable for the fresh market, storage or both, but mainly storage. Type is not hardy enough to overwinter outdoors, except in mildest coastal areas. White cabbages for storage must be cut and stored *before* there is any frost damage and can then be marketed in late winter.

January King varieties   Early selections may be cut before Christmas but are valued for January–February; best sorts are very frost hardy. Few outer leaves,

with purple cast, prominent veins and producing heads rather under 1 kg. Leaves somewhat wrinkled and classed as a savoy elsewhere in Europe.

Savoys   These may be distinguished by their deeply wrinkled foliage and dark green colour. Savoys are more frost hardy than other forms of cabbage, so are especially suitable for growing in colder areas. Regarded as of higher quality than hardy cabbage; flavour improves after exposure to frosts. Varieties have wide range of maturity, varying from 140 to over 200 d after transplanting.

Savoy × winter white hybrids   These have dense heads like white parent, with some of the hardiness of the savoy; they are high yielding. Grown as January Kings.

Varieties are listed in Vegetable Growers Leaflet No. 2, *List of Cabbages* (NIAB).

Soil, cultivations and establishment

Winter cabbages are grown on a wide range of soils but soil must be free draining. Deep soils are best. Avoid light sands or shallow soils. Plant on early, medium and late fields to secure succession. Tilth as for previous types but allow ample time for preparation after clearing previous crop. Plant into moist soil. If soil remains dry for few days after transplanting, irrigation is essential.

For sowing and transplanting dates *see Table 5.5*. Crops may also be direct drilled on good tilths. Aim for 3–4 plants/m^2 in rows 600 mm apart, 400–600 mm within rows.

Manuring (Table 5.3)

Work in base dressing during tilth preparation, including all P_2O_5 and K_2O. Adequate N is essential for healthy growth but excess damages winter hardiness of crops of savoy and winter cabbage to be left after Christmas; excess also impairs quality of storage cabbage. Direct drilled crops and transplants on light soils should not receive more than 150 kg/ha in seedbed; top dress remainder within one month of crop emergence or transplanting. Additional N may be required as top dressing where rainfall greatly

Table 5.5 Winter cabbage, types and times of sowing, transplanting and cutting

Type of cabbage	Sow	Method of raising	Plant out	Cutting period	Direct drilled
Winter white cabbages	late April	O	early July	Should be cut and stored by November	mid May
January King and savoy × winter white hybrids	late May	O	early July	November onwards	June–early July
Savoys	late April–mid May	O	late June	early varieties September to December, late varieties January to March	mid–end May

O = raised in open.

exceeds transpiration within two months of base application.

Storage cabbage

This is largely marketed from January to April, for which storage is essential. Cut before frost damage and when the crop is dry. Leave outer leaves in the field, do not trim, handle heads with care. Storing in clamps, barns or cold store should be completed by late November, cut early in the day, to lose field heat. Cabbages store well until late March. Maximum depth: barns 2.4 m, clamps 1.2–1.5 m. Smaller cabbages not given too much N store best. Aim for a temperature of 0 °C in store where possible.

Marketing from field

The earliest crops of winter cabbage and savoy sometimes overlap late crops of summer and autumn cabbage. Winter cabbage and savoy mature over a long period and regular cutting secures firm, compact heads of uniform size and quality. Modern F_1 hybrids mature more evenly.

Yield

Storage cabbage:
Average   30–32 tonnes/ha
Good   45 tonnes/ha

Winter hardy cabbage/savoy:
Average   20–22 tonnes/ha
Good   30 tonnes/ha

Cattle cabbage

Cattle cabbage produce large-hearted cabbages for autumn and winter use but are of little importance nowadays, more suited to small farms where massive productivity from small areas is essential. Under really fertile conditions yields of 150 tonnes/ha are obtained but 50–75 tonnes/ha is more common. Grown as a maincrop, it has a high labour demand and its feed value is well below that of marrow stem kale.

Devon Flatpoll, with purple leaves, is the main variety. Seed is usually sown in nursery beds in mild areas in August or early September at 1.25 kg seed, dressed against flea beetle, 12 mm deep in 230–300 mm rows on 0.125 ha. This provides plants for 1 ha of crop. Transplant to final cropping position in April or early May in rows 760 mm–910 mm apart, 910 mm between plants.

Cattle cabbages need very fertile soils and do well on deep, rich medium-to-heavy soils, liberally manured with farmyard manure 50–60 tonnes/ha ploughed in. Work 125–200 kg/ha N, 125 kg/ha each of P_2O_5 and K_2O into the seedbed, giving the lower rate of N after applying a heavy dressing of dung. The seedbed must be deep and loose enough for the transplanter or planting by hand with a dibber. Cabbages must be firmly planted.

Cauliflowers

Cauliflowers, which are produced all the year round from different areas in the UK, may be divided into two main groups:

(1) *winter cauliflowers*, marketed from early November to late May or June according to variety; and
(2) *summer and autumn cauliflowers*, cut from early May to November or even early December in mild open years.

A comprehensive list of varieties of all types suited to culture in England and Wales is issued by the National Institute of Agricultural Botany—Vegetable Growers Leaflet No. 1, *List of Cauliflowers*. The two groups are quite distinct in their varieties and cultural methods and must be treated as separate crops.

Winter cauliflowers

Grown as farm crops in relatively frost-free areas with mildest winters: south west England, the south coast, South Wales and the Channel Islands. To avoid frost damage to curds, varieties heading in mid winter should be grown only in most favoured situations, usually close by sea coast. Only varieties resistant to ring spot (*Mycosphaerella brassicicola*) of, or derived from, Roscoff type (as recommended by the NIAB) should be grown in the south west. More winter hardy sorts, heading from March to June include varieties of Angers, English Winter and Walcheren types. These are only half grown when winter sets in and are more widely cultivated.

Soil

Choose free draining friable soil, pH 6.5–7.0 on arable farms, 6.5 on holdings with livestock. Crop highly susceptible to clubroot (*P. brassicae*) and must not be grown more than once in five years except on strongly alkaline light soils near Penzance in west Cornwall. Maintain good structure with regular use of leys or farmyard manure.

Plant bed preparation and sowing

Winter cauliflowers are unsuitable for direct drilling and are always raised in plant beds. Select land *known* to be free from clubroot; arrange for soil test by ADAS if in doubt. Some growers use permanent seedbed. Keep soil pH above 7.0. Plant bed may also be sited on newly ploughed grass but treat with approved insecticides against wireworm and leatherjackets. Plough before end of December to allow settling, frosting and preparation of fine firm weedfree soil in spring. Incorporate approved insecticide into soil to control cabbage rootfly. Avoid excess N in plantbed. Very fertile soils need no fertiliser in plant bed. Otherwise work in

	N (kg/ha)	P_2O_5 (kg/ha)	K_2O (kg/ha)
After grazed/clovery ley	nil ⎫	55	55
Arable rotation	25 ⎭		

Sow Roscoff types during first week in May. Sow others late May and during June, all to be ready for transplanting in July. Sowing too early results in subsequent losses from transplants being too large, buttoning, wirestem and mildew.

Allow 280–420 g of seed to produce enough transplants for 1 ha of crop. Drill in beds, rows 225–250 mm apart, seed 20 mm deep or cut into soil moisture. Seedrate in bed 28 g/91 m of row to give 33 plants/m of row. Place dressed seed 18 mm apart with precision drill. Plants are lifted from seedbed when required; protect against sun and transplant immediately. Select sturdy plants 180–230 mm height with good root systems. Reject weak, deformed or blind plants, or those damaged mechanically or by pest or disease. Use routine calomel root dip in areas where clubroot is prevalent. (Use 49 g 100% calomel in 1 litre water dipping roots in bunches of 25–30 plants; this will treat 600–1000 plants.) Protect transplants against aphids and cabbage root fly.

Field preparation and transplanting

Plough before end of March at latest to allow adequate time for preparation of well settled fine moist tilth to receive transplants. Always have land waiting for plants, not other way round. Mark out land with spaced tine implement across line of transplanting to ensure adequate population. *Transplant* late varieties in first half of July and earliest during third week July.

Finish planting by end of July at very latest. Complete gapping up failed plants or misses within one week of planting. *Banking up* plants with earth is traditional in windy areas but must be completed early to avoid severe damage to roots.

Plant population and spacing

Aim for 23 500 plants/ha for varieties heading after early March and 20 000 plants/ha for earlier varieties, which require row widths 650–750 mm and interplant spacings 550–600 mm.

Irrigation

Irrigate or water in when transplanting during a dry period. Plant *firmly*.

Manuring

Base fertiliser, including all P and K is worked in during soil preparation (*Table 5.6*). Roscoff types cut in February in frost free areas are top dressed in late autumn; crops for cutting after mid February are top dressed in mid February. Other types for spring cutting are top dressed between January and March. Avoid excess N applications.

Cutting

Inspect crop frequently and once production starts cut every other day to minimise losses from frosting or yellowing of curds.

Cut every good head *before* it has reached full size. During frosty periods cut only in middle of day, when temperature is rising. Cutting at other times damages growing crop and keeping quality of heads; also cut harder during periods of sustained frost although heads will be smaller. Cut into trailers or tractor mounted bins, one man usually taking three rows. Cutters trim while harvesting, leaving sufficient leaves

Table 5.6 Manuring of cauliflower (kg/ha)

N, P or K index	N			P_2O_5	K_2O
	Early summer, late summer, early autumn, late autumn	Winter		All cauliflower types[1]	
		Roscoff types	Winter hardy types		
0	250	75	75	200	300
1	200	40	40	125	250
2	125	Nil	Nil	60	200
3	—			60	125
4	—			Nil	60
N top-dressing[2]		60–125	125–200		

ADAS figures
[1] When winter cauliflowers follow crop leaving large residues reduce P_2O_5 by half and K_2O by 60 kg/ha.
[2] For transplants on light soils or direct drilled crops top dress N in excess of 150 kg/ha; apply residue within one month of emergence or transplanting.

and length of leaf above top of curd to protect during transit.

Quality, packing and yield

Heads when cut should be firm, compact and of good white colour. Once seriously yellowed by sunlight or blown by overstanding, they are unwanted. Discard also badly damaged heads. 'Ricey' and 'bracted' curds are of low quality.

All cauliflowers must normally be graded according to EEC regulations (*see* MAFF leaflet EEC Standards for Fresh Cauliflowers)—Extra class and Classes 1, 2 and 3. Most good production usually falls into classes 1 and 2. Extra is too exacting for most commercial winter production; grade 3 is of low quality and usually only justifies local marketing. Packing is best carried out under cover for workers' comfort and efficiency. Produce for wholesale markets is packed in non-returnable containers of 12, 16, 24 or 30 heads. Cauliflowers will usually keep for a week in a cool store; only store top quality curds cut as soon or immediately before they are quite ready.

Yield

	(Crates/ha)
High	625
Average	525
Low	425

Summer and autumn cauliflowers

Crops are grown to provide succession of cuttings throughout summer and autumn until November and are grouped as follows.

Early summer cauliflowers Sown in late September or early October, pricked out into frames or Dutch lights and transplanted as 'peg' plants into cropping position in March or early April. Alternatively sown in gentle heat in February or early March and transplanted four to five weeks later, as soon as ground conditions permit. Undue delay involves increased risk of 'buttoning'. Maturity (50% of marketable heads cut—NIAB) varies from approximately 80–110 d after transplanting according to variety. Use of peat blocks or pots minimises root disturbance.

Late summer cauliflowers Sown under protection in March or in open ground in April transplanted mid–late May, maturing about 60–80 d after transplanting.

Early autumn cauliflowers Sown in late April or early May and transplanted mid to late June. Mature 70–85 d after planting.

Late autumn cauliflowers Sown in mid May, transplanted late June or early July, maturing 70–130 d after transplanting.

Soil type and cultivations

These are more specialised crops and require greater care in soil selection and management than winter cauliflowers. *Plants must never receive a check through lack of moisture or nutrients.*

Choose deep rich moisture retentive soils, such as alluvial silts, medium to heavy loams, brickearths and lower greensands. Avoid soils prone to drying out. Soil should have good structure and be liberally supplied with organic matter, either by leys or heavy dressings of farmyard manure. Cultivations should be deep and thorough as for winter cauliflowers. Allow adequate time after harvesting previous crop for preparation of high quality moist weed free seedbed. Loose soils result in crop failures and low quality curds. Firm with roller but follow with harrows prior to transplanter.

Crop also highly susceptible to clubroot—control pH to 6.5–7.0.

Plant raising methods

Block or pot raised plants do not suffer from root disturbances, grow away and mature earlier. Some growers raise the bulk of their plants this way, especially on heavier bodied soils. Seed grown in boxes and pricked out individually when two true leaves start to expand breaks taproot, so giving well branched root system, but technique is laborious. Discard undersized, damaged or untypical plants. Sowing two seeds direct into blocks or pots and removing surplus seedlings by hand is quicker but plants are less evenly developed. Where plants are raised in plantbeds and transplanted, management as for winter cauliflowers. Insecticide to control cabbage root fly should be incorporated in block or pot medium: 100 g of seed is sufficient for 12 000 plants.

Plant population and spacing

This depends on maturity group and size of curd required, head size increasing with lateness of maturity.

Type of crop	Plant popula-tion/ha	Spacing (mm)
Earliest crops (from frame raised transplants)	37 000 to 30 300	600 × 450 to 600 × 550
Late summer and early autumn crops	27 700	600 × 600
Late autumn crops	23 800	600 × 700

Manuring

A steady supply of N is essential. If deficient, plants make poor growth and start to button, whereas excess leads to undesirable features such as browning, scorching, bracting, hollow stems and loose curds. Dressings over 150 kg/ha N should be split; top dress within one month of transplanting. Potash is said to increase head firmness. Very heavy dressings of muriate of potash can damage plant, so apply well before planting or part before ploughing. Some growers prefer sulphate of potash, which is essential under glass (*see Table 5.6*).

Irrigation

Summer and autumn cauliflower make heavy demands on soil moisture; do not allow soil to dry out. In dry periods water immediately after planting, which can suppress buttoning. Soil should be well irrigated in early stage of curd development. Regular watering with 10 mm when needed gives best results.

Quality, packing and yield

Only top quality high yielding crops justify production. Quality and packing as for winter cauliflower. Yield increases with lateness of maturity; late autumn crops tend to have large heads and give high proportion of crates of '12's.

Type of crop (good yield)	Crates/ha
Early summer	1300–1500
Late summer and early autumn	1500–1750
Late autumn	2000–2250

BRASSICAS GROWN PRIMARILY FOR FODDER

Fodder radish

Although fodder radish produces sizeable bulbs or roots, it is sown thickly to produce an abundance of foliage and used as a green forage crop. Initial growth rate is extremely rapid and yields of 75 tonnes/ha of green material are recorded, but 37–50 tonnes/ha is more usual. Unfortunately the crop runs to seed in 6–11 weeks according to weather and variety and is then highly unpalatable and neglected by stock. Foliage is not frost hardy and the crop must be cleared by early autumn. Period of use is inflexible and the crop is not popular.

Resistant to mildew, it is virtually immune to clubroot and does not harbour beet cyst nematode (*Heterodera schachtii*), hence it may be of special value on sites heavily infested with clubroot or as a catch crop in a sugar-beet rotation. It is highly sensitive to poor seedbed conditions and adverse climatic conditions, as in the north of England and Scotland.

Varieties

These include early and late cultivars but only late sorts should be sown, e.g. Neris, Slobolt. Early varieties flower much too quickly.

Seeding

Drill 9 kg/ha in 180 mm rows or broadcast 13.5 kg/ha in June or July (August in southern England only) and consume before flowering starts. Seedbed and manuring as forage rape.

Kale

Widely grown, kales are unimportant in Wales and north-east Scotland. Planted mainly as early sown catch crops or late-sown maincrops, they may follow early harvested crops, Italian ryegrass or grass for hay, silage or grazed.

Types and varieties

Marrowstem Gives heavy yield 60–70% stem, which is very thick, fleshy with low plant densities, when it lodges. Not very frost hardy and unsuited after the new year. Stems become woody quickly and are then neglected by stock. Grown mainly for dairy cows. Green and purple stemmed varieties exist, the latter of little importance. Resistant to clubroot.

Thousand head Yield mostly as leaf and tender sideshoots arising in late winter and early spring. The stem becomes hard, woody and neglected by stock. It is really frost hardy and well suited to feed in late winter. Both tall (stem height approximately 1.1 m) and dwarf types (0.7–0.9 m) available, the latter, including Canson, most frost hardy and least liable to lodge. Reistant to clubroot. Mainly used for sheep.

Hybrid varieties

These are somewhat similar in appearance to marrowstem, are shorter, less liable to lodging and retain their digestibility longer, stems being slower to become woody. Varieties for autumn (Merlin), and winter (Bittern) use are available. Maris Kestrel and Proteor may be used throughout autumn and are hardy enough to stand until mid March, when flowering may begin. Hybrids are slightly less resistant to clubroot than marrowstem.

Hungry gap and rape kales See Rape.
Details of kale varieties are given in Farmers Leaflet No. 2, *Recommended Varieties of Green Fodder Crops* issued by the NIAB.

Soil, cultivations and sowing

Kale is very tolerant of soil type but avoid heavy or poorly drained soils owing to liability to severe poaching when utilised. Direct drilling (destruction of surface vegetation by herbicide and drilling seed into undisturbed surface with special drill) is highly suitable under the right conditions, as it minimises poaching, annual weeds and loss of soil moisture and

Table 5.7 The kales-dry matter content, D value, seeding and approximate period of utilisation

Crop	Dry matter content of whole plant (%)	D value of whole plant	Approximate range of fresh yields (t/ha)	Usual row width (mm)	Seedrate (kg/ha)		Approximate range of sowing dates	Approximate period of utilisation
					Drill	Broadcast		
Marrow stem varieties	12	66–67.5	50–75 Very heavy crops 100 Late sown S. England 25–40	180–350 up to 550 with conventional cultivations	OD 3–4.5 PD 1–2.5	7	N 15 April–31 May	N Aug–Nov
Early hybrid varieties	12	69–70					S 1 April–31 July (according to district)	S Aug–Dec
Frost hardy hybrid varieties	12	69–72.5	25–70 according to sowing date	180–350	OD 2–3.5 PD 1–2.5	5	As marrow stem	N Aug–Dec S Aug–Mar according to variety and sowing date
Thousand head Tall	15	Tall 65–66	Tall 35–50	250–550	OD 1–1.7	2.5–3.0	N May–June	N Oct–Dec (not common)
Dwarf	14	Dwarf 69–70	Dwarf 25–40		PD 0.6–1.0		S June–July successional sowings pointless	S Jan–Mar

Key:
N = Scotland, north of England, hill, upland and other colder areas.
S = South, south western England, west Wales and other areas with mild autumn and winter.
PD = Precision drills—use *graded seed* and check suitability of belt or wheel.
OD = Other drills. Seed may require bulking with *inert* diluent, e.g. mini slug pellets. Do not use fertiliser. Includes direct drills of non precision type.

requires much less labour. With conventional cultivations ensure a fine free moist seedbed; drying out of soil is a particular hazard when growing as a second crop. For sowing details *see Table 5.7.*

Manuring

Kales require liberal application of N, of which direct drilled crops require an additional 25 kg/ha N. Relatively unresponsive to P and K unless deficiency exists (*see Table 5.8*).

Table 5.8 Manuring of kales and summer turnips (kg/ha)

N, P or K index	N	P_2O_5	K_2O
0	125	100	100
1	100	75	75
2	75	50	50
3 and over	—	25	50

ADAS figures.
Increase N by 25 kg/ha for direct drilled crops. Above recommendations assume crop is grazed.

Utilisation

Kales may be fed *in situ* or zero grazed. Early sowings provide excellent part replacement for grass in August and September for cows. Successional sowings of suitable varieties can provide green fodder until mid March in south. Thousand head or winter hybrids

may be sown in alternating blocks with swedes. It is unsafe to feed more than 25–28 kg or 18–20 kg/d to large (Friesian, etc.) or small (Guernsey, Ayrshire) breeds of cow, respectively. A phosphorus and iodine rich mineral supplement should be fed. Calcium rich foods such as sugar beet pulp should not be given with kale.

Rapes

Rapes grown for fodder include forage rape and so-called rape kales. Oilseed rape is a completely different crop.

Forage rape

Forage rape is a quick growing, green forage, ideal for catch cropping, especially in cereal stubbles. Subject to mildew in dry areas if sown early, it does best in the west, north and Wales. Not really frost hardy, it can only be relied on to provide feed to Christmas, except in mild winters. All varieties, except Nevin, are highly susceptible to clubroot; Nevin contracts the disease but is tolerant and does not collapse. It is inadvisable to catch crop with rape in rotations containing maincrop brassicas, or grow more than once in five years, because of susceptibility to clubroot. Soil should be maintained at pH 6.0–6.5. Initial growth is quicker than the kales but final yield only higher in a short growing season of 8–12 weeks.

Varieties

These are divided into Giant and Dwarf types, although in practice they form continuous series.

Giant varieties. These are quicker growing, heavier yielding and always preferred to dwarf types for stubble catch cropping. Nevin is lower yielding and only appropriate to situations where clubroot is present, when it excludes other rapes. Continental turnips are then a preferable alternative. Details of varieties of forage rapes are given in Farmers Leaflet No. 2, *Recommended Varieties of Green Fodder Crops* issued by NIAB.

Dwarf varieties Broad leaved Essex types have slightly higher digestibility, recover better from grazing but give low yields and are no longer recommended.

Swedelike or rape kales

These include two well known varieties, *Hungry Gap kale* and *rape kale*, are really rapes and possess many features of rape, notably susceptibility to mildew and clubroot and capacity for rapid initial growth. Cultivated almost entirely in the south, where they are used for stubble sowings for keep in the hungry gap period during the following March and April, both are frost resistant, although the variety rape kale is somewhat hardier but lower yielding. Both varieties lose a large amount of leaf around Christmas; the bulk of the keep is produced from sideshoots which develop in spring.

Sowing

Forage rape and rape kales require a fine firm seedbed, when drilling in close rows is usually more reliable than broadcasting; crops may also be direct drilled, depth 20 mm (*see Table 5.9*).

Manuring

Rapes require ample lime and high fertility. They cannot be grown without fertiliser, especially in cereal stubbles (*see Table 5.10*).

Table 5.10 Manuring of forage rape and stubble turnips (kg/ha)

N, P or K index	N	P_2O_5	K_2O
0	100	75	100
1	75	50	75
2	50	25	50
3 and over	—	Nil	50

ADAS figures.

Utilisation

Grown largely for sheep, rape is almost invariably fed *in situ*. Stock, especially cattle, must be introduced gradually to it with clean roughage (hay or sweet straw), and with grass, available at all times. Gorging with rape should never be allowed. If properly managed, rape is second to none for fattening lambs in the autumn.

White mustard

Crop now only of localised importance and for feeding purposes is largely superseded by continental turnips or fodder radish. Sometimes grown as a catch crop for green manure or flushing ewes after early potatoes or barefallow. It must be fed or ploughed in before flowering, seven to eight weeks after sowing and is not frost hardy. Drill in 180 mm rows at 11–17 kg/ha or broadcast 23–25 kg/ha on fine firm seedbed mid April to mid August. Fertiliser is unnecessary after well manured potatoes or barefallow. Give 75–90 kg/ha N to stubble sown crops. Green yield 25–50 tonnes/ha.

Turnips and swedes

Swedes may be readily distinguished from turnips by the presence of a 'neck' from which leaves arise; these

Table 5.9 Dry matter content, D value, seeding and approximate period of utilisation of forage rapes

Crop	Whole plant (% DM)	Whole plant D value	Range of fresh yield (t/ha)	Usual row width (mm)	Seedrate Broadcast (kg/ha)	Drilled (kg/ha)	Approximate range of sowing dates	Approximate period of utilisation
Forage rape, giant varieties	12	68	30–40 (average at 8–12 weeks)	180 (or broadcast)	7–8	4.5	N 30 May–15 July S 15 June–15 Aug	N 1 Sept–30 Nov S 1 Sept–31 Dec
Swede-like kales—rape kale and hungry gap kale	12–13	—	25–40	180 (or broadcast)	6	3.5	S only 15 July–15 Aug	S only late Mar–mid April

N = Scotland, northern England, hill, upland and other colder areas.
S = South and southwest England, western Wales and other areas with a mild autumn and winter.

are usually bluish or dark green and smooth. With turnips the 'neck' is so small that it is almost absent and leaves are coarsely hairy. Both may have white or yellow flesh and are used for (a) human consumption, or (b) animal fodder.

Varieties

Turnips   These mature more quickly than swedes, give a higher yield in a shorter growing season and can be sown much later, although they have a lower dry matter content and feeding value. Types of turnip differ widely in time of maturity, keeping quality and frost hardiness.

Human consumption—garden turnips   Grown for their small bulbs, white or yellow flesh, in several skin colours. Shape mainly round or flat. Varieties numerous. A very quick growing specialist crop.

Animal fodder

Varieties grown primarily for leaf or forage

White fleshed 'continental stubble turnips'   Mainly of Dutch origin, have amazing initial growth capacity and produce heavy crops ready for folding eight to nine weeks after sowing. Under ideal conditions 110 tonnes/ha of fresh material can be produced in 11 weeks. This is a quick-growing, green forage crop rather than a root crop, for 50% of the yield is leaf and the roots deteriorate rapidly. It is essential to consume continental varieties whilst still growing actively, as the bulbs become unpalatable, woolly and finally hollow as they approach maturity. Earliest sowings in late April or May (summer turnips) produce heavy crops capable of fattening lambs weaned off grass from late June onwards; sowings can be made throughout the summer for a succession of feed to early January. There are many varieties but yield differences are relatively small; although all varieties are early maturing there are still marked differences. Earliest varieties include Civasto and Debra, later varieties Ponda, Taronda and Vobra. Many, including the above, show good resistance to clubroot.

Choose varieties with good resistance to bolting for early spring sowing; Civasto, Marco, Taronda.

Hybrid grazing turnips   They produce 80% top with only a small bulb. Heavy yield of leaf and will produce regrowth. Root anchorage is superior to continental turnips. Varieties: Appin, Tyfon.

Varieties grown primarily for root production

Traditional British white fleshed varieties   These are not resistant to clubroot, make slower initial growth, but produce heavier crops of roots which are sweeter, more palatable and last longer than continental types. Except for Hardy Green Round, these varieties are

not frost hardy and mature early; best varieties include Imperial Green Globe (syn. Green Globe, Norfolk Green Globe), Hampshire Hardy Green Round, Lincolnshire Red Globe, Purple Top Mammouth and Red Tankard.

Yellow fleshed turnips   They are seriously underestimated; easier to grow than swedes, they grow on poor land and can be sown later. *Soft yellow* varieties (Centenary, Fosterton or Dales Hybrid, Grampian) are comparable to white types, being early maturing and less frost hardy. *Hardy yellows* are akin to swedes in hardiness and keeping quality and are preferable where fertility is low. Varieties Aberdeen Green Top Yellow (syn. Green Top Scotch, Green Top Yellow Bullock), Aberdeen Purple Top Yellow (syn. Aberdeen Purple). The Bruce (syn. Tammie Mackie) and The Wallace and Champion Green Top Yellow. Latter three show resistance to clubroot, which is excellent in Brimond and Findlay.

Swedes

Frost hardiness and keeping quality, in which swedes are generally superior to turnips, increase with dry matter content. Few varieties apart, swedes have yellow flesh and are classified by skin colour. There are no distinct varieties for market and livestock feed; market, fodder or dual purpose crops may be grown. *Market* requires swedes 0.5–1.0 kg each with small top and neck, yellow flesh and purple skin. Varieties Acme, Devon Champion, Marian. Manchester market prefers green skinned varieties.

Purple skinned varieties

These are by far the most widely cultivated, are divided into light and dark skins, the latter having the higher dry matter content and best frost hardiness and are most important.

Bronze skinned varieties

These have comparable characteristics to the purple but are unsuited for market sale.

Green skinned varieties

They are rarely grown in England and Wales; roots are smaller but have a higher dry matter content, flesh is very hard, skin tough and pulping is advisable to avoid damaging teeth of stock. Longest keeping, these varieties will store into May in Scotland.

Details of varieties of turnips and swedes are given in Farmers Leaflet No. 6, *Varieties of Fodder Root Crops* issued by the NIAB. Sow *only* varieties resistant to clubroot on infected land (e.g. Marian; varieties resistant to powdery mildew are preferable for early sowing, especially in the south.

Climate and soil

Apart from continental and grazing varieties of turnip, which can be grown in most areas, swedes and turnips grown for their bulbs are best suited to cooler moister parts of the UK, including Scotland, Wales and that part of England lying north and west of a line drawn from Hull to Weymouth; to the south east of this line they are of little importance, being highly susceptible to dry weather and giving low yields. Market crops of swedes are common in Devon, especially on red soil. Crops are grown on a wide range of soils provided they are well drained but retentive of moisture or where rainfall is ample, soil pH at least 6.0, preferably 6.5.

Rotation, cultivations and sowing

Turnips for green forage or leaf They are grown as catch crops, either early, as with summer turnips or after a silage cut or, in the south, a winter cereal. Do not grow as catch crops in rotation containing swedes. May be sown in narrow rows, up to 180 mm, or broadcast on fine firm conventional seedbed or direct drilled. Use dressed but ungraded seed (*Table 5.11*).

Swedes and turnips for bulbs These are grown as maincrops or very early sown catch crops. Sowing date according to area and variety. In Scotland and the north swedes and yellow turnips are sown early as traditional root break between cereal crops, whereas in the south they are grown as late sown maincrops after spring fallow or early bite or as an early sown catch crop following early harvested crops, say early

potatoes. White turnips are usually grown as catch crops. Clear first crop early enough to allow ample time for seedbed preparation and timely sowing.

Nowadays all crops for bulbs are drilled to a stand on a fine firm level seedbed; singling is obsolete. Precision drill *graded* seed of high germination capacity, treated with combined insecticidal/ fungicidal seed dressing, into the soil moisture, usually 10–20 mm deep and leave well firmed. Watch emerging crop for flea beetle attack and treat if required. An appropriate residual herbicide will be necessary. Seed is usually spaced 125–175 mm in row, according to size of bulb required. Numerous small bulbs are wanted for market and are preferable for feeding direct, being less liable to frost, splitting and disease and keep better. For cultural details, *see Table 5.11*.

Manuring

Turnip for green forage or leaf Where a very heavy crop is expected, e.g. summer turnips, apply fertiliser as for kale (*Table 5.8*), stubble catch crops and others, apply as for rape (*Table 5.10*). Give 25 kg/ha extra for direct drilled crops. Crop is not very responsive to P or K.

Manuring swedes and turnips for bulbs Nitrogen applications depend on level of soil fertility (*Table 5.12*), rainfall and method of cultivation. Crops grown with conventional cultivations should receive low dressings of N especially in high rainfall areas; excess produces large tops and necks with reduced yields of bulbs of poor keeping quality. Swedes and bulb

Table 5.11 Row width, spacing (precision drill), seedrate and range of sowing dates for turnips and swedes

Crop	Usual row widths (mm)	Spacing between seeds (precision drill) (mm)	Seed rate Broadcast (kg/ha)	Drilled (kg/ha)	Approximate range of sowing dates
Continental turnips					
Summer crops					
Bulbs and top (sheep)			2–3	1.5–2	
Tops only (cows)	180	—	4.5–7	4.5	1 April onwards
Stubble crops			4.5–7	2.5–4.5	S only, mid July–mid Aug
White turnips (British) for bulb production	500–550	100–150	—	0.5–1.1	N 20 May–15 June S 20 June–31 July
Soft yellow turnips for bulb production	500–550	100–150	—	0.5–1.1	N 15 May–15 June S 15 June–15 July
Hardy yellow turnips for bulb production	500–550	100–150	0.6 yellow turnips + 23–28 Italian ryegrass	0.5–1.1	N 1–31 May S 1–30 June
Swedes (fodder)	500–550	125–175	—	0.5–1.1	N 20 Apr–31 May S 1 June–25 June
Swedes (market)	375–550	100–150	—	0.5–1.5	S 15 May–5 July

Key:
N = Scotland, northern England, hill, upland and other colder areas.
S = south and south west England, western Wales and other areas with a mild autumn and winter.

Table 5.12 Manuring of turnips and swedes for bulb production (kg/ha)

N, P or K index	N	P_2O_5	K_2O
0	75	150	125
1	50	125	100
2	25	75	60
3	—	50	60
Over 3	—	50	60

ADAS figures.

turnips are very responsive to phosphate and potash, especially at low soil levels. In high rainfall areas with soil pH below 6.5 ground mineral phosphate is a suitable alternative to water soluble phosphates. On soils known to be deficient in boron apply solubor or use boronated fertiliser.

Harvesting and utilisation

Catch crops and turnips grown for leaf are normally eaten *in situ*. Turnips and swedes grown for roots are similarly treated but the practice is wasteful in very heavy crops, where it is generally best to clear lanes at regular intervals for feeding elsewhere and fold the residue. Harvesting and feeding of roots is fully mechanised; they may be pulped and fed mechanically into mangers, or whole roots may be stored and self-fed indoors.

Yield and dry matter content

See Table 5.13.

BRASSICAS GROWN FOR SEED PROCESSORS

Oilseed rape

A useful break crop for intensive cereal growers, oilseed rape does not harbour pests and diseases of cereals. Once established foliage produces tall dense cover; used in conjunction with suitable herbicides crop gives good control of annual and perennial grassy weeds. Handled by same basic machinery as cereals, it has a low labour requirement and integrates well on intensive cereal farms. An excellent preparation and re-entry for winter wheat, although spring sown rape may be harvested too late in the north.

All varieties are highly susceptible to clubroot (*P. brassicae*) and canker (*P. lingam*) so do not grow on land that has carried brassicas in previous five, preferably six, years. Also, maintain soil pH above 6.0, preferably 6.5. Adequate isolation necessary in areas producing seed crops of other brassicas. Crop is subject to Beet Cyst Nematode Order 1977. Avoid fields known to be infested with charlock, runch or white mustard, which cannot be controlled by selective herbicides or contamination and rejection of crop will result.

Uses of oilseed rape

The only outlet for rape seed is for crushing, so it is advisable to grow only on contract to a merchant or cooperative. Crushers are not usually interested in small tonnages. Rape oil is used for:

Table 5.13 Approximate fresh yield, dm content and periods of utilisation of turnips and swedes

Crop	Approximate range of DM (%)	Approximate range of fresh yield (t/ha)	Approximate period of utilisation
Continental turnips Summer Stubble	leaf 9–11	50–110 20–50	June to Dec according to sowing date. Later use possible but deterioration can be severe. Allow 50–60 d and 80 d before feeding summer and stubble crops respectively
White turnips (British) for bulb production	bulb 6–8	40–75	N, Aug–Oct S, Oct–Dec
Soft yellow turnips for bulb production including hardy low dm varieties	bulb 7–8	40–75	N, Sept–Nov S, Nov–Dec. Hardy varieties, e.g. Champion Green Top Yellow usable till late Feb
Hardy yellow turnips for bulb production—(mainly grown in Scotland)	bulb, 8–9	40–75	N, Oct–Feb S, Dec to Jan or Feb
Swedes—purple and bronze skin varieties	bulb 8–11	N, 75–115 S, 40–50	N, throughout winter, Oct–April, partly stored S, Jan, Feb and Mar
Swedes—purple, market crops		S, 30–37	Oct–Mar
Swedes—green skin varieties (mainly grown in Scotland)	9.5–12	N, 75–105	Late winter, early spring; partly stored

Key:
N = Scotland, north of England, hill, upland and other colder areas.
S = south, south western England, west Wales and other areas with mild autumn and winter.

(1) human consumption, e.g. margarine, cooking oils and shortening, when it is extracted only from varieties with a low erucic acid content,

(2) industrial purposes, e.g. lubricating oils and detergents, when oil with a high erucic acid content is appropriate.

The by-product from crushing, rapeseed meal, is included up to 10% in concentrates for animal feed; the meal must then be low in glucosinolates which are toxic to pigs and poultry.

Varieties

Varieties of oilseed rape behave quite differently from forage rape varieties and the two are not interchangeable. Only the swede rape (*Brassica napus*) is now grown in Europe and UK; turnip rape (*B. campestris*) has been displaced.

Winter rape This is more widely grown as it gives about 0.5 tonnes/ha more seed than spring rape and 3–4% higher oil content. Harvested in early August in south; also liable to pigeon damage in late winter.

Spring rape This may not reach contract standard of 40% oil content in most seasons and is rarely harvested before mid September, when weather conditions are becoming difficult. Seed losses from sprouting and shattering can be severe.

A list of varieties, Farmers Leaflet No. 9, *Varieties of Oilseed Rape* is issued by NIAB. Varieties are selected for yield of seed, content and quality of oil, early maturity and resistance to lodging and shattering. Breeders are paying attention to edible qualities of rapeseed, notably erucic acid and linolenic acid contents; the latter may reduce shelf-life of oil and cause undesirable flavours. Varieties with very low erucic acid and glucosinolate contents are referred to as 'double zero' varieties.

Soils, cultivations and sowing

Oilseed rape is grown over a wide range of soils in arable areas, giving highest yields on heavier and moisture retentive soils if well drained; avoid soils liable to dry out in early spring. Crop well suited to direct drilling (i.e. agricultural method—drilling with little or no prior cultivation), which reduces loss of seedbed moisture and halves time taken to establish crop, compared to conventional cultivations. But surface must be trash free, a good stubble burn being desirable, there must be no compaction or panning and surface water must drain off freely. Shed cereals are often a problem. With conventional cultivations a fine firm seedbed is essential for quick even germination. Avoid uneven cloddy seedbeds. *Sow seed treated against flea beetle and canker. Winter rape* is sown between mid August and mid September; later drilling results in reduced yields, if not poor establishment. *Spring rape* is sown from early March to mid April but it is essential to wait for good seedbed conditions; heavy soils can be difficult. Late sowings more liable to powdery mildew in dry years.

Drill 15–20 mm deep in rows 90–180 mm apart; these give less lodging, earlier maturity and slightly higher yield than wider rows, which in winter rape are also more susceptible to pigeon damage. Narrow rows provide better support for swathes where windrowing is used at harvest. Target population for *winter rape*, which branches freely is 100–110 plants/m^2; lower populations compensate by branching if evenly distributed. Number of seeds/kg 200 000–220 000; drill 6–8 kg/ha; use higher rates for poor seedbeds. *Spring rape* does not branch as freely so 120–130 plants/m^2 are necessary, but seed is smaller, 230 000–240 000 seeds/kg so 6 kg/ha is usually adequate.

Manuring

Fertiliser recommendations are shown in *Table 5.14*. Unless soil is deficient in P and K, (index 0), oilseed rape rarely shows any response to these elements. Applications for indices 1 and 2 are to maintain soil reserves. Ample N is of vital importance. *N top dressing: winter rape:* apply all N in February or early March; no advantage in splitting dressing. *Spring rape:* apply all N by early May. Split N dressing on

Table 5.14 Manuring of oilseed rape (kg/ha)

N, P or K index	N								P_2O_5	K_2O
	Winter oilseed rape				Spring oilseed rape				All crops and soils	
	Sandy soils and shallow soils on chalk		Other soils		Sandy soils and shallow soils on chalk		Other soils			
	Seed bed	Top dress	Seed bed	Top dress	Seed bed	Top dress	Seed bed	Top dress		
0	60	225	60	200	50	150	150	Nil	75	75
1	50	200	50	175	50	125	125	Nil	40	40
2	40	175	40	150	50	100	100	Nil	40	40
3	—	—	—	—	—	—	—	—	40	Nil

ADAS figures.

sandy soils to avoid damaging germination and losses from leaching. All N must be applied by early May.

Harvesting and drying

Crop may be either direct combined or windrowed and combined from swath. Whichever method is used, seed must be fully ripe when threshed from pod. Fit vertical cutter bar. Unevenly ripened crops require windrowing, which is normal for winter rape. Direct combining is quicker, but liable to serious losses in strong wind. Combining from swath gives cleaner, more uniform seed with lower moisture content but persistent rain on windrow will result in sprouting and even loss of entire crop.

For windrowing, cut when seeds in pods at middle of stem are turning brown. Leave a stubble 200 mm long to keep pods clear of soil. Crop is left in windrow for one to two weeks and then combined. *Direct combining*: seed should be black in majority of pods. Use of desiccant, e.g. Diquat is especially valuable in wet harvests, in weedy or late maturing crops, especially spring rape in the north, giving cleaner samples with lower moisture content and reduced sprouting. Spray when majority of pods at centre of stem are yellow and seed is chocolate brown and pliable, usually some three days after windrowing would have occurred. NB desiccants must *only* be used for crop desiccation—they do not hasten crop maturity. Seal all gaps in trailers, lorries or combine to prevent heavy seed losses.

Immediate drying is essential, since seed heats rapidly and oil goes rancid, losing value. Pre-clean trashy crops. Dry to maximum of 8% moisture content for long-term storage. Allow maximum temperature of 66 °C, which is safe for oil quality; do not exceed 50 °C for crops intended for seed. Use two-stage drying for very wet seed. After drying blow *cold* air through to reduce temperature as low as possible to make conditions unfavourable for development of mites and insects.

Yield

Winter rape 2600 kg/ha, spring rape 1900–2000 kg/ha. Good crops produce more. Except in seasons of severe litter shortage, *rape straw* is valueless to the grower and does not burn readily, unless fired as soon as combine has finished and is very dry, so may be best chopped and ploughed in or collected into heaps and burnt; it may have a fuel value if baled dry.

Mustard grown for condiment

Like oilseed rape, mustard is a valuable cereal break crop and an excellent preparation for wheat. Also a useful cleaning crop for wild oats. No special labour or machinery is required. Mustard is grown on arable farms in eastern and south eastern England, mainly Lincolnshire, Norfolk and Cambridgeshire; also in Essex and Suffolk.

Varieties

Two distinct types are grown for manufacturing: Brown mustard (*Brassica juncea*) var. *Newton* and white mustard (*Sinapis alba*) var. *Kirby*, both bred by Colmans of Norwich.

Soil, cultivation and sowing

Crop does best on free working medium loams but yields are satisfactory on light or sandy soils if rainfall and fertiliser are adequate. It is often difficult to obtain a good seedbed on heavy soils, so if mustard must be grown, white mustard is rather safer.

A fine firm seedbed is essential. Plough in autumn to give a good frost mould, work to a fine tilth and consolidate with roller if necessary. Use cage wheels to avoid compaction.

Sow as soon as a good seedbed can be obtained in March or early April. Sowing after end of April may result in failure. Precision drill seed in rows 340–510 mm apart, with 40–50 mm between individual seeds to give average of 70 mm between plants. Seed requirement 1 kg/ha Newton or 2.25 kg/ha Kirby; this allows for some seeds failing to establish. For non-precision drills allow 1–1.5 kg/ha Newton and 3–3.5 kg/ha Kirby. Do *not* sow thicker and try to thin. Seed is drilled shallowly, deep enough to place in soil moisture and left rolled if soil is dry.

Start tractor hoes as soon as rows are visible from tractor seat and continue until mid May. Standard herbicides for wild oat control may be used pre-drilling or during early growth of crop to continue rotational wild oat control programmes.

Manuring

On fertile soils, ADAS N index 1–2, e.g. best silts after peas or potatoes, apply up to 65 kg/ha N; on less fertile soils or after a cereal crop give 200 kg/ha N. High N usage is essential for Newton and Kirby varieties. Kirby shows no response to phosphate but with low P index (0–1) give 30 kg/ha P_2O_5 for Newton. Neither variety responds to K, which is not recommended for any situation. All fertiliser should be applied to seedbed; any form of N may be used. The above varieties tolerate pH down to 5.6 but it is wiser to maintain pH 6.5.

Harvesting

Provided crop is clean, direct combining is cheapest and best method. Mustard is usually fit to combine in early September, when seed in all pods is fully coloured and hard to bite. Straw brittle below lowest pod, although it may still be green at base. Ripe mustard dries out very quickly while standing; allow adequate drying before combining. Seed moisture content usually: Newton 8–15%, Kirby 10–16%. Neither variety sheds seed readily and may be left one to two weeks after full ripeness but beware hailstorms or gales. Swath *only* if there is much green in crop as seed may be discoloured and heavy losses can occur. Desiccant sprays not recommended. Combine leaning crops one way (i.e. with the lean). Gaps in trailers, combine or lorries should be sealed to prevent loss.

Drying

Seed over 15% moisture content tends to heat rapidly, especially in bulk or where rubbish collects in pockets. Dress seed to remove cosh, weed seed and rubbish if necessary. Air temperature for drying should not exceed 65 °C; that of seed 52 °C. Dry down to max. 10% MC or as required by buyer.

Pests and diseases

Flea beetle   Seed is dressed with insecticide; attacks after emergence controlled with γ HCH. *B. juncea*

var. *Newton*: inspect daily from bud formation, in last half of May, to start of flowering in early June for pollen beetle, swede seed weevil and stem weevil and treat if infestation develops. Give second application, if necessary, before flowers open; do not treat later to avoid killing bees. *S. alba* var. *Kirby* does not need treatment. *Beet Cyst Nematode Order 1977* applies to seed mustard crops. Do not grow mustard on land affected with clubroot (*P. brassicae*).

Pollination

Newton mustard is self fertile. Kirby is not but wind and insects are effective enough to preclude the necessity of introducing hives of bees; mustard flowers are, however, a valuable source of honey.

Yield

On fertile soils 2750 kg/ha, light soils 1900 kg/ha. The only outlet is for condiment manufacture, so always grow on contract.

Graminae—cereals, maize and cereal herbage seed production

The group includes:

(1) major cereals grown primarily for grain: wheat, barley, oats;
(2) rye, grown for early bite and, in UK, on very limited scale, for grain;
(3) silage maize; little maize is grown for grain in UK as climate is too cold;
(4) seed production of cereals and the major grasses; clover seed is included for convenience of treatment.

CEREALS GROWN PRIMARILY FOR GRAIN

General

Choice of seed and seed treatment

Always sow high quality certified seed of high vigour. Seed should be treated with combined fungicide/insecticide to protect seedlings against certain soil-borne fungi and to control certain seed-borne diseases. While surface acting fungicides are adequate for surface-borne diseases, systemic fungicides are necessary for diseases carried inside grain (*see* Chapter 9). Seed dressings are best applied by merchant, as application is much more even and thus effective.

Recent failures to dress seed have resulted in the reappearance of certain seed-borne diseases that have been controlled by seed dressing for many years, e.g. Bunt or stinking smut of wheat (*Ushlago tritici*). Do *not* feed surplus treated seed to livestock, as it is dangerous.

High vigour seed is especially important when it is sown in adverse conditions or if such conditions occur shortly after sowing, e.g. early sown spring crops, cold wet soil. Improvements in plant establishment, early growth, ear size and yield of grain can then be very substantial.

Choice of variety

Sow only varieties included in Farmers Leaflet No. 8 *Recommended Varieties of Cereals* issued annually by NIAB. Choose varieties suitable for the purpose for which they are being grown. The following features are common to all cereals.

Winter and spring varieties

Winter varieties of wheat, barley and oats usually give higher yields than spring sowings as they have a longer growing season and get their roots down before the onset of dry weather. On heavy land adverse soil conditions usually preclude drilling of spring crops early enough for maximum yield. Winter cereals, especially barley, ripen well before spring varieties, thus allowing an earlier start to harvesting, when weather conditions are good, and better utilisation of combine capacity. Earlier maturity of winter barley allows more time for stubble cleaning and makes crop especially suitable to precede winter oilseed rape or stubble catch crops which must all be drilled early for best results.

Winter hardiness

Winter wheat and six row winter barleys are generally frost hardy enough for UK conditions but winter wheats with low vernalisation requirements (which can be sown later in spring) and two row winter barleys are less hardy. Winter oats are not normally grown north of the Humber owing to insufficient hardiness.

Vernalisation requirement

Winter wheat and winter barley require adequate cold to cause them to come into ear. Sowing too late in spring gives insufficient cold for vernalisation and crop stays in vegetative state. For winter wheat complete sowings before date given in NIAB leaflet No. 8 for latest safe sowing date for each variety. Winter oats may be sown up to 7 March but thereafter yield is reduced and maturity is later than that of spring oats. Winter varieties of barley *must not* be sown in spring.

Earliness of ripening

Choose early ripening varieties where earliness of harvesting is important, e.g. with late sowing of spring crops, in cold late areas or where cereal is to be followed by catch crop or late summer sown ley.

Disease resistance

Choosing resistant varieties is the most economical way of reducing crop losses. Ascertain which diseases are most troublesome in your area and select varieties which have a high degree of resistance (i.e. low risk); most diseases have a characteristic geographical distribution (*see* NIAB Farmers Leaflet No. 8). Avoid growing large areas of a single variety, as it may become vulnerable to new races of disease, e.g. mildew in barley and yellow rust in wheat. Grow several varieties and use Farmers leaflet No. 8, *Variety Diversification Schemes* (NIAB). There are no varieties of wheat or barley resistant to Take All (*Ophiobolus graminae*) while the incidence of Eyespot (*Pseudocercosporella herpotrichoides*) depends on intensity of cereal growing. Choose resistant varieties for second or third crops and continuous wheat growing. Loose smut can be controlled by seed treatment.

Thousand grain weight

This is a measure of grain size and thus of the genetic potential of a variety but is influenced to a very high degree by growing conditions; it must be ascertained where it is desired to control plant population by seed numbers.

Specific weight

This is a measure of the weight of a given volume of grain, stated as kg/hl (formerly bushel weight). Affected by variety but growing conditions exert a large influence.

Lodging

Lodging (i.e. collapse of straw) can occur early or late; the former may result in near total loss, the latter is not usually serious if combine can get under ears. Most likely in crops with long straw, in exposed or windy positions and wet summers on high fertility soils; also occurs with eyespot infections. Barley is more susceptible to lodging than wheat so prefer wheat on highly fertile sites and select shortest stiffest strawed varieties. Avoid excessive seedrates and applications of N. Adjust N level to ADAS N index. Ensure timing of application suits soil fertility. Where there is serious risk of lodging or heavy dressings of N are given, use recommended straw shortener (e.g. chlormequat on wheat and oats) at appropriate growth stages. Spring grazing of proud winter cereals reduces risk of lodging. Grow eyespot resistant varieties. Modern varieties of wheat and barley are much more resistant to lodging than older cultivars.

Straw shorteners (growth regulators, e.g. chlormequat)

These reduce effect of lodging by increasing internode diameter and reducing straw length. Yield is increased where lodging is a problem; straw shorteners should be used:

(1) where lodging is a problem, and
(2) to allow maximum use of N without lodging, especially on longer or weaker strawed types or varieties. Some varieties of wheat are not worth treating.

Hardly worthwhile on lower yielding crops. Check for which cereals a particular straw shortener is suited; also recommended growth stage of crop before applying product.

Grazing winter cereals

Spring grazing may be obtained from winter cereals:

(1) specially grown for the purpose, usually forage rye,
(2) planned dual purpose cereals, mainly winter barley, very occasionally grown for this purpose in mild south west England to provide extra sheep keep, and

(3) to remove excess vegetative growth from winter proud crops, which is only a problem with early sowings in mild winters and districts; also in case of feed shortage.

Grazing shortens straw, reducing incidence of lodging and perhaps disease but *only* winter proud crops benefit; grazing frequently reduces grain yield and is not commonly practised. Growth stage for grazing is critical and *must be completed before* growing point at centre of ensheathing leaves has started to elongate. Avoid grazing on wet ground, as soiling of crop results in rejection and consequently uneven ripening. Top dress with N immediately after grazing is complete, not before. Apply normal N recommendations for particular cereal.

Growth stages of cereals

Wherever practicable all recommendations for stage of crop growth at which N fertilisers, growth regulators, herbicides, fungicides and insecticides are applied to cereals are now given in terms of Zadok's decimal code for the growth stages of cereals (*see Table 5.15*).

Production systems

Highest grain yields do not necessarily give maximum profits. Aiming for highest yields frequently increases total cost of inputs such as fertilisers, fungicides and pesticides without generating a corresponding increase in cash return. Before embarking on high cost production be sure that quality of site, especially soil type, depth, structure and drainage, has potential to achieve high yield; if not, it is much safer to reduce costs with lower inputs. Various 'blue print' systems, mainly of continental origin, have been tried which use high plant densities and high inputs of fertiliser and fungicide but these are unnecessarily rigid. Although such systems frequently give very high grain yields, profit is often well below those obtained from lower cost production. Each individual cost increase must thus be justified on its own merit on each particular site.

Highest yields

Highest yields, e.g. 10 tonnes/ha depend not only on a suitable site and factors such as high fertility levels, adequate plant population, right variety and efficient disease control, but *on careful attention to detail in all factors*. Numerous apparently less important items can each produce a *small* yield reduction *but added together they can reduce yield considerably.*

Rotation

Cereals are grown under many different systems, which may be broadly divided into:

(1) *intensive cereals*, i.e. 50% or more of arable land in cereals, including continuous cereals or cereal sequences with occasional breakcrops, or
(2) *mixed systems* with under 50% of arable land in break crops.

Situations include mixed cropping, where cereal is used to cash in on built up fertility, e.g. ley or permanent grass or as a change crop from more valuable non cereal crops.

Wheat (mainly winter) is traditionally grown as first crop after a break crop, e.g. potatoes, beans or ley or up to four successive crops after fertile old grass. Lucerne is an ideal precursor for two crops of wheat. Many wheat crops are now included in intensive cereal rotations, when highest N applications are needed. *Barley and oats* are most accommodating and can be grown in any position in rotation but prefer wheat wherever fertility is high, it is less liable to lodge and cash returns are better. *Oats* is very susceptible to cereal cyst nematode (*Heterodera avenae*); laboratory cyst count should be taken by ADAS before inclusion in intensive cereal rotations.

Disease

Only soil- or residue-borne pests and diseases are controllable by rotational means, e.g. Eyespot (*Pseudocercosporella herpotrichoides*), Take All (*Ophoiobolus graminae*) and cereal cyst eelworm (*Heterodera avenae*). First two can heavily damage wheat and to lesser extent, barley; heavy infestations of cereal cyst nematode (*H. avenae*) can damage all three but oats most susceptible. Use resistant varieties and fungicides to control leaf-borne diseases (*see* Chapter 9).

Soil-borne diseases and pests are mainly a problem with intensive cereal situations, although some grasses can act as alternative hosts and serious outbreaks can occur after old pasture, e.g. take-all (*O. graminae*), Eyespot (*P. herpotrichoides*), primarily attacks wheat, and is worst with early sowing, mild winters and wet springs. Use resistant varieties with continuous cereals or second and subsequent white straw crops. There are no varieties resistant to take-all (*O. graminae*); level of infestation increases to around fourth successive crop of wheat or barley (*Figure 5.1*) and then declines; yield decreases until maximum infestation is reached, then rises and may attain higher level than at first—phenomenon known as 'Take-All (TA) Decline'. Intensive cereal grower has two major choices for containing effect of take-all (*O. graminis*):

(1) to go through TA decline, minimising losses culturally and then to continue to grow cereals; grow barley in most susceptible years, or
(2) to introduce non susceptible break-crops, e.g. beans, oilseed rape, mustard, oats, but these should be grown in two *consecutive* years.

Table 5.15 Zadoks' decimal code for the growth stages of cereals

Code		Code	
0	GERMINATION	5	EAR EMERGENCE
00	Dry seed	50	—
01	Start of imbibition	51	First spikelet of ear just visible
02	—	52	
03	Imbibition complete	53	$\frac{1}{4}$ of ear emerged
04	—	54	—
05	Radicle emerged from seed coat	55	$\frac{1}{2}$ of ear emerged
06	—	56	—
07	Coleoptile emerged from seed coat	57	$\frac{3}{4}$ of ear emerged
08	—	58	—
09	Leaf just at coleoptile tip	59	Emergence of ear completed
1	SEEDLING GROWTH	6	FLOWERING
10	First leaf through coleoptile	60	—
11	First leaf unfolded	61	Beginning of flowering (not easily detectable in barley)
12	2 leaves unfolded		
13	3 leaves unfolded	62	—
14	4 leaves unfolded	63	—
15	5 leaves unfolded	64	—
16	6 leaves unfolded	65	Flowering half-way
17	7 leaves unfolded	66	—
18	8 leaves unfolded	67	—
19	9 or more leaves unfolded	68	—
2	TILLERING	69	Flowering complete
20	Main shoot only	7	MILK DEVELOPMENT
21	Main shoot and 1 tiller	70	—
22	Main shoot and 2 tillers	71	Seed coat water ripe
23	Main shoot and 3 tillers	72	—
24	Main shoot and 4 tillers	73	Early milk
25	Main shoot and 5 tillers	74	—
26	Main shoot and 6 tillers	75	Medium milk ⎫
27	Main shoot and 7 tillers	76	— ⎪ Increase in solids of liquid endo-
28	Main shoot and 8 tillers	77	Late milk ⎬ sperm visible when crushing the
29	Main shoot and 9 or more tillers	78	— ⎪ seed between fingers
3	STEM ELONGATION	79	— ⎭
30	Pseudostem erection (winter cereals only) or stem elongation	8	DOUGH DEVELOPMENT
31	1st node detectable	80	—
32	2nd node detectable	81	—
33	3rd node detectable	82	—
34	4th node detectable	83	Early dough
35	5th node detectable	84	—
36	6th node detectable	85	Soft dough (Finger-nail impression not held)
37	Flag leaf just visible	86	—
38	—	87	Hard dough (Finger-nail impression held, head losing chlorophyll)
39	Flag leaf ligule just visible		
4	BOOTING	88	—
40	—	89	—
41	Flag leaf sheath extending	9	RIPENING
42	—	90	—
43	Boot just visibly swollen	91	Seed coat hard (difficult to divide by thumb-nail)
44	—		
45	Boot swollen	92	Seed coat hard (can no longer be dented by thumb-nail)
46	—		
47	Flag leaf sheath opening	93	Seed coat loosening in daytime
48	—	94	Over-ripe, straw dead and collapsing
49	First awns visible	95	Seed dormant
		96	Viable seed giving 50 per cent germination
		97	Seed not dormant
		98	Secondary dormancy induced
		99	Secondary dormancy lost

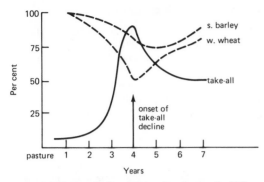

Figure 5.1 Yield and take-all in continuous cereals. (After Lester, E. Disease problems in continuous cereal growing, *Agricultural Progress*, 44, 1969. Reproduced by courtesy of the editor and publishers)

Good soil conditions, especially drainage, structure and lack of compaction, ample N and stubble sanitation are valuable for reducing effects of take-all (*O. graminae*). Although not controllable by rotation, leaf diseases can build up to high levels in intensive cereal crops; crop sanitation and protective treatment are very important.

Weeds

Broad leaved weeds, annual and perennial, are most readily controlled by suitable modern herbicides in intensive cereal rotations (*see* Chapter 8). A few, e.g. forgetmenot (*Myosotis arvensis*), are difficult. Grassy weeds readily build up to infestation level, especially when a run of winter crops is taken; proper preventive measures, according to species, are essential and include stubble treatment and use of herbicides. Modifying rotation, e.g. following winter barley with spring barley allows ample time for stubble cleaning to control couch; spring crops are much less subject to autumn germinating grassy weeds, e.g. blackgrass (*Alopecurus myosuroides*) but yield is lower.

Cultivations

All cereals may be grown by *conventional cultivations*, with or without the mouldboard plough, to varying depths or by *direct drilling*. The latter involves placing seed in soil without prior cultivation or, where necessary, with minimum precultivation, using special drills and coulters of various designs. Provided there are no pans or compaction, deep seedbeds are not necessary and working to 150 mm deep is normally adequate. Subsoil where necessary to break up pans or compaction. With omission of ploughing NB direct drilling, presowing control of surface vegetation is essential; contact or translocated non residual herbicide used according to weed problem (*see* Chapter 8).

Conventional cultivations

Winter cereals Prepare seedbed early, firm but not compacted with ample crumb to receive seed but not too fine on top, i.e. some surface clod to reduce capping. Delay drilling until some rain falls if excessively dry but do not work final tilth too far ahead of drilling; in wet autumns drill close behind minimal working. Late wet cloddy seedbeds impair plant establishment. Mouldboard plough has now been superseded on many clayland and other farms by heavy cultivators or chisel ploughs. On heavy soil start shallow and work gradually deeper or tilth will be excessively cloddy. Herbicide may be required to control grass weed with no ploughing, especially in wet autumns.

Spring cereals

(1) Early sown crops (say up to late March) plough before Christmas for well settled frosted tilth, especially on heavy soils. Do not travel on land before it is fit, whether minimally cultivated or ploughed. With earliest sowings use minimal number of passes, consider cage wheels and do not leave too fine a surface.
(2) Late sowings (early April onwards). When sowing cereals after late cleared crop (e.g. folded roots, winter cauliflower) wait until land is dry enough to plough a crumbly furrow slice then work down *quickly* tight behind plough to secure fine seedbed and conserve moisture. Leave crop cambridge rolled after drilling; loose seedbeds are fatal to crop. Except earliest sown crops, aim for fine firm even and level seedbed, especially malting barley and long leys undersown.

Direct drilling

All cereals may be direct drilled either occasionally or sequentially; long-term direct drilled sequences may include suitable combine harvested break crops, e.g. ideal for oilseed rape.

Soil type and conditions are critical for technique, especially sequential direct drilling. Major problem areas include:

(1) *drainage*: install proper drainage system, mole drain and/or subsoil before starting; ponding must not occur on surface;
(2) *lack of tilth*: localised tilth is required to cover seed and allow penetration of soil by rootlets; drilling into compacted soil fails to cover seeds, gives smearing by coulters, causes ponding around seeds and heavy losses. Crop establishment is better after stubble burning, which improves surface tilth, keeps soil drier for a while and disposes of trash which can produce toxic residues; predrilling cultivations and tine coulters on drill also improve tilth;

(3) *topsoil compaction is produced by any heavy machinery* including combine harvesters and grain trailers or tractors if soil is wet, when structural strength is minimal; keep off land or if essential cut travelling to minimum on wet soil, reduce implement weight and use wide or flotation tyres. Problem is reduced on quick draining soils with high organic matter content and surface mulching. Soil texture is most important, e.g. non calcareous clays, silts, sandy clays and low organic matter sands showing little resistance to topsoil compaction.

Classification of soils according to suitability for sequential direct drilling of cereals

Group 1

Soils give similar yields from direct drilling and conventional cultivations for winter and spring cereals. These are well drained permeable soils of good structure; loams, including some clays with high natural lime and humus rich soils.

Group 2

Similar yields from direct drilling and conventional cultivations, with good management, for winter cereals only. Spring cereals risky. Imperfectly or moderately well drained soils but relatively impermeable layer at moderate depth. Correctable by drainage.

Group 3

Risk of loss of yield from direct drilling, especially spring sown crops. Low organic matter sands, silts, wet alluvial soils, clayey soils and clay loams which return to field capacity before 1 November.

Soils in groups 2 and 3 may be direct drilled with cereals provided: drainage is improved, shallow pre-drilling cultivations are given, excess traffic is avoided and *only* winter cereals are planted, drill before mid October.

General precautions for direct drilling cereals

Ensure clean weed and trash free site, burn stubble where practicable, drill in good time into friable tilth and adopt high standard of husbandry throughout. Maintain adequate pH in top 25 mm of soil, give extra N, drill slug pellets and watch for slug damage. Do *not* drill in adverse soil conditions or too deeply and use suitable coulters. Check on soil group, drain and precultivate if necessary. Some annual grassy weeds are much more troublesome in sequentially direct drilled cereals than in conventionally cultivated crops where small seed is buried by the plough, so ensure appropriate herbicidal control in direct drilled crops; a good stubble burn is helpful with some species. Follow the guidelines (ICI/Plant Protection).

Plant population, yield and seedrate

Yield of grain is the product of number of ears (fertile tillers) × number of grains/ear (ear size) × weight of individual grains. (Thousand Grain Weight, i.e. TGW). Ears arise from vegetative tillers which in turn arise from individual plants, each produced by a single grain on germination. It is therefore necessary to ensure sowing an adequate number of seeds to secure a population large enough to utilise fully the resources of the area of land they occupy.

Cereal types differ greatly in their tillering capacity. Barley and winter wheat tiller profusely, while spring oats and spring wheat produce relatively few tillers. Types such as winter wheat and winter barley, which are capable of creating high tiller numbers are especially flexible or plastic in their ability to compensate for low plant densities and give similar yields of grain from widely differing plant and ear populations. Large varietal differences in tiller production have also been noted. If initial population is inadequate, yield potential is not reached; once plant numbers are adequate for the site, yield remains on a broad plateau until it declines through overcrowding (*Figure 5.2*), quite apart from excessive seed cost; ear size is also reduced. Initial plant density and seedrate are not critical in high tillering winter wheat and winter barley varieties, but control of seedrate is an essential first step in helping to give a dense enough plant population to provide a high yielding crop structure. Drill a seedrate high enough to ensure an adequate margin of safety—there is no point in exceeding this.

Plant establishment and subsequent development are highly dependent on soil conditions, weather and management. Early sown crops on good seedbeds, especially in autumn, tiller vigorously, so less seed is

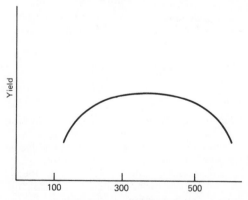

Figure 5.2 A diagrammatic representation of the relationship between yield and plant population. (After Hayward, P. R. *et al., Developments in the business and practice of cereal seed trading and technology*, p. 149, 1978. RHM Gavin Press. Reproduced by courtesy of the publishers)

then required. With crops drilled late in cold, wet soil (e.g. November), seedling mortality is high and more seed is needed but this does not compensate for retarded crop development, especially winter barley. Ear numbers and size are determined by many other factors. Farmer can control choice of variety, quality of seed and N application; he cannot control emergence factor, winter survival nor summer rainfall, which much affects N utilisation and consequent crop performance.

Always sow

(1) modern high yielding variety recommended by NIAB and suited to your circumstances,
(2) high vigour seed,
(3) seed treated with appropriate fungicidal /insecticidal dressing,
(4) an appropriate seedrate, and
(5) drill winter cereals early, spring crops as soon as soil conditions permit.

Drilling a standard seedrate or weight of seed is a poor way of controlling number of seeds sown, as it takes no account of differences in seed size and hence number of seeds/kg between varieties and different lots of seed; where required population and TGW are known, it is more accurate to calculate seedrate as follows:

$$\text{Seedrate (kg/ha)} = \frac{\text{target population/m}^2 \times \text{TGW (g)}}{\text{expected \% establishment}}$$

e.g. target population = 250 plants/m^2 established, TGW 50 g, expected establishment 80%, e.g.

$$\text{Seedrate (kg/ha)} = \frac{250 \times 50}{80} = 156 \text{ kg/ha}$$

Formula applies only to high laboratory germination, 95% and over, when allowance for percentage germination makes little practical difference; *avoid sowing low germination seed*. Have home grown seedlots tested for germination by Official Seed Testing Station, NIAB. Field establishment is *very difficult to forecast* and can vary from 60% or less with winter crops sown in bad conditions, about 80% in good conditions and up to 90% in spring grown crops in very good conditions. Standard seedrates and target populations, as appropriate, are given with individual crops.

Drilling and plant distribution

Well distributed populations are much better placed to compensate for low plant numbers than are patchy stands; plant counts can then be unreliable. Operate drill carefully, avoid high speed drilling, using wider or an additional drill if necessary. Buy a reliable drill and maintain it properly. Sow seeds as shallowly as covering, tilth and cutting into soil moisture permit, usually 25–50 mm.

'Tramlines'

These are paired continuous plant free strips, designed to allow passage of tractor with sprayers and fertiliser distributors without damaging standing crop and may be formed by:

(1) shutting off coulters at drilling, or
(2) making *accurate* continuous wheelings in emerged crop.

Tramlines are an integral part of modern high input cereal production (*cf.* bed system in vegetable crops) and necessary to allow numerous passes through standing crops, especially at later stages of crop growth, e.g. fertiliser application, and spraying fungicides and aphicides. It is essential to match widths of drill, fertiliser distributor spread and spray boom.

Manuring of all cereals grown for grain production, P_2O_5 and K_2O

Liberal applications of P, and on many sites, especially lighter soils, of K, are necessary on newly reclaimed deficient soils; also lime. On cultivated land, including well managed temporary grassland, past applications of P and K have built up reserves of these nutrients, so that cereals rarely give an increased yield from fresh P or K fertiliser. Only where soil indices are low (e.g. K on light and chalky soils) or where little P and K has previously been applied, are large responses likely. Responses are rare where P and K indices are over 2 (*Table 5.16*). It is advisable to maintain soil reserves of P and K, which are applied annually, or, with continuous cereal production, P may be given at regular intervals during the rotation. Levels of maintenance dressing depend on yield, as heavier crops remove more nutrients. Allow for straw removal which removes some K but little P. Light soils tend to

Table 5.16 Phosphate and potash manuring of all cereals for grain production (kg/ha)

P or K index	Straw ploughed in or burnt		Straw removed	
	P_2O_5	K_2O	P_2O_5	K_2O
Farm yield—level average or less[1]				
0	75	75	75	75
1	40	40	40	60
2	40	40	40	60
3	40	Nil	40	Nil
Over 3	Nil	Nil	Nil	Nil
Farm yield—level consistently above average[1]				
0	75	75	75	100
1	60	60	60	80
2	60	60	60	80
3	60	Nil	60	Nil
Over 3	Nil	Nil	Nil	Nil

ADAS figures.
[1] Yields exceeding 5 t/ha of wheat or oats or 4 t/ha barley.
Apply all P and K to seed bed.

be low in K and even at index 2 maintenance dressings of K are wise. Most soils with high clay content can release enough K and maintenance dressings are unnecessary at index 2. Be sure to maintain adequate levels when rotation includes P or K demanding crops.

Combine drilling slows down work considerably and such drills are more expensive and laborious to maintain than ordinary drills. P and K should be combine drilled at index 0; if broadcast, increase each by 30 kg/ha. There is no response from combine drilling at high P and K levels.

Nitrogen manuring

See individual cereal crops.

WHEAT

Grain quality

Grain of suitable quality is essential for human consumption, when it is milled and blended into a wide range of flours for bread, confectionery or biscuit making; remainder is used for animal feed, mainly by feedstuff compounders. Suitability of sample for milling and bread or biscuit making is essentially a varietal feature but modified by growing, harvesting and storage conditions.

Milling quality

Varieties may be grouped as 'hard' or 'soft' in milling texture. In good milling (i.e. hard) varieties of wheat, bran is readily separated from endosperm or flour; the latter is granular and so is free running, passing through sieves quickly without blockage. Conversely, with poor (i.e. soft) varieties, bran does not come away cleanly and flour is irregular and lumpy, so milling has to be slowed down to prevent blockage of sieves. Extraction rate also varies, plump well filled samples giving highest percentage extraction. Other than plumpness of sample, these are essentially varietal characteristics, little affected by cultural methods.

Bread making

'Hard' wheats are best for bread making because they give a flour which absorbs more water in the dough, making more loaves from a given weight of flour and which take longer to go 'stale'. Good dough is also elastic, thereby trapping the carbon dioxide bubbles from the yeast giving large, soft, well piled loaves. Suitable quality is indicated by:

(1) *Hagberg Falling Number (FN) test*, which indicates alpha amylase activity. A high FN is essential, showing that the grain has a low activity of the enzyme alpha amylase. A high activity, indicated by a low FN, results in degradation of starch during baking, giving a sticky crumb structure of loaf; loaves are then difficult to slice mechanically. Where crop is to be sold for bread making, choose only varieties with a high Hagberg FN in NIAB recommended list and good resistance to sprouting; once germination starts, alpha amylase is produced and Hagberg FN is reduced.

(2) *Protein (i.e. gluten) quality*, which determines elasticity of dough. A 'strong' gluten gives a loaf of large volume which is light and spongy in texture but effect is also determined by *quantity* of gluten present. Gluten *quality* is largely a varietal feature, unaffected by husbandry techniques.

(3) *Protein content* is largely determined by environmental conditions, especially late application of N; an extra 40 kg/ha N may be considered for application between flag leaf and ear emergence (GS 37–39 to 51) to increase protein content of grain.

Biscuit making

Good biscuit flour produces a weak dough which combines maximum extensibility with minimum elasticity, which can be rolled into a thin, even sheet; the biscuits cut from it retain their shape. Good biscuit varieties are soft milling in texture, yet reasonably free milling and have suitable protein quality. Samples of intermediate protein content are required.

Varieties

Choose varieties suitable for the required purpose. For *bread making* select varieties with good scores for bread making and ease of milling, with hard endosperm texture. Minimise likelihood of rejection by selecting those with good scores for Hagberg falling number and resistance to sprouting. Prefer varieties with naturally high grain protein content. For biscuit making choose high yielding soft wheats of suitable quality. For crops grown only for feed, high yield is the main criterion.

Climate and soil

Wheat is widely grown but is more reliable in drier climates of south and eastern Britain. Crop is suited to a varying range of soil types but deep rich moisture retentive soils with good structure are essential for highest yields. Well drained fertile medium heavy loams, clay loams and alluvial soils are especially suitable. Avoid thin, infertile and badly drained soils, as yields are usually low.

Sowing date

Winter wheat

Best yields are usually obtained from sowing during first two weeks of October; complete drilling before end of month, as November sowings invariably give reduced yields. Drilling should begin in September on large areas where a later start may result in November completion. September sowing is also beneficial in cold areas where winter begins early. Spring sowing of winter varieties should be finished before latest safe sowing date recommended by NIAB, Farmers leaflet No. 8.

Spring wheat

Sow as soon as soil is fit to drill, the earlier the better. Range 1 February–15 April.

Plant population and seedrate

Winter wheat

Winter wheat is extremely flexible in its ability to compensate for variations in plant density, similar grain yields being obtained over a very wide range of initial plant and ear populations. Experiments have shown that on evenly distributed stands, yield increases only by 5% when population is increased from 100–200 plants/m²; established populations of 200–300 plants/m² are adequate for highest yields. When calculating seedrate aim for 250 plants/m² to allow for high safety margin and to produce minimum of 500 ears/m². Thresholds below which yield declines are probably around 125 plants/m² on good seedbeds on heavy soils, 150 plants/m² on light soils. Evidence suggests that provided thin crops can be liberally top dressed with N, it is not worth redrilling with spring barley unless established density is below 50–60 plants/m². There are large differences between some varieties in tillering capacity. When calculating seedrate, allow 80% establishment for early sowing, good seedbed, reduce to about 60% for late sowing,

adverse conditions; these allowances are only approximations.

If TGW is not known, sow 140 kg/ha for early sowings, choice seedbed; 160 kg/ha average conditions and 180–200 kg/ha November–December sowing, cold rough seedbeds. High seedrates for early sowing on good tilth may reduce yield, e.g. by favouring disease growth in resultant dense crop.

Spring wheat

Tiller production is low and high seedrates are essential. Sow 190–220 kg/ha, according to TGW and seedbed conditions, average 200 kg/ha.

Manuring

Nitrogen (P & K, see Table 5.16)

Nitrogen supply is a prime factor in production of high grain yields but timing of applications is most important.

Total application of N fertiliser This depends largely on soil index; *Table 5.17* provides an *approximate* guide to rates for winter and spring wheat. Increase or decrease according to condition of crop and soil and expected economic response on site.

Winter wheat

Timing of N application

Nitrogen stimulates tillering and subsequent vegetative growth and encourages tiller survival and duration of green leaf after flowering. *Adequate but not excess N is essential for each phase of crop development but amounts required vary greatly. Nitrogen uptake is small from sowing until tillering is completed about end of March. Deficiency during this phase results in a crop whose frame is inadequate to achieve high yield; it cannot be remedied by later applications. After completion of tillering, vegetative growth and

Table 5.17 Nitrogen manuring of wheat (kg/ha N)

N index	Sandy soils and shallow soils over chalk/limestone	Fen medium and heavy silts, heavy clay soils	All other mineral soils	Fen light peats, loamy peats	All other peats and organic soils
Winter wheat
0	175[1]	150	150	50	90
1	125	75[2,3]	100	Nil	45
2	75	Nil	40	Nil	Nil
Spring wheat
0	150[1]	125	125	40	70
1	100	50[2]	75	Nil	35
2	50	Nil	30	Nil	Nil

ADAS figures.
[1] Response limited by drought in some years.
[2] Increase N by 25 kg/ha in poor soil conditions, e.g. after sugar beet.
[3] Only 50 kg/ha N needed on deep silty soils and brick earths.

N uptake of winter wheat accelerate rapidly, the latter being maintained until about end of June. Nitrogen requirements are heavy during this period but N uptake and N response can be seriously reduced by drought.

Seedbed applications

No response with conventional cultivations; give 30 kg/ha N to direct drilled crops only, in addition to main recommendation.

Early top dressing (GS 20–23)   About mid March, during tillering. Split dressings, giving a maximum of 40 kg/ha of total N dressing at this stage, are advisable in the following cases:

(1) direct drilled crops,
(2) crops on lighter soils, N index 0,
(3) crops with low populations, weak tillering or poor soil conditions, and
(4) if main top dressing is delayed.

Do not apply N to vigorous crops on N rich soils at this stage, e.g. after lucerne, intensively managed grass or on peaty soils or yield may be reduced.

Main top dressing (GS 30/31)   Mid-late April. Nitrogen uptake is now rapid and restricted N supply adversely affects dry matter production. Translocation of N to ear is more effective now, as opposed to earlier application. Spread remainder of total N recommendation now and normally total recommendation for soil N indices 1 and 2.

Very late N applications (GS 37–39 to 51)   Flag leaf emergence and later. Applications during this period increase N content of grain but usually fail to give economic response in yield; 40 kg/ha N may be considered to raise N content of grain for sale for baking or possibly on sites of highest potential yield when soil moisture is adequate and late application of fungicide is given. Given as an *addition* to main application and is an integral part of some continental wheat systems.

Spring wheat

Split applications of N may increase yield on light chalky and sandy soils, when half is worked into seedbed in compound fertiliser and rest is topdressed at GS 30, mainly with early sowings. Elsewhere all N is incorporated in seedbed. Do not combine drill more than 75 kg/ha N. Crop is less responsive to N than winter wheat (*Table 5.17*).

Harvesting

Wheat must be dead ripe when combined. Given a choice, harvest wheat in preference to barley, as it is slower to dry out in wet weather. Combine wheat crops for breadmaking, first, then biscuits and finally feeding if fit.

Drying and storage

Dry wet crops at once. Temperatures should not exceed 43 °C for seed wheat and 65 °C for milling wheat, when excessive temperatures 'perish' gluten and loaves will not rise. Dry grain to 16% moisture content (MC) for sack and 14% MC for bulk storage.

Yield

Grain

Winter wheats normally outyield spring sowings considerably. Well managed winter crops give around 7 tonnes/ha, best reaching 9 tonnes/ha or a little over. Average to moderate crops and spring sowings usually range from 3.75–5.0 tonnes/ha but badly grown or drought stricken crops can yield less.

Straw

Yield 2.5–3.75 tonnes/ha.

Specific weight

Average 77 kg/hl.

BARLEY

Major use of barley is as an animal feedstuff, either consumed at home or sold to animal feed compounders. Top quality barley used for malting, 'green grain' alcohol production, a small market for pearling and other miscellaneous uses, e.g. coffee substitute.

Malting

A good malting sample has germination as near 100% as possible, grain plump, even, a bright pale lemon colour, finely wrinkled skin, total absence of moulds or mustiness and free from broken grains or impurities. In cross section it is white and mealy, not grey and steely, the latter denoting high nitrogen content. A good sample will not contain over 1.5% N, although up to 1.6% is acceptable by maltsters; more in bad seasons. Malting quality is dependent on choice of variety and environment: a dry season with enough moisture at the end to mellow the grain, a fine, firm, level seedbed, avoidance of lodging from excessive seed rates or too high N application, harvesting when fully mature, not drying above 43 °C and dry storage. A late top dressing increases the N content of the grain.

Feeding

Good feeding barley should be clean, dry (*see* storage) and free from mould. Plump well filled sample is best, as thin poorly filled grain with high fibre or husk content has lower feeding value. Fibre content is largely determined by growing conditions, although six row barleys do have a somewhat higher fibre level than two rowed sorts. High protein content of grain is beneficial, as it reduces the amount of protein concentrate needed to balance a ration.

Varieties

Recommended varieties are listed in Farmers Leaflet No. 8, *Recommended Varieties of Cereals* (NIAB). Winter varieties include two and six row types, the latter being very high yielders but unsuitable for malting. For malting select variety with high score for malting grade, although this is no guarantee of a malting sample. For feeding go for high yield provided variety shows adequate resistance to foliar diseases which are troublesome locally. Rhychosporium is most frequent in wet seasons and around south coast so choose resistant variety for winter and early spring sowings; brown rust can be especially serious in the south west. Mildew is widespread. Use the NIAB Varietal Diversification Scheme to reduce spread of mildew in spring barley and consider routine fungicidal treatment for winter and late sown spring crops; it is essential for susceptible varieties. Choose earliest maturing varieties for late spring sowings. Winter varieties are totally unsuitable for sowing after mid February.

Climate and soil

Barley is grown in most areas. Traditionally regarded as a light land crop, it is well suited to a wide range of soils. Good drainage and soil pH 6.0 or over are essential; usually pH 6.5 is safer.

Sowing date

Winter barley

Early drilling necessary for top yields, especially in colder more exposed areas, e.g. Cotswolds. Start mid September and complete by mid October, particularly direct drilled crops. Early November drilling seriously reduces yield if seedbed is cold and wet.

Spring barley

Range 15 Jan–30 April, according to soil and rainfall. The soil should be dry and friable and it is better to wait for the right conditions than force a tilth. Sow as soon as conditions permit but generally sow in March and in the southwest, mid March to mid April.

Plant population and seedrate

Winter barley

Given even distribution and adequate density, high yields can be achieved over a wide range of ear populations. Precise level is unimportant. For safety aim to establish 250–300 plants/m² by early spring with general target of 900 ears/m² at harvest; with six row varieties 800 ears/m² is adequate, two row varieties 1000 ears/m² is preferable. Optimal seedrate depends on TGW, percentage, date of sowing and condition of soil, assuming seed of high percentage germination. With early sowing, good seedbed expect 80% establishment; November sowing, poor seedbed, 60% or less and poorly developed late tillered crop. Calculate seedrate or, if TGW is not known, usual rate is 160 kg/ha, range 140–190 kg/ha, according to above conditions.

Spring barley

Spring barley tillers vigorously and high seedrates are unnecessary. Usual range 125–160 kg/ha, former for really good seedbed conditions. Reduce to 100 kg/ha if undersowing with slow establishing long ley.

Manuring

Nitrogen (P and K, see Table 5.16)

Total N recommendations are shown in *Table 5.18*, winter barley and *Table 5.19*, spring barley. *Do not* reduce total amount of N for *malting barley* as this reduces yield and there is no guarantee of suitable sample or compensatory price. Late topdressing increases N content and reduces size of grain, so for malting apply all N early in life of crop, according to type, i.e. winter or spring barleys.

Table 5.18 Nitrogen manuring of winter barley (kg/ha N)

N index	Sandy soils	All other mineral soils	Fen light peats, loamy peats	Other organic soils
0	160[1]	140	50	90
1	125	100	Nil	45
2	75	40	Nil	Nil

ADAS figures.
[1] Drought will limit response to lower level.

Time of application

Winter barley

Development of crop is parallel to that of winter wheat but stages occur earlier; principles are similar and N is again key factor for high yields.

Seedbed   Nitrogen is usually not worthwhile except for direct drilled crops, when 25 kg/ha N should be

Table 5.19 Nitrogen manuring of spring barley, winter and spring oats and rye for grain

N index	Sandy soils	Fen light and medium silts	All other mineral soils	Fen light peat, loamy peat and warp soils	Organic soils over chalk	Other organic soils
0	125	125	100	40	70	60
1	100	60	60	Nil	35	30
2	50	30	30	Nil	Nil	Nil

ADAS figures.
Figures represent kg/ha N.

given, in addition to recommended dressing. All P and K applied to seedbed.

Spring top dressing   All N is normally applied in spring; finish application to winter barleys in March for malting crops to minimise adverse effects of N fertiliser on grain size and N content. All N may be applied as a single dressing in March but split application of two or three dressings may be given and appear to give best response, i.e. (a) February and again in late March or early April, or (b) early February, mid March and again late March–early April. Flexibility of approach to N applications is essential and rate and timing should be governed by plant population, state of crop and soil and weather conditions: e.g. N requirements may be less after exceptionally dry winters.

Spring barley

Apply all N to seedbed for malting barley; otherwise N may be applied at any time from shortly before sowing to GS 13 (three leaf stage). Split applications are usually only worthwhile for very early sown crops on light sandy soils liable to leaching.

Harvesting and storage

Barley should be combined when dead ripe and with the ear turned right down. Further delay results in shedding and necking, to which some varieties are very prone. The grain should be dried down 16% MC for storage in sacks or 14% in bulk, the latter cooled with dry, cold air to prevent weevil infestation. Maximum drying temperature for seed and malting crops: MC 22% and over 43 °C MC under 22% 49 °C. Grain for stock feed needs no drying if treated with propionic acid or stored in airtight silos.

Yield

Grain

Winter barleys outyield spring sowings. Heavy crops exceed 6 tonnes/ha, highest yields 8–10 tonnes/ha. 'Average'–moderate yields 4.0–5.0 tonnes/ha.

Spring barleys good over 5.0 tonnes/ha, average 3.75–5.0 tonnes/ha, poor under 3.75 tonnes/ha.

Straw

Yield 1.8–2.3 tonnes/ha, longer strawed winter varieties may give more.

Specific weight

Average 70 kg/hl.

OATS

Importance of oats has declined greatly in last 25 years. Grain is grown mainly as a livestock feed, especially for horses. Market is limited as feed compounders are not interested on account of high husk content. Minor market for best samples for production of oatmeal and porridge oats. Traditionally some crops have been cut immature and fed unthreshed in wetter parts of Scotland and north west England. Many years ago, when oats were cut immature by binder, oat straw was considered equivalent to average hay but today mature combined oat straw is of less value than that of barley. Oats are now of minor importance.

Varieties

Suitable varieties are listed in Farmers Leaflet No. 8, *Recommended Varieties of Cereals* (NIAB).

Climate and soil

Oats require ample moisture for straw production and grain filling and cannot tolerate drought. Crop is best suited to cooler and moister regions of Scotland, Wales and north and western England. Oats may be grown in south and eastern England but winter oats are more reliable and then only on moisture retentive soils. Good drainage is essential. Crop may be grown on a wide range of soils provided they are not liable to drying out. Oats tolerate some degree of acidity but are extremely susceptible to manganese deficiency (grey speck) on overlimed soils and peats of pH 6.0 and over.

Rotation

Oats are not susceptible to takeall (*Ophiobolus graminae*) or eyespot (*Pseudocercosporella herpotrichoides*) and are used as breakcrops by some cereal growers. As they are extremely susceptible to cereal cyst nematode (*Heterodera avenae*), which wheat and barley can carry to a much higher level without showing adverse effects, cyst (soil) counts by ADAS are necessary before inclusion in an intensive cereal rotation. Oats can also carry stem and bulb nematode (*Ditylenchus dipsaci*) which affects several other unrelated crops. Oats will tolerate any position in rotation but choose shortest stiffest strawed varieties for rich conditions, e.g. after heavily stocked ley or permanent grass.

Sowing date

Winter oats

These can be sown from 1 October to early November, optimum first half of October. Complete direct drilling by mid October. Not only are winter oats higher yielding but they are less affected by frit fly and dry summers, have a higher kernel content and ripen earlier than spring oats. May be sown in early spring up to 7 March; thereafter yield is reduced and maturity is very late.

Spring oats

These are often less reliable than winter oats, with sowing dates 15 Jan–15 April, optimum as soon as soil is fit in late February, early March. Late sowings are more susceptible to frit fly damage, so barley is frequently prefered for April sowing.

Seedrate

Sow according to seedsize, sowing date and quality of tilth, using lower rates for good conditions.

Winter oats 150–170 kg/ha.

Spring oats 180–200 kg/ha, on account of low tillering capacity.

Manuring

Total N recommendations are shown in *Table 5.19*, P and K recommendations in *Table 5. 16*.

Winter oats

On cultivated land all N is applied as a top dressing, usually in April when crop is 150–200 mm high; pale crops may require some earlier N. Direct drilled crops should receive 20 kg/ha N at drilling time in addition to total recommendation and may require an early top dressing. If grazed, crop should be top dressed immediately *after* grazing. Recommended straw shortener should be used if lodging is likely, as crop is susceptible in wet areas.

Spring oats

Nitrogen is usually applied to seedbed in compound fertiliser.

Harvesting

Grain is extremely susceptible to heating and moulding in store if at all immature or damp. Crop *must* be fully ripe, i.e. straw completely dead and all grains hard, when it is combined and dried immediately. Drying and storage as for barley.

Yield

Grain 2.5–5.0 tonnes/ha, according to soil and season; yield seriously reduced if drought occurs when grain is filling. Good winter crops can yield considerably more.

Straw 2.5–3.8 tonnes/ha.

Specific weight

Average 52.4 kg/hl.

RYE

Rye may be grown as a green forage crop or for grain; each must be treated as a distinct crop, requiring totally different varieties and management.

Forage rye

Rye starts growth at lower soil temperatures than Italian ryegrass; so is especially suitable for earliest spring bite for dairy cows, ewes and lambs. Well drained soil, which warms up early in spring, is essential. Crop grows away quickly, soon becoming stemmy and unpalatable; therefore start grazing early, before stem formation or waste is excessive. Start cows when rye is 150 mm high, sheep earlier.

Varieties

Use leafy long strawed forage varieties (e.g. Lovaszpatonai, Rheidol); do not leave grazed crops to set grain as yields are usually very low.

Sowing

Early sowing, August in the North and late August to mid September in the south, is essential; later sowings do not give the massive flush of early spring growth from many tillers. Rye may be sown after grass or other cereals and direct drilled or drilled conventionally on a fine firm seedbed. Graze surplus growth in early November. Drill 220 kg/ha seed for pure stands. Where several grazings are required, reduce rye to 100–125 kg/ha and include 25 kg/ha leafy tetraploid Westerworths (milder areas) or hardy Italian ryegrass.

Manuring

Rye is almost insensitive to acidity. Give 40–50 kg/ha N, 40–50 kg/ha P_2O_5 and 40–50 kg/ha K_2O in the seedbed. Top dress with 75–90 kg/ha N as soon as spring growth shows signs of starting; give two dressings of 50–65 kg/ha N if two shorter grazings are taken.

Grain rye

Grain rye suits poor, light, sandy, acid soils and is grown as second or subsequent straw crop or used for reclamation on acid land. Grown on contract for rye crispbread and straw for thatch, it is of little importance compared to the other cereals.

Varieties

Spring and winter varieties occur but winter varieties only are sown in the UK. Recommended varieties are listed in Farmers Leaflet No. 8, *Recommended Varieties of Cereals* (NIAB).

Sowing

Direct drilled or drilled on conventionally prepared moderately fine, firm seedbed during September and as late as mid October according to district. Sown after mid February, the crop may never come into ear. Drill 125–190 kg/ha of seed 50 mm deep with the higher rate for late sowing.

Manuring

Nitrogen recommendations are shown in *Table 5.19* and PK recommendations in *Table 5.16*.

Harvesting

Rye required for thatch is cut by binder when a little green under ear and grain can be broken cleanly over the thumbnail. Stook for 7–10 d and cart. Rye for combine harvester should be dead ripe. Straw is very tough and durable and unsuitable for feeding.

Yield

Grain 2500–4400 kg/ha; straw 3750–5000 kg/ha.

Specific weight

Average = 68.7 kg/hl.

MAIZE

Maize is grown world wide in subtropical areas and warmest parts of temperature zone for grain, silage, green forage, sweet corn and other uses. There are numerous types. Sweet corn (corn on the cob) is a luxury vegetable requiring distinct varieties and specialised production and marketing and is not considered here. Maize has been grown from grain in warmest parts of south eastern England but even there climate is neither warm nor reliable enough, so grain is not fit for combining until early winter, if at all. Maize has been grown for provision of green fodder in early autumn but wastage of 25% has been experienced with grazing *in situ*; kale is cheaper and more reliable.

Forage maize

Only forage or silage maize can be grown in the UK with any degree of reliability and then its use must be confined to low lying, warm, sheltered or south facing sites, mostly south east of a line drawn from the Wash to the Severn; even then, crop may not reach desired state of maturity before it is killed by first sharp frost of autumn. Earliest maturing varieties may be considered for most favoured sites north of this line. Maize is not a difficult crop to grow but a very high standard of cultivation and attention to detail is essential.

For highest yield and dry matter content of silage maize, grain must reach cheesy ripe stage at cutting, usually late September to mid October. To succeed crop needs long warm growing season and no exposure to strong winds. Very early sowing is impracticable as germinating seeds cannot tolerate cold soil and many rot. Growth is terminated by onset of cold weather on first sharp frost in autumn. Crop cannot reach necessary growth stage or yield with late springs and cold short summers.

Variety

Suitable varieties, listed in Farmers Leaflet No. 7, *Recommended Varieties of Forage Maize* (NIAB) are essential. Avoid late varieties for all sites in UK; although potentially high yielding they fail to reach required degree of maturity. Medium and medium late varieties (NIAB maturity classes 5 and 6) have highest yield potential but only achieve it on good sites in warm summers. For greatest reliability, choose early or medium early (NIAB maturity classes 8 and 7) and always on less favoured sites. High scores for early vigour indicate varieties with rather better tolerance of colder spring conditions and ability to compete with any weed growth in early life. Best varieties have good resistance to lodging.

Soil

Maize is *totally dependent* on good soil conditions, requiring first class seedbed on fairly deep well drained and structured topsoil. Also moisture retentive subsoil, as crop is particularly susceptible to drought during tasselling and silking period (i.e. male and female flowers). Deep, free working loams are best; avoid shallow or infertile soils, those subject to drying out, i.e. light sandy soils or soils overlying gravel or shale and thin chalks. Cold heavy clays are late to warm up, give poor seedbeds and result in difficult harvesting in wet autumns.

Seedbed preparation

Although maize seed is large it is highly sensitive to poor seedbed conditions. Rough seedbeds give stunted growth, immature cobs and failure of soil-applied herbicides to control weeds. Land should be ploughed before Christmas to obtain a good frost mould. Fine seedbed, loose enough to allow minimal working. Drill marks help rooks to find the seed and should be removed by rolling or chain harrowing immediately after drilling. Atrazine should be sprayed on to the soil surface when work is completed but shallow incorporation of herbicide improves weed control in dry periods.

Sowing date

Sowing period is short and critical. Drill as soon as soil temperature reaches 10 °C, but not before; occasionally in late April on best sites, more often early to mid May. Sowings made from late May onwards germinate and grow rapidly in warm soil, producing tall vigorous dark green plants but these are very deficient in cob. Result is an extremely immature crop at harvest which gives only a low quality watery silage.

Plant population and seedrate

Maximum yield of dry matter compatible with high proportion of cob, one cob per plant, is required. Excessive populations reduce proportion of cob while inadequate plant numbers give lower yields of dry matter. Aim for 110 000 plants/ha (i.e. 11 plants/m²). Seed size varies greatly between and even within varieties. Usually 2600–4000 seeds/ha, so it is normally sold in packs calculated to give 110 000 plants/ha after allowing 10% for losses after drilling. Crop should be precision drilled 50 mm deep. Merchants usually state number of seeds/kg and size grade of seed, indicating belt or wheel number required; suitable belts or wheels should be ordered well in advance of drilling. Drilling in 760 mm rows requires interseed spacing of about 100 mm to achieve requisite population. Check that seed has been dressed with appropriate fungicidal–insecticidal seed dressing.

Bird damage

Rooks can devastate a promising crop in hours just at emergence and are difficult to control. Phorate granules, used to protect crops against frit fly, are useful repellents. Heavy duty black nylon thread, strung about 1 m above ground on canes 14–18 m apart in each direction, is valuable on smaller areas, together with banger sound scarers. Dead rooks shot on dawn patrols should be strung up.

Manuring

Maize gives a poor response to fertiliser applications and recommendations for P and K shown in *Table 5.20* are largely for maintenance; much more K is usually removed: e.g. a 12 tonnes/ha dry matter. Crop will remove around 150 kg/ha K_2O which must be replaced during rotation unless regular applications of slurry are being given. When liberal dressing of slurry is given prior to sowing, there is no need to apply P or K for a maize crop. Some 40 kg/ha N should still be applied unless dressing was very heavy (i.e. 50 m³/ha or more of undiluted cow slurry) and spread shortly before sowing. All N should be applied to seedbed; avoid top dressing as fertiliser granules are trapped in leaf funnels, resulting in scorch. Never combine drill, as it may seriously damage germination.

Table 5.20 Manuring of silage maize (kg/ha)

N, P or K index	N	P_2O_5	K_2O
0	60	75	75
1	40	40	40
2	40	40	40
Over 2	—	Nil	Nil

ADAS figures.

Harvesting

Feeding value of the whole plants stays constant over harvest period, with average D value of around 70 and crude protein content 9%. Aim to harvest when dry matter content of whole plant is between 20 and 25% for clamp and 25–30% for sealed towers. Crop is fit to harvest up to 25% MC, the grain being the consistency of hard dough and tough to handle; in the south this stage is reached by early and medium-early varieties in late September or early October. The later stage (around 30%) is reached when grains are hard and difficult to crack with the finger nail and husks and lower leaves are papery.

Frosting does not harm mature crops but increases dry matter content. Harvesting must be within 10 d of frosting or deterioration sets in, and it is essential to use precision chop harvester, either a special single row machine or one with one or two row adaptors. Unadapted flail mowers result in much waste of cob, uneven chopping and poor fermentation, and animals pick out best parts. Properly chopped maize ensiles without difficulty and is rich in soluble carbohydrates; additives are unnecessary.

Yield

Well grown crops of silage maize usually produce 11–14 tonnes/ha of dry matter or 52–62 tonnes/ha of fresh material at 21% dry matter, which should produce from 44 to 53 tonnes/ha of silage.

SEED PRODUCTION

Seed certification

The aim of seed certification is clean, healthy stocks of seed, free from weed seed, pests, diseases, alien varieties and other cultivated species. Seed must produce a crop true to type as produced by breeders, of the stated variety, guaranteed at a minimal standard and of high germination. All seed offered for sale in EEC must be certified, which includes seed

(1) produced under statutory certification scheme,
(2) certified seed imported from EEC,
(3) from third country having equivalence as certified seed.

For details of producers and requirements for producing seed for certification in England and Wales, consult MAFF or elsewhere in UK, Department of Agriculture, Scotland or Ministry of Agriculture, Northern Ireland. *National lists* of varieties are published by National Institute of Agricultural Botany, who also publish a range of seed growers leaflets.

Production of certified seed is carried out exclusively under the control of the appropriate certifying authorities. Only authentic seed of varieties included in the *national list* for each crop or group of crops (e.g. cereals, grasses, potatoes) is eligible for certification; to quality for this a variety must be distinct, uniform, stable and of acceptable agricultural value. *Recommended lists* have no statutory status. All crops being grown for certified seed are inspected solely by *licensed inspectors* and must reach minimal standard for each grade of seed. Processing and sale of seed is also controlled; only licensed *seed merchants* may sell seed, only *licensed processors* may treat, clean and label basic or certified seed while samples for processor must be taken by a *licensed sampler*.

Seed production and marketing is a highly complex business which may be divided into three stages: growing, storage and processing. The latter is entirely outside the province of the grower.

Seed growing

Production of certified seed is a specialised job requiring excellent husbandry and management. Meticulous care, good planning, and very high standards of training and supervision of staff are indispensable. Farmers who cannot satisfy these essentials should not attempt seed growing.

It is necessary to meet the following basic requirements.

Authentic seed

Crops eligible for certification may only be planted with authentic seed of a grade of high standard and of a variety approved by the certifying authority; this normally includes foundation seed in sealed sacks straight from breeder and other seed with an appropriate certificate. Be sure there is adequate seed to finish the field: once-grown seed must not be used.

Adequate isolation

Isolation from crops of the same species prevents cross pollination and contamination of the crop with disease or unwanted seeds of other varieties. Cereals, normally self pollinated (i.e. wheat, oats, barley), need a physical barrier, hedge, fence, or an area of fallow or non-cereal crop at least 2 m to prevent admixture. A distance of 50 m from potential sources of loose smut infection (e.g. crops not planted with certified seed) is recommended to prevent cross infection; six row barleys are prone to cross fertilising and should be similarly isolated from two row barleys. Crops normally cross fertilised require much larger minimum distances, e.g. rye: 250 m. Grass seeds, same spp. or IRG and PRG: fields up to 2 ha 100 m, fields over 2 ha 50 m. Leys and permanent pasture are included in minimum isolation distance: if they adjoin seed stand, keep topped or grazed to prevent heading during flowering periods of seed crop, unless they contain *only* same variety. Be careful in selecting fields on farm boundary or adjoining grass verges of roads. Brassicas require at least 200–1000 m according to species and grade of seed.

Clean land

Seeds likely to contaminate seed crops lie dormant in the soil for a long time so a cleaning rotation is necessary. Cereals crops for seed should be preceded by two years of non-cereals, unless the previous crop was the same variety and grade. Similarly it is necessary to use the same grass or clover variety in ordinary leys as that grown for seed. It is also desirable, if possible, to keep the same field for the same variety and species of grass. The field must be free from noxious weeds or those likely to contaminate the crop. Do not feed hay on seed-producing leys or apply fresh farmyard manure to land used for seed production.

On arrival at the buyer's premises seed must be cleaned. Modern seed cleaning machinery removes impurities provided they are substantially different in size, weight or shape from seed crop. The closer the form of the impurity to the seed crop, the more difficult to remove and the greater the amount of crop seed removed with the impurity during cleaning. If seeds of crop and impurity are identical or very similar, separation is impossible and the contaminated crop is worthless. Thus, some weeds and other crop plants often disregarded in normal crops are highly undesirable in a seed crop.

Clean machinery and equipment

These are essential to prevent contamination. Drills or broadcasters must be thoroughly cleaned on waste land. Combine harvesters should be run empty and then thoroughly cleaned out with a compressed air line and industrial vacuum cleaner. Use clean sacks. Trailers, augers, elevators, conveyors and bins must be emptied and thoroughly cleaned before use. To avoid admixture of stocks, produce of each crop should be kept separate and be identifiable by variety, reference number of stock, field OS number and name. Bins should be kept locked and sacks clearly labelled, with stacks of sacks separated by a solid partition.

Farm suitability

Cereals and herbage seed production are best suited to large farms, in a ring fence. Small farms or those spread out find it impossible to exert control over adjoining cropping needed to obtain the required isolation. It is advantageous for farmers to group together to plan cropping of boundary fields, especially for grass seed crops. Large cereal farms have the advantage of complete cereal growing equipment, combines and drier under their own control. Contractors usually fail to clean out combines between farms and fields; contamination from other crops is soon transferred to the sample. Seed production is thus better suited to larger farms.

CEREALS

Cultivation for seed crops is dealt with under the appropriate cereal crops. Inspection by a qualified inspector should be preceded by rogueing when other species, varieties or weeds can be removed from the crop; dirty crops cannot be rogued. Avoid herbicides which stunt but do not kill all wild oat plants (*Avena* spp.). Drying needs care, as high temperatures damage germination capacity; drying at 49 °C is normally safe for grain up to 24% MC. Dry down to a maximum of 15% MC and store in a cool, dry place, protecting from vermin. (*See also* Farmers Leaflet No. 1, *Seed Quality* (NIAB) and Seedgrowers Leaflet No. 1, *Growing Cereals for Seed*; also MAFF publications, Sectional list No. 1.)

HERBAGE SEED PRODUCTION

Leys for herbage seed production are valuable break crops; the roots restore structure of worn out arable soils and provide large amount of organic matter. In addition, they provide out-of-season grazing and additional livestock can be kept. The stover or straw, often sold as threshed hay is useful fill-belly. Principal crops include grasses (perennial and Italian ryegrass, cocksfoot, timothy, meadow and red fescues) and clovers (red and white clover).

(*See also* Seed Growers Leaflet No. 5, *Growing Grasses and Herbage Legumes for Seed* (NIAB) and Welsh Plant Breeding Station, Aberystwyth, Bulletin *Principles of Herbage Seed Production*.)

Crop establishment–general

Seedbed

Plough early for good frost mould. Fine, firm, level weed-free seedbed is essential; knobbly seedbeds give weak plants. Shallow drilling 12–18 mm on rolled surface is best in dry areas, otherwise broadcast on rolled surface. Place seed in soil moisture for rapid germination and vigorous growth, drilling as shallowly as possible. Small seeded species (white clover, timothy) 8–12 mm, larger seeded sorts (ryegrasses, cocksfoot, meadow fescue 20–25 mm deep).

Use of cereal cover crop

Undersowing or sowing seeds without a cover crop after cereal harvest gives income from cereal during year of establishment, but in some slower developing species, notably timothy, tall fescue and cocksfoot, yield in first harvest year is heavily reduced (20–70%). Sowing bare (without cover crop) eliminates this loss but it is necessary to balance loss of income from cereal crop against this increase. Practice depends on species and circumstances; major risk of undersowing is total loss of seed crop under lodged or dense cereal. To minimise adverse effects, drill grass and cereal on same day, reduce cereal seedrates and N applications (*see* section on lodging in cereals, p. 126). For sensitive species grown in wide rows and undersowing either

(1) block off every second or third row in cereal drill, according to grass crop row width, and drill seed crop after cereal emergence, or
(2) fit dividers into seed box and sow cereals and grass in same operation.

Row width

Two major methods include

(1) drilling in narrow rows, 180 mm or less, or broadcasting, and
(2) drilling in wide rows 450–600 mm apart.

Method 1 is perfectly satisfactory for all ryegrasses, red and white clover but in cocksfoot, timothy and meadow fescue results in excessively dense tiller populations and consequent reduction in seed yield, especially in high tillering pasture varieties and in second harvest year.

Sowing date

Undersown crops are sown simultaneously with cereal in March or early April. Early drilling of *all* crops ensures adequate soil moisture for establishment in dry areas as well as full tiller development in time for vernalisation the following winter. If sowings without a cover crop are intended late in the year, speed of establishment and tillering rate are determining factors in deciding how late a species can be sown with safety; it also depends on season and soil temperature. In mild southern areas Italian ryegrass, hybrid ryegrasses and early flowering perennials can be sown up to mid September, perhaps even up to end of September but the later the sowing the greater the risk of failure. It is usually inadvisable to delay sowing other grasses beyond end of July at latest. Red clover performs very poorly from late summer or autumn sowings.

Seedrate

Recommendations depend on soil conditions. With choice seedbed and consequent high tillering, low seedrates are best; increase if conditions are not so good. Excessive seedrates result in very high tiller densities, which suppress seed yields. Aim for high quality seedbeds with reduced seedrates, to give low population of vigorous plants. Increase seedrates with closer spacing or broadcasting, according to seed size and establishment rate of species. Ideal population appears to be one mature plant/150 mm of drill run, as a general guide.

Establishment of individual species

Perennial ryegrass

Prefer to undersow in spring cereal but can be sown bare after early harvested winter cereal. Drill in rows 100–200 mm apart, diploid varieties 9 kg/ha, tetraploids 17 kg/ha or broadcast 13 kg/ha diploids, 22 kg/ha tetraploids; 0.5–1 kg/ha wild white clover may be added to crops of pasture varieties to improve grazing; large leaved white clovers may cause difficulty in harvesting.

Italian ryegrass

This grass establishes rapidly from seed and may be undersown in a spring cereal or drilled bare after cereal is removed up to mid September. Sow in drills 100–200 mm apart, diploid varieties 13–14 kg/ha, tetraploids 16–17 kg/ha.

Tetraploid hybrid ryegrasses

As for Italian ryegrass but drill 16–22 kg/ha.

Cocksfoot

For highest seed yields in first harvest year drill in spring without cover crop in rows 400–600 mm apart. Sow 2–5 kg/ha, lower amount for superfine seedbed and good conditions.

Timothy

Drill bare in spring or up to late July in rows 500–600 mm apart, 2–4 kg/ha. If undersowing, reduce row width to 350 mm and increase seed to 5–8 kg/ha.

Meadow fescue

Drill in spring under cereal crop in rows 150 mm apart with 12–18 kg/ha seed.

Tall fescue

Drill in spring without cover crop in wide rows 600 mm apart, seedrate 5–7 kg/ha, after cleaning crop.

Red fescue

Drill in spring under cereal crop in rows 100–200 mm apart at 10–15 kg/ha.

White clover

Broadcast, or in dry areas drill in narrow rows, under cereal crop in early spring. Large and small leaved varieties may be sown pure at 2–4 kg/ha or mixed with 3–5 kg meadow fescue or pasture perennial ryegrass, whose seeds are easily cleaned from white clover; avoid timothy which is difficult to separate. Inclusion of grass improves grazing; sward then forms pasture when seed production is finished.

Red clover

Undersown in cereal crop in spring; avoid late summer sowings as establishment is then unreliable. Prefer drilling in narrow rows in dry areas. Best sown alone, diploid varieties 8–10 kg/ha, tetraploids 11–13 kg/ha. Some growers mix 3–5 kg/ha timothy or perennial ryegrass in broadcast stands.

Row indicator plants

Grasses grown in wide rows germinate slowly and cannot be seen early enough for inter-row cleaning. Include 0.6 kg/ha mustard, or on charlock (*Sinapsis arvensis*) land 0.3 kg/ha lettuce with the crop. These seeds emerge early, indicating the rows and are subsequently killed by herbicides.

Manuring

Establishment year

Total applications shown in *Table 5.21*.

Table 5.21 Manuring of herbage seed crops establishment year (kg/ha)

N, P and K index	*Direct (bare) sown grasses*			*Undersown grasses and clovers (autumn application immediately after harvest)*		
	N	*P_2O_5*	*K_2O*	*N*	*P_2O_5*	*K_2O*
0	80	120	120	0–50	100	120
1	60	80	90	0–40	80	90
2	60	50	60	0–40	50	60
Over 2	—	30	30	—	30	30

ADAS figures.

Direct sown grasses   Phosphorus and potassium may be applied in previous autumn or to seedbed; if crop is cut for forage give extra 40 kg/ha P_2O_5 and K_2O. Backward crops may need extra N but control weeds first.

Undersown grasses and clovers   If there is a P or K deficiency, broadcast fertiliser for cereal and increase these nutrients; do not combine drill or stripping occurs. Otherwise all additional P and K required is applied immediately after cereal harvest, with N; clovers do not require N.

Production years

Annual applications of P and K are shown in *Table 5.22*, N in *Table 5.23*. Phosphorus and potassium may be applied in spring, according to N index. Spring N may be given in one or two dressings and timed according to species and variety. For Italian ryegrass, early perennial ryegrass, cocksfoot and meadow fescue, apply in early spring; apply N to medium and late perennial ryegrasses and timothy in late spring. Rate of application also depends on weather, management and condition of sward, older, weak or backward swards need rather more N. Conversely, excessive N application, e.g. direct combined late

Table 5.22 Manuring of herbage seed crops production years P and K requirements—all crops and soils

P and K index	*P_2O_5*	*K_2O*
0	100	120
1	80	90
2	50	60
Over 2	30	30

ADAS figures.

varieties of perennial ryegrass, can cause considerable harvesting difficulties from lodging.

Autumn applications, of N, given as compound fertiliser immediately after harvest are designed to produce vigorous leaf growth of grasses but amount is dependent on age and condition of sward. Up to 50 kg/ha N may be needed for backward crops, especially cocksfoot and red fescue. Also apply up to 50 kg/ha N if sward is to be grazed or cut for forage in autumn.

Sward management and defoliation

Herbage seeds are grown under such a wide range of soil, climatic and farming conditions that it is impossible to lay down precise details of sward management. Some crops are grown on stock farms but most are grown on large predominantly arable farms, many with no stock at all. Surplus grass may be removed by stock or mechanically.

Appropriate extent and timing of removal of excess herbage vary not only between species but between types and varieties within same species. Obtain appropriate advice when growing a new variety for first time. Appropriate defoliation technique is also modified greatly by amount and distribution of rainfall and water holding capacity of soil. Species differ greatly in their dates of flower formation and head emergence; in general, grazing should be completed well before critical stage of flower formation; defoliation at this stage or later reduces seed yields drastically in most species. The earlier flower formation occurs, the earlier must grazing be completed; late varieties or species can be grazed later. Effect of spring grazing is much influenced by soil moisture. In dry years and areas spring grazing where appropriate

Table 5.23 Manuring of herbage seed crops. Production years

Time of N application	*Soil type*	*Perennial and Italian ryegrasses*	*Cocksfoot*	*Timothy and meadow fescue*	*Red fescue*	*Red and white clover sown alone*
Spring	Sandy soils	125–150	170	100	Nil	Nil
	Other soils	100–125	120	80	Nil	Nil
Autumn	All soils	0–50	0–50	0–50	0–50	0–50

ADAS figures.
N requirements (kg/ha).

should be early and light, omission is often preferable as late grazing in these circumstances can reduce yield heavily. Always aim to complete grazing *quickly*— then move on.

Perennials and Italian ryegrasses

These produce the bulk of their fertile tillers in spring. Autumn or early winter grazing, October to end January, is beneficial to all varieties, as it encourages tillering and removes excess herbage, which suppresses tillering and causes winter burn. Value of spring grazing, other than supplying feed, is much less certain and can reduce seed yield; properly carried out, grazing shortens straw and helps to reduce lodging. Adequate N is essential to prevent loss in yield.

Finish grazing early varieties of perennial ryegrass, e.g. S24 by first week in April, intermediate, e.g. S321 by second week in April and late varieties, e.g. S23 by third week in April; the latter allow heavier grazing and need higher N levels.

Italian ryegrass

Spring grazing shortens straw and reduces lodging risk. Crop produces ample fertile tillers in spring; provided adequate N is given in early March, crop can be grazed up to end of April, according to variety, without loss of seed yield. If crop is not grazed reduce N. Additional N does not mitigate effects of grazing too late.

Hybrid ryegrasses

Growth pattern permits considerable flexibility of management, which is similar to that of Italian ryegrass. Removal of flowering tillers in some varieties does not reduce yields, as they are replaced, provided grazing is not too late and ample N is given. Check on varietal requirements.

Cocksfoot, timothy and tall fescue

Majority of seed bearing tillers are produced in autumn and different grazing pattern is required. In general graze after autumn growth has stopped but before spring growth starts, quickly and once only. Removal of excessive leaf, which restricts tiller growth is beneficial but avoid grazing too tightly.

Cocksfoot best grazed leniently in December; danger of heavy yield suppression if grazed early February onwards. Do not graze backward crops in establishment year. Least susceptible species to lodging; generous N application increases number of fertile tillers and size of infloresene. *Timothy*, management similar to cocksfoot. Graze leniently once only in period October–December; late grazing disastrous. *Meadow fescue*, graze once during autumn or early winter and never later than early March. Very susceptible to lodging. *Tall and red fescues*, graze once

in autumn or early winter and complete by mid February.

Clovers

Grazing after removal of cover crop is beneficial in first autumn; graze thin crops *lightly* with sheep, lush crops may be grazed fairly hard to produce short even sward; do not overgraze or bare tight.

Red clover

Broad red clover is traditionally first cut for silage or hay and the seed crop taken from the aftermath to prevent heavy leaf and straw production. Late or heavy cuts reduce yield severely; do not take more than a light early cut of silage before laying up. *Late flowering red clover* should be left untouched or a *very short* silage cut (10 cm) taken in spring. Research indicates that highest seed yields may be obtained by omitting early cutting and or grazing and taking seed cut first. A recommended desiccant can deal with heavy greenery if properly applied.

White clover

Swards sown with companion grass should be grazed *hard* when grasses are growing actively in early spring, to encourage clovery sward. Clover is grazed in late spring to remove excess growth which interferes with flower production and gives harvesting difficulties. Cease grazing *before* significant number of flower buds are taken (i.e. when first flower buds appear): medium to large leaved varieties finish by mid May; small leaved or wild white varieties, cease end of May or first week of June. In dry areas or seasons, finish two weeks earlier, under wetter conditions grazing may be extended by a week.

Post harvest treatment

Mow over after harvest and remove dead stubble and trash; stubble removal is most important if grazing is required, otherwise stock reject grass. Cocksfoot stubbles should be burned if cocksfoot moth (*Glyphipterix simpliciella*) is a problem. Red fescue swards are best gang mown. Apply fertiliser to cleared stubble. If inter-row cultivating wide drilled crops do not work too close to drills.

Harvesting

The binder, ideal for cocksfoot, has largely disappeared. Crops cut loose may be put on *tripods*, a safer practice, producing better samples in wet seasons but too laborious and now little used. Alternatively cut and cure in the swath and thresh with a combine and pick-up attachment. Reasonably long stubble is left, according to species. Moving weathered crops around to dry results in heavy seed losses: if grass has grown through swaths, undercut with mover, allow to dry, and combine. Swathing is best suited to unevenly ripened crops, those containing greenery and late flowering or pasture varieties which ripen unevenly. *Direct combining* is most suitable for early ripening grass varieties and red clover: the latter may require use of chemical desiccant. Crop must be ripened evenly and free from green material. This method has lowest labour requirement but shedding is very serious if combining a fully ripe crop is delayed in wet windy weather.

Double combining

This is when crop is threshed and then re-threshed after an interval of a few days, is suitable for unevenly ripened crops, especially timothy.

Drying, conditioning and storage

Damp or trashy parcels must be dried, immediately or, if impracticable, spread thinly on a clean airy floor or hessian; never leave in bags, even overnight, for rapid heating ruins germination. Maximum air temperature reaching the seed should not exceed:

	Normal seed (°C)	*Very wet seed* (°C)
Timothy and clovers	37	26
Other grasses	48	32

Store in cool, dry, weatherproof permanent building and periodically check for condition. Method and time of harvesting, target yield and duration of the stand are given in *Table 5.24*.

Table 5.24 Harvesting of herbage seed crops

Species	Type/variety (E: early, M: medium, L: late flowering)	Ready for swathing	Usual/best method of harvesting	Target yield (kg/ha)	No. of harvest years
Italian ryegrass Hybrid rye grasses	—	Very variable—mid July to mid Aug	Swathed, then combined. Can be direct combined but only advisable in badly laid crops	875–1250	1
Perennial ryegrass	(E) S24	Early to mid July		750–1000	2
	(M) S321	Mid to late July		800–1250	2
	(L) S23 or S101	Late July (good years), mid Aug (late wet summers)		600–1000	2
Cocksfoot	(E) S345, S37	End June to early July— 14 d later in bad year	Best swathed, then combined. Stands well but only direct combined in emergency	450–1000	
	(L) S26, S143	Mid July to 14 d later in bad year			2[1]
Meadow fescue	(E) S215	Early to mid July	Swath and combine or direct combine if lodged, with little greenery	525–750	2[1]
	(L) S53	Mid July		350–500	
Timothy	(E) Scots, S352	Mid Aug	According to variety: swathing double- or triple-combining for varieties difficult to thresh, e.g. S48	375–650	3
	(L) S48	Late Aug to mid Sept		375–650	
Red fescue	S59	Early July	Swath, then combine. Does not shed readily	375–500	4–6[2] or even longer
Red clover[3]	Broad red varieties		Seed does not shed readily when ripe. Direct combining. Green crops require chemical desiccant	<500	1
	Late flowering red varieties	Sept onwards according to season			
White clover[3]	Giant White, S100, Kersey	Late July to early Sept according to weather	Tricky except in dry weather; choose dry spell. Swath crop and combine from swath: put into windrows first with light crops	375	2[1]
	Wild White, S184, Kent			<375 old pastures 80	2[1] indefinite if old pastures

[1] Three crops can be taken but two are usually more economic.
[2] Sward may be sold profitably as fine turf at end of its life with landlord's agreement.
[3] Yields highly dependent on weather.

Leguminous crops

This group includes beans

(1) *Vicia faba* type, i.e. field and broad beans,
(2) *Phaseolus* spp., i.e. dwarf or french beans; also includes the runner and little grown navy beans; peas, lucerne, lupins, sainfoin and vetches or tares.

Grown variously for mature and immature grain or pods for human and animal consumption, according to variety; lucerne, sainfoin and vetches (tares) grown for ruminant fodder only.

Rotation

All are valuable break crops in cereal rotations, although amount of atmospheric N they 'fix' varies greatly. Lucerne and sainfoin leys, down for several years leave rich residues (ADAS N Index 2) but peas and *V. faba* beans leave less (ADAS N Index 1). *Phaseolus* spp. do not fix their own N but considerable plant residues can still be ploughed in after harvest.

For reasons of pest and disease control, peas, field, broad and dwarf beans, lupins and vetches should be treated as a single crop. Peas, broad and field beans, lupins and vetches are hosts to pea cyst nematode (*Heterodera goetingiana*) and are also hosts to some soil-borne diseases attacking peas and dwarf beans. None of these crops should be grown more frequently than once in five years; if previous crops have been attacked, a longer interval is advisable.

LEGUMINOUS SEEDCROPS FOR HUMAN AND/OR ANIMAL CONSUMPTION

Beans

Field beans

Field beans are a valuable break crop in cereal rotations as cereal diseases are not perpetuated; it has been estimated that the crop is capable of fixing 200 kg/ha N from atmospheric nitrogen, leaving good reserves for the next crop. Only normal cereal equipment is needed and labour requirement is low. Grain is valuable when protein prices are high.

Varieties

Winter and spring varieties are used. The latter includes (a) tick beans, small seeds (TGW under 560 g) and grown to sell for pigeon feeding where they command a premium over the larger (b) horse beans (TGW over 560 g), which are only suitable for livestock feeding or protein extraction. Winter varieties are all of the horse bean type. Winter beans ripen two to four weeks earlier than spring beans and usually give about 600 kg/ha more grain. Suitable varieties are listed in Farmers Leaflet No. 15, *Varieties of Field Beans* (NIAB).

Climate and soil

Winter beans are more frost tolerant than spring varieties (the most frost resistant will stand up to $-15\,°C$) but are rarely grown in Scotland and the north. Both types do best in areas where relatively mild winters are followed by warm dry springs and medium summer rainfall. Yield of spring beans is badly reduced by prolonged summer drought. Still sunny weather during flowering increases yields.

Soil pH should be above 6.5. Soils should be well drained but not liable to drying out in summer, to which spring beans are especially susceptible. Clay or chalk/clays are ideal. Avoid very light soils or those high in humus. Spring beans are grown on a wide range of soils; light soils permit early drilling.

Sowing

Beans require a seedbed similar to wheat, of adequate depth but not too fine. Complete *winter* sowings by the end of October, so that plants have three to four leaves before onset of frost.

Spring beans should be sown as soon after 15 February as possible to ensure rapid development before dry soil conditions occur. Drilling after mid March usually results in serious yield loss. Drill at least 75 mm deep if simazine is used, which must be applied within 7 d of sowing. With other herbicides 25–50 mm may be adequate.

Row width may vary from 150 to 500 mm and has little effect on yield of beans; many growers prefer 360 mm drills as this permits tractor working if inter-row cultivation or post emergence sprays are needed. There is little evidence as to optimum plant populations: for winter beans which produce several productive branch stems at their base, 45–50 plants/m² appear adequate while for spring beans, which do not branch, 50–60 plants/m² are needed. Sow 225 kg/ha small seeded/tick varieties, 250–275 kg/ha winter beans and 275–300 kg/ha spring horse beans, according to size.

Manuring

Beans of *V. faba* species, give no response to N, which should not be given in any circumstances, even with poor seedbeds. Rates of application for P_2O_5 and K_2O are given in *Table 5.25*. Broadcast dressings should be given after ploughing and cultivated deeply into the seedbed. As beans are more responsive to placed fertiliser optimum rates of K_2O are higher

Table 5.25 Manuring of field beans (kg/ha)

P or K index	P_2O_5	K_2O		
		Broadcast	Placed	Combine drilled
0	75	120	120	$50^1 + 70$ broadcast
1	40	50	80	50^1
2	40	40	40	40^1
Over 2	Nil	Nil	Nil	Nil

ADAS figures.
[1] Applies to rows 180–200 mm apart. Cut down proportionately for wider rows to give same rate per row, e.g. 360 mm rows combine drill maximum of 25 kg/ha. Exceeding these amounts may impair germination.

than for broadcasting. Placing fertiliser close to but not in contact with the seed is most efficient method of application but cannot be used with narrow rows.

Harvesting

Field beans are direct combined when the pods are black and dry and the haulm shrivelled. Shedding is not a problem; the crop combines easily. Desiccate with diquat where crops for animal or pigeon feed are weed infested, unevenly ripened or spring crops are still green in late September. Use slow drum speed to avoid damage. Dry at low temperature (< 40 °C) and ensure proper drying to below 15% MC before storage.

Diseases and pests

Winter beans are susceptible to chocolate spot (*Botrytis fabae* and *B.cinerea*). Spray with benomyl as soon as first symptoms appear. *Leaf spot* (*Aschochyta fabae*) exhibits similar symptoms—control by using only healthy seed checked for absence of disease.

Black bean aphid (*Aphis fabae*) is a serious pest of beans and can ruin spring sown crops. Inspect regularly before and during flowering. Spray with recommended aphicide as soon as first aphids are seen. Spray crops in flower during evening to avoid injury to bees.

Yield

Average: 2.5 tonnes/ha. Good crops 3.8 tonnes/ha or more. Straw 3.2 to 5.0 tonnes/ha.

Broad beans

Broad beans are grown for picking green for market or for processing, i.e. freezing or canning. As with peas, picking green is relatively unimportant. Harvest of broad beans for processing must be integrated with other crops in freezing and canning programmes. Maximum yields are obtained from early sowings but sowing dates must be arranged so that broad bean crops are ready between pea and dwarf bean harvests or before pea harvest.

Types and varieties

Uses of broad beans may be differentiated by their flower colours. Varieties with pigmented flowers include:

(1) *garden varieties*, winter sorts—Seville Longpod and Aquadulce Claudia and spring sorts—Windsor and Longpod varieties. Large seeded and used for fresh market or slicing young;
(2) *freezing varieties*, Ipro, Minica, Pax, Primeur and Primo. *Dual purpose varieties* have non pigmented (i.e. white) flowers and are suitable for *freezing* or *canning* and include: Polar (Autumn), Archie, Beryl, Feligreen, Rowena and Threefold white.

Flower colour is important, as coloured flowers indicate presence of leucoanthocyanidins, which cause discoloration of beans on canning or cooking. Winter sown varieties of broad beans tend to outyield spring sown sorts, but their simultaneous maturation with vining pea crops presents problems to processors unless they are geared to harvesting and processing as a separate operation. Varietal selection to suit agronomic and marketing requirements is thus of prime importance.

Distribution, soil, seedbed and sowing date

Apart from crops for picking green, must be in easy range of factory. Avoid extreme soil types. After ploughing use minimum cultivations necessary to produce an 'open' type of seedbed. Avoid very fine overworked seedbeds and compaction. Drill seed into soil moisture, usually 50 mm deep, but increase depth to 75 mm to avoid damage if simazine is used. Owing to large seed, not all drills are suitable. Consult manufacturer.

Broad beans are sown to reach required maturity immediately after last crops of vining peas, except where overwintered crops are harvested first. For processors requiring young freezing beans start sowing spring varieties in late April; for more mature crops for canning start sowing in early April. Successional sowings will then give supply of beans of correct maturity for rest of season. Overwintered varieties come to harvest just before vining peas. Sow all winter varieties in first half of October and Spring varieties for picking green in late February and throughout March, according to soil conditions. Seed should be dressed with approved fungicide against pre-emergence damping off.

Plant population and seedrate

Optimum population is 18 plants/m², drilled in rows 450 mm apart and 125 mm between seeds; may be reduced to 14 plants/m² if seed is very dear or on fertile silt or organic soils which are likely to produce extra strong vegetative growth. Use higher density on poorer soils, e.g. sands. Seed varies greatly in size, so it is essential to use formula to adjust seedrate:

$$\text{Seedrate} \atop \text{(kg/ha)}} = \frac{10\,000 \times \text{required plants/m}^2}{\text{No. seeds/kg}}$$
$$\times \frac{100}{\% \text{ germination}}$$
$$\times \frac{100}{100 - \text{field loss}}$$

Field loss usually 5%.

Manuring

Broad beans largely fix their own N and response is only obtained to N fertiliser on N deficient soils; crop very responsive to P and K in deficient situations (*Table 5.26*).

Table 5.26 Manuring of broad beans for market and processing (kg/ha)

N, P or K index	N	P_2O_5	K_2O
0	60	250	250
1	25	200	200
2	Nil	125	125
3	—	75	60
4	—	40	Nil
Over 4	—	Nil	Nil

ADAS figures.

Harvesting

Beans are windrowed with pea cutter, usually wilted for 12–24 h to ease threshing; then vined with mobile viner. Efficient vining and cleaning with minimal delay before processing is essential.

Pests and diseases

Largely similar to those of field beans.

Yield

Canning and freezing 3.0–5.0 tonnes/ha.

Dwarf beans (green or french beans)

Dwarf beans are now a large scale farm crop grown on contract for processing, i.e. freezing, canning or artificial drying. Before introduction of efficient mechanical harvesters they were grown solely as a market garden crop and picked by hand for fresh sale.

Varieties

'Stringless' varieties, slender, smooth skinned and straight, now predominate. Varieties are numerous. Choice of variety, which is a most important factor influencing yield, quality of produce and efficiency of mechanical harvesting, is usually made by processor.

Seed

Seed is extremely fragile; skin is tender and easily cracked and attachment between embryo and cotyledons is slender and easily broken. Avoid rough handling of seed, e.g. dropping bags on hard floor, or germination will be adversely affected.

Climate, soil and cultivations

Dwarf beans are best suited to warmer areas of south and east England; further north only warm sheltered south facing slopes are suitable. Crop is very sensitive to soil conditions and best grown on deep loamy soils with good structure and ample organic matter. Avoid heavy, wet, or overdrained soils. Do not overcompact soil when preparing seedbed. To conserve moisture use minimum cultivations necessary and leave level surface for efficient harvesting. Drilling is delayed until soil temperature has risen to 10 °C and is usually completed in first half of May.

Sowing

Rows 140 mm apart give the highest yields but row width is determined by type of harvester available, many machines working on 450 or 600 mm row widths. Different harvesters are required for narrow rows. Optimal plant densities are 36 and 42 plants/m² for wide and narrow rows respectively, but varied according to cost of seed.

The following formula should be used to calculate seedrate:

$$\text{Seedrate} \atop \text{(kg/ha)}} = \frac{10\,000 \times \text{required plants/m}^2}{\text{No. seeds/kg}}$$
$$\times \frac{100}{\% \text{ germination}}$$
$$\times \frac{100}{100 - \text{field loss}}$$

Field loss usually 10%.

Seed dressed with an approved fungicide is drilled into the soil moisture; 25–40 mm is usually adequate but a depth of 50 m may be necessary in some situations.

Irrigation

Irrigation before flowering has little effect on yield, but water applied during flowering and early pod growth increases yield substantially.

Manuring

Present position is unsatisfactory, as adequate experimental evidence is lacking. Dwarf beans do not as yet appear to have a satisfactory nodule forming bacteria and so cannot fix their own N like peas; crop is therefore highly responsive to N, especially arable soils index 0. There are good responses to P and K on soils index 0–1, but where P and K indices are 2 or over, a 3:2:1 fertiliser appears to be adequate.

However P level can be increased at index 2 to build up soil P reserves (*Table 5.27*). All fertiliser is usually applied to seedbed, but a top dressing of 25 kg/ha N may help if N deficiency develops during growth.

Table 5.27 Manuring of dwarf beans for market and processing (kg/ha)

N, P or K index	N	P_2O_5	K_2O
0	150	300	200
1	100	250	125
2	75	125	60

After ADAS.

Pests and diseases

Aphids (*Aphis* spp.) These cause distortion and stunting of plants and pods; treat with approved aphicide.

Anthracnose (*Colletotrichum lindemuthianum*) Produces angular brown spots in which holes appear, on leaves, and sunken pink bordered brown lesion on pods. Most severe disease attacking dwarf beans. Apply approved systemic fungicide at full flower stage.

Grey mould fungus (*Botrytis cinerea*) Thrives in high humidity attacking dying or damaged tissue, e.g. scratched tips of pods touching soil. To prevent apply approved systemic fungicide at full flower stage of crop.

Halo blight (*Pseudomonas phaseolicola*) Produces brown spots in centre of large yellow spots on leaves, the halo, spots on pods and wilt and death of plants in wet humid weather. Remove affected plants in bags from field, burn and apply copper oxychloride or colloidal copper spray at once and at 10 d intervals until cured.

Yield

Average crop 7.5 tonnes/ha.
Good crop 12.5 tonnes/ha or more.

Peas

Peas are mainly grown in the drier eastern counties, where they make an excellent cereal break crop. Crop is grown for human consumption; also grain and forage for stockfeed.

Types and varieties

Peas for human consumption

Peas for human consumption may be grown for:

Picking green for market Much the least important outlet. Area has declined greatly. Little demand as quality is unreliable, due to flats, maggots, staleness and previous use of early low quality varieties. Production costs are high. Varieties are of 'garden' type with white flowers with round or wrinkled seed; peas with the latter type of seed are of much higher quality. Should be picked while young and tender. Pods on these varieties become ready to pick over a period of about 10 d.

Varieties may be divided into *first earlies*: (a) round seeded types, suitable for sowing in autumn or February but of moderate quality: Clipper, Feltham First, Meteor, (b) wrinkle seeded types less hardy but higher quality; also used for late crops: Kelvedon Wonder, Laxtons Progress, Thomas Laxton. *Second earlies*: Early Onward, Kelvedon Triumph, Little Marvel. *Maincrops*: Kelvedon Monarch, the Lincoln, Onward; varieties with a short haulm 0.3–0.6 m preferred.

Contract growing for vining Shelled green by mobile viners while young and tender and transported rapidly to factory for processing into frozen, canned or dehydrated *garden peas*. High quality of frozen product and absence of need to shell substantially responsible for decline of picked types. Provided it is clean, pea haulm makes useful silage. This type of pea is entirely confined to the eastern counties, as all processing plants are in the east. Processors require minimum blocks of 120 ha for vining. Factories require an even flow of high quality peas for a long period during growing season and consequently contract with numerous growers to exploit climate and soil type with a succession of sowing dates of varieties of differing maturity.

Special varieties of garden type with white flowers and wrinkled seed are used but pods are carried at top of plant and mature simultaneously. For high quality, peas must be harvested at correct stage of maturity, which is decided by means of a *tenderometer*. For freezing a tenderometer reading (tr), 100–105 is usual, but canning peas are harvested more mature, (tr) about 110–115; yield is then slightly higher. The ideal variety is of upright habit with haulm of medium length (about 0.6 m), resistant to diseases such as downy mildew (*Peronospora viciae*) and fusarium wilt (*Fusarium oxysporum*) and herbicides to be used. Pods must mature uniformly and be thin to allow shelling without excessive beating and damage to peas. Peas should not be prone to splitting and be of uniform size, moderately dark but bright green in colour, with good fresh pea flavour. Texture should be smooth, even, skin not tough. Varietal lists are published by Processors and Growers Research Organisation (PGRO), The Research Station, Thornhaugh, Peterborough, together with other technical literature.

Varieties may be divided into: *first earlies* which include earliest peas that can be successfully grown and harvested. Short straw, relatively low yielding but allow early start to vining: Avola, Banff, Sparkle, Sprite. *Second earlies*: Galaxie, Visto. *Early maincrops*:

Scout, Skinado. *Maincrops*: Dark Skinned Perfection, Hurst Green Shaft, Puget. Extra late varieties have been bred to extend pea harvest into August; others to produce higher yields of *petits pois* (small green peas).

Processing peas   Grown to maturity and combine harvested as dry grain. Sold for packeting as dry peas or canned as *processed* peas. These types have white flowers, green or blue seed and are not sweet. They include:

(1) *Marrowfats* (Harrisons Glory type): large seed, slightly flattened and suited to a wide range of soils. Varieties: Maro, Progreta, Sleaford Dryad.
(2) *Large blues*: seed round, smaller and bluish green with very short straw, the highest yields obtained only on fertile soils. Varieties: Rondo, Dik Trom. Allround and Finale are resistant to pea wilt.
(3) *Small blues* (Lincoln or Prussian small blues): seed round, smallest of all, bluish green seed with relatively long straw, suited to lighter soils. Varieties: Polaris, Vedette.
(4) *White peas*: white round seed. A small acreage of variety Birte is grown. Yield similar to Maro.

Peas for livestock feed

Forage peas   Include varieties selected for their ability to produce a high yield of haulm. Widely cultivated in Europe, have been grown commercially in UK since 1975. Flowers variegated, pink, maroon, red or mixed with white, seed mottled, or brown. May be sown alone or with cereals. Suggested as a nurse crop for Italian ryegrass, but may form a dense 'smother' and kill young seeds before harvest under 'growthy' conditions. There are numerous varieties. Early varieties ready for cutting in 12 weeks, late varieties take 15 weeks, giving higher yield in longer growing season.

Peas for grain   Include Maple and Dun types. Maple has purple variegated flowers, mottled seed and is not sweet. Dun is similar but has brown seed. These types may also be used for forage, hay or silage.

Soil and cultivation

Peas suit a wide range of soils provided they are well drained, of good structure and texture, friable, and with depth of mould to allow easy root penetration. Cold wet conditions, raw soil or capping result in failure. Avoid very light sands and heavy wet clay. A good mouldy seedbed is produced by ploughing early, giving maximum weathering by frost; prepare seed bed with minimal working—a single stroke of the harrows is frequently adequate. Use of pre-emergence residual herbicides requires reasonably fine seedbed.

Sowing

Peas for human consumption

Sowing date   Round seeded varieties for *picking green* may be sown in October or February; other sowings are made until end of April, according to variety. Sowing date of *vining peas* is decided by processor, in order to achieve succession of crops for harvesting on planned dates; product deteriorates rapidly if 'over-stood'. *Dried peas* must be sown early for maximum yield. Sow late February or by end of first week in March; thereafter yield falls by 125 kg/ha per week of delay so do not wait for ideal conditions. Early sown peas suffer less from pests and mature earlier, usually in better weather.

Seedrate   It is essential to achieve the right population. Optimum depends on type of pea and varies according to cost of seed and value of produce; high value produce and low cost seed justify higher plant populations. Lower densities can be used where vigorous vegetative growth is likely; for poor growth, increase density. Aim for the following target populations:

	Plants/m²
Vining peas	90
Marrowfats	65
Large blues	70
Small blues	100

Row width should not exceed 200 mm for maximum yield. Closer rows are suitable but do not increase yield further.

Correct seedrate depends on required plant production, seedsize, percentage germination and expected field losses, as determined by field conditions and is obtained from the following formula:

$$\frac{\text{Seedrate}}{\text{(kg/ha)}} = \frac{10\,000 \times \text{required population/m}^2}{\text{No. of seeds/kg}}$$
$$\times \frac{100}{\%\ \text{germination}}$$
$$\times \frac{100}{100 - \text{estimated field loss}}$$

Probable field losses are shown in *Table 5.28*.

Peas for livestock feeding

Sowing date   Forage peas are best sown in late March or early April. July sowings give faster initial growth but poor autumn bulk. Sow peas for grain as early as seedbed conditions permit.

Seedrate   If requisite information is available, use formula for peas for human consumption, otherwise *drill forage peas*: small seeded varieties 150 kg/ha,

Table 5.28 Expected field losses (PGRO) on free draining soils[1]

Sowing time	Expected field losses (%)			
	Vining peas	Marrow fats	Large blues	Small blues
Very early (Feb)	20	15	18	20
Early (Mar)	15	10	12	15
Mid-season (Apr)	10	—	—	—
Late (May–June)	5	—	—	—

[1] For heavy soils add 5% to losses.

large seeded varieties 175 kg/ha *or* 90 kg/ha peas with 60 kg/ha oats or barley; do *not* exceed given rate for cereal *grain*. Drill 125–160 kg/ha according to size of pea grain. Drill both types 30–50 mm deep with corn drill in rows 200 mm apart or less; do *not* broadcast.

Manuring

Peas do not respond to N, which is unnecessary even on poor seedbeds. Rates of application for P_2O_5 and K_2O are shown in *Table 5.29*. Optimal rates of K_2O are higher for placed fertiliser than for broadcasting. Placement close to but not in contact with seed is most efficient but is impracticable with close drilling.

Table 5.29 Manuring of peas for vining, drying and forage (kg/ha)

P or K index	P_2O_5	K_2O		
		Broadcast	Placed	Combine drilled
0	50	150	150	50[1] + 70 broadcast
1	25	40	100	50[1]
2	Nil	Nil	50	50[1]
Over 2	Nil	Nil	Nil	Nil

ADAS figures.
[1] Applies to rows 180–200 mm apart. Cut down proportionately for wider rows to give same rate per row, e.g. 360 mm rows combine drill maximum of 25 kg/ha. Exceeding these amounts may impair germination.

Harvesting

Vining peas

Date of cutting is decided by contracting firm on tenderometer reading. Most peas now harvested by minimum intake viner; previously cut into windrows with pea windrower and shelled in the field by mobile viner; the peas are then rushed to the freezing plant, since they perish rapidly once shelled. Peas must be free from maggots and impurities like poppy heads and mayweed, which cannot be removed, and toxic berries such as black nightshade and bryony. Efficient weed control is thus essential.

Dried peas

Good colour and freedom from defects as these largely determine price; both are much influenced by harvesting technique. Dry packet peas should be dark green. Canners for sale as processed peas of a lighter but even colour.

Conditioning peas in windrow is risky in all but driest areas and years. Wet seasons result in severe discoloration and shedding from constant turning. Peas may be safely left on tripods or fourpoles for some weeks before threshing; colour of peas is retained but few farmers now have adequate labour.

Direct combining is safest. Combine as soon as grain will thresh without damage; usually under 30% MC. Weedy or unevenly ripened crops should first be treated with a desiccant (e.g. diquat) but not before peas on least mature plants are starchy and can be marked with fingernail without splitting. Top pods will be wrinkled, lower pods in parchment stage. Combine 7–14 d after spraying, once all is desiccated. Spraying immature crops results in pod splitting.

When *drying* great care is needed to avoid damage. Do not exceed the following drier temperatures:

	Moisture content of peas	
	< 24%	24% and over
Seed peas	43 °C	38 °C
Consumption peas	49 °C	43 °C

When double drying allow at least 2 d between dryings for MC to even out. For storage in bulk or closely stacked bags dry down to 15% MC for winter storage; 17% MC is adequate for storage up to four weeks; 1% higher MC is permissible for storage in bags fully exposed to air, in bulk ventilated by forced draught or turned frequently.

Forage peas

Cut early varieties at podding, when dry matter yield reaches peak (about 12 weeks) or senescence and leaf loss occur, reducing yield and quality. Late varieties flower and produce pods over a longer period and do not suddenly senesce; date of cutting is less critical. Lodging likely by harvest; avoid setting cutter too low or soil contamination of produce may result.

Pests, diseases and disorders

Pea weevil (*Sitona lineatus*) needs treatment if large numbers appear when plants are small on cloddy seedbeds in slow growing conditions. *Pea aphids* (*Acyrthosiphon pisum*) do severe damage in large numbers; control before build up in warm weather.

Pea moth (*Cydia nigricana*) causes maggoty peas; larvae feed on peas in pod and severely reduce value. Use 'Oecos pheromone' pea moth trapping system to assess need and timing of treatment. Consult ADAS for further information.

Fungus diseases include pea wilt (*Fusarium oxysporum*) and downy mildew (*Peronospora viciae*); sow resistant varieties. Foot rot (*Aschochyta pinodella*) worst on heavy soils. Improve drainage, avoid compaction. Leaf and pod spot (*A.pisi* and *Mycosphaerella pinodes*): sow disease free seed.

Marsh spot caused by deficiency or non availability of manganese on organic or alkaline soils, yellowing around leaf margins and between veins appears. If untreated, brown spots appear in centres of peas, reducing value. Apply 11.4 kg/ha manganese sulphate HV with wetting agent on appearance of symptoms, again if they reappear: to prevent spot formation apply also at full flower and again 7 d later.

Yield

	tonnes/ha
Picking green (market) Early	7.5
Vining	
First early	3 to 6
Second early	3.5 to 6
Early maincrop	4 to 9
Maincrop	3.5 to 7
Dry	
Marrowfat	2 to 5
Small blue	2 to 4
Large blue	2 to 5
Whites	3 to 5
Forage peas (cut green)	
Dry matter	3.5 to 6
Maple and other	2.5 to 3

Lupins

Lupins have been grown at times for green manuring in England and there have been recent trials in production of the protein rich grain when varieties with a low alkaloid content of seed (i.e. sweet lupins) were used. Species extensively tried included white lupin (*Lupinus albus*), large seeds with high protein content and some oil. Grown mainly in USSR and eastern Europe. Narrow leaved lupin (*L. angustifolius*) produces smaller seeds and less oil, mainly grown in Australia. Neither species proved of commercial value in UK and trials were discontinued.

Yellow lupins (*L. luteus*) seeds have a high oil content but plants are initially weak with poor yields of seed. Grown in eastern Europe, not worth cultivation in UK.

LEGUMES GROWN FOR FODDER ONLY

Lucerne

A native of Asia Minor, lucerne is now cultivated world wide; known as alfalfa in the USA. A deep taprooted legume, it is highly drought resistant and fixes large amounts of atmospheric nitrogen to provide heavy yields of protein rich fodder and rich nitrogenous residues; well managed crops yield up to 10.5 tonnes/ha or even more of dry matter. A perennial crop, it normally occupies the land for four years.

Essentially a cutting crop, it has been grown widely for drying and makes excellent silage if it is wilted first. Difficult to make into hay by conventional methods as stem is difficult to dry and leaf shatters readily; well suited to barn hay drying. Grazing only taken on a limited basis. Lucerne requires very different treatment from grass and clover crops. Management is rather specialised and new growers should take advice first. A sun loving plant, it is especially suited to drier warmer summers of southern and eastern counties but withstands hard winters well. May grow satisfactorily in higher rainfall areas but grass is easier to manage. The crop tolerates soil types ranging from light sand to clay but cannot stand poor drainage or acidity, requiring a pH of 6.5–7.0 in the top soil and a subsoil minimum pH of 6.0 to at least 30 cm deep. Soil profile must allow deep root penetration; useless on unfissured rock or where pans exist but excellent on chalky soils.

Varieties and seeds mixtures

The once popular midseason type *Provence* is now superseded by varieties of the early or *Flanders* type, which give the earliest spring growth, longest growing season and highest yield. Variety Europe has consistently given highest yields but where either *verticillium* or bacterial wilts or stem eelworm (*Ditylenchus* spp.) are known to occur, resistant varieties should be chosen, even if of lower yielding capacity. Varieties are listed in Farmers Leaflet No. 4, *Varieties of Herbage Legumes* (NIAB). Lucerne varieties are not mixed together in seeds mixtures.

Lucerne may be grown pure or sown with a companion grass; not grown with clovers. Pure stands are preferable for drying or where grassy weeds are controlled chemically. Otherwise the inclusion of a companion grass helps to suppress weeds, gives variety to the herbage and is said to make hay and silage making easier. Lucerne does not thrive in a dense grass sward so the grass chosen should be a minimum space demander and a similar growth rhythm to lucerne. Cocksfoot is only suitable in close drilled stands on light dry soils or dry areas and requires hard winter grazing to control it. Meadow fescue is preferable for general use. Do not mix with lucerne seed but cross drill to avoid close competition with the young lucerne plants. Drill 13 kg/ha lucerne seed 12 mm deep in rows 100–180 mm apart; cross drill 1.2 kg cocksfoot or 3.5–4.5 kg meadow fescue if required.

Seed inoculation

Lucerne fixes atmospheric nitrogen only if effective

nodules essential for vigorous growth are formed on the roots by the appropriate nodule bacterium *Rhizobium meliloti*; this is usually not present where lucerne has not been grown before. Inoculation of seed with correct nodule bacterium is then essential and may only safely be omitted if field has recently grown lucerne. Methods include a slurry peat inoculum in water applied on the day of sowing; seed is allowed to dry before drilling. Inoculation may also be done by pelleting up to two weeks before drilling.

Establishment and seedling management

Lucerne is best drilled without a cover crop on a superfine firm level seedbed after soil has warmed up and several crops of weed seedlings killed by harrowing, between late April and mid July; midsummer sowings are usually best, as growth is very rapid in warm soil. Avoid early spring sowings, as germination and growth are slow and weed competition severe. Crops may also be undersown in cereals, when a silage crop removed in July is preferable to a grain crop; use reduced seedrates of stiff strawed barley variety, 100 kg/ha.

Young plants each develop initially from a slender single stem and do not form a cover dense enough to suppress vigorous early weed growth, so treatment with a suitable herbicide is normally advisable. If weeds are mown, set cutter *above* growing tips of young lucerne. Allow flowering to begin before removing first crop or allow growth to die back in autumn.

Management of established stands

Allow a large bulk to develop before cutting. First cut, about mid May, should be taken as soon as flower buds appear, subsequent cuts not later than opening of first flowers. Further delay seriously reduces crop digestibility and retards next cut. Cutting too early weakens stand; overfrequent cuts kill it. On strong stands second or third cuts can be taken before recommended stage without damage in favoured areas. If grazed, *under no circumstances* must young regrowth be eaten; always use a back fence close to the grazing.

Autumn management is critical. Rest for eight weeks in autumn from about late August to end of October to allow build up of root reserves. Crops may then be grazed or cut while green or allowed to die back. Companion grasses are grazed heavily in winter, especially cocksfoot, but remove stock before spring growth starts. If needed mechanical treatments loosen a compacted surface but complete while lucerne is dormant.

Herbicides are more reliable for controlling heavy growth of weed grasses. Paraquat is effective for controlling creeping grasses and checking green growth.

Manuring

Check soil pH and apply lime if necessary.

Seedbed dressings (Table 5.30)

This should be worked in prior to sowing. Nitrogen is only worthwhile for grass mixturees or soils of ADAS N index 0. If recent analysis is not available use highest rates of P_2O_5 and K_2O.

Maintenance dressings (Table 5.30)

Lucerne gives no response to fertiliser N once established but demands a liberal supply of K_2O. Deficiency must not be allowed to develop for it is difficult to cure. A crescent of white or yellow spots around tips of leaflets, yellowing of lower leaves and stunting of the plants, often in patches, indicates K_2O deficiency. A single cut if young lucerne, 19 tonnes/ha fresh material, removes some 120 kg/ha K_2O, while a crop of total production 10.5 tonnes/ha dry matter removes 190–225 kg/ha K_2O.

Sainfoin

Also known as St Foin, Cockshead or Holy Grass, sainfoin was once grown extensively on chalk and

Table 5.30 Manuring of lucerne

	Nitrogen index			Phosphorus index				Potassium index			
	0	1	2	0	1	2	Over 2	0	1	2	Over 2
Seedbed—no cover crop pure lucerne	25	—	—								
Seedbed—no cover crop lucerne with companion grass	50	25	—	100	75	50	25	125	75	30	30
Rate of application per cut	—	—	—	100	80	50	40	120	100	60	30

ADAS figures.

limestone soils, especially in the Cotswolds, Hampshire and around Newmarket; now rarely grown. Like lucerne, sainfoin has a deep taproot, is highly drought resistant and leaves valuable N residues for subsequent cereal crops. Digestibility, palatability, protein content and high feeding value, long appreciated, have now been confirmed by modern research. Well made sainfoin hay has been much in demand for race horses: aftermath is invaluable for fattening lambs. Yield of British stocks, about 7500 kg/ha dry matter is well below that obtainable from lucerne; some European stocks of sainfoin appear to do considerably better. Cut for seed, 625–750 kg/ha of seed in husk is obtained.

Sainfoin is very suitable for hay or silage making, with aftermath grazing, on light dry soils, provided they are well supplied with lime and have pH 6.5. Primarily suited to well drained soils in warmer dry climates of the south, sainfoin fails utterly on cold wet soils.

Varieties

There are two main types, *Common* and *Giant*. Common sainfoin is truly perennial and once remained down 15–20 years, later shortened to about four years according to weed incidence. Flowering in late May or June gives a single cut of hay but remains prostrate thereafter and is best suited to aftermath grazing. Giant sainfoin establishes quickly but only persists for about two years, giving two cuts annually.

Seedbed, sowing and seedrate

Sainfoin replaces red clover in the rotation and is undersown in spring barley. Clean land and fine firm seedbed are essential. Drill at right angles across the rows of barley in rows 100–180 mm apart and 12 mm deep from March to early May; deep drilling leads to poor establishment; ground must be rolled after drilling.

Seed may be obtained in husk (unmilled) or husk removed (milled); the latter contains fewer hard seeds, giving better germination and is free from empty husks and weed seeds. Only fresh coloured light brown plump seed of high germination should be bought; black shrivelled seed is old or harvested badly.

Seedrate 65 kg/ha for milled and 125 kg/ha for seed in husk. Giant sainfoin is sown as a pure stand while common may be sown pure or mixed with grass, the latter giving higher yields and less weed. Either 7 kg/ha meadow fescue or 3.5 kg/ha timothy may be included, each with 1 kg/ha giant white clover. Cocksfoot at 1 kg/ha may be considered on very poor soils but is very aggressive unless heavily grazed in winter.

Manuring

Little UK experimental evidence exists for recommendations. Overseas work indicates that sainfoin is able to utilise P_2O_5 of very low availability and does not respond to applications of either P_2O_5 or K_2O. Up to 125 kg/ha N given in first harvest year has been shown to increase dry matter yield. As a precaution, give 60 kg/ha each of P_2O_5 and K_2O in seedbed and an annual maintenance dressing of 40 kg of each. Where the sward contains substantial proportions of grass or clover, higher applications of K_2O appear desirable. A little N early in the life of the plant increases nodulation.

Management and conservation

Cut at flowering bud stage or quality is seriously reduced. Like lucerne, sainfoin hay requires careful handling in later stages and is well suited to 'barn drying'. Crimping immediately after cutting bruises the fleshy stems and speeds up drying. Start grazing *before* flower buds appear. Quick defoliation is best, as in folding; use a back fence and avoid overgrazing. Allow adequate autumn regrowth for plants to build up carbohydrate reserves in roots.

Vetches or tares

Winter and spring types are listed but little is known of differences between them; winter varieties are hardier. Winter varieties are sown in September or early October, spring varieties from February to April, using 190 kg/ha of seed for pure stands. Seedbed and manuring are as for peas. Winter varieties are mixed with rye for early spring sheep feed or oats or beans and oats for silage, including 35–70 kg/ha in mixture. Of less importance than previously since the seed is expensive, dry matter production per unit area is less than well manured cereals. Vetches grown for silage rot in the base if left too long in wet weather. However, they produce heavy yields of green protein rich material. Grown alone, vetches produce 1500–2250 kg/ha of seed.

Root and bulb crops

This group includes:

(1) *Grown primarily for human consumption, direct or processed*: carrots, onions (dry bulb), potatoes, sugar beet.

(2) *Grown for livestock feeding*: fodder beet, mangels.

Carrots

Carrots are a specialised crop, grown on contract for processing, i.e. canning small whole or sliced, dicing, freezing or dehydration or grown for market and prepacking as earlies or maincrops. Only clean straight undamaged roots of required diameter and type are suitable; reject the rest and feed to livestock. Crops are usually washed and must have deep orange flesh and core colour, be free from green shoulders, be well graded and of bright attractive appearance.

Types and varieties

Varieties may be grouped into the following types:

Amsterdam Forcing

Early maturing small tops, roots small to medium size, slender cylindrical stump rooted. Grown for early 'bunching' market and prepacks.

Autumn King ('Flakkee' type)

Late maturity, large vigorous tops, roots very large, tapering stump rooted. Grown mainly for late-season fresh market and dicing.

Berlicum ('Berlikum')

Rather late maturity, medium foliage, roots large, cylindrical, stump rooted. Grown particularly for fresh market and prepacks, quality good. Also slicing and dehydration.

Chantenay

Late maturity, medium foliage, roots medium size, conical stump rooted, good core and flesh colour. Widely grown for fresh market and processing. Crop often graded: small roots canned whole, medium roots sliced or fresh market and large roots diced for processing or for fresh market.

Nantes

Medium maturity, medium foliage and root size, cylindrical stump rooted shape. Widely grown for prepacks, also market and canning.

The latest recommended varieties are listed in Vegetable Growers Leaflet No. 4, *Varieties of Maincrop Carrots* (NIAB).

Soil

Shape is very important; misshapen roots go for stockfeed. Soils must be of such a texture and structure as to allow easy root penetration and even unrestricted root expansion in lateral and vertical directions, to ensure good shape and easy harvesting. Best carrot soils are well drained, deep, stone free sands or light and loamy peats, having minimum water holding capacity of 38 mm/300 mm topsoil and overlying moisture retentive subsoil. Avoid soils with high silt or clay contents or overlying gravel subsoil.

Irrigation

Carrots can be grown without irrigation on sandy soils in years with normal rainfall. If irrigating wait until four-leaf stage and then apply water at 25 mm soil moisture deficit on sand overlying sand, if drying out. If large deficit is allowed to build up, subsequent irrigation encourages splitting.

Rotation

Carrots may follow cereals but ensure that soil is free from couch grass (*A. repens et al.*) and other perennial weeds. Stubble clean and/or apply recommended herbicide pre-sowing. Frequent cropping with carrots leads to a build up of carrot cyst eelworm (*Heterodera carotae*) and violet root rot (*Helicobasidium purpurea*) in the soil, which can only be controlled by rotational means; ensure a minimum of five years between carrot crops, preferably longer.

Manuring

Ideal pH is 5.8 on peats and 6.5 on sands. Avoid overliming, as crop is sensitive to deficiencies of boron, manganese and copper on sands and manganese and copper on peats. (Treat respectively with: Bo—incorporate 22 kg/ha borax or 11 kg/ha solubor pre-sowing, boron index 0–1; Mn—foliar spray 9 kg/ha manganese sulphate; Cu foliar spray 2.2 kg/ha copper oxide or oxychloride or incorporate 60 kg/ha copper sulphate pre-sowing.)

Response to N is minimal and N is only applied on deficient soils (*Table 5.31*). Crop responds well to P and K and on sandy soils; 400 kg/ha salt (150 kg/ha Na) should be ploughed in or worked deeply into the soil before drilling. All fertiliser should be applied and worked in at least one month before sowing. Use cage or double wheels and wheel track eliminators.

Table 5.31 Manuring of carrots (kg/ha)

N, P or K index	Maincrop and processing				Early bunching		
	N		P_2O_5	K_2O	N	P_2O_5	K_2O
	Fen peats	Other soils	All soils				
0	Nil	60	300	200	60	400	200
1	Nil	25	250	125	25	300	125
2	Nil	Nil	200	100	Nil	250	100
3	—	—	125	Nil	—	150	—
4	—	—	60	Nil	—	125	—

Seedbed preparation and cultural methods

Plough at any time during winter but complete by end of February. Subsoil previously if pan exists. Prepare a fine firm level, clod-free seedbed. Good carrot soils work freely and seedbed preparation is easy but take care to avoid moisture loss by overworking or badly timed cultivations. Avoid compaction by using double or cage wheels where feasible with minimal number of passes. Chemical weed control is essential for all crops (*see* Chapter 8).

Many row arrangements and widths may be used but correct choice depends on type and width of harvester, grade of carrot required and soil texture. Systems vary from single, double or triple rows on 380–510 mm centres to scatter rows and multi-row beds with at least 460–510 mm for wheel spacings between them; width of bed is determined by width of share on the harvester. A wide range of row widths and arrangements is possible within the bed. Particular advantages of the bed system are the elimination of soil compaction and improved yields from higher plant densities at close row spacings. Choice of row width within beds is limited by the proximity it is possible to achieve between drill units; double or treble scatter rows from each unit are popular solutions.

Plant population and drilling

Within a given variety, size of produce is decided jointly by plant density and sowing date; size of root is also affected by water supply, (i.e. timing and amount of rainfall and irrigation) and soil fertility. It is thus impossible to do more than predict that a particular combination of sowing date and density is *likely* to give a high proportion of roots within the required grade.

As a general guide aim for 325–375 plants/m² for canners, prepacks and market and 65–85 plants/m² for large carrots for slicing or dicing. Precise density will be varied to suit variety and major grade required. Crops are frequently graded into several classes for market and processing.

Sowing date

At 325–375 plants/m² maximal yield and proportion of canning size occurs some 16–20 weeks after sowing; thus, for small carrots, sow in March, April, May and early June for harvesting end of July, end of August, end of September and October and later, respectively. Irrigation should be available for crops sown on sands after late May; complete all sowings by 20 June at latest. Crops grown at low densities for large roots are best sown in April and May.

Drilling

Drill as shallowly as possible but place seed in moist soil at an even depth. Seed must be dressed; it is *essential* at drilling to apply a recommended insecticide to the soil to protect against first generation carrot fly (*Psila rosae*).

Seedrate

This must be adjusted to obtain plant density required. Rate for a given density depends on seed size, laboratory germination and field conditions. Use formula:

$$\text{Seedrate (kg/ha)} = \frac{10 \times \text{required population/m}^2}{\text{No. seed/10 g } (10^3) \times \text{field factor} \times \% \text{ laboratory germination}}$$

Numbers of seeds/10 g varies from 5500 to 16 000. *Field factor*, which takes account of likely establishment in varying field conditions, should be applied as follows: cold soil and poor tilth 0.5; average conditions 0.6; good conditions 0.7; ideal conditions 0.8.

Pest control

Carrot fly (*Psila rosae*)

Presents the main problem. First generation must be controlled by soil application of persistent approved insecticide at drilling. Crops to be lifted *after* end of

September *must* be treated for control of second generation carrot fly, according to manufacturers instructions. Untreated crops are almost invariably heavily damaged and unmarketable. It is also necessary to control *carrot willow aphid* (*Cavariella aegopodii*) with approved insecticides.

Harvesting

Carrots are harvested with a top lifting harvester or an elevator digger type. With the former the roots are loosened with a small share and twin spring-loaded belts grip the tops and lift the roots, which are topped on the machine. It gives high rates of work, and soil, stones and tops are all left behind in the field. The machines are limited to row widths and cannot be used after tops have died down.

With elevator digger types it is generally necessary to remove tops before harvesting begins. They can operate after the tops have died down. Elevator diggers are also suited to a wider range of growing systems but dirt and stones may be carried over with the roots unless sorted on the machine. System and row width must be selected to suit the harvester.

Storage

Canning carrots are generally cleared early. Carrots may be stored in the field by the following methods:

(1) Earthing over: the rows, minimum width 510 mm are covered with at least 150 mm of earth by plough (or ridger) but this is hardly practicable in wide beds. Used for crops lifted from January onwards.
(2) Strawing: straw may be laid loose 300 mm deep in late November to early December. Very costly; only justified with low value straw and high crop returns.
(3) Clamping in the field: carrots store satisfactorily in clamps till March.
(4) In mildest areas carrots may be left as grown in field but there is serious risk of total loss in hard winters.

Controlled temperature storage is only justified if higher prices are obtained to cover the extra cost. Optimum storage temperature is 1 °C at 95% relative humidity.

Yield

Yield varies greatly according to soil, weather and other variables.

Average: mineral soils 40 tonnes/ha, peaty soils 48 tonnes/ha.

Some high density systems can produce 50–100 tonnes/ha; exceptional crops can produce yields in excess of 100 tonnes/ha.

DRY BULB ONIONS

Onions are grown as (a) *salad* or *spring* onions, which are specialist crops, usually grown on a market garden scale to provide a succession for bunching, and (b) *dry bulb onions*. These may be grown for ware for fresh household use or by caterers (small ware bulbs 25–40 mm diameter; other ware 40 mm and over. Caterers prefer bulbs over 45 mm), for processing for soups, canning and dehydration or for pickling when bulbs under 25 mm are required.

Good appearance and quality of bulbs is essential to command top prices and compete with high quality imports, especially Spanish. Skin bright golden brown, may vary from straw colour to brown, *must not* be stained. Necks thin and bone dry; bull necked onions must be removed as they will not dry. Shape varies from flat to globe, according to variety; globes fetch best prices.

Types and varieties

Onions for dry bulb production are classified as

(1) *autumn sown*, which stand the winter and are harvested in June and July, and
(2) *spring sown*, harvested in September.

Both open pollinated (normal pollination between selected plants) and F_1 hybrids are available.

Varieties for autumn sowing

Japanese varieties Only bolting resistant varieties mainly of Japanese type or bred from these in Europe, are suitable. Spring varieties must never be sown in autumn as they invariably bolt. Open pollinated and F_1 hybrid varieties are available. Maturing in June and July, bulbs are generally rather flat or irregular.

Appearance is much inferior to Rijnsburger types and bulbs do not keep well; crop should be sold by 31 August at latest. Valuable for earliest crops.

Varieties for spring sowing

Rijnsburger varieties Globe shaped with skin colour varying from pale straw to brown. Good keepers and croppers. Favoured by large scale growers, e.g. Robusta, Balstora.

North European F_1 hybrid varieties Globe shaped, appearance very similar to Rijnsburger varieties from which they are usually bred, but more uniform and usually higher yielding, maturity mainly midseason. Good keeping quality, e.g. Hyper, Hygro Hydeal. Over 90% of UK spring grown crop is drilled with Rijnsburger and North European F_1 hybrids.

North American F_1 hybrid varieties Early maturing but low yielding. Bulbs high shouldered with dark copper coloured skins, e.g. Elba Globe, Granada.

Other UK varieties   Generally outclassed by Rijnsburger and North European hybrids, which are superior in yield and far better keeping quality, e.g. Ailsa Craig, Bedfordshire Champion.

Pickling varieties   Are sown thickly to produce small onions; they include brown pickling varieties and smaller white skinned varieties, e.g. Barletta and Paris silverskin.

Climate and distribution

Warm sunny climates are best to promote quality of produce and easy harvesting. Mainly grown in eastern counties.

Soil and cultivations

Choose only best soils on the farm, which must be well drained, deep, free working and not subject to capping after heavy rain. Select moisture retentive soils holding more than 38 mm/300 mm depth. Avoid stony and heavy soils and those likely to dry out. Onions are very sensitive to acidity; soil pH should be between 6.3 and 7.0 on mineral soils and 5.5 to 7.0 on peaty soils. Best soils include well drained silts, brick earths and medium loams. Peats are particularly suitable. Sandy soils are used extensively and are especially suitable for autumn sown crops as they are very free draining; yields are lower with spring sown crops.

   Seedbed must be fine, firm, level and clod free. Plough in autumn to allow weathering and settling. To avoid compaction during seedbed preparation use cage or double wheels, with minimum number of passes. Bed system of cultivation is highly suitable. Prepare seedbed on day of drilling.

Rotation

Previous crop must be cleared in ample time for seedbed preparation for autumn sown onions and for autumn ploughing for spring sown crops. Onions and leeks should never be grown closer than one year in six as both are susceptible to white rot (*Sclerotium cepivorum*), which is cheaply controlled by rotational means; crops are also attacked by stem eelworm (*Ditylenchus dipsaci*) whose wide range of host crops includes peas, beans, clover and other legumes, oats, carrots, parsnips and sugar beet.

   Brassicas, potatoes, wheat and barley are not affected and are suitable for rotations which include onions.

Manuring

Onions are highly responsive to P and K, but response to N is usually poor (*Table 5.32*). Keeping quality is improved by K but reduced by N. Autumn sown

Table 5.32 Manuring of dry bulb onions (kg/ha)

N, P or K index	N Autumn sown		N Spring sown		P_2O_5 All soils	K_2O All soils
	Fen peat	Other soils	Fen peat	Other soils		
0	50	100	60	100	300	300
1	25	80	30	80	250	250
2	Nil	60	Nil	60	200	200
3	—	—	—	—	125	125
4	—	—	—	—	60	60

ADAS figures.

onions require top dressing in spring; apply 100 kg/ha N (50 kg/ha N on fen peats) in late February. After exceptionally wet winters top dressings of 150 kg/ha N may be necessary (100 kg/ha N on fen peats) but split dressing between January and normal time. Spring sown crops may require a top dressing of 50 kg/ha N on mineral soils. With heavy applications of K, part should be applied before ploughing. Avoid overliming peaty soils or manganese deficiency is likely. If lower reaches of soil are acid, plough in some lime.

Plant population seedrate and drilling

Total yield of bulbs increases with plant density but size of individual bulbs is reduced. Optimum density depends on size of onion required: *Ware* allow 70–80 plants/m² (range 65–85 plants/m²); *picklers* allow 320 plants/m².

   When calculating seedrate, seed size (range 2400–2900 seeds/10 g), percentage germination and field factor (0.5 cold soil, poor tilth; 0.6 fair conditions; 0.7 good conditions; 0.8 ideal conditions) must be taken into account.

Seedrate (kg/ha)
$$= \frac{10 \times \text{required population/m}^2}{\text{No. seed/10 g } (10^3) \times \text{field factor} \times \% \text{ laboratory germination}}$$

No. of plants/m of row
$$= \frac{\text{mean row spacing (mm)} \times \text{plant population/m}^2}{10 \times \text{field factor} \times \% \text{ laboratory germination}}$$

Row width and arrangement

This must fit share width of harvester to be used. Ideal mean row width is 300 mm, using single rows or paired rows 50–75 mm apart; the latter can only be obtained by certain makes of drill. Row width should never exceed 450 mm for ware and 350 mm for pickling onions. Beds of four rows 300 mm apart and 1500 mm wide have proved satisfactory. Drill into soil moisture, just sufficient to cover seed, 12–25 mm deep; autumn sowings may require slightly deeper sowing to contact soil moisture.

Irrigation

Autumn sown crops may require watering at sowing time to ensure rapid germination, as timing is critical for success. Otherwise it is preferable to grow onions on soils where irrigation is not needed, as uneven application results in uneven ripening. Irrigation is usually only required for spring sown crops on sandy soils in a hot dry summer: apply 2.5 cm in the post crook stage *before* mid July. Multiseeded peat block crops require irrigation immediately after planting out.

Methods of culture and sowing dates

Onions may be grown commercially in three distinct ways:

(1) *overwintered crops*, sown in August,
(2) *spring sown crops* and
(3) *multi-seeded peat block onions*.

Onion 'sets' are of little commercial importance to growers.

Overwintered crops

Use only appropriate varieties. Sowing date is critical, as it is essential to obtain a strong plant to enter winter, with several lateral roots to prevent frost lift, without bolting. Sowing too early causes bolting, while late sowing results in weak plants and serious losses. Sow 9–15 August in north England, 15–31 August in south England. A really free draining site is essential, as is efficient chemical weed control. Quick marketing is important as potential quality of Japanese types does not justify expensive conditioning techniques. Improved varieties are available.

Spring sown crops

The major part of the UK onion area is of this type. Sow as soon as a fine seedbed can be obtained in late February or early March; if not, it is better to delay until conditions are suitable but complete as early as possible. Sow by mid April at latest, as delay results in serious loss of yield. Crop must be harvested as early as possible in September.

Multi-seeded peat blocks

All too frequently soil conditions do not permit early spring sowing, with resulting loss of yield or even of crop and lower quality; problem is particularly serious in north England. With multi-seeded peat blocks a crop is assured, with larger bulbs, higher yield and improved quality; also harvesting occurs in August, when good conditions are more likely. Rijnsburger types and hybrids, of better quality and appearance than Japanese varieties, should be used. Problem is greatly increased costs, but these can be justified by much higher yields and prices obtainable with crops harvested in late summer. Multi-seeded peat blocks are likely to be less attractive once high quality overwintered varieties are bred according to circumstances.

Method is not critical. Allow six to eight seeds per block, covering lightly with silversand in gentle heat (18 °C) using cubed peat blocks; sow 38–40 mm blocks late January; 27 mm blocks mid February. Once germinated reduce temperature to 15 °C for four weeks, then withdraw heat and harden off. After six weeks NK feed at weekly intervals, soaking blocks with solution of 15 g potassium nitrate + 15 g ammonium nitrate in 50 ℓ of water.

Plant out end March to early April in beds, 300 mm between rows and 230 or 380 mm between blocks for small prepack and large ware onions respectively. Irrigate immediately and give 60 kg/ha N within one to two weeks of planting at latest. Method gives widest choice of herbicide for weed control. Crop ready for harvest in August.

Pre-harvest treatments

Apply sprout suppressant, maleic hydrazide, while tops are still erect; it is not absorbed once tops have fallen over. Effect is to kill sprouts when dormancy breaks, usually in March but also in wet autumns. Use foliar desiccant only as a last resort and then only in conjunction with sprout suppressant.

Harvesting

Peak yields are attained in first half of September. Harvest as soon as tops have gone down. If too early, yield is lost; if too late quality suffers and skins split.

Traditional methods

Crop is undercut and windrowed as soon as tops go down and then left to dry in field for a maximum of 7–10 d in dry weather. Crop may be loaded into drying room earlier if weather is difficult, but ensure adequate airflow or crop will be ruined. Wet weather can result in skin discoloration and also root growth in crops untreated with sprout suppressant. Lift and load into store or sell on completion of drying period; stack up to 3 m deep, not more or crop will be damaged by compression.

Do not lift while dew is on crop or excessive soil will be brought into store. Remove soil and small, bull-necked or diseased onions before storage.

Direct harvesting methods

No field drying period is allowed. As soon as tops have gone down they are removed by flail machine, leaving 80 mm length of neck. Some 3 h are allowed for stem debris to dry off. Crop is then lifted mechanically into trailers and put into store up to 3 m deep on same day that tops are removed. Surface moisture must be removed within 3 d of harvesting. Direct harvesting results in improved skin colour and retention, but choose varieties with retentive skins and long keeping characteristics.

Drying

Process is carried out in two stages.

Removal of external moisture

Start blowing air heated to 10–15 °C above ambient as soon as store loading begins. Ensure rapid uniform continuous drying for 3 d or longer. Even a short period of wetness results in discoloration and moulds. Stage one is complete when onions on top of heap are dry. Skins should then be bright golden brown.

Removal of neck moisture

Blow intermittently until *all* other batches have completed stage one. Then blow continuously at reduced airflow, using heat if required. Stage two is complete when necks of onions on top are raffia dry, usually middle or even end of October. Do not cool until necks are quite dry, then reduce temperature as low as possible without freezing.

Normal ventilated storage may be used until end of March. Cold air is blown at night, using differential thermostat, when ambient is 3 °C below store. Frost guard set at − 2 °C. *Refrigerated storage* is necessary for storing onions from April to June.

Diseases

Neck rot (Botrytis allii)

This can cause serious losses in stored onions if not controlled. Sow only seed treated with recommended fungicide. Previous onion crops and dumps are a serious source of infection, so grow new crops well away from these areas. Handle onions carefully at harvest.

White rot (Sclerotium cepivorum)

Disease is soil-borne and best controlled by rotational means (*see* Rotation); also by fungicidal treatment. Less important diseases include *downy mildew (Peronospora destructor)* and *leaf spot (Botrytis squamosa; B.byssoidea; B.cinerea)* which are controlled by appropriate protective spraying.

Pests

Stem eelworm (*Ditylenchus dipsaci*) is the most serious pest of onions (*see* Rotation). Onion fly (*Delia antiqua*) is much less serious and can be controlled with an appropriate seed dressing.

Yield

	tonnes/ha
Mineral soils	
Poor	30 or lower
Very good	40
Peat	37
Multiseeded blocks	40–45

POTATOES

Potatoes are grown for many purposes, each market having its own requirements and needing a different type of tuber. Hence there are several distinct types of production.

Maincrop 'ware' or table production

By far the largest outlet, requiring heavy yield of saleable, medium to large sized potatoes; larger sizes preferred by chip fryers. Crop is lifted at full maturity and earliness is unimportant, provided crop is ready for lifting by early October. Tubers for pre-packing must be free from blemishes or losses in grading are heavy. Local markets and demands vary.

Earlies

Tubers are lifted in immature state to provide a succession of new potatoes retaining firm waxy or soapy texture after cooking through the summer. First, then second early varieties are used. Earliness is of prime importance for earliest crops; they are planted very early and lifted as soon as marketable, Earliest areas like Cornwall and Pembroke start lifting about 20 May. Price falls sharply as other districts start lifting and yield increases. Later planted first and second earlies, giving higher yields, continue to provide a supply of new potatoes until maincrops are ready.

Potatoes for processing

Generally grown on contract and used to manufacture wide range of products. *Potato crisps* require uniform, medium-sized tubers with shallow eyes, dry matter content over 20% and reducing sugar content below 0.25%. *Chips or french fry* require medium to large tubers, shallow eyes, regular shape, white creamy flesh, dry matter over 20% and reducing sugar under 0.4%. With *dehydrated* potato for instant mash, maximum yield of dry matter is needed at minimum cost. Requirements are large, uniform, mature tubers with minimal peeling loss, over 20% dry matter, reducing sugars under 0.5%, white creamy flesh, free from taints and after-cooking discoloration, the starch granules and cells separating readily during processing. *Canned new potatoes* should give high yield of small uniform tubers 18–38 mm with creamy flesh colour, dry matter under 19%, remaining waxy and solid after steam peeling and canning, with clear, sludge-free brine. *Canned potatoes* for dicing need large blocky tubers over 50 mm, white flesh, waxy texture, below 10% dry matter; cubes must retain shape after cutting.

Seed production

Objective is to grow high yield of small tubers within range 32 to 60 mm, producing crops true to type, free from undesirable variations, other varieties, virus and certain other diseases. Seed production is a specialised job mainly carried out in cool areas of low aphid incidence such as north and east Scotland, Northern Ireland, designated areas in north, north west and west England and Wales. Many early growers produce their own seed for planting earliest crops.

General quality requirements for cooking potatoes

Smooth even tubers with shallow eyes for minimal wastage and labour in peeling, without coarse or misshapen tubers. Damage from second growth (glassy, cracked tubers), greening, pests (slugs, wireworm), disease (blight, *Phytophthora infestans*, scab, *Streptomyces scabies*) must be absent, whilst mechanical damage during lifting, storing, grading and transportation must be minimal. Cooking quality is partly a varietal feature but is substantially modified by soil type, rainfall and soil moisture, manuring and storage temperature. Well matured tubers lifted in early autumn before soil has cooled down and become wet give best quality; wounds also heal more readily.

Boiling tubers

These must not disintegrate during boiling nor discolour or blacken after cooking. Maincrop tubers may be waxy or floury after cooking.

Roasting or frying

Product must be golden brown outside, not dark brown or black due to presence of reducing sugars. Chips should be crisp, floury inside and absorb minimal fat.

Varieties

Essential varietal features include ability to produce commercial yield, saleability, good cooking quality and absence of deep eyes and poor shape, with adequate resistance or at least lack of extreme susceptibility to diseases (e.g. late blight, *P. infestans*, dry rot, *Fusarium caerulum*). Varieties differ much more than in many other crops in suitability to cultural and climatic conditions and response to management.

Particular attention is needed for varietal requirements, including storage temperature, sprouting regime, seed rate and spacing.

Varieties are classified according to maturity group.

First earlies

Produce the earliest crop of new potatoes. They only produce dwarf foliage, thus give lower yields than maincrop if left to maturity.

Second earlies

Are slower bulking than first earlies but give high yields of new potatoes for follow on digs.

Maincrop varieties

Tubers bulk later than previous types, so larger foliage is built up before growth is slowed down by tuber production, hence heavier yields. Except for a few crops for very late new potatoes, these varieties are only lifted when mature in October.

Recommended varieties are listed and described in Farmers Leaflet No. 3, *Recommended Varieties of Potatoes* (NIAB).

Seed certification

It is only possible to maintain satisfactory yields of potatoes when seed is free from severe virus and certain other pests and diseases, e.g. potato cyst nematode (*Globodera rostochiensi* and *G. pallida*) and blackleg (*Erwinia carotovora* var. *atroseptica*). It is essential that seed is of variety stated, free from rogues, undesirable variations such as bolters and semi-bolters, producing a crop true to type. Certified seed is produced under supervision and control of Agricultural Departments in the UK. Certified seed is only produced from approved parent material and growing crop inspected by officers of the appropriate certifying authority.

Grades of seed

Certificates are issued by the Agricultural Departments in the UK, i.e. *England and Wales*—Ministry of Agriculture, Fisheries and Food (MAFF); *Scotland*—Department of Agriculture (DAS); and *Northern Ireland*—Minstry of Agriculture (MANI). Certificates refer only to health and purity of crops at time of inspection. Grades fall broadly into two groups

(1) basic seed intended primarily for further multiplication as seed; and
(2) seed intended mainly for production of ware crops.

England and Wales

Basic seed intended mainly for multiplication as seed.

VTSC (Virus tested stem cuttings) an elite stock raised initially in aphid proof greenhouses.

FS (Foundation seed) derived from VTSC or FS stocks. A number, 1–4 indicates year of multiplication since VTSC.

SS (Stock seed) for producing further seed crops.

AA Prime quality, commercial seed, suitable for producing once grown seed or ware crops.

Certified seed intended for production of ware crops only. Unsuitable for once grown seed.

CC   Healthy commercial seed.

Scotland (Scot)

Basic seed intended mainly for multiplication as seed.

Virus Tested Stem Cuttings

FS Foundation stock
AA1 (*cf.* SS but SS no longer issued in Scotland).
AA as England and Wales
CC not issued.

Northern Ireland (Nor. Ir.)

Basic seed intended mainly for multiplication as seed.

Virus Tested Stem Cuttings

Special Stock Seed
Foundation Stock
Stock Seed
Grade A (*cf.* AA in England and Wales)
CC not issued.

The letters NI after the grade indicate that variety is non-immune to wart disease (*Synchitrium endobioticum*). Particulars of certificates will be supplied by Department of appropriate country on application.

Source of seed

Growers may

(1) purchase total requirement every year of basic or certified seed from seed producers, or
(2) they may purchase enough basic seed every year to grow on to produce enough once-grown seed to plant the whole crop the following year. Purchased seed is grown in isolation, given appropriate insecticidal treatment and rogued;
(3) seed may be taken from the ware crop and new seed purchased every other year or as required. Practice varies according to type of production.

Earlies

'Once-grown' seed produces an early crop for lifting 10–14 d earlier than one grown from imported seed. Early growers produce their own once-grown seed from grade SS or AA1 imported seed, or purchased seed from a grower producing seed of identical and predictable performance, which involves controlling planting date of the seed crop.

Maincrops

In areas of high aphid incidence it may be necessary to purchase new certified seed every year. Elsewhere, stocks remain healthy for longer periods, depending on the number of aphids, date of migration, and freedom from virus of surrounding crops. Inexperienced growers should change stocks annually, but experienced growers will know how long to keep stock before renewal.

Seed size

Seed size is of less importance than seed rate. As seed size increases, the number of mainstems (or individual plants) produced from the seed tubers increases, so large seed produces more stems and tubers than small seed, but average tuber size is smaller. This effect is modified by widening spacing as size of seed increases, up to certain limits, i.e. by modifying seed rate. Small seed gives better yields than large seed planted at the same seed rate but optimum size is determined by type of production and variety.

First earlies

Small seed under 32 mm with less food reserves is slower bulking, tubers are lifted later and show poor recovery from frosting. Seed must be size graded and planted at intervals from 200 to 350 mm according to size; do not use chats.

Maincrops

Modern planters work better with closely graded seed. If not already size graded, this should be done on arrival. Seed graded 32×60 mm should be split into two or sometimes three grades according to size, distribution and shape of consignment. All sizes between 32 and 60 mm are satisfactory if planted at appropriate spacing and similar seedrate. Tubers over 50 mm are less efficient especially with varieties or on sites where numerous small tubers are formed. Chats (i.e. under 32 mm) from healthy stock are perfectly satisfactory if planted at appropriate seed rates.

Seed production

Chats under 32 mm are unsuitable for producing seed potatoes; otherwise, all sizes may be used if planted at appropriate seed rates. Seed producers in Scotland plant ware tubers over 50–60 mm, 230–300 mm apart, as they are worth less than seed size and give a good yield of seed. Chats may be used to produce ware for re-planting for seed.

Seed rate and spacing

Seed rate

This is determined by row width, spacing within row and average weight per set of number of tubers per 50 kg. Optimum seed rate varies greatly, depending on type of production (i.e. size of produce required), variety, conditions of growth (soil type and fertility, water supply), and cost of seed. To increase tuber numbers and reduce size, increase seed rate with larger seed and/or closer spacing; to reduce tuber numbers and increase size reduce seed rate by using smaller seed and/or wider spacing. Management of seed to produce larger number of sprouts per set also increases stem and tuber numbers and *vice versa*. Seed rates required are shown in *Table 5.33* and spacings are shown in *Table 5.34*.

Row width

In high rainfall areas and where tubers are lifted immature (i.e. first earlies) close rows 540–660 mm give more uniform distribution and higher yields: Maincrops have traditionally been grown in 710 mm rows but wider rows, 760, 800 or 900 mm allow more room for tractor tyres and more soil for a ridge, resulting in similar yields with few green tubers. Canning potatoes and seed are grown in 660–760 mm rows.

Seed treatment

Pre-sprouting or chitting

Seed is placed in chitting trays, four trays per 50 kg for earlies, 3.5 trays for maincrops; these are stacked

Table 5.33 Guide to seedrates for potatoes

Type of production	Seedrate (kg/ha)	Remarks
Maincrop		
King Edward and Red King	1500–2300	Very small seed 1600/50 kg plant 1500 kg/ha. Larger seeds require higher seed rate. On sites which form large numbers of small tubers reduce seedrate.
White varieties producing large tubers and Desirée	3200–4500	These varieties produce fewer stems and tubers; too low a seedrate, especially on fertile sites, results in large coarse unsalable tubers. Use high seedrate with very fertile conditions.
Earlies	3200–4500	Single sprouted seed required for earliest crops; multiple sprouted seed satisfactory for later lifting and second earlies. First early seed should be tightly graded and planted 200–360 mm apart according to size.
Canning	Optimum 7500 Economic range 2500–5000	High stem density 160–220/m² required. Use multisprouted seed spaced 150–230 mm apart according to size. Size grade seed. Optimum seedrate depends on cost of seed and value of produce.
Seed production		
Home produced seed, e.g. Scotland	5000–7500	High seed rates are fully justified where cheap seed is available; home produced ware size, which is of lower value than seed, is planted.
Purchased seed	3800–5000	High seed rates are very expensive if seed is bought at high price, e.g. early growers producing once grown seed. Use lower rate and multisprouted seed.

NB When growing new variety check on suitable seedrate—there are large differences.

Table 5.34 Spacing of potatoes according to seed rate

	Required seed rate												
kg/ha (× 100)	17.5	20.0	22.5	25.0	27.5	30.0	32.5	35.0	37.5	40.0	45.0	50.0	75.0
No. of sets/ 50 kg	Spacing between sets (mm × 10) in 760 mm rows												
400	94	84	74	66	61	56	51	48	43	41	37	33	21
500	76	66	58	53	48	43	41	38	35	33	29	26	17
600	63	56	48	43	41	38	33	30	30	28	24	21	15
700	53	48	43	38	35	30	28	28	25	23	21	19	12
800	48	41	38	33	30	28	25	23	23	20	19	16	11
900	43	38	33	30	28	25	23	20	20	18	16	15	10
1000	38	33	30	28	25	23	20	18	18	18	15	14	9
1100	35	30	28	25	23	20	18	18	15	15	14	12	7
1200	30	28	25	23	20	18	18	15	15	13	12	11	6

in greenhouses or suitably lit buildings to produce sprouts 12–18 mm long before planting. Pre-sprouting accelerates cycle of growth; emergence, tuber initiation, bulking and maturity are brought forward 10–14 d. Effect on tuber yield depends on circumstances. Sprouted seed invariably increases yield of early crops, preferably always use sprouted seed. With maincrops increases from sprouting are less certain, depending on conditions of growth. With late planting, or in very cold, late areas, well sprouted seed gives up to 5 tonnes/ha more ware, but when unsprouted seed is planted early in warmer areas and blight is controlled, there is usually no increase in yield. Even so, sprouting some seed allows an earlier start on harvesting and seed which does not sprout properly can be picked out; misses from skin spot (*Oospora pustulans*) are also eliminated. Full sprouting is costly and heavy on labour, and as yet sprouted seed is unsuitable for bulk handling or in fully automatic planters.

Sprout size

This is controlled by varying storage temperature: below 4 °C no sprout growth; 5.5 °C fast sprouting varieties grow; 7 °C all varieties start, fast sprouting varieties grow rapidly; 10 °C all varieties grow rapidly, fast sprouting varieties grow excessively. Effect of temperature depends on level and duration; thus 7 °C maintained over long periods is too high for a fast sprouting variety. Provided the store is frostproof, the difficulty is to keep the temperature of fast sprouting varieties low enough to prevent excessive growth. A fan and ducting fitted with differential thermostat for blowing cold (but not frosty) night air into the store to control temperature is usually necessary; refrigeration equipment may be needed in mild areas but is expensive. Ideally sprout length should be restricted to 19–25 mm for all types of production. Varieties differ in speed of sprout growth and need different management; two differing varieties should not be sprouted in the same store, e.g. Home Guard (fast), Ulster Prince (slow). Sprout growth must be watched and adjusted accordingly; to speed up sprout growth, raise temperature; to slow down, lower temperature of the store.

Number of sprouts

This is controlled by varying time at which sprouting starts. If tubers are sprouted at 16 °C in the early autumn, a single sprout develops which suppresses growth of all other sprouts; excessive growth is prevented by lowering temperature once the single sprout is formed.

Multi-sprouted seed is produced by delaying start of sprouting until the New Year or later, or removing the first sprout and resprouting; the number of sprouts produced increases with size of seed but small seed still produces relatively few sprouts. Single-sprouted seed is essential for earliest crops while seed for canning and seed crops need as many sprouts as possible.

Mini-chitting

Seed is managed to produce mini sprouts 2–3 mm long at planting. Labour required is low, trouble from skin spot (*Oospora pustulans*) eliminated and possible increase from sprouting maincrops largely obtained; cold storage facilities are essential. Seed may be treated in bags, boxes of up to 500 kg capacity or in bulk. On arrival seed is cured at 13–16 °C for 10 d to heal wounds; temperature is then reduced and held at 3 °C (30°F) to stop sprout growth and then raised to 7–10 °C to obtain the required sprout development three weeks before planting; management must be precise and is critical.

Storage of unsprouted seed

Where cold store facilities are not available bags should be stacked, maximum 1.8 m high, on boards or straw with ample airspace between in a frostproof building; the shallower the better, for heating is less likely. If necessary, cover sacks with straw in frosty weather but keep as cool as possible as spring approaches by opening doors.

Cutting seed

Small whole tubers always give higher yields than cut pieces of the same weight. Cutting is now rarely done and is only worthwhile with very large tubers and expensive seed. Unsprouted seed should be cut lengthwise and then heaped and covered by moist bags at 13–16 °C for 10 d to allow healing. Cut surfaces must not dry out nor be treated with disinfectant. Lime and soot are worthless. Seed for sprouting may be cut for two-thirds of way from apical end towards heel, the two sides left pressed together until planting, then pulled apart; each side sprouts well.

Date of planting

First earlies

These when fully sprouted, are planted as soon as soil conditions permit. In earliest areas, e.g. Cornwall and Pembroke, the crop is planted in February; other areas follow in early March.

Unsprouted maincrop

This seed is planted as early as practicable, certainly by mid April. In the south, planting occurs in late March and first week in April; in the north, early April. Automatic planters speed planting and hence may increase yield of unsprouted seed.

Sprouted maincrop

This seed may be planted later, without loss of yield; if possible plant by mid April in the south, end of April in the north. Precise results of late planting depends on season; in dry years heavy losses result and tops of some varieties may never meet across the rows. By contrast, with ample later rainfall there is relatively little effect provided blight is controlled and planting is not excessively late.

Climate

Earliest crops

These are grown in areas close to the sea where winters are mild and very early planting is possible, e.g. Cornwall, Pembroke. As climate and site become progressively colder, early lifting is increasingly delayed, late first earlies merging into second early crops.

Maincrops

Maincrops are grown over a very wide range of climatic conditions; for these soil texture and structure are more important than climate.

Soil

Good potato soils are well drained, deep and allow deep cultivation. Deep loams, sandy loams or alluvial soils are admirable but avoid stiff clays, harvesting being very difficult in wet autumns. Peat soils give heavy yields but tuber dry matter is usually lower. Potatoes are almost insensitive to acidity and lime is only necessary on the most acid reclaimed land, after bracken.

First earlies

These require light soils, light loams or sandy loams with southerly aspect, which warm up early.

Maincrops

They are best on deep moisture retentive soils; avoid soils prone to drying out unless irrigation is freely available.

Cultivations and seedbed

Deep friable moist mouldy clod free seedbed is required. Plough flat broken furrow slice 200–300 mm deep, according to topsoil depth, with reversible plough well before Christmas to allow weathering and settling. To avoid compaction and clod production use cage or double wheels and minimum number of passes in spring; power harrows are especially useful for reducing number of passes. Most planting is done mechanically; where done by hand, ridges are opened and closed quickly to avoid moisture loss. In dry areas ridges may be rolled down after planting.

After cultivations

Where weeds are controlled by cultivation, ridges may be pulled down with chain harrows, followed by periodic inter-row cultivations after crop emergence. This needs doing carefully, not running too close to the plants to avoid damage and loss of yield. Crop should be finally ridged up when 200–250 mm high and before roots have started spreading between rows. Weed control in early crops and high rainfall is best effected by use of chemicals; elsewhere practice varies (*see* Chapter 8).

Prevention of potato blight (Phytophthora infestans)

At least one spraying with a MAFF approved fungicide must be given to all maincrops before haulm meets across rows. Thereafter, spraying frequency depends on weather. In warm, humid, *blight* weather, i.e. in the south west or in wet years, weekly spraying is necessary when using contact type of fungicide. In dry periods and years (other than in the south west) a second spraying can be delayed until blight warning; spraying is then advisable at 7–10 d intervals. Burning off the haulm in autumn with MAFF approved desiccant prevents spread of blight to tubers. Essential preventive measures include good earthing up, destroying groundkeepers and old clamps and avoiding very susceptible varieties.

Manuring

Dung, when used, should be well rotted and ploughed in during autumn. Fertilisers ploughed in give inferior results to spring applications. Recommendations in *Tables 5.35* and *5.36* are for broadcasting on the flat, in front of the planter. Placing fertiliser close to but not in contact with the seed gives more efficient utilisation of nutrients but there is risk of damage if fertiliser comes too close to the seed. Risk is greatest with sprouted seed, high rates of fertiliser and on soils liable to dry out. When fertiliser is broadcast over ridges or band placed, reduce recommendations by 25%. Fertiliser attachments to planters slow down rate of working. Lime is *only* applied before potatoes on *extremely acid* soils such as bracken reclamation land in upland areas and then only in small quantities.

Nitrogen

Raising level of N application increases tuber yields up to a certain point but reduces dry matter or specific gravity. Excess N delays tuber initiation, and maturity of foliage and tubers, which being immature are watery and of low cooking quality. Susceptibility of haulm and tubers to blight is also increased. Level of N should be reduced for very fertile soils (e.g. after heavy applications of dung or rich turf or in areas of high summer rainfall) whereas an increase is necessary for low rainfall areas with arable rotations.

Table 5.35 Manuring of maincrop potatoes (kg/ha)

N, P or K index	All mineral soils			Fen light peats loamy peats			All other peat and organic soils, including moss and warp		
	N	P_2O_5	K_2O	N	P_2O_5	K_2O	N	P_2O_5	K_2O
0	220	350	350	130	350	350	180	350	350
1	160	300	300	90	300	300	130	300	300
2	100	250	250	50	250	250	80	250	250
3	—	200	150	—	200	150	—	200	150
Over 3	—	100	100	—	200	100	—	200	100

ADAS figures.

Table 5.36 Manuring of early, seed and canning potatoes (kg/ha)

N, P or K index	Earlies			Seed and canning		
	N	P_2O_5	K_2O	N	P_2O_5	K_2O
0	180	350	180	180	350	250
1	130	300	150	130	300	200
2	80	250	120	80	250	150
3	—	250	60	—	250	75
Over 3	—	200	60	—	200	75

ADAS figures.

Phosphorus

This encourages early tuber growth, important for early crops, as well as earlier maturity, harder skins and lower blight incidence. Potatoes respond well to P in terms of tuber yield but level of response is lower at high P indices. Use only water soluble forms of P. First early potatoes grown on slightly acid soils of the west give very large responses to P.

Potash

The more mature the crop at lifting, the better the response. Maincrops and second earlies are highly responsive to K first early crop much less so. The use of sulphate of potash instead of muriate (chloride) increases tuber dry matter somewhat and may reduce incidence of after cooking blackening; sulphate form never used for new potatoes. Heavy dressings of muriate form reduce dry matter content of tubers and consequently reduce susceptibility to internal bruising.

Harvesting and storage

Ware and seed crops

Crops must be completely mature, haulm dead, tubers with hard skins, before lifting; crops should never be lifted with immature or blighty foliage. Allow at least 10 d for skins to harden after burning off foliage, around 21 d for crops with large vigorous foliage. Certified crops for seed are burnt off by order of certifying authority by specified date, to prevent aphid infestation and introduction of virus; burning off also reduces size of seed tubers. Relatively early harvest, from mid September to mid October, gives drier conditions, cleaner, easier lifting, less dirt, better healing of wounds, better cooking quality. Careful implement setting and handling are essential to prevent damage during harvesting and transport.

Storage

Potatoes are now mostly stored in buildings. Indoor storage gives independence of weather conditions for sorting and grading which can then be carried out when other work stops, so farm organisation is improved. Working under comfortable conditions results in better quality grading and 60–100% higher output. Rotting is minimised and temperature control improved. Against this, should rotting occur, many tons of tubers may need moving quickly to reach the trouble, which could prove disastrous if not immediately checked. Capital cost need not be high and a wide range of buildings can be adapted cheaply. General purpose umbrella buildings are ideal, provided tubers are properly protected.

Indoor storage

Clean (no loose earth), dry, sound tubers are required, with minimal damage or disease. Buildings should have:

(1) strong walls, reinforced if necessary, as a stack of potatoes exerts side thrust and weak walls collapse; temporary walls of straw bales should be well reinforced;
(2) proper insulation to combat frost and weather;
(3) adequate ventilation, including ducting to let air through the stack, and gable end ventilation with ample headroom to prevent condensation and keep the potatoes dry, with up to 600 mm of loose straw on top of the heap;
(4) proper siting for ease of access;
(5) correct management is essential.

Potatoes may be stacked up to 1.8 m with only convection ventilation; stacking beyond this is likely to require forced draught ventilation.

Sprout suppressants

Potatoes keep satisfactorily until January or February, according to district, with only normal ventilation. Thereafter they show increasing tendency to rapid sprouting as temperature rises. Refrigeration is unsatisfactory and uneconomic for ware. Late storage thus requires a sprout suppressant approved by MAFF to be used. Follow maker's instructions closely.

Yield

Earlies 7.5–10 tonnes/ha according to current price up to 25–30 tonnes/ha according to earliness of lifting.

Canning potatoes 7.5–10 tonnes/ha or tubers 18–38 mm with up to 10 tonnes/ha oversized tubers.

Maincrops average 25–30 tonnes/ha, good crops 45–50 tonnes/ha: exceptional crops may exceed this.

SUGAR BEET, MANGELS AND FODDER BEET

Rotation

Sugar beet, mangels and fodder beet make excellent cereal break crops, especially on deep light or medium soils; they should *not* follow potatoes owing to danger of groundkeepers. Also avoid following crops where persistent herbicides were used. In order to control beet cyst nematode (*Heterodera schachtii*) the British Sugar Corporation (BSC) contract requires that no sugar beet, fodder beet, mangels, red beet, spinach beet or 'excluded' crop, i.e. all brassica crops including brassica catch crops, shall not be grown in the two years preceding the beet crop. A longer break (e.g. six years) between beet crops is much safer, especially where weed beet has appeared (*see* Chapter 8). A sound rotation is essential, as it is the only satisfactory way of controlling beet cyst nematode (*H. schachtii*).

Sugar beet

Grown only on contract to British Sugar Corporation, who have a monopoly of seed supplies and purchase all roots. Sugar beet is confined mainly to the eastern counties and east and west midlands, within easy reach of factories; elsewhere the area of beet is small.

Varieties

Suitable varieties are listed in Farmers Leaflet No. 5, *Recommended Varieties of Sugar Beet* (NIAB); trial results are published annually in summer editions of *British Sugar Beet Review*. When selecting choose at least two varieties; one of them may suit individual farm conditions better than the other. Also buy bolting resistant varieties for March sowings; sow tolerant or resistant varieties in areas where virus yellows or downy mildew are a problem. Choose smaller topped varieties for Fenland or high fertility situations, larger topped varieties for livestock feeding or where obtaining full ground cover is a problem.

Soil and cultivations

Beet requires deep well drained stone free soil that is not acid. Shallow soils, plough or chemical pans or acidity result in fangy roots, making lifting difficult and with high dirt tare. Stony soils cause excessive wear on implements. In practice beet is grown on a wide range of soils from sand to better structured clays or marls but good medium textured soils regularly give the highest yields. A high standard of management is required to provide a well structured soil free from compaction.

A smooth level seedbed ensures even drilling. An excessively fine surface should be avoided, to prevent cupping after rain but seed must be surrounded by fine firm moist soil to ensure rapid even germination. Compaction prevents root penetration and seriously reduces yield. Seedbed preparation starts in autumn, with stubble cleaning and subsoiling if required. Dung and certain nutrients, according to soil type and analysis (*see* manuring), are applied prior to ploughing to avoid compaction. Plough a coarsely broken level furrow slice. Deep ploughing is unnecessary on most soils; 200–250 mm is usually adequate. Late ploughing of medium or heavy soils when wet causes pans and poor seedbeds. Autumn or winter post ploughing cultivations for levelling allow the production of the spring seedbed with a single pass, but must *only* be undertaken when the soil is dry or well frosted. Chisel ploughing is used on some farms. Type of ploughing and winter work have a profound effect on quality of spring seedbed. Cage or double wheels on tractor should be used wherever possible to avoid compaction.

Wait until soil is dry enough before starting spring seedbed preparation. Work must be shallow with minimum number of passes to avoid loss of moisture and compaction. Deep working brings up lumps and gives loose cloddy seedbeds. To conserve moisture in dry weather, avoid working down a larger area than can be sown the same morning or afternoon. Precise choice of method depends on soil type and growers experience. Always select harrows with *straight* tines to prevent bringing up clod and avoid overworking.

Sowing

Early sowing is essential for maximum yields of sugar. Drilling should start as soon as possible after 20 March and be completed by 10 April, subject to satisfactory ground conditions. After this date yield of sugar declines; in dry years there is inadequate soil moisture for satisfactory plant establishment and activity of residual herbicides. Very early sowing, i.e. before 20 March, is inadvisable as it may give a marked increase in the number of bolters (*see also*

Weed beet in Chapter 8) and poor establishment and uneven stands; germination is slow and erratic at soil temperatures below 5 °C.

Full evenly distributed plant population is needed for maximum yield of sugar. Aim for a final stand of at least 75 000 plants/ha. Up to 100 000 plants/ha produce no harvesting difficulties. If under 62 000 plants/ha, yield is seriously affected. Irregular distribution may reduce yield, seriously at low populations, and increases harvesting difficulties. Closer seed spacing does not compensate for irregular establishment, as topping mechanisms cannot cope adequately with large variations in size of beet and many small beet remain unharvested.

Row width may need to be adjusted to suit other crops but yield is reduced if row width exceeds 500 mm. If wider rows are contemplated a bed system may be considered if a suitable harvester is available.

Only monogerm seed of high germination capacity and in pelleted form, usually graded to 3.40–4.75 mm diameter, is now supplied by BSC to growers; this gives an emergence rate at least equivalent to the former rubbed and graded seed. A standard inter-seed spacing of 175 mm is normally satisfactory, provided that seedbed conditions and standards of drill operation are good enough to achieve an establishment rate of at least 70%. In practice many growers fail to achieve this level. For the earliest sowings, difficult conditions or where problems have been experienced in obtaining satisfactory populations, inter-seed spacings down to 155 mm should be considered.

Failure to achieve satisfactory standards of maintenance and operation of precision drills is frequently the prime cause of poor establishment. Coulters must be sharp, depth of sowing should be checked and sowing mechanisms serviced well before drilling. Store units under cover. In operation do not exceed forward speed of drill of 5 km/h, when each unit should cover 1 ha in 8 h. For faster ground coverage, increase number of drill units, not forward speed. Match number and spacing of drill units to other operations, e.g. spraying, tractor hoeing and harvesting.

Manuring

Adequate N is essential for rapid foliage development and establishment of full leaf canopy but all too frequently excess is applied, increasing leaf growth with very little extra weight of root. Sugar content and consequently financial returns are reduced (*Table 5.37*).

Beet does not respond well to P and yield increases are not likely above ADAS index 1. Applications suggested for index 2 and over are solely to maintain soil P reserves and may be given at any stage of rotation.

Sodium

Sodium is an essential nutrient for sugar beet and should always be given except on fen silts and peaty soils; it reduces need for K and application of 150 kg/ha elemental sodium (Na) (400 kg/ha agricultural salt) is assumed in K recommendations in *Table 5.37*. If salt is not applied, increase potash applications by at least 100 kg/ha K_2O except on fen silts and peaty soils; this treatment is more expensive than a combination of sodium and potassium. When tops are carted off soil is depleted by about 150 kg/ha K_2O, which should be applied subsequently in rotation.

Magnesium deficiency

Magnesium deficiency, to which crop is highly susceptible, is common in light soils where farmyard manure is not given. At Mg index 0 apply 100 kg/ha Mg, index 1 apply 50 kg/ha Mg. Kieserite at 600 kg/ha will supply 100 kg/ha Mg and is more available than calcined magnesite but is less persistent in soil. If lime is also needed, use magnesium limestone where available. Deficiency symptoms cannot be controlled by foliar application.

Manganese deficiency

Manganese deficiency, causing speckled yellows, may occur on peaty soils pH over 6.0 and on sandy soils pH over 6.5. Avoid overliming and when symptoms occur

Table 5.37 Manuring of sugar beet (kg/ha)

N, P or K index	N					P_2O_5	K_2O
	Light and medium fen silts chalk soils	Lincolnshire limestone soils	All other mineral soils	Peaty loams and sandy peats	Light peats and loamy peats	All soils	
0	125	140	100	75	50	100	200
1	100	100	75	50	25	75	100
2	75	75	50	25	Nil	50	75
3	—	—	—	—	—	50	75
Over 3	—	—	—	—	—	Nil	75

ADAS figures.

spray with 10 kg/ha manganese sulphate in low volume. Up to three applications may be required. On land regularly affected, consider use of sugar beet seed pelleted with manganous oxide, which delays need for foliar applications.

Boron deficiency

Boron deficiency, causing heart rot occurs on alkaline sandy soils. Apply 10–40 kg/ha borax or 7–10 kg/ha solubor to soil; solubor may also be mixed with aphicide and applied to leaves.

Fertiliser application

Germination can be seriously damaged by N, potash and Na if applied in quantity shortly before drilling. Agricultural salt (sodium) is best applied before ploughing in autumn except on very light soils where leaching may occur; it is then applied on top of ploughing. All P_2O_5, K_2O and Mg may also be ploughed in during autumn except where soil analysis shows one or more to be deficient, when part should be applied on top of ploughing at least two weeks before drilling, i.e. ADAS index 0 or 1. Nitrogen is always applied in spring but if applied too early may be lost by leaching; it can also damage germination. Band application of N between rows immediately after drilling is very effective but if applicator is not available, with dressings of 100 kg/ha N or more, broadcast one-third of dressing at or just after drilling and two-thirds as soon as all plants have emerged.

Lime

Aim for pH 6.5–7.0 except on sandy (pH 6.0–6.5) or peaty (pH 6.0) soils. Lime requires adequate time to penetrate soil and is best applied to crop which *precedes* beet. If lime is applied just before beet crop, plough in half and apply remainder after ploughing. Application of whole dressing after ploughing acid soil results in acid layer about 100 mm below surface and consequent fanging. On sandy soils apply lime little and often to avoid trace element deficiencies. In areas near beet factories, use waste factory lime.

Pests and diseases

Apart from beet cyst eelworm (*H. schachtii*) which can only be controlled satisfactorily by sound rotation and epidemic attacks of virus yellows, pests and diseases do not usually reduce beet yields seriously.

Seedlings are so small that damage is often fatal. Millipedes (*Class Diplopoda*), symphalids (*Class Symphyla*), spring tails (*Order Collembola–Onychiurus* spp.) and pigmy mangel beetle (*Atomaria linearis*) can all cause serious loss of plant. Routine treatment with approved granular insecticides, other than for control of early aphid attacks and docking disorder caused by *Trichodorus* and *Longidorus* spp., is only worthwhile where soil-borne pests have previously caused serious losses. Mere 'blanket' treatment of all crops puts up cost without increasing yield, may kill beneficial species and induce resistance to chemical in some species. Slugs (*Derocreras reticulatum*) and leather-jackets (*Tipula* spp.) can also cause severe damage. Wireworm (*Agriotes* spp.) can be serious when beet closely follows grass in the rotation; prior treatment with an approved insecticide is essential.

Field mice (*Apodemus sylvaticus*), which live on insects and weed seed, can cause loss of beet seed. Ensure all seed is well covered or mice will eat seed on surface and then work along rows by smell. Inspect field the day after drilling and place approved poison bait (under cover, e.g. section of drainpipe) at three points/ha, according to makers' instructions.

Virus yellows (Beet yellows virus BYV or Beet mild yellowing virus BMYV) carried and spread by peach potato or green aphid (*Myzus persicae*), is the most damaging disease of beet. Extent of losses depend on earliness and severity of attack. As a rough guide, 3% of potential yield may be lost per week of infection before mid October. Losses are greatest in East Anglia and in bad years may be as high as 40–50%. Reduce incidence and severity of disease by controlling aphids: deal with overwintering sites of aphids, e.g. residues of beet crop, ground keepers, plants at cleaner loader sites, overwintered green crops and steckling beds (beds of young beet plants for planting out for seed crop). Local gardens and allotments are often a serious source of aphids. Early sowing gives quicker total ground cover which is less attractive to aphids than partially bare ground. Routine seed furrow applications of approved insecticide are of value in areas of highest infestation in East Anglia but are of questionable economic value elsewhere. Routine foliar treatment, according to incidence of aphids is advisable. First foliar application of approved aphicide should be given as soon as an average of one aphid can be found on four plants or on receipt of BSC spray warning card. A second application may be economically viable in bad years. Aphids have now built up resistance to some organophosphorus insecticides and use of others, e.g. carbamates, is often wise.

Powdery mildew (*Erysiphe betae*) can reduce yield in dry years from August onwards. Spray with wettable sulphur in early stages of disease and not later than end of August.

Irrigation

Sugar beet ultimately produces a deep tap root, about 1.8 m deep by September but before this it is vulnerable to water shortage, especially in early stages of growth. Crops on sands and loamy sands suffer most. Irrigation in a dry June, applying 250 mm as water deficit approaches 500 mm, is particularly beneficial in helping to accelerate root extension and early completion of leaf canopy. Irrigation is also worthwhile in dry weather in July and August to prevent wilting, when closure of stomata stops sugar production. Prevent deficit exceeding 500 mm; irrigate to near field capacity. There is no response to irrigation in September, as the tap root has penetrated deep enough to draw adequate water from the soil.

Harvesting

Correct setting and operation of harvester is essential to minimise field losses of beet. Aim to finish by early December. When clamping ensure that beet are clean, correctly topped and free from trash. Concrete rafts are well worth their cost and provide by far the best clamping site. To allow maximum ventilation and reduce heating, use long narrow clamps in early portion of harvesting period. Later on, when damage from frost is likely, use wider clamps. Do not exceed height of 2.5 m. Protect beet from frost as factory will reject frosted beet once thawed, as it cannot be processed and may bring refining to a halt. Frosting also reduces sugar content seriously. Use baled straw to protect tidy well made clamp in frosty weather; in mild weather covering impedes ventilation and allows temperature to rise excessively. Use thermometers at side and in centre of clamp to monitor temperature. Keep clamp as cool as possible without freezing, i.e. below 5 °C. Respiration and consequent sugar losses rise sharply above 10 °C; temperatures over 15 °C are too high. Load beet for factory in same sequence as it is harvested.

Method of payment for beet

Under EEC regulations beet is paid for under 'A' and 'B' quotas. If total white sugar yield for contract year does not exceed BSC's basic white sugar quota, contract for all beet delivered to BSC by grower is paid at consolidated 'A' price. Only when national total white sugar quota is exceeded does 'B' quota start to operate and then on those growers exceeding their 'A' quota. Price for 'B' quota = (Price/tonne of beet in 'A' quota minus levy to cover cost of exporting surpluses).

Beet delivered is then paid for on basis of minimum price per tonne of washed and correctly topped beet at a standard sugar content of 16%; precise price paid being adjusted for actual sugar content by addition or subtraction of stated sum for each 0.1% by which sugar content is above or below 16%. Increasing sugar content raises proportion of the total yield of sugar which can be extracted by factory and the grower consequently receives a higher price per tonne of sugar. It is therefore essential to manage crop to maximise sugar content, e.g. avoid excess N, obtain high plant population.

Adjustments

(1) *Additions*: (a) early and late delivery allowance, (b) hand unloading allowance, and (c) transport allowance.
(2) *Deductions*: (a) growers' contribution due to Sugar Beet Research and Education Committee, (b) growers' contribution to cost of National Farmers' Union representative at factory.

Full details of contract are set out in booklet *Sugar Beet Contract Season (Year), Conditions* issued by BSC and revised annually.

Yield

Yield is extremely variable, some growers achieving very poor results.

Washed roots 30–40 tonnes/ha, very good crops 50 tonnes/ha. 'Good average' attainable yield of sugar 5.5 tonnes/ha. Yield of sugar/ha =

$$\frac{\text{(weight of washed beet/ha} \times \% \text{ sugar)}}{100}$$

By-products

In addition to cash return, growers benefit from by-products.

Tops (leaves and crowns)

Tops give fresh yield of 20–30 tonnes/ha and are equal in feeding value to marrow stem kale; 1 ha beet tops gives yield equivalent to 0.5 ha kale at 50 tonnes/ha. Tops must be kept free from soil and wilted for a few days before feeding, since fresh tops contain oxalic acid; 1 ha of tops should last 250 ewes one week. Dairy cows should be limited to 19 kg/d. Tops may be ensiled if free from dirt; harvesters should be specially chosen if tops are to be saved.

Pulp

Pulp, of which an allocation is made to growers, is the residue left from sliced roots after sugar extraction; it is available as beet pulp or molassed beet pulp. Dry pulp or nuts are equivalent in feeding value to oats.

Mangels and fodder beet

These crops are capable of giving very heavy yields of highly digestible dry matter. When succulent roots are included in right proportion in a ration, they can increase the total dry matter intake of cattle and sheep. Provided they are protected against frost, roots may be stored over a very long period, sometimes until early June. Mangels are an invaluable feed for ewes penned at lambing time or in times of hard frost.

Varieties

These form a continuous series from mangel to fodder beet but may be grouped as follows:

Mangels, low dry matter (10–12% DM)

Include traditional English varieties, globe or intermediate in shape, with a red, orange or yellow skin. Varieties include Prizewinner, Wintergold, Yellow and Red Intermediates. These varieties sit well on top of the ground and are easy to lift and clean by hand; they give the highest yield of roots, over 200 tonnes/ha having been recorded. Flesh is soft and much less likely to damage teeth of stock than roots of high DM fodder beet. Tops are small, so yield of DM/ha is some 10% below that of fodder beet, although fresh root yields are 20–30% higher.

Mangels, medium dry matter (12–15% DM)

Mainly of continental origin, usually giving higher yields of dry matter, they have larger tops but roots are smaller.

Fodder beet, medium dry matter (15–17% DM)

Mainly of continental origin, roots are intermediate to pear shaped, sitting reasonably well on top of the ground and fairly easy to lift with swede harvester or by hand. Tops large, with high yield of DM. Suitable for sheep and cattle.

Fodder beet, high dry matter (17–20% DM)

Of continental origin, roots grow mainly below ground level and are very similar to sugar beet in shape; require lifting with sugar beet harvester. Yields of fresh roots are low but these varieties can give very high yields of DM; however, if the yield of DM is to be utilised fully, it is necessary to feed the tops, which constitute some 25% of total DM yield. Excessive dirt, especially where fanging occurs, causes serious problems with feeding.

Varieties of mangel and fodder beet are listed in Farmers Leaflet No. 6, *Varieties of Fodder Root Crops* (NIAB).

Seed

Plant breeders have developed a range of monogerm varieties, mostly in the fodder beet group, which produce over 90% of single plants. Cultivation of these types can be readily mechanised, as with sugar beet, so that handwork is eliminated. Traditional varieties of mangel and fodder beet are multigerm, giving two or more plants per seed; some handwork is still necessary.

Climate and soil

Mangels and fodder beet require warm, sunny climates and produce heaviest yields in the southern half of Britain; they are little grown in Scotland. Crops are grown on a wide range of soils provided they are well drained and of high fertility. Heavy soils need careful management to produce a tilth and cause problems at harvesting.

Cultivations and seedbed

Mangels and fodder beet usually follow cereals. Stubbles should be cleaned and, with mangels and medium DM fodder beet, receive a heavy dressing of well rotted farmyard manure, 50 tonnes/ha or more together with 400 kg/ha agricultural salt (150 kg/ha Na) before ploughing. P and K may also be applied before ploughing, on similar conditions to sugar beet. Remaining fertiliser is applied during spring seedbed preparation, so that it is well incorporated into soil (*Table 5.38*). Where salt is omitted, increase K dressing by 100 kg/ha K_2O except on fen silts and peats.

Table 5.38 Manuring of mangels and fodder beet (kg/ha)

N, P or K index	N	P_2O_5	K_2O
0	125	100	200
1	100	75	100
2	75	50	75
3	—	50	75
Over 3	—	Nil	75

ADAS figures.

Land must be ploughed before Christmas to allow production of a good frost mould, which secures the *stale* seedbed required and gives a fine, firm, level tilth containing adequate moisture. These pre-conditions are essential for precision drilled crops.

Sowing, plant population and spacing

Drill in second half of April in the south of England, as soon as weather permits in the North. Later sowings may yield well, but potential DM yield is reduced and drier soil conditions give poor results from soil applied herbicides. Varieties susceptible to bolting (NIAB Farmers Leaflet No. 6) should not be sown very early, especially in long-day conditions of the north.

Most crops are now sown with a precision drill; monogerm varieties are drilled to a stand, multigerm varieties may be chopped out subsequently. Aim to achieve final population of 75 000 plants/ha. Cut seed into soil moisture 12–18 mm deep in rows 500–600 mm apart. Drill pelleted monogerm 175–200 mm apart rubbed and graded precision seed (2.75–4.25 mm) 100–125 mm apart in rows respectively.

Harvesting

Mangels must mature or they will not keep. Maturity occurs in late October to November, according to district, evidenced by withering of outer leaves. Root or bulb must not be cut during harvesting and varieties with low dm need gentle handling.

Fodder beet is harvested like sugar beet with tops kept clean for feeding. Both crops can be harvested mechanically with an adapted swede harvester (not long types of fodder beet), an adapted sugar beet harvester, or multi-purpose harvesters. The crop is drilled in rows wide enough to accommodate tractor tyres; 600 mm width is the minimum required for mangels lifted by swede harvesters pulled by tractors with standard tyres. Stored roots need protection from frost by at least 0.75 m of straw or fine hedge trimmings and should be earthed up or stored under cover in cold situations.

Yields

These vary considerably.

Mangels average 75–115 tonnes/ha of roots to 7.5–11.5 tonnes/ha DM. Stored mangels should not be fed until after Christmas.

Fodder beet 50–75 tonnes/ha of roots to 7.5–11.5 tonnes/ha DM. Tops 30–50 tonnes/ha, 5–8 tonnes/ha DM, which must be fed to achieve full crop potential; wilt before feeding.

6

Grassland

J. S. Brockman

Grassland occupies a major part of the agricultural land of the UK, as shown in *Table 6.1*. Even if rough grazings are excluded, the total area of grass is still double that devoted to cereal production. The following definitions apply to the three types of grassland mentioned in *Table 6.1*:

(1) rough grazing (uncultivated grassland)—wholly unenclosed or relatively large enclosures found on moorlands, heaths, hills, uplands and downlands;

(2) permanent grass—grassland in fields or relatively small enclosures and not in an arable rotation;
(3) rotational grass (temporary grass)—grassland that is within an arable rotation.

Table 6.2 shows the extent of UK grassland in relation to the other eight members of the EEC. This table indicates that 40% of grassland in the EEC is in France, with the UK, Italy and West Germany accounting for most of the remainder. France has

Table 6.1 Grassland in the UK (10^6 ha)

	England and Wales	Scotland	N. Ireland	UK total	% of UK total	
Grass—rough grazing	1.8	4.5	0.2	6.5	35	⎫
—permanent	4.0	0.4	0.5	4.9	27	⎬ 74
—rotational	1.4	0.7	0.2	2.3	12	⎭
Cereals	3.2	0.5	0.1	3.8	20	
All other crops	1.0	0.1	0.02	1.12	6	
Total	11.3	6.2	1.0	18.5	100	

Table 6.2 Grassland and grass-dependent stock in nine EEC countries

	Grass[1]		Total cattle[2]		Total sheep	
	10^6 ha	%	10^6 head	%	10^6 head	%
West Germany	5.5	13	13.9	18	1.1	3
France	16.3	40	22.5	30	10.2	24
Italy	5.7	14	8.8	12	7.8	18
Netherlands	1.3	3	4.1	5	0.6	1
Belgium/Luxembourg	0.9	2	2.9	4	0.06	0.1
Denmark	0.8	2	2.8	4	0.06	0.1
Eire	3.3	8	6.4	9	2.8	7
United Kingdom	7.2	18	13.5	18	19.6	47
Total	41.0		74.9		42.2	

[1] Excludes rough grazings.
[2] Includes cattle of all ages.
Source: NEDO, 1974 with permission.

most cattle and the UK most sheep. Also notable is the fact that the Netherlands has only 3% of EEC grassland and 5% of EEC total cattle, in both cases about half of that found in Eire.

DISTRIBUTION AND PURPOSE OF GRASSLAND IN THE UK

Grassland is not distributed evenly over the UK, although every county within the UK contains a good proportion of farmed grass. For example, predominantly arable counties such as Norfolk and Kent have 18% and 42% respectively of their agricultural land in grass. Most grassland occurs in the western half of Great Britain and in Northern Ireland, where rainfall is high and the soils are heavier, and so less suited to arable cropping, than in the east. (For more details of crop distribution *see* Coppock, 1976.)

In some mainly arable counties grass is grown either as a break from regular arable cropping or on land that is unsuited for other crops. In the truly grassland areas, grass is grown as a highly specialised crop, often on farms growing little but grass.

Most of the grass grown in the UK is eaten by the ruminant animals, dairy cows, beef cattle and sheep, and the proportion of total farm income derived from these three animal enterprises varies considerably, ranging from 100% down to a very low proportion indeed. It follows that there is a great diversity in the way in which grass is farmed. This chapter outlines general principles that underlie economic grassland production, irrespective of the type of grassland farming being practised.

Use of grass

Grass as such is rarely a cash crop, unless sold as hay or dried grass. On nearly every farm, grass is processed by ruminant animals which are adapted to digesting the cellulose found in grass and also synthesising proteins from simple nitrogen compounds (*see* Chapter 13). Whilst it is possible to measure grassland production as weight of grass produced (and research results are often expressed in this way), it is important to realise that growing grass is only a part of the production process. Apart from producing grass, the farmer is concerned about the proportion of the grass that is grown which is eaten by stock, i.e. its degree of utilisation. Also, ruminants are fed material of non-grass origin, e.g. cereals. The diagram illustrates the essential steps:

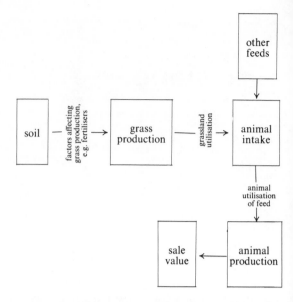

Although the ruminant can extract most of the energy and protein it requires from grass, it has been estimated that in the UK some 40% of the energy used by ruminants comes from cereal feeds, as shown in *Table 6.3* (NEDO, 1974). In pastoral countries, such as New Zealand, where animal products have a low cash value, virtually all ruminant production is derived from grass: this illustrates that the relative proportion of grass and other feeds in the ruminant diet is related to the economic value of the animal product. Thus it is possible that grass could contribute more significantly to ruminant nutrition in the UK in the future; such a move would depend on improvements in grass production and utilisation, or other feeds becoming relatively more expensive. On the other hand it would be possible for grass to become even less significant than now if alternative feeds became more attractive—either in monetary terms or convenience to grow and feed.

Table 6.3 Contribution of grass and cereals to energy input of UK ruminants

	%
Permanent and rotational grass	52
Rough grazing	5
Forage crops	4
Other feeds (mainly cereals)	39

Source: NEDO, 1974 with permission.

Rough grazings

These are uncultivated, natural grasslands where little or no attempt is made to improve herbage production or alter the botanical composition of the sward. The level of production from such natural vegetation is restricted by the environmental limitations present, mainly altitude and aspect. Generally rough grazings are used for summer grazing only. The Hill Farming Research Organisation (HFRO), financed by the Agricultural Research Council and based in Scotland, is devoted to studying production systems from rough grazings (HFRO, 1980).

The main types of rough grazing are as follows.

Mountain grazing

This is land over 615 m in altitude, where high rainfall, low soil temperatures and extreme acidity restrict the plant species that will grow and limit their growth. Winters are very cold in these areas, farms are isolated and the stock carrying capacity of the land is very low.

Moors and heaths

Moors and heaths are found at lower altitudes and are often dominated by heathers (*Calluna* spp.) in relatively dry areas and purple moor grass (*Molinia caerulea*) or moor mat grass (*Nardus stricta*) in wet areas. Heather can be utilised by sheep and these areas can be improved by liming, selective burning and application of phosphates, grass will be encouraged, often the indigenous sheeps' fescue (*Festuca ovina*).

Stock will eat moor mat grass in spring, but as the season progresses it becomes unpalatable, especially to sheep. Most natural species occur in patches and clumps, their occurrence reflecting the environment and the selective grazing of animals. Normally cattle will graze moor mat grass and purple moor grass more readily than sheep, and some improvement in overall production can be encouraged by regular grazing together with the use of lime and phosphate. Under these circumstances, encouragement will be given to bents (*Agrostis* spp.), sheeps' fescue and red fescue (*Festuca rubra*). Although these three grass species have a low production potential compared with grass species that are sown in lowlands, they have a greater value than *Molinia* and *Nardus*.

Some rough grazings occur on lime-rich soils, such as the downs in the south of England, where the soil is too shallow or too steep for cultivations: here sheeps' fescue and red fescue dominate the sward.

Permanent grassland (long-term grassland)

As shown in *Table 6.1*, permanent grassland occupies a very significant area of the UK. Also, very many different types of agricultural situation are represented by permanent grass. From 1971–1980 a special Permanent Pasture Group at the Grassland Research Institute studied permanent grassland (Forbes *et al.*, 1980), defining permanent grassland as 'enclosed

grass not normally in an arable rotation'. This definition distinguishes permanent grassland from rough grazings and also indicates that much permanent grass is found on land that is not suited to arable cropping. It has been estimated that 30% of permanent grass in the UK suffers from serious physical limitations that impede use of machines: the main factors are rough and steep terrain, stones and boulders on the surface and very poor drainage. Farmers can improve drainage and such a step can lead to a marked increase in output from this type of land. Much useful information on permanent grass is given by Hopkins and Green (1978).

Several surveys have classified permanent grass on the basis of its botanical composition, an account of the classification being given by Davies (1960). In simple terms, swards can be placed in order of probable agricultural value as follows (bearing in mind that grassland is rarely homogeneous in composition, and most natural species occur in irregular patches in the field):

rough grazing
—heather (*Molinia, Nardus*)   poor
—areas with red fescue   quality
—*Agrostis* with red fescue
   (with rushes and sedges on wet soils)

permanent grass
—mainly *Agrostis*
—*Agrostis* with a little perennial ryegrass
   (PRG) (less than 15% PRG, often with
   rough-stalked meadow grass, meadow
   foxtail, cocksfoot and timothy)
—second-grade PRG (15–30% PRG with
   Agrostis) (often with crested dogstail,
   sheeps' fescue, cocksfoot and tufted hair
   grass)
—first-grade PRG (over 30% PRG) (often   good
   with white clover, rough-stalked meadow quality
   grass, *Agrostis*, cocksfoot, sheeps' fescue
   and timothy)

Table 6.4 gives a summary of surveys conducted by Baker (1962) and Green (1974): it shows that *Agrostis* dominates the permanent grasslands of UK, although the proportion of ryegrass swards increased slightly from 1962 to 1973.

Table 6.4 Classification of permanent grass on botanical composition

	% of fields containing	
	(Baker, 1962)	(Green, 1974)
15% or more of PRG	20	26
Mainly *Agrostis*	59	56
Fescue dominant	7	10
Others	14	8

Rotational grass (temporary grass)

This is defined as grass within an arable rotation, and as such it has one or more definite functions, quite apart from its essential role in supporting an animal enterprise. These functions include:

(1) essential part of rotation to

    (a) break disease and/or pest cycle,
    (b) improve fertility,
    (c) break weed cycle,
    (d) improve soil structure,

(2) grow grass species best-suited to animal enterprise, giving

    (a) better fertiliser response than indigenous grass,
    (b) high yields of grass and animal product,
    (c) good seasonal distribution of grass production.

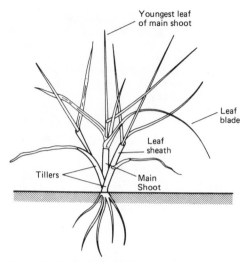

Figure 6.1 Vegetative development in grasses

GRASSLAND IMPROVEMENT

The botanical composition of permanent grass and older rotational grass is related to the nature of the environmental factors operating there. The main items are:

(1) climate,
(2) soil acidity,
(3) soil nutrient status,
(4) drainage,
(5) frequency and severity of plant defoliation.

Only item 1 is outside the farmers' control. Items 2 and 3 are easily controlled and as drainage schemes progress, so item 4 is corrected. Often the implications of the fifth point, concerning defoliation management, are not widely appreciated or practised. The following section gives a brief account of basic grass physiology and shows how the farmers' management of grass can have a marked influence on the type of sward produced.

Nature of grass growth

Figure 6.1 illustrates a plant of perennial ryegrass (PRG) in a vegetative state. All the while the plant is not producing seed heads, the region of active plant growth (or stem apex) remains close to the ground and below cutting or grazing height. Growth of the plant continues in two possible ways:

(1) Increase in size of existing tillers. As each leaf grows (*Figure 6.2*), cells close to the stem apex divide and elongate, so that the youngest part of

the leaf is at its base and the oldest at the tip. As each leaf develops, its base surrounds the apex, so the stem apex is continually sheathed by developing leaves. This method of vegetative growth maintains the stem apex at ground level and protects it from damage caused when the oldest part of the leaves may be removed during cutting or grazing.

(2) Tiller production. Each tiller bud contains a replica stem apex and further tillers will develop as the plant grows, giving rise to vegetative

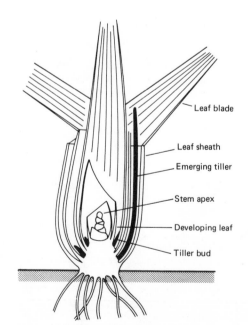

Figure 6.2 Position of stem apex and development of leaves and tillers in grass

reproduction. These young tillers are also well-protected from damage and rapidly develop a root system and an identity of their own—to further continue the process of vegetative growth and development.

When grasses have the ability to tiller freely, and are encouraged to do so, they can spread laterally over the ground surface and form a thick or 'dense' sward, characterised by a large number of grass tillers per unit area.

During inflorescence development (sexual reproduction) the pattern of grass growth is quite different in two important ways. When inflorescence development is triggered, by a mechanism based on temperature and day length, two changes alter the normal vegetative growth:

(1) new tiller development is suppressed and existing tiller buds start to form the inflorescence,
(2) the stems of these inflorescence buds elongate rapidly, to carry the developing ear well above the ground.

When ear-bearing stems are defoliated, they are incapable of regrowth as the inflorescence was their apex, and plant regrowth can only occur from further developing tiller buds at the base of the plant.

When a grass plant bears visible ears, much of its impetus for vegetative growth is lost, and the longer the plant is left bearing ears, the slower will be the subsequent development of vegetative organs. Thus removal of a heavy, tall grass crop with ears emerged will leave a thin or 'lax' sward, with few tillers per unit area and a slow propensity to regrow. However if regular frequent defoliation is then practised, the sward will become more dense, particularly if there is a high population of PRG, as this species tillers freely in the vegetative state. The effect of cutting date is well-illustrated in *Table 6.5*, where a delay in taking the first silage cut gave a high silage yield but seriously delayed speed of regrowth.

Tillers and plants are in a state of continual change, the life of an undefoliated leaf in PRG being about 35 d in summer and up to 75 d in winter. If grass is left undefoliated, there is a rapid burst of vegetative (tiller) development in spring which is suppressed in May/June by tall inflorescence-bearing stems that restrict further vegetative development until autumn, followed by drastic tiller restriction during winter. Frequent defoliation in spring will restrict ear development, maintain tiller formation and lead to a dense sward. On soils of low–medium fertility, application of fertiliser N will further encourage overall plant growth and tiller production.

The timing and frequency of defoliations will have a marked influence on the seasonal distribution of grass growth, as shown in *Figure 6.3*. It is important to realise that there is always an uneven pattern of grass growth during the growing season, but this pattern is exaggerated when spring defoliations are infrequent.

Most varieties of PRG can tiller freely compared with less desirable 'weed' species of grass, so that regular frequent defoliation enables PRG to compete favourably for space in the sward, and become dominant. Also most varieties of PRG are comparatively prostrate in growth habit compared with weed grasses, and so defoliation close to the ground will further encourage development of a PRG dominant sward.

The practical importance of the basic physiology described above was demonstrated 50 years ago in classic experiments by Jones (1933). He showed on a relatively fertile lowland site that provided there was at least 15% PRG in the sward, an *Agrostis* dominant sward could become PRG dominant in two years if subjected to regular, tight grazing. Similarly, he showed that a newly-sown PRG sward would rapidly degenerate in composition if subjected to irregular high defoliation.

Some other plants, like white clover and creeping bent (*Agrostis stolonifera*), can propagate vegetatively by stolons: under some circumstances close defoliation will favour these plants as well.

Methods of improvement

The farmer has three improvement strategies available:

(1) replace—plough or chemically destroy the previous crop and reseed;

Table 6.5 The effect of date of first cut on yield and regrowth (perennial ryegrass swards, adequate fertiliser and moisture)

| Date cut | Tonnes/ha DM at cut | Delay in regrowth[1] (d) | Regrowth to reach (d) | | Tonnes/ha DM on 20 July |
			2.5 tonnes/ha	4.0 tonnes/ha	
May 20	4.2	8	32	46	5.5
May 30	5.4	11	35	49	4.1
June 10	6.6	15	39	53	2.6
June 20	7.8	20	44	58	1.1

Note: the range in total yield up to 20 July was only 8.9–9.7 tonnes/ha DM.
[1] Compared with uninterrupted growth.
Source: Wolton, 1980.

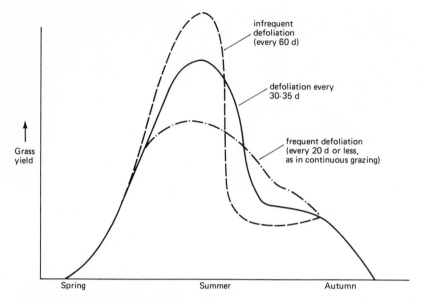

Figure 6.3 Effect of defoliation frequency on seasonal distribution

(2) renovate—introduce new species into the old sward, with or without partial chemical destruction;

(3) rejuvenate—improved defoliation technique, often with attention to fertilisers and drainage.

Checklist for action

(1) *Why is field in its present state?* bad drainage, steepness, stones, altitude, climate, aspect, acidity, low fertility, weeds, low stocking rate, poor grazing control?

(2) *Which improvement method?*

	replace	*renovate*	*rejuvenate*
speed of action	very rapid	quick	slow
cost	expensive	moderate	very cheap
stability	can revert easily	adjustable to acceptable level	improvement based on farmer's ability
special consider-ations	other limita-tions must be corrected	topography may limit	no real physical limits

Methods of grassland establishment

Grassland is established by sowing seeds either as part of a replacement/renovation procedure where grass follows grass or as part of a rotation involving other crops. The timing and method of sowing the seeds will be influenced by the purpose of the crop,

soil texture and structure, climate and farm type. Whatever method of sowing is used, it must be remembered that the grass seed has a relatively small food reserve and so must have an adequate supply of available nutrients near the surface of the soil. The biggest single demand is for phosphate and except on high phosphate soils a fertiliser supplying 80 kg/ha of P_2O_5 should be applied at seeding time. Also grass seeds should be sown no deeper than 25 mm below the soil surface; where sowing is preceded by cultivations care should be taken to ensure that the seedbed is level and fine. After sowing, grass establishment is enhanced greatly if the seedbed is well-consolidated; the firmer the better, provided soil structure is not impaired or surface percolation rate reduced.

Undersowing

This is the most common single method of establishment, and can be a very successful technique provided a few simple rules are observed. The most important rule is to prevent the 'nurse' crop, usually spring cereals, from overshadowing the establishing grasses and thus depleting the density of the sward before the nurse crop is removed. Where cereals are undersown, an early-maturing, stiff-strawed variety should be used. Whilst farmers successfully undersow wheat, barley and oats, and the choice often depends on rotational considerations, barley is the most suitable, having short straw and an early harvest. Seed rate should not exceed 125 kg/ha and fertiliser N rate should be about two-thirds that which would have been used had the crop not been undersown, to avoid lodging. The seeds mixture may be broadcast or drilled, with drilling always being preferred if conditions are likely to be dry. The grasses may be sown immediately after the spring cereals are sown and the ground then harrowed and rolled: alternatively, some farmers prefer to wait until the cereal is established before sowing the grass seeds. If the cereals are sown in February, it is a good idea to delay sowing the grasses until March. When the cereals are taken through to harvest, the straw must be removed as soon as possible and often the grasses will benefit from a dressing of complete fertiliser, say 400 kg/ha of a 15:15:15 compound, followed by grazing three to four weeks later to encourage tillering.

Some farmers do not take the cereal to harvest, but cut the crop for silage when the grain is still soft. Removal of the cereal at this earlier stage benefits grass establishment and is good practice, provided there is a use for the 'whole-crop' cereal silage that is made. Other nurse or cover crops are used sometimes, including forage peas.

Whenever crops are undersown, it must be remembered that herbicide use on the cover crop may be restricted, particularly if the seeds mixture contains clover.

Seeding without a cover crop

Here the seed mixture is sown as a crop in its own right, the nature and timing of operations being adjusted to the needs of the grass crop. In the past this technique has been known as *direct seeding* but this term is not encouraged now as it can be confused with the modern (and different) *direct drilling*. It is advisable to sow without a cover crop in either spring (mid-March to end of April) or late summer (mid-July to mid-September). Outside these periods it is risky, because of cold conditions in winter and drought in summer. Location has an important bearing on correct sowing dates—for example in the south west of England any time in the periods mentioned may be suitable, but in East Anglia, where summers are more dependably dry, farmers would not sow grass seeds in the spring unless undersowing.

Where grass is sown without a cover crop, establishment must be rapid both to minimise weed competition and reduce the period that land is unproductive. Thus adequate fertiliser must be applied to the seedbed; 80 kg/ha of both P_2O_5 and K_2O and 60–80 kg/ha of N if no clover is sown. If clover is in the mixture, it is important that no more than 25 kg/ha of N is used, otherwise clover establishment will be seriously impaired. An early grazing will benefit the establishment by encouraging tillering, but a September sowing should not be grazed before winter as it will drastically reduce the sward's ability to build up root reserves during the winter.

Cocksfoot and meadow fescue establish more slowly than ryegrass, with tall fescue the slowest of all.

Seeding without a cover crop is best suited to ryegrass mixtures, particularly straight Italian and hybrid mixtures with no clover.

Direct drilling

Grass and clover seeds can be direct-drilled successfully provided:

(1) the drilling date is within the periods given in the preceding section,
(2) the previous crop has been thoroughly killed off using a herbicide such as paraquat (if annual weeds are present) or glyphosate (if perennial weeds exist) at 6 ℓ/ha. If there was a dense mat of previous vegetation and paraquat is used then two half-rate sprays at three to four week intervals should be made. After spraying, *at least* two weeks should elapse before sowing where paraquat was used, and four weeks where glyphosate was used,
(3) the seeds should be placed 10–20 mm deep in mineral soil. Where there has been a previous mat of vegetation and/or where the ground is rough from poaching by grazing animals, it is not easy to ensure all the seeds are sown where they are required. Direct-drilling grass into grass can result in failure, principal reasons being insufficient time between spraying and sowing seeds into the surface mat and not into the soil proper.

Normal direct-drilling practice includes application of fertiliser at similar rates to the preceding section and use of 8 kg/ha of mini slug bait. Autumn sown crops are particularly susceptible to damage from frit fly and in some localities routine use of an insecticide is advised.

Slit-seeding

This technique is used for sward renovation, the principle being that at least some of the original sward is allowed to survive. If the old sward contains a reasonable proportion of PRG, but is lax and beginning to fill with annual meadow grass (*Poa annua*) then slit-seeding may be the quickest and cheapest way of renovating the sward and boosting production. Also some farmers use this technique to introduce Italian ryegrass into old swards in an attempt to extend the grazing season and increase production. Where there are no real problems from either grassy or broad-leaved weeds, then no herbicide treatment is needed. Otherwise, a selective chemical such as dalapon will have to be band-sprayed ahead of the slit-seeding of the grass.

There are several types of slit-seeder available, but the basic technique is to drill rows of grass seed into an existing sward with no previous cultivation. Some machines can apply a band of herbicide ahead of the drill; some cultivate the narrow band of the drill before the seed is sown; some can apply fertiliser and slug bait at the time of sowing; and some push the slit of old sod to one side so it cannot fall back over the seeds. Two basic rules are:

(1) ensure aggressive weeds in the old sward are killed (particularly stoloniferous species), otherwise they will 'strangle' the newly emerging young plants;
(2) use the narrowest width of drill feasible and cross-drill if possible, to get the most rapid knitting together of the sown species.

The fertiliser rates and sowing periods are similar to the preceding sections. In some cases, slit-seeding may be tried on very old swards and on fields not suited to normal cultivations. It is essential to check soil pH and also ensure that bad drainage will not cancel out any good that may arise from sowing new grasses.

Scatter and tread

On an opportunist basis, it is possible to renovate grassland by simply scattering the seeds on the soil surface under moist conditions. This should be followed immediately by turning in some grazing stock (preferably sheep) for 2–3 d to tread in the seeds. This technique is a gamble, as its success depends on a thorough, quick tread and then some two weeks of continuing moist, but not wet, weather to allow the seeds to establish. The technique is cheap, and can be used to patch-up worn areas in grazed paddocks.

CHARACTERISTICS OF AGRICULTURAL GRASSES

In this section, the important characteristics of the commonly-used grass species are reviewed, as a precursor to information on the formulation of seeds mixtures. All the information on the performance of grass species and on seeds mixtures is based on information from NIAB and ADAS. An essential publication is NIAB Farmers' Leaflet No. 16 (revised annually) (*see Table 6.6*).

Table 6.6 Grass species used in agricultural seeds mixtures in UK (1978/79)

	% by weight of seed sold	
Perennial ryegrass	57 ⎫	
Italian ryegrass	25 ⎪	91
Hybrid ryegrass	8 ⎬	
Westerwolds ryegrass	1 ⎭	
Timothy	6	
Cocksfoot	2	
Meadow fescue	0.7	
Tall fescue	0.3	

Perennial ryegrass (PRG)

Perennial ryegrasses are very successful under British conditions; not only do they account for over half the

grass seeds sown, but also the best permanent pastures have an appreciable PRG content. Provided they are defoliated regularly and grown under conditions of medium-to-high fertility, they will give high yields of digestible herbage over a long season. PRG has the following valuable characteristics:

(1) tiller well and so form dense swards under good management;
(2) fairly prostrate and ideal for grazing (but must be protected from prolonged overshading by taller species on cut swards);
(3) recover well from defoliation, both because of high tiller numbers and good rate of replacement of root reserves;
(4) persist well and are suited to long-term grassland.

PRG is classified into three main groups, based on date of ear emergence in spring. The following heading dates are for central–southern England and will be approximately one week later in northern England.

Very early and early group	10–25 May
Intermediate group	26 May–1 June
Late and very late group	2–15 June

In general, when early heading varieties of PRG are compared with late heading varieties, the early heading varieties:

(1) grow earlier in spring, but not as much earlier as the difference in heading date: for example a three-week earlier heading date would give 7–10 d earlier spring growth;
(2) are more erect;
(3) tiller less freely;
(4) are easier to cut for conservation, both because of erect growth and good stem elongation at heading.

Compared with early heading varieties, late heading varieties are:

(1) more prostrate;
(2) tiller well and are more persistent;
(3) give good midseason growth (June–August).

PRG has a limited temperature tolerance compared with other grasses, being unable to withstand very hot summers where temperatures exceed 35 °C (not found in the UK but can occur in some temperate Continental areas) or very cold winters where temperatures persist below about − 10 °C. Lack of winter hardiness is a problem with PRG in the north of Scotland, where only the most winter-hardy varieties are recommended. Winter survival in PRG is enhanced if little or no autumn nitrogen is applied and the plants are defoliated to at least 100 mm before the onset of winter.

Recommended varieties of PRG come from both British and continental breeders, with little difference in general characteristics, except that continental varieties can be slightly more winter hardy.

Breeders have produced *tetraploid* varieties of PRG. Compared with diploid varieties of the *same heading date*, tetraploid varieties:

(1) appear higher yielding and may have a greater yield above, say 80 mm, but total DM yields are generally the same. Cattle may be able to graze tetraploids more easily under some conditions;
(2) have a higher soluble carbohydrate content which may enhance palatability when grazed;
(3) are more erect, with leaf of higher moisture content: the higher moisture content can render them less suitable for conservation;
(4) tiller less freely and are less persistent;
(5) more winter hardy and more drought resistant;
(6) have larger seeds and so require a higher seed rate.

Italian ryegrass (IRG)

These are grasses with an expected duration of two to three years. They are very erect and although capable of tillering, they rarely tiller enough to prevent a steady decline in sward density from establishment onwards. IRG will grow for a long season, starting growth in spring before PRG and continuing well into the autumn under conditions of high fertility. However, the bulk of growth from IRG is in the April–June period, making them particularly useful for early spring grazing and a heavy conservation cut. During the summer they tend to run into seed-head every 35–40 d and this can lower their feeding value at this time. Most varieties of IRG are even less winter hardy than PRG. The spring heading date of IRG coincides with the early group of PRG, and IRG are classified simply into two groups based on their persistence. IRG is specially useful as a rotation grass, where its two to three year duration is not a disadvantage. *Tetraploid* varieties are important, as tetraploidy emphasises the natural advantages of IRG in terms of erectness, leaf size and soluble carbohydrate content.

Hybrid ryegrass (HRG)

During recent years plant breeders have attempted to combine the vigorous pattern of growth found in IRG with the greater tiller density and persistence found in PRG. The New Zealand hybrid Grassland Manawa (formerly known as H1) has been available for many years, but is only suited to mild areas as it is not winter hardy. The first of a new series of hybrid ryegrasses coming from the Welsh Plant Breeding Station was recommended by NIAB in 1974 and more varieties

are coming forward. In 1968 only 2% of grass seeds used in UK agriculture were HRG compared with 8% in 1978. All of the new HRG varieties coming from the WPBS are tetraploid, and the varieties at present available are best regarded slightly more persistent types of IRG that are also rather more leafy in summer than IRG.

It is likely that other hybrid grasses, not necessarily involving a ryegrass parent, will become available as breeders perfect new techniques for developing useful crosses between grass species.

Westerwolds ryegrass

These are annuals. When sown in the spring or summer they flower in the year of sowing and do not persist over winter. They can be sown in autumn in mild areas and can then provide very early spring growth. They are only of real value where high yields of grass are required for a short period and they are best regarded as catch crops, otherwise IRG should be used, as it will give equivalent yields and longer duration for a similar seed cost.

Timothy

Timothy is a persistent and winter hardy grass that will grow well in cooler, wetter parts of Britain. Also, it has survived well in drought years in other areas. Varieties of timothy head 1–25 June—as late and later than the latest heading of the PRG varieties. Timothy is very palatable and grows well in summer and autumn, but when grown alone it does not give a dense sward or very high yields. It is an ideal companion grass in seed mixtures and historically was included with meadow fescue and clover. Today it is used with PRG and is found in most mixtures sown in cool, wet parts of the country. Timothy seed is some two and a half times lighter than ryegrass seed, so its percentage use by weight of 6% shown in *Table 6.6* is an underestimate of its real importance. Timothy is very winter hardy and is widely used in Scotland, where sometimes it is grown alone for heavy hay cuts followed by sheep grazing.

Cocksfoot

Cocksfoot is a native grass that is very responsive to nitrogen and, with a very well developed rooting system, is useful on light soils in low rainfall areas, where it shows drought resistance. It grows and regrows rapidly, is erect in habit and has relatively broad leaves for a grass, the leaves also being coarse and hairy. Cocksfoot heads in early to mid May and has an aggressive growth habit that makes it competitive in swards that are not closely defoliated during the main growing season. Over recent years cocksfoot has declined in popularity as it was found that its digestibility was below that of PRG and IRG. Breeders are producing new varieties that will correct this deficiency, and also are seeking to breed varieties that are less coarse and hairy, thus increasing its palatability.

Meadow fescue

This is a very adaptable grass that tolerates a wider range of climatic conditions than ryegrass. It does not tiller as freely as PRG and so forms an open sward that is more prone to weed invasion, particularly as it is not aggressive. Meadow fescue makes an ideal companion grass in mixtures, for example with timothy and/or cocksfoot: also, it remains leafy during summer under conditions of fairly low fertility, unlike PRG which demands higher fertility if it is to remain leafy. As fertiliser use on grass has increased over the last 20 years, the importance of PRG has risen and the popularity of meadow fescue has declined, except where the winters are too severe for PRG—for example in Scandinavia. It heads mid to late May.

Tall fescue

This is one of the earliest grasses to grow in spring, it is winter hardy, drought resistant and very persistent under regular cutting (which PRG is not). It is very slow to establish, taking up to one season to become fully productive. Also it has a very rough, stiff leaf that is extremely unpalatable to grazing stock, although this unpalatability does not apply to silage made from the grass. Tall fescue's main role is on grass drying farms in eastern England, although a few farmers on light land have used tall fescue in areas they cut regularly for silage.

HERBAGE LEGUMES

Legumes have root nodules that can be inhabited by *Rhizobial* bacteria which can fix atmospheric nitrogen (N) and make it available to the host plant (discussed further on p. 186). Herbage legumes are grown either on their own ('straight') or mixed with grasses where their ability to acquire fixed N can also benefit associated grasses.

The important herbage legumes used in UK are shown in *Table 6.7*.

Table 6.7 Legumes used in UK (1978/79)

	% by weight of seed sold
Red clover	33
White clover	46
Alsike	7
Lucerne	6
Others	·8

Red clover (RC)

There are three main types of red clover:

(1) *Single cut*   Traditionally this was used for one hay cut in rotational farming where the clover break lasted one year. It is slow to grow in spring and lower yielding than other types.
(2) *Broad red* (also Double-cut or Early Red)   These give early growth and high yields over two cuts and the possibility of autumn grazing. It is not very persistent and is best grown for two years only, either straight or in mixtures with erect high-yielding grasses such as IRG.
(3) *Late flowering*   These flower two to three weeks later than Broad Red and are more persistent, being suited to medium-term leys where a red clover constituent is required. Such leys should be cut periodically as red clover does not persist well if regularly defoliated more frequently than every 35 d.

Single cut clover is scarcely used now and NIAB do not recommend varieties. In both other groups there are recommendations for both diploid and tetraploid varieties. Tetraploid red clovers do not necessarily give a higher yield than diploid counterparts and choice of variety can depend on selection for disease resistance. Red clover can suffer severely from two diseases:

(1) clover rot (*Sclerotinia trifoliorum*),
(2) clover stem eelworm (*Ditylenchus dipsacci*).

Both diseases can be problems in areas where red clover has been commonly grown, e.g. in the arable areas of East Anglia. Where either disease is present, resistant varieties must be grown; tetraploids offer best resistance to clover rot but are often less resistant to stem eelworm than diploids. Tetraploids give low seed yields and have bigger seeds than diploids, so tetraploid seed is expensive.

White clover (WC)

White clover is scarcely ever grown straight in the UK: even white clover seed is produced from suitable defoliation management on a grass/clover sward. Unlike red clover, white clover is persistent under grazing, often developing well when associated grass competition for light is reduced by constant defoliation. White clovers are classified according to their leaf size:

(1) *Small leaved*   These are very prostrate and have been selected from cultivars existing in old pastures (e.g. Kent Wild White). They are very hardy, but always remain small even when under

favourable growing conditions: thus they do not stand up to grass competition when fertiliser N is used or when a conservation cut is taken. Their real value is on tightly grazed swards, conditions of low fertility and at high altitudes.

(2) *Medium leaved* This is the largest group used in the UK and NIAB split them into 'medium small' and 'medium large'. In recent years breeders have produced bigger types of WC that can persist better under high N usage and in conserved swards. Leaf size is not a consistent characteristic in the field: if a medium-large clover is used in a sward that is regularly and tightly grazed, its leaf size will be quite small. Also, there is a general tendency for persistence to decrease as leaf size increases, but this is not a simple relationship, e.g. under a high N system even a medium-large will not persist.

(3) *Large leaved* (Ladino clover) These are not suited to the UK.

It is important to select the most appropriate type of white clover for the expected management conditions: often it is advisable to sow a range of types in the mixture.

Alsike

This is rather more persistent than red clover under grazing and is also more tolerant of wet acid conditions and is more winter hardy and disease tolerant. It is often used with late flowering red clover in mixtures and this accounts for its relatively high use.

Trifolium (crimson clover)

An annual, its main use is to provide a heavy crop of green material in May following late summer sowing. It can be sown into cereal stubbles, either straight, or with rye (corn) or IRG. Seed rates are 25 kg/ha if alone, or 10 kg/ha with 125 kg/ha of rye, or 15 kg/ha with 10 kg/ha IRG.

Trefoil (black medick)

Very useful for heavy yields of catch crop on thin calcareous soils. A mixture of 6 kg/ha with 20 kg/ha IRG sown in August will give excellent production the following spring for early grazing or conservation.

Sanfoin

See Chapter 5.

Lucerne

See Chapter 5.

BASIS FOR SEEDS MIXTURES

Seeds mixtures should be made up from specific varieties that are recommended for their individual merits. There are very many seeds mixtures available to farmers in the UK and the following outlines a way in which the suitability of mixtures can be examined and appropriate mixtures formulated. A golden rule is to keep the mixture as simple as possible commensurate with its function, also ensuring that sufficient of each ingredient is present to confer its expected characteristic.

Purpose of mixtures

Spread growth pattern

A single variety will exaggerate the spring peak in production and use of several varieties with varying heading dates can spread this peak to some extent, e.g. a mixture of early, intermediate and late PRG.

Speed up regrowth

When grasses are defoliated for conservation, regrowth is slow from grass plants that were fully headed. If a mixture is used it is possible that some grass plants are relatively vegetative and will give rapid growth from tillers when the crop is cut. *However*, the date of cutting such mixtures is critical, as if cutting is delayed, not only will the later heading varieties start to head, but also the feeding value of the early-heading varieties will fall rapidly.

As a general rule:

(1) *for cutting areas*, have grasses with well-matched heading dates;
(2) *for grazing areas*, a greater spread of heading dates can give a greater spread of growth over the season than a single cultivar.

Easier grazing management

With a mixture the timing of grazing is less critical than with single-variety swards. Also, if a range of varietal types is present, there is a good chance that varieties well-suited to the particular management will develop. However there is always a danger of incursion of weed species if this management is *too* lax, or if some of the varieties in the mixture are too short-lived for the expected duration of the ley.

Cheaper seed cost per hectare

This is possible as merchants can buy seed under contract or in bulk. The individual components of a mixture should be checked, as sometimes mixtures include non-recommended varieties.

Lessen disease risk

Grasses can suffer from a number of diseases and pests such as crown rust and ryegrass mosaic virus, but so far research has not shown the economic merit of controlling these. It is likely that some varieties are less susceptible to pests and diseases than others and a mixture spreads the risk.

Types of mixture

Mixtures can be classified into 12 possible groups, based on expected duration and intended use, as follows:

very short term (1–2 years)
short term (2–3 years)
medium term (3–5 years) $\times$ mainly for cutting / mainly for grazing / general purpose
long term (over 5 years)

Table 6.8 gives an outline of the basis for formulating seeds mixtures in ten of these groups and suggests the other two areas are not appropriate.

Seeds rates

The correct seed rate will vary to some extent with the climatic and soil conditions at sowing and expected during the establishment period. Under *ideal* sowing conditions, experiments have shown that a PRG seed rate of 4 kg/ha will establish a good sward. Farmers always use much higher rates than this because field conditions are not ideal, but even so it should be remembered that although a very high seed rate may give a rapid green cover, it will not lead to higher yields of grass as plant competition kills out many of the young plants within three months of sowing.

IRG (diploid)	35 kg/ha	PRG (diploid)	25 kg/ha
IRG (tetra-ploid)	40 kg/ha	PRG (tetra-ploid)	30 kg/ha
Meadow fescue	25 kg/ha	Cocksfoot	25 kg/ha
Timothy	15 kg/ha[1]	Tall fescue	40 kg/ha
White clover	2 (+grass)	Red clover	10 (with IRG) 4 (in leys)

[1] But rarely sown alone.

Table 6.8 Basis for seeds mixtures

Main use	Very short term (1–2 years)	Short term (2–3 years)	Medium term (3–5 years)	Long term (over 5 years)
Cutting	35 of 1a/b (diploid)	20 of 1a+	10 of 1a/c+	not applicable except
	40 of 1a/b (tet)	15 of 2a	10 of 2a+	35 of 6a
	40 of Westerwolds	20 of 1c+	10 of 2b	for grass drying
	20 of 1a/b+	15 of 2a	15 of 2a+	
	10 of 8a		10 of 2b+	
			4 of 7c	
Grazing	40 of 1a/b (tet)	20 of 1a+	12 of 2a+	10 of 2b+
		15 of 2a	6 of 2b+	10 of 2c/d+
	40 of 1c	20 of 1c+	4 of 3a/b+	4 of 3c+
		15 of 2a/b	2 of 7b/c	2 of 7a/b
	20 of 1a/b (tet)+	15 of 2a+	15 of 5a/b+	20 of 2c/d+
	15 of 1c	15 of 2b+	5 of 3a/b+	4 of 3c+
		4 of 7c	4 of 7b/c	2 of 7b
General purpose	not appropriate	not appropriate	6 of 2a+	6 of 2b+
			6 of 2b+	6 of 2c+
			6 of 2c+	4 of 3c+
			4 of 3b/c+	4 of 4b+
			2 of 7b/c+	2 of 7b+
			2 of 8c	2 of 8c
			6 of 2a/b+	6 of 2b/c+
			4 of 3c+	4 of 3c+
			4 of 5a/b+	4 of 5b+
			6 of 4a+	4 of 4a+
			2 of 7b/c+	2 of 7b+
			2 of 8c	2 of 8c

All rates are in kg/ha.

Note: Several varieties may be used within each species group, but individual components should not be less than 4 kg/ha (timothy and white clover 2 kg/ha).

Key to species groups:

Italian ryegrass	*Perennial ryegrass*	*Timothy*	*Cocksfoot*
1a = high persistence	2a = early	3a = early	4a = early/medium
1b = low persistence	2b = intermediate	3b = intermediate	4b = late
1c = hybrids	2c/d = late and v. late	3c = late	

Meadow fescue	*Tall fescue*	*White clover*	*Red clover*
5a = early	6a	7a = small-leaved	8a = broad red
5b = late		7b = med-leaved	8b = single-cut
		7c = med/large	8c = late flowering

Seed rate also depends on the number of seeds present where mixtures of the above are used, the quantities should be amended *pro rata* as in *Table 6.8*.

Check-list for studying mixtures

The main points to note are:

(1) rate/ha of individual components, at least 4 kg/ha for grass (2 for timothy), at least 2 kg/ha for clover;
(2) use of recommended variety, if not recommended, is there a reason? e.g. suited to locality, cheapness, personal preference;
(3) diploid or tetraploid, bearing in mind characteristics of each;
(4) heading dates, generally little spread for spring cutting swards;
(5) persistence, particularly important in medium- and long-term mixtures.

Table 6.9 illustrates the way in which a seeds mixture can be analysed. The medium-term grazing mixture has a spread of heading dates, uses one outclassed and one non-recommended variety and two tetraploids, but generally it seems suited for its purpose. The long-term ley (main use unspecified) also has a wide spread of heading dates which might lead to problems if laid up for spring silage: two varieties are outclassed and, strangest of all, three varieties (making up over half of the mixture) have low persistence: conclusion = not a very good mixture for a long-term ley.

Table 6.9 Examples of seeds mixtures and their analysis (mixtures available commercially in 1980)

kg/ha	Variety	NIAB recc.	Tet	HD[1]	Pers
Medium-term cow grazing					
9	IRG sabalan	✓	✓	56	2
5	IRG combita	0		57	2
5	PRG S.24	✓		42	5½
5	PRG reveille	✓	✓	49	4½
5	PRG S.321	X		62	5
3	PRG melle	✓		72	7½
2	WC blanca	✓		—	—
Long-term ley					
5	IRG lema	0		57	3
9	PRG S.24	✓		42	5½
5	PRG melle	✓		72	7½
5	PRG S.23	✓		66	7
2	Tim S.51	0		76	5½
2	Tim S.48	✓		84	8
2	WC huia	✓		—	—
2	WC S.184	✓		—	—

[1] Heading date = d after 1 April in central–southern England.

EXPRESSION OF GRASSLAND OUTPUT ON THE FARM

Farmers seldom weigh grass, and the most convenient way to assess grassland production is by estimating the quantity of grass eaten by the ruminant stock on the farm. In practice, the grassland appetite of all ruminant animals can be expressed as 'cow equivalents', using the relationship of Baker (1964):

$$y = 0.0234x + 0.32$$

where y = DM intake in kg/d, x = liveweight of animal in kg.

The average weight of Friesian cows has been found to be 550 kg (Forbes *et al.*, 1980) so the standard Friesian cow will eat 13.19 kg/d DM of grass or grass products. *Table 6.10* gives the cow equivalents (CE) for other stock of varying weights: note that ewes and beef cows have been found to consume less than expected and have a separate column.

Table 6.10 Cow equivalents of other ruminant stock

Liveweight (kg)	CE		Liveweight (kg)	CE	
10	0.04	growing sheep			
20	0.06				
30	0.08				
40	0.10		40	0.08	ewes
60	0.13	growing cattle	60	0.10	
80	0.17		80	0.13	
100	0.20				
150	0.29				
200	0.38	heifers			
250	0.47				
300	0.56	Channel Island cows			
350	0.64				
400	0.73		400	0.56	beef cows
450	0.82		450	0.61	
500	0.91	Friesian	500	0.68	
550	1.00		550	0.75	
600	1.09	Friesian dairy cows			
650	1.18				

From Forbes *et al.*, 1980.

Estimation of stocking rate

Where grass is the only contribution to bulk feed, and it is all grown on a known area of the farm, then it is easy to express stocking rate as 'CE/ha of grassland'. On many farms, however, other bulk feeds may be used and grass may be brought in from other areas of the enterprises not in the nominal grassland sector.

A quick guide to stocking rate is:

(1) area of grassland on the farm   ha (A)

(2) number of cow equivalents on the   (B)
     farm

(3) forage crops used for stock (ha ×
     1.5)
     hay and feeding straw fed to stock
     from area *not* in box A (tonnes ×
     0.25)
     brewers' grains, feed pots and
     silage *not* from area in box A
     (tonnes × 0.05)
                               Total   (C)

(4) deduct box C from box B   (D)
(5) stocking rate on grass as
     CE/ha = D/A

Normal range for stocking rate is 1.0–3.0: if outside
this range check calculations and/or data used.
Note: The coefficients 0.05, 0.25 and 1.5 are based on
average crop yields and animal appetites.

FERTILISERS FOR GRASSLAND

As with any crop, the soil is able to supply some or all
of the essential nutrients needed for growth. Fertilisers
are used to make up the difference between those
which the soil can supply and the quantity required
by the particular crop. In grassland the nitrogen (N)
supply is so important that it has the effect of
controlling the level of grassland yield.

Research results are nearly always expressed as
'kg/ha of dry-matter yield' (DM), that is, the weight
of *oven-dried* material produced. Dry-matter yield has
no direct practical significance. However, if it is
assumed that utilisation of grass remains constant,
the following gives a general guide for conversion of
DM yield into practical terms (note these are
approximate equivalents)

DM yield (kg/ha)	Stocking rate (CE/ha)	Cow grazing days (CE d/ha)
2000	0.4	160
4000	0.9	320
6000	1.3	480
8000	1.8	640
10 000	2.2	800
12 000	2.6	960

Nitrogen supply controls grassland production and
the requirement for other nutrients is related to the
yield of grass produced and hence the N supply.
Because grass has to be utilised through an animal
system to achieve monetary value, not all farmers seek
the highest possible grass yields, but rather to balance
the level of grass yield with the requirements of their

particular stock enterprises. The skill of economic
grassland production lies in balancing production
with requirement very carefully.

It follows that farmers are justified in operating at
a wide range of N inputs on their grassland, depending
on the yields needed. As the next section shows, the
quantity of N available from non-fertiliser sources
can vary, and so the amounts of fertiliser N used to
'top-up' N supply will range from nil up to more than
400 kg/ha of N (equivalent to 1160 kg/ha of ammonium nitrate fertiliser). It is important the rate of
fertiliser N applied is adjusted for the level of
production required and the other sources of available
N.

Sources of available nitrogen

Soil nitrogen

Very large quantities of N are present in the organic
matter in the soil, but most of this N is present either
in living matter or complex organic compounds and
is not available to plant roots. However, biochemical
activity does release some N in available forms each
year, the quantity depending on:

(1) organic matter (OM) content of the soil,
(2) biochemical activity of that organic matter,
(3) length of time in grass.

There can be a very wide range in available N
supply from the soil, from 20 kg/ha on soils low in
OM (e.g. old arable and light-textured soils) to
150 kg/ha on soils with large quantities of active OM
(e.g. old grassland). In the UK there is no accepted
method of soil analysis for predicting available soil N
under grass and estimates have to be based on the
above factors.

Animal recirculated N

Grazing animals retain only some 5–10% of the N
they eat in herbage, voiding about 70% of returned N
in urine and 30% in dung.

Potentially, grazing animals can recirculate a high
proportion of herbage N, but both dung and urine
occur in relatively small patches in the field, affecting
only a small proportion of the sward in one grazing
season. Recirculation of N by animals is not very
effective unless there is a high stocking rate, in which
case there will be a high density of dung and urine
patches. As a guide, for stocking rates up to 1.3
CE/ha, recirculation should be ignored. For 1.4–
1.9 CE/ha, a recirculation of 10–15% can be assumed
and for 2.0 CE/ha and over the recirculation can be
20–30% depending on actual stocking rate (SR).

Where slurry (and farmyard manure) is applied to
grassland there is an effective recirculation of animal
N, the value depending on the rate of application and
the N-value of the material.

Clover nitrogen

Rhizobial bacteria in clover root nodules will fix atmospheric N and this N is used by the clover plant. As the clover plants grow, old roots die and some fixed N in these roots becomes available to grass roots: this is termed *underground N transfer*. Also clover foliage contains fixed N and if the clover is grazed, then some of this N is returned to the soil in dung and urine: this is termed *overground transfer*.

The value of clover N depends on the quantity of clover in the sward. Where there is a very large amount of clover, likely to average 30% or more of the weight of herbage grown in the year, then clover N fixed can be 200–250 kg/ha of N under British conditions and the quantity of this transferred to grass some 60–80 kg/ha (and up to 120 kg/ha where swards are grazed at a high SR).

Even with skilful management the application of fertiliser N will decrease the clover content, so that fertiliser N and clover N are not additive. As a rule of thumb, application of every 2 kg/ha of fertiliser N will decrease clover N by 1 kg/ha. This means that if 200–250 kg/ha of fertiliser N is applied over the season to a grass/clover sward, then clover N contribution will be nil (and *pro rata* for lower fertiliser N rates).

Fertiliser nitrogen

This should be applied to make up the difference between the total available N supply from other sources and the required N level. The following gives a guide to probable fertiliser N rates for many situations found on farms:

Stocking rate (CE/ha)	kg/ha of N needed from fertiliser	
	Good clover and/or soil N value	Poor clover and/or soil N value
1.0–1.3	none	50–100
1.4–1.7	40–100	100–160
1.8–2.2	80–150[1]	160–300
over 2.2	150–300[1]	over 300

[1] High soil contribution, as clover value will be small at these SRs.

Efficient use of fertiliser N

The target annual rate of fertiliser N application can be obtained from the above. The seasonal use of this N can have a marked influence on the quantity of grass produced and the economic value of the N.

As shown in *Figure 6.4*, grass takes up N much more rapidly than it grows in response to that N. Not only is N uptake always more rapid than growth response, but also N uptake can occur when temperature and moisture limit growth. The greenness of grass is associated with N content (i.e. N uptake) and not grass growth, and it does not follow that very green grass following N application is growing rapidly. Research has shown a definite relationship between the quantity of N grass can use for growth and the number of days of active growth available to the grass. From this work, a *maximum* N rate of 2.5 kg/ha for every day of active growth is suggested.

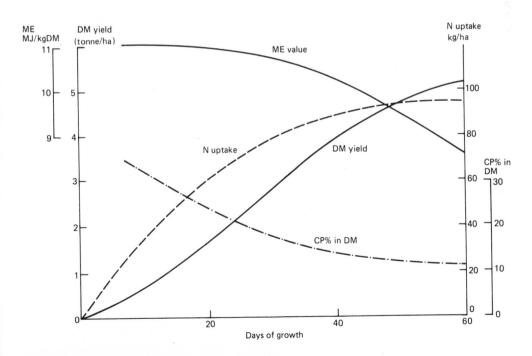

Figure 6.4 Relationship between days of grass growth, DM yield, N uptake, crude protein content and ME value

Thus *maximum* individual dressings of fertiliser would be:

for 21-d paddock grazing, $21 \times 2.5 = 52.5$ kg/ha
monthly application in continuous
grazing, 30×2.5 $= 75$ kg/ha
for silage grown 40–50 d,
40–50×2.5 $= 100$–125 kg/ha

Because grass takes up N so rapidly and completely, there is very little residual N left following defoliation: where high annual rates of fertiliser are needed, some N should be used for each growth period rather than relying on infrequent heavy dressings.

Grass grows most rapidly in spring (May and June) and it gives the biggest responses to N in this period. However very high rates of N application in the spring may deplete tiller production and limit valuable mid-season growth. It is suggested that in grazing systems the maximum daily N rate should never exceed 2.5 kg/ha, even in spring. Where silage cuts are taken, a lack of big N responses in summer should be recognised and a guide is 120 kg/ha of N for spring silage, 100 for July cuts and 80 for September cuts. From September onwards the day-length is decreasing rapidly and grass growth slows appreciably: it is recommended that no N dressing should exceed 40 kg/ha after mid-September.

Phosphate and potash

The soil can supply some or all of the phosphate and potash needed by grass, fertiliser being required to bridge the gap between supply and demand. As a guide, phosphate demand is 60–80 kg/ha of P_2O_5 and that for potash is 150–300 K_2O, depending on the level of grass production (and hence available N supply). For both phosphate and potash soil analysis can be a useful guide to the quantity of fertiliser needed (*see Table 6.11*).

Table 6.11 Basis for fertiliser recommendations on grassland

Stocking rate (CE/ha)	P_2O_5 *soil index*					K_2O *soil index*			
	0	1	2	3+	Util.	0	1	2	3+
1–1.3 } 1.4–1.7 }	60	40	20	0[1]	cut	100	80	40	0
					grazed	80	40	20	0
1.8–2.2 } over 2.2 }	80	60	30	0[1]	cut	200	120	90	45
					grazed	80	60	30	0

kg/ha of nutrient to apply.
[1] Some may be needed every three to four years to maintain level in soil.

Grazing animals

These return a substantial proportion of both the phosphate and potash eaten, virtually all the phosphate being in dung and most of the potash in urine. Dung is very unevenly distributed over the field and broken down very slowly to release the phosphate: as a result animal recirculation of phosphate is not assumed. Urine covers a much greater area of the sward than dung and all the potash in urine is in a readily available state: thus allowance is made for potash recirculation on grazed swards.

Plant uptake of phosphate is steady and is well-related to plant needs: it follows that the annual fertiliser requirement for phosphate can be applied in one annual application if desired, without causing nutrient imbalance. However, uptake of potash by the plant is related to the quantity of potash available to the roots and *not* to the plant needs: thus grass can take up in a single growth period as much as twice the potash it needs for adequate growth. This excessive uptake is called 'luxury uptake' and can have two harmful consequences:

(1) Mg content is inversely related to K content in grass, so the unnecessarily high level of K leads to very low Mg content and a greater risk of hypomagnesaemia in stock eating the herbage;
(2) growth following defoliation of this herbage may be restricted by lack of available potash.

Where high annual rates of potash are needed, some potash should be applied for each growth period, except none should be applied in spring to swards that will be grazed.

Phosphate and potash for grassland establishment

Young developing grass and clover plants have a high demand for phosphate and potash, particularly phosphate. As the young plant has a very small root system, it is necessary to ensure that fertiliser supplies large quantities of available phosphate and potash.

For undersowing use normal cereal recommendation for phosphate and potash at sowing: if soil is index 0 or 1, apply 50 kg/ha of P_2O_5 and/or K_2O immediately after harvest.

For other methods of establishment apply the following before or at time of sowing:

Soil index	*kg/ha of* P_2O_5				*kg/ha of* K_2O			
	0	1	2	3+	0	1	2	3+
	100	80	60	40	100	80	40	nil

Other nutrients

Calcium

Calcium is applied in lime and maintenance of a pH of 5.8 or above will ensure that acidity does not limit the growth of grasses and clovers *and* that adequate calcium is present in the soil.

Magnesium

Magnesium deficiency can occur in grazing animals as hypomagnesaemia. This is a problem that cannot be cured reliably by applying magnesium to the soil; the best prevention of the disease is to place magnesium compounds in the animals' food or water. However, where soil magnesium is low, it is sound practice to use a magnesium-containing form of lime when lime is applied.

Grassland in the UK suffers rarely if ever from other nutrient deficiencies, although occasionally grazing stock may suffer from lack of minerals such as copper. The safest rule is to supply the deficient element directly to the animal rather than through the herbage: this not only ensures that the animals concerned receive a correct dosage but also ensures that areas of grassland with adequate levels of minerals are not enhanced to reach toxic levels.

PATTERNS OF GRASSLAND PRODUCTION

The basic pattern of grassland production is shown in *Figure 6.5*, and shows that grass can be available for grazing over a long period of the year, with conservation as hay or silage removing surplus grass and making this grass available for winter feeding. As conserved grass has a lower feed-value than fresh grass, other feeds have to be used with hay and silage, normally a cereal-based concentrate.

Generally over 50% of total annual grass production has occurred by the end of May and some management practices can emphasise the spring peak even more than this. For example, heavy use of N during spring with much lower rates later in the year can result in 70% of annual production by the end of May: where one very heavy conservation cut is taken in early June, this cut can account for as much as 85% of annual production.

Spring peak in production will be *emphasised* by:

(1) very high N use in April–May, particularly as this can limit July–September production below that which would otherwise be obtained;
(2) taking one very heavy cut in spring, particularly as this will deplete tiller numbers and subsequent regrowth;

(3) use of a single grass variety, particularly an early-heading one;
(4) low summer rainfall reducing production from June onwards.

Spring peak will be *minimised* by:

(1) restricting N use in April–May, although this will lower annual grass yield as grass is most responsive to N during this period.
(2) very frequent defoliation in the May–June period.
(3) using a mixture of grasses with different spring growth periods.

Every farmer has slight differences in his detailed grassland management, but the underlying strategy is the same, that is to utilise by grazing and conservation a high proportion of the grass grown, and at a stage of growth where its feeding value is suitably high for the system concerned.

GRASS YIELD AND FEEDING VALUE

Ruminant animals depend on grass as a major source of energy and protein, and also obtain from grass some essential minerals and vitamins. The protein value of grass is expressed normally as crude protein (CP), this being calculated by multiplying the total N content of the material by 6.25. Rumen bacteria can synthesise amino acids from simple N substances and so not all the protein needed by the ruminant has to enter its mouth in protein form.

The total energy value of the food eaten by the animal is its *gross energy* (GE). Some of this energy is passed out in faeces and the remainder is *digestible energy* (DE). Some of the DE is lost as methane from the rumen and some is passed out in urine; that

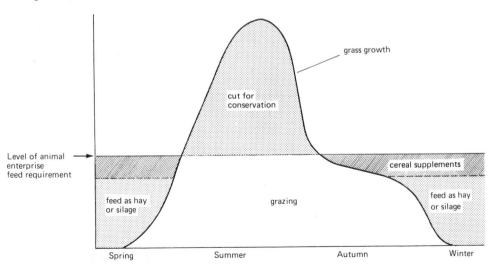

Figure 6.5 Pattern of grass production and use

Table 6.12 Effect of cutting frequency on grass yield and quality

Three cut system			Two cut system		
Date	DM yield (tonnes/ha)	ME (MJ/kg DM)	Date	DM yield (tonnes/ha)	ME (MJ/kg DM)
(a) *Field data*					
late May	4.6	10.6	early June	7.8	9.6
early July	3.2	9.8	mid-August	3.7	9.0
mid-August	1.8	9.6			
Total	9.6 Mean	10.0	Total	11.5 Mean	9.2
(b) *Feeding data*					
concentrate feeding	low	high		low	high
concentrate intake (kg/d)	5.0	9.3		5.0	9.3
silage intake (kg/d DM)	11.2	9.0		10.3	8.4
milk yield (ℓ/d)	19.9	20.8		17.3	19.7
land required for 180-d winter (ha/cow)	0.25	0.21		0.20	0.16

Source: Leaver and Moisey, 1980.

remaining is termed *metabolisable energy* (ME). Some ME is lost as heat and the remainder, *net energy* (NE), is used for maintenance and production. Whilst NE is the nearest assessment to the production potential of a feed, it has been found that ME is well-correlated with production and is much easier to assess on a routine basis than NE.

In the 1950s workers at the Grassland Research Institute found that the *digestibility* of grass (i.e. the percentage of grass eaten that was digested by the animal) was well-correlated with the animal intake of grass and with the subsequent animal production from it. Since that time much advice and literature refers to digestibility. By convention, the term 'D-value' (i.e. the use of the capital letter D) should refer specifically to *the digestible organic matter as a percentage of the dry matter*. There is a relationship between D-value and ME in grass, such that

$$ME = 0.15D$$

Figure 6.4 shows the effect of increasing grass growth period (maturity) on yield, N uptake, CP content and ME value. As grass matures it increases in yield but decreases in % CP and ME value: it is important to define the quality of grass required and defoliate it at the appropriate stage of growth. Generally high grass quality is associated with low yields and *vice versa* (*see Table 6.12*). Not only does a feed low in ME provide a low feed value, but also the animals eat less of such a feed. *Table 6.13* shows the double-action of low ME value on intake and production (Brockman and Gwynn, 1961).

Clearly the target ME-value of grass must depend on the productive potential of the animals concerned and on whether the grass is grazed or conserved: the latter point being influenced by the inevitable loss of feed value when grass is conserved and the need for at least a moderate yield to justify the costs of conservation. Guidelines for minimum ME levels for good production from grass are:

Utilisation	Minimum ME (MJ/kg DM)
Dairy cow grazing	11.5
Grazing other growing stock	10.5
Grazing dry cows	10.0
Grazing dry ewes	9.5
Silage and barn-dried hay	9.5
Field hay	9.0

Table 6.13 Effect of grass ME value on intake and production in dairy cows

Grass ME value (MJ/kg DM)	Grass DM intake (kg/d)	ME intake (MJ/d)	Milk yield (ℓ/d)
11.3	12.8	145	14.6
9.5	12.0	114	11.8
8.4	10.6	89	10.0

GRAZING SYSTEMS

Many areas of grass, particularly those used for beef and sheep, are grazed on a very extensive basis that is not systematic in nature. On the other hand most dairy farmers have a very definite basis to their grazing policy and an increasing number of the more successful beef and sheep farmers are adopting grazing strategies similar in principle to dairy systems. This section will deal with the major methods used to graze dairy cows; some of the methods outlined are appropriate for other stock.

Whatever the grazing method or stock used, the following basic principles should apply:

(1) *for stock*, obtain a high intake of grass at above minimum ME value;
(2) *for grass*, maintain good utilisation commensurate with good intake.

Two sward system

Here one area is regularly cut for conservation and the other regularly grazed. It is difficult to accommodate the seasonal growth pattern in the grazed area, but the system has advantages where part of the grass area is inaccessible to grazing (e.g. on a split-farm) or where the cutting grass is in an arable rotation and does not justify fencing and a water supply.

Otherwise the integration of cutting with grazing is a powerful tool in smoothing out the availability of grass for grazing, because whereas only one-third of the total area may be needed for grazing in May, two-thirds may be needed in June–July and the whole area from August onwards.

Set-stocking

This implies a given number of stock on a fixed area for a long period, often the whole season. This method is advised under very extensive conditions only at low stocking rates, as there is a tendency for undergrazing in spring, leading to poor quality mature herbage, followed by overgrazing in late summer and autumn resulting in low animal performance. One advantage is that provided the perimeter of the area is stock-proof, fencing costs and water supply problems are minimal. The term *set-stocking* is sometimes used erroneously for *continuous (or full) grazing (see below)*

For	Against
simple and cheap	does not match grass growth
good for large groups	possible disease build-up

Continuous grazing (or full grazing)

This refers to systems where the stock are allowed to graze over a large area for a fairly long period, say two to three months. Many farmers use this system for dairy cows and associate it with a high stocking rate, high and regular fertiliser N usage and a willingness to feed cereals if grass supplies become inadequate. As rate of grass growth slackens during the season, farmers can both reduce stock numbers as cows dry off and increase the grazing area by adding some silage aftermaths. Thus part of the area may be grazed for the whole season, but neither the total area nor the stock numbers are fixed, and supplementary feed is used as a buffer against periods of low grass availability. This system depends on grass growing as rapidly as possible over the whole season and is best associated with high N application, with monthly applications of about 70 kg/ha of N from March to August.

Overall there should be 1 ha of available grazing for every three cows: at turnout in spring the whole area may be needed but as growth picks up the area for spring grazing will be about 1 ha per five cows; following spring silage cuts the area may need extension to 1 ha per four cows, with the whole area in use from August onwards, depending on the season.

For	Against
reduces poaching as stock dispersed	no visible assessment of grass growth
saves cost in fencing and water	regular check on performance essential
increases sward density and clover	cow collection can take time
eliminates daily decision on grazing area allocation	willingness to feed supplementary feed at grass essential

Three-field system

This is one that has much to commend it if the whole grass area can be grazed conveniently (this is often difficult with cows that have to walk to and from milking and for this reason this method is more often used in beef and sheep systems). Basically this method formalises a continuous grazing system into three periods in the season and in a very simple way adjusts to the expected pattern of grass production. The grassland area should be split by stock-proof fencing into two areas, one area being approximately double the size of the other. In spring the stock graze the smaller area and the larger area is cut, preferably for silage as this will provide more reliable aftermaths than where hay is taken. As soon as the aftermaths are available, the stock are switched and the smaller area is taken for a conservation cut. Once the smaller area has available aftermaths, the whole area is grazed.

For	Against
simple self-adjusting to seasonal growth	inflexible
suitable for large animal groups	based on *expected*, not *actual* growth
good parasite control from 'clean' grass	often unsuited to dairy grazing

Block grazing systems

These are ones in which the grazing area is divided into a number of fairly large blocks with the aim of grazing on a rotational basis. Thus one block of grass is grazed with a large number of animals for a short period, usually 1–7 d, and then they are moved to other blocks whilst the grazed block regrows. After about one month or less, the block can be grazed again. Blocks that are surplus to grazing are cut for conservation. This basic principle is used for all classes of grazing stock, with the following common amendments.

For dairy cows Farmers often like cows to have a fresh allocation of herbage each day, so an electric fence can be used to ration the grass. This is formalised in the *Wye College* system where cows are given one-seventh of a block each day and one block lasts one week.

For ewes and lambs   The fence separating adjacent blocks can have spaces wide enough for the lambs to pass but not the ewes. Thus lambs can obtain the pick of the next block to be grazed by the ewes (*forward creep grazing*) or creep into an area never grazed by ewes (*sideways creep grazing*).

For cattle rearing   Young stock may be grazed one paddock ahead of older cattle so that the younger ones obtain the pick of the grass and the older cattle clean up, ensuring good utilisation. This system is called *leader follower grazing* and has been used particularly in heifer rearing.

For dairy cows   Occasionally farmers split their herd and do a leader/follower with the cows, putting the high yielders in the leader group, or putting dry and nearly-dry cows as followers behind the main herd.

For	*Against*
efficient grass utilisation	good fencing round each
good parasite control for	paddock essential
young stock	regular stock-moving de-
areas large enough for	cisions needed
conservation	many water points
	needed

Paddock grazing

This represents a very formal method of rotational grazing, where the grazing area is divided into some 21–28 permanently-fenced and equal-sized paddocks. The aim is that one paddock is grazed each day and the rotation around the area is completed as soon as the first paddock is ready for grazing again. Surplus paddocks can be taken out for conservation but usually the operation is restricted by the small size of the paddocks. Paddock size must be related to the number of grazing animals, as the stock density must be sufficient to ensure that the grass is efficiently utilised during the short grazing period. A guide is to allow 100–125 cows/ha daily, depending on stocking rate.

Thus for a highly-stocked herd of 100 cows and where 25 paddocks are selected, each paddock should be about 1 ha in size. Every paddock must have a water point and independent access on to a track to avoid poaching.

For	*Against*
easiest rotational system	high fencing and water
to manage	costs
gives objectivity to graz-	small areas for
ing plan	conservation
	some wasted land in
	trackways

Strip-grazing

This involves the use of an electric fence to give a fresh strip of herbage once or twice daily. Ideally, stock should be confined to the daily strip only, to prevent the regrazing of young regrowth, but this is done rarely, because of problems with animal access and water supply. Although this is the most sensitive system to allow adjustment for fluctuations in grass growth, it is not common because:

(1) it requires daily labour to decide on area allocation and move the fence;
(2) there is a tendency for grass to become over-mature ahead of grazing in large fields;
(3) there is a risk of serious poaching along the fence line in wet weather.

Many of the best features of strip-grazing are achieved more easily in block grazing.

Zero grazing

This implies that grass at grazing stage is cut and transported to stock that are either housed or kept in a 'sacrifice' area. Zero-grazing is practised widely in some countries where grass is inaccessible for grazing (e.g. in 'strip field' systems) or where a range of crops need to be grown to ensure continuity of supplies (e.g. in semi-arid areas). In temperate climates there is research evidence (Marsh, 1975) that stocking rates can be higher under zero grazing because:

(1) the sward does not suffer the deleterious physical effects of treading and selective grazing,
(2) herbage can be harvested at the ideal stage for the stock concerned, and
(3) utilisation can be high as there is no field refusal of grass (although this implies the stock eat all the grass carted to them).

Both milk and meat production per hectare have been shown to be greater from zero grazing than from other grazing systems. However, zero grazing is not used commonly in the UK except for specific opportunist reasons such as:

(1) reduce poaching in early spring and autumn;
(2) provide 'grazing' from distant or inaccessible fields;
(3) use other crops during crisis periods, as in drought.

The reason for the lack of popularity of zero grazing despite its technical excellence for high production per hectare are:

(1) high labour and machinery costs for feeding *and* for removing slurry;

(2) complete dependence on machinery, with risk of breakdowns;

(3) output per animal tends to be depressed (but high SR gives high output per hectare);

(4) problems of refused grass at feeding face, particularly as wet grass heats rapidly when heaped.

CONSERVATION

Conserved grass is used as the basis of ruminant feeding when grass for grazing is not available. Grass for conservation is either grown specifically for this purpose or taken as a surplus in a grazed area: often both situations apply in any one year on an individual farm. Where grass is grown for conservation, it is possible to tailor the crop to fit the particular requirements of the unit, in terms of herbage varieties, yield and feeding quality. Where grass arises as a grazing surplus, an over-riding factor is the rapid removal of the crop in order to allow regrowth for a further grazing.

When green crops are cut biochemical changes occur and if these are allowed to continue unchecked, degradation of the material will take place, releasing heat and effluent (water that contains much of the soluble cell contents). There are four methods by which this progressive degradation can be restrained, but at present only the first two methods are used in commercial practice and only these two methods will be discussed in detail.

Dehydration

Degradation ceases when the moisture content of the material falls below 15% (85% DM: grass in the field averages about 20% DM).

Dehydration can take place under natural conditions in the field and is *haymaking*. Sometimes hay that is almost dry is placed in a building and subjected to forced-draught ventilation to complete dehydration and is *barn hay drying*. On some specialised units, grass at or near field moisture is carted to a high-temperature drier and dehydrated very rapidly; this is *green crop drying*.

Acidification

Degradation ceases when the pH of the material falls below 4.2–5.0 (depending on the moisture content) *provided* the material is anaerobic (oxygen-free). This material is *silage*, and can be made by heaping herbage in as near oxygen-free conditions as possible and allowing the natural process of biochemical change to produce acids which can effectively 'pickle' the material. Additives can be used to both accelerate the process and give a more reliable end-product.

Preservation

Some chemicals can prevent the process of degradation and so preserve the material in almost the same chemical state as at cutting.

At present research is seeking a suitable chemical that will preserve herbage economically and yet not restrict animal intake, affect rumen fermentation or be carried into the animal products.

Freezing

Freezing will inhibit completely chemical breakdown in the material.

For many years research institutes have been using frozen grass in feeding experiments as a normal means of preserving fresh grass and all its intrinsic value. At present the cost of freezing and storing herbage on a farm-scale is prohibitive.

Haymaking

Haymaking is still the most popular method of conservation, involving the reduction of moisture from fresh grass at 80% to about 20% when the product can be stored. Successful haymaking should follow these principles:

(1) Grass should be at the correct stage of growth when cut. As grass is allowed to mature its total yield increases and its moisture content falls, so it may be tempting to allow a very heavy, mature crop to develop before cutting. *But* digestibility falls at a rate of one-third to one-half of a D-value unit per day once the seedheads have formed; mature hay has low feed value and low intake characteristics, even though it may be well made.

(2) Losses should be kept to a minimum. Losses can arise in the following ways:

(a) respiration losses will occur in the field as the herbage continues to respire after cutting. Rapid drying will minimise these losses, which can amount to 1–5% of the DM yield.

(b) Mechanical losses occur if herbage is fragmented by machinery into pieces too small to be picked when the hay is collected for storage. Mechanical losses tend to become greater as the herbage becomes drier and when the action of the machines is abrasive. A mechanical loss of 10% is acceptable.

(c) Leaching of nutrients can take place when the cut material is exposed to rain. Ideally hay is made and removed from the field without rainfall, when this loss is zero, but long periods of heavy rain can lead to a DM loss of up to 15% and a soluble nutrient loss far in excess of this.

(d) Some of the hay may itself be inedible due to dust and mould. Both these factors arise when hay is made under adverse conditions: the loss can be up to 15% of the DM and it emphasises the importance of making hay under good climatic conditions.

When grass is cut and a wide area of swath exposed to wind and sun, there is a rapid initial loss of moisture, as external moisture is lost and water from the outer cells of the leaves and stems. At this stage drying will take place without much sun provided the atmosphere has a low relative humidity. Also at this stage of drying the material can be treated quite roughly by machines. As drying proceeds it becomes progressively more difficult to remove water and the rate of moisture loss declines, together with an increasing risk of mechanical damage. For continued drying sun is necessary to provide heat and wind is valuable for removing water vapour from within the swath. The rate of drying can be speeded up if the grass is cut with a flail mower or passed through a crimper/conditioner immediately after cutting: however, grass treated in this way suffers more in bad weather, and flail mown material can suffer mechanical losses as high as 50% if subjected to prolonged wet weather.

Field drying rate is maximised if the cutting machine leaves the largest possible leaf area exposed to sun and wind, and not tight swaths. For this reason drum mowers have acquired a deserved reputation. If bad weather threatens during the drying process, the material should be winrowed to present the smallest area to the rain. Also it is sound practice to swath-up the grass at night to minimise the effect of dew, ensuring the ground and surface of the swath are dry before the material is again spread next morning.

The rate of drying depends on the weather, proper use of machinery, rates of N used, varieties in the sward, stage of maturity at cutting, bulk of grass present, time of year and desired moisture content at transporting from the field. Most hay is baled and it should not be baled until it is fit for storage, i.e. with a moisture content of about 20%. Frequently hay is baled at a higher moisture content than this and then the bales are left in the field to 'cure'. Very little further moisture loss can occur in the field from the bale, and yet the bales will acquire considerable moisture if rain falls. As a general rule, baling should be regarded as the first step in the process of transport into storage, and grass should not be baled until the herbage is dry and transport into storage is organised.

Barn hay drying

This is a very useful technique for reducing the risk of bad weather, minimising losses and making hay from younger material (at a higher feed value) or where higher N rates have been used. The material is cured as far as possible in the field, certainly down to at least 30% moisture. It can then be baled, taking care not to over-compress in the bale, and the bales carefully placed over a grid or ducts through which air can be blown. There must be no gaps between the bales, otherwise air will take the line of least resistance and fail to pass through the bales. If the outside humidity is low and only some 5–8% of moisture needs to be removed, then cold air blown for up to a week will suffice, often ceasing to blow at night when humidity might be high. During periods of sustained high humidity or when considerable moisture must be removed, then the air must be heated. The air can be heated by either a flame or electric heaters, but in any event the heating of air is *very expensive*. Also, all barn-drying installations cannot deal with a sudden large volume of material to cure. Barn-drying is best regarded as a means of producing some very high quality hay either from the rest of the hay being made or from within a silage system.

Success in barn-drying depends on:

(1) justifying the extra costs involved, rather by the higher quality of the hay than by a salvage operation on a mediocre crop;
(2) allowing moisture content to fall to below 30% (and certainly 35%) before baling;
(3) using bales at a low-moderate packed density;
(4) stacking the bales carefully to avoid cracks through which the majority of the air can pass;
(5) having sufficient fan capacity to obtain a good flow of air, with heating available if humidity is high or bales are wet.

Hay additives

Hay additives are based on either propionic acid or ammonium bispropionate. Both chemicals have a strong anti-fungal activity and will reduce the rate of decomposition of hay (and control heating) when the material is still too wet for immediate storage. However there are four snags to hay additives:

(1) It is very difficult to apply these chemicals evenly to the herbage. Attempts to apply within the bale chamber or spray on the swaths before baling have both failed to obtain the necessary even application (in contrast to the addition of silage additives in a forage harvester).
(2) Although both the chemicals mentioned give good initial control of decomposition, they are broken down gradually and give protection for some two to three weeks only. Thus the hay must be dehydrated to a proper storage moisture content soon after baling and storage.
(3) Under warm conditions at baling some 50–75% of the additive may be lost by volatilisation.
(4) Rate of application should depend on the moisture content of the hay—and this will vary from swath to swath and throughout the day, for example:

% moisture	kg of additive/tonne of hay
28	7
30	9
32	11
34	13

Development of a successful hay additive is proving difficult and at present it should not be regarded as a means of storing wet hay but rather as an aid to delaying the need to get to a safe storage moisture content.

Hay facts and figures

Quality in hay can be judged by its colour, which should be bright green/yellow, by its sweet smell and an absence of dust. Also the feeding value of hay can be determined by its analysis: typical figures for a range of types of hay are given in *Table 6.14*.

Yields

Yields of hay vary considerably, and high N rates should not be used to grow a hay crop because it aggravates the curing problem: a maximum N rate of 80 kg/ha for the growth period is recommended.

light crop   = 2–3 tonnes/ha of made hay
medium crop = 4.5 tonnes/ha of made hay
heavy crop   = 5 tonnes/ha and over

A standard bale of hay measures approximately 0.9 m × 0.45 m × 0.35 m and weighs 20–30 kg depending on type of material and density (33–50 bales/tonne). A big round bale of hay measures about 1.5 m high and 1.8 m in diameter and contains some 25 times more material than a standard bale, weighing 500–600 kg. A big rectangular bale is approximately 1.5 m × 1.5 m × 2.4 m with the same weight as the round bale.

Approximate storage volumes are:

	m^3/tonne
loose medium-length hay	12–15
standard bales	7–10
big rectangular bales	10

Green crop drying

When green crops are passed through an efficient high temperature drier, they are rapidly reduced to a stable moisture content with little or no loss of nutrient value. Also artificially dried green crops are very palatable to stock. In the UK the majority of green crops dried are either grass or lucerne; both are traded under the general term 'dried grass'. Less than 1% of the grassland area of the UK is used for dried grass production, and most of the drying is done in very specialised large units where the size of operation justifies the use of the large, very efficient high-temperature triple pass drier, the smallest of which can produce some 3000 tonnes of product per year, needing about 400 ha of adjacent land to provide material. Most green crop driers are members of the British Association of Green Crop Driers, who supply specialist information to members on a variety of topics related to the industry. Although dried grass has a very high reputation as a supplement to silage in cattle and cow rations, it is unlikely that green crop drying will expand in the foreseeable future because it demands a high input of fossil fuel—equivalent to about 200 ℓ of oil per tonne of material produced, quite apart from the large equipment needed to cut and transport the grass to the drier and the power needed by the mill and cuber to package the material after drying.

In an efficient grass-drying unit, dry-matter losses are only about 3–5%, and in this respect the technique is far superior to hay and silage making. Some feed analysis figures for dried grass and lucerne are given in *Table 6.14*.

Silage making

The process of ensilage consists of preserving green forage crops under acid conditions in a succulent state. When such green material is heaped, it respires until all the oxygen in the matrix is exhausted: during respiration carbohydrates are oxidised to carbon dioxide with evolution of heat. Continued availability of oxygen, as in a small outside heap of grass, will lead to enhanced oxidation and overheating.

Assuming the oxygen supply is restricted, bacteria can control the fermentation process. These bacteria, present on the crop, the machinery and in soil contamination, fall into two categories—desirable and undesirable.

The desirable bacteria are ones which can convert carbohydrate into lactic acid and are mainly of the *Lactobacillus* and *Streptococcus* species. These are anaerobic bacteria and their even distribution and activity throughout the grass in the silo is encouraged by mechanical chopping of the grass, rapid consolidation and exclusion of air. Lactic acid is a relatively strong organic acid and its rapid production within the ensiled grass leads to a low pH and conditions which inhibit the lactic acid producing bacteria and *all* other bacteria as well. The pH at which this 'pickling' occurs depends on the moisture content of the grass: the wetter the grass the lower the pH needed and the greater the quantity of lactic acid that has to be produced. Thus wet silage with a DM% of about 20 will need a pH of around 4.0 for stability but 30% DM grass will be stable at a pH of 4.6–4.8. Silage with a good lactic acid content is light brown in colour, has a sharp taste and little smell: it is very stable and can

Table 6.14 Typical analyses of some common grass and forage materials

Feed	Nutritive value on DM basis					Chemical analysis on DM basis		Digestibility coefficients	
	DM (%)	ME (MJ/kg)	DCP (%)	D-value	GE (MJ/kg)	Crude protein (%)	Crude fibre (%)	Crude protein (%)	Crude fibre (%)
Grasses									
Set-stocked pasture	20	12.1	22.5	75	18.6	26.5	13.0	85	81
3 week rotation pasture	20	12.1	18.5	75	18.9	22.5	15.5	82	81
4 week rotation pasture	20	11.2	13.0	72	18.5	17.5	22.5	74	82
Extensive grazing (spring)	20	10.0	12.4	64	18.0	17.5	20.0	71	65
Extensive grazing (autumn)	20	9.7	10.1	63	18.1	15.5	22.0	65	59
Legumes									
Red clover (flowering)	19	10.2	13.2	65	18.3	17.9	27.4	74	58
White clover (flowering)	19	9.0	15.2	57	18.1	23.7	23.2	64	60
Lucerne (early flower)	24	8.2	13.0	54	17.6	17.1	30.0	76	44
Lucerne (in bud)	22	9.4	16.4	62	18.2	20.5	28.2	80	50
Peas (early flowering)	17	8.5	14.0	56	18.8	20.6	35.3	68	50
Sanfoin (early flowering)	23	10.3	14.3	65	18.5	19.6	20.9	73	45
Sanfoin (full flowering)	25	8.4	11.6	54	18.3	17.5	28.5	69	49
Trefoil	20	9.0	12.1	57	18.5	17.5	28.5	69	49
Vetches (in flower)	18	8.6	12.3	56	18.2	17.8	29.4	69	45
Clamp silage									
Grass (high D)	25	10.2	11.6	67	18.1	17.0	30.0	68	81
Grass (mod. D)	25	9.3	10.7	61	18.2	17.0	30.5	63	76
Grass (low D)	25	8.8	10.2	58	18.2	16.0	34.0	64	73
Maize	21	10.8	7.0	65	18.8	11.0	23.3	64	68
Pea haulm	21	8.7	9.5	51	16.9	15.7	29.0	57	56
Rye	18	8.3	7.1	55	18.5	12.3	33.8	58	60
Sugar beet tops	23	7.9	6.5	50	13.4	10.4	14.8	62	73
Vetch and oats	27	9.6	8.2	60	18.3	12.6	29.3	65	58
Barley (whole-crop)	30	9.6	5.0	62	18.0	9.5	25.0	53	53
Oats (whole crop)	30	8.6	6.1	57	18.2	7.9	32.8	60	60
Wheat (whole crop)	30	8.4	3.6	55	18.2	7.8	30.0	47	43
Tower silage									
Grass (v. high D)	40	10.4	12.1	68	18.1	17.0	31.3	71	80
Hay									
Red clover (v. good)	85	9.6	12.8	61	18.4	18.4	26.6	70	50
Red clover (good)	85	8.9	10.3	57	18.5	16.1	28.7	64	47
Red clover (poor)	85	7.8	6.7	50	18.4	13.1	34.0	51	40
Grass (high D)	85	10.1	9.0	67	17.7	13.2	29.1	68	76
Grass (mod. D)	85	9.0	5.8	61	17.6	10.1	32.0	57	70
Grass (poor D)	85	8.4	3.9	57	17.7	8.5	32.8	46	61
Grass (v. poor D)	85	7.5	4.5	51	17.8	9.2	36.6	49	56
Lucerne (before flowering)	85	8.3	14.3	54	18.2	19.3	32.1	74	42
Lucerne (half flowering)	85	8.2	16.6	55	17.9	22.5	30.2	74	48
Lucerne (full flowering)	85	7.7	11.6	51	18.1	17.1	35.3	68	45
Sanfoin (in flower)	85	9.0	11.5	58	18.1	15.8	33.5	73	42
Trefoil	85	8.8	13.9	56	18.4	18.4	29.2	76	44
Vetches (full flower)	85	8.0	11.3	52	17.9	17.1	30.6	66	50
Vetches and oats	85	8.1	7.7	52	17.9	13.8	28.8	56	51
Dried grass and lucerne									
Grass (very leafy)	90	10.8	11.3	70	17.3	16.1	21.7	70	83
Grass (early flower)	90	9.7	9.7	64	17.6	15.4	25.8	63	75
Lucerne (bud stage)	90	9.4	17.4	60	17.7	24.4	19.8	71	53

All values taken from ADAS Paper LGR 21 'Nutrient allowances and composition of feedingstuffs for ruminants' (Crown copyright, reproduced with permission).

be kept for years if necessary provided nothing is done to permit oxygen to enter the material.

The undesirable bacteria are:

(1) obligate anaerobes of the *Clostridium* species that can ferment carbohydrate and lactic acid to form butyric acid;
(2) aerobic species of bacteria that can oxidise carbohydrate to carbon dioxide and water;
(3) *Clostridium sporogenes* which can break down amino acids to ammonia and amines, some of the latter being toxic to stock.

Butyric silage is olive-green in colour, has a rancid smell and is unpalatable to stock. Also it has a higher pH than lactic silage, is unstable and will not keep for more than a few months.

Clostridial activity in silage can be inhibited by:

(1) reducing moisture content of the grass, as this will lessen the quantity of acid needed to prevent decomposition;
(2) ensuring adequate carbohydrate is present for lactic acid bacteria, or applying an acid to assist in lowering pH;
(3) avoiding soil contamination.

Intrusion of air *during* fermentation will delay or even prevent the achievement of a stable pH and will lead to an excessive amount of carbohydrate being used, thus lowering the nutritional value of the silage. Intrusion of air *after* the silage has reached a stable condition will lead to secondary respiration and a further progressive loss of carbohydrate. Secondary respiration can shorten the storage life of well-made silage and can occur when the silo is opened for use if the exposed feeding face is too large for the rate of silage removal.

Silage-making can result in considerable loss of material during the wilting, fermentation and feeding periods. Field losses can range from 0–10% of the DM yield depending on degree of wilting in the field: fermentation losses are inevitable and commonly range from 10–20% of the DM yield: not all of the material present in the silo is suitable for feeding, due mainly to side and top waste, and in clamp systems this 'visible' loss can be 5–15% of the DM yield. Thus in a really good silage system, only 80% of the weight of grass cut will be available for feeding as silage and often the figure is as low as 65–70%.

For the best fermentation the crop should have a high carbohydrate content (to provide ample substrate for fermentation) and a low moisture content (to reduce the volume of material requiring acidification). For these reasons emphasis is given to cutting crops when their carbohydrate (soluble sugar) content is high and when they can be wilted quickly. ADAS advice is that sugar content should be at least 3% at time of cutting and grass should be wilted from its normal moisture content of about 80% down to 70–75%. Often these two desirable conditions cannot be met and farmers can use additives to help alleviate the problem.

Additives fall into three main types:

(1) *Sugars*   By adding extra carbohydrate, e.g. molasses, the crop is better able to produce a lactic fermentation. Some additives contain material designed to stimulate the *lactobacilli*.
(2) *Acids*   In some countries inorganic acids are used at high rates in the silo at time of filling to create a pickled effect. At present in the UK acids such as formic and sulphuric are applied at 3–5 ℓ/tonne of grass as the grass is picked up from the field. These acids reduce the quantity of lactic acid needed to reach a stable pH and consequently reduce the time taken for stability to be reached in the silo.
(3) *Preservatives*   Some acid additives also contain chemicals that should suppress unwanted biochemical reactions: examples of such chemicals are formalin and sodium metabisulphite.

Storage of silage is in *silos*, and these fall into two basic types, one where grass is added and removed horizontally (clamps) and one where it is done vertically (towers).

Clamps are found in a variety of forms, e.g. walled or unwalled, roofed or open, on the surface or in pits.

Towers are made on concrete or galvanised/vitreous enamelled steel: material for ensiling is always added to the top but, depending on type, silage is removed from either the top or the bottom.

Making silage in clamps

Assuming the material to be ensiled is either high in sugars and low in moisture or having an additive applied, the main principle during the filling process is to eliminate as much air from the matrix of herbage as possible and keep the material airtight. Polythene sheeting is an essential feature of silage-making and is available for this purpose in 300 or 500 gauge and in widths up to 10 m. The use of polythene sheets can be taken to the ultimate in the production of *vacuum silage*, where the crop is stacked on a sheet laid on the floor of the silo: then another sheet is placed over the heap and the two sheets joined together at ground level by a plastic 'strip-seal', after which the whole mass is evacuated by a vacuum pump. Such complete removal of oxygen leads to excellent fermentation, but the process is laborious and difficult to carry out on a large scale. Polythene sheeting is easily punctured and on some farms polythene in any form is nibbled by rodents. Some farmers have tried to make silage in plastic bags or polythene sausages: these techniques are again useful in making small quantities of high quality silage, but are not suited to the rapid filling rate of 100–200 tonnes of grass per day that is necessary on the medium-to-large farm.

Farmers have found that many of the advantages of vacuum and similar techniques can be obtained more simply by making silage in a walled pit using a *wedge-filling* principle (sometimes called *Dorset wedge*) (*see Figure 6.6*).

On the first day of filling the cut crop is stacked at one end of the silo, against an end wall (*ad*) that is either solid or has a polythene sheet lining. If the side walls are not solid, they too must have a polythene lining. The slope *bc* is used to load the material into the silo using a push-off buckrake. The buckraking tractor maintains the slope at the angle *dcb* and when the material has reached the maximum intended height *ad* the slope is progressed forwards, leaving a fixed height of material. One principal objective of the wedge system is to prevent warm air rising out of the silo, as this will encourage oxygen-rich cold air to come in at the bottom and sides: so a polythene sheet is placed over the grass each night, and when one section of filling is complete, the sheet is left in place, so that the silo is gradually wrapped in polythene sheets. If the technique is carried out correctly, the oxygen in the air in the silo is soon used up, lactic acid fermentation proceeds, and the crop consolidates under its own weight, often resulting in a drop in crop height to two-thirds to three-quarters of the original. Because of this shrinkage the polythene sheet should not be fixed rigidly, but rather covered with a flexible and convenient material such as old tyres, sand bags or even a net. If the sheet is not held down it will flap in the wind, allowing in more air and eventually tearing the sheet.

Long material can be made into good silage in clamps, but chopped material is easier to handle and consolidates better, with less oxygen trapped in the matrix. Fairly wet grass can be placed in a clamp, but it will produce effluent. ADAS figures show that grass ensiled at 80% moisture will release an average of 200 ℓ of effluent per tonne of grass ensiled, and the quantity of effluent decreases progressively until material ensiled at 72% moisture should give no effluent.

Silage effluent

This is a real problem if it enters a water course, as it has a high Biological Oxygen Demand (BOD), and will kill many oxygen-demanding organisms in the water, including fish. Many farmers wilt their grass simply to avoid or minimise the effluent problem. As a precaution it is advisable to construct an effluent tank adjacent to the silo, taking care to ensure that *only* silage effluent can enter it (i.e. no surface or rainwater). Silage effluent is very acid and all effluent-conducting channels and ducts must be coated with acid-resisting material. Silage effluent contains some plant nutrients (say 2, 1 and 1.5 kg/1000 ℓ of N, P_2O_5 and K_2O respectively) but as it is very acid it is very phytotoxic. It can be applied to arable land by tanker or a slurry irrigation system, but care is needed to ensure it does not get into land drains and hence to a water course.

Making silage in towers

The crop has to be blown into the top of the tower and at feeding time the silage is removed by mechanical means: thus it is vital that the ensiled material is well-chopped and sufficiently dry to remain friable after compaction in the tower during storage. For this reason, grass going into a tower must be below 65% moisture and many tower operators prefer 45–55% moisture (sometimes called *haylage*). Very wet material must *never* be placed in a tower as the effluent

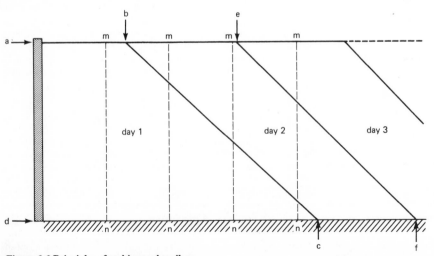

Figure 6.6 Principles of making wedge silage

from it will cause very serious corrosion to the tower structure leading to possible collapse of the tower. Because the material going into a tower must be well-chopped, and because a tower is almost airtight, compaction in a tower is good and excellent fermentation is assured. Towers have a deserved high reputation for producing good silage with minimum in-silo losses of 5–10%. (But field losses during the necessary prolonged wilting period will be 10–15%.) Some farmers have shown that application of 'tower' techniques in terms of wilting and chopping the grass, but then placing the material in a polythene-lined clamp produce silage equivalent to the tower: even so the tower is a first-class starting point for mechanised feeding and on some farms is justified for this reason alone.

Feeding clamp silage

Feeding methods can be split into two types—self-feeding and mechanical feeding.

Self-feeding

This takes place when the animals are allowed to help themselves to the silage. The settled height of the silage must not be higher than the animals can reach and usually the animals' access to the face is restricted by some physical means, such as an electric fence, so that they cannot climb on the silage or pull out big lumps of silage and waste it by treading it under foot. Where animals are given 24-h access to the face, then some 20 cm of face width should be allowed per cow equivalent. If access time is limited, the width per beast must be greater. The rate of feeding can be controlled by the distance the electric fence is moved each day, and by this means the silage can be rationed to last the whole winter. Self-feeding is a very cheap and effective method of feeding silage and is very popular with farmers. The only real justifications for using other feeding techniques for clamp silage are where several groups of animals are to be fed from one silo face, or where a balanced feeding programme is planned in which silage must be premixed with other feeds.

Mechanical removal of silage from clamps can be done by fore-loader, grab, block-cutter or a cylindrical silage-cutter. Where the silage is greatly loosened during removal, it must be eaten within 12 h or it will deteriorate from re-fermentation: in this respect the block-cutter is good as it removes blocks weighing some 0.6–0.7 tonnes without losing much of the original density of the silage.

Crops for ensilage

Many green crops can be ensiled and also some arable by-products such as *sugar beet tops* and *pea haulms*. By far the most common crop is *grass* (including *grass/clover herbage*) and grass should be ensiled when its digestibility is at least 65, and for high quality silage, the D-value at cutting must be 67–70.

The crop will lose about 2 units of D during the ensilage period even where there is a good fermentation, and as much as 5 units of D can be lost if fermentation is poor.

Whole-crop cereals

These can be made into silage, the cereals being cut about two weeks after full ear emergence. Yields of 8–10 tonnes/ha of DM can be obtained from this crop, but digestibility is often low at around 62–63D, so that animal intake and performance are fairly low: also it is possible for whole-crop barley to have an intake below expectation due to the physical dislike of the barley awns in the silage. Oats have the highest feed value of the cereals, but in practice other considerations may dictate choice of crop, such as the need to sow a quick-growing cereal in spring to meet an emergency situation, in which case barley will be the choice.

Maize

Maize can make excellent silage provided it is well-chopped before ensiling. A crop of maize with grain at the 'pasty' stage has a high carbohydrate and low protein content, a D-value of 63–65 and a moisture content of about 70% as it stands in the field. In the silo, this crop ferments well without additives and it is very well suited to mechanical handling. Provided its low protein value is recognised at feeding time, it is excellent material for silage and a good contributor to stock nutrition. The main snag with maize is the relatively low yield obtained in the UK coupled with the late harvesting in October or even November. A good crop of maize should produce 30 tonnes/ha of 30% DM material in mild, sunny parts of southern England, but in practice yield of only 16–20 tonnes/ha have been recorded.

Legume silage

This usually based on red clover or lucerne, can have a high protein content, but it is more difficult to get a good fermentation than with grass, cereals or maize because of the low sugar content in legumes: an additive is very often essential. Also legumes tend to have fibrous stems when cut for silage and the material compacts much better in the silo if a precision-chop harvester is used.

Special mixtures

These are sometimes grown for arable silage, usually based on the traditional oats-legume combination: this has the advantage of combining the high protein value of the legume with the good carbohydrate

Table 6.15

Cutting period	Fertiliser (kg/ha)	Yield (tonnes/ha)		
		DM cut	Grass cut at 25% DM	Silage fed at 25% DM
May	120–150N + 30 P_2O_5 and K_2O on index 0, 1 soils	4–6	16–24	12–18
July	100–125N + at least 50 K_2O on soils below 3	3–4	12–16	9–12
Aug/Sept	80–100N + at least 40 K_2O on soils below 3	2	8	6
	Average high yield from 3 cuts	12	48	32

Density of silage
clamp silo unchopped 20–25% DM = 720–800 kg/m^3
            chopped   20–25% DM = 800–850 kg/m^3
tower   chopped   35–45% DM = 500 kg/m^3

Feeding values
Typical values for a range of silages are given in *Table 6.14*.

content of the cereal. Also this type of mixture requires less fertiliser N than straight cereals or grasses. Examples of such mixtures are:

	125 kg/ha oats	125 kg/ha oats
	35 kg/ha vetches	45 kg/ha forage peas
or	35 kg/ha beans	

Silage—facts and figures

Yields

High yields of crop are essential for silage to justify the machinery labour and fuel costs involved. *Table 6.15* is a guide to the fertiliser rates and expected yields of grass silage cut at three different times in the season.

GLOSSARY OF GRASSLAND TERMS

For a full explanation of these and other terms *see* Hodgson (1979) and Thomas (1980).

Anthesis: flower opening.
Biomass: weight of living plant and/or animal material.
Browsing: the defoliation by animals of the above-ground parts of shrubs and trees.
Canopy: the sward canopy as it intercepts or absorbs light.
Canopy structure: the distribution and arrangement of the components of the canopy.
Closed canopy: a canopy which either has achieved complete cover or intercepts 95% of visible light.
Cover: the proportion of the ground area covered by the canopy when viewed vertically.
Crop growth rate (CGR): the rate of increase in dry weight per unit area of all or part of a sward.

Crop (standing): the herbage growing in the field before it is harvested.
Crown: the top of the tap-root bearing buds from which the basal leaf rosette and shoots arise (appropriate to clovers but *not* grasses).
Culm: the extended stem of a grass tiller bearing the inflorescence.
Date of heading (or date of ear (inflorescence) emergence): for a sward, the date on which 50% of the ears in fertile tillers have emerged.
Defoliation: the severing and removal of part or all of the herbage by grazing animals or cutting machines.
Density: the number of items (e.g. plants or tillers) per unit area.
Foliage: a collective term for the leaves of a plant or community.
Forage: any plant material, except in concentrated feeds, used as a food for domestic herbivores.
Forage feeding: the practice of cutting herbage from a sward (or foliage from other crops) for feeding fresh to animals.
Grassland: the type of plant community, natural or sown, dominated by herbaceous species such as grasses and clovers.
Grazing: defoliation by animals.
Grazing cycle: the length of time between the beginning of one grazing period and the beginning of the next.
Grazing period: the length of time for which a particular area of land is grazed.
Grazing pressure: the number of animals of a specified class per unit weight of herbage at a point of time.
Grazing systems:
  Continuous stocking: the practice of allowing animals unrestricted access to an area of land for the whole or a substantial part of a grazing season.
  Set stocking: the practice of allowing a fixed number of animals unrestricted access to a fixed area of land for a substantial part of a grazing season.

Rotational grazing: the practice of imposing a regular sequence of grazing and rest from grazing upon a series of grazing areas.

Creep grazing: the practice of allowing young animals (lambs or calves) to graze an area which their dams cannot reach.

Mixed grazing: the use of cattle and sheep in a common grazing system whether or not the two species graze the same area of land at the same time.

Harvesting: defoliation by machines.

Harvest year: first full harvest year is the calendar year following the seeding year.

Herbage: the above-ground parts of a sward viewed as an accumulation of plant material with characteristics of mass and nutritive value.

Herbage allowance: the weight of herbage per unit of live weight at a point in time.

Herbage consumed: the herbage mass once it has been removed by grazing animals.

Herbage cut: the stratum of material above cutting height.

Herbage growth: the increase in weight of herbage per unit area over a given time interval due to the production of new material.

Herbage mass: the weight per unit area of the standing crop of herbage above the defoliation height.

Herbage residual: the herbage remaining after defoliation.

Inflorescence emergence: the first appearance of the tip of a grass inflorescence at the mouth of the sheath of the flag-leaf.

Leaf: in grasses = lamina + ligule + sheath; in clovers = lamina + petiole + stipule. Note leaf is *not* lamina.

Leaf area index (LAI): the area of green leaf (one side) per unit area of ground; *Critical LAI* is LAI at which 95% visible light is intercepted: *Maximum LAI* is the greatest LAI produced by a sward during a growth period: *Optimum LAI* is the LAI at which maximum crop growth rate is achieved.

Leaf area ratio: total lamina area divided by total plant weight: *specific leaf area* is the lamina area divided by lamina weight.

Leaf burn (or scorch): damage to leaves caused by severe weather conditions, herbicides, etc. This contrasts with *leaf senescence* which is a genetically predetermined process where leaves age and die, usually involving degradation of the chlorophyll in the leaves.

Leaf emergence: a leaf in grass is fully emerged when its ligule is visible or when the lamina adopts an angle to the sheath. In clovers the leaf is emerged when the leaflets have unfolded along the midrib and are almost flat.

Flag leaf: in grass this is the final leaf produced on the flowering stem.

Leaf length to weight ratio: lamina weight divided by lamina length.

Net assimilation rate (NAR): defined as $1/A \cdot dW/dt$ where A = total leaf area, W = total plant weight, t = time, units = $(g\ m^2)/d$

Palatable: pleasant to taste.

Plastochron: the interval between the initiation of successive leaf primordia on a stem axis.

Preference: the discrimination exerted by animals between areas of a sward or components of a sward canopy, or between species in cut herbage. *Preference ranking* is based on the relative intake of herbage samples when the animals have a completely free choice of the materials.

Pseudostem: the concentric leaf sheaths of a grass tiller which perform the supporting function of a stem.

Regrowth: the production of new material above the height of defoliation after defoliation, often with initial regrowth at the expense of reserves stored in the stubble.

Rest period: the length of time between the end of one grazing and the start of the next on a particular area.

Seed ripeness: the stage at which the seed can be harvested successfully.

Seeding year: the calendar year in which the seed is sown.

Seedling emergence: a seedling has emerged when the shoot first appears above the ground.

Selection: (by animals) the removal of some components of a sward rather than others.

Shoot bases: the part of the sward below the anticipated height of defoliation. This becomes *stubble* after defoliation.

Sod: a piece of turf lifted from the sward, either by machine or grazing animals.

Spaced plant: a plant grown in a row so that its canopy does not touch or overlap that of any other plant.

Standing crop: the herbage growing in the field before it is harvested.

Stem: the main axis of a shoot, bearing leaves.

Stocking density: the number of animals of a specified class per unit area of land actually being grazed at a point in time.

Sward: an area of grassland with a short continuous foliage cover, including both above- and below-ground parts, but not any woody plants.

Sward establishment: the growth and development of a sward in the seeding year. *A primary sward* is one that has never been defoliated. *A mixed sward* is one that contains more than one variety or species. *A pure sward* is one that contains a single stated variety or species.

Simulated sward: an assemblage of plants intended to represent in convenient form a 'normal' sward.

Tiller: an aerial shoot of a grass plant, arising from a leaf axil, normally at the base of an older tiller. *An aerial tiller* is one that develops from a node of an extended stem.

Tiller appearance rate (TAR): the rate at which tillers become apparent to the eye without dissection of the plant.

Tiller base: the part of a growing tiller below the height of defoliation.

Tiller stub: the part of a tiller left after defoliation.

Turf: the part of the sward which comprises the shoot system plus the uppermost layer of roots and soil.

Winter burn: leaf burn in winter.

Winter hardiness: the general ability of a variety or species to withstand the winter.

Winter kill: death of plants in winter.

References

BAKER, H. K. (1962). *Proceedings of the Sixth Weed Control Conference, Brighton* **1**, 23–30

BAKER, R. D. (1964). *Journal of the British Grassland Society* **19**, 149–155

BROCKMAN, J. S. and GWYNN, P. E. J. (1961). *Journal of the British Grassland Society* **16**, 201–202

COPPOCK, J. T. (1976). *Agricultural Atlas of England and Wales*. London: Faber and Faber

DAVIES, W. (1960). *The Grass Crop*. London: Spon

FORBES, T. J., DIBB, C., GREEN, J. O., HOPKINS, A. and PEEL, S. (1980). *Factors Affecting the Production of Permanent Grass*. Hurley: Grassland Research Institute

GREEN, J. O. (1974). *Preliminary Report on a Sample Survey of Grassland in England and Wales, Report 310*. Hurley: Grassland Research Institute

HFRO (1980). *Science and Hill Farming: 25 years work at the HFRO*. Edinburgh: Hill Farming Research Organisation

HODGSON, J. (1979). *Grass and Forage Science* **34**, 11–18

HOPKINS, A. and GREEN, J. O. (1978). Changes in Sward Composition and Productivity, *Occasional Symposium No. 10*, 115–129. Hurley: British Grassland Society

JEWISS, O. R. (1979). *Grass Farmer* **4**, 3–7

LEAVER, J. D. and MOISEY, F. R. (1980). *Grass Farmer* **7**, 9–11

MARSH, R. (1975). *Pasture Utilisation and the Grazing Animal, Occasional Symposium No. 8*, 119–128. Hurley: British Grassland Society

JONES, M. G. (1933). *Empire Journal of Experimental Agriculture* **1**, 43–47; 122–128; 361–367

NEDO (1974). *Grass and Grass Products*. London: National Economic Development Office

THOMAS, H. (1980). *Grass and Forage Science* **35**, 13–23

WOLTON, K. M. (1980). *AgTec*, Summer 1980, 10–12: Felixstowe: Fisons Limited Fertiliser Division

Further reading

HOLMES, W. (1980). *Grass—its Production and Utilisation*. Oxford: Blackwells Scientific Publications for the British Grassland Society

7

Trees on the farm

R. D. Toosey

Trees play little part in the economy of most farms today, while in some parts of the country there are large areas of derelict or unmanaged woodland. With the ever increasing shortage and rising price of timber, judicious planting and the rehabilitation of existing woodland must be a first class long-term investment.

Trees grown for timber have an extremely important part to play in rural land management as they achieve high productivity on land such as steep banks, waste places and on the poorer soils, with little or no agricultural value. Managed woodlands, even where small, can also be a valuable source of farm timber, fencing posts and firewood. Trees for shelter are often neglected; the provision of windbreaks and shelterbelts is a means of increasing productivity of exposed farms. Well managed woodlands also improve the amenity and sporting value of the countryside.

CHOICE OF SITE AND SPECIES

Once planted, a tree crop is unlikely to be changed for 50–100 years unless it fails completely. Also it takes many years for mistakes to become apparent, so that a wrong choice of tree can result in wasted time and serious financial loss. As it is essential to marry the tree successfully to the site, ecological factors override all other considerations, e.g. degree of exposure, frost risk, rainfall, surface soil and underlying rock formation, drainage, susceptibility of the site to drought, local topography, elevation and aspect. Only when species of trees suited to the particular site have been determined can a final selection be made on the basis of marketability and other considerations. When two or more species are equally suitable for the site, it is usual to select the species with the highest potential volume production, as indicated by General Yield Class (GYC) value.

It is vital to make a thorough initial inspection of the site, which must be capable of growing trees; fissured rock is satisfactory, but unfissured rock is useless. Ease of access for routine operations and timber extraction must be checked. Shape and size of the plantation affect fencing costs, square shaped and large plantations giving the lowest cost per unit area.

When selecting species it is advisable to consult with local experience and inspect woodlands on similar sites. Natural plant associations give a very useful guide to site conditions, e.g. grass/herbs usually indicates a moist fertile site, with a wide choice of species; purple moor grass (*Molina caerulea*) indicates a wet site suited to the spruces (not Sitka spruce in frosty hollows), while heather (*Erica* spp.) or ling (*Calluna vulgaris*) indicates a leached compacted soil of low fertility, suited to the pines according to area (*see Forestry Practice*, p. 138, Forestry Commission Bulletin No. 14, London: HMSO).

Careful inspection of the soil and subsoil is required and the nature of the underlying rock should be ascertained. Compacted soils do not permit satisfactory root penetration and ploughing or other pre-planting cultivation will usually be required. Depth of soil and the state of the drainage should also be noted. Soil analysis is only of limited value. Species differ greatly in water requirements, and consequently suitability to high or low rainfall areas, and in their ability to withstand exposure; the latter is more important than altitude. *Every ecological factor must be examined carefully as the neglect of only one can result in failure.*

On fertile, sheltered ground, especially old woodland, a wide choice of species is possible. The deciding factor is then the GYC value of the species and its likely uses or markets. Conversely, on infertile soils and exposed situations only one or two species are likely to thrive and choice is then severely restricted.

Most common timber trees are fully resistant to winter cold, but the new spring growth of some species, notably Douglas fir, is particularly susceptible to damage by late spring frosts. Only frost hardy species should be planted in hollows where cold air collects, i.e. frost pockets. *Light demanding* trees can only be grown in full light and require earlier thinning. *Shade tolerant* trees thrive initially in partial shade cast by a tall pole crop of larger trees and are suitable for underplanting, which is particularly advantageous on frost prone sites.

Coniferous trees, the so-called softwoods, generally grow most rapidly, produce the highest volume of timber, need the shortest *rotation* (i.e. number of years from planting to clear felling) and, except for larch, are evergreen. Most prefer acid soils and some will tolerate exposed hill situations. As a group they give by far the best yield of timber and financial return.

Broadleaved trees, the so-called hardwoods, are generally slower growers, lower volume producers and more suited to less acid fertile lowland conditions. With their lower volume production, longer rotation and low-priced thinnings, they are much less profitable than conifers. While plantings are minimal, hardwoods should not be overlooked as they have great amenity value, can provide valuable firebreaks and good hardwood timber is becoming increasingly scarce and valuable.

Hardwoods may be either planted as a pure stand, when they should be grown on small areas, carefully tended and high pruned to eliminate knots, or planted in a matrix of slow growing conifers which act as 'nurses', e.g. Japanese larch or Scots pine. The latter method is best in more exposed situations.

MEASUREMENT OF GROWTH RATE OF TREES

Whilst it is physically possible to measure the growth of trees in terms of weight of fresh or dry material, height or volume, volume is the most meaningful, as timber is normally sold by volume. *Measurable volume* is described by convention as stemwood over 70 mm overbark.

For practical management purposes it is necessary to measure and predict the rate of growth of a plantation, i.e. the annual volume increment of timber produced per unit area of land, expressed as cubic metres per hectare (m³/ha). *Current Annual Increment* (CAI) is the volume added in any single year, while *Mean Annual Increment* (MAI) is the *total* volume produced to date divided by the *age of the stand*, i.e. number of growing seasons since planting.

Yield class is a classification of growth rate in terms of the potential maximum MAI of volume greater than 70 mm top diameter, irrespective of tree species or age of culmination. These measurements only apply to even aged stands.

The MAI and therefore the yield class (or maximum potential MAI) are determined jointly (a) by the species and (b) by the quality of the site.

As shown in *Table 7.1* the yield class varies greatly from species to species and at the same time there is a wide range within species, according to the quality of the particular site. A high volume producing species on a good site will thus produce highest yields while a low volume producer on a poor site will produce the lowest (*Table 7.1*).

Table 7.1 General yield class (GYC) and length of rotation of major timber species

Species	GYC (m³/ha per annum)		Approximate length of rotation[1] (years)
	Range	Average	
Conifers			
Pines—Corsican	6–20	12	60–70
Lodgepole	4–14	8	65–80
Scots	4–14	8	
Spruces—Norway	6–22	12	55–60
Sitka	6–24	14	
Douglas Fir	8–24	14	
Western red cedar	12–24	14	45–55
Western hemlock	12–24	14	
Larches—European	4–14	8	
Hybrid	4–16	8	35–45
Japanese	4–16	8	
Broadleaved trees			
Oak	4–8		120
Beech	4–10		80–100
Ash	4–12		50–70
Sycamore	4–12		
Poplar	4–14		35–40

[1] Varies according to GYC, as influenced by species and site.

There is a good correlation between top height per hectare of largest diameter at breast height (dbh) 1.3 m above ground level and total volume production of a stand of timber. This relationship is used to avoid detailed measurement of total volume production.

In any even aged plantation it is thus possible to obtain the General Yield Class (GYC) which is adequate for most purposes, by calculating the top height of a small representative sample of largest diameter trees at dbh and plotting it against the age of the stand in years on the GYC curves in Forestry Commission Booklet No. 34, *Tables (Metric) Forest Management*, London: HMSO. These graphs will also indicate the time of first thinning and age of maximum MAI according to GYC.

MEASUREMENT OF FELLED SAWLOGS

The simplest and quickest method is to measure the length of the log in metres and the top diameter, underbark, in centimetres and obtain the volume of the log (m³) from the Forestry Commission Booklet

No. 31 *Metric Top Diameter Sawlog Tables*, London: HMSO. This method is more convenient than the long established method of determining volume of a log from quartergirth and length.

LENGTH OF ROTATION

Although some areas are managed as continuous cover forest, most commercial woodlands in the UK are managed on a rotational basis, i.e. clearfelled and then replanted. The *length of the rotation* is the number of years between planting and clearfelling the mature crop. Maturity is not easy to define.

Physical maturity indicates the age at which a stand is physically mature for felling—the best saw timber needs large trees from a long rotation while trees for small poles, chipboard and pulpwood only need a short rotation or are taken from thinnings. *Financial maturity* is the age at which a stand will give the best financial return, which in turn depends on the value of the timber at each stage of growth or maturity.

The main financial problem in forestry is the long period during which capital is locked up and consequently high accumulated interest charges. Major costs such as land clearance, fencing and planting occur at the beginning of the rotation, and attract interest from the day they are incurred, but it is many years before a cash return is available for repayment of capital and accumulated interest. In order to reduce the interest liability it is necessary to:

(1) delay all expenditure until the latest feasible stage, and
(2) speed up the process and obtain a quicker return by planting fast growing species and by shortening the rotation where possible (*Table 7.1*).

Although the larger trees produced in a long traditional rotation will be more valuable and may give a higher income per hectare than those in a shorter rotation, there is so much capital locked up that the percentage return is seriously reduced. Thus, even though a plantation in a shorter rotation may yield less cash when clear felled, it is usually more profitable:

(1) on account of the lower interest charges, and
(2) because the rate of growth of timber falls off once maximum MAI is reached; the best financial return is often given by felling just before this but in practice it is usually recommended to clearfell ± five years from the point of maximum MAI shown in the Forestry Commission GYC curves for the appropriate species.

The precise timing of felling depends on current timber demand and value. The new crop following replanting will give a better growth rate and return than the ageing stand it replaced.

There are no hard and fast rules on length of rotation, which is jointly determined by:

(1) rate of growth, being shorter with fast growing conifers and on good sites, longer with slower growing species and on poor sites. Very slow growing broadleaved trees such as oak and beech require such a long rotation that the interest charge alone becomes excessive.
(2) nature and extent of the market and size of plantation. Where there is a good local market for small logs and thinnings and on small areas a short rotation is possible; with a poor market for such material and on large areas, a longer rotation may be preferable.

Total production of a rotation

$$T = Y \times D$$

where T = total production (m^3/ha)
        Y = GYC (m^3/ha per annum)
and D = duration of rotation in years.
e.g. Douglas Fir (very good site)
$$Y = 18; D \quad 50; T = 900 \ (m^3/ha)$$
Larch (satisfactory site)
$$Y = 10; D \quad 60; T = 600 \ (m^3/ha)$$
Oak (satisfactory site)
$$Y = 4; \quad D \ 120; T = 480 \ (m^3/ha)$$

TIMBER QUALITY

The best timber trees are grown under forest or woodland conditions. Thick planting allows a high degree of selection for the final stand trees and the 'crowded' conditions so produced draw up the trunks long and straight and limit side branch development and hence knots. Valuable hardwoods and sometimes the final stand trees in conifer plantations may be highpruned to eliminate side branches. Hedgerow and parkland trees are generally unsatisfactory for quality timber, as the trunks tend to be short and the open conditions allow heavy branching with the consequent development of large knots. The lower part of the trunk frequently conceals old staples, nails and wire, and has to be discarded.

Good quality saw timber comes from trees which are long, straight, knot free and with the minimum taper. Knots, wide rings and cross grain all tend to cause a rough surface finish. Large knots and rings of knots cause weakness. A variable growth rate of the tree gives unevenly spaced rings, which may result in cupping or twisting of the sawn timber. Homegrown timber is frequently inferior to the imported article in these respects.

Wood is used for a very wide range of purposes. The higher the price, the more stringent are the

Table 7.2 Value of major species of conifers for timber and pulpwood

Species of tree	Alternative trade names for timber type	Building strength class[1]	Durability of heartwood[2]	Ease of impregnation[3]	Suitability for pulpwood
Pines—Corsican	Redwood, red or yellow deal, Baltic redwood	S2	ND	MR	Yes
Lodgepole				R	Yes
Scots				MR	Yes
Spruces—Norway	Whitewood, white deal, Baltic whitewood	S3	ND	R	Yes
Sitka				R	Ideal
Douglas fir	Columbian or Oregon pine	S1	ND	R	Yes
Western hemlock	—	S2	ND	R	Ideal
Western red cedar	—	S3	D	R	Not normally used
Larch spp.	—	S1	MD	R	No

[1] Building strength class, S1 strongest, *see* BSCP 112.
[2] Durability of heartwood under conditions suitable for decay. Sapwood is non-durable; ND, perishable, D = durable.
[3] Ease of impregnation of heartwood with preservative under pressure.
Moderately resistant (MR) satisfactory under 2–3 h pressure; resistant (R) is difficult and requires a longer period. Sapwood normally permeable.

quality requirements. Only the pick of the clean, straight, knot-free large diameter logs are suitable for veneer, which is a highly specialised job. This is followed by trades where a fine finish is required, notably furniture manufacture, high-grade joinery and boatbuilding. Good sawn building timber used for structural purposes needs to be strong in bending and in compression—Douglas fir is ideal in this respect. Although building timber (*Table 7.2*) does not need to be as fine as the joinery grade, it should be clean, straight, even-grained and free from large knots. Thinnings and the lower grades are used for a wide variety of work such as turnery, fencing, fencing stakes, garden furniture, rustic work, general estate work, mine timber, chipboard, fibreboard, wood wool, packing cases and pulpwood, but command lower prices than sawlogs, minimum top diameter 18 cm. There are also various local markets for timber.

The quality of timber is jointly influenced by the method of manufacture, including seasoning, the species of tree and conditions of growth. Not only is it essential for a timber grower to select the right species for the site, but source or provenance of his transplants is extremely important: a poor type will produce a very unsatisfactory timber tree, especially in the cases of European larch, Corsican and Lodgepole pines, and Norway and Sitka spruces.

SPECIES OF TREE TO PLANT

The conifers (softwoods). (I). Major timber species

Scots pine (*Pinus sylvestris* L.)

Origin: British Isles and Northern Europe.

Growth rate and volume production rather low. Windfirm. Very frost hardy; a strong light demander. A very safe tree to plant, succeeding over a wide range of conditions. Suited to dry heather sites, light or sandy soils and low rainfall areas. Suitable as a 'nurse' when establishing hardwood species, e.g. beech. Avoid wet or soft ground, moorlands in high rainfall areas, chalk and limestone soils and high elevations although it grows well in glens up to 450 m altitude in north east Scotland.

Uses

Good general purpose timber with wide range of uses: joinery, furniture, building, flooring, pitwood, fibre board, wood wool, chipboard, telegraph poles. A valuable shelterbelt tree.

Corsican pine (*Pinus nigra* var. *maritima* (Ait) Melville)

Origin: Corsica.

Grows faster than Scots pine but is difficult to establish. Probably best planted in April or use container raised trees for amenity purposes. Plant only at low elevations in south and east England. Well suited to areas near the sea. Does well on light sandy soils and heavy clay in low rainfall areas. Preferable to Scots pine on chalky soils.

Uses

Slightly lower strength than Scots pine. Lower quality. Uses include building, box manufacture, pitwood, fencing, fibreboard, chipboard. Especially suited to wood wool slab manufacture. A productive shelterbelt tree in suitable areas.

Lodgepole pine (*Pinus contorta.* Doug. ex Loud.)

Origin: Western North America.

A very slow grower, often producing an even lower volume than Scots pine. Thrives on the worst sites; probably the best 'pioneer' crop and is now widely planted. Should be confined to the poorest heaths, peats and even some sand dunes, where it is about the only timber tree to survive. Withstands exposure well. Quite unsuitable on fertile soils, where it produces a very coarse timber. Essential to obtain plants from a suitable origin.

Uses

As for Scots pine.

Norway spruce (*Picea abies* (L) Karst.)

Origin: Northern Europe.

Only suited to moist sites with adequate rainfall—over 875–1000 mm; grassy or rushy soils and less acid peats. Suited to old woodlands and heavy clays. Much more sensitive than Sitka spruce to exposed situations and liable to windblow on badly drained sites. Avoid dry heather sites and low rainfall areas of UK. Liable to frost check. A valuable timber tree and a high volume producer on the most suitable sites but less adaptable than Sitka spruce.

Uses

The only species suitable for Christmas tree production. A good general purpose whitewood type timber. Especially suitable for building, as it is stable over a wide range of humidity. Not as strong as Douglas fir. Works and nails well. Also joinery and kitchen or whitewood furniture, packing cases, chipboard, wood wool, fibreboard, pitwood. Not suitable for outdoor use, except for small treated sapwood posts or poles, as the heartwood is resistant to preservative.

Sitka spruce (*Picea sitchensis* (Bong.) Carr.)

Origin: Western North America.

Only suited to high rainfall areas, minimum 875–1000 mm where the summer is cool or on soils with high water table. Very windfirm and suited to exposed sites. Widely planted in Scotland and north west England, especially on wet moorlands, also wetter parts of south west. A very high volume producer on good sites.

Uses

Strictly a timber and productive shelterbelt tree, not for amenity planting, as lower branches die off. Timber of whitewood type, unsuitable for high grade joinery; uses otherwise as for Norway spruce. An excellent pulpwood. Unsuitable for Christmas tree production.

Douglas fir (*Pseudodotsuga menziesii* (Franco) Mirb.)

Origin: Western North America.

One of the highest volume producers but great care is essential in choice of site. Needs sheltered situations with deep moderately fertile well drained soil. Ideal for valley slopes; is fairly drought resistant once established. Avoid exposed situations, as it readily loses its leading shoot in high winds and is susceptible to windblow in soft ground; also avoid heather ground, wet or shallow soils and frost pockets. Very susceptible to spring frost damage when young. One

of the most valuable and productive species on suitable sites. Widely planted in UK.

Uses

Timber especially suited for construction work, as it has a high strength to weight ratio in bending and compression. Stronger than the pines or redwood. Harder to work and nail than most other softwoods. Used for flag and telegraph poles, joinery, flooring, pitwood, chipboard and fibreboard manufacture.

Western hemlock (*Tsuga heterophylla* (Raf.) Sarg.)

Origin: Western North America.

A high volume producer. Grown over a wide range of climatic conditions but is particularly suited to moisture retentive soils and to the higher rainfall areas of the west. Does well on acid mineral soils and better peats. Slow to start on heathland; unsuited to dry loose sands. Withstands heavy shade well and is especially suited to underplanting, when it is usually easier to establish than in the open. Preferable to Douglas fir for replacing coppice in less sheltered places and on heavier soils.

Uses

A good general purpose softwood intermediate in quality between pine and spruce. Properly graded material suitable for joinery and building. Also used for estate work and pitwood.

Western red cedar (*Thuja plicata* D. Don.)

Origin: Western North America.

A high volume producer which needs moisture. Most vigorous in mild moist regions on moderately fertile soils or shallow soils over rock. Elsewhere suited to heavy moist base rich soils. Also succeeds on drier sites overlying chalk and on some peats. Avoid very dry or acid situations and deep loose sands.

Uses

Stands shade well and is ideal for underplanting. Stays well clothed to the ground, so is suitable for single row windbreaks and shelter belt margins. Stands wind and frost well when established; younger trees may turn purplish colour in cold weather but recover. Unsuited to exposed situations and not now recommended for hedges, owing to susceptibility to keithia disease, caused by the fungus *Didymascella thujina*. May be raised from cuttings. Florists value foliage.

A lightweight timber with a very durable heartwood suitable for a wide range of outdoor uses and estate work; portable buildings, weather boarding, wooden greenhouses, 'cedar' roofing shingles and seed boxes.

The larches

These comprise: European larch (*Larix decidua* Mill.), Japanese larch (*L. leptolepis* Murr.) and Hybrid larch (*L. eurolepis* Henry).

Growth rates, canker resistance and site requirements differ but grade for grade, the *timbers* are of similar quality.

Uses

Especially suitable for outdoor work or where a stronger or more durable timber than ordinary red or white deal is required. Larch is heavier than most other softwoods. Heartwood naturally durable but sapwood needs preservative for outdoor use: fencing, gates, estate and rustic work, garden furniture, telegraph poles, chipboard and pitwood. Suitable parcels also for boat building and vat manufacture.

General cultural remarks

Larch is a strong light demander so the first thinning should be carried out early—about 12 years on good sites. Very suitable as a 'nurse' for hardwood establishment. Site must be chosen with care. Tends to corkscrew on very fertile soils. Avoid dry sites, especially in low rainfall areas (under 750 mm/annum).

European larch

Site requirements most exacting, doing best on moderately fertile, moist but well drained soils. Avoid wet, badly drained or dry sites, frost pockets, poor sands or peats. Will not stand exposure. Susceptible to larch canker so now much less commonly planted. Produces a better timber form than Japanese larch.

Japanese larch

Commonly planted, especially on private estates. Suited to a much wider range of conditions than European larch, grows faster and shows good resistance to larch canker. Does well in wetter humid regions of the north and west. Tolerates poorer soils than European larch, including grass and heather slopes but avoid sites susceptible to drying out or low fertility. Needs side protection in high uplands, as it leans and twists badly in very exposed situations.

Hybrid larch

Produced by crossing Japanese with European larches. Combines the vigour and canker resistance of the Japanese parent with the superior timber form of the European. Superior to Japanese larch under poor and exposed conditions; on European larch sites is free from dieback and grows faster. Use only first or second generation transplants.

The conifers. (II) Other species, not widely planted for timber or suitable only for shelter or amenity

Austrian pine (*Pinus nigra* var. *nigra Harrison*)

Origin: Austria.

Its heavy branching habit results in a coarse knotty trunk of low timber value; is only worth planting for shelter or amenity where the site rules out other conifers, as with exposed seaside situations or shallow dry soils such as chalk, sand or limestone. Corsican pine is better on wet or heavy soils. Very windfirm, frost hardy and withstands aerial pollution. A strong light demander.

Bishop pine (*Pinus muricata* Blue form D. Don.)

Origin: Californian coast.

Selection similar to Monterey pine but tolerates poorer soils and is rather hardier inland. Avoid calcareous soils. Timber quality similar to Monterey pine. High yields (GYC 6–24 m^3/ha per annum). Use *only* blue form, not green.

Maritime pine (*Pinus pinaster* Ait.)

Origin: Mediterranean.

Grows quickly and valuable for shelter belts in exposed coastal sites. Frost tender and only suitable to sandy soils in south west England. Timber of very low value, e.g. pitwood, etc.

Monterey or Radiata pine (*Pinus radiata* D. Don.)

Origin: California.

Grows extremely rapidly in early life, producing a very large tree in 30 years. Windfirm and resists salt laden winds but is not frost hardy. May be considered for quick growing shelter belts in mild coastal areas only, on well drained light or sandy soils. Prefer small seedlings for planting on exposed sites. Timber of very low quality, coarse and knotty.

Mountain pine (*Pinus mugo* Turra.)

Origin: mountains of Central Europe.

Not a timber tree and only worth planting on very worst sites, where it will withstand extremes of soil and climate. Frost hardy and very windfirm. Suggested as a margin for the most exposed shelter belts. Very slow growing.

Lawsons cypress (*Chymaecyparis lawsoniana* (A. Marr) Parl.)

Origin: Western North America.

Rarely planted as a timber tree; used mainly for single row screens or windbreaks. A prime favourite for evergreen hedges, for which it is safer than western red cedar. Foliage valued by florists.

Site requirements are not exacting but prefers a deep fertile soil with a fairly sheltered position. Avoid dry or heather sites. An excellent shade bearer and suited to underplanting. Tends to fork and liable to snow-break.

Monterey cypress (*Cupressus macrocarpa* Hart.)

Origin: California.

Once widely planted in the milder coastal regions of south west England, as it withstands severe exposure to sea winds. Very susceptible to frost, many specimens being killed in the hard winter of 1976. A very untidy grower, tending to die back in the middle. Timber coarse, knotty and of little value. Useless for hedges. Not recommended for any purpose.

Leyland cypress (*Cupressocyparis leylandii* Dall.)

A hybrid from the Nootka and Monterey cypresses; can be reproduced only from cuttings and is thus much more expensive than Lawsons cypress. One of the fastest growing conifers, showing a high degree of vigour under a wide range of conditions. Particularly valuable for quick growing screens; also used for hedges but may require very frequent cutting. The species may have some timber value.

European silver fir (*Abies alba* Mill.)

Origin: Central Europe.

Not now safe to plant, owing to liability to severe damage from the insect *Adalges nusslini*. Timber quality and uses as for Norway spruce.

Grand fir (*Abies grandis* Lindl.)

Origin: Western North America.

Best suited to deep moist but well drained soils. Avoid exposed sites, frost pockets and poor acid soils. Shade tolerant and well suited to underplanting. Resists *Fomes annosus* better than larch, western red cedar and Norway and Sitka spruces. Source of seed of critical importance. Produces a large volume of whitewood timber on suitable sites, which is of similar quality to the spruces (GYC 14–30 m^3/ha per annum).

Noble fir (*Abies procera* Rehd.)

Origin: Western North America.

Likes deep moist but well drained soils and is more frost resistant than other silver firs. Stands exposure well and suited to moderately acid soils. Avoid poor dry soils. Shade tolerant and suitable for underplanting. A high volume producer and a useful shelterbelt tree. (GYC 10–22 m^3/ha per annum, average 15.) Little demand for timber.

Serbian or Omorika spruce (*Picea omorika* (Pancic) Purkyne.)

Origin: Yugoslavia.

Timber quality and site requirements similar to Norway spruce but tolerates poorer soils and is less exacting in its water requirements. More frost hardy and suited to frost pockets where other spruces fail. Slower growing than other spruces but very windfirm (GYC 10–20 m^3/ha per annum).

Californian redwood (*Sequoa sempervirens* (D. Don.) Endl.)

Origin: California.

One of the highest volume producers but only suited to deep fertile soils in sheltered high rainfall areas. Avoid shallow or dry soils and frosty sites. Best established under tall cover. Timber naturally durable and useful for estate work (GYC 16–30 m^3/ha per annum).

Wellingtonia (*Sequoiadendron giganteum* (Lindley) Buch.)

Origin: California.

Habit and volume production similar to Californian redwood but is hardier; while tolerating greater exposure and drier and more acid soils, it still needs deep fertile soils in relatively sheltered situations (GYC 16–30 m^3/ha per annum).

The broadleaved trees (hardwoods). (I) Major timber species

Oaks

The main species include the *pedunculate oak* (*Quercus robur* L.) and the *sessile oak* (*Quercus petraea* (Matt.) Lieb.)
Origin: British Isles and Europe.

The two have frequently inter-crossed so that a high proportion of oak trees found naturally in Britain are hybrids between the two. Oaks are long lived, windfirm trees but very slow growers, requiring deep fertile well drained soils; they do well on stiff clays and marls. Avoid any infertile, shallow, badly drained or exposed sites. The sessile oak is the faster growing of the two and will tolerate less fertile soils, as in Wales and the north. Both are strong light demanders but are difficult to establish in open or exposed situations.

Uses

Oak is one of the most erratic timbers as regards quality, varying from the first class to rubbish. Good quality oak is hard to come by and very expensive. Oak has been used with profligacy in the past and frequently the only way nowadays to obtain good oak is to buy second-hand material. The heartwood is hard, resistant to abrasion and naturally durable but the sapwood needs preservative treatment if used outdoors. Top quality oak is used for floors, veneer and furniture, etc. Lower grades have a wide range of uses, including weather boarding, engineering and sawn mining timber. Round oak may be used for chipboard and hardwood pulpwood. (GYC 2–8 m³/ha per annum, average 4—much higher has been achieved by free growing oak.)

Red oak (*Quercus borealis* Michx.)

Origin: Western North America.

Less demanding on the soil than common oak, growing well on acid light or sandy soils of moderate to high fertility. Avoid infertile or lime rich soils. Of special value on sites which are marginal for the better hardwoods; also suitable for use in mixtures. Hardy and more shade tolerant than common oak but the timber is less valuable and can only be classed as a useful general purpose hardwood, which takes preservative well. A useful amenity tree, the foliage turning bright red in the autumn. (GYC 4–10 m³/ha per annum.)

Turkey oak (*Q. cerris* L.)

This is not a good forest tree as the timber lacks durability and working qualities. Grows into a large tree and is wind resistant, thriving on a wide range of soils. Mainly of value for shelterbelt and amenity purposes.

Beech (*Fagus sylvatica* L.)

Origin: Southern UK and Europe.

Likes fertile well drained loams of all types, chalks and limestones. Avoid poorly drained or leached soils and frost pockets. Very shade tolerant and ideal for underplanting. When older it casts a dense shade, suppressing growth in the understorey. Very windfirm and a useful shelterbelt tree but it should not be allowed to become dominant, as old stands of beech become open and draughty and are difficult to regenerate. Usually planted as a mixed stand in open situations, when larch or Scots pine are suitable companions. Very susceptible to squirrel damage.

The ornamental *copper* (var. *cuprea*) and *purple* (var. *purpurea*) beeches are similar in growth and in site requirements to the main type.

Uses

Planted in shelterbelts, for hedges, ornament and timber, which is employed almost entirely for indoor work, where it has a wide range of uses according to grade; it is strong, gives a good finish and takes stains well. Used for furniture, veneer, bentwood, flooring turnery work and pulpwood. (GYC 4–10 m³/ha per annum, average 6.)

Ash (*Fraxinus excelsior* L.)

Origin: British Isles and Europe.

One of the most exacting species for good growth and quality timber production, requiring deep moist fertile soils; deep chalk and limestone soils also suitable. Ash will grow on a wider range of soils but will not usually produce good timber. Not suitable for widespread planting, as only small localised sites are normally suitable and must be selected with great care. Avoid dry, shallow, heavy clay or badly drained soils, heaths and moorlands.

Uses

The timber is springy, with a high resistance to shock and is thus used for tool handles, oars and other sports equipment; also turnery, furniture and pulpwood. (GYC 4–10 m³/ha per annum, average 5.)

Sycamore (*Acer pseudoplatanus*)

Origin: Central Europe.

A very dependable tree; like ash prefers deep fertile soils, but is fairly adaptable on all soils which permit deep rooting. Avoid shallow, dry or badly drained soils, heaths and moorlands. Withstands severe exposure but not salt laden winds. Frost hardy, will tolerate some shade in early years and is well suited for admixture with conifers in shelterbelts. Can be ruined by squirrels.

Uses

A valuable shelterbelt tree. Timber is white and widely used where it comes in contact with food, e.g. butchers' blocks, breadboards, etc. A useful turnery and pulpwood. Figured sycamore is valued for veneer and furniture. (GYC 4–12 m³/ha per annum, average 5.)

Sweet chestnut (*Castanea sativa*. Mill.)

Origin: Mediterranean.

Suited only to deep fertile soils and mild climates. Avoid frosty, poorly drained or exposed sites and less fertile or heavy clay soils..

Uses

When grown as coppice the timber is used for cleft fencing and hop-poles. Owing to the risk of shake, trees for sawn timber should not be left to grow too large. Valuable for furniture, etc. (GYC 4–10 m³/ha per annum, average 6.)

The poplars or cottonwoods

Comprising *black hybrids* (*Populus* × *euramericana* Dode Guinier)—various—*P. robusta, P. serotina*, etc.

Origin: Europe.

Balsam poplars and hybrids (*Populus trichocarpa* Torr. and Gray and *P. tacamahaca* × *Trichocarpa* hybrids).

Origin: North America.

Black hybrids

Poplar production is a highly specialised job: requirements for timber poplars are extremely exacting, so that the number of good sites is very limited. Ample moisture is necessary but avoid stagnant water or badly drained soils. Rich alluvial or fen soils or deep loams are ideal, on sheltered sites. Avoid exposed or acid peats. (GYC 4–14 m³/ha per annum.)

Balsam poplars

Only clones resistant to bacterial canker should be planted. Stand slightly higher acidity and better suited to cooler and wetter areas than the black hybrids. (GYC 4–16 m³/ha per annum.)

Uses

Suitable large timber is peeled for matches and chip baskets. Poplars are the most widely planted shelter trees in the temperate regions of the northern hemisphere and are valued for windbreaks or single row screens, when shrubs may also be necessary to provide low shelter. Owing to the certainty of severe damage by the extensive root system, poplars should be kept well away from drains or buildings. When planted for shelter other poplars may be considered: *(Grey poplar* (*P. canescens* (Ait.) Sm.) is hardy and wind resistant; prefers moist fertile soils. Very vigorous and suckers profusely. *The aspen* (*P. tremula* L.) is very frost hardy; tolerates more difficult soil and climatic conditions, growing well on northern uplands. Avoid dry or waterlogged soils. Suckers abundantly.

The broadleaved trees (hardwoods). (II) Other species, not widely planted or only suitable for shelter or amenity

Alders
Common or black alder (*Alnus glutinosa* L.)

Origin: British Isles and Europe.
   Grows naturally along watercourses and in low lying places too wet for other species except willow. A hardy and adaptable species. Avoid very acid peats, badly aerated or dry acid soils. Grows rapidly when young and is planted for shelter on soils with a high water table; may alternate with poplars. Withstands cutting or pollarding and coppices well. Produces straight stemmed trees. Timber is used for turnery work, localised industrial uses, clog soles, hat blocks, and pulpwood. (GYC 4–12 m³/ha per annum.)

Grey alder (*A. incana* L.)

Has similar site requirements to black alder but grows successfully over a wider range of soils. Avoid dry, shallow or infertile sites. Usually the most suitable species for windbreaks. Very hardy, suckers freely, somewhat resistant to browsing of livestock and fast growing. Timber of little commercial importance. Mainly a pioneer species.

Italian or cordate leaved alder (*A. cordata* Desf.)

Grows rapidly on drier chalk and limestone sites in the south of England. A very adaptable tree but avoid acid sites. Stocks are mixed and costly. Rather frost tender but tolerates some shade. Mainly a pioneer species.

Birch
Origin: British Isles and Europe.
   Silver or warty birch (*Betula pendula* Roth.) and *common or white birch* (*B. pubescens* Ehrh.) are very hardy wind firm and frost resistant trees adapted to a wide range of soils including acid situations. Silver birch is the more successful on drier sites while common birch does better on peaty or waterlogged sites and tolerates some shade. Not usually worth planting for their own sakes but they are ideal pioneers and soil improvers. Valuable as 'nurses' for frost tender conifers, beech or oak. Silver birch is widely planted as an ornamental tree, as is the paper birch (*B. papyifera* Marsh.). The demand for birch timber is very limited unless log sizes suitable for veneer or furniture can be produced.

Elms
Origin: England, introduced.
   The common *English elm* (*Ulmus procera* Salisb.) and other field elms were very valuable hedgerow and parkland trees until they were cut back by Dutch elm disease; they are now quite unsafe to plant. Mature timber is very tough and resistant to splitting but is not naturally durable and needs treatment with preservative. Takes nails more easily if they are oiled first. Wide range of uses according to grade, including veneer, furniture, turnery, boarding and anywhere where a tough split proof timber is required. The *wych elm* (*Ulmus blabra* Huds.) thrives under woodland conditions and is an indicator of fertile soils. Does well in fertile valleys in the west and north and extends further into cooler areas than the field elm. Abounds in Scotland on the better soils. Also susceptible to Dutch elm disease. The timber is even tougher than English elm; the best grades are suitable for furniture and boat building.

   Suitable replacements for elm: limes and Norway maple.

Gean or wild cherry (*Prunus avium* L.)

Origin: Britain and Europe.

One of the few trees to produce showy blossom (white) and timber, although not usually grown for profit in Britain. Good logs in demand for furniture and turnery work. A valuable tree for the margins of woodlands and shelterbelts and for amenity purposes. Does best on fertile woodland sites and calcareous soils. Avoid any infertile soil, high elevations and cold exposed areas.

Bird cherry (*Prunus padus* L.)

Also produces showy blossom. An upland species indigenous to Scotland, northern England and mid Wales. Suckers freely producing a bush or tree 6–9 m high) and suited to a wide range of soils. Valuable as a margin for shelterbelts, especially conifers. Best logs suitable for veneer.

Horse chestnut (*Aesculus hippocastanum* L.)

Origin: Albania and Greece.

Primarily an amenity tree for parks and avenues, producing showy white candles of blossom. The timber is soft and fine grained but with limited uses. Only suited to fertile soils and rarely lives more than 80 years.

Limes (*Tilia* spp.)

Origin: British Isles and Europe.

The lime species (*T. cordata* Mill., *T. platyphyllos* Scop. and *T. europea* Henry) are frost hardy, windfirm and long lived. They are suitable for shelter, especially single row windbreaks, screens and avenues and may replace elm as a hedgerow tree. Expensive to buy, they are shade tolerant and coppice readily. Growth is fairly rapid. Deep moist soils are best but limes are adaptable provided fertility is good. Avoid all infertile sites. Of limited value as a timber tree but good for turnery and in demand for carving and pulpwood.

Norway maple (*Acer platanoides*)

Origin: Northern Europe.

A hardy windfirm species demanding conditions similar to sycamore but tolerates somewhat poorer and drier soils. A strong light demander. Avoid all infertile situations. A useful amenity tree, the leaves turning an attractive golden colour in autumn. Well suited for inclusion in shelterbelts, replacing elm in hedgerows or avenues and in mixtures on soils not really suited to sycamore. Grows faster than sycamore initially but does not achieve the same size, maximum height 23 m. Produces a strong hard smooth textured timber; the best logs are suitable for carving and furniture.

Southern beech (*Nothofagus* spp.)

Origin: South America.

There are numerous species but the best timber producers *N. procera* and *N. obliqua* are natives of Chile. Both are very fast growers, giving average MAI of 10–12 and 12–14 m^3/ha per annum respectively; the best sites do more. These species show considerable promise but are still in the experimental stage. Not exacting in their soil requirements, thriving on a wide range of soils from sand to clay. Avoid frosty or exposed sites and altitudes over 300 m. Both species are light demanders; early thinning is essential to avoid long but excessively thin trunks. *N. procera* produces the best timber form and is suited to the wetter west, annual rainfall 750–1500 mm. *N. obliqua* gives a more branched form and suits shallower and poorer soils and the drier east, annual rainfall under 750 mm. Quality of mature imported timber comparable to beech but with 20% less bending strength; also suitable for pulpwood. Immature wood is of lower quality. There is not yet enough wood available to gauge the quality of home grown timber.

Willows (*Salix* spp.)

Only the cricket bat willow (*S. Alba* var. *coerulea* Sim.) is of any importance, used for cricket bats, chip baskets and pulpwood. Grown only on deep, fertile land, free from frost and with an abundant supply of moving water, it increases in height by 1 m/annum and is ready for marketing in 16 years. Production is highly specialised. Several other species of willow are used for shelter purposes, including the common white willow (*S. alba* L.) and the crack willow (*S. fragilis* L.); the latter suffers from high winds. Most willows are suitable for single-row windbreaks or shelterbelt margins as they throw coppice or pollard shoots readily. Moist, fertile sites are required and with adequate moisture; roots readily from cuttings.

PREPARATION OF SITE

Trees must form a canopy quickly to suppress all unwanted vegetation early in the life of the crop. Thorough preparation of the site is thus essential to provide an environment in which the young trees can establish and grow rapidly.

The work requires clearance of unwanted vegetation, cultivation if required by ploughing, provision of drainage for surface water, fencing for control of farm livestock, rabbits and deer, control of fungal diseases and insect pests and the provision of adequate plant nutrients.

Clearing unwanted vegetation and cultivation

It is essential to provide the young trees with

satisfactory soil conditions for rapid root development and to control vegetation on and immediately around them.

Failure to do so results in the suppression of the young tree, which is robbed of water, light and nutrients or even smothered by dense vegetation. Site clearance is generally expensive, the amount of work necessary depending on the quantity and type of vegetation or dead material present and on whether it is likely to hinder subsequent operations. The approach depends on whether the land to be planted is bare land (i.e. land which has not carried trees previously) or former woodland.

Bare land

Ploughing

Wherever practicable ploughing should be used when cultivation is required, as it is quick and relatively cheap. Forestry ploughs, which can work up to 0.9 m deep, have two types of mouldboard assembly:

(1) for digging or disturbing the soil and breaking up pans, or
(2) drainage ploughs for producing a clean drainage furrow.

Ploughing is *essential* on soils with a pan or compacted layer and on waterlogged soils where it can make all the difference between success and failure. On compacted heaths or on soils where iron pans exist ploughing opens up drainage passages in the soil, allowing water to reach the lower levels and young roots to penetrate. On waterlogged soils, notably peats and peat gleys, it is essential to plant on a site raised above the original soil level, so that the soil around the roots of the young trees is well drained. The drainage plough produces a strip of inverted raised turf for planting. On old pasture or in relatively fertile situations ploughing is not usually necessary provided appropriate steps are taken to control vegetation.

Where ploughing is not necessary, the use of pre-planting total herbicide treatments should be considered (*see* Chapter 8). Overall or strip applications 1.0 m wide, in the centre of which the young trees are to be planted, not only removes direct competition but also results in the decomposition of unwanted vegetation which provides the new crop with additional nutrients.

Handwork

Handwork is slow and expensive, and is mainly used on small or remote areas and difficult terrain.

Screefing

Screefing, i.e. the clearing of a thin square of surface vegetation 300 to 450 mm with a spade or mattock may be used for each young tree on grass or heather ground.

Turf planting

Inverted turves are used where the vegetation is thick and screefing would leave a deep hollow in which water might collect. A square of turf with sides 300–450 mm and 150 mm thick is cut and replaced upside down in the original hole to receive the young tree. *Raised turves* are preferable on wet ground where an elevated planting site is necessary. Again, turves 300–400 mm × 150 mm thick are cut and inverted, but are laid alongside the original hole. Alternatively, on steep slopes, *hinged turves*, cut only on three sides, may be swung out from the original hole to prevent them from sliding or washing downhill. The rotting faces of the inverted turves are believed to provide some plant nutrients, but cut and invert the turves well in advance of planting to allow time for settlement and rotting, or failure will result. On waterlogged sites it is also necessary to provide drainage channels.

Former woodland

This situation poses problems very different from those on bare land. Cultivations are more difficult. The net cost depends on the condition of the site and the realisation value of the timber. Much of the subsequent weed growth is of woodland origin and woodland fungi and insect pests are often waiting to attack the new crop, especially where the replanting succeeds or adjoins coniferous plantations. Treating the young trees with an insecticidal dip is frequently advisable. It may also be necessary to repair the fencing and return the drainage to a state adequate for young trees. Each side presents different problems, whose nature depends on a very wide range of ground conditions.

Former cleared woodland

Coniferous woodland, especially under acid conditions at higher elevations and in Scotland, generally produces little regrowth and, provided excessive lop and top does not have to be dealt with and drains or fences made good, presents few problems; it is then relatively cheap to replant. *Old broadleaved woodlands*, especially in the south of England, soon become overgrown with scrub and woody weeds and can then be difficult and expensive to clear. The problem includes previously unmanaged woodlands, disused coppice, e.g. hazel, oak, chestnut and scrub, including silver birch and rhododendron.

Methods of clearance and restocking

Methods of clearance and restocking include:

(1) clearfelling and replanting,
(2) retention of part of the existing crop, if of adequate quality,
(3) allowing regeneration to occur from seedlings, or
(4) thinning to leave temporary cover followed by underplanting.

The choice of method depends on species to be planted and on condition of the site. The decision is complex and it is usually advisable to obtain expert advice. Some of the above methods may also be used in combination.

Clearfelling

Clearfelling is essential before planting any light demanding species. It is simple and easy to supervise but exposes the site to insolation and frost; the complete removal of cover frequently causes vigorous regrowth of coppice and weeds of all kinds, often greatly increasing weeding costs.

Clearfelling very large areas in one operation removes shelter from the young trees, making re-establishment more difficult and often ruining, at least temporarily, any existing amenity value. Clearfelling should be planned carefully with these aspects in mind. Contractors will often undertake a difficult job at relatively low cost where a good return is available from the sale of timber for pulpwood, mining timber or box wood. With scrub, a total kill may be obtained, using a suitable herbicide (*see* Chapter 8); cost is low if immediate replanting can occur. Rhododendron and dense thorn or briar should always be completely cleared.

Partial clearance

This may be worthwhile where existing material is good enough to produce a crop of satisfactory quality. Areas of 0.05 ha or more are suitable. Select clean straight young trees; in coppice leave only one or two straight vigorous shoots per stool. Avoid large individual trees, which occupy too much space and are difficult to clear later.

Regeneration from seed

The technique requires careful judgement, as the quality of the resultant crop is only as good as its parent trees, which, if it is to be successful, must be of high genetic potential. Suitable species which regularly set ample viable seed include Scots pine, oak and sycamore; beech is unsuitable as it masts very irregularly. Rapid weed growth, which soon smothers the young seedlings, is a major problem and it is usually safer to use transplants.

Thinning to leave temporary cover with underplanting

Only suited where highly shade tolerant species are to be planted, e.g. Western hemlock (*Tsuga heterophylla*) and Western red cedar (*Thuja plicata*). Does not destroy the woodland environment and preserves its amenity value. Trees left for temporary cover protect the young trees from frost and insolation and reduce weed growth; their transpiration also helps to prevent potentially wet soils from becoming waterlogged.

Method

Thin out to leave a tall pole crop which gives a dappled shade on the ground. Where possible cut out or kill all low or heavily branched trees and brash all that are left to a height of 1.5 m. Light crowned trees, e.g. silver birch make an ideal cover crop.

Cover is often removed in two stages: half after two years and the rest after four years; six years is an absolute maximum. Removing cover too early increases the frost risk and weeding costs while excessive delay results in reduced tree growth and higher cost of removal. As an alternative to felling, the cover trees may be killed while standing by ring barking or basal bark treatment and left to disintegrate; this reduces costs but the dead trees can be unsightly and the method is unsuited to prime amenity situations. Large trees are better felled.

Drainage

The prime function of woodland drains is to remove surface water and prevent surface waterlogging, which kills trees and results in serious windblow later on. Once fully established, the trees themselves draw water from the soil, which they transpire and this forms a very efficient drainage system.

Open drains, i.e. ditches or ploughed furrows, are the rule in forest or woodland, as tree roots invariably block unsealed land drains, which they enter at the joints. Salt glazed, solid plastic, concrete or similar pipes, where the joints can be sealed, are necessary for farm drains passing through or near tree roots. Except in high rainfall areas, most of the water comes from higher lying land above a plantation or from springs within it.

Before starting a drainage scheme *within* the plantation, it is necessary to intercept any external water entering the plantation from above and lead it away in long *cut off* or interceptor ditches. These run across the slope, approximately at right-angles to the water flow. Once external water is controlled, drainage within the plantation may proceed.

Drainage is expensive and one should not overdrain. The system must last for the whole rotation. Choice of method depends on the circumstances—soil type, rainfall, topography and the size of the catchment area. Plantations on heavier soils and in high rainfall areas require more drains which are drawn closer than elsewhere. Wet peats and impacted or panned podsols need overall ploughing; in the latter case, busting the pan allows vertical percolation of the water and an immediate improvement; some soils then drain quite freely. Springs should be picked up and led into the drainage system.

On *unploughed land* water falling within the plantation is collected by contour drains, with a small fall running across the slope, at right-angles to the flow. Short feeder drains may be herringboned in. Each drain leads eventually to a main stream or, on agricultural land, a main ditch. Drains running downhill on moderate or steep slopes collect little water and scour badly.

The layout, spacing and size of drains should vary with soil and local conditions. A common size of drain is 0.3–0.4 m deep, 0.6 m wide and 0.2 m across the bottom. Main ditches are usually 1.0 m deep and 0.4 m across the bottom.

On low lying land with little fall, deep ditches may be dug. Trees are planted on the spoil thrown on to the banks and as rooting depth is increased the water table is lowered.

A plan of the drainage system should be drawn to scale and kept in the office for reference. The system should be regularly maintained. Operations such as brashing and thinning are especially likely to result in blockages. When clearfelling takes place, restitution of the drains should be written into the sale contract.

Fencing

Fencing is expensive and for productive forest large square plantations, which give a relatively low cost per unit area, are the general rule, but for small farm woods and shelterbelts, fencing forms a very high proportion of the total cost of establishment. Sound stock-proof fences are necessary where livestock come into contact with woodland but rabbit- and deer-proof fences are only essential where these are a problem.

Specifications depend on needs, e.g. rabbits, sheep, cattle or a combination of them. Deer fences should be 1.8 m high with at least eight strands of wire barbed and plain; sheep netting must be used at the bottom against roe deer. Specifications are obtainable from the Forestry Commission. CCA treated timber and stakes should always be used (Forestry Commission Bulletin No. 14, *Forestry Practice*, p. 138, London: HMSO).

PLANTING
Plant supply

Unless a large area is to be planted annually, the highly skilled job of raising trees from seed or even bought in seedlings is not worthwhile. Plants should be purchased from a professional *forest tree* nurseryman. Names and addresses can be obtained from the Royal Forestry Society of England, Wales and Northern Ireland. Under the Forest Reproductive Material Regulations cuttings and plants of the listed species intended for the production of wood may not be sold unless obtained from sources approved and registered by the Forestry Commission in Great Britain or other appropriate authority elsewhere in the EEC.

Sturdy transplants with well developed roots should always be bought. Small trees recover more quickly from transplanting than large ones, but are more likely to be smothered by weeds or grass if not kept clean. Larger transplants are preferable where rapid weed growth is likely but such sites are usually fertile and relatively sheltered and can support a larger tree. Smaller trees are much safer on poor or exposed sites as they are less liable to damage from windrocking and easier to protect. Trees over 600 mm require staking. Age is normally described by number of years in the seedbed and transplant lines respectively, e.g. a 1+1 tree has spent one year in each. Trees may be undercut in the nursery bed, i.e. mechanical severance of the taproot 75–100 mm below the surface, to encourage vigorous development of fibrous roots. Hardwoods generally benefit most.

Plants may be offered straight from the seedbed but transplants or a similar size of undercut plant are normally used, owing to their balanced top and root development, With *conifers* 2+2 and 2+1 are suitable, but 1+1 or a similar size in undercut plants is most popular as these are cheaper. Plants 300–380 mm tall are suitable for average conditions but smaller plants, 150–250 mm in size are preferable for exposed situations, wet peat or dry heath provided initial weed competition is eliminated by ploughing or a suitable herbicide. Fertile sheltered positions and sites subject to strong weed growth need larger plants, up to 600 mm tall.

With most *hardwoods*, two-year seedlings, undercut in the nursery bed or 2+1 transplants are best. Well grown one year seedlings may also be used. Poplars and willows are supplied either as unrooted or rooted two to three year cuttings or sets.

Conifer seedlings grown in containers (i.e. tubed or paper pot planting stock) and raised in controlled conditions in polyhouses are now used to plant considerable areas. From seed to planting in the forest usually takes from 12–16 weeks. The plants are small and it is essential to have a well prepared weed free site. A special tool is necessary for planting paper pot stock.

Amenity plants may also be purchased in polythene containers which must be removed before planting. These are many times more expensive than transplants and cannot be justified for plantings for timber. Suitable for trees difficult to establish as transplants, e.g. Monterey pine, valuable amenity trees and late spring or summer plantings. Care should be taken in planting and the plants should be watered as needed during the establishment period.

Plant requirement

Spacing

Optimum spacing, usually on the square, depends on initial growth rate of the species and on quality of the site. Fast growing species on a good site need relatively wide spacing while slower growing species on poor or very exposed sites may be set more closely. Spacing also depends on length and type of bole required, length of rotation and thinning technique. Wider spacings than those used traditionally are now employed, as plant quality, soil preparation and weed control have all improved greatly in recent years. As a general guide approximately 2×2 m is suitable for conifers and mixed stands in most situations. Closer settings may be preferred on poor or very exposed sites but it is questionable whether the extra cost is justified. Pure stands of hardwood are no longer planted at the traditional very close spacings. Poplars, willows and pruned free grown oak whips are not thinned and are planted at their final cropping positions. *Table 7.3* gives an approximate guide to spacing; *Table 7.4* shows the number of trees required.

Table 7.3 Spacing of trees

Species and situation	Approx. spacing (m) (planting on the square)
Conifers	
Christmas trees	0.75–1.75
Slow growers (e.g. Scots, Lodgepole pines on poor exposed sites)	1.5–1.75
All conifers on average to good sites	2.0
Hardwoods	
Pure stands	4.0–5.0
Alder, single row windbreak	1.0–2.0
Poplars—single row windbreak	1.75
Lombardy type	5
Bushy top types	6–10
Willows and pruned free grown oak whips	10

Table 7.4 Tree requirements (number of trees required per hectare planted on the square)

Spacing (m)	No. trees/ha	Spacing (m)	No. trees/ha
0.75	17 778	2.75	1322
1.00	10 000	3.00	1111
1.25	6400	4.00	625
1.50	4444	4.50	494
1.75	3265	5.00	400
2.00	2500	6.00	277
2.25	1975	10.00	100
2.50	1600		

Time of planting

Planting is done when the plants are dormant, from October until early April. Sites in colder areas and at higher altitudes may be planted with cold stored (i.e. refrigerated) plants in late May.

Spring planting is commonest, as autumn planted trees suffer from winter gales and hard frosts. Late planting frequently leads to failures. Planting starts as early as possible, provided there is no frost. Larch, which flushes early, is planted in autumn or first in spring, as also the hardwoods. Other conifers, notably western red cedar and Lawson cypress, are probably best planted in the spring. Excellent establishments have been obtained in mild areas by planting in early autumn whilst the soil is warm.

Ordering

Trees should be ordered well in advance. Nurseries usually pack plants in polythene bags, which generally obviates the need for heeling in. On no account must bags be exposed to direct sunlight, for the trees heat up and wither, and newly arrived trees should be protected against frost. Trees should be heeled in when not in polythene bags. A trench should be dug in well prepared ground before arrival of plants, the roots placed in the trench, covered and firmed. The soil should be well packed around the plants to prevent damage by drying winds or frost. In frosty weather the plants should be covered with straw or bracken.

Planting

When trees are taken out for planting the roots must not be allowed to dry out—even a few minutes exposure to hot sun or wind can kill. When planting in drying wind, the roots should first be dipped in water. Trees are best carried in plastic or plastic lined bags slung on the planter's shoulder and withdrawn as required. On no account should the trees be 'lined out' prior to planting.

Planting at regular intervals in straight lines is essential to expedite subsequent operations. The requisite inter-tree spacing may be obtained initially by using a planting stick cut to the appropriate length, but more experienced planters pace the distance. On *unploughed land* straight lines are achieved by furnishing each planter with two long straight sighting sticks, one for each end of the planting area. The planter aims for his stick at the opposite end, reaches it and then offsets it to the required row width. He then aims back to his other stick. With more than one planter, each man has his own colour of cloth tied to the top of his sticks. On *ploughed land* the ridge or furrow, as appropriate, are followed to plant each tree in a similar position.

Methods of planting

Notch planting

Notch planting is normal for large planting areas, except on loose ploughed land. It can be done with a garden spade but a straight bladed notching spade is quicker; on rocky ground a mattock is best. Drive the tool to full depth, withdraw it and make a notch at right-angles to the first. The pattern is like a letter T or L. Lever the ground up for a crack to open. Insert the tree into this crack, draw the tree through until the old ground level mark (or collar) of the tree coincides with the surface of the disturbed ground. Withdraw the tool and allow both earth and tree to fall back, level with the surrounding ground. Finally, stamp everything firm with the heel.

On *ploughed land* choice of position on ridge or in the furrow depends on soil and conditions, planting in the furrow bottom for dry areas but in heavier soil at the side of the furrow. On wet land ridge planting is the rule but not in peaty ground, the peat cracking and shrinking, but rather on the sides. Planting on the side of the ridge away from the prevailing wind should be practised on the wettest sites with over 890 mm rainfall.

Turf planting

Allow time for turf or furrow to settle before planting. Drive a spade through it, making a single cut or side notch from centre to outside, allowing the roots to be inserted and stamped firm. The roots lie between two surfaces of turf which rapidly decompose. Air pockets can cause failures. The side notch which is inclined to dry out and open up should be closed if seen. Cutting a T notch, not breaking the sides of the sod, is a safer but slower method of establishment.

Pit planting

Pit planting is used only for poplars, willows and other valuable specimen trees. A hole large enough to take the spread-out roots is dug, the tree held at the right level, crumbled earth returned around it and finally stamped firm. Trees of any size require staking. In grass, turves are returned upside down to the bottom of the pit, a mulch of leaves, cut grass or farmyard manure used as filling, this method being applicable on stiff soils lacking aeration, and on old woodland sites. Not used for normal forest planting.

Beating up

Beating up is the replacement of young trees that failed on newly planted land, a few trees invariably failing from drought or severe frost. In woodland and wide shelterbelts beating up is not profitable unless failures exceed 25% in a pure stand, provided they are evenly distributed. Beating up is advisable where failures occur in groups, a wide spacing has been used or, in a mixed stand, where numerous trees of the species which is to give the final crop have failed. For shelterbelts, the deciding factor is whether losses are likely to reduce effectiveness. Failures must be replaced in narrow belts and single-row windbreaks. Beating up must be carried out the year after planting to be fully effective, best quality plants used, at least as large as the originals, and the work done carefully. Slower growing species are likely to be suppressed by the original planting.

FERTILISER APPLICATION

Poor, acid, wet and peaty soils are usually deficient in one or more of the major elements, nitrogen, phosphorus and potassium. Lime is undesirable for conifers. Unless the necessary fertilisers are applied the trees may 'take' but then remain 'in check' for many years (i.e. make little or no growth). Phosphorus is by far the most frequently deficient element, followed occasionally by K. Nitrogen is rarely required. Where in doubt, a dressing of P should be given.

Phosphorus is best applied as URP (Unground Rock Phosphate) which is less dusty and much pleasanter to handle than GMP (Ground Mineral Phosphate); there is little to choose between them in cost and effectiveness. Water soluble phosphates, i.e. Superphosphate and Triple superphosphate are dearer and soon become unavailable to plants on acid soils. Potassium is applied as muriate of potash (KCl) and N as prilled urea.

	kg/ha	*g/tree*
Apply URP (or GMP)	375	150
and *if* needed		
muriate of potash	200	80
prilled urea	350	140

Fertiliser is normally applied as a top dressing soon after planting. Large areas, especially on rough or hilly country, are most cheaply treated from the air but staff must be trained to cooperate in loading and marking the area to be covered. Small plantations are top dressed by hand in the area immediately surrounding each tree.

WEEDING AND CLEANING

Weeding means the suppression of vegetation in the immediate vicinity of the young tree and is essential for three to four years after planting. Neglect allows weeds to rob the young trees of nutrients, water and light, checking growth severely; if weed growth is allowed to collapse on the young trees in the autumn they may be smothered. Vigorous weed growth may follow cultivations which often encourage germination of weed seeds of unexpected species, e.g. gorse (*Ulex* spp.), spearthistle (*Cirsium lanceolatum*).

Cleaning refers to the removal of woody growth, which includes:

(1) fast growing wild tree species, e.g. birch (*Betula* spp.), sallow or goat willows (*Salix caprea* L.) and coppice shoots from hardwood stumps,
(2) harmful climbers, e.g. old man's beard (*Clematis vitalba*), ivy (*Hedera helix*) and honeysuckle (*Lonicera periclymenum*), and
(3) other species gorse and brambles (*Rubus* spp.).

Most of these can restrict growth rate or even suppress the planted species if not controlled up to the time of the first thinning. Weed control may be effected by manual, mechanical or chemical means or by a combination of methods (*see* Chapter 8).

PREPARATION FOR THINNING

It is usual to cut *inspection racks* or paths during the thicket stage to allow the state of the plantation to be assessed; these should be adequate in number and are made by cutting off the lower branches of two adjacent rows with a pruning saw or lightweight chain saw, tight to the trunks.

Brashing

This is the removal of the branches on the lower 2.0 m of the trunks with a *sharp* saw: either a pruning saw or lightweight chain saw, but flush with the trunk without damaging the bark. Traditionally whole plantations were brashed but the work is expensive. Today, with selective thinning, only enough trees are brashed to allow inspection and marking of those to be felled. With line thinning, brashing is largely eliminated. Species differ in the ease with which their lower branches are removed: spruces are very tough, larch branches break off readily, while pines are intermediate between the two.

Pruning

Pruning sometimes called high pruning, is undertaken to a higher level than brashing, usually to a maximum of 6 m. The object is to produce knot-free timber. The work is expensive and mainly confined to hardwoods.

Free grown oak should be inspected annually and pruned as required. Conifers are not usually pruned but final crop trees in top quality stands may justify the work to improve the appearance and price of the standing timber. Pruning saws with extendable handles will work up to 6 m; two to four prunings are usually given. Pruning wounds can be protected by applying approved fungicide/sealants (BS 3998: 1966 Recommendations for tree work).

THINNING

Conifers are normally planted relatively thickly, around 2500 trees/ha to:

(1) draw the trunks up straight and suppress branching and hence the development of large knots, and
(2) provide a wide selection for the final crop.

Inter-tree competition begins as soon as the dense thicket stage is reached, so that many trees overtopped by their dominant neighbours ultimately die. If left unthinned, the process will continue throughout the life of the plantation, so that no income is obtained from thinnings and the final crop consists mainly of small undersized trees of low value.

The object of thinning is to maximise income from the crop over the whole rotation. This includes income from thinnings, each increasing in value over the previous one, together with the final crop of large saw timber. The income from a single thinning cannot be considered in isolation, as earlier thinnings allow the remaining trees to grow to a greater trunk diameter, thereby becoming more valuable. Frequently the first thinning barely pays for itself. A substantial proportion of the total production, up to 80–90% of the number of trees and 50–60% by volume may be taken as thinnings without reducing total volume production over the whole rotation.

The *thinning intensity* is the annual yield from thinnings or the rate at which volume is taken as thinnings from a plantation: i.e. V/C where V = volume removed at a single thinning, and C = number of years in the thinning cycle.

Marginal thinning intensity is the maximum without reducing subsequent volume production. Higher thinning intensity raises yield from thinnings and increases diameter and value of the final crop but, if too much volume is taken at any single thinning, the marginal thinning intensity is exceeded and subsequent volume production is lost. Marginal thinning intensity is thus the most profitable rate of thinning but must not be exceeded.

Age at which the first thinning is taken is determined jointly by species and quality of site. In general, the higher the yield class (GYC) the earlier the first thinning should occur. Species differ considerably. On a good site Japanese and hybrid larches may need thinning 13–14 years after planting, whereas other species may wait until 16–20 years. Poorer sites are thinned later. Detailed information on date of first thinning, according to GYC is given for the main timber species in the Forestry Commission Booklet No. 34, *Forest management tables (metric)* 1971.

When selecting trees to be taken out at the first thinning the following must be included for removal: *wolf trees*—vigorous but deformed trees of planted species which suppress the growth of good adjacent trees unless removed; *whips*—trees with long whiplike stems, which flay about in wind and damage the tops of adjacent trees. *Suppressed* trees should be removed where economically worthwhile.

The *thinning cycle*, i.e. the frequency or number of years between thinnings varies from three to six years for young or fast growing crops to ten years for older or slow growing crops. Quick growers on good sites, notably Douglas fir, Sitka spruce and Japanese and hybrid larches, may need thinning every three years, while for Norway spruce, Scots pine and most hardwoods a five-year cycle is quite adequate. Thinning tends to be delayed where very large areas are being managed or with low prices for thinnings. Lower frequency of thinnings with more volume removed are less costly in labour, but may result in increased incidence of windblow.

Sound selection of trees needs considerable experience. The aims are to maintain, as far as possible, an unbroken canopy so that growth is uninterrupted and to improve steadily the quality of the remaining trees. The final crop trees should be the best, having long, clean straight trunks with a large diameter, little taper and free from heavy branches or forking; the crowns, which manufacture the timber, should be large and well developed. Remove trees which are coarse or leaning, or have a wide spread, a weak unhealthy crown or spiral bark, which frequently indicates spiral grain in the wood.

The *thinning type* denotes the position in the canopy from which the thinning is taken. With a *crown thinning* the bulk of the trees removed are dominant or co-dominant, having their crowns in the upper storey of the canopy. Conversely, with a *low thinning*, the trees removed are mainly sub-dominants and suppressed trees, coming from the lower story of the canopy. The most productive trees come from the upper storey of the canopy, so *low thinning* is generally the rule, although some crown thinning is usually needed to maintain the balance of the upper canopy. A guide to the correct number of trees to be taken at each thinning may be obtained from production management tables (Forestry Commission Booklet No. 34, *Forest management tables (metric) 1971*).

Methods of thinning

There are two basic approaches to thinning:

(1) purely selective thinning, the traditional method, and
(2) line thinning, where the simplest form is to remove in its entirety every fourth row at the first thinning.

Subsequent thinnings should be selective for maximum production. Line thinning simplifies mechanical extraction, largely eliminates the need for brashing and reduces costs, but potentially good trees are lost in the cleared lines and susceptibility to windblow is increased.

HARVESTING

In order to safeguard the national reserve of standing timber, it has proved necessary to control felling and to require re-planting, where appropriate, of cleared land. These controls are exercised by the Forestry Commission under the Forestry Act 1967. Applications for licences should be made to the Commission's District Office.

Modern timber harvesting, which includes the operations of thinning and clear felling, is essentially a specialist's job, requiring efficient planning and organisation coupled with a high degree of skill on the part of the work force. Chain saws and other forest machinery are extremely dangerous in untrained hands and much money can be lost by relying on an inexperienced work force. The work may be undertaken by:

(1) a timber merchant who has bought the standing trees,
(2) specialist timber handling contractors, or
(3) most suited to larger estates, the timber owner's own employees.

MARKETING

The most profitable outlet for poles and timber grown on a farm or estate is generally some use on the property itself. Usually there is a surplus of material or it is of unwanted size or some of it is of special quality. Larger estates or landowners with experience of the trade and their own equipment, saw mills transport and a skilled labour force can profitably process their timber to some finished form, except for the highest grades such as logs for veneer or furniture, where the buyers require to do their own processing. Otherwise the best course is to sell the timber *in the round* to a merchant, who will have the skilled labour to handle it and can place *all* grades, each on a suitable market, unlike many manufacturers who will handle only the best. Merchants usually handle pulpwood and pit props. Timber may be sold *standing* in the woods, or *at stump* after felling, or *at roadside* after hauling out with tractor, the price rising considerably with each step in preparation and transport.

Most reputable timber merchants wish to return in a few years to purchase another parcel. They may bargain keenly over price, but will agree to and observe all reasonable conditions in the contract of sale. These should include: dates for entry, final removal of all material and purchaser's equipment, and payment; method of measurement, routes to be used for hauling, repairs to gates, fences, roads or bridges damaged, clearance of lop and top, fire precautions, and arbitration in case of dispute. The more stringent the conditions, the less the merchant will pay.

If only a small quantity of timber is involved, a sale by private treaty is often satisfactory, but for a large or valuable parcel it is better to invite several merchants to submit sealed tenders to be opened on a stated date. As a safeguard against selling for a price far below true value, it is often worthwhile having an independent valuation by a professional adviser. Auction sales are only worthwhile for large parcels of timber. Prices fluctuate widely according to the type and quality of the timber.

Conversion of timber on small saw-benches driven by a power take-off is an excellent way of meeting home needs, but such equipment cannot compete in accuracy of sawing and speed of output, with more specialised plant.

TIMBER PRESERVATION

A wide range of preservatives against fungal and insect attack both indoors and outdoors are available. Use treated timber for construction work, but the processes are highly specialised and expert advice should be sought. Only treatment of farm or estate timber is considered here.

Seldom is timber durable when exposed to alternating wet and dry conditions favouring growth of wood rotting fungi. The heartwoods only of oak, sweet chestnut, larch, western red cedar, Lawson cypress and yew possess natural durability; small logs and *all sapwood* do not. Untreated non-durable wood used for fencing stakes lasts barely four to five years whereas properly treated timber has an almost indefinite life.

Preparation of timber

Timber must be seasoned before treatment; freshly felled timber is useless. Seasoning involves storing in free moving air, under cover to lower the water content to at least 30%, preferably 25%, according to species. Logs or stakes are peeled, stacked criss-cross or with spacers to allow the passage of air; supports are used to keep the bottom layer off the ground. Two to six months' seasoning is usually necessary, depending on the month of felling. Seasoning is not a preservative treatment. Home grown planks and scantlings required for indoor work need seasoning for six months in a dry airy building, small sticks between them allowing air circulation. Seasoned timber is lighter, slightly stronger and much less liable to shrinkage or change in shape.

Preservatives

The effect of any wood preservative depends on concentration, amount retained in the wood and depth of penetration achieved. Factory treatment, though more expensive, gives far better control and a really durable product.

Brush applications

These are ineffective for stakes and only appropriate for thin boards up to 12 mm ($\frac{1}{2}$ inch) thick, above ground level and reached on both sides as in a board fence.

The CCA process (copper chrome arsenate)

This has superseded creosote. Water soluble, it gives a high degree of penetration and forms a totally insoluble chemical complex with the conifer wood resins which remain permanently in the timber.

The process requires a specialist plant operated under factory conditions; some local sawmills provide the service. Stakes require de-barking, pointing and kiln drying before treatment.

Factory treated CCA timber has a life of 30 years or more and is a once and for all treatment. CCA treated stakes and timber for outside use replace all others.

Creosote

Prior to the introduction of CCA, creosote was the main treatment; the main problem is to secure adequate penetration. Only suited for timber *above* ground not otherwise treated.

Pressure creosoting secures deep penetration, as with telegraph poles, and is highly effective, but few, if any, plants now operate.

PROTECTING WOODLANDS

Fire risk

The danger of fire is always present, but degree of risk varies greatly. On the average wooded estate, where varied kinds and ages of tree are intermixed and broken up by fields, it is never so high as in a large expanse of young conifer plantations. Everyone owning woods should take simple precautions:

(1) Have available telephone numbers and addresses of fire brigade and police; if in doubt dial 999.
(2) Inform fire brigade of location and type of woods.
(3) Make sure all parts of woods are accessible for fire fighting; locked gates, broken culverts or fallen trees can cause delay, allowing minor outbreaks to become major disasters.
(4) Have adequate, if simple, fire fighting tools at strategic points; include spades, buckets, axes,

bill hooks and birch brooms. Only birch brooms can safely be left out in the woods; site stacks in full view near gateways.

(5) Clear away inflammable vegetation from points of high risk; for example, clear gorse along roads or railways.

(6) Brash young conifers early, especially near roads or footpaths.

(7) Ensure that all staff and contractors observe fire precautions and are trained in fire drill.

(8) Insure against fire risk.

Few private owners have woods big enough to merit a regular patrol, but in dry weather both owner and woodmen should keep alert for signs of fire. The most dangerous period is usually from March to May; afternoon and early evening are the most dangerous times of day. Fires may arise at any time when herbage is dry.

Wind damage

Avoidance of serious damage by gales is mainly a matter of correct choice of species for site, proper maintenance of drains to prevent soft ground conditions and proper and timely thinning. In particular, the spruces, Douglas fir and larch are very susceptible on shallow clay soils; shallow peats affect all species. Avoid long periods between thinnings. Line thinned plantations are more susceptible. With best management, however, no forester can stop the occasional gale or gust from bringing down groups of trees. Windblows are rarely a total loss; any tree big enough to be blown down is also big enough to sell. *Windsnaps*, where the tree breaks off part way up the trunk, cause considerable loss, as split wood has to be discarded, and lengths are shortened.

INSECT PESTS AND FUNGI

Insects

When clear felled conifers are replanted or conifers are planted in close proximity to existing conifer plantations, the young trees are liable to be attacked by either the large *pine weevil* (*Hylobius abietis*) or the *black pine beetle* (*Hylastes* spp.)

Pine weevils

Pine weevils breed in conifer stumps. The adults emerge and feed on the bark of all species of newly planted conifers, which die as soon as they are ring barked by the weevils. The traditional method of control was to leave the site bare for three or four years after felling and trap the insects, but this method wasted time and allowed heavy weed infestations to develop. Control at planting time may be achieved by

dipping the aerial parts of the bundles into a liquid dip of gamma HCH. If control is incomplete or has not been undertaken prior to planting, a topspray of gamma HCH is given to the young trees. Neglect of control measures can result in total loss of the planting.

Black pine beetles

Black pine beetles also breed in conifer stumps and then attack the young transplants, but below the region of the collar. Pines are especially liable to damage, also spruces. Root dipping of the bundles in a liquid HCH dip as far as the collar gives good control. Alternatively, control of both pests may be achieved from a total dip of the bundles.

A number of other pests may attack woodlands and forests, some causing considerable damage, but they usually appear sporadically and are very localised in their occurrence.

Fungus diseases

Trees are subject to attack by many kinds of fungi. The most serious for the ordinary grower are those causing heart rot or butt rot. The fungi concerned, such as *Fomes annosus*, gain entry to the heartwood of the tree either through some deep wound, such as the breaking of a branch, or through the roots. They cause decay in the centre of the trunk, which may be slight or so serious as to make it both dangerous to passers-by and worthless as timber. Unfortunately, only when infection is well advanced is it revealed by fructifications—toadstools or brackets—usually at the foot of the tree, but sometimes higher up the trunk.

Control of heart rot in hedgerow or woodland trees is seldom possible once there is cause to believe trees are affected. Some should be felled and the trunks examined. If rot is prevalent, it may be advisable to clear the crop before sound trees deteriorate in quality and value, but the occasional unsound tree need cause no general alarm. Few stands of maturing timber are wholly free, while the timber merchant can usually cut sound lengths of timber from all but the worst affected logs.

As a rule it is quite safe to replant ground on which heart rot has developed; but where it has been severe it is best to change the kind of tree grown. When thinnings are carried out in young conifer woods routine flood painting of the stumps with a fungicide, which includes a coloured indicator, is strongly advised. A concentrated solution of urea *applied within 15 min* is ideal. The treatment prevents fungi from entering roots and spreading through root crossings to other trees.

Honey fungus (*Armellaria media*) can be a problem in old broadleaved woodlands, affecting both broadleaves and conifers. It spreads from old stumps through the soil and kills young trees. Losses are sporadic and rarely affect a whole wood. There is no economic treatment. Other problem diseases include Dutch Elm Disease (*Ceratocystis ulmi*) and beech bark disease (*Nectria coccinea*), both carried by insects.

Details of insect pests and fungus diseases and their control may be obtained from Forestry Commission publications.

MISCELLANEOUS
Shelterbelts

Grown to protect crops, orchards or livestock, shelterbelts often produce timber as well. On arable land shelter can depress yields of crops nearby, because of shading and the roots drawing water and nutrients from the soil. Increased yields over a much larger area some distance away are noted and an overall benefit is secured. Provision of shelter is especially important with orchards, the shelter being preferably planted several years before the trees. In some areas production of vegetables and flowers is very difficult without shelter. On exposed livestock farms shelter makes the crucial difference between land worth improving and rough grazing. Shelter must be available for outlying stock, ewes and lambs.

On level ground a shelterbelt gives appreciable shelter over a distance equal to 20 times its own height; thus a belt 18 m high shields a field 360 m wide. When siting new belts advantage should be taken of existing natural features—woods, ridges, etc.—to provide a system giving protection from several directions. Belts are best placed at right-angles to the prevailing, most damaging wind, but local features call for some variation.

The width varies from single-row windbreaks upwards, depending on space available, value and productivity of land. If the value of agricultural production exceeds that from forestry, width should be the minimal, 9 m normally being ample for arable situations. Where the value of agricultural production is low, wider belts can be considered, varying from 20–40 m or more and having a productive interior. On hill grazings, shelterblocks to protect stock from all quarters are necessary. Three-armed plantations of conifers are highly effective, allowing stock to move around to the lea side, according to wind direction.

Belts with more or less vertical edges and an irregular profile are most efficient. The latter can only be obtained with mixed stands. With narrow belts, designed primarily for shelter, the mixing of species within rows and planting alternating rows of different species is normal practice. With arable and horticultural crops less dense shelterbelts are necessary, thereby filtering the wind to prevent formation of eddies; dense belts result in substantial local crop damage.

Amenity

In special amenity situations the fringes of commercial woodlands and forest can be landscaped with single rows, small groups or single specimens of attractively growing trees, e.g. western red cedar, beech, copper beech or Norway maple.

Game birds, such as pheasants and partridges, benefit from good woodland management, for this ensures certain woods (though not always the same woods) are at different stages of growth, thus meeting needs of both birds and sportsmen. Recently felled and newly replanted woods give open feeding grounds and low cover for nesting; older plantations provide ample safe roosting places; maturing timber leads to high flying pheasants at shooting time. Keep woods at various stages evenly spread over an estate.

Coppice

Coppice or underwood is only profitable for limited production in a few districts. Sweet chestnut is suitable in districts where cleavers buy it by the hectare for fencing. Hazel is largely unprofitable and little hazel coppice is now worked. It usually pays best to replant, after clearance, with a quick growing conifer to smother young coppice shoots, and involves a minimum of weeding; Japanese larch is recommended. Birch coppice, also unprofitable, is best thinned out to one shoot per stump and then underplanted with a shade bearing tree such as western hemlock or beech. Oak coppice can be partially cleared for firewood and the ground planted with Douglas fir.

Firewoods

Branchwood from felled broadleaved trees is best used or sold as firewood or charcoal wood; but coniferous timber, even if of small dimensions, commonly fetches a better price for pulpwood or chipboard material. A simple portable saw bench drive from the power take-off is adequate for firewood cutting. Oak, ash, birch and beech are excellent firewoods; chestnut burns well, but is apt to spark; elm is useless when wet but burns well when really dry. Poplar is very poor. All the conifers burn well but are liable to spark badly, especially if wet; they are satisfactory in closed woodburning stoves. Fruit trees give good firewood. Seasoning for a few months, preferable under cover, improves the burning qualities of fresh felled timber but the loss of weight is a disadvantage when selling.

AGENCIES FOR ASSISTING WOODLAND OWNERS

To employ a full time forester ensures maximum returns in the case of several hundred hectares of woodland, but a common difficulty with small acreages is that need for thinning, felling or replanting arises only at intervals of several years. The help of a forestry adviser, in constant touch with local conditions, particularly markets and ruling prices, is worthwhile. Numerous landowners now let out the management of their woodlands to contract.

The *Forestry Commission* gives free technical advice to woodowners and is the main source of grant aid for new plantings, but timber production must be one of the primary objectives. Current schemes include the *Small woods scheme* for planting or restocking small detached woodlands of less than ten hectares.

The Dedication Scheme—Basis III only applies to areas of ten hectares or over. Previous grant schemes, e.g. Dedication Basis I and II, the Approved Woodlands Scheme and a Small Woods Planting grant scheme are now closed to new applicants, but grants are still paid to those remaining in the schemes. Applications for grants should be made to the local Conservator of Forests whose address will be found in the booklet *Grants for Woodland Owners*. Grants may also be obtainable for amenity plantings from the *Countryside Commission* and for shelterbelts on lowland farms with development plans or horticultural holdings through the *Ministry of Agriculture, Fisheries and Food*.

Cooperative forestry societies

These are independent associations operating throughout Scotland and in many districts of England and Wales. In return for a small subscription, members have the services of a professional forester for advice on planting or marketing of produce, and most societies undertake the actual operations for a fee or commission. Societies also arrange bulk buying of young trees, tools, and materials at favourable rates.

Planting contractors

The planting and subsequent maintenance of woods for a stated period are undertaken by firms of planting contractors for an agreed charge. Some timber merchants maintain a planting service which can be employed to re-stock land cleared by felling.

Forestry consultants

These are professionally qualified men who advise on woodland management, or actually undertake it, for an agreed fee. Many land agency firms offer a similar service and can advise on taxation and other legal problems appertaining to woodlands.

Forestry societies

The Royal Forestry Society of England, Wales and Northern Ireland, 102 High Street, Tring, Herts HP23 4AH, and The Royal Scottish Forestry Society, 26 Rutland Square, Edinburgh, EH1 2BU, issue quarterly journals, organise excursions and conferences at which forestry problems are discussed and can advise on legal and other matters.

LEGAL RESPONSIBILITIES

If a tree growing on land bordering a highway falls and causes damage to passers-by, the landowner may be held liable, but only if it was apparent to a prudent, observant person that the tree was dead, diseased or otherwise unsafe, before the accident happened.

Local authorities have powers to remove trees that have blown down across a road, or are otherwise obstructing or endangering passers-by; they may do so without prior notice and are entitled to recover costs from the owner of the trees.

Branches extending over a neighbour's property may be cut back without notice, but only as far as the legal boundary line. The severed branches must be offered to the owner of the tree.

Should the roots of a tree extend on to a neighbour's land and cause damage, the tree owner may be held responsible. The owner may also be held liable if poisonous foliage, such as yew, extends on to a neighbour's field and harms cattle.

Where land is leased for farming, it is usual practice to reserve 'all timber and timber-like trees' to the owner. In the absence of such a clause, local custom may allow a tenant to use timber or underwood, for such purposes as fence repairs or firewood, but only on the leased property.

Bodies possessing special powers with regard to trees, mainly in the interests of keeping open lines of communication, include: the Post Office (with regard to telegraph lines); electricity supply undertakings; railways; air navigation authorities; and land drainage boards.

ADDRESSES

Main offices of the Forestry Commission

Headquarters of the Forestry Commission:
231 Corstorphine Road, Edinburgh, EH12 7AT (031 334 0303)
London office:
25 Savile Row, London, W1X 2AY (01 734 4251)
Senior Officer for Wales:
Churchill House, Churchill Way, Cardiff, CF1 4TU (0222 40661)

Director of Research and Development:
Alice Holt Lodge, Wrecclesham, Farnham, Surrey
GU10 4LH (042 04 2255)
Northern Research Station:
Roslin, Midlothian, Scotland, EH25 9SY (031 445
2176)

Other organisations

Timber Grower's Organisation, Agricultural Centre,
Kenilworth, Warwickshire, CV8 2LG
Scottish Woodland Owners Association, 6 Chester
Street, Edinburgh
The Countryside Commission:
England: John Dower House, Crescent Place, Chel-
tenham, Glos. GL50 3RA
Wales: 8 Broad Street, Newtown, Powys, SY16 2LU

The Countryside Commission for Scotland: Battleby,
Redgorton, Perth, PH1 3EW
Nature Conservancy Council, 19/20 Belgrave Square,
London, SW1X 8PY, and 12 Hope Terrace, Edin-
burgh, EH9 2AS
Agriculture:
Addresses of Regional Offices of MAFF, ADAS and
DAFS may be found in the telephone directory.

Forestry Commission publications

A catalogue is obtainable from the Publications
Officer, Forest Research Station, Farnham, Surrey,
giving an up-to-date list of titles in print. These may
be purchased through any Government Bookshop or
through most booksellers.

8

Weed control

R. D. Toosey

Weeds are injurious or harmful to growing crops, and if not controlled reduced yields from crops and grass result, harvesting and other operations are hindered, produce contaminated or taints imparted and the product rendered unfit for sale. A number of weeds are poisonous.

Most weeds of significance are *annuals* or *perennials*. *Annuals* complete their life cycle within a year and reproduce only from seed, of which they set an abundance. Growth is rapid, some species producing two or more generations in a year. *Perennials* also produce seed but do not die after seeding and, as they also reproduce from vegetative storage organs, serious infestations can build up rapidly if not controlled. *Biennials* only live two years. They make vegetative growth in the first and set abundant seed in the second, after which they die; they reproduce only from seed. Few are important, e.g. ragwort (*Senecio jacobea*) and spear thistle (*Cirsium lanceolatus*), which infest badly managed grassland. For convenience of weed control, weeds may also be classified as grasses, broadleaves, rushes and sedges and woody weeds.

OCCURRENCE OF WEED PROBLEMS

Grassy weeds

Annuals

A serious problem in cereals, especially wild oats (*Avena* spp.), blackgrass (*Alopecurus myosuroides*) and *sterile brome* (*Bromus sterilis*), but can be controlled chemically; not usually a serious problem in broad-leaved crops. Only annual meadow grass (*Poa annua*) occurs in grassland.

Perennials

A widespread problem in grassland and they occur on poorly managed arable land, especially cereals, where they cannot be controlled chemically while crop is growing actively; stubble treatment or spraying shortly before harvest is usually necessary. Good chemical control in young trees, no problem later.

Broadleaves (non-woody)

Annuals

Except for mouse eared chickweed (*Cerastium vulgatum*) these occur almost entirely on arable land, where there are good chemical controls in most crops. Elsewhere only a problem following soil disturbance, i.e. on new leys or poached grassland.

Perennials

Sprayed cereals are an excellent cleaning crop for these, where they are readily controlled by translocated herbicides. Cannot usually be controlled chemically in broadleaved arable crops. Some are difficult to control chemically in grassland, e.g. docks (*Rumex* spp.) unless ploughed prior to treatment. Adequate controls in young trees.

Rushes and sedges (perennial)

Found on poorly drained and managed grassland— sign of low fertility. Should not be an arable weed.

Woody weeds

Mainly a problem in young woodland. Can be readily controlled by chemicals in establishing conifers. May establish in undergrazed grassland, e.g. gorse (*Ulex* spp.), brambles (*Rubus* spp.), etc.

GENERAL CONTROL MEASURES— APPLICABLE TO ALL SYSTEMS

Prevention of spread of weeds by seed

Preventive sanitation is necessary to stop the spread of seed of a wide range of weeds, e.g. wild oats (*Avena* spp.), docks (*Rumex* spp.) and gorse (*Ulex* spp.). Sow clean seed, burn straw infested with weeds and avoid buying weedy hay or straw or transporting onto land destined for seed production. Allow farmyard manure to heat up in a neat pile before spreading. Prevent spread of weed seeds from hedgerows or banks, e.g. sterile brome (*Bromus sterilis*) and waste places, e.g. docks (*Rumex* spp.), ragwort (*Senecio jacobea*), and seeds and vegetative organs from headlands.

Cultural and rotational methods

See also major cropping systems—Arable Crops (Chapter 5), Grassland (Chapter 6) and Woodland (Chapter 7).

PRINCIPLES OF CHEMICAL (HERBICIDAL) WEED CONTROL

The effect of any herbicide or chemical weedkiller depends on the manner and circumstances in which it is used, namely the herbicidal treatment. If recommendations are to be successful they must be given in relation to the whole treatment. The three basic components of herbicidal treatment are the crop, type of weed, and type and dosage rate of herbicide. Herbicidal treatments can be:

(1) *total or non-selective*. Applied with the object of killing all vegetation, either before planting the crop or in non-crop situations, e.g. paths, around buildings and waste land;
(2) *selective*. Designed to suppress weeds without damaging the crop. However, all treatments become total if a large excess of herbicide is applied.

Herbicides may be applied either to the foliage (foliar application) or to the soil (residual application).

Foliar application

Foliar applied herbicides may be of two types.

Contact types

These kill only parts of the plant they contact. Single applications are suitable only for use against annual weeds.

Translocated types

These are transported within the plant to parts remote from the point of application. They can be used for perennial weeds with well protected storage organs.

Applications may be made at these stages of crop growth:

(1) *pre-sowing* or *pre-planting* before the crop is sown or planted;
(2) *pre-emergence*, after sowing but before emergence;
(3) *post-emergence*, after crop emergence.

Each may consist of contact, translocated or residual applications or a combination of two or more of these.

Herbicides may be applied *overall*, when the whole crop area or soil is covered. Most spraying for fallows, cereals or grassland is so treated.

Alternatively the application may be *directed*, where only part of the area is sprayed. Special equipment is needed, hence the acreage must justify the capital cost. With *band spraying*, where a band of chemical about 180 mm wide is applied along the rows, say for sugar beet (pre- or post-emergence), some two-thirds of the cost of herbicide is saved. Weeds between the rows are controlled by rowcrop tackle. Chemical hoeing, where the herbicide is applied to weeds or soil between the rows of an emerged crop, the crop being protected by metal guards on either side of the row, is used only on a limited scale. Directed applications may also be made from *tractor-mounted spraylances, knapsack sprayers* or *granule applicators*. Directed applications are widely used in woodlands and forests.

Selective treatments

Contact applications

These are used almost entirely for controlling annual weeds but kill only emerged weeds, being applied to the seedling or young plant. Contact herbicides usually have little or no residual effect and are thus ideal for application immediately before sowing. A high degree of operator efficiency is necessary to achieve adequate spray penetration and coverage. Higher pressures, smaller droplets and medium or high volumes are required.

Uses

(1) *Pre-sowing or pre-planting*. For (a) control of emerged annual weed seedlings, a 'stale seedbed' technique; (b) sward desiccation prior to direct drilling or ploughing (paraquat); (c) stubble cleaning, desiccation of stubble for burning and couch control (paraquat). Can be combined with a residual herbicide to control weed seedlings not emerged at the time of application. Several residual herbicides have some contact action.

(2) *Post-emergence*. Specific herbicides used on brassicas, sugar beet, mangolds and similar crops. Generally used in mixture with a translocated herbicide for cereals.

Translocated applications

These are used for controlling annual and perennial weeds. These usually act more slowly than contact herbicides but a much lower degree of cover is necessary, provided there are no misses. Damage to susceptible crops from drift is a serious hazard. High pressures and small droplets should not be used. Usually applied in low or low–medium volume. Many translocated herbicides need 6 h of fine weather but thereafter unaffected by rain; others need longer.

Uses

(1) *Pre-sowing*. This is for (a) control of couch and other grassy perennials in stubbles or fallows; (b) control of late starting perennials in stubbles and seedlings germinating after a first spraying; (c) control of perennials for direct drilling, provided chemical has little residual effect.
(2) *Post-emergence or post-planting*. By far the greatest proportion of land sprayed is treated in this way, including most of the cereals treated for broadleaved weeds and all permanent and temporary grassland where selective weed control is required. Also used on clovers, lucerne, peas and in forestry.

Soil or residual applications

These remain active in the soil for a variable period, depending on rate of application and speed with which herbicide is dissipated from the soil by rainfall and bacterial activity. Cropping restrictions exist according to the individual herbicide and dosage rate and the time necessary between application and sowing a susceptible crop. Degree of activity, effectiveness and requisite dosage rate of residual herbicides depend on soil type. Some are of limited value in certain soils, e.g. organic soils and peats.

Uses

(1) *Pre-sowing*. These herbicides are usually incorporated into the soil and those with high volatility need immediate incorporation. The ideal tool is a rotary cultivator and some are fitted with spray bars. Alternatively, two passes of heavy harrows at right-angles to each other may be used. Some herbicides are also formulated as granules, obviating need for incorporation. They control difficult perennial weeds like couch, or, if applied shortly before sowing, annual broadleaved weeds, wild oats or blackgrass in a wide range of annual crops, cereals, peas, brassicas and sugar beet.

(2) *Pre-emergence*. Treatment is highly dependent on adequate moisture and a smooth, high-quality tilth, and is almost useless in dry conditions and rough seedbeds. Some herbicides in this group are mixed with a contact herbicide to give immediate control of annual weed seedlings. Several have some contact properties of their own used to control annual grassy and broadleaved weeds in a wide range of broadleaved and cereal crops.
(3) *Post-emergence*. Largely used for directed application to perennial crops, orchards, soft fruit, and forestry; of little importance in farm situations.

A number of herbicides are used for crop desiccation prior to harvesting, as for instance the desiccation of leafy red clover seed crops, potato haulm and green material in a cereal crop ready for harvesting. They may be used for root and shoot destruction on potato and mangold clamps or old clamp sites.

CHOICE OF HERBICIDE

A very wide range of herbicides is on the market. New herbicides require time to be fully proved and some are shown unreliable. It is generally safest, except perhaps for special situations, to buy only chemicals listed by the Agricultural Chemicals Approval Scheme in the booklet *Approved Products for Farmers and Growers*, issued annually and obtainable from HM Stationery Office or any bookseller. Brand names of herbicides and their manufacturers are given for each chemical, with cross references. Other information includes sale, use and application of chemicals, first aid measures, chemicals listed in the Agriculture (Poisonous Substances) Regulations and those subject to the Poisons Act, together with a crop guide.

When making a choice, it is also *essential* to consult the *manufacturers' literature to ensure that the herbicide(s) under consideration is suitable for the crop and purpose in hand in every respect*.

Choice of herbicide is governed by the following factors.

Crop

Only crops and varieties for which a herbicide is specifically recommended should be treated; others are liable to be severely damaged or even destroyed. Even where several herbicides are recommended for a particular crop, they may vary in their degree of crop safety.

Weeds to be controlled

The occurrence of weed species and the nature of weed problems vary greatly between different types of crop and from farm to farm; species which are very troublesome in one situation may not even be present nearby. In some cases a weed infestation may consist of a single species, which may be dealt with by a simple herbicide, but the problem is usually much more complex, involving a wide spectrum of weeds, some of which are resistant to the simple herbicide. Successful weed control, at reasonable cost, involves a combination of the following approaches.

Broad spectrum

These treatments generally contain a mixture of two or more (sometimes four) herbicides given as a single application. These can be very successful for the control of a wide range of weeds, which have reached simultaneously an appropriate growth stage at the time of application, especially for the late spring foliar treatment of broadleaved weeds in cereals. Herbicides must be miscible. The same principle is used also for residual applications.

Herbicidal programmes

Single applications are only partially effective on mixed weed floras which develop over a relatively long period, e.g. winter cereals, where the crop may first be infested by autumn germinating annual grassy and broadleaved weeds, and then by spring germinating annual grass and by a wide range of broadleaved weeds. An autumn application, including a residual herbicide, may be highly effective for much of the winter but may lack persistency to give adequate control of spring germinating wild oats and broadleaved weeds. Conversely, if application is delayed until the spring, the autumn weed germination may well have reached infestation proportions. A *sequence* of applications, designed to deal specifically with the weeds being encountered, is therefore given at appropriate times in the life of a particular crop and is referred to as *a herbicidal programme*. The principle is now used in a very wide range of crops—sugar beet, root and vegetables, cereals, perennial crops and forestry.

Integration of herbicide use with those used on other crops

Crops differ greatly in the ease with which certain weed types may be controlled selectively during the life of the crop and in the post-harvest opportunities which they offer for chemical or mechanical cleaning. Cereals are an excellent crop for cleaning broadleaved weeds with herbicides—a sequence of two or three cereal crops sprayed with translocated herbicides will control most, if not all, perennial broadleaved weeds.

A winter cereal stubble allows ample time for cleaning grassy perennials before a spring sown crop. Conversely, it is not often possible to control selectively broadleaved perennials in broadleaved crops

such as sugar beet and potatoes; these weeds usually present serious problems if not controlled in the preceding cereal crop. Grassy annuals, which are frequently expensive to control in cereals, especially early sown winter varieties, can be readily controlled in spring sown broadleaved crops, where they should not present any problem. Integration of herbicidal treatments over a run of crops pays dividends.

Stage of growth of crop and weed

Treatment at the wrong stage of crop growth is likely to result in damage to the crop, which can be very severe or even lethal, while application of a herbicide to weeds at an unsuitable stage results in unsatisfactory control. Foliar applied contact herbicides used to control annual weeds need very early application while translocated types, especially for control of perennials, need later treatment. A wide degree of crop or weed tolerance is especially valuable when climatic or weather conditions restrict the period of application (e.g. late spring applications to winter cereals) when a more expensive herbicide or one with a narrower spectrum may be justified. Avoid, where possible, herbicides with very narrow tolerances of the stages of growth of crop or weed between which application must be made.

Suiting the environment

Wherever possible herbicides least harmful to animal life should be selected. Highly toxic substances may only be justified where the situation demands such use and no satisfactory alternative is available.

Herbicides differ in their reaction to low temperatures. Hence suitability must be checked before applications are made very early or late in the season. The same basic herbicide is frequently formulated as a number of different compounds differing in type of activity. Choose the most suitable for the purpose in hand. Residual herbicides should not be used on unsuitable soils.

Farming system and cropping programme

Use of sophisticated herbicidal treatments requires considerable expertise and equipment, both generally available on intensive arable farms. On many livestock farms, however, neither the equipment nor the experience exist and simple treatments, applied by a contractor, are preferable. Herbicidal treatments should be planned in conjunction with the cropping programme and must take probable residual effects into account.

Cost of treatment

This must be considered in relation to type and degree of weed infestation controlled. The value of crop and saving in labour and cultivations must also be taken into account. Heavy infestations of wild oats, black-grass or couch grass in wet conditions justify expensive herbicidal treatment. Light infestations or conditions where cultivations are cheap and effective seldom do. Herbicides which result in complete mechanisation of row crops or permit use of narrow rows giving higher yields, as with carrots, are well worth while. High value crops like sugar beet, potatoes or carrots permit the use of sophisticated herbicidal treatment. Less valuable fodder crops seldom justify heavy expenditure on herbicides. Where labour is freed by the use of herbicides it must be saved or used for other profitable enterprises. When labour is in short supply use of herbicidal treatment makes the difference between growing a crop or not growing it.

Spray application

Success or failure of herbicidal treatment depends largely on efficient application. A high standard of maintenance is required and sprayers should be overhauled and worn parts replaced before winter storage. Sprayers should be calibrated before use and output checked periodically. Use correct pressure, volume rate and nozzle type for each job. Good marking and correct height of spray bar are essential; fans or cones of spray should meet a few centimetres above the top of the crop being sprayed.

The manufacturer's latest instructions on dosage rate, mixing, application, and use of protective clothing must be followed. Use clean tap water, for dirty water blocks filters. Establish a proper spraying routine using gloves, face shield, and protective clothing when handling dangerous concentrates. Empty drums must be rinsed out into a spray tank, the empty drum being disposed of properly or burned to avoid contaminating water supplies or streams. Wash out the sprayer thoroughly, using a synthetic detergent, after each day's work or when changing types of chemical or crop, on waste land away from water courses. *Even small traces of herbicide can severely damage susceptible crops.*

Field conditions, stage of crop growth and weed must be right, for spraying at the wrong stages gives poor control or severely damages the crop. Best results are obtained when weeds are growing vigorously, not in cold or droughty weather. Spraying immediately before heavy rain is expected should be avoided. *Drift* is a major hazard and does irreparable damage to susceptible crops. Even slight traces of growth regulator herbicides ruin glasshouse, fruit and many market garden crops. Spray only in calm weather, never up-wind with susceptible crops, and avoid unnecessarily high pressure or small droplets.

Always possess adequate insurance cover against third-party claims from damage by spray drift. Cover needs to be substantial for glasshouse crops involve very large sums of money.

Nearly all herbicides are applied as a spray, water being the normal diluent. The volume of water used may be:

Volume	ℓ/ha
v. low	< 90
low	90–200
medium	201–700
high (rarely used)	> 700

ULV (ultra low volume) water is *not* used. Special formulations required (mainly forest application) 5–20 ℓ/ha.

Low or low-medium volume applications of 280 ℓ/ha or under require less labour and a larger acreage can be sprayed in a given time.

ARABLE CROPS AND SEEDLING LEYS

Once established, a heavy leaf canopy suppresses further weed growth but individual crops differ in density and duration of the canopy. Crops like kale produce a heavy, lasting canopy but cereals and potatoes may 'open up' late in the season allowing weed growth to re-start. The problem of weed control is thus twofold:

(1) To establish a dense leaf canopy as quickly as possible and maintain it throughout the life of the crop. All factors which promote rapid growth and increase yield, such as free drainage, good soil structure, adequate soil moisture, ample fertiliser, high quality seed and freedom from pests and diseases, allow crops to form a thick, well maintained leaf canopy. Conversely, drought or poor plant population favour weed growth.

(2) Control measures to suppress weeds until a complete leaf canopy is formed or where the crop canopy is poor. These measures include preventive sanitation, rotations, cultivations and chemical weed control.

Rotations

A well balanced rotation prevents build up of certain weeds by changing the environment from one year to another. If one crop is grown continuously, management and timing of cultivations remain constant and particular types of weed increase, e.g. blackgrass in winter cereals, wild oats in all cereals. Rotations including three year leys and roots give much less trouble. The alternation of cereals with broadleaved crops also allows herbicides to be rotated and used much more efficiently. Build up of couch and other perennial grasses can be prevented by routine stubble cultivations.

The rotation may also be adapted to deal with other weed situations. Wild oats die out if left undisturbed under a long ley (seven to eight years), while two to three year perennial ryegrass leys, if *grazed hard* and not mown, usually eradicate couch.

Cultivations

Annual weeds

(1) *Preparation of 'false seedbed' prior to sowing crop.* The seedbed is prepared early, but not too fine, to encourage weed seeds to germinate. They are then killed by harrowing, another crop of weeds is allowed to grow and then killed; the process may be repeated several times. Method results in a very clean seedbed and is suitable where time allows, e.g. summer seedbeds for lucerne, direct sown leys, swedes or kale, for general annual weed control. Only suitable for the control of weeds whose seed will germinate at the time of year cultivations are carried out.

(2) Thus, the similar technique, *stubble cleaning* or early autumn cultivation of cereal stubbles for annual weed control is highly effective against autumn germinating weeds, e.g. blackgrass, (*Alopecurus myosuroides*) or sterile brome (*Bromus sterilis*), but is useless against weeds which germinate mainly in spring, e.g. spring wild oat (*Avena fatua*). The latter may be killed by delaying the sowing of spring crops to allow harrowing in March and April.

(3) *Delayed sowing of crop*, combined with repeated harrowing, is an effective method of controlling some weeds, e.g. delaying sowing of winter cereal until 5 November gives good control of blackgrass and seedling grasses but usually reduces yield; also loss of yield from delaying spring cereal sowings.

(4) *Inter-row cultivations for crops grown in wide rows*, e.g. sugar beet. Hand hoeing obsolete. Weeds within rows most effectively controlled by herbicides but overall application expensive. Use of band sprays (directed band of spray 90 mm each side of row), with inter-row cultivations saves two-thirds of chemical cost at each application.

(5) *Green smother crops*, e.g. cereal silage may be grown to control wild oats, which ensile satisfactorily and prevents seeding. May be followed with a catch crop or undersown if not planted too thickly.

Perennial weeds

Cultural methods for controlling perennial weeds are nowadays restricted to grassy species, couch (*Agropyron repens* and *Agrostis gigantea*) and others, in cereal stubbles in hot dry weather. Broadleaved perennials are most economically controlled in a run of sprayed cereal crops.

Fallows may be used but where possible stubble cleaning is cheaper. Costs are minimal where routine stubble cleaning is employed as a *preventive* measure. Regular attention should be given to the *headlands* where infestations frequently start.

There are two approaches to controlling infestations:

Desiccation

This process relies on hot dry weather, when it is cheapest and very effective. The ground is loosened by chisel plough or heavy cultivator, the couch rhizomes worked to the surface and desiccated in the sun by frequent shallow cultivations. The rhizomes should not be ploughed in during autumn and the seedbed for the following cereal should be prepared by shallow cultivations.

Exhaustion

Normal growing weather is ideal. Soil and rhizomes are broken up, usually with a rotary cultivator, to encourage fresh growth. This is then destroyed by further cultivations when shoots are 50 mm long or have an average of one and a half to two leaves each. The process is repeated several times to exhaust the rhizomes and buds. A *bare fallow*, which covers a full year, is rarely used. The *bastard fallow*, the land being broken up with heavy cultivators or ploughed in late summer and then baked in the clod, is excellent for dirty, worn-out leys or old pastures prior to a winter cereal.

General chemical weed control

Pre-sowing sward desiccation

Used prior to direct drilling, minimal cultivations or to kill grass swards before ploughing to give grass free seedbed. Only herbicides with minimal residual effect suitable. Use (1) *paraquat* (foliar applied contact) for general purpose where couch grass (*Agropyron repens*) and rhizomatous grasses and perennial broadleaved weeds are absent, or (2) *glyphosate* (foliar applied translocated) for couch and other rhizomatous grasses and perennial broadleaved weeds.

Pre-sowing control of grassy perennials

Treatments are shown in *Table 8.1*. Herbicides are more reliable than cultural methods for heavy infestations in wet weather. Routine annual treatments around headlands should be considered.

WEED CONTROL IN CEREALS

Cereals not undersown

Winter cereals

These occupy the land for some ten months of the year and when they follow another winter cereal little time is available for cleaning. Crops are now sown earlier to increase yield (September and early October) and are then more exposed to infestations of autumn germinating *annual and other weed grasses*, including blackgrass (*Alopecurus myosuroides*), sterile brome (*Bromus sterilis*), annual and rough stalked meadow grasses (*Poa* spp.) and winter wild oat (*Avena ludoviciana*).

Cultural and rotational controls include: prevention of seeding from banks and hedgerows and control on headlands where infestations of sterile brome (*B. sterilis*) often start, stubble cultivations and delaying sowing until 5 November, which usually reduces yield somewhat, or changing to a spring sown crop (incomplete control of *A. ludoviciana*).

Herbicial treatments

Those for autumn application are shown in *Table 8.2*. If needed, failure to treat at this stage results in a heavy infestation by spring. Some pre-emergence herbicides used for grass seedlings also control several species of broadleaved annual weeds, but may not control wild oats and do not persist long enough to control spring germinating broadleaves; post-emergence applications for the latter two situations are frequently necessary. A full programme for an early sown winter cereal crop may thus include as many as four applications:

(1) stubble application for perennial grasses (*Table 8.1*);
(2) autumn applied residual for annual grasses and broadleaved weeds (*Table 8.2*);
(3) spring post-emergence application for wild oats (*Table 8.2*);
(4) spring post-emergence application for broadleaved weeds (*Table 8.3*).

One of the problems of spring applications of broadleaved herbicides to winter cereals is to complete the job before the cereals have passed out of a safe growth stage. Select a herbicide with a *wide application tolerance*—several broad spectrum herbicides are now available for winter cereals which can be applied up to the two node stage (Zadok scale 32, *see Table 5.15*, p. 128). Autumn germinating grassy weeds are unlikely in late sown winter cereals, but these crops are subject to spring germinating wild oats and broadleaved weeds.

Spring cereals

Spring cereals (sown some six months after winter crops) allow ample time for stubble cleaning; residual effects from herbicides applied in the previous season are much less likely. The weed situation is much simpler and, in the absence of wild oats, only a single application for broadleaves is necessary.

Herbicides for broadleaved weeds

Perennials require treatment with *translocated* herbicides—contact types are useless. The 'simple' herbicides, e.g. mecoprop, MCPA and 2,4-D give satisfactory control in cereals provided adequate leaf area is allowed to develop before spraying. Do not spray too early—wait as long as possible. Control is enhanced by a series of sprayed cereal crops. More complex mixtures containing translocated herbicides are only necessary where the presence of other difficult annual weeds requires them.

A wide range of annuals is controlled by simple herbicides but some species are resistant and more complex herbicides are necessary (*Table 8.3*). Choose according to crop and weeds present. First choice herbicides are marked '* or mixture given'. Check stages of crop and weed growth, etc. suitability from manufacturers' literature.

The more expensive 'broad spectrum' herbicides are not usually necessary every year—cheaper simple herbicides or those with a more limited weed spectrum may be considered for intervening years, but in no circumstances is it advisable to omit the annual cereal spray for *broadleaved* weeds—even if only MCPA is used—otherwise, docks and thistles soon develop.

Undersown cereals and seedling reseeds containing clover

Use only herbicides to which clover is tolerant. General purpose herbicides such as MCPB or MCPB + MCPA control many weeds but a number are resistant (*see Table 8.4*). Selection should be made on this basis. Sprays which control chickweed should always be chosen where the weed is known to be a problem, as it can rapidly suffocate young leys. *All* undersown and seedling leys should be sprayed as dock seedlings, which are usually present and inconspicuous, are highly susceptible in the seedling stage to all translocated herbicides—adult docks are much more difficult to control.

ROOT CROPS, ROW CROPS AND OTHER BROADLEAVED CROPS

The principles of weed control are similar in all these crops—the main differences lie in the herbicides used.

Weed problems

Unlike in cereals, perennial weeds are generally difficult or impossible to control selectively during the growing period of these crops. Grassy perennials (couch *Agropyron repens* and others) are best controlled in cereal stubbles or pre-sowing by the herbicides listed in *Table 8.1*. Exceptions include couch in *maize*—use atrazine (high rate) but crop must be grown for two successive years. *Potato*—EPTC, worked in pre-sowing.

Broadleaved perennials

These should be controlled by translocated herbicides applied to preceding cereal crops wherever possible or, if feasible, by pre-sowing application of glyphosate. Exceptions include creeping thistles (*Cirsium arvense*) in *peas*—apply MCPB post-emergence (NOT beans or MCPB+MCPA) and *sugar beet*—3,6-dichloro-picolinic acid.

Grassy annuals

Grassy annuals (i.e. blackgrass *Alopecurus myosuroides* and wild oats *Avena* spp.) may be controlled by herbicides listed in *Table 8.5*.

Broadleaved annuals

For broadleaved annuals consult *Table 8.6*. Selection should be made according to the weeds present or expected to be present and with residual herbicides on soil type; on peaty soils the range is very limited.

Herbicidal programmes

These are necessary in broadleaved crops which germinate slowly and take a long time to form a complete leaf canopy. Such crops are extremely vulnerable to weed competition when there are successive germinations of different weed species. Sugar beet is an excellent example, where programmes, which are constantly changing, have become sophisticated and costly.

Individual applications may include two or more herbicides. Band application allows the more expensive treatments which may last longer to be used, but if soil incorporation is required, overall application is necessary. Incorporation usually increases herbicidal activity although in a wet spring it may reduce both persistence and activity.

The programme must be planned well in advance, taking account of previous experience, the likely proportions of early and late germinating weed species and manufacturers' limitations on soil type. Water solubility of herbicides should be matched to average rainfall for the area, e.g. lenacil, which has a low solubility, is unreliable in dry situations or years but is excellent where there is adequate moisture. Conversely PCF mixtures are preferable in dry situations but lack persistency and may check the crop in wet

conditions. Chloridazon is intermediate between the two. Ensure miscibility of products if tank mixing and also that each product to be used in the programme has the approval of manufacturers of subsequent products. Alternatively, follow a single manufacturer's programme, if suitable.

Where wild oats are expected use a pre-drilling soil-incorporated treatment; volatile substances must be incorporated within 20 min of application.

All programmes for sugar beet should include a pre-emergence application to control weeds which emerge early in the life of the crop. Even where resistant species grow successfully, subsequent applications give better control than where a pre-emergence herbicide is omitted. Sole reliance on post-emergence control gives more inferior results.

Post-emergence herbicides should be applied, weather and crop growth stage permitting, when the most difficult weeds to control are at their most susceptible stage, which may last only for a few days. Timing of the application is of great importance. Choose a herbicide whose weed spectrum compensates for deficiencies in previous treatments. Trifluralin, which is cheap, controls some later germinating species and weed beet but requires inter-row incorporation.

WEEDS FROM SHED CROP SEED OR GROUNDKEEPERS

Problems arise from weed beet, shed oilseed rape, potato and beet groundkeepers and, when oilseed rape or stubble brassicas are to follow, shed cereal grains. Wherever possible prevention is better than cure; methods include preventive sanitation, adequate length of rotation, cultural controls as for annual weeds and herbicides (*see Table 8.7*).

Weed beet

Weed beet is an annual form of beet which produces no harvestable root and sheds large quantities of seed in the year of germination. Although indistinguishable from sown beet in the seedling stage, weed beet may be detected as they occur in patches. Any beet out of place between or within sown rows must be treated as weed beet; they grow from seed shed by bolters in beet crops, which may come from varieties with low bolting resistance and weed beet or groundkeepers elsewhere. Bolting must be prevented as sugar beet varieties are genetically unstable when allowed to cross pollinate freely in an uncontrolled environment and soon degenerate into the annual weed beet form.

Control

Good isolation and tight control of cross pollination by seed growers is essential. Root growers should walk their fields before row cropping, looking for patches of seedlings outside rows or in cereal crops and identify the problem. *Prevention of seeding is better than cure.*

Measures

Pull *all* bolters before flowering (i.e. yellow pollen visible); if pulled after flowering remove from field and destroy to prevent shedding of viable seed. Destroy all weed beet at cleaner loader sites or on wasteland to prevent cross pollination with bolters in root crops. If hand labour is not available control bolters by mechanical cutting at least twice in July. Obtain advice from the British Sugar Corporation field staff as to correct stage for cutting. Drill only after 20 March to avoid bolting, use only *non-bolting varieties* (NIAB Farmers leaflet No. 5).

Additional measures (*once weed beet are identified*): lengthen rotation with non-beet crops and use effective herbicides in other crops (e.g. bromoxynil mixtures in cereals). Spread beet over whole farm. Tractor hoe frequently, close to rows to remove about 90% of weed beet seedings and use trifluralin for late inter-row weed control. Alternatively use weed wiper with glyphosate when beet is above crop. If seed of weed beet has been shed, leave it on the surface; avoid ploughing in, which preserves it until brought to the surface by later ploughings. Use direct drilling or minimal cultivations for cereals on suitable soils. Only relatively few fields are as yet seriously affected but weed beet poses a serious menace to the whole British home grown sugar industry.

HERBICIDE APPLICATION

General key to Tables 8.1 to 8.7

Key to type of application

Co Herbicide applied to foliage with contact type of action.

Fo Herbicide where type of foliar action is not distinct, is not known, is of secondary importance or is not particularly strong.

So Herbicide applied to soil, having a residual action.

Tr Herbicide applied to foliage with translocated action.

Joint activity is stated, e.g. CoSo = Herbicide with contact and soil activity.

1 Pre-sowing (of crop) application.
2 Pre-emergence (of crop) application.
3 Post-emergence (of crop) application.

Thus Co1 = Contact pre-sowing application, etc.

A High mammalian toxicity—also dangerous to livestock. Suggest less toxic chemicals used wherever possible.

B Varietal limitations in one or more crops.

C Limitations in choice of subsequent crop.

D Minimum period between application and harvesting of edible crops ( ) weeks.

E Other crops may not be sown within ( ) months of application—or (weeks) if stated.

F Autumn or winter crops only.

G Stage of growth of weeds or crop and/or condition of crop at time of spraying critical.

H Allow minimum of (mm) length of foliage to develop before spraying.

I Specified time interval must elapse between treatment and sowing.

J Take extreme care to avoid drift.

K Limitations on mixing with other chemicals; do *not* mix glyphosate with any other chemical.

L Do not cultivate before treatment.

M Only available or applied in mixtures.

N Requires incorporation in soil within ( ) minutes—if no brackets consult manufacturers' label; drill crop within ( ) days of application.

O Limitations on application according to growth stage, weather conditions or district.

P Limitations on soil type to which residual herbicides may be applied; not usually organic/sandy soils.

Q Apply within ( ) months of manufacture; *see* manufacturers' label.

R Do not use treated straw in glasshouse culture.

S Do not use in undersown cereals.

T Suitable for use in undersown cereals.

U Temporary leaf scorch, leaf discoloration or chlorosis may occur after use.

V Limitation on permissible sowing dates.

W Minimum depth of settled soil after drilling (mm) is required.

X Heavy rain or frost after application may cause damage.

Y Temperature limitation at time of spraying. Maximum temperature (°C).

Z Limitation on application of previous or subsequent herbicides.

a Plough at least 150 mm deep for subsequent crops.

b Do not use on potatoes during or after prolonged dry periods or damage to tubers may occur. Consult manufacturer after a dry period.

c Only some products.

d Do not apply in frosty or very cold weather.

Note: These tables do not include *all* herbicides included in *Approved Products for Farmers and Growers.* Inclusion or omission of any chemical does not signify a recommendation or otherwise on the part of the author or the publishers.

Table 8.1 Pre-sowing control of perennial grassy weeds—stubble cleaning and elsewhere

Herbicide	Type of application	Species controlled	Limitations of herbicide
Aminotriazole	Tr1	Couch (*Agropyron repens* and *Agrostis gigantea*) and many grasses. Broadleaved perennials	CH(125 mm)I
Dalapon	Tr1	Couch, perennial and annual grasses, volunteer cereals	CH(125 mm)I
Glyphosate	Tr1	Couch, most grasses and broadleaved perennials	H(125 mm)IJKL
Paraquat	Co1	Annual and stoloniferous perennial grasses: *A. stolonifera* and *P. trivialis*. Couch *only* if repeated application technique used—NOT single application	H(75 mm)I
TCA	So1	Couch and other grasses	CIN

Key: to table, *see* p. 233.

Table 8.2 Selective weed control in cereals annual/seedling grasses

Herbicide	Type of application	Grasses controlled				Suitable cereals	Notes
		Blackgrass (*A. myosuroides*)	Meadow grasses (*Poa spp.*)	Wild oats (*Avena spp.*)	Some annual broadleaved weeds		
Barban	Tr3	Yes		Yes		Wheat. Barley	B
Benzoylpropethyl	Tr3			Yes		Wheat	KT
Chlorfenpropmethyl	Tr3			Yes		Barley. Spring oats and wheat	BT
Chlortoluron	TrSo2 or 3	Yes	Yes	Yes	Yes	Winter barley. Winter wheat	BO
Diclofopmethyl	Fo3			Yes		Barley. Wheat	O
Difenzoquat	Tr3			Yes		Winter and spring barley. Wheat, rye	BKT
Flampropmethyl	Tr3			Yes		Wheat	KT
Isoproturon	TrSo2 or 3	Yes	Yes		Yes	Winter barley. Winter wheat	O
Methabenzthiazuron	TrSo 2 or 3	Yes	Yes		Yes	Winter barley, oats, wheat, rye. Spring barley, annual meadow grass and grass seed crops	O
Metoxuron	TrSo3	Yes	Yes		Yes	Winter barley. Wheat	BO
Metoxuron + Simazine	TrSo3	Yes	Yes		Yes	Winter barley and wheat	BOE
Nitrofen	So2	Yes			Yes	Winter wheat	P
Pendimethalin	So2	Yes	Yes		Yes	Winter barley	OP Annual meadow grass
Terbutryne	So2	Yes	Yes		Yes	Winter wheat and winter barley	O
Tri-allate (Granule/liquid)	So1 or 2	Yes		Yes		Barley or wheat, rye	ON

Key: to table, *see* p. 233.

Table 8.3 Selective control of annual broadleaved weeds in cereals *not* undersown

Herbicide or predominant herbicide in mixtures and type of action. (*Mˣ = mixtures* only suit-able or available)	Suitable mixtures	General weed control	Combinations difficult to control with simple herbicides								Suitable cereals	Limitations of herbicide
			Black bindweed (*P. convolvulus*) and Redshank (*P. persicaria*)	Common chickweed (*S. media*)	Cleavers (*G. aparine*)	Corn marigold (*C. segetum*)	Knotgrass (*P. aviculare*)	Mayweed spp. (*Anthemis etc.*)	Speedwell spp. (*Veronica spp.*)	Spurrey (*S. arvensis*)		
Bentazone MˣCo3	a	Yes		a	a	a		a		a	All cereals	
Bromofenoxim MˣCo3	b	Yes	b	b		b					Spring cereals	Annual weeds only
Bromoxynil MˣCo3	cdefg	Yes	*			c	*	*	*	fg	All cereals	(c) Annual weeds only
Cyanazine MˣFoSo3	hi	Yes	*	*	k		*	*			All cereals	PQ
Dicamba MˣTr3	jklmn	Yes	*	mn			*	n		*	All cereals	
2,4-D		Yes									Not spring oats	
Dichloropicolinic acid MˣTr3	op	Yes	*	*	*	*	*	*		*	All cereals	
Dichlorprop Tr3	qrs	Yes	*	*	r						Not rye, not p on spring oats	
Dinoseb (amine and ammonium salts) CoSo3				(*)	(*)	(*)		(*)			All cereals	A
DNOC Co3				(*)	(*)	(*)					All cereals	A
Ioxynil MˣCo3	tu	Yes	st	st	*			*	*		s not rye	
Linuron CoSo2						(*)					Spring cereals	Annual weeds only
MCPA Tr3		Yes									All cereals, not rye	
Mecoprop Tr3	vw	Yes		*	*							
2,3,6-TBA Mˣ Tr3	xy	Yes	wx	wx	wx		wx	*		wx	All cereals	GR
Trifluralin MˣSO₂	z	Yes	z	z			z	z	z	z	Winter cereals. Annual weeds only. Some grasses. Inspect crop in spring	
In above mixtures only: Benazolin Fenoprop Terbuthylazine												

Key:
* Denotes a first choice chemical for stated weed. Additional weeds controlled to first choice standard by a mixture shown by appropriate letters.
( ) Brackets denote not usually first choice for this particular weed.
Italics denote herbicide also or solely used by itself.

NB: There is a great difference between herbicides in crop tolerance of application at different growth stages; selection is often made on this point. Consult manufacturers' literature.

Herbicide mixtures

a	Bentazone + dichlorprop
b	Bromoferoxim + terbuthylazine
c	Bromoxynil + ioxynil
d	Bromoxynil + MCPA
e	Bromoxynil + ioxynil + dichlorprop
f	Bromoxynil + ioxynil + mecoprop
g	Bromoxynil + ioxynil + dichlorprop + MCPA
h	Cyanazine + MCPA
i	Cyanazine + mecoprop
j	Dicamba + MCPA
k	Dicamba + mecoprop
l	Dicamba + dichlorprop + MCPA
m	Dicamba + mecoprop + MCPA
n	Dicamba + benazolin + dichlorprop
o	Dichloropicolinic acid + mecoprop
p	Dichloropicolinic acid + dichlorprop + MCPA
q	Dichlorprop + 2,4-D
r	Dichlorprop + MCPA
s	Dichlorprop + mecoprop
t	Ioxynil + mecoprop
u	Ioxynil + dichlorprop + MCPA
v	Mecoprop + 2,4-D
w	Mecoprop + fenoprop
x	2,3,6-TBA + dicamba + MCPA + mecoprop
y	2,3,6-TBA + dichlorprop + mecoprop
z	Trifluralin + linuron

Other uses of chemicals on cereals:
Straw shortening, e.g. chlormequat—wheat and oats, including undersown.
Desiccation of laid crops and attendant weed growth, e.g. diquat.

See also general key, p. 233.

Table 8.4 Selective control of annual broadleaved weeds in undersown cereals; seedling leys (grass and clover, lucerne and sainfoin)
(a) Established legumes and seed crops

Herbicide or predominant herbicide in mixtures and type of action. (M^x = mixtures only suitable or available)	Suitable mixtures	General weed control	Combinations difficult to control with simple herbicides								Suitable crops	Limitations of herbicide
			Black bindweed (P. convolvulus) and Redshank (P. persicaria)	Common chickweed (S. media)	Cleavers (G. aparine)	Corn marigold (C. segetum)	Knotgrass (P. aviculare)	Mayweed spp. (Anthemis etc.)	Speedwell spp. (Veronica spp.)	Spurrey (S. arvensis)		
Benazolin M^xTr3	a	Yes	*	*	*		*				Undersown cereals, direct sown leys. *Not* lucerne or sainfoin	U
Bentazone M^xCo3	b	Yes		*	*	*		*			Undersown cereals; *not* undersown lucerne	
Bromoxynil M^xCo³	c	Yes	*			*	*	*			Undersown cereals	Annual weeds only
2,4-DB Tr3		Yes	*				*				Lucerne—undersown cereals, direct sown leys	
	d	Yes	*				*				Seedling red and white clover, undersown cereals, direct sown leys not lucerne	Not established clovers
	e	Yes	*				*				Undersown cereals, seedling leys not lucerne	
Dinoseb, amine and ammonium salts CoSo3				(*)	(*)	(*)		(*)			Clover, lucerne, sainfoin undersown cereals	A
MCPB Tr3		Yes									Sainfoin, seed clovers undersown leys, *not* lucerne	⎫ Not alsike for seed or yellow trefoil
Herbicides used only in above mixtures: 2,4-D Ioxynil MCPA	f	Yes									Sainfoin, leys and undersown cereals, *not* lucerne	⎭

Table 8.4 (b) Selective weed control in established legumes and herbage seed crops

Herbicide and type of action	Purpose	Suitable crops	Limitations of herbicide
Legumes			
Carbetamide FoSo3	Winter control of annual grasses, volunteer cereals, chickweed and speedwells	Clover, lucerne and sainfoin	P
Dalapon *Glyphosate* *Propyzamide* *TCA* }	Couch grass; also other grassy weeds	Lucerne, pure stands and winter application only	K P-stands at least 1 year old
Grass for seed/pure grass stands			
Difenzoquat	Wild oats *see Table 8.2*	Ryegrasses	U
Ethofumesate TrSo2/3	Meadow grasses, wild oats	Ryegrasses. Tall fescue. Wide range of pure grasses	P
Methabenzthiazuron	Meadow grasses *see Table 8.2*	Perennial ryegrass	
Also 2,4-D, dicamba mixtures *MCPA, mecoprop*, TBA mixtures	Broadleaved annual and perennial weeds—*see Table 8.3*	All pure grass swards	

Key:
* Denotes a first choice chemical for stated weed. Additional weeds controlled to first choice standard by a mixture shown by appropriate letters.
( ) Brackets denote not usually first choice for this particular weed.
Italics denote herbicide also or solely used by itself.

Herbicide mixtures
a   Benazoline + 2,4-DB + MCPA
b   Bentazone + MCPB
c   Bromoxynil + ioxynil
d   2,4-DB + MCPA
e   2,4-DB + 2,4-D + MCPA
f   MCPB + MCPA

Desiccation of clover seed crops:
Dinoseb in oil; diquat.

See also general key p. 233.

Table 8.5 Selective control of wild oats and blackgrass in broadleaved field and field vegetable crops

Herbicide or predominant herbicide in mixtures Type of action given individually (M = mixtures only suitable or available)	Field crops										
		Beet			Brassicas						
	Beans, field	Sugar beet	Mangolds and fodder beet	Seed crops/ stecklings	Kale	Mustard	Rape, forage	Rape, oilseed	Swede/turnip	Maize	Potatoes
Blackgrass and wild oats											
Barban	Tr3	Tr3									
+Phenmedipham[1]		TrCo3									
Diallate		Sol			Sol		Sol	Sol	Sol		
+Chloridazon[1]		Sol									
Triallate (granule or liquid)	Sol/2	Sol/2	Sol/2		Sol/2	Sol/2	Sol/2	Sol/2	Sol/2	Sol/2	
Blackgrass only											
Dinitramine[1]					Sol				Sol		
Metoxuron[1]											
Wild oats only											
Benzoylprop ethyl	Tr3					Tr3		Tr3			
Cycloate M[1]		Sol	Sol								
Diclofopmethyl	Fo3	Fo3			Fo3	Fo3	Fo3	Fo3	Fo3		Fo3
Difenzoquat										Fo3	
Propham[1]		Sol									
TCA		Sol			Sol		Sol	Sol			

Key: to table, *see* p. 233.

[1] Also some broadland annual weeds.

Field vegetables												Limitations of herbicide
Bean			Brassicas brussels sprouts, cabbage, cauliflower or broccoli									
Broad	French	Beetroot	All brassica crops	Brussels sprout/cabbage only	Spring cabbage only	Carrot	Leek	Onion	Parsnip	Pea	Sweet corn	
Tr3										Tr3		D(6)G
												DG
		Sol	Sol	Sol	Sol							N
												NP
Sol/2		Sol/2	Sol/2	Sol/2	Sol/2	Sol/2	Sol/2	Sol/2	Sol/2	Sol/2		N liquid
												I
												K N (24 h)P
												13
						TrSo3						
												Culinary mustard K
		Sol										E(4)N(15)(d21)PV
Fo3			Fo3	Fo3	Fo3	Fo3		Fo3	Fo3	Fo3		U
												K
										Sol		
										Sol		

Table 8.6(a) Selective control of broadleaved annual weeds in broadleaved field and field vegetable crops

Herbicide or predominant herbicide in mixtures. Type of action, given individually (M = mixtures only suitable or available)	Field crops										
	Beans, field	Beet			Brassicas					Maize	Potatoes
		Sugar beet	Mangolds and fodder beet	Seed crops/stecklings	Kale	Mustard	Rape, forage	Rape, oilseed	Swede/turnip		
Atrazine										So2 FoSo3	
Aziprotryne											
Bentazone											
— +MCPB											
Carbetamide	FoSo3			FoSo3			FoSo3				
Chlorbromuron											CoSo2/3
Chloridazon		So2/3									
— +chlorbufam											
Chlorpropham											
— +diuron		So2									
— +fenuron	So2/3	So2/3									
Cyanazine											
— +MCPB											
— +linuron											So2
2,4-DES										So2	
Desmetryne					Co3	Co3					
3,6,-Dichloropicolinic acid		Tr3									
Dinoseb salts	CoSo3										
— +monolinuron											CoSo2
Ethofumesate M											
— +lenacil		TrSo2	TrSo2								
— +chloridazon		TrSo2	TrSo2								
— +PCF mixtures		TrSo2	TrSo2								
— +phenmedipham		TrCo3	TrCo3								
Ioxynil											
— +linuron											
Lenacil		So1	So1								
— +cycloate		So1	So1								
— +phenmedipham		CoSo3									
Linuron											CoSo2
— +metoxuron											
— +monolinuron											CoSo2
MCPB											
Metamitron		TrSo 1/2/3									
Methazole											
Metribuzin											TrSo2
Monolinuron											CoSo2
— +paraquat											CoSo2
Nitrofen					TrSo2		TrSo2	TrSo2	TrSo2		
Pentanochlor											
— +chlorpropham											
Phenmedipham		Co3	Co3								
Prometryne											TrSo2
Propachlor											
Propham		So1									
— +chlorpropham+ fenuron (PCF mixt.)		So2	So2								
Simazine	So2/3									So2/3	

Vegetables

Beans		Beetroot	Brassicas (brussels sprouts, cabbage, cauliflower or broccoli)									Limitations of herbicide
Broad	French		All brassica crops	Brussels sprout/cabbage only	Spring cabbage only	Carrot	Leek	Onion	Parsnip	Pea	Sweet corn	
											So2	E(7)
											FoSo3	couch E(18)
		Co3	So3				So3	So3		So2		CPW (2.5 cm peas) X
										CoTr3		BUY (max 21 °C)
					FoSo3							E (according to crop) P
						CoSo2/3			CoSo2/3			P
												PY (max 21 °C) Z also approved mixtures
							So2/3	So2/3				P
So2						SoFo3	SoFo3	SoFo3		So2		P
So2/3	So2/3	So2/3				So2/3	So2/3	So2/3	So2/3	So2/3		P
										FoSo2		BGPW (2.5 cm) peas
										FoSo3		P
											So2	O
				Co3								BD (4 weeks)
												BGZ
CoSo3								CoSo3		CoSo3		AB
	CoSo2											AGP
		TrSo2										Pa
		TrSo2										Pa
		TrSo2										Pa
												Pa
							Co3	Co3				GPYa
							CoS3	CoS3				U
		So1										PZN (organic soil)
		So1										PZ
						CoSo2/3			CoSo2/3			GPY (21 °C) Z
						TrSo3						
										Tr3		B
												E(4)P
							TrSo3	TrSo3				BE(4 +)GP
							CoSo3					B (potatoes) EPU
	CoSo2											B (french bean)—G
				TrSo2								E (4 weeks) P
						Co2/3			Co2/3			
						CoSo2/3			CoSo2/3			GY (21 °C) Z
		Co3										
						TrSo3	TrSo3			TrSo2		D(6)E (6 weeks) P
							So2	So2				U
			So2							So1		
		So2										
So2/3											So2/3	E(7)W beans (7.5 cm)

(cont.)

Table 8.6(a)—*(cont.)*

Herbicide or predominant herbicide in mixtures Type of action, given individually (M = mixtures only suitable or available)	Field crops										
		Beet			Brassicas						
	Beans, field	*Sugar beet*	*Mangolds and fodder beet*	*Seed crops/ stecklings*	*Kale*	*Mustard*	*Rape, forage*	*Rape, oilseed*	*Swede/turnip*	*Maize*	*Potatoes*
Sodium monochlor- acetate (SMA)					Co3						
Terbutryne M											
+terbuthylazine	So2										So2
Trietazine M											So2
+linuron											
+simazine	So2										
Trifluralin	So1	So3			So1	So1	So1	So1	So1		
General pre-emergence control of annuals; paraquat Co2											

Key: to table, *see* p. 233.

(b) Potato haulm destruction

Potato haulm destruction: dinoseb in oil[b]
diquat[b]
metoxuron[cd]

Key: to table, *see* p. 233.

Vegetables												Limitations of herbicide
Beans			Brassicas brussels sprouts, cabbage, cauliflower or broccoli									
Broad	French	Beetroot	All brassica crops	Brussels sprout/ cabbage only	Spring cabbage only	Carrot	Leek	Onion	Parsnip	Pea	Sweet corn	
				Co3			Co3	Co3		So2		P potatoes W BE (12 weeks) peas (2.5 cm) E (12–14 weeks) P
So2												
So2										So2		BE (12–14 weeks) P W peas (2.5 cm)
Sol	Sol		Sol			Sol			Sol			{ E(5): E(12) after autumn application. NPUZ At least 3 d before emergence on very sandy soils

Table 8.7 Control of volunteers or groundkeepers, in certain crops

Volunteer or groundkeeper	Herbicide or predominant herbicide in mixtures — Type of action given individually (M = mixtures only suitable or available)	General			Beet		Brassicas						Limitations of herbicide
		pre-sowing	pre-sowing or pre-emergence	Bean, winter field	Sugar beet	Stecklings	Cabbage, spring	Rape, oilseed	Seed crops (some)	Stubble crop (forage)	Carrots	Cereals	
SELECTIVE CONTROLS													
Beet, weed	Bromoxynil + ioxynil + mecoprop M											Fo3	
Beet, weed	Glyphosate				Tr3								Glyphosate using weed wiper only
Cereals volunteer	*Carbetamide*		FoSo3				FoSo3						E (according to crop) P
	Dalapon				Tr3		Tr3			Tr3	Tr3		
	Propyzamide		So3		So3		So3	So3	So3				E (10–40 according to crop) FP
	TCA						So2						E (according to crop) FP
Oilseed rape	Bromoxynil + ioxynil + mecoprop M											Fo3	
Potato groundkeepers	Linuron + metoxuron										TrSo3		
UNSELECTIVE CONTROLS													
Beet, cereals and oilseed rape	*Paraquat*	Co 1/2											*See Table 8.1*
Potatoes	*Glyphosate*	Tr1											*See Table 8.1* K

Key: to table, *see* p. 233.

Key to Table 8.8

A Improve drainage.

B Apply lime.

C Increase fertility. Apply N, P and K as required; a base dressing of 190 kg/ha insoluble P_2O_5 given as basic slag or ground mineral phosphate on phosphate-deficient pastures in high rainfall areas or wet situations.

D Avoid poaching; keep cattle off grassland in winter.

E Avoid overgrazing in winter, spring or early summer.

F Avoid undergrazing in summer.

G Increase stocking rate, mow or top over after grazing.

H Change management to include close cutting, using forage harvester to cut silage.

I Cut twice annually for two or three years.

J Heavy grazing with sheep reduces infestation.

K Heavy trampling with cattle but unstock cattle on bracken to avoid bracken poisoning.

L On ploughable land plough bracken deeply with digger plough when fronds are three-quarters open in spring or early summer. Cut up rhizomes with several passes of disc harrows and consolidate. Grow pioneer crop, e.g. rape or turnips, graze in autumn, disc harrow in December; do not plough. Follow with re-seeded ley or potatoes; re-seed must be intensively stocked and managed or bracken returns.

M Prevent flowering and seed production by spudding, cutting, flail or swipe mower.

N Several perennial weeds are difficult to control with inexpensive herbicides in established grassland but are easily controlled by such herbicides in the presence of cultivations. Where practicable, ploughing and cultivations, followed by either (a) re-seeding with Italian ryegrass and spraying with MCPA, 2, 4-D or mecoprop as soon as possible, followed by two further sprayings during the summer, or (b) growing a run of two or more sprayed cereal crops.

O Undersown leys and reseeds must be sprayed with an MCPB or 2, 4-DB mixture (*see Table 8.4*) to kill perennial broad leaved weeds in the susceptible seedling stage. Once established these weeds become much more resistant to herbicidal treatment.

P Spray in spring or early summer before flowering.

Q Spray in autumn.

R Spray in at least two successive years.

S Spot treatment—wet clumps to run-off point.

T Spray when plants have made maximum growth.

Note. When pasture contains a very large amount of weed and little or no valuable grass or clover, reseeding should be considered, especially if intensive stocking is intended.

Table 8.8 Control of weeds in established grassland

Weed	Situations where especially troublesome	Preventive cultural and management controls	Most effective herbicidal treatments and herbicides
Bent, creeping (*Agrostis stolonifera*)	Poorly managed lowland pastures, overgrazed in winter and spring, undergrazed in summer	G, E, F, G	Asulam used for docks gives some useful effect
Bracken (*Pteridium aquilinum*)	Heavily understocked sites, upland, moor and hill lands, poisonous	B, C, I, K, L	Asulam
Buttercup, bulbous (*Ranunculus bulbosus*)	Common in most permanent grassland. Poisonous except in hay. *R. bulbosus* occurs in drier conditions	If very serious N otherwise CE	QR (2, 4-D[1]) (MCPA[1])
Buttercup, creeping (*R. repens*)		A, C, E	P2, 4-D[1] MCPA[1] MCPB and mixtures P MCPA[1]
Buttercup, tall (*R. acris*)		A, C, E	MCPB and mixtures
Chickweed (*Stellaria media*)	Poached grassland along with other annual weeds	D	*Benazolin mixtures* mecoprop[2]
Daisy (*Bellis perennis*)	Overgrazed poor permanent grass and lawns	C, E R if necessary	2, 4-D[1] MCPA[1]
Dandelion (*Taraxcum officinale*)	Common in poorly managed grassland	C, E, F	2, 4-D[1] MCPA[1]
Docks, broad and curly leaved (*Rumex obtusifolius* and *R. crispus*)	Universal on land stocked only with cattle; also treated with slurry and mown	J, M, N	O *Asulam* or *Dicamba mixtures[1]*
Gorse (*Ulex* spp.)	Grossly undergrazed acid permanent grass and waste places	B, C, D, G, H, M	2, 4, 5-T
Horsetails (*Equisetum* spp.)	Soils with wet subsoil; remains poisonous in hay and silage	A, C, E, F, G	RT (2, 4-D)[1] (MCPA[1]) (MCPB mixtures)
Meadow grass, annual (*Poa annua*)	Very common in reseeds and poor, open or poached grassland. May suddenly appear/disappear	D, E	Ethofumesate[1] *if* cost is justified
Mouse-eared chickweed (*Cerastium holosteoides*)	Universal in grassland	C, D	(2, 4-D[1]) (MCPA[1]) (MCPB mixtures)
Ragwort (*Senecio jacobea*)	Neglected and overgrazed old pastures and wasteland, hedgerows. Remains poisonous in hay and silage	C, D, E, J, M Undersow new leys	2, 4-D[1] MCPA[1] Once sprayed keep stock out until ragwort has disappeared
Rush, hard. (*Juncus inflexus*)	Badly drained wet land of low fertility; establish from seed in poached or open sward	A, C, D, E, G, I Reseed new leys *without* cover crop	Resistant to selective herbicides
Rush, soft, (*Juncus effusus*)		A, B, C, D, G, E, I Reseed new leys *without* cover crop	P, 2,4-D[1] MCPA[1]. Cut 4 weeks after spraying
Sorrel, common (*Rumex acetosa*)	Poor permanent grass, usually damp	B. C. G	(2,4-D[1]) (MCPA[1])
Sorrel, sheeps (*Rumex acetosella*)	Indicates acidity on poor permanent grass	B, C, G	(2,4-D[1]) (MCPA[1])
Stinging nettle, great (*Urtica dioica*)	Mainly on grazed grassland, often with loose structure; encouraged by surface litter	H, K N if serious	PQRS 2,4, 5-T and mixture with 2,4-D[1]
Thistle, creeping (*Cirsium arvense*)	Universal in undergrazed grassland—rarely seeds	C, E, F, G N if serious	P QR 2,4-D[1] MCPA[1] *MCPB mixtures*
Thistle, spear (*C. vulgare*)	Universal. Establishes readily from blown seed	M	O 2,4-D[1] MCPA[1] MCPB mixtures
Tussock grass (*Deschampsia caespitosa*)	Badly managed wet grazings, establishes from seed	A, C, D, E, H, M	No selective chemical control
Yorkshire fog (*Holcus lanatus*)	Wet, acid or badly managed lowland grassland	A, B, C, E, F, G	Asulam used for docks gives useful effect

[1] Clovers killed or severely checked.
[2] Mecoprop is effective against chickweed but kills clover. Use only on grass seed crops, pure grass swards and in emergency, when chickweed (*S. media*) is likely to smother sward or oversowing.
( ) brackets denote limited effect.
Italics denote first choice herbicide.

WEED CONTROL IN ESTABLISHED GRASSLAND

The presence of infestations of weeds or weed grasses is generally an indication that all is not well with management and growing conditions. Weed infestations can also arise in leys, where the species originally sown lacked persistency and faded out, leaving the field to be colonised by weeds.

The first step is to decide whether the sward justifies retention and subsequent improvement or whether to kill all vegetation and reseed or grow a run of one or more arable crops before reseeding. Retention is often best on moist fertile soils or in wetter situations as a more poaching resistant sward is maintained; costs are lower and no grass production is lost. Conversely, where a temporary sward is worn out, in dry situations or where a poor sward is infested with perennial weeds which are difficult to control by herbicides while the field remains down to grass, destruction of the sward has much to commend it. A herbicide which gives a total kill if properly used (e.g. glyphosate) is a valuable first step (*see Table 8.1*). The majority of grassland weeds will not withstand well managed cultivations and arable crops for long, especially where the arable break starts with cereal crops sprayed with translocated herbicides.

The main weeds of grassland and their appropriate controls are listed and summarised in *Table 8.8*. Where weedy grassland is to be retained, improvements in management, growing conditions and fertility levels are essential. Herbicides are often an essential part of the programme but cannot maintain any improvement by themselves.

WEED CONTROL IN FOREST AND WOODLANDS

Adequate weed control is essential for the satisfactory establishment of young trees and their subsequent growth. The nature of the problem and weed spectrum is much more variable than with normal agricultural crops where much of the land is reasonably level and the weed spectrum is confined largely to weed grasses and broadleaved herbaceous species. In woodland a much wider weed spectrum occurs and terrain is often difficult. The situation varies from sites where little or no weeding is necessary to those where weed growth is vigorous and rapid and may cause total failure of the young trees; the commonest cause of smothering is the collapse of tall vegetation on top of them in the winter. Frequently the control of one type of weed allows another to develop, e.g. grasses often follow cutting down of brambles (*Rubus* spp.) while ploughing temporarily controls some grass species but may result in heavy growth of thistles (*Cirsium* spp.) or gorse (*Ulex* spp.). It is necessary to be prepared for these changes. Total eradication of weeds over the whole site is unlikely to repay the cost; generally the suppression of weed growth in the immediate vicinity of the young tree is adequate, provided that tall woody broadleaves are not allowed to develop. Considerable time and money can be saved by knowing when to intervene and when intervention is unnecessary.

Weed types

For practical purposes weeds may be grouped as follows:

(1) perennial grasses, including 'soft' and rhizomatous types, rushes (*Juncus* spp.) or mixtures with herbaceous broadleaved weeds;
(2) bracken (*Pteridium acquilinum*);
(3) heather (*Erica* and *Calluna* spp.);
(4) woody broadleaved weeds, i.e. fast growing trees, shrubs and harmful climbers, either seedling or coppice.

Weeds in groups (1)–(3) and low growing woody weeds like gorse (*Ulex* spp.) must be controlled until the young trees are about 2 m high (roughly four to five years after planting) thereafter the crop trees should suppress them even if weed growth looks vigorous. Fast growing trees such as birch (*Betula* spp.) and sallow or goat willow (*Salix* spp.) can also cause severe damage much later in the life of the crop.

Methods of control

Control measures must be carefully integrated to suit the weed flora, topography and general conditions of the site. Each operation should be carried out as part of the overall establishment plan and not in isolation. Methods can be grouped into the following categories: (a) hand, (b) mechanical, and (c) herbicidal.

Hand methods

The traditional method is to walk down a line of plants, trimming away vegetation around each plant for a sufficient distance to prevent smothering, using a curved grass or reaping hook. *Do not make any cuts until the plant is located.* A light stick is carried to push weed growth away and if of the same length as the planting distance within the row, location of small trees is much quicker. Work is much easier where rows are quite straight and trees are evenly spaced. Some foresters save labour by treading down vegetation around the trees. For bramble and gorse a slasher and gloves are required. The effect of hand weeding is only temporary and it is expensive, specially if two passes are required in a season. Handwork also creates a summer labour peak, as it is only effective in the growing season. Handweeding and cleaning have been substantially superseded by herbicides in the forest, but may be considered for very mixed weed floras where herbicides cannot be used for environmental reasons and on small areas. The method is also simple and needs little staff training.

Mechanical weeding

Machines can be tractor mounted, pedestrian controlled or hand held. Cutting mechanisms include crushing rollers and reciprocating knives for grass or soft weed or rotating flails, chains or saw blades for tougher material. Tractor mounted machines with a suitable attachment can give a high output at low cost on soft weed or deal with considerable woody growth, but cannot operate on rough or very steep terrain. Pedestrian controlled machines are rather more flexible regarding terrain, while hand held machines are suitable for steep or rough ground and some woody growth but output of both types is low and costs are high. Machines are useful on very mixed weed floras on level land or where vegetation is out of hand, but involve capital expenditure and staff training for work and maintenance. Row widths and spacing may need adjustment for tractor working.

Herbicidal treatments

These have now displaced handwork on large areas of forest as they are normally cheaper, last for a whole season or more and save labour. A wide range of well tried herbicides for most weed situations is now available. Some methods require no diluent (e.g. ULV and granules) and are well suited to steep or rough sites or where water supply is difficult. The main problems lie in persuading some woodland owners to use herbicides and in staff training.

No single herbicidal treatment can be relied on to control the very broad spectrum of weeds that may develop in the early years of a plantation—some weeds are resistant and different herbicides may be needed later. Some treatments have a short period of use in a growing crop, while timing in relation to growth stages of trees and weeds is important; if wrong, weed control may fail or the crop may be damaged. Careful forward planning is therefore essential to integrate the herbicidal programme with all other operations during the establishment period—from the initial preparation of the site until the trees have formed a dense thicket.

Herbicidal application

Equipment

Steep or rough terrain and tree growth largely preclude the use of normal farm equipment for woodland. The following are suitable, according to herbicide, species and type of application.

Knapsack MV sprayers

Directed applications around base of young trees—some herbicides require use of *tree guards* (i.e. to cover tree or spray jet). NB glyphosate, paraquat. Also stump and basal bark treatments.

Mistblowers LV

Knapsack or tractor mounted for overall foliar applications. The latter machine only of limited value in lowland forest.

Ultralow volume (ULV)

Overall foliar application by incremental spraying.

Distributors for granular herbicides

Airflow type distributors are essential for accurate granule distribution. Handwork is too laborious, inaccurate and wasteful of expensive herbicides; overdosing can occur. Suitable for overall or strips 1–2 m wide; flow can be cut off with wide tree spacings for spot application. ULV and granules require no diluent and are ideal for areas where water supply or carriage is difficult.

Tree injectors

These are for injecting undiluted herbicide into standing trees at (a) breast/waist height—cut made with axe, or (b) base of tree with chisel bit for penetration.

Types of treatment

Pre-planting

These treatments are given to control unwanted vegetation prior to planting, either overall or in strips, in the centre of which trees are subsequently planted; selectivity at this stage is not needed but some herbicides require an appropriate interval between application and planting. Avoid drift onto adjacent susceptible crops.

Post-planting

Post-planting treatments if given *overall must be selective* in the species concerned, as determined by age, size, and growth stage of the trees, time of year and herbicidal activity; otherwise *directed* applications must be given, using *tree guards* where necessary (e.g. glyphosate, paraquat, 2,4,5-T applications in susceptible spp.). *Foliar* applications are dependent for success on the active growth of *weeds* at the time of application, e.g. with broadleaved woody weeds apply before leaf senescence. *Selectivity* is often obtained by application when the crop is not making rapid growth or is dormant. Timing is all important for control of weeds and crop safety, as application

248

Table 8.9 Selective weed control in forests, woodland, shelterbelts and other tree situations (except fruit, ornamentals)

Weed problem[1]	Herbicide[2]	Type of application[3]	Timing of application		Specific limitations of herbicide[5]
			Pre-planting	Post-planting[4]	
Perennial grasses—'soft' grasses and rhizomatous rushes (*Juncus* spp.) and broadleaved herbaceous weeds and mixtures	*Altrazine* Gs	FoSo	Yes	Yes	Conifers only AB Feb–May F
	Chlorthiamid Gr, He	So	Yes	Yes	Wide weed spectrum—ADEFG granules
	Dalapon Gr	Fo	Yes	Yes	D Mar–early April E K
	Dichlobenil GrHe	So	Yes	Yes	Similar to chlorthiamid ADEFG granules
	+Dalapon GrHe	So	Yes	Yes	Coarse grasses, rushes ADEFG granules
	Glyphosate GrHe	Fo	Yes	Yes	Wide weed spectrum ACDHJKL
	Hexazinone GrHe	FoSo	No	Yes	Wide weed spectrum. Conifers only ABC
	Paraquat Gs	Fo	Yes	Yes	Shortgrass. D. early summer, autumn or winter, J
	Propyzamide Gr(He)	So	Yes	Yes	*Deschampsia* spp. BD winter only F granules or wettable powder
Bracken (*Pteridium aquilinum*)	Asulam	Fo	Yes	Yes	May check young trees. L
Heather (*Calluna vulgaris* and *Erica* spp.)	2, 4-D (ULV only)	Fo	Yes	Yes	Conifers only A, BD late summer
Woody broadleaved weeds	2,4, 5-T (also ULV)	Fo	Yes	Yes	ABCDN
	2,4, 5-T	V, WorX	Yes	(Yes)	Spray only when dry DE. Best applied pre-planting
	Undiluted 2,4, 5-T or 2,4-D	Z	Yes	Yes	
Mixed herbaceous and woody broadleaved weeds;	Mixtures of 2, 4-D and 2,4, 5-T	Fo	Yes	Yes	ABCDN
Brambles (*Rubus* spp.)	Glyphosate	Fo	Yes	Yes	ACDHJKL
Woody weeds resistant to 2,4, 5-T	*Ammonium sulphamate* (AMS)	V	Yes	No	EI
		X Y	Yes	Yes	Use dry AMS crystals 15 g/notch
Rhododendron (*Rhododendron ponticum*) and Laurel (*Laurus* spp.)	*Ammonium sulphamate* (AMS)	T or V	Yes	No	EI
	2,4, 5-T heavy dosage	T or V	Yes	(Yes)	D-only in crop's dormant season

Key:

1
'Soft' grasses include: *Agrostis stolonifer*, *Deschampsia* spp., *Festuca* spp., *Holcus* spp. and *Poa* spp.
Ammonium sulphamate (AMS) is highly toxic to young trees. Use *only* for spp. resistant to 2,4,5-T or any future replacement. AMS readily absorbed through tree roots and is used selectively for *spot treatment only: Do not apply overall.*

2
Italics denote herbicide also or solely used by itself.
( ) Brackets denote herbicide not first choice or limited effect on weed type.
Gs = Controls 'soft' grasses only
Gr = Controls couch (*Agropyron repens*) and other perennial grasses.
He = Controls herbaceous broadleaved weeds.

3
Fo   = foliage absorbed
So   = root absorbed—i.e. soil acting or residual.
T   = Spray woody regrowth
V   = Cut stump application
W   = Basal bark treatment
X   = Frill girdling
Y   = Notch application   carried out on standing trees
Z   = Tree injection

4
( ) Brackets denote pre-planting treatment is preferable.

5
Restrictions on use—consult manufacturers' label or literature.

A   Some species of tree excluded
B   Age or size of trees.
C   Growth stage of trees.
D   Only certain times of year—stated.
E   Minimum period required between treatment and planting.
F   Not organic soils or heavy trash.
G   Only established grass—avoid contact with tree roots.
H   Do not mix with other chemicals.
I   Apply only to freshly cut stumps within 24 h of cutting.
J   Extreme care needed to avoid drift—use tree or spray guards.
K   Adequate leaf area and active growth at time of application essential.
L   Apply at least 24 h before rain.
M   Apply fertiliser in spring prior to spraying.
N   Gorse and broom killed by winter application. Not hardwoods or larches or some pines unless *directed* by knapsack, application.

Table 8.10 Weed control (unselective) in non crop situations

Herbicide[1]	Perennial weed situations for which herbicide is particularly recommended					General use, i.e. path, drives, roads, yards, around buildings, etc.	Approximate period of persistency of initial dose in soil and consequent control of seedlings of susceptible species[2]	Mixtures available (see Key)[3]	Specific limitations of herbicide[4]
	Bracken (Pteridium aquilinum)	*Couch and other rhizomatous grasses (Agropyron repens etc.)*	*Docks (Rumex spp.)*	*Great stinging nettle (Urtica dioica)*	*Woody weed species*				
Foliage absorbed herbicides (see also Table 8.1)									
Aminotriazole		Yes				Yes	LP	abcdef	C
Dalapon		Yes				Mixtures only	MP[2]	gh	B
Glyphosphate	Yes	Yes	Yes			Yes	None	None	FJ
Paraquat						Yes	None	*See* other herbicides	CF
Foliage and/or root absorbed herbicides									
Ammonium sulphamate (AMS)				Yes			MP[2]	None	D—but *see Table 8.9*
Picloram			Mixtures			Yes	VP	ij	AEFH
Sodium chlorate						Yes	MP[1]	kl	DI
Root absorbed herbicides (residual)									
Atrazine						Yes	VP	n	K
Bromacil		Yes				Yes	VP	o	D
Chlorthiamid	Yes	Yes	Yes			Yes	P		E only granules
Dichlobenil	Yes	Yes	Yes			Yes	P	p	E only granules.
Diuron						Yes	VP	q	D
Simazine						Yes	VP	*See* other herbicides	
TCA		Yes					MP[2]		*See Table 8.1*
Other herbicides with limited use or spectrum									
Asulam	Yes	Yes					LP		
MCPA						Yes	LP		AE
2,4,5-T				Yes	Yes		LP		AE

Key:

1
Italics denote herbicide also or solely used by itself.

2
Persistency of chemicals is determined to a large extent by the dosage rate, (increased dosage rate = longer persistency) soil type and rainfall. The following times are *only approximate* and apply to the *straight herbicide only*. Additives may increase persistency greatly:

VP (very persistent) a season or longer
P (persistent) a season
MP[1] (moderately persistent) 3 months to a season
MP[2] (rather less persistent than MP[1]) 3 to 4 months
LP Low persistency. Up to 8 weeks
None No persistency—sowing can follow soon after treatment if required.

3
Herbicide mixtures
a Aminotriazole + atrazine
b Aminotriazole + atrazine + 2,4-D
c Aminotriazole + atrazine + diuron
d Aminotriazole + dichlorprop + diuron + MCPA
e Aminotriazole + diuron
f Aminotriazole + simazine
g Dalapon + MCPA
h Dalapon + diuron + MCPA
i Picloram + bromacil
j Picloram + 2,4-D
k Sodium chlorate + atrazine
l Sodium chlorate + atrazine + 2,4-D
m 2,3,6-TBA + MCPA
n Atrazine + sodium monochloracetate
o Bromacil + diuron
p Dichlobenil + dalapon
q Diuron + paraquat

3 and 4
A = Grasses resistant to straight herbicide
B = Broadleaves resistant to straight herbicide
C = Safe to use in close proximity to established trees or ornamentals but check restrictions on species.
D = Avoid use in close proximity to any trees or ornaments to be retained, as a lethal dose may be taken up by their roots.
E = Avoid use near hops or in or around glasshouses or near any other susceptible crops—consult label.
F = Requires *extreme* precautions—guards etc. to avoid drift.
G = Soil type restrictions for residual herbicides.
H = *Only* non agricultural grass areas where there is no botanical interest.
I = Any material impregnated with solution becomes highly inflammable when dry. Do not light matches, smoke or expose contaminated clothes to naked flame.
J = Do not mix with other herbicides.
K = Also available as granules.

NB There is a need for great care when using herbicides in the proximity of glasshouses, orchards, ornamentals and any susceptible crops. Avoid drift at all times. Many of the above herbicides are quite unsuited for amateur use.

periods are frequently *very limited*. *Surface* or soil applications, granular or liquid depend on adequate rainfall for activity. Restrictions exist on timing of applications.

TREATMENT OF INDIVIDUAL LARGE TREES

Stump treatment

This treatment is given to freshly cut stumps (within 24 h of felling) of broadleaved trees or woody weeds to prevent coppice regrowth. Give as a *pre-planting* treatment. With 2,4,5,-T saturate cut surface and remaining bark; ammonium sulphamate (AMS) cut surface only. Allow recommended intervals for each herbicide between treatment and planting. Failure to treat usually results in heavy regrowth of coppice shoots.

Unwanted standing trees

Basal bark spray

Saturate bottom 30 cm (thin bark) or 45 cm (thick bark) of whole circumference of trunk to run off with 2,4,5-T in paraffin—water is useless; take great care with spring or summer post planting application.

Frill girdling

A ring of overlapping downward cuts encircling the trunk close to ground level is first made with a light axe or bill hook, to penetrate the cambium and preferably the sapwood; herbicide is then sprayed onto bark just above cuts and runs into them. Suitable for 2,4,5-T in paraffin or AMS for trees less than 15–20 cm diameter breast height.

Notching

This is for applying solid AMS crystals to larger trees; cut a ring of steps with an axe, maximum 10 cm edge to edge apart at base of tree, floors of steps sloping slightly inward to retain AMS crystals, 15 g per step.

Tree injection

Tree injection requiring a special tool for injecting undiluted herbicides into the translocation systems of a tree, may be carried out safely at any time of the year.

Choice of herbicidal treatment

A sequence of treatments is likely to be necessary in the early life of the crop; the treatments for various weed situations are summarised in *Table 8.9*. Particular attention should be paid to the requirements and limitations of each herbicidal treatment.

WEED CONTROL (UNSELECTIVE) IN NON-CROP SITUATIONS

Frequently the object is to keep ground free from vegetation for a prolonged period. If temporary control only is needed and cropping is to follow, select herbicides with only short residual effects, e.g. glyphosate (*see Tables 8.1* and *8.10*). Avoid using herbicides which may be taken up by the roots of adjacent crops, trees or ornamentals, e.g. sodium chlorate.

Initially it is essential to kill established vegetation, including all parts above and below ground. Maintenance dressings to keep the ground clear are then applied at appropriate intervals. Leaf and soil acting herbicides are frequently mixed to give the dual effect of immediate kill of foliage and a subsequent residual effect at the first application: 'follow up' applications are still required. Mixtures are also employed to broaden the weed spectrum. Effects of residual herbicides are much affected by ground conditions. Above average dose levels are needed for adsorptive surfaces e.g. clay, peat or ashes. With less adsorptive but highly penetrable conditions where leaching is rapid, e.g. sand or gravel, lower rates of chemical are adequate but more frequent maintenance applications are necessary. Resistant weeds may require an additional herbicide or a higher dosage rate. Advice should be obtained before treating slopes or hard impenetrable surfaces where run-off may affect wanted vegetation, streams or watercourses. Some suitable herbicides are summarised in *Table 8.10*.

NB While every precaution has been taken to ensure the correctness of this chapter, the author and publishers cannot accept any responsibility whatsoever for losses arising from using chemicals mentioned herein. Selection for field use should be made from *Approved Products for Farmers and Growers*, published annually by the Ministry of Agriculture, Fisheries and Food (Agricultural Chemicals Approval Scheme), obtainable from HMSO or through any bookseller. *Manufacturers' literature and labels must always be consulted before use.* Detailed recommendations are also given in *Weed Control Handbook* Vol. II Recommendations, edited by Fryer J. D. and Makepeace R. J., published by Blackwell Scientific Publications, Oxford.

9

Diseases of crops

G. Moule

Recent changes in farming practice have resulted in changes in the disease problems of crops. The increase and intensification of the cereals and oilseed rape cropping in recent years, coupled with the trend towards increased autumn sowings, has resulted in the widespread incidence of many air-borne foliar diseases. Many seed-borne diseases however, are now quite rare due to the general use of the various seed treatments available. Similarly disease problems in the future are only likely to change rather than disappear totally.

The cost effectiveness of any crop protection treatment is dependent upon many factors which may vary widely with the region, season, the chemical used and the individual crop and variety concerned. In the short term the cost of the treatment has to be set against the increased value of the crop and there are likely to be large differences between the treatment of high value crops such as field scale vegetable compared with fodder crops for livestock. In the long term it is likely to be preferable to undertake uneconomic measures with an initial infection of a persistent problem (e.g. long-lived soil-borne disease or pest) to prevent or reduce recurring losses in future crops. Many problems of this nature are best controlled by strict preventative measures in the first instance.

Crop diseases can be divided into the following general groups,

(1) air-borne;
(2) soil-borne;
(3) seed-borne;
(4) vector-borne diseases.

Many diseases have two or more distinct phases of attack and may belong to more than one of these groups, e.g. canker (*Leptosphaeria maculans*) of brassicas may be seed- and soil-borne as well as having a very distinct and destructive air-borne phase. How-ever each disease has been allocated to the most appropriate group for control measures.

AIR-BORNE DISEASES

These quickly establish in the crop and mostly attack the foliage and stems producing an overall blanket-type field infection, i.e. virtually all plants are affected to the same degree. They are all of fungal or bacterial origin and produce spores which are easily dispersed by wind and air currents. Spore production may be vast and rapid during suitable environmental conditions and long distance spread can occur in a short period of time. Many are associated with wet weather conditions particularly when the air temperatures are above 10 °C, i.e. April to September. Important diseases included in this group are potato blight, (*Phytophthora infestans*) the yellow rusts (*Puccinia striiformis*) of wheat and barley, *Septoria* spp. of wheat, *Rhynchosporium* leaf blotch of barley, rye and ryegrasses, crown rust (*Puccinia coronata*) of oats, canker (*Leptosphaeria maculans*) of brassicas and various leaf and pod spots (*Aschochyta* and *Botrytis* spp.) of peas and beans. Their activity is often curtailed by hot dry weather. Others tend to be more prevalent during the warm dry spells of summer and early autumn, e.g. powdery mildews (*Erysiphe* spp.) of various crops and brown rusts (*Puccinia* spp.) of wheat and barley. The powdery mildews, however, are more independent of specific climatic conditions for spread than most other diseases.

Many air-borne diseases are very host specific and produce large numbers of asexual spores giving rise to specific races of the fungus with the ability to attack only a very few species or even several varieties (cultivars) of one crop. This extreme specificity can be utilised to advantage when choosing varieties of certain cereals (*see* NIAB Diversification Schemes).

Control

Disease epidemics are only likely to result if

(1) a susceptible host is sown,
(2) a virulent race of the pathogen occurs, and
(3) the environment is favourable for disease attack and spread.

If any of these three factors is limiting in any way crop damage is not likely to be severe. Wherever possible all forms of control measures should be used to give a fully integrated control programme (*Figure 9.1*). If, however, a resistant variety is used or climatic conditions have not been favourable for disease spread, chemical control can safely be reduced or omitted on economic grounds.

Cultural

Crop rotations are not very useful in controlling air-borne diseases as they cannot prevent wind-blown spores coming in from neighbouring infected areas. Similarly efficient stubble clearing after harvest drastically reduces carryover of spores onto the following crops but at best, like crop rotation, is only likely to delay subsequent re-infection. Both are very much more useful for soil-borne diseases and some types of weed control. Burning of stubble followed by deep ploughing to bury the remainder is preferable to the use of heavy cultivators with their poor burial ability. The date of sowing may affect the length of the intercrop period considerably and the subsequent disease carryover. Many diseases are very common

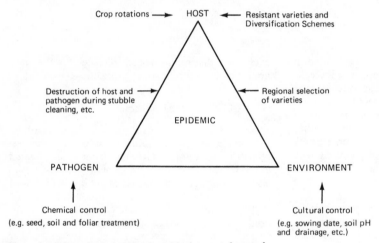

Figure 9.1 A hypothetical epidemic and its integrated control

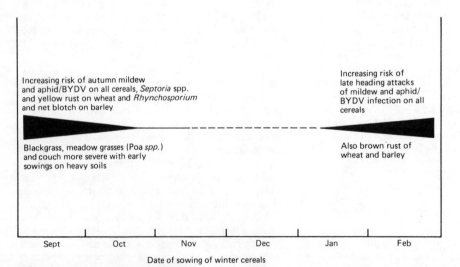

Date of sowing of winter cereals

Figure 9.2 The relationship between the date of sowing of winter cereals and associated disease, pest and weed problems (BYDV—barley yellow dwarf virus)

on late tillers and volunteers consequently early sown autumn crops are particularly prone to disease carryover from the previous crop. The seedlings of these crops often become greatly infected with various seed-borne diseases from the later germinating shed corn after stubble cleaning (*Figure 9.2*).

September sowings of winter cereals are often very high yielding but seldom give as high returns as October sown crops due to the increased costs incurred for the early autumn disease, pest and weed control required. For these reasons spring sown crops may be useful as cleaning crops in a sequence of autumn sown crops although returns tend to be considerably lower.

Air-borne diseases are best controlled by the use of resistant varieties. Where these do not exist or break down due to new pathogen races chemical control should be used. All variety choice should be made with consultation to the National Institute of Agricultural Botany (NIAB) recommended leaflets paying particular attention to those diseases common in the region. Where such schemes exist, use should also be made of the NIAB Variety Diversification Schemes to help reduce the spread and severity of those air-borne diseases which exist in many physiological races. The scheme for winter wheat variety choice in 1982 has been taken from the NIAB Farmers leaflet No. 8 *Cereals*, and is reproduced in *Figure 9.3*.

Farmers should grow, preferably, several varieties chosen from different diversification groups and not just one or two popular varieties that may often belong to the same diversification group and thus the same pattern of susceptibility. Diversification in time may also be useful, e.g. where wheat follows wheat it would be advisable if the second wheat crop variety was chosen from a different diversification group to the first wheat variety.

Chemical control

Seed treatments

These are useful in controlling many seed- and soil-borne diseases and give good control during the first half of the plant's life. Few are persistent enough to give effective control of most air-borne diseases of later critical stages of growth such as flag leaf emergence and heading of cereals. Those that are persistent are often more expensive than similar foliar treatments and, in years of relatively low disease levels, may prove to be unnecessary.

Foliar treatment

This treatment is considerably more flexible than seed treatment particularly with regard to the number and choice of chemicals available. The correct chemical(s) can be chosen accurately for the specific disease(s) as and when they occur. Protectant fungicides with little or no eradicant action must always be applied before disease build up. As a general rule even when using systemic fungicides with good eradicant activity the best economic responses are obtained when applications are made at the first sign of disease particularly if this coincides with weather conditions that favour disease development (*Figure 9.4*).

Winter Wheat

VARIETAL DIVERSIFICATION SCHEME TO REDUCE SPREAD OF YELLOW RUST AND MILDEW IN WINTER WHEAT 1982

Severe infections may result if yellow rust or mildew spreads from an adjacent winter wheat crop into a variety with a low level of resistance. This risk can be reduced by choosing varieties with high levels of resistance. Further benefit can be obtained by sowing adjacent fields with varieties chosen from different diversification groups as this reduces the spread of disease from one to the other. Similarly, varieties to be grown in the same field in successive years or in a seed mixture should be chosen from different diversification groups.

Diversification groups (DG) of currently recommended winter wheat varieties are:

DG 1B	DG 1F	DG 3B
Avalon	Rapier	Norman
Bounty		
Fenman	DG 2B	DG 4C
	Hustler	Armada
DG 1E	Mardler	
Aquila	Maris Huntsman	DG 6B
Flanders	Virtue	Brigand

Choosing varieties to grow together

1. Decide upon first-choice variety and locate its DG number.
2. Find this number under 'Chosen DG' down left hand side of table.
3. Read horizontally across table to find the risk of disease spread for each companion DG.
4. Ensure that chosen varieties are not all susceptible to another disease.

Chosen DG	Companion DGs						
	DG 1B	DG 1E	DG 1F	DG 2B	DG 3B	DG 4C	DG 6B
DG 1B	m	+	m	m	m	+	m
DG 1E	+	m	m	+	+	+	+
DG 1F	m	m	m	m	m	+	m
DG 2B	m	+	m	ym	m	+	m
DG 3B	m	+	m	m	ym	+	m
DG 4C	+	+	m	+	+	ym	+
DG 6B	m	+	m	m	m	+	ym

+ = good combination; low risk of spread of yellow rust or mildew

y = risk of spread of yellow rust

m = risk of spread of mildew

Figure 9.3 Scheme for winter wheat variety choice (Reproduced by permission of the National Institute of Agricultural Botany)

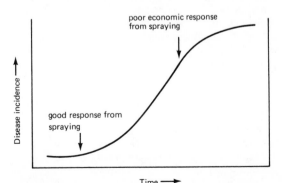

Figure 9.4 The effect of time of spraying on the likely yield response

It is normally more cost effective to use a fungicide in high risk situations, for example

(1) susceptible varieties,
(2) high disease levels in nearby crops,
(3) suitable weather conditions for disease spread,
(4) pre-disposing factor(s) operative.

Broad-spectrum fungicides generally give better responses than specific types particularly if the variety is susceptible to more than one disease and/or two or more diseases are likely to attack the crop. With many crops there are often critical growth stages that benefit from good disease protection, e.g. flag leaf/ear emergence in cereals and rapid bulking up of potatoes in August. Generally crops should not be sprayed with any crop protection chemical if they are suffering from stress. Fungicides used during periods of drought often give rise to quite significant yield reductions.

SOIL-BORNE DISEASES

The main symptoms usually occur on the roots and stem bases which often give rise to wilting and stunted plants. Affected plants occur in patches of varying sizes in the field and may be confused with poor drainage, shallow soil depth and various mineral deficiencies. These diseases generally have very limited powers of mobility and are commonly dispersed in contaminated soil on machinery wheels, animals' feet, clothing and footwear. Severe wind and water erosion of soil may be significant in certain areas also.

The organisms generally exist in a limited number of races and either attack a wide range of crops and are short lived without a host, e.g. 'Take-all' (*Gaeumannomyces graminis*) of cereals and grass or possess a narrower host range coupled with long-term survival in the soil as a resting spore, e.g. potato wart (*Synchitrium endobioticum*), onion white rot (*Scerotium cepivorum*) and pea wilt (*Fusarium oxysporum*). Club root (*Plasmodiophora brassicae*) of brassicas is a particularly difficult disease as it has a wide host range including most agricultural/horticultural brassicas and many cruciferous weeds such as charlock (*Sinapis arvensis*), shepherds purse (*Capsella bursa-pastoris*) and field pennycress (*Thlaspi arvense*). Its soil-borne resting spore is also capable of surviving for at least eight years between susceptible hosts. Once established eradication of these diseases is extremely difficult and often quite costly, therefore the main form of control measure should be preventative using a sensible cropping sequence. Overcropping should be avoided and extreme care should be taken with contaminated fields to prevent further spread.

Often the most severe effects of the disease can be alleviated by various cultural measures such as generous fertiliser application, drainage and liming where appropriate. Some specific crops can be sown on infected soils provided resistant varieties are used. As a result of legislation (The Wart Disease of Potatoes (Great Britain) Order 1973) requiring the use of immune varieties to the UK races of potato wart since the 1920s this particular disease has been virtually eliminated. Chemical control is considerably more difficult, expensive and generally less effective than with air-borne diseases. Often it is only worthwhile with high value arable crops.

SEED-BORNE AND INFLORESCENCE DISEASES

Symptoms are most likely to be seen in the seedling and young plant stages followed later by further major attack at the end of the crop's life on the flowers, seed pods and seeds in general. May cause a systemic infection of the plant causing no visible external symptoms during the vegetative phase and then appear quite dramatically on the ears, e.g. loose and covered smuts (*Ustilago* spp.) of cereals, bunt (*Tilletia caries*) of wheat and maize smut (*Ustilago maydis*). Others cause severe attack on the seedlings and foliage during the early stages of growth and relatively little damage until the heading stage, e.g. various seedling blights of cereals caused by *Fusarium*, *Pyrenophora* and *Septoria* spp. This latter group is often associated with untreated seed such as shed corn as a result of poor stubble cleaning. The diseases develop on the volunteer plants then infect the newly sown autumn crops. This is particularly common with early sown winter cereals.

Seed-borne diseases are potentially very serious indeed and can cause considerable yield reduction. In the UK most of them are controlled by the general use of healthy seed produced in various Seed Certification Schemes and such fungicidal seed dressings as organomercury, benomyl, carboxin, iprodione and thiram. Seed treatments are an inexpensive form of disease control and should always be used where the health of a seed sample is in doubt. Bunt of wheat and the covered smuts of barley and oats are now exceedingly rare as a result of the standard organomercury seed dressings used today. Partly as a result of this effective treatment research into breeding resistant varieties has been limited in recent years. However, strains of seedling blight on oats (*Pyrenophora avenae*) have developed which are resistant to organomercury dressings and more research may be necessary in the future.

VECTOR-BORNE DISEASES

The important diseases in this group are all viruses and the main vectors are either aphids or nematodes. Aphid-borne diseases may show a superficially similar blanket pattern of distribution to air-borne diseases in the field. However not all plants are infected and the appearance and severity of symptoms is more variable. Crop attack and appearance of symptoms are directly dependent upon aphid activity initially and therefore seldom occur much before May in spring and continue until October in the autumn. Yield reductions are related to the age of the plant at the time of infection. Seedling infection, as often occurs with early sown autumn and late sown spring crops, can cause severe reductions in yield in individual plants whereas late attacks on mature plants may cause negligible loss. The yield loss is therefore directly related to the numbers of individual plants affected and their age at infection. Most true seed is naturally virus-free at planting and will give rise to healthy seed despite being infected during vegetative growth. However, where vegetative 'seed' is used, e.g. potato seed tubers, once infected these will give rise to infected seed. Consequently to maintain virus-free stock for commercial growers, potato seed tubers are produced in areas such as Scotland and Northern Ireland with low aphid populations and sold subject to statutory restrictions (Prevention of Spread of Pests (Seed Potatoes) (Great Britain) Order 1974 and Seed Potatoes Regulations 1978) on purity and health.

Similar schemes also exist in horticulture for production and sale of virus free stocks of strawberries, raspberries and top fruit that are typically produced by vegetative means.

Nematode-borne viruses occur in patches in the field similar to most other soil-borne pathogens. Attack by these free-living types of nematodes and subsequent virus infection is much more common on lighter sandy soils where nematode populations are naturally higher. Many crops and weeds may be attacked by these nematodes but potatoes, sugar beet and raspberries are at greatest risk.

Once infected with viruses, plant yields can be severely reduced and chemical control is not possible. Control should be aimed at preventing or delaying infection for as long as possible by the use of healthy seed, isolation from infected sources of material and then chemical control against the vectors concerned. The control of aphid vectors tends to be generally less expensive than for nematodes but prevention of virus infection is variable and dependent upon virus type.

Lastly it should not be forgotten that plants are often attacked by several organisms simultaneously and virus infection may occur alongside or be confused with fungal diseases and in particular mineral deficiencies.

DISEASES OF WHEAT

Leaf and stem diseases

Powdery mildew (Erysiphe graminis)

Superficial grey-white fungal pustules on leaves, stems and ears, particularly at heading. Pustules darken with age. Attacks wheat, barley, oats, rye and grasses but cross-infection unlikely. Common in May/June (ears) and Oct/Nov (seedlings) and favoured by warm dry conditions. Air-borne disease surviving on stubble, late tillers, volunteers and early sown winter wheat.

Cultural control

(1) Use NIAB Farmers leaflet No. 8 for variety choice and Diversification Schemes.
(2) Destroy stubble and preferably plough soon after harvest.
(3) Avoid early autumn and late spring sown crops.
(4) Avoid close proximity of winter and spring crops and wheat after wheat.

Chemical control

Foliar treatment Use of one of the following fungicides alone or in mixtures at first sign of disease in spring and up to full ear emergence (Zadoks scale—ZCK 59) (*Table 5.15*, p. 128).

    ethirimol
    fenpropimorph
    prochloraz
    propiconazole
    triadimefon
    tridemorph

Seldom economic to apply fungicides in autumn or more than one well timed application at start of epidemic in spring.

Yellow rust (Puccinia striiformis)

Orange-yellow pustules occurring in stripes on mature leaves or groups on leaves of young plants. Air-borne diseases favoured by cool moist conditions in May/June and wet summers in general. Survives on volunteers, late tillers and early sown winter wheat. Occurs also on barley and rye but cross-infection from barley is unlikely. Very common in eastern England.

Cultural control

(1) Use NIAB Farmers leaflet No. 8 for variety choice and Diversification Schemes.
(2) Destroy stubble and preferably plough soon after harvest.
(3) Avoid early sowing of winter wheat.
(4) Avoid close proximity of winter and spring crops and wheat after wheat.

Chemical control

Foliar treatment   Use of one of the following fungicides alone or in mixtures at first sign of disease in spring and up to full ear emergence (ZCK 59):
    benodanil
    fenpropimorph
    propiconazole
    triadimefon
    tridemorph and 'Polyram'

Brown rust (*Puccinia recondita*)

Orange-brown pustules randomly scattered or grouped in patches on leaves. Air-borne disease favoured by hot dry conditions in June and July. Seldom severe until after ear emergence, therefore less common or as damaging to yield as Yellow Rust. Survives on stubble, late tillers, volunteers and early sown winter wheat in mild winters. Also occurs on rye and barley but a different species is involved on barley and, therefore, cross-infection is not possible. Generally of infrequent occurrence.

Cultural control

(1) Use NIAB Farmers leaflet No. 8 for variety choice.
(2) Destroy stubble and preferably plough soon after harvest.
(3) Avoid late sown and late maturing crops.
(4) Avoid close proximity of winter and spring crops and wheat after wheat.

Chemical control

Foliar treatment   Use of one of the following fungicides alone or in mixtures at first sign of disease in spring and up to full ear emergence (ZCK 59):
    benodanil
    fenpropimorph
    propiconazole
    triadimefon

Glume blotch (*Leptosphaeria nodorum* syn. *Septoria nodorum*) and Leaf spot (*Mycosphaerella graminicola* syn. *Septoria tritici*)

Septoria spp. cause brown, often irregular shaped lesions on leaves and purple/brown glume blotch phase on glumes at heading. Difficult to diagnose unless leaves and head are still green. Associated with high rainfall areas, wet summers and high humidity at heading. Rain splash/air-borne disease surviving on infected stubble and seed.

Cultural control

(1) Use NIAB Farmers leaflet No. 8 for variety choice.
(2) Destroy stubble and preferably plough soon after harvest.
(3) Avoid early sown winter wheat and wheat after wheat.

Chemical control

Seed treatment   The following fungicidal seed treatments will give partial control of *Septoria nodorum*.
    benomyl + thiram
    guazatine
    organomercury

Foliar treatment   Most cost effective when applied to protect flag leaf and ear (ZCK 39–59):
    benomyl
    captafol
    carbendazim
    chlorothalonil
    prochloraz
    propiconazole

Barley yellow dwarf virus (BYDV vectors—cereal aphids)

Seedling infection can result in stunted plants of wheat, barley, oats, rye and grasses. Infection in early summer results in purple-red leaves in wheat, bright canary yellow in barley and pink leaves in oats. Spread by two main cereal aphids, bird-cherry (*Rhopalosiphum padi*) and grain aphid (*Sitobion avenae*), during warm dry weather in autumn and May/June. Most common in late sown spring crops. Survives in late tillers, volunteers and grasses.

Cultural control

(1) Winter wheat most tolerant. Spring barley most susceptible to yield reductions.
(2) Destroy stubble and preferably plough soon after harvest.
(3) Avoid early sown winter and late sown spring crops.
(4) Avoid close proximity of cereals and grasses.

Chemical control

Foliar treatment

(1) Delay infection to as late in the plants' life as possible by avoiding high aphid populations.
(2) September and May sown cereals crops are high risk for aphid/BYDV and mildew infection. Dual low cost treatment of both may be worthwhile.

Root and stem base diseases

Eyespot (Pseudocercosporella herpotrichoides)

Dull brown indefinite eyespot at base of stem eventually causing lodging. Attacks wheat, barley, oats and rye. Common in intensive cereals on heavy damp soils. Soil- and stubble-borne surviving two to four years between susceptible crops. Favoured by long wet cold periods in winter and spring.

Cultural control

(1) Use NIAB Farmers leaflet No. 8 for variety choice.
(2) Deeply plough stubble soon after harvest.
(3) Avoid three or more years of continuous wheat and/or barley especially on heavy land.
(4) Use three year break from wheat or barley for disease reduction.
(5) Avoid early sown lush crops.

Chemical control

Foliar treatment   Use of one of the following fungicides alone or in mixtures at late tillering stage (ZCK 30–31):

   benomyl
   carbendazim
   prochloraz
   thiophanate-methyl

Take-all (Gaeumannomyces graminis)

Dead, stunted and thin patches of varying sizes in fields of wheat and barley. Black fungus kills roots causing empty bleached 'white heads' and premature ripening. 'White heads' often attacked by sooty moulds at harvest. Oats and rye are usually resistant to wheat/barley race of fungus. Disease is favoured by light alkaline soils and above average rainfall in winter and spring. Survives on couch (*Agropyron repens*) and stubble for two years. Common in intensive cereals.

Cultural control

(1) Use two year break from wheat and/or barley for eradication.
(2) Prevent build-up by avoiding more than three years of continuous wheat or barley.
(3) Avoid wheat after barley.
(4) Direct drilled crops often less affected than traditional sown crops.
(5) Oats, maize and grasses are effectively resistant in most commercial situations.
(6) No resistant varieties of wheat or barley.

Chemical control

(1) No economical chemical control.
(2) 40 kg/ha N extra top dressing at first sign of disease in spring may alleviate most damage.

Sharp eyespot (*Pellicularia filamentosa* syn. *Rhizoctonia solani)*

Attacks stems and stem bases causing numerous and clearly defined 'eyespot' lesions. Soil-borne disease with wide host range (including barley, oats and rye) and, therefore, not easily controlled by rotation. Most common on light sharp soils of neutral/acid nature during cold dry conditions. More prevalent in rotation with grass leys, peas and root crops. Less prevalent in intensive cereals.

Cultural control

(1) No resistant varieties, oats and rye most susceptible, barley least so.
(2) Avoid late sown winter and early sown spring
crops.

Chemical control

No economic chemical control.

Brown foot rot and ear blight (*Fusarium* spp.)

Ill-defined brown rotting of stem base, roots and seedlings especially on cold, heavy, poorly drained acid soils in wet autumns and winters. Also occurs on other cereals.

Cultural control

(1) No varietal resistance known at present.
(2) Avoid late sown winter and early sown spring
crops in cold soils.

Chemical control

Seed treatment   Use of a seed dressing containing one of the following will give partial control:
  benomyl + thiram
  guazatine
  organomercury

Ear diseases

Loose Smut (*Ustilago nuda)*

Ears only visibly affected. All grains destroyed and replaced by black spores. Conspicuous in early June, at the start of heading. Not serious in UK. Seed-borne disease.

Cultural control

(1) Use of NIAB Farmers leaflet No. 8 for variety
choice.
(2) Use clean certified seed.

Chemical control

Seed treatment   Use of a seed dressing containing one of the following may be necessary for many seed certification schemes:
  benomyl
  carboxin

Bunt (*Tilletia caries)*

Ears small and stunted. Internal contents of grain replaced by black fishy-smelling spores but seed coat intact initially. Now very rare in UK. Seed-borne disease.

Cultural control

(1) Use clean certified seed.
(2) Avoid untreated home saved seed.

Chemical control

Seed treatment   Use of a seed dressing containing one of the following:
  benomyl
  carboxin
  guazatine
  organomercury

Black/sooty moulds (*Cladosporium herbarum/Alternaria* spp.)

Black sooty appearance on overipe/late harvested crops in wet seasons and/or on prematurely ripened grains as a result of other disease attacks, e.g. Take-all, foot rots. Cosmetic damage only.

Cultural control

(1) Avoid late harvesting of crops where possible.

Chemical control

Foliar treatment   Chemicals used for Septoria glume blotch control at heading will give incidental control of sooty moulds. Otherwise not economic to spray.

DISEASES OF BARLEY

Leaf and stem diseases

Powdery mildew (Erysiphe graminis)

Superficial grey-white fungal pustules on leaves and stems particularly during stem elongation. Pustules darken with age. Attacks wheat, barley, oats, rye and grasses but cross-infection unlikely. Common in May/June (adult plants) and Oct/Nov (seedlings and shed corn). Favoured by warm dry conditions. Air-borne disease surviving on stubble, late tillers, volunteers and early sown winter barley. Very common and serious on spring barley.

Cultural control

(1) Use NIAB Farmers leaflet No. 8 for variety choice and Diversification Schemes.
(2) Destroy stubble and preferably plough soon after harvest.
(3) Avoid early sown winter and late sown spring barley crops.
(4) Avoid close proximity of winter/spring crops and barley after barley.

Chemical control

Seed treatment   Use of a seed dressing containing one of the following:
  ethirimol
  triadimenol + fuberidazole
  thiophanate − methyl
  triforine

Foliar treatment   Use of one of the following alone or in mixtures at first sign of disease in spring and up to full ear emergence (ZCK 59).
  ditalimfos
  fenpropimorph
  nuarimol
  prochloraz
  propiconazole
  triademifon
  tridemorph
  triforine

Seldom economic to apply fungicides in autumn or more than one well-timed application at start of epidemic in spring.

Yellow rust (Puccinia striiformis)

Distinct form from that on wheat and cross-infection unlikely. Very similar in all other respects and full account given under 'Wheat'. Control similar to that on wheat with the addition of triadimenol + fuberidazole seed dressings which may control infections on young plants. A later foliar application is also likely to be needed with this particular seed dressing.

Brown rust (Puccinia hordei)

Distinct species from that on wheat and cross-infection is not possible. Very much more common than wheat brown rust and likely to cause greater crop damage. Yields and quality are likely to be reduced. Most common in low rainfall areas of southern and central England during July at heading time, especially in hot, dry seasons. Control is similar to that for brown rust of wheat with addition of triadimenol + fuberidazole seed dressing which may give control on young plants although these early infections tend to be of irregular occurrence. Only likely to be of use with late sown spring barley.

Leaf blotch (Rhynchosporium secalis)

The disease causes blotches with purple-brown borders on the leaves and stems of barley, IRG, PRG and rye.

Blotches initially pale grey-green, often diamond-shaped or at base of leaf blade. Air-borne foliar disease often very severe in high rainfall areas, coastal regions and generally widespread in wet summers. Yield and quality reduction can be very severe when weather is cool and wet in May/June particularly. Survives on stubble, late tillers, volunteers and early sown winter barley crops. Also sometimes seed-borne.

Cultural control

(1) Use NIAB Farmers leaflet No. 8 for variety choice.
(2) Destroy stubble and preferably plough soon after harvest.
(3) Avoid early sowings of both winter and spring barley.
(4) Avoid close proximity of winter/spring crops and infected rye grasses.
(5) Avoid barley after barley.

Chemical control

Seed treatment　Seed dressings may give control of early infections:
　imazalil + thiophanate-methyl
　triadimenol + fuberidazole

Foliar treatment　At first sign of disease in spring and up to full ear emergence (ZCK 59):
　benomyl
　captafol
　carbendazim
　fenpropimorph
　prochloraz
　propiconazole
　thiophanate-methyl
　triademifon

Barley yellow dwarf virus (vectors—cereal aphids)

Extremely damaging on yield where infection occurs in seedling stage on early sown winter or late sown spring barley. Later infection at heading time in May/June has minor yield effect. Barley and oats are much less tolerant of attack than wheat. Control measures are similar for those on wheat. Aphicidal control is likely to be more cost-effective on barley particularly when combined with powdery mildew control which often accompanies BYDV infection. Aphicide control after symptoms appear on plants will only reduce further losses occurring and will not affect initial infection damage. A full account is given under 'Wheat'.

Net blotch (Pyrenophora teres)

The disease causes dark brown net-like blotches on seedlings of volunteers in particular. Symptoms spread to young crops in autumn and attack mature plants in June. Favoured by warm wet conditions in Sept/Oct and May/June. Disease may be seed-borne or from crop debris and volunteers left in field.

Cultural control

(1) Avoid untreated seed.
(2) Avoid early sown winter barley.
(3) Avoid winter barley after spring barley.
(4) Destroy stubble and preferably plough soon after harvest.

Chemical control

Seed treatment
　guazatine + imazalil
　organomercury
　thiophanate-methyl + imazalil
　　triadimenol + fuberidazole

Foliar treatment　Spray to protect flag in May/June with one of the following:
　benomyl
　carbendazim
　fenpropimorph
　prochloraz
　propiconazole
　triforine + mancozeb

Leaf stripe (Pyrenophora graminea)

Long brown stripes on leaves usually running entire length. Leaves often split and shred later. Most common on untreated seed in cool wet weather.

Cultural control

(1) Avoid untreated seed.
(2) Avoid early sowings.
(3) Avoid barley after barley.

Chemical control

Seed treatment Easily controlled by most seed dressings.

Halo spot (*Selenophoma donacis*)

Small angular spots with dark margins and pale centres. Most common on top leaves and awns in May/June during cool moist conditions. Seldom damaging in UK. Seed- and stubble-borne but may also survive on wheat, rye, cocksfoot and timothy.

Cultural control

(1) Avoid untreated seed.
(2) Avoid barley after barley.

Chemical control

Foliar treatment If symptoms are severe spray foliage with one of the following:
  benomyl
  dichlofluanid
  thiophanate-methyl
  zineb

Root and stem base diseases

Eyespot (*Pseudocercosporella herpotrichoides*)

Similar but generally less important than that on wheat. Cross-infection is likely and taken into account for crop planning. Chemical control measures are unlikely to be as necessary or as economic compared with disease control on wheat. A full account is given under 'Wheat'.

Take-all/white heads (*Gaeumannomyces graminis*)

Similar but generally less important than that on wheat. Cross-infection occurs but barley is inherently more tolerant of attack and there is less time for the disease to spread in the soil with spring sown barley. Spring barley usually shows a lower incidence of attack than winter barley due to the longer intercrop period. A full account is given under 'Wheat'.

Sharp eyespot (*Pellicularia filamentosa* syn. *Rhizoctonia solani*)

Rarely economic on barley. *See* under 'Wheat'.

Brown foot rot and ear blight (*Fusarium* spp.)

Seldom recorded on barley probably due to the majority of barley being spring sown and on lighter free draining soils which discourage disease development. *See* under 'Wheat'.

Ear diseases

Loose smut (*Ustilago nuda*)

This disease is a distinct form from that on wheat and cross-infection does not occur. Otherwise very similar in all other respects and for further information *see* under 'Wheat'.

Covered smut (*Ustilago hordei*)

Very similar to bunt of wheat, for control *see* under 'Wheat'.

Black/sooty moulds (*Cladosporium herbarum/Alternaria* spp.)

Identical to those on wheat causing cosmetic damage to diseased and/or later harvested crops. Chemicals used for Septoria glume blotch control on wheat are likely to be uneconomic for sooty mould control solely. Many are also used for control of various leaf diseases of barley.

DISEASES OF OAT

Leaf and stem diseases

Powdery mildew (Erysiphe graminis)

Grey-white superficial fungal pustules on leaves and stems. Pustules darken with age. Often very severe on late sown spring oats during warm dry weather in May/June and on winter oat seedlings in Oct/Nov.

Also occurs on wheat, barley, rye and various grasses but cross-infection unlikely.

Cultural control

(1) Use NIAB Farmers leaflet No. 8 for variety choice.
(2) Avoid early sown winter and late sown spring oat crops.
(3) Destroy stubble and preferably plough soon after harvest.
(4) Avoid close proximity of winter/spring oat crops.
(5) Avoid oats after oats.

Chemical control

Seed treatment   None available at present.

Foliar treatment   Use of one of the following at first sign of disease in spring and up to full ear emergence (ZCK 59):
ethirimol
fenpropimorph
triadimefon
tridemorph

Crown rust (Puccinia coronata)

Orange-brown pustules on leaves and sometimes stems. Attack cultivated and wild oat species as well as several important grasses (PRG and IRG). Cross-infection from grasses is unlikely. Air-borne disease favoured by cool wet conditions especially during May/June and Oct/Nov. Survives on stubble, late tillers, volunteers and early sown winter oats. Most severe in high rainfall areas.

Cultural control

(1) Use NIAB Farmers leaflet No. 8 for variety choice.
(2) Avoid early sown winter oats.
(3) Destroy stubble and preferably plough soon after harvest.
(4) Avoid close proximity of winter/spring oats.
(5) Avoid oats after oats.

Chemical control

Foliar treatment   Use of one of the following alone or in mixtures at first sign of disease in spring and up to full ear emergence (ZCK 59):
triademifon

Leaf stripe and seedling blight (Pyrenophora avenae)

Seedlings discoloured and stunted, adult lower leaves with short purple-brown stripes and upper leaves with spots. Seed- and stubble-borne disease, most damage occurs on seedlings during cold wet weather.

Cultural control

(1) Avoid untreated seed.
(2) Destroy stubble and preferably plough soon after harvest.
(3) Avoid very early autumn sowings of winter oats.
(4) Avoid sowing in cold wet soils.
(5) Avoid oats after oats.

Chemical control

Seed treatment   Resistant strains to organomercurial seed dressings have become common. Therefore where thought necessary use a seed dressing containing guazatine+imazalil.

Foliar treatment   Very seldom necessary to use fungicides on adult infections.

Barley yellow dwarf virus (or red leaf)

Similar to virus infection on other cereals. For further details *see* under 'Wheat'.

Oat mosaic virus (fungal vector)

Plants stunted in patches in field. Few tillers formed and leaves have dark green mosaic appearance. Fungus vector—*Polymyxa graminis* present in infected soil causes virus infection. Most common in winter oats during cool wet springs and summers. Infested soils remain so for long periods and thus crop rotations are not very effective. No chemical treatment economic on a farm scale.

Root diseases

Take-all (*Gaeumannomyces graminis*)

Almost immune to 'Take-all' that affects wheat and barley and often used as a break crop where appropriate. However it is susceptible to a distinct oat strain occurring in several western and northern parts of the UK.

Control measures for oat strain of Take-all are similar to those used for wheat and barley.

Ear diseases

Oat smuts (*Ustilago hordei and Ustilago avenae*)

Similar to those on wheat and barley. Rarely economic on the oat crop. For more details *see* under 'Wheat'. Most seed treatments are effective including organomercurials which are likely to be the cheapest available.

DISEASES OF RYE

See under 'Wheat diseases' for the following diseases of rye: powdery mildew, yellow and brown rust, eyespot, sharp eyespot, take-all, brown foot rot, bunt and sooty moulds.

See under 'Barley diseases' for leaf blotch of rye and barley.

DISEASES OF FORAGE MAIZE AND SWEET CORN

Stalk rot (*Gibberella/Fusarium* spp.)

Base of stalks rot, plant wilts and often lodge especially in July/Aug. Some strains may also attack wheat. Control includes the use of resistant varieties found in NIAB Farmers Leaflet No. 7 and prevention of lodging such as avoiding windy exposed sites and late harvesting. No chemical treatments are available.

Take-all (*Gaeumannomyces graminis*)

Roots attacked mainly but seldom damaging in UK. May act as an alternative host of the disease for following crops of wheat or barley. No recommendations for control of the disease on maize. Further details *see* under 'Wheat'.

Maize smut (*Ustilago maydis*)

Typically causes black galls on the cob or elsewhere. Spectacular in appearance but of little economic importance at present in the UK. Soil- as well as seed-borne survival. No chemical control measures can be recommended. Growers should avoid maize growing more than one year in four and avoid known infected sites.

Damping-off (*Pythium* spp.)

Poor emergence and slow early growth especially under cold wet conditions. Control measures include the use of treated seed and avoidance of cold early sowing especially in poorly drained and/or low lying fields. Chemical seed treatment with thiram is recommended.

DISEASES OF POTATO

Leaf and stem diseases

Potato blight (*Phytophthora infestans*)

Grey-brown leaf and stem lesions spreading rapidly to kill haulm and infect tubers. Very common and serious on maincrop potatoes after long periods of wet windy weather in July, August and September. Very common in high rainfall areas and the main potato growing regions of UK. Yield and quality losses with first and second earlies may be slight but the plants may act as a source of infection for nearby maincrops. Infected tubers may cause very high storage losses as a result of secondary bacterial infection.

Cultural control

(1) Use NIAB Farmers leaflet No. 3 for variety choice.
(2) Use clean pre-sprouted seed tubers.
(3) Avoid late sowings.
(4) Avoid close proximity of maincrops with earlies or previous potato fields.
(5) Maintain stable and well earthed up ridges to reduce tuber blight.
(6) Destroy groundkeepers and potatoes on dumps and in clamps.
(7) Isolate potato dumps from growing areas.
(8) Defoliate and harvest early.

Chemical control

Foliar treatment
    Protectant fungicides
(1) Start spraying maincrops approximately every 10 d just before haulm meets in rows or at first sign of disease if earlier.
(2) Vary spraying interval rate and chemical used depending upon incidence of wet weather and age of crop.

Young plants
    captafol
    chlorothalonil
    cufraneb
    mancozeb
    maneb
    'Polyram'
    propineb
    zineb

Mature plants and tuber blight control
    captafol
    chlorothalonil
    fentin acetate + maneb
    fentin hydroxide

If an acceptable yield has been formed stop spraying and kill haulm where 5% disease level occurs on varieties susceptible to tuber blight. Haulm destruction may be delayed in those varieties showing good tuber resistance. If the crop is required for long-term storage do not lift for at least 10 d after the haulm is completely dead.

Potato leaf roll virus (main aphid vector—peach potato aphid—*Myzus persicae*)

Symptoms vary with variety of potato, strain of virus and time of infection. Plants stunted, leaflets with margins rolled upwards and inwards often with purple tinges. Foliage is hard and leathery; tubers are smaller, fewer in number and yield is much reduced. Mother tuber seldom rots during season and remains hard till harvest. Infection after flowering produces no visible symptoms till after planting next year. Very common in southern England with home saved seed.

Cultural control

(1) Use certified virus free seed.
(2) Use NIAB Farmers leaflet No. 3 for variety choice.
(3) Isolate new seed from home saved seed.
(4) Isolate potato dumps and clamps from growing fields.
(5) Destroy groundkeepers and potatoes on dumps.
(6) Isolate growing fields from allotments, market gardens and housing areas.
(7) Rogue virus infected plants early in season.

Chemical control

Prevent aphid infestations at all times in chitting houses, stores, potato fields and dumps using appropriate aphicides. Spread in field can be drastically reduced by efficient use of aphicides (*see* under Chapter 10).

Potato virus Y (severe mosaic and leaf drop streak—main aphid vector—Myzus persicae)

Symptoms vary with variety of potato, strain of virus and the time of infection. Plants stunted, leaves small, crinkled, showing mosaic, mottling and necrosis. Some varieties show leaf drop streak symptoms with certain strains of Y. Mosaic symptoms usually more severe than Virus X.

Cultural control

As for leaf roll virus.

Chemical control

Prevent aphid infestations at all times in chitting houses, stores, potato fields and clamps. Spread in the field is *not* easily reduced by aphicides due to non-persistent nature of the virus in the aphid.

Potato virus X (Mild mosaic)

Typically symptomless or a very mild mosaic. No reduction in size of leaflets and yield losses usually slight but may accentuate effects of other viruses. No vector involved, mechanically transmitted by machinery, footwear, clothing and general physical contact of contaminated and healthy plants. Physical method of spread coupled with general lack of symptoms makes roguing difficult and spread can be rapid in susceptible crops. No visible symptoms in tubers.

NB   Several other viruses of generally minor importance can also cause mild mosaic symptoms. Best controlled by the use of certified virus free seed.

Cultural control

(1) Use certified virus free seed.
(2) Use NIAB Farmers leaflet No. 3 for variety choice.
(3) Isolate new seed stocks from home saved seed.
(4) Destroy groundkeepers.

Chemical control

None.

Black leg (Erwinia carotovora var. atroseptica)

Soft black bacterial decay on lower parts of stem and heel end of tuber. Plants may wilt and die and stems are easily pulled out. Storage losses greater than in field. More common under wet conditions in field and store. Mainly seed-borne.

Cultural control

(1) Certified virus-free seed is generally safer than home saved seed.
(2) Avoid cutting or damaging seed planting.
(3) Avoid poorly drained fields.
(4) Avoid putting wet tubers in store.
(5) Lift early under clean dry conditions thus reducing tuber damage.

Chemical control

None

Spraing (Tobacco rattle virus—vectors—nematodes)

Leaves show yellow mottle, lines or rings with distorted leaf margins. Locally common on light dry sandy soils with high populations of free living nematodes. Tubers unmarketable due to brown concentric rings or arcs internally. Symptoms not usually visible till tuber cut. Yield reductions negligible compared with loss of tuber quality. Most of the progeny from spraing infected tubers are virus free.

Cultural control

(1) Use NIAB Farmers leaflet No. 3 for variety choice.
(2) Efficient weed control in preceding crops may be beneficial in lowering nematode populations.
(3) Avoid home saved seed in virus prone areas.

Chemical control

Use of one of the following nematicides may be useful on known infected sites:
    carbofuran
    oxamyl
    Also phorate when used for wireworm control.

Spraing (Potato mop-top virus—vector—powdery scab fungus)

Symptoms and control similar to spraing caused by tobacco rattle virus. The powdery scab fungus vector (*Spongospora subterranea*) is most common in the wetter west and northern parts of the UK. Mop-top spraing is not associated with any particular soil unlike tobacco rattle spraing. Mop-top virus is more readily transmitted in infected tubers than tobacco rattle so that they should not be used for seed purposes.

Tuber and storage diseases

Gangrene (Phoma exigua var. foveata)

An increasingly troublesome disease of storage maincrop and seed potatoes especially from northern Scotland and Northern Ireland. Dark coloured round or oval shallow depressions in tubers one to two months after lifting. Soil-borne fungus enters damaged areas at lifting especially under cold late conditions. Often secondary invasion by dry rot and black leg organisms occur also. Large dry hollow cavities usually form in the centre of the tuber in store. Well sprouted but infected tubers usually produce normal plants.

Cultural control

(1) Use of NIAB Farmers leaflet No. 3 for variety choice.
(2) Early lifting of maincrops (i.e. before 10 Oct).
(3) Avoid damage at lifting/riddling time.
(4) Curing period of 10 d at 13–16 °C after lifting and riddling will help wound healing.

Chemical control

(1) Fumigation with 2-aminobutane 14–21 d after lifting controls gangrene and skin spot.
(2) Thiabendazole dips, dusts and ULV mists at harvest or within three weeks after also controls gangrene, silver scurf and skin spot.

Dry rot (Fusarium solani—f.sp. caeruleum)

Soil- and seed-borne disease causing wrinkled concentric rings with pink, white or blue fungal pustules on tubers within one to two months after storage. Eventually tuber dries out, shrinks and mummifies. Bacterial wet roots may also invade under damp storage conditions. Less important than gangrene.

Cultural control

(1) Use of NIAB Farmers leaflet No. 3 for variety choice.
(2) Early lifting of maincrops (i.e. before 10 Oct).
(3) Avoid damage at lifting and riddling time.
(4) Curing period as for gangrene.

Chemical control

(1) Use of sprout suppressant tecnazene (TCNB) dust at lifting.
(2) Thiobendazole treatments as under gangrene.

Common scab (Streptomyces scabies)

Very common superficial skin blemishing disease especially on light/gravelly alkaline soils low in organic matter. Seed-/soil-borne disease affecting selling quality but not yield.

Cultural control

(1) Use of NIAB Farmers leaflet No. 3 for variety choice.
(2) Avoid liming potato or preceding crop.
(3) Avoid low organic matter soils.
(4) Irrigate crop especially during June.

Chemical control

No chemical control available.

Powdery scab (Spongospora subterranea)

Very uncommon except on wetter soils of the west and north. Most common in wet seasons and low lying areas of field. Long lived soil- and seed-borne disease.

Cultural control

(1) Use NIAB Farmers leaflet No. 3 for variety choice.

Chemical control

No chemical control available.

(2) Avoid use of known infected sites for at least five years after last potato crop.

Silver scurf (*Helminthosporium solani*)

Very common superficial skin blemishing disease affecting appearance but not yield. Silvery grey lesions develop during storage under high temperature and humidity conditions, causing loss of fresh weight in storage. Mostly seed-borne.

Cultural control

Avoid planting infected tubers.

Chemical control

(1) Treatment of seed tubers before planting with benomyl or thiabendazole.
(2) Post-harvest treatment with thiabendazole or benomyl mists and dusts.

Skin spot (*Polyscytalum pustulans*)

Superficial skin blemishing disease—common and important. Purple spots appear during storage and can affect crop emergence if used for seed. Most common on cold dry or heavy loam soils.

Cultural control

(1) Presprout seed before planting and discard affected tubers.
(2) Avoid late cold wet harvesting.
(3) Store tubers in boxes under dry ventilated conditions.

Chemical control

(1) Treatment of seed tubers before planting as for silver scurf.
(2) Fumigation of seed tuber after lifting with 2-aminobutane.

Pink rot and watery wound rot (*Phytophthora and Pythium* spp.)

Soil-borne diseases causing wilting in field and pink rot and/or rapid rot in store. Very sporadic occurrence. Locally common in hot summers on heavy badly drained soils. Commonly spreads in store.

Cultural control

(1) Avoid damaging tubers at lifting/riddling time especially under damp soil conditions.
(2) Extend period between potato crops to eight years or more and prevent infection of clean fields.
(3) Improve drainage.

Chemical control

No chemical control available.

Wart disease (*Synchytrium endobioticum*)

Large external warts and deformities on tubers and stolons. Now very rare due to Government legislation and use of field immune varieties. Common in Europe. Soil-borne disease. More common in wet north and west.

Cultural control

(1) Use of NIAB leaflet No. 3 for selection of immune varieties.
(2) Avoid spread on infected implements and dung from animals fed with infected tubers.
(3) Rest field for 30 years. Outbreaks must be reported to MAFF (notifiable disease).

Chemical control

Not permitted.

Black scurf and stem canker (*Thanatephorus cucumeris*)

Superficial black scurfy skin blemishes and brown or white girdling of stem bases. Black patches easily removed from skin. Young sprouts destroyed on seed tubers causing delayed emergence. Soft leaf rolling and wilting may occur (*cf.* virus leaf roll). Control includes the use of well-sprouted healthy seed. Avoid early/deep plantings in cold dry conditions and it is generally less severe if the crop is harvested early.

DISEASES OF BRASSICAS

Leaf and stem diseases

Powdery mildew (*Erysiphe cruciferarum*)

Silvery white patches on most brassicas causing eventual defoliation. Very common on swedes, turnips and brussels sprouts in dry summers. Air-borne disease.

Cultural control

(1) Use of NIAB Farmers leaflets Nos. 2, 6 and 9 and Vegetable Growers leaflets No. 1, 2 and 3 for variety choice.
(2) Delayed sowing of spring and summer sown brassicas may be beneficial.

Chemical control

Foliar treatment At first sign of disease use one of the following (on swedes):
  benomyl
  dinocap (brussels sprouts also)
  fluotrimazole
  tridemorph
  triadimefon

Canker (*Leptosphaeria maculans* /*Phoma lingam*)

Beige leaf spotting in autumn followed by stem cankers and lodging in spring and eventually pod and seed infection. Very important disease of oilseed rape and all brassica seed crops may be affected. Seed-, stubble- and debris-borne disease favoured by wet weather.

Cultural control

(1) Use of NIAB Farmers leaflets Nos. 2, 6 and 9 and Vegetable Growers leaflets No. 1, 2 and 3 for variety choice.
(2) Isolate crop from other brassica crops and previous oilseed rape fields.
(3) Use treated seed.
(4) Chop/burn and deeply plough oilseed rape stubble soon after harvest.
(5) Use at least four year break between brassica crops.

Chemical control

Seed treatment Use seed treated with one of the following:
  benomyl + thiram
  iprodione
  thiabendazole

Foliar applications If infection occurs in autumn or spring on foliage use of benomyl, carbendazim, iprodione or thiabendazole may be required.

Light leaf spot (*Pyrenopeziza brassicae/Cylindrosporium concentricum*)

Pale green/bleached areas on leaves and passing onto stems and inflorescences especially during wet weather. Common on most brassicas especially seed crops. Seed- and debris-borne.

Cultural control

(1) Isolate crops from other brassica crops and previous stubble.
(2) Use treated seed.
(3) Chop/burn and deeply plough stubble soon after harvest.
(4) Use at least four year break between brassica crops.

Chemical control

As for canker control on oilseed rape.

Dark leaf spot (*Alternaria* spp.)

Small dark leaf spots on foliage and seed pods especially during wet weather. Very common on all brassica seed crops and rape and stubble turnips in general. Seed- and debris-borne.

Cultural control

(1) Isolate crop from other brassica crops and previous stubble.
(2) Use treated seed.
(3) Chop/burn and deeply plough stubble soon after harvest.

Chemical control

Use of thiram seed soak treatment or dry seed treatment with iprodione.

Downy mildew (Peronospora parasitica)

Yellowing of lower leaves with white fungal growth on undersurfaces especially on autumn sown seedlings and young plants during winter months, and wet weather. Not usually important on mature plants or during summer period. Air-borne disease.

Cultural control

(1) Isolate crops from other brassica crops and previous stubble.
(2) Chop/burn and deeply plough stubble soon after harvest.
(3) Avoid very thick plant populations and low lying cold wet fields.

Chemical control

Not normally necessary but foliar applications on seedlings and young plants with dichlofluanid or zineb may be beneficial at times.

Cabbage ringspot (Mycosphaerella brassicola)

Brown/black concentric ringspots on foliage of brassicas especially cauliflowers, brussels sprouts, kale and cabbages in south west England during winter months. Seed- and stubble-borne.

Cultural control

(1) Isolate crop from all other brassica crops and previous stubble.
(2) Chop and deeply plough stubble soon after harvest.

Chemical control

No chemical treatment normally necessary. Seed infection is deep-seated requiring hot water treatment and thiram dressing.

Foliar treatment   Benomyl as required.

Cauliflower mosaic virus (aphid vectors—Myzus persicae and Brevicoryne brassicae)

Mottling and vein clearing of leaves, plants stunted and yields severely reduced. Very common on all brassicas especially cauliflowers where curd production is affected.

Cultural control

Isolate crops from all other brassica crops and previous stubble.

Chemical control

Control aphid populations at all times and delay virus infection to late in plant's life.

Foliar treatment   Use of one of the following aphicides at first sign of aphid attack will reduce spread:
demephion
demeton-S-methyl
dimethoate
formothion
malathion
mevinphos
pirimicarb
thiometon

Clubroot (Plasmodiophora brassicae)

Stunting of plants as a result of tumour growths on and later rotting of crop roots. All brassica crops may be affected but particularly severe on summer crops though rarely damaging on kale. Swedes and turnips on cold wet acid soils may be very severely affected. Soil-borne disease occurring in patches in fields. Resistant spores may survive for more than eight years between susceptible crops in affected soils. Potentially very serious on oil-seed rape crop.

Cultural control

(1) Use of NIAB Farmers leaflets Nos. 2, 6 and 9 and Vegetable Growers leaflets No. 1, 2 and 3, for variety choice.
(2) Avoid overcropping of brassicas (not more than one year in five as a preventative measure).
(3) Avoid transport of infected soil on boots, etc.
(4) Avoid feeding of infected roots on fields planned for future brassica production.
(5) Improved drainage and liming may reduce severity of attack.
(6) Use a break of at least eight years from brassicas after infection.

Chemical control

Seed treatment Sterilisation of seedbed for high value brassica crops with dazomet.

Transplant application Pre-plant dip treatments with one of the following:
   calomel
   benomyl
   carbendazim
   thiophanate-methyl

DISEASES OF SUGAR BEET, FODDER BEET, MANGOLD

Virus yellows (aphid vectors—peach-potato—Myzus persicae. Black bean aphid—Aphis fabae)

Leaves turn yellow and may be more susceptible to fungal attack. Yield and quality reduced depending upon earliness of infection. Aphids spread the two viruses concerned after feeding on infection plants. Beet yellow virus (BYV) carried only for a few hours in aphids but Beet Mild Yellowing Virus (BMYV) carried by aphids for most of its life. Reduce spread in spring to young crops from clamps, seed crops and any over-wintered beet, plants or remnants left in field. Many aphicides now less effective due to aphid resistance in south and east.

Cultural control

(1) Use of NIAB leaflets No. 5 and 6 for variety choice.
(2) Sow early.
(3) Avoid very low plant populations.
(4) Use aphicides and delay virus infection to as late as possible in the season.
(5) Avoid close proximity of root crops to steckling beds and previous crops.
(6) Inspect crops regularly and spray if one plant in four has aphids.

Chemical control

Seed-furrow treatment Using one of the following:
   aldicab
   carbofuran
   oxamyl
   thiofanox
followed by

Foliar treatments of
   acephate
   demephion
   demeton-S-methyl
   ethiofencarb
   pirimicarb

Downy mildew (Peronospora farinosa)

Lower leaves yellow with white mycelium underneath especially during winter six months on seed crops and early sown root crops in spring. Root yields and juice quality can be seriously affected in wetter seasons and low lying areas.

Cultural control

(1) Use of NIAB Farmers leaflets No. 5 and 6 for variety choice.
(2) Avoid close proximity of seed and root crops.
(3) Avoid early sowing of root crops on heavy soils.

Chemical control

Foliar treatment Spray autumn and spring seedling plants with one of the following:
   copper oxychloride
   maneb
   zineb

Powdery mildew (Erysiphe spp.)

Powdery white mycelium on foliage especially during dry weather in late summer. Sugar yield may be significantly reduced in south and east. Spray foliage with wettable sulphur at first sign of disease in summer.

Blackleg (Pleospora bjoerlingii)

Important fungus causing damping off and poor seedling growth leading to low plant populations. Seed-borne disease more important in crops drilled to a stand. Use seed treated with EMP (ethylmercury phosphate).

DISEASES OF PEAS AND BEANS (FIELD, BROAD/DWARF, NAVY)

Damping off, foot rot and seed-borne diseases (Pythium, Phytophthora, Ascochyta and Colletotrichum spp. but not halo blight)

Poor germination and early growth.

Cultural control

(1) Use disease free or treated seed.
(2) Avoid sowing into cold wet soils.
(3) Chop/burn and preferably plough stubble soon after harvest.
(4) Avoid poorly drained or low lying fields.
(5) Use appropriate fungicides on growing crops.

Chemical control

Seed treatment
Peas
  benomyl
  drazoxolon
  thiram
Field/broad beans
  benomyl
  drazoxolon
  thiram
Dwarf beans
  benomyl
  drazoxolon

Leaf, stem and pod spots (Ascochyta pisi—peas. Ascochyta fabae—field and broad beans. Botrytis fabae—chocolate spot of field and broad beans. Colletotrichum lindemuthianum—anthracnose of dwarf/navy beans)

Various types and sizes of brown/grey coloured spots of leaves, stems and pods. Cause partial defoliation, collapse of stems and discoloration of pods and seeds especially during wet periods in May/June, July. Chocolate spot is most severe on winter beans; seed- and stubble-borne.

Cultural control

As above for seed-borne diseases.

Chemical control

Foliar treatment   At first sign of disease in spring spray with benomyl repeated three weeks later if necessary. OR a single application at flowering may be sufficient.

Pea wilt (Fusarium oxysporum)

Rapid wilting of plants within patches in fields during late May and June. Foliage turns grey then finally goes yellow starting at base of plant then upwards. Very persistent soil-borne disease severely affecting yields in the affected patches. May be seed-borne. Mostly controlled by rotation and resistant varieties.

Cultural control

(1) Use disease free or treated seed.
(2) Use NIAB Pea leaflet for variety choice.
(3) Use at least a four year break after a *healthy* crop of peas *or* beans. Longer if unhealthy.

Chemical control

No chemical control available.

Halo blight (dwarf, navy beans) (Pseudomonas phaseolicola)

Small lesions surrounded by a yellow halo and greasy spots on pods. A bacterial disease favoured by wet windy weather causing reduction of pod quality in the main. Seed- and stubble-borne.

Cultural control

(1) Use disease free seed.
(2) Isolate crops from other dwarf bean crops and previous stubble.
(3) Chop/burn and deeply plough stubble soon after harvest.
(4) Avoid poorly drained or low lying fields.

Chemical control

Foliar treatment   If occurrence is likely spray every 10–14 d from emergence to pod set with copper oxychloride.

Pod rot of beans (Botrytis cinerea)

Lesions with grey fungal growth on pods especially those damaged or in contact with soil. Very common and severe in wet seasons causing serious reduction in quality.

Cultural control

(1) Chop/burn and deeply plough stubble soon after harvest.
(2) Avoid poorly drained or low lying fields.

Chemical control

Foliar treatment   Spray at flowering with benomyl, carbendazim or thiophanate-methyl to protect pods.

DISEASES OF CARROTS

Damping-off and leaf blight (Alternaria dauci)

Causes damping-off and leaf blight on wet soils and during wet seasons. Seed- and soil-borne.

Cultural control

(1) Use disease free or treated seed.
(2) Avoid poorly drained or low lying cold fields.
(3) Avoid early sowings.

Chemical control

Seed treatment   Thiram seed soak for 24 h.

Violet root rot (Helicobasidium purpureum)

Roots covered with purple fungal growth at lifting time especially on cold poorly drained soils. Whole host range including sugar beet, beetroot, parsnips, potatoes and weeks such as docks and dandelions. Brassicas are resistant. Soil- and stubble-borne disease. Yield reduced in store and quality markedly reduced.

Cultural control

(1) Avoid susceptible root crops being grown more than one year in five.
(2) Chop and deeply plough crop debris soon after harvest.
(3) Practice good weed control.

Chemical control

No chemical control available.

Black rot (Stemphylium radicinum)

Large black sunken lesions on mature roots and rotting in store. Also considerable seed losses can occur in seed crops and damping-off of seedlings in the field. Seed-borne disease mainly.

Cultural control

(1) Use disease free or treated seed.
(2) Chop and deeply plough crop debris soon after harvest.

Chemical control

Seed treatment   Thiram seed soak for 24 h.

Carrot motley dwarf virus (vector—willow carrot aphid)

Yellow mottling of leaves and stunting. Also affects parsnips and celery.

Cultural control

(1) Isolate carrots field from other aphid hosts and previous cropped fields.
(2) Large fields are less prone to overall attack than small fields.
(3) Reduce aphid populations at all times.

Chemical control

Seedbed treatment   Disulfoton or phorate granules at crop emergence.

Foliar application   At first sign of aphids use one of the following:
    demephion
    demeton-S-methyl
    dimethoate
    oxydemeton-methyl
    thiometon

Storage rots (Sclerotinia sclerotiorum, Botrytis cineria)

Grey and white fungal mycelium on roots in store especially when damaged and stored under damp conditions.

Cultural control

(1) Avoid late lifting under cold wet conditions.
(2) Avoid damaging roots at harvesting.
(3) Provide adequate ventilation to keep roots dry and cool in storage.

Chemical control

Use of benomyl on roots before storage.

DISEASES OF ONIONS

Downy mildew (Peronospora destructor)

Pale oval lesions on leaves and die back on tips. Spreads extensively within field under cool wet conditions. Fungus overwinters in bulbs and in soil for many years, causing further infection. Common on autumn sown onions during winter months. Attacks all onions and shallots.

Cultural control

(1) Isolate crops from other onion crops and previous onion stubble.
(2) Avoid using known infected fields for at least five years.
(3) Avoid low lying cold or poorly drained sites.
(4) Practice good weed control to help air circulation within crop.

Chemical control

Foliar treatment   Apply zineb at first sign of infection and repeat every 14 d.

Smut (Urocystis cepulae)

Leaves blister and rot to release black powdery fungal spore-masses. Can be very common and serious on seedlings and young plants of onions, leeks, shallots, chives and garlic. Fungus penetrates in seedling stage only. Therefore if raised in disease free seedbed transplants cannot become infected later. Fungal spores survive in soil for at least ten years.

Cultural control

(1) Avoid contaminating soil with infected debris, or soil on machinery, boots, wheels, etc.
(2) Burn all infected plants immediately. Do *not* bury in soil.
(3) Encourage fast germination and early growth. Avoid sowing early in cold wet soils.
(4) Avoid using infected fields for at least ten years.

Chemical control

Seed treatment

(1) Either for each sowing treat seed with thiram + methyl cellulose sticker for dry powder application,
(2) or for each sowing treat seed furrow with 40% formalin solution while drilling.

Thiram treatment more suitable under high rainfall conditions at sowing and/or lightly infected soils.

White rot (Sclerotium cepivorum)

Plants yellow, stunted/wilted with rotten base and covered with white fungal growth. Soil-borne fungus surviving in soil for many years.

Cultural control

(1) Avoid contaminating soil with infected debris or soil on machinery, boots, wheels, etc.
(2) Burn all infected plants immediately. Do *not* bury in soil.
(3) Avoid any infected soils for at least eight years.

Chemical control

Seed treatment   Use of one of the following will give good control:
  benomyl
  calomel
  carbendazim
  iprodione
  thiophonate-methyl

Foliar application Spray overwintered crops in March, or spring sown 14 d after emergence with iprodione. Followed by further applications as necessary at three week intervals.

Neck and storage rots (Botrytis allii)

Onions soften and rot internally while in store. Discoloration and rotting of neck occurs after several weeks in store. Very common in December and January. Mostly seed-borne but crop debris and onion dumps can be an important source of infection. Infected seedlings and plants appear healthy in field.

Cultural control

(1) Use disease free or treated seed. Care is needed in using home saved seed.
(2) Deeply bury crop debris soon after harvest.
(3) Completely cover old onion dumps with soil.
(4) Avoid damage at harvest and ensure adequate curing/drying occurs.

Chemical control

Seed treatment Dry or slurry applications of benomyl + thiram.

Foliar application Several applications of benomyl to the plant foliage at three to four week intervals may be beneficial particularly to machine harvested crops during wet seasons.

DISEASES OF GRASSES AND HERBAGE LEGUMES

Crown rust of rye grasses (Puccinia coronata)

Crown rust causes orange fungal pustules on the leaves of rye grasses, fescues and cultivated oats, but cross-infection is unlikely. Most common on late summer/autumn silage cuts of rye grasses (especially IRG). Associated with hot weather and cool dewy nights coupled with low nitrogen application. Reduces yield, palatability and digestibility of forage. Disease is more severe in a pure stand of rye grass cultivar especially early heading types. Very common in south and west England. Less frequent in north. Air-borne disease surviving on established leys.

Cultural control

(1) Use NIAB Farmers leaflet No. 16 for variety choice.
(2) Increase defoliation by cutting or grazing more frequently.
(3) Change management to all grazing in late summer through to early winter.
(4) Increase nitrogen levels to 250 kg/ha or more.

Chemical control

No chemical control economically feasible at present but the following will give good control:
  benodanil
  nickel sulphate + maneb
  oxycarboxin
  triadimefon

Leaf blotch of ryegrasses (Rhynchosporium secalis and Rhynchosporium orthosporum)

Leaf blotch causes dark brown blotches with light centres on barley, cocksfoot, couch, timothy and rye grasses (especially IRG). Cross-infection is possible but generally restricted. Most common under cool moist conditions on spring IRG silage crops. Yield, palatability and digestibility are reduced. Air-borne disease surviving on established plants.

Cultural control

(1) Use NIAB Farmers leaflet No. 16 for variety choice.
(2) Increase defoliation by cutting or grazing more frequently.
(3) Change management to all grazing in early spring and summer months.

Chemical control

No chemical control economically feasible at present but the following chemicals should give control:
  benomyl
  carbendazim
  thiophanate-methyl
  triadimefon

Powdery mildew (*Erysiphe graminis*)

Grey/white superficial fungal growth on leaf surface. Attacks a wide range of grasses/cereals but cross-infection unlikely. Common during and after dry periods. More conspicuous under high soil nitrogen conditions. Reduces yield, palatability and digestibility but generally less damaging than crown rust and leaf blotch. Air-borne disease surviving on established plants.

Cultural control

(1) Use NIAB Farmers leaflet No. 16 for variety choice.
(2) Increase defoliation by cutting or grazing more frequently.

Chemical control

No chemical control economically feasible at present but benomyl, tridemorph or triadimefon should give control.

Barley yellow dwarf virus

Attacks a wide range of grasses mostly without showing symptoms. Very important disease of cereals and although grasses can be severely affected main importance in ryegrasses is as a reservoir of infection for neighbouring cereals. The virus is transmitted by various cereal/grasses aphid species.

Cultural control

(1) Isolate cereal and grass crops from each other, particularly early sown winter barley.
(2) Destroy cereal stubble and volunteer as soon as possible after harvest.
(3) Avoid early sown winter cereals.

Chemical control

Chemical control unlikely to be economically feasible on grass crops alone. Aphicides may be worthwhile on undersown cereals and grass seed crops.

Rye grass mosaic virus (Mite vector—*Abacus hystix*)

Causes mottling and streaking in mild strains and a dark brown leaf necrosis with severe strains. Disease is widespread and severe in the south in seed crops and conservation and grazed leys (especially IRG). Undersown spring sown crops are more severely affected than straight autumn grass reseeds. Yield, palatability and digestibility can all be seriously affected.

Control

Hard autumn grazing reduces mite populations and the following spring virus infection. Chemical control not economically feasible.

Clover rots (*Sclerotinia trifoliorum*)

Very serious disease affecting trefoil, lucerne, sainfoin, white and particularly red clover. Soil-borne disease killing large patches of plants in high rainfall areas during autumn and winter months. Often damaging on first year lucerne crops but once established lucerne is generally resistant. Responsible for clover sickness with continuous clover cropping.

Cultural control

(1) Use NIAB Farmers leaflet No. 4 for variety choice.
(2) Use healthy seed free from fungal sclerotia.
(3) Maintain sward in short condition during autumn but do not poach in winter months.
(4) Use at least five year break between susceptible crops.

Chemical control

Foliar treatment. Fungicide treatments using benomyl and quintozene have been partially effective but remain uneconomic.

Verticillium wilt (*Verticillium albo-atrum* and *Verticillium dahliae*)

Yellowing and wilting of lucerne followed by poor re-growth, stunting and finally death of large patches in crop. Plants shrivel from the base upwards particularly in late summer and autumn. Very serious disease of lucerne after three to four years' cropping on infected soil. Seed- and soil-borne surviving for long periods in soil.

Cultural control

(1) Use NIAB Farmers leaflet No. 4 for variety choice.
(2) Restrict lucerne crops to maximum of three years' duration.
(3) Cut/graze healthy crops first.
(4) Use as long a break as possible on infected sites.

Chemical control

Seed treatment
(1) Thiram seed treatment to prevent disease introduction on new sites.
(2) No foliar treatment possible.

NB   While every precaution has been taken to ensure the correctness of this chapter, the author and publishers cannot accept any responsibility whatsoever for losses arising from using chemicals mentioned herein. Selection for field use should be made from *Approved Products for Farmers and Growers*, published annually by Ministry of Agriculture, Fisheries and Food (Agricultural Chemicals Approval Scheme), obtainable from HM Stationery Office or through any bookseller. *Manufacturers' literature and labels must always be consulted before use.* Detailed recommendations are also given in *The Pest and Disease Control Handbook*, edited by Nigel Scopes assisted by Michael Ledieu (1979) BCPC Publications, Croydon and various ADAS advisory leaflets covering specific diseases.

10

Pests of crops

D. J. Iley

Farm crops are subject throughout their growth to attack by pests belonging to various animal groups, the most important being insects, mites, nematodes, slugs, birds and mammals. Different pest species vary in their mobility, host-specificity, period of peak abundance, regularity of occurrence and response to climatic conditions.

It is well known that problems such as the development of resistance in pests, the killing of beneficial organisms and the entry of persistent chemicals into food chains have resulted from the use of pesticides. Because of these problems, and the increasing cost of chemical control, a great deal of effort is now being directed towards the development of techniques aimed at improving the effectiveness of pesticide usage so that smaller quantities may be used in a more selective manner. Examples of such techniques are:

(1) establishing economic thresholds so that growers can be advised more precisely on pest population levels at which control is justified;
(2) forecasting the abundance of pests so that growers can be advised whether or not preventive control measures are necessary;
(3) employing biological and ecological information so that other methods of control can be used to replace or enhance chemical control;
(4) employing equipment which is capable of applying pesticides accurately in smaller quantities.

An increasing body of information and advice based on these principles is now becoming available.

The growing of 'new' crops such as oilseed rape and field grown vegetables and the development of techniques of crop production such as direct drilling and sowing to a stand have created new pest problems.

In addition, there may be legal constraints or recommended codes of practice associated with the control of a pest or the use of a pesticide. All the factors mentioned above should be considered when contemplating prevention or control of a pest and additional advice should be sought, if necessary, from appropriate sources such as ADAS, growers' organisations, technical representatives or agricultural consultants.

The pests are dealt with under 'host-crop' headings but it must be remembered that many of the general feeders are associated with a wide range of crops. It is not possible to include every pest which might injure a crop and those covered have been selected somewhat arbitrarily on the basis of their actual or potential economic importance or on their frequency of occurrence. Chemicals named are given as examples. The omission of an insecticide, nematicide, acaricide, molluscicide or other crop protection chemical does not necessarily imply ineffectiveness. The MAFF book *Approved Products for Farmers and Growers*, published in February every year gives complete lists of officially approved products, together with much useful information on their use.

Note: The information on each pest is organised, with variations where appropriate, as follows.

The first entry gives relevant information on the life cycle and biology of the species. Sections headed

A. deal with circumstances in which the pest is most likely to be a problem.
B. deal with recognition and injurious effects.
C1. deal with non-chemical (cultural) control methods if applicable.
C2. deal with chemical control methods and ways of ascertaining when control is justified.

CEREAL PESTS

Many pests whose normal hosts are grasses infest cereals either directly via mobile winged adults or indirectly when the feeding stages migrate from ploughed or desiccated grass to a following cereal crop. Directly drilled cereals are especially prone to the latter.

Wheat bulb fly (*Delia coarctata*)

A serious pest of winter wheat in eastern Britain; winter barley and rye and early sown spring wheat may also be attacked. Adult fly lays eggs in cracks and crevices in bare soil in July–September, eggs hatch and larvae invade susceptible cereal plants in following January–March; hatching delayed by cold conditions. Larvae feed inside base of main shoot or tillers and migrate, as they grow, to infest more shoots.

A. Host crops drilled in fields which were fallow or sparsely covered in previous July, e.g. roots. Late sown winter wheat, untillered, is especially susceptible. ADAS forecasts, based on egg counts of soil from sample fields, are issued before winter sowing.
B. Patches of dead or damaged plants in January–March, expanding in April–May. Damage shows as 'deadhearts', centre leaf dies and turns yellow, outer leaves remain green. The larva, a typical legless fly maggot, is revealed on peeling outer leaves away from base of shoot.
C1. Sow winter wheat before end of October at a shallow depth. Sow spring wheat and barley after mid-March.
C2. (1) Seed dressings of gamma HCH, carbophenothion or chlorfenvinphos; the latter two must not be used on seed sown after 31 December because of risk to wild birds. Control less effective when seed is deep drilled.
    (2) Fonofos granules or chlorfenvinphos sprays mixed into top 50 mm of soil at, or immediately before, drilling give better protection but cost more; use when ADAS forecast indicates a high risk.
    (3) Sprays of chlorpyriphos, chlorfenvinphos or pirimiphos methyl applied at egg hatch where soil conditions permit; advice on timing from local ADAS.
    (4) Where earlier treatments have been omitted, or have failed to stem an attack, spray with dimethoate, formothion or omethoate if damage is observed; not later than mid-March.

Frit fly (*Oscinella frit*)

Grasses, especially ryegrass, are the natural host plants of the frit fly whose larvae feed inside the bases of the shoots. There are three generations, sometimes four in southern counties, of flies in a year and larvae overwinter in the shoots of their host plants. Cereals may be attacked in two different ways.

Migration from sward

A. Autumn sown wheat, barley, oats and rye drilled over ploughed grass or grassy stubbles, or directly drilled into desiccated swards may be invaded by larvae migrating from the decomposing grass.
B. Slight angular bend above coleoptile where larva has penetrated at single shoot stage of growth, followed by 'deadheart', i.e. central leaf dies and turns yellow. Legless, transparent larvae 2–5 mm long, inside base of shoot. Young plants usually killed.
C1. Attack greatly reduced by leaving an interval of at least four weeks between ploughing/desiccating and drilling. Late sown crops suffer less because lower soil temperatures reduce migration.
C2. Spray with omethoate, triazophos or chlorpyriphos immediately at first sign of attack, or at crop emergence on advice from local ADAS.

Direct oviposition

Female flies lay eggs directly on spring oat and maize plants, the eggs hatch after a few days and the larvae enter the shoots.

A. Later sowings of spring oats are at greater risk of attack by first generation flies because the latter are attracted only to plants or tillers with fewer than five leaves. Maize is often at risk because of its late sowing date but attacks are unpredictable.
B. First generation attack results in 'deadhearts' and death of young oats and maize plants or excessive tillering of older plants. Second generation flies lay eggs in spring oat spikelets, the larval feeding producing blindness or shrivelled and damaged grain.
C1. Sow spring oats before mid-March in south, late March in Wales and north Britain.
C2. For spring oats, spray with chlorpyriphos if crop has not reached the four leaf stage by first half of May in south, late May in Wales and north. Or apply triazophos immediately damage is observed. Economic yield response likely to be small unless attack is severe. For maize, granules of phorate or carbofuran in the seed row as a precaution. Or spray with chlorpyriphos, fenitrothion, pirimiphos methyl or triazophos at crop emergence.

Wireworms (*Agriotes* spp.)

The larvae of click-bettles; natural habitat is the soil under permanent grass where they feed on the underground parts of plants.

A. Only likely to be a problem in the first or second year after permanent grass. All cereals are affected but wheat and oats are most susceptible. Soil can be sampled by ADAS prior to cultivation; treatment depends on numbers present.

B. Main damage occurs in autumn and spring when wireworms feed on underground stems and hypocotyls of cereals. Seedlings may be completely severed and turn yellow and die while still remaining upright in the ground. Several plants in a row may show progressive symptoms. Damage often occurs in patches where other factors are contributing to poor growth, e.g. disease, poor drainage. Careful excavation of soil around injured plants reveals stiff, smooth, yellow larvae up to 20 mm long.

C1. Increase seed rate in high risk situations. Assist recovery by rolling and nitrogen top dressing if soil conditions are suitable.

C2. Seed treatments based on gamma HCH are cheap and are recommended when wireworm population is less than 1.25 million/ha. To reduce the risk of plant injury the seed should be sown soon after treatment, moisture content should not exceed 16% and the dressing should be applied uniformly at the prescribed rate. For higher populations apply gamma HCH* as a spray to the soil when the seedbed is being prepared.

Leatherjackets (*Tipula* spp.)

The larvae of daddy longlegs or craneflies. The flies lay eggs in grassland mainly in September. The eggs hatch in 10–14 d and the leatherjackets feed in the soil on the underground parts of plants but will come to the surface in dull, moist and relatively warm conditions to feed at ground level. Activity is reduced in cold and dry conditions.

A. Cereals following grass or grassy stubbles are at risk especially when preceding September–October has been wet. All cereals suffer; winter cereals may be attacked in early winter but the greatest damage normally occurs on spring cereals in April and May. ADAS monitors populations in most regions and forecasts are issued or are available on request. Rooks are attracted to cultivated fields with large numbers of leatherjackets and may be seen searching for them by turning over surface clods.

B. Young plants may be chewed below or at ground level and spring cereals in particular may be completely severed or roughly grazed down.

Leaves with ragged holes. Damage often in patches coinciding with wet areas of fields. Confirm by examining soil in vicinity of affected plants for presence of legless, grey-brown, fleshy grubs up to 50 mm long, with tough, wrinkled skins and a number of small, pointed protuberances at tail end.

C1. Plough or desiccate grassland before September; kill grassy stubbles immediately after harvest. Rolling and top dressing with nitrogen when soil conditions permit will encourage the crop to grow away from an attack.

C2. Examine crop daily for signs of attack in risk situations because damage can occur very rapidly and insecticide must be applied immediately to avoid excessive loss. Chemical control is justified when a *total* of 15 or more leatherjackets are found on examining ten separate 30 cm lengths of row selected at random diagonally across the field. This figure applies when the rows are 17.5 cm apart, at narrower spacings a total of ten or more is the critical level. The soil within 3–4 cm of the plants should be carefully searched.

Insecticides may be applied in these circumstances as sprays, granules or poison baits, the latter normally giving best results. In all cases the application is best made when the leatherjackets are active on the surface, i.e. in the evening in damp, warm weather.

Poison baits

Mix gamma HCH*, fenitrothion or DDT with bran moistened to a crumbly consistency. Thorough mixing to ensure even distribution is essential, and the bait should then be distributed as evenly as possible over the field. Bran plus gamma HCH pellets are available commercially.

Sprays

Gamma HCH*, chlorpyriphos, triazophos and DDT†. These may also be applied during the preparation of the seedbed when leatherjacket populations are known to be high.

Granules

Apply chlorpyriphos using an applicator capable of distributing small quantities/ha.

---

* Gamma HCH should not be used if potatoes or carrots are to be planted within 18 months because of risk of taint.
† Rye and some barley varieties are damaged by DDT sprays (not baits)—consult ADAS, NIAB lists, or latest *Approved Products for Farmers and Growers* for susceptible varieties.

Field slug (*Deroceras reticulatum*)

This is the most important of several species of slugs which may feed on cereals, and is the only one which regularly feeds above or close to the soil surface. Slug numbers vary considerably according to soil type, previous cropping and climate; their activity is also affected by temperature and humidity but, in general, their numbers are greatest in autumn and late spring.

A. Autumn sown cereals, especially wheat, are most likely to suffer economic loss; damage may be seen on spring cereals but this is not normally important. Populations are highest in undisturbed soil, especially if it has good moisture retaining properties and where there is dense ground cover. Direct drilled cereals or cereals following grass, oilseed rape, peas, beans and other crops with bulky residues on silt or clay soils are most at risk.
B. Winter wheat and rye may be severely damaged by slugs feeding on the germ of the seed shortly after drilling, giving the impression of total or partial seed failure. Young shoots of all cereals may be eaten as they germinate, or grazed at ground level and their growing points destroyed. The leaves of older plants may be shredded longitudinally but this damage, though conspicuous, is not normally of great importance.
C1. Prepare a fine, well consolidated seed bed which restricts movement of slugs through the soil.
C2. Test bait prior to sowing by placing small handfuls, about ten per field, of slug pellets protected by slates, fertiliser bags, etc. on the ground. These should be examined every 3 d and when an average of four or more slugs per site is found regularly, treatment is recommended. To prevent seed attack on wheat or rye apply baits containing metaldehyde or methiocarb to the seedbed and leave it undisturbed for 3–4 d before drilling. Alternatively, 'mini' pellets can be mixed and drilled with the seed. After brairding, apply baits to soil surface when damage is seen to be increasing. Test and treatment baits are only likely to be effective when slugs are active on the surface, i.e. in moist, mild weather.

Cereal cyst nematode (*Heterodera avenae*)

Mainly a problem on spring oats. Winter oats, maize, wheat, barley and rye are progressively more resistant to damage. Present in small numbers in grassland which increase rapidly when cereals are grown frequently. Small, less than 1 mm long, lemon shaped cysts, the bodies of dead females, containing approximately 400 eggs when young may persist in the soil for many years their contents declining slowly in the absence of suitable hosts. Minute larvae emerge from the cysts in spring in response to substances exuding from growing cereal roots and invade and feed inside the rootlets. Females swell and form young cysts in July which may be seen attached to roots of affected plants, white at first and darkening as they age.

A. Mainly confined to chalky and light soils in southern England. Oats following several successive crops of wheat and/or barley are most at risk but other cereals may suffer when populations are high. Suspect fields can be sampled by ADAS to determine population levels.
B. Damage normally shows up as patches of pale, stunted plants whose roots are short and much branched in comparison with those of healthy plants.
C1. Reduce high populations by sowing a ley. Avoid growing oats after several successive crops of wheat or barley. Resistant varieties may be grown where cyst counts are high and are useful in reducing nematode populations; the spring barleys Tyra and Tintern, the winter oat Panema and the spring oat Trafalgar all exhibit resistance to the two principal pathotypes (races) of cereal cyst nematode.
C2. Chemical control is not economically justified at present.

Cereal aphids

Three species are important; like all aphids they feed by sucking plant sap causing direct damage to their host plants, they may also transmit cereal virus diseases. Correct identification, and careful observation of population trends wherever possible, is important because different species may require different treatment. Special care must be taken when deciding whether or not control measures are justified because spraying is costly in both economic and ecological terms on a crop which occupies such large areas of land.

Grain aphid (*Sitobion avenae*)

A. Mainly a problem on winter wheat causing direct damage, but may occur on all cereals. Winged females migrate from grasses in late May–early June in southern Britain, later in the north. Numbers build up in hot, humid conditions. Losses greatest when heavy infestations occur when flowering heads are developing.
B. Largish aphids, 2–3 mm long, colour varying between reddish brown and green, two black tubes (siphunculi) projecting from upper surface at rear of body. Early arrivals feed on leaves but transfer to flowering heads as latter emerge and feed on rachilla and developing grain; resulting

grain is light and shrivelled. This species may also transmit milder strains of barley yellow dwarf virus (BYDV) from grasses to cereals.

C1. No cultural control recommended.

C2. Spray winter wheat when, at the beginning of flowering (GS 10.5.1.), an average of five or more aphids per ear is found on at least 50 ears selected at random from all parts of the field except the headlands, and when weather is conducive to aphid build up. Chlorpyriphos, demeton-S-methyl, dimethoate, formothion, heptenphos, oxydemeton-methyl, phosalone, pirimicarb and thiometon are all effective and all are available in formulations cleared for aerial application. Local beekeepers should be warned of impending spraying and, where there is a risk to bees, pirimicarb should be the insecticide of choice.

Rose grain aphid (*Metapolophium dirhodum*)

A. Winged females disperse from brambles and wild roses to cereals and grasses late May–early June. Rapid build up in warm, humid conditions.

B. Light green aphids, 2.25–3.0 mm long, with darker green stripe down middle of back. Normally on undersurfaces of lower leaves but will colonise upper leaves and surfaces and stems as numbers build up. Only likely to cause economic loss when flag leaves are heavily infested.

C1. No cultural control recommended.

C2. In winter wheat and spring barley spray when average number of aphids per flag leaf exceeds 30 at any time between flowering and milky ripe stage (ZCK 75). Insecticides and precautions as for grain aphid.

Bird cherry aphid (*Rhopalosiphum padi*)

Mainly important as a transmitter of virulent strains of BYDV (*see* Chapter 9) from grasses to cereals and then within the cereal crop. Overwinters naturally as eggs on bird cherry but, in areas with a mild winter climate, can also overwinter as small colonies of wingless adults and juveniles on grasses and early sown winter cereals; may then build up and spread rapidly in the spring.

A. Main risk of virus transmission is in Wales and south west England on early sown, September–early October, winter cereals. Late sown spring cereals may also become infected by aphids migrating in May. Cereals following grass, in close proximity to grassland, and in fields with high hedges are more likely to suffer.

B. Small, 1.5–2.3 mm long, brown to greenish-brown aphids with rust red patches at rear of body; found on all parts of the plants. Cereals infected in the autumn do not normally show BYDV symptoms until the following spring when patches of affected plants can be seen. Infected plants in spring sown crops normally become conspicuous at ear emergence.

C1. Careful ploughing under of grass before drilling cereals in Wales and south west.

C2. Prevent spread of virus within early sown winter crops by spraying in November when 5% of barley or 10% of wheat plants are found to be infested. Local specialist advice should be sought on the need to treat spring cereals.

Wheat blossom midges (*Contarinia tritici* and *Sitodiplosis mosellana*)

A. Injurious attacks are sporadic and difficult to predict but widespread outbreaks have occurred in the north of England in recent years and chemical control has effected economic yield responses.

B. Lemon wheat blossom midge (*C. tritici*), so-called because of the colour of the eggs, larvae and adults. Eggs are laid in groups on the ears shortly after emergence and the larvae feed in the spikelets rendering them sterile. Up to ten larvae per spikelet may be present.

Orange wheat blossom midge (*S. mosellana*)—the orange red larvae normally feed singly on the milky grain which is light and shrivelled at harvest.

The small, 3 mm long, fragile adults of both species are best observed at dusk at the time of ear emergence.

C1. No cultural control.

C2. Ground or aerial sprays based on fenitrothion between ear emergence and flowering.

POTATO PESTS

Pests of potatoes fall into two main categories—those that reduce the productivity of the plants by direct feeding damage or by transmitting virus diseases and those that disfigure the tubers thereby reducing their market value. In both cases the most effective control measures are normally applied before or at the time of planting.

Peach potato aphid (*Myzus persicae*)

Overwinters as eggs on peach trees and as small colonies of adult and juvenile females on a wide range of host plants especially in mild winters and in protected situations such as glasshouses and chitting houses. Winged females arrive on potato plants in spring and feed by sucking the sap, colonies of wingless aphids build up rapidly in hot, dry conditions, movement within the crop is responsible for transmission of virus from infected plants. Mainly important as a vector of leaf roll virus and rugose mosaic (virus Y) into and within home grown 'seed' crops and in the chitting house. Early migration from overwintered colonies and rapid build up in early summer may create a virus problem in main crop potatoes in some years.

A. More likely to be a problem following a mild winter and early spring and when calm, warm weather prevails in summer.
B. Small scattered colonies of green to pinkish aphids, relatively inactive even when disturbed.
C1. Destroy volunteers, discards, clamp site debris and other potential sources of potato viruses. Plant certified healthy 'seed'. Grow 'seed' crops in isolated sites and rogue virus infected plants as soon as symptoms appear. Burn off foliage early.
C2. Routine chemical control is recommended only for home grown 'seed' production and is more effective against leaf roll virus which, in contrast to virus Y, is not transmitted instantly by aphids. Granular applications of insecticide to the seedbed or planting furrow will generally protect the crop until mid June, thereafter one or more sprays should be applied until haulm destruction. Alternatively, spray at 80% plant emergence and repeat at intervals. On main crops, spraying is normally justified when an average density of three to five aphids per leaf is observed in the first half of July. This will also control the potato aphid (*Macrosiphum euphorbiae*), a larger species which may form dense colonies in hot, dry summers causing direct damage and leaf curling in some varieties. If large numbers of aphid predators (ladybird and hoverfly larvae) and parasites (indicated by the presence of swollen, brown aphids on the leaves) are present do not spray.

Chemical control is complicated by the increasing prevalence of strains of aphid which are resistant to varying degrees to commonly used organophosphorous, carbamate and pyrethroid insecticides. This pattern of resistance is constantly changing and growers are advised to obtain local advice on the best choice of insecticide. In chitting houses, pirimicarb smokes may be used to kill aphids of several species, including *M. persicae*, on the sprouts.

Potato cyst nematodes (*Globodera rostochiensis* and *G. pallida*)

The most important pests of potatoes in the UK—*G. rostochiensis*, which forms yellow cysts on the roots is present in all potato growing districts; *G. pallida*, with white cysts, is more restricted in its known distribution but is probably present at low population densities even in areas where it is not considered to be a threat. Life cycles are similar to that of the cereal cyst nematode except that potatoes are the only important field hosts; tomatoes are also susceptible.

A. Yields are reduced when potatoes are grown in soil containing viable cysts, and cyst numbers increase rapidly when potatoes are grown frequently in infected fields. The number of eggs/g of soil can be estimated by ADAS and the information should then be used to plan a cropping and control programme in consultation with specialist advisers.
B. Infestations normally show up first as patches of stunted plants which wilt readily. Their root systems are shortened and much branched and small spherical cysts may be seen attached to the rootlets.
C1. The main objective of control is to contain the cyst/egg population to an acceptable level and is best achieved by combining lengthened rotations, resistant varieties and nematicide treatment. EEC regulations require that potatoes sold as 'seed' can only be grown in fields which prove to be eelworm free on sampling. Where initial egg counts are high potatoes should not be grown for several years; thereafter the rotation can be shortened. Destroy volunteer plants. The early variety Pentland Javelin and the maincrop varieties Maris Piper and Cara are resistant in the sense that the sex ratio of the invading nematodes is altered and few cysts are formed, but plants suffer some degree of injury as a result of the invasion. Four distinct pathotypes of *G. rostochiensis* and one of *G. pallida* have been identified in the UK and resistance extends only to pathotype RO1 of *G. rostochiensis*, which predominates in some of the major ware producing areas including the Isle of Ely. It is unfortunate that, when RO1 resistant varieties are grown frequently in these areas, populations of *G. pallida* will almost certainly increase. However, varieties resistant to *G. pallida* are undergoing selection but are not likely to be available until 1985–86.
C2. Incorporate granules of aldicarb, carbofuran or oxamyl in the seed bed. Rates of application may vary according to soil type and whether an early or maincrop variety is to be grown. Effectiveness depends on thorough mixing with the soil to the correct depth. Autumn application of dazomet, metham-sodium or dichloropropene may also be used, special applicators may be required.

Free living nematodes (*Trichodorus* and *Paratrichodorus* spp.)

Minute, less than 1.00 mm long, soil inhabiting nematodes. Migrate through the soil and feed externally, sucking the contents of surface cells of tubers and roots.

A. Associated with coarse, sandy soils.
B. Loss of yield negligible. Important as vectors of tobacco rattle virus (TRV) causing spraing which renders tubers unmarketable (*see* under potato disease, Chapter 9).
C1. The variety Record, recommended only for crisping, is resistant to spraing.
C2. Aldicarb granules thoroughly incorporated into the seedbed. Phorate granules used primarily for wireworm control also bring about some control.

Garden slug (*Arion hortensis*)

The principal slug species affecting potatoes, it lives almost entirely underground only coming to the surface in numbers during rainy weather in July and August. The keeled slug, *Milax budapestensis*, may also contribute to tuber damage.

A. Numerous in wet, heavy soils especially those with high organic matter content. Damage likely to be greatest during a wet autumn following a mild, wet summer; also in irrigated crops.
B. Slugs penetrate tubers through relatively small, rounded entrance holes and excavate wider tunnels and cavities thereby reducing the marketability of the crop. The garden slug is black with a yellow sole. Millipedes may be present as secondary feeders.
C1. Improve drainage. Harvest as soon as possible. Maris Piper and Cara are highly susceptible and should not be grown in slug risk situations. Pentland Dell and Pentland Ivory are among the least susceptible varieties.
C2. Surface baits containing methiocarb can be useful when applied in wet conditions in late July and August when the slugs come to the surface. Test baiting, as for slugs in cereals, can be used to determine the best time for treatment. In this case an average of one or more slugs per bait site is the criterion.

Wireworms (*Agriotes* spp.)

See wireworms in cereals p. 279. Relatively low populations can cause serious economic loss.

A. Numbers are likely to be high in the two years following permanent grass but damaging populations, above 75 000/ha, may persist even in arable land. Numbers are estimated by soil sampling. Wireworm activity increases in autumn. Unlikely to be a problem on early lifted varieties.
B. Small round holes in the tuber leading into tunnels of the same diameter, wireworms may be found on cutting tubers open. Entry of slugs, millipedes and other soil pests is encouraged. Loss in yield is negligible but crop may be rejected or market value greatly reduced.
C1. Lift maincrops as soon as possible after maturation.
C2. Work aldrin dust or spray into the soil prior to planting when population levels merit chemical control; higher rates are needed on peaty soils. Alternatively, apply phorate granules in the furrow at planting, this will also reduce spraing and early aphid attack. On no account should gamma HCH be used.

Cutworms

The caterpillars of a number of related species of nocturnal moths, which damage a wide range of crops. The most important species in recent years has been the turnip moth (*Agrotis segetum*). Eggs are laid on foliage and stems of many crop and weed plants from May–July. The small caterpillars hatch and feed on the leaves for two to three weeks, during which time they are highly susceptible to wet conditions, and then descend into the soil feeding just below ground level although they may come to the surface at night.

A. Prevalent when weather has been warm and dry during early feeding stages, especially on light, well drained soils and weedy fields. Irrigated crops are unlikely to suffer.
B. The fleshy, smooth, greenish-brown caterpillars, up to 5 cm long when fully grown, excavate irregular shallow pits in the surface of the tubers, especially those close to the surface.
C1. Control weeds.
C2. Spray with chlorpyriphos, triazophos or pyrethrins when the newly hatched caterpillars are feeding on foliage prior to descending into the soil.

SUGAR BEET PESTS

The seedling is the most vulnerable growth stage of any plant, a fact of particular relevance to the sugar beet crop which is now almost entirely sown to a stand with pelleted monogerm seed. This results in a low seedling population and pests feeding on seed or seedlings can damage or destroy a high proportion of the plants very quickly even though they themselves are present in relatively small numbers. Effective weed control increases the problem in the case of general feeders. It is essential, therefore, that protection should be applied in advance against predictable pests and that the crop should be examined every day from sowing until it is well established so that control measures can be applied in time to be effective.

Seed furrow treatments with granular pesticides, extensively used against aphids, will also control or reduce the injurious effects of most seedling pests. In general, they are preferable on ecological grounds to post-emergent spray treatments against seedling pests which may kill important predators and parasites of aphids.

Sugar beet pests also affect fodder beet and mangels but the cost of control on these crops may only be justified in exceptional circumstances.

Woodmouse (*Apodemus sylvaticus*)

The mice live in burrows in the open fields hence their alternative name of long-tailed field mice. They dig up and break open the husks and feed on ungerminated seed, working their way along rows. Each individual has a large foraging area. Populations vary enormously from year to year during the relatively short time that seed is at risk.

A. Damage is greatest on early sowings in dry soils and where some seed is exposed on the surface. BSC field staff monitor populations during the spring.

B. Soil disturbed along rows. Seed pellets fragmented.

C1. Avoid spillage and exposure of seed.

C2. Place proprietary mouse baits or breakback traps at a density of three to four per hectare in the open fields away from hedgerows as soon as damage is observed. Baits and traps should be placed in pipes or other suitable containers to protect non-target mammals and birds. Traps must be reset and baits replenished until the seed germinates.

Millipedes

Body composed of many similar segments, examination with a hand lens reveals the presence of two pairs of legs per segment. The spotted snake millipede (*Blaniulus guttulatus*), cylindrical in section with a line of reddish-brown spots along each side, is the most injurious but flat millipedes including *Polydesmus angustus* may also contribute to the damage.

A. Commoner in well structured and heavy soils, often recurring in fields where previous damage has been evident.

B. Seedling roots and stems tunnelled or rasped causing collapse of plants; large numbers may congregate around individual seedlings in May. Damage often patchy.

C1. No cultural control.

C2. Work gamma HCH into seedbed before sowing *or* apply aldicarb, oxamyl or carbofuran granules in seed furrow, the latter must not come into direct contact with the seed. Granule treatments give some control of seedling foliage pests. When damage is seen on seedlings application of gamma HCH in large volumes of water in bands along the rows gives some protection; potatoes and carrots should not be grown in soil treated with gamma HCH until at least 18 months have elapsed.

Symphylids (*Scutigerella* spp.)

Small, primitive arthropods with weak mouthparts. Found in similar situations to millipedes and springtails, especially in fissured soils. They become active as soils warm up in spring and feed on soft plant tissue. Control with gamma HCH or carbofuran as for millipedes.

Springtails (*Onychiurus* spp.)

Minute, white, soil-inhabiting insects with weak mouthparts; numerous in heavier soils with high organic content. They are active at lower temperatures than symphylids and may kill newly germinated shoots in the soil before emergence. Roots of older seedlings are damaged causing stunting and reduced vigour. Control as for symphylids.

Pygmy beetle (*Atomaria linearis*)

Small, 2 mm long, beetles which disperse, by flying when temperatures exceed 15 °C and wind speeds are below 7 km/h, during mid-April–June from previous year's beet fields into fields carrying the current year's crop where they feed on the seedlings. Populations build up during the year but little harm is caused to plants beyond the seedling stage. Adults overwinter in soil and under clods and plant debris on the surface.

A. Numerous in intensive beet growing areas, almost certain to be a problem when beet follows beet in the rotation because the beetles are already in the field when the seeds of the second crop are germinating.

B. Blackened feeding pits in the hypocotyl below, or just above the soil surface which may bring about complete severance and death. At a slightly later stage beetles feed in the young heart leaves which become very distorted and misshapen as they unfurl and grow.

C1. No cultural control.

C2. Methiocarb, normally present in the seed pellets, gives some protection against early attack. When beet follows beet, or in areas where damage occurs regularly, control as for millipedes. Thiofanox used for aphid control is also effective.

Beet flea beetle (*Chaetocnema concinna*)

Small, 2.5–3.5 mm long, shiny bronze beetles active in dry sunny weather. Adults overwinter in hedgerows and other sheltered situations close to the fields in which they have been feeding. Emerge on first warm days of spring and disperse to fields of seedling beet to feed on cotyledons and leaves. Seedlings beyond the cotyledon stage are normally able to withstand an attack.

A. Outbreaks sporadic, more likely in dry areas with abundant shelter and in cold dry springs when beetles are active but seedling growth is slow.

B. Small circular pits, later developing into holes on the cotyledons—'shothole' damage. In bad attacks the cotyledons are completely consumed and the seedlings killed. Beetles may be seen on the cotyledons, jumping when disturbed.

C1. Early sown crops are more likely to be past the susceptible stage at the time of attack.

C2. Carbofuran granules in the seed furrow are effective—*see* millipedes. Spray with gamma HCH if cotyledon damage is increasing, smaller quantities are required if the spray is applied as a band along the rows.

Wireworms (*Agriotes* spp.)

See under cereal pests, p. 279. On sugar beet, relatively low populations can lead to poor establishment.

A. Most likely in the second or third year after permanent grass, but damaging populations may persist longer. Damage more serious on early drilled crops. Control is justified when population is assessed at over 250 000/ha.

B. Hypocotyl chewed below ground level, seedlings die.

C1. No cultural control recommended.

C2. Seed furrow treatments of aldicarb, carbofuran and thiofanox (but not oxamyl) used against other pests give reasonable control. Otherwise, work gamma HCH into the seedbed, using higher rates on soils with a high organic content.

Leatherjackets (*Tipula* spp.)

See under cereals pests, p. 279.

A. A problem in wetter, western beet growing areas and on fields with a high water table; also when beet is grown after grass or weedy stubbles. Much lower populations are injurious to beet than to cereals.

B. Leatherjackets feed in spring, cutting off the seedlings at ground level. Grubs in soil near damaged plants.

C1. Plough grassland before September.

C2. Apply gamma HCH as a spray or bait when damage is seen. Do not grow potatoes or carrots within 18 months of this treatment.

Cutworms

See under potato pests, p. 283. Sporadic attacks by a number of species occur on sugar beet. In the fens, the caterpillars of the garden dart moth (*Euxoa nigricans*) cause occasional damage to beet seedlings, grazing and severing them at ground level between April and mid-June. Control, when damage is observed, by band sprays of trichlorphon or DDT applied in large volumes of water in the late afternoon in moist conditions when cutworms are more likely to come to the surface. Effectiveness is increased if the insecticide is then hoed into the surface soil alongside the seedlings. Other species may feed later in the season on the tap roots but control is not normally justified.

Skylark (*Alauda arvensis*)

Birds of the open fields, skylarks have emerged as seedling pests since the rapid increase in sowing to a stand. They feed on the cotyledons and leaves of the plants throughout the spring and early summer and the defoliation can cause death or reduction in vigour. Other birds, notably partridges, may cause similar damage. Control is difficult and uncertain but aldicarb, used as a seed furrow treatment against other pests, may give some protection.

Peach potato aphid (*Myzus persicae*)

See also under potato pests, p. 281. Sugar beet is one of the numerous host plants of this species which is of major importance as a vector of beet yellows virus (BYV) and beet mild yellowing virus (BMYV)—*see* section on sugar beet diseases, p. 270. Viruses are not transmitted via the true seed of sugar beet, so every plant starts life free from infection but may become infected by aphids migrating into the crop after having fed on other infected sources, including mangels, further spread then occurs within the crop as aphid populations increase and disperse.

A. Early migration and, therefore, early virus infection follows a mild winter. Crops in which a lot of bare soil is exposed are more attractive to aphids. BSC monitors local aphid populations.

B. Aphids may be seen on the plants. Reduction in yield and quality is proportional to the earliness of virus infection; it is estimated that for each week that the crop shows symptoms there is a 3% reduction in yield.

C1. Control is directed at breaking the cycle of virus transmission by eliminating overwintering sources of virus and by reducing aphid numbers and movement within the crop, particularly in its early stages of growth.

Sow early and encourage rapid early growth. Sow virus resistant varieties in recognised 'yellows' areas. Plough in plant residues on previous year's beet fields, destroy debris on cleaner-loader and mangel clamp sites before aphid migration gets under way in April. The seed crop is an important overwintering 'bridge' for virus and aphids—stecklings should be grown in isolation and/or under cover crops, insecticidal treatment of seed crops is specified in BSC contracts.

C2. The development of resistance in aphid populations to many widely used organophosphorous and some carbamate insecticides, especially in areas where aphids and virus transmission are recurrent problems, is a major threat to the beet industry. It is therefore very important that remaining and newer insecticides are used only when absolutely necessary in order to prolong their useful field life.

Seed furrow granule treatments using aldicarb on light soils and thiofanox on heavier soils will protect plants up to the six to eight leaf stage, carbofuran and oxamyl are not so long lasting. These treatments should only be used specifically for aphid control, following mild winters, in areas where virus infection is a regular problem. Sprays may be necessary when this treatment is omitted or as a supplement in bad years. Apply ethiofencarb, pirimicarb, demephion or demeton-S-methyl, preferably in bands along the rows, on receipt of BSC warning cards or when aphids can be seen on more than one plant in four.

Black bean aphid (*Aphis fabae*)

This aphid can seriously reduce the yield and quality of beet crops as a result of direct feeding damage. It can transmit BYV but is relatively unimportant in this respect. Overwintering occurs almost entirely as eggs on spindle trees (*Euonymus europaeus*) from which winged females migrate to summer hosts, including beet, in May and June. Dense colonies of wingless females grow rapidly in hot weather and more plants become infested as the season progresses. Numbers decline in September as a result of emigration and build up of predators and parasites.

A. Injurious in hot dry seasons especially on late sown crops and when plant growth is retarded by drought. Loss in yield occurs when average number of aphids per leaf exceeds two.

B. Conspicuous black aphids often co-existing with green peach potato aphids. Damage most important on heart leaves which become distorted and brown.

C1. Early sowing.

C2. Normally controlled together with peach potato

aphid and the same treatments and insecticides are recommended. Sprays should be used when numbers exceed two per leaf and hot weather prevails.

Beet cyst nematode (*Heterodera schachtii*)

The life cycle is similar to that of the potato and cereal cyst nematodes, dealt with under their respective headings. There are, however, some special features which are significant when considering control measures. The development time from the emergence of the larvae from the cysts to the formation of the next generation of cysts is short and up to three generations may be completed in a single growing season, resulting in a much greater increase in cyst population when a host crop is grown. This species also has a wide range of hosts including all beets, various weeds and most brassicas; the latter, although efficient hosts producing many cysts, are relatively unaffected by the invading larvae. Since the relaxing of statutory controls under the old Beet Eelworm Orders of 1960 and 1962 the problem has increased in certain areas. The Beet Cyst Nematode Order 1977 replaced the older Orders and provides powers to impose control of cropping if infestations should become serious in any particular area.

A. Injurious populations are more likely to occur in light sandy or peaty soils. Frequent cropping with sugar beet, fodder beet, red beet, spinach, mangel and brassicas (except radish and fodder radish) will lead to a rapid increase of cyst populations in fields which are infected.

B. Causes 'beet sickness' in fields where no rotational control is carried out. Patches of stunted, unhealthy plants whose outer leaves wilt readily in dry conditions, and turn yellow and die prematurely. Tap roots poorly developed with numerous fibrous side roots. Lemon shaped cysts visible on roots.

C1. The only reliable control is to adopt long rotations covering all the crops listed in 'A', combined with weed control and prompt destruction of any host crop residues. Length of rotation is related to the cyst population found by soil sampling. On 'clean' fields the minimum interval, stipulated in BSC contracts, between host crops is two years unless the first crop followed grass of at least three years' duration. On sandy or peaty soils a longer interval is preferable. In fields known to be, or suspected of being, infected the interval should be not less than three years on heavy soil and five years on sandy or peaty soils.

C2. Chemical treatment is not recommended as a means of regulating cyst populations because of the difficulty of controlling the second and third generations.

Free living nematodes (docking disorder) (*Trichodorus* and *Longidorus* spp.)

Similar to the species which attack potatoes.

A. A problem on light, sandy soils only, particularly when plants are under stress from other causes.
B. Patchy damage in crop, associated with sandier areas. Individual healthy plants often conspicuous among surrounding stunted plants. Attacked roots distorted, fangy or with lateral extensions.
C1. Provide good growing conditions.
C2. Seed furrow applications of aldicarb, carbofuran or oxamyl as for millipedes, *or* pre-sowing soil treatment with dichloropropene.

OILSEED RAPE PESTS

This relatively new crop has raised a number of new pest problems and as the area and intensity of cropping increases it is possible that more will arise. Pests which affect the developing pods, flowers and the flowering stems are of particular importance and in this connection it is necessary to distinguish between winter and spring sown crops because their susceptible growth stages occur at different times and, since the pests normally have relatively fixed times of maximum abundance, they are subject to a different range of pests.

Bees are strongly attracted to rape flowers and, although they make little impact on seed set in this self-pollinating crop, sprays should not be applied to crops in flower. This is not normally necessary, but if it should be, e.g. when flowering is uneven, the sprays should be applied in the early morning or the evening using phosalone or endosulfan which are *relatively* safe to bees. Local beekeepers should be given adequate warning of any intention to spray.

Spray booms should be set high to promote even distribution when advanced crops are being treated. At the present time only phosalone is cleared for aerial application.

Field slug (*Deroceras reticulatum*)

See also under cereal pests, p. 280.

A. A problem mainly in the autumn and early winter on winter rape directly drilled into cereal stubbles on wet soils and in cool conditions which retard germination and plant growth.
B. Seedlings grazed.
C1. No cultural control.
C2. Drill slug pellets containing metaldehyde or methiocarb with the seed *or* apply pellets as soon as damage is seen.

Brassica flea beetles (*Phyllotreta* spp.)

Life cycle and injurious effects similar to beet flea beetle. Mainly a problem in spring sowings. All seed is dressed with gamma HCH as a protection but it may be necessary to band spray the seedling rows with gamma HCH in the face of sustained attacks when seedling growth is slow. Potatoes and carrots should not be grown for 18 months after such treatment.

Cabbage stem flea beetle (*Psylliodes chrysocephala*)

A pest of winter rape at present mainly in parts of Northants, Bucks, Hunts and Cambridgeshire and a potential pest in other areas. Adults migrate in September from fields of the current year's crops of rape and other brassicas into fields where autumn sown seedlings are emerging. The adults cause shothole damage on the seedlings but this is normally of little importance. Females lay eggs in the soil in October–November, these hatch at temperatures above 5 °C. The larvae burrow into the stems of the plants via the petioles and accumulate there as hatching continues. When feeding is completed they emerge to pupate in the soil in April and May giving rise to adults in May and June.

A. The presence of large number of the small, dark blue beetles with thickened hind legs in the harvested seed of the current year's crop and/or shothole damage in autumn seedlings indicate that damage is likely to occur in the winter or early spring.
B   Attacked plants collapse as their stems are hollowed. Small, white grubs with dark heads and six small legs may be seen on opening affected stems.
C1. No cultural control.
C2. Kill adults by spraying with gamma HCH when shothole damage is seen *or* kill hatching larvae before they enter the plants by a similar spray in late winter when soil temperatures start to rise. Neither treatment is likely to be entirely successful because egg laying and hatching continue over a long period.

Pigeon (*Columba palumbus*)

Winter crops are attractive to flocks of pigeons which can cause extensive damage as a result of their grazing activities especially on late sown crops of small plants with a high proportion of bare soil. Control by sowing early and encouraging rapid germination and growth. Dalapon and TCA applied to control grass weeds and volunteer cereals render the crop unpalatable for a few weeks.

Cabbage stem weevil (*Ceutorhynchus quadridens*)

Affects spring crops. Adults migrate from fields of previous year's crops, feed on seedling leaves and lay eggs. Larvae hatch and tunnel in petioles and stems in May and June causing parts of the plants above where they are feeding to wilt and collapse. Larvae white with darker heads and legless. Can be important when flowering stems are affected. Control is difficult and not normally directed specifically against this pest. Seed dressings for brassica flea beetles and the sprays applied to control blossom beetles on spring rape at the early green bud stage give some protection.

Cabbage aphid (*Brevicoryne brassicae*)

Occasionally a problem, especially on spring crops in hot, dry seasons when dense colonies of the mealy grey aphids form on the flowering stems. Specific control is necessary only when an appreciable proportion of the stems is colonised, care must be taken to avoid killing bees if insecticides are applied.

Blossom beetles (*Meligethes* spp.)

Also known as pollen beetles. A problem mainly on spring crops but may affect backward winter crops. The adult beetles hibernate in sheltered situations such as hedges, copses, etc. and fly to rape crops, assembling in increasing numbers during hot, sunny periods in April and May. Eggs are laid in the early green flower buds and the larvae, which hatch after a few days, feed on and destroy the reproductive parts of the flowers. Feeding by adults and larvae continues as the flowers open. Large numbers of beetles cause a significant reduction in the number of pods formed.

A. Common in most districts especially where there is an abundance of winter shelter. Most damage occurs when peak numbers coincide with the green bud stage of development.
B. Look for small, 1.5 mm long, oval, shiny black beetles with conspicuous knobs on the ends of their antennae.
C1. No cultural control.
C2. Spraying is most effective at the green bud stage when adults are present in numbers but before many eggs have been laid. To assess the need for control select 20 plants at random through the field, avoiding headlands, and count the number of beetles per plant—move carefully to avoid disturbance. On spring rape, spray when average population reaches three to five per plant. On winter rape, spray when average reaches 15–20 per plant. Use azinphos-methyl mixtures, endosulfan, gamma HCH, malathion or phosalone. A second application may be required if numbers build up again in a slow growing crop, but

treatments from the late yellow bud stage onwards are not justified and are dangerous to bees.

Cabbage seed weevil (*Ceutorhynchus assimilis*)

A major pest in both winter and spring crops. The adult weevils overwinter in hedgerows and other sheltered situations and migrate in spring to rape crops. They fly actively in hot, calm weather and rapidly colonise the fields. Some time is spent feeding on the leaves and other parts of the plants but this causes little or no loss. Eggs are laid only in the soft young green pods shortly after flowering. Normally only one egg is laid per pod and the damage is caused by the larvae feeding on and destroying the young seeds. The larvae emerge when fully fed and pupate in the soil, giving rise to a second generation of adults which move to other brassica crops in the area in July–August.

A. Highest populations develop in areas of intensive oilseed rape production where the crop has been grown for many years.
B. Small, 2–4 mm long, grey beetles with long thin 'snouts' are seen on the plants, flying actively in sunny weather. White, legless grubs with brown heads in the seed pods.
C1. No cultural control.
C2. On spring crops maximum populations normally assemble on the plants before flowering and the same sprays used against blossom beetles should be applied when, at the early yellow bud stage, an average of one adult weevil per plant is found. Use the same counting technique as for blossom beetles.

On winter crops, which are usually more advanced, peak numbers often coincide with mid-flowering. In this case it is necessary to make the count when the crop is in flower and then wait until after petal fall, when the crop presents an overall green appearance, before spraying with phosalone or triazophos if the threshold population was exceeded. These insecticides kill the young larvae in the pods, phosalone should be used if bees are still working late flowers.

Brassica pod midge (*Dasyneura brassicae*)

Sometimes known as the bladder pod midge because of the swellings characteristic of pods which contain the small white magggots. Affected pods ripen and shed seed prematurely. The midge is associated with the seed weevil because the adults, which are fragile and weak, make use of the weevils' entrance and exit holes to penetrate the pods for egg laying. Control of seed weevil usually protects the crop against the midge but when exceptionally large populations occur throughout the field a specific treatment using endo-sulfan or triazophos (winter rape only) may be required.

KALE PESTS

The main pests of kale, grown as animal fodder, are those which affect the establishment of the seedlings.

Brassica flea beetles (*Phyllotreta* spp.)

Life cycle and injurious effects similar to beet flea beetles.

A. Mainly a problem in hot, dry conditions in April and May on spring sowings when the seedlings are in the cotyledon stage and growth is slow. Plants are much more tolerant of injury after the first true leaves develop.
B. 'Shothole' damage to cotyledons and stems, leading to complete loss in severe attacks. Small, 2–3 mm, beetles some with a yellow stripe down each wing case, jumping actively when disturbed.
C1. Sow early and encourage rapid early growth by providing a fine, well fertilised seedbed.
C2. Dress seed with gamma HCH, usually combined with a fungicide. This may have to be supplemented by a treatment of gamma HCH as a spray or dust in the face of sustained attacks. Potatoes and carrots should not be grown within 18 months of this treatment.

Cabbage stem flea beetle (*Psylliodes chrysocephala*)

See under oilseed rape pests, p. 287. Mainly a problem on late sown overwintering kale crops. Early sown, well grown crops are more tolerant of injury. Chemical control as on oilseed rape.

PEA AND BEAN PESTS

Apart from the general pests which disrupt the normal growth of the crops, growers of peas in particular have to contend with a number of pests which feed within the pods and affect the appearance of the peas. A relatively small percentage of damaged peas can reduce the market value of the crop and may even lead to rejection by the processor.

Pea and bean weevil (*Sitona lineatus*)

Hibernating adults become active from mid-April onwards and move into pea fields, where they feed on the leaves. Later, they lay eggs and the larvae burrow in the soil and feed on the root nodules.

A. Leaf injury is only likely to be important when young plants are heavily attacked at a time when plant growth is slow.
B. Characteristic U-shaped notches in the leaf edges or on the folds of unopened leaves. Seedlings may be killed by defoliation and plant growth may be reduced later as a result of nodule destruction.
C1. Encourage rapid early growth.
C2. Spray, when most seedling leaves show symptoms, with azinphos-methyl mixtures, fenitrothion or triazophos. Crop inspection may indicate that only the periphery merits treatment.

Pea cyst nematode (*Heterodera goettingiana*)

Life cycle similar to cyst nematodes of cereals, potatoes and sugar beet but with peas, beans and vetches as hosts. Peas are seriously damaged but beans, although efficient hosts, are little affected by larval invasion.

A. Cyst numbers build up rapidly in fields cropped too frequently with peas and beans.
B. Patches of stunted, prematurely senile pea plants are normally the first indication of trouble, lemon shaped cysts can be seen on the short, fibrous roots which bear few root nodules.
C1. Lengthen the interval between host crops. Cysts are very persistent and expert advice on the length of the interval should be sought.
C2. Reduce damage by mixing oxamyl granules in the seedbed before drilling. This must be considered to be a supplementary treatment and is not an acceptable alternative to lengthening the rotation.

Pea aphid (*Acyrthosiphum pisum*)

Winged females migrate from a wide range of leguminous plants into pea crops during the summer months. These found colonies of wingless aphids and further spread occurs within the crop.

A. Colonies build up rapidly in hot, settled weather reaching a maximum in June and July.
B. Colonies of large, long-legged green aphids on the growing points and spreading to the pods when numerous. Extraction of plant sap results in poor growth and badly filled pods. Plants may be covered with sticky honeydew. Also transmits pea virus diseases.
C1. No cultural control.
C2. Sprays used to control pea moth and pea midge also control the aphid. Where no such control has

been applied spray with any of a large number of aphicides when there is an average of one or more colonies on every fifth plant. The choice of aphicide may be limited if the crop is to be harvested shortly after spraying.

Pea moth (*Cydia nigricana*)

The moths emerge from the previous year's pea fields and congregate in the current year's crop, the flight period lasting from early June to mid-August with a peak round about mid-July. Eggs are laid on the leaves and the minute caterpillars crawl to and penetrate the soft young pods and feed inside for three weeks. They emerge after this and overwinter in the soil.

A. Common in all pea growing areas and particularly prevalent in East Anglia and Kent. Especially important on peas for freezing and canning because a relatively low infestation may result in rejection. All peas which flower in the main flight period are at risk.

B. Peas burrowed into and fouled by the excrement of the small, pale caterpillars with black heads.

C1. Early picking peas which flower before the main flight period escape serious attack.

C2. Insecticide treatments must kill the caterpillars in the relatively short time that they are exposed outside the pods, therefore correct timing is crucial to success. Warning systems are operated by the local ADAS and commercial advisers but there may be considerable variation in moth abundance and flight periods even in neighbouring fields. Growers can use commercially available pheromone (sex-attractant) traps combined with temperature data calculators to predict moth abundance and the optimum spraying time in their own crops. Sprays containing azinphos-methyl mixtures, fenitrothion, carbaryl, deltamethrin, permethrin or triazophos, all in large volumes of water are effective—the first two are also cleared for aerial application. A second treatment may be required on dry harvesting peas. Observe recommended intervals before harvesting.

Pea midge (*Contarinia pisi*)

Mainly a pest of vining peas. Midges migrate from previous year's pea fields and lay eggs near the developing flower heads; the small pale, jumping maggots feed in the developing buds, on the leaves and sometimes inside the pods.

A. When midges assemble in numbers about a week before flowering. Small, fragile, yellowish flies concealed within the terminal leaf clusters may be found by careful examination.

B. Flower heads sterile; terminal parts of flowering stems shortened and leaves distorted to form 'nettleheads'.

C1. No cultural control.

C2. Spray warnings are issued in some ADAS regions and by some processors. Otherwise, spray immediately when an average of one midge per five flowering stems is found when flowers are in the early green bud stage. Azinphos-methyl mixtures, demeton-S-methyl, dimethoate, dimethoate with triazophos and fenitrothion are recommended.

Black bean aphid (*Aphis fabae*)

See also under sugar beet pests, p. 286. Mainly a problem on spring sown field beans.

A. Serious infestations more likely on late sown spring crops and in southern England. Colonies build up rapidly in hot, dry weather.

B. Dense colonies of black aphids on stems and growing points. Pods distorted, poorly filled, contaminated by honeydew and moulds. Bean virus diseases transmitted.

C1. Sow early and encourage rapid early growth.

C2. Beans are bee pollinated and aphicides should be applied just before flowering when forecasts issued by ADAS in conjunction with Imperial College suggest that this is necessary. In areas where the risk is lower, treatment may be restricted to the headlands. Granules and pirimicarb sprays are the least harmful to bees and should be used if it becomes absolutely essential to treat the flowering crop.

FIELD VEGETABLE PESTS

Uniformity of appearance and freedom from blemish are the main requirements of vegetables grown for human consumption and very low pest populations, which have a negligible effect on yield, may therefore be a cause of considerable financial loss. Standards of control must be exceptionally high and often involve routine and intensive use of insecticides. This has created problems of increasing resistance to many of the chemicals which have been successfully used in the past. The choice of insecticide may be further limited by the short interval between final treatment and harvest, and other factors such as soil type.

Anyone growing such crops is recommended to take full advantage of forecasts, spray warnings and advice provided by the local ADAS and commercial organisations. The principal pests of the main vegetable crops are:

(1) brassicas—flea beetles, cabbage root fly, cabbage caterpillars (various species), cabbage aphid;

(2) broad and french beans—black bean aphid, bean seed flies;

(3) carrots—carrot fly, willow-carrot aphid;

(4) onions—onion fly, stem and bulb nematode.

SCHEMES, REGULATIONS AND ACTS RELATING TO THE USE OF INSECTICIDES

Virtually all the insecticides in use are potentially hazardous, some more than others, to users, consumers, wildlife and beneficial organisms and there are many schemes and legal requirements governing their purchase, use, application, storage and disposal. The principal schemes and Acts are as follows:

Pesticides Safety Precautions Scheme,
Agricultural Chemicals Approval Scheme,
Deposit of Poisonous Waste Act 1972,
Rivers (Prevention of Pollution) Acts 1951 and 1961,
Control of Pollution Act 1974,
Health and Safety (Agriculture) (Poisonous Substances) Regulations 1975,
The Poisons Act 1972.

These are summarised, and sources of further information are given, in *Approved Products for Farmers and Growers*, published every February for MAFF by HMSO.

11

Grain preservation and storage

P. H. Bomford

Grain is stored after harvest so that it can be marketed, or used, in an orderly manner. Prices are generally at their lowest at harvest time, and can be expected to rise with length of storage, until the next harvest approaches. If livestock are to be fed with home-produced grain, careful storage can ensure a year-round supply of high quality feed.

A successful storage system will preserve those qualities of the product which are important for its proposed end-use. These qualities may include a high germination percentage, good baking or malting properties, ease of extraction of constituents such as starch, oils or sugars, freedom from impurities, discoloration or taint, and the nutritive value of the grain to livestock.

Grain storage alternatives

Grain is generally preserved in good condition by controlling its moisture content and temperature so that the organisms which cause deterioration cannot develop. Grain stored in this way can be sold into any appropriate market, or fed to livestock on the farm.

Where the grain is definitely to be used as stockfeed on the farm, there are advantages in storing it at the harvested moisture content. In this case, harmful organisms are controlled by the use of a chemical preservative, or by storing in an oxygen-free environment in a sealed container.

CONDITIONS FOR THE SAFE STORAGE OF LIVING GRAIN

The main agencies which cause deterioration in stored grain are moulds, insects and mites. All occur naturally on cereal crops, and so are likely to be brought into the store with the crop. Residual populations of insects and mites may be present in crevices or crop residues in the 'empty' grain store, but thorough cleaning and insecticide/miticide treatment of the store before it is filled can greatly reduce this hazard.

The grain is protected from attack by a tough outer skin or pericarp and impermeable seed coat or testa. Unfortunately, the severe threshing treatment to which the crop is subjected at harvest can damage these layers, allowing easy access by mould organisms to the starchy endosperm or the embryo. In one study between one-fifth and one-third of all wheat grains were found to have threshing injuries to the skin of the embryo area. The embryo contains sugars, fats and proteins and is thus an excellent source of nutrients for the moulds. Damage to the embryo will impair the viability of the seed.

The relationship between grain condition and pests (over a 35-week period) and storage conditions is illustrated clearly in the classic diagram (*Figure 11.1*) which has been produced by research workers at the MAFF Slough Laboratory.

In the figure, *line A* defines the conditions under which mites will breed, and therefore infestations can build up. Breeding can be prevented by keeping the temperature below 2.5 °C, or the grain moisture content below 12%. Neither of these measures is practicable and adequate control is generally achieved under British farm conditions by thorough store hygiene to eliminate residual populations of mites.

Line B covers the risk of loss of germination due to embryo damage, and also the deterioration of baking (wheat) and malting (barley) qualities. Grain stored under conditions to the *left* of the line may be expected to retain all these qualities and thus be suitable for sale into premium markets.

Line C defines the conditions under which moulds will grow on and subsequently in the grain. In addition

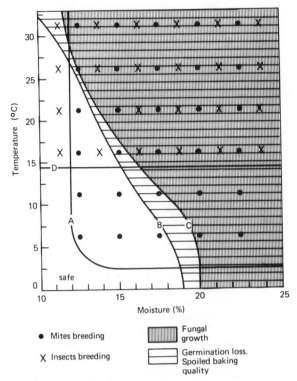

Figure 11.1 Combined effects of temperature and moisture content on grain condition and grain pests over a 35 week storage period (Crown copyright. Reproduced with permission)

● Mites breeding

X Insects breeding

▓ Fungal growth

▭ Germination loss. Spoiled baking quality

to the 'invisible' forms of damage mentioned in the previous paragraph, mould growth can lead to a loss of crop weight as the mould organisms feed on the grain, discoloration due to visible mould growth, and the unpleasant health hazards of mycotoxins (poisons produced by the mould organism which can kill or debilitate livestock or even humans who eat contaminated foodstuffs) and mould spores (a very fine dust which can cause farmers' lung, a pneumonia-like affliction, when inhaled). It is essential to wear an adequate *organic dust respirator* if working with grain (or hay) which is suspected of being mouldy.

Moulds, like other living organisms, produce heat as they break down food with oxygen, liberating *water* and carbon dioxide. Thus, any large concentration of moulds (or insects or mites) will make the grain around it warmer and more moist. As can be seen from the diagram, this will have the effect of making conditions even more favourable for their own development, and may also encourage the development of other pests or even allow the grain to germinate, which will happen if the local moisture content increases to 25–30%.

Severe mould infestation, if not detected and controlled, will result in the grain being caked together, particularly at or just below the surface and

against the walls of the store, and sprouted at the surface.

This can reduce or eliminate the market value of the corn, and may also render it hazardous to livestock and to the farmer himself.

Line D shows that insects such as grain weevils and beetles, which are naturally present in small numbers on the harvested grain as it comes into store, can generally be prevented from becoming active and breeding if the temperature of the grain is kept below 15 °C. In fact, the cooler the grain, the less active the insects and the greater the margin of safety, even down to freezing point. Reducing grain moisture content has no effect in controlling insect activity.

Insect activity in the grain will lead to loss of dry matter and reduced germination, contamination with insect remains and faeces, and as above, to an increase in temperature and moisture levels which can lead to moulding and even sprouting of the grain.

Where an insect problem has been found to exist and it is not possible to control it by cooling the grain, the insects may be controlled by fumigation of the grain store with substances such as methyl bromide or hydrogen phosphide. These substances are toxic and their successful use involves sealing the grain store and the deployment of specialised control and safety equipment. This should be left to specialist pest control contractors.

The grain may also be treated with an insecticide such as pirimphos-methyl as it is conveyed into store, if insect problems are expected.

It can be seen from *Figure 11.1* that (ignoring mites) the preservation of living grain in top condition for a 35-week storage period may be achieved by keeping conditions within the area below line D and to the left of line B. If the grain is intended for use or sale as livestock feed, then the right hand boundary moves out to line C, allowing a higher moisture content to be accepted.

Generally, the crop is safest when temperature and moisture content are low, and at increasing risk as either factor rises.

THE TIME FACTOR IN GRAIN STORAGE

As may be expected, insect, mould or mite infestations take time to build up. This means that grain, too moist or too warm to be stored safely for seven months without drying or cooling, can be held for several weeks with little or no harm. The length of time such a parcel of grain can be held without loss of quality decreases with increasing temperatures or moisture content. *Table 11.1* shows the maximum recommended moisture content for bulk grain stored at various temperatures and for various periods. (It must be remembered that storage for any extended period at temperatures over 15 °C entails the risk of insect build-up *however dry the grain*.)

Table 11.1 Moisture content at which bulk grain can be stored for various periods at 15 °C

Storage period	% moisture content	
	Feed	Seed/malting/baking
Up to 4 weeks	17	16
To April	15	14
Beyond April	14	13

GRAIN MOISTURE CONTENT AND ITS MEASUREMENT

Grain moisture content is expressed as *wet-base* (WB) percentage, or the ratio of the weight of water contained in a sample of the grain, to the total weight of the sample, including the water.

In a laboratory, the moisture content of a sample of grain is found by drying the weighed sample in an oven until no water remains, and then re-weighing to find the weight of the water which has been driven off. The moisture content (wet base) is found from the expression:

$$\text{MC\%} \, (\text{WB}) = \frac{\text{Loss of weight on drying}}{\text{Original weight of sample}} \times 100$$

This is the only completely accurate method of measuring the moisture content of a sample of grain, and is always used in scientific research, or when moisture meters are to be calibrated. However, the complete oven-drying of even a milled sample of grain will take 2 or 4 h. Faster moisture measurement systems have been produced to satisfy the need for an 'immediate' evaluation of grain moisture content.

Most grain moisture meters measure an electrical property of the grain such as its resistance or capacitance, which have been found to vary with moisture content. A good quality instrument, well maintained, can be expected to give a reading which is correct to within $\pm 1\%$ of the true moisture content of the sample, and this reading can be available within a few minutes of the sample being taken.

Some instruments require the grain to be ground, which increases the time needed to make a measurement. However, a whole-kernel instrument may give false results with grain that has just passed through a drier, or with very cold grain since in these cases the condition of the exterior of the grain may not reflect the internal situation.

When measuring the moisture content of a sample of grain, it is most important to ensure that the sample is truly representative of the bulk of grain from which it has been drawn. This can be best achieved by systematically extracting several small samples from different points and depths in the load or store, mixing them together and then taking the final sample from this mixture.

Without great care in the taking of samples, even the best of moisture meters and the most careful of operators will fail to produce satisfactory results.

DRYING GRAIN

Where incoming grain is at too high a moisture content for long-term storage, it must be dried at the beginning of the storage period. In the process of drying, heat is provided to evaporate the moisture from the grain, and a stream of air picks up the water vapour and carries it away into the atmosphere.

The more a batch of grain has to be dried, due either to a high initial moisture content or to a low final moisture content, the greater will be the cost both in terms of energy usage and drying equipment operating time.

When grain is dried the resulting product will weigh less than it did before drying, by the amount of water which has been removed. It is often important to know what this weight loss will be, either for the purpose of estimating drying costs or for converting harvested grain weights to equivalent weights of dry grain. The following formulae are helpful.

Where
    X is the weight loss on drying,
    M_1 and M_2 are initial and final moisture contents (% WB),
    W_1 and W_2 are weights of grain before and after drying.

$$X = \frac{W_1(M_1 - M_2)}{100 - M_2}, \qquad X = \frac{W_2(M_1 - M_2)}{100 - M_1}$$

In addition to the weight loss on drying, a further minimum weight loss of 1–1.5% can be expected over the storage period, due to respiration of the grain itself, and to the activities of storage pest organisms.

Drying systems range from those which are capable of drying grain within 2 or 3 h of its arrival from the field to those where complete drying can take two weeks or more.

Because of the air temperatures which are used, high-speed drying systems may be classified as 'high-temperature driers', while the slowest are known as 'low-temperature', or 'storage' driers, the latter term because drying takes place after the grain has been put into the store, and the store does double duty as a drier.

High-temperature driers

High-temperature driers can extract moisture from grain very rapidly because a high temperature allows moisture to diffuse rapidly out from the interior of each grain, and the very hot air which is blown through the grain has a very great attraction for water vapour and a very large water vapour carrying capacity.

As the hot air passes through the grain, it gives up some of its heat to evaporate water, and is thus cooled as it picks up water vapour. The distance that the air travels through moist grain must be limited so that there is no risk of it being cooled to the point where it becomes saturated (i.e. can no longer carry all its water, and water is deposited back on the grain by condensation). In high-temperature driers the thickness of the grain layer rarely exceeds 250 mm, while in low-temperature storage driers it may be as much as 2.4 m.

The higher the temperature at which a given drier is operated, the faster is the rate of moisture extraction, the higher is the grain throughput of the drier, and in many cases the lower is the energy consumption per unit of water removed. Hence operating a drier at a higher temperature can reduce drying costs in terms of both overhead and fixed costs per tonne dried.

If the temperature of the drying air is too high, grain temperature will rise to a level high enough to cause damage to germination, baking or malting properties, extraction of oils or starch or even to the livestock feeding value of the grain.

Table 11.2 The recommended maximum air temperatures for high-temperature grain driers

Grain types	Max. air temperature (°C)
Grain for stock feed	80–105
Wheat for milling	65
Malting barley or seed corn up to 24% MC	50
Malting barley or seed corn above 24% MC	43
Oily seeds	45

Table 11.2 shows the recommended maximum air temperatures at which to dry grain for various markets. These recommendations date from the early 1940s. New developments such as concurrent flow driers may make them obsolete. New recommendations are expected in 1984 or 1985.

Unless the grain is completely dry, grain temperature will always be lower than air temperature due to the cooling effect of evaporating water; the greatest cooling effect will be where the grain is moist, and the evaporation rate is high. Even at the end of the drying process, grain temperatures will still be 10–20 °C lower than drying air temperature.

High-temperature driers may be classified as 'continuous', where there is a constant flow of grain through the machine and into store, or 'batch', where a quantity of grain is loaded into the machine where it remains while it is dried, and is then unloaded into store.

Continuous flow driers

Most continuous grain driers are of the crossflow type, where the grain moves in a layer 200–300 mm thick while the drying, and later cooling, air is directed through the grain at right angles to its direction of travel. In order to give an equal drying effect to all grains, the grain should be turned or mixed several times as it passes through the drier.

The crossflow principle is popular due to its simplicity, but other airflow systems can offer advantages, and a few have appeared on the market. In a counterflow drier system, drying air flows in the opposite direction to the grain, so that the hottest and driest air meets the driest corn. This gives faster and more efficient drying, but there is a greater risk of damaging the grain by overheating, and this method is in fact only used in the *cooling* section of some driers where the temperature gradient is in the reverse direction and this system gives a more gentle temperature reduction and the advantage of maximum heat transfer can be fully utilised.

In a concurrent flow drier, grain and drying air move in the same direction although the air moves at a much faster rate. Thus the hottest air always meets the wettest corn, where the cooling effect of evaporation will be greatest. By the time the air reaches the driest corn, its temperature will have been much reduced, minimising the temperature stress on the grain. This may permit air temperatures above the recommended maximum, without excess grain temperatures. There is a faster throughput from a given drier size, and more efficient use of heat.

The properties of mixed-flow driers are intermediate between the three types.

Not all the drying is done in the drying section of the drier; a final 1 or 1.5 percentage points of moisture are removed while the grain is being cooled, utilising the heat energy remaining in the grain from its passage through the drying section. The cooling section occupies one-third to one-quarter of the drier, and is generally capable of cooling grain to near ambient temperature before it is discharged from the drier.

If there is cooling available in the store, drier capacity can be increased since the whole of the drier can be used for drying the grain, which is cooled in bulk later.

The drying capacity of a continuous-flow drier is generally quoted in tonnes of grain/h, dried by five percentage units of moisture (often 20–15%) at a specified drying-air temperature (often 65 °C). Output of the drier will be increased if a higher air temperature can be (safely) used, or if the grain requires drying by less than five percentage units, and *vice versa*.

The amount of drying done to each grain depends on the length of time for which that grain is exposed to the stream of hot air in the drying section of the drier, and this is controlled by the rate at which grain moves through the machine. This rate is adjustable, and adjustments are made according to the moisture content of the incoming grain and the resulting final moisture content of the dried product. Both of these values should be checked regularly. Several manufacturers offer automatic drier control units, which can monitor final grain moisture content (or grain temperature at the end of the drying section, which is a related value) and regulate the flow of grain accordingly.

It should be remembered that grain can take 3 h or more to pass right through a continuous-flow drier, so that any adjustment made to the machine will take this long to come into full effect.

Energy consumption by continuous driers

Continuous-flow and other high-temperature driers, which evaporate water at a very rapid rate, consequently consume energy at a rapid rate (*Table 11.3*).

Table 11.3 Furnace capacity of continuous driers

Tonnes/h of wet corn to be dried from 20–15% MC at 65 °C air temperature	Drying capacity					
	2.5	5.0	10.0	20	30	60
Evaporation rate (kg water/h)	150	300	600	1200	1800	3600
Furnace capacity (kW)	220	450	900	1860	2700	5300

Such a high heat requirement clearly rules out many sources, and at present the available types are fired by diesel (gas oil) or LP gas (propane). Potential alternative fuels are natural gas (if a piped supply is available) or coal.

Thus, 2.45 megajoules (MJ) of energy are required to convert 1 kg of water to vapour at 20 °C. Tests at NIAE have shown an average energy use of 6.7 MJ/kg of water evaporated, for a range of continuous driers. Clearly, significant amounts of energy are used up in other ways. Apart from the energy used for the operation of fans and conveyors, and the proportion which leaves the drier in the form of latent heat locked up in water vapour, most of the 'wasted' energy is lost in the form of heated, unsaturated air either through the lower part of the drying section or from the cooling section of the drier. Research work and experience in the field have shown that a 10% reduction in fuel consumption can be achieved by recovering the heat from the cooling section. Several driers at present on the market recover this heat and use it to pre-heat the drying air which is drawn inwards through the cooling section, and then passes on through the fan and heater to the drying section. There is also considerable evidence that a further 25% fuel saving can be achieved by returning the hot unsaturated exhaust air from the lower part of the drying section for a second pass through the machine.

In operating the drier to save energy, the most important factor is to *avoid over-drying*. Every kg of water that is removed from the grain adds to the cost of the operation and slows output. Reducing the drying air temperature will *not* save energy, and will certainly increase the cost of drying a tonne of grain.

High-temperature batch driers

The high-temperature batch drier carries out the same functions as the continuous drier, but it operates in a series of separate steps instead of a flow process. The sequence of events can be controlled manually, or automatically so that the drier can be left to run unattended. Each batch spends 2–4 h in the drier.

Batch driers can be installed on a permanent base, or some models can be moved on their own wheels from one site or one farm to another. Portable driers can be easily re-sold. The unit may be rated by the amount of grain in a batch, or by the number of tonnes of grain that can be dried by five percentage points in 24-h operation. Capacities range from 38–135 tonnes/24 h. The capacity of a drier in 24 h should be equal to the daily harvesting capacity.

Storage, or low-temperature, driers

Where grain is to be stored on the farm after it has been dried, it is possible to reduce costs by combining the functions of drying and storage. *Tables 11.4* and *11.5* refer to the storage volumes of various grain crops. Capital cost is less than for a separate drier and a store, and typical energy consumption is 3.5 MJ/kg of water evaporated. Against this, the system gives a slow drying rate, 0.5% per d or less, and its successful operation demands a high level of management.

Table 11.4 Crop weights and volumes

	kg/m^3	m^3/tonne
Wheat	785	1.28
Barley	689	1.42
Oats	513	1.95
Beans	817	1.20

Table 11.5 Round bin capacities (wheat)

Bin diameter (m)	Height (m^3/m)	Height (tonnes/m)
2.4	4.5	3.5
3	7.1	5.5
3.7	10.8	8.4
4.3	14.5	11.3
4.9	18.9	14.8

Measurement and control of air relative humidity

The operation of any low-temperature drier relies on the fact that there is an equilibrium relationship between the moisture content of grain and the relative humidity (rh) of the air which surrounds it (*Table 11.6*). Grain is brought to a safe moisture content by ventilating it for two weeks or more with air at the corresponding equilibrium relative humidity. Any deviation in air quality will affect both the rate of drying and the final moisture content, so it is clear that the accurate measurement and control of relative humidity must be a prime requirement for successful drier management.

Table 11.6 The equilibrium between air relative humidity and grain moisture content at 15 °C

Air rh (%)	50	57	65	72	77	82	86	87	88
Grain MC (%)	12	13	14	15	16	17	18	19	20

Note: As temperature rises, the equilibrium shifts so that rh increases in relation to MC, and the air gains water from the grain. The reverse happens if the temperature falls. At 38 °C, air at 65% rh reaches equilibrium with grain at 11.8% MC, while at 5 °C the grain MC would be 14.8%.

Relative humidity is measured with a wet and dry bulb hygrometer, which consists of a pair of identical thermometers. The bulb of one thermometer is covered with a wet cloth sleeve. After exposure to the air stream until readings stabilise, the two thermometers are read and the relative humidity is found from a slide rule or chart supplied with the instrument. All other instruments for measuring or controlling relative humidity must be calibrated regularly against the basic, and inexpensive, wet and dry bulb instrument.

If the relative humidity of the air is too high, it may be reduced by heating the air. Equipment which will raise the air temperature by 6 °C is adequate to deal with any weather situation. Electrical heater banks are commonly used, because of their ease of installation, operation and control, but oil or gas fired heaters are available, or a stove may be fuelled with wood or straw. Diesel-engine driven fans apply the waste heat of the engine, which can raise the air temperature by up to 4.5 °C. So 1.1 kW of heater banks are needed to raise the temperature of 1 m³/s of air by 1 °C, and thus reduce the relative humidity by 4.5 units.

Heater output must be adjusted with every change in ambient conditions, if the maximum drying rate is to be maintained without unnecessary use of heat. An automatic humidity control (humidistat) which will regulate the output of a series of electric heaters to maintain the drying air at the desired relative humidity will soon pay for itself in energy savings.

The low-temperature drying process

As the drying air moves upwards through the grain mass it picks up water and is cooled. Grain at the bottom of the store dries first, and a 'drying front' then moves slowly upward through the grain until all the grain is dry after two weeks or longer. During drying, the grain above the drying front remains as wet (or wetter) than when it was harvested. This portion of the grain is preserved before drying by the cooling effect of the air which has dried the lower layers of grain.

If grain at a moisture content above 22% is to be dried, it must not be loaded to a depth of more than 1 m. This reduces the drying time, and it also prevents the soft grains being squashed by the weight of a full 2.4 m depth of corn. If grains at the bottom were crushed, this would reduce the air spaces between them and restrict the flow of drying air.

Progress of the drying front towards the grain surface can be followed by extracting samples from different depths with a sampling spear. The depth of the transition between dry and wet grain indicates the position of the drying front.

Air supply and distribution

The drying air is distributed over the base of the grain mass by means of a system of ducts, or through a fully perforated floor. An adequate quantity of air must be provided to make the drying period as short as possible. An airflow rate of 0.1 (m³ s)/m² of floor area, or 0.05 (m³ s)/m² of grain, whichever is greater, is recommended. Since not all the grain in the store is being dried at once, the recommendations are applied to one-half to three-quarters of the store according to geographical region. Adequate fan capacity is a cornerstone of successful crop drying.

FANS FOR GRAIN STORES

A fan moves air by creating a pressure behind it. The air output of the fan decreases as the back pressure or static pressure against which it is working increases. Back pressure builds up as the air is forced along ducts, round bends and through the grain. The greater the air speed, the higher the back pressure.

The pressure generated by a fan is measured with a water manometer (wg = water gauge), and fan output can be determined from this reading by reference to manufacturers' performance curves. Working pressures for crop drying fans are 60–150 mm wg.

A fan is 40–85% efficient. The remaining 60–15% of its power consumption is lost as heat, which warms the air by up to 3 °C. This may be enough heat to dry a crop, but it is a disadvantage when the fan is used for cooling. Cooling systems sometimes operate under suction with the fan at the outlet; this way, the fan's heat does not enter the crop.

Some fans absorb more power as back pressure drops and air output increases, and the motor may become overloaded and overheat. Only non-overloading fans are suitable for crop drying.

Centrifugal fans

Centrifugal fans develop pressure by accelerating the air in the rotor, and then converting its kinetic energy into pressure energy in the diverging housing or volute. They are quiet in operation, and bulky. A form of centrifugal fan with backward-curved (b-c) blades on the rotor is non-overloading, and is frequently used in grain stores. Pressures up to 300 mm wg are possible, according to rotor diameter and speed.

Axial-flow fans

Pressure is created by the rotation of an airscrew-like impeller in a close fitting circular housing. Air passes in a straight line through the fan, giving high peak efficiencies. The fan alone can only produce a pressure of 80–120 mm wg, but stationary intake guide-vanes can increase this by 20–60%. Two fans contra-rotating in the same tubular housing can increase the pressure up to three times. The fan is non-overloading, and more compact than a centrifugal fan of the same output. It is noisy, particularly when guide vanes are fitted. Silencers are available, or the fan can be shielded with bales to absorb the noise.

Table 11.7 shows the output of various crop drying fans against a range of static pressures.

Table 11.7 The output of various crop drying fans against a range of static pressures

Type and power of fan	Static pressure (mm wg)					
	50	75	100	125	150	175
	Air output (m³/s)					
3.8 kW b-c	4.6	3.9	3.0	—	—	—
7.5 kW b-c	8.2	7.7	7.0	5.9	4.4	1.8
18.7 kW b-c	13.7	13.1	12.4	11.5	10.5	9.4
30 kW axial	10.0	8.8	7.1	4.7	1.8	—
52 kW axial	18.8	17.7	15.9	14.1	11.8	9.4

Ducts and drying floors

After being pressurised by the fan, and heated if necessary, air reaches the grain via a main duct and a series of lateral ducts spaced at 1–1.2 m. To keep back pressure as low as possible, ducts should have a cross-sectional area of 1 m² for every 10 m³/s of air that passes. This will give a maximum air speed of 10 m/s in the ducts. Similarly, the size of the openings through which the air passes from the ducts into the grain should ensure that air speed at this point does not exceed 0.15 m/s. Lateral ducts do not normally exceed 10 m in length because the output of air along the duct becomes uneven beyond this length. A better distribution of air to the grain lying between the ducts is achieved when air escapes from the sides rather than the top of the duct.

Above-floor ducts convert any sound, level concrete floor into a drying floor, and can be removed if the building is needed for some other use. However, the operations of filling and emptying the store are made more difficult by the presence of above-floor ducts. Many stores are equipped with permanent below-floor ducts which give a flat floor and are strong enough to support the weight of trailers or bulk lorries. Although the flexibility of use of the store is curtailed, moving the grain in and out of store is greatly facilitated.

Completely even distribution of drying air over the floor area can be achieved if the entire floor is perforated, and air passes up from a plenum chamber below the floor, through the perforations and into the grain. This system is used in some grain bins. A further refinement to this type of floor is an arrangement of louvres which directs jets of air to 'sweep' the last few tonnes of grain out through an unloading chute using the power of the drying fan. This eliminates the dusty task of sweeping or shovelling the last corn out of a flat-bottomed bin.

Uniform air flow throughout the grain mass also demands that the grain itself be uniform. Any differences in depth or compaction, or any local concentrations of rubbish will affect uniformity of air flow. Pre-cleaning the grain before it enters the store will remove most of the rubbish. Even layers of grain should be built up over the whole floor area by the careful use of conveyors, spreaders or grain throwers, to prevent rubbish, or a particular batch of wet grain, from being concentrated in one spot.

If air flow to any part of the store is restricted, the grain in that part of the store will dry slowly or not at all. Air flow through the grain can be measured at many points on the surface using an inverted funnel type of airflow meter, and any shortfall can be corrected. The instrument costs less than the price of a tonne of corn, and is simple and quick to use.

To ensure that moisture-laden air can escape easily from the roof space above the grain store, exhaust openings of at least 1 m²/(2.5 m³ s) of air flow must be provided at eaves, ridge or gable ends.

GRAIN CLEANERS

It is desirable to remove chaff, dust and impurities by pre-cleaning grain before it is dried or put into store. This avoids the expense of drying valueless material. The presence of impurities in a bulk store can block the spaces between the grains, and inhibit air flow. Concentrations of such impurities can build up under the discharge of stationary conveyors.

Grain may also be cleaned or graded when it is removed from store prior to sale; the ability to do this is particularly valuable where the grain is to be sold at a premium for seed, malting or baking, or in a situation of over supply where it is necessary to produce a good-looking sample.

Dual-purpose grain cleaners are available from many sources. For pre-cleaning, machines can accept 4–40 tonnes/h; this output is halved for cleaning after storage when more time is available and a more thorough job is required. Most machines combine one or two aspirations (separation by airblast) with a series of vibrating sieves. Air blast and sieve sizes are adjusted to suit the crop being cleaned. Inexpensive aspirator pre-cleaners are available with outputs to 12 tonnes/h.

Cleaners with reciprocating sieves should be mounted on a rigid base so that no damage will result from their constant vibration.

MANAGEMENT OF GRAIN IN STORE

Monitoring grain temperature

After the grain has been dried and placed in store, careful management and monitoring of its condition is essential if the full value of the crop is to be retained. Grain temperature is normally monitored, since heating is a symptom of most of the problems which affect stored grain. Any local increase in grain temperature must be detected and the cause corrected before there is time for grain damage to occur.

Many manufacturers offer permanently installed systems of grain temperature sensors, with a central monitoring station where all the measurement points can be checked at least weekly. Some systems will themselves perform the checking, and sound an alarm if a pre-set temperature is exceeded. At the other end of the scale, a temperature sensing spear can be pushed into the grain and the temperature read off in a number of positions and depths in the grain store to build up a composite picture of grain temperature.

Aeration

Any grain store over 20 tonnes in capacity should be fitted with aeration equipment. Storage driers will have this equipment anyway, but where the grain has been dried elsewhere and then put into store, small fans and air distribution ducts must be provided. The required airflow is 0.5 to 0.7 m^3/min for each m^2 of floor area, or 10 m^3/(h tonne) of grain. A 0.5 hp fan unit can aerate up to 50 tonnes of grain at a time. To give even air distribution, ducts should not be spaced more widely apart than twice the depth of the grain.

Aeration helps to maintain grain quality in the following three ways.

General cooling

The cooler the grain, even down to freezing point, the better it will keep (*see Figure 11.1*). Without refriger-

ation, grain can only be cooled to ambient temperature. The ability to force cold air into the grain as the season cools, and to take advantage of cold nights in order to reduce the temperature of the grain is a valuable aid to the management of the store.

Where aeration equipment is available in the store, the ouput of a continuous drier can be uprated by converting its cooling section into additional drying space, and cooling the dry grain in the store.

In addition to the conditioning of dry grain, aeration has been used to preserve undried grain of up to 21% MC which is to be used on the farm. The success of this system is dependent on the season; cold nights during harvest, and a cool dry autumn are ideal; a mild, moist autumn might not permit the grain to be cooled quickly enough to prevent mould development. As with other low-cost storage systems, careful management and a thorough understanding of the principles involved, are essential for success.

In the 1960s, a number of stores were fitted with refrigeration equipment for chilling the grain, but very few are still in operation.

Elimination of hot-spots

When localised heating of the grain is detected the traditional solution is to 'turn' the grain to break up the hot spot and allow it to cool down by exposure to the air. When aeration equipment is fitted, cool air can be blown through the grain to reduce spot temperatures without the need to move the grain.

Maintaining an even temperature throughout the grain

When the grain goes into store, it is all at approximately the same temperature as the outside air. As the season progresses, the layers of grain in contact with the walls of the store will cool, while the grain in the centre of the store will remain at its original temperature. Sunshine will heat the grain at the south side of a store. Any temperature inequalities of this sort will set up convection currents in the air which fills the spaces between the grains. This will result in moisture removal from the warmer grains and condensation on the cooler grains, particularly just below the surface at the centre of the store. This process can result in mould development, which may be followed by heating, insect activity and even sprouting.

If the grain is cooled in line with the cooling of the season, temperature will be uniform throughout the grain mass and moisture movement will not occur. This eliminates the need for costly over-drying which gives a 'safety margin' to allow for the effects of moisture movement. As well as minimising any losses due to deterioration of the grain, the correct use of aeration equipment will permit drying costs to be reduced.

HANDLING GRAIN

Most grain storage installations rely on conveyors for moving the grain. The characteristics of the common types of handling equipment are summarised below.

Auger conveyors

The grain is moved by a screw, or 'flight' which rotates inside a steel tube. Grain is drawn in at one end, and is normally discharged at the other end although intermediate discharge points are possible.

Augers are made in diameters of 90, 115, 150, 200 and 300 mm. Outputs of dry wheat, when the conveyor is inclined at 45 degrees to the vertical are as follows: 114 mm–25 tonnes/h; 150 mm–35 tonnes/h; 200 mm–58 tonnes/h; 300 mm–120 tonnes/h. Output is reduced with other grains, or if the grain is damp, or if the angle is steeper, and is increased if the angle is flatter. Portable augers are made in lengths up to 15 m: large units have a wheeled tripod stand.

The auger is a versatile conveyor, which will operate in many positions without adjustment. There is considerable churning and abrasion of the conveyed material, particularly if the auger is only partially full. The churning effect may be used to advantage in mixing a preservative with grain, but augers must not be used for conveying fragile materials such as rolled barley or pelleted feeds, because of the risk of damage.

An auger may be used to meter the flow of, for example, an ingredient into a mixer by regulating its speed of rotation. This gives very accurate control of output.

The flight of the auger projects beyond the tube at the intake end. This portion must be guarded at all times with a welded-mesh screen to avoid the risk of serious personal injury.

Chain and flight conveyors

The grain is moved along a smooth wooden or steel trough by an endless chain fitted with horizontal scrapers or 'flights'. The grain can be fed into the trough at any point, and is discharged at the end of the conveyor or at intermediate discharge points if these are fitted. Some models are reversible.

Chain and flight conveyors are normally horizontal, although some can work at small slopes. They are quiet in operation, long-lasting, economical in power, and do little damage to the grain. Lengths up to 60 m are available; outputs range from 15 tonnes/h to more than 100 tonnes/h. They are widely used as top conveyors in bin-type grain stores. 'Double-flow' models can supply grain to a continuous drier, and also take excess grain back to store.

Belt conveyors

The grain lies on an endless fabric-reinforced rubber belt which slides along a flat bed. Grain can be loaded onto the conveyor at any point, through a hopper which concentrates the grain in the centre of the belt. Discharge is over the end of the belt, or at a movable discharge point where the belt passes in an S-shape over a pair of rollers. The grain shoots over the upper roller and is deflected into a side discharge trough to right or left.

A motorised discharge can move slowly back and forth along a conveyor distributing the grain evenly along the whole length. In conjunction with a grain thrower to project the grain across the store, a level layer of grain can be built up over the complete area.

The belt conveyor is quiet, and very gentle with the material being conveyed. It is found as top and bottom conveyors in bin-type grain stores. Conveyors with cleated belts can operate at steep angles, but the typical grain version has a smooth or textured surface without cleats, and operates horizontally.

Belt conveyors are available in lengths up to 45 m and capacities from 20 to over 50 tonnes/h. Power requirement is greater than for the chain and flight machine.

The bucket (or belt) elevator

This is a device for raising grain vertically. A series of steel or plastic scoops or buckets are bolted to a flat endless belt which moves vertically between a pair of pulleys. The motor and drive pulley are at the top, while the belt passes round a tensioning idler pulley at the bottom. Grain is fed into the conveyor through a regulating orifice up to 1 m above its lowest point, and is discharged from a spout up to 1 m below its highest point. For this reason, the elevator usually stands in a hole at least 1 m deeper than the bottom of the receiving pit, and is often subject to flooding in winter when the water table is high. The top of the conveyor is generally accommodated in a small extension of the grain store roof where access for maintenance may be less than good.

No damage can occur to the grain while it is in the buckets, but careful operation of the intake slide, and correct design of intake and discharge points are necessary if damage is to be avoided entirely.

Bucket elevators are available in capacities from 5 to 120 tonnes/h, and heights to 25 m. 'Double-leg' models combine two separate conveyors with a single drive unit. The two legs may be used separately; for example, one leg may load wet grain into a continuous drier while the other takes dry grain from the drier and raises it to the top conveyor of the store. When a high throughput is required, for example when loading a lorry, both legs can be used to raise grain from the bottom conveyor into the reversed top conveyor which carries it to a discharge chute outside the store.

Multiple valving devices can be used to direct the grain from the conveyor's discharge to alternative destinations. Controls at ground level simplify the operation of such valves.

Pneumatic conveyors

The grain is carried along a closed tube 130 to 210 mm in diameter by a stream of air moving at 70 to 100 km/h. The air stream is produced by a narrow, large-diameter fan driven by electric motor or tractor power take-off.

Intake may be by a flexible suction spout, which is very convenient for emptying flat bottomed bins or clearing floor stores. Alternatively, the grain may be injected into the air stream through a metering hopper. Discharge is from the far end of the system, which may be up to 200 m from the intake if sufficient power is applied, at outputs of 5 to 40 tonnes/h. The grain does not pass through the fan but considerable abrasion can occur as it moves along the pipes and round bends; this shows itself in the large amount of dust which is generally produced at the discharge end.

This is a very versatile conveying system, as the pipework can be joined together to accommodate any storage arrangement, and the power unit is generally portable and can be used at several locations during its season of work. Against this must be set a high initial cost, and the power consumption of 1–2.5 kW/(tonne h) is high, particularly for those models with a suction intake. Fragile materials such as rolled grain or pelleted diets should not be handled in this way.

The grain thrower

A floor store can be filled in a series of flat layers if a grain thrower is available. The grain is metered onto a fast moving conveyor belt or a rotating paddle-wheel which throws it up to 12 m horizontally, and to a height of up to 5 m. Larger models stand on the floor, smaller units may be mounted at the discharge of other types of conveyors. Outputs range from 5–120 tonnes/h, at a power consumption of 1 kW per 4–7 tonnes/h. There is little damage to the grain, although some dust is produced.

A further advantage of loading a floor store in this way is that any rubbish present in the grain is distributed evenly throughout the store and will not have any local effect on the drying process.

Grain weighers

A very important instrument in the grain producer's armoury is the weigher. It permits him to determine yields at harvest time, to control his stock of grain, and to weigh the grain as it comes out of store for sale or for livestock feed on the farm.

The weigher consists of a counterbalanced hopper which is positioned in the flow path of the grain. When the hopper has received a pre-set weight of grain, the flow is cut off, one increment of weight is recorded on a counter, the hopper is dicharged and the cycle is repeated. Very accurate weighing is possible at rates from 6 to over 100 tonnes/h according to model.

Loading lorries

If the crop is to be sold in bulk, it may have to be discharged into lorries of 25 tonnes or more. If conveying equipment of 20 tonnes/h is used, which is adequate for most other functions in the grain store, the lorry will be tied up for more than an hour and this may make the product less attractive to a potential customer. High capacity conveyors are available, but would only be fully utilised for a few hours per year.

To overcome this problem some grain stores are equipped with covered V-bottomed hoppers of 30-tonne capacity with several large discharge ports, mounted high enough for a bulk lorry to be driven underneath. The hopper can be filled slowly with the store's existing conveyors, allowing plenty of time for the grain to be passed through cleaning equipment if required, and yet the lorry can be filled in 10 min or so with the minimum of effort or dust. In the likely event of there being a buyer's market for grain, it is this type of detail, plus a quality product, that will ensure a sale in a competitive situation.

STORAGE OF GRAIN IN SEALED CONTAINERS

Where grain is to be processed into livestock feed on the farm, there are many advantages to storing it without drying. Apart from the fact that there is no drying cost, the grain can be handled into store at a very fast rate so there are no harvest bottlenecks. Moist grain can be rolled easily, and produces little dust when ground; this can improve palatability to stock and improve working conditions for the stockman.

When stored under airtight conditions, the oxygen concentration is quickly reduced to less than 2% by the respiration of the grain and other aerobic organisms such as moulds. The grain is killed and most destructive organisms die or become dormant. Some fermentation may take place if the moisture content is above 24% and this will taint the grain.

As long as the container remains sealed, the condition of the stored grain is stable. When grain is removed from the store, care must be taken to allow only the minimum of air to enter, and to reseal as soon as possible.

Once exposed to the air, the moist dead grain will deteriorate rapidly especially if the weather is warm. Normally, only enough for one or two days' use should be withdrawn at a time.

The container for airtight storage may be a plastic or butyl bag, or the more durable enamelled steel tower silo. Capacities are from 15 to 800 tonnes/silo. Unloading is by means of augers, a sweep arm which can traverse the circular bottom of the silo, and a fixed arm which brings the grain from the centre of the silo to discharge outside through a resealable spout.

If grain above 22% MC is to be stored in a tower silo the lower layers are likely to be compressed by the weight above, and may bridge and be difficult to unload. A few tonnes of dry corn at the bottom of the silo will facilitate this initial stage of unloading, as will the use of a gentle method of filling such as a bucket elevator rather than the more convenient but damaging blower.

USE OF CHEMICAL PRESERVATIVES TO STORE GRAIN

An alternative method of controlling moulds and pests in undried grain is to treat the grain with a chemical preservative such as propionic acid, at a rate which increases with the moisture content of the grain.

After treatment, no special storage structure is required; the grain must be kept dry to avoid dilution or leaching of the preservative so floor, walls and roof must be waterproof. Since propionic acid is highly corrosive, the materials of the store must resist corrosion, or be coated with plastic or a chlorinated rubber paint for their protection. Any equipment which handles either the preservative, or grain which has been freshly treated, must be thoroughly washed out with water after use, to minimise corrosion.

The preservative itself or its fumes can cause irritation or injury to the skin, eyes, mouth or nose. Protective garments should be worn, and a supply of water kept close by in order to wash off splashes. The operator should stand up-wind of working application machines or piles of recently treated grain.

The applicator generally consists of a pump, flow regulator and nozzle which draw the chemical from a container and spray it onto the crop at the intake of a short (1.5 m) length of auger. The tumbling effect of the auger is very effective in distributing the chemical over every grain, an important requirement of this system. Charts are provided so that the chemical flow rate can be adjusted to match the grain moisture content (up to 30% and more) and the (measured) throughput of the auger.

The cost of the chemical is relatively high, so in order to be viable the handling and storage parts of the system must be inexpensive.

Since the grain is killed and tainted by the preservative, there is no market for the grain other than for stockfeed. Weed seeds are also killed, which may be an advantage.

Because the chemical is absorbed into the grain, protection continues after the grain is removed from store, so quite large batches of feed can be prepared at a time even in warm weather, without risk of deterioration.

Part 2

Animal production

12

Animal physiology

R. M. Orr

Profitable animal production involves the efficient conversion of feed into animal products in the form of meat, milk, eggs and wool. Consequently an understanding of those physiological activities that are part of these processes is a prerequisite.

In this chapter those main aspects of animal function and their physiological control that are of particular relevance to the animal productionist are given attention. The chapter is divided into five main subdivisions:

(1) regulation of body function;
(2) reproduction;
(3) lactation;
(4) growth;
(5) environmental physiology.

A further area that is clearly relevant to this chapter, that of 'digestion and metabolism' is covered in Chapter 13, 'Animal nutrition'.

REGULATION OF BODY FUNCTION

In order that animals can survive their environment and changes in that environment they must possess the necessary coordinated mechanisms to maintain their own internal environment in a steady state. These physiological reactions that maintain the steady states of the body have been designated 'homeostatic' and are achieved by the combined action of the various organ systems of the body. This coordinated action is carried out by the nervous and endocrine systems. These systems must initiate the necessary adjustments that enable animals to respond to external environmental changes such as the availability of food, temperature and light.

The nervous system

The nervous system in mammals is an extremely complex collection of nerve cells which plays an essential part in the functioning and behaviour of the animal. The basic units of the nervous system are the nerve cells or neurons and the associated receptors.

The ability of animals to maintain homeostasis is dependent on the nervous system receiving and responding to information from specialised receptor cells and to this end mammals have evolved receptors that are sensitive to a wide range of physical (e.g. temperature), chemical (e.g. specific components of the blood) and mechanical (e.g. muscle stretch) stimuli. When any response to information received at receptors is necessary the information is carried from receptors to the cell or tissue (effector) where appropriate adjustments are made. This is the function of the neurons.

There are three different types of neurons:

(1) sensory or afferent neurons which carry information from receptors to the spinal cord and brain;
(2) motor or efferent neurons which carry information from the spinal cord and brain to the effector organ; and
(3) interneurons or association neurons which make connections between sensory and motor neurons within the spinal cord and brain.

Neurons consist essentially of a cell body with an elongated projection, the axon, and numerous shorter projections, the dendrites (*Figure 12.1*). For our purposes the cell body can be thought of as performing two functions: firstly, it receives information from the axons of other cells both directly onto its surface membrane and also via the dendrites and, secondly, it acts to sum all the effects of inputs, which may be

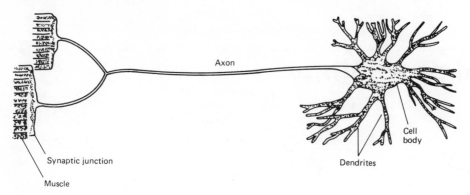

Figure 12.1 A nerve cell

either excitatory or inhibitory, that it receives. Should this sum of inputs exceed a critical level then an impulse is triggered which is then carried along the axon to the synaptic region which in turn passes the neuronal signal to the next stage in the nervous system, either another neuron or, alternatively, a muscle or gland. Propagation of signals along an axon is dependent on changes in the electrical potential between the inside and the outside of the axon membrane which results from changes in the permeability of the membrane to sodium, potassium and chlorine ions. Passage of neuronal signals at synaptic regions involves the release of a chemical transmitter substance into the small gap between the synaptic region and the membrane of the next neuron. The nature of the transmitter substance determines whether the influence on the next neuron is excitatory (e.g. acetylcholine and noradrenaline) or inhibitory (e.g. γ-amino butyric acid). These transmitter substances released by the synapse are received by specialised sites on the dendrites of the next neuron which have the effect of changing the electrical potential of the neuron.

It is the nerve cells—the basic building blocks—from which the various components of the nervous system of mammals are built (*Figure 12.2*).

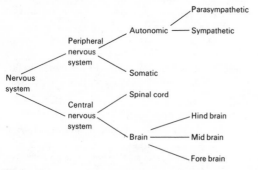

Figure 12.2

Division of the nervous system between central and peripheral is made simply on the basis of whether the

fibres or cell bodies lie outside the spinal cord. The peripheral nervous system can be further subdivided into the somatic nervous system and the autonomic nervous system. The former is involved in the control of the skeletal muscles in the body whereas the latter controls specific target organs in the body such as the heart, lungs, blood vessels and intestines. The autonomic nervous system has two subdivisions, the sympathetic and the parasympathetic. In many cases these two divisions have opposite effects on the target organs and glands. The sympathetic nervous system acts to increase the level of activity or secretion in an organ whereas stimulation of the parasympathetic will have the opposite effect.

The central nervous system consists of the spinal cord and brain. The spinal cord is contained in a continuous channel within the vertebrae, which form the backbone. At each vertebra there are two openings in the base through which nerves can pass in and out of the spinal cord. Sensory nerves enter through openings on the dorsal surface, motor nerves leave through openings on the ventral surface. The brain itself is a vastly complex organ consisting of billions of nerve cells, interconnected by neurons. However, certain areas of the brain (*Figure 12.3*) have been found to have specific functions:

(1) the hind brain: this consists of the medulla and the cerebellum. The medulla is involved in the regulation of the heart, breathing, blood flow and posture. The cerebellum is involved in the coordination of muscle activity,

(2) the fore brain: this consists of the thalamus, the hypothalamus and the cerebrum. The thalamus plays an important role in the analysis and transmission of sensory information between the spinal cord and the cerebral cortex. Lying directly beneath the thalamus is the hypothalamus which, despite its small size has been shown to be involved in the control of basic behaviours such as hunger, thirst and sexual behaviour. The hypothalamus also has close connections with the pituitary—a hormone secreting gland—which,

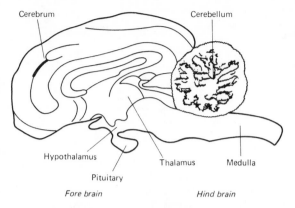

Cerebrum

Cerebellum

Hypothalamus

Thalamus   Medulla

Pituitary

Fore brain   *Hind brain*

Figure 12.3 Midline section of the brain showing the location of major structures

under the control of the hypothalamus, secretes hormones into the blood stream. In mammals, the cerebrum constitutes by far the major part of the brain. Definite areas of the cerebrum have been shown to control specific motor and sensory function.

The endocrine system

Like the nervous system, the endocrine system acts as a means of communication in the regulation of the physiological and biochemical functions of the body. Chemical regulation by hormones is known to be involved in reproductive activities, control of growth, regulation of intermediary metabolism and adaptation to external factors such as light intensity and temperature.

The endocrine system consists of various endocrine glands, the locations of which are shown in *Figure 12.4*. These glands produce chemical substances (hormones) which are carried by the blood to some target tissue or organ where they exert their effect. These effects may be either excitatory or inhibitory. Certain hormones can exert both excitatory and inhibitory effects depending on their concentration. In addition to acting on distant target tissues hormones may also act at the site of hormone production to inhibit further hormone secretion. This mode of control is termed a negative feedback mechanism.

Although certain hormones are probably secreted continuously in small amounts, the secretion of hormones is under the control of either the nervous system, other hormones or changes in the chemical composition of the blood. Specific examples of how hormonal secretion is controlled will be discussed in other parts of this section and other sections as appropriate.

The action of hormones on target tissues at the cellular level is not well understood. It is thought that certain hormones may act by influencing the production of a particular enzyme whilst others may act by way of altering the permeability of cell membranes to specific substances.

The hormones produced by the various endocrine organs, their chemical nature, target tissue and chief physiological effects are listed in *Table 12.1*. As can be seen from the table some endocrine organs produce more than one hormone. Where this occurs each hormone is produced by a different type of cell within the organ.

Since the endocrine system plays such a fundamental role in the control of animal activities there has been considerable activity in recent years in the use of natural and synthetic hormone preparations in animal

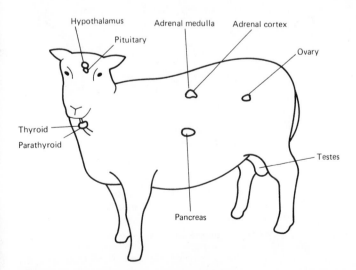

Hypothalamus   Adrenal medulla   Adrenal cortex

Pituitary

Ovary

Thyroid

Parathyroid

Testes

Pancreas

Figure 12.4 Location of major endocrine glands

Table 12.1 Endocrine structures and hormones

Endocrine structure	Hormone	Chemical nature	Major activity
Hypothalamus	Growth hormone releasing hormone (GHRH)	Unknown	All are neurohormones regulating the release of the pituitary hormone indicated
	Prolactin releasing hormone (PRH)	Polypeptide	
	Luteinising hormone releasing hormone (LHRH)	Polypeptide	
	Follicle stimulating hormone releasing hormone (FSHRH)	Polypeptide	
	Thyrotropin hormone releasing hormone	Peptide	
Anterior pituitary	Growth hormone (GH)	Protein	Stimulates growth
	Adrenocorticotrophic hormone (ACTH)	Protein	Stimulates release of hormone of the adrenal cortex
	Thyrotropin (TH)	Protein	Stimulates release of thyroxine
	Follicle stimulating hormone (FSH)	Protein	Regulates development of ovarian follicles in females and spermatogenesis in the male.
	Luteinising hormone (LH)	Protein	Triggers ovulation and stimulates progesterone and testosterone release.
	Prolactin	Protein	Stimulates production of milk by mammary gland
Posterior pituitary	Oxytocin	Peptide	Stimulates uterine contraction and involved in milk release
Adrenal gland	Vasopressin	Peptide	Reabsorption of water by the kidney tubules
Cortex	Aldosterone	Steroid	Regulates water and electrolyte balance
	Cortisone and corticosterone	Steroid	Regulator of carbohydrate metabolism
Medulla	Adrenaline and noradrenaline	Amino acids	Regulator of carbohydrate metabolism in muscle and liver, constriction of peripheral vessels and contraction of smooth muscles
Thyroid	Thyroxine	Amino acid	Stimulates oxidative metabolism
	Calcitonin	Polypeptide	Regulates calcium levels in body fluids ($\downarrow$)
Parathyroid	Parathyroid hormone	Polypeptide	Regulations calcium levels in body fluids ($\uparrow$)
Pancreas	Glucagon	Polypeptide	Raises blood glucose levels
	Insulin	Polypeptide	Lowers blood glucose levels
Digestive tract Duodenum	Secretin	Polypeptide	Stimulates release of pancreatic enzymes
	Cholecystokinin	Polypeptide	Stimulates release of pancreatic enzymes and regulates the gall bladder
Stomach	Gastrin	Peptide	Stimulates secretion of HCl and pepsin by gastric mucosa
Testes	Androgens (e.g. testosterone)	Steroids	Stimulates development of secondary sex characteristic in male; maintains accessory sex organs
Ovary	Oestrogens	Steroids	Stimulates development and maintenance of female secondary sexual characteristics
	Progesterones	Steroids	Stimulates uterus in preparation of ovum implantation; development of mammary gland for lactation
	Relaxin	Polypeptide	Relaxation of pelvic tissues at parturition

science, particularly in the areas of promoting increased growth rate and the regulation of the reproductive process.

REPRODUCTION

The process of reproduction occupies a position of central importance in animal systems; calving indices in dairy and beef herds, lambing percentages in sheep and number of piglets born per year in sows all greatly influence the economics of their related enterprises.

The reproductive organs

The primary reproductive organs in males are the testes and in females the ovaries. The male reproductive system of farm animals consists of two testes

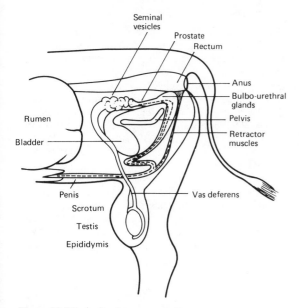

Figure 12.5 Reproductive tract of bull

contained in the scrotum, accessory organs including ducts and glands, and the penis. The anatomy of the reproductive tract of the bull is illustrated in *Figure 12.5*. The testes produce the male gametes (or spermatozoa) and testosterone, the male sex hormone. The scrotum provides an environment of a lower temperature than the rest of the body suitable for sperm production whilst the ducts, including the epididymis and vas deferens, are concerned, along with the penis, in providing a means by which the spermatozoa may reach the site of fertilisation in the female reproductive tract. The accessory glands, including the seminal vesicles, prostate and bulbo-urethral glands function in the production of seminal

fluid which, whilst not essential to successful fertilisation, does produce a suitable environment for sperm transport.

The female reproductive system of farm animals consists of two ovaries, two uterine or Fallopian tubes, the uterus, the vagina and the vulva. The anatomy of the reproductive tract in the cow is illustrated in *Figure 12.6*. The ovum (or egg) is expelled from the ovary and received by the infundibulum and carried to the Fallopian tube. Fertilisation normally occurs during the passage of the ovum from the ovary to the uterus. Within the uterus the fertilised ovum develops into an embryo and then a fetus and finally passes out of the uterus through the vagina and vulva as a newborn animal at parturition.

The development of the reproductive system

The growth and development of the reproductive tract is a gradual process and it is some considerable time after birth before the young animal is sexually active. Puberty is the time when the animal becomes able to reproduce, spermatogenesis and ovulation occurring in response to changed levels of two hormones, follicle stimulating hormone (FSH) and luteinising hormone (LH), produced by the anterior pituitary gland which itself is regulated by neurohormonal secretions from the hypothalamus. The increased secretion of these two hormones or gonadotrophins results in the development and eventual rupture of one or more follicles in the ovaries which marks the point of puberty in the female animal. Thereafter she will exhibit a regular cyclical pattern of reproductive behaviour. In the male animal no such accurate identification of puberty is possible since levels of spermatogenesis sufficient to effect conception are achieved gradually but as in the female the onset of puberty results from increased levels of FSH

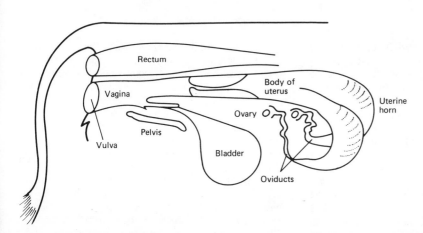

Figure 12.6 Reproductive tract of cow

and LH. The latter is sometimes referred to as interstitial cell stimulating hormone (ICSH) since its action in the male is on the interstitial cells of the testes where spermatogenesis occurs.

The attainment of puberty in the female tends to be a function of both age and nutrition in the pig, whereas in cattle and sheep the age of puberty is markedly affected by nutrition and is more closely associated with weight than age. However, in sheep the effect of nutrition is complicated by the time of year when lambs are born, since the pituitary in sheep is particularly responsive to decreasing day length.

The oestrous cycle

Once puberty has been attained in domestic female animals oestrous cycles follow at fairly regular intervals unless interrupted by pregnancy. The oestrous cycle is controlled directly by hormones from the ovaries, oestrogen and progesterone, and indirectly by hormones secreted by the anterior pituitary, FSH, LH and prolactin. The oestrous cycle may be divided into several well-defined phases called pro-oestrus, oestrus, metoestrus and dioestrus, the basic pattern being the same in all domestic animals although species differences, particularly with regard to duration, are found in different parts of the cycle.

Pro-oestrus

Under stimulation, mainly of FSH secreted from the anterior pituitary, the ovaries increase their secretion of oestrogens which cause the development of the ovarian follicle with its enclosed ovum. At the same time there is increased vascularisation of the uterus and vagina in preparation for oestrus and possible subsequent pregnancy.

Oestrus

This is the period of sexual receptivity (heat) largely determined by the levels of circulating oestrogens; it is during this period or shortly after this time that ovulation occurs. The stimulus to this phase is a reduction in the FSH levels in blood and an increase in LH levels which, in addition to causing the follicle to swell and mature, initiates the production of progesterone from the developing follicle. Oestrus terminates about the time that rupture of the ovarian follicle occurs, the ovum being expelled from the follicle to pass into the upper part of the Fallopian tube. Should the follicle fail to break and release the egg it will continue to grow and a condition known as *cystic ovaries* may result with the animal appearing to be 'on heat' much of the time. These cysts can be palpated through the rectum.

Metoestrus

With the release of pressure following rupture the follicle walls collapse and in the cavity remaining there develops a yellow coloured body termed the *corpus luteum*. Unless pregnancy intervenes this body has a highly predictable life, secreting progesterone which through its effect on the pituitary inhibits the development of further follicles and hence the occurrence of further oestrous periods for a fairly specific number of days before its life is terminated. Progesterone is necessary for implantation and nourishment of the fertilised ovum and for the development of the mammary gland should pregnancy result. Sometimes a corpus luteum persists when an animal is not pregnant, the continued secretion of progesterone preventing the next phase in the oestrous cycle. Manipulation of the ovaries via the rectum by a veterinary surgeon will usually result in resumption of normal cycles.

Dioestrus

Dioestrus is the relatively short period of inactivity between oestrous cycles. In ewes oestrous cycles are limited to a particular time of year, termed the breeding or sexual season, at the end of which, should they remain non-pregnant, they go into a period of anoestrus until the following breeding season. The period of anoestrus is decided by seasonal changes in length of day, ultimately affecting the release of neurohormonal factors from the hypothalamus which in turn effects the release of hormones from the anterior pituitary.

In sheep the duration of the sexual season differs among breeds and is related to their latitude of origin, northern breeds such as the Blackface having a much more restricted season than breeds such as the Merino which originated in Mediterranean latitudes.

Further information is given in *Table 12.2*.

Semen production and quality

The semen is composed of two parts, the spermatozoa and the seminal plasma. The production of sperm takes place in the seminiferous tubules that comprise most of the testicular mass. Between the tubules there are interstitial cells which secrete the male sex hormone testosterone which is essential for spermatogenesis, maintains the secondary sex organs and is responsible for the secondary male characteristics. Storage of sperm takes place in the epididymis and the seminal plasma is contributed by the secretory fluids produced in the accessory sex organs when the semen passes the vas deferens and urethra during ejaculation. The seminal plasma acts as a buffer due to its content of protein, citrate and bicarbonate. These buffers protect the sperm against the acidic conditions prevailing in the female genital tract.

Table 12.2 Some data on normal female reproduction

(a)

Animal	Onset of puberty	Age first service	Length of cycle	Duration of oestrus
Cow	8–12 months	15–18 months	21 d (18–24 d)	18 h (14–26 h)
Ewe	4–12 months (first autumn)	First autumn or second autumn	17 d (14–20 d)	36 h (24–48 h)
Sow	4½–10 months	7–10 months	21 d (18–24 d)	48 h (24–72 h)

(b)

Animal	Time of ovulation	No. of ova shed	Length of gestation	Optimum time for service
Cow	10–15 h after end of oestrus	1	282 d (277–290 d)	Mid to end of oestrus
Ewe	Near end of oestrus	1–4	149 d (144–152 d)	16–24 h after onset of oestrus
Sow	Middle of oestrus	10–20	114 d (111–116 d)	15–30 h after onset of oestrus

The average volume and sperm concentration of each semen ejaculate varies greatly with species, but in general volume and density are inversely related. Typical amounts are:

bull   4–8 cm³ containing 1 million sperm/cm³
ram   0.5–2 cm³ containing 3 million sperm/cm³
boar   150–400 cm³ containing 100 000 sperm/per cm³

A knowledge of semen quality is desirable where the fertility of a male used for natural mating is being checked and where semen for artificial insemination is being diluted and stored. Microscopic examination of semen can provide information on concentration, motility and appearance of sperm.

Fertilisation and implantation

Fertilisation of ova shed during or shortly after oestrus has ended generally occurs in the upper part of the Fallopian tube. Normally, natural service takes place during the heat period prior to ovulation so that sperm reach the site of fertilisation some time before the ova. This is desirable because spermatozoa require a period in the tract before they are capable of penetrating the shell membrane of the ova (zona pellucida) and the egg membrane (vitelline membrane). This phenomenon is referred to as capacitation.

Typical viability times for the ova of cows, sows and ewes are 6, 15 and 24 h respectively. The sperm of domestic animals are viable within the female genital tract for 24–36 h. With both ova and sperm, loss of viability is not sudden but fertilisation involving ageing gametes frequently results in embryonic death.

Fertilisation involves penetration of the sperm into the egg, the breakdown of the membrane of the spermatozoon to form the male pronucleus which unites with the female pronucleus to form the zygote or new individual.

After fertilisation the embryo passes down the Fallopian tube, reaching the uterus in a few days. For implantation to take place the uterus must be in precisely the correct condition. The stimulus to this comes from the changes in the levels of oestrogen and progesterone that occur during the oestrous cycle. Recent developments in embryo transfer have highlighted this necessity for the uterus to be in an appropriately synchronous physiological state.

Development of the fetus and parturition

In the initial stages of development the nutritive requirements of the embryo are met by the diffusion of nutrients secreted by uterine glands. But as it increases in size the placenta, a vascularised arrangement of membranes, develops such that nutrients from the dam can reach the fetus and, in turn, waste products from the fetus be transferred to the dam and excreted. By the fourth to sixth week after fertilisation the embryo has differentiated into its major components and both the fetal and maternal placenta are growing rapidly. Placental attachment does not occur until about one-third of the way through pregnancy.

Growth of the developing young is slow in the early stages of pregnancy and little affected by nutrition, but towards the end growth of the fetus is rapid and related to the nutritional status of the dam and begins to outstrip the growth of the placenta. This imposes a stress on the developing young and at around this time a change in the levels of circulating oestrogen and progesterone occurs. Progesterone production which has been responsible for the maintenance of pregnancy begins to decline whilst the levels of oestrogen produced by the placenta increase. This change increases the susceptibility of the uterus wall to the action of oxytocin, contractions begin to occur and labour commences. In addition to the effects of oxytocin a further hormone, relaxin, which is formed by the ovaries and placenta towards the end of pregnancy causes relaxation of the cervix muscles and pelvic ligaments allowing the fetus to be expelled from the uterus.

Developments in reproductive physiology

The successful development of artificial insemination (AI) and its widespread use in animal breeding has been one of the great advances in animal production in recent times. Techniques of semen dilution (one ejaculate from a bull may provide sufficient semen for up to 1000 inseminations) and preservation allows enormously extended use of superior sires. To date AI has found greatest application in dairy cows and to a lesser extent sows since the husbandry of these species allows the early recognition of oestrus in individual animals. This is necessary in order that the timing of insemination be correct. However, even in these species 'standing' oestrus may be of short duration and at least four separate checks are required each day to detect the maximum number of oestrus in cows.

To overcome difficulties of oestrus recognition, techniques of oestrus synchronisation have been developed. In sheep fairly successful procedures based on the administration of progestagens in intravaginal sponges for a period of about 15 d may be used. These substances inhibit ovulation and, on withdrawal, oestrus follows after a fairly predictable time. In cattle treatments based on natural or synthetic progestagens have been less successful but procedures based on the use of a synthetic analogue of the hormone prostaglandin $F_2\alpha$ offer an alternative. This latter substance is a potent luteolytic compound in cattle and will synchronise oestrus in animals that are in the luteal phase of the oestrous cycle. In addition to offering the advantage of timeliness of AI, oestrous synchronisation also offers potential benefits in the control of calving and lambing patterns.

Whilst AI has for many years provided a means by which the animal breeder can exploit the genetic potential of the male it is only in relatively recent years that methods by which superior females could be exploited have received attention. Although the ovaries of breeding females have a very large number of ova present in them only a very small number of these will be fertilised when natural breeding methods are used. However, techniques of super-ovulation induced by the injection of FSH from sources such as pregnant mares' serum (PMS) provide a means by which a greater number of eggs may be shed from the ovaries at ovulation.

Following fertilisation these eggs may be obtained by non-surgical means and transferred to appropriately synchronised recipient females. In cattle, where the technique has found widest implementation, prostaglandin treatments in both donor and recipient animals has been used to synchronise breeding cycles for transfer purposes.

This technique of ova transfer is also being directed towards obtaining twin calves.

In considering these valuable developments it should be noted that their widespread use to breeding is still full of procedural difficulties which are the subject of continuing investigation.

LACTATION

The growth and development of the mammary gland and the subsequent manufacture and secretion of milk represents an important phase of the reproductive cycle in mammals. In this section an attempt will be made to give coverage to our knowledge of the structure and development of mammary tissue, of the processes involved in the manufacture and secretion of milk and of the involvement of hormones in lactation.

Structure and development of the mammary gland

The mammary gland (*see Figure 12.7*) consists of numerous lobules made up of clusters of rounded alveoli. It is the epithelial cells lining the lumen of these alveoli, the secretory cells, which are the site of milk synthesis, the milk secreted passing into small ducts which unite with ducts from neighbouring alveoli to form the large ducts or lactiferous tubules. The micro-anatomy of the mammary gland where these large ducts converge on the mammary papilla or teat differs in the various species. In the cow and ewe the large ducts lead into a single udder cistern. Prior to milking about 40–50% of milk is held in the gland cisterns and large ducts, the remainder being held in the small ducts and alveoli. Continuous with the gland cistern is the teat cistern and the streak canal through which the milk is eventually secreted. The streak canal is surrounded by a sphincter composed of circular smooth muscle fibres. In cows that tend to leak milk this sphincter is not tight enough whilst the opposite is the case in cows that are slow milkers.

The mammary gland of the sow differs in that each teat contains two separate streak canals each leading to a teat cistern and gland cistern.

The mammary glands develop along the so-called milk line of the embryo. In the sow the normal number of teats is seven pairs distributed from the sternum to inguinal regions; in the cow there are two pairs in the inguinal region, the rear two accounting for about 60% of total milk production; and in the ewe a single pair of glands again in the inguinal region. Extensive

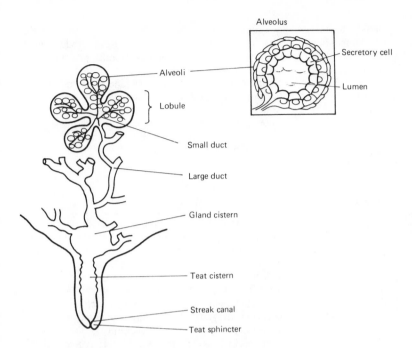

Figure 12.7 Diagram showing structure of the mammary gland of the cow

development of the mammary gland is associated with puberty and is largely due to increases in levels of the ovarian hormones. Oestrogen is involved with the development of the larger ducts, whilst progesterone stimulates alveolar development. In addition to these ovarian hormones the full development of the lactating gland requires prolactin, growth hormone, insulin, thyroid hormone and adrenal corticoids, the secretions of which are under either indirect or direct control of the pituitary. In other words successful lactation requires an appropriate hormonal environment.

Mammary development during pregnancy is characterised by the growth of small ducts and extensive proliferation of the alveolar structures under the influence of the ovarian hormones. These large amounts of oestrogen and progesterone also keep the mammary glands in their unresponsive state during pregnancy and it is not until near parturition when their levels decline that the lactogenic hormone, prolactin, in combination with growth hormone and adrenal corticoids, can act directly on the epithelial cells lining the alveoli and initiate milk secretion.

Milk biosynthesis and secretion

Whilst the composition (*Table 12.3*) and quantity of milk secreted by the dams of the different domestic species varies considerably the milk from all species has the same basic qualitative composition. Milk essentially consists of two phases; an aqueous phase

Table 12.3 Approximate average composition of milk of different species (% by weight)

	Total solids	Fat	Protein	Lactose	Minerals/water soluble vitamins
Cow	12.6	3.7	3.6	4.6	0.7
Sow	20.0	8.0	6.0	5.2	0.8
Ewe	17.0	5.5	6.2	4.5	0.8
Rabbit	35.0	18.3	14.0	2.0	0.7
Human	12.4	3.8	1.0	7.0	0.6

in which are partitioned proteins in colloidal suspension and lactose, minerals and water-soluble vitamins in solution; and a lipid phase containing true fats, phospholipids, sterols and fat-soluble vitamins. The lipid in milk is mainly triglyceride and exists in the form of globules ranging from $3-5\,\mu m$ in diameter surrounded by a complex membrane which is acquired at the time of secretion from the lactating cell.

Because of the commercial importance of cow's milk as a food more is known about it than about the milk of other mammals and the mechanisms of synthesis of milk have been much investigated. In this section, therefore, synthesis will be described with reference to the cow.

The site of milk biosynthesis is in the vast number of epithelial cells which line the alveoli, each cell in an alveolus discharging its milk into the lumen or hollow part of the structure. The alveoli of the mammary gland are well supplied with blood capillaries and it is from the arterial supply that the metabolites used for milk synthesis are drawn. These include acetate, triglycerides, glucose, amino acids and proteins. There is still little known about the mechanisms involved in the transport of these metabolites from the capillaries into the lactating cells.

Once inside the lactating cells these nutrients may be used in the milk synthesis process. In the case of the milk proteins, the total free amino acids absorbed are apparently sufficient for synthesis but since the balance of absorbed amino acids is not the same as that present in milk proteins (80% casein; 20% lactalbumin and lactoglobulin) a certain amount of amino acid synthesis must occur. Protein synthesis involves the assembly of amino acids into a polypeptide chain through peptide bonds. The sequence of the amino acids in the various milk proteins is dictated by the usual method of transcription by ribonucleic acid on the ribosome of the cell.

Lactose is a disaccharide consisting of one molecule of glucose and one molecule of galactose. The precursor used for synthesis is glucose. Some absorbed glucose is converted to galactose within the lactating cells.

The fatty acids present in the triglycerides in milk fat range from 4–18 carbon atoms each and in comparison with other fat sources are characterised by having a relatively high proportion of shorter-chain fatty acids. Those fatty acids with chain lengths up to and including 16 carbon atoms are mainly synthesised within the lactating cell, the main precursor utilised being acetic acid. The feeding of diets low in roughage leads to the production of low proportions of acetic acid in the rumen and consequent reductions in the levels of fatty acid synthesis. The longer chain fatty acids are obtained from the glycerides of the blood. Non-ruminants appear to use glucose for the synthesis of milk fat. In all species the metabolism of glucose by the lactating cell would appear to be the main energy source for these synthetic processes.

The secretion of milk from lactating cells is a continuous process, but the rate of secretion is not constant and is inversely related to pressure in the alveoli and ducts. Immediately after milking when pressure is lowest the rate of secretion of milk is at a maximum. The actual mechanism by which such large particles as the protein micelles and fat globules are secreted from the lactating cell into the alveoli without destroying the integrity of the cell membrane is still the subject of investigation.

The length of time over which the lactating cells will continue to synthesise and secrete milk varies between species and also individuals. Certain minimum levels of prolactin and cortisol are particularly essential for the maintenance of lactation. Although secretory activity must eventually decline and cease under the influence of hormones produced by the ovaries the major causes of decline in secretory activity are the start of another pregnancy along with, under natural conditions, a reduction in the frequency of suckling by the young as they begin to consume solid food. Both these events result in a reduction in secretion of the lactogenic hormone, prolactin and further to this in the pregnant animal the fetus competes with the mammary gland for certain of the hormones and nutrients associated with milk secretion. A further consequence of a reduction in suckling frequency is the retention of milk and increase in pressure in the alveoli which has an inhibitory effect on secretory activity.

The milk ejection mechanism

As a result of secretion, milk accumulates in the alveoli and ducts of the mammary gland. This milk is ejected from the alveoli and ducts into the gland cistern by contraction of the smooth muscle fibres (myoepithelial cells) which surround these tissues. This process is termed milk 'letdown' and is under the control of the pituitary hormone, oxytocin. A number of signals will stimulate oxytocin release. In the situation of natural suckling the butting of the mammary gland prior to suckling is the primary stimulus to oxytocin release. This is relayed to the pituitary via the hypothalamus. In machine milked dairy cows the normal routine carried out by the cowman prior to milking can stimulate the secretion of oxytocin by conditioned reflexes.

The milk ejection reflex can be disrupted by emotional disturbance. It is likely that this is due to the vasoconstrictive action of adrenaline reducing oxytocin entry to the gland.

GROWTH AND DEVELOPMENT

Meat production is essentially the growth of body tissues and an understanding of the processes involved and of the factors that may influence growth is essential in the quest for increased efficiency of production.

In this section the biological nature of growth and development is defined and an attempt made to elucidate the complex relationships between growth potential, body composition and efficiency of feed utilisation.

The nature of growth and development

'Growth' in animals may be simply defined as an increase in size and/or weight of the whole animal or part of the animal. 'Development' refers to the changes in proportions of the various organs and tissues that make up the animal as it grows from conception to maturity.

If we express graphically the increase in weight or size that occurs in the whole animal from conception

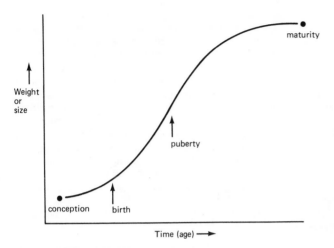

Figure 12.8 The sigmoid growth curve

to maturity a sigmoid curve results (*Figure 12.8*), there being a steadily increasing rate of growth from conception up to puberty after which the growth rate starts to decline at an increasing rate until maturity is reached.

The liveweight growth of farm animals is the gross expression of combined changes in the carcass tissues, organs and viscera and gut fill. In studies with farm livestock the liveweight growth of the body is primarily a reflection of the growth of the carcass tissues comprising the muscle, bone and fat. It is the relationship between these three tissues that the animal productionist aims to control and alter.

The temporal growth of each of these components of the body also follows a sigmoid curve similar to that of liveweight, but different growth curves are not in phase one with another. This is due mainly to the fact that the different tissues vary in priority for the available nutrients. The bone tissue reaches maximal growth rate prior to maximal muscle growth with adipose tissue being the latest of the body tissues to attain peak growth intensity (*Figure 12.9*). Thus carcass composition in terms of the proportions of muscle, fat and bone changes as an animal grows.

The chemical composition of the body changes with age in a way which clearly substantiates the phasic development of the tissues. The most marked changes with age are the decrease in the proportion of water in the body and the increase in the proportion of lipid. The almost inverse relationship between the water and fat content of the body reflects the lower water content of fat (10%) compared with that of muscle tissue (75%). The percentages of protein and mineral matter decline only slowly as growth proceeds mainly as a result of changes in lipid content; in fact if chemical composition is expressed on a fat-free basis the chemical composition of the body remains remarkably constant during growth reflecting that the bones and muscles of the limbs and trunk remain in proportion to one another.

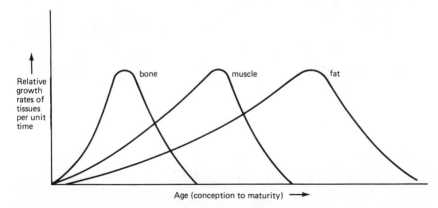

Figure 12.9 Diagram indicating relative growth rates of body tissues from conception to maturity

Factors influencing growth

The patterns of tissue growth and consequent changes in chemical composition of the body are influenced by several interrelated environmental and genetic factors. Genetic differences occur in that animals of the same species vary in their mature size and weight, a feature which is reflected in differences in carcass composition. Generally animals which reach the asymptote of the growth curve at an early age have lower mature body weights. The converse also applies. This is exemplified in the Angus and Friesian breeds of cattle and largely reflects differences in the timing of adipose tissue growth. Although it is often implied that animals which have a large mature body weight are more efficient because of their lower maintenance requirements when expressed on a unit weight basis, there is little evidence that this is the case and the main determinant of efficiency appears to be the choice of slaughter weight.

In addition to these differences in total body size and body composition there are genetic influences on the proportions of muscle and bone tissue. These are most noticeable in comparisons of dairy and beef breeds of cattle, the latter generally having more muscle relative to bone when comparisons are made at equal levels of fatness.

Differences in the growth and composition of body tissues occur between the sexes. In cattle and sheep, if comparison of body composition are made at equal weights or ages, entire males have less fat than castrates, and castrates have less fat than females. The situation differs slightly in pigs where gilts have less fat than castrates at equal body weights. These differences may be largely explained by differences in the levels of testosterone found in males, females and castrates.

Nutrition is generally the dominant factor influencing the expression of growth potential in farm livestock. Both quantitative (plane of nutrition) and qualitative (diet composition) variation in nutrition influence growth. The general influence of plane of nutrition on growth develops from the priorities that exist for available nutrients as an animal grows (*Figure 12.9*). On a high plane of nutrition the growth curves are telescoped together, whereas on a low plane of nutrition the sequence is extended. The most profound effects are on the deposition of fat. The main feature of diet composition which can influence growth is the energy and protein content of the diet. At any specified level of energy intake, animals increase in fatness if their muscle growth is restricted by the amount, and in the case of non-ruminants the quality, of protein in the diet. Thus a balance between the energy and protein content of the diet is necessary for the production of carcasses with the desirable level of fatness. In ruminant animals the situation is further complicated by the fact that the nature of the diet influences the ratios of volatile fatty acids (VFA) arising from rumen fermentation. The overall efficiency of utilisation of these sources of energy for growth and fattening is dependent on the proportions of the different acids present (*see* p. 330).

A further nutritional influence that requires mention here is that of 'compensatory growth'. Animals whose liveweight growth has been retarded by restriction feeding exhibit this phenomenon when introduced to a high plane of nutrition, and there is a temptation to assume that 'compensatory growth' is an inevitably more efficient process. In biological terms the issues involved are complex but in economic terms the phenomenon can be put to good use. Under UK conditions the growth rate of cattle is frequently restricted during the winter (store period) with 'compensatory growth' being achieved on summer grass.

The control and manipulation of growth

The endocrine system is a major regulator of animal metabolism and probably all hormones either directly or indirectly influence animal growth. Those considered to have specific effects on growth are the hormones of the anterior pituitary, the pancreas, thyroid, adrenals and gonads. These hormones affect body growth mostly through their effects on nitrogen retention and protein deposition. The exact modes of action of the individual hormones is only partly understood and is complicated by the fact that in some cases at least their effects depend on the sex, species of animal and balance of other hormones. However, manipulation of the endocrine system particularly by the use of anabolic agents with properties similar to the sex steroids has met with some success in the stimulation of growth and improved feed conversion efficiency through the partition of nutrients from fat to protein deposition.

The use of anabolic agents has been adopted most widely in cattle and is based on the hypothesis that a balance of both androgens and oestrogens is required for maximum growth to be realised. Thus best responses have been obtained by the exogenous administration of androgens to females, oestrogens to entire males and a combination of androgenic and oestrogenic agents to castrates. The main substances currently being used are trenbolone acetate (a synthetic androgen) and the oestrogenic substances hexoestrol and zeranol. To date the use of such agents has been arrived at on an empirical basis. Perhaps when the endocrine control of growth is more fully elucidated more effective agents may be developed.

ENVIRONMENTAL PHYSIOLOGY

'Environment' is clearly an all-embracing term but in the context used here refers to climatic conditions, both macro- and micro-, in which animals live thus omitting day length, stocking density and other aspects of the animal environment. This section is concerned with the animal's physiological reaction to its local environment and with the effect of environmental factors on animal production. Such knowledge is clearly a prerequisite in defining the optimum environment for animals and is essential information in the design of animal housing.

The most important direct effect of climatic factors is the exchange of heat between the animal and its environment; it is related to animal productivity as the amount of heat energy losses from the animal will directly affect the amount of dietary energy retained by the animal.

Maintaining thermal balance

Domestic animals maintain a high (37.5–39 °C) and relatively constant body temperature, despite large fluctuations in the environment. Body temperature is regulated by a combination of behavioural and physiological responses, the anterior hypothalamus playing a central role in controlling and coordinating these activities.

The maintenance of a constant body temperature requires that there is a balance of heat production within the body and heat loss from the body. Heat is produced within the body as a by-product of the metabolism of ingested nutrients for maintenance and the various productive processes. Heat transfer between the body and the environment occurs by the methods of conduction, convection, radiation and evaporation. Since generally the animal body is at a higher temperature than the environment heat may be lost by all these methods. The partition of losses between the various routes of heat exchange varies according to the air temperature with evaporative losses being high at high air temperature whereas radiation and convection are the main avenues of loss at low and intermediate temperatures. In addition to air temperature affecting heat loss, other interacting environmental factors—wind, precipitation, humidity and sunshine—are determinants of heat transfer. Wind, precipitation and humidity particularly affect evaporative losses; wind also affects convection losses by increasing air movements around the animal's body thus destroying the insulating layer of air trapped in the animal's hair or wool; and sunshine especially affects the net radiation exchange between the animal and its environment in animals kept outdoors.

The mechanisms by which animals can regulate body temperature fall into two categories: either the rate of heat production can be altered, or the rate of heat loss can be altered. The animal has a number of mechanisms by which the rate of heat loss can be altered—by varying the degree of vasodilatation,

sweating, piloerection and numerous behavioural responses such as changes in posture, activity and food intake—and it is these mechanisms which come into play first of all in the regulation of body temperature. The range of temperature within which they can regulate body temperature is termed the zone of thermoneutrality. The air temperature at which these mechanisms are no longer successful in dissipating sufficient heat to prevent a rise in body temperature is termed the upper critical temperature; the temperature at which the animal must increase its rate of heat production through increased involuntary activity (shivering) is termed the 'lower critical temperature'. These critical temperatures are not fixed and are especially dependent on the animal's metabolic body size and level of production since size of animal affects maintenance costs and waste heat production and level of productivity also influences heat production. Thus a small slow-growing animal will be less able to cope with a temperature deficit and will have to resort to diverting feed energy from productive processes to extra-thermoregulatory heat production at a higher temperature than a large fast-growing animal. Conversely small slow-growing animals will more easily cope with high environmental temperatures.

Environment and animal production

To date much of the precise work on the effects of the environment on the animal has been concerned with intensively kept pigs and poultry and has primarily studied their interaction with the nutrition of the animal. Energy is the main nutrient affected, higher environmental temperatures releasing more food energy for productive purposes but secondary effects on the animals' protein metabolism, food intake and carcass composition may be produced. In the light of these effects it is clear that where quantitative data on environmental effects are available they may be taken into account in the planning of feeding programmes, the design of buildings, and optimising of investment in insulation and supplementary heating.

In the outdoor situation experienced in the UK most interest on environmental effects revolves round the combined effects of extreme cold, wind and rain. The effects on adult animals are primarily discomfort and loss of production but in the young and newborn such weather conditions are a threat to life. Surveys have shown perinatal lamb mortality can vary from 5% to 45% on hill farms and that at the higher mortality rates extreme cold is the main contributor. In extreme conditions, shivering thermogenesis is often insufficient to maintain body temperature in the neonate. These animals may increase heat production by non-shivering thermogenesis, a process mediated by the effect of cold on the sympathetic nervous system and particularly by the action of noradrenaline on brown adipose tissue. The breakdown of this special type of adipose tissue results in much higher levels of heat production than are produced from other catabolic processes and serves to help prevent hypothermia in the neonates of many species. It appears likely that variations in the amounts of brown adipose tissue present in the body at birth may explain in part the genetic variation in the ability of newborn lambs to survive adverse conditions.

13

Animal nutrition

P. J. Davies

In this chapter the study of farm animal nutrition is presented in five sections. The first three describe the constituents of food, the changes they may undergo during digestion and the functions fulfilled by major nutrients within the animal body. The next section is devoted to the terminology used in assessing nutritive value and the final section reviews the factors that influence the nutritional requirements of farm livestock. The study is designed to provide an outline of the principles on which practical livestock feeding is based.

FOOD CONSTITUENTS

The constituents of animal feedstuffs may be conveniently illustrated by the following diagram.

some foods protein predominates. The composition of the animal body contrasts with that of its diet in containing very little carbohydrate but a relatively large proportion of fat.

Carbohydrates

Carbohydrates are organic compounds containing the elements carbon, hydrogen and oxygen. The formula $C_xH_{2y}O_y$ may be used to represent any carbohydrate and indicates that the ratio of the number of hydrogen atoms to oxygen atoms is always 2:1 as in water. As a group carbohydrates range from simple compounds of low molecular weight to compounds having large and complex structures. Being widely distributed in

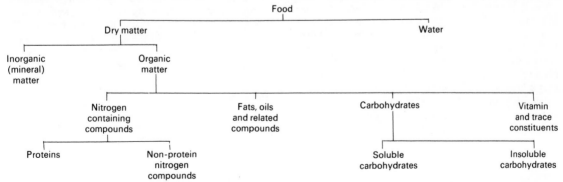

Whilst the proportion of each constituent varies with different foods certain broad generalisations can be made for the large number of foods used in livestock feeding. Organic constituents together constitute over 90% of the dry matter. Fat or oil content of the dry matter seldom exceeds 7%. The predominant fraction of most foods is the carbohydrate fraction whilst in

plants, carbohydrates normally constitute the major proportion of the dry matter consumed by farm livestock.

From a nutritional viewpoint it is convenient to consider the carbohydrate of plant material in two parts. The first being the carbohydrate within the plant cells, the other being the carbohydrate of the cell wall. The latter is often referred to as structural carbohydrate or 'fibre'. The distinction is important because it is generally regarded that carbohydrate within the cell is readily available and utilised by all farm animals whilst fibre is only degraded and utilised, to a large extent, by ruminant animals.

The various carbohydrates that exist in feeding stuffs are classified chemically as monosaccharides, disaccharides and polysaccharides. Monosaccharides are the basic structural units of all carbohydrates. Whilst only very small quantities of monosaccharides as such are found in foods they are produced in quantity from more complex carbohydrates during digestion. Monosaccharides of significance fall into two groups namely *pentoses* (formula $C_5H_{10}O_5$ and *hexoses* (formula $C_6H_{12}O_6$). The more important pentoses are arabinose, xylose and ribose whilst the hexoses of importance include glucose, fructose and galactose. That different compounds have the same formula is explained by the fact that the internal arrangement of atoms within the molecule is different. Compounds sharing the same molecular formula but exhibiting different properties are termed *isomers*.

Disaccharides are formed when two monosaccharide molecules unite chemically with the elimination of water. The reaction is termed condensation and is the chief type of reaction employed in the animal body and in plants for the synthesis of large molecular structures from relatively small ones. These conversions take place only when the appropriate conditions and enzymes are present. Whilst they may occur readily within living cells they generally cannot be reproduced outside cells. The equation

$$C_6H_{12}O_6 + C_6H_{12}O_6 \rightarrow C_{12}H_{22}O_{11} + H_2O$$

illustrates the combination of two monosaccharides (hexose) units to give a disaccharide. The actual disaccharide formed will depend on the constituent monosaccharides that combine and the points of linkage between them. Nutritionally the most important disaccharides are sucrose and lactose. Sucrose is synthesised by the condensation of one molecule of glucose and one molecule of fructose. Root crops are major examples of foods containing significant quantities of sucrose. Lactose is synthesised by the condensation of one molecule of glucose and one molecule of galactose. The process occurs in the cells of the mammary gland and so lactose is unique to milk, representing over 95% of the total carbohydrate present.

Disaccharides are converted into the constituent monosaccharides by reversal of condensation. Such reactions involve the addition of water to the molecule, a process known as hydrolysis. During digestion this process is of great significance and is facilitated by the appropriate enzymes. Hydrolysis of most food constituents is readily brought about in the laboratory by boiling with dilute acid.

Polysaccharides result from condensation of large numbers of monosaccharide units. Normally each polysaccharide is constructed from one of the monosaccharides. Several polysaccharides are widely distributed in plants and exist either as storage polysaccharides within the cell contents or as structural components of the cell wall. The most important storage components are starch and fructans. Starch is a condensation product, or polymer, of glucose whilst fructans are polymers of fructose. Starch is most abundant in seeds, the cereal grains being the major source to farm animals, with tubers and roots also rich sources. Fructans are the principal storage polysaccharides of the grasses and as such are consumed in significant quantities by ruminants.

On hydrolysis by the enzyme amylase starch produces maltose, a disaccharide composed of two glucose molecules. Glucose is not produced without the presence of the enzyme maltase which is necessary for hydrolysis of maltose.

Quantitatively the most important structural carbohydrates of plants consumed by animals are cellulose and the xylans (sometimes referred to along with other minor components as hemicelluloses). Cellulose is a polymer of glucose whilst xylans are polymers of xylose. The proportion of cellulose in cell walls increases as the plant matures. It is extremely resistant to hydrolysis by chemical agents and only certain species of bacteria produce enzymes capable of bringing about its hydrolysis. The presence of large numbers of those organisms within the digestive system of ruminant animals explains their ability to utilise fibrous foods. Within the cell wall cellulose is associated with lignin. The ratio of lignin to cellulose increases with maturity and this is thought to inhibit the efficiency of enzymic breakdown. In consequence the digestibility of the diet declines. Cereal straws are examples of highly lignified tissue. The nutritional improvement of straw by treatment with sodium hydroxide is based on the disintegration of the complex architecture of the cell wall thereby increasing the susceptibility of cellulose and other plant constituents to enzymic degradation.

Proteins

Proteins are constituents of living cells, being synthesised from structural units termed amino acids by the process of condensation. Some 21 acids, listed in *Table 13.1*, exist in typical dietary proteins.

All amino acids, and therefore proteins, contain the elements carbon, hydrogen, oxygen and nitrogen. The nitrogen content of most proteins varies between 14 and 18% with an average of 16%. Measurement of protein content of foods is based on estimation of nitrogen content. Traditionally protein content is expressed as crude protein (CP) where $CP = N \times 100/16$. Three amino acids, cystine, cysteine and

Table 13.1 Major amino acids occurring in food proteins

(1) Alanine	(12) Leucine
(2) Arginine	(13) Lysine
(3) Aspartic acid	(14) Methionine
(4) Cysteine	(15) Phenylalanine
(5) Cystine	(16) Proline
(6) Glutamic acid	(17) Serine
(7) Glycine	(18) Threonine
(8) Histidine	(19) Tyrosine
(9) Hydroxylysine	(20) Tryptophan
(10) Hydroxyproline	(21) Valine
(11) Isoleucine	

methionine also contain the element sulphur. The sulphur content of proteins therefore is more variable than the nitrogen content and is a reflection of the proportion of the three sulphur containing amino acids present, e.g. wool has a high sulphur content reflecting the high cystine content of wool protein.

The many proteins present in animal and plant tissues are normally classified into two major groups, simple and conjugated proteins. Simple proteins, when hydrolysed, yield a mixture of amino acids only, examples being albumins, globulins, glutelins and scleroproteins. Conjugated proteins when hydrolysed yield compounds in addition to the amino acids and include phosphoproteins, lipoproteins, chromoproteins and nucleoproteins.

Lactalbumin and lactoglobulin are simple proteins which occur in milk. Scleroproteins are typical of skeletal structures and protective tissues, and include ossein of bone and collagen of connective tissue. Casein, the chief protein of milk, is a phosphoprotein. Chromoproteins are proteins combined with a coloured group. Haemoglobin of blood is protein globin combined with haematin. Nucleoproteins, characteristic of cell nuclei, consist of protein combined with nucleic acids.

During hydrolysis of protein, intermediate compounds are formed before the constituent amino acids are released. These intermediate compounds are termed peptides, chain-like molecules composed of condensed amino acid units. These range in size from dipeptides, where two such units are condensed, to polypeptides where several hundred units may be present. Peptides differ essentially from protein in being structurally simpler.

Proteins are very large molecules and range in shape and size. Solubility in water varies from completely soluble to completely insoluble. Chemically proteins are amphoteric, i.e. they function as bases or acids and their presence in tissue helps to keep a fairly constant pH.

The nutritional value of any protein to simple stomached (monogastric) animals such as pigs and poultry will depend principally upon the ease of hydrolysis and on the amino acid composition. To the ruminant animal protein value will be influenced by the degree to which rumen micro-organisms can degrade the protein. Nutritional value may be modified by the action of heat and certain chemical agents. For example, excessive heat treatment in the manufacture of fish meal can seriously impair its value to pigs and poultry. Treatment of silage with formaldehyde can reduce the extent of protein degradation by rumen micro-organisms.

Fats

Fats occur in all living organisms. In plants fat is normally present in liquid form (oil), in animals chiefly in solid form. Most animal feedstuffs contain 2–6% oil in the dry matter whilst the animal body may contain more than 30% fat.

Naturally occurring fats are mixtures of triglycerides, compounds produced by condensation of glycerol (one molecule) and fatty acids (three molecules). Thirteen fatty acids occur in natural fats (*see Table 13.2*). Differences between fats reflect differences in the proportions of various fatty acids present. Fatty acids are conveniently subdivided into two groups, saturated and unsaturated, terms referring to features in molecular structure. Animal fats are characterised by a high proportion of saturated fatty acids, plant fats (oils) by a high proportion of unsaturated acids. Three acids—palmitic, stearic and oleic—constitute the bulk of fatty acids present in most fats. Three other acids—butyric, caproic and caprylic— are normally found only in butter fat.

Table 13.2 Common acids of natural fats

Saturated		Unsaturated
Butyric	Myristic	Oleic
Caproic	Palmitic	Linoleic
Caprylic	Stearic	Linolenic
Capric	Arachidic	Arachidonic
Lauric		

Minerals

Minerals are elements, other than carbon, hydrogen, oxygen and nitrogen, required by animals. The essential elements are designated as either major or trace elements depending on their concentration within the animal body. these elements are shown in *Table 13.3*.

Many other elements are present in animal tissue but there is no conclusive evidence that they are essential to the animal or that a deficiency results in loss of production or ill health. Such elements include boron, fluorine, lead and strontium. Many elements, including some of the essential trace elements, are toxic to animals when ingested in large quantities.

Table 13.3 The essential elements

Major elements	Trace elements
Metals	Metals
calcium	iron
potassium	zinc
sodium	copper
magnesium	manganese
	cobalt
Non-metals	molybdenum
phosphorus	selenium
chlorine	
sulphur	Non-metal
	iodine

Within the body minerals occur in both inorganic and organic combination and have several functions:

(1) They form part of structural components, e.g. the skeleton.
(2) They help to control the acid/base balance within tissues and fluids.
(3) They contribute significantly to the osmotic properties of all body fluids.
(4) They form an integral part of many enzyme systems.

Enzymes are protein in nature but their catalytic action is usually dependent on the presence of other compounds. Many of the trace elements function primarily in the role of enzyme activators.

Animal feedstuffs vary widely in mineral composition. The concentration of an element in a single food can vary appreciably depending on the soil and fertiliser practice used in its production. All classes of farm livestock can be affected by deficiencies involving mineral elements, such deficiencies may arise due to inadequate intake of an element or to unavailability of the element in the diet. Unavailability may be due to the chemical form in which the element is present or the presence of other elements interfering with its absorption.

Calcium

Some 98% of the calcium in the animal body is present in bones and teeth, the element is therefore particularly important in the diet of young growing animals where skeletal development is taking place. The remaining small fraction of calcium is distributed throughout the body soft tissues and fluids where it has roles in controlling muscle and nerve activity. These roles are so crucial that the concentration of calcium in the blood plasma is very closely regulated and so is kept within very narrow limits.

When dietary calcium intake is low the animal absorbs a high proportion of the intake. Where the intake is high the proportion absorbed declines. These changes in absorption permit the animal to meet its requirement over a wide range of intakes without accumulating excess within the body. Where calcium intake is inadequate the animal is able to mobilise some of the substantial reserves of calcium that exist in the bones to supplement that absorbed from its food and thus maintain again a normal plasma concentration. The high yielding dairy cow at peak lactation is almost certainly in this situation, the bone calcium reserves being replenished at a later period when milk yield declines. The mobilisation and replenishment of bone calcium is largely under the control of hormones secreted by the parathyroid glands.

Milk fever in cows and lambing sickness in ewes normally occurs just after parturition and are characterised by hypocalcaemia and usually hypophosphataemia, i.e. low concentrations of calcium and phosphorus in the blood plasma. The initiation of milk secretion just prior to parturition causes a substantial and sudden increase in calcium requirement. The condition is thought to be largely hormonal in origin, with bone calcium not mobilised quickly enough to meet the drain on blood calcium made by the mammary gland. Older cows are more susceptible than younger, and evidence suggests the Jersey breed is most susceptible. Feeding practice can modify the inherent susceptibility of the animal to milk fever. For example, cows fed high calcium diets in the weeks prior to calving are more prone to the condition since such diets have a detrimental effect on the cow's ability to mobilise bone calcium and increased absorption from the gut cannot occur quickly enough to meet the new demand. Recent recommendations for the prevention of this condition in susceptible cows rely on giving diets of low calcium content for some weeks prior to calving then increasing the intake about one week before the expected date of calving, i.e. matching the calcium intake more closely to requirement. Animals affected by the condition are injected subcutaneously or intravenously with a calcium borogluconate solution (usually containing some 8–12 g Ca). In most cases recovery is rapid but some cases require prolonged treatment.

Calcium is widely distributed in foods, the quantity varying considerably with the food. Green crops and milk are rich sources, roots and cereal grains poor.

Potassium

Potassium is a major ionic constituent of body fluids and in particular intracellular fluid. It contributes significantly to the osmotic properties of these fluids and is concerned with muscle control and nerve activity. Potassium is not stored within the body, excess being excreted in the urine.

Deficiency of potassium is highly unlikely in practice since the content of most plants (foods) is high and intake normally more than adequate.

Sodium

Sodium, like potassium, is a major ionic constituent of body fluids and the major cation present in extracellular fluid and blood. Concentration of sodium in blood plasma is held within narrow limits, excess excreted principally as sodium chloride. When intake is inadequate excretion is sharply reduced to conserve sodium, there being no reserves.

Deficiency of sodium results in reduced performance. Milk yield of cows is adversely affected by diets low in common salt and, therefore, sodium. Cows on such diets go to extreme lengths, licking soil and urine patches, in attempts to increase intake.

Supplementation of diets by salt is generally necessary to offset the relatively low concentration of sodium and chlorine in most foods. The major ingredient of most mineral supplements and licks is salt. Many cases are reported of salt poisoning in young pigs, brought about by dehydration resulting from lack of drinking water. Increasingly sodium bicarbonate is used to replace part of the sodium chloride in many dairy concentrate diets. Apart from supplying sodium this salt also helps in the neutralisation of acids within the rumen and so maintains a more constant pH regime.

Magnesium

About 60% of magnesium in the animal body is in bones and only about 1% in body fluids. The efficiency of exchange between bone and blood magnesium is not as great as exists for calcium. Consequently blood concentration of magnesium is a reasonable index of magnesium intake. In cattle and sheep the normal range of magnesium in blood serum is between 20 and 32 mg/ℓ, values below this indicating hypomagnesaemia. This may exist as a sub-clinical condition but in severe cases nervous symptoms known as staggers or tetany are shown. Injection of magnesium sulphate may cure but in many cases the condition is rapidly fatal. The condition may affect dairy cattle, suckler cows, calves and sheep but major outbreaks occur during grazing of spring grass, a time frequently coinciding with high production levels. The magnesium of such grass is largely unavailable and high application rates of fertiliser N and K tend to aggravate the situation.

Preventative measures are based on increasing magnesium intake, achieved by feeding concentrated mixtures fortified with commercial magnesium oxide (calcined magnesite) or adding magnesium salts to drinking water. A magnesium 'bullet' may be administered which is a relatively insoluble, dense form of magnesium compound which remains in the reticulum and releases magnesium slowly over the danger period. This method of control is attractive where stock do not receive regular hand feeding. It suffers from the drawback that occasionally the bullet is passed out in the faeces.

Phosphorus

Some 80% of the phosphorus in the animal body is in the skeleton. Phosphorus also occurs in many proteins and is necessary for efficient utilisation of carbohydrate. Blood serum concentration of inorganic phosphorus is an index of phosphorus intake.

Deficiency of phosphorus results intially in reduced animal productivity. Serious deficiency results in infertility and bone disorders.

Whilst phosphorus is a major plant nutrient certain plant species contain insufficient amounts to meet high animal requirements. Many pastures are deficient, hays and straws a poor source, but cereal grains, their by-products and materials such as meat and bone meal and fish meal, are all good sources.

Chlorine

Chlorine as chloride is the major anionic constituent of body fluids. It is a constituent of hydrochloric acid secreted in the gastric juice. Chlorine deficiency is usually associated with sodium and is prevented or cured by salt.

Sulphur

Sulphur contained in body and food is largely present in the form of protein, since sulphur is a constituent of the three amino acids cysteine, cystine and methionine. Proteins present in hair, wool and feathers are rich in these amino acids.

Sulphur deficiency, therefore, is reflected in a shortage of the sulphur amino acids, particularly important in pig and poultry diets where these amino acids are often limiting, depending on the constituents of the diet. Deficiency in the ruminant is unlikely since rumen micro-organisms utilise inorganic sulphate, plentiful in most foods, for the synthesis of the sulphur amino acids.

Iron

More than half the iron of the animal body is present as a constituent of the protein haemoglobin, found in red blood corpuscles and responsible for oxygen transport to tissues. Most of the remaining iron is a constituent of the brown coloured protein ferritin. This protein, containing up to 20% iron, is found in the liver, spleen, kidneys and bone marrow and provides a store of iron. Iron is also a constituent of certain enzyme systems.

The absorption of dietary iron from the digestive system is relatively small. Much iron is liberated, as a result of breakdown of red blood corpuscles, and utilised in the synthesis of new haemoglobin and so dietary requirement is small. Deficiency of iron gives rise to nutritional anaemia, characterised by lowered blood haemoglobin values, which results in unthriftiness. Animals most susceptible are suckling piglets for concentration of iron in milk is low and unaffected by the iron intake of the dam. Consequently it is routine practice to inject intramuscularly an iron preparation designed to liberate the element slowly over the suckling period. Piglets reared out of doors are unaffected since they acquire sufficient iron by rooting in soil. The relative growth rate of young pigs is greater than calves or lambs. They double their birth weight more quickly, which undoubtedly contributes to susceptibility. Survey work on blood composition of dairy cows indicates haemoglobin levels are lower in winter than summer, but the reasons are unknown. It is unlikely that simple iron deficiency is the cause.

Iron is widely distributed and usually present in adequate quantities in all solid foods, with the possible exception of cereal grains. Almost all proprietary mineral supplements contain iron salts.

Zinc

Zinc is distributed widely in the animal body and in feedstuffs. Its major role within the animal appears to be as a constituent of certain enzyme systems. Deficiency of the element is rare in ruminant stock but a skin disorder in growing pigs termed parakeratosis is often alleviated when the dry meal they consume is fortified with added zinc.

Copper

Copper occurs in minute quantities in all body tissues. It is particularly concentrated in the liver which functions as a storage organ. Copper is present as a constituent of certain pigments found in hair and wool. Haemoglobin formation is dependent on copper, although it does not contain the element. This suggests a role as part of an enzyme system. Certain other enzymes are known to be dependent on copper supplies.

Various conditions affecting ruminant stock are known to result from shortage of copper. They occur in well defined geographical regions of the country. Suckler calves are particularly prone, becoming stunted and suffering from anaemia. Blood serum copper concentration is also below normal limits. Cure is based on injection of organic copper preparations which release copper slowly over an extended period. Although the dams of such calves rarely show symptoms of copper deficiency it is likely they have small reserves. Survey work on blood copper in dairy herds showed that the lowest levels occurred in dry cows and heifers during winter, a reflection of the copper content of the roughage diet. Heifers showed poor growth rates and had high incidence of infertility. A condition in lambs known as swayback is associated with copper metabolism. Affected animals are unable to walk properly, suffering uncoordination of muscle activity in the hind legs. Post-mortem examination reveals damage to the spinal cord. The condition is not the result of a simple copper deficiency but tends to occur in specific regions and outbreaks are sporadic. Preventative measures include injection of pregnant ewes with organic copper preparations. Copper deficiency has little practical significance in pig nutrition.

Whilst copper is an essential trace element it is also highly toxic when consumed in quantity. Excess copper is stored in the liver but above a certain storage limit, liver damage occurs and is followed by death unless diagnosed and treated. Sheep are particularly susceptible to copper poisoning, and pigs relatively resistant. Diets for growing pigs normally include added copper sulphate, the concentration of copper in the final diet being of the order of 250 mg/kg of food. Addition of copper improves growth rate and food conversion efficiency. Diets for breeding stock should never contain added copper.

Copper is widely distributed in foods but is likely to be low in roughage foods.

Manganese

This element occurs in most tissues of the body and is important as an enzyme activator. Deficiency is uncommon in farm livestock. In parts of south west England infertility in dairy cows has been associated with low intakes of manganese. The element is widely distributed but only a very small proportion of dietary intake is absorbed.

Cobalt

The chief role of cobalt is as a constituent of the vitamin B_{12} molecule. It may have an independent role as enzyme activator.

The major source of vitamin B_{12} to ruminants results from bacterial synthesis within the recticulum and rumen. In certain areas the cobalt content of herbage is so low that intake is insufficient to permit necessary production of the vitamin. In such cases animals suffer from a condition known as pine or illthrift. These names suggest the animals suffer loss of appetitie, have poor growth rates and are anaemic. Prevention in these areas depends on oral administration of cobalt, either as soluble salts or for long-term protection by a cobalt 'bullet', a dense form of cobalt oxide meant to remain in the reticulum and release cobalt slowly. Bullets may sometimes be lost in faeces or become coated with layers of calcium phosphate and thus protection is lost.

Deficiency of cobalt is of little practical significance in pig nutrition.

Molybdenum

Molybdenum is an essential component of certain enzyme systems. Deficiency is rarely encountered but well known symptoms occur when intake is excessive, including scouring and unthriftiness. The condition is prevalent on the so-called teart pastures in Somerset, where herbage molybdenum levels are greatly elevated. Control is by oral administration of copper sulphate.

Selenium

This element has only recently been considered essential in animal nutrition. Its precise function is unknown but it can in some cases prevent symptoms of vitamin E deficiency arising. Selenium is toxic to livestock if ingested in quantity, but this occurs only in isolated regions where plants take up large amounts from the soil.

Iodine

The main role of iodine in the body is as a constituent of thyroxine, a hormone secreted by the thyroid gland. Thyroglobulin, a protein found in this gland, contains thyroxine which is liberated when necessary. The chief role of thyroxine is to control metabolic rate. Iodine deficiency results in decreased thyroxine production and enlargement of the thyroid gland, symptomatic of goitre. In practice this is most likely to occur in newborn animals which may be born hairless.

Iodine is present in small amounts in most foods, the quantity largely dependent on soil levels of the element. Fish meal is one of the best sources. Several common foods, notably kale and rape, contain goitrogenic agents which cause goitre if fed in large amounts; they interfere with thyroxine production even though there is apparently sufficient iodine in the diet. Their influence is nullified by feeding extra iodine; this is most easily achieved by inclusion of sodium iodide in the diet.

Vitamins

Vitamins are organic compounds required in extremely small quantities for health and productivity.

Some 15 vitamins are recognised as essential. They are unrelated chemically, having separate and independent functions in metabolism. The principal role is as integral parts of various enzyme systems. Deficiencies, therefore, reveal a variety of disorders and symptoms resulting from disruption of orderly and synchronised chain reactions which characterise metabolism.

Vitamins are classified into two groups according to solubility; the important vitamins and their classification are shown in *Table 13.4*.

All vitamins have chemical names but many continue to be known by letters of the alphabet which provided a means of identification prior to a knowledge of their chemical identity.

Table 13.4 Vitamins important in animal nutrition

Vitamin	Chemical name
Fat-soluble vitamins	
A	retinol
D_2	ergocalciferol
D_3	cholecalciferol
E	tocopherol
K	phylloquinone
Water-soluble vitamins	
B_1	thiamine
B_2	riboflavin
—	nicotinamide
—	pyridoxine
B_6	pantothenic acid
—	biotin
—	folic acid
—	choline
B_{12}	cyanocobalamin
C	ascorbic acid

The subdivision of vitamins according to solubility has important implications. Vitamins that are soluble in fat and oil are stored in the fat components of the body; short periods of deficiency may therefore be unimportant since the animal utilises stored reserves. Water soluble vitamins, however, cannot be stored and must be in the diet on a regular basis.

The metabolic machinery of farm animals requires a supply of all the vitamins in *Table 13.4*. This does not imply, however, that animals must receive every vitamin in the diet. A dietary source of vitamin C is not essential for cattle, sheep or pigs since the material is synthesised within the tissues from a substance, glucuronic acid. Ruminants differ from non-ruminants in respect of dietary requirements of B vitamins since the bacterial population of the rumen synthesises B vitamins as part of their own protoplasm. When later digested, the vitamins are released and become available to the host. Very young ruminants and pigs do not benefit from this source. Significant amounts of B vitamins may, however, be produced by bacteria in the large intestine of all animals. In some cases this provides an adequate supply to meet the requirement of the animal, whilst in others a dietary intake is essential to avoid deficiency symptoms.

Intensive systems of animal production often necessitate supplementation with synthetic forms of certain vitamins to ensure necessary intakes are achieved. Vitamin supplements are often combined with mineral supplements, thus achieving even distribution when mixing rations.

The following are notes on vitamins, with special reference to role and function, consequences of deficiency, and distribution in foods.

Vitamin A

Vitamin A may have several roles. It maintains mucous membranes in a moist and healthy condition and cells in the retina contain a vitamin A compound. The vitamin also has a function in bone formation.

At birth the newborn may be dead, blind, deformed or weak if the diet of the dam has been grossly deficient during pregnancy. In growing animals growth rate is depressed by sub-optimum intakes of the vitamin. In adults, scaly skin and rough coat may indicate vitamin deficiency, whilst blindness may also occur. In breeding females it is thought infertility results from shortage. It must be stressed that some of these symptoms can result from causes other than vitamin A shortage.

Most livestock diets contain very little vitamin A as such. Plant sources normally contain precursors (substances that are converted into the active vitamin within the body), the most important of which is the yellow pigment β-carotene. This substance is converted to vitamin A in the intestinal wall and in the liver. In practice five units of dietary β-carotene are equivalent to one unit of vitamin A. Green foods are rich in carotene whilst roots and cereals are poor. Diets rich in cereals must be fortified with synthetic forms of the vitamin.

The extent to which cattle convert carotene to the vitamin depends on breed. Channel Island breeds are less efficient than others in this respect and in consequence their milk fat and body fat are more yellow.

Vitamin D

Two forms of the vitamin are important, namely D_2 and D_3. These have roughly equal potency for cattle, sheep and pigs, but D_3 is considerably more potent than D_2 for poultry. The role of vitamin D is chiefly concerned with calcium and phosphorus metabolism and is involved with absorption of these elements from the gut and their deposition in bone.

Deficiency of the vitamin chiefly affects young animals. Bone development is affected, giving rise to the condition known as rickets. D vitamins as such are not found in growing plants. The bulk of animal requirement is normally met by sunlight acting on a precursor present in the fat layers under the skin. This produces the active vitamin D_3. Clearly, housed animals cannot benefit from this source and are dependent on diets supplemented with synthetic vitamin. Sun-dried herbage, such as hay, contains small amounts of vitamin D_2 resulting from the effect of radiation on another precursor present in plant tissue. Colostrum contains about ten times as much of the vitamin as normal milk.

Vitamin E

Vitamin E has many roles in metabolism, chiefly as part of enzyme systems. The most common ailment resulting from its deficiency is muscular dystrophy, a term describing skeletal and heart muscle degeneration.

Green foods and cereal grains are amongst the best sources of the vitamin.

Vitamin K

This vitamin is thought to be a part of several enzyme systems in the animal. It has a particular role in the formation of prothrombin, a protein produced by the liver and involved in the clotting of blood.

Deficiency of the vitamin in practice is not known since it is widely distributed and bacterial synthesis occurs in the digestive system.

B vitamins

The major role of all these vitamins is as integral parts in body enzyme systems. Adult ruminants have no dietary requirement under normal conditions. The special case of vitamin B_{12} has already been discussed (*see* p. 324). In practice it is rare to encounter deficiencies of these vitamins in pigs since they are widely distributed and some intestinal synthesis occurs. Some pig diets, however, are fortified with riboflavin since cereals, the major component of pig diets, are relatively deficient in this vitamin.

Vitamin C

Vitamin C, or ascorbic acid, plays a vital role in certain oxidation–reduction systems in the body. Although farm animals may consume vitamin C from green crops it is not essential, since the substance can be synthesised in tissues from glucuronic acid. This synthesis does not take place in man and so a dietary requirement exists.

Water

Water is contained in all plant and animal cells, the water content of foods varying widely. Young succulent spring grass contains 85% water and 15% dry matter (DM), whereas dried grass contains only 10% water. When comparing food intakes it is essential that comparison is made on the basis of DM intakes. The water content of the animal body is also variable. In general, water content is highest in young, lean animals and least in mature, fat animals.

Water is obtained by drinking and from food. Additionally, water is made available from oxidation of organic compounds within the body. This is termed metabolic water and for every kg of carbohydrate oxidised 0.7 kg of water is made available, whilst for every kg of fat oxidised approximately 1.1 kg water is produced. Animals lose water through faeces and urine, water vapour in expired air, and through the skin. Milk secretion represents a major loss to lactating animals, whilst the pregnant animal deposits quantities of water within the uterus particularly in late pregnancy.

Within the animal body water is the major component of all body fluids and internal secretions. The high specific heat and latent heat of water aids maintenance of a constant body temperature, whilst chemically the reaction of water with certain food constituents under the influence of enzymes (hydrolysis) is the major reaction in digestion.

Water intake is related to DM intake, DM composition, especially in relation to the salt content, and the environmental temperature. All stock need adequate supplies of water at all times; insufficient water means reduced DM intake with resultant depression of productivity.

DIGESTION

Most organic food constituents cannot be utilised directly by the various body tissues. The process of digestion produces from these constituents relatively simple compounds subsequently absorbed across the wall of the digestive tract into the blood and lymph. The digestive system or gut of farm animals is conveniently considered under three regions, the stomach, the small intestine and the large intestine. The processes involved in digestion may be grouped into physical, chemical and microbial. The physical processes include chewing and the muscular action of the stomach and intestines which result in the breakdown of food particles and their transport through the system. The chemical action is brought about by enzymes contained in the various secretions coming into contact with the food. Microbial digestion is brought about by enzymes produced by microorganisms, chiefly bacteria, which inhabit the tract.

For convenience of study farm livestock are subdivided into (a) simple stomach or monogastric animals, the pig being the chief example, and (b) ruminants, which possess a compound stomach having four regions, cattle and sheep being the principal examples.

Digestion in the pig

Dry food is first wetted by saliva which acts as a lubricant and enables mastication and swallowing of the material via the oesophagus to the stomach. Where pigs are fed meal with water, rate of consumption is increased and the role of saliva is less significant. In the stomach the food becomes impregnated with gastric juice which is produced by cells in the wall of the stomach. Due to the presence of hydrochloric acid in this fluid the pH of stomach contents is low, normally near a value of 2. The semiliquid digesta passes from the stomach to the small intestine where it comes into contact with a series of alkaline secretions which cause the pH of the digesta to rise. The secretions are bile, pancreatic juice and intestinal juice and the enzymes contained in the latter two secretions are functional at the pH values encountered in the small intestine. Absorption of products of digestion occurs principally from the small intestine even though changes continue to take place as the digesta moves to the large intestine. In this latter region changes take place under the influence of bacterial action and whilst some absorption of products takes place it is considered small in comparison to that taking place from the small intestine. Large quantities of water are however absorbed from the large intestine. Waste material, or faeces, voided from the tract contain along with water, undigested food residues, bacteria, remains of secretions that have entered the tract, cells eroded from the tract wall and products of bacterial action not absorbed.

The fate of the major food constituents is now considered.

Carbohydrate

Carbohydrates are broken down to monosaccharides by the action of specific enzymes. Since starch is normally the predominant carbohydrate of pig diets, glucose is the chief end product of digestion. Saliva contains the enzyme *amylase* capable of splitting starch to the disaccharide maltose. It is doubtful whether this enzyme has much significance since its action will be fairly quickly inhibited by the low pH of the stomach contents. There is no enzyme in gastric juices capable of bringing about the hydrolysis of carbohydrates. In the small intestine a series of enzymes ensures the total hydrolysis of susceptible material. *Amylase* brings about maltose production whilst *maltase* converts maltose into glucose. The enzyme *sucrase* converts sucrose to glucose and fructose and *lactase* converts lactose to glucose and galactose; lactose is present only when diets contain milk, skimmed milk or whey and *lactase* activity is maximal in the suckling animal, the activity declining as milk makes a declining contribution to the diet. Raw potato starch is relatively unaffected by the enzyme systems produced by the pig and so it has a low nutritive value. Cooking or steaming however renders the starch susceptible to hydrolysis and therefore greatly enhances the value of potatoes. The monosaccharides produced by digestion are largely absorbed from the small intestine and transported in the portal blood to the liver.

Protein

Within the stomach some protein is partially hydrolysed to polypeptides by the enzyme *pepsin*, one of the very few enzymes to function at low pH. Entering the small intestine further breakdown of protein to peptides occurs due to the enzymes *trypsin* and *chymotrypsin*. Peptides are finally converted to amino acids by the action of enzymes known as *peptidases*. Amino acids on absorption from the small intestine are transported to the liver.

Fat

Dietary fats and oils are not chemically altered until they reach the small intestine. Here fat is firstly emulsified under the action of bile, this action greatly increases the surface area of material and so enhances the activity of the enzyme *lipase* which brings about the hydrolysis of fat to glycerol and fatty acids. Only a portion of the fat intake is so hydrolysed, the products of hydrolysis being absorbed into the wall of the small intestine where they re-combine to form neutral fat. This, with very fine particles of unhydrolysed fat, is absorbed into the lymph system and ultimately into the blood as minute fat droplets.

Digestion in the ruminant

In ruminant animals the stomach is subdivided into four compartments, reticulum, rumen, omasum and abomasum. The abomasum resembles, in form and function, the stomach of the monogastric animal. In newborn ruminants the reticulum and rumen are relatively undeveloped but grow proportionally faster than remaining compartments. In adults they occupy over 80% of the total stomach volume with the rumen itself occupying more than 60% of the total volume.

The ruminant is adapted to consume and digest large quantities of coarse food, i.e. food containing large amounts of cellulose. Since no enzyme capable of splitting cellulose is produced by the animal the ability of ruminants to deal with such diets centres on the vast population of micro-organisms, chiefly bacteria, inhabiting the reticulum and rumen. Some species of bacteria produce an enzyme, *cellulase*, capable of bringing about hydrolysis of cellulose. Other species are capable of hydrolysing starches whilst others effect the breakdown of proteins. Distribution of various organisms within the total population is influenced by the nature of the diet. Diets rich in cellulose promote a high proportion of bacteria capable of hydrolysing cellulose, whilst diets rich in starch bring a decline in number of these organisms but a corresponding increase in those capable of hydrolysing starch.

The rumen is anaerobic and chemical transformations occurring within it are referred to as fermentation. Rate of fermentation depends on the nature and composition of the diet. Coarse roughage ferments slowly, whilst concentrated non-fibrous foods ferment rapidly. Rumination whereby food is regurgitated to the mouth for re-chewing is responsible for the physical breakdown of the food and this results in a vastly increased surface area for bacterial action and a consequent increase in rate of breakdown.

The end products of bacterial activity include large quantities of organic acids. The acidity of the rumen contents is kept within fairly narrow limits since saliva, produced in copious quantities, contains sodium bicarbonate which partially neutralises the acids. Additionally, partially neutralised acids are absorbed across the wall of the rumen. Large quantities of gas are produced by fermentation which is expelled by eructation. In the omasum much water is absorbed from the digesta, leaving it drier to enter the abomasum. Acidity of the gastric juice produced here quickly inhibits bacterial action. Bacterial protoplasm together with undigested food residues undergo digestion from this point onwards in a manner similar to that of the pig.

The material arriving in the fourth stomach is thus quite different from that fed; by contrast material entering the stomach of a pig is very little different from the food consumed.

Carbohydrate

The large variety of carbohydrates in ruminant diets are hydrolysed by the extracellular enzymes of the micro-organisms to produce monosaccharide sugars. These are absorbed into the cells of micro-organisms where the sugars are utilised (a) as raw material to synthesise polysaccharide and other cell components, and (b) as metabolic fuel. Because conditions within the rumen are anaerobic complete oxidation of sugars to carbon dioxide and water is not possible due to lack of oxygen. The major end products of carbohydrate fermentation are the three volatile fatty acids (VFA) acetic, propionic and butyric and the gases carbon dioxide and methane. The acids are present dissolved in the rumen fluid and the production in a lactating cow can amount to some 3 kg daily. Acetic acid is normally present in the greatest quantity, followed by propionic acid. The molar ratio of acetic:propionic in the rumen is not constant. It is influenced by the type of carbohydrate fed and the manner of feeding. Fibrous diets containing a high proportion of cellulose tend to produce a wide ratio whereas diets containing high levels of concentrate foods, which have high starch content, tend to narrow the ratio. Animals having continuous access to food consume on a little-and-often basis, this helps to maintain a more even rate of fermentation which in turn promotes a wider acetic:propionic ratio compared with feeding situations where the same food is consumed in fewer meals.

The acids produced during fermentation are partially neutralised by sodium bicarbonate in the saliva before being absorbed from the rumen and transported to the liver. Increasingly sodium bicarbonate is incorporated in diets where high rates of VFA production are envisaged such as in dairy cows consuming large quantities of concentrate foods.

Rumen acidosis can develop where ruminants unaccustomed to diets containing high levels of readily fermentable carbohydrate, suddenly consume large quantities of such food. The clinical symptom of this condition is the almost complete loss of appetite. Thus care is necessary in feeding such diets and certain minimum levels of roughage are essential for normal rumen function. Again high frequency of feeding can help to prevent violent fluctuation in rumen pH.

Since the VFA are incompletely oxidised products of bacterial action they are a major source of energy to the host animal. On most diets little glucose is produced in and absorbed from the lower gut.

Protein

Within the reticulum and rumen a large but variable proportion of dietary protein is degraded by bacterial activity. Amino acids produced by hydrolysis serve as raw material for some species of bacteria for synthesis of their own protein but the bulk of the amino acids produced are further degraded to produce organic acids, carbon dioxide and ammonia. This ammonia is the major form whereby rumen bacteria acquire nitrogen for synthesis of their own protein. Thus the protein entering the abomasum will consist of a mixture of bacterial protein and undegraded dietary protein, i.e. protein that has escaped breakdown by bacterial action.

The efficient activity of the rumen micro-organisms is thus dependent on an adequate supply of degradable protein. Any shortage of such protein in the diet would result in a reduced rate of breakdown of other dietary ingredients and an associated reduction of food intake. Excessive quantities of degradable protein in the diet result in ammonia production surplus to the requirements of the micro-organisms. Ammonia in excess of bacterial needs is absorbed from the rumen and transported to the liver where it is converted to urea. This urea is then excreted in the urine, but a proportion is recirculated to the rumen via the saliva.

Ammonia needed by rumen micro-organisms may be derived from sources other than dietary degradable protein. Thus it is possible to partially replace protein in a diet or supplement certain low protein diets with compounds that provide ammonia directly or indirectly. Most commonly used is synthetic urea, a compound that occurs naturally in saliva. Urea is hydrolysed by enzyme action to carbon dioxide and ammonia, the rate of hydrolysis is particularly rapid and so it is essential to control intake strictly, to achieve high efficiency of conversion into bacterial protein.

The protein leaving the rumen is thus a mixture of microbial protein and undegraded dietary protein and the mode of digestion in the abomasum and small intestine will resemble that in the pig. Owing to the considerable transformations that occur within the rumen the amino acid composition of the protein mass entering the abomasum can be markedly different from the composition of the dietary protein. Therefore relatively little attention is currently given to the amino acid composition of proteins in ruminant diets.

Fat

Certain bacterial strains within the rumen produce enzymes capable of hydrolysing fat to glycerol and fatty acids. Glycerol is fermented within the rumen to propionic acid. Certain unsaturated fatty acids are modified within the rumen either to become less unsaturated or totally saturated. Fatty acids are not broken down at this stage but pass to the small intestine where they are absorbed. Within cells of the intestinal wall they recombine with glycerol to produce fat. This fat is considerably harder than dietary fat.

METABOLISM

Metabolism refers to all chemical processes within tissues and organs of animals. Here emphasis is laid on ways in which animals utilise major end products of digestion to sustain life and support production.

Metabolic processes can be considered under one of two headings: anabolism refers to reactions of a synthetic nature, i.e. reactions producing large molecular structures from small ones; catabolism refers to reactions whereby large structures are simplified. Catabolic changes are often oxidative in character and involve oxidation of a compound to provide energy. This energy is needed to support physical work and also to allow synthetic reaction to proceed. The compound oxidised is referred to as metabolic fuel. Within the body several compounds are utilised as fuels, although usually one predominates. Relative contribution of various compounds to total energy production is determined by level of feeding and composition of diet.

The chief aspects involved in the metabolism of carbohydrate, protein and fats in pigs and ruminants are now summarised.

Carbohydrate

In the pig the major end product of carbohydrate digestion is glucose and on absorption from the small intestine this is carried to the liver. Although the supply of glucose reaching the liver is variable depending on the diet and the time that has elapsed following a meal, the concentration of glucose in systemic blood is held within narrow limits under the influence of the hormones insulin and glucagon produced by the pancreas. Glucose surplus to immediate needs is converted to the insoluble polysaccharide glycogen in the liver and stored. Glycogen is reconverted to glucose in response to utilisation of blood glucose through one or more of the following routes:

(1) by the muscle for glycogen synthesis,
(2) by the placenta for the developing fetus,
(3) by the mammary gland for synthesis of lactose,
(4) by the liver for fat synthesis,
(5) by all tissues and organs as metabolic fuel.

In ruminants little glucose as such is absorbed from the gut. Blood glucose is derived therefore from other compounds by reactions in the liver called gluconeogenesis. Major sources of glucose are propionic acid, some amino acids and glycerol. In high production situations, as encountered in ewes carrying multiple fetuses in later pregnancy and cows in early lactation, blood glucose is at a premium to the female ruminant. When food intake does not support the productivity blood glucose concentration drops (hypoglycaemia). This condition in cows is often associated with a reduction in the solids-not-fat content of milk and reduction in conception rate. In extreme cases the condition of acetonaemia develops (*see* under Fat Metabolism, p. 331).

Acetic acid, the major fatty acid absorbed from the rumen is the only VFA present in significant quantities in blood. It has a major role as a metabolic fuel and, in the lactating ruminant, is a precursor of fatty acids synthesised in the udder which are subsequently incorporated into milk fat.

In the dairy cow it has been shown that relative proportions of volatile fatty acids produced during rumen fermentation have an influence on the composition of milk. The lower rumen pH and lowered ratio of acetic:propionic acid observed with diets rich in concentrates is associated with milk of lowered fat content.

In growing and fattening cattle the efficiency with which digested nutrients are utilised in the tissues is affected by the pattern of rumen fermentation. Highly digestible concentrate based rations which favour propionic acid production result in superior performance to diets supplying equivalent energy derived largely from fibrous foods.

Protein

Amino acids resulting from protein digestion are not stored in the liver but enter the pool of amino acids existing in blood and other body fluids. Amino acids are utilised in the synthesis of new proteins, as part of body tissues or present in products such as milk and wool. All enzymes and some hormones are protein by nature and require a source of amino acids for synthesis.

In the pig efficiency with which dietary protein is utilised depends on the amino acid composition.

Certain amino acids are synthesised from others within the liver by transamination. Such acids are termed non-essential amino acids. If there is a relative shortage of such an amino acid in diets, protein synthesis within the animal is not necessarily retarded since the deficit is corrected. By contrast, certain other amino acids cannot be synthesised in this way. These are termed essential amino acids, their presence in the diet being vital to protein synthesis. Ultimately a shortage of a single amino acid, termed the first limiting amino acid, may be responsible for a reduction in protein formation. In most pig diets the acid most likely to be limiting is lysine. Hence considerable attention is given to ensuring adequate lysine levels in practical pig diets. Generally proteins of animal origin have a higher proportion of essential amino acids than vegetable proteins. These are termed high quality proteins, smaller quantities being needed to support a given level of protein production.

The utilisation of amino acids for protein synthesis is never 100% efficient and amino acids not utilised for protein synthesis are converted by the liver in a process known as deamination to keto acids and ammonia. The keto acids, compounds of carbon, hydrogen and oxygen only, are utilised by the animal either as metabolic fuel or in certain circumstances in the formation of fat.

Ammonia, a highly toxic substance, is quickly converted in the liver to non-toxic urea which is excreted in urine.

In the case of ruminants, amino acid composition of the dietary protein is not so critical as in pig nutrition. The requirement of the tissues for amino acids is met largely by the supply made available by digestion of bacterial protein. The quantity of bacterial biomass entering the abomasum is governed by the quantity of organic material fermented in the reticulum and rumen. It is therefore related to the energy intake of the animal. In certain situations of high productivity, in particular the very high yielding dairy cow and the rapidly growing young calf and lamb, the supply of amino acids made available from digestion of the biomass is, of itself, insufficient to meet the requirement of the tissues. In these situations the content of undegradable protein in the diet is critical if the full potential of the animal is to be achieved.

Fat

The fat content of the animal body is largely derived from nutrients other than fat. Fat composition is determined primarily by species but can be modified by feeding practice; if a pig diet includes large levels of oil this modifies the composition of the body fat. The glycerol and fatty acids of butter fat are largely synthesised within the udder, but some of the fat is of dietary origin. Most diets contain little fat but addition of extra fat is increasingly practised to produce high energy diets for certain intensive livestock enterprises. A minimum level of fat in diets is essential since some unsaturated acids cannot be synthesised by the body. These are called essential fatty acids. Additionally dietary fat acts as a carrier for fat soluble vitamins.

Within the body fat acts as an insulator and reserve source of energy. Fat is a major metabolic fuel and is superior to carbohydrate and protein, yielding some two and a half times as much energy as an equal weight of carbohydrate. At any one time the body normally oxidises a mixture of carbohydrate, protein and fat. The contribution of each of these materials to the fuel mixture is not constant, varying with food intake and composition. The proportion of fat oxidised increases as animals call on reserves at times when energy intake is inadequate. For cows such situations arise at peak lactation, whilst in ewes late pregnancy is a time of severe metabolic stress. At these times the female ruminant is particularly prone to a metabolic disorder thought to result from an excessive rate of fat breakdown and utilisation in relation to carbohydrate utilisation. The disorder, known as acetonaemia, or ketosis, results in high levels of acetone in the blood. In sheep the disorder is commonly referred to as twin lamb disease, since affected animals invariably carry multiple fetuses. Prevention lies in ensuring adequate food intake at times of great need.

NUTRITIVE VALUE OF FOODS

The term 'nutritive value' is broad in its implication, and it is customary in evaluating foods to consider nutritive value under four main headings, energy, protein, minerals and vitamins.

Energy

The energy content of a food is represented by the very large number of organic compounds it contains. Energy in this form is referred to as chemical energy. Normally over 90% of food dry matter is composed of organic matter, the remaining inorganic (mineral) fraction not contributing to energy content. Organic compounds are combustible, i.e. they may be oxidised and energy in the form of heat liberated in the process. The energy released from a given weight of pure compound is a constant and is referred to as heat of combustion. The total energy released by the combustion of unit weight of food is usually referred to as gross energy and this can be readily determined in an apparatus known as a bomb calorimeter.

There is good reason why energy value should be considered initially in food evaluation since it is energy intake that primarily determines the level of performance an animal can sustain. Diet formulation then becomes primarily a matter of ensuring an adequate intake of the other nutrients so the animal functions as efficiently as possible on any particular energy intake.

Whilst food represents the sole source of energy to the animal there are several routes of energy loss, namely:

(1) faeces,
(2) urine (soluble organic compounds, e.g. urea),
(3) methane produced as one of the end products of fermentation in the rumen and to a much lesser extent in the large intestine, and
(4) heat, which is constantly lost from body surfaces.

The fate of the gross energy provided by a diet may therefore be summarised in the flow diagram:

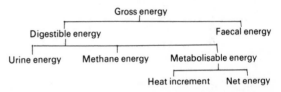

From the flow diagram four possible ways are seen in which energy value of food may be expressed. These, together with other related expressions, are discussed below. The SI unit of energy is the joule (J) and this is currently used to describe the energy value of foods. Since the use of the joule involves very large numbers it is customary to use the kilojoule (kJ) or more usually the megajoule (MJ).

$$1 \text{ MJ} = 1000 \text{ kJ} = 1\,000\,000 \text{ J}$$

For ease of comparison it is usual to express the energy value on a dry matter basis so that the effect of variable moisture content of foods is eliminated.

Gross energy value

This is measured by burning a known weight of the food in a bomb calorimeter and recording total heat produced. The gross energy value of any food is the sum of the energy values of its constituents. Carbohydrate, the dominant fraction of most foods, has an energy value of approximately 17.5 MJ/kg of dry material. Proteins and fats have energy values of about 24 MJ/kg and 38 MJ/kg, respectively. It would be expected therefore that gross energy value would be higher for foods rich in protein and fat. In practice most farm animal foods contain very little fat and carbohydrate is usually the dominant fraction. In consequence the gross energy value of a wide range of foods varies little from an average value of some 18 MJ/kg of dry matter. By itself gross energy concentration is the least satisfactory assessment of energy value, since no account is taken of any losses associated with feeding the food.

Digestible energy value (DE)

This is determined from the results of a digestibility trial. Individual animals are held in specially designed crates permitting accurate assessment of food intake and faecal output. Recordings are carried out over a period of some 7–10 d following a preliminary period of acclimatisation, usually of some 10–14 d. The example below illustrates the calculation of the digestible energy value of a sample of dried grass fed to sheep.

DM intake   = 0.90 kg/d
gross energy of diet   = 18.40 MJ/kg DM
DM output (faeces)   = 0.28 kg/d
gross energy of faeces   = 18.80 MJ/kg DM

Thus

energy intake   = 0.90×18.40 = 16.56 MJ/d
energy output   = 0.28×18.80 = 5.26 MJ/d
energy digested = $(16.56 - 5.26)$ = 11.30 MJ/d
DE value $= \dfrac{11.30}{0.90}$ = 12.55 MJ/kg DM

D value

D value is defined as the percentage of digestible organic matter (OM) in the dry matter of the feed. It is determined from results of a digestibility trial and knowledge of organic matter content of feed and faeces. The latter is determined by burning a sample at 450°C when all organic matter is destroyed. The loss of weight represents the weight of organic matter in the original sample. In the above example the D value would be calculated as follows:

DM intake   = 0.90 kg/d
OM content of diet   = 92.0%
DM output   = 0.28 kg/d
OM content of faeces   = 90.0%

Thus

OM intake $= 0.90 \times \dfrac{92}{100}$ = 0.83 kg/d

OM output $= 0.28 \times \dfrac{90}{100}$ = 0.25 kg/d

OM digested $= (0.83 - 0.25)$ = 0.58 kg/d

D value $= \dfrac{0.58}{0.90} \times 100$ = 64.4

The D value is now used widely by advisory services to describe the energy potential of roughage foods, particularly hay and silage.

Determination of D values using live animals is time consuming and not suitable for routine estimation of large sample numbers. In practice evaluation is carried out by an *in vitro* or test tube method developed at the Grassland Research Institute or is predicted from chemical analysis.

There is a close relationship between the D value and the DE concentration. Little error is involved in converting one to the other by the equation

$D = 5.4 \times DE$ (MJ/kg of DM).

Factors affecting the digestibility of energy

Digestibility is mainly a function of the food, being very little influenced by the consumer. For a given ration, animal factors such as age, size, breed and sex have very little, if any, influence. Sheep and cattle digest foods to very similar extents but pigs cannot digest fibrous rations.

Foods having fairly constant chemical composition have fairly constant digestibility. Cereal grains, harvested at the same stage of growth each year, differ relatively little in composition or digestibility. By contrast, grassland is conserved and utilised at varying stages of growth and varies widely in composition and D value. As the crop matures the proportion of crude fibre increases, chiefly due to increase in proportion of cellulose and lignin in the cell walls of the plant. This has the effect of reducing the rate and degree of digestion. Digestibility of poor quality roughage diets may be further reduced by their low crude protein content which limits the activity of rumen micro-organisms.

An increase in level of feeding generally results in a faster rate of passage of material through the digestive system. The food is thus in contact with bacterial and enzyme action for a shorter period, resulting normally in a slight reduction in the proportion of energy digested. Digestible energy and related values should therefore be determined at a specified intake. The most convenient is that necessary to provide sufficient energy for maintenance.

Finally, the way foods are prepared for feeding influences efficiency of digestion. Cereal grains are rolled or milled to prevent the passage of intact whole grains through to the faeces. Cooking or steaming potatoes is necessary before feeding to pigs but unnecessary when feeding cattle.

In summary, the expressions DE and D give assessments of energy value of foods based on the degree to which the energy is digested, but since the efficiency with which digested energy is utilised in the tissues of the animal is not taken into account it is possible that two diets may provide equal quantities of DE yet not produce the same performance in the animal.

Metabolisable energy value (ME)

From the flow diagram (p. 331) it can be seen that

Metabolisable = gross energy–(faecal energy + urine
  energy   energy + methane energy)
or   = digestible energy – (urine energy
                                     + methane energy)

Since further losses of energy associated with feeding are taken into account, clearly ME values of food are more accurate assessments of energy value than DE values. The estimation of ME values depends, however, on the measurement of methane production, which requires sophisticated experimental equipment and conditions. The routine evaluation of large numbers of samples is therefore precluded. In ruminants some 5–10% of gross energy intake is lost as methane, whereas in the pig it is probably less than 1%. The ME value of a diet depends, therefore, on the species to which it is fed.

Much research work has indicated that ME values of ruminant rations can be predicted from their DE values with little error using the equation $ME = 0.81 \times DE$ but if this equation is used it must be realised that ME values are no better assessments of energy value than the DE values from which they are predicted. Clearly ME values are influenced primarily by the digestibility of the food. High ME values are only possible with foods that are highly digestible. As with DE values the evaluation of ME values should ideally be conducted when the diet is fed at the maintenance level of nutrition. Values are influenced by protein content of the diet. Excessive dietary protein involves additional loss of energy in excreting surplus ammonia as urea.

Net energy value (NE)

The ME provided by a diet may be looked upon as representing the nutrients made available to the tissues of the body for their various metabolic processes. Utilisation of ME is never 100% efficient— a fraction will be utilised (the net energy), and the remaining fraction is lost to the animal in the form of heat (the heat increment).

The NE provided by a diet may be utilised in one of two ways:

(1) maintaining the animal in energy balance, i.e. meeting the maintenance requirement;
(2) formation of a product such as liveweight gain, milk or fetus.

The energy utilised for maintenance is used mainly to perform work and will thus be converted eventually to heat, leaving the body along with the heat increment. On a maintenance ration, therefore, the total heat output from the animal is numerically equal to the metabolisable energy intake. The energy utilised for production is found in the product being formed. In theory NE values represent the true energy worth of foods, since all losses associated with feeding are taken into account. The calculation of NE value of a diet may be illustrated from the following data which shows the food being used for liveweight gain (LWG).

	For main-tenance	For maintenance + LWG
Dry matter intake (kg)	3	5
Gross energy intake (MJ)	57	95
Energy of faeces (MJ)	17	30
Energy of urine and methane (MJ)	8	13
Heat production (MJ)	32	43
ME provided (MJ)	32	52

Increasing the DM intake by 2 kg results in an increase of 20 MJ in ME intake and this results in an increment of 11 MJ in the heat production. In other words 55% of the ME intake is lost as heat whilst 45% (9 MJ) is utilised and retained within the animal as liveweight gain. The net energy value of this diet for liveweight gain is thus $9/2 = 4.5$ MJ/kg DM.

Note that two levels of feeding are necessary to calculate the answer. The animal would be kept in a respiration calorimeter to allow measurement of its heat production.

In practice there are difficulties involved in using NE for precise ration construction. The two most serious of these are:

(1) The NE of very few foods has been determined directly, owing to the great cost and time involved.
(2) A single food may possess several different NE values. This arises from the fact that the efficiency of utilisation of ME differs with the various physiological functions being sustained; it is known that energy is most efficiently utilised for maintenance and that lactation is a more efficient process than fattening.

As the determination of net energy values is difficult and time consuming current practice related to diet formulation for ruminants assumes that little error is involved in using a factor of 0.72 to convert ME values to NE values for maintenance (NE_m), thus

$$NE_m = ME \times 0.72 \quad (1)$$

i.e. for a range of diets there is little variation in the degree of utilisation of ME for maintenance around 72%.

For lactation little error is involved in predicting net energy value (NE_l) by the equation

$$NE_l = ME \times 0.62 \quad (2)$$

For growth and fattening the efficiency of utilisation of ME varies widely according to diet (between 35% and 55%) and this precludes the use of a simple equation as (1) and (2) above when dealing with diets for growing calves and lambs. Thus

$$NE_g = ME \times K_g \quad (3)$$

where K_g is the efficiency of utilisation of ME for growth which is determined separately by the equation

$$K_g = 0.0435 \times ME \text{ value of diet} \quad (4)$$

Therefore for a diet (A) having an ME value of 8 MJ/kg DM

$$K_g = 0.0435 \times 8$$
$$= 0.348$$

$$NE_g = 8 \times 0.348$$
$$= 2.78 \text{ MJ/kg DM}$$

For a diet (B) having an ME value of 12 MJ/kg DM

$$K_g = 0.0435 \times 12$$
$$= 0.522$$

$$NE_g = 12 \times 0.522$$
$$= 6.26 \text{ MJ/kg DM}$$

Although diet B has 50% more ME/kg than diet A it has 125% more NE/kg when fed to growing/fattening ruminants. In other words, the relative value of foods for maintenance are not the same as their relative values for growth and fattening. It is thought this difference is due to concentrate diets producing mixtures of fermentation products within the rumen that are better utilised for fattening than those produced by roughages. When utilised for maintenance there is a much smaller difference in the degree of utilisation of the mixtures produced.

Protein

Whilst many nutrients are interchangeable as sources of energy the animal has a specific requirement for amino acids. These acids are utilised for the synthesis of protein within the body and of proteins in products such as milk and wool which leave the body. The amino acids must therefore be supplied at the site of synthesis in sufficient quantity and in the correct proportions to allow the potential of the animal to be achieved.

The value of foods in meeting these requirements is expressed in different ways, the more common expressions used in farm animal nutrition being the following.

Crude protein (CP)

Protein (and amino acids) are distinguished from other major nutrients by the fact they contain nitrogen. Most nitrogen in foods is present as protein. Thus if the nitrogen content of a food is determined it gives an assessment of protein content. Two assumptions are made in calculating protein content from nitrogen content. First, all nitrogen in food is present as protein, and second, all proteins contain 16% nitrogen. Since neither of these assumptions is absolutely correct the protein content is expressed as crude protein thus:

$$\%CP = \%N \times (100/16),$$

or

$$\%CP = \%N \times 6.25.$$

Percentage CP gives no indication of how efficiently the protein is digested or utilised.

Digestible crude protein (DCP)

The DCP content of a diet is determined from the results of a digestibility trial and is a more realistic measure of protein value than CP content.

Assume the DM of the diet shown in the example on p. 332 has a crude protein content of 180 g/kg (18%) and that the faecal DM contains 160 g/kg crude protein then

DM intake	= 0.90 kg
CP content of DM	= 180 g/kg
DM output	= 0.28 kg
CP content of faecal DM	= 160 g/kg

Then

CP intake $= 0.90 \times 180 = 162$ g/d
CP output $= 0.28 \times 160 = 45$ g/d
CP digested $= (162 - 45) = 117$ g/d
DCP content $= \dfrac{117}{0.9} = 130$ g/kg

The DCP system of assessment is used widely in formulating ruminant diets. However the values do not give a true guide to the quantities of amino acids absorbed from the small intestine. When calculating DCP content determination is made of nitrogen intake and nitrogen lost in the faeces. The difference between these figures is the nitrogen apparently digested. In the ruminant nitrogen absorbed from the digestive tract can be in the form of ammonia absorbed from the rumen as well as N in the form of amino acids absorbed from the intestines. Thus two diets may provide an equal quantity of absorbed nitrogen or DCP yet one might provide a greater proportion of nitrogen as amino acids than the other.

In pig feeding, it is thought that DCP values are not significantly better than CP values, since the degree of digestibility of protein is considered fairly constant.

Protein quality

Although diets may have equal crude and/or digestible crude protein content the performance of animals consuming the diets often differs. Such differences are attributed to differences in protein quality. An initial assessment of protein quality may be given by its apparent biological value (ABV). This is determined by a nitrogen balance experiment which measures nitrogen intake and nitrogen losses in the faeces and urine and is defined as the proportion of digested nitrogen retained by the animal.

N digested = N in diet − N in faeces

N retained = N digested − N in urine

$$\text{ABV} = \frac{\text{N retained}}{\text{N digested}} \times 100$$

In pig nutrition the efficiency with which digested nitrogen is utilised at a given energy intake is largely determined by the supply of essential amino acids; as long as these are provided in sufficient quantities the animal is able to express its potential for protein synthesis. The economics of production, however, demand that protein should not be supplied in excessive quantity and it is therefore a matter of reducing protein content to the point below which production would be impaired. It is likely that impairment of productivity would in the first instance be due to lack of one essential amino acid. This acid is termed the limiting amino acid and in pig diets it is normally lysine. In ration formulation it is therefore imperative that the lysine content of the diet meets the minimum level necessary to achieve acceptable performance criteria.

In ruminant nutrition nitrogen retention is not affected significantly by differences in amino acid composition of the protein since bacterial action within the rumen results in considerable transformation of protein composition. Normally a large fraction of the protein reaching the abomasum is composed of bacterial protein and this has a fairly constant value. Of greater significance is the relative contribution of rumen degradable protein (RDP) and undegradable protein (UDP) contained in the diet. Simply

CP = RDP + UDP

or

RDP = CP × dg

(where dg is the coefficient of degradability), RDP is that protein from which micro-organisms derive nitrogen to sustain their total activity, UDP is that protein which escapes fermentation and is available for digestion along with microbial protein within the abomasum and small intestine.

Taking an extreme example, if dietary protein were totally degradable (dg = 1) then the protein entering the abomasum would be entirely microbial. Any nitrogen in excess of bacterial needs would be absorbed across the rumen wall and largely excreted as urea. If the protein was not totally degradable then a fraction would be available for digestion directly in the abomasum and small intestine. Provided that the nitrogen requirement of the bacteria was still satisfied the animal would derive a greater amino acid supply from the small intestine and lose less nitrogen as urea thus giving a potentially greater utilisation of dietary protein.

The degradability coefficient of a protein is currently assessed in the laboratory by incubation with rumen fluid.

Estimates vary for nominally similar foods from different sources and therefore the values currently used must be considered provisional. It is clear that the degradability values for certain proteins are modified by the treatment to which they are often subjected. For example the protein of fresh grass is highly degradable. Drying and pelleting the material reduces the degradability probably because of reduced particle size and increased rate of passage through the rumen. The longer the retention time of material within the rumen the more likely is the extent of degradation to increase. Conservation as silage increases degradability which might be expected since considerable degradation of protein occurs during ensilage. The addition of formalin to grass at the time of ensilage protects the protein resulting in a reduction in the degradability.

The protein concentrates differ markedly in degradability. Some fish meal proteins are particularly resistant to breakdown whilst soya bean meal that has been heat treated to inactivate trypsin inhibitors appears to be less degraded than other solvent extracted oil seed meals such as ground nut and cotton seed meal.

Minerals

It is usual to express mineral content of foods as the total quantity present. Availability of dietary minerals varies greatly and is influenced by many factors. This, therefore, precludes a statement of available mineral contents.

It is important to view mineral concentration in relation to the energy concentration of the food. Diets of high energy value need high concentrations of minerals to achieve the necessary mineral intakes, since less food is required to provide a given energy intake.

Vitamins

Vitamin concentration of foods is usually expressed as milligrams or micrograms per kilogram. However, the levels of vitamins A and D are normally expressed as International Units (IU), defined as follows:

1 IU vitamin A = 0.300 µg vitamin A alcohol,

1 IU vitamin D = 0.025 µg crystalline vitamin D_3.

NUTRIENT REQUIREMENTS

Since the nutritive value of foods has been treated under the headings of energy, protein, minerals and vitamins, it is sensible to treat the requirements of animals under the same headings. In the case of energy and protein it has been seen there are several ways in which the values can be expressed, and so in dietary formulations it is essential that provision and requirement are stated in exactly the same terms.

The fact that individual animals vary in response to various nutrients and variation exists between samples of a food means great precision is not possible in calculating requirements or specifying responses. The provision of tables of food values and nutrient requirements is thus to guide the livestock feeder and does not detract from the art of the stockman, nor does it guarantee success to the novice.

Nutrient requirements or recommended nutrient allowances may be expressed in two ways, either as a quantity or as a proportion of the diet. Thus the lysine requirement of a growing pig may be either 12 g/d or 0.8% of the diet, on the assumption that the pig is consuming 1500 g/d of diet.

The remainder of this chapter is devoted to a survey of the factors influencing the requirements of animals under practical conditions.

Requirements for maintenance

Animals given no food call upon body reserves to supply the nutrients essential to life and so lose weight. The maintenance ration is that quantity of a diet which allows animals to maintain body weight and composition constant.

Energy

The energy requirement for maintenance is proportional to the size or body weight, W, of the animal, but it is not directly proportional. Energy requirement does not double as body weight doubles. A Friesian cow of 700 kg does not require the same energy as two Jersey cows each weighing 350 kg. For any species, requirement is more closely related to body surface area than to body weight. Since this is difficult to assess it is of academic interest. Requirement is, however, closely related to a function of body weight, namely $W^{0.73}$, termed the metabolic weight of the animal. The following calculations show the necessary steps in calculating the metabolic weight of a 500 kg cow and a 50 kg ewe.

	Cow	Ewe
W(kg)	500	50
log W	2.6990	1.6990
0.73 × log W	1.9702	1.2403
antilog of (0.73 × log W)	93.37	17.39
Metabolic weight (kg)	93.4	17.4

The metabolic weight of the cow is therefore approximately 5.4 times greater than that of the ewe, hence its energy requirement is greater by this proportion.

The energy requirement of an individual animal of given weight does not necessarily remain constant. Two factors can materially modify the estimate,

namely the activity of the animal and climatic conditions. Clearly animals that have to forage for food expend more energy than housed animals. The extra energy expenditure involved is debited to the maintenance requirement. The influence of climate is chiefly mediated through the effects of temperature, wind speed, and rain. For housed animals temperature effects are overriding.

All farm animals have characteristic and constant body temperatures. This implies that the rate of heat production is exactly equal to the rate of heat loss when measured over a period of time. Under 'normal' environmental conditions this equilibrium is readily achieved, but under extreme climatic conditions the animal may have to produce extra heat in order to maintain body temperature. This clearly implies an increased maintenance requirement. The environmental temperature at which the animal is forced to increase heat output is known as the effective critical temperature. This is not a constant temperature but is influenced by such factors as wind speed and whether the animal is wet or not. Subcutaneous fat and coat or fleece act as effective insulating materials and help reduce rate of heat loss. Ruminant animals have a lower critical temperature than pigs, largely due to the heat of fermentation produced from the rumen. The animals most susceptible to the effects of low temperature are the newborn, particularly in the first hours of life before the temperature regulating mechanisms of the animal are fully operative.

Protein

Proteins of body tissues are constantly being renewed. Amino acids resulting from breakdown of tissue protein are not re-utilised with 100% efficiency for synthesis of new protein. Thus the diet must provide protein in sufficient quantity to keep the animal in nitrogen balance. The quantity of protein necessary depends on its digestibility and efficiency of utilisation of the products of digestion within the tissues.

Minerals and vitamins

The quantity of any mineral necessary in the diet must ensure no net loss occurs from the animal. This requires an estimate to be made of the availability of the mineral. The dietary vitamin requirement is that intake which ensures prevention of any deficiency, clinical or sub-clinical.

Requirements for growth and fattening

The commercial measure of growth in farm animals is liveweight gain (LWG). The term performance is often used but this has wider implications and relates to the efficiency with which food is converted into LWG. Under any given set of conditions animals have a certain genetic potential for LWG. Whether this potential is achieved depends largely on nutrition.

Energy

The LWG made by an animal may be looked upon as a deposition of energy within its carcass. The quantity of energy deposited in unit weight of LWG depends on the chemical composition of the LWG and it is convenient to consider this composition in terms of water, bone, protein (flesh) and fat. The energy content of the gain is almost entirely represented by the protein and fat fractions. The composition, and therefore energy content, of LWG is not constant throughout the growth phase of an animal. Whilst it is considered all four components are present in the gain of animals of all ages, the proportions can vary widely. In general, gain made by young animals has a low energy content since it contains a high proportion of water and bone with little fat. In older animals, energy content of the gain is considerably greater, largely reflecting the high proportion of fat present. This explains why growth of older animals is so often referred to as 'fattening' and it therefore follows that older animals produce fatter carcasses at slaughter than young animals.

The greater the rate of LWG the greater the fat and, hence, energy content of the gain. In other words, animals grown quickly to a given slaughter weight have fatter carcasses than those grown slowly to the same weight.

In specifying energy requirements for growth it is necessary, therefore, to indicate the rate of growth anticipated. There are many instances where animals are required to grow at their maximum potential. This is normally achieved in practice by allowing *ad libitum* access to foods of high energy concentration, giving the so-called intensive systems of production. Conversely, restriction of energy intake reduces growth rate and also reduces the fat content of the carcass at slaughter, a principle applied in the production of pigs for bacon where over-fatness of carcass is severely penalised.

Protein

Growth rate is primarily determined by the energy intake. At a given growth rate there is a specific requirement for the various amino acids to permit the synthesis of new tissue protein. The actual protein intake necessary (or the protein content of the diet) for a particular level of production in pigs will be governed by the digestibility and amino acid composition of the protein fed. For ruminants the proportion of degradable to undegradable protein is important particularly where rapid rates of LWG involving high rates of protein deposition are involved. Certain minimum levels of undegradable protein are essential to augment the amino acids made available from

digestion of microbial protein. These considerations explain the observations that equal intakes of protein do not necessarily sustain equal rates of productivity.

Minerals and vitamins

Criteria for the adequacy of minerals and vitamins are similar to those described under 'Requirements for maintenance'. Most investigations on mineral requirements of growing animals have centred on the need for calcium and phosphorus. This is natural since these two elements are required in greater quantities than any others and symptoms resulting from their deficiencies are well known. Along with vitamin D these elements are concerned largely with skeletal development and the condition of rickets is due to inadequate and improper bone development. In ruminants retardation of growth is often caused by deficiency of the elements copper and cobalt, whilst in young pigs deficiency of iron is most likely to cause growth depression.

Food conversion ratio (FCR)

FCR is a practical measure of the efficiency with which food intake of growing animals is converted to LWG

$$FCR = \frac{\text{Weight of food consumed}}{\text{LWG produced}}$$

Clearly a ratio of small magnitude indicates high efficiency and *vice versa*. The ratio achieved in any system of production depends on various factors, discussed below.

Nutrititive value of the diet

The energy concentration of the diet has a major bearing since the LWG is more a response to energy intake than to food intake. High energy diets thus promote superior FCR, whilst those deficient in protein result in a reduction of rate of LWG and in turn produce gain of high fat content. Both effects tend to result in inferior FCR. Likewise sub-optimal intakes of a mineral or vitamin must have the effect of reducing the efficiency of the animal.

Energy intake or level of feeding

High energy intake promotes a high rate of LWG which in turn produces high efficiency of food utilisation. This is explained by the fact that as level of production increases so the proportion of total food intake utilised for maintenance decreases. This is illustrated by a theoretical example of a 200 kg steer fed quantities of a diet designed to produce daily liveweight gains of 0, 0.5 and 1 kg/d.

	LWG		
	0 kg/d	*0.5 kg/d*	*1 kg/d*
Daily food intake (kg)	3.2	4.2	5.5
Daily LWG (kg)	0	0.5	1.0
FCR	∞	8.4	5.5
Percentage of food used for maintenance	100	76	58

Notice that although the second half-kilogram of gain requires more food than the first, overall efficiency increases since the proportion of energy used for maintenance declines. However, a point may be reached where the last increment of LWG is so expensive to achieve in terms of food input that the FCR will actually begin to get worse.

Much experimental work with fattening pigs has shown that the FCR is improved where a degree of food restriction is practised compared with *ad libitum* feeding.

Slaughter weight

Other factors being equal, FCR deteriorates as slaughter weight increases. This reflects the high energy requirement for fat production which forms an increasing proportion of the LWG as the animal gets heavier. Thus 'heavy' pigs (slaughter weight 120 kg) have inferior ratios to pork pigs (slaughter weight 60 kg).

Breed

Genetic differences between breeds (and between strains within a breed) can result in significant differences in efficiency of food utilisation as measured by the FCR. Breeds differ in their mature body size, so if animals of different breeds are slaughtered at the same weight it follows that this represents a smaller proportion of mature body weight in the large breeds compared with the smaller. At this weight carcasses of the potentially large animals will contain more lean and less fat. Because less food is consumed to produce such gain potentially large animals have a superior FCR. Although over-fat carcasses are undesirable it must be appreciated that a certain degree of fat cover of 'finish' is essential to a high grade carcass. In practice, potentially large animals are often slaughtered at heavier weights than potentially small animals.

Sex

Entire males, castrates and females of a breed or cross normally differ in body composition at equal body weights. Bulls and steers are leaner than heifers of the same weight. In view of the high food requirement to produce fat it follows that males have a superior FCR. The corollary is that heifers are normally slaughtered at lower weights than steers. Similarly, male fat lambs

are leaner and superior food converters to females. In the case of pigs, gilts are leaner than hogs at slaughter weight and it follows they are superior food converters, more likely to produce better grading results when carcasses are evaluated for bacon.

Clearly all the effects of these factors cannot be taken into account in publishing requirements for growing animals, but they do indicate why responses to the guide figures may turn out to be quite variable.

Requirements for reproduction

Factors influencing the requirements of female livestock just before and during pregnancy are discussed below.

Energy

The period immediately before mating is considered important since conception is more likely to occur when animals are in good and improving condition. Further, the period immediately following conception is critical in that faulty or inadequate nutrition at this time jeopardises the successful attachment of the fetus to the uterine wall. The female may then have to be re-mated or the number of eventual offspring may be reduced. It is common to put ewes on to good grazing some weeks before and during tupping, a practice known as flushing.

Once the fetus is successfully attached to the uterine wall it is thought to have very little influence on the nutritional requirements of the dam for the first two-thirds of pregnancy. Thereafter, by virtue of its accelerating growth, it makes an increasing and appreciable demand on the dam, and hence materially increases her requirements for energy and other nutrients. With young females pregnant for the first time assessing requirement is further complicated by the fact that the animals are themselves still growing; there is therefore a requirement for this growth.

With dairy cows it is practice to terminate a lactation some eight weeks before the birth of the next calf and to feed supplementary concentrates for up to six weeks before calving. This practice is referred to as steaming-up and ensures the cow is in good condition to commence her next lactation; it has relatively little influence on the birth weight of the calf.

The requirements of the pregnant ewe during the last six to seven weeks of pregnancy are largely influenced by whether she is carrying one or more lambs. Double and triple fetuses produce a burden some 50% greater than a single fetus. Failure to provide ewes carrying multiple fetuses with an adequate and consistent energy intake during this time can result in reduced lamb birth weights and, more seriously, predispose the ewe to pregnancy toxaemia. In lowland flocks it is customary to feed supplementary concentrates at an increasing rate as lambing approaches. In hill and mountain flocks often no supplementary food is provided but then the incidence of multiple births is not as high as in lowland flocks. Over-generous nutrition in mid pregnancy is not to be recommended. It results in ewes becoming fat and as such can often further predispose them to pregnancy toxaemia.

Pregnant sows are known to be more efficient at utilising their food than their non-pregnant counterparts. In one experiment pregnant and non-pregnant sows were given the same total quantity of food (225 kg) over the period of pregnancy. Whilst the allowance resulted in a small gain of weight (4 kg) in the non-pregnant sows the pregnant sows were 20 kg heavier post partum than at service.

Although it is customary to restrict food intake of pregnant sows to prevent unnecessary fatness and inefficiency of food utilisation, it is desirable the food allowance permit them to make a net increase in body weight over pregnancy. Much of this gain will be lost in the ensuing lactation. The level of feeding may be increased in the last third of pregnancy, but in many systems sows are fed a constant daily intake throughout pregnancy with impunity.

Protein

Rapid fetal growth in late pregnancy creates an increased requirement in the dam. Since this increase is roughly in proportion to the extra energy required, no special consideration is given to increasing the percentage of protein in the diet.

Minerals and vitamins

Deficiencies of minerals or vitamins must be severe to cause death of a fetus, but fetal abnormalities may often be traced to nutrient deficiency. Vitamin A deficiency is probably most often implicated in such conditions, but copper, iodine and riboflavin deficiencies are known to produce classic symptoms in the newborn.

Requirements of breeding males

Requirements of adult males are not precisely known. The eye of the stockman is therefore imperative to assess that animals are in good condition without being over-fat. There is some evidence to indicate the maintenance energy requirement of entire males is slightly higher than that of females and castrates of comparable weight.

Requirements for lactation

Energy

The lactating animal has a given genetic potential for milk production. In the dairy cow the attainment of

this potential depends primarily on energy intake. With animals suckling their young, milk yield is considerably influenced by the number of offspring suckling, as well as by the plane of nutrition. The energy requirement for milk production depends on (a) energy value of the milk, and (b) yield of milk, since these together determine energy output.

Energy value of milk is determined directly by bomb calorimetry or predicted from its compositional analysis. The energy yielding components of milk, i.e. fat, protein and lactose, are considered to have energy concentrations of 38.5, 24.5 and 16.5 MJ/kg respectively, thus allowing computation of the total energy value. However, the constituents of cows' milk most commonly determined on a routine basis are fat and solids-not-fat. With a knowledge of these values, fairly accurate predictions of milk energy value can be made using the equations of Tyrrel and Reid: either

$$E\ (MJ/kg) = 0.386F + 0.205SNF - 0.236 \quad (5)$$

or

$$E\ (MJ/kg) = 0.406F + 1.509 \quad (6)$$

In the typical lactation of a dairy cow milk yield increases after calving until a peak yield is established normally within four to eight weeks; thereafter there is a gradual decline until the cow is dried off prior to the next calving. The precise pattern of production throughout the period of lactation as exemplified by the lactation curve will vary between individuals; it is generally more flat in heifers than in older cows and it can be modified by the level and type of nutrition. In the first weeks following calving it is unlikely that the cow can consume sufficient energy to match the high energy output as milk. The output is sustained because energy is made available by the mobilisation of body reserves, chiefly fat, resulting in a loss of body weight and condition. Dry matter appetite normally rises to reach a peak some weeks after peak milk yield and this enables the cow to replenish reserves and put on weight since energy intake exceeds that necessary for the immediate milk output. The situation can be illustrated by *Figure 13.1*.

It is important to realise that the partitioning of available energy between milk yield and body tissue will vary between individual cows, thus cows of high potential may be described as those individuals having high dry matter appetite and which partition available energy more in favour of milk than body tissues when compared with average individuals. Additionally the composition of the diet affects the partition of energy. It is known that diets containing very high proportions of concentrates increase the proportion of energy diverted to body tissue at the expense of milk production.

All recent experimental evidence indicates the desirability of feeding cows of high potential generously in early lactation. This not only encourages the attainment of high peak yield but ensures that body

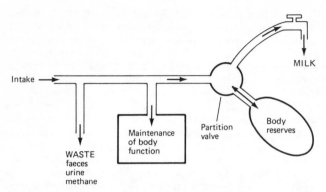

Figure 13.1 Model cow (After J. B. Owen, with permission from the publisher Farming Press Ltd, Ipswich)

reserves are not over depleted. Appropriate energy inputs have of course to be ensured over the whole lactation if the potential productivity is to be secured.

Protein

Calculations of protein requirement are based on the protein content of the milk and efficiency of utilisation of dietary protein. For the high yielding dairy cow it is important to ensure that the diet contains an adequate quantity of undegradable protein (UDP) in addition to an adequate supply of degradable protein (RDP) which is essential for efficient microbial activity within the rumen. Thus the perfect diet would meet both these requirements exactly.

If the diet should provide adequate UDP but insufficient RDP then the cheapest supplement is likely to be a source of non-protein nitrogen such as urea. If on the other hand the diet provides an adequate or more than adequate quantitity of RDP but insufficient UDP then some substitution of ingredients is required, i.e. some constituents having a high UDP content such as fish meal should replace ingredients having a lower UDP content. During early lactation when cows are losing weight the UDP requirement is likely to be increased since inadequate protein is being mobilised from tissues to match the energy mobilised from fat deposits. Again, these considerations explain the observations that diets of equal energy and crude protein content do not necessarily sustain equal rates of milk production.

Minerals and vitamins

Most information on mineral and vitamin requirements is obtained from studies with dairy cows. Calcium and phosphorus have been most studied and allowances per kg of milk (4% butter fat) are put at 2.8 g calcium and 1.7 g phosphorus. Such milk contains about 1.2 g calcium and 1.0 g phosphorus. These allowances take into account long-term requirements. It is considered normal for dairy cows in early lactation to be in negative calcium and phosphorus balance, indicating a drain on body reserves, and to be in positive balance in later lactation when milk yield is lower. Another element that can become deficient in high yielding cows is sodium. This is prevented by provision of salt, either in concentrate foods or as the major component of proprietary salt licks.

Dietary vitamin requirements for ruminants are confined to the fat soluble vitamins. Since these vitamins are stored in the body it is difficult to estimate requirements with accuracy. Additionally, prolonged shortage of vitamin A results in reduction in concentration of the vitamin in milk. Concentrate diets for cattle are often fortified by addition of synthetic forms of both vitamin A and vitamin D. These are also added to diets for lactating sows along with vitamin B_2, considered the only water soluble vitamin likely to be deficient in normal diets.

14

Animal feedstuffs

R. M. Orr

INTRODUCTION

Between 65 and 70% of total agricultural output in the UK is in the form of livestock production. The nutrient requirements for this output is met through the feeding of grass, conserved grass and other bulky fodders along with the use of both home grown and imported concentrate foods.

Whilst pig and poultry diets consist almost entirely of concentrated feedstuffs the requirements of ruminant livestock are met by a combination of roughages and concentrates, the contribution of concentrates to total requirements being greatest in dairy cow feeding (38%) and least in the feeding of sheep and suckler cows (less than 10%).

In meeting the nutrient requirements of livestock a wide range of raw materials may be utilised. In this chapter an overview of the role of the various feeds and feedstuffs in meeting nutrient requirements is presented and particular characteristics of individual feedstuffs, which impose limits on the contribution they can make to diets, are given attention.

CLASSIFICATION OF FEEDSTUFFS

In discussing feedstuffs it is useful to distinguish between those feedstuffs which are relatively high in fibre and/or water content (bulky feeds) and those which are low in fibre and water (concentrated feeds). *Figure 14.1* indicates into which of these major groupings different feedstuffs fall.

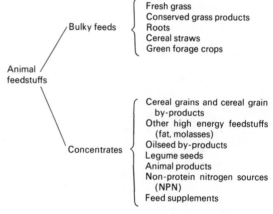

Figure 14.1

COMPOSITIONAL ANALYSIS AND FEED QUALITY

In feeding animals a basic consideration is the nutritional value of available feedstuffs. This information has been accumulated over many years and is being continually updated as more precise measures of nutritive values are obtained. Details on composition and nutritive values of the more widely used feedstuffs for the different classes of livestock are given in *Table 14.1*. More comprehensive information may be obtained from the following publications: *Nutrient Allowances and Composition of Feedingstuffs for Ruminants* (1976), MAFF: LGR21; *Nutrient Allowances for Pigs* (1978), MAFF: Advisory Paper No. 7 (LPR22); Bolton, W. and Blair, R. (1974), *Poultry Nutrition*, MAFF Bulletin 174: HMSO London.

Table 14.1 Guide to the chemical composition and nutritive values of some commonly used feedstuffs

Feedstuff name	Dry matter content (%)	Chemical composition					Nutritive value		
		CP	EE	CF	NFE	Ash	Metabolisable energy		Digestible energy
							Ruminants (MJ/kg DM)	Poultry (MJ/kg A-F)*	Pigs (MJ/kg A-F)
		(% of feedstuff DM)							
(1) Pasture grass									
(2) Conserved grass products	Details of typical nutritive values and chemical composition of some common grass and forage materials are given in Chapter 6, Grassland								
(3) *Roots*									
Cassava (dehydr. ground)	90.0	2.8	0.3	8.7	84.8	3.4	11.8	14.43	15.3
Fodder beet	22.0	6.8	0.3	5.9	79.1	7.7	12.2	—	2.9
Mangels	13.0	9.2	0.8	6.2	76.9	6.9	12.4	—	1.97
Swedes	12.0	10.8	1.7	10.0	71.7	5.8	12.8	—	1.7
Turnips	9.0	12.2	2.2	11.1	66.6	7.8	11.2	—	1.4
Sugar beet pulp (dried)	90.0	9.9	0.7	20.3	65.7	3.4	12.7	—	12.9
(4) *Cereal straws*									
Barley (spring)	86.0	3.8	2.1	39.4	49.3	5.3	7.3	—	—
Barley (winter)	86.0	3.7	1.6	48.8	39.2	6.6	5.8	—	—
Oat (spring)	86.0	3.4	2.2	39.4	49.3	5.7	6.7	—	—
Wheat (winter)	86.0	2.4	1.5	42.6	47.3	6.2	5.7	—	—
(5) *Green forage crops*									
Kale-marrow stem	14.0	15.7	3.6	17.9	49.3	13.6	11.0	—	—
Rape	14.0	20.0	5.7	25.0	40.0	9.3	9.5	—	—
Green legumes	Details of typical nutritive values and chemical composition of these materials are given in Chapter 6, Grassland								
Green cereals									
(6) Cereal grains and cereal by-products									
Barley	86.0	10.8	1.7	5.3	79.5	2.6	13.0	11.13	13.2
Wheat	86.0	12.4	1.9	2.6	81.0	2.1	13.5	12.18	14.3
Oats	86.0	10.9	4.9	12.1	68.8	3.3	12.0	11.05	11.4
Maize	86.0	9.8	4.2	2.4	82.3	1.3	14.2	13.22	14.9
Maize (flaked)	90.0	11.0	4.9	1.7	81.4	1.0	15.0	—	15.3
Sorghum	86.0	10.8	4.3	2.1	80.1	2.7	13.4	12.98	14.3
Brewers' grains (dried)	90.0	20.4	7.1	16.9	51.2	4.3	10.3	7.91	7.6
Distillers' grains (dried)	90.0	30.1	12.6	11.0	44.3	2.0	12.1	9.37	—
Bran	88.0	17.0	4.5	11.4	60.3	6.7	10.1	8.45	8.9
Fine wheat middlings	88.0	17.6	4.1	8.6	65.0	4.7	11.9	11.80	11.8
(7) *Other high energy feeds*									
Fat and oil	98.0	—	100	—	—	—	—	—	29.0
Molasses	75.0	9.0	0	0	84.1	6.9	12.5	7.91	
(8) *Oilseed by-products*									
Ex soya bean	90.0	50.3	1.7	5.8	36.0	6.2	12.3	10.67	13.6
Ex. rape seed	90.0	41.3	3.4	10.4	36.6	8.2	10.9	7.36	12.2
Ex. groundnut (decort.)	90.0	34.3	2.1	27.3	31.6	4.7	9.2	11.4	13.7
Ex. sunflower	90.0	42.3	1.1	18.1	31.2	7.2	10.4	7.5	12.6
(9) *Legume seeds*									
Field beans									
—spring	86.0	31.4	1.5	8.0	55.1	4.0	12.8	10.42	12.2
—winter	86.0	26.5	1.5	9.0	59.1	4.0	12.8	10.00	12.2
(10) *Animal products*									
Fish meal (white)	90.0	70.1	4.0	0	1.8	24.1	11.1	11.46	11.8
Herring meal	90.0	76.2	9.1	0	4.4	10.2	—	13.35	13.5
Meat and bone meal (medium protein)	90.0	52.7	4.4	0	1.7	41.2	7.9	—	9.6
Dried skim milk	95.0	37.2	1.1	0	53.2	8.5	14.1	12.26	15.6

* A-F = as fed.

Sources of information:
Bolton, W. and Blair, R. (1974). *Poultry Nutrition*. MAFF Bulletin 174 (HMSO London).
Ensminger, M. E. and Olentine, C. G. (1978). *Feeds and Nutrition—complete*. Clovis, California: Ensminger Publishing Co.
McDonald, P., Edwards, R. A. and Greenhalgh, J. F. D. (1973). *Animal Nutrition*. Edinburgh: Oliver & Boyd.
Nutrient Allowances and Composition of Feedingstuffs for Ruminants (1976). MAFF, LGR21.

| Digestible crude protein | | | Amino acid composition | | | | | | Mineral composition | |
| Ruminants (g/kg DM) | Pigs (% A-F) | Poultry (% A-F) | Glycine | Lysine | Methionine + Cysteine | Iso-Leucine | Threonine | Tryptophan | Calcium | Phosphorus |
			(% of total feed A–F)						(% of total feed DM)	
—	1.7	—	—	0.09	0.03	0.05	0.05	0.01	—	0.03
47.0	0.7	—	—	0.035	—	0.024	0.036	—	0.22	0.22
58.0	0.8	—	—	—	—	—	—	—	0.22	0.22
91.0	0.9	—	—	—	—	—	—	—	0.42	0.33
73.0	0.9	—	—	—	—	—	—	—	0.44	0.33
59.0	3.1	—	—	0.69	—	0.37	0.37	0.09	0.96	0.09
9.0	—	—	—	—	—	—	—	—	0.42	0.08
8.0	—	—	—	—	—	—	—	—	0.39	0.08
11.0	—	—	—	—	—	—	—	—	0.24	0.09
1.0	—	—	—	—	—	—	—	—	—	—
123.0	—	—	—	—	—	—	—	—	2.14	0.31
144.0	—	—	—	—	—	—	—	—	0.93	0.42
82	7.6	9.0	4.8	0.35	0.34	0.35	0.30	0.15	0.05	0.38
105	9.8	8.8	3.8	0.31	0.39	0.40	0.31	0.12	0.03	0.40
84	7.3	8.5	0.55	0.38	0.34	0.36	0.34	0.12	0.09	0.37
78	6.7	6.7	0.32	0.26	0.29	0.32	0.32	0.08	0.02	0.27
106	9.0	—	—	0.29	0.33	0.36	0.36	0.10	—	0.29
87	7.2	8.4	0.3	0.22	0.31	0.38	0.30	0.08	0.03	0.68
145	13.5	20.8	0.3	0.52	0.55	0.94	0.55	0.21	0.32	0.78
214	—	—	—	—	—	—	—	—	0.31	0.33
126	10.4	11.1	0.87	0.60	0.50	0.55	0.45	0.25	0.16	0.84
129	12.2	15.0	0.41	0.64	0.40	0.49	0.49	0.22	0.13	0.91
—	—	—	—	—	—	—	—	—	—	—
31	—	—	—	—	—	—	—	—	—	—
453	39.0	42.8	1.94	2.94	1.35	2.52	1.81	0.60	0.23	1.02
343	31.0	26.1	0.43	2.04	1.26	1.36	1.58	0.44	0.59	0.94
316	46.2	37.8	2.53	1.67	1.25	1.67	1.36	0.52	0.12	0.51
381	31.0	30.3	2.39	1.60	1.49	1.71	1.56	0.50	0.41	1.33
248	21.6	21.1	1.15	1.78	0.52	1.25	1.05	0.25	0.16	0.66
209	18.2	17.5	1.00	1.50	0.36	0.93	0.84	0.21	0.19	0.67
631	56.8	59.0	5.64	4.40	1.26	2.46	2.46	0.61	7.93	4.37
—	64.4	66.6	4.50	5.62	1.62	3.10	2.91	0.58	3.26	2.22
411	42.2	—	—	2.61	1.48	1.26	1.55	0.29	11.44	6.00
350	34.7	27.5	6.07	2.63	1.26	2.42	1.58	0.47	1.17	0.96

Nutrient Allowances for Pigs (1978). MAFF Advisory Paper No. 7 (LPR 22).
Whittlemore, C. T. and Elsley, F. W. H. (1976). *Practical Pig Nutrition*. Farming Press.

Explanation of the terminology and methodology of feed evaluation is given in Chapter 13, Animal nutrition.

The compositional and nutritive details given in these publications merely represent the averaging of data from a number of sources. Since considerable variation may be found in the nutrient content of feedstuffs it is important that producers and feed compounders recognise the value of an adequate feed analysis programme. In fact compounded feedstuffs offered for sale must be accompanied by a statutory statement giving particulars of the contents of fat, fibre, protein and the protein equivalent of any urea or other non-protein nitrogen compound in the formation.

The most widely used routine methods of evaluating the nutrient content of feedstuffs are based on chemical procedures. Such analyses are only of value, however, if the feed sample analysed is representative of the entire batch and to this end it is important that appropriate sampling procedures be followed. Using appropriate sampling equipment a number of core samples should be taken from different areas of the feedstuff concerned, these samples thoroughly mixed together and a suitable size of subsample taken for analysis. It is important that samples for analysis are appropriately packaged, labelled and stored prior to analysis. For example, silage samples not analysed immediately should be frozen.

Analysis of concentrate feeds

The most widely used routine analysis is the proximate analysis in which feeds are broken down into six fractions (*Table 14.2*).

overestimated whilst that of others may be underestimated. It is likely that the use of proximate analysis will decline in the future as new techniques are used more widely.

When the proximate analysis of a feed is known it is possible to estimate the energy value. For ruminant feeds an estimation of the ME value of a food may be made by applying appropriate factors to the amounts of digestible nutrients it contains. Workers at the Oskar Kellner Institute at Rostock have proposed the following equation for this purpose:

$$ME \ (MJ/kg \ DM) = 0.0152 \ DCP + 0.0342 \ DEE \\ + 0.0128 \ DCF + 0.0159 \ DNFE$$

where

DCP   = digestible crude protein
DEE   = digestible ether extract all as
DCF   = digestible crude fibre g/kg DM
DNFE = digestible nitrogen free extractives

Details of digestible proximate constituents for feeds are given in MAFF Paper LGR 21. In the absence of precise information on formulation, digestibility coefficients of 0.8, 0.9, 0.4 and 0.9 may be assumed for protein, ether extract, fibre and NFE respectively.

Analysis of forage feeds

While proximate analysis serves a useful purpose in predicting the nutritive value of concentrate feeds it is less effective in predicting the nutritive value and quality of forages.

Table 14.2 The fractions of proximate analysis

Fraction	Components	Procedure
(1) Moisture	Water and any volatile compounds	Heat sample to constant weight in oven
(2) Ash	Mineral elements	Burn at 500 °C
(3) Crude protein ($= N \times 6.25$)	Proteins, amino acids, non-protein nitrogen	Determine nitrogen by Kjeldahl method
(4) Ether extract (oil)	Fats, oils, waxes	Extraction with petroleum spirit
(5) Crude fibre	Cellulose, hemicellulose, lignin	Residue after boiling with weak acid and weak alkali
(6) Nitrogen-free extractives	Starch, sugars and some cellulose, hemicellulose and lignin	Calculation (100 minus sum of other fractions)

Individual mineral and vitamins can be assayed but are usually only done when a specific problem exists.

Whilst most of the data on feed composition at the present time is reported in terms of proximate analysis it is generally recognised that it has shortcomings particularly with respect to the measurement of how much indigestible material there is present in the feed such that the nutritive value of some feeds may be

A procedure which has achieved wide consideration as a substitute for the conventional crude fibre estimation has been developed by Van Soest and associates at Beltsville, Maryland. Although full analysis according to the Van Soest method is both too costly and time consuming for routine analysis one of the analytical procedures involved, the Acid Detergent Fibre determination, serves as the basis of the routine analysis of herbage samples carried out by

MAFF laboratories. Acid detergent fibre (ADF) is obtained by refluxing the forage with acid detergent for 1 h the residue left representing recovery of indigestible residues including cellulose, lignin, cutin, tannin–protein complexes, heat damaged protein, plant silica and soil minerals. In the MAFF procedure, a modification of the acid detergent fibre (ADF), modified acid detergent fibre (MADF) is used. Developed at the Grassland Research Institute and differing from the Van Soest method in that the length of boiling and acid strength are increased this gives a good correlation with D-values and ME values of herbage.

Using appropriate regression equations it is possible to predict the D-value and ME value (MJ/kg DM) of forages:

e.g. for grass silages:

$$\text{D-value} = 91.4 - (0.81 \times \text{MADF})$$
$$\text{ME value (MJ/kg DM)} = 14.6 - (0.13 \times \text{MADF})$$

For grass hays:

$$\text{D-value} = 104.8 - (1.27 \times \text{MADF})$$
$$\text{ME value (MJ/kg DM)} = 17.1 - (0.221 \times \text{MADF})$$
(where MADF is calculated as % of DM)

Further information on forage composition and quality is also given by the following chemical procedures:

(1) dry matter or moisture percentage;
(2) pH in silages;
(3) crude protein.
From this digestible crude protein (DCP) estimates can be made using appropriate prediction equations:
e.g. for grass silages
$$\% \text{DCP} = (0.82 \times \text{CP}) - 2.4$$
for grass hays
$$\% \text{DCP} = (0.92 \times \text{CP}) - 3.8$$
(where CP is expressed as % of DM)
(4) Ammonia–nitrogen in silages. This gives an indication of the level of proteolysis that has occurred. In well preserved silages values of <5–8% NH_3-N (expressed as % of Total N) are encountered whilst in poorly preserved materials the value may be in excess of 20%.

TYPES AND ROLES OF FEEDSTUFFS

Pasture grass

Grazed herbage provides a major source of nutrients for ruminant animals. As a feedstuff, however, it has the disadvantage of being variable in nutritive value and in addition its efficiency of utilisation is difficult to control. Aspects of this are covered under Chapter 6, Grassland, the salient points being that whilst a number of factors such as climate, soil fertility and botanical composition of the sward may affect nutritive value, the most important factor is stage of growth; thus since nutritive value, stage of growth and yield of material are inter-related, systems of utilisation must take into account these variables together with animal production targets and other management constraints, e.g. amount of conserved herbage required for winter use. A relatively high degree of control of these factors may be exerted on lowland pastures but in the case of hill pastures the restricted growing season and difficulties in controlling herbage utilisation leads to a very cyclical pattern of nutrition in grazing animals. Although during the growing season (May to October) grass in hill pastures will be in the range of 65–55 D during the winter months D-values may be as low as 40 with little or no DCP present.

Conserved grass

A feature of all conserved grass is that the nutritive value is dependent on the quality of grass used and the efficiency of the conservation process. Thus the aim in all conservation processes is to minimise the losses of nutrients from the grass utilised and to produce a feedstuff that fulfils the production objectives of the livestock to which it is to be fed.

The main methods of conserving grass are as hay, silage and as artificially dried grass. Most grass is conserved as hay, with conservation as silage not far behind. Because of costs of production artificially dried grass is produced on a relatively small scale for specialised purposes. Details of these processes and factors affecting the efficiency of the processes are given in Chapter 6, Grassland.

As the nutrient value of conserved hays and silages can vary considerably it is advisable that appropriate samples be analysed so that most prudent use of conserved forage and supplementary concentrate feeds may be made. Since grass is generally cut at an earlier growth stage for silage than for hay the nutritive value is generally higher. Thus when compared with the best quality hays good quality silages have approximately 10% more ME and up to 50% more DCP. Retention of the vitamins and minerals in grass during hay and silage making processes is variable but in general the mineral content of both materials reflect levels in the crop unless effluent losses or leaching losses are high; silages are good sources of vitamin A but contain negligible amounts of vitamin D whereas the opposite situation pertains with sun-cured hays.

Whilst the nutritive properties of conserved forages may be defined to an extent by chemical analysis one of the main problems in feeding roughages when they are a major component of the diet is to predict how much animals will consume. This is more of a problem with silages where factors such as digestibility, dry matter content, chop length and the fermentation pattern are operating. Intakes tend to be enhanced by high dry matter, high digestibility and reduced chop length. Whilst intakes of silages with high pH and high ammonia levels resulting from poor fermentation are low, at the other end of the scale high levels of acidity may also restrict intake. At present there is insufficient information to predict accurately roughage intake by different classes of livestock.

The ensilage of whole crop cereals, particularly maize gives a product with a higher energy value than grass silage, but lower levels of DCP. Such silages may be usefully supplemented with NPN compounds.

Artificially dried grass is a product which by virtue of its relatively low fibre content occupies an intermediate position between the bulky feeds and the concentrates. Due to the rapid removal of water dried grass has a similar feeding value to the original grass when expressed on a dry matter basis. Artificial drying is an expensive process and only young leafy material with a high D-value and high crude protein content is used. To conform to the definition of 'dried grass' under the terms of the Fertilisers and Feedingstuffs Regulations it must contain at least 13% crude protein at 10% moisture content. It is generally only included in the diets of high producing ruminant animals, particularly dairy cows during early lactation or occasionally in monogastric diets.

Cereal straws

All cereal straws are of relatively low nutritive value but differences occur between the different cereal types and also between spring and winter-sown cereals.

Since the levels and digestibility of the protein in straw is negligible and since there are relatively few minerals and vitamins present the only nutrient which straw can supply is energy. The availability of this energy, however, is very low mainly due to the high levels of lignin protecting the fibrous carbohydrates from microbial breakdown. In addition when straw is a major component of the diet the low levels of nitrogen available to the rumen micro-organisms may be a further factor in limiting its digestibility. Considerable improvements in the utilisation of the gross energy in cereal straw can be brought about by treatment of the straw with alkali (e.g. sodium hydroxide). This has the effect of disrupting the association between the fibrous carbohydrates and lignin. When in addition to alkali treatment the straw is ground and pelleted improvements in D-value from 40 to 60 can be achieved. As with the feeding of untreated straw, however, it is important that the remainder of the diet supplements the low levels of protein and nitrogen present. The use of ammonium hydroxide as the alkali agent helps overcome this problem.

Due to their increased digestibility the intake by animals of treated straw is much higher than that of untreated straws allowing much wider inclusion in the diets of ruminant animals.

Root crops

The root crops include turnips, swedes, mangolds and fodder and sugar beet. All are characterised by having a high water content (75–90%) and low crude fibre (5–10% of the DM). The main component of the dry matter is sugar. Thus roots are highly digestible and the principal nutrient they supply is energy. The digestible crude protein content is in the range 40–120 g/kg DM, a fairly high proportion of this being in the form of NPN. As suppliers of vitamins and minerals (potassium excepted) roots are poor and diets containing a high proportion of roots require appropriate supplementation.

Whilst roots have been fed to pigs and may be particularly useful where a wet feeding system is used they find greatest use in the diets of ruminant animals either being fed *in situ* to sheep or carted and fed to housed animals. Where large quantities of roots are to be fed to housed animals recognition should be made of the large quantities of urine that will be excreted.

Sugar beet is not primarily grown for animal consumption but sugar beet pulp, the residue left after sugar has been extracted at the factory, is widely used in ruminant diets where in energy terms it can substitute directly for barley. The higher fibre content compared with barley makes it a useful component of the concentrate diet when cows go out to grass in spring but account must be taken of its lower DCP value when it substitutes for cereals.

In recent years, cassava, a tropical root crop from which tapioca is made, has been increasingly used as an animal feed ingredient.

Green forage crops

The main green forage crops fed are kale, rape and cabbage. These brassica crops can provide a useful source of succulent feed during autumn and winter. Composition varies between the different brassica species and varieties, being particularly affected by the leafiness. As a generalisation they compare favourably in feeding value with good quality silage.

In feeding these brassicas account has to be taken of certain agents that may prove harmful when brassicas are fed in excess. The presence of goitrogens which affect iodine utilisation and of certain organic sulphur compounds which can produce haemolysis mean that their use in diets should be limited. It is recommended that intake of these brassica forage crops be limited to a maximum of 30% of total dry matter intake and particular attention paid to the health of animals. Additionally, it should be remembered that the high levels of calcium present in brassicas may upset the Ca:P ratio in the diet, necessitating phosphorus supplementation.

Cereal grain and cereal grain by-products

Cereal grains are the main ingredients of rations for pigs and poultry and provide the major source of energy in compound feeds fed to ruminant animals.

The main cereal grains used in the UK are barley, wheat and oats which are mainly homegrown along with imported maize and sorghum.

All cereal grains are rich in carbohydrate which is mainly in the form of starch. Starch accounts for about 70% of the seed, varying between grain types. Thus grains are a real source of energy. The crude protein content of feed grain is relatively low, ranging from 8–12%. This crude protein is of relatively poor quality being low in lysine, methionine and tryptophan (maize) and containing 10–15% of the nitrogenous compounds in the seed as NPN.

The digestibility of cereal grains is in general high, with maize, wheat and sorghum having the highest content of digestible nutrients and with barley having slightly less. The relatively high fibre content of oats considerably reduces its digestibility.

As regards the fat content of grains, oats and maize have higher levels than the other cereals. This fat is particularly rich in linoleic and linolenic acids, the oxidation of which may cause rancidity during prolonged storage.

Of the other nutrients present all cereal grains are poor sources of calcium and whilst the phosphorus content is fairly high much of this is unavailable through its being present in the form of phytic acid. Of the vitamins most cereals are good sources of vitamin E although under moist storage conditions much of this may be destroyed. Except for yellow maize, cereal grains are low in carotene and vitamin A and likewise all cereals are poor sources of vitamin D and most of the B-vitamins.

Cereal preparation and processing

The aim in all cereal processing methods is primarily to increase the efficiency of utilisation of the nutrients. Such improvement may result simply from an increased nutrient availability but other factors such as changes in palatability and nutrient density may also contribute towards improvements in performance.

Most grain processing methods have as their main objective improvement in the availability of the starch present. Some of the techniques involve solely physical change, others chemical and some a combination of both physical and chemical; in addition, some processes are carried out 'wet', others 'dry'; some involving heat treatment, others under cold conditions.

Mechanical alterations in the grain are the most widely used and involve physical disruption of the grain such that the starch is made more available. Grinding, rolling and crushing are the most common processes employed. It is worth noting, however, that such treatments have little or no effect on the nutritive value of barley or wheat offered to sheep.

Processing procedures which bring about chemical changes through the gelatinisation of the starch include such techniques as steam flaking, micronising, popping and pelleting. The 'flaking' process has been long applied to produce 'flaked maize' in a process involving steaming of the grain either at atmospheric pressure or in a pressure chamber followed by rolling. Micronising and popping are both 'dry heat' processes. In the former grain is passed under gas-fired ceramic tiles, the radiant heat from which produces rapid heating within the grain causing it to soften and swell. It is then crushed in a roller mill which prevents reversal of the gelatinisation process. Popping is the exploding of grain through the rapid application of dry heat. Popped grain is usually rolled or ground prior to feeding. Popping has been shown to be particularly effective in processing sorghum grain.

When feeding concentrate mixtures to certain livestock classes it is common practice to produce it in the form of pellets. From a management point of view it reduces wastage, prevents selection of ingredients and makes for easier storage and handling. Additionally, there are in some instances nutritional advantages to pelleting, this being due in the main to improvements in the available energy content.

Recent studies involving the treatment of grain for ruminant consumption with alkali appear to have potential as a means of increasing the availability of energy.

Cereal by-products

When cereal grains are processed for human consumption a number of by-products are produced which are used extensively as livestock feeds. The main sources of by-products are the flour milling industry and the brewing and distilling industries.

When wheat is milled to produce flour for human consumption about 28% of the grain becomes available as by-products of the process. These arise from the removal of the outer layers of the kernal and are usually classified on the basis of decreasing fibre as bran, coarse middlings and fine middlings. The coarser fractions are usually fed to ruminant species, the finer fractions being useful in formulating diets for pigs and poultry. Since the protein in the wheat grain is concentrated mainly in the outer layers, crude protein levels in these residues are higher than in the whole grain. Likewise wheat offals contain a high proportion of the B-vitamins.

The main by-products arising from the brewing and distilling industries are brewers' grains and distillers' grains, ('draff'), which are essentially the part of the grain that remains after the starch has been removed in the malting and mashing processes. They may be purchased without being dried and fed either fresh or after ensiling or alternatively may be purchased after being dried. Since both feeds have a relatively high fibre content their use is limited to ruminant rations where they particularly supply useful amounts of protein. In the case of distillers' grains the relatively high lipid content (80–90 g/kg DM) is known to interfere with the cellulolytic action of the rumen microflora but this can be overcome to an extent by the addition of suitable amounts of calcium.

Other by-products of these industries, namely malt culms, dried brewers' yeast and dried distillers' solubles may also be fed to livestock.

Other high energy feedstuffs (fats and molasses)

Whilst feed grains are the main energy supplying concentrate feeds, other feeds are routinely used to supply energy to livestock. Of particular value are fats and molasses.

Fat is of particular value in increasing the energy density of the ration, since its energy value is more than twice that of carbohydrates. The inclusion of fats in animal diets has found greatest application in milk replacers for suckling animals and in the diets of pigs and poultry where energy density is a factor controlling total energy intake. The inclusion of fat in the diets of ruminant animals has been limited until recent years by the adverse effects of fat on the activity of the cellulolytic bacteria in the rumen. However, the development of the 'rumen protected fat system', in which the fat is emulsified with non-degradable protein and thus protected from bacterial breakdown in the rumen but can be digested in the small intestine has led to the feasibility of increasing fat inclusion in ruminant diets. Such inclusion may be of benefit as a means of increasing the energy intake of high producing animals, e.g. dairy cows in early lactation and may also offer a means of altering the fatty acid composition of either milk or body fat.

Molasses is a by-product of sugar production, the main constituents of which are sugars. It is particularly used in ruminant feeds where it may be of particular value as a pellet binder or as a component of feeds which include NPN sources.

Oilseed by-products

Several oil-bearing seeds are grown to produce vegetable oils for human consumption and industrial purposes. The residues that remain after extraction of the oil are rich in protein and of great value as livestock feeds.

Among such high protein feeds are soya bean meal, coconut meal, groundnut meal, linseed meal, sunflower seed meal, rapeseed meal, cottonseed meal, sesame meal and palm kernal meal.

Oil is extracted from these seeds by hydraulic pressure (expelled) or solvent (extracted). Most oilseeds are now subjected to the latter treatment, the efficiency of oil extraction being much higher, leaving little oil ($<1\%$) in the residue whereas expeller methods leave up to 6% of the oil in the residue. The amount of oil left in the residue affects the energy value of the feed.

Fibre levels will also affect the energy value of the feed and in some cases, e.g. groundnuts, fibre levels will vary according to whether the seeds have been decorticated (removal of husk) prior to processing.

Of those oilseeds used widely, soya has the best quality of protein followed closely by sunflower and rapeseed. As a generalisation, the main essential amino acids in deficit in oilseed proteins are the sulphur-containing amino acids and lysine.

In addition to nutritive values a number of other factors have to be taken into account when feeding certain oilseed residues. For example, the presence of toxic substances called gossypols in cottonseed and aflatoxins in groundnut meal and the presence of an anti-nutritional factor (glucosides) which may affect growth and reproduction especially in pigs and poultry in rapeseed meal may restrict their inclusion in feedstuffs. The presence of trypsin and urease inhibitors in soya beans is overcome by adequate heat treatment during processing.

Legume seeds

The main legume seed used in the UK is field beans. Although they tend to be regarded primarily as a source of relatively good quality protein, being high in lysine but low in methionine, they are additionally useful sources of energy.

Because of the low levels of methionine the inclusion of home-grown field beans as a supplementary protein source in the diets of pigs and poultry must be limited, but for ruminant animals levels up to 50% of the total ration have been fed without adverse effects.

Some species of beans contain components with anti-nutritional properties, e.g. trypsin inhibition but this would not appear to be a problem in varieties grown in the UK.

Animal products

Feedstuffs of animal origin are without exception principally included in diets as supplemental sources of high quality protein which can remedy deficiencies in the essential amino acid composition of the rest of the diet. Although this role has for long been recognised in the feeding of pigs and poultry current thinking on the protein nutrition of ruminants indicates high quality protein that is resistant to degradation in the rumen may be of value.

Protein supplements of animal origin are derived from slaughterhouse wastes as meat meal and meat and bone meal; from milk and processed milk chiefly in the form of dried milk powders; and from fish and processed fish as fish meals.

The composition of slaughterhouse wastes is variable. However, to meet the Fertiliser and Feedingstuff Regulations (1973) meat meal must contain at least 55% protein whilst meat and bone meal must contain not less than 40% protein. In addition, there must be no inclusion of hoof, horn or feathers in either of the products. As would be expected the protein quality of these products is good. The fat content varies according to the efficiency of the rendering processes from about 3–13% and can considerably affect the levels of available energy. Ash content will also affect energy concentrations, there being clearly more minerals present in meat and bone meal.

Fish meal is generally manufactured from either demersal species (such as cod) or pelagic species (such as herring, pilchard and anchovy). The main compositional differences of meals from these sources is that the former are low in fat (2–6%) whilst the latter contain fairly high levels of fat (7–13%). Although the fat contributes energy to the diet, levels of inclusion of these high fat fish meals are usually limited because of the possibility of rancidity, taint, or 'soft' body fat developing. Such problems occur due to the presence of the polyunsaturated fatty acids found in these meals. The main contribution that fish meals can make to diets is high quality protein, ranking close to milk proteins. Additionally, fish meals are usually good sources of B-vitamins and high in most of the mineral elements, particularly calcium and phosphorus.

The main milk product used in animal feedstuffs is dried skim milk. Whilst skim milk powder is more than an adequate substitute for fish meal or meat meal in diets in terms of its protein quality it must be remembered that it is a much poorer source of minerals and certain vitamins. Dried skim milk is widely used in milk replacers.

All of these animal products are expensive and are included in animal diets primarily to supplement the essential amino acids in the rest of the diet. To this end the high levels of lysine found in all animal proteins and of methionine and tryptophan in fish meal are particularly useful in complementing the amino acid profile of dietary ingredients.

The methods of processing used to prepare these products may have an effect on feeding values. In all cases heat treatment is involved and the extent of this may markedly affect the digestibility of the protein.

The desirability of particular protein digestibility characteristics may depend on whether it is being fed to monogastric or ruminant species. Whilst extensive protein denaturation and reduction in solubility is undesirable if the protein is being fed to pigs or poultry the effect of heat treatment in reducing the rumen degradability of high quality protein fed to ruminants is likely to be desirable and necessary if its inclusion in the diets of adult ruminants is to be justified.

Non-protein nitrogen sources (NPN)

Feedstuffs which contain nitrogen in a form other than protein are termed non-protein nitrogen. Such compounds can serve as useful components of ruminant diets in that they provide a source of nitrogen for rumen micro-organisms to synthesise microbial protein. Although a wide variety of compounds can be used as NPN sources the market is dominated by urea. When urea is fed it is initially broken down to ammonia and carbon dioxide by microbial urease. This ammonia may then be utilised along with appropriate oxo-acids in the synthesis of microbial protein. The efficiency of urea utilisation is particularly dependent on the availability of oxo-acids and of energy to meet the needs of protein synthesis. These are affected by the amount and type of dietary carbohydrate. Starch appears to be the best source. Failure of micro-organisms to incorporate ammonia rapidly into microbial protein leads to a loss of nitrogen through urinary excretion and in extreme cases of ammonia production outstripping utilisation, toxic levels of ammonia in the blood may result. A variety of factors must be considered in utilising urea in feeds. These may be summarised as follows:

(1) The diet to which urea is being added must be suitable in terms of its energy, protein and mineral status to allow efficient use of NPN.
(2) Diets containing urea must be introduced gradually to allow rumen micro-organisms to adapt.
(3) Urea should not be used in pre-ruminant diets or where the level of NPN in the diet of adult ruminants is already fairly high, e.g. silage.
(4) Levels of inclusion should be appropriate to the class of stock being fed. For example, levels exceeding 1.25% in the concentrate ration will affect production in dairy cows, particularly in early lactation whilst for beef cattle and suckler cows levels should not exceed 2.5% with restricted feeding or 3% with *ad libitum* feeding.

Feed supplements (nutrients)

In formulating a ration the primary aim is to fulfil the animals' requirements for energy and protein and sometimes fibre. Should the ration prove to be lacking in micronutrients then additions of the appropriate minerals, vitamins or amino acids may be made. Nutrient supplements are commercially available which allow the addition of small amounts of specific nutrients without changing the general make-up of the initial formulation.

Supplementation of rations with specific amino acids is mainly of concern to non-ruminants where cereal based diets may be sub-optimal in such amino acids as lysine, methionine and tryptophan. Although it is not current practice to supplement ruminant rations with individual amino acids some research has indicated a potential use in the future.

Mineral supplementations may be provided in the form of licks or feeding blocks which allow the animal free access to a suitable combination of minerals or alternatively those specific minerals in deficit in a ration may be added to the ration in the form of a powder. In using mineral supplements the interrelationships among minerals must be recognised since excessive amounts of one mineral can cause a deficiency of another. Many proprietory supplements containing mixtures of macro- and trace minerals, appropriate to particular production situations are available commercially.

As with minerals, any vitamins in deficit in a formulation must be made good by supplementation. For ruminants only fat-soluble vitamins need be considered whilst for pigs and poultry both the water- and fat-soluble vitamin content of the diet may require supplementation. A variety of balanced vitamin premixes formulated for particular circumstances and usually containing the vitamins in a chemically pure form such that only very small amounts are required are available commercially. In assessing the need for vitamin supplementation it is important to recognise the variability in the vitamin content of feedstuffs and also that vitamins are easily destroyed by agents such as heat, light and oxidation.

15

Cattle

J. Kirk

Domesticated cattle are nearly all descended from two major species: *Bos taurus*, which includes the European types, and *Bos indicus* to which the Zebu cattle belong. Man has selected from these and has developed a number of well defined breeds. These breeds vary from types used primarily for milk production to those developed for beef production.

In some breeds an attempt has been made to combine the desirable qualities of both dairy and beef cattle to provide dual-purpose cattle.

During the last two decades a number of exotic breeds have been imported, mainly from Europe. The rationale behind this importation of breeding stock has been to improve beef production although some are dual-purpose breeds in their country of origin.

Most of these immigrants have been large bodied, fast growing cattle producing lean carcasses. Characteristics particularly valuable for satisfying consumer demand for lean meat (*Tables 15.1–15.4*).

Table 15.2 Slaughtering of fatstock in the UK in 1979

Cattle	Nos ($\times 10^3$)
Steers and heifers	2859
Cows and bulls	990
Calves	147
Total	3996

Table 15.3 Sources of beef and veal in the UK in 1979

	Tonnes ($\times 10^3$)	£ ($\times 10^6$)
Produced in UK	1029.2	
Imported into UK	271.9	
	1301.1	321.0
Exported from UK	92.4	
	1208.7	145.0

Table 15.1 Cattle numbers in the UK (June)

	$\times 10^3$		
	1975	1977	1979
Cows and heifers in milk			
dairy herd	2903	2935	2972
beef herd	1605	1432	1297
Cows in calf (not in milk)			
dairy	339	330	316
beef	294	249	238
Heifers in calf			
dairy	664	634	684
beef	239	189	180
Bulls	97	94	90
Others			
2 yr+	987	1011	1030
1–2 yr	3560	3220	3107
6 month–1 yr	2062	1918	1831
Under 6 months	1919	1803	1766
Calves for slaughter	50	39	32
Total (rounded)	14719	13854	13543

Source: (Meat and Livestock Commission 1980) with permission.

Table 15.4 Household meat consumption and expenditure in UK

	Consumption (kg/person per week)	Expenditure (pence/person per week)
Beef and veal	0.235	55.76
Mutton—lamb	0.122	22.78
Pork	0.103	19.01
Total carcass meat	0.460	97.55

Source: MLC 1980 with permission.

PRINCIPAL BREEDS (*Tables 15.5–15.10*)

Dairy breeds

Ayrshire

The breed is brown and white in colour and has long horns that curve upwards and slightly backwards. The breed is hardy and the udder conformation is excellent. The fat globules are small and this makes its milk very suitable for cheese making.

Purebred Ayrshire male calves are not suitable for most forms of beef production but if the cost of the calves is low enough some are utilised for veal. Crossing with Charolais, Hereford, Friesian or Beef Shorthorn bulls produce beef type calves.

The number of Ayrshire dairy cows has declined throughout the UK due to the higher milk yield and superior beef qualities of the Friesian.

Friesian

Originating in the Netherlands the breed is known in the USA as the Holstein–Friesian and in Canada as the Canadian Holstein. These latter two types are larger and more extreme dairy types than the British Friesian. All three types are normally coloured black and white although there are Red and White Friesians which are homozygous for the recessive red factor.

The dramatic change in the breed structure of the UK since the Second World War has been due to this breed. The reasons for the change have been the high milk yields of the British Friesian and the value of the pure and crossbred Friesian in producing beef.

Late maturity made the purebreed particularly suitable for intensive (barley beef) systems of beef production, thus providing a valuable by-product from the dairy herd. Most beef breeds crossed with

Table 15.6 Milk yields and butterfat of recorded herds in England and Wales 1978–79

Breed	Yield (kg)	Butterfat (%)
Ayrshire	4863	3.94
Holstein	6218	3.76
Friesian	5441	3.79
Guernsey	3941	4.65
Jersey	3776	5.14
All breeds	5335	3.83

After MMB (1980) with permission.

the Friesian are suitable for producing calves for the more extensive systems of beef production.

Friesians account for more milk and beef than any other breed in the UK.

Guernsey

The breed originated from the island of that name. Fawn in colour with white markings, slightly smaller than Ayrshires. As the pure breed is unsuitable for beef its existence depends on premiums for high quality milk and higher stocking rates due to its size. Guernsey cows crossed with extreme beef type bulls, e.g. Charolais, produce useful beef cattle.

Jersey

Again the breed originates from the island which bears its name. They are found in a variety of shades of yellow some with white markings and all have a grey ring of hair round a black muzzle. The breed is the smallest of the important dairy breeds and has a lower body weight and lower milk yield with a higher fat content than the Guernsey. Purebred it produces poor beef carcasses. By virtue of its small body size it is capable of high stocking rates and high output of milk solids per unit of grazing land.

Dual purpose breeds

The traditional dual purpose breeds have lost popularity since the early 1950s and the Friesian breed has taken over their role.

Dairy Shorthorn

During the present century this breed has changed from the major milk producer to a minority breed. It has been largely replaced by the Friesian with its higher yield, higher fat percentage and faster growth rate. Red, white and roan are the accepted colours.

Red Poll

The breed was developed in Norfolk and Suffolk and

Table 15.5 Dairy herd size distribution

Herd size (cows)	England and Wales (1979)	Scotland (1978)	Northern Ireland (1979)
Herds as a % of total herds			
3–19	17.1	1.6	50
20–39	26.3	13.1	26.4
40–69	29.5	35.3 ⎫	
70–99	14.7	25.8 ⎬ 21.4	
100–199 ⎫	12.4	⎧ 22.0 ⎫	
200+ ⎭		⎩ 2.2 ⎭	2.2
Cows as a % of total cows			
3–19	3.5	0.3	17.7
20–39	13.8	5.2	26.9
40–69	27.7	24.3 ⎫	
70–99	21.6	27.3 ⎬ 45.0	
100–199 ⎫	33.4	⎧ 35.7 ⎫	
200+ ⎭		⎩ 7.2 ⎭	10.4

After MMB 1980 with permission.

as suggested by the name the cattle are red in colour and without horns. Milk yield does not compare favourably with Dairy Shorthorns. Pure and beef cross Red Poll cattle are useful beef animals. The number of cattle in this breed is now small. Poll character is dominant and transmitted to progeny.

South Devon

The largest bodied breed of cattle native to the UK. Horned and brown/red curly coat, now a minority breed, no longer the dominant breed in its own locality, having much lower milk yield than Friesian and fat content lower than the Channel Island breeds. Now kept for beef production, cows produce milk yields greater than any other beef breed, resulting in high daily liveweight gain. They are also able to utilise rough as well as good grass. The breed often carries the gene(s) for muscular hypertrophy.

Welsh Black

These are black, horned cattle. They are hardy and thrive in inclement weather, therefore well suited to high rough ground. They have a higher growth potential than the Highland cattle. Many herds of Welsh Black cows are used to single-suckler calves.

Beef breeds

Aberdeen Angus

Famed for its carcass characteristics. The breed is black and polled and both of these characters are dominant. An important feature when animals are sold as stores because the identity of one parent can be easily recognised. The Aberdeen Angus is early maturing with a low mature body weight and slow growth rate. Often used as a crossing bull on Friesian dairy heifers in an attempt to prevent dystokia (difficult calving).

Hereford

An important beef breed throughout the world. A red breed with a characteristic white face and other white markings. The white face is dominant and is therefore transmitted to crossbred progeny, colour marking the calves. The breed is moderately early maturing and has a relatively good growth rate, especially if compared with Aberdeen Angus. Hereford bulls are suitable for crossing with most breeds.

The pure or crossbred Hereford performs well at grass but is also suitable for intensive systems of beef production. When crossed with a Friesian it produces an excellent suckler cow giving more milk than the pure Hereford and is hardier and faster growing than a pure Friesian.

Table 15.7 EEC dairy and beef cow numbers ($\times 10^3$) in 1978

	Dairy cows	Beef cows
Belgium	967	182
Denmark	1054	73
France	7387	2750
West Germany	5397	153
Irish Republic	1598	498
Italy	2692	755
Luxembourg	68	8
Netherlands	2291	NA
UK	3278	1584
Total	24 732	6003

After MLC (1980) with permission.

Galloway

The Galloway is a black, dun or belted breed. The Belted Galloway has a white belt running round the middle of the body. All types are very hardy, with thick coats and are polled. Later maturing than the Aberdeen Angus but earlier than Highland with a rather slow growth rate. Popular for crossing with other breeds for the production of commercial suckler cows.

Highland

Their characteristic appearance with long shaggy coat and wide sweep of horns gives them their picturesque appearance. Found in a variety of colours from yellow (fawn) to dark brown. Very slow growing and late maturing. Their hardiness makes them suitable for poor quality moor and hill land.

Luing

This new breed is the result of crossing the Beef Shorthorn and Highland cattle to provide a breed to suit adverse conditions in Scotland. They combine the early maturity of the Beef Shorthorn with the hardiness of the Highland.

North Devon

Commonly called a Ruby Devon because of the

Table 15.8 Average size of dairy herds and average milk yield/dairy cow (ℓ) in EEC (1978)

	No. of cows/herd	Av. milk yield/cow
Belgium	15.8	3850
Denmark	21.1	4871
France	13.6	3451
West Germany	11.2	4305
Irish Republic	12.4	3257
Italy	6.5	3250
Luxembourg	20.0	3750
Netherlands	29.9	5050
UK	46.3	4791

Source: MMB (1979) with permission.

distinctive colour. A hardy breed having the ability to withstand heavy rainfall, severe winters and sparse pastures. These factors combined with its docility and good mothering ability have made it a desirable mother in suckler herds. The Devon Bull has been used on a variety of other breeds with excellent results in terms of weight gain and carcass quality of progeny.

Sussex

Very similar to the Devon in many respects—colour, size, rate of maturity and growth rate. A hardy animal with an excellent reputation as a grazing animal.

Lincoln Red

A large breed with the reputation for rapid growth and early maturity. Its ease of calving has been of positive importance in suckler herds and coupled with a high milk yield produces fast growing calves.

Beef Shorthorn

A true beef breed found mainly in Scotland and northern England. Smaller than the Hereford and the red beef breeds. The growth rate is not particularly high and there is a tendency to lay down fat unevenly. The breed has been used in the past at home and abroad to upgrade breeds. It is useful to cross with dairy breeds, e.g. Ayrshire to produce calves for beef. White Shorthorn bulls crossed with black Galloway cows produce the Blue Grey cattle. Cows of this cross are polled, hardy and good mothers. Blue Grey cows are often crossed with Hereford or Aberdeen Angus bulls to produce good beef calves.

Imported breeds

Blonde D'Aquitane

A French breed resulting from a fusion of types. Large cattle producing big calves which develop into good beef cattle.

Charolais

The first of the French breeds to be imported into the UK. The cattle are horned and usually white although some are cream in colour. It has a high liveweight gain and when mature is one of the largest breeds. Many of these animals carry the muscular hypertrophy factor and this accounts for heavy development of muscle particularly on the rump and the high proportion of lean meat in the carcass. The Charolais bull has proved useful as a sire on dairy breeds and in suckler beef herds. One problem resulting from its use has been an increase in dystokia.

Table 15.9 UK dairy herd breed distribution (%)

Breed type	England and Wales (1978–79)	Scotland (1978)	Northern Ireland (1976)
Ayrshire	3.4	35.5	2.4
Friesian	88.6	36.6	91.3
Holstein	1.4	0.4	—
Guernsey	2.4 ⎫	0.4	Included in
Jersey	2.0 ⎭		others
Others	2.2	27.1	6.3

After MMB (1980) with permission.

Chianina

An Italian breed which has the distinction of being the tallest in Europe. A white breed with short horns, the calves are born tan in colour but turn white at a young age. These animals produce lean carcasses with little fat cover. Although the mature animals are large the calves are small at birth and calving difficulties are less frequent than in other breeds, e.g. Charolais.

Limousin

Another French breed which is fast growing and has a large mature size. The breed has a red body coat. Carcasses are noted for their exceptionally high muscle/bone ratio.

Maine–Anjou

Another of the large French breeds. Colouring is red or red on white. The Maine–Anjou is a fast growing, early maturing breed with good carcass qualities. Used as a choice sire for terminal cross or for replacement suckler cows.

Simmental

This breed originated in Switzerland and is now popular throughout Europe including the UK. It is

Table 15.10 Approximate mature weight of cows

Breed	Weight (kg)
Ayrshire	500
British Friesian	600
Guernsey	365
Jersey	385
Dairy Shorthorn	565
Red Poll	520
South Devon	660
Welsh Black	545
Aberdeen Angus	500
Devon	590
Galloway	475
Hereford	545
Highland	545
Shorthorn	570

It should be stressed that these are only approximate and individuals may vary quite considerably from the average.

red and white with a white face and a large proportion of calves are colour marked by a white face in the same way as a Hereford. It is a dual purpose breed, the cows yielding above the average pure beef breed and the calves a high weight gain, being only slightly less than the Charolais; thus proving useful where a breed is needed to improve both the beef and milk characteristics in producing a good suckler cow.

DEFINITIONS OF COMMON CATTLE TERMINOLOGY

At birth
male—bull calf; bullock calf if castrated.
female—heifer calf; cow calf.

First year
male—yearling; year old bull
female—yearling heifer

Second year
male—two year old bull; steer, ox, bullock
female—two year old heifer

Third year
male—three year old bull; steer, ox, bullock
female—three year old heifer, becomes cow or bearing calf

Heifer—usually applied to a female over one year old which has not calved. An unmated animal is known as a maiden heifer and a pregnant one as an in-calf heifer. In some areas the term first-calf heifer is used until after the birth of a second calf. A barren cow is either barren, cild or farrow and when a cow stops milking she is said to be yeld or dry.

Stirk—limited to males and females under two years in Scotland. It is usually applied to females only in England, the males being steers.

Store cattle—stores, are animals kept usually on a very low level of growth for fattening later.

Veal—is the flesh from calves reared especially for this purpose and normally slaughtered at about 16 weeks of age.

Bobby veal, stirk veal—flesh from calves slaughtered at an early age, often only a week or two old. These animals tend to be of extreme dairy type, thus making them undesirable for rearing as beef cattle.

Cow beef—beef from unwanted cows, often used in processing.

Bull beef—beef from entire male animals.

AGEING

The development of the incisor teeth can be used as an indication of the age of cattle.

The dental formula for a full mouth is:

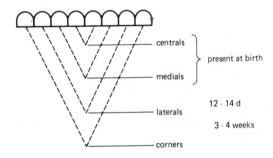

Permanent molars	Temporary molars	Incisors	Temporary molars	Permanent molars	
$\dfrac{3}{3}$	$\dfrac{3}{3}$	$\dfrac{0\mid0}{4\mid4}$	$\dfrac{3}{3}$	$\dfrac{3}{3}$	$= 32$

There are four pairs of incisor teeth. These are found in the lower jaw, the upper jaw has no incisors but is a hard mass of fibrous tissue known as the dental pad. Starting in the middle of the jaw the pairs are known as centrals, medials, laterals and corners. The times at which the temporary incisors are shed and replaced by the permanent incisors are important.

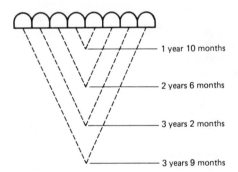

Individual animals will vary from these figures. Breed, management and feeding all have an influence.

CALF REARING

The foundations of the future health and well being of the calves are laid by good feeding and management throughout the first three months of age.

Common to all the systems of rearing is the need for an adequate supply of colostrum (the secretion drawn from the udder at the time of parturition).

At birth the calf is virtually free of bacteria but quickly becomes infected with organisms from its surroundings. The blood of the newborn calf contains no antibodies until the calf has received colostrum.

Table 15.11 Composition of colostrum (first 24 h after calving) and milk

	Colostrum	*Milk*
Fat (%)	3.6	3.5
Solids-not-fats (%)	18.5	8.6
Protein (%)	14.3	3.25
Casein (%)	5.2	2.6
Albumin (%)	1.5	0.47
Immune globulin (%)	6.0	0.09
Calcium (%)	0.26	0.13
Magnesium (%)	0.04	0.01
Phosphorus (%)	0.24	0.11
Carotenoids (g/g fat)	25–45	7
Vitamin A (g/g fat)	42–48	8
D (ng/g fat)	23–45	15
E (g/g fat)	100–150	20
B_{12} (g/g fat)	10–50	5

From Roy, J. H. B. (1980).

As can be seen in *Table 15.11* colostrum contains a high percentage of proteins especially immune globulins with their attendant antibodies, minerals, vitamins, especially the fat soluble A, D and E and carotene which is the cause of the yellow coloration.

It is vitally important that the calf receives colostrum during the first 24 h. The calf's ability to absorb the antibody protein is greatest during the first 6–12 h of life and then declines with the result that little can be absorbed after 24 h of life. Hence, to obtain the maximum protection the calf should be fed colostrum during the first 6 h of life.

The globulins are only able to pass unchanged into the blood stream during this time.

Colostrum feeding should continue for 4 d, at the rate of 3.4 ℓ/d in three equal feeds. Some farmers achieve this by leaving the calf to suckle, others remove the calf at birth. If the calf is removed at birth it is easier to teach it to drink from a bucket. Frequently a greater quantity of colostrum is produced by the cow than is needed by the newborn calf. Any excess colostrum can be diluted with water at the rate of two parts colostrum to one of water and fed to older calves in place of milk. Alternatively colostrum may be frozen and kept in case of an emergency and fed to newborn calves.

The composition of colostrum changes to milk during the first 4 d of milking. If pre-calving milking is practised this change takes place before the calf is born.

The zinc sulphate turbidity (ZST) test enables the concentration of circulating antibodies in the calf's blood to be determined. There is a high correlation between the results of this test and calf health, showing that it is essential calves get adequate amounts of colostrum. This is undoubtedly one of the major factors involved in ensuring that disease and ill health are kept under control.

Calves should be kept on the farm on which they were born for at least 7 d. Newly purchased calves should be allowed to rest for a few hours after a journey and then fed a glucose feed (100 g glucose in 2 ℓ of warm water). They should be receiving their full quantity of milk substitute within 48 h of arrival; however it is imperative to avoid overfeeding. Numerically the most important method of rearing calves is 'by hand'. Artificial milk replacers are used instead of cows' milk because the cost per unit of milk replacer is less than the equivalent amount of milk. Calves destined as replacements in the dairy herd are almost always reared by hand. The method requires more specialist accommodation and is more exacting in terms of labour but enables economies in food costs to be achieved. Milk substitutes basically consist of skim milk powder and added fat and vitamins and minerals; whey powder is sometimes included.

Two systems of rearing are practised.

Late weaning

After 4 d of age colostrum is replaced by milk substitute. Water, hay and rearing concentrates are offered after three weeks of age.

Early weaning

In this method the calf is weaned onto a diet of dry food by five weeks of age. The early introduction of concentrates, hay and water at 7 d encourages the development of the rumen. To achieve the desired intake of concentrates it is essential that they be palatable and fresh. Calves should be eating 0.75–1.00 kg of concentrates daily with a liveweight of 65 kg by weaning at five weeks. The advantages of early weaning are that concentrates are much cheaper than milk replacer, the labour requirement up to 14 weeks is much less and after weaning, group housing can be used thus saving house space. At 14 weeks there should be no difference in liveweight between animals reared on the two systems.

'Milk' feeding can be practised in several forms. Once or twice daily feeding of milk substitutes is the common practice on farms. *Ad libitum* feeding of milk substitute was first made possible with the advent of high fat milk substitutes and automatic dispensing machines. The introduction of 'acid' milk replacers has enabled *ad libitum* feeding of cold milk replacer to be practised without the need for sophisticated machines. Acid milk replacers are classified according to two types:

(1) medium acid based on skimmed milk. These have a pH of 5.5–5.8 and a protein content of 24–26% with 17–18% fat;
(2) high acid based on whey from cheese manufacture. These have a pH of 4.4–5.8 with 19–20% protein and 12–15% fat.

The acidity has a positive effect in helping to reduce the incidence of digestive upsets. The pH of the abomasum prior to feeding is 2.0–2.8 but after feeding conventional milk replacer of pH 6.2–6.5 the abomasum pH rises to between 4.5–6.2. It then declines to pre-feeding levels after 3–5 h. Satisfactory digestion depends on enzyme action and the optimum pH for this to occur is between 3 and 4. The conclusion therefore is that by feeding high acid milk replacer the abomasal pH is kept near optimum levels for enzyme activity and below the pH at which most organisms can survive.

Acidification of milk replacer therefore helps not only in the preparation of milk replacer, allowing 3 d feed supply to be mixed at the same time, but is also important in aiding the normal digestive process of the calf.

The sucking action of milk through a teat by the calf ensures correct closure of the oesophageal groove allowing milk to enter the abomasum without spilling into the developing rumen, where it can ferment and possibly cause digestive troubles. The argument that cold feeding is bad for calves probably grew out of the bucket feeding system where a calf can suffer a physiological shock when consuming large amounts of cold milk in a short time. With an *ad libitum* feeding system the 'little and often' effect of food consumption ensures that the small amount of milk taken in at one time is rapidly warmed up to body temperature.

There are three important stages if the use of *ad libitum* feeding acidified milk is to be applied correctly:

(1) Ensuring that calves are consuming enough milk. One of the easiest ways of achieving this is to individually feed the calves for the first 5–10 d by bucket and teat. Any initial difficulty with calves not drinking may be overcome if the milk is fed warm, later the calves readily consume the milk replacer if it is fed at cooler temperatures until cold feeding is practised.
(2) Calves should remain on *ad libitum* acid cold milk for three weeks. At no time should the milk be allowed to run dry as excess consumption will take place when replenished and this may lead to scours.
(3) After three weeks the intake of milk may be reduced by substituting cold water for replacer during the night. This encourages the consumption of concentrate food. Calves must have access to highly palatable concentrates. These should preferably be sited near the teats where milk is drunk thus encouraging their consumption.

Criticism of *ad libitum* feeding is usually made due to the over consumption of milk linked to the cost of feeding more milk replacer. Commercial farms claim that although it may cost more to rear a calf that animal has experienced less stress and fewer health problems and as a result is a better calf. Added to this is the reduction in labour time per animal coupled with a more flexible work routine.

Calves are usually housed in individual pens until weaning. A pen 1.8 m long and 1 m wide will accommodate calves up to eight weeks old. Less pen space is needed by calves reared in groups, an area of 1.1 m² per calf should be sufficient up to eight weeks, this area being increased to 1.5 m² per calf up to 12 weeks.

When reared in groups and machine fed care should be taken to ensure that individuals know how to suck and that there is no bullying so all have an opportunity to feed. Evenly matched batches and close observation for the first few days are essential.

Whatever system is used warmth, a dry bed and particularly, prevention of draughts are important to the well being of young calves.

The three most commonly practised systems of natural rearing are single suckling, double suckling and multiple suckling.

Single suckling

This is by far the most popular and is mainly carried out on hill and marginal farms, pedigree beef breeders, lowland farms with inaccessible grassland and on some arable farms utilising grass as a break crop. In this method the cow rears her own calf, milk being the calf's main source of nutrients. The calf remains with the dam until weaning at approximately six months of age. Cows are normally calved in either autumn or spring. The advantage of autumn calving is that calves are old enough to make full use of the grass in spring and are heavier at weaning in the following autumn. The disadvantage is that cows and calves often have to be housed during some of the winter period. Spring calving has the advantage that cows can be overwintered outdoors, the disadvantages are that the young calf cannot make as good a use of spring grass, and spring grass may cause a flush of milk in the cows resulting in the calves scouring. Concentrates should be fed before weaning so eliminating any loss in condition later.

Double suckling

The objective is for each cow to rear two calves together. After calving a second calf should be introduced, and allowed to suckle with the cow's own calf. Care and attention must be practised at first until the cow will willingly accept the new calf. The success of the system depends on a readily available supply of newborn calves when required, cows having an adequate supply of milk (often a cull dairy cow or a Friesian cross cow) and good management.

Multiple suckling

This necessitates high yielding nurse cows and a supply of suitable calves. Each suckling period usually lasts about ten weeks. In the first ten weeks of lactation four calves are suckled, in the second ten weeks three, in the third period three and for the last period two, giving a total of 12 calves reared during lactation. Cows are often removed from the calves between feeds.

Veal production

For veal production calves capable of high rates of gain are required, Friesian bull calves normally being used. These are reared on an all milk diet and slaughtered between 140–150 kg liveweight at 12–16 weeks of age. Friesian bull calves will have a killing out percentage of 55–60%.

Correct feeding is critical if an adequate return on capital is to be achieved. Milk replacers of the high fat type with at least 15% and up to 25% fat are required for maximum gains. The aim is for a daily liveweight gain to slaughter of 1 kg/d or more. Good stockmanship with particular attention being paid to hygiene and observation of animals for ill health is of paramount importance.

Calves are normally housed in buildings with a controlled environment and kept in individual pens on slatted floors. New systems of rearing veal calves involve loose housing in barns with Yorkshire boarding sides, floors bedded in straw and the animals fed *ad libitum* milk from machines.

MANAGEMENT OF BREEDING STOCK REPLACEMENTS

Rearing policy from weaning depends on two main factors, the season in which the replacement was born (autumn or spring) and the age at which it is to be calved. Age at calving is important as it is related to conception, dystokia, milk yield in first lactation, overall lifetime milk production and the herd calving pattern.

Conception is largely influenced by liveweight, oestrus being associated with weight. The following table (*Table 15.12*) gives the target liveweights for various breeds.

Table 15.12 Target liveweights

Breed	Wt at service (kg)	Calving wt (kg)
Jersey	230	340
Guernsey	260	390
Ayrshire	280	420
Friesian	330	500
Hereford × Friesian	320	500
Aberdeen Angus × Friesian	290	430

Dystokia problems require particular consideration and help to determine the appropriate weight and age for first calving. Calving problems are greater in younger heifers, especially if calved before 22 months of age. The size of the calf and of the heifer also being important. Feeding influences both the size of the heifer and the calf. Condition scoring (*see Appendix 15.1*, p. 374) can be a valuable management aid and heifers should be of condition score between 3 and 3.5, six to eight weeks prior to parturition. If the condition of heifers is correct at this time restricted feeding during the last weeks of pregnancy will not affect heifer size but will help to reduce calving difficulties by minimising the growth of the calf.

Choice of bull can also affect calving difficulties, but the choice may also be influenced by the value of the calves in term of herd production and the required number of replacements to be retained for the breeding herd. The large beef breeds cause the greatest problems in Friesian heifers, rather than calves sired by Herefords and Aberdeen Angus. As Friesian bulls tend to cause more problems than either of these two smaller beef breeds their use on heifers is not recommended unless there is need for a large number of replacements. Another advantage in favour of both the Hereford and the Angus is that both breeds colour-mark their calves thus readily identifying them and adding value to them. Whatever the choice the individual variation within breeds means that some sires can be identified as having a greater incidence of difficult calvings.

First lactation yield is lower in heifers calved early, milk yield being closely related to liveweight at calving. In subsequent lactations differences in milk yield are minimal, and as evidence suggests early calved heifers (two years) are kept in the herd to the same age as later calved (three years) they average one lactation more and thus their lifetime yield is increased, besides providing the extra calf.

Heifer rearing should be planned so that replacements enter the herd to fit the intended calving pattern. This is one of the main determinants in maintaining a system of block calving. Once the age and the month at which the heifer is to calve has been decided then growth rates to achieve the necessary target liveweights at service and calving can be calculated.

Differences between growth rates for autumn and spring born heifers are largely due to the higher weight gains achieved at grass (*Table 15.13*). Maximum use of grass means economical rearing and the advantage of compensating growth.

Autumn-born calves should be weaned at five weeks of age when consuming 0.8 kg/d of concentrates. The concentrate should be palatable and contain 17% crude protein. Between 5 and 12 weeks calves should be fed concentrates containing 15% crude protein *ad libitum* up to a maximum of 2 kg/d and hay to appetite. Silage as a partial substitute for hay can be fed from

Table 15.13 Target weights for two year calving

	Weight (kg)
Autumn born heifer	
Birth	35
Turnout (6 months)	150–170
Yarding (12 months)	275–300
Service (15 months)	330
Turnout (18 months)	375–400
Calving (24 months)	500–520
Spring born heifer	
Birth	35
Turnout (3 months)	80–100
Yarding (6 months)	140–160
Turnout (12 months)	250–280
Service (15 months)	330
Yarding (18 months)	400–420
Calving (24 months)	500–520

Source: MLC/MMB joint publication. Reproduced with permission.

six weeks of age. From three months to turnout in the spring the concentrates fed can be cheapened by reducing the crude protein to 14%. Hay or silage should be fed *ad libitum*. By six months each calf will have consumed 50 kg of early weaning concentrates, about 300 kg of rearing concentrates and 330 kg of hay.

To achieve the desired gains during the first grazing season (6–12 months of age) a continuous supply of good quality grass and control of parasitic worms are essential. Calves should be vaccinated against husk before turnout unless clean pastures are available; clean pastures being those that have been free of cattle since the previous mid-summer. Supplementary feeding of concentrates (1–2 kg/d) after turnout prevents any check in growth which might otherwise occur. A change to clean silage aftermath after dosing against stomach worms in mid-summer is generally recommended. At this time cereal feeding (9% protein plus vitamins plus minerals) may be introduced when grass becomes scarce or very wet and lush.

The grazing system should be integrated with the conservation area; 0.25 ha/animal can be divided into three sections. One section is grazed until the end of June, whilst the other two are cut. After this time the two conserved areas are grazed and the other one cut, finally grazing all three.

The second winter is best sub-divided into two halves. From yarding to service at 15 months and from service to turnout at 18 months. Cattle benefit from dosing against internal parasites at yarding and from the use of a systemic insecticide against warble fly. A daily liveweight gain of 0.6 kg/d is important to ensure a target weight of 330 kg and a good conception rate. Conserved forages form the basis of the ration (25 kg silage or 6 kg hay/d) and this is supplemented by 2.5 kg/d of concentrates (12% protein). The concentrates can be based on barley and the protein

content and quantity of concentrate adjusted according to the quality of roughage.

Identification of bulling heifers is often found to be a problem. Careful observation and heat detection devices may prove to be a valuable aid. After service the liveweight gains may be reduced to 0.5 kg/d, this allows the concentrate levels to be reduced. The overall concentrate use during 12–18 months should be about 250 kg of concentrates and 4–4.5 tonnes of silage or 1.25 tonnes of hay.

During the second grazing season target weight gain should be about 0.7 kg/d. With good grassland management no concentrates are necessary until steaming-up takes place in late summer. Excess steaming-up should be avoided to prevent overstocking of the udder prior to calving. The best guide to the level of feeding during late summer is body condition score. This should be between 3 and 3.5 six weeks before calving.

SPRING BORN HEIFER CALVING AT TWO YEARS OLD

The management and feeding of the newborn calf until weaning is the same as for autumn born calves. As young calves are too small to make efficient use of grass, concentrate (16% crude protein) feeding should be continued after turnout. Hay should also be available during this time (*Table 15.14*). If growth rate from grass alone falls below 0.5 kg/d concentrate feeding should be restarted. The target stocking rate should be about ten calves/ha. This may be achieved by a similar system of grassland management as for autumn born calves. It is important to ensure that grazing is clean as very young calves are extremely susceptible to parasites. At yarding calves should weigh between 140–160 kg and have consumed 100 kg of concentrates and 125 kg of hay from weaning.

From yarding at six months to turnout in the following spring at 12 months a growth rate of 0.6 kg/d should be maintained. At yarding animals should be dosed against internal parasites and dressed against warble fly during the winter. The basis of the ration will be either silage or hay, the amounts fed being 18 kg or 5 kg/d, respectively. The level of concentrate feeding depends on the quality of the conserved forages, a guide being 2–3 kg/d of 16% crude protein. If these growth rates are not maintained the target service weight of 330 kg at 15 months of age in the

Table 15.14 Approximate feed quantities used

	Autumn born	*Spring born*
Milk substitute (kg)	13	13
Concentrates (kg)	700	920
Silage (tonnes)	5.5	7.25
and hay (kg)	100	
or hay (tonnes)	1.9	2

spring will not be achieved and conception rates will be poor.

The second grazing season growth rates from turnout to service will be 0.8 kg/d. After service target growth rates can be reduced to 0.6 kg/d. Supplementation with mineralised cereals should commence in early autumn.

During the second winter the aim should be to produce a Friesian heifer calving down at about 500 kg. This can be achieved by feeding 2 kg/d of concentrates and up to 30 kg/d silage. This will take approximately 4.5 tonnes of silage or 1.25 tonnes of hay and 360 kg of concentrates. Steaming up prior to calving should be practised and the heifers' body conditions should be between 3 and 3.5 six weeks prior to calving.

Heifer replacement rearing enterprises compete for resources with the main milk producing animals. The two major economies that can be made in the resources required in the production of replacement heifers are, the reduction in the number required and the reduction of the age at which they calve. Both these enable considerable economies in land, labour and capital invested in buildings and livestock to be made. However heifers grown well enough to calve at two years of age need a higher plane of nutrition. This necessitates the feeding of greater quantities of concentrates thereby increasing the cost of concentrate feed for animals calving at two rather than three years of age. As far as total feed costs is concerned heifers calving around two years cost nearly as much as those calving a year older. Thus intensification of heifer rearing with a greater reliance on concentrates has important repercussions in that the land and labour saved can be made available for milk production or other more profitable enterprises.

In addition to these direct savings in resources the heifer calving at the younger age will have a longer herd life with a greater total lifetime milk production (*Table 15.15*).

Table 15.15 Milk production according to age at first calving

	Age at first calving (months)				
	23–25	*26–28*	*29–31*	*32–34*	*35–37*
Herd life (lactations)	4.00	4.03	3.84	3.81	3.78
Lifetime yield (kg)	18 747	18 730	17 964	17 991	17 657

Source: Wood, P. D. P. (1972).

These higher lifetime yields make up for a slightly lower first lactation yield; 4.300 kg for two year old heifers compared with 4.500 kg for three year olds.

In beef suckler herds calves have a slightly slower growth rate than the calves out of five year old cows (*Table 15.16*).

These slower calf growth rates are largely due to the early calved beef heifers having a slightly lower milk yield. Over their lifetime beef heifers calved first at

Table 15.16 Effect of age at calving on calf growth. (Percentage difference in 400 d weights compared with calf out of a five year old cow)

		Percentage difference
Heifer	first calving at 2 years	−8
	2½ years	−5
Cow	second calf	−3
	third calf	−2
	fourth calf	0

Source: MLC/MMB joint publication. Reproduced with permission.

two years of age produce more calves over their lifetime than those calving at three years.

As far as total feed is concerned it appears that heifers calving at two years of age costs nearly as much as those calving a year older. The financial saving in grazing and forage for the younger calving animals is substantially eroded by the need for a higher concentrate input to maintain growth rates. This is of especial importance during the winter periods unless the diet comprises ample good quality forage.

Even if there is no great saving in the cost of producing younger calving heifers there is a saving in capital investment in young stock because of the quicker turnover in animals and a reduction in the number of followers kept. Indirect benefits may result from the successful adoption of such a system of rearing, the intensification required having repercussions throughout the whole dairy enterprise, especially with respect to better grass production, conservation and utilisation. The effects that the more rigorous discipline involved in rearing heifers to calve at two years imposes are likely to extend right across the rearing enterprise and benefit the whole farm economy.

BEEF PRODUCTION

Traditionally beef cattle were slaughtered between two and a half to three years of age. They were kept during the winter at very low levels of feeding, often with no increase in weight and were then expected to grow during the spring and summer.

Recent developments in beef production systems, using imported breeds, better grassland management and more efficient use of concentrate feeds have resulted in greater liveweight gains and a move from grass to yard finishing.

Baker (1975) suggests four factors that need to be considered when choosing a system for beef production.

(1) Financial resources. These include cash flow requirements and the capital availability.

(2) Physical resources, e.g. the area and quality of grassland available, field structure and the availability of water.

(3) Date of birth of calves—autumn or spring.

(4) Type of cattle, pure dairy, dual purpose or beef, or crosses.

Growth is usually measured by the change in liveweight of an animal. As this includes changes in the weight of feet, head, hide and the internal organs including the contents of the gut besides carcass tissue it is not always a reliable indicator of the final amount of saleable beef. Liveweight gain follows a characteristic sigmoid curve (*see Figure 15.1*). The rate of growth being influenced by nutrition, breed and sex.

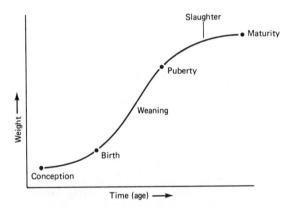

Figure 15.1 Typical sigmoid growth curve

An important phenomenon in many beef systems is compensatory growth. When animals have been fed on low quantities and/or low quality feed growth will slow down or cease—this is known as a store period—when full levels of feeding are resumed these animals will eventually catch up animals that have been kept on full feeding levels; this is compensatory growth. It is exploited in beef systems so winter feeding can be kept at as low cost as possible and when spring feeding is started compensatory growth takes place thus taking advantage of relatively cheap grazed grass.

The development of the animal varies due to tissues maturing at different rates; nervous tissue first, followed by bone, muscle and finally fat. If at any time the energy intake of the animal is in excess of that needed for the growth of earlier maturing tissues bone and muscle then fat will be deposited. In time the fattening phase is reached when fat grows fastest. Fat is deposited in the body in a certain order: subperitoneal fat (KKCF)* first, then intermuscular fat, subcutaneous fat and finally intramuscular fat (marbling fat).

* Kidney knob and channel fat.

In the later maturing breeds, which have characteristically high growth rates, fat deposition occurs at a greater age and weight than in the early maturing breeds when fed on similar diets. Sex effects are also important, bulls grow faster than steers and these in turn grow faster than heifers. In parallel, bulls are later-maturing than steers and steers later than heifers. As a result heifers of early maturing breeds quickly reach an age when they are depositing fat and are then slaughtered at a young age. Bulls of the late maturing breeds need high levels of feeding before fattening commences.

The amount of fat in the animal influences killing-out percentage. As the animal gets heavier and fatter the killing-out percentage increases, dressing percentage being the yield of carcass from a given liveweight. Systems of describing carcasses have been developed in many countries. The purpose of beef carcass classification is to describe carcasses by their commercially important characteristics. The development of the classification system sought to improve the efficiency of marketing throughout the whole industry. The description enables wholesalers and retailers to define their requirements. Producers should be able to obtain higher returns by producing animals to match these market requirements. The result is a flow of information from the consumer through the retailer to the farmer. Producers can then assess the economics of providing one type of carcass from others and plan their systems of management accordingly.

The British scheme devised by the Meat and Livestock Commission (MLC) is based on a grid system which describes fatness on the horizontal scale and conformation on the vertical scale (*see Figure 15.2* and *Table 15.17*).

Table 15.17 Composition of steer carcasses (ex KKCF)

MLC fat classes	1	2	3L	3H	4	5
Lean (%)	68	64	61	59	56	52
Fat (%)	7	9	10	11	13	14
Total saleable tissues (%)	75	73	71	70	69	66
Carcass residue (%)						
Bone and waste	18	18	17	16	15	14
Fat trim	7	9	12	14	16	20

Source: MLC Handbook *Changing Britain's Beef*. Reproduced with permission.

Fatness is described from leanest (1) to fattest (5) with Z for overfat carcasses. Currently 50% of carcasses fall into class 3 and this has been divided into 3L and 3H. In carcasses where the distribution of fat is unusual the classification is more clearly defined by: P—patchy fat, V—excessive udder or cod fat and K where there is excessive kidney knob and channel fat.

Conformation is based on the scale 1–5, carcasses from extreme dairy types being 2.

Fatness and conformation are determined by visual appraisal. Conformation considering shape and taking into account carcass thickness, blockiness and the

Fat class

		Leanest					Fattest
		1	2	3L	3H	4	5
Very good	5						
	4						
	3						
	2						
Poor	1						

Conformation class

Figure 15.2

fullness of the round. Fatness is always referred to before conformation, e.g. a carcass falling in fat class 2 and conformation 3 would be described as 23. The scheme also includes information on weight, sex and sometimes age. The MLC undertake many demonstrations of the system throughout the UK every year and anyone interested should contact the nearest MLC Regional Fatstock Offices.

Selecting cattle for slaughter

The two most important aids to the selection of cattle for slaughter are weight and fat cover. Knowledge of the weight of animals and their previous performance means that decisions can be based on fact. Once animals reach their minimum weight the decision to sell should be based on finish. Excess fat is wasteful and has to be trimmed off carcasses besides which feed turned into excess fat is wasted and is very expensive. The best way of assessing carcass fat cover is to handle the cattle.

The diagram (*Figure 15.3*) shows the five key points for handling:

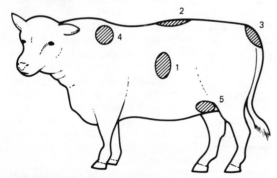

Figure 15.3 Key handling points

(1) over the ribs nearest to the hindquarters,
(2) the transverse processes of the spine,
(3) over the pin bones on either side of the tailhead and the tailhead itself,
(4) the shoulder blade,
(5) the cod in a steer or udder in a heifer.

If producers are doubtful about the fat levels in animals they have selected for slaughter they should check their classification reports or follow the carcasses through the abbattoir.

BEEF PRODUCTION SYSTEMS
Cereal beef

Cereal beef more commonly known as barley beef is so named because of the system of production. Dairy bred animals are fed an all concentrate ration—usually based on barley, this encourages rapid liveweight gain and animals are slaughtered at 10–12 months of age weighing 380–410 kg. These will yield a carcass of 205–240 kg.

Calves are reared on an early-weaning system. Weaning takes place at five weeks of age onto a concentrate ration containing 17% crude protein. At 10–12 weeks of age a diet based on rolled barley supplemented with protein, vitamins and minerals giving a mix of 14% crude protein is fed *ad libitum*. From six to seven months the protein level of the mix can be reduced to 12% thus reducing cost (*Table 15.18*).

The system is best suited to late maturing animals as early maturing animals become overfat at light weight. The Friesian is most commonly used because of its ready supply from the dairy herd. As bulls are later maturing than steers and have better food

Table 15.18 Liveweight and protein levels (%) of diet

Period	Liveweight (kg)	Protein content of diet (%)
5–12 weeks	100	17
3–6 months	250	14
6 months–slaughter	380–410	12

conversion efficiencies (FCE) there has been an increasing number of entires kept on this system. Of great advantage to the keeping of bulls is the fact that animals are housed throughout. Their faster and leaner growth, as compared to steers, means that bulls can be slaughtered at the same age (10–12 months) weighing 450 kg (*Table 15.19*).

Table 15.19 Cereal beef targets (Friesians)

Period	Gain (kg/d)		Economy of gain kg feed : kg gain	
	Bulls	Steers	Bulls	Steers
0–5 weeks	0.45	0.45	—	—
6–12 weeks	1.0	0.9	2.7	2.8
3–6 months	1.3	1.2	4.0	4.3
6–slaughter	1.4	1.3	6.1	6.6
Overall	1.2	1.1	4.8	5.5
Slaughter age (d)	338	345		
Slaughter weight (kg)	445	404		
Carcass weight (kg)	231	211		
Feed inputs	(kg)			
Milk powder	13			
Calf concentrate	155			
protein supplement	225			
Barley	1500			

Source: Allen, D. and Kilkenny, D. (1980).

Practical problems that may be encountered include respiratory diseases and bloat. As animals are housed throughout building design is of paramount importance both to the handling and physical management of stock and to reducing respiratory diseases. Good ventilation and draught-free buildings significantly reduce the incidence of pneumonia. Bloat or rumen tympany is a greater problem when animals are kept on slats rather than in bedded pens where they can consume some roughage. If 1 kg/d of hay is fed this should prevent any incidence of bloat.

Cereal beef production is sensitive to the relative prices of calves, barley and beef. Its main advantages as an enterprise are that it makes no direct use of land as all feeds can be bought-in and it is not seasonal, hence an even cash flow can be established once a regular throughput of animals is established.

Maize silage beef (*Tables 15.20* and *15.21*)

Continental producers have developed this system of fattening dairy bred calves and it is now a common system for bull beef production. British interest has

Table 15.20 Maize silage beef production (Friesians)

	Bulls	Steers
Gain (kg/d)	1.0	0.9
Slaughter weight (kg)	490	445
Slaughter age (months)	14	14
kg feed DM/kg gain	5.6	6.0
Protein concentrate (kg)	500	500
Maize silage DM (tonnes)	1.75	1.65
Stocking rate (cattle/ha assuming 10 tonnes DM/ha)	5.7	6.1

Source: MLC (1978). Reproduced with permission.

Table 15.21 Slaughter weights for maize silage beef

	Slaughter weight (kg)	
	Bulls	Steers
Friesian	480–500	430–450
Charolais, Simmental, Blonde d'Aquitaine	510–540	460–480
Limousin and South Devon × Friesian		
Devon, Lincoln Red and Sussex × Friesian	480–500	430–450
Hereford × Friesian	450–480	400–430
Angus × Friesian	420–450	380–410

Source: Alan, D. and Kilkenny, B. (1980).

been aroused as maize silage systems are fully mechanised and the crop provides a useful arable break crop.

Calves are reared to 12 weeks in the same way as for cereal beef previously described. Maize silage is then introduced and fed *ad libitum*. A protein supplement must be fed to bring the overall crude protein of the diet to 16%. The main problem with maize silage being its low protein content.

As with the cereal beef the system favours animals of late maturing type.

Maize silage can also be used for finishing suckled calves. It is important to ensure that the silage has a high dry matter content. Silages with low dry matter levels will result in lower feed intake and lower liveweight gains.

15 month grass/cereal beef (*Table 15.22*)

Calves born in the winter or early spring are reared on the early weaning system (five weeks). They are then grazed during the summer and fattened on a diet comprising mostly concentrates to 13–17 months and weigh 400–450 kg. The system therefore uses grassland for grazing and only very little grass conservation is needed.

Late maturing breeds such as the Friesian are better suited to the 15 month grass/cereal production system than earlier maturing types, e.g. Hereford × Friesian, because of the problem of early maturing types being overfat at too light a weight. Again entire bulls fit the

Table 15.22 MLC targets for 15 month grass/cereal beef

	Calf	
	Winter	Spring
Gain (kg/d)		
Birth to turnout	0.7	0.7
Grazing	0.7	0.6
Finishing winter	1.1	1.2
Overall	0.8	0.9
Stocking rate (animals/ha)		
Turnout to mid season	9.9	14.8
Mid season to yarding	4.9	7.4
Overall (no conservation)	7.4	11.1
Total concentrates (tonnes)	1.28	1.37
kg concentrates/kg gain in finishing winter overall	3.5	4.0

Source: MLC data summaries on Beef Production and Breeding, 1978. Reproduced with permission.

system well because of the heavier weights they may be taken to without becoming overfat.

The success of the system largely depends on a high level of grassland management during the summer to provide good grazing.

As little conservation is needed the control of the early flush of grass and the provision of clean grazing aftermath is made difficult. The high cost of concentrates has stimulated interest in the use of forage crops as substitutes for barley in the winter fattening period. However, the cost of the winter feed is offset to some extent by the compensatory growth which is exhibited during the winter after the summer grazing.

18 month grass/cereal beef (*Table 15.23*)

Autumn born calves are reared on a conventional early weaning system, until they are five weeks old when they are weaned onto a palatable high quality early weaning ration and good quality hay.

Friesian and Hereford × Friesian calves are commonly used. The autumn born steer should weigh 180–190 kg when turned out in the spring. Calves born during the winter and early spring weigh less and have lower weight gains at grass. The daily gain of heifer calves is 20% poorer than steers and finish at lighter weights. Bulls grow more rapidly and produce carcasses about 10% heavier than steers (Baker, 1975).

Table 15.23 Targets 18 month grass/cereal beef

Period	Gain (kg/d)	Period	Stocking rate (animals/ha[1])
To turnout	0.75	Turnout to mid season	8.5
Grazing	0.8–0.9	Mid season to yarding	5.0
Finishing winter	0.8–0.9	Overall (inc. conserv.)	
		450 kg at slaughter	4.0
		500 kg at slaughter	3.2

Source: MLC 1978. Reproduced with permission.
[1] Based on 230 kg/ha of fertiliser N.

During the grazing season the aim is to achieve a daily liveweight gain of 0.8–0.9 kg. To achieve this aim grass must be managed to provide a continuous supply of high quality herbage. Supplementary feeding with rolled barley may be practised to maintain the target growth rate, this will usually be necessary for the first few weeks after turnout in the spring and after late August when herbage quality and quantity begins to decline.

The animals should weigh about 320–350 kg at yarding. The target gain should be maintained at 0.8–0.9 kg/d. Winter feeding is usually based on silage which is supplemented with mineralised rolled barley. The amount of barley fed will depend on the quality of the silage and the desired weight gains. Cattle are slaughtered at 15–20 months of age weighing 400–520 kg.

20–24 months grass beef (*Table 15.24*)

Calves born between October and April are most suited to this system. Early maturing types, e.g. Hereford × Friesian or Angus × Friesian, fit more easily into grass finishing than Friesian. For the same reason heifers are more suitable than steers or bulls.

Table 15.24 Targets 20–24 month grass beef

Period	Gain (kg/d)	Period grazing (summer)	Stocking rate (animals/ha)
To turnout	0.7	First	10.0
First summer	0.7	Second (a)	3.5
Winter	0.5	(b)	3.0
Second summer	1.0		
Overall	0.7		
		(a) Slaughter at 450 kg (H × F)	
		(b) Slaughter at 500 kg (F)	

Source: MLC 1978. Reproduced with permission.

The calves are reared on the early weaning system and turned out to grass when conditions in the spring are suitable. Barley is fed as a supplement to the grass during the first grazing seasons for as long as the animals will eat it. Supplementary feeding can be beneficial in the late autumn to achieve the desired 0.7–0.9 kg gain/d.

The winter feeding programme is based on silage or hay and supplemented with concentrates to achieve a daily gain of 0.5 kg.

Animals are slaughtered during the latter half of the second grazing period at 20–24 months weighing 430–530 kg.

Suckled calf production

Suckled calf production is practised under a varied range of environmental conditions, from hill land to lowland. From the earlier descriptions of the various cattle types it can be seen that beef suckler cows also vary from the larger and milkier types to the smaller and hardier types. Suckler herds are most likely to be kept for one of the following reasons:

(1) the utilisation of hill and upland grazing,
(2) the utilisation of inaccessible grass on lowland farms,
(3) the utilisation of the grass break crop and arable by-products,
(4) the production of pedigree beef cattle.

The number of purebred beef calves has declined, crossbreds becoming more prevalent, and this introduces the heterosis effects of greater fertility and calf viability.

The seasonality of calving is affected by the availability of buildings and winter grazing, the bulk of calvings taking place in either the autumn or late winter/early spring. A short calving season enables the cows to be fed as a group without over or underfeeding of individuals and also gives a more uniform batch of calves to be sold or fattened.

The use of condition scoring has added some precision to the management of suckler herds. The body condition of breeding cows at service and calving is particularly important for reducing barren cows.

Herds that have a condition score of above 2, with the optimum being 2.5–3, have the better calving intervals and the greatest number of calves reared (*Tables 15.25* and *15.26*).

Table 15.25 Relationships between body condition and reproductive performance (beef cows)

Cow scores[1]	Calving interval (d)	Herd average	Calves weaned/100 cows served
1–2	418	1–2	78
2	382	2	85
2–3	364	2–3	95
3+	358	3+	93

Source: MLC 1978. Reproduced with permission.
[1] Scores on the scale 1 (very thin)–5 (very fat).

Table 15.26 Body condition score targets

Stage of production	Target score	
	Autumn calving	Spring calving
Mating	2.5	2.5
Mid pregnancy	2	3
Calving	3	2.5

DAIRYING

Milk production is the largest enterprise in UK agriculture with an annual net sum received by producers of some £1756 million, this accounts for about 22% of the total value of all agricultural output.

The number of milk producers reached a peak in 1950 but since then has declined. In contrast there has been a slow increase in the number of dairy cows kept in the UK. The result is that the average herd size has increased from 18 cows in 1955 to 52 in 1979 and it is probable that the trend towards larger herds will continue.

Herd replacements

The number of replacements required for a herd each year will depend on

(1) the size of the herd,
(2) the rate of replacement,
(3) whether the herd size is increasing, decreasing or being held static.

Replacement rate is related to culling policy. In the national herd culling rate is around 20–25%, the average cow having a lifetime production of four or five lactations. Only about 25% of cows are culled for genetic reasons, i.e. low milk yield and poor conformation. A large proportion of cows are culled for reproductive and mastitis problems and only 11% for reason of age. If there were a reduction in the number of animals culled, for reasons other than genetic, this would enable either fewer replacements to be reared or a greater selection pressure to be exerted on the heifers retained for the herd.

When planning a replacement policy it should be assumed that on average for every 100 cows mated only 83 live calves will be produced owing to conception failures, abortions and calves being born dead. Of these 83 only half will be female and with the inevitable deaths and reproductive problems in the heifers only 35 animals will be available as replacements. Assuming a replacement rate of 25% this only leaves ten surplus heifers. With better management and fewer losses more heifers could be reared so increasing selection or more cows could be crossbred to provide calves for sale to beef producers. Alternatively fewer animals would need to be reared and the surplus grass could be used for milk production. Dairy replacements compare unfavourably with dairy cows in terms of gross margin per forage hectare. Contract rearing schemes enable breeders to follow a planned programme of genetic improvement. At the same time the buying in of disease is minimised and the dairy herd can fully utilise the land and maximise output. The two commonest forms of contract rearing are as follows.

(1) The breeder sells the calf to the rearer at an agreed price. The calf is transported by the rearer, it becomes his responsibility and he bears all the expenses and losses. The breeder has first option

on the heifers which he may then buy back at approximately £400 + the cost of the calf for Friesians eight weeks prior to calving and pays for their transport.

(2) Alternatively the breeder retains ownership of the calf and is charged expenses by the rearer. Approximate charges would be

From 2 weeks old   £12–£14/month
From 6 months   £10–£12/month

The rearer supplies transport and pays for veterinary and medical expenses and pays for losses.

Most contract rearing schemes are based on one of these lines. The final alternative to rearing replacements is to buy in all requirements. This is only a reasonable alternative if yields are maintained and there is no increase in replacement rate. In practice this is often not the case as purchased animals tend to be of an unknown quality and fail to yield above average and there is little possibility of planning a programme of genetic improvement. There is always a danger of introducing disease, thus increasing replacement rates.

Lactation curves

If the milk yield of cows is plotted as a graph a lactation curve is produced. The standard lactation is 305 d and the annual cycle can be conveniently split into four distinct parts of early, mid and late lactation and a dry period (*Figure 15.4*).

After calving the milk yield will rise for a period of four to ten weeks when peak milk yield will be achieved. The time taken to reach this peak yield varies depending on breed, individual variation and yield.

Once a cow has reached peak yield the subsequent decline is approximately 2.5% per week or 10% per month. (Many producers are now beating this performance.) The decline in yield of heifers is 7–8% per month.

The daily peak yield of cows is approximately 1/200th of total (305 d) yield. Thus a cow giving 30 kg at its peak will have a total yield of approximately (30 × 200) 6000 kg.

The heifers' peak yield will be approximately 1/220th of total (305 d) yield. Thus a heifer giving 18 kg at peak will yield approximately (18 × 220) 3960 kg. A useful 'rule of thumb' guide is that two-

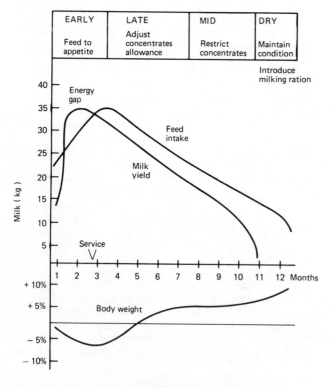

Figure 15.4 Feed intake, milk yield and body weight lactation relationships for a dairy cow producing 35 kg milk daily at peak (From the Scottish Agricultural Colleges (1979) *Feeding the Farm Animal—Dairy Cows***, publication no. 42. Reproduced by courtesy of the publishers)**

thirds of the total yield will be produced in the first half of the lactation.

It is clear that the height of the curve at peak milk yield has a great influence on the total lactation yield. The factor that is most likely to limit the level at which cows reach their peak lactation is nutrition.

Feeding

The nutrition of dairy cows can be divided into two distinct requirements: first to provide the means of maintaining life (maintenance requirement); secondly to provide nutrients for growth, the development of the unborn calf and milk (production requirement).

The metabolisable energy (ME) required for maintenance can be calculated from

$$Mm = 8.3 + 0.091 \, W$$

where Mm = maintenance allowance in megajoules (MJ), W = liveweight in kg. The energy allowance for milk production is calculated from

$$M_L = 1.694 \, EV_L$$

where $EV_L = 0.0386 \, BF + 0.0205 \, SNF - 0.236$, BF = butterfat (g/kg), SNF = solids-not-fat (g/kg).

The metabolisable energy required for 1 kg liveweight gain is 34 MJ of ME. A 1 kg liveweight loss spares 28 MJ of ME in the diet.

To formulate rations for dairy cows it is necessary to estimate dry matter intake (DMI). Intake is influenced by many factors including body size, milk yield, stage of lactation, digestibility of food and the number of feeds per day. The following is a useful estimate of DMI for cows in mid and late lactation.

$$DMI = 0.025 \, W + 0.1 \, Y$$

where DMI = dry matter intake (kg/d),
W = liveweight (kg), Y = milk yield (kg).

In early lactation the appetite is reduced by 2–3 kg/d.

The methodology of calculating diets and the energy requirements of dairy cows with worked examples is explained in Technical Bulletin No. 33 *Energy Allowances and Feeding Systems for Ruminants*, London: HMSO.

Protein allowances are as follows: 320 g of digestible crude protein (DCP) are needed for every 1 kg of liveweight gain and 50 g of DCP for 1 kg of milk produced.

Accurate feeding of dairy cows involves long-term planning as the requirements and allowances at any one time can be influenced to a large extent by previous nutrition.

The practice of feeding prior to calving is a good example of the nutrition in one period affecting the production in another. During the last two months of pregnancy—the latter part of the dry period—the unborn calf grows rapidly and there is extensive growth and renewal of mammary tissue. The practice of steaming-up is designed to fulfil these two needs, as well as providing reserves of body tissue which the cow can catabolise during early lactation and accustom the rumen to consuming large quantities of concentrates.

Broster (1971) reviewing experiments on pre-calving feeding concluded that a bodyweight gain of up to 0.5 kg can give a response of 10–15% in the following lactations yield. Above 0.5 kg there is no extra benefit in milk yield.

Steaming-up is usually started some six to eight weeks prior to calving, beginning with a small quantity of concentrates which is then increased each week, until about one-half to three-quarters of the amount that it is expected will be needed at peak lactation, is fed in the week before parturition. The amount of concentrates fed will depend on the body condition of the animals, the expected level of milk yield at peak lactation and the quality of bulk fodders. For efficient and economic feeding bulk fodder should be analysed for their nutrient value. Heifers are not usually given more than 3 kg of concentrates per day unless their body condition is very poor.

As the date of calving becomes close the animal may have a temporary additional loss of appetite, this should have fully recovered within 3–4 d after calving. As the appetite increases the cow is commonly fed for the quantity of milk being produced and an extra 1 kg/d of concentrates in an attempt to increase future production—this is usually termed 'lead-feeding'. However, too rapid an increase in concentrate feeding can lead to digestive upsets. It is important to realise that feeding dairy cows should not be viewed entirely in terms of concentrate nutrients as the bulk food part of the ration can contribute a substantial part of the total nutrients.

Correct feeding pre- and post-calving is essential to prevent excess weight loss in the early part of the lactation. Some weight loss is inevitable as the cow's appetite will not reach its maximum until after peak lactation. This difference in the time taken for an animal to reach peak lactation and peak DMI means that a nutrient gap occurs between the two. This deficiency in nutrients is made up by the cow catabolising her own body fat, and a daily liveweight loss of at least 0.5 kg can be expected. A too great a loss in weight can be conducive to a higher incidence of ketosis (acetonaemia). The manipulation of body weight (condition) in dairy cows has become necessary to sustain high milk yields. The pattern in *Table 15.27* is commonly suggested.

Frood and Croxton (1978) showed that the ability of a cow to reach a predetermined level of milk was closely related to its condition at calving.

Cows whose condition score (*Appendix 15.1*) was below 2 at calving did not achieve their predicted milk yield, those whose score was above 2.5 yielded

Table 15.27

Week no.	Liveweight change (kg/d)	Change during 10 weeks (kg)	Net effect on liveweight (kg)
0–10	−0.5	−35	−35
10–20	0.0	0	−35
20–30	+0.5	+35	0
30–40	+0.5	+35	+35
40–52	+0.75	+63	+98

Source: HMSO (1975). (Crown copyright, reproduced with permission.)

more than their predicted yields. Animals with a high condition score at calving, i.e. having ample body reserves, gave a higher earlier peak milk yield than animals in poor condition whose body reserves could not furnish enough nutrients during the time of low appetite and high energy demands of the cows.

The condition at calving and subsequent weight loss can have repercussions on conception rates at service. To ensure good conception rates cows should have a condition score of at least 2.5 at service. Below this score fertility is adversely affected. Cattle that are 'milking off their backs' must have their diet supplemented with minerals and protein additional to that normally included in the diet.

Once peak milk yield has been achieved and production starts to decline concentrate use can be reduced and intake of bulk foods increased. The level of concentrates fed will be determined to some extent by cow condition. Animals that have lost much weight will need liberal feeding until after they are served and are in calf. Concentrate levels should still be fed slightly in excess of milk production in an attempt to prevent the decline in yield. Liveweight during the period of mid lactation should be stable.

During late lactation cows have a maximum DMI relative to milk yield. The aim is to maintain milk production and restore the body condition of the cow. Maximum use of good quality forage with little or no concentrates should achieve the aims.

Suggested condition scores for dairy cows are as follows:

Calving	3.5
Service	2.5
Drying off	3

Flat rate feeding

With the traditional method of feeding differing amounts of concentrates are fed according to stage of lactation and yield. On a flat rate feeding system all cows are fed the same daily amount of concentrates throughout the winter regardless of their individual yields. Bulk food is usually silage because of its better feed value; this must be of a good quality (at least 10 MJ/kg DM) and must be fed *ad libitum*. If there is not enough silage available other good quality bulk

foods can be fed in its place. On most farms the optimum level of concentrates per cow will be between 6–10 kg/d, this does depend on the yield of the cows and the quality of the forage fraction of the diet.

The system is most effective and easier to manage if the herd has a tight calving pattern.

Complete diets

In this system all the dietary ingredients are mixed in such a manner that individual ingredients cannot be selected out from the rest and the mixture is offered *ad libitum*. The whole system can be mechanised and does away with the need for equipment designed to feed concentrates both in and out of the parlour. It is a feature of larger dairy herds where cows can be split into at least three groups according to the stage of lactation and appropriate rations can be fed to each group.

Self-fed silage

This system became popular after the Second World War with the advent of loose housing. The silage depth should not exceed 2.25 m for Friesians or they will not be able to reach the highest part of the clamp. If the depth of silage is greater than 2.25 m the upper part will have to be cut and thrown down, otherwise there is the tendency for cows to tunnel into the clamp and animals can be trapped when the overhang collapses, this is especially dangerous if tombstone barriers are used as the cows can have their necks trapped between the uprights of the barriers. The width of the feeding face should be 150 mm/cow. It is unwise to have a greater width than this or the silage will not be consumed quickly enough causing spoilage and secondary fermentation.

The commonest barrier to keep the cows from the silage face is either an electrified wire or an electrified pipe; the latter being less likely to break than wire and providing a better electrical contact. Consumption can be controlled by the distance between the barrier and the feed face. The layout of buildings is important as cows need ready access to the silage. The width of the face is relative to the number of animals that will feed from it and one of the problems encountered with such a feed system is that any herd expansion is difficult because of the inability to increase the width of the silo face.

Milk quality

Milk is a major source of protein, fats, carbohydrates, vitamins and minerals in the human diet. All these are present in milk in an easily digestible form. The greater part of milk is water. The remainder is solids which can be divided into two main groups, the milk

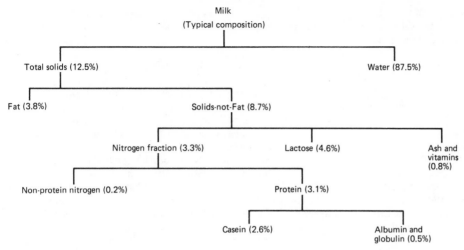

Figure 15.5

fat (butterfat) and the solids-not-fats (SNF) (*Figure 15.5*).

The total solids content of milk varies. Milk fat is a mixture of triglycerides and contains both saturated and unsaturated fatty acids. Casein is the main protein contained in milk.

There is generally an inverse relationship between milk yield and the milk fat and protein percentages. The higher the yield the lower the percentage composition of these components.

The breed of cow is an important factor affecting milk composition. Friesians, one of the heaviest milk yielding breeds, have some of the lowest fat and protein contents. Jerseys on the other hand have low yields but high fat and protein content (*see Table 15.6*). Individual variation within a breed can also cause a large effect in milk composition. Old cows tend to produce milk with a lower total solids content because yields increase up to the fourth lactation and udder troubles increase with age, these factors have a depressing effect on fat and solids-not-fats. A regular intake of heifers into the herd will help to maintain milk quality.

Day to day variations in milk quality can be caused by incomplete stripping of milk from the udder. The first milk to be drawn from the udder at milking contains low levels of fat when compared with the last milk to be drawn off, solids-not-fats change very little during the milking period.

Fat percentage is usually lower after the long period between milkings, part of this is due to the higher udder pressure which causes lower fat secretion.

The fat content of milk often drops when cows are turned out to lush grass in the spring. This is a function of corresponding increase in yield and because of low fibre levels. This problem can be mitigated to some extent by feeding 2 kg of hay or straw before the cows go out to grass and by restricting the grass so that the more fibrous stem fraction is eaten as well as the leaf. Long fibre is necessary for the rumen fermentation to produce acetic acid, acetate being the main precursor of milk fat.

Progressive underfeeding and poor body condition are responsible for cows producing milk with low solid-not-fat. If energy supply is deficient either during the winter or as a result of grass shortage during the summer then solids-not-fats will drop. Sub-clinical mastitis can be the cause of a reduction in solids-not-fats.

If the above factors are not the cause and there is still a milk quality problem then the answer might be to change the breed or the problem can be rectified to some extent by a progressive breeding programme using a bull selected for his high milk quality characteristics.

Oestrus detection

Oestrus (bulling or heat) is the time during which the cow will stand to be mounted by the bull. This lasts on average 15 ± 4 h. During the winter the period may be greatly reduced, a shorter period of 8 h is also common with heifers. The average interval between oestrus is 20–21 d.

Failure to detect animals on heat can be a major problem especially with the increasing size of herds. Poor conception rates are often blamed on poor artificial insemination techniques, bull infertility, disease, etc. while the actual problem is poor management.

The first management factor necessary to improve oestrus detection and conception is good cow identification. There are a number of methods by which cows can be identified clearly, but freeze branding is probably the best. In conjunction with identification should be the keeping of good records. It is essential that simple, accurate and complete records are available whilst cows are being observed for signs of oestrus. Detecting cows in oestrus is a part of good stockmanship as cow behaviour is the best indication of heat.

Signs indicating oestrus are:

(1) mounting other cows,
(2) sore, cuffed tail-head and hip bones,
(3) mud on flanks,
(4) mucus from vulva,
(5) restlessness,
(6) steaming animals.

Stockmen must be allowed adequate time for observation of the herd and the most productive times are after the animals are settled after milking and feeding. Mid morning and late evening are particularly productive. Observation should ideally take place three or four times per day and the herd should be watched for at least half an hour at a time, and any signs of oestrus should be written down and not left to memory. The aim should be to detect 80% of all cows in oestrus. Various aids may be used, including vasectomised bulls, pads or paint placed on the cows rumps have proved useful in some cases.

The timing of service is important. If cows are served too early or too late in the heat period poor conception rates may result. Ideally, cows should be served in the middle of the heat period between 9 and 20 h after the start of heat giving the best results.

The aim should be for a calving interval of 365 d. If the cow is to have a lactation of 305 d and a dry period of 60 d and pregnancy lasts from 278–283 d cows should conceive 81–86 d after calving. To achieve this in practice service should take place 40 d after calving. If service is delayed until the first heat after 60 d—a common practice—the result will be a slightly higher conception rate but a longer calving interval.

Oestrus can now be controlled and synchronised by the use of prostaglandins. The technique is useful where a group of heifers need to be inseminated. The animals will come into heat at the same time and the result will be a tighter calving pattern. In older cows the technique can be used on individual animals rather than the herd as a whole. The technique is not cheap and is not a cure for infertility and will only bring a cow on heat that is cycling normally—but at a selected time.

The practice of calving the herd over a period of about ten weeks is not very common despite the simplification of management routine. All the main tasks, e.g. calving, oestrus detection, service, drying off occur during a specific time period for the entire herd. Coupled to this is the fact that the feeding management of the herd can be simplified by having uniform groups of animals at similar stages of lactation. Heifer rearing becomes simpler as the animals have a smaller range of age. Not all farmers are attracted to such a routine but the regimentation

can create conditions more conducive to the achievement of better results.

CATTLE BREEDING

The aim of selecting breeding animals is to produce a future generation with improved performance.

Improvement is brought about by increasing the frequency of desirable genes and by creating favourable gene combinations. The genotype of an animal is its genetic constitution. The phenotype is the sum of the characteristics of the animal as it exists and is the result of both genetic and environmental effects. It is possible for animals to have the same genotype but different phenotypes owing to environmentally-produced variation, i.e. how it is fed, housed and managed. Environment does not affect all characters to the same extent. For example, the normal homozygous black coat colour of the Aberdeen Angus is always dominant to the recessive red of the Hereford. However, the Angus bull will not cover up the dominant white head of the Hereford, thus all Angus × Hereford cattle are black with a white head—environment has no effect as these characters are controlled by the presence of one pair of genes. The expression of many characters of economic importance—body growth, milk yield, milk quality, fertility—is controlled by many genes and environment also exerts an effect. Characters in which there is a close resemblance between parent and offspring, whatever the environment, are characters of high heritability (*Table 15.28*).

Table 15.28 Heritability of breeding stock characters

Trait	Heritability
Birth weight	0.4
Weaning weight	0.2
Daily gain, weaning to slaughter	0.3–0.6
Food conversion	0.4
Mature weight	0.4
Wither height	0.5–0.8
Heart girth	0.4–0.6
Dressing percentage	0.6
Points of carcass quality	0.3
Cross section of eye muscle	0.3
Bone percentage	0.5
Lactation yield	0.2–0.3
Butterfat content	0.5–0.6

After Johansson, I. and Rendel, J. (1968), *Genetics and Animal Breeding*, Edinburgh: Oliver and Boyd. Reproduced with permission.

Greater progress can be expected if the breeding programme is concentrating on characters of high heritability.

As the sire often serves many cows a great deal of attention needs to be placed on his selection. The two main methods of evaluating bulls are:

(1) performance testing,
(2) progeny testing.

Performance testing involves measuring an animal's individual performance—growth rate, feed conversion efficiency—and comparing them with animals from a comparable group which have been subjected to similar conditions of feeding and management. The advantage of performance testing is that it is much cheaper and quicker than progeny testing. The disadvantages are that it is only of value where the characters being measured are of a high heritability and thus likely to be passed on to its progeny. It can be used only to measure characters in the live animal, this used to be a disadvantage in assessing carcass composition as that used to necessitate slaughtering, however, new techniques such as ultrasonics have overcome this particular difficulty to a large extent.

Progeny testing involves the examination of an animal's offspring. The characters are measured and compared against the progeny of others. Again progeny should be kept under similar environmental conditions. It is particularly suitable for testing dairy bulls as milk production is only measurable in the female and the characters in question are often of low heritability. The method is used by the Milk Marketing Board for evaluating its dairy bulls for AI purposes. The bull's progeny can be compared with daughters sired by other bulls in the same herds. The records should then reflect the differences attributable to the bull's genetic constitution and is referred to as a contemporary comparison. These contemporary comparisons can provide guides when selecting a bull for AI. The higher the contemporary comparison the greater the chances are that the bull will pass on genes resulting in improved daughters. The greater the number of daughters used to evaluate a bull in this way results in a more reliable comparison figure, the number of daughters used is referred to as a 'weighting' which appears with the contemporary comparison figure.

MILK PRODUCTION

Milk production and the premises under which milk is produced are controlled by the Milk and Dairy Regulations 1959, a requirement being that all dairy farms are registered by the Ministry of Agriculture. Premises, stock and methods of production are open for inspection at any time.

The milking process

Milk ejection or 'let-down' is controlled by the hormone oxytocin. Milk let-down is a condition reflex in response to a stimulus. Quiet handling of animals prior to milking is essential; if cows are nervous or frightened milk let-down will be inhibited. The natural stimulus for cattle is the calf suckling, in dairy herds a substitute stimulus in the form of feeding and/or udder washing is used. Besides providing this stimulus udder washing removes dirt which may contaminate the milk as it leaves the teat. Methods of udder washing include sprays, or buckets of water, the warm water often includes an antibacterial agent and cloths or paper towels are used to clean and dry the udder and teats. Disposable paper towels are preferable to cloths as there is less chance of infection being passed from cow to cow.

Before or after udder washing fore-milk (the first milk from the udder) is removed by hand in to a strip cup to reveal any signs of mastitis in the form of clots, flakes or watery milk. This is a useful indication of clinical mastitis.

Teat cups should be applied as soon after washing as possible. Maximum rate of milk flow is reached after about 1 min, later the flow rate declines quite rapidly. Once milking is started it should be accomplished as rapidly as possible. Overmilking should be avoided as this may damage the udder, thus predisposing to mastitis. Stripping—the removal of the last milk—can be achieved by applying downwards pressure to the teat cup with one hand and the udder massaged downward with the other, the forequarters being done first. In parlours the stripping of milk and automatic removal of teat cups, when the milk flow slows to a predetermined rate, is becoming more common.

Teat disinfection by dipping the teats in a cup containing an approved iodophor or hypochlorite solution helps prevent the spread of bacteria causing mastitis.

Efficient milking is largely dependent upon a good milking routine. This should be simple and consistent providing a 'let-down' stimulus for the cow and a routine series of operations which the milker has to perform on each animal.

After milking, the milk will be at an ideal temperature (37 °C) for the growth and multiplication of most bacteria. As the milk is collected once a day by bulk tankers some must be kept overnight. It is important to cool the milk quickly to prevent its deterioration. Cooling takes places in a refrigerated bulk tank and the Milk Marketing Board require milk to be cooled to below 4.5 °C by 30 min after milking. Before milk is allowed in the bulk tank it is usually passed through a sock filter.

Milk is routinely tested for keeping quality by direct or indirect tests and penalties are imposed for failure to meet the required standards. Stringent penalties are also incurred if antibiotics are detected in milk, these usually occur from the intramammary treatment of mastitis.

Poor keeping quality can be caused by inadequate cleaning of the milking equipment giving a build up of residue. Hand washing is normally used for bucket milking plants. A cold water rinse of approximately 10 ℓ of water can be drawn through the clusters of each unit, this removes the film of residue left after milking. The equipment is then dismantled and washed once in either detergent or a detergent and sterilant solution at 50 °C, then final rinsed in clean water also containing a sterilising agent.

With pipeline systems where it is impractical to dismantle the equipment, cleaning is done *in situ* by circulation cleaning. Here the success of the operation relies on the properties of the chemicals and heat for the disinfectant effect. Again the basic process starts with a rinse of cold water, this is followed by the circulation of a detergent solution. The initial temperature needs to be fairly high (80–85 °C) as the solution will be cooled during the first cycle round the plant. A final rinse with cold water which may contain sodium hypochlorite for sterilisation is circulated.

The commonest sterilising and disinfecting agents are sodium hypochlorite, bromates and iodophors, they should be used in accordance with the manufacturers' recommendations.

Another method of cleaning equipment *in situ* is by the use of acidified boiling water. Here hot water (96 °C) is flushed through the plant for 5–6 min and allowed to run to waste, this pre-rinses and warms up the plant. Nitric or sulphamic acid is mixed with the water, this has no disinfecting effect but removes the milk deposits from the equipment. The aim is to heat the plant to 77 °C for at least 2 min to achieve disinfection.

The circulation cleaning method involves the use of expensive chemicals and takes more time (15 min) whereas the acidified boiling water system needs more water (13–18 ℓ/unit) at a much higher temperature thus using more energy, but only takes 5–6 min.

To prevent a build up of scale on equipment it may be necessary to use a milkstone remover once per month.

Bulk tanks can be cleaned by hand, using a long handled brush, or by a mechanical spray. As tanks are cooling mechanisms and often contain an ice bank at the base of the tank a cold system of cleaning is normally used. This uses an iodophor or bromate cleaning agent at mains water temperature. The tank is rinsed with cold water immediately after emptying by hand and the cleaning agents applied. In automatic systems the rinse is sprinkled into the tank; this is then followed by a solution containing the chemicals. The inside is then rinsed before milking.

Particular attention should be paid to the outlet, paddle, dipstick and underneath the lid and bridge of the tank.

MILK AND MILK PRODUCTS

Milk is a food important in the human diet and has been described as nature's most perfect food. Milk varies in composition and some of the main factors affecting milk composition have been dealt with

earlier. Milk is tested for its compositional quality when purchased from the farm and at point of sale to the consumer.

The price paid to the producer is based upon its hygienic and compositional quality. Hygienic quality is measured and failure to reach the necessary standard results in price penalties. The presence of antibiotics, sterilants and cleaning agents are measured by the triphenyltetrazolium chloride (TTC) test, this compares the growth of *Streptococcus thermophilus* BC in the milk sample with its growth in a sample of milk free from inhibitory substances. Compositional quality is measured by the Gerber and density hydrometer tests; bonus and penalty payments are made above and below a defined level. The Gerber Test determines the fat content of milk and the hydrometer test the density, this is then related to the known fat content and a formula

$$\% \, SNF = 0.25 \, D + 0.22 \, F + 0.72$$

where D = hydrometer degrees, F = fat percentage, gives the solids-not-fats percentage.

A freezing point test is used where milk is suspected of containing added water. Milk freezes within a range of -0.52 to -0.56 °C, any added water brings the freezing point of milk nearer 0 °C, so samples freezing about -0.53 °C are assumed to have extraneous water.

Taints and flavours

Feed taints may be caused by certain foods, e.g. kale, turnips, wild garlic. These are easily controlled by reducing or eliminating the food source. Milk can also absorb odours from the atmosphere, e.g. silage, disinfectants, fuel oils. Most of these absorbed taints are volatile and are removed by heat treatment of the milk. Ill health may cause taints, mastitis gives milk a salty taste and acetonaemia gives a sweetish taste. The enzyme lipase causes a fat breakdown resulting in a rancid taste. Finally, bacterial taints may be caused by contamination, but clean hygienic milk production systems can avoid such problems.

Milk processing

Pasteurisation

The objective in pasteurising milk is to kill any pathogenic organisms that may be present, without affecting the nutritional value and palatability of the milk. Two methods of heat treatment are commonly used. The original method was to heat milk to 63–66 °C for 30 min then cooling to 10 °C or below. The second method involves heating to a temperature not less than 72 °C for at least 15 s, before again cooling to at least 10 °C. Rapid cooling is necessary to prevent the growth of heat resisting organisms. Two tests are used on pasteurised milk. The Phosphate Test indicates whether adequate heat has been used to pasteurise effectively the milk and the methylene blue test measures the keeping quality of pasteurised milk.

Ultra-heat treatment (UHT)

As in the second method of pasteurisation UHT involves the flow heating of milk but it is heated to at least 132 °C for 1 s before cooling. If it is then packaged under sterile conditions into a sterile container it will keep for four to six weeks if unopened. A colony count test is used to check that UHT milk has been properly heat treated.

Sterilisation

Milk is heated in the sealed airtight container in which it is sold. The temperature is usually in the range 105–115 °C for around 30 min. The treatment should kill all organisms present in the milk with the result that it will keep for 12–16 weeks without refrigeration.

The milk is required to pass the turbidity test to check the efficiency of the sterilisation process. The prolonged exposure to heat changes the protein structure and causes caramelisation of the lactose, thus altering the flavour and colour of the milk.

Homogenisation

This process results in a more uniform composition of the milk with no cream line at the top of the bottle. Milk fat globules have on average a diameter of 4 μm, homogenisation is achieved by forcing the milk through a small hole. This results in the fat globules having a size of about 1 μm in diameter.

Colour codes for bottle caps

The statutory colours for bottled milk are specified by the Milk and Dairies (Milk Bottle Caps) (Colour) Regulations 1976 and are as follows (*Table 15.29*):

Milk powder

Milk powder is produced by removing water from milk resulting in a solid with less than 5% moisture. The two main milk powders are dried whole milk and dried skimmed milk. The latter contains very little fat as this is removed with the cream. Three methods are used in the manufacture of dried milk. Freeze-drying produces a high quality milk powder but is expensive to produce because of the high energy needed in the process. With roller drying milk is distributed onto rotating hot rollers and the water is evaporated and the powder is scraped off. Spray drying involves atomising the milk and allowing it to pass through a stream of hot air which evaporates any water and the powder falls to the floor of the chamber.

Butter and cream

Cream is separated from milk by centrifugal action at about 6000 revolutions/min. As cream is lighter it stays in the centre and the skimmed milk is thrown outwards and separated off. The fat content of the milk remains in the cream and most of the protein in the separated milk.

Cream is sold as liquid, clotted or churned into butter. Cream sold fresh is pasteurised. The Cream Regulations 1970 specify the butterfat content of the varying types.

	Butterfat
Single cream must contain no less than	18%
Double	48%
Whipping cream and whipped cream	35%
Clotted cream	55%

Cream for buttermaking contains 35–40% butterfat. The skim milk is often dried to milk powder.

Cheese

Cheese type depends on the method of manufacture and cheeses are classified as follows:

(1) hard pressed cheese, e.g. Cheddar, Cheshire, Leicester, single and double Gloucester, Lancashire, Derby, and Scottish Dunlop;

Table 15.29

Type of milk	Colour of cap	Lettering
Pasteurised	Silver	Black
Pasteurised homogenised	Red	Black or silver
Pasteurised Channel Island and South Devon	Gold	Black or silver
Untreated	Green	Black or silver
Untreated Channel Island and South Devon	Green with single gold stripe	Black or silver
Sterilised	Blue	Black or silver
Ultra-heat treated	Pink	Black or silver

(2) blue veined, e.g. Stilton, Blue Wensleydale, Blue Vinney, Roquefort and Gorgonzola;

(3) soft cheese, e.g. Camembert, Pont L'Eveque.

Basically all cheese making processes involve lactic acid fermentation and the production of curd, leaving whey as the by-product. Cheeses are classified by their water content. The blue veined cheeses are stored in a very humid atmosphere which encourages mould growth, *Penicillium roquefort*.

Yoghurt

Other fermented products include yoghurt which may be made from whole skimmed or dried milk. A culture of *Lactobacillus bulgaricus, Streptococcus thermophilius* and occasionally *Lactobacillus acidophilus* are added and incubated at 43 °C. Yogurt can be sold either in its plain form or flavoured.

References

ALLEN, D. and KILKENNY, B. (1980). *Planned Beef Production*. London: Granada

BAKER, H. K. (1975). *Livestock Production Science* **2**, 121

BROSTER, W. H. (1971). *Dairy Science Abstracts* **33**, 253

HMSO (1975). *Energy Allowances and Feeding Systems for Ruminants*. Technical Bulletin No. 33

JOHANSSON, I. and RENDEL, J. L. (1968). *Genetics and Animal Breeding*. London: Oliver and Boyd

MLC (1978). *Beef Improvement Services*. Data summaries on beef production and breeding

MLC (1980). *UK Meat and Livestock Statistics*

MLC HANDBOOK. *Changing Britain's Beef*

MLC/MMB JOINT PUBLICATION. *Rearing replacements for beef and dairy herds*

MMB (1979). *UK Dairy Facts and Figures*

MMB (1980). *UK Dairy Facts and Figures*

ROY, J. H. B. (1980). In *The Calf*. London: Butterworths

SCOTTISH AGRICULTURAL COLLEGES (1979). *Feeding the Farm Animal—Dairy Cows* Publication No. 42

WOOD, P. D. P. (1972). MMB Better Management No. 7: 1

APPENDIX 15.1

Condition scoring of cattle

Condition scoring is a means of estimating the fatty tissue under the skin at the tailhead and loin and is a means of quantifying body reserves.

Scores range from 0 (very poor) to 5 (grossly fat), with half scores giving an 11 point scale.

The loin and tailhead areas are scored independently along the 11 point scale. The tail score is then adjusted by half a score, if it differs from the loin by one or more points.

Score 0 (very poor)

Tailhead area:	Deep cavity under tail and around tailhead. Skin drawn tight over pelvis with no fat in between.
Loin area:	No fat detectable. Shape of transverse processes clearly visible.

Score 1 (poor)

Tailhead area:	Cavity present around tailhead. No fatty tissue felt between skin and pelvis but skin is not draw tight but supple.
Loin area:	End of transverse processes sharp to touch and upper surfaces are easily felt. Deep depression in loin.

Score 2 (moderate)

Tailhead area:	Shallow cavity lined with fatty tissue felt at base of tail. Some fatty tissue felt under the skin. Pelvis easily felt.
Loin area:	End of transverse processes feel rounded, but upper surfaces felt only with pressure. Depression in loin visible.

Score 3 (good)

Tailhead area:	Fat easily felt over the whole area. Skin appears smooth but can be felt.
Loin area:	Ends of transverse processes can be felt with pressure, with thick layers of fat on top. Slight depression visible in loin.

Score 4 (fat)

Tailhead area:	Folds of soft fat present. Patches of fat apparent under skin. Pelvis felt only with firm pressure.
Loin area:	Transverse processes cannot be felt. No depression in loin.

Score 5 (grossly fat)

Tailhead area:	Base of tail buried in fat. Skin distended. No part of pelvis felt even with firm pressure.
Loin area:	Folds of fat over transverse processes. Bone structure cannot be felt.

16

Sheep

R. A. Cooper

INTERNATIONAL PICTURE

There were 1055 million sheep in the world in 1978. Details of their distribution are given in *Table 16.1*. Since then there has been a tendency for numbers to increase, particularly in the major producing countries. One per cent increases have been seen in Australia, the USSR, France and Spain, with increases closer to 3% in New Zealand and the UK

Table 16.1 World distribution of sheep (× 10⁶) (1978[1])

Africa 168.5		North America 22	
South Africa	31.5	USA	12.3
Ethiopia	23.0	Mexico	7.8
Sudan	15.5	Canada	0.3
Morocco	14.0		
Algeria	10.0		
South America 103		Asia 300	
Argentina	34.0	China[2]	101
Uruguay	18.0	United Arab	
Brazil	17.0	Emirates	43.5
Peru .	14.0	India	40.5
		Iran	33.0
		Afghanistan	22.5
		Philippines	22.0
Europe 129		Oceania 192	USSR 141
East Europe	42.6	Australia	131.5
UK	29.6	New Zealand	60.3
Spain	15.5		
France	11.0		
Italy	8.5		
Greece	8.0		

Data taken from *FAO Production Yearbook 1978* with permission.
[1] Total 1055 million.
[2] Includes Mongolia.

(1980 figures). Production of sheepmeat from the major producers is shown in *Tables 16.2* and *16.3*. In Europe, output from sheep flocks is largely as meat, but in south western France (Aveyron and Basses–Pyrenees) and in Italy the most important output is milk for cheese. In Italy lambs are allowed to suckle for some six weeks, and are then slaughtered as milking of the ewes begins. This leads to average carcass weights of only 9 kg compared with 15 kg for the rest of Europe.

Table 16.2 Sheepmeat production forecast for 1980 (10³ tonnes)

USSR	865	Australia	503
New Zealand	580	North America	140
EEC	549		

Data taken from *MLC International Market Survey 1980 No. 2* with permission.

Table 16.3 Sheepmeat production and consumption of major EEC countries—forecast for 1980 (10³ tonnes)

	Produced	Consumed
EEC	549	812
UK	260	405
France	168	218
Italy	38	73
Ireland	38	NA
West Germany	19	60

NA—not applicable.
Data taken from *MLC International Market Survey 1980 No. 2* with permission.

SHEEP IN THE UK

The total sheep population of the UK is some 29.6 million head (*Table 16.1*). Data for breeding animals are given in *Table 16.4*, with an approximate breakdown by flock size shown in *Table 16.5*. *Table 16.4* also indicates the importance of sheep in hill and

375

Table 16.4 Numbers and distribution of breeding ewes in the UK (10³ head)

	Hill flocks	Upland flocks	Lowland flocks
England and Wales	3303	1422	4292
Scotland	2225	656	341
Northern Ireland	249	85	160
Totals	5777	2163	4794

Data taken from *MLC Sheep Facts 1978* with permission.

Table 16.5 Distribution of breeding ewes by farm size (%)

Flock size	England and Wales		Scotland	
	Holdings	Ewes	Holdings	Ewes
1–49	32.2	4.4	27.6	3.2
50–99	20.8	8.7	23.3	7.4
100–199	21.1	17.5	19.6	12.3
200–299	10.0	14.1	8.6	9.3
300–499	8.7	19.4	8.8	15.2
500–999	5.6	22.0	8.1	24.6
1000	1.6	13.9	4.0	28.0
	100	100	100	100
Average flock size	170		222	

Data taken from *MLC Sheep Facts 1978* with permission.

upland areas, which produce about 50% of UK output of sheep and wool. It can be seen that whilst a majority of holdings carry less than 100 breeding ewes, most ewes are found in flocks of 300 head or bigger.

The National sheep flock is made up of some 50 'pure' breeds and over 300 crosses, but few of the pure breeds, and even fewer of the crosses, are of major significance except in localised circumstances. Details of individual breeds are available from many sources, in particular the publication *British Sheep* published by the National Sheep Association. Flock categories,

important characteristics and major breeds are given in *Table 16.6*.

Because of the variations in conditions under which sheep are kept, and because of the multiplicity of breeds and crosses available, sheep production systems in the UK are many and diverse. There is, however, one basic and fundamental thread linking the various sectors of the industry. This is the very substantial crossbreeding programme generally known as stratification. An outline of the concept of stratification is given in *Figure 16.1*, with the picture in rather more detail in *Figure 16.2*. *Figure 16.1* also illustrates the interdependence of different production systems. Under many hill situations ewes cannot thrive under the very harsh conditions for more than

Table 16.6 Breed types, important ewe characteristics and main breeds

Type	Desirable characteristics	Main breeds
Hill breeds	Regular lambing Milking/mothering ability Ability to rear 100% crop Hardiness Good wool weight and quality	Blackface Cheviot (North and South) Speckleface Swaledale Welsh Mountain
Upland breeds	Milking ability High fertility	Devon Closewool Kerry Hill Clun Forest
Longwools	High fertility and prolificacy Milking ability Growth rate	Bluefaced Leicester Border Leicester Teeswater
Down breeds	Fertility Growth rate Carcass quality	Suffolk Dorset Down Hampshire Down

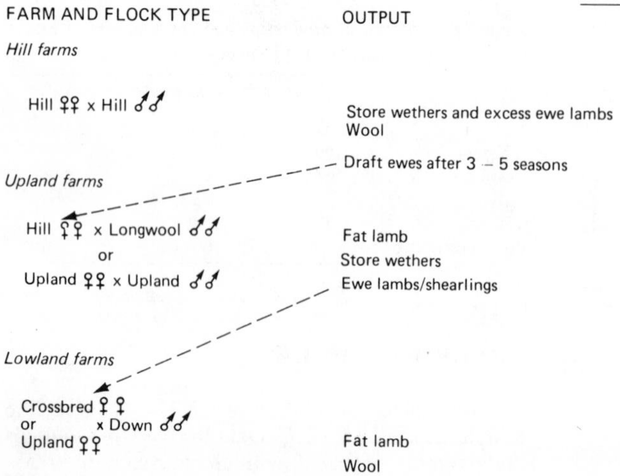

FARM AND FLOCK TYPE   OUTPUT

Hill farms

Hill ♀♀ x Hill ♂♂   Store wethers and excess ewe lambs
                                Wool

                            — Draft ewes after 3 – 5 seasons

Upland farms

Hill ♀♀ x Longwool ♂♂   Fat lamb
    or   Store wethers
Upland ♀♀ x Upland ♂♂   Ewe lambs/shearlings

Lowland farms

Crossbred ♀♀
  or   x Down ♂♂
Upland ♀♀   Fat lamb
                                Wool

Figure 16.1 An outline of the stratification of the British sheep industry

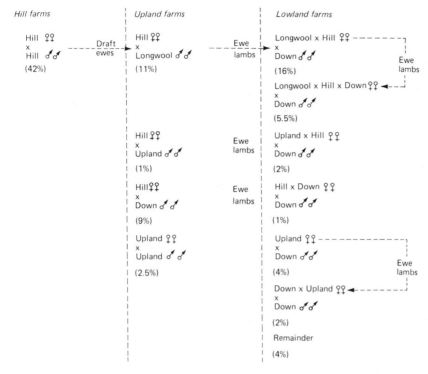

Figure 16.2 Estimated structure of national flock (by breeding policy)

four or five seasons, thus the hill farm retains many of its ewe lambs as breeding stock and the sale of draft ewes provides an important part of the hill farm income (15–20% of flock gross output at 1979 prices). Drafted into better conditions these ewes have several years of productive life left and given better climate and nutrition are capable of lambing at levels well beyond the 80–90% they have previously achieved.

On the upland farms, to which these draft ewes go, there is often little scope for fattening lambs satisfactorily. Put to a longwool ram, however, these ewes produce a crossbred lamb, the females of which are noted for their high fertility, milkiness and general vigour. There is, therefore, a strong demand for such animals for lowland breeding flocks. Five important crossbred types can be recognised. These are listed in *Table 16.7*. Forty-five per cent of all ewes tupped under lowland conditions are derived from longwool × hill matings.

FACTORS AFFECTING FLOCK PERFORMANCE

Given the diverse nature of sheep production in the UK it is difficult to discuss flock performance in general terms. Many flocks are now recorded in detail by the Meat and Livestock Commission and the standards which such flocks are achieving are shown

Table 16.7 Important crossbreeds and their derivation

Cross	Sire breed	Dam breed
Greyface	Border Leicester	Blackface
Masham	Teeswater, Wensleydale	Dalesbreed, Swaledale, Rough Fell
Mule	Blue-faced Leicester	Swaledale, Clun, Blackface, Speckleface
Scottish halfbred	Border Leicester	Cheviot
Welsh halfbred	Border Leicester	Welsh Mountain

in *Figures 16.3* and *16.4*. In each case target figures are given in parenthesis alongside the achieved values.

It is apparent from the data in these figures that there is room for considerable improvement in both lowland and hill flocks. The factors which contribute to this situation are discussed below.

Oestrus

The sheep is a species in which there is generally an annual rhythm of breeding activity, the onset of the breeding season corresponding to a period of decreasing daylength and the cessation occurring as daylength increases in spring. There are, however, wide variations within as well as between breeds, and a much

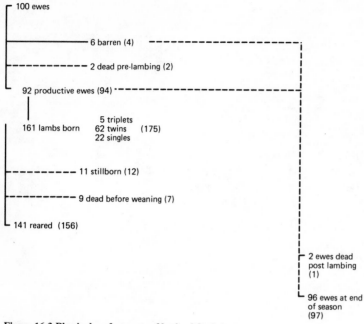

Figure 16.3 Physical performance of lowland flock (target values in parentheses)

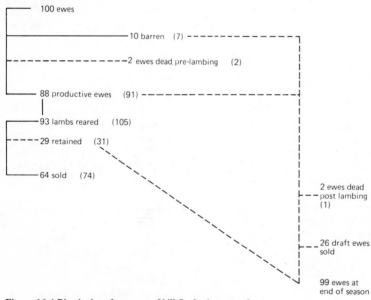

Figure 16.4 Physical performance of hill flocks (target values in parentheses)

reduced cyclicity is evident at lower latitudes. Some idea of the range of dates involved is given by *Table 16.8*. These values can be taken as no more than guides because of the interactions of many other factors as well as daylength. Some of these other factors are as follows.

Temperature

Very little direct effect in absence of daylength change although there is some indication that high temperatures in July and August may delay onset and this may produce year-to-year variations within the same flock.

Table 16.8 Breeding seasons for some British sheep breeds

Breed	Onset		Cessation	
	Mean	Range	Mean	Range
Blackface	25 Oct	26 Sept–10 Nov	19 Feb	17 Jan– 3 Apr
Border Leicester	9 Oct	24 Sept–13 Oct	10 Feb	18 Dec–11 Mar
Dorset Horn	24 July	15 June–21 Aug	2 Mar	22 Jan–22 Apr
Romney	4 Oct	16 Sept–19 Oct	2 Mar	20 Jan–23 Mar
Suffolk	3 Oct	12 Sept–23 Oct	17 Mar	7 Feb–20 Apr
Welsh Mountain	25 Oct	15 Oct–11 Nov	17 Feb	1 Feb–25 Feb

Adapted from Hafez, E.S.E. (1952), *Journal of Agricultural Science* **42**, 189 with permission.

Breed

Much variation within as well as between breeds. Obvious examples of breeds having a short or non-existent seasonal anoestrus are the Finnish Landrace, Dorset Horn (and polled Dorset) and some strains of Merino.

Breeding history

Post-lambing there occurs a period of anovulation known as lactational anoestrus. The duration of this period is difficult to ascertain since it usually runs into the period of seasonal anoestrus. It does, however, tend to be longer in late-lambing animals although nutritional effects may confound this.

Ram introduction

The sudden introduction of rams during the period immediately before the onset of oestrus can stimulate a more synchronised onset. This technique, often involving vasectomised rams, can be used to obtain a tighter lambing pattern.

Ovulation

One or more silent ovulations often precede the onset of behavioural oestrus. Ovulation occurs approximately every 17 d (range 14–19 d) throughout the breeding season. Oestrus lasts for about 30 h (range 24–48) with ovulations occurring towards the end of oestrus. One of the prerequisites for a good lambing percentage is a high ovulation rate, and subsequently a high fertilisation rate with minimal embryo mortality. Genotype is obviously going to have an effect here, although nutritional interactions may confuse the situation. Within an individual flock ovulation rate will be influenced by the following factors.

Nutrition

It has been known for many years that putting ewes onto a rising plane of nutrition for two to three weeks prior to tupping has a beneficial effect. Traditionally ewes were weaned onto hard conditions, in part to assist drying off, in part to put them into lean condition before 'flushing'. This 'dynamic' effect gives some 6–8% more twins compared with the performance of ewes mated whilst on a maintenance diet, and 12–16% more twins compared with ewes mated whilst fed below maintenance. More recently it has become apparent that a 'static' effect is also present, this being dependent upon the level of body reserve in the ewe; ewes with greater body reserves having a higher ovulation rate. It has been concluded that, for ewes on a level plane of nutrition, each extra 3.5 kg bodyweight at mating represents an extra 6% twins (comparisons within breed and within farm). In recognition of the impracticability of weighing ewes on commercial farms, the Hill Farming Research Organisation has produced a simple, rapid and accurate method of judging the body condition of ewes by palpating the spinal column in the loin area, immediately behind the last rib. The degree of fat cover over and around the spinous processes is converted into a score on a scale which runs from 0–5. Scoring is possible to the nearest half. The descriptions attached to each of the six scores are given in *Appendix 16.1*, p. 390.

Body condition scoring may be used at any time of year as a guide to the adequacy of the nutrition of the flock. It is particularly useful in the pre-tupping period when it can be used to divide the flock into groups according to their needs in terms of the duration and degree of flushing they require. Lowland ewes should have a condition score of at least 3.5 at tupping; poorer conception rates can be expcted from over-fat and very thin ewes. In the normal flushing period (approximately three weeks), ewes having a score of 3 or higher can achieve their target score weight easily. Those scoring 2.5 or less, on the other hand, will need at least six weeks on improved grazing, or supplementary concentrates, if lambing percentage is not to suffer. The relationship between score at tupping and lambing percentage is shown in *Table 16.9*. Thus a scoring at least six weeks before tupping can identify those animals in need of preferential treatment and a second, three weeks later, can be used to estimate the response. An improvement in condition from condition score 2.5 to a score of 3.5 is equivalent to a gain of some 6 kg liveweight.

Age

Ewe lambs may attain puberty in their first year.

Table 16.9 Relationship between body condition and lambing %

Breed	Average score at tupping	Live lambs/100 ewes to ram
Scotch halfbred	3.8	184
	2.6	120
Welsh halfbred	3.5	158
	2.7	133
Masham	3.4	178
	2.2	127
Clun	3.6	156
	2.0	131

After MLC 1980, Sheep Improvement Services, Body condition score of ewes (Reproduced with permission).

There is a nutrition/date of birth interaction and with lightweight or late-born lambs oestrus may not be attained. Although most animals which do attain puberty will rear one lamb at best, there is some evidence that animals which are bred as lambs may be more prolific in later life. Subsequently mean ovulation rate will continue to rise for five to six years and may then stay at this level for several more seasons. Hence the draft ewe from the hill flock has only just achieved her potential when she is sold.

Season

Many ewes begin the breeding season with a 'silent' oestrus. Ovulation rate then continues to rise for three to four cycles and likewise will drop again at the very end of the season. On average therefore lambing percentages will be higher in later-lambing flocks, and will tend to be low in out-of-season lambings. The Finnish Landrace, and its crosses, do not appear to be affected to the same degree. The relationship between tupping date (and hence lambing date) and lambing percentage is shown in *Table 16.10*.

Under normal conditions all lowland ewes will ovulate and produce at least one ovum; the majority should produce two or more ova. Genotypically there is no reason why this should not apply to hill breeds, the lower ovulation rates experienced in hill flocks being largely the result of environmental effects.

If all ova produced are to be fertilised, rams must be fertile and active. At the beginning of the breeding season all rams should be examined to make sure that

Table 16.10 Effect of tupping date on mean lambing %

Lambing date	Lambing %	
	MLC data[1]	WSCA data[2]
January	148	137
February	155	145
March 1–15	160	
March 16–31	167	159
April	168	—

[1] *MLC sheep facts 1978*. Reproduced with permission.
[2] West of Scotland College of Agriculture Research and Development Publication No. 7.

they are healthy, sound in legs and feet and that they have no obvious abnormalities, especially of the testes. Microscopic examination of a semen sample, obtained by electro-ejaculation, is a valuable additional check. Correct ewe to ram ratio will depend on circumstance. A strong ram should be able to cover 50–60 ewes adequately, but under extensive conditions, or following attempts to synchronise oestrus this may be reduced to 25 ewes per ram or even lower. The inexperienced ram lamb may be used (carefully) in his first year, but at a low ewe-to-ram ratio and only with mature ewes. It is particularly important such rams do not have to compete with more experienced rams at this time.

The practice of applying ochre to the breastbone of rams (keeling or raddling), or the use of the Sire–Sine harness, is a useful management tool. Rams treated in this way leave a mark on all ewes mated, and this can give a good indication of how the breeding season is going, as well as forming the basis for subsequent subdivision of the flock at lambing time. In larger flocks this can lead to a significant saving in concentrate feeds. With such a system the colour of the raddle should be changed every 16 d.

A significant percentage of all fertilised ova is lost in early pregnancy, especially in the first 30 d. Losses as high as 40% have been suggested but the mean is about 25%. Some loss of abnormal embryos is normal and to be expected, but environmental factors appear to be important. In particular nutrition during the immediate post-tupping period is critical and should be held at an above-maintenance level. Embryo mortality also increases with increasing ovulation rate, so that ewes which have been flushed pre-mating are likely to be more sensitive to post-mating drops in feed intake. On the other hand embryo mortality is reported to be heavier in ewes mated early in the breeding season, and also in younger animals. In many countries a further factor is likely to be heat stress, which can cause very high embryo losses if it occurs at the early cleavage stage.

During the middle part of the pregnancy few fetal losses appear to occur. The ewe is able to stand periods of under-nutrition, hill ewes often losing 5–10% of bodyweight during this period, without significant effects. Such losses may reduce lamb birthweights and increase pre- and perinatal mortality. Perinatal mortality (including stillbirths and abortions) is a major source of loss in many sheep flocks. Surveys suggest that as many as four million lambs may be lost in the UK every year. MLC estimates put the causes of these losses as: abortions and stillbirths 30–40%, starvation and exposure 20–30%, infectious diseases 15–20%, misadventure 5–10%, congenital defects 5–10%. Organisms of the *Pasteurella*, *Toxoplasma*, *Brucella*, and *Rickettsia* groups may be implicated in so-called 'abortion storms' in sheep. Vaccination can help avert the problem. Similarly, appropriate vaccines (e.g. against

Clostridia) given to the ewe pre-partum, can help reduce post-natal mortality.

The nutrition of the ewe in the six weeks pre-lambing is extremely important. During this period the fetus makes 70% of its growth. Under-nutrition of the ewe at this time will lead to reduced birthweight and an increased susceptibility to starvation (*see Tables 16.11* and *16.12* and *Figure 16.5*) or may even lead to the death of some, or all, fetuses *in utero*. Additionally, correctly-fed ewes are much less likely to suffer from pregnancy toxaemia (twin lamb disease), a metabolic disorder which may kill both ewe and lambs. There is some evidence to suggest that reducing worm burden by dosing in-lamb ewes with a suitable anthelmintic six weeks before lambing may also show considerable benefit in terms of the number of lambs reared, and that, for housed ewes, shearing at housing may be beneficial.

Table 16.11 Main causes of lamb mortality (% of total mortality)

Ante-natal loss	30	Mummified fetus	8
		Fetal death	22
Post-natal loss	70	Starvation	29
		Trauma	13
		Infection	17
		Other	11

Adapted from MAFF EHF Results 1973–76.

Table 16.12 Intensive care routine at lambing

(1) Check lamb breathing: clear mucus if necessary
(2) Draw milk from both teats of ewe
(3) Treat navel with chloromycetin or other suitable material
(4) Individually pen ewe and lambs
(5) Put lamb to teat to initiate suckling
(6) Use stomach tube to administer colostrum if necessary

Once the lamb is born, careful shepherding can do much to ensure the lamb's survival. Critical points are the lamb's ability to withstand the challenges of climate and infection. Particularly at risk are lambs from very young ewes, very small lambs, and lambs that are weak at birth. Heavy lambs, from ewes in good condition, are able to survive for some time without food in warm weather, and have enough fat reserves to meet the challenge of a cold, wet environment if necessary. The small lamb may quickly burn up body reserves in bad conditions and thence be too weak to suckle properly. Hence it does not replenish its energy stores, nor does it obtain sufficient colostrum to confer adequate passive immunity. Careful shepherding is therefore very important. Factors involved in an intensive care routine at lambing are given in *Table 16.12*.

MEAT PRODUCTION FROM SHEEP

Growth and nutrition

Under UK conditions the main output of the sheep farm is in the form of lamb—either fat or store. The money made for each lamb sold on any date is dependent upon its weight, either live or carcass, and its quality. This section discusses the important factors affecting this output.

As can be seen from *Figure 16.6*, the development of the components of the conceptus is very variable, with most placental development having finished by six weeks prior to lambing, whilst most fetal development is yet to take place (*see Table 16.13*). The importance of correct ewe nutrition during this period has already been discussed in relation to the survival of the newborn lamb; it is equally important in terms of the subsequent growth of that lamb, especially in the case of multiple births.

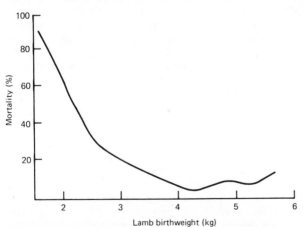

Figure 16.5 Relationship between lamb birthweight and mortality. (From Maund, B. 1974, Drayton EHF results)

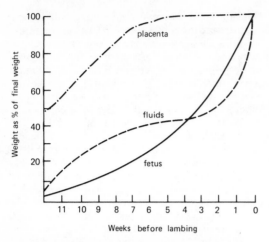

Figure 16.6 Differential development of components of conceptus during last 12 weeks of pregnancy. (From Robinson, J. J. *et al.*, *Journal of Agricultural Science* **88**, 539. Reproduced by courtesy of the editor and publishers)

Table 16.13 Growth of fetus during twin pregnancy

Stage of pregnancy/(d)	Fetus wt (g)
28	1
56	89
84	1000
112	4250
140	10 000

After Wallace, L. R. (1948), *Journal of Agricultural Science* **38**, 93

From a daily requirement of 12–18 megajoules (MJ) in late pregnancy, the ewe's energy demands will increase to 20–30 MJ at peak lactation. Especially in the flock lambing before grass becomes available this energy demand may not be met and the ewe will begin to 'live off her back'. Ideally the ewe will lamb down with a body condition of at least 3–3.5 if she is going to milk satisfactorily. Ewes suckling more than one lamb respond by producing more milk; some 40% more if suckling twins, 50–55% more for triplets. There is, therefore, a case to be made for running ewes suckling twins/triplets separately and feeding them accordingly. Notwithstanding this extra milk, lambs reared as multiples are unlikely to perform as well as those reared as singles. Some average daily milk yields are given in *Table 16.14*. For the lowland ewe the average of 1.6 kg milk produced/d in early lactation supports a daily gain of about 300 g in a single lamb. When producing 2.54 kg milk ewes can support some 460–500 g daily gain, but this is only 230–250 g/lamb. By the third month the relevant figures are 220 and 130 g respectively. The effects of this on lamb liveweight are shown in *Figure 16.7*. *Figure 16.7* also indicates that for the first eight to ten weeks of life growth rate is fairly constant, but drops off subsequently. This is consistent with a situation in

Table 16.14 Average daily milk yields during first 12 weeks of lactation (kg)

Type of ewe	No. of lambs	Yield/d		
		First month	Second month	Third month
Hill	1	1.34	1.29	0.95
	2	2.24	1.85	1.26
Lowland	1	1.60	1.55	1.24
	2	2.54	2.10	1.43

Taken from MAFF Technical Bulletin No. 33 with permission

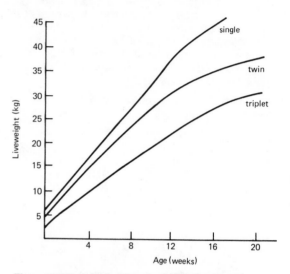

Figure 16.7 Typical growth curves for lambs reared as singles, twins or triplets. (From Spedding, C. 1970 *Sheep Production and grazing management.* London: Bailliere. Reproduced by courtesy of the publishers)

which the lamb obtains most of its nutrient requirement from the ewe during this period, and that milk production after 12 weeks is greatly reduced. There appear to be few advantages in leaving lambs with their dams beyond this time, and indeed, in situations where grass is in short supply, there may be distinct advantages in weaning lambs early. A second factor affecting daily gain beyond the 10–12 week period is the shift, in terms of the components of any gain, away from lean tissue and towards fat. The characteristics of growth of the lamb are very similar to those for other species, namely that initially bone development predominates, followed by lean tissue and finally fat. Animals in which the waves of growth for bone, lean and fat are steep, and close together, are called early-maturing. In the sheep the tendency is for smaller breeds, such as the Southdown and Ryeland, to be early maturing, and to start to lay down fat at relatively low carcass weights, and for heavier breeds to be late maturing and therefore capable of producing, given time, a heavier carcass at any given level of fatness. The current trend is towards a demand for leaner carcasses. The Meat and Livestock Commission have developed a grading scheme for lamb carcasses which can be used to assess, and describe, any carcass in terms of its fatness and overall conformation (*see Figure 16.8*). On that basis current demand is for a carcass falling into fat class 2 or 3L. Many butchers now pay premiums for lambs falling into categories 2E and 3E, and demand penalties for those in fat classes 4 and 5.

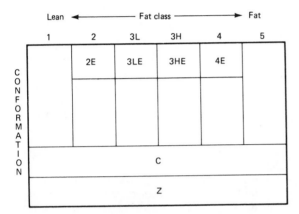

Figure 16.8 MLC system of lamb carcass classification

It is possible to predict with a reasonable degree of accuracy, the ideal carcass weight for any breed or cross of lamb, using the formula

$$\frac{(WD/2 + WS/2)}{2} \times \text{estimated killing-out } \%$$

(where WD = mature weight of dam breed and WS = mature weight of sire breed). Thus, for example, using a Suffolk ram (90 kg) on a Scotch halfbred ewe (90 kg) will produce lambs whose carcasses should weigh 19.5–22.5 kg at a fat score of 2–3L. Using a Southdown ram (60 kg) on similar ewes will produce an 18.5 kg carcass. Generally, however, the heavier lamb will remain on the farm longer and this may be inconvenient. Choice of sire breed is a matter of individual judgement. It is essential that the breed of ram chosen for use on a fat lamb flock suits the lamb fattening system. For example, lambs aimed at the early market need rapid growth and early maturity, and will tend to be sold at lower carcass weights, while lambs to be fattened on roots as late lambs or hoggets (animals sold after 1 January) need to be capable of producing a heavy carcass without laying down much fat.

Systems of lamb production

If lambs are categorised according to time of sale, then early lambs, sold before the end of May, account for only 3% of total output; lambs fat off grass 52%;

lambs fattened on forage crops 22% and hoggets 23%. There are many systems of production used to achieve these ends. Important points of the main ones are as follows:

Early lamb

Aimed particularly at the Easter market, prices peaking between mid March and late April as hoggets' numbers decline. Dorset Horns and their crosses are able to lamb regularly in the period October to December, as are carefully selected strains of other breeds. The system is characterised by high lamb prices and high stocking rates for weaned ewes, but these points tend to be offset by lower lambing percentages and higher concentrate usage. Lambs should be offered creep concentrates with about 170 g/kg DCP from three weeks of age. Weaning can take place any time after six weeks provided the lamb weighs 18 kg or over, with no need for milk substitutes. After weaning, protein levels may be reduced to 140–150 g/kg. On these diets a FCE of 3–3.5:1 should be aimed for, with lambs going off at 34–36 kg liveweight at 12–14 weeks of age. This type of system fits particularly well on arable farms where grass may be limiting, and where 'off-peak' buildings and labour are available. It may also be used in a split-flock situation on larger sheep units.

Lambs off grass

These systems should aim for high lambing percentages (170%+), high stocking rates (15–18 ewes/ha) and as many lambs sold fat off the ewe as possible (75%). Lambs not sold before weaning may then either be finished on silage aftermaths or sold as stores. Two such systems, which have been well documented, are the 'Grasslamb' system developed by Grassland Research Institute (*see* GRI Farmers' Booklet No. 2), and the 'Follow-N' system of the National Agricultural Centre. The main difference between these two systems is that the GRI system uses rotational grazing and forward creep grazes the lambs, while the NAC set stock, do not creep graze and put greater reliance on the use of anthelmintics to reduce worm burden in lambs. They both use high levels of nitrogen (upwards of 250 kg/ha). These figures compare with average stocking rates on traditional systems of only 10 ewes/ha at a nitrogen level of 98 kg/ha.

Fattening lambs and hoggets on forage crops

The animals available for fattening will be wethers of the hill breeds, weighing only 23–30 kg, wethers of the hill crosses (Mule, Masham, etc.) at 25–35 kg, and the remains of fat lamb flocks at 30–40 kg. Feeds available are likely to be grass, catch-crop roots, main crop roots and arable by-products. It is important that the correct type of lamb is purchased for the feeds available, and that stocking rates are carefully controlled. For example, maincrop roots are expensive to grow and the aim must be for high yields and high stocking rates (60+lambs/ha). Catch crops generally yield less well, but should support 35 lambs/ha, with a supplement of 10–12 kg barley/head. Arable by-products, such as sugar beet tops and brussels sprouts, have the advantage of a very low cost, and should carry 35 lambs/ha with a supplement of 17–20 kg barley/head. On a daily ration of 5 kg roots, 0.25 kg barley and 0.25 kg hay (providing approximately 12 MJ, 85 g DCP and 1.0 kg DM) a 35 kg lamb should gain at 160–170 g/d (rather more than 1 kg/week). Hill wethers will go off at up to 35 kg liveweight while Down crosses may be kept on to 50–55 kg. The market at the top end of this range is extremely specialised. Extreme care also needs to be taken if over-fat carcasses are to be avoided.

Fattening on concentrates

As an alternative to forage feeding the intensive feeding of lambs indoors may be considered. In this case a cheap concentrate based on barley and having about 150 g/kg DCP will be used. Lambs should be wormed and vaccinated against pulpy kidney prior to the commencement of concentrate feeding. Concentrates should be offered at grass and animals not housed before they are eating 0.7 kg daily. On housing 0.5 kg (hay head)/d should be offered and concentrate feeding can be increased until it is *ad libitum*. A FCE of 5–5.5:1 should be achieved, with lambs fattening in five to six weeks. Careful supervision and stockmanship are essential and 'poor-doers' must be removed early. Lambs need 0.5 m² bedded floor area in well-ventilated buildings. *Pasteurella* pneumonia may be a problem and routine vaccination should be considered.

An all-concentrate diet may also be considered for lambs from birth in situations such as those associated with frequent lambing systems or for rearing one lamb of twins in the hill environment. On such a system it is essential that the lamb receives adequate colostrum in the first 24 h of life, it can then be introduced to teat-fed milk substitute. A training period of 3–4 d, during which time each lamb is introduced to the teat three to four times/d, may be necessary. Concentrates (12.5 MJ, 17.5% CP) should be introduced from one week of age. Milk substitute should be offered *ad libitum* for three to four weeks, followed by one week of restricted access, before weaning, to encourage concentrate uptake. A feed conversion of 1:1 (DM basis) is possible on the milk part of this diet and of 4:1 on concentrates. Total consumption to slaughter will be some 10 kg milk powder, 100 kg concentrates and a little hay, and overall gains of 300 g/d are possible.

WOOL

For many years wool was the main source of income for the sheep farmer. Today income from wool accounts only for some 10% of gross return per ewe on lowland farms, but up to 20% in many hill situations. Yield of wool varies greatly between breeds, with longwools such as the Teeswater and Devon longwool producing 5–6 kg/head, while hill breeds such as the Swaledale and Herdwick will only yield 2 kg.

Wool is produced by follicles in the skin. Two distinct types of follicle are identifiable. Primary follicles, normally associated with an erector muscle and a sweat gland, produce coarser, medullated fibres of hair or kemp. In improved breeds of sheep many of these coarser fibres are shed soon after birth. Secondary follicles, associated only with a sebaceous gland, produce the finer wool fibres. The ratio of secondary:primary follicles determines the fineness of the fleece and varies between breeds. For example, the fine-woolled Merino may have 25 secondaries per primary, while most British breeds will have no more than eight.

Fineness is the major criterion in wool grading, the standard measure of fineness being the Bradford Count. Bradford Count is the number of hanks of yarn 510 m long that can be spun from 450 g of wool. Values will vary from those of the Merino at 80+ to those of the Blackface at 27–30+. There are also within-fleece differences, with breech wool being coarsest and shoulder wool finest. Breech and neck wools generally contain a higher percentage of kemp. A second quality factor in wool is its crimp; that is the number of corrugations per 25 mm length. Crimp values vary from 8 to 28 and are closely correlated with fineness.

The growth of wool is cyclical and photoperiodic, 80% of growth taking place between July and November. Wool fibres are almost entirely keratin—a protein with a very high cystine content (12–14% compared with 2–3% in microbial protein). Notwithstanding this the main relationship between nutrition and wool growth is in terms of energy input. Wool continues to grow even when the sheep is in negative energy balance. It does however grow more slowly and is finer under such conditions. Whilst wool which is uniformly fine along its length is desirable, the development of a thinner area in an otherwise thicker fibre, a condition known as tenderness, is to be avoided, since it reduces the value of the fleece and may even cause wool to be shed prematurely. Tenderness develops in conditions of underfeeding, for example in the ewe suckling twins or triplets, or during bad attacks of parasitic gastro-enteritis or liver fluke. Along similar lines, work has been done on the use of a drug—cyclophosphamide, which, when given by mouth, produces a short-term check on cell division and makes it possible subsequently to pull the fleece off without the need for conventional shearing.

At shearing, care should be taken to preserve the quality of the fleece. Double cuts, which reduce staple length, should be avoided, organic matter such as straw and faeces should be kept out of the wool, and the fleece should be carefully wrapped and stored in the dry. Considerable penalties are incurred for badly presented fleeces. Details of grades, prices and penalties are published annually by the Wool Marketing Board as their Wool Price Schedule.

No advantage accrues from shearing sheep more frequently than once a year, in terms of wool yield. Recently, however, the shearing of ewes immediately prior to winter housing has been examined. This technique appears not to affect adversely total wool yield and produces advantages in the form of higher feed intakes, heavier birthweights and cleaner fleeces.

FLOCK REPLACEMENTS

Flock replacement costs are a major element in the profitability of a sheep enterprise. In 1979 the costs averaged £6.1/ewe in MLC recorded flocks.

Keeping these costs down is a question of minimising the number of ewes to be replaced each year and of keeping down the cost of each replacement.

On most hill farms some ewes will be lost and some will need to be culled. The majority of ewes leaving the farm will be drafted out at four to five years of age simply because they are no longer able to cope with conditions. Thus on average some 25% of the flock will need to be replaced each year. Set against this is the fact that many of these ewes will be sold as draft breeding stock rather than culls, thus making a substantial contribution to the output of the flock (17% for MLC recorded flocks in 1979/80). Traditionally, and of necessity, replacement ewes will be retained from within the lamb crop. This is essential if the ewe lamb is to become a productive member of the flock. Within a hill flock, groups of sheep become territorially organised or 'hefted'. This tendency is encouraged by careful shepherding, for it reduces the need for fences and assists the survival of the sheep since they know of, and will seek out, areas of shelter, and sources of food during bad weather. From a husbandry viewpoint one of the major problems created by the need to keep ewe lambs as replacements is what to do with them over their first winter. If they stay on the hill then they add to feed requirements at a difficult time, and some indeed may not survive the winter. On the other hand, the traditional 'tacking' or 'agistment' of ewe lambs onto lowland farms has been made more difficult by a reduction in the number of farmers willing to cooperate and by escalating costs, both of agistment and of transport. One compromise which may become more common is the provision of housing for these animals on the home farms.

Under lowland conditions most ewes being sold off will be culls. Criteria taken into account when deciding on which animals to cull may include condition of udder, feet and mouth, previous history, condition and temperament. Many of these factors are influenced, to a greater or lesser extent, by management. It could be argued, for example, that culling because of bad udder or bad feet should only occur as an extreme measure, and that if many sheep are involved it may be that flock management is suspect. Culling on teeth condition will depend on two main management factors: stocking rate and winter feed policy. The need for a ewe to be 'sound' in mouth relates to her ability to feed. As stocking rates increase so competition for grass increases and the ewe needs to be able to graze closer to the ground, a facility made more difficult if the ewe is 'broken-mouthed'. Similarly, if winter feeding involves the use of 'hard' roots, such as swedes or turnips, the broken-mouthed ewe will be unable to compete and will tend to lose condition. Additionally, the feeding of hard roots tends to increase teeth loss. By four and a half years of age up to 75% of ewes may still be full-mouthed if fed on hay and concentrates; in root-fed flocks the figure may be as few as 35%.

In addition to the above factors, opportunism plays a part in determining culling rate. In years when replacements are expensive culling levels tend to be lower than in years when they are relatively cheap. The cost of each replacement can also be influenced by the age of the animal at purchase; ewe lambs being appreciably cheaper than shearlings. Having opted for the cheaper ewe lambs, the farmer must next decide whether to put them to the ram in their first year or not. Puberty in sheep is influenced by age, bodyweight and daylength, older heavier animals being more likely to attain puberty than younger or lighter ones. The advantages of breeding from ewe lambs are that it increases lifetime lamb production per ewe and decreases replacement cost, but against this must be set the likelihood of more mis-mothering problems and difficulties in getting their lambs away fat. In the end the decision may well depend on the source of the lambs. For homebred animals, whose management has been aimed at producing an animal suitable for tupping at seven to nine months, the system can work quite well. For the producer who is using longwool crosses, and who of necessity has to purchase all his replacements, it may be much less viable.

FEEDING THE EWE
Theory

As has been emphasised, the nutrition of the ewe is a major factor in determining the performance of the flock. Feed requirements, and the means of meeting those requirements vary with ewe liveweight, environment and the production system used. Details of these requirements and suggested systems to meet them, are discussed in the MLC publication *Feeding the Ewe* and MAFF Technical Bulletin 33. What follows is a brief outline of the salient points. (*See also Tables 16.15* and *16.16.*)

Table 16.15 ME allowances (MJ/d) for pregnant ewes outdoors

Liveweight (kg)	No. of fetuses	Weeks before lambing		
		8	4	0
30	1	5.1	6.3	7.7
	2	5.1	6.8	9.2
40	1	6.1	7.4	9.1
	2	6.1	8.2	11.0
50	1	7.0	8.6	10.5
	2	7.1	9.6	12.8
60	1	8.0	9.8	11.9
	2	8.1	10.9	14.7
70	1	8.9	10.9	13.4
	2	9.2	12.3	16.5
80	1	9.9	12.1	14.8
	2	10.2	13.7	18.3

Taken from MAFF Technical Bulletin No. 33 with permission

Table 16.16 ME allowances (MJ/d) for lactating ewes

Live-weight (kg)	No. of lambs	Stage of lactation					
		Hill			Lowland		
		Month 1	Month 2	Month 3	Month 1	Month 2	Month 3
40	1	16.3	14.9	13.2			
	2	23.3	20.2	15.6			
50	1	17.3	16.9	14.2	19.3	18.9	15.7
	2	24.3	21.2	16.6	26.6	23.2	18.0
60	1	18.3	17.9	15.2	20.3	19.9	16.7
	2	25.3	22.2	17.6	27.6	24.2	19.0
70	1				21.3	20.9	17.7
	2				28.6	25.2	20.0

Taken from MAFF Technical Bulletin No. 33 with permission

Voluntary feed intake

Dry matter intake will vary with stage of pregnancy/lactation and with diet, as well as with bodyweight. Intakes will normally be within the range 60–70 g/kgW$^{0.75}$ (where W$^{0.75}$ = metabolic bodysize), but may be as high as 100 g/kgW$^{0.75}$ for finely ground diets. On silage based diets, however, intakes as low as 45 g/kgW$^{0.75}$ have been recorded. Appetite is depressed during late pregnancy, but recovers quickly after lambing to reach a peak some two to three weeks after peak milk yield is achieved. The increase in intake between late pregnancy and peak may be as high as 60%.

Energy requirements

Energy requirements vary according to bodyweight, stage of pregnancy and milk yield, as well as being influenced by housing. Maintenance requirements, based on Mm = 1.8 + 0.1 W, will normally fall within the range 5–10 MJ/d. During pregnancy, requirements will increase significantly, being some 20% higher six weeks before lambing, and up to 150% more immediately before lambing. This increase is dependent on the number, and weight, of fetus(es) carried and is of the order of 1.5 MJ/kg fetus/d.

During lactation energy requirement varies with milk yield. Average milk yields for the first 12 weeks of lactation are given in *Table 16.14*. Energy requirement/kg milk produced is 7.8 MJ.

Protein requirements

Protein requirements also vary according to the physiological status of the ewe. Maintenance levels are generally between 50 g and 70 g/d DCP. During tupping and the first month of pregnancy these requirements increase by about 50%, but during mid pregnancy maintenance levels are again adequate. During the last eight weeks of pregnancy requirements increase to between 75 g and 150 g DCP depending on ewe weight and number of lambs carried. For the lactating ewe it is normally suggested that 16% CP in the diet is desirable although levels as low as 12% appear not to affect milk yield. Recently, however, it has been suggested that for ewes in negative energy balance (i.e. losing weight) protein level may be more critical. On a per-ewe basis requirements will be between 150 g and 300 g/d DCP.

Feeding in practice

In practice it may be neither practical, nor economically desirable, to meet the ewes' requirements in full at all times. Most flocks contain ewes with a range of bodyweights, or varying ages and carrying or suckling differing numbers of lambs, and requirements therefore vary. During the pre-lambing period, when most ewes will be at pasture, the aim may well to be to allow a moderate degree of under nourishment. A loss of 5% bodyweight is acceptable. This should not produce any unacceptable reductions in performance. During late pregnancy it is essential that the ewe be brought up in body condition. Failure to do this may lead to reduced birthweights, higher perinatal mortality and lower milk yields. Under lowland conditions the best way of achieving this aim is to feed all ewes as if they were carrying twins. Using body condition scoring as a guide, concentrates should be offered some six to eight weeks pre-lambing at 100 g/head per day, and this amount should be gradually increased to 500–750 g/d at lambing. Hay should be available more or less *ad libitum*. In many hill situations such feeding is economically more difficult to justify and practically often very difficult, not only because of the difficulty of reaching the sheep, but also because it is undesirable to interrupt the normal grazing behaviour of the flock.

Under such conditions the judicious use of feed blocks may be considered. Such blocks should be sited in strategic points, allowing one block/30 ewes per week.

After lambing the amount of hand feeding will depend largely on date of lambing. Where grazing is not immediately available the continued use of concentrates may be necessary. Otherwise root crops such as kale or swedes may be utilised. Some examples of rations suitable for in-lamb ewes are given in *Table 16.17*. Using such feeding scales it should be possible to restrict concentrate usage to about 45 kg/ewe per year.

Table 16.17 Suggested rations for a 75 kg lowland ewe in late pregnancy

	Amount fed (kg/d)					
	Silage	Hay	Swedes	Kale	Concen-trates[1]	Barley
Ration 1		0.6	4.0		0.6	
Ration 2		0.7		4.0		0.5
Ration 3	4.0				0.6	
Ration 4 (sheep indoors)		1.0			0.6	

[1] 13 MJ/kg DM, 130 g/kg DCP, DM

GRAZING MANAGEMENT

Good grazing management is one of the most important factors influencing profitability. On the one hand variable costs will be affected by the extent to which the flock needs supplementary feeding, and by the forage variable costs, while on the other gross margin/ha is very dependent on stocking rate. Since stocking rate and the physical output of a flock tend to be inversely related the aim should be to increase stocking rate as far as is possible without affecting output (*see Figure 16.9*). A comparison of MLC 'Average figures with those for the top one-third of producers suggests that 65% of the difference between them, in gross margin/ha terms, can be accounted for by difference in stocking rate, with only 12% being due to differences in number of lambs produced. In 1979 average stocking rate was 11.1 ewes/ha overall, with the top one-third of producers averaging 13.5 ewes/ha.

One of the problems associated with the allocation of forage costs, and with the whole question of overall stocking rate, is that of grassland recording. The simplest way of standardising these figures is to use the Livestock Units system (*see* Chapter 6, *Table 6.10*).

The use of ewes/ha as a measure of stocking rate can be misleading, since it takes no account of ewe size. In the MLC publication *Prime lamb from grass* an attempt has been made to overcome this problem by interrelating ewe bodyweight, prolificacy and fertiliser usage rather than ewe numbers. Represent-ative values, for farms on average-quality land, are given in *Table 16.18*, but individual farm circumstan-ces may markedly affect these values. Although there is a general trend towards higher stocking rates at higher levels of nitrogen application, many farmers fail to exploit fully the opportunity to increase stocking rates which increased nitrogen application offers.

The times of the year when problems of high stocking rates are likely to be most apparent are in early spring and in the latter half of the grazing

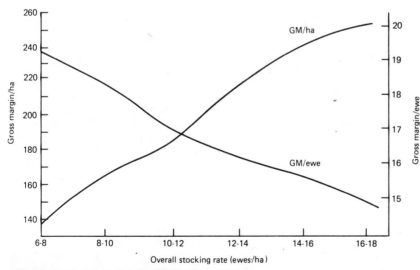

Figure 16.9 Relationship between stocking rate and gross output/ewe and /ha (1979 data). (From *MLC Commercial Sheep Production Yearbook, 1980*. Reproduced by courtesy of the publishers)

Table 16.18 Relationships between ewe weight, nitrogen usage and ewe breeds on average quality land

Target	Nitrogen use (kg/ha)		
	0–75	*75–150*	*150–225*
Wt of ewe carried (kg/ha)	750	900	1050
Wt of lamb produced (kg/ha)	600	750	850
Stocking rates to achieve above (ewes/ha)			
Welsh halfbred	13	16	18
Mule	11	13.5	15.5
Scotch halfbred	10	12.5	14.5

After *MLC Prime Lamb from Grass*. Reproduced with permission.

season. In early spring the provision of fresh grass soon after lambing is important if a good start to milk yield and lamb growth are to be obtained. It is easier to obtain this early grass if off-wintering facilities are available.

Where the flock is wintered at grass the effect on subsequent yield will be dependent on the weather and on the use to which the grass is subsequently put. In a wet year poaching may reduce early yields by up to 40%, but much of this reduction will have been made up by 1 June. Wherever possible, therefore, ewes should be overwintered on fields destined for hay.

In-wintering of ewes is expensive and cannot normally be justified on the basis of increased ewe outputs. Whatever the advantages of housing it can only be justified economically if it is used as a means of increasing grassland utilisation and stocking rates.

In the latter half of the grazing season the problem is competition between ewes and lambs, which leads to reduced growth rates and higher levels of parasitic gastro-enteritis in the lambs. This situation is most difficult on farms where sheep are the only grazing enterprise, since on such farms there is less scope for using hay or silage aftermaths or for alternating grazing between species.

THE FUTURE

From the data already presented it is obvious that there is room for improvement in the performance of most flocks. These improvements may come either from improving management within traditional systems or through the introduction of more sophisticated techniques.

Under hill and marginal land conditions improvements are likely to be carried out within the traditional framework. Areas to be considered include the winter nutrition of pregnant ewes, closer attention at lambing and improved stocking rates. On many hill farms winter-carrying capacity is low, and limits flock size, even where summer grazing is plentiful. The use of fencing, the application of fertilisers and lime and the improvement of swards can all produce benefits. The

major problem is that such improvements are expensive and may be difficult to finance.

Under lowland conditions improvements within existing systems are possible in both lambing percentage and stocking rate. The data in *Table 16.19* suggest that in terms of gross margin/ha there is more potential in increasing stocking rate. It must be borne in mind, however, that these two factors are interrelated, and that an increase in one may be to the detriment of the other.

Table 16.19 Effects of lambing percentage and stocking rate on gross margins (1979 data)

	Average lambing %	
	High	*Low*
High stocking rates		
Lambing %	145	122
Stocking rate (ewes/ha)	12.6	13.0
Conc. use/ewe (kg)	62	57
Lamb sales/ewe (£)	35.2	30.3
Gross margin/ewe (£)	17.0	14.9
Gross margin/ha (£)	214.2	193.7
Low stocking rates		
Lambing %	1.52	1.27
Stocking rate (ewes/ha)	9.6	9.1
Conc. use/ewe (kg)	53	52
Lamb sales/ewe (£)	38.5	32.5
Gross margin/ewe (£)	21.2	17.6
Gross margin/ha (£)	203.5	160.2

Source *MLC Commercial Sheep Production Yearbook (1980)*. Reproduced with permission.

More recently considerable interest has been shown in alternative systems which increase lamb output. Two broad approaches are possible; the use of highly prolific breeds or crosses lambing conventionally, or the development of systems which involve increased frequency of lambing.

Increased prolificacy

The development of a flock of highly prolific ewes will normally involve the use of animals with some Finnish Landrace blood (e.g. the Finn–Dorset) or the newly-developed Cambridge ewe. Where such animals are used as breeding females then lambing percentages in excess of 250% can be expected, with mature ewes having prolificacy of 2.25–3.0 lambs per ewe. Where suitable prolific breeds, for example the ABRO Dam line, are used on hill ewes, instead of the more traditional longwool ram, then mature prolificacies of 2.0–2.25 may be expected but flock avarages are unlikely to be above 200%.

The management of the prolific flock is fairly standard, but involves close attention to detail and higher levels of stockmanship. The main additional problem is the need to restrict suckling to two lambs per ewe and to artificially rear surplus ram lambs if necessary. Chief features of the management of such flock are as follows:

(1) careful organisation of mating, to facilitate organised lambing,
(2) generous feeding in late pregnancy,
(3) closely supervised lambing, normally indoors in March,
(4) restriction of suckling to two lambs/ewe,
(5) weaning by mid July to allow ewes to regain condition prior to October tupping.

Increased frequency of lambing

Theoretically it is possible for ewes to lamb down every six months, and some ewes may do this. However, physically ewes may not recover from one lambing in time to be mated again so quickly, and practically the problems of operating such a system are many. In particular low conception rates following the mating of lactating ewes necessitate very early weaning, and any failure of ewes to conceive to first or second service greatly increases barren percentage or spreads subsequent lambing to an unacceptable degree. More commonly ewes lamb down every eight months, and often two flocks will be run, four months out of synchronisation, so that ewes failing to conceive in one flock may be 'slipped' into the other and so be given a second chance (*see Figure 16.10*).

In a frequent breeding system a variety of breeds may be used, but although exogenous hormones or daylength manipulation will generally be used, highly prolific ewes having a very long breeding season are to be preferred. In this respect ewes of the Finn–Dorset type are probably ideal.

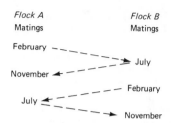

Figure 16.10 Organisation of split flock to allow 'slipping' of ewes failing to conceive

Although daylength may be manipulated to increase breeding frequency (*see Figure 16.11*) this method is not commercially viable, and most units will use exogenous hormones for out-of-season breeding. Generally an intravaginal sponge, impregnated with progesterone, will be used, followed by an injection of 750 units of pregnant mares' serum (**PMS**) at sponge withdrawal. A detailed calendar describing such a system is given in *Appendix 16.2*.

Critical to such a system are the management of the rams, the close supervision of lambing, the abrupt weaning of lambs at one month of age and the careful nutrition of the ewe. Inevitably these, and in particular the high concentrate usage, increase variable costs. It has been suggested that an annual production rate in excess of 2.5 lambs/ewe must be achieved if such a system is to be economically viable. Up to now few units have consistently achieved this target.

A detailed discussion of this aspect of sheep production may be found in *A study in high lamb output production systems* (Scottish Agricultural Colleges, Technical Note No. 16).

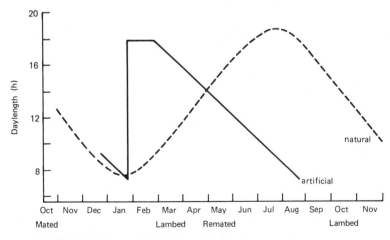

Figure 16.11 Manipulation of day length to control oestrus. (From Robinson, J. J. *et al.* 1975 *Annals Biol. Anim. Biochem. Biophys.* 15, 345. Reproduced by courtesy of the editor and publisher)

APPENDIX 16.1 MLC CONDITION SCORING CLASSIFICATIONS

0 Extremely emaciated and on the point of death. Not possible to detect any tissue between skin and bones.

1 Spinous processes prominent and sharp. Fingers pass easily under ends of transverse processes. Possible to feel between each process. Loin muscles shallow with no fat cover.

2 Spinous processes prominent but smooth; individual processes felt as corrugations; transverse processes smooth and rounder, possible to pass fingers under the ends with little pressure. Loin muscle of moderate depth but little fat cover.

3 Spinous processes have little elevation, are smooth and rounded, and individual bones can only be felt with pressure. Firm pressure required to feel under transverse processes. Loin muscles full with moderate fat cover.

4 Spinous processes just detected with pressure, as a hard line; ends of transverse processes cannot be felt. Loin muscles are full with thick covering of fat.

5 Spinous processes cannot be detected, even with firm pressure; depression between layers of fat where processes would normally be felt. Loin muscles are full with very thick fat cover.

APPENDIX 16.2. CALENDAR OF EVENTS FOR FREQUENT LAMBING FLOCK

July
14   Insert sponges.
27   Withdraw sponges. Inject 750 units PMS.
29   Mating. Allow adequate rams (1:10) and rotate between paddocks every 8 h or use AI.

August
15   Mate repeats.
20   Remove rams.

December
23   Begin lambing.

January
 9   'Repeats' lamb.
27   Wean first lambs. Insert sponges into all ewes.

February
10   Withdraw sponges. Inject 750 units PMS. Wean 'repeats'.
12   Mate.
28   Mate 'repeats'.

March
 3   Remove rams.

July
 9   First ewes lamb.
25   'Repeats' lamb.

September
20   Wean all lambs.

November
 5   Introduce rams (normal mating, no hormonal treatment).

April
 1   First ewes lamb.

June
14   Wean all lambs.

July
14   Insert sponges.
     Etc.

17

Pigs

P. H. Brooks

THE CONTRIBUTION OF THE PIG TO THE WORLD MEAT PROVISION

Introduction—the global view

The last 30 years have seen a rapid increase in the world production of meat, as livestock producers strive to fulfil the demands of a burgeoning human population. This increase has been particularly marked in the pigmeat sector where slaughterings in 1979 were some 98% greater than the average for the period 1960–1967 (*Table 17.1*).

Table 17.1 World production of meat[1]

| | 10³ head | | % change |
	1966–70	1979	1979 / 1966–70
Beef and veal	35 904	46 469	+29.4
Mutton and lamb	6149	7442	+21.0
Pigmeat	25 824	51 214	+98.3

[1] Derived from FAO statistics.

In less developed countries the growth in pig numbers has resulted from an attempt to maintain or improve the nutritional standards of the human population. In developed countries expansion has been fueled by the twin demands of an increasing population and a greater per capita consumption of meat products.

Production has not increased uniformly around the world (*Table 17.2*). In the Americas and Oceania where red meat is still in abundant supply the percentage increase has been small. In the African continent the large percentage increase is deceptive as the pig population is comparatively small. The potential of the pig in helping to meet the protein needs of the human population cannot be exploited

Table 17.2 World pig numbers[1] (10³ head)

	1960/69	1978	% of total (1978)	% increase 1978 / 1960–69
World	570 227	735 021	—	+28.9
North America	76 687	78 831	10.7	+ 2.8
South America	70 345	68 000	9.3	− 3.3
West Europe	68 319	94 803	12.9	+34.8
(EEC)	53 094	72 083	9.9	+35.8
East Europe	47 874	68 416	9.3	+42.9
USSR	55 993	70 300	9.6	+25.6
Africa	5692	8583	1.2	+50.8
Asia	240 759	343 308	46.7	+42.6
Oceania	2558	2780	0.4	+ 8.7

[1] Derived from USDA and FAO statistics.

because of religious strictures against the eating of pork.

Asia has almost half the pigs in the world, but although the increase in numbers here has been dramatic the effect has been largely negated by the expansion of the human population.

In Europe and in particulary in the EEC the situation is somewhat different. In addition to an increased population there has also been a notable increase in per capita consumption of pigmeat (*Table 17.3*). The reasons for this are complex but two of the most important factors are undoubtedly the relative competitiveness of pigmeat when compared with other meats and its suitability for manufacture into a vast range of 'convenience' food products.

Pigmeat production in the EEC

In 1979 the EEC produced 9.75 million tonnes of pigmeat, over 50% more than the average production in the 1960–69 period. This level of production meant

Table 17.3 Meat consumption in the EEC[1] (kg/head per annum)

	1960–69	*1978*	*% change*
Beef and veal	23.3	25.7	+10.3
Mutton, lamb and goatmeat	3.4	3.0	−11.8
Pigmeat	26.4	35.7	+35.2
Poultrymeat	7.7	13.4	+74.0
Offal	4.8	6.0	+25.0
Total	67.8	87.0	
Pigmeat as % total meat consumed	38.9	41.0	

[1] Derived from MLC statistics.

that the community was self sufficient for pigmeat. Over the same period there was an increase in the per capita consumption of all meats, with the exception of sheepmeat (*Table 17.4*). Although the biggest percentage increase was for poultrymeat the actual per capita consumption of poultrymeat increased by only 5.7 kg/head per annum compared with an increase of 9.3 kg/head per annum in the case of pigmeat.

Table 17.4 EEC production of meat[1] (10³ tonnes)

	1960–69	*1979*	*% change 1979 / 1960–69*	*% self sufficiency*
Beef and veal	5146	6680	+29.8	98.9
Mutton and lamb	475	510	+ 7.4	64.8
Pigmeat	6340	9750	+53.8	101.2
Poultrymeat	1841	3419	+185.7	105.4

[1] Derived from MLC statistics.

Although the EEC is self sufficient in pigmeat production approximately 16% of pigmeat consumed in the community was produced by 'third' countries while a similar quantity was exported. In addition to this trade there is considerable inter-community trade, as member states differ greatly in both their per capita consumption and their degree of self sufficiency (*Table 17.5*). The UK has the dubious distinction of having

the second lowest per capita consumption of pigmeat while also being the least self sufficient. The reasons for this are complex having their roots in historical trade agreements and more recently reflecting the inequality of opportunity afforded the UK pig industry following entry to the EEC.

Member states differ considerably in the size of their national herds (*Table 17.5*). West Germany has the largest national sow herd but due to the prodigious level of pigmeat consumption in the country is still a net importer of pigmeat. Denmark, the Netherlands and Belgium all have sow herds in excess of one million and export pigmeat as bacon to the UK, manufacturing meat to West Germany and canned products throughout the EEC and to countries outside the community.

The average pig carcass weight varies considerably too. The highest being in Italy (99 kg) where large manufacturing carcasses are required for the production of Parma ham and the lowest being in the UK (64 kg) where lightweight pork pigs are still favoured by the fresh meat trade. The UK, the Irish Republic and Denmark still cure a large percentage of whole carcasses for bacon. This process generally demands the production of a lighter weight carcass than is favoured for manufacturing purposes in the remainder of Europe. However, as new processing techniques develop and the demand for bacon declines it seems likely that average carcass weights will tend to converge throughout the community.

STRUCTURAL CHARACTERISTICS OF THE UK PIG INDUSTRY

The UK pig industry does not exist in isolation from other agricultural enterprises. It is true that some pig production units may be of an intensive nature and have limited land attached to them. Nevertheless they still rely upon land as a source of feed, particularly cereals, even though this need may be met through a third party, such as a feed compounder. Pig units almost always need land as an outlet for the vast quantities of dung and urine they produce. This does

Table 17.5 Characteristics of pig production in the nine EEC countries (1977)

	Sows for breeding (10³ head)	*As % of EEC population*	*Pigmeat production (10³ tonnes)*	*Consumption (kg/person per annum)*	*Self sufficiency*
Belgium	628	7.5	587	36.9	175.3
Luxembourg	14.6	0.2			
Denmark	1053	12.6	741	41.7	353.8
West Germany	2471	29.5	2928	52.6	87.9
France	1260	15.1	1689	35.6	84.6
Irish Republic	104.8	1.3	130	27.9	150.4
Italy	893	10.7	904	20.1	75.6
Netherlands	1020	12.2	967	35.3	221.5
UK	912	10.9	905	24.8	64.8

Source: MLC (1979) EEC Statistics Vols 1 & 2. Reproduced with permission.

not imply the need to own the land area but it does necessitate access to it. For this reason pig units are generally situated on and form an integrated part of mixed farming enterprises often utilising farm grown cereals and straw for bedding and returning manure. Pig units are of many different types, they may be intensive, i.e. with the pigs permanently housed, or extensive with sows living and breeding outdoors either on land expressly reserved for them or as part of a rotation. Units may both produce and grow piglets to slaughter or may concentrate on one of the two functions. Finally, units may be involved in seedstock production, its multiplication, or in the production of slaughter pigs only.

The changing nature of the UK pig production industry

During the 1970s far reaching changes occurred in the UK pig industry. Between 1970 and 1977 the number of holdings with pigs in the UK fell by more than a half (from 84 400 to 40 800). This fall in numbers was compensated by an increase in the size of remaining units indicating an increasing specialisation by the remaining producers. Of the 40 800 units

77.6% had breeding sows. There has been a similar fall in the number of holdings with breeding stock (*Figure 17.1*) again with a corresponding increase in unit size. In 1970 only 24% of sows were to be found in herds of 100 plus but by 1979 this percentage had increased to 54.4%. A continuation of this trend will lead to the domination of the industry by around 3000 units of 200 plus sows in the next few years.

During the same period there were also geographical changes in pig distribution. Pig numbers declined in all regions except the eastern and northern regions where the populations increased by 18.0 and 20.1% respectively. These two regions now account for 50% of the total UK pig population. Results obtained from the Meat and Livestock Commission's Feed Recording Scheme reflect the changes and trends in pig performance which have occurred over the last decade. Sow productivity has increased, mainly due to the increasing number of units weaning piglets at younger ages and to a lesser extent as a result of a small reduction in piglet mortality, probably as a result of the increased use of specialist buildings for farrowing (*Table 17.6*).

In the feeding herd there has been a tendency for average slaughter weight to be reduced and a substantial improvement in food conversion efficiency (*Table 17.7*). The latter trend is due to the

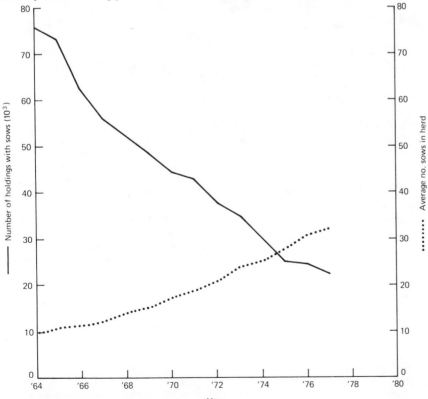

Figure 17.1 Holdings with sows and average herd size (England and Wales) (MAFF statistics)

Table 17.6 Trends in performance and feed costs in the breeding herd

	Breeding herd									
	1970	1971	1972	1973	1974	1975	1976	1977	1978	1979
No. sows and gilts in herd	63	70	84	93	102	107	112	136	147	157
Sow performance										
No. litters/sow per annum	1.8	1.8	1.8	1.9	1.9	1.9	2.0	2.0	2.1	2.2
Live pigs born/litter	10.2	10.3	10.3	10.3	10.2	10.3	10.4	10.3	10.3	10.3
Mortality of pigs born alive (%)	15.2	15.4	14.4	14.6	14.7	14.1	14.1	14.5	13.7	13.6
Pigs reared/litter	8.7	8.7	8.8	8.7	8.8	8.8	8.9	8.9	8.9	8.9
Pigs reared/sow and gilt/year	15.5	15.9	16.2	16.4	16.3	16.5	17.5	18.1	18.7	19.1
Weight of pigs reared (kg)	25.2	25.5	26.2	24.9	22.4	21.3	20.7	19.6	19.8	19.0
Feed usage										
Quantity of feed/sow and gilt/year (tonne)	1.59	1.58	1.63	1.62	1.61	1.62	1.63	1.57	1.64	1.65
Quantity of feed/pig (kg)	107	104	105	98	102	98	94	94	92	86
Kg feed/kg weaner produced	5.1	4.9	4.9	5.0	5.2	5.1	4.7	4.8	4.6	4.5
Feed costs										
Ration cost/tonne (£)	36.0	42.2	39.6	56.0	76.0	87.8	90.6	117.7	112.2	124.3
Feed cost/pig (£)	3.8	4.3	4.1	5.2	7.1	7.8	8.5	11.1	10.5	11.4
Feed cost/kg pig produced (p)	17	20	19	29	36	39	41	56	53	76

Source: MLC (1980) *Commercial Pig Production Yearbook 1979*. Reproduced with permission.

Table 17.7 Trends in performance and feed costs in the feeding herd

	Feeding herd									
	1970	1971	1972	1973	1974	1975	1976	1977	1978	1979
Average herd size (no. pigs)	526	590	620	633	656	672	688	788	862	887
Average weight of pigs sold (kg)	81	80	79	78	78	79	79	80	79	75
Average weight of pigs at start (kg)	22	21	21	21	19	20	20	18	17	16
FCE (kg feed/kg gain)	3.7	3.5	3.5	3.5	3.4	3.4	3.4	3.2	3.1	3.0
Feed costs										
Ration costs/tonne (£)	32.2	37.4	36.4	43.8	73.8	78.2	85.0	112.8	105.8	119.7
Feed cost/kg gain (p)	11.7	13.0	12.8	15.4	24.7	26.5	28.9	36.3	32.9	37.0

Source: MLC (1980) *Commercial Pig Production Yearbook 1979*. Reproduced with permission.

increasing use of stock of superior genotype, improvements in nutrition and housing and most recently to increasing percentage of entire male animals being reared for slaughter.

It appears likely that the majority of the trends observed in the 1970s will continue into the 1980s as most are a result of management changes inspired by the need to *increase* productivity in order to *maintain* profitability.

Breeds and breeding

Breeds

In recent years the UK pig industry has become dominated by two white breeds. The prick-eared Large White, which is a native breed, and the lop-eared British Landrace, which was originally developed from Scandinavian stock. Over 75% of slaughter pigs result from some cross between these two breeds and over 90% of slaughter pigs have one or other breed as a parent.

The reasons for the popularity of these breeds is not difficult to find. They have been selected successfully for fast growth, efficient food conversion, and for the production of carcasses with low levels of subcutaneous fat. The carcasses of both breeds are white skinned. This is an important feature in meat marketing as consumers react adversely to coloured skin on their pork or bacon. This consumer resistance is reflected in the price paid by meat traders for pigs with pigmented skin. An additional advantage of the Large White is that it produces pigmeat of a quality unsurpassed by any other pure breed.

Only two other breeds, the Welsh and the British Saddleback, are of numerical significance. The Welsh, is in many respects similar to the Landrace and can be regarded as an alternative to it for crossing with the Large White.

The British Saddleback owes its continuing survival as a commercial breed to its popularity with outdoor pig producers who place a high value on its docility, good mothering abilities and hardiness.

Breeds such as the Tamworth, Middle White, Large Black and Gloucester Old Spot, though visually attractive, are outclassed in terms of their performance characteristics and must now be regarded largely as 'fanciers' breeds. A number of imported breeds, in particular the Hampshire and Duroc breeds from North America and the Pietrain from Belgium, have gained some support although none has yet made a significant impact on the National Herd. With renewed interest now being shown in the development of specialist sire and dam lines, the importance of some of the more recent imports may increase. However, this pattern of breeding is as yet unproven in pigs and it will be some years before the approach can be critically appraised.

Crossbreeding

Crossbreeding is an essential part of modern pig production. It is estimated that over 80% of sows on commercial units are crossbreds and that the proportion of crossbred slaughter pigs is even greater. Crossbreeding is practised primarily as a means of improving the reproductive characteristics of the pig. When two breeds are crossed the litter productivity (numbers born and weaned and piglet birth and weaning weight) exceeds that of the better parent by 5–15%. This is due to these low heritability traits exhibiting heterosis (hybrid vigour). The greatest advantage is gained when first cross sows are used to produce the slaughter generation. If second cross sows are used much of the hybrid vigour is lost again. For this reason the ideal commercial production system probably consists of using a crossbred sow (in the UK this would usually be a cross between a Large White and a Landrace) and back crossing it with a male of one of the parent breeds, e.g.

Grandparent generation	Breed A ♂ × Breed B♀ (or B ♂ × A ♀) ⟶	AB♂ also♀ rejected for breeding slaughtered
Parent generation	A or B ♂ × A B ♀ (F1 cross)	
Slaughter generation	Male and female progency slaughtered for meat production	

Alternatively a male with a totally different genetic make-up could be used to sire the slaughter generation. Such a male could be a purebred animal of a third breed or a 'synthetic' line, that is to say a male which was himself a crossbred or a genetically stable mixture of several breeds, e.g.

Parent generation	C or (C × D) or 'synthetic' ♂ × AB♀	
Slaughter generation	Male and female progeny slaughtered for meat production	

Some commercial producers breed their own crossbred ('parent generation') female stock by buying in purebred ('grandparent') females of one breed and crossing them with boars of a second breed. This has the advantage of limiting stock importations to a herd but the disadvantage that genetic progress may be slower. Also if relatively few purebred females are maintained it is often difficult to produce just the right number of gilts at the appropriate times.

Other producers prefer to buy in their 'parent' gilts. Crossbred or 'hybrid' gilts are readily available from both independent breeders and from breeding companies and probably around 50% of sows derive from these sources. The term 'hybrid' has no particular legal definition and cannot be taken to imply that the animals in question have any particular characteristics, health status or genetic merit. Producers have to ascertain precisely what any potential supplier of 'hybrid' gilts implies by the term.

The practice of buying in all female replacements simplifies management considerably as the producer has only one type of pig to manage and can procure the number of replacement gilts required when they are wanted. A further advantage is that it is easier for the producer to keep up with genetic progress if all his gilts are bought in from a supplier who is in the vanguard of genetic improvement. The main disadvantage of such an approach is that the importation of such a large number of females each year does represent a considerable health risk if stringent precautions are not undertaken. (It should be remembered that many units replace 35–45% of their sows each year.) Nevertheless few problems should arise if the producer follows three basic rules, namely

(1) always purchase replacement stock from the same source,
(2) quarantine imported stock for three weeks (during which they should be inspected by a veterinary surgeon), and
(3) allow a three-week contact period with some old cull sows before the gilts are introduced to the main herd in order to give them an opportunity to develop some resistance to diseases already present in the herd.

Some producers have established herds with a very high health status usually by producing or buying hysterectomy derived stock. The health status of such herds is jealously guarded and elaborate precautions taken to ensure that disease is not introduced either by stock or human visitors from other units. Many 'nucleus' units producing purebred 'grandparent' stock have such high health status. In units of this type the only means of introducing new genetic material is by artificial insemination or by deriving further 'disease free' piglets by hysterectomy. The high costs of establishing and maintaining such units are often prohibitive for normal producers marketing meat animals but are more easily met by breeders who can sell at a premium animals with both a high health status and high genetic merit. Indeed high health status is important if a breeder wishes to be able to trade with a wide spectrum of customers and is often essential if the breeder wishes to export bloodstock to other countries.

Genetic improvement

The rate of genetic improvement in the National Herd has been and continues to be impressive. Comparisons which have been made between pigs from MLC Nucleus herds and Control herds (in which no genetic selection is practised) indicate that between 1974 and 1979 the annual rate of genetic progress was worth 53.5 p/slaughter pig in the case of Large White pigs and 60.9 p/pig for animals of the Landrace breed (Steane, 1980, personal communication). This rapid rate of improvement may be attributed to three factors. First, the characteristics of the end product desired by the meat trade (most particularly the bacon industry) have been clearly defined. Secondly, the more important production characteristics (FCR, growth rate and carcass traits) can be easily and objectively measured and, as they are moderately heritable, can be selected for successfully. Thirdly, the existence of a National Pig Improvement Scheme (initially operated by the Pig Industry Development Authority and subsequently by the Meat and Livestock Commission) has enabled the efforts of individual producers to be coordinated. In addition it has provided independent evaluation and comparison of bloodstock from different sources and played an important part in assisting commercial producers to improve the genetic merit of their stock.

The Pig Improvement Scheme has four main herd categories—nucleus herds, breeding companies, reserve nucleus herds and nucleus multiplier herds.

Nucleus herds

These are normally closed populations which practise rigorous selection of stock for economically important characteristics. Superior males from such herds are either retained, sold to other nucleus herds or purchased by artificial insemination studs. Above average boars not needed within the herds are offered for sale to commercial producers. Female stock may be sold to commercial producers but are more often subjected to a multiplication stage (often involving crossbreeding to produce the FI 'parent' female) before sale to commercial producers. MLC Nucleus herds have access to Central Boar Testing facilities, whereby they can ascertain the relative merit of their stock by comparison with contemporary boars from other nucleus units.

Reserve nucleus herds

These are herds which aspire to Nucleus status and are testing stock in the hope of entering the Nucleus category.

Breeding companies

These companies may have their own breeding and selection procedures or may be part of the MLC scheme. In addition other companies exist completely outside the scheme. A number of breeding companies take part in the MLC's Commercial Product Evaluation (CPE) Scheme. Boars and gilts being offered for sale by participating companies are purchased by the MLC and evaluated at a central testing station. Test reports are published and from these commercial producers are able to assess the relative merits and suitability of company stock for their own particular purposes.

Nucleus multiplier herds

These exist to multiply or 'bulk up' superior stock for sale to commercial producers. In addition these herds frequently undertake the crossing programme needed to produce first cross 'parent' females. Though not tested as rigorously as Nucleus herds care is taken to ensure that they keep up with genetic progress by using only top quality boars and by maintaining a rapid generation turnover in both their male and female stock.

All stock offered for sale has to exceed minimum standards set within individual herds.

The genetic improvement of pigs may thus be envisaged as a breeding pyramid (*Figure 17.2*). The most rigorous selection and most rapid progress being made at the top of the pyramid in the Nucleus herds and usually being passed down to the commercial producer either directly through the sale of 'grandparent' purebred boars and gilts or by way of multipliers buying 'grandparent' stock and selling 'parent' generation crossbred gilts to the commercial producer.

Feeds and feeding

The pig is omnivorous and can utilise a wide variety of food materials including roots, human food waste (swill), milk by-products, grass and silages and vegetable wastes. The use of bulky foods is restricted by the pig's appetite capacity, additionally fibrous foods are poorly digested and tend to reduce both the voluntary intake and overall digestibility of the diet. The use of bulk foods is further limited by the practical difficulties encountered in collecting, storing and presenting the food to the pig and by controlling intake levels with accuracy. Nonetheless, where these problems can be overcome the use of bulk foods is often the basis for profitable production.

The majority of pigs are fed complete diets formulated to provide all the animals' nutrient requirement. The major factors affecting the choice of materials are fibre level, the content and availability of essential amino acids and the relative cost of the

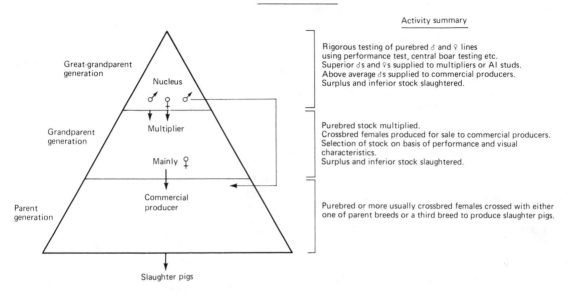

The breeding pyramid

Activity summary

Great-grandparent generation — Nucleus

Rigorous testing of purebred ♂ and ♀ lines
using performance test, central boar testing etc.
Superior ♂s and ♀s supplied to multipliers or AI studs.
Above average ♂s supplied to commercial producers.
Surplus and inferior stock slaughtered.

Grandparent generation — Multiplier — Mainly ♀

Purebred stock multiplied.
Crossbred females produced for sale to commercial producers.
Selection of stock on basis of performance and visual characteristics.
Surplus and inferior stock slaughtered.

Parent generation — Commercial producer

Purebred or more usually crossbred females crossed with either one of parent breeds or a third breed to produce slaughter pigs.

Slaughter pigs

Notes (1) Nucleus and multiplier units may be totally independent (e.g. within MLC scheme) cooperating (e.g. in a cooperative breeding company or organisation) or totally integrated and dependent (e.g. in a breeding company organisation).

(2) Multipliers and commercial producers may be either independent, part of a cooperative or integrated (e.g. a commercial producer may be his own multiplier).

Figure 17.2 The breeding pyramid

nutrients they provide. Reliability of supply and consistency of a raw material are also important as these factors reduce the expenditure on laboratory analyses and the need for frequent reformulation of diets.

In the UK barley, wheat, and wheat 'offals' are usually the major component in diets and supply much of the dietary energy. In 'high energy' diets (i.e. diets with a high dietary energy yield in a given mass) fats and oils may also be used as energy sources.

The most common protein sources are extracted soya bean meal, fishmeals and to a lesser extent meat and bonemeal. However in most diets the cereal grains also make an important contribution to the protein. In diets for young pigs, e.g. creep foods and early weaning diets, dried milk products are used extensively. The energy and protein provided by materials such as dried skim milk are well utilised by the immature digestive tract of the young pig. In some circumstances lysine, usually the first limiting amino acid in pig diets, may be provided more economically by the addition of synthetic lysine than by increasing the proportions of proteinaceous foods in the formulation.

In formulating diets careful attention has to be paid to the provision of minerals and vitamins. The natural contribution of the major minerals (calcium, phosphorus, sodium and chlorine) has to be supplemented and balanced by the use of mineral supplements such as limestone flour, dicalcium phosphate and common salt. It is not unusual for the total requirement for trace elements and vitamins to be met by supplementation with refined or manufactured sources. The reasons for this are that the natural contribution from other ingredients in the diet is extremely variable and that the availability and stability of naturally occurring micronutrients is uncertain.

Feed ingredients for pigs are usually ground. This produces a greater surface area for enzyme action and increases digestibility. It also results in a more homogeneous mixture and reduces the likelihood of separation of ingredients during handling and feeding. Some feed ingredients, particularly cereal grains, may be cooked (i.e. micronised, steam flaked or expanded) prior to milling. The cooking process ruptures the starch grains and increases energy digestibility. Such cooked products are often used in diets for very young pigs where enzyme activity levels are low and may limit energy utilisation. The process of cubing or pelleting often involves the use of steam and pressure which generate high temperatures, this too can have a beneficial effect on digestibility. However it can also have deleterious effects on vitamins and some proteins unless great care is exercised. Cubing is less frequently practised by home-mixers than by the compound feed trade.

It is difficult to ascertain the proportions of feeds purchased as compounds and home mixed. *Table 17.8*

Table 17.8 Proportions of breeding and feeding herds using different types of diets (MLC Feeding Recording Scheme 1979)

	Breeding herds %	Feeding herds (%)
Compound diets	57	42
Home-mixed diets		
using proprietary concentrates	22 ⎤	28 ⎤
	⎬ 39	⎬ 51
using straights	17 ⎦	23 ⎦
Others	4	7

Source: MLC (1980) *Commercial Pig Production Yearbook 1979.* Reproduced with permission.

shows the proportions of different types of diets used by MLC feed recording herds. Because this sample tends to be drawn from larger units it is likely that the proportion of herds using compound feeds is somewhat higher than this on a national basis.

THE SOW AS A PRODUCTION UNIT

The breeding female pig fulfils only two functions. First, to produce and support piglets to an age at which they are capable of being reared successfully and economically without her assistance, and second to transmit to her offspring desirable genetic characteristics. Therefore the sow should be regarded as the basic unit of production. In order to understand the factors which affect the productivity of a breeding herd it is first necessary to examine the characteristics of the individual animal which may influence her capacity to function effectively as a production unit.

The reproductive cycle of the sow

Puberty

Puberty in the pig is characterised by the occurrence of the first oestrous period (heat) and generally marks the stage at which the animal first becomes competent to reproduce.

The age at which puberty is attained varies greatly from as early as 135 d of age to in excess of one year. Age at puberty is influenced to a limited extent by genotype, nutrition and season. However in recent years it has been recognised that management events often 'trigger' puberty and account for much of the observed variation in pubertal age. Indeed it seems likely that some external stimulus is essential in facilitating the transition from the non-breeding to the breeding condition. The two most potent stimuli discovered to date are a change of environment and first contact with a mature male. If prepubertal gilts in excess of 140 d of age are subjected to either, or preferably both, of these stimuli the attainment of puberty is encouraged. It has been found that the response is most predictable if first contact with boars

occurs at 165 d of age. If the gilts have had no prior contact with males and if large, sexually mature boars are used, 70% or more of gilts will attain puberty within 10 d. At present this technique provides the only readily available means of exerting some control over puberty attainment and of reducing the variation in pubertal age within a commercial unit.

The oestrous cycle

Following puberty the gilt usually adopts a pattern of regular oestrous cycles. These are characterised by periods of oestrous activity, when the gilt is receptive to the male, interspersed by periods of quiescence. The duration of oestrous cycles can vary considerably but the majority of normal animals would have cycles in the range 18–23 d. In gilts a number of the oestrous periods may be 'silent' that is the gilt may ovulate without any outward manifestation of oestrus. In addition some oestrous periods may be poorly expressed, the animal showing some sign(s) of oestrus but being unwilling to accept the boar.

Weaned sows also adopt a pattern of oestrous cycles if not successfully mated at the post-weaning oestrus.

The oestrous period

The most significant part of the oestrous cycle from the producers' point of view is the oestrous period. It is during this period that the female is receptive to the male and must be mated if conception is to occur (*Figure 17.3*).

In the period immediately preceding oestrous, follicles in the ovary develop rapidly towards maturity under the influence of follicle stimulating hormone (FSH). These follicles produce oestrogen which primes the uterus and may also produce characteristic changes in the external genitalia. In gilts the vulva usually becomes engorged and oedematous and shows a marked change in coloration to a deep reddish purple hue immediately prior to oestrus. In sows these changes are often completely absent. The circulating oestrogen also produces behavioural changes. The pro-oestrous female shows a general increase in locomotor activity and, if given the opportunity, will seek out and keep close company with a boar. During oestrus the female becomes receptive to the attentions of the boar adopting a passive stance inviting mating when approached by the male. This 'standing reaction' can sometimes be elicited by a stockman applying pressure to the back of the animal. If such a response is elicited it is a positive indication that the animal is 'on heat'. Unfortunately the response is not shown by a majority of sows, so a negative reaction cannot be assumed to indicate that the animal is not in oestrus.

Towards the end of the oestrous period luteinising hormone (LH) is released which results in ovulation. The duration of the various phases of oestrus show considerable variation between individuals with gilts

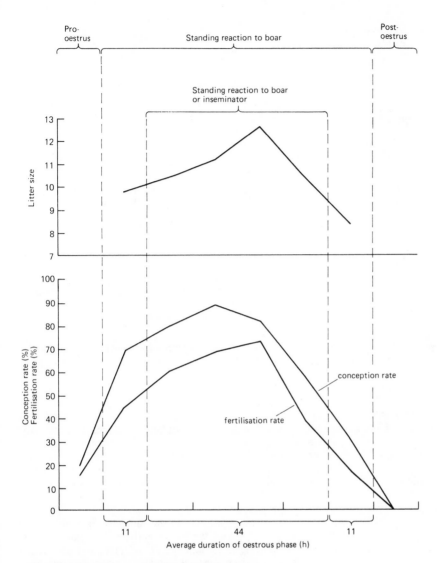

Figure 17.3 Effect of time of mating during the oestrous period on conception rate and litter size. (After Willemse, A. H. and Boender, J. (1967) *Tijdschr. Diergeneesk* **92, 18–34. Reproduced by courtesy of the editor and publisher)**

usually having shorter oestrous periods than sows. Season also has an effect on the duration of oestrus (*Table 17.9*).

Table 17.9 Seasonal variations in duration of oestrus

Season	No. oestrous periods observed	Average duration of oestrus (h)
Winter	365	53.98
Spring	369	54.75
Summer	515	58.85
Autumn	439	56.83

Source: Signoret, J.P. (1967), *Annls. Biol. anim. Biochim. Biophys* **7**, 407. Reproduced with permission.

Ovulation occurs at the end of the oestrous period. As ovulation rate sets the upper limit to litter size a high ovulation rate is desirable. Ovulation rate shows heterosis when dissimilar breeds or strains of pigs are crossed. This in part explains the increase in litter size in crossbred pigs. Ovulation rate increases with the age of the female. In young gilts low ovulation rates may limit litter size and because of this mating of gilts is frequently delayed until the second or third heat period. Although this may increase the size of the first litter it does not affect lifetime piglet production and has an adverse effect on food utilisation (*Table 17.10*).

Ovulation rate is favourably influenced by high

Table 17.10 Effect of mating at puberty, second or third heat on the performance of sows over three parities

	Heat mated		
	1	*2*	*3*
Pigs weaned (first litter)	7.8	8.3	8.6
Pigs weaned (litters 1–3)	26.5	26.4	26.9
Total weight weaner produced in first 3 litters (kg)	280.7	282.4	284.8
Weight of gilts at mating	88.1	98.2	115.1
Weight of sows at mating for fourth litter	165.5	168.9	165.8

Source: MacPherson, R. M., Hovell, F. D. De B. and Jones, A. S. (1970). *Anim. Prod.* **24**, 333. Reproduced with permission.

energy intakes during the follicular stage of the oestrous cycle; for this reason the food intakes of gilts and of weaned sows are usually increased for 7–14 d prior to mating.

Pregnancy

The pig is a polytocous (litter bearing) animal and its reproductive tract is well adapted to this function. The uterus is divided into two horns which connect near the cervix. This enables ova to pass from one horn to the other thereby increasing the opportunity for ova to attach at a suitable site for implantation and development.

The ova commence attachment to the uterus about 10 d after ovulation, and implantation is complete by day 25. Each piglet forms its own membranes and has a placenta discrete from its litter mates. In the later stages of pregnancy the membranes tend to fuse and at parturition the interconnections rupture to form a tube through which the piglets can be born.

Not all the ova shed at an ovulation develop to produce piglets. Some fail to become fertilised, others fail to implant or develop due to genetic defects, endocrine disturbances, nutritional inadequacies or disease. Some indication of the extent of losses between ovulation and weaning is given in *Figure 17.4*.

Although ovulation rate increases with reproductive experience there is not a proportional increase in litter size. In older animals a number of potentially viable embryos fail to develop because they are in excess of the animal's capacity to support them. These losses were originally thought to be due to insufficient 'space' in the uterus, but in recent years it has been shown that most of the losses occur before day 25 of pregnancy, that is at a stage when the physical size of the conceptus is of little significance. In gilts ovulation rate may be lower than the litter carrying capacity. In this case increasing ovulation rate may result in more piglets being born. In sows ovulation rate is rarely a limiting factor.

Pregnancy terminates with parturition. The duration of pregnancy varies more widely than is generally

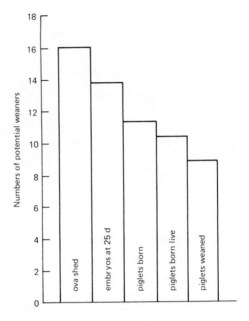

Figure 17.4 Losses of potential weaners through pregnancy and lactation

appreciated. Although the mean duration is just over 115 d 'normal' pregnancies may be between 108 and 122 d in duration, with larger litters being carried for less time than small litters. Despite this range about 99% of farrowings occur with 4 d of the mean, i.e. on days 111 to 119. It follows that wherever possible sows should be settled into farrowing accommodation 7 d before the average due date, i.e. by day 108 of pregnancy.

Prior to parturition the sow becomes more active and, if given the opportunity, will indulge in nest building behaviour, gathering litter and chewing it up. This is very noticeable in sows kept outdoors where the sow will select and defend a site chosen for farrowing. With confined sows this activity is less obvious and may be displaced by ground pawing and bar biting.

A few hours prior to farrowing it becomes possible to express milk manually from the sow's teats due to the release of oxytocin into the bloodstream. (It is only in this limited period around parturition that sows can be milked by hand.)

The duration of farrowing may vary from 1–24 h in 'normal' births. Farrowings tend to be shorter in duration if the stock are young and have had plenty of exercise. Birth order has a marked effect on piglet survival. Later born piglets stand a greater chance of being born dead particularly if farrowing is protracted. The deaths are mainly due to anoxia caused by disruption of the piglets' oxygen supply while in the uterus.

Lactation

The pig does not normally exhibit reproductive behaviour or have oestrous periods during lactation. The release of hormones involved in reproduction being suppressed by the circulating lactogenic hormones. This suppression may be overcome in rare circumstances, e.g. if the sow ceases lactation prematurely or if the piglets are withheld from the sow for some of the day thereby reducing the suckling stimulus. It has also been shown that if sows are kept in groups with a boar and fed very generously (e.g. with food on free access) they may resume oestrous activity while lactating. Although such a system may appear attractive the additional building, boar, food and labour costs negate any potential benefit to be gained from the more rapid rebreeding of the sow.

The quantity and composition of milk produced by the sow is affected by a vast range of factors but most particularly by stage of lactation and by nutrition. Sow milk yield increases rapidly after birth and peaks at about the third week of lactation (*Figure 17.5*). The composition changes considerably over this period (*Table 17.11*). The protein content is high initially due to the presence of immunoglobulins. The total solids and fat content is higher than that of cows' milk. This is important for the piglet has only a small gut capacity and needs to receive its nutrients in a concentrated form.

Females selected for breeding will normally possess 12–14 well spaced mammary glands. The glands are anatomically separate and produce differing quantities of milk, the anterior glands producing more milk than the posterior ones. The size of teat, spacing and presentation also influence their suitability for and accessibility to the piglet. As piglets rapidly establish a teat order after birth, and thereafter tend to suckle the same teat, the 'quality' and productivity of the gland obtained by the piglet influences its preweaning growth (*Table 17.12*).

In commercial units the duration of lactation is determined by the stockman not the sow. Given the correct management piglets can be artificially reared from birth. However, rearing colostrum deprived piglets is an extremely difficult and expensive opera-

tion not normally attempted on normal commercial units. The age at which pigs are weaned commercially varies from as early as 10 d post partum to in excess of eight weeks. The factors influencing choice of weaning age will be discussed in a later section.

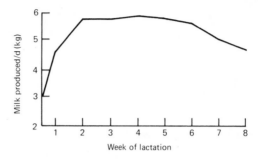

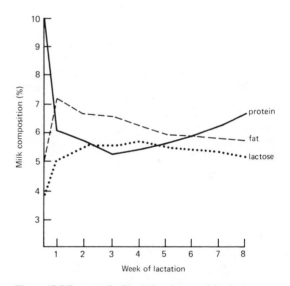

Figure 17.5 Pattern of milk yield and composition in the sow (After Salmon-Legagneur, E. (1961) *Ann. Biol. anim. Biochem. Biophys.* 1, 295–303. Reproduced by courtesy of the editor and publisher)

Table 17.11 Changes in composition of sow's colostrum and milk with time post partum

	Time after parturition (h)						(Wk)
	0	6	12	15–24	27–48	72–120	2–8
Total solids (%)	30.2	26.6	20.8	19.6	21.2	21.8	21.2
Fat (%)	7.2	7.8	7.2	7.7	9.5	10.4	9.3
Protein (%)	18.9	15.2	10.2	7.2	6.9	6.8	6.2
Lactose (%)	2.5	2.9	3.4	3.7	4.0	4.6	4.8
Ash (%)	0.63	0.62	0.63	0.66	0.72	0.77	0.95
Calcium (%)	0.05	0.05	0.06	0.07	0.11	0.16	0.25
Phosphorus (%)	0.11	0.11	0.11	0.12	0.13	0.14	0.15

Source: Perrin (1955). *J. Dairy Res.* **22**, 103. Reproduced with permission.

Table 17.12 Weight of piglet in relation to teat number suckled

	Teat position	Piglet weight (kg)	
		Birth	6 weeks
Front	1	1.41	10.7
	2	1.29	9.7
	3	1.31	9.2
	4	1.32	8.7
	5	1.34	8.8
	6	1.31	8.2
Back	7	1.25	8.5

Source: English, P., Smith W. and Maclean, A. (1977). *The sow improving her efficiency.* Ipswich: Farming Press. Reproduced with permission.

The weaning to remating period

Following the removal of the piglets at weaning the mammary glands rapidly become engorged with milk. This build up of milk in the gland triggers a feedback mechanism which inhibits the lactogenic hormones. This in turn removes the inhibition of hypothalamic releasing factors and these reinitiate cyclic breeding activity. The time taken from weaning of the sow to the occurrence of the first post-weaning oestrus varies considerably. Some sows can return to oestrus within 2–3 d of weaning others may take several weeks (*Figures 17.6*). It follows that producers must anticipate sows returning to oestrus on *any* day post-weaning and plan their oestrous detection management accordingly.

A number of management factors can influence the pattern of post-weaning oestrus. Gilts generally take 3–5 d longer to return to oestrus than older sows. Nutrition also plays a significant role. The practices of reducing feed intake in the last week of lactation and withholding food and/or water for 24–48 h after weaning have now been discredited. Indeed these practices tend to inhibit rather than encourage a rapid return to service. In gilts a generous food allowance post-weaning has been shown to increase both the number of gilts returning to oestrus and the mean interval (*Table 17.13*). In sows high feed levels post-weaning do not have such dramatic effects but do tend to increase the percentage of sows exhibiting oestrus within 10 d of weaning (*Table 17.14*).

Table 17.13 Effect of post-weaning feed level on return to service following the first lactation

	Feed intake (kg/d)		
	1.8	2.7	3.6
% weaned gilts mated			
within 7 d	14	22	17
within 10 d	29	33	88
within 42 d	66	75	100
% weaned gilts farrowing	58	75	100
Mean interval weaning to oestrus (d)[1]	21.6	12.0	9.3
Mean litter size	9.4	10.1	11.6

[1] Gilts returning within 42 d of weaning
Source: Brooks and Cole (1972). *Anim. Prod.* **15**, 259. Reproduced with permission.

Table 17.14 Effect of post-weaning feed level on return to service in sows

	Food intake (kg/d)			
	1.8	2.3	3.6	4.5
% weaned sows mated				
within 7 d	86	84	89	98
within 10 d	89	91	93	98
Mean interval to mating	6.70	6.96	6.50	6.13
Conception rate (%)	93	95	86	100
Mean litter size	11.9	12.3	11.7	11.8

Return to oestrus is further stimulated by housing weaned sows in close proximity to boars. Ideally they should have the opportunity to see, hear and smell boars continuously and should have physical contact each day until remated.

There is considerable argument about the effect of weaning age on the interval to remating. Oestrus is less well exhibited in early weaned sows so the detection rate may be reduced giving the impression of increased intervals to remating. However with good oestrous detection the interval is little affected by the weaning age over the range 18–35 d. However when lactation length is reduced to less than 18 d the interval to remating tends to increase.

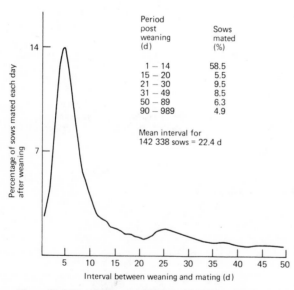

Period post weaning (d)	Sows mated (%)
1 – 14	58.5
15 – 20	5.5
21 – 30	9.5
31 – 49	8.5
50 – 89	6.3
90 – 989	4.9

Mean interval for 142 338 sows = 22.4 d

Figure 17.6 Weaning to remating interval in sows (After Legault, C., Dagorn, J., and Tastu, D. (1975). *Jour. Rech. Porcine en France*, INRA-ITP. Ed Paris XLIII–LI. Reproduced by courtesy of the Editor and publisher)

Sow herd productivity

Productivity is measured in terms of physical output and while it may have a strong relationship with profitability within an individual unit, this may not be the case when comparisons are made between units. If it is considered that

Profitability = Value of product(s) − Cost of production,

it is apparent that profit can be improved by increasing the number or value of products or by reducing cost of production. Similarly, if two units produce a similar value of product but have different production costs their profitability will differ.

Within a unit the costs of housing, feeding and tending individual sows vary little but the output of those sows may differ greatly. Consequently, identifying sows with suboptimal performance and either eliminating them or correcting their deficiencies can have a greater impact on overall herd performance and profitability than attempting to effect, marginally, improvements in the production of all individuals in the herd.

It should also be noted that producers generally show a desire to maximise rather than optimise output. Attempting to increase profits by improving the output of animals already performing close to their biological limits may increase the vulnerability of the herd. On the other hand a reduction of production costs effected by improved management and elimination of substandard performance incurs little risk and frequently minimal investment. The difference between 'average' producers and those in the 'top third' category is usually a combination of attention to detail and a degree of luck.

The three most important factors determining piglet output from a herd are the frequency with which litters are produced, the number of pigs born in the litters produced and the survival rate of those piglets. These factors are considered below.

Frequency of parturition

The interval between successive farrowings is determined by the length of lactation, the interval from weaning to effective service and the duration of pregnancy. Although there is some natural variation in the duration of pregnancy this is unaffected by management except in one circumstance, that is when parturition is induced by the use of prostaglandins. Even then farrowing interval is reduced by no more than a couple of days. As discussed previously the interval from weaning to effective service can be significantly affected by management. Within a unit, monitoring average 'empty days', (that is the number of days from weaning to effective service) can help to pinpoint any deficiencies in post-weaning management. In recent years producers have increasingly adopted management systems involving earlier weaning (i.e. shorter lactation lengths) as a means of increasing farrowing frequency. The theoretical difference in piglet production which can be derived from weaning pigs at different ages is illustrated in *Table 17.15*.

If these theoretical calculations are compared with results obtained in recorded herds (*Table 17.16*) some interesting observations can be made. First y, the majority of herds fall well short of theoretical performance targets particularly in respect of farrowing frequency. Secondly, there are very large differences between the performance of average and 'top third' herds at each weaning age. When it is remembered that the 'average' values include the 'top third' producers it may be concluded that the performance of some herds is very far below potential. Thirdly, it should be noted that the 'top third' of producers weaning at 40+ d actually rear more pigs/sow per annum than the 'average' producer weaning at 25 d or less. Comparisons of food utilisation are a little confusing as piglets are transferred from the breeding herds at different weights. However, if feed costs are compared for a subsample of the recorded herds, transferring pigs with a similar weight range, it can be seen that earlier weaning on average results in a lower feed cost/pig transferred.

It is apparent from the foregoing that while earlier weaning affords opportunities to increase sow output the potential productivity of many units is not realised in practice. Two factors which are frequently implicated in this failure are first, inability to maintain a short average weaning to remating interval due to

Table 17.15 Theoretical productivity of sows having different lactation lengths

	Lactation length					
	14	*21*	*28*	*35*	*42*	*56*
Pregnancy (d)	114	114	114	114	114	114
Lactation (d)	14	21	28	35	42	56
Minimum weaning to remating interval (d)	5	5	5	5	5	5
Farrowing interval (x)	133	140	147	154	161	168
Litters/sow per annum (365/x)	2.74	2.61	2.48	2.37	2.27	2.17
Piglets/sow per annum (assuming 9.5 pigs reared/litter)	26.0	24.8	23.6	22.5	21.6	20.6

Table 17.16 Performance of average and top third producers weaning piglets at different ages

| | Weaning age | | | | | | | |
| | 25 d or less | | 25–32 d | | 33–39 d | | 40 d and over | |
	Average	Top third	Average	Top third	Average	Top third	Average	Top third
Av. no. pigs born alive/litter	10.2	10.5	10.4	10.8	10.3	10.7	10.5	11.0
Av. no. pigs reared/litter	8.9	9.4	9.0	9.6	8.8	9.4	9.1	9.7
Av. no. litters/sow + gilt/year[1]	2.24	2.39	2.16	2.27	2.06	2.18	1.99	2.10
Av. no. reared/sow + gilt/year[1]	19.9	22.3	19.4	21.8	18.2	20.4	18.1	20.4
Total feed consumption/kg pig produced (kg)	6.7	6.6	6.2	5.6	5.8	5.4	5.6	5.0
Total feed cost/kg pig produced (p)[2]	86	83	80	73	76	70	70	63
Feed cost/pig transferred (av. 20.1 to 34 kg) (£)[2,3]	14.48	13.30	15.39	14.85	15.10	13.78	15.67	15.21

[1] Adjusted for unserved gilts
[2] Costs at March 1980
[3] Sub-sample of survey population
Source: MLC Feed Recording Scheme. Reproduced with permission

inadequate oestrous detection and poor mating management and secondly reduced productivity resulting from high culling rates.

Oestrous detection and mating management

In a free living population of pigs the 'boar-seeking' activity of the female obviates the need for oestrous detection. Within commercial pig units oestrous detection presents a considerable challenge to the management skills of the stockman. It is often wrongly assumed that if females are allowed to cohabit with boars the problem will take care of itself. With sufficient circulation space and a sufficiently small sow to boar ratio it may. However, if matings are not supervised it is impossible for the stockman to determine whether all females have been mated, whether the matings have been satisfactory or when they occurred. If this information is not obtained the unit may end up carrying a number of unproductive animals and will be unable to make efficient use of available buildings. Such a system may be tolerable on a low cost, outdoor system but even on such a system a proportion of unmated animals can prove expensive.

In most commercial units sows and boars are confined separately. In such circumstances oestrous detection assumes great importance. It will be apparent from the previous section that while a number of physical and behavioural changes may be indicators of oestrus none can be relied upon for the detection of oestrus in all females. The detection of oestrus can only be effectively carried out by a boar. It is essential that all potentially oestrous females are presented to the male once or preferably twice each day. The oestrous check is best carried out by introducing individual females to the boar either in his home pen or in a service area in which the boar habitually operates. This arrangement limits the amount of extraneous non-mating activity by the boar and so

saves time. Matings should be supervised to ensure intromission and to minimise the chances of physical damage to either the sow or the boar. Both conception rate and litter size can be influenced by the timing of mating relative to ovulation (*Figure 17.3*). As it is difficult to anticipate when ovulation will occur it is usual to repeat mate the sow, ideally at 12 h intervals, throughout the standing period. Such a practice will ensure that viable sperm are available in the female reproductive tract at the time of ovulation.

In order to ensure that an adequate concentration of viable sperm are deposited in the tract it is important that boars are not overworked. There is evidence to suggest that boars should be permitted only five to six services/week (*Table 17.17*). This number should be reduced further in the case of young animals whose willingness to serve frequently exceeds their capacity to produce adequate viable sperm.

Table 17.17 Effect of number of ejaculations on reproductive performance

| | Number of ejaculations in previous 6 d | | | | |
	0–1	2–3	4–5	6–7	8–9
No. of litters	547	433	193	62	28
No. born alive	10.1	10.2	10.0	9.4	8.6
No. stillborn	0.9	0.9	0.9	0.8	0.8
Return rate (%)	15	12	11	9	15

Source: Reasbech, N.O. (1969). *Br. Vet. J.* **125**, 599

Number of pigs born and reared

In addition to the factors already mentioned the most important influence on numbers born and reared is the genotype of the sow. Litter size is a trait with low heritability (10% or less) which generally shows little response to selection. However, in common with other low heritability characteristics it does exhibit heterosis ('hybrid vigour') with crossbred sows producing 4–8% more pigs per litter and 6–10% more pigs per year.

Litter size is also affected by sow age. Numbers born increase over the first two to three litters and then remain relatively stable for litters four to eight after which they may decline. However, although older sows may farrow more pigs they do not necessarily succeed in rearing them. There is an inverse relationship between numbers born and reared. In addition older sows may be more clumsy and have fewer fully functional teats (due to previous infection and/or damage) and as a consequence are less competent to rear all the piglets they produce.

Increasing the number of liveborn pigs which survive to weaning remains one of the greatest challenges to pig producers.

Data from recording scheme herds show that few units consistently have a pre-weaning mortality rate of less than 10% and that normally rates in excess of 20% are far from uncommon. The majority of these piglet losses occur in the first few days of life. Generally over 50% of losses occur in the first 2 d after farrowing and 75% in the first week. The most significant causes of piglet mortality are starvation and crushing. It is not difficult to see why mortality is high in the first few days of life. The piglet is born at a young physiological age, is extremely small in relation to its dam, has very limited body energy reserves and virtually no passive immunity.

The limited body energy reserves and poorly developed temperature control mechanisms of the newborn piglet make it vital that it achieves an adequate suckle soon after birth. The newborn piglet rapidly dissipates its energy reserves to maintain body temperature. This problem is exacerbated in piglets of low birthweight. The smaller the piglet the greater its surface to volume ratio and hence its rate of heat loss. In addition the smaller the pig the more limited its body reserves. The small piglet may find it difficult to obtain and retain a teat particularly if there is excessive competition from larger litter mates and at best will have to settle for one of the less favoured and less productive posterior teats.

In addition to supplying nutrients the first suckle also provides the piglet with immunoglobulins which provide the piglet with some degree of passive immunity to the disease organisms which it is likely to encounter in its environment. Failure to obtain sufficient immunoglobulins renders the pig more liable to disease and death at a later stage of lactation.

It will be apparent from the foregoing that attempts to reduce piglet mortality will involve the consideration of a large number of interacting factors. The simplest solution might appear to increase piglet birthweight. Unfortunately, this is not easily achieved. With the exception of crossbreeding, which tends to result in the production of larger piglets, there is little that can be guaranteed to increase birthweight. Increasing the intake of food or changing the nutritional composition of food in late pregnancy (e.g. by adding lipid) can result in the production of larger piglets but the response is unpredictable and the economics dubious.

Greater attention to the management of the newborn pig and the environment into which it is born are more likely to pay dividends. The provision of accommodation which affords the piglet protection from overlying and a suitable climatic environment is essential. Establishment of the piglet at the udder can be improved by the stockman assisting weakly piglets to a teat and by reducing competition by cross fostering piglets to ensure that sufficient functional teats are available to enable all the piglets in the litter to suckle. Finally the disease challenge can be reduced by providing accommodation constructed of materials which can be thoroughly cleaned and sufficient units of accommodation that this is feasible.

The significance of herd age profile and culling rate

Unlike some other livestock enterprises commercial pig units operate on a continuous throughput basis. This means that once established a unit will consist of a population of breeding females of different ages. It has already been noted that the reproductive performance of the female pig alters with age, therefore the age profile of the herd will influence its productivity. In a newly established or expanding unit it is inevitable that the percentage of young females will be high. In an established unit the herd profile will be determined by the culling and replacement rate. Estimates of annual culling rate vary widely but most fall in the range of 35–45%. The results of two recent surveys of reasons for culling are summarised in *Table 17.18*. It is interesting that in both surveys only a little over a third of sows were culled for poor performance or old age. These may be regarded as 'voluntary' cullings. The decision to cull in the remainder of cases was forced on the producer by the inability of the sow

Table 17.18 Reasons for culling sows

	UK survey[1]	French survey[2]
Not holding to one or more services	12.3 ⎫	31.0
Not in pig	17.2 ⎭	
Not showing oestrus	2.6	5.4
	32.1	36.4
Poor performance	13.9	8.4
Age (lowered productivity)	24.5	27.5
	38.4	35.9
Lameness	11.8	8.8
Abortion	5.6	2.8
Milk failure	0.6	2.3
Other causes	11.5	13.8
	100.0	100.0

[1] MLC Pig Improvement Services Newsletter No. 14. March 1980
[2] Dagorn, J. and Aumaitre, A (1979). *Livest. Prod. Sci.* **6**, 167

to make any further useful contribution to the herd. Reproductive failure accounted for about one-third of cullings and lameness for approximately 10%. In the two surveys 40 and 49% of sows were culled prior to their fourth litter, that is before the stage at which they might be expected to reach maximum productivity.

The high incidence of 'involuntary' culling has two important effects. First, it increases the proportion of less productive young females in the herd and this in turn lowers herd productivity. For example increasing culling rate from 30–50% would lower herd productivity in the order 1.5–2.0 pigs/sow per year. Secondly, it limits 'voluntary' culling, producers being less able to discriminate between sows which are capable of breeding because it would further increase their replacement rate.

Finally it should be noted that herd productivity is reduced when sows remain in the herd for a prolonged period between weaning their last litter and culling. Surveys have shown that unproductive sows are frequently retained in the herd for 50–100 d before disposal. These sows are still eating expensive food and taking space in expensive buildings while making no contribution to herd output. Better recording and the use of pregnancy detectors could significantly reduce this source of inefficiency in many units.

Feeding the reproducing female

The nutrient requirements of the sow are determined by a number of interdependent factors. These include the weight of the sow, her reproductive state and her genetic makeup. The requirement of an individual animal is modified by her social and thermal environment and by her previous nutritional history. Thus an allowance for an individual sow needs to take into account not only her weight and stage of pregnancy or lactation but also the temperature of her surroundings and the state of her body reserves. Unfortunately although some 80% of sows are individually fed (either because they are housed in stalls or fed in individual feeders) only a very small proportion are individually rationed. On the majority of units a basal feed level is given in pregnancy and any deviation from that level represents remedial action to correct previous errors in feeding. There is no doubt that the productivity of sows could be greatly improved and their herd life extended if feeding regimes on commercial units were designed to cater for sows as individuals rather than making the unwarranted assumption that all sows were not only created equal but remained equal throughout their working life.

General aims of a sow feeding regime

The most important aims in feeding the sow are that she should produce and rear a large number of piglets

and that she should have a long herd life. The sow has a great capacity for overcoming nutritional abuse and will continue to reproduce in surprisingly adverse circumstances provided these adverse conditions are of limited duration and opportunity for recovery is not long delayed. However, the aim should be to match the sow's requirements at each stage of the cycle. Pigs have not achieved their mature body weight when mated for the first time so some provision must be made for growth during the first three to four pregnancies. Given the opportunity the sow tends to eat in excess of her energy requirements in pregnancy and deposit large amounts of body fat which are then mobilised for milk production in the following lactation. This leads to marked fluctuations in body-weight and body condition throughout the cycle which have been shown to be undesirable.

In general the aim should be to allow the sow to gain 10–15 kg between successive matings for at least the first three pregnancies. It is likely that in the next few years as methods of measuring sow fat deposits (either directly or using some form of condition scoring) become available it will be possible to replace this crude weight recommendation with a target fat reserve for different ages of animal and stages of reproduction. In the interim the general aim should be to minimise weight loss in lactation as far as possible while allowing sufficient gain in pregnancy to meet the target gain of 10–15 kg/reproductive cycle.

Feeding the pregnant sow

Suggested energy allowances for pregnant sows of different weights derived using a factorial approach are given in *Table 17.19*. These allowances assume the sows are not in a cold environment. In a cold environment the maintenance requirement might increase from 0.5 to 0.75 MJ.DE/kg metabolic body weight or higher. This would increase the energy requirement considerably.

The dietary protein requirement of the sow is less variable than the requirement for energy and a daily intake of 210–250 g crude protein/d (providing 7.5–10 g lysine) would be appropriate. In general protein is oversupplied in commercial pregnant sow diets as producers prefer to feed a single diet to their sows throughout the cycle. Consequently as the protein energy ratio is higher in lactation pregnant sows may receive more protein than they require.

Suggested micronutrient allowances are given in *Table 17.20*. A number of micronutrients have a marked effect on reproduction therefore even though they represent only a tiny percentage of the diet their importance cannot be overestimated.

Individual feeding of housed sows is vitally important as competition for food can lead to bullying and the failure of timid animals to obtain their share of the food. There is little to be gained from altering feed rates over the pregnancy period. Higher levels of

Table 17.19 Energy allowances for pregnant sows

Sow mating wt (kg)	Equivalent metabolic body wt (kg^0.75)	Sow maintenance requirement (MJ.DE/d)	Requirements for products of conception (MJ.DE/d)	Allowance for sow wt gain (MJ.DE/d)	Total daily DE requirement (MJ/d)
85	28.0	14.0	2.0	6.0	22.0
100	31.6	15.8	2.0	6.0	23.8
115	35.1	17.6	2.0	6.0	25.6
130	38.5	19.3	2.0	6.0	27.3
145	41.8	20.9	2.0	6.0	28.9
160	44.9	22.5	2.0	6.0	30.5
175	48.1	24.1	2.0	6.0	32.1
190	51.2	25.6	2.0	6.0	33.6
205	54.2	27.1	2.0	6.0	35.1
220	57.2	28.6	2.0	6.0	36.6

Notes:
(1) Maintenance allowance of 0.5 MJ.DE/kg metabolic body wt assumes a thermoneutral environment.
(2) Requirement increases with stage of pregnancy but can be met with a level allowance throughout without detriment to performance.
(3) Energy cost of maternal gain approximately 27 MJ.DE/kg gain. This allowance would be equivalent to a gain of approximately 25 kg over the pregnancy.

Table 17.20 Suggested allowances of minerals and vitamins in breeding pig diets

Minerals		Vitamins	
Calcium (%)	0.7–1.0	Vitamin A (units/kg)	12 000–15 000
Phosphorus (%)	0.55–0.65	Vitamin D (units/kg)[1]	2000
Salt (NaCl) (%)	0.45–0.55	Vitamin E (mg/kg)	10–25
Iron (mg/kg)	25–50	Vitamin K (mg/kg)	2–5
Zinc (mg/kg)	80–120	Thiamine (B1) (mg/kg)	1–2.5
Copper (mg/kg)	5–10	Riboflavin (B2) (mg/kg)	3–6.5
Manganese (mg/kg)	40–80	Nicotinic acid (mg/kg)	10–20
Iodine (mg/kg)	1–2	Pantothenic acid (mg/kg)	6–12
Selenium (mg/kg)	0.1–0.15	Pyridoxine (B6) (mg/kg)	1–6
		Vitamin B12 (µg/kg)	10–20
		Folic Acid (mg/kg)	0–2
		Biotin (µg/kg)	50–250
		Choline	75–150
		Vitamin C[2]	(1 g/d/late pregnancy?)

Notes: with the exception of calcium and phosphorus and salt the values indicate the range of micronutrient supplementation (that is additional to any natural contribution from raw materials)
[1] Most diets supplemented with legal maximum of 2000 units/kg
[2] Vitamin C may be required in late pregnancy but not evidence of requirement at other stages of the cycle.

feeding in late pregnancy can sometimes lead to the production of heavier piglets but the level of feeding needed is usually too great to be cost effective. An excessive energy intake early in pregnancy should be avoided as this tends to increase embryo mortality and lead to reduced litter size.

Feeding the lactating sow

Energy allowances for lactating sows are summarised in *Table 17.21.* Sows suckling large litters are likely to be producing up to 7 kg/d milk. Sows with small litters and sows weaned early are more likely to be producing at the lower level (5 kg/d milk). For sows producing 5 and 7 kg milk respectively, appropriate crude protein allowances would be 730 and 940 g/d CP. The DE:CP ratio is thus 1 to 11–13 hence a 12.5 MJ.DE/kg diet would need to contain 14–16% crude protein.

It is very easy to overfeed the sow in early lactation and underfeed her later. In the early stage the piglets' consumption of milk is low and overfeeding may encourage mastitis through producing an overstocked udder. Consequently the feed allowance should be increased gradually over the first 10 d of lactation then maintained until weaning. Appetite may be a problem in lactating sows particularly if they have been overfed in pregnancy and are kept in very hot farrowing houses. This problem can be partly overcome by more frequent feeding and using a more dense diet and of course by lowering the house temperature.

Table 17.21 Energy requirements of lactating sows

Sow farrowing weight (kg)	Equivalent metabolic body weight (kg$^{0.75}$)	Sow maintenance requirement (MJ.DE/d)	Total daily energy requirement at two levels of production			
			Producing 5 kg milk		Producing 7 kg milk	
			Body weight constant	10 kg weight loss in 35 d	Body weight constant	10 kg weight loss in 35 d
110	34.0	17.0	61.2	47.8	78.8	65.4
125	37.4	18.7	62.9	49.5	80.5	67.1
140	40.7	20.4	64.6	51.2	82.2	68.8
155	43.9	22.0	66.2	52.8	88.2	70.4
170	47.0	23.5	67.7	54.3	85.3	71.9
185	50.1	25.1	69.3	55.9	86.9	73.5
200	53.2	26.6	70.8	57.4	88.4	75.0
215	56.2	28.1	72.3	58.9	89.9	76.5
230	59.1	29.6	73.8	60.4	91.4	78.0
245	61.9	31.0	75.2	61.8	92.8	79.4

Notes:
(1) Sow maintenance allowance 0.5 MJ.DE/kg W$^{0.75}$ assumes thermoneutral environment.
(2) Requirement for milk production assumed to be 8.93 MJ.DE/kg milk. Therefore 5 kg milk requires 44.2 MJ.DE, 7 kg milk requires 61.8 MJ.DE
(3) Loss of liveweight assumed to be fatty tissue. This loss spares 47 MJ.DE/kg fat lost. 10 kg wt loss in 35 d = 470 MJ.DE total = 13.4 MJ.DE/d.
(4) If a sow fed for 5 kg yield at '0' weight loss was producing 7 kg milk she would lose 374 g/d or 13 kg in 35 d lactation.

Feeding the weaned sow and the replacement gilt

In the past sows were frequently fasted at weaning in the belief that this hastened return to oestrus. This practice has now been discredited and demonstrated to have adverse rather than beneficial effects. Weaned sows should receive a generous feed allowance (40–45 MJ.DE/d) from weaning to reservice. The high intake should be continued until 24 h after the end of oestrus, by which time ovulation should be complete, and the allowance then reduced to the pregnancy level.

Replacement gilts also benefit from high energy intakes prior to mating. Ideally their energy intake should be increased to 40–45 MJ.DE/d for 14 d prior to mating. In practice if gilts are to be mated at the second heat period feed intakes can be increased 7 d after the first heat period is observed. Individual feeding of gilts is not common but is very advisable as competition for food can result in some gilts in a group becoming overfat (and unproductive) and others being underfed and showing delayed oestrus and poor subsequent litter size.

THE PRODUCT—PIGS FOR SLAUGHTER

The end product of the pig production industry is pigs for slaughter. These animals provide principally meat for human consumption but in addition various offals, and glands may be utilised by the pharmaceutical industry, skins may be cured for leather and bristles may also be utilised for commercial purposes. However, it is meat which is the overriding consideration and it is towards the satisfaction of this market that the breeding, feeding and management of the growing pig is aimed.

The markets for slaughter pigs in the UK

In the UK pigs can be marketed over a very wide weight range but are usually slaughtered in the range 50–130 kg liveweight. Four categories of market are usually distinguished, pork, bacon, cutter and manufacturing (heavy hog). An indication of the weight range of pigs sold within these categories is given in *Table 17.22*. However it should be noted that different slaughterhouses and different contracts may stipulate that carcasses within any category fall in a much narrower weight bond than that illustrated.

Table 17.22 Market outlets for pigs[1]

	Liveweight (kg)	Deadweight (kg)	Killing out (%)
Pork pigs	49–93	36–68	73
Cutters	80–107	60–80	75
Bacon	81–94	61–70.5	75
Manufacturing	108+	81+	79

[1] Values represent extreme ranges of weight bands. Individual contracts will pay premiums for pigs falling into more narrowly defined weight bands and penalise over- and underweight pigs.

At one time it was possible to distinguish clearly between the different categories in terms of the use to which the carcasses were put. Pork pigs were used exclusively for the fresh meat market, bacon pigs were cured as whole sides for bacon and ham. Cutter and manufacturing pigs were used for a variety of cooked, processed and comminuted products in addition to being used for fresh pork. However as processing and marketing methods change the distinctions between the categories become less precise. For example there is a trend for fewer whole sides to be turned into bacon using the Wiltshire cure. While some 34% of carcasses are sold on the basis of length measurement (i.e. on a bacon contract) only about 14% of the total kill is processed as Wiltshire sides. Fifty per cent of bacon is now produced as a result of curing part carcasses using more rapid and less expensive techniques than that employed for Wiltshire curing. Although the demand for bacon has declined somewhat in recent years some 500 000 tonnes of bacon and ham were available for consumption in 1978 of which 42% was home produced. The majority of the balance came from Denmark (44%) with signficant imports from the Netherlands (8%), Poland (8%) and the Irish Republic (4%). Whereas home production accounts for less than half of the bacon consumed in the UK some 96% of the 626 000 tonnes of pork consumed in 1978 was home produced.

The pork market has the greatest future potential primarily because of the multitude of attractive 'convenience food' meat products which can be prepared from pig meat. This diversity of product lines has undoubtedly played an important part in the increased demand for pig meat in recent years.

Carcass quality and classification

Consumers are very reluctant to buy fat meat. Fat is unpalatable to the majority of consumers and is often perceived as a health hazard. Consumer resistance to fat in pig carcasses is particularly marked, first because the pig has clearly defined layers of subcutaneous fat and second because many cuts of fresh pig meat and bacon from traditional Wiltshire cured bacon is sold with the skin (rind) still on. Thus the consumer can readily assess fatness and reject products considered overfat. This problem can be and is overcome in two ways. Either excess carcass fat can be trimmed off before presentation to the consumer or alternatively fewer fat carcasses can be produced. Trimming is costly both in terms of labour and loss in edible meat yield. The production of leaner carcasses is a preferable approach but is long term, involving changes in breeding and management.

Producers are encouraged to supply suitable pigs for slaughter by the operation of a classification system and contracts which give premiums for superior pigs. The most important feature of pig carcass classification is the measurement of subcutaneous fat thickness. For pork, cutter and manufacturing pigs two measurements are made 4.5 cm (P1) and 8 cm (P3) from the midline of the back at the last rib. For bacon pigs backfat thickness is measured 6.5 cm (P2) from the midline at the last rib (*Figure 17.7*). These measurements of fat thickness correlate very well with the total lean content of the carcass and are often used as the basis for classifying carcasses both as a means of describing the carcass for potential buyers and as a basis for payment to the producer.

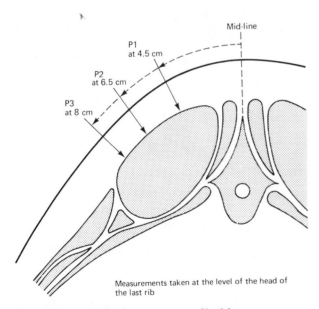

Figure 17.7 Positions for the measurement of backfat thickness on the pig carcass
Note the P1, P2 and P3 probe measurements are taken 4.5, 6.5 and 8.0 cm respectively from the mid-line at the level of the head of the last rib. On carcasses these measurements are taken using an optical probe. On live animals measurements may be taken at similar points using an ultrasonic probe

Other additional constraints may be built into the payment scheme. For example grading schemes for bacon carcasses often use additional fat measurements at the midline usually maximum shoulder and minimum loin and also include a minimum carcass length constraint.

The majority of pigs slaughtered are now marketed on the basis of dead weight and grade and irrespective of the market, pressure is applied to the producer to produce leaner carcasses. This approach has undoubtedly been effective (*Figure 17.8*). Nevertheless there is still enormous variation in the carcasss fat thickness of pigs offered to the different markets (*Figure 17.9*). In 1979 the Meat and Livestock Commission estimated that the UK pig industry was still producing 140 000 tonnes of unwanted fat per annum. This is equivalent to a wastage of almost half a million tonnes of cereal grains each year. The reduction of this waste remains one of the great challenges to producers in the 1980s.

Factors affecting growing pig performance

The performance of the growing pig is ultimately limited by its genotype, hence the attention which has been given to improving the genetic merit of the National Herd (*see* earlier section on Breeds and Breeding). However, whether the pig achieves its genetic potential or not will depend upon its management. Environmental factors act to modify or limit the expression of genetic potential. These can conveniently be divided into two groups: 'internal factors' such as age, sex and health status and 'external factors', in particular nutrition and external environment.

Internal factors modifying performance

One of the most important factors influencing performance of the pig is its age. As the pig ages its voluntary food intake and its growth rate increase.

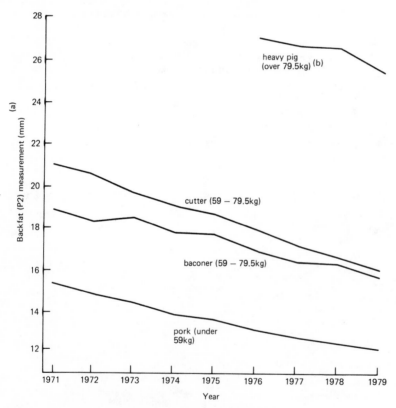

Figure 17.8 Trends in fat depth in pigs marketed at different weights (After MLC (1980) *Commercial Pig Production Yearbook 1979.* **Reproduced by courtesy of the publisher)**
　Note (a) where P1 and P3 measurements taken P2 approximated by (P1 + P3)/2
　(b) Data in earlier years insufficient to examine trends

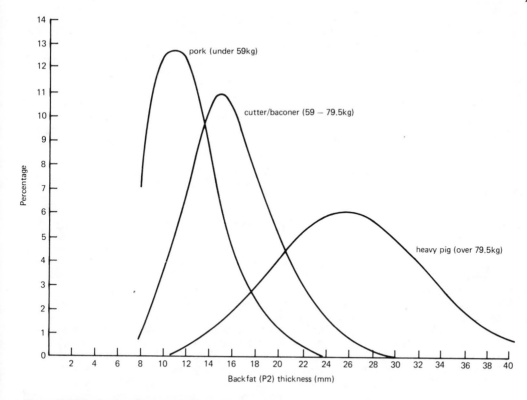

Figure 17.9 Distribution of backfat thickness (P2 measurement) in pigs of different weights (1979) (After MLC (1980) *Commercial Pig Production Yearbook 1979.* **Reproduced by courtesy of the publisher)**

However, the faster growth increasingly consists of fat tissues rather than lean. If fat deposition is excessive it reduces the value of the carcass. Furthermore as a unit of fat growth requires considerably more energy than a unit of lean growth food conversion ratio gets worse as the animal ages and the cost of producing each extra kg of weight increases. *Table 17.23* illustrates the relative performance characteristics of pigs of high genetic merit slaughtered at different weights. It can be seen that the pigs slaughtered at higher weights converted their food to lean tissue much less efficiently than pigs slaughtered at lower weights. These increased costs of production have to be related to the value of the carcass produced in determining which forms of production are most likely to generate an acceptable return on investment.

Another factor which has a significant effect on production characteristics is the sex of the animal (*Table 17.24*). For animals with the same breeding the entire male has a greater potential for lean tissue growth than the female. Castration removes this advantage; indeed castrates have a lower lean growth potential than their female litter mates. The meat trade has shown considerable reluctance to accept entire male pigs for meat production as a small

Table 17.23 Average performance of pigs slaughtered at different weights (MLC commercial pig evaluation test 1975–1978)

	Slaughter weight		
	61 kg	*87 kg*	*116 kg*
Feed conversion ratio	2.63	2.87	3.35
Daily gain (g)			
restricted scale	597	653	630
ad libitum	781	835	843
Daily feed intake on *ad libitum*			
feeding (kg)	2.02	2.38	2.76
Killing out %	73.0	75.7	79.1
P2 fat thickness (mm)	11.9	16.0	21.0
% lean in carcass	51.6	49.6	46.8
Lean tissue conversion ratio	7.00	7.71	9.10
Feed cost/kg lean (p)	95.2[1]	97.9[2]	108.3[3]

[1] Feed at £136/tonne
[2] Feed cost £127/tonne
[3] Feed cost £119/tonne

Source: MLC (1980) *Commercial Pig Evaluation Sixth Test Report.* Reproduced with permission

percentage of entires slaughtered at heavier weights have 'boar taint'. That is the meat gives off an offensive 'piggy' smell when cooked. However, meat traders are accepting more entire boars particularly at the lower slaughter weights. In 1979 12% of all pigs

Table 17.24 Effect of sex on performance and carcass characteristics of lightweight pork pigs (27 to 58 kg liveweight)

	Sex		
	Boars	Castrates	Gilts
FCR	2.67	3.46	3.22
Growth rate (g/d)	717	653	635
Killing out %	76.1	76.4	77.6
Carcass composition			
Lean	46.7	43.9	46.0
Fat	24.3	27.9	26.8
Bone	11.0	10.7	10.2
Skin	4.2	3.9	3.8

Source: Lodge, G. A. and Day, N. (1967). *Rep. Sch. Agric. Univ. Nott. 1966–67,* 62

classified (i.e. males and females) were entire males most of which weighed less than 62.5 kg carcass weight.

A discussion of the effects of health status on performance is outside the scope of this chapter. Suffice it to say that disease at clinical or subclinical levels adversely affects the growth rate of the animal and reduces the efficiency with which food is converted to saleable meat. Consequently, every effort is taken when establishing units to eliminate disease by using high health status foundation stock. In existing units appropriate preventive measures can reduce the adverse effects of disease. Preventive medicine procedures can have a very considerable cost benefit if carefully planned and successfully implemented.

External factors modifying performance

The most important and expensive single input into growing/finishing pig units is food. As this is the case and as virtually all other external factors exert their influence directly or indirectly through the utilisation of the food input it is appropriate to consider the nutrition of the pig as the central theme and examine how nutrition is influenced by other factors.

Nutrition of the growing pig

The nutrient requirement of the pig is influenced by a multiplicity of factors including its genotype, age and sex, its climatic and its social environment. The requirement is further modified by the production objectives determined for the animal. It must be recognised that there is often a discrepancy between a nutrient 'requirement' which would enable a pig to express its maximum biological potential within a given system, and the appropriate 'allowance' which will optimise profit. The optimisation of profit additionally involves other non-nutritional factors such as, availability and cost of buildings and labour, and demand for working capital. As these factors vary both between units and with time, generalisations can be misleading. Because of the dangers inherent in considering any one factor in isolation producers are relying, to an increasing extent, on computer models to predict the outcome of proposed changes in management. The following section touches briefly on some of the more important considerations.

The food consumed by the pig has first to satisfy its requirement for maintenance; any surplus is then available for growth. It follows that a faster growing pig should use its food more efficiently as the total maintenance requirement is reduced due to the shorter growing period. Pigs in a thermoneutral environment require approximately 0.5 MJ.DE/kg metabolic body weight per day for maintenance. The lower limit of

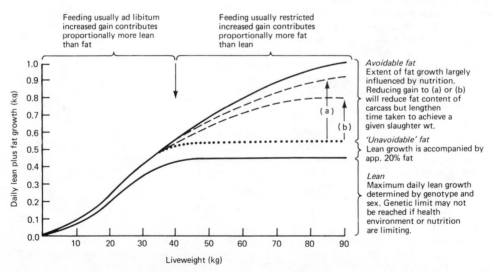

Figure 17.10 Generalised pattern of lean and fat growth in the pig

the thermoneutral range is known as the Lower Critical Temperature (LCT) at environmental temperatures below the LCT the pig has to divert dietary energy for the maintenance of body temperature. The LCT is not a fixed value but ranges from as high as 23 °C in a newly born pig to as low as 10 °C in a 100 kg pig. Pigs kept in groups on deep straw lose less body heat than pigs kept in cold, wet or draughty conditions. If pigs are kept on cold wet floors in a draught of 0.5 m/s their effective temperature may be as much as 8 °C below air temperature. For each 1 °C that the effective temperature of the pig falls below the LCT the pig requires an extra energy intake of 0.017 MJ.DE/kg metabolic body weight. For an average pig this would represent about 2 kg feed/°C 'cold' over the growing period.

Energy in excess of the requirement for maintenance is used for lean and fat tissue growth. The pig requires about 15 MJ.DE/kg lean deposited and about 50 MJ.DE/kg fat laid down. The pig does not lay down lean and fat in equal proportions throughout its life (*Figure 17.10*). As it gets older its daily liveweight gain increases and so does the percentage of fat in each 100 g of gain. Consequently as the deposition of a unit weight of fat requires almost four times as much energy as a similar weight of lean the Food Conversion Ratio (FCR) gets worse as the animal gets older.

The rate at which the animal deposits lean depends on its genotype, sex and its nutrition. An 'average' pig can probably deposit around 450 g/d lean, while gilts and boars of genetically 'improved' strains may have lean deposition rates of 530 and 590 g/d respectively. Castration of the male reduces the lean deposition rate by as much as a third. Lean deposition is always accompanied by at least 20% of its weight as fat. Thus 'improved' pigs depositing lean at 560 g/d will, in addition, deposit at least 140 g/d of 'unavoidable' fat giving a total daily gain of 700 g/d.

The appetite of the pig is such that from about 50 kg onwards its intake would support daily gains considerably in excess of these values. However, as the extra daily gain is as fat which is both expensive to produce and an undesirable component of the carcass it is usual to restrict energy intake during this period to limit fat growth. Such restrictions increase the length of time taken to achieve a given slaughter weight but normally the additional housing and labour costs entailed are more than compensated by a higher realisation price for the leaner carcass produced.

Conversely it is important to maximise growth rate prior to 50 kg. During this period lean growth rate is usually below its potential so any improvement in daily gain will be predominantly in the form of lean tissue. Estimation of protein requirement is complicated by the changing maintenance requirement (which declines from about 0.12 to 0.05% of body weight as the protein mass increases) and by the digestibility and biological value of the protein fed. These frequently have to be estimated from rather limited published information. An indication of energy and protein allowances for growing pigs is given in *Table 17.25*. The allowances are based on anticipated daily gains of 600 and 700 g/d (assuming daily lean tissue gains of 480 and 550 g/d respectively) and assume a thermoneutral environment.

The amount of nutrients the pig receives is determined by the quantity of food which it is given and the concentration of nutrients in that food (nutrient density). Providing that the necessary intake level does not exceed the pig's appetite limit diets of very different nutrient density can be used to meet the animal's requirement by making appropriate adjustments to the feed allowance. Thus the requirements of a pig weighing 60 kg growing at 700 g/d could be met by feeding 2.05 kg/d of a 'high energy' diet (13.7 MJ/DE and 159 g CP/kg) or 2.26 kg/d of a 'medium energy' diet (12.5 MJ/DE and 14.4 g CP/kg).

Table 17.25 Suggested energy and protein allowances for growing pigs[1]

| | Approximate growth rate | | | |
| | 600 g/d | | 700 g/d | |
Liveweight (kg)	Energy requirement (MJ/DE)	Protein requirement[2] (g/Cp)	Energy requirement (MJ/DE)	Protein requirement[2] (g/CP)
20	12.6[3]	185[3]	12.6[3]	185[3]
30	18.8	255	18.8[4]	255[4]
40	22.0	275	24.5	300
50	23.9	285	26.5	310
60	25.5	300	28.2	325
70	27.0	315	29.8	340
80	28.4	325	31.2	355
90	29.8	335	32.6	365

[1] Derived from ADAS (1978). *Nutrient allowances for pigs.* Advisory Paper No. 7, MAFF. Reproduced with permission
[2] Assumes protein digestibility 80% biological value 65. (Equivalent to an amino acid contribution in the protein of lysine 5.1% methionine + cystine 2.8% threonine 3.3% and tryptophan 1.0%)
[3] Growth rate 400–500 g/d
[4] Growth rate 600 g/d

Table 17.26 Suggested vitamin allowances for growing pigs

	Weight of pig		
	Up to 20 kg	25–55 kg	55–120 kg
Vitamin A (units/kg)	12 000–20 000	10 000–15 000	5000–10 000
Vitamin D (units/kg)	1500–2000	1500–2000	1000–2000
Vitamin E (mg/kg)	12–30	10–15	5–15
Vitamin K (mg/kg)	2–5	1–4	1–4
Thiamine (B1) (mg/kg)	1–3	0–2	0–2
Riboflavin (B2) (mg/kg)	3–6	3–6	3–10
Nicotinic acid (mg/kg)	20–25	10–20	10–15
Pantothenic acid (mg/kg)	10–15	10–15	5–15
Pyridoxine (B6) (mg/kg)	2.5–6	1–4	0–4
Vitamin B12 (µg/kg)	10–30	10–30	10–20
Folic acid (mg/kg)[1]	0.8	0.5	—
Biotin (µg/kg)[1]	200	150	100
Choline (mg/kg)	150	150	100
Vitamin C (mg/kg)[1]	20	10	—

[1] Where only a single value is presented there is a paucity of published evidence on which to suggest likely allowances.

Table 17.27 Suggested dietary allowances of minerals

	Liveweight		
	Up to 20 kg	20–55 kg	55–120 kg
Calcium (%)	0.6–0.75	0.6–0.90	0.6–1.0
Phosphorus (%)	0.65	0.62	0.60
Salt (NaCl) (%)	0.35–0.60	0.35–0.50	0.35–0.5
Potassium (%)	0.22	0.22	0.22
Iron (mg/kg)	50	50	50
Magnesium (mg/kg)	—	—	—
Zinc (mg/kg)	160	160	160
Copper (mg/kg)	180	180	180
Manganese (mg/kg)	80	80	80
Iodine (mg/kg)	1.5	1.5	1.5
Selenium (mg/kg)	0.2	0.2	0.1

Note: with the exception of calcium, phosphorus and salt values indicate the suggested level of micronutrient supplementation.

Because of the wide range of nutrient density in diets offered to producers and the modifying effects of environment no attempt is made here to translate allowances into feed scales.

To meet the pig's requirements over the whole growing period two or more diets differing in protein percentage would be required in order to fulfil these requirements precisely. However, some producers, particularly those using automated feed systems and relying on bulk deliveries of food find it convenient to use only one diet and accept that in the latter stages of growth protein is oversupplied. Such a decision depends upon the relative cost of the energy and protein components of the diet and the value placed by the producer on 'convenience'.

Suggested mineral and vitamin allowances are summarised in *Tables 17.26* and *17.27*. These suggested allowances include sufficient safety margin to be applicable over the normal range of nutrient density encountered. The higher vitamin allowances are suggested for animals of superior genotype.

18

Poultry

John I. Portsmouth

Poultry are kept in the UK primarily for meat and egg production. The two sectors are economically different, but naturally have physiological similarities, e.g. meat chickens are hatched from eggs produced by specially bred parent stock and commercial eggs are produced by specialist bred parents. The distinction is thus in the selection of meat or egg production characteristics. The utility fowl of the 1950–60s no longer exists. Chickens are specifically bred, managed and marketed to meet a very definite market requirement of size, quality, quantity and price.

MEAT PRODUCTION

For a summary of UK poultry meat production, *see Table 18.1*. Total output increased by 12% in the eight-

year period with turkey meat expanding by 22% and chickens by 9.7%. Waterfowl decreased by 5.9%. Between 1972–75 total output decreased but increased in 1975–79. A period of slower growth is anticipated in the early 1980s as different sectors of the poultry meat industry concentrate on developing new markets for newer products as saturation point approaches for whole carcass sales. Such products are rolls, sausages, portions and meat loaves.

Figures in *Table 18.2* show how poultry meat consumption has increased by 40% in 14 years—an annual increase of some 2.8%. Turkey meat consumption on the other hand has increased by 63% (4.5% per annum).

Further increases in poultry meat consumption are likely to occur from the 'further-processing' of whole birds, i.e. portions, chicken/turkey roll, smoked

Table 18.1 UK poultry meat production 1972–79

	1972	1973	1974	1975	1976	1977	1978	1979
Numbers (a) (10^6)								
Fowls—over 6 months	46.1	39.1	45.6	44.0	39.1	39.7	41.3	41.7
Fowls—under 6 months	326.4	331.5	320.0	324.2	352.2	364.7	362.8	371.0
Total fowls	372.4	370.6	365.6	368.2	391.3	404.4	404.1	412.7
Ducks	7.7	7.9	6.9	4.9	6.5	6.4	6.9	7.1
Geese	0.1	0.1	0.1	0.1	0.1	0.1	0.1	0.1
Turkeys	17.3	18.9	18.4	16.9	18.8	19.5	22.2	22.5
Total poultry	397.5	397.6	390.9	390.1	416.7	430.4	433.3	422.5
Meat (b) (10^3 tonnes)								
Fowls—over 6 months	69.4	59.0	68.8	66.4	59.6	61.0	63.5	63.1
Fowls—under 6 months	482.6	488.6	472.0	480.1	518.0	538.2	540.7	556.1
Total fowls	552.0	547.2	540.8	546.5	577.6	599.2	604.2	619.1
Ducks	16.6	17.1	14.8	12.8	14.0	13.8	14.9	15.5
Geese	0.4	0.7	0.5	0.5	0.6	0.5	0.5	0.5
Turkeys	86.1	94.1	91.0	84.2	93.3	95.4	99.8	110.3
Total poultry	655.2	659.5	647.1	644.0	685.5	708.9	719.4	745.4

Source: after MAFF official statistics. Reproduced with permission.

Table 18.2 Poultry meat consumption (kg/head per year)

	Total	Turkey meat	Chicken
1966	8.09	0.73	7.36
1967	8.59	0.77	7.82
1968	9.68	0.86	8.82
1969	10.09	1.00	9.09
1970	10.45	1.23	9.22
1971	10.68	1.27	9.41
1972	12.05	1.55	10.50
1973	11.77	1.68	10.09
1974	11.64	1.62	10.02
1975	11.40	1.52	9.88
1976	11.50	1.65	9.85
1977	11.40	1.66	9.74
1978	11.80	1.80	10.00
1979	13.50	1.97	11.53

Source: British Poultry Meat Association. Reproduced with permission.

poultry meat and sausages. Waterfowl meat consumption has changed little in the last five years and no change in this trend is envisaged.

Broilers

Of total poultry meat produced in the UK 75% is derived from broilers. A broiler is a young bird, of either sex, marketed between 1.45 kg and 2.75 kg liveweight. It is slaughtered between 40 and 56 d (approximately) depending on the range of body weights needed and the most economic time of production. The broiler is a specifically bred hybrid meat bird derived from breeds and strains originally developed in the USA. Basic leading UK breeding

companies are: Cobb (UpJohn), Ross (Imperial Foods), Shaver (Cargill), Hubbard (MSD), Marshall (name in bracket indicates major shareholding company).

Depending on killing age a broiler will convert food into meat at about 2:1. Optimum economic performance depends on correct nutrition and husbandry conditions. Work by ADAS, at Gleadthorpe, EHF, shows that broilers respond optimally to high nutrient density (HND) feeds when housed at 21 °C and killed at 47–49 d. *Tables 18.3, 18.4, 18.5, 18.6* and *18.7* summarise the nutritional needs of broilers.

It is anticipated that by 1982 the figures quoted in *Table 18.5* will be achieved in 2 d less and by the end of the present decade a 2 kg liveweight will be reached in 35 d.

Anticoccidial drugs are an essential part of an overall disease prevention and control programme. Without these drugs modern broiler/turkey production under intensive husbandry methods would be

Table 18.4 Broiler feed programmes (kg/1000 birds)

	40–47 d kill	48–53 d kill	+53d kill
Starter	1000	1000	1000
Grower	1500	1250	1000
Finisher 1	to finish	to finish	2000
Finisher 2	—	—	to finish

The amount of each feed is fed until completely used.
Source: Peter Hand (GB) Ltd—technical information. Reproduced with permission.

Table 18.5 Broiler performance guide

	Days at killing			
	42	45	52	59
Av. liveweight (kg)	1.59	1.77	2.13	2.45
FCR	1.85	2.0	2.10	2.18
Livability %	97.5	96.5	96.0	95.5

Source: Peter Hand (GB) Limited—technical information

Table 18.3 Recommended broiler ration specifications

Nutrient	Feed type			
	Starter	Grower	Finisher 1	Finisher 2
Protein %	23	20	19	17.5
Lysine %	1.30	1.15	1.05	0.95
Av. lysine %	1.20	1.04	0.95	0.85
Methionine (M) %	0.60	0.54	0.46	0.40
M + cystine %	0.95	0.82	0.76	0.70
Energy ME kcal/kg	2995	3060	3100	3080
Energy ME MJ/kg	12.51	12.80	12.97	12.88
Calcium (Ca) %	0.90	0.90	0.90	0.90
Phosphorus (P) %	0.70	0.70	0.70	0.65
Av. phosphorus %	0.45	0.45	0.45	0.42
[1] mEq/100 g	22–23	22–23	22–23	22–23
Linoleic acid % (EFA)	0.80	0.80	0.60	0.60

[1] Sum of sodium + potassium − chloride ions to obtain satisfactory acid–base balance.
Source: Peter Hand (GB) Limited—technical information. Reproduced with permission.

Table 18.6 Micronutrients recommended for inclusion in broiler diets

Broiler supplements	Broiler starter	Broiler finisher
Usage (kg/tonne)	5	5
Calcium (%)	—	—
Vitamin A (m units)	12	12
Vitamin D3 (m units)	4	4
Vitamin E (g)	10	5
Vitamin K3 (g)	2	2
Vitamin B1 (g)	1	1
Vitamin B2 (g)	5	5
Nicotinic acid (g)	30	20
Pantothenic acid (g)	8	6
Vitamin B12 (mg)	10	10
Vitamin B6 (g)	3	1
Choline (g)	250	200
Folic acid (g)	1	0.5
Biotin (mg)	150	100
Ethoxyquin (g)	10	10
Manganese (g)	100	70
Zinc (g)	60	60
Copper (g)	10	8
Iodine (g)	1	1
Cobalt (g)	0.5	0.5
Iron (g)	25	20
Selenium (g)	0.15	0.15
Molybdenum (g)	0.5	0.5
Zinc bacitracin (g)	20	20

Source: Peter Hand (GB) Ltd, with permission.

impossible and the disease, coccidiosis, would cause great economic loss.

Antiblackhead drugs are an important part of blackhead disease control. Their use in continuous preventive medication is, however, relative to specific disease situations rather than in 'blanket' medication.

Growth promoting drugs such as the antibiotic zinc bacitracin are used in about 95% of the UK broiler feeds. It is generally acknowledged that their presence probably enhances growth and improves feed conversion efficiency between 2 and 5%.

COMMERCIAL EGG PRODUCTION

The UK laying flock has been declining since 1968. This is due to a fall in demand for eggs and an increase in egg production per bird due to technological advancement. *Figures 18.1* and *18.2* illustrate laying flock number and egg output. Total number of poultry farms with laying birds has fallen from 125 258 in 1971 to under 70 000 in 1978. It is currently estimated to be below 60 000. The egg industry concentrates in fewer and larger units with over 50% of the country's eggs produced from about 450 flocks averaging 20 000 or more laying birds each. The trend to fewer and

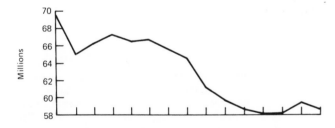

Figure 18.1 Laying flock (From *Poultry World*, 17 July 1980. Reproduced by courtesy of the editor and publisher)

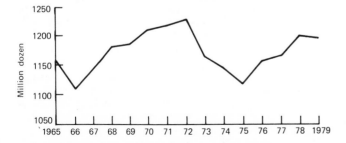

Figure 18.2 Egg output (From *Poultry World*, 17 July 1980. Reproduced by courtesy of the editor and publisher)

Table 18.7 Additives commonly used in poultry production without veterinary prescription

Additive	Stock type	Age (weeks)	Level of active ingredient in final feed (ppm)
Growth and egg promoters			
Avotan	Broilers	0–16	7.5–15 avoparcin
Eskalin	Broilers	0–9	0–20 virginiamycin
Flavomycin	Broilers	0–16	2–5 flavomycin
	Turkeys	0–26	2–5 flavomycin
	Laying hens		2–5 flavomycin
Payzone	Broilers	0–16	10 nitrovin
	Turkeys	0–12	10 nitrovin
Zinc bacitracin	Broilers	0–4	5–50 zinc bacitracin
	Broilers	4–16	5–20 zinc bacitracin
Albac	Turkeys	to 26	5–20 zinc bacitracin
	Laying hens and breeders		15–100 zinc bacitracin
ZB100	Broilers	0–4	5–50 zinc bacitracin
	Broilers	4–16	5–20 zinc bacitracin
	Turkeys	to 26	5–20 zinc bacitracin
	Layers		15–100 zinc bacitracin
Anti-coccidials[1]			
Elancoban	Broilers, pullets	0–16	100–120 monsensin
Pancoxin[1]	Broilers, pullets	0–16	50–100 amprolium
	Broilers, pullets	0–16	100 amprolium
	Pullets	0–6	80 amprolium
	Pullets	6–16	65 amprolium
	Pullets	12–16	50 amprolium
Supacox	Broilers	0–16	100 amprolium
Stenerol	Broilers	0–16	3 halofuginone
Arpocox[1]	Broilers	0–16	60 arpinocid
Stenerol[1]	Turkeys	until 5 d before slaughter	3 halofuginone
Anti-blackhead medications			
Emtryl	Turkeys		125 dimetridazole
Salfuride	Turkeys		50 nifursol

[1] Where an additive contains more than one active ingredient only the major constituent is shown.
Source: *Poultry World*, 1980 Feed Facts. Reproduced with permission.

Table 18.8 Selection of egg producing strain

Name and code	Feather colour	Company
Babcock B300V	White	Babcock
Dekalb XL–Link	White	Dekalb
Hisex White	White	Euribrid
Nick Chick	White	H & N
Hubbard White Leghorn	White	Hubbard
Ross White	White	Ross Poultry
Shaver 288	White	Shaver
Ross Tint	Flecked	Ross Poultry
Babcock 380	Brown	Babcock
Dekalb–Amber Link	Brown	Dekalb
Hisex Brown	Brown	Euribrid
Golden Comet	Brown	Hubbard
Ross Brown	Brown	Ross Poultry
Shaver 585	Brown	Shaver
Tetra SL	Brown	Tetra
CT 404	Brown	Thornber
Warren	Brown	ISA

Source: Peter Hand (GB) Limited—technical information, with permission

larger units will continue unless action by welfare activists moves the industry away from the battery cage and back to the less intensive management systems where husbandry practice favours the smaller unit.

In the late 1950s and early 1960s the laying flock was made up of predominantly white feathered birds producing white eggs. In 1975 64% of the flock was brown feathered producing brown eggs. In 1979 alomost 90% were brown egg layers and in 1980 this figure has reached 95%.

The demand for brown eggs (including tinted) has caused breeding companies to intensify efforts into increasing feed efficiency, which currently favours white egg layers, because of their smaller size and lower energy needs for maintenance. *Table 18.8* shows the main 'breeds' and companies involved in breeding commercial laying stock in the UK.

Tables 18.9 and *18.10* show egg yields for hens housed in different husbandry systems and also per capita egg consumption since 1960.

Table 18.9 Egg production/hen per annum by management system

Management system	1960/61	1972/73	1973/74
Free range	166.5	181.2	172.9
Deep litter	187.9	238.5	200.9
Battery cage	206.2	214.0	232.3
	1974/75	1976/77	1978/79
Free range	175.7	192.1	196.8
Deep litter	207.0	224.5	227.8
Battery cage	236.4	245.4	249.3

Source: Eggs Authority, with permission

Table 18.10 1960–79 annual egg consumption in UK (no. eggs/person)

	Shell products (egg equivalent)		Total
1960	237	21	258
1961	243	21	264
1962	243	21	264
1963	238	22	260
1964	249	24	273
1965	250	21	271
1966	246	24	270
1967	249	26	275
1968	250	23	273
1969	248	23	271
1970	251	24	275
1971	254	19	273
1972	253	20	273
1973	244	18	262
1974	240	16	256
1975	232	14	246
1976	237	11	248
1977	236	12	248
1978	238	12	250
1979	239	12	251

Source: Eggs Authority, with permission

Table 18.11 Recommended intake of certain essential nutrients to produce 55 g egg output

Nutrient	Minimum daily intake	% Nutrient in ration at specified average daily intake		
		100 g	110 g	120 g
Protein	18.0 g	18	16.4	15
Methionine	400 mg	0.40	0.36	0.33
Methionine + cystine	640 mg	0.64	0.58	0.53
Lysine	860 mg	0.86	0.78	0.72
Calcium	3.8 g	3.8	3.45	3.20
Av. phosphorus	0.35 g	0.35	0.32	0.30
Linoleic acid (EFA)	1.25 g	1.25	1.14	1.04
[1]Xanthophylls (Roche Fan 10)	14 mg/kg	14	12.75	11.7
Calories MJ/d	1.19	1.19	1.31	1.42

[1] Recommended level of xanthophylls ensures a good yellow–orange yolk colour.
Source: Peter Hand (GB) Ltd—technical information, with permission

In 1960–61 birds housed in cages on average produced 18.3 eggs more than deep litter flocks, whilst in 1978–79 this widened to 21.5 eggs/bird. The average of 249 eggs/bird is in fact insufficient to cover the production costs of 1980. Profitable flocks need to produce 260 eggs on low feed intakes.

Reverting to the deep litter management system could, at today's costs and returns, bankrupt the whole industry. The consumer would have to pay almost twice as much to compensate for the loss in efficiency and higher capital costs of housing and land required by alternative management systems.

To obtain high and economic production certain minimum amounts of nutrients must be provided. These are shown in *Table 18.11*.

Feed programme for laying hens

Laying hens are invariably fed *ad libitum*. The adjustment of feed intake to changes in ration energy level is not precise but sufficiently accurate to make *ad libitum* feeding the most economic system. Restricted feeding, whereby a set allowance is allocated once or twice daily, is hazardous. The variation in feed consumption within a flock may be as much as 40%, thus any physical restriction of feed will penalise the small appetite to the great detriment of egg output and profit. Free choice feeding, whereby cereals fed whole with a pelleted protein/vitamin/mineral concentrate allows the individual bird to select its own protein (amino acid) needs according to rate of lay, has shown nutritional/physiological advantages. The system needs management perfection before finding practical use.

Feed restriction for replacement pullets

Excess body fat adversely affects subsequent rate of lay. Consequently, replacement pullets and replacement broiler breeding stock are fed controlled amounts of feed to reduce excessive energy consumption. With commercial pullets the mid-term restriction during rearing is more important than early and late restriction. This means that the severest level of restriction occurs in the 9–15 week period and to a lesser extent in the 6–8 and 16–20 week periods. Compared to conventional full-term restriction the severe mid-term restriction gives good laying house performance and greatest economic returns.

Feed control with broiler breeders is vital if excess energy consumption is to be prevented. Broiler breeder companies have recommended programmes. These should be closely followed for maximum economic performance. Care is necessary to ensure that 'non-energy' nutrients are adequate. If they are not, both peak lay, hatchability and chick quality will be unsatisfactory.

Table 18.12 details the essential micronutrients such

Table 18.12 Micronutrient recommendations for commercial layers, replacement and breeding stock

	Poultry breeder	Chick and grower	Poultry layer
Usage (kg/tonne)	2.5	2.5/2	2.5
Vitamin A (m units)	12	10	6
Vitamin D3 (m units)	3	3	3
Vitamin E (g)	10	6	4
Vitamin K3 (g)	2	2	2
Vitamin B1 (g)	1	0.5	0.5
Vitamin B2 (g)	8	5	3
Nicotinic acid (g)	25	20	8
Pantothenic acid (g)	10	8	2
Vitamin B12 (g)	6	6	3
Vitamin B6 (g)	4	0.5	1
Choline (g)	150	100	—
Folic acid (g)	2	0.5	—
Biotin (mg)	100	—	—
Ethoxyquin (g)	10	10	10
Manganese (g)	70	65	70
Zinc (g)	60	40	40
Copper (g)	8	8	6
Iodine (g)	1	1	1
Cobalt (g)	0.5	0.5	0.5
Iron (g)	20	15	20
Selenium (g)	0.1	0.1	0.1
Molybdenum (g)	0.5	—	—
Methionine (g)	450	—	800

Source: Peter Hand (GB) Limited—technical information, with permission

as vitamins and trace elements which are used to supplement the normal ingredients of poultry rations.

Environment

Table 18.13 shows the ventilation rates needed for intensively housed poultry. The optimum temperature for adult layers is 21 °C and for broilers, after initial brooding, 20–21 °C depending on sex and ration nutrient density.

Ventilation rate is also expressed according to the amount of feed consumed. Thus, the optimum minimum for broilers is 2 mstd (cubic metres of air per second/tonne feed/day).

Maximum ventilation is ten times minimum (i.e. 20 mstd).

NB. 1.0 mstd is approximately equal to 1.0 cfm/(lb d) (cubic feet of air per minute/poundfeed/day). This standard therefore relates closely to the metric m^3/s/tonne/d, 25 cfm/lb/d being equivalent to 25 mstd, and 2 cfm matching 2 mstd.

Lighting

Because nearly all modern poultry units employ windowless and controlled environment houses artificial lighting, both in extent and intensity is extremely important. In laying birds changes in artificial daylength stimulate or depress the secretion of gonadotrophic hormones. Increasing daylength stimulates and *vice versa*. Change in daylength can be used to manipulate age at sexual maturity. Extending the 'day' in rearing advances sexual maturity. With laying birds daylength is generally increased from about 8 h at 20 weeks to a maximum of 16–17 h achieved by weekly increases of 30 min each. Daylength is then held at this level to the completion of laying. Rate of lay can be manipulated by so-called ahemeral lighting patterns where daylength may be more or less than 24 h cycles. With a longer daylength, i.e. 26–28 h, egg weight is increased, whilst with cycles less than 24 h rate of production can be increased at the expense of egg weight. Ahemeral patterns are currently not widely used in practice but may find application in selection of breeding stock in the future.

Table 18.13 Ventilation rate requirements for intensively housed poultry

Stock	Weight (kg)	Max. needed		Min. needed	
		m^3/h bird	cfm/bird	m^3/h bird	cfm/bird
Layers	1.2	10	6.0	1.0	0.6
	2.0	12	7.0	1.2	0.75
	2.5	14	8.0	1.5	0.9
	3.0	14	8.0	1.7	1.0
	3.5	15	9.0	2.0	1.2
Broilers	0.05			0.1	0.06
	0.4			0.5	0.3
	0.9			0.8	0.45
	1.4			0.9	0.50
	1.8	10	6.0	1.3	0.75
	2.2	14	8.0	1.7	1.0
Turkeys	0.5	6	3.5	0.7	0.4
	2.0	12	7.0	1.2	0.7
	5.0	15	9.0	1.5	0.9
	7.0	20	12.0	2.0	1.2
	11.0	27	16.0	2.7	1.6

Source: MAFF, Gleadthorpe EHF. Reproduced with permission.

In broiler production the most popular lighting system is 24 h for the first 48–72 h, followed by a 23-h day to slaughter. The 1 h dark period is used to accustom the birds to darkness in the event of power failure.

Intermittent lighting programmes are also practised. After the continuous 24 h for the first 72 h, the programmes may be 2 h light followed by 2 h darkness, or 1 h light and 3 h darkness. Intermittent programmes are thought to improve efficiency of digestion through the enforced rest of the dark period.

Light intensity

Under intensive housing systems low light intensity is essential to avoid feather pecking and cannibalism. Additionally, birds are quieter, easier to manage and less active, thus using less energy for maintenance. Young chicks should be given 20–30 lx for the first 72–96 h to encourage early feeding. Thereafter, 6–7 lx is sufficient to allow normal feeding, optimum activity and ease of operations. For laying birds, the optimum intensity is 5–7 lx. In the dark period one-tenth of the optimum figure is recommended.

TURKEYS AND WATERFOWL

Turkeys

Figure 18.3 shows how turkey production has increased three-fold since 1965. Disastrously low selling prices due to overproduction in 1973 saw drastic cutbacks in 1974 and more severely in 1975. Since then better organised marketing in the Easter, Whitsun and August Bank Holiday periods have led to steady expansion. The market prospects for Christmas 1982 and the first half of 1983 are good with production and demand in balance. However the situation is affected by expansion in the EEC, particularly Italy and France, and the prospect of imports affecting the UK market is a serious matter.

In 1971, 2300 flocks produced the 14 million turkeys, whilst in 1978 this number had fallen to about 1064 for 1.6 times the output (22.5 million). Thus, the trend of fewer and larger is seen in the turkey industry. *Table 18.14* shows how in 1978 some 86% of all turkeys were produced by units with in excess of 100 000 birds compared to 7% produced by flocks of 5000 to 10 000 birds!

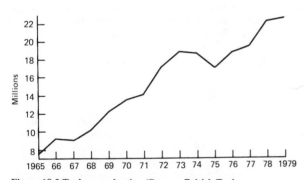

Figure 18.3 Turkey production (Source. British Turkey Federation, with permission)

Table 18.14 Number of turkeys in England and Wales and proportion of flock by size of flock

Size of flock	1976		1977		1978	
	× 10³	%	× 10³	%	× 10³	%
1–25	3.8	0.1	3.3	0.1	3.1	0.1
26–99	6.8	0.1	5.3	0.1	5.6	0.1
100–499	45.0	0.8	32.9	0.8	35.2	0.7
500–999	53.9	1.0	36.3	0.8	43.7	0.9
1000–4999	371.2	6.6	290.3	6.0	278.4	5.5
5000–9999	425.5	7.6	351.1	7.3	360.3	7.1
100 000 and over	4720.2	83.8	4110.8	85.0	4370.1	85.8
Totals	5626.4	100.0	4830.0	100.0	5096.5	100.0

Source: MAFF statistics. Reproduced with permission.

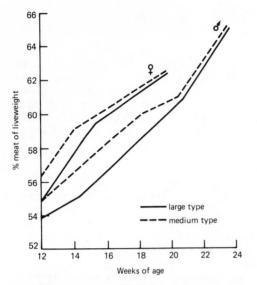

Figure 18.4 Relationship between % meat/liveweight and turkey age (From *Poultry World* 26th June 1980. Reproduced by courtesy of the Editor and publisher)

Table 18.15
Turkey carcass analysis

	kg	%
Weight	5.2	100
Breast meat	1.634	31.42
Leg meat	1.31	25.20
Visceral fat	0.007	0.14
Total bone	1.493	28.71
Total meat	3.277	63.02
Total meat/skin	3.70	71.15

Source: MAFF, Gleadthorpe EHF. Reproduced with permission.

Turkeys are into a relatively new phase of marketing with expansion into the cut-up, portioned and further processed markets. Turkey boned rolls, sausage, paté, burgers are all examples of markets being pursued with success. This market needs a large bird with a high meat-to-bone ratio.

Such birds are 20–26 weeks old and may weigh 12 kg to 15 kg compared to the whole bird market which demands a 4–6 kg oven-ready bird.

Table 18.15 shows the carcass analysis of 5.2 kg male turkeys (from MAFF, ADAS, Gleadthorpe 1980).

Figure 18.4 shows the relationship between percentage meat of liveweight and turkey age. For a 22 week old medium stag some 63% of total liveweight is meat. In the data of *Table 18.15*, the 5.2 kg bird is 14 weeks old and such birds have a 55–56% meat yield of total liveweight. Thus, the larger the bird the greater its meat yield and usefulness in further processing.

Consumption per capita of turkey meat is shown in *Table 18.2*. The average rate of increase is 4.5% per annum. In the last five years this has been due to a

Table 18.16 Management standards—Turkeys. Growth rate and food conversion

Age (weeks)	Live weight (kg)	Total food consumed (kg)	Cumulative FCR
1	0.13	0.11	—
2	0.27	0.29	1.07
3	0.45	0.60	1.33
4	0.75	1.13	1.51
5	1.18	1.86	1.57
6	1.68	2.77	1.65
7	2.27	3.86	1.70
8	2.84	5.13	1.80
9	3.47	6.59	1.90
10	4.18	8.20	1.96
11	4.77	10.0	2.09
12	5.54	11.8	2.13
13	6.22	13.8	2.21
14	6.86	15.9	2.31
15	7.5	18.2	2.42
16	8.2	20.7	2.52
17	8.77	23.1	2.63
18	9.32	25.4	2.72
19	10.0	28.2	2.82
20	10.5	31.0	2.95

Source: Peter Hand (GB) Ltd—technical information, with permission.

greater consumption of turkey meat at times other than Christmas, e.g. Easter and Whitsun.

Table 18.16 shows growth rate, feed consumed and feed conversion for turkeys from 1 to 20 weeks of age. These figures are at best a guide to performance as strain differences, etc., are wide.

Tables 18.17 and *18.18* deal with the nutritional requirements of turkeys of all ages.

Waterfowl

Duck and goose meat production is small in comparison to turkeys and chicken. *Table 18.19* shows the growth in the duck market from 1970 to 1979 in the UK and other EEC countries. By far the largest output is from France with almost 40% of the total. The UK produces nearly 17% of EEC duckling production.

Based on 1979 figures consumption per capita is less than 0.25 kg. Duck meat is still regarded as a luxury. The majority of the UK duck production is from the world's largest producer, Cherry Valley Farms Limited, Lincolnshire.

The two mains breeds are Aylesbury and Pekin. Plumage of both is white. The Pekin hybrids commonly used in commercial production, have better reproductive qualities than the Aylesbury.

The growth rate of duckling is far superior to chicken and turkey, as *Table 18.20* shows.

For a given age under eight weeks, the duckling outgrows all other poultry. Its food conversion however, is inferior. This is due to the higher feed

Table 18.17 Nutritional requirements of turkeys

	Pre-starter	Starter	Rearer	Early finisher	Late finisher	Grower	Heavy stag
Protein %	29	27	24	21	15	18	14
Methionine (M) %	0.63	0.61	0.56	0.52	0.36	0.40	0.32
Lysine %	1.80	1.70	1.45	1.32	0.76	0.90	0.70
M + cystine %	1.15	1.10	1.02	0.95	0.65	0.70	0.60
ME MJ/kg	12.0	12.0	12.0	12.0	11.8	11.6	11.5
Calcium %	1.20	1.20	1.20	1.30	1.40	1.20	1.40
Phosphorus %	0.85	0.80	0.80	0.75	0.72	0.80	0.72
Av. phosphorus %	0.70	0.65	0.65	0.60	0.55	0.65	0.55
Salt %	0.36	0.36	0.36	0.36	0.42	0.42	0.42
Na %	0.16	0.16	0.16	0.16	0.17	0.17	0.17

Source: Peter Hand (GB) Ltd—technical information, with permission.

Table 18.18 Micronutrients recommended for inclusion in Turkey diets

Turkey supplements	Turkey starter	Turkey grower/ finisher	Turkey breeder
Usage (kg/tonne)	5	5/4	5
Vitamin A (m units)	16	12	16
Vitamin D3 (m units)	5	5	4
Vitamin E (g)	15	10	20
Vitamin K3 (g)	4	3	2
Vitamin B1 (g)	2	1	1
Vitamin B2 (g)	10	8	12
Nicotinic acid (g)	60	45	50
Pantothenic acid (g)	15	12	15
Vitamin B12 (mg)	8	6	8
Vitamin B6 (g)	4	2	2
Choline (g)	500	300	200
Folic acid (g)	1	0.5	3
Biotin (mg)	120	80	50
Ethoxyquin (g)	10	10	10
Manganese (g)	80	70	80
Zinc (g)	65	60	70
Copper (g)	15	10	8
Iodine (g)	1	0.5	1
Cobalt (g)	0.5	0.5	0.5
Iron (g)	50	40	20
Selenium (g)	0.15	0.15	0.10
Molybdenum (g)	—	—	0.5
Zinc bacitracin (g)	20	—	—

Source: Peter Hand (GB) Limited—technical information, with permission.

Table 18.19 Duck hatchings ($\times 10^3$) in EEC countries

	1970	1971	1972	1973	1974	1975	1976	1977	1978	1979
West Germany	4760	4839	4233	4227	3710	2301	2943	3779	4176	4103
Netherlands	4026	2983	3814	3704	3519	2672	3314	3019	3141	5915[1]
Belgium/Luxembourg	291	255	333	337	350	293	268	259	232	256
Italy	2899	3568	2900	2986	2048	2518	3368	4476	4264	4440[1]
France	6106	6399	10 214	13 672	12 988	13 888	14 085	14 797	15 492	17 538
Denmark	na	na	na	2989	2829	2542	2809	2769	2717	2858
Eire	na	289	403	594	514	626	1191	1206	1236	1263[1]
UK	6400	6000	7500	7700	7400	6400	6803	7202	7667	7569[1]
EEC	na	na	na	36 249	33 358	31 232	34 780	37 507	38 924	43 942

[1] Jan-Nov
na = not available
Source: After MAFF and AGRA Europe. Reproduced with permission.

Table 18.20 Comparative growth performance

	Broiler strain	Duckling	Turkey
Age (d)	47	47	47
Liveweight (kg)	1.8	3.1	2
FCR	2.0	2.8	1.75
Feed consumed per bird (kg)	3.6	8.68	3.5

Source: Peter Hand (GB) Limited—technical information, with permission.

Table 18.21 Comparison of cooking losses in ducks, turkeys and broilers

	Duck	Turkey	Broiler
(1) Liveweight (kg)	2.7	3.6	1.8
(2) Bled/plucked weight of liveweight (1)	82%	90%	87%
(3) Eviscerated weight of (1)	73%	70%	75%
(4) Cooking loss of (1)	37%	31%	22%
(5) Field of edible cooked meat of (1)	25%	34%	27%

Source: After Hollows, (1978). *Productivity of Ducks*. Society of Feed Technologists. Reproduced with permission.

consumption per unit of body weight gain and it is this which makes duckling a relatively expensive meat to produce.

The yield of edible cooked meat is lower for duck compared with turkeys and broilers. *Table 18.21* compares the cooking losses.

GEESE

Goose production is the poor relation of the poultry industry. Few accurate records are available to show how many are produced annually. Many are raised on general farms and are unrecorded. Sales mainly occur at festive occasions, particularly Christmas. Being extremely good grazers and converters of grass into meat, the most profitable management and feeding system utilises this phenomenon.

Most popular breeds are Embden, Toulouse, Roman and hybrids involving crosses of these major breeds. Weights vary depending on feeding system and age at killing with Embdens reaching 14–15 kg by six months, whilst the small Roman may only achieve 6 kg. The Toulouse is slightly smaller than the Embden.

After feeding on a good proprietary chick pellet for three to four weeks after hatching, the growing goose is capable of surviving and growing on good quality grassland and small supplements of feed, which can be dispensed with after eight weeks. A satisfactory finish can be achieved by grain feeding in the last month of fattening. It is usual to restrict the range area at this time to conserve energy requirement. Penning the geese in straw yards is sometimes used in East Anglia where straw forms a cheap wall and bedding.

The selling of day old or part grown goslings can be a profitable sideline. Breeding geese often mate for life and this strong 'pairing bond' is established in the autumn prior to a laying season in early spring. Normally one gander pairs with three to four geese. For best fertility the gander should be one year or more older than the geese. Young geese show reproductive improvement with age through to ten years or more.

Egg production varies, but may range from 30 to 75 eggs/goose per year. The eggs are best hatched artificially as geese make indifferent incubators and mothers.

19

Horses

M. Ponting

The horse is now re-established as an important income earning animal on the farm as more farmers are now deriving some of their production from equine pursuits of various kinds. These enterprises, such as riding stables, livery yards and studs can often be run in conjunction with normal farming operations and provide a useful income supplement.

Since the lapse of interest in horses, with the advent of the motor, many wishing to keep horses have little or no experience of them. It is timely, therefore, to restate the principles of good horse management. Horses, like any other class of stock, respond best if the standard of management and stockmanship is high and many of the rules relating to other farm stock can be applied to the horse.

BREEDS

The horse as a species shows an enormous degree of variation between breeds and types, ranging, for example, from the small Shetland pony standing 8 hands high, to the Shire horse at 17 hands (*see Figure 19.1*). Accordingly, horses can be classified into heavy draught, light draught, riding horses and ponies. As well as purebreds many crossbreds are produced, combining the features of the parent breeds.

Heavy draught breeds

Shire, Clydesdale, Suffolk Punch and Percheron are

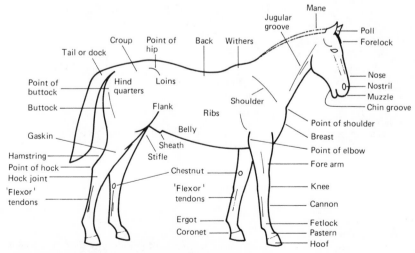

Figure 19.1 Points of the horse. The horse's height is measured at the highest point of the withers. 1 hand = 10.16 cm = 4 in (e.g. 15 hands = 152.4 cm)

425

heavy draught breeds. Having declined to very low numbers, all these breeds are now increasing in popularity and an important export market exists.

Shire

The largest of the heavy draught horses in this country, with a mature stallion standing 17 hands high and weighing 863–1016 kg. Very large feet and an abundance of hair or 'feather' on the legs and fetlocks typify this willing and majestic animal.

Clydesdale

Originally developed in Scotland, the Clydesdale is often recognised by a white blaze on its face and legs white to the knees. The mature stallion stands 16.2 hands and the mare 16 hands high.

Suffolk Punch

Originating from the eastern counties this horse is invariably chestnut in colour and clean-legged. Average height is about 16.1 hands.

Percheron

This French breed was introduced into Britain shortly after the First World War and is typically grey in colour. The stallion stands from 16.2 to 17 hands and mares average 16.2 hands.

Light draught breeds

Cleveland Bay

A large breed standing 16.0–16.2 hands high and bay in colour, developed as a coach horse. Size and good temperament have led to extensive use for crossing with thoroughbreds to produce hunters and 'event' horses. The former role as a harness horse is causing renewed interest due to the popularity of driving as a sport.

Hackney

A harness breed standing 15 hands with a characteristic, exaggerated leg action. Used mainly in harness in the show ring.

Riding horses

Thoroughbred

The English Thoroughbred is the world's most famous horse. A quality animal, with a fine appearance and possessing great speed. Used extensively as a racehorse but contributing to all equine riding sports and widely used for crossing to improve the quality of native stock.

Arab

A smaller breed than the Thoroughbred, not usually standing more than 15 hands high. A quality animal used as a high class hack and also for crossing.

Hunter

Not a breed but a type of animal capable of crossing country at speed, carrying considerable weight. The thoroughbred sire is often used on a light draught mare to produce a horse with substance, quality, speed and endurance.

Ponies

The native British breeds of ponies are types developed in certain geographic areas in response to the nature of those areas, and are typically hardy and strong for their size. Due to their usual good temperament they are used widely as children's ponies.

Welsh Mountain

An extremely popular pony now enjoying world-wide distribution as children's mounts and quality show ponies. Their smart action earns them a place as driving ponies. Several sections exist, namely:

Section A: up to 12 hands
Section B: 12–13.2 hands
Section C: up to 13.2 hands—Welsh pony of Cob type
Section D: the Welsh Cob, 14–15 hands approximately.

New Forest

A breed standing 12.2–13.2 hands high and possessing considerable speed.

Dartmoor

A small pony standing about 12.2 hands high, with a good temperament.

Exmoor

This breed is typified by its mealy-coloured muzzle and stands 11.2–12.2 hands high.

Fell

A large pony, 13.0–14.0 hands high and fairly thick set.

Dale

A very similar breed to the Fell, probably standing a little higher. Both Dale and Fell breeds have considerable feather on the fetlocks and abundant manes and tails.

Shetland

A very small hardy breed, 7.0–10.0 hands high. Very popular for small children and widely exported.

Highland

A fairly variable breed standing 13.0–14.2 hands high. A very sturdy animal, sure footed and docile. Many attractive dun colours exist and a marked dorsal stripe is often visible.

AGEING THE HORSE

As with other classes of stock, it is possible to estimate the age of horses by examining the teeth, although this method requires considerable practice. The incisor teeth in particular are used for ageing and there are 12, six in the upper and six in the lower jaw (*see Figure 19.2*).

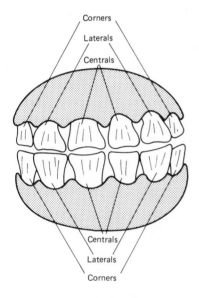

Figure 19.2 Front view of the horse's teeth

The foal is born with the two temporary central incisors already erupted. The temporary lateral incisors erupt at about four weeks and the temporary corner incisors by nine months. These temporary incisors differ from the permanents, possessing a more distinct neck and being smoother, whiter and smaller. The permanent incisors erupt at the following times:

Two permanent centrals, 2.5 years
Two permanent laterals, 3.5 years
Two permanent corners, 4.5 years

Ageing after 4.5 years becomes more difficult and relies on changes taking place on the grinding surface or table of the incisors and on the angle at which the upper and lower incisors meet.

BREAKING-IN AND HANDLING

Horses respond best to gentle but firm handling. Sudden actions or a noisy approach serve to make a potentially nervous animal definitely so. A creature of habit, it readily comes to accept and indeed enjoy a set daily routine, this satisfying the need to feel secure.

A foal should be handled from as early an age as possible in order to accustom it to being haltered and led with the mare. Early handling and gradual approach help to make the actual breaking-in a much less traumatic experience.

Breaking-in commences at two years old and initially involves getting the horse accustomed to having a bit in its mouth. This is followed by lungeing or long-reining which, in conjunction with words of command, teach the horse the basic movements, e.g. walk-on, trot-on, bear left, bear right, canter, halt. At the end of this process the horse understands the use of the bit and the words of command.

Saddling is the next step and again this is easier if during its early life the horse has been used to having objects placed over its back. When the horse is mounted for the first time or backed as it is called, the rider should be as light as possible. Work must commence slowly and should not be excessive before the horse is four years old. Fairly intensive schooling follows to complete and polish the horse's education.

BUILDINGS

Since the horse is a grazing animal, living at pasture is probably the simplest method of management. Horses engaged in hard, regular work or kept in harsh climates require housing. This in its simplest form may be a field shelter with a large, permanently open doorway which horses at grass can use at will. Thoroughbreds cannot be kept successfully at grass all the year round in the UK and need stabling in winter. Horses were formerly housed commonly in individual stalls, tethered via a head collar in such a way that they were able to lie down. Modern opinion is in favour of loose housing in individual boxes and most horses are now stabled in this way. *Table 19.1* gives the necessary dimensions for loose boxes. When constructing stables, adequate provision must be made for drainage, dung disposal and ventilation. Doorways must be wide enough to prevent injury to animals passing through, and if stabled animals can see activity, boredom is minimised.

The type of bedding most commonly used is wheat straw, but peat, woodshavings and shredded newspaper are also used. Droppings and soiled bedding must be removed twice daily and sufficient clean bedding introduced to provide a deep, elastic bed. Insufficient bedding leads to capped hocks and elbows and seriously interferes with the horse's usefulness.

Provision must be made in stabling for feeding and watering and the receptacles provided must not have sharp or protruding edges liable to damage the horse. Glass windows should be well protected with bars or mesh and all doors should have outside catches which cannot be opened by the horse.

Table 19.1 Dimensions for loose boxes (m)

Thoroughbred size	
Width	3.66
Depth	3.66
Height to eaves	2.23
Height to ridge	3.20
Cob and pony size	
Width	3.02
Depth	3.66
Height to eaves	2.23
Height to ridge	3.20
Foaling box	
Width	4.88
Depth	3.66
Height to eaves	2.23
Height to ridge	3.20
All doorway widths	1.065

FEEDING

The horse needs to eat large amounts of bulky foods to satisfy its nutritional needs and digestive processes. However, unlike the ruminants it does not possess a large stomach, and most of the digestion of cellulose takes place in the caecum and colon. Thus in nature, the horse spends much of its time feeding. This time is reduced by using more concentrated foods, but because of the small size of the stomach, feeding little and often is a good basic rule.

Horses at grass and especially ponies tend to become over-fat in summer, and this predisposes them to laminitis. Hence removal from the pasture for several hours each day may be necessary. Conversely, many ponies at grass are underfed during the winter and supplements of hay are necessary when the weather is bad. The two basic foods for horses, apart from grazing, are hay and oats. It is imperative that the hay is of good quality and free from mould growth. Fairly coarse hay can be used for horses but it obviously does not provide the same amount of nutrients. Oats, often mixed with wheat bran, are the traditional concentrate food for horses in the UK. However, barley, maize and more recently, proprietary horse and pony nuts also have a part to play.

The amounts of food required depend on the size of the animal and the work it has to do, but the estimates given in *Tables 19.2* and *19.3* are a rough guide. They are only general rules and the person responsible for feeding must keep a close watch on the animal's condition, and increase or decrease amounts accordingly.

Table 19.2 Maintenance requirements of hay

Variety of horse	Liveweight (kg)	Amount of hay (kg/d)
Small child's pony	254	4.3
Child's pony	356	5.4
Hunter or thoroughbred	508	6.8
Heavy hunter	610	7.9

Table 19.3 Oats requirement for work

Work	Amount of oats (kg/d)
Light	3.2
Medium	5.0
Heavy	7.3

Table 19.4 Amount of cereals equivalent to unit weight of oats

Cereal	Amount of cereals equivalent to unit weight of oats
Bran	1.32
Barley	0.83
Maize	0.77
Linseed cake	0.80

Oats may be replaced by other cereals and *Table 19.4* shows the weights of cereals equivalent to unit weight of oats. Whilst bran is a useful food to mix with oats, aiding efficient digestion, it is worth remembering, as *Table 19.4* shows, that 1.3 kg of bran are needed to replace 1 kg of oats.

WATERING

Horses require a sufficient supply of clean, fresh water and this is best provided by continual access to a watering device. If this is not possible, then water buckets should be regularly re-filled. The horse should not be allowed to drink large quantities of water immediately after a period of heavy exercise or a large corn feed.

REPRODUCTION

In the UK the horse is seasonally polyoestrous, showing a peak reproductive activity during spring and summer when daylength is long. In more favourable climates, or where management, particularly nutrition, is good, the horse can be bred successfully at other times of the year.

Puberty

The age at which horses, both male and female, become capable of reproducing is extremely variable and is affected by many environmental factors such as climate, nutrition and the season of their own birth. However, some fillies are bred from as two or three year olds and stallions may be used at three or four years old.

Oestrous cycle

The oestrous cycle of the mare is 16–24 d long (average 22 d) with oestrus or 'heat' lasting 2–11 d (average 6 d). Ovulation takes place 1–2 d before the end of heat, therefore matings should be concentrated around that period.

Foal heat

The 'foal heat' or post-partum heat occurs 5–15 d after foaling. As this is constant, it has been common practice to mate mares at the foal heat. However, as uterine involution is far from complete at this time, conception rates are poor and owners are better advised to mate at subsequent heats. Signs of heat to watch for are:

(1) change in temperament, often becoming more excitable;
(2) increase in frequency of urination;
(3) 'winking' of the vulva to expose the clitoris;
(4) interest in and acceptance of the stallion.

Mating management

The best breeding performance is achieved under natural conditions, such as occur when stallions of the pony breeds run out with their mares. However, due to the high value of stallions of the larger breeds, particularly thoroughbreds, hand mating is commonly practised. Competent handlers for both the mare and stallion are required, and if possible, the mares should be tried with a teaser (a male of low value) before introduction to the stallion to ensure she is on heat. Effective teasing helps to reduce injuries to the stallion, and trying boards or gates which are solid are also very useful. Highest fertility is achieved if the mare is mated every other day during oestrus as this ensures that mating coincides closely with ovulation.

Pregnancy and parturition

Average length is 336–340 d. During pregnancy mares should be only lightly worked; the rider should not be heavy and cantering, galloping or jumping should not be allowed. It is a good general rule not to work the mare at all after the fifth month of pregnancy. Signs of approaching parturition in the mare are:

(1) swelling of the teats and udder;
(2) a waxy material develops as drops on the teats 1 or 2 d before foaling.

Parturition or 'foaling' in the mare is relatively short, often lasting not more than 1 h, and abnormalities of presentation are uncommon. The afterbirth or 'cleansing' is normally passed within a few hours of birth. Professional veterinary assistance should be called if:

(1) visible straining lasts for more than 1 h;
(2) presentation is abnormal;
(3) afterbirth is retained for longer than 12 h.

GENERAL MANAGEMENT (GROOMING, SHOEING, ETC.)

Grooming is necessary for appearance and also for the maintenance of good skin hygiene. The removal of the waste products of scurf, secretions and loose hairs keeps the skin clean and reduces incidence of skin diseases. Stabled animals should be groomed daily. However, animals running at grass should not be groomed, as the natural oils present in the coat are important in protecting the horse from inclement weather.

Animals stabled and worked hard greatly benefit if the coat is clipped during the winter. This reduces sweating when working but necessitates the use of clothing (blankets and rugs) when at rest.

Washing of horses is a controversial subject, but there are times when it is extremely beneficial. Probably the most important point is that drying should be thorough and circulation stimulated by mild exercise or brisk grooming.

Shoeing is necessary to protect the foot from the unnatural degree of trauma and abrasion that it meets when worked, especially on hard surfaces such as roads. Frequency of shoeing depends on rate of horn growth and degree of work on abrasive surfaces but is usually necessary every four to six weeks.

DISEASES

The good horseman must develop a sound knowledge of normal equine behaviour and use this as a foundation on which to build an understanding of disease. Very often, the early signs of abnormality are very slight and will only be spotted by the most observant person. Early diagnosis and prompt treatment are needed to avoid unnecessary suffering and excessive loss of usefulness. Therefore continual vigilance and good observation are important in maintaining the horse in the most healthy condition.

Diseases of the horse can be classified as infectious and non-infectious. The former are caused by infectious agents or pathogens such as bacteria and viruses whilst the latter have a variety of causal agents such as nutrition, management and trauma.

Infectious and parasitic diseases of the horse

Virus diseases

Some of the most important virus diseases of the horse are those affecting the respiratory system causing coughing and respiratory difficulty. These lead to massive loss of use of horses and periodically spread widely amongst the equine population particularly when horses meet at various events such as hunting and pony club camps.

Equine influenza

This is also called the cough or Newmarket cough and is a highly infectious disease caused by a myxovirus of which two types are known, A/equi 1 (Prague Strain) and A/equi 2 (Miami Strain). The symptoms are a raised temperature, 39–41 °C, loss of appetite, a persistent, distressing cough and a watery nasal discharge. Secondary bacterial infection is possible, thus treatment often incorporates antibiotics. Animals infected must not be worked or exposed to other horses.

Fortunately, vaccines are available to help prevent this disease and their use in horses entered in many competitive sports is now compulsory. Two injections with an interval of four to six weeks between the injections are required for basic immunisation followed by a booster six to nine months after the second injection. Further boosters at yearly intervals are required although this interval might well be shortened if epidemics occur or horses are exposed to high levels of risk. Mares should be given boosters in the last month of pregnancy to produce high levels of colostral immunity for the foal and foals should receive their first injections at two to three months old.

Colds and catarrh

These are also called stable cough and are mild virus diseases of the upper respiratory tract. Several viruses have been incriminated as causing this condition in which a profuse nasal discharge develops which is at first watery and becomes pus-like. It is common in younger horses and may produce coughing and spread rapidly through a stable. Although the virus infection itself is fairly mild, secondary bacterial infections may develop as a sequel, producing more serious respiratory diseases. Antibiotics and mucolytic agents are used in treatment and it is very important that

infected animals are not worked or exposed to other horses.

Bacterial diseases

Tetanus or lock-jaw

This is an extremely unpleasant disease caused by the bacterium *Clostridium tetani*, an anaerobic, spore-forming organism found in the soil. In the anaerobic conditions found deep in soil-contaminated wounds the organism grows and divides to produce a toxin which is absorbed into the bloodstream to cause the characteristic symptoms: general stiffness and difficulty in moving with ears and tail erect, hindered jaw movements and swallowing. External stimuli may produce violent spasms. The third eyelid or nictitating membrane is commonly flicked over the eyeball. Treatment is usually unsuccessful and humane destruction advised.

Tetanus is now only rarely seen due to the use of a highly successful vaccine which is given as an initial course of two injections four to six weeks apart followed by a booster a year later. Subsequent boosters are given at two to three yearly intervals but this period is frequently reduced especially in horses suffering from injuries and wounds when boosters may be given at the time of injury to raise circulating antibody levels.

Mares should be given boosters in the last month of pregnancy to produce high levels of colostral immunity for the foal and foals should receive their first injections at three months old.

Strangles

This is an infectious disease caused by the bacterium *Streptococcus equi*. It occurs particularly in young horses and is common after exposure to infection, especially at sales. It is typified by high temperature and swelling of lymph glands under the lower jaw, which eventually burst to release pus. The affected animal lacks appetite, has a profuse nasal and ocular discharge and may develop a cough. Other body organs may be involved although this is rare. A course of antibiotic treatment must be given promptly.

Contagious equine metritis

A bacterial disease caused by the organism *Haemophilus equigenitalis* was first seen in England in 1977 causing a genital infection, spread mainly venereally, with a vaginal discharge developing some days after mating. In 1977 this infection reached serious levels causing serious disruption and financial loss particularly in thoroughbred studs. The Horserace Betting Levy Board in Great Britain set up a scientific committee of enquiry to devise control measures and advise on a 'code of practice' to prevent the spread of this disease. Vigorous application of this code which includes routine swabbing of mares prior to service has resulted in good control and the condition is now described as uncommon.

Fungal infection

Ringworm

The fungus responsible for this disease survives on bedding, saddlery and stabling, hence eradication is difficult. The hair falls out in characteristic patches which spread and coalesce, leaving reddened areas which scab over. All materials, such as grooming kits, in contact with the horse should be thoroughly disinfected and not used on other horses. Antifungal preparations should be applied to the lesions with care, as man can also become infected. It is advisable to wear rubber gloves when one is handling infected horses.

Treatment of ringworm with the drug griseofulvin given by mouth as a feed additive over a 10 d period is now considered to be the most successful treatment for ringworm. In addition external applications of antimycotic washes will aid recovery and reduce cross-infection rates.

Parasitic diseases

Parasitic diseases of the horse are common and particularly relevant where horses graze the same pasture continually. In such situations levels of worm infestation can build up to very high levels and cause serious problems particularly in young horses.

Redworms

These roundworms (nematodes) of the horse are the most important parasites of the horse. Several species are involved and they parasitise the intestines. The life history is as follows.

Females in the intestine lay eggs which pass out in the dung onto pasture. These eggs hatch and become infective larvae on the pasture. As pasture is grazed, these infective larvae are consumed and undergo development inside the host. With some nematode species, these developments involve penetration of the gut wall and migrations within the abdominal cavity. In the case of *Strongylus vulgaris* the larvae enter the blood vessels supplying the small intestine and can cause permanent damage. The larvae then enter the intestine to become adults which mate and the females commence egg production. Symptoms of disease caused by redworms can be divided into those caused by adult worms in the intestines and those caused by migrating larvae. In the former case adults suck blood from the intestines causing haemorrhage, diarrhoea, colic and anaemia. Complete rupture of the intestines by heavy infestations has been reported in young horses. Larvae during their migrations can cause a number of pathological lesions involving peritonitis, aneurysms, colic and general wasting.

Treatment and prevention involves the regular use of anthelmintics. Choice of drug to be used and frequency of administration should be the subject of advice from the veterinary surgeon. Regular removal of droppings from the pasture, particularly where horses are confined to small paddocks, is also a useful method to reduce the amount of infection taken in by horses.

Whiteworms

These large roundworms (nematodes) of the horse have the following life cycle. Females in the intestine lay eggs which pass out into the dung onto pasture. Infective larvae develop on the pasture and are consumed as the pasture is grazed. The larvae burrow through the intestinal wall into the blood stream which carries them to the liver, heart and lungs. From the lungs they travel up the windpipe to the throat and are swallowed down into the stomach and intestines.

Symptoms of infection which are commonest in young horses involve wasting, lung damage and occasionally rupture of the gut.

Treatment and prevention as for redworms.

Lungworms

The species of lungworm affecting horses and donkeys is *Dictyocaulus arnfeldi*. These worms inhabit the lungs causing irritation and inflammation of the airways. Bronchitis, coughing and often pneumonia are symptoms of lungworm infestation and veterinary advice should be sought if cases are suspected. As donkeys can carry large numbers of lungworms without necessarily showing any symptoms they should not be allowed to graze with horses unless they are free of infestation.

Non-infectious diseases

Lameness

A horse is said to be lame if there is interference with normal locomotion. It renders an animal unable to perform useful work and therefore is of great concern to the owner. If a horse is lame on a front leg, the horse nods his head down as the sound leg is put to the ground. As the lame leg is put to the ground, the horse jerks his head upwards with the effect of reducing weight taken by the affected limb. With hind leg lameness, the hip on the affected side is raised as the lame leg meets the ground and the affected leg has a swinging appearance.

Lameness is best detected at the walk and particularly the trot and on a hard surface. The causes of lameness are many and varied and in most cases veterinary advice must be sought.

Foreign body penetration

Most lameness originates in the foot. A large proportion of cases involve penetration of the soft structures of the foot by a foreign body. Nails and sharp stones are commonly incriminated and if the horse has recently been shod, a shoeing nail driven too deeply might be suspected.

Laminitis

Laminitis, also known as fever in the feet, is a disease of bad management, seen particularly in ponies receiving too much to eat and too little exercise. The front feet in particular are involved, becoming hot and painful, making the animal markedly lame in the acute stage. If the disease is prolonged then chronic changes take place in the anatomy of the foot, causing the pedal bone to drop and the hoof wall to become concave and ridged. Prompt veterinary attention is required if permanent damage to the foot is to be avoided.

Navicular disease

This condition is seen in the front feet of horses subjected to excessive concussion caused by work on hard surfaces and jumping. Upright pasterns also lead to a higher incidence. The navicular bone is a small bone situated in the foot near the joint formed by the pedal bone and the second phalanx. The deep flexor tendon runs behind the navicular to insert on to the pedal bone and in cases of this disease inflammation develops at this site, resulting in pathological changes in the bone and tendon.

The horse develops a typical 'pottery' gait, taking very short steps and tending to take weight on the toes rather than the heels.

Anti-inflammatory drugs are used in treatment and more recently the use of anti-thrombotic drugs such as warfarin have been successfully employed. These materials however, carry great risk in use and can only be contemplated where strict veterinary supervision is available.

Another long-term treatment is to remove the nervous sensation to the foot by an operation known as neurectomy.

Exostoses

This means the growth of new bone and often occurs in horses as a result of inflammation of bones or joints.

Spavin

This is an exostosis on the internal side of the hock joint and results in hind-leg lameness, seen particularly if the hock is flexed acutely. The animal takes a shortened stride and may wear the toe excessively. Treatment mainly involves rest and anti-inflammatory drugs, but in some cases surgery is possible.

Ringbone

This is an exostosis on the pastern bones and may be called high or low, and is usually caused by trauma. A hard swelling may be felt and seen externally on the pastern and the horse is markedly lame in the acute stage. Recovery depends on rest but anti-inflammatory drugs may also be used.

Sidebone

This is the result of ossification of the lateral cartilages within the foot, seen particularly in heavy horses. Slight lameness may be caused.

Synovial distensions

Joints and tendons are lubricated by a sticky fluid called synovia which is held in sacs called bursae. If these bursae become inflamed the resultant bursitis can be seen as swellings externally. These swellings or synovial distensions are an indication that the horse has worked hard and they usually present no permanent problems.

Thoroughpin

This is swelling of the bursae just above and behind the hock joint, seldom causing lameness.

Windgall

This is a swelling seen above and behind the fetlock joints in both front and back legs, rarely causing problems. Common in older horses.

Capped elbow

This is when a bursa protecting the point of the elbow becomes enlarged when the elbow is exposed to excessive trauma. This happens when insufficient bedding is supplied or the elbow is repeatedly rubbed by the heels of the shoes when the animal is lying down.

Capped hock

Bruising of the back of the hock may result in swelling of the bursa at that site. It may cause trouble when fresh but although unsightly is rarely a lasting problem provided the bruising does not recur.

Sprained tendons

The superficial and deep flexor tendons on the front legs bear enormous strains when the horse is galloped and jumped and are prone to physical damage. Tendon fibres can be actually ruptured and when large numbers of fibres are so damaged, the situation is extremely serious. The horse is said to be 'broken down' when this occurs.

Old treatments such as blistering and firing are viewed as unsatisfactory by many veterinarians and more modern treatments such as carbon fibre implants are being extensively tested. In all cases however, long periods of rest are essential if the horse is to recover satisfactorily.

Soundness of wind

An efficient respiratory system is an obvious pre-requisite for a healthy horse. Non-infectious conditions affecting the horse's wind are an important cause of respiratory troubles in the UK.

Broken wind or emphysema

This is a condition in which the alveoli in the lungs are damaged, so interfering with normal expiration of air from the lungs. Broken winded horses show a characteristic double-lift of the flank muscles during expiration and have a cough heard at both exercise and rest. Dusty conditions and mouldy hay are incriminated in this disease. Affected animals should be bedded on peat moss rather than straw and the hay fed damped. Mouldy hay must never be fed to horses.

Roaring and whistling

These are sounds produced in the larynx during inspiration as a result of partial paralysis of the vocal cords. As the airway is thus partially blocked, maximal respiratory flow is impossible, so affecting the horse's performance. The condition can be relieved by what is known as Hobday's operation.

Colic

This means pain in the abdomen and is a common condition in horses. Several types of colic are known, but no attack should be treated lightly and veterinary help must be sought.

Symptoms are variable but a horse with abdominal pain refuses to eat, sweats excessively, anxiously looks round at its flanks, paws the ground, lies down, and rises repeatedly. In severe cases the animal may roll violently and be dangerous to handle.

If the colic is due to impaction of the alimentary tract liquid paraffin and purgatives may be given by stomach pump. If tympany due to excessive fermentation is causing pain, relief is obtained by giving anti-spasmodic drugs. In cases of twists in the gut itself only prompt surgery will give any hope of successful treatment. In all cases of colic the veterinary surgeon is likely to sedate the animal in order to prevent injury to horse or handlers.

20

Game conservation

J. Collins and P. B. Beale

INTRODUCTION

It is not possible to produce a full practical guide to the requirements of, and methods used in, game conservation in this chapter. Its aim is to provide an awareness of the potential benefits of a game crop, to the owners and tenants of farms and small estates. The game crop can be propagated, husbanded and harvested as any other crop could be.

In order to give adequate coverage to habitat requirements and management of game, this chapter will only attempt to cover pheasants and grey partridges.

The grey partridge (*Perdix perdix*) is a native species, whilst the red-legged or French partridge (*Alectoris rufa*) was successfully introduced into Suffolk, from Europe around 1770. The pheasant too, is an introduction and it is considered that the 'old English' pheasant (*Phasianus colchicus*), which lacks a white ring round the neck, was probably introduced from near the Caspian Sea by the Romans between AD 350 and 400. The Chinese ring-necked race (*Phasianus colchicus torquatus*) was probably introduced in 1785. The Japanese pheasant (*Phasianus versicolor*) is also considered to have been an eighteenth century introduction. They may not, in fact, be separate species, but different races of the same species. The pheasant of the twentieth century combines the hybrid characteristics of many races or sub-species. Partridges and pheasants are particularly associated with lowland farming areas, where agricultural changes are reducing the area and quality of the habitat these birds require.

Modern and intensive agriculture puts pressure on as great an area of the farm to become productive as possible. Areas which were regarded as agriculturally unproductive in the past are now, due to economic pressures and greatly improved technology, being brought under cultivation, or are being influenced by it. Modern machinery, improved crops, herbicides and insecticides, make almost the ultimate in cultivation possible; habitats which are vital to game are inevitably affected, both directly and indirectly.

Landscapes of the prime game conservation areas of lowland Britain are being changed due to losses of hedgerows, copses, small woodlands, wetland areas and others, because, very often, these areas no longer serve the function for which they were created. It has, however, been shown that the hedgerow, which is a vital part of a balanced game habitat, is cheaper to maintain as a means of stockproofing than barbed wire fencing (Brooks, 1975). If the hedgerow has a function on stock and mixed farms, a function can be found for woodlands and other unproductive areas as well. Since these areas form part of the cover which is vital to the successful rearing and holding of pheasants and partridges, a viable game enterprise could restore their function. As a spin-off these features will also provide a habitat for a wide range of wildlife species.

The shooting industry is currently worth an estimated £100 million annually in Great Britain. This figure is made up from lettings and sale of goods. An estimated £30 million worth of food, in the form of game, is produced and over £40 million is earned in foreign revenue from sales of equipment and selling of shooting to foreign sportsmen. The ancillary trades within the industry employ over 25 000 people and about 6000 full-time gamekeepers are employed in Great Britain. Over one million people shoot regularly and this number is increasing.

The quality and diversity of shooting available in Britain is the envy of many foreign countries and we must do our utmost to ensure it remains so.

The countryside is subject to many pressures, modern farming included, but the pressures of recreational use are increasing rapidly. Many farmers and landowners have already developed enterprises to meet the demands of people seeking recreation. Given a business-like approach and correct management of habitats to meet the requirements of the birds, a game conservation enterprise can become a valuable source of income.

A farmer who enjoys his shooting can produce a first class shoot from a relatively small area. A 150 ha farm, with a mixed cropping rotation and at least 5% of the area as woodland and hedges, interspersed with some cover crops to create the right habitat conditions, can produce a first class shoot. All that is needed other than these requirements is plenty of enthusiasm and a degree of sympathetic farming.

It is, however, most important to carry out a reasonable degree of management to ensure success. Whether a full-time gamekeeper is employed, or a keen tractor driver who is prepared to undertake part-time keeping duties, is not important, so long as management is done. Maintenance and improvement of habitat, control of vermin, integration of game into the farming system, winter management of birds, a rearing programme if needed to supplement wild birds, correct release of reared birds and the prevention or control of diseases, are all important and necessary aspects of management.

Owners of estates will already be aware of the excellent work which is being done by the Game Conservancy at Fordingbridge in Hampshire. Whether you require a formal day of driven shooting, if you are interested in selling a day's shooting to help finances or if you are more concerned with creating or improving a shoot for yourself and your friends, advice is always desirable. Anyone seriously interested in improving game conservation on his farm would be well advised to refer to the Conservancy's publications, or to obtain advice from them.

AGRICULTURE AND GAME

Undoubtedly the factors which have had the most detrimental effect on game are.

(1) the destruction of habitat (loss of hedgerows, woodlands and a general increase in field size), and
(2) the recent monumental change in the techniques and patterns of agriculture.

Gone are the halcyon days for the shooting man of weed infested crops, abundant insect life, small fields bounded by hedgerows, the patchwork quilt of varied crops and rotations. These were the golden days for game, wild stocks were abundant and shooting was a matter of harvesting the surplus.

Today, due to economic necessity, a farmer cannot afford to be anything but intensive and efficient, less mixed farming occurs and many hedgerows have

disappeared as a consequence. In addition, chemical use has often replaced rotations. These factors when combined make it very difficult for game birds to compete.

A farm which can provide varied cropping that will give a continuation of food and cover throughout the year, ideally still undersowing cereals, and that has a greater proportion of area down to arable crops rather than grass, will have the advantage over a farm practising monoculture. Hedgerows will provide ideal territorial nesting sites for both partridges and pheasants and if autumn stubbles can be left as long as possible before ploughing this will again help provide a food and cover source. The recent increase in the acreage of winter cereals has helped.

Farming operations do cause physical losses to both adult and young game birds.

The whole spectrum of modern chemicals all, either by direct or indirect means, influence game. Herbicides used unselectively kill the weeds which are often the food source or host plant for a number of insects which chicks, especially the grey partridge chick, need during the first few weeks of life. Recent findings have also indicated that fungicides have a detrimental effect on insect life. Fortunately the present situation regarding poisonous chemicals is improving. Gone are the persistent chlorinated hydrocarbon seed dressings and standards of safety are continually being improved, although we still have a long way to go. With commonsense and a few precautions, the risk to wildlife can be reduced, especially if fewer chemicals are used for merely cosmetic effects. The benefits to maximum crop production from use of chemicals are obvious, and use of some chemicals could be deemed as beneficial to game; for example, the combination of paraquat and direct drilling.

The grass crop, if cut for silage, can be an absolute 'killer' for pheasants and partridges, especially if late cuts of silage and early cuts of hay are made. If there is no better alternative nesting cover then it is likely that these fields will hold many nesting birds, especially between mid May and mid June. Not only can eggs be destroyed and young chicks killed, but also a large number of sitting hens can end up in the silage trailer. Little can be done to prevent these annual losses, but perhaps a total disaster may be avoided if:

(1) better alternative nesting cover is provided;
(2) early grass strains for any early silage cut are used;
(3) hay, especially a late cut, is conserved instead of silage, but this, of course, is impractical in many instances;
(4) by leaving the last few cuts of each field overnight, returning the next day to finish the job. This will allow birds to move to safer cover, or for you to 'dog' these last strips;
(5) by cutting the field from the centre outwards, this may push any early broods out of the field. Modern

forage conservation machinery travels so rapidly in the crop that flushing bars are no longer practical.

Other crops such as vining peas also present a danger.

Stubble burning, now an essential operation on many arable farms, can, if done carelessly, cause damage not only to game, but also to habitat. Other farming operations such as irrigation can also cause high mortality to young birds through chilling.

VERMIN CONTROL

Just as a garden needs weeding, a degree of vermin 'weeding' is an essential task. As game increases in an area, it is only a matter of time before their predators increase accordingly. Controlling vermin will ultimately benefit game, and also the many other creatures that appear in the predator's diet. The sheer cost of rearing pheasants should be sufficient encouragement for the shooting man to invest in a few spring traps.

The most critical time of year to make a concentrated attack on vermin is during the spring, when a fox or stoat will easily kill a sitting hen bird, so you not only lose breeding stock, but also a potential brood. This is particularly devastating to wild stocks of game.

Mammalian vermin which can legally be controlled and that should appear on the shoot's 'wanted list' should include the following.

Mink (Mustela vison)

This is fast becoming the most serious of our predators, as it is now widespread throughout the UK. Being extremely versatile, it swims as well as an otter, can climb like a squirrel. It will kill fish up to 1.8 kg and birds the size of turkeys. Often the first clue indicating the presence of mink is the sudden disappearance of waterfowl, especially moorhens, and closer observation may reveal the characteristic five-toed star-shaped footprint. It can be cage trapped quite successfully along riverbanks, ditches or culverts. The cage trap should be constructed out of 2.5 cm × 2.5 cm welded, galvanised, 14 gauge square mesh—use any smaller gauge and the mink will be able to escape from the cage. Camouflaging the trap will encourage the inquisitive mink to enter it, whilst also hiding it from passers-by. The trap can be baited, using fish heads, and should be checked regularly.

Fox (Vulpes vulpes)

Undoubtedly the fox can do tremendous harm to a shoot, especially if it breaks into a release pen. The fox will also account for a number of rats and rabbits during its life, thus doing good, so perhaps one can be tolerated in an area, but if things get out of hand, then measures should be taken. A snare in the right hands can be extremely efficient and humane, but used indiscriminately can cause pain and suffering. If snaring is practised it is essential that the snare is checked regularly, at least twice daily, that the wire has a minimum breaking strain of 254 kg. A 'deer leap' should be placed over the snare if deer are present in the area. Shooting and gassing also have a role to play in controlling foxes.

Stoat (Mustela erminea) and Weasel (Mustela nivalis)

These can be controlled by tunnel trapping and shooting.

Hedgehog (Erinaceus europoeus)

This can account for many pheasant and partridge eggs and can be controlled by tunnel traps.

Rat (Rattus norvegicus)

Every attempt should be made to control rats throughout the year due to the serious damage they cause. Annually they soil and eat tonnes of valuable foodstuffs, as well as eating many eggs of wild birds. They are also a potential disease risk to humans, transmitting leptospirosis and salmonella. The anticoagulant poison warfarin is most commonly used against rats, but recently resistance to this poison has built up in some areas. This is mainly caused by incorrect baiting, usually by using too little bait or not having sufficient baiting points. This allows the rats to eat a sub-lethal dose and slowly become resistant. It is essential that poisoning should be carried out until the bait is no longer taken.

Grey squirrel (Sciurus carolinensis)

Grey squirrels can cause severe damage to forestry, and to a lesser extent predate on game eggs. It can be cage trapped easily, especially if pre-baited, or controlled by dray poking.

Avian vermin such as crows, rooks, jackdaws, magpies and jays are all able egg stealers and may need controlling. Shooting and the use of funnel and letterbox type cage traps are effective.

It must be remembered that badgers, otters and *all* birds of prey are protected and under no circumstances should they be controlled. One should be familiar with the Protection of Birds Act 1954 and 1967 and the Badger Act of 1973, as exceptions do occur.

Tunnel trapping

A series of tunnel traps situated on vermin highways, i.e. hedgerows and ditches, operating if time allows all the year round, will be the best line of defence against ground vermin. A tunnel made from wood (measuring 15 cm × 15 cm × 45 cm will take a Fenn Mk IV) or a tunnel made from natural materials is legally required under the Spring Traps Schedule. The trap is placed within the tunnel, firmly tethered by the chain, the entrance being staked to prevent long necked birds from being caught. A number of spring traps are available, but it should be remembered that:

(1) only ministry approved spring traps can be used, namely, the Imbra trap Mks I and II, the Fenn Vermin trap Mks I–IV, the Fenn Rabbit trap—which can only be used for catching rabbits and set below the ground in a rabbit hole—the Juby trap, the Fuller, Sawyer and Lloyd trap;
(2) they must be placed within a natural or artificial tunnel;
(3) they must be checked at least once a day.

It is illegal to use pole traps, and, since 1958, it is also illegal to use the gin trap. It is advisable to be aware of the legal implications concerning vermin control.

WINTER FEEDING

Supplementary winter feeding is necessary for several reasons:

(1) to hold birds in an area to prevent straying;
(2) to maintain the birds when natural food is insufficient for the population;
(3) to enable birds to winter well and enter the breeding season in reasonable condition.

Ideally pheasants need to be fed on a strawed feed ride, situated in a warm covert, with low shelter close at hand. Wheat straw is the best material, being cheap and free draining it will remain unrotted longer. (Good leaf litter will also do.) Wheat, the best quality available, is the most favoured grain.

Where possible the rides should be hand fed twice daily, morning and afternoon, ensuring that one is regular both with timing and quantity. The grain being scattered on the straw will keep the birds occupied while they scratch for it. The birds will soon become accustomed to a call during feeding and once the call is used, birds will travel a considerable distance to the feed ride.

The feed ride should be as long as possible, but should not continue to the end of a covert, as this will tend to let the wind in. The ride should also have several kinks along its length, since this will help to prevent a dominant cock bird from ruling the feed ride and from driving off other birds.

With reared pheasants and partridges, it is essential to provide them with water; wild birds will fend for themselves. An inexpensive drinker can be made from inverting a filled five gallon drum of water into a shallow tray. This will automatically fill as birds empty the tray. Dusting shelters and feed straddles are useful to keep the birds contented on the feed ride.

Where daily hand feeding is not possible, hopper feeding should be undertaken. A five gallon drum with slits 6 cm × 5 mm cut on the bottom of the drum and hung 40 cm from the ground will enable pheasants to pack grain from the hopper. Small birds and squirrels become expert grain robbers, so a sparrow guard is advisable (these are available from the Game Conservancy). Sufficient hoppers must be made available for the birds, about one hopper per 16 birds feeding is a guide. Each hopper should have a rat baiting point close by—a length of tile drain pipe makes an excellent poison dispenser.

COVER CROPS

If insufficient natural cover is available, or a greater variety of drives are required, careful planning and positioning of cover crops can be used to increase the potential of a shoot.

Cover crops can also be used to:

(1) increase the number of drives on a shoot day;
(2) provide nesting cover;
(3) extend hedgerows, spinneys and woodlands;
(4) provide a food supply, also insects;
(5) increase an areas game 'holding' capacity;
(6) provide return drives;
(7) make the best use of the natural topography of the ground to show quality birds.

Only a small area of the farm need be set aside, and if strips of kale or other fodder crops are grown, these may well be utilised by livestock after the shooting season.

Probably the most popular cover crop grown is kale (*Brassica oleracea*), which can hold birds well if there is no better natural alternative, although if the crop is too wet, then birds may not be found in it. Kale should not be driven too late in the day, and, as with all drilled cover crops, it should be drilled in the direction of the drive. Kale, in this instance, can be left for a second year.

Maize (*Zea mays*) provides not only good cover, but if cobs are knocked to the ground, a food supply for the birds.

A problem with both these crops can be large flushes of birds towards the end of the drive, since the birds tend to run along the rows and flush all together. This can be partly overcome by transversing the strip at intervals with sewelling or by undrilled patches.

Artichokes (*Helianthus tuberosus*) also provide cover and food for pheasants, as they will often peck at any exposed tubers. This is a perennial crop which needs periodic thinning, as it does tend to become extremely thick. It is not recommended for exposed sites.

Mustard (*Brassica alba*) can be successfully 'oversown' in standing spring cereals, and once the main crop is removed, will rapidly produce some excellent early season cover for both pheasants and partridges.

Game mixtures have been developed, which include maize, sunflower, rape, kale, etc., but drilling and establishment can be a problem because of varying seed size.

For additional nesting cover narrow strips of cocksfoot (*Dactylis glomerata*) or canary grass (*Phalaris tuberosa*) can be sown, especially useful along headlands or hedgerows. Pheasants and partridges will readily nest in such areas, and if one can make as many separate nesting territories as possible, then more potential nests should result.

Many other crops are available for game. The value of cereal stubbles left well into the autumn is considerable, and if undersowing can fit into the farm cropping plan this will also benefit game. This will give a continuation of cover throughout the winter, and it will also encourage greater insect life.

REARING

The need to actually 'restock' annually with reared game would be in question if the farm is suited to wild pheasants and partridges. Time and money spent on encouraging and protecting wild birds will, in many cases, be more beneficial than rearing.

Unfortunately the ideal conditions for wild birds are no longer common and one must consider adopting a rearing programme to maintain numbers, especially if regular shooting is demanded.

Deciding which rearing policy to adopt can be a problem, as consideration has to be given to capital expenditure, labour, and other factors. If the right habitat prevails, then perhaps all that is needed is an initial stocking of adult breeding birds, and thereafter, to manage them as wild stock. Perhaps the purchase of six week old poults from a game farm would fit the situation better, and this can be considered.

The main rearing policies available to the shooting man are:

(1) the purchase of eggs from a game farm. Although it is cheaper to purchase eggs rather than day-old chicks or poults, the cost and risks of incubation and rearing still have to be faced;
(2) the purchase of day-old chicks. Obviates worries of incubation and is more likely to produce a guaranteed number of birds;
(3) the purchase of six weeks old poults. The cost will be approximately seven times that of an egg, but someone else has had the risks and mortalities of the first six weeks of rearing;

(4) the purchase of adult stock. Very expensive, but no rearing equipment involved;
(5) custom hatching by a game farm in return for eggs;
(6) the catching-up and retaining of pheasants for 'home' produced and incubated eggs. (Allowing for mortalities one hen should produce 10–12 six weeks old poults.)

If time and labour allow, the last method is the most rewarding, as you have direct control over all the stages of rearing.

The rearing procedure is quite straightforward, and a beginner can achieve excellent results. However, the job must be done well, since short cuts and skimping will lead to trouble. Hygiene is essential at all stages and equipment must be well washed and disinfected. Diseases and behaviour such as feather pecking can arise, so a high degree of stockmanship is essential.

Hopefully, a strong and well hardened bird will eventually be released onto the shoot, providing a testing quarry and without the slightest hint of anything artificial.

More people are becoming interested in the rearing and release of red legged partridges. However, these have not been covered in this chapter due to the greater level of expertise required. Those interested are advised to seek expert advice, initially.

Two excellent publications are available from the Game Conservancy on rearing pheasants and partridges (*see* Further reading).

RELEASING

The importance of correct release and acclimatisation of reared birds into the wild cannot be over emphasised. High mortalities can occur during and just after release; such mortalities lead to inflated rearing costs. In 1981, it costs about £10 per pheasant to put over the gun; this is allowing for an average return of 38%.

A release pen is essential for safe acclimatisation, as reared birds have become dependent on someone supplying all their creature comforts. As soon as they are released, they are expected to act as wild birds, so the initial security of a pen is necessary to enable this transition to take place with minimal losses.

The ideal release pen would be situated within a sheltered wood, and within the pen approximately 30% of the area should consist of low ground cover, 30% with medium height trees (so birds learn to roost), and 40% of open ground (so birds can sun themselves). A feed ride (or hopper) and watering points should be provided in the pen, so the birds will associate them with food once they are released. Hand feeding should also start to accustom birds to the whistle, or rattle of bucket, during feeding; this familiar sound will draw birds to the feed ride once they become independent.

The pen should be constructed of 3.125 cm wire netting, with the bottom 23 cm dug into the ground or turned out for 45 cm to prevent foxes digging under the wire. The top of the wire should have a floppy overhang to stop predators climbing the fence. As a guide, 1 m of perimeter fence per bird will give an adequate area, the circular being the better shape of pen. At 40 m intervals, fox proof re-entry grids will allow birds to walk back into the pen during the early stages of release.

Only well 'hardened' poults, at least six weeks of age, should be released into the pen. To prevent birds leaving the pen prematurely the juvenile flight feathers outside the most recent blood quills should be cut. Within a few weeks these juvenile feathers will moult and be replaced by adult feathers. The birds will then be full flighted and will be able to fly out of the release pen.

Even when great care is exercised, it is quite possible to lose 25% of the released birds before the shooting season.

WOODLAND

Areas of woodland (preferably small blocks well scattered throughout the shoot, to enable birds to be driven from one wood to the next) are extremely valuable for two reasons, as habitat and to increase the height of driven birds. The pheasant makes the most of the right kind of wood, the most critical area of which is where the field meets the wood; the pheasant is a bird of this 'edge', where nesting sites and close cover are available.

Problems can arise when attempts are made to shoot large areas of woodland, since much time can be spent chasing birds from one area to another. Even when these are flushed they may present only indifferent 'tree top' targets. Large areas of closely planted softwoods which prevent sunlight penetrating the canopy and are continually dark and damp hold no attraction for game birds.

When consideration is being given to planting up an area of woodland for game, or to improving an existing wood, several important factors should be considered:

(1) It is essential that a wood should be kept as warm as possible. This is best done by planting a shelter hedge with a thick bottom around the perimeter of the wood, to slow down ground wind. Suitable species would be quickthorn, especially when cut and laid, shrub honeysuckle (*Lonicera nitida*) or tree species such as Norway spruce (*Picea excelsa*). Lawsons cyprus (*Chamaecyparis lawsoniana*) are good if they are 'topped' at the right height. This perimeter hedge also creates ideal nesting sites. Within the hedge several rows of intermediate height softwoods should be planted to provide a medium height windbreak.

(2) For a wood to be successful, it must have a proportion of ground cover. This can be encouraged by felling timber, thereby allowing more sunlight to reach the woodland floor. A chain saw is often the best tool for woodland improvement.

(3) If possible, a 'flushing' point within the woods should be planted. This consists of a small area of ground shrub cover e.g. *Lonicera nitida* or common laurel (*Prunus laurocerasus*) towards which birds can be guided during a drive and then gently flushed. There should be an area of relatively clear canopy between the flushing point and the guns so that the birds are able to rise freely through the trees.

(4) Rides within a wood, especially if shooting is to take place within the wood, need to be as wide as possible.

(5) A mixture of tree species within the wood is desirable, ideally at least 50% deciduous and 50% softwood.

(6) The planting of cover shrubs within the wood to provide warm ground cover is essential. Suitable species would include shrub honeysuckle, snowberry (*Symphoricarpus rivularis*), box (*Buxus sempervirens*). The cotoneaster species provide berried shrubs, some of which are eaten by pheasants. If possible, do not plant or encourage rhododendrons, as they soon take over the woodland floor; cherry laurel needs regular cutting back to keep it under control.

CONCLUSION

Conservation of game ultimately benefits our wildlife and landscape, and gives a function to areas not under cultivation. With high capital investment in land it is increasingly important that every hectare should pay. With today's demand for, and value of shooting, a farmer should consider the contribution which a game crop can make to overall farm income.

Further reading

BROOKS, A. (1980). *Woodlands: a practical conservation handbook*. British Trust for Conservation Volunteers

COLES, C. (1975). *The Complete Book of Game Conservation*. Barrie and Jenkins

Game Conservancy Advisory Booklets:

Game and Shooting Crops
Wildfowl Management on Inland Waters
Partridge Rearing and Releasing
Pheasant and Partridge Eggs: Production and Incubation

Diseases of Gamebirds and Wildfowl
Farm Hazards to Game and Wildlife
Pheasant Rearing and Releasing
Game Records
Grouse Management
Feeding and Management of Game in Winter
Forestry and Pheasants
Woodlands for Pheasants
Predator and Squirrel Control
Roe Deer
Red-legged Partridge
Grey Squirrels

The European Woodcock
Rabbit Control
Game Conservancy Annual Review No. 11—Published May 1980

Reference

BROOKS, A. (1975). *Hedging: a practical conservation handbook.* British Trust for Conservation Volunteers

21

Fish farming

J. S. Goddard

INTRODUCTION

Fish farming in the British Isles, formerly based on a few units producing fish for restocking sporting fisheries, has expanded considerably in recent years. Whilst the number of fish produced for restocking has continued to grow in response to the increasing pressures of modern angling, the major expansion has been in the intensive rearing of fish for food. Fish are poikilotherms (cold-blooded) and in our temperate climate many species exhibit slow growth. Attention has thus focused on those species with a sufficiently high market value to offset the cost disadvantages of prolonged husbandry. The culture of salmonids (fish of the trout and salmon family) has proved most successful.

The production of rainbow trout (*Salmo gairdneri*) far exceeds that of all other species, and is the fastest growing sector of this small industry (*Figure 21.1*).

The farming of Atlantic salmon (*Salmo salar*) is now established at a number of sites on the west coast of Scotland. Total production in 1980 was an estimated 700 tonnes, making this the second most important farmed species. Salmon farming, and a variety of development projects, which include the rearing of eels (*Anguilla anguilla*), turbot (*Scophthalmus maximus*), and Dover sole (*Solea solea*), have attracted considerable investment from large corporations with interests in the food industry. In contrast, the majority of the estimated 400 trout farms in Great Britain are small, privately funded businesses (Goddard, 1980).

The basic principles of rainbow trout farming are considered in detail in this chapter with reference, where appropriate, to other farmed species. The production of fish for restocking angling waters is described under a separate heading.

FISH FARM SITES

If the current phase of expansion in trout farming is to be maintained, many new sites will have to be exploited. Considerable expansion is taking place in the west of Scotland, partly as a result of direct stimulus from the Highlands and Islands Development Board, whilst in England and Wales more sites are being developed outside the main trout farming areas of Wessex, North Yorkshire and Lincolnshire.

Trout are a demanding species in terms of water requirement and this governs the availability of suitable sites. In general, water supplies must be free from pollution, or risk of pollution, and saturated with dissolved oxygen. Alkalinity (measured as $CaCO_3$) should lie between 20–200 mg/ℓ and pH within the range 6.5–8.0. Flow rate requirement for trout production varies in accordance with water temperature. Less oxygen is carried per unit volume of water at higher temperatures, and flow rate must be increased or stocking density reduced in order to compensate. A general estimate, often applied to water supplies, is that a continuous flow rate of 50 ℓ/s will support the production of approximately ten tonnes of rainbow trout per annum.

Fish growth rates increase with rising temperature. Assuming that low oxygen levels do not become limiting, maximum growth rates of rainbow trout occur within the temperature range 14–18 °C. Production cycles, from egg to marketable portion size (220–280 g), thus vary significantly in duration between geographical areas. These may range from a minimum of ten months in southern England to 18 months in northern England and Scotland. This effect of higher water temperature on growth rates confers a natural advantage on farms sited in southern England.

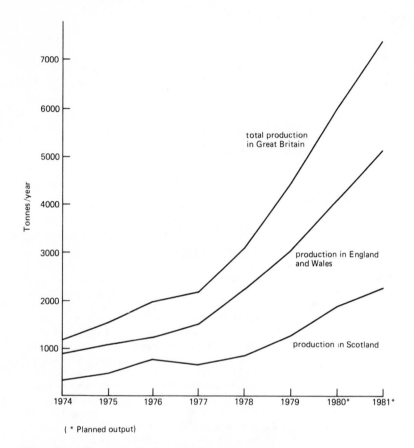

(* Planned output)

Figure 21.1 Estimated production of rainbow trout for direct consumption in Great Britain 1974–1981. (After Lewis, 1980) (* = planned output)

Whilst some farms operate their own hatcheries to supply their own young fish, many purchase fry or fingerlings (young fish up to 10 g) from specialist producers for growing on. Hatcheries are generally sited on natural springs or boreholes which provide pure water at constant temperature. Farms producing table trout generally abstract water from river systems. Alternatively young rainbow trout, reared in fresh-water hatcheries, may be transferred to floating net cages moored at sheltered marine sites or in large freshwater lakes. Salmon are reared for one to two years in fresh water then are transferred to sea water for a further period of two years.

Other important site factors which must be taken into account include any risk of flooding, security against vandalism and theft and proximity to markets. Land requirements are relatively small and depend on the type of farm construction. The most extensively planned farm, producing for example 50 tonnes of trout per annum, would not normally require in excess of 5 ha, including land for buildings and services.

In the earliest stages of planning a farm, Regional Water Authority officials can generally offer valuable advice relating to any particular site. Information should be collected which includes details of water flow rates, temperatures and water quality. It is vital to record details of any seasonal fluctuations in water flow or changes in water quality since it is the minimum flow rate which sets the limits for the production capacity of the farm.

Planning and consents

Fish farmers have to meet several sets of Local Authority and Regional Water Authority regulations. Although farming fish for food is now generally recognised as an agricultural practice, the status of fish rearing is not yet entirely clear and is currently under review. This may affect, where applicable, planning consents, grant applications, rating and licensing for water abstraction and discharge. Advice from the Local Authority and Regional Water Authority should be sought at the earliest opportunity for an assessment of any particular site. The fish farming committees of the National Farmers Unions of England and Scotland can offer advice with regard to the planning and legal aspects of fish farming. Grant aid may be available from within the EEC Agricultural Fund (FEOGA) or the MAFF Agricultural and Horticultural Development Scheme (AHDS). Enquiries should be made in the first instance to the MAFF concerning grant aid under either of these schemes. Grant aid may be available from the Highlands and Islands Development Board for farms proposed within that region.

Local authority regulations

If the farm proposed is intended for the exclusive production of fish for food then planning permission is not generally required from the Local Authority. Conditions may be imposed however, depending upon the nature of the enterprise and the siting of the farm. Consents are required for the construction of any dam or weir and for any excavation work (Town and Country Planning Act, 1971).

Regional water authority regulations

Creation of lakes or ponds

A licence must be obtained from the Water Authority before constructing a dam or weir to impound water (Water Resources Act, 1963).

Abstraction of water

If the fish farm is producing fish exclusively for food, and is abstracting from a surface water contiguous with the farm, no abstraction licence is required. Abstraction from underground strata does however require a licence (Water Resources Act, 1963).

Discharge of water

Water discharged from a fish farm is classified as a trade effluent and requires consent from the Water authority for discharge (Rivers (Prevention of Pollution) Acts, 1951–1961). Standards for the quantity and quality of effluent are determined individually for each discharge and are designed to protect the receiving water. Biochemical oxygen demand (BOD), suspended solids content, ammonia content and turbidity are regularly monitored by Water Authority officials and must fall within the standards set.

Introduction of fish

It is an offence to introduce fish or fish eggs into inland water, including fish farms, without the consent of the Water Authority (Salmon and Freshwater Fisheries Act, 1975). Live salmonids or coarse fish species may not be imported from abroad and imports of salmonid eggs require a licence from the MAFF. Exotic ornamental fish do not require an import licence.

Those intending to rear fish at coastal marine sites must, in addition to satisfying Local Authority regulations, obtain consents from the following:

(1) Crown Estate Commissioners—seabed rights;
(2) MAFF (England and Wales), DAFS (Scotland)—permission to anchor cages at sea (Dumping at Sea Act, 1974);
(3) Local Coastguard or Department of Trade (Marine Division)—(Coast Protection Act, 1949).

FARMING SYSTEMS

A variety of holding systems are used for growing fish. The choice is generally dictated by the topography of the site and the nature of the water supply. Capital costs and husbandry techniques vary considerably between the different systems.

Excavated earth ponds

This is generally the cheapest method of construction although it is restricted to those sites where the sub-soil contains sufficient clay to seal the ponds. Earth ponds are widely used throughout Europe for the production of rainbow trout. They are operated at a slower rate of water exchange than other holding facilities and hence are stocked less densely. Pond dimensions vary considerably but those used in trout production are typically of the order 30 m long, 10 m wide and up to 2 m deep, with sloping sides.

Prefabricated tanks

These are most commonly circular and constructed from galvanised steel or fibre-glass set into a concrete base. They may be set above or below ground level. Their efficient operation rests on maintaining a sufficiently rapid rate of water exchange. The circular flow pattern of water, entering tangentially and leaving through a central drain pipe, maintains solids (faeces and uneaten food) in suspension. This results in a degree of self-cleaning and aids general hygiene. Tank dimensions commonly range from 2–10 m diameter and from 0.5–2 m deep.

Raceways

These are long rectangular channels lined with concrete or reinforced block walls. In common with circular tanks, raceways operate efficiently with high water exchange rates. This, in turn, enables high stocking densities to be maintained. Raceways are of various lengths, sub-divided by screens. Their width rarely exceeds 5 m and depths range from 0.5 m–2 m. Establishment costs are generally highest for farms based on a raceway system.

Floating net cages

Net cages are used for growing fish at sheltered coastal sites and in large freshwater lakes. They consist basically of a net bag suspended from a floating framework. The shape of the net bag is maintained both by the floating framework and by weights suspended from each of the bottom corners. Shapes and dimensions vary but nets are usually between 4–6 m deep and have an enclosed volume within the range 100–300 m^3. Water exchange through cages is maintained both by natural currents and by the swimming movements of the enclosed fish. Cages may be moored singly or in lines or rafts. Factors influencing marine site selection include exposure to wind, tidal range and water depth, current speeds, salinity and access. Careful account must be given to the siting of fish cages in freshwater lakes or gravel pits since in the absence of sufficient water exchange pollution problems are likely to arise.

For any particular site, each of the above holding systems may offer distinct advantages. Of prime importance in the design of any land based farm is the facility for emptying tanks or ponds completely and also for isolating individual ponds. Fish can then be more easily graded and harvested and, should disease occur, be isolated from other stocks. Stocking densities for rainbow trout range, in practice, from figures of approximately 10 kg/m^3 of water volume in earth ponds, to figures exceeding 50 kg/m^3 in raceways. It is however impractical to discuss stocking densities in any detail since it is water flow rates which effectively dictate stocking policy.

FARM CAPITAL COSTS

Capital costs of fish construction vary widely in accordance with the characteristics of the chosen site and water supply, and the availability of existing buildings, labour and machinery. A recent survey of trout farms by Lewis (1979) gives some indication of capital costings for the most commonly used systems, ponds and tanks. The survey showed that the average cost of establishing a 50 tonne per annum production unit for rainbow trout required a capital investment of £36 000, if based on ponds, and £45 000 if based on tanks. These figures include farm buildings, holding systems, water supply and services but exclude land costs. Hatchery and early rearing facilities generally account for only a small proportion of the total capital investment.

ANNUAL OPERATING COSTS

The major operating costs in table trout production are food and labour. Food currently costs in the region of £300/tonne for grower diet. Trout are carnivores and require a high protein/low carbohydrate ration. In general, food conversions of between 1.5–2.0 units of dry food: 1 unit of wet fish flesh are achieved. Taking an average food conversion ratio of 1.7:1 food costs in producing each tonne of trout will be in the region of £500.

As farm size increases percentage food costs increase whilst percentage labour costs decrease. With reduced labour costs, and other economies of scale, total operating costs on the larger farms, per tonne of trout produced, are substantially lower than those of small units. The survey by Lewis (1979) showed that on large farms, producing in excess of 100 tonnes per annum, average production cost per tonne was £846. On small farms, producing 10 tonnes or less per annum, the average production cost rose to £1586/tonne.

Other major variable costs to be accounted for include the purchase of eggs or young fish, packaging, transport and power. Fixed costs include administrative and maintenance costs, insurance, depreciation and interests on capital. There may be a period in excess of 18 months from stocking the farm initially to first sales. Full operating costs during this period must be accounted for. Any continuous requirement for pumping or artificially aerating water will add substantially to operating costs.

SALES AND PROFITABILITY

Capital and operating costs for small units are considerably higher, per tonne of trout produced, than for larger units. Small units, which may include ancillary enterprises on large agricultural holdings, must therefore have access to the most highly priced sales outlets. These include farm-gate sales, sales to local hotels and restaurants and direct to retailers. Current prices at these outlets range from £1500–£2250/tonne.

Large farms, in contrast, sell the bulk of fish produced to wholesale markets and for processing. Wholesale prices are currently within the range £1250–£1650/tonne, depending on seasonal availability.

Most farms operate a variety of sales outlets and a meaningful figure for percentage return on capital can only be assessed for any planned unit when detailed sales outlets are known. In the survey by Lewis (1979) the larger farms, with their more uniform wholesale outlets showed an average profit margin of about £300/tonne. This figure did not take into account charges for land or capital.

In general all units must be planned with their markets clearly defined. Farmed fish sales account for less than 1% of total fish sales in this country and there remains considerable scope for developing local markets. Trade is however becoming increasingly competitive in some areas and more home-produced, processed, frozen trout is becoming available. This may affect local sales of fresh fish if prices become competitive.

PRODUCTION OF FISH FOR RESTOCK-ING ANGLING WATERS

The growth of 'put and take' angling waters, particularly in new gravel pits, reservoirs and lakes, has led to an increasing demand for live fish. Rainbow trout, brown trout (*Salmo trutta*), and brook trout (*Salvelinus fontinalis*) are the principal salmonids, whilst carp (*Cyprinus carpio*) is the only coarse fish species reared on any scale in the British Isles.

Rearing stock fish is generally conducted in earth ponds at low stocking densities. This helps to ensure fish of good visual appearance. Most restocking farms keep their own broodstock fish and conduct limited forms of selective breeding to improve strains. Restocking farms must also undertake or arrange for live fish deliveries.

Prospects are particularly good for the rearing of several species of coarse fish, including carp, tench (*Tinca tinca*) and roach (*Rutilus rutilus*), since live fish imports are now forbidden. The Regional Water Authorities rear fish for restocking their own waters. In some areas this restricts the market open to the private sector.

In common with the planning of a table fish-farm, markets for stock fish must be established at the outset. Large stock fish are grown over a period of several years and this necessitates detailed planning of production.

FISH FARM MANAGEMENT

Sound husbandry and management are the greatest single factors influencing profitability. The main husbandry skills relate to maintenance of fish health, optimising food use, and extending production through as great a part of the year as possible. Interruption of water supplies, disease and water pollution are the main causes of fish loss. Farms must be monitored continuously if such losses are to be avoided. Advice on various aspects of fish husbandry is available from MAFF (England and Wales) and DAFS (Scotland). Both these organisations offer disease diagnosis facilities and will advise on treatments. Advice may also be sought through certain of the regional Veterinary Investigation Centres. The NFU has produced a useful booklet entitled *A Code of Practice for Farmers rearing Fish for the Table and Restocking Markets*. Several of the Regional Water Authorities also offer useful guides.

In general, experience is vital and in many cases it is advisable to operate a pilot scheme before committing large amounts of capital to a fish-farming venture.

Further reading

BARDACH, J. E., RYTHER, J. H. and MCLARNEY, W. O. (1972). *Aquaculture: The Farming and Husbandry of Freshwater and Marine Organisms*. New York: Wiley Interscience.

DRUMMOND SEDGEWICK, S. (1976). *Trout Farming Handbook*. London: Seeley, Service and Co.

Fish Farmer. Surrey, England: Agricultural Press Ltd (bi-monthly).

Fish Farming International. London: Arthur J. Heighway Publications Ltd (quarterly).

HUET, M. (1970). *Textbook of Fish Culture: Breeding and Cultivation of Fish*. Surrey, England: Fishing News (Books) Ltd.

References

GODDARD, J. S. (1980). *Journal of the Farm Management Association* **4** (1), 19

LEWIS, M. R. (1979). University of Reading, Dept. of Agricultural Economics and Management, Miscellaneous Study 67

LEWIS, M. R. (1980). University of Reading, Dept. of Agricultural Economics and Management. Miscellaneous Study 68.

22

Animal health

I. Fincham

The productive potential of an animal can only be achieved by maintaining it in good health. This first requires the birth of strong, viable stock from dams themselves in fit condition to achieve this.

Ill health is not necessarily due to specific diseases alone; it is often initiated by poor husbandry, such as exposure to cold, wet conditions, indoors or outside, resulting in stress, particularly to young stock and pregnant animals. This lowers resistance to diseases, as also do digestive disorders and sudden changes of environment, even travel itself.

Intensive husbandry methods can provide a controlled healthy environment and lowered risk of illness; mismanaged, they present increased risk of disease. The stockman's experience has to ensure the right balance, always with the welfare of the animals in mind—summed up as 'tender, loving, care'. Early observation of abnormal behaviour hastens diagnosis and early corrective measures or treatment.

DIAGNOSIS

This may require clinical examination by a veterinary surgeon, supported by laboratory tests of suitable specimens, e.g. milk samples in mastitis; faeces, for suspected intestinal worm infestations and bacterial diseases. So-called 'indirect tests' can reveal unsuspected disorders before clinical symptoms show, e.g. tuberculin tests, blood serum tests for brucellosis, leptospirosis, virus infections. Biochemical blood tests may indicate disturbances in body metabolism, commonly associated with dietary imbalances.

Health records

Weight gains, milk output, reproductive cycles, lambing percentages, litter sizes: changes in any of

these may be the first indication of inapparent ill health affecting profitability. Routine veterinary inspections also detect problems early, such as non-conception, mastitis, liverfluke disease. Some 'herd health schemes' are available as computer programs.

HYGIENE

Diseases can be spread in different ways, as discussed in their respective sections, but there are certain fundamental control measures.

Airborne

Most respiratory and certain other bacterial and viral diseases are spread by airborne droplets or dust particles: correct siting of buildings reduces windspread between buildings. Pens should have solid walls and animals of the same age kept in small groups, and out of contact with older animals, which themselves may be immune but may be carrying infections. The principle of 'all in, all out' enables pens or buildings to be thoroughly cleansed, disinfected and rested empty between batches for several weeks.

Enteric diseases

Enteric diseases are spread by faeces; drains should not run from pen to pen. Manure should be put on arable fields, or grass fields saved for silage or hay. Slurry storage for two months will kill many pathogenic organisms, but dressed pasture should not be grazed for three months—longer, with some infections; rain reduces risks by dilution and wash off.

Flies and biting insects

Flies and biting insects can also transmit certain diseases, e.g. wound infections, summer mastitis, redwater (babesiosis), tick pyaemia (in sheep).

Surgical instruments

Surgical instruments can carry infective material, so castration knives, glass syringes, and needles should be washed in cold water, then boiled for 20 min. Disposable syringes and needles may be used, preferably with a new needle per animal; dispose of them safely afterwards.

Quarantine

Quarantine each group of animals and birds for at least four weeks, to reduce risks of apparently healthy carrier animals introducing disease. The incubation period (delay between infection and evident symptoms) can be long e.g. Johne's disease, six months to two years; rabies up to six months; some 'slow virus' diseases even two to three years. Delivery lorries should unload or load away from stockyards.

Imported animals are compulsorily quarantined in approved lairages, under supervision by veterinary inspectors; they may be subjected to certain tests before release to UK farms.

Isolation

Isolation of sick animals, individually or in small groups, avoids spreading infection. Healthy animals should be attended before sick ones, by separate attendants if possible, or at least wearing separate protective clothing, washed and disinfected for each group.

Medication

Modern medicaments are expensive and some are only effective for certain specific diseases. Bacteria may develop drug resistance, particularly to antibiotics; laboratory tests demonstrate this and aid selection of effective treatments. Correct dose rate and length of treatment are important. 'Preventive treatments', so-called, use low-level inclusion of medicaments in either food or water, but these levels are seldom adequate to treat actual outbreaks of disease.

Preventive inoculation

Vaccines are available for some diseases, such as tetanus, blackleg, 'husk'; immunity may not develop for several weeks, and often needs boosting by a second dose and re-inforcing annually. Antiserum gives an immediate protection of short duration,

perhaps two to three months, but is useful when a disease outbreak occurs in flock or herd.

Disorders and diseases of cattle, sheep and pigs described represent some of those of major importance in the UK. Nutritional inadequacies are dealt with in Chapter 13. Surgical procedures are omitted—elementary ones are described in Stockman's Manuals, and practical instruction is often available, e.g. through the Agricultural Training Board. It is convenient to classify diseases for the principal age group of animal which is usually affected, although many diseases do not respect age. Diseases may also be ascribed to the body system principally affected, e.g. respiratory, alimentary; when the whole body is affected the disorder is 'systemic', e.g. bacterial septicaemia.

SCHEDULED DISEASES

Statutory control

Certain highly contagious diseases, e.g. foot and mouth disease (FMD) are compulsorily controlled by the State Veterinary Service (SVS), under legislation by the Animal Health Act (1981) and various orders. These scheduled diseases if suspected, must be reported by the person in charge of the animal (or by the visiting veterinary surgeon) to the Divisional Veterinary Office or the Local Authority—usually a police officer. The suspect animal must be kept isolated, and all stock movements suspended pending instructions of a veterinary officer.

EEC regulations also require certain diseases to be compulsorily notifiable, within Member States.

At present (1981) absent from the UK are: cattle plague (rinderpest), contagious bovine pleuro-pneumonia, *foot and mouth disease, *rabies, *swine fever and *African swine fever, *Teschen disease. In equines, epizootic lymphangitis, glanders, parasitic mange. Sheep-pox.

Those occurring in the UK (1981) are: anthrax, bovine tuberculosis, enzootic bovine leucosis, sheep scab, maedi-visna, swine vesicular disease.

Anthrax (Bacillus anthracis)

World-wide, more prevalent in tropical countries, it can affect all mammals including man. In the UK it occurs sporadically in cattle and pigs, rarely in horses and sheep. Human contact with infected material should be reported to a doctor or District Community Physician.

Anthrax bacilli form spores which can live in soil for 20 years—a source of infection over burial places of affected carcasses. Discharges from infected animals, animal products (meat and bone meal), knackers' dried meat, tannery effluent, flies. Infection enters

*These present hazards to UK stock are described below.

by contaminated food, skin wounds, inhalation of dust, and perhaps bites of blood-sucking insects.

Symptoms

Incubation period is usually short, 1–3 d, up to 14 d. Often an animal is found dead, bleeding from nostrils, mouth and anus. Otherwise there is high fever, diarrhoea, swollen throat or neck, difficulty in swallowing. Mild cases can respond to treatment. In milking cows, mild cases may only show slight drop in milk yield, with mild fever: temperatures of in-contact animals should be taken twice daily.

Treatment

Treatment is by antibiotic injections. Vaccines and serum are used as protective measures in localised areas where anthrax is more prevalent.

Control

Control is by carcass disposal, etc. on the premises, as guided by the veterinary inspector. Any sudden death is suspect and must be reported. Do not move carcasses but cover with sacks soaked in disinfectant, also the contaminated area, and temporarily fence off. Healthy, in-contact animals should be isolated but ailing stock should not be moved except as instructed.

Bovine tuberculosis (*Mycobacterium bovis*)

Tubercle bacilli of several types affect man, animals and birds. They can survive for three to six months in dung, buildings and sheltered pasture, and resist disinfectants for longer than many organisms. Infection is spread by sputum, dung, dust and in food (e.g. milk, meat from infected animals). Milk pasteurisation kills infection. An eradication programme for bovine tuberculosis made the UK an attested area in 1966; a few cases still occur.

Symptoms

Incubation period varies, up to six months. Cattle lose body condition, develop a chronic cough, and show swollen throat glands. Tuberculous mastitis causes slow hardening (induration) of the udder, perhaps internal nodulation, swollen udder lymph glands: milk, at first normal, later becomes whey-like with clots.

Control

There is no satisfactory treatment. Suspect cases must be reported and will be slaughtered, as will reactors identified by tuberculin tests, also any dangerous in-contact animals. The person in charge of the suspect case must isolate the animal, sterilise the milk before discarding it, and cleanse and sterilise milking equipment before further use. A veterinary officer will advise on disinfection of buildings and subsequent re-stocking. On some farms with infected cattle, badgers also are infected, but not necessarily so. If badger infection is demonstrated, and suspected as a source of tuberculosis for cattle, control will be advised by the veterinary and pests control officers.

Enzootic bovine leucosis (EBL) (virus)

An infectious disease of cattle which causes multiple tumours in body tissues, sometimes leukaemia. EBL is present in most European countries and North America. It was first recorded in the UK during 1978 when blood reactors were identified, mainly in Canadian Holstein cattle imported since 1967, as inapparent carriers of virus. EBL is not known to affect man: experimentally, sheep and goats can be infected. To date (1981) no clinical disease in the UK has been seen. A sporadically occurring condition, sporadic bovine leucosis (SBL) is not infectious, and usually affects only a single animal in a herd, under three years of age.

The virus can be spread by contact with bovines of any age, particularly by a calving cow, and by urine, faeces and milk. Pre-natal infection affects about one in five calves born from infected dams (who may still be clinically normal). Dissemination to other herds may readily occur through sales of reactor breeding stock. Infection could also spread by ticks, biting insects, ear punches, dehorning and hypodermic needles (use a fresh sterile needle for each animal).

Symptoms

The incubation period is four to five years, very rarely seen in animals less than two years old. Only few animals (2–5% of a herd) may ever show clinical disease, but many become blood reactors. Cattle lose condition and appetite, become anaemic and weak, then lymph glands are visibly enlarged. Eyes may show staring appearance. Death follows within a few weeks, due to tumorous involvement of internal organs. Diagnosis (via the local Divisional Veterinary Officer) can only be confirmed by laboratory tests of blood and tumour samples. Tumours in slaughtered animals must be reported to the DVO, who will examine the carcase.

Control

Within a reacting herd the DVO serves notice restricting affected animals to the farm until they die, or are licensed for slaughter. These animals should be isolated from non-reactors, isolation premises then cleansed and disinfected before re-occupation, particularly buildings wherein they calve. Afterbirth and discharges are dangerous—dispose of these immediately, hygienically. Remove the calf at once, do not feed dam's colostrum or milk from any known reactors (these should be milked last), all equipment then thoroughly cleansed with a dairy detergent.

Herd control includes prevention of introduction of infected animals into a clean herd (even borrowed or hired stock could introduce disease).

Purchase tested stock; isolate and re-test four months later. Voluntary eradication is encouraged, e.g. Holsteins in particular should be tested at four month intervals and reactors slaughtered under licence.

The SVS are tracing animals moved from infected herds, to identify possible spread to other herds; in this, all movement records will be necessarily invaluable.

Sheep scab (Psoroptes ovis)

Specific to sheep, the tiny mange mites bite the skin, and are easily spread by contact with affected sheep or their rubbing posts. It re-appeared in England and Wales in 1973 for the first time since eradication 20 years ago.

Symptoms

Symptoms may be delayed for some months, since mites remain dormant during the summer around the tail, groin and head. They become active during October–March particularly, causing skin irritation. Sheep show restlessness, nibbling and biting affected area (shoulders, back, sides and tail), causing extensive wool loss. Severe cases show extensive areas of very thickened scabby skin. Mites may be seen, under a magnifying glass, at edges of scabby skin lesions, especially on the head and tail-base. Severely affected lambs may die.

Control

Sheep scab is subject to an eradication programme involving annual compulsory dipping within designated control areas. On instruction from SVS officers and supervised by Local Authority staff, the whole flock must be treated with a Government approved dip (usually containing benzene hexachloride or an organophosphorus compound), repeated as instructed. Sheep must be thoroughly soaked, for 1 min, plunged to wet the skin and head and ears, ducked at least once. Effectiveness is quite rapidly depleted as more sheep are dipped, also by dirt. Proper dippings should give protection for six to eight weeks. Drain old dip and replace by a fresh mix if large numbers of sheep are to be dipped. Follow manufacturer's instructions when mixing the dip and subsequent replenishments, and as regards disposal afterwards of empty cans and waste dip, which is a hazard to fish and other wildlife. Follow strictly the personal precautions, especially when handling the concentrate dip.

Maedi–visna (slow virus)

These are two diseases, caused by the same virus, affecting sheep, occasionally also goats. Up to 1981 no clinically affected sheep had been seen in the UK, but infected sheep have been identified in some flocks, by virus isolation and by blood tests. These cases usually related to imported breeds of sheep, from Europe, e.g. Texel, and their offspring. Maedi occurs in most sheep rearing countries, apart from Australia and New Zealand.

Symptoms

May not be seen until four to five years after infection enters a flock. The virus soon dies outside its animal host. Maedi produces a slowly developing progressive pneumonia, causing panting, listlessness, emaciation, later coughing and nasal discharge. Death ensues after three to eight months.

Visna is rarely seen outside Iceland. Again, it affects only sheep over two years old, causing unsteady gait and progressive paralysis of the hind legs. Illness may last up to a year.

Virus spreads most readily between infected and non-infected sheep by close contact, e.g. housing in winter. Virus in milk is probably a major source of infection for young lambs. Other spread is probably by biting insects, which can transfer infected blood, whilst needles of vaccination syringes could also do so—use a new one for each animal.

Control

There is no treatment, nor vaccine. Pre-import tests are now required by UK. Infected flocks should blood test every six months, slaughtering reactors and their offspring, and not introducing new stock until tests are clear. Other measures should be discussed with your own veterinary surgeon.

Swine vesicular disease (SVD) (virus)

First seen in the UK in 1972, introduced from abroad by infected animal material. Like foot and mouth disease, it can resist chilling, and can remain viable in meat unless heated to over 68 °C. It is infective in manure and the ground for at least six months. Unlike foot and mouth disease virus, airborne spread is not known to occur.

Symptoms

Symptoms are indistinguishable from foot and mouth disease. Only pigs are susceptible. Incubation period is 3–7 d, sometimes longer if pigs are exposed to small amounts of virus. During active disease large amounts of virus are spread from pigs even before symptoms are evident. Lameness, with tender feet and arching of the back, may not be very obvious unless suspect pigs are walked on concrete. Blisters rupture to form raw ulcers above the coronary band of the feet; discoloration under the horn and separation of this from the hoof; snout and mouth ulcers; fever. A severe outbreak can affect nearly all pigs in a pen, but milder outbreaks may slightly affect only a few, with few foot ulcers, and pass unnoticed. Sows and older pigs often suffer lameness from other causes. Few pigs die.

Control

Control is by eradication through compulsory slaughter under instructions from a veterinary officer. This depends on prompt reporting of suspect cases and thorough cleansing and disinfection (specific disinfectants are listed for SVD) of premises and vehicles. Swill must be properly cooked, as infection is readily spread by untreated animal waste. (Swill feeders have to be licensed in UK, and the boiling plant approved.)

The following diseases are not normally present in the UK, but are scheduled as they present possible risk through importation.

Foot and mouth disease (FMD) (virus)

This is caused by seven main types of virus, with other sub-types of varying virulence. All types are resistant to cold, even surviving in chilled meat or other animal products. All cloven-footed animals are susceptible, including deer and occasionally coypu; outbreaks in the UK have usually affected cattle or pigs, occasionally sheep.

The highly infective virus can be transmitted in saliva, discharges, milk from infected animals, even before diagnostic symptoms are shown. Contamination of vehicles, personal clothing, birds, can also distribute infection, as may contaminated straw, hay, feed bags, troughs. Raw meat or bones must not be fed to livestock but always properly cooked: swill users must be licensed and their boiling plant approved.

Birds flying from the European continent may bring in infection, which can also be spread by non-susceptible but carrier animals such as dogs, cats, fowl.

Virus can be windborne, under certain meteorological conditions.

Symptoms

Incubation period is usually 2–4 d, but may be 12 h to 12 d. Symptoms are drooling saliva, smacking lips (due to mouth ulcers), lameness due to ulcers between claws of feet, fever. Calves may die before mouth blisters develop. Body condition is lost rapidly and milk yield falls—blisters can occur on teats. Mild infections can remain unnoticed, but are a major hazard in spread of disease.

Sheep usually show lameness, rapidly spreading through a flock.

Pigs show foot lesions, also snout blisters and in mouth, virtually identical with swine vesicular disease.

Control

Treatment is not allowed in UK. Government policy is to eradicate FMD whenever an outbreak occurs. Suspected herds are subject to immediate standstill on livestock movements on or off the premises, pending examinations and diagnosis by a Veterinary Officer. Confirmed cases are slaughtered, buried or burned on the farm, as are dangerous in-contact animals (including indirect contacts via markets, vehicles, or human movements). Food and fodder may also be destroyed.

Movements of livestock and vehicles within the controlled area are strictly controlled. Compensation is payable. In other European countries vaccination is practised but so far has not been found necessary in the UK.

Rabies (virus)

This is not present in the UK, but occasionally occurs in imported dogs and cats undergoing compulsory six months' quarantine. The virus can infect any domestic or wild mammal (which then widely spreads disease), also man. Rabies is world-wide and is being spread westwards through Europe, mainly by foxes. Infection is introduced by a bite from a rabid animal, by infected saliva.

Symptoms

Dogs change their behaviour, either hiding, perhaps paralysed (so-called 'dumb rabies'), or aggressive, howling, eyes staring, excessive salivation, convulsions. Cattle and other farm livestock may also show aggression, or the passive 'dumb' symptoms. Death rate is very high.

Control

It is essential to keep rabies out of the UK by import regulations; disregard of these results in heavy fines and imprisonment. At present imported carnivores are quarantined for six months, and vaccinated (preferably before importation). If disease should become imminent, control measures would be announced by the State Veterinary Service.

Swine fever (virus)

Highly infectious, specific to pigs, causing more severe disease in younger animals. This virus survives normal meat processing, including salting and freezing. It is spread by contact with affected pigs or their excretions, via troughs, ground or contaminated foodstuffs, e.g. unboiled swill. Also spread by contaminated boots, clothing, sacks, brooms, utensils, or vehicles.

Symptoms

Incubation period of 5–10 d, sometimes several weeks. Dullness, inappetance and shivering (high temperature). Purple rash may develop on the skin of ears, hocks and belly; diarrhoea. Incoordination and convulsions can occur. Acute disease causes emaciation and unthriftiness, pneumonia, fewer deaths; pregnant sows may abort or produce weak, trembling piglets.

Other diseases of pigs may produce some symptoms similar to swine fever, e.g. swine erysipelas.

Control

In the UK an eradication policy eliminated this disease and would be applied to any new outbreaks: compensation is paid for any pigs compulsorily slaughtered. Any suspicion of swine fever must be reported to the Divisional Veterinary Office or Local Authority. At any time, pigs moved from a market or dealer's premises must be kept separate from all other pigs for 28 d after arrival (legal requirement). A pig owner can prevent introduction of this and other diseases by not visiting other piggeries to see sick pigs, nor permit visitors to his own premises without due hygienic precautions, e.g. providing washable waterproof coats, disposable overshoes. Swill food must be boiled for at least 1 h before use; this is also a legal requirement. Preferably, a separate person should deal with the raw swill, not the pig herdsman.

African swine fever (virus)

Outbreaks have occurred in continental European countries, but not in the UK.

Symptoms

Symptoms are almost identical with classic swine fever, but with fewer acute deaths. Recovered animals may carry infective virus for life, and virus could live for several years, during cold weather, in ground and farm buildings. It requires special disinfection procedures.

Control

Control in the UK would be by eradication, as for swine fever, and early suspicion of disease must be reported.

Teschen disease (virus)

This disease has spread into some European countries, but has not yet (1981) appeared in the UK. It was

scheduled to comply with an EEC Directive. The virus affects only pigs, quickly spreading through a herd, mainly by infected faeces or indirectly by inadequately cooked swill.

Symptoms

Symptoms are loss of appetite, fever, irritability and nervous signs such as hind-leg paralysis, stiffness, tremors, eye twitching, convulsions and death in 3–4 d. (Some of these symptoms are caused by the virus of Talfan disease, which is present in the UK.)

Control

If the disease should occur in the UK, control measures would be instituted by the State Veterinary Service.

ZOONOSES

These are diseases which affect both man and animals, spreading either by direct contact with infected animals, or their faeces or milk, or indirectly by consuming infected food, perhaps imperfectly cooked.

Animal attendants should always observe strict personal hygiene, washing and disinfecting hands and protective clothing and footwear, particularly after handling sick animals; avoid splashing and inhalation of infected discharges, pus, faeces, urine, womb discharges. It may be necessary to wear disposable gloves and face masks.

The Zoonoses Order 1975 (a Diseases of Animals Act) schedules two specific diseases as *reportable*, and can add others as necessary. These are *brucellosis* (which can cause undulant fever in humans), and *salmonellosis*, where bacteria cause food poisoning, fever and diarrhoea, sometimes fatal. Both diseases are described fully later. The person who submits the animal specimen found to be thus infected has the responsibility of reporting the case to his local Divisional Veterinary Officer. In the event of any possible risk to human health, both veterinary and medical advice are given.

Veterinary Inspectors have authority to fully investigate cases, if necessary taking samples, require disinfection and apply restrictions on movements of animals and birds.

Medical Officers of Environmental Health (Local Authorities) may advise on human risks, via a general medical practitioner.

Other infections include *psittacosis* (from parrots, budgerigars, pigeons); *leptospirosis* (infectious jaundice—from dogs, rats, cattle); *anthrax* (contact with carcasses so affected); *Q fever* (from sheep, possibly untreated milk); *tuberculosis* (milk, or infected cattle, dogs); *Campylobacter* (an enteritis probably contracted from affected dogs, perhaps milk); *trichinosis* (a parasite, from pig meat, rare in the UK); *Toxocara* (a parasite from dogs); *ringworm* (skin infection, from dogs, cats, cattle, horses). Some conditions are uncommon, others present occupational hazards, e.g. to stockmen. If suspected, consult your medical practitioner.

Diseases of cattle

In cattle, particularly, different systems of management influence the risk and types of illness; susceptibility also varies with age. The completely self-contained dairy herd, well fed on a balanced ration, is at lowest risk from infectious diseases, although high yielders may present more cases of so-called production disorders such as ketosis, fatty liver syndrome. This health harmony may be disturbed if newly purchased animals introduce disease.

By comparison, an outwintered beef suckling herd, on a low cost system fed only home grown fodder and grain, is liable to dietetic deficiencies reflected in difficulties in calf-rearing, even muscular dystrophy and sudden deaths.

At high risks are calf rearing units relying on continual turnover of batches of calves bought via markets and dealers, which may introduce a variety of infections; their resistance can also be lowered by the travelling involved early in life. If the system does not allow for cleansing, disinfection and depopulation of pens between batches, and does mix different age groups within a building, then new batches frequently succumb to infections carried by older, recovered animals.

Comfortable housing remains important; even a new, modern building can encourage illness if badly designed, draughty, incorrectly sited and improperly drained.

It is convenient to consider disorders in the broad age groups of calves, up to six months old; older growing stock; adults, e.g. in-calf heifers and older.

DISEASES OF CALVES

Colostrum, taken not later than 6 h after birth, is essential to provide vitamin A and antibodies produced by the dam—although some cows are inadequate in this respect. *Weakly or stunted calves* may be born, die soon or are unthrifty and difficult to rear. Poor nutrition of the dam may be responsible, including specific deficiencies of iodine, copper, selenium, vitamin A. Infections during pregnancy may affect the fetus yet pass unnoticed in its mother, e.g. the virus of bovine mucosal disease. Occasionally, faults may be inherited.

Navel or joint-ill (bacterial)

The unhealed navel is readily invaded by bacteria from dirty surroundings. Frequently *E. coli* bacteria enter the newborn calf's blood stream through this route, causing septicaemia and rapid death, or infection may cause abscesses, in swollen limb joints or liver and other organs. Lameness results, or death later.

Treatment

Treatment is by early medication, usually antibiotics.

Prevention

Prevention is by dipping the navel-cord into antiseptic, tying off with sterile tape. Good hygiene in the calving area is essential.

Digestive system

Nutritional errors commonly initiate indigestion, inflammation of the stomachs and intestines: overfeeding on milk; incorrect mixes of milk substitute; too rapid changes of food, e.g. before the rumen has developed to proper rumination.

These may cause excess gas, showing as bloat, in the left flank region, or diarrhoea of yellow-white, or greyish colour.

Rest from food for 24 h, but with water or fluid replacement solutions available to combat dehydration. Re-introduce slowly a correct feeding regime. Simple intestinal astringent medicines will assist—antibiotics should not be necessary.

Specific diseases are as follows.

Enteritis

Bacteria and viruses can infect the intestinal tract, causing diarrhoea—in the young calf often known as 'white scour'. Perpetual diarrhoea itself can lead to body dehydration and death. Faulty bucket feeding, poor hygiene, excess milk, are contributory factors.

E. coli bacteria comprise numerous types, some pathogenic, others normal inhabitants of the intestines.

Salmonella bacteria also with many types, can be highly infective and fatal; these are readily introduced by newly purchased calves, or by 'carrier' cows, sometimes by vermin, occasionally in foodstuffs of animal origin. They can affect cattle of all ages.

S. dublin is common and fairly specific for cattle, whilst *S. typhimurium* will affect a variety of animals and birds, also causing food poisoning in humans (*see also* 'Zoonoses'). Salmonella can survive for many months in dung; allow a six month interval between applying this to pasture and subsequent grazing.

Symptoms

Diarrhoea, dark-coloured or bloody; fever, depression, septicaemia, sometimes pneumonia. Older cattle show same symptoms, or abort though seeming healthy. *S. dublin* can exist in a herd for years, with few clinical cases apparent. Pneumonia can also be caused by *Salmonella*.

Treatment

Medicaments for *E. coli* and *Salmonella* infections include nitrofurans, sulphonamides, antibiotics, given by mouth or injection as necessary; laboratory tests help to guide the best choice, as some types become drug resistant. Simple intestinal astringent medicines control the diarrhoea; fluid loss must be compensated by intake of suitable electrolytic fluids.

Prevention

Vaccines and sera are available, not infallible due to the variety of types of *Salmonella* and *E. coli*.

Virus infection by rotavirus and corona virus are also associated with diarrhoea in calves, from a few days of age up to six to eight weeks. These viruses can live for nine months in faeces, but carrier adult cattle are the likely source of infection, spread indirectly also by stockmen and contaminated equipment; severe illness also occurs in suckled beef calves kept at grass.

Symptoms

Diarrhoea, rapidly spreading among calves a few days old, but even four to six weeks of age also. Illness can be mild, or fatal. Cases may occur every year in new calves, but some immunity is given by adequate colostrum.

Treatment

Intestinal astringent mixtures and ample saline (electrolyte) intake to combat dehydration. There are no medicines specific for these viruses, nor vaccines, but supportive treatments may control co-existing bacterial infections.

Coccidiosis is caused by minute parasites which invades the intestinal lining. Some types of coccidia are non-pathogenic. Adult cattle may carry infection, which is multiplied by susceptible calves and spread by faeces, particularly in overcrowded housed animals where humidity fosters the life cycle of the parasites.

Symptoms

Affects usually the weaned calf, with sudden onset of blood-stained faeces (dysentery), the animal straining. Blood loss may cause anaemia and weakness. Recovery, within a week, often is followed by unthriftiness.

Treatment

Clinically affected animals should be housed, treated, and given adequate fluid replacements. The companion group should have mass medication, to suppress development of clinical disease; this should be prescribed by a veterinary surgeon.

Respiratory system

Specific bacterial and viral infections are involved but the respiratory system is very susceptible to adverse environmental factors such as draughts, chilling, e.g. in vehicles, markets; overcrowding in houses especially when humidity is high (inadequate proper ventilation). Calves can adjust to poor weather if accustomed slowly. Even outdoors, poor hygiene and crowding in yards is debilitating. Unless these faults are avoided, respiratory illness will be widespread, sometimes fatal, and expensive medicaments will be wasted.

Virus pneumonia

Many different viruses affect the respiratory system.

Symptoms

Some are mild, rarely noticed; others cause severe changes (lesions) in lungs, with rapid breathing, cough and some fever and death within a few hours. Various superimposed infections by different bacteria usually exacerbate symptoms, causing mucopurulent discharges from nose and eyes: e.g. *Pasteurella, Mycoplasma, Chlamydia, Corynebacterium pyogenes*.

Treatment

Treatments include antibiotics and other drugs, which control bacterial invaders, but there are none which directly affect viruses. Treatment should be administered for 3–5 d, otherwise illness relapses occur. Recovery may be slow and some animals may fail to grow normally and remain unthrifty.

Prevention

The complex combination of viruses and bacteria precludes a universal vaccine. Vaccines can be effective in some cases, but immunity may be of only short duration.

Good calf management is the prerequisite.

Poisoning by lead

Calves are highly susceptible to poisoning by lead substances, e.g. flaking paint on walls, doors and woodwork, or from car batteries: calves, being inquisitive, often lick these objects.

Symptoms

Symptoms include staggering, fits, blindness, bellowing, prostration and death. Early diagnosis and treatment is essential, as there are other disorders causing similar symptoms, e.g. *hypomagnesaemia, cerebrocortical necrosis,* some *brain diseases.*

Treatment

Treatment is specific—consult your veterinary surgeon.

There are many potentially poisonous substances commonly used on farms, especially the concentrate chemicals before dilution, e.g. herbicides. Medicines, and external applications, should always be carefully used according to manufacturers' instructions. Discarded containers present a real risk—dispose of them safely, remembering also their human hazards.

DISORDERS OF OLDER CATTLE

Older calves and yearlings share some diseases such as salmonellosis, pneumonia, but are subject to other conditions when first turned out to pasture. Since some affect all ages it is convenient to consider these in alphabetical order.

Bacterial diseases

Actinomycosis and actinobacillosis (lumpy jaw and wooden tongue)

Similar swelling of affected tissues is caused by these different bacteria, both penetrating through small wounds in the mouth, e.g. caused by foreign bodies, changing of teeth. Actinomycosis causes swelling of the jaw-bone with ultimate suppuration. Actinobacillosis causes swelling of the tongue and neighbouring soft tissue, with purulent changes in neck glands; it can also affect the stomach and other internal organs, sometimes the udder.

Symptoms

Symptoms are salivation, difficulty in masticating food, swollen tongue, or abscessed throat glands which later burst. Infections of internal organs give indefinite symptoms, sometimes progressive loss of bodily condition.

Treatment

Iodine preparations given orally or by injection;

sulphonamides, antibiotics. Early treatment is essential, but jaw swellings respond less well than actinobacillosis. Abscesses must be drained and locally treated. Destroy the pus hygienically; it is highly infective.

Prevention

No vaccines. When cases are frequent the infection is probably being spread via communal food or water troughs, ponds, scratching posts.

Blackquarter (Clostridium bacteria) (blackleg, felon)

On some fields spores of *Clostridium* bacteria are long-lived in soil and infect small or larger wounds. Young cattle, 6–30 months old, are chiefly affected.

Symptoms

Sudden deaths are common in younger animals without previous symptoms. Others show lameness, limbs swollen, hot and painful, fever. The swellings are doughy at first, later sometimes ballooned by gas which crackles when the skin is pressed: if burst open, there is a rancid odour.

Treatment

Temperatures of all in-contact animals should be taken and feverish cases treated with antiserum and antibiotics (penicillin).

Prevention

Vaccinate young stock every six months until two to two and a half years old. Burn affected carcasses.

Note that sheep of all ages suffer from a similar disease, especially after difficult lambing (more often when manually aided), or when dipping or shearing causes injuries: the multiple-type *Clostridium* vaccines may include protection against blackleg.

Foul-in-the-foot (bacterial)

Mixed infections usually. Common on some farms.

Symptoms

Lameness. Heat and swelling between claws and above hoof, even higher up the leg. (Abscesses may develop and penetrate deep to tendons and bones.)

Note that foot and mouth disease should always be considered when lameness occurs.

Treatment

Early injections of sulphonamides or antibiotics. 'Proud flesh' may need surgical removal; in bad cases, claw amputation. The healing wounds must be protected by bandaging.

Johne's disease (Mycobacterium johnei)

These bacteria are related to tuberculosis and similarly

survive on shady pasture, stream and pond margins and muddy gateways for up to 12 months. Infection is by mouth, from dung-contaminated forage, causing bowel infection and later excretion of bacteria in faeces.

Symptoms

Long incubation period; calves up to six months old are most susceptible but symptoms may not develop for one to two years, often after calving or some stress. Infection between adult cattle is unlikely. Diarrhoea becomes profuse, foul smelling and bubbly. Rapid loss of condition is followed by death. Such affected cows may infect their embryo calves so offspring should not be kept as herd replacements but considered as a potential danger, intermittently shedding infection before symptoms are shown.

Diagnosis

Suggested by symptoms, but confirmed by laboratory tests on dung and blood samples; certificates for export of healthy stock often require these tests and a Johnin skin test.

Treatment

Treatment aims first to alleviate diarrhoea and weight loss until slaughter: isolate meanwhile in a separate loose box. Use dry food and intestinal astringent medicines. Few cases fully recover, and these remain a permanent risk to other cattle.

Prevention

Bucket-feed calves from birth to prevent early infection by dung on the udder. Rear out of contact with adults or fields used by them. Put dung on arable, not grazing, fields. Improve watering facilities and field drainage. Slaughter calves from affected dams.

Vaccination

Use of Johne's disease vaccine needs the approval of the divisional veterinary officer and is limited to herds with a known problem.

Lameness

This commonly begins with damage to the sole of the feet, often hind-feet, such as by stony tracks, rough stubble fields, kale stumps. Cubicle bedding may be too coarse, concrete too rough. The heels can be bruised. Excess slurry in yards, silage clamp areas, dung passages, and around water troughs, gateways, etc. foster heavy bacterial contamination which enters the wounds.

Symptoms

Heat and swelling between claws and above hoof, even higher up the leg. (Abscesses may develop and penetrate deep to tendons and bones.)

Note that foot and mouth disease should always be considered when lameness occurs.

Treatment

Early injections of sulphonamides or antibiotics. 'Proud flesh' may need surgical removal; in bad cases, claw amputation. The healing wounds must be protected by bandaging.

Laminitis

Laminitis is attributed to excess concentrate or grain feeding, not generally common except when growing stock are 'forced', e.g. for show. At times, dairy cattle are affected, uterine infections (metritis and retained afterbirth) coinciding.

Symptoms

Symptoms are acute lameness, often in all feet, with tangible heat in the hoof, pain, sweating, uneasy movements and preference to lie down.

Treatment

Treatment must begin quickly, otherwise permanent hoof damage remains. The medicines need prescription, after differential diagnosis by your veterinary surgeon. Cold water footbaths relieve the pain, and intermittent walking exercise is beneficial.

Any feeding errors or excesses need correction; laminitis is best prevented by only slow introduction to heavy feeding regimes.

Leptospirosis (Leptospira bacteria)

Various types of leptospira affect most domestic and wild animals (e.g. rats, voles), also humans. Sometimes no clinical illness ensues, but different effects result from the several serotypes as follows.

Symptoms

Dairy herds may exhibit *mastitis*, with overall drop in milk yield. *Abortions* can occur, also some fever. *Jaundice*, and death, is seen, usually in young calves on farms where rats are numerous.

Treatment

Mastitis is usually temporary, treatment not needed, nor do aborting cows remain ill. Jaundice requires differential veterinary diagnosis and appropriate treatment and control measures—which include vermin destruction.

There is always a risk to humans when animal infection occurs and medical advice should then be sought.

Mastitis (bacterial)

This is an 'inflamed' udder, but mastitis can be classified as clinical, showing obvious swelling, pain, and milk clots; or sub-clinical, with no visible changes but an established bacterial infection and higher cell count detected by laboratory tests. Older cows become more susceptible and there seem to be inherited tendencies. The risk is greater in the first month of lactation.

Infection is due to a variety of bacteria, perhaps also viruses (not yet identified). It is spread by milk from infected udders directly, or indirectly by contaminating milking equipment, udder cloths, hands, teat and udder skin and broom handles. Bacteria multiply in teat and udder sores and wounds.

Streptococcus agalactiae infection is contagious, readily spread by infected milk via hands, and can be eradicated from a herd. Less contagious bacteria are commonly present on teats, udders and body openings of most cows, with complete eradication not possible. They include *Staphylococci, Streptococcus dysgalactiae* and *Streptococcus uberis, E. coli. Corynebacterium pyogenes* causes summer mastitis, affecting even dry cows and heifers; often spread by flies. Environmental organisms, e.g. *E. coli, Pseudomonas*, some moulds, can cause severe mastitis, general illness and death; these infections occur during winter, accumulating in cubicles and yards, especially in sawdust bedding.

Symptoms

With the acute type, heat, swelling, pain in affected quarters, clots in milk, fever and inappetence. *Summer mastitis* is acute, with secretion of much foul-smelling pus; affected quarters seldom recover—the cow may die. With sub-acute type, less severe general upset, but still marked udder and milk changes.

The chronic type is either sub-clinical, or quarters become enlarged and firm, alternatively showing obvious shrinking. *Tuberculous mastitis*, now rare, is chronic, usually with detectable nodulation in the udder tissue.

Pre-disposing causes

Injuries and sores (some infectious) especially to teat-tip. Teat chaps, milking machine faults from poor maintenance, milking technique, e.g. overmilking, can damage the teat canal.

Treatment

Clinical cases are treated with various antibiotics injected hygienically up the teat canal: afterwards, discard milk for several milkings according to prescribed instructions. An ill cow may need other general treatment. *Streptococcus agalactiae* responds well to a programme of milk testing; treatment and very good hygiene eradicate infection from a herd. Plan this with the veterinary surgeon. *Staphylococcal*

and other bacteria are often resistant to some antibiotics: choice may need laboratory tests. Herd eradication is not possible as re-contamination occurs from skin and environment. Summer mastitis cases may need stripping-out, hygienic pus disposal, and thorough disinfecting and washing of hands and utensils before handling other cows. NB Phenolic disinfectants can taint milk

Prevention

No fully effective vaccines yet. Control aims at reducing level of herd infection by sound hygiene and good milking routine. Teat-cups are best cleaned between each cow by running hot water, 85–90 °C (185–200 °F), through each milk tube and cluster; cold water is less effective, cluster dipping in disinfectant even less. Separate paper towels should be used for each cow, or failing this two or three cloths used in turn, soaking in udder disinfectant. Sterilise between milkings. After milking and dressing teat sores with germicidal cream, dip teats in test disinfectant, hypochlorite, iodophor, chlorhexidine types, diluted as instructed.

With dry cow therapy, after the final milking, antibiotic is inserted into each teat to provide long duration contact with any latent infection.

Milking machines should be regularly serviced, the vacuum controller checked daily and cleaned weekly if necessary.

Cows with chronic mastitis should be culled—they carry infections, do not milk to capacity, and interrupt milking routine.

Parasitic diseases

Parasitic diseases can be due to internal parasites, e.g. stomach and intestinal worms, or external flies, lice ticks. Younger animals are more liable to suffer from parasitic worms, but adult stock (partially immune) often still carry sufficient parasites to infest pastures. Some parasites have life cycles necessitating passage through another intermediate host animal, e.g. liver fluke (snail and cattle, sheep), tapeworms (dog, cattle, sheep), redwater (tick, cattle, sheep, wild animals): their control requires attention to each host. Climatic conditions influence development of eggs of liverfluke and stomach worms, and some parasites are seasonal (warble fly, tick).

Control of most parasites necessitates knowledge of each life cycle, then good stock management to avoid seasonal risks and build-up of infestation in pastures or houses.

Treatments are applied to animals at strategic times, to prevent illness rather than cure. New medicaments are continually being evolved and choice and usage are best discussed with a veterinary surgeon. Different programmes may be needed on farms, varying with weather conditions. More common parasitic diseases are outlined below.

Lice

Both blood-sucking and biting lice live on the skin of calves, causing irritation, restlessness, unthriftiness and sometimes anaemia. The skin of neck and back becomes very scurfy, with lice and their eggs seen on the hairs.

Treatment

Louse powders, often containing hexachlorocyclohexane (HCH) should be dusted into the hair, repeated as instructed, usually twice.

Note: do not use horticultural preparations of HCH—some contain higher concentrations and are toxic to calves when licked off the hair.

Liverfluke disease (Fasciola hepatica) (liver rot, bane, cord)

Liverflukes affect cattle, sheep, rabbits, occasionally horses and donkeys, and man. The adults live in the livers of affected animals, the eggs excreted in dung, hatching on the ground within 2–20 weeks under suitable ground and weather conditions; larvae then require to enter the intermediate host, a mud-snail (note—not a permanent water snail). In this they multiply, after six weeks leaving the snail and forming cysts on grass, to infect the grazing animal.

Symptoms (see also 'Diseases of Sheep')

The disease in cattle is seldom as acute as in some sheep outbreaks. Flukes cause chronic inflammation and thickening of the bile ducts, with intermittent digestive disturbance and diarrhoea, bottle jaw, unthriftiness, and lowered milk yield in dairy cows. Abattoir surveys show that fluke-damaged livers are widespread; in some years over 70% may be wholly condemned. Laboratory tests on dung samples confirm diagnosis.

Treatment

Various medicaments are available, administered by mouth drenches, as granules in food, or by injection: some are not usable for cows in milk. Discuss choice and timing of treatment with the veterinary surgeon, in relation to type of animal and the annual 'fluke forecast': this is based on summer weather conditions and regular monitoring of mud-snails on observation farms. It gives general guidance to earliness or lateness of fluke infestation on herbage, also possible severity, and timing of the animal treatment. Generally, cattle are treated twice, at beginning and middle of winter:

a third dose before spring turn-out may be recommended.

Low-level feeding, or other debilitating factors (stomach worms), increase the ill effects of fluke disease. Poor weather in a year produces low quality fodder and ground conditions ideal for fluke build-up.

Prevention

Regularly treat cattle and sheep. Do not graze the wetter, known snail habitats from October to December, but improve drainage, fence off stream margins and 'flushes' beside springs. Chemical molluscicides can be applied to such local areas to kill snails during their active periods, especially wet spells in June–July.

Lungworms (husk, hoose)

Different species affect cattle, sheep, horses and pigs; usually younger animals, but adults are not entirely immune, often maintaining pasture infestation as a hazard to youngsters grazing after adults. Lungworm larvae hatching from dung live in damp grass, are eaten, then migrate through the body to the lungs.

Symptoms

The slender, thread-like lungworms live in air passages of the lungs, causing bronchitis, husky cough, sometimes nasal discharges. Animals are unthrifty and may die from secondary pneumonia due to bacteria. Laboratory dung examinations may demonstrate lungworm larvae, but not in the early stages.

Treatment

Choice of medicament depends on diagnosis of stage of infection. Injections and oral treatments are available, some of which also kill stomach and intestinal worms, which are common simultaneous infestations. Treat, then move to 'clean' pastures, say hay aftermath.

Prevention

The hazard is created when calves in spring acquire lungworms by grazing areas contaminated by older cattle. This group may not suffer disease, or only slightly, but can so further multiply ground infestation that a following group of calves suffer severe disease. Preferably, turn out all youngsters at the same time, on fields left unstocked all winter.

Calves can be given a live lungworm larva vaccine by mouth; two doses are essential, four to six weeks apart, before turning out: the veterinary surgeon will arrange this. *See also* 'Diseases of Sheep' p. 463, and 'Diseases of Pigs' p. 471).

Mange (Psoroptes and Sarcoptes)

Minute mites rarely parasitise cattle, but when present mainly affect the tail-base or neck region. Skin irritation results, with scurf, scaliness and loss of hair due to rubbing.

Treatment

A variety of washes, sprays, and creams containing hexachlorocyclohexane, organophosphorus, or other suitable insecticide are available. Use as directed, with personal precautions especially when mixing.

Redwater (Piroplasmosis, Babesiosis)

Caused by microscopic parasites entering and destroying red blood cells. The common ticks (*Ixodes ricinus*) are intermediate hosts, whose infected bites spread infection. Dormant ticks in ground cover remain infective for several years. Calves under six months old are seldom ill but acquire immunity. The greatest risk is to adult cattle brought in from redwater-free districts.

Symptoms

Listlessness, high fever (41–41.7 °C (106–107 °F)), diarrhoea then constipation, death if not treated early; severe anaemia. Urine may become reddish-brown due to broken-down red blood cells. Recovered animals remain a source of infection, via ticks, and stress may cause illness to recur.

Treatment

Early injections of specific remedies rapidly cure but may be ineffectual after constipation is shown. House, give light laxative diet, perhaps purgatives, and ample water.

Prevention

Eliminate ticks by destroying shelter, i.e. scrub and rough pasture. Insecticide sprays of persistent type (e.g. HCH) on cattle during tick season, April–May and again August–September; avoid grazing tick areas, especially during these periods, and do not introduce susceptible adult cattle. Sheep can be run over tick-infested land, then dipped.

Ringworm

Highly contagious fungus infection of the hair, more common in younger, yarded cattle, but affecting also cows even at grass. In stockmen causes *dairyman's itch*, with lesions on arms and face.

Symptoms

Grey scabby circular patches, with hair loss, mainly on face and neck.

Treatment

Skin treatments can be sprayed or painted on, more effective if heaviest scabby areas are first scrubbed off using old sacking and hot water and soda. Some fungicides are given by mouth and act by excretion through skin.

Prevention

Difficult, as ringworm spores live for many years in old woodwork. Scorch carefully with blow-lamp, then creosote.

Stomach and intestinal worms

Adult and immature stages of the different types of parasitic worms live in either the stomach or intestine. Their eggs pass out in dung and develop into infective larvae on the pasture, but only in moist, warm weather: this takes 7–21 d. Young animals at first grazing are most susceptible. Adults acquire some immunity, but still carry light infestations of worms, to re-infect pasture.

Symptoms

Symptoms are diarrhoea, inappetence, loss of condition, sometimes severe anaemia and death. Diagnosis is aided by laboratory tests on blood and dung, although very acute cases occur before worm eggs are excreted in dung.

Treatment

Cattle 6–18 months old suffer from various types of worm, but particularly from *Ostertagia* species of stomach worm. These are picked up from grass in June and treatment is recommended in late June or early July, followed by moving to 'clean' grazing, e.g. aftermath. A second early treatment may be necessary in severe cases, and is advisable when housing the following winter. Good feeding and ample water are needed to counteract the anaemia and loss of weight. Recovery may be slow, due to permanent damage to the stomach lining. Choice of medicament should be discussed with the veterinary surgeon, there being a wide choice.

Prevention

Avoid overstocking: the handy calf paddock is a menace. Graze calves on pastures kept free of older stock during winter and spring.

Strategic treatment

Treat in late spring if weather conditions and tight grazing produce a hazard. Treat and move in July. Risk is increased if stock are poorly fed. Disease can occur indoors if litter is inadequate and wet.

Tapeworms

Tapeworm segments, resembling cooked rice grains may occasionally be seen in dung, but cattle seldom carry enough adults in the intestine to warrant treatment. They might contribute to unthriftiness due to poor diet and other diseases.

Tick-borne fever

Another disease of the blood. Minute organisms invade the white blood cells and again are carried from affected to healthy cattle by tick bites.

Symptoms

Fever (41–41.7 °C (106–107 °F)), sudden fall in milk yield, but no urine reddening. Seldom fatal; cattle recover within a week when treated with antibiotics, sulphonamides (i.e. different drugs from redwater treatment).

Ticks (Ixodes ricinus)

In the UK the grey sheep tick is distributed through western counties. Seldom present in large enough numbers to cause anaemia, it is an intermediate host and transmits important diseases to cattle (*see* 'Redwater', 'Tick-borne fever' and Sheep ticks, p. 469).

Warble fly (Hypoderma)

Warbles on the back are the swellings inhabited by the maggot of the warble fly. These appear from January to June, usually after migration through the body tissues beginning the previous summer. Warble flies lay eggs on the hairs of the legs during May–September.

Symptoms

Warbles are obvious—a growing swelling under the skin, with a breathing hole (sometimes a little pus is present). Internally, migrating larval maggots damage the gullet and flesh over the back, and ruin the hide. *Gadding*, with loss of milk, is provoked by the flies, although these do not actually bite.

Control

The Warble Fly Order 1978 makes it compulsory to treat cattle affected with warbles, during 15 March to 31 July. Then, a Declaration of Treatment form must accompany the cattle when moved off the farm. An Amendment in 1981 further requires treatment in the autumn of all cattle over 12 weeks old on premises where there were warble affected cattle during the spring. This intensifies the MAFF eradication campaign, commenced at the request of farmers.

Treatment

Treatment should be 'strategic', to prevent damage. 'Pour-on' insecticides (often organophosphorous compounds, OP) should be used once over the back during September–October (never after mid-November). The absorbed chemical kills the larvae in migration. This can be used when warbles appear, but the damage is done.

Note that OP compounds are poisonous to man and animals: strictly follow manufacturers' instructions. New parasiticides, avermectins, are injectable and may be used during any time of the year, but not in lactating animals. Fly repellent sprays to legs and flanks, during summer, help deter the warble fly.

Diseases of production

Diseases of production are disorders due mainly to imbalance between uptake of nutrients, including major minerals and trace elements, and the animals' requirements for maintenance, production and reproduction. They can be grouped into three main categories, *infertility, metabolic disorders* and *ill-thrift* or *pine*.

Infertility

Failure of female animals to conceive or male animals to promote conception. Herd infertility is usually temporary and curable. Individual animals may suffer from permanent, incurable changes in the reproductive system, sometimes inherited. Diagnosis requires veterinary consultation and is aided by careful records of heat periods and service data.

The bull-borne diseases, *vibriosis* and *trichomoniasis*, have been largely controlled by artificial insemination.

Energy (carbohydrate) deficiency is now a common cause of infertility, particularly in high yielders inadequately rested between lactations. An inadequate intake during the first 100 d of lactation causes depletion of a cow's reserve and loss of body weight ('milks off her back'); fall in blood sugar levels can be shown by laboratory tests. Milk shows low, solids-not-fat (SNF) due to low milk protein. *Ketosis* (acetonaemia) may occur in extreme cases.

Contributory factors

Contributory factors are poor grazing due to cold weather, drought, or over-estimating the feeding value of hay, silage and autumn pasture (low in starch). Cows lying out and in cold weather have a higher energy requirement; ill-thrift due to mineral deficiencies or parasitic diseases.

Treatment

Supplementary feed of 1.8 kg cereal for nine weeks, commencing six weeks after calving. Preferably, first analyse fodder, then balance total ration with a bias to energy value, as in lead-feeding. Often quality and quantity of milk improves.

Trace elements deficiency

Especially of manganese, can cause anoestrus and infertility. Manganese becomes unavailable to herbage uptake after over-liming has increased pH to over 6.5; this may not be rectified for several years after last liming. At pH 6.0 or under, full grass production and fertiliser response are also possible.

Treatment

Ensure a daily supplement includes 1 g of manganese for nine weeks, commencing six weeks after calving.

Metabolic disorders

Acetonaemia (ketosis)

This usually affects high yielding cows up to three weeks after calving, especially in autumn and winter, but also in late summer on energy deficient pasture. It occurs when the cow is in negative energy balance, short of glucose precursors, especially when she is drawing on bodily reserves and using ketones instead of glucose for energy. Also associated with starvation, stress, excess protein in the ration and silage high in butyric acid. A secondary ketosis can follow acidosis due to feeding excess cereals in the diet. (Ketosis in pregnant ewes is known as *pregnancy toxaemia* or 'twin lamb disease'.)

Symptoms

Inappetence, lower milk yields, slowed rumen movement, constipation or diarrhoea, nervous signs and recumbency. Breath has characteristic pungent sweetish odour, also detectable in urine and milk (and confirmed by laboratory tests, also in blood samples).

Treatment

It may resolve spontaneously as milk yield falls. Injection of hormone preparations such as anabolic steroids (under veterinary direction); oral dosage with glyceral, propionates (i.e. glucose precursors) vitamin B injections to improve appetite. Acetonaemia is better prevented, by ensuring ample digestible energy (lead feeding). It rarely occurs on good grass in spring and early summer.

Milk fever (hypocalcaemia)

This is not a fever, but due to a temporary drop in blood calcium within 24–48 h after calving. It is associated with a sudden demand for extra calcium due to the onset of lactation, so the udder should only be relieved, not milked out. It is most prevalent in the autumn and there appears to be an age incidence (over second calving), also individual and breed susceptibility.

Symptoms

Symptoms are uneasiness, 'paddling' of hind-legs, perhaps excitement and violence, then the cow lies down either on its side or characteristically with head turned round into the flank. If untreated, coma and death may follow. Complications are absence of rumination, defaecation and urination.

Treatment

Injections of soluble salts of calcium (preferably with phosphates, glucose and sometime magnesium). A laxative diet of after-calving aperient helps the absorption of calcium from the bowel.

Prevention

The pre-calving diet should be well balanced, including essential minerals, especially phosphates, but not excess calcium! High dosage of vitamin D for 5 d before calving helps. The rate of 'steaming up' should not be too high. Avoid over-dressing pastures and conservation fields with potash or poultry manure. (Herbage potassium values of 3% and over may interfere with the utilisation of calcium.)

Grass staggers (hypomagnesaemia, Hereford disease)

This is associated with a subnormal level of magnesium in the blood. It can affect older calves fed only milk (normally deficient in magnesium), outwintered beef animals and dairy cows on lush grass in spring or autumn. (Ewes are also susceptible.) The diet may actually contain inadequate magnesium, leading to slow depletion in the body, as on poor winter keep. An acute illness (e.g. dairy cows) is associated with high potash (3% and over) and nitrogen content in the grass, which appears to interfere with utilisation of herbage magnesium. Spring grass may have adequate magnesium (0.16%) for a 18 ℓ cow, but a level of 0.20–0.25% would be necessary under high potash or nitrogen levels. Clovers contain more magnesium than grass.

Symptoms

Altered behaviour, nervousness, muscular tremors, twitching eyeballs. Sometimes there is only a sudden onset of tetanic muscular spasms, perhaps death, but this acute form is more common in the cow on lush grass. Blood tests confirm diagnosis.

Contributory factors

Contributory factors are sudden onset of cold, wet weather, or other stresses.

Treatment

Prompt injections of magnesium salts, usually with calcium. Keep patient warm and quiet. For calf, then include daily 15 g calcined magnesite in the milk or concentrate.

Prevention

Avoid excessive potash and nitrogen fertilising (also poultry manure in excess). Include daily, per head, 60 g magnesium oxide (calcined magnesite) in the ration—e.g. 3.2 kg ration would need 18 kg of magnesium ozide/tonne concentrate. Also available as self-help liquid feed, magnesium 'bullets', or is dusted on pasture or silage just before feeding.

Pine or ill-thrift

This can be associated with sub-optimal intake of trace elements, particularly copper or cobalt, often aggravated by general malnutrition and (or a heavy burden of) internal parasites. Over-liming depresses herbage uptake of both elements. High soil levels of molybdenum, made more available by liming, are usually the cause of copper deficiency (*hypocupraemia*). Cobalt in food is essential for normal rumen digestion, and aids red blood cell production. This deficiency disorder more commonly affects sheep than cattle.

Symptoms

Symptoms are not distinctive, just similar to any debilitating disease: poor growth, listlessness, inappetence, rough coat, anaemia. Diagnosis is confirmed by blood tests and herbage analyses.

Treatment

Cattle respond to a single dose of 0.5–1.0 g cobalt sulphate. Prevent by top-dressing selected grazings with cobalt salts at 0.5 kg/ha. Normal concentrate rations include enough cobalt.

Copper deficiency

Marked illness (*hypocupraemia*) occurs in certain specific districts, typically on peat land, i.e. Somerset, where illness is called teart. These areas contain high levels of molybdenum, which interferes with utilisation and liver storage of copper, especially at high pH values.

Symptoms

Symptoms range from some unthriftiness to marked stunted growth of very young cattle during first year outside. Black hair fades to brownish, especially round eyes; teart cases have severe diarrhoea, infertility. Blood tests confirm diagnosis.

Treatment

Teart cake provides high copper levels, adequate lower levels are included in other mineral supplements, dairy cake. Copper compounds can be injected. On a high cobalt intake, store cattle can tolerate some degree of copper deficiency.

Note that excess copper can be poisonous: sudden release from liver storage causes jaundice, anaemia and collapse.

Reproductive disorders

Infertility problems have already been outlined. Normal conception should result in straightforward birth of a normal calf, usually single, after 278–283 d gestation period. Allow the cow time to calve by herself; too early interference can be unnecessary, harmful, and cause difficulties if professional veterinary assistance is ultimately necessary. If the calf presents hind-legs first, early help is needed, and cows bearing twins also are likely to require aid. A retained afterbirth should be attended to if it fails to cleanse properly after 3 d.

Occasionally, developmental defects cause birth or abortion of abnormal calves but all abortions or premature births require veterinary consultation.

Abortion and stillbirth

Abortion is the premature birth of a dead calf at any stage of pregnancy. Stillborn calves are delivered dead at full term. Some calves which are born alive prematurely may live and yet carry certain infections. The dam herself may have become infected with bacteria or viruses during pregnancy, not necessarily showing obvious illness at the time yet the fetus or womb can be damaged. Abortion may follow, or birth at full term of an abnormal calf; e.g. virus mucosal disease (also known as bovine virus diarrhoea, BVD) can damage a fetal brain, resulting in either a blind calf, or spastic—lacking normal balance.

Brucellosis (Brucella abortus)

Eradicated in major cattle producing countries, the UK also has concluded an eradicating programme, after terminating a national vaccination scheme which reduced the heavy economic and livestock losses caused by this highly contagious disease. Also, *human brucellosis* (undulant fever), caused by *Brucella abortus*, will henceforward be uncommon—it can be acquired through direct contact with infected fetal, placental, and womb discharges, also via infected milk (if not pasteurised).

Symptoms

The calf may be aborted (slipped), often at the fifth to seventh month of pregnancy, or born alive but weakly after the eighth month, often soon dying. Some are stillborn at full term; other calves may appear normal and healthy. Any calves which survive may spread infection in the dung for several months. The afterbirth is frequently retained and this and any discharges are highly infective. Udder infection (even after normal calving) continues over several lactations.

Spread of infection

Spread of infection is by aborted calf, afterbirth, discharges and milk. These contaminate pasture, food and water, either directly or indirectly by flies, faeces, scavenging birds, also by the hand, boots and clothing or use of unsterilised calving ropes. Bulls may be infected, usually orally or by the respiratory or ocular route (flies) and many become chronic carriers without becoming sterile as sometimes occurs.

Diagnosis

Laboratory tests may be made on freshly aborted calves, afterbirth, blood or milk samples.

Treatment

No effective treatment in cattle. Cows usually only abort once, but often carry the infection in the womb and udder for the rest of their lives.

Control

The UK herds are now accredited under the Brucellosis Eradication Scheme. Abortions must be reported to the local Divisional Veterinary Officer, who will arrange for appropriate diagnostic testing of the calf fetus and dam; since there are other possible causes of abortion, it is essential to submit to the veterinary investigation laboratory as wide a variety of specimens as possible, e.g. the fetus itself, with some placenta; milk; vaginal swab; a number of blood samples will be required to also check for other infections.

Salmonellosis

Usually due to *Salmonella dublin*, this causes sporadic abortions, sometimes small outbreaks within the herd, the cows not necessarily being ill, although they can be severely affected. (*See also* 'Salmonellosis'.)

Leptospirosis (various leptospira bacteria)

Similarly causes abortions and may be diagnosed only by blood tests, infection being inapparent but at times causing mastitis and general illness.

Mycotic abortion

Certain moulds which spoil hay, silage, or straw, can cause abortion, probably by inhalation of the profuse spores produced. This type of abortion is more prevalent in winter and spring, and in the wetter counties of the UK. There is no specific treatment, but the cows' reproductive potential is not permanently damaged.

Fetal goitre

During the last three months of pregnancy the calf fetal thyroid gland develops twice the activity of the dam's thyroid, and requires sufficient iodine for proper function. Some brassica and legume foods may increase the requirement for iodine (due to goitrogenic factors). Dairy cake is adequate in iodine, except in certain areas where the iodine level in the grass is abnormally low.

Symptoms

Iodine deficiency results in late abortion, stillbirth, or calves born weakly which soon die; their thyroid glands are enlarged (goitrous). First calf heifers are particularly prone.

Prevention

Feed an iodised supplement, or incorporate in concentrates, to ensure a daily intake of 10 mg iodine by each pregnant lactating cow, especially during the last three months of pregnancy.

Virus abortion

Virus infections may cause many abortions which at present are not specifically diagnosed in the UK (about 75%), as indicated by USA research. In the UK research suggests that the viruses of bovine virus diarrhoea (BVD) infectious bovine rhinotracheitis (IBR), and enteroviruses may be responsible. Diagnosis usually required blood tests.

Genital vibriosis and trichomoniasis

Diseases transmitted by the bull used to cause widespread infertility and early abortion, but are now only very rarely diagnosed.

Miscellaneous cattle disorders

Bloat

Bloat is distention of the rumen by gas or frothy green food, seen as bulging of the left flank. In calves only individual cases occur, when milk or substitutes ferment in the wrong stomach (rumen). Access to good quality hay and careful feeding are helpful.

In older animals, bloat follows obstruction of the

gullet, or a lush diet of grass, kale, clover, or frosted foods. Pressure can cause respiratory distress. Some acute cases die unless immediately relieved by puncture of the rumen at the highest point (in emergency use a knife, but preferably trochar and cannula). Frothy bloat required oral dosage by silcone-type medicines; a simpler dose is 30 ml turpentine in 0.75 ℓ of raw linseed oil.

Cerebro-cortical necrosis (CCN)

A nervous disorder, symptoms resembling lead poisoning, which affects older calves, sometimes adults. Vitamin B utilisation is interfered with, probably by enzymes produced by bacteria on grass and fodder. There is good response to early treatment with vitamin injections.

Muscular dystrophy

Single suckled calves, up to three months, are susceptible when dams are fed a limited diet of turnips or swedes and straw, as are yearling cattle fed indoors on grain and straw or hay. These diets are deficient in vitamin E and often in selenium; research suggests propionic acid additives may be a contributory cause.

Symptoms

Symptoms often occur on turning out to grass; difficulty in standing, sometimes heart failure, also brown coloured urine, due to muscle degeneration. Vitamin E and selenium supplements prevent illness, and can cure less severe cases.

Mycotoxicosis

Some fungal infections (mycoses) of foodstuffs, including cereals, concentrates and fodders, can produce harmful products in the spoiled food. These are called mycotoxins, are of many types and may produce a variety of different illnesses, ranging from general inflammation of the alimentary tract, causing diarrhoea, and irreparable damage to the liver, sometimes nervous symptoms, including skin irritation. Blood clotting may be reduced, affected animals suffering haemorrhages under the skin or in any body tissue, sometimes fatal.

Diagnosis

Diagnosis can be difficult, requiring a range of chemical and microbiological tests on the suspect foodstuff.

Treatment

Often none specific, but requires veterinary consultation.
  Note: some fungi in foodstuffs present a real hazard to humans handling them, through inhalation of the spores; e.g. 'farmer's lung': appropriate approved airflow face masks should be worn.

Virus mucosal disease

Several viruses can invade the digestive tract and cause mild diarrhoea and transient fever, or severe disease, mainly in animals 2–18 months old, with high mortality. In a dairy herd illness may last several months, with no deaths but many cows affected and lowered milk yields. Diarrhoea can be severe and blood-stained; some animals are lame due to ulcers between the claws; emaciation is rapid.

Treatment

Treatment is seldom helpful. Vaccines may aid prevention.

Diseases of sheep

Young lambs tend to be at greatest risk from various disorders. The viability of the newborn relates closely to its birthweight, as body heat is more quickly lost by lightweight lambs under adverse weather conditions. Good birthweight requires adequate feeding of the in-lamb ewe, especially when carrying multiple fetuses.
  The flock lambed indoors requires only simple airy buildings for weather protection, but overcrowding and poor hygiene can induce serious illness by build-up of various bacterial infections, especially if lambings follow rapidly in the same pens.

On the other hand, intensive indoor sheep husbandry seems to present fewer hazards than met in intensively reared cattle; properly managed and adequately dosed and vaccinated, the major sheep hazards—parasitic gastro-enteritis and clostridial bacterial diseases—are minimised.

SCHEDULED DISEASES

These have been described earlier. In sheep they include anthrax, foot and mouth disease, maedi-visna, rabies, sheep scab, sheep-pox.

DISEASES OF LAMBS

Managemental defects which affect young lambs are similar to those in calves (*see* 'Diseases of Calves').

Chill-starvation

Loss of body-heat, especially in lightweight newborn lambs, results in apathy and failure to suck colostrum, death following in 2–4 d. This is a common disorder during severe weather conditions—wet, with high speed cold winds. Under such conditions, lamb crates can save many lambs.

Navel or joint-ill (bacterial)

The unhealed navel cord is readily invaded by opportunist bacteria, often *E. coli*, lambs being more prone when ewes lamb-down in the same fields over several years, or in overcrowded lambing pens.

Symptoms

Sudden death at 2–4 d old, or lameness due to joint abscesses forming after several weeks (*see also* 'Tick Pyaemia').

Treatment

Early injection of suitable antibiotics.

Prevention

Lamb ewes hygienically. Dip the navel-cord of newborn lamb into iodine solution, chlorhexidine, or other skin antiseptic, preferably tying the cord with sterile tape.

Border disease (virus)

The term 'hairy shakers' is an apt description of the unsteady, quivering lamb, which may carry noticeably excess hairy areas in place of wool.

This virus, which also causes bovine virus diarrhoea, can affect ewes inapparently during pregnancy, resulting in birth of affected lambs, also some abortions.

Other neonatal disorders follow outbreaks of abortion (*see* p. 465), causing early mortality, or weakly lambs which later die.

Colibacillosis (white scour of lambs)

E. coli bacteria invade the navel at birth to cause septicaemia and death (*see* 'Navel Ill'). Multiplication of some *E. coli* types in the intestine can cause diarrhoea, dehydration and death in lambs up to two to three weeks old. Infection builds up particularly in unhygienic indoor lambing pens.

Treatment

Oral dosage with sulphonamides or antibiotics; dehydration must be prevented by dosage with suitable salt solutions and lambs not overfed by heavy-milking ewes.

Prevention

Specific *E. coli* antiserum at birth may be helpful.

Lamb dysentery (Clostridium welchii type B)

Occurs in northern England, north Wales and southern Scotland. Lambs 7–10 d old either are found dead or suffer diarrhoea (often blood-stained): up to 50% soon die.

Treatment

Vaccinate ewes, use antiserum on lambs at birth.

Pulpy kidney (Clostridium welchii type D)

Older lambs are usually affected (at 6–16 weeks), sometimes also older sheep. Calves may also be affected. The best thriving animals are found dead, rarely seen ailing. Lambs sucking full-milking ewes, or grazing lush grass, most commonly suffer. Folded sheep are susceptible when moved to fresh fold crops.

Ewe vaccination gives colostral protection to lambs for a few weeks, when the lambs need vaccination. In an actual outbreak, give lambs serum and vaccine (repeat latter in four to six weeks time).

Swayback

When ewes show low blood levels of copper, their lambs may be born showing swayback, or develop symptoms up to six weeks of age. The exact cause is still unknown. All breeds are susceptible. Swayback is more prevalent following milder winters, and occurs sporadically throughout the UK.

Symptoms

Vary from complete inability to stand, perhaps blindness, to slight weakness in hind-quarters, apparent when lambs are hurried. Severe cases die, mild ones may recover. Ewes are not affected.

Treatment

A few cases respond to injections of copper salts.

Prevention

It is essential to first confirm by blood tests of pregnant ewes that a susceptible flock has low blood copper levels, as misuse of copper dosage is readily fatal to sheep. Include in the ewes' concentrate food a copper-containing mineral supplement, or provide mineral licks, or inject ewes with copper compounds.

Tetanus (*Clostridium tetani*)

This infects principally deep wounds. The bacterial spores commonly exist in soil, especially in the presence of organic matter such as manure. Lambs are infected at castration or tailing, more rarely via the navel at birth.

Symptoms

These may be delayed for one to three weeks after infection, and include limb stiffness, head thrown back, sometimes muscular spasms. Mildly affected cases may recover.

Prevention

Vaccinate regularly. Antiserum may be injected at the time of operations, but it may be ineffective when symptoms have developed. Risk is reduced by hygienic operations in clean pens.

Note: *Horses* are affected more severely, usually following deep pricked wounds in the feet or stake injuries while jumping. *Cattle* are less prone to infection, less severely affected. *Pigs* suffer similarly to lambs. *Humans* are also susceptible, mainly through deep or pricked, dirty wounds. Vaccination of farm workers is a wise precaution.

Tick pyaemia

In tick districts, tick bites of young lambs result in limb joint and internal abscesses, symptoms resembling joint-ill. Staphylococcal bacteria are often involved.

Treatment

Antibiotics may help, but infection is usually too widespread before a case is seen.

Prevention

Regular sheep-dipping to control ticks. Newborn hill lambs can be dusted with insecticide preparations; preferably dip the hill ewe flock during the last week of pregnancy (their lambing often coincides with the spring flush of ticks).

DISEASES OF OLDER SHEEP

Reproductive disorders

Infertility

In sheep, a sub-fertile or sterile ram is a more likely cause of failures to conceive than faults in the ewe, although her nutritional status and body condition can affect the lambing percentage. Very poor feeding can promote resorption of fetuses, and a 'barrener'. Adequate diets help ewes with the genetic potential to bear twins. Inadequate rams can often be identified by thorough examination of genitalia; semen tests may offer useful supporting evidence but single tests are not always reliable.

Abortion

Diagnosis of the several types of sheep abortion is necessary to decide control measures. Laboratories require aborted lambs and afterbirths, also blood samples.

Bacterial

Two main bacterial causes are *Campylobacter* (previously *Vibrio*) (different strain from that affecting cattle), and *Salmonella abortus ovis*. The former is widespread in England and Wales; the latter is more localised, now uncommon, principally in south west England. Both types are introduced by carrier animals, themselves immune, which contaminate pasture and water supplies. Mating may further spread infection.

Symptoms

Campylobacter causes abortions from the second month of pregnancy. Abortions due to *S. abortus ovis* begin about six weeks before lambing is due; 10–40% ewes may abort, even 60–70% at times. A proportion of ewes die.

Prevention

No treatment is known. Once affected, ewes remain immune, but some will spread infection to susceptible animals which may be bought-in, or start infection in a new flock if sold. After abortion, a flock should be kept 'closed', and only a few further abortions are likely in subsequent years.

Other types of Salmonella, can cause abortion, sometimes severe with mortality among the ewes.

Enzootic abortion (*Chlamydia*)

This infection of the placenta causes abortion about 10–14 d before lambing is due, or dead or weakly lambs may be born at full-term. The virus survives only a few days outside the animal. Once mainly occurring in lowland flocks in south west Scotland and north east England, but now prevalent in England and Wales: 10–15% of ewes may die.

Prevention

Vaccines are available, but must be used over several seasons before full benefit is apparent. Since infection is spread during lambing or abortion, crowding of ewes should be avoided: newly infected sheep will abort during the next lambing season. Aborted lambs and all (even healthy) afterbirths should be destroyed (by burning, burying, after temporary collection in tanks of disinfectant). Affected ewes should be isolated, especially from ewe hoggs, for four to six weeks after abortion; they should only foster male lambs.

Toxoplasmosis

A further cause of abortion is due to organisms larger than bacteria. Abortions are not always numerous but in some flocks it may result in severe abortion 'storms': live births may be weakly and soon die. The disease within a flock is self-limiting, and aborting ewes should be retained.

Border disease (virus)

See p. 464; this is a less common cause of sheep abortions.

Bacterial diseases

Clostridium infections

This large group of bacteria are commonly present in the intestines of sheep (and cattle), also on the ground where they develop resistant spores. Some can affect man (gas gangrene, tetanus). Wounds can be invaded (blackleg), others multiply rapidly in the intestines producing fatal toxins (enterotoxaemia, as in pulpy kidney disease). Differentiation of the bacteria requires laboratory tests on fresh specimens.

The diseases are much less common since *preventive vaccines* are available for each type; often, multiple vaccines cover several diseases, being used particularly to inoculate ewes *twice*, with a booster dose annually. *Antiserum* against some types is used during disease outbreaks, or when suspected, to give immediate immunity but of short durations.

Black disease (Clostridium oedematiens)

These bacteria multiply in the sheep liver, producing necrosis and toxins. It is activated by liverfluke larval invasions, so is an extra hazard where fluke disease occurs, but not on all farms. Death is usually sudden.

Prevention

Vaccinate all sheep on known flukey and black disease farms, especially when forecasts predict a bad fluke season.

Blackleg ('Gas gangrene') (Clostridium chauvoei)

These bacteria invade the body of ewes through injuries, often at lambing, shearing or dipping; lambs are infected at castration or docking.

Symptoms

These show within 2–4 d, mainly stiffness of the legs,

swelling of the hind-quarters, pain and distress. Death follows in a few hours.

Prevention

Vaccinate ewes three to four weeks before lambing. Antiserum can be given at the time of lambing or dipping, shearing, or tailing and castration. Cleanly operations greatly reduce risk. Combined vaccination is available against braxy and blackleg.

Braxy (Clostridium septicum)

The stomach lining is invaded, causing severe inflammation, usually following indigestion due to frosted foods. It causes sudden deaths of yearling and older sheep on hill grazings, mainly in north England, Scotland and Ireland.

Diagnosis

Diagnosis is only certain by post-mortem examination within 1 h of death.

Treatment

Vaccinate in early September.

Lamb dysentery

See diseases of lambs.

Pulpy kidney

See diseases of lambs.

Struck (Clostridium welchii type C)

This is not to be confused with 'fly-strike'. Sheep over one year old are found dead, rarely showing prior symptoms, e.g. dullness. The condition occurs in late winter and spring, mainly on Romney Marsh, Kent, sometimes north Wales and elsewhere; losses are often confined to certain fields.

Prevention

Two injections of vaccine are given, 14 d apart.

Foot rot (Spherophorus nodosus)

The causal bacteria live in damp soil, but die out if pasture is rested from sheep for at least two weeks. Secondary infections often aggravate the lesions in the feet and cause septicaemia.

Symptoms

Lameness in one or more feet, sometimes severe as evil-smelling discharges under-run the wall and sole of the foot. Pain and difficulty in grazing cause loss of body condition; septicaemia can cause death.

Treatment

Several types of foot dressing may be used, but none is likely to be effective unless the affected feet are properly pared absolutely clear of the diseased horn. After paring, foot-baths of 5% formalin or 8% copper sulphate solutions can be used, or these can be well brushed in by hand dressing. Antibiotic sprays are also available, but require the same thorough preliminary paring of the feet. Affected sheep should be grazed separately from healthy, inspected every 2–4 d, and diseased feet again pared and re-treated. Healthy sheep should graze pastures rested from sheep for at least 14 d.

Note: foot and mouth disease can also cause lameness, usually spreading rapidly: a Notifiable Disease.

Prevention

Foot-rot vaccine is available; after two primary injections, booster doses are needed every six months. Timing in relation to lambing is important—follow the prescribed instructions.

Scald

Also causes widespread lameness, especially in lambs. Wet ground conditions and abrasions (e.g. stubble grazing) soften the coronary skin and allow bacterial infection. Dry grazing and antibiotic sprays promote recovery.

Strawberry foot rot

Severe encrustations affect the legs, covering ulcers, due to bacteria resembling mycotic dermatitis (*see below*).

Mastitis (garget)

Caused by a variety of bacteria, including staphylococci. More common in folded flocks on arable land and during wet, cold weather. Disease occurs from a few days to a few weeks after lambing.

Symptoms

Inflammation rapidly develops, followed by blackening and gangrene of the affected half of the udder. Death may take place early, or recovery follow sloughing of the affected tissue. This may be helped by amputation of the teat or udder.

Treatment

Antibiotics must be injected promptly. Injections of toxoid in late pregnancy may increase resistance to infection.

Mycotic dermatitis ('Fleece rot', 'Rain scab')

Fungal-like bacteria present in some flocks multiply on the skin usually during warm, wet weather conditions. Breeds with close-woolled fleeces are more susceptible than more open, long-woolled types.

Symptoms

Patches of 'broken' fleece first occur. The skin show inflammation and scabbiness, particularly over loins and rump. Discoloured bands of scabs rise with wool growth. Severe infections cause fleece loss. The ears can also be affected, especially lambs.

Treatment

Dips or sprays containing suitable antibacterial agents are used after shearing, when there is enough wool-growth to hold the dip.

Pneumonia (Pasteurella bacteria)

Lambs are prone to this infection, in both late spring and in early autumn; ewes also are susceptible often acting as carriers of latent infection. Stress factors such as hot weather, overcrowding, overexertion, e.g. gathering for dipping, and travel often initiate the illness.

Symptoms

Nasal discharges, heavy breathing, coughing, fever. Virulent infections cause septicaemia or death. Intestinal infection shows as diarrhoea.

Treatment

Isolate and move sick animals. Medication, e.g. with antibiotics, may help but is not always successful. Avoid further stresses. There are many different serotypes of *Pasteurella*, which account for lack of full effectiveness of vaccines, but vaccination is advised prior to the expected risk period, Concurrent virus infections of the lungs may be added complications, as may *Corynebacterium* bacteria which cause purulent abscessation.

Metabolic disorders

Milk fever (lambing sickness)

This resembles milk-fever in cattle, i.e. the level of blood calcium falls below normal, but the condition is more rapidly fatal if not treated early.

Symptoms

It may occur after lambing, or before when pregnant ewes are exposed to exertion and fatigue, a dazed appearance, i.e. excitement, is rapidly followed by stiffness of the limbs, coma and death. Symptoms may easily be mistaken for louping ill, and may resemble pregnancy toxaemia.

Treatment

Injections of calcium salts are effective; addition of glucose is also useful to counteract any tendency to pregnancy toxaemia. Careful inflation of the udder also remains an effective cure.

Pregnancy toxaemia (twin-lamb disease, ketosis)

In-lamb ewes are affected during late pregnancy when their diet during the previous two to three months has been inadequate for the demands of the growing fetuses (especially when twins or triplets are present). Ewes are often in fat condition, but either are short of readily available carbohydrates or have suffered sudden deprivation of food, e.g. as in sudden inclement weather. Lack of exercise may be a contributory factor.

Symptoms

Dullness, slow or staggering gait, either unusually low or high head position, possibly blindness. Temperature is normal, and the condition remains unchanged for some days, although the sheep may be unable to rise. Unless ewes lamb within a few days, or are treated, mortality is heavy. Unlike 'lambing-sickness', pregnancy toxaemia does not occur after lambing.

Treatment

Injections of glucose solutions and dosage with glycerine but these are not always successful. Caesarean operation is helpful if not delayed. The flock should be given concentrate foods and good quality hay. During the last two months of pregnancy a rising level of nutrition is essential.

Staggers (hypomagnesaemia)

This is associated with a sub-normal level of magnesium salts in the blood. The diet may actually be deficient in magnesium, e.g. on poor grassland; or high potash and nitrogen levels in good grass can interfere with the ewe's utilisation of magnesium. Clovers contain more magnesium than grasses.

Symptoms

Ewes are affected after lambing, sometimes a week or more. Deaths may occur suddenly, or ewes may exhibit general incoordination, muscular tremors or spasms, nervousness, prostration.

Treatment

Injections subcutaneously or intravenously of soluble magnesium salt solutions, usually with calcium and glucose also.

Prevention

Prevention is by provision of magnesium salts in the diet, e.g. calcined magnesite at 7–10 g/head daily.

Nutritional disorders

Sheep are susceptible to two specific nutritional deficiencies; *cobalt* and *copper*. They are also subject to *selenium* and *vitamin E* inadequacies (causing *muscular dystrophy*). *Cerebro-cortical necrosis* can also occcur, the initiating cause not yet identified.

Cobalt pine ('vinquish')

Caused by cobalt deficiency in the diet. Granite-derived soils, and some sea–sand grazings, are low in cobalt. Heather can take up more cobalt than companion grasses, so supplies some moor-grazing sheep.

Symptoms

Evident by poor lamb growth after weaning, severe cases showing dullness, anaemia and a poor fleece, and emaciation. Blood and liver tests confirm diagnosis, supported by analysis of herbage or soil.

Treatment

Vitamin B_{12} injections give temporary remedy. Cobalt is not stored within the body, so it must be continually available. Solutions of cobalt sulphate or chloride can be given weekly, by mouth, or sprayed in strips across selected grazing areas. Cobalt 'bullets' given orally are long-acting, but may be coughed up or inactivated in the stomach by surface deposits. Mineral licks and mixtures with cobalt supplement can be offered *ad libitum* in boxes sheltered from rain.

Copper (swayback)

Ewes which do not assimilate enough copper may produce 'swayback' lambs (*see* section on diseases of lambs).

Selenium (muscular dystrophy)

Selenium is only required in trace amounts, but may still be deficient on some farms. When the diet is entirely home-grown, the selenium deficiency can result in muscular weakness, so-called muscular dystrophy or white muscle disease. Vitamin E is also involved in this condition; its potency disappears on storage of grain.

Symptoms

Growing lambs are affected, most commonly between three and six weeks of age, but even at birth or up to one year old. They walk stiffly, back arched, but may only seem abnormal when driven. Swayback causes similar symptoms, so careful diagnosis is essential.

Treatment

Injection of selenium and vitamin E preparations, with care as to dose rate since excess can be toxic.

Prevention

Suitable mineral and vitamin supplements to the diet.

Parasitic diseases—external

Blow-fly strike

Blow-flies are active from May until the end of September; they prefer shelter of woodland, hedges, bracken, etc. Attracted by soiled fleece, e.g. around breech after diarrhoea, they lay eggs on these areas: maggots hatch in 12 h, feed on the skin for 2–3 d, then drop off to pupate on the ground. Further flies are attracted by the smell of the first maggots.

Symptoms

Unless sheep are inspected daily maggots can be well-grown before the typical irritation is shown by restlessness and tail wriggling. Heavy 'strikes' can be fatal—large areas of skin are damaged.

Prevention

Modern persistent insecticide dips maintain a fly deterrent effect for several weeks. They can also be used to hand-dress affected areas, but insecticide creams act also as germicides. 'Crutching' of wool around tail and breech, and clipping off soiled wool, reduces fly-attraction. Treatment for stomach worms reduces diarrhoea (also caused by rich feeding). Sheep penned closely for dosing or foot-paring should be released as soon as possible.

Lice and keds

These parasites cause irritation, ragged fleece, sometimes loss of weight. Persistent insecticide dips and sprays are again effective.

Ticks

The common sheep tick is *Ixodes ricinus* (described in the section on cattle). Their main importance in sheep is as carriers of several diseases, e.g. tick pyaemia, louping-ill (*see* notes on these diseases). Control is by insecticide dips, washes or sprays, used according to instructions, especially noting the strength of the 'topping up' solution. Excessive dirt will reduce dip efficacy.

Parasitic diseases—internal

Coccidiosis

Certain species of this protozoon parasite affect only sheep. They invade the intestinal lining of grazing lambs then are shed with dung and develop further on moist ground to become infective by ingestion to further grazing lambs.

Symptoms

The practical importance of coccidiosis in lambs is not yet determined; experimental infections do not always cause illness, but symptoms of diarrhoea and unthriftiness can occur. On the farm, coccidiosis usually accompanies infestations by stomach and intestinal worms.

Treatment

Symptoms of diarrhoea should first prompt treatment to control stomach worms. If this is not fully effective, your veterinary surgeon would test dung samples and treat for coccidiosis, if necessary, usually by sulphonamide drugs.

Gid

Tapeworm cysts pressing on the brain cause symptoms of nervous disorder in older sheep, e.g. 'circling' or other abnormal carriage of the head. Lambs may show symptoms of meningitis resembling other diseases.

Prevention

These cysts are an intermediate stage in the life cycle of a tapeworm inhabiting the intestine of dogs, whose faeces contaminate the pasture. Sheep carcasses re-infect the scavenging dog. Sheepdogs should be treated for tapeworms at least four times each year. There is no treatment for affected sheep.

Liverfluke disease ('liver rot', 'bane', 'cord')

The life cycle is described in the section on cattle, also control of the intermediate snail host. In bad fluke seasons sheep can suffer from *acute fluke disease* due to massive infestations by immature flukes; sudden deaths can occur as early as July and August. *Chronic disease* is due to mature flukes causing thickening of the bile-ducts visible as 'cording' of the liver. Affected sheep lose condition, weaken, are anaemic and may show 'bottle jaw'.

Treatment

Annual 'fluke forecasts' for England and Wales guide timing of treatments, i.e. whether beginning in early or late autumn. A variety of medicaments includes some given by injection, others combining stomach worm treatments; some are more efficient than others in killing the young, immature flukes. A usual programme could be two to three treatments at monthly intervals, beginning in August or later (*see* annual forecast), then a dose in late spring to kill flukes carried over winter.

Note: *see also* black disease, an additional hazard on some farms.

Lungworms

Species different from the cattle lungworm affect sheep and goats, but similarly inhabit the bronchial air-passages in the lungs, causing chronic coughing, nasal discharges, sometimes pneumonia (this may be aggravated by bacterial infection).

Treatment

Specific medicaments, some injectable. Some compounds combine treatment for both lungworms and stomach and intestinal worms.

Note: as yet, there is no vaccine for *sheep* lungworms.

Stomach and intestinal worms

Infestation by these causes parasitic gastro-enteritis (PGE), e.g. inflammation of the stomach and intestines. The growing lamb and hogg is particularly susceptible; some heavy infestations can kill adult sheep quite suddenly, usually when overstocked. Ewes are the source of pasture infection when lambs begin to graze over their faecal contamination; the worm eggs hatch in wet, warm weather to produce infective larvae on the grass.

Nematodirus types differ in that their life cycle is from one year's lamb crop to next year's lambs, the eggs and infective larvae over-wintering on the ground; greatest risk is from April to June.

Symptoms

Unthriftiness, diarrhoea, sometimes severe anaemia. Inapparent low level infestations may only produce a lowered growth rate.

Treatment

Dosage of lambs may begin at one to two months of age, usually two doses four to six weeks apart in spring, repeated in early autumn, but more frequently if a farm problem is known to exist, viz due to heavy stocking. There are many medicaments available, with varying ranges of effectiveness on the different types and stages of the parasitic worms. Good feeding and adequate water are necessary to combat the dehydration and debility.

Control

Avoid overstocking. Graze treated ewes with their lambs on 'clean' pasture, i.e. rested from sheep overwinter. Forward—or sideways—creep grazing allows lambs a clean bite. Folded sheep should not have more than 2 d run-back, as this area can build up heavy infestations.

Nematodirus—do not graze lambs where last year's crop experienced this infection; at least avoid suspect fields during the initial months April–June. Adult sheep can safely graze here.

Tapeworms

Many lambs carry intestinal tapeworms, usually stomach worms also. Unless very numerous, when bowel irritation causes diarrhoea, specific treatment is seldom necessary (medicaments for stomach worms do not kill tapeworms).

Virus diseases

Louping-ill ('trembles')

A nervous disease due to a virus carried by ticks, occurring more severely during the tick seasonal activity, April–June and September–October. Cattle, horses, pigs and humans may also contact louping-ill.

Symptoms

Acute fever develops 6–18 d after infection, with dullness and lack of appetite. Five to ten days later nervous symptoms develop, e.g. excitability, tremors of head and neck muscles, unsteadiness (if made to move a sheep may fall), eventually paralysis. When driven by a dog, some sheep jump jerkily or show high-stepping walk with head held high. Deaths are more common in younger sheep (perhaps 4% in yearlings). Sheep introduced from tick-free areas may suffer 60% mortality. *Symptoms* may seem to resemble lambing sickness, but this is only seen in ewes in late pregnancy or lactation, and responds readily to treatment, whereas there is no specific treatment for louping-ill. Shelter, quiet, and good nursing enable some to recover.

Prevention

A vaccine may be used on all but very young lambs. Tick control is essential.

Orf

A well-known disease causing pustules, ulceration, and scab formation on the skin, usually of the lips and mouth, teats and udder, coronet of the feet, also the genitalia. The causal virus is long-lived in the dried scabs. Lambs up to one year old are more commonly affected: sucklers may die from inability to suck, but older sheep usually recover and remain immune for about six months.

Treatment

Various antibiotics and antiseptic ointments help to prevent secondary infections by bacteria.

Prevention

Ewes can be vaccinated four to six weeks before lambing, as can lambs over four weeks old, even at birth in high-risk flocks.

Note: humans can contract this disease and workers should take hygienic precautions when handling affected sheep or using the live vaccine.

Scrapie

A virus infection with a long incubation period, up to three years. Most cases occur in sheep one and a half to two years old or over, more commonly in Border Leicester and half-breds, Suffolk × half-bred, and Cheviots, but all breeds are susceptible.

Symptoms

At first, only slight change in appearance of fleece. Fine muscle tremors of head and neck, then 'nibble-reflex' when back is scratched. Intense skin irritation develops causing rubbing and marked raggedness of the fleece. Nervous symptoms increase, with head held high, high-stepping action: convulsions are induced by excitement or fright. Death is inevitable, often after six months' illness.

Prevention

There is no treatment. This disease can be reduced in a flock by a selective breeding policy. First, keep no ewe lambs for two years. By then, affected breeding ewes should have shown symptoms and should be slaughtered and their offspring. Breed from the healthy ewes mated only to rams which have not contacted infected ewes. Thus a healthy flock is built up. Flock records and ear tagging are, of course, essential.

Note: *sheep scab* (Scheduled disease) also causes skin irritation, but more widespread. Diagnosis should be made by your veterinary surgeon.

Tick-borne fever

Not strictly a virus disease, but due to rickettsial organisms (smaller than bacteria) which infect red blood cells. Carried by bites of infected ticks. Most sheep survive, are immune but still infective to others via ticks. Lambs infected soon after birth are immune.

Symptoms

Fever for only 2–3 d, or becoming normal slowly over two to three weeks. Temporary listlessness and inappetence, but susceptible pregnant ewes (e.g. brought into the flock) may abort.

Treatment

Sulphonamides and antibiotics, but treatment may interfere with immunity. As yet, no fully effective vaccines. Tick control is essential.

Diseases of pigs

Generally, the profitability of pig units is influenced more by managemental and nutritional factors than by outbreaks of fatal diseases, although when these do occur in susceptible groups of pigs losses can be high. Certain diseases of low virulence can lower output, recurring in successive litters or groups, e.g. enzootic pneumonia can retard growth, as does atrophic rhinitis.

Sows possessing good potential for producing large litters may suffer from inapparent virus infections during pregnancy, causing resorption of embryo piglets, or mummification and abortion, or just small litters. Such infections, and others, can be introduced by new stock animals such as boars, also by human visitors' shoes, etc.; maintain strict hygiene, and locate a delivery area (for stock and food vehicles) out of contact with the actual pig units.

Regular records of performance (e.g. sow fecundity, pig weights at different ages, food consumption) may indicate where to investigate sub-optimal productivity.

Since different diseases may show similar symptoms, early consultation with your veterinary surgeon is essential—and he will be aided by well-presented records.

SCHEDULED DISEASES

Scheduled diseases (except Aujeszky's disease) have been described previously, e.g. anthrax, foot and mouth disease, swine fever and African swine fever, swine vesicular disease, Teschen disease, tuberculosis. Aujeszky's disease of swine is described under virus diseases.

DISORDERS OF PIGLETS

Developmental defects

Some piglets may be born lacking a functional anus, *atresia ani*. Correction by surgery may be possible, perhaps uneconomical, and such survivors should not be kept for breeding.

Scrotal hernia

This is not uncommon and should be carefully looked for before castration is carried out; operation at that time can rectify this inherited fault.

Hermaphrodite pigs

These pigs display varying sexual features of each sex, both internally and externally. Testes, one or both, may be present as well as an apparent vulva with functional urethra. Some pigs will have one testis and one ovary. Needless to say, such pigs are fit only for fattening.

Trembles

Also well described as 'shakes', affected piglets, from birth, shiver and may be unable to suck. This can be an inherited trait, or caused by various virus infections. Piglets may lose these symptoms, as they grow, provided they are helped to the sow for suckling.

Splay leg

Again, piglets are born affected, unable to stand properly, usually sitting splay-legged. This can be inherited, but similar symptoms can occur if pregnant sows eat grain contaminated with *Fusarium* mould producing a mycotoxin (F2, zearalenone).

Bacterial diseases

Atrophic rhinitis (Bordetella bronchiseptica bacteria)

Rhinitis is inflammation of the nasal passages, visible as snuffling and sneezing in the young suckler. This may subside after a few weeks, without complications, but leaving 'carrier' animals. Atrophy refers to the malformations of the turbinate bones in the nose, which can follow the initial rhinitis. *Bordetella* infection alone may not be the sole cause; inclusion body rhinitis virus and mycoplasma may be involved. Various other factors have been blamed, e.g. genetic (certain strains of Landrace appeared to be more susceptible); nutritional; management and housing—always important in any respiratory disease. The infection can be passed by the sow or by other piglets.

Symptoms

After sneezing, sometimes with bloody nasal discharge, the growing piglets show progressive deformity of the snout, due to uneven growth of nasal bones, particularly the turbinates. The upper jaw shortens, wrinkling the skin, and may curve sideways. Tear-stains run from the eyes. Body growth may be checked, but mild disease may not be apparent unless the carcass snout is sectioned.

Treatment

Without good husbandry, reliance only on medication will fail. Piglets can be given preventive doses whilst suckled by the sow, who herself could have had treatments just before farrowing. Choice of treatment should be discussed with your veterinary surgeon.

Control

Keep a closed herd, or buy from selected sources. Rear litters up to nine weeks old in isolation, and do not mix different age groups in the same building. Maintain warm, ventilated houses, without overcrowding—especially of sows and litters. Do not breed from affected pigs; cull sows which have infected their piglets.

Colibacillosis (E. coli)

Neonatal diarrhoea occurs when newborn piglets acquire certain pathogenic *E. coli* infections, which produce toxins.

Symptoms

Illness can commence within 12 h after birth, then death within 2 d without any diarrhoea, which may develop in survivors. Piglets become emaciated and dehydrated, tails and anal region being encrusted with faeces.

Treatment

Various antibiotics and other medicaments may be given orally, or offered in food or water. The choice will depend on laboratory tests of the sensitivity of the particular *E. coli*, as some strains show resistance to certain antibiotics.

Prevention

Thorough cleanliness of the farrowing sow and pen will reduce the numbers of bacteria. The sow herself can have pre-farrowing vaccinations, to induce good levels of antibody in her colostrum.

Meningitis

Usually caused by streptococcal infection, specific types affecting pigs but also occasionally humans, e.g. *Strep. suis* type 2.

Symptoms

Young pigs (6 d to eight weeks old) are susceptible. They become uncoordinated, unbalanced, walk unsteadily perhaps in circles. They may lie on their sides, paddling legs, with occasional convulsions. Body temperature is raised, and joints may become swollen.

Treatment

Treatment with selected antibiotics is effective in less severe cases. Effective vaccines have not yet become available, but if this disease recurs then preventive medication is recommended, for the particular age groups usually affected.

Prevention

Try to buy pigs from herds without a history of this disease. Isolate sick piglets, and thoroughly cleanse and disinfect their pens before restocking: ensure good ventilation. Stress appears to provoke outbreaks of illness among pigs which may be carrying latent infection.

Greasy pig disease (Staphylococci)

Also known as exudative epidermitis, this bacterial disease can rapidly affect whole litters, 5–35 d old sometimes with high mortality. Skin abrasions, e.g. from rough concrete floors, exacerbate losses. Infection is readily carried from one area to another.

Symptoms

These range from peracute, with death within 3–5 d, to a more slowly developing condition. A moist, greasy exudate affects ears, eye area, and skin over the whole body; slower development causes the skin to wrinkle, the piglet is off food, dehydrates and becomes emaciated. Older skin lesions become crusty and have a foul odour. Occasional mild cases may only exhibit reddish-brown spots on the ears, and dandruff flakes on the skin.

Other conditions (mange, lice, nutritional deficiencies) cause some similar symptoms, so careful differential diagnosis is wanted.

Treatment

Antibiotic treatment is effective in less severe cases, but the particular organisms should be cultured and checked for its drug sensitivity. Skin debris should be washed off, and antibacterial ointments applied. If the disease has spread internally to the kidneys, the prospect of recovery is remote. To prevent spread of disease, affected litters should be isolated, in good, clean, dry and hygienic pens.

Virus diseases

Transmissible gastro-enteritis (TGE virus)

Yet another disease causing gastro-intestinal symptoms, requiring differentiation from coli-bacillosis, salmonellosis, epidemic diarrhoea, for example. As per title, it is very readily transmitted, by direct and indirect contact, via birds, dogs, vermin, people, also lorries.

Symptoms

Sudden onset of diarrhoea in young piglets, killing most of those under two weeks old, and affecting the whole herd (all ages of pig) within 2–3 d. Some pigs vomit, even sows. The diarrhoea is watery, foul smelling, perhaps of greenish colour.

Fattening pigs seldom die but scour, whilst sows and boars may only be ill transiently. Survivors may remain unthrifty.

Laboratory tests on blood samples or pig carcasses may be necessary to confirm diagnosis.

Treatment

Since this virus is not susceptible to antibiotics or other medicines, there is no specific treatment (but medicines may help if other infections are complicating TGE). The dehydration must be corrected by readily accessible fluids, e.g. electrolyte solutions.

Control

By good pig management and hygiene, do everything possible to prevent spreading infection to highly susceptible newborn piglets. Gilts and sows, however, *should* contact infective material at least three weeks before farrowing. Their colostrum then will protect their piglets.

Note: this procedure is *contradicted* for other pig diseases, so should first be discussed with a veterinary surgeon. Isolate sows due to farrow within the next fortnight, to *prevent* their infection.

Prevention of spread to other herds, or its introduction into your own, requires basic precautions which are recommended as routine on all pig farms. Stop selling pigs, except direct for slaughter, and avoid visiting markets and other pig farms. Bury dead piglets quickly. Stop visitors and vehicles at the farm gate and place boot disinfectant baths, replenished at least once daily, using either lysol 2.5%, formalin 2%, hypochlorites with not less than 0.5% available chlorine. Nobody should casually look around your pigs.

Lorry loading points are always better arranged well away from the pigs themselves: clean and disinfect the area after use.

Keep down vermin, and make buildings bird-proof. When TGE is suspected, please inform your veterinary surgeon and local veterinary investigation centre;

a general press and radio warning is then issued locally, to warn pig keepers to redouble their disease precautions.

Nutritional disorders

Suckling, fattening, and adult pigs can suffer from various deficiencies in their diet, either of major or minor minerals, vitamins, or amino acids. They may display specific symptoms, or merely poor growth and food conversion rates. Nutritional requirements are dealt with elsewhere in this book. A few nutritional deficiency diseases are outlined below.

Anaemia

This is the prime nutritional defect of piglets reared indoors, but so readily preventable that it is much less common. Iron and copper salts are essential for the adequate development and balance of the blood system. The sow's milk cannot meet the growing piglets' demand for these; natural access to soil, in outdoor farrowing systems, overcomes this, but at the same time may contribute parasitic (large round-worms) or bacterial diseases.

Symptoms

At two to three weeks of age, piglets become listless, grow poorly, have rough hair growth, drooping ears and tail; mucous membranes of eye and mouth, also the skin itself, appears pale. Breathing may become laboured, jerky. Death may follow. Diarrhoea is common, often due to bacterial infections of the debilitated pigs.

Treatment

A variety of treatments include pastes squeezed into the mouth, and iron compounds injected into the muscles, during the first 7–10 d of life. Suitable 'paints' can be applied to the sow's teats.

Iron toxicity can occur if excess is given. Some piglets are susceptible to correct doses, if their dam's diet is low in vitamin E and selenium: General drowsiness and difficult breathing may occur, progressing to coma and death, as quickly as 3–12 h after the iron administration.

Rickets

Rickets due to inadequacy in quantity and relative ratios of calcium and phosphorus are only likely to be seen in the fattening pig, when its ration lacks these essential elements, combined with deficiency of vitamin D. The faster growing pig is more susceptible, but 'rickets' (more properly osteomalacia) may affect heavily lactating sows.

Symptoms

Reduced growth rate, excitability, and multiple fractures of bones. A sow usually exhibits posterior paralysis, is lame and stiff, in the second half of her lactation.

Treatment

Immediate correction of the diet by inclusion of correct amounts of salts of calcium and phosphorus, with vitamin D.

Sodium salt poisoning

The weaned pig's normal ratio should provide correct levels of salt, as sodium chloride. If excess is consumed, often it derives from other foods such as bakers' waste, swill, occasionally from access to salt licks, buttermilk. Provided that ample water is always available and consumed, salt levels higher than normal can be tolerated; 'water deprivation' is an alternative description of this condition.

Symptoms

Thirst and constipation. Later, skin irritation, and some pigs become blind and deaf, wander aimlessly, pushing against walls. Epileptic type fits occur, with collapse and coma. Mortality varies with the degree of poisoning.

Treatment

Remove the offending source of the sodium salt. Offer *small* amounts of water—*not* excess, which can be harmful. Recovery is often spontaneous.

DISORDERS OF OLDER PIGS

This section deals with pigs from weaning age to adults, although some diseases can affect all ages of pig. Weaning time itself is critical, often involving changes in both feeding and housing, which can be stressful and precipitate illnesses.

Bacterial diseases

Colibacillosis (E. coli bacteria)

Some serotypes of *E. coli* are specific pathogens for pigs. These serotypes multiply in the intestinal tract and produce toxins, causing death. Less severe cases exhibit diarrhoea, as a result of enteritis, and *oedema disease* is another manifestation. The bacteria are spread in faeces, and become prevalent if there are dirty food and water troughs, and housing systems which cannot be kept hygienic and warm. Newly introduced pigs can bring in new serotypes of *E. coli*.

Symptoms

Weaners and fatteners are susceptible, rarely adult pigs (*see also* 'Disorders of piglets'). Acute infection may show as sudden deaths; less severe, diarrhoea, with dehydration, inappetence and lethargy. If prolonged pigs lose weight and look scruffy.

The *oedema disease* type of colibacillosis commonly coincides with change of ration, e.g. at weaning. The eyelids swell, also face. Pigs are unsteady, may tremble or develop spasms and paralysis. Although a quick succession of cases may occur in one group, disease does not seem to spread. Some may die suddenly, others recover aided by treatment; purgatives may assist.

Treatment

Orally administered antibiotics and other drugs control further multiplication of the intestinal *E. coli*. Water is essential, but reduce the quantity of food, then increase slowly over 4–7 d.

Coli vaccines given two weeks before the expected risk period are helpful in some herds; oral vaccines, too, have some success. The colostral-derived immunity, if any, wanes rapidly—hence the poor resistance of weaners. Drug-resistance occurs in *E. coli*; laboratory tests may aid correct selection of treatment.

Erysipelas (*Erysipelothrix rhusiopathiae*)

This bacterial disease usually affects store pigs and fatteners, causing septicaemia. (Many other species of animals and birds may carry these organisms, sometimes becoming ill, e.g. sheep, with post-dipping lameness; turkeys, septicaemic; humans, with 'erysipeloid'—skin lesions).

Symptoms

Acute disease—sudden death. In-contact ill pigs show high fever (40–41 °C), poor appetite, staggers, diarrhoea or constipation, inflamed areas of 'diamond skin'.

Chronic disease causes lameness due to joint infection, heart valve disease (and sudden death of apparently healthy pigs). In epidemic, death rate may be 80%.

Treatment

Usually favourable response to penicillin and antiserum, except in chronic illness. Annual vaccination is necessary on farms where disease recurs.

Pig paratyphoid (*Salmonella bacteria*)

An enteric and septicaemia infection caused specifically in pigs by *Salmonella cholerae suis*. Other serotypes, e.g. *S. typhimurium*, also affect pigs as well as other animals (including rats and mice) and birds.

Some serotypes originate in animal protein in foodstuffs, but a Protein Processing Order 1981, now requires certified heat treatment of these, so eliminating risk from those sources.

Symptoms

In acute disease, sudden death. Less acute cases are feverish (41 °C), off food, show diarrhoea, perhaps purple discoloration of the ears. Mortality can be high.

Note: this disease may resemble swine fever, now eradicated from the UK, causing similar ulcers in the bowel.

Treatment

A variety of different medications may limit losses, but radical action is necessary to curtail an outbreak. Slaughter all chronically ill animals; send all healthy fat pigs to the abattoir. Cleanse and disinfect pens before restocking. These bacteria can remain infective for six to nine months in soil in shaded areas.

Swine dysentery (*Treponema hyodysenteriae*)

This infection of the intestines affects fatteners principally, also adults and sometimes sucklers. It is easily spread in faeces, either from ill pigs or inapparent carriers, directly or indirectly via dirty boots, vehicles. Apart from deaths, insidious infection in a herd causes long-term losses due to poor growth rate.

Symptoms

Diarrhoea characteristically contains mucus and blood, or is almost black ('black scours'), after beginning as yellowish or grey coloured. Fever may occur, not always. Arched back indicates abdominal pain, and dehydration, thirst, and weakness lead to emaciation. Disease may spread only slowly, rarely causing acute mortality. Other diseases may resemble this one, and differential diagnosis is important.

Treatment

Water medication is used long-term, several drugs being available, but illness may recur if treatment is withdrawn. No vaccine is yet available. Hygiene of pens, and manure disposal, has to be of very high standard to reduce spread of infection.

Pasteurellosis (*Pasteurella*)

Respiratory disease can be caused, but usually *Pasteurella* complicates enzootic pneumonia (*see* later, virus diseases), causing bronchopneumonia and sometimes death. Hence appropriate treatments are given to control this.

Virus diseases

Aujeszky's disease of swine (scheduled)

This virus disease was made notifiable by an Order in 1979. Owners, or pig keepers, or veterinary surgeons, who suspect that pigs are affected, must report this to the local police or divisional veterinary officer of Ministry of Agriculture, Fisheries and Food.

Unlike some other Scheduled Diseases, only a few temporary restrictions on movement are enforced, e.g. until a Ministry Veterinary Officer has examined the pigs and advised the owner: there is no compulsory slaughter policy, but a control policy is under continual review. Generally, this disease only appears sporadically in the UK in localised areas.

Because some symptoms resemble those of rabies, Aujeszky's disease is also misleadingly called pseudorabies, but rabies is in no way related to Aujeszky's disease. The pig is the natural host animal but the disease can be transmitted to dogs, cats, sheep, cattle and rodents, in which species it is always fatal. It does not affect man.

Symptoms

All age groups can be affected. Acute disease can kill many young piglets, up to two weeks old. Month old pigs usually will recover, after nervous and respiratory symptoms, whilst infection in sows may only show as abortions, stillbirths, or mummified fetuses. In small herds, once the acute outbreak is over, after about one month, further illness is seldom seen, but latent infection persists in carrier sows. Piglets may vomit, show diarrhoea, then lethargy and nervous symptoms such as trembling and incoordination, then death within 2–3 d. Fatteners at first may cough, possibly vomit; constipation, nervous symptoms and convulsion follow, perhaps death after a week. Sows affected early in pregnancy will abort. Later infections may kill only a proportion of fetuses, and sows may farrow these and healthy ones at full term.

Treatments

None specific, vaccination not permitted. Control measures will be discussed with the Veterinary Officer, and will depend on the size and type of the pig enterprise, e.g. purely fattening for direct slaughter, as distinct from a breeding herd from which carrier sows are best disposed of, to produce a 'clean' herd. Fortunately, the virus itself survives outside its host for only six weeks, and is susceptible to most disinfectants.

It is essential to isolate affected groups from susceptible pigs (especially pregnant sows—but barreners and maiden gilts will acquire useful immunity by contact).

Keep other animal species away from infected animals and slurry, and apply the usual hygienic security to stockmen and vehicles.

Enzootic pneumonia (Mycoplasma)

Strictly, mycoplasma is a very small bacterium, but the disease is commonly called virus pneumonia. *Mycoplasma hyopneumoniae* affects the lungs; related organisms cause arthritis and polyserositis. The infection is widespread in pigs, usually contracted by baby piglets from their carrier mothers. Purchased 'coughers' also introduce infection, so symptoms may first become obvious when different litters and sources of weaners are mixed together.

Symptoms

A slight chronic cough, often after exercise. Appetite is retained, but growth rate can be depressed. Mortality is low unless other bacterial infections (*Pasteurella*) invade the lungs.

Treatment

None is fully satisfactory and the long-term needed is uneconomic. Secondary infections may be controlled and minimise overall effects.

Control involves the usual good hygiene, warmth, lack of stress, and food feeding. Since older sows lose their infectivity, subsequent breeding gilts should be selected from their litters—but these must be reared in isolation: cough shown by any piglet would indicate an infected litter. The extent of infection in a herd can also be assessed by veterinary examination of samples of slaughter carcasses. 'Specific Pathogen Free' (SPF) piglets—hysterectomy derived—offer another source of healthy stock.

Epidemic diarrhoea (virus)

Characterised by sudden onset and rapid spread through the pig herd, this disease resembles transmissible gastro-enteritis, but differs in not affecting unweaned pigs. Most pigs recover spontaneously, few die.

Symptoms

Commencing in adults, these are off food, with profuse diarrhoea for 2–3 d, but no fewer. Nursing sows lose their milk so some baby piglets may become ill and scour. Sows recover within a week, but illness occurs in other groups within about a fortnight.

Control

There is no specific treatment, and it is better to encourage spread of disease within the whole herd (unlike advice for other diseases). Some growth check will occur, perhaps delaying slaughter weight by one to two weeks. Breeding pigs suffer no long-term effects.

Pig pox (virus)

Infection is spread by contact, the virus being still viable in scabby material for up to one year: skin injuries, e.g. by biting lice, predispose.

Symptoms

A mild disease usually of younger pigs, up to four months old. Papules progress over 7–10 d to small black scabs, on the thin skin areas of the belly especially. Slight fever at first may cause inappetence and some lethargy. Scabs fall off and pigs recover after three to four weeks, unless secondary skin infections occur due to dirty pens. Death rarely occurs, then only in young piglets.

Treatment

None needed, nor is vaccination. Recovered pigs are immune. Lice should be controlled, and good hygiene maintained.

Transmissible gastro-enteritis
(*see* Disorders of piglets)

In contrast to epidemic diarrhoea, note that it is more severe in piglets and mortality can be high.

Parasitic diseases

External lice (Haematopinus suis)

Size up to 6 mm. These are common, and specific to pigs, seen in skin folds around the neck, jowl, ears, and inside the legs. They live entirely on their hosts, only surviving for 2–3 d off them.

Symptoms

Mild to severe irritation, depending on numbers of lice, caused by the skin punctures. Heavy infestations cause anaemia, restlessness, lowered growth rate. Lice can carry pig pox virus.

Treatment

Several insecticides are available, often either hexa-chlorocyclohexane or organophosphorus compounds. These are applied by spraying (knapsack type for a few animals), at 2–3 ℓ/sow, less for smaller pigs—which could be individually washed or bathed. Note the manufacturers' recommended precautions when handling the concentrate, also the careful disposal of run-off solutions—toxic for fish. Two or three treatments are necessary, to kill later hatched lice. Bedding should be destroyed, and pens cleaned.

Mange (Sarcoptes scabei, var. suis)

These microscopic mites burrow into the skin, where they live and reproduce. Infestations are common, often persisting in the ears, and mites survive off their hosts for up to two weeks on posts, walls, utensils, especially in a service area if a mangy boar is used.

Symptoms

Itchiness, causing rubbing and scratching. Young piglets, six to eight weeks old, show a red rash on ears, face, down their legs. Later, crusty scales develop, then thickening and folding of the skin, with exudate and characteristic odour. Severe cases are unthrifty and may die.

Treatment

Similar skin lesions can be caused by other conditions ('greasy pig disease', nutritional deficiencies), so correct diagnosis is essential before selecting treatment, which can be hexachlorocyclohexane or organophosphorus compounds. These can be used as washes or baths, for small pigs, more usually by knapsack sprayers, or power sprayers for many larger and badly scabbed pigs. Dressing must wet the whole body especially underneath and the ears.

Three applications, at 10–15 d interval, will kill mites hatching later.

Sows should be treated twice a year, pre-farrowing, boars also but three times at 10 d intervals. Ears can be separately dressed with benyl benzoate emulsion. New entrant pigs should be isolated until dressed three times.

Precautions as for lice dressings also apply, concerning handling and disposal of the solutions.

Lungworms (Metastrongylus)

Adult lungworms live in the lungs, their eggs being coughed up, swallowed, excreted in faeces and viable for a year on shaded pasture. Earthworms, the necessary intermediate hosts, consume these eggs, which develop into infective larvae ready to develop into adults when pigs eat the earthworms. Indoor housing on concrete, or slatted floors, has reduced the occurrence of lungworm infestations, which are now uncommon in the UK.

Symptoms

Mild infestations may be harmless. Large numbers of lungworms can cause severe coughing, respiratory distress, inappetence. They aggravate co-existing respiratory disease, such as enzootic pneumonia, and may actually carry swine influenza virus.

Treatment

Several anthelmintics are available, some as additives to food or water.

Grazing paddocks should be left fallow, in sunshine, before ploughing and reseeding.

Roundworms (Ascaris suum)

This large roundworm of the pig can grow to 25–30 cm in length and large numbers may obstruct the small intestine. Eggs passed in faeces in large numbers are infective direct to other pigs, wherein larvae hatch, migrating via liver and lungs before reaching the intestine to become adults. Eggs can remain viable for several years. Infections sometimes pass into sheep, evident at slaughter by liver damage, rarely as intestinal worm burdens.

Symptoms

Migrating larvae may cause slight coughing, during lung passage, and so-called 'milk spots' can downgrade livers at slaughter. Piglets can acquire heavy infestations whilst suckling, from sow's faeces, then become unthrifty and stunted, with some diarrhoea, and may die. Adult pigs are rarely ill.

Treatment

Several medicaments are available, as for lungworms. It is essential to treat sows before farrowing, then thoroughly scrubbing them clean and moving to clean farrowing quarters. Pigs are often treated when weaned at five to six weeks old, repeated later if ascarid risk is high. Pasture grazing should be rotated, stocked only by pigs *after treatment* in pens or yards readily cleansed, *not* out on the fields.

Stomach worms (Hyostrongylus rubidus)

Very small, 4–9 mm, this parasite can cause serious stomach ulceration, with subsequent unthriftiness—affecting adult and older sows, but is not commonly seen in the UK under indoor husbandry.

Coccidiosis (Protozoa)

These microscopic organisms parasitise the lining cells of the intestinal tract of young pigs, causing unthriftiness, perhaps diarrhoea, but few deaths. This condition may be more common in the UK than thought previously, as firm diagnosis requires special examinations of sections of intestine from killed, suspect animals.

Treatment

Sulphonamide drugs are used for treatment.

Tapeworms

Although pigs do not suffer from adult tapeworm infestations, the larval stages of some tapeworms of man and dogs do live in pig muscles ('measly pork'— the uncooked meat being infective for man), or on the liver ('bladder worms', which infect dogs eating uncooked offal). Virtually no harm ensues for pigs, the importance being the public health aspect: one human excretor can result in 'measly pork'. Dogs should be frequently (every three to four months) treated for tapeworms. Neither parasite is common in pigs in the UK.

Trichinella

Almost unknown in the UK, this minute, 1–4 mm, intestinal worm causes little ill health in the pig, but the larval stages in uncooked pig meat pass to humans, e.g. via raw sausage meat, who may be affected. Strict meat inspection procedures eliminate this risk.

Nutritional disorders

The composition of pig foods and rations for different ages are described elsewhere: if adhered to, there should not be any dietary inadequacies. A few specific deficiencies are outlined below; there are others which might occur if formulations were incorrect or unevenly mixed, or spoiled ingredients were incorporated.

Thin-sow syndrome

After farrowing, failure to recover normal body weight is usually due to plain underfeeding, no specific deficiency, underestimating the demands of the litter. It also is advisable to have dung samples examined for evidence of gastro-intestinal parasitic worms, which can cause similar chronic debility and anaemia, and can be eliminated by treatment.

Osteomalacia

See also Rickets, in 'Disorders of Piglets'. Sows in heavy milk production have a high requirement for calcium, suitably in balance with phosphorus and essential vitamin D (D2 and D3). Deficiency leads to decalcification and softening of bones. Excesses can also be harmful.

Symptoms

Lameness, stiffness, posterior paralysis, due to bone fractures, e.g. of femur, pelvis, or spine, exacerbated by over-exertion or slippery floors.

The condition is best prevented, not treated.

Copper deficiency

Copper is necessary for the blood system; together with iron (*see* Anaemia, Baby Piglets), but its deficiency can also impair development of the central nervous system.

Symptoms

Uncoordinated gait, with back-swaying motion, or posterior paralysis. Severe cases will not respond to treatment; again, preventive measures are best, ensuring adequate intake in food.

Zinc deficiency

Among several causes of skin diseases, inadequate zinc causes gross thickening of the skin (parakeratosis) commencing on the abdomen and inside thighs, eventually spreading over the whole body. The crusty lesions appear to cause little irritation, unlike mange, from which it must be differentiated, also from 'greasy pig' (which affects younger pigs).

Soyabean protein, and high calcium, in the diet enhance the requirement for zinc, easily remedied by correct inclusion in the food.

Selenium deficiency

Trace amounts of selenium, in combination with vitamin E, are particularly important for normal development of muscle, both skeletal and cardiac. Liver necrosis and 'mulberry heart' disease are other manifestations of deficiency, as may be reproductive malformation, including infertility. Cereals in store can lose vitamin E potency, and may be low in selenium if grown on deficient ground. Adequate mineral—vitamin supplementation is necessary when compounding them into a pig ration. Stale, possibly rancid, ingredients must be avoided.

Symptoms

More often affect growing pigs: sudden death, due to heart failure. Stiffness, general lassitude, reluctance to move.

Vitamin A

When inadequate in pregnant sows' diets the litters may be stillborn, or born weak and blind with minute eyes, whilst the sows are normal.

Poisons

Apart from careless access to poisonous chemicals (herbicides, fertilisers, paint, etc.) pigs do sometimes suffer from inadvertent excesses of products normally added to fattening rations, e.g. *arsenicals* and *copper salts*, even *sodium chloride*. Ensure correct inclusion levels—that little extra is more harmful than beneficial.

Symptoms

These often resemble a variety of other disorders; veterinary diagnosis is recommended if pigs show nervous disorders, jaundice, undue thirst.

Mycotoxicoses

Any mouldy food is potentially of poor feeding value, possibly harmful. Certain moulds, not always easily visible, produce potent toxins (mycotoxins) on foods and fodders. Their identification requires detailed culture tests and chemical analyses, and much further research is still needed.

Symptoms

These vary from unthriftiness to sudden deaths, e.g. due to body haemorrhages. Sows may exhibit vulval enlargement.

Reproductive disorders

A sow's profitability directly relates to her ability to produce 2–2.4 litters, totalling 18–24 healthy pigs/year. Failure to achieve this is evidence of poor 'fertility', either in sow or boar, from actual sterility (e.g. due to abnormalities of genital organs), or anoestrus, return to service, abortions, stillbirths, mummified piglets, small litter size, and low viability: from birth to weaning, the highest percentage mortality occurs within the first week of age, often due to faulty feeding and management of sow during pregnancy, at farrowing and immediately afterwards. Pig feeding and management are dealt with in Chapter 17.

Specifically venereal diseases are rare, but infections of the sow by either bacteria or viruses do cause abortions and stillbirths.

Anoestrus and sub-oestrus

Complete inactivity of ovaries, or inapparent signs of heat despite ovarian cyclicity, can result from early weaning systems; the sow's uterus may not reach normality for three weeks after farrowing, or the sow may still be too thin, or conversely overfat. She must receive her correct ration of minerals and vitamins. Infertility can be a hereditary characteristic.

Returns to service

Up to 21 d post service is a highly sensitive period for fertilised eggs, not yet implanted in the uterine wall, readily upset by stress of any kind; hot weather, bullying, movement. After 21 d, embryos plus total contents of the uterus may be resorbed. Bacterial or viral infections are possible causes, perhaps certain mycotoxins in foodstuffs.

Abortion

This is expulsion of partly developed, dead fetuses, from 30–110 d post service, seldom showing obvious pathological changes, and the placentae also seem normal. Outbreaks are usually only sporadic in individual pigs, or affecting particular groups of sows or gilts.

Bacterial infections of the sow herself, inducing fever, can cause abortion, e.g. swine erysipelas, *Pasteurella*, leptospira, salmonella, *E. coli*, streptococci, corynebacteria. These would be controlled by appropriate medications and/or vaccinations.

Mummification

Fetuses dying after 40 d conception can remain in the uterus until near to normal term of pregnancy. Rarely are all the fetuses mummified (dried, due to resorption of their body fluids), and they will probably vary in size.

Viral infections of susceptible, non-immune gilts and young sows are common causes; fetuses are killed, whereas their dams may not visibly be affected. In the UK so-called SMEDI parvoviruses are involved, and further losses are prevented by ensuring that young breeding stock, before mating, mix well with older affected animals, gaining a degree of immunity.

Note, however, that when *other* infections are diagnosed quite the opposite may be recommended, e.g. complete hygienic isolation of affected from unaffected animals, otherwise widespread disease may be encouraged. No vaccines or specific treatments are available.

Stillbirths

Five to seven per cent of piglets are born dead, about full-term, even more in large litters particularly in older sows. Infections are rarely responsible, many stillbirths being due to death of piglets at or just before the act of farrowing, if the placenta separates too soon, or a short umbilical cord still within the uterus.

Full-time attention to the farrowing sow could save many piglets which would otherwise die as a result of prolonged parturition.

Low viability

Live piglets companion to some stillbirths may soon die. So, too, may any undersized piglet, perhaps born to inadequately fed mothers, e.g. when thin, or lacking certain essential minerals (manganese, calcium, phosphorus) or vitamins (e.g. A, B complex). Such piglets are even more vulnerable than the normal newborn piglet to chilling—their thermoregulatory mechanism is not adequate at birth, and body reserves of glucose (as quickly available energy) are also low. Losses are greater during winter months. A proportion of dead piglets are diagnosed as due to crushing, overlain by the sow, but this probably results from the inactivity of weakly piglets.

Post-farrowing disorders

These include *agalactia*, failure to let-down milk at all or for only a few hours. Usually gilts are affected. Unless these are promptly treated by hormone injections, the starved piglets will die, but can be saved by immediate hand feeding. *Farrowing fever*, often with *metritis*, can result from infections such as *E. coli*.

Symptoms

Symptoms occur 2–3 d post-farrowing, with fever up to 41 °C, inappetence, uterine discharge.

Antibiotic treatments, in some infections vaccination, are effective.

Mastitis can be acute, within 3–4 d of farrowing, with swollen, hot, and painful glands, and some fever. Unless treated, some cases may die. A chronic form is due to deepest seated infections, less severe, but reducing the number of lactating glands. *E. coli* bacteria are commonly involved.

Miscellaneous disorders

Porcine stress syndrome

Certain genetic lines of pigs have an inherited stress susceptibility, unable to adapt quickly to stresses such as excitement, fighting, or particularly handling for loading into transport vehicles. They over-respond to energy demands, by rapid glycogenolysis of muscles, causing lactate acidosis and fever, often death.

Symptoms

Early symptoms are tremors of the tail and muscles, heavy breathing; rest and quiet will lead to uneventful recovery. Further stress causes severe respiratory difficulties, blotching of the skin, high temperature, purpling, then total collapse in spasms and death in terminal shock—when no treatment succeeds.

Treatment

Tranquillisers, corticosteroids, possibly bicarbonates to counteract the lactate acidity. Prior to stressful situations, tranquillisers can limit the problem for known susceptible pigs.

Heat stroke

Pigs cannot properly sweat, lose less heat via lungs than other animals, and carry thick fat as insulation. In hot weather, heat stroke can occur when pigs are overcrowded in confined pens lacking sufficient cool air flow and ventilation; in absence of shade, sunburn is a further complication, as also are handling and other stress situations, also inadequate water and limited intake of sodium chloride.

Symptoms

Pigs seek shade and water. Breathing is rapid, becoming difficult, fever may reach 42 °C, and frenzy is followed by coma and death.

Treatment

Pour cold water on the floor (not over the whole pig—swab legs, belly and head only). Ice packs can be applied to head and legs and fans used to circulate air. Tranquillisers are helpful, also corticosteroids, on veterinary advice.

Part 3

Farm equipment

23

Environmental services

R. P. Heath

SERVICE DUCTS

When planning any buildings or structures it is difficult to anticipate all future likely uses. As such it is unreasonable to put a prediction value on installing all services at construction. It seems reasonable to allow for future services by leaving ducting. Below ground these should withstand hydraulic and soil loadings and any anticipated direct load from above. Movement prevention is a must.

Deterioration of any service by entrainment of effluents, dissolved salts or the like into a duct must be countered when the duct is constructed. Where there is a high water table, water exclusion should be a priority for choice of materials for the duct.

RAINWATER GOODS

The installation and size of rainwater goods must satisfy the Building Regulations.

A storm is deemed to be 75 mm/h rain intensity. This is found to occur over a 5 min period once in four years and over a 20 min period once in 50 years, over most of the UK. Rainwater goods are matched to this storm requirement.

The maximum roof drip into a gutter should not exceed 50 mm. Falls on gutters should accommodate this. Valley gutters require special attention to allow for snow loads, human access, overall fall and volume. The construction of downspouts and gutters is related to anticipated precipitation. The length of gutter either side of a downspout should not exceed 6 m. Downspouts are particularly prone to blockage and mechanical damage. The choice of material and installation location should accommodate this problem.

The roofing trade works in 'squares' for roof area. The old standard for a square of 100 ft² (9.29 m²) is likely to change to 10 m². The sizes and falls of gutters can be estimated by reference to *Building Research Station Digest* Nos. 188 and 189.

Roof water, clean but not pure, may be stored separately from foul drainage and used for yard washing or fire defence. Whilst a useful source of dilution for animal slurries, volumes can be excessive if all drainage is to be stored.

Many different materials may be used for gutters and downspouts. Downspouts within a structure should be sealed at joints. Roof water may be discharged directly into a water course.

DRAINAGE

Reference should be made to BS5502:1978 and Section 3.1 is particularly useful. Waste disposal, pollution and river acts and local bye-laws must be observed.

Natural water may be disposed of directly into a water course, or into a soakaway, should subsoil conditions permit. The soakaway should be away from a building boundary. A perforated wall may be used to contain the porous fill, but a solid cover is needed. The volume is taken as 13 mm over the drained area.

Waters containing waste products, contaminants or pollutants, must be dealt with appropriately. Livestock and vegetable wastes, organic matter, cleansing, disinfecting or chemical agents and fuel oils must be assessed in terms of biochemical oxygen demand, toxicity and physical handling properties.

No waste should be disposed of through areas of human habitation, food preparation or storage areas.

Static and dynamic loadings, internal and external, from stored solid or liquid or vehicle passage, must be accounted for when constructing stores.

Animal access should be restricted and provisions under the Health and Safety at Work (1974) Act observed with respect to humans. Should access be necessary appropriate precautions must be taken.

Where workers occupy a building for 4 h or more daily, lavatory accommodation is required. Human waste must be disposed of separately from all other wastes. Septic tank capacity of 0.1 m³/person daily is required. CP 301:1971 and BS 5502:1978 will be of value.

Effluent movement and disposal are centred round adequate dilution with water to enable suitable transportation, subsequent treatment and subsequent disposal. Air, land and water pollution risks must be catered for when designing a disposal system. Back-flow under flood or blockage occurrences must be allowed for by valving and external access points provision.

Stackable solids should be retained as such and structures designed to suit.

Silage effluent removed from a settling sump can be diluted with water and spread on land. Care should be taken with surface run-off or drainage to a water course because of the massive biological oxygen demand (BOD). Spreading is the only satisfactory disposal method.

Waters contaminated with dairy products should collect in a 'settling' tank, unless a liquid manure store is available when it can be admixed.

Blind ditches or oxidation ditches should not be used for waste contaminated with dairy products.

Diseased animal effluent or peculiar risk wastes should be kept and disposed of separately from other wastes.

No store should be within 150 m of domestic premises.

Drains should be of stoneware, glazed, with no less than 150 mm of concrete surrounding them. This may prove insufficient where ground movement is possible and a flexible pipe may be needed. An adequate fall is necessary but the old 'rules of thumb', of 1 in 40 for 100 mm drain and the like, are now unsatisfactory.

Clean water velocity should not fall below 0.75 m/s and it rarely exceeds 3 m/s. Foul water may have drain sizes based on Colebrook and White formula. The maximum velocity of 1.5 m/s allows for cleaning to be maintained. In excess of this figure the liquid fraction is likely to leave the solid behind and blockages result. BS 5502:1978 contains tables from which soil pipe sizes may be determined.

The pipes should be laid socket-end uphill, and joints made watertight. Caulking the joint and masking with an even volume cement–sand mix can achieve this objective and the joints should not interfere with flow.

The use of rubber and plastic gaskets and steel, plastic or pitchfibre pipe can overcome the brittle nature of the solid jointed glazed clay or iron pipes.

Drains should be laid in straight lines. Inspection chambers should be at all joints and rodding eyes and at changes in direction.

No part of a drain may be more than 45 m from an inspection chamber and there must be either a chamber or a rodding eye at the head of the drain.

No drain should pass under a building or interfere with foundations. Where such an event is unavoidable then likely ground movement must be catered for.

Drains adjacent to buildings, where the bottom of the trench is lower than the foundation of a wall, should be 'filled in' with concrete. If the distance to the side of the trench is less than 1 m from the wall footing the concrete must be to the level of the underside of the foundation. No continuous length of fill concrete should exceed 9 m.

Resultant shear loads on the pipe and toppling moments of a wall, foundation or slab must be calculated and strengthening provided. Inspection chamber covers should comply with the current British Standard for the loads and location envisaged. Chamber covers may be sealed by grease, or a grease–sand mix at the locating slot. Any screw fixing should be protected from corrosion.

Ventilation of drains should be provided and free air passage along the pipe length should not be impeded.

Discharges to foul water drains or stacks must in all cases pass through suitable 50 mm seal water traps. Only air vents to positions likely not to cause a nuisance or health hazard are permissible.

Interceptor traps should include rodding eyes wherever possible. Drains should be rodded after installation to ensure free egress for liquids and ensuing wastes.

A permanent record of installations should be kept. Where a public sewer is involved then all detail such as size, position, likely dilution and discharge should be submitted to the Authority. Approval must be gained prior to work.

WATER

The final use will determine the quality, delivery method, storage method, flow rate and pressure. Statutory requirements of the Water, and Public Health Authority must be followed. The current somewhat rigid water byelaws are expected to be modified under pressure from the EEC.

Any installation must be adequately supported, frost proofed and corrosion proofed. Electrical continuity for earthing aspects must be ensured. When burying pipes external loadings must be taken into account. Outside buildings they should be buried at least 800 mm deep. Under paved areas or under floors of buildings a suitable depth should be sought.

Galvanized steel pipe, asbestos (tested to twice normal operating pressure) and copper pipe may be used universally. PVC and polyethylene pipe may be used for cold water; 18/10 stainless steel may be used where copper tube (BS 2871) would be specified (*see* BS 4127 and *Building Research Establishment* No. 83) (*Tables 23.1, 23.2* and *23.3*). Wall thickness and type of pipe will determine the pressure of liquid they can withstand. The local water authority will have details of the class and type of pipe for a particular service and their guidance should be sought.

Table 23.1 Copper tube for water, gas and sanitation

Nominal diameter (mm)	(in)	Wall thickness (mm) Table X	Table Z	Maximum working pressure, MN/m^2 (N/mm^2) Table X	Table Z
10	($\frac{3}{8}$)	0.6	0.5	7.7	7.8
15	($\frac{1}{2}$)	0.7	0.5	5.8	5.0
22	($\frac{3}{4}$)	0.89	0.59	5.1	4.1
28	(1)	0.89	0.59	4.0	3.2
35	($1\frac{1}{4}$)	1.185	0.685	4.2	3.0
42	($1\frac{1}{2}$)	1.185	0.785	3.5	2.8
54	(2)	1.185	0.885	2.7	2.5

Table X tube—half hard, light gauge ⎱ BS 2871, Part 1, 1971
Table Z tube—hard drawn, thin wall ⎰
Table Z tube should not be formed into bends but jointed to change direction

Table 23.2

Light steel tubes		Heavy steel tubes	
Nominal bore (mm) (in)	Thickness (mm)	Nominal bore (mm)	Thickness (mm)
15 ($\frac{1}{2}$)	2.0	50	4.5
20 ($\frac{3}{4}$)	2.35	80	4.85
25 (1)	2.65	100	5.4
40 ($1\frac{1}{2}$)	2.9	125	5.4

Other sizes available from 150 mm in light, medium and heavy weight tubing

Table 23.3 Availability of polythene pipe to BS 1972 and BS 3284 (low and high density)

Pipe class and colour code	B, Red		C, Blue		D, Green	
Working pressure head	60 m		90 m		120 m	
Density type—32 (low) and 50 (high)	32	50	32	50	32	50

Nominal size (in)	Permitted range of outside diameter (mm)						
$\frac{3}{8}$	17.0–17.3	×	×	×	✓	✓	✓
$\frac{1}{2}$	21.2–21.5	×	×	✓	✓	✓	✓
$\frac{3}{4}$	26.6–26.9	×	✓	✓	✓	✓	✓
1	33.4–33.7	✓	✓	✓	✓	✓	✓
$1\frac{1}{4}$	42.1–42.5	✓	✓	✓	✓	✓	✓
$1\frac{1}{2}$	48.1–48.5	✓	✓	✓	✓	✓	✓
2	60.1–60.5	✓	✓	✓	✓	✓	✓

Other sizes used in agriculture 3, 4 and 6 in.

Storage vessels should have dust covers and allow access for cleaning. A number of suitable materials are available (*Table 23.4*).

Table 23.4 Small water tanks and cisterns to BS 417 (galvanised steel), BS 2777 (asbestos cement) and BS 4123 (polythene and polypropylene)

Galvanised steel	Asbestos cement	Polythene	GRP
SCM45/18	AC27M(6)/17	PC4/18	5/23
SCM70/36	AC45M(10)/28	PC8/36	—
SCM110/68	AC90M(20)/62	PC15/68	15/68
SCM135/86	AC114M(25)/80	PC20/81	—
SCM180/114	AC136M(30)/114	PC25/113	30/136
SCM270/191	AC227M(50)/188	PC40/182	—
SCM320/227	AC273M(60)/227	PC50/227	50/227
SCM360/264	AC364M(80)/300	PC60/273	65/295
SCM570/423	—	PC100/455	—
SCM680/491	AC591(130)/473	—	105/477

BS ·reference numbers and capacity in litres (except for *GRP* where a manufacturer's range is quoted) in near equivalent sizes (gallons and litres)

Ball valves of two main sorts are fitted to control the delivery to the cistern; the traditional piston and sealing washer type, known as the Portsmouth valve, and the more recently designed diaphragm and plunger valve. The first one is usually supplied in brass and all parts are 'wet', the nozzle is short and the orifice size can be varied for action against different pressure heads. It suffers from corrosion and sediment which impede its working. The other style of valve has a flat flexible disc, which is pushed up against a protruding nozzle from the supply pipe by a plunger, which is on the dry side of the disc to seal off the incoming flow. This is usually manufactured in plastic; the nozzle and diaphragm may be changed to cope with different pressure heads. An advantage of this over the Portsmouth valve, is the outlet, which may be above the body so eliminating the possibility of back syphoning.

Overflow pipes should be fitted. For tanks up to 4.5 m³ capacity the overflow may be positioned to give warning. Above 4.5 m³ a warning pipe of 25 mm diameter at 50 mm below the invert of the overflow pipe should be fitted.

No pipe work should pass into or through, any ash pit, manure pit, sewer, drain, cesspool, refuse area or such like, which may cause pipe deterioration or egress of foul material.

Every building served by water should have its own stop valve. This should be fitted as near as possible to the point of entry of the pipe on the inside. Each storage vessel of >0.02 m³ should be fitted with a stop valve close to the vessel on each of the exit pipes.

Water supplies to vessels from which stock are allowed access should not be supplied directly from the main, unless the inlet is >150 mm above the free surface of the water and out of direct reach of stock; this should prevent contamination.

Stock should not be allowed to vandalise plumbing fixtures so guards may be needed.

Cattle troughs are usually made from galvanised steel or concrete, measuring 450 mm wide by 400 mm deep, holding 50 ℓ/300 mm of length, from 1.2 m to 3 m. A separate service box may be provided, equipped with the ball valve on one end of the trough. Water bowls can be either gravity filled from a separate control tank equipped with ball valve or they can be constant pressure supplied and fitted with a tongue- or snout-operated integral supply valve. All troughs and bowls should be securely fixed at an appropriate height. As a rough guide, one 1.8 m trough is usually sufficient for 30 cows and one bowl is adequate for 15 adult pigs.

Pipes buried beneath wall decorations or surface finishes can cause problems (cracking, condensation and access problems may be experienced).

Water for dairy use comes within the bounds of the Milk and Dairies Regulations and must be clean and pure. Vegetable washing water follows similar constraints.

Private sources must yield water of an adequate quality and quantity. Storage of a volume to enable repair to a pumping failure should be considered.

In any situation where low temperature is going to affect the service insulation or electric heater pipe wraps should be used.

Exhaust water from dairy heat exchangers may prove a useful source for stock.

Water demands for domestic, stock and washing vary considerably. Average data can considerably underestimate the actual demand (*see* CP 310: 1965) (*Table 23.5*).

Table 23.5 Daily water use for farm production and related storage

	Water required (ℓ)
Cow consumption, 22ℓ milk/d	70
Cow consumption, dry	35
Cow allowance, dairy cleaning	20–50
Cow allowance, milk cooling	60–120
Beef animal, drinking	25–45
Calves (up to 6 months), drinking	15–25
Sows drinking, in milk	18–23
Sows drinking, in pig	5–9
Boars drinking	9
Pigs drinking, growing, fattening	2–9
Sheep drinking, growing, fattening	2.5–5
Sheep dipping, all sizes/dip	2.5
Poultry drinking (100 birds), layers	20–30
Poultry drinking (100 adult birds), fattening	13
Turkeys drinking (100 adult birds), fattening	55–75
Storage capacity allowance for sink	90
Storage capacity allowance for WC	90
Storage capacity allowance for hot water system	130
Storage capacity minimum for farm office/canteen	225
Storage capacity minimum for farm workshop	450

LIGHTING

Daylight is extremely variable in intensity but offers an attribute to working man that non-daylight cannot provide. Artificial light can give adequate shadow-free levels of illumination. The level of illumination should allow for safe egress, work and social needs. The stroboscopic effect near rotating machinery due to the frequency of discharge should be noted.

Quartz–halogen lights can lead to snow blindness and offer an explosion hazard in dust laden atmospheres. It will be found that some light sources make more efficient use of electricity and provide a better output in the useful visible range.

Flexibility of switching can aid waste reduction.

Providing glare is eliminated, the better the illuminance, the better will be visibility (*Table 23.6*). The level of illuminance is measured in lux (lx).

Table 23.6 Illuminance intensity

Area or task	Intensity (lx)
For selection by colour or detailed assembly	500–1000
For 'close' inspection	300
Farm workshop (general level)	100
Milking premises and passages	100
Others	50

After BS 5502:Sect 3.5:1978

Daylight may be added to the artificial provision. The intensity is variable so that it may be considered more as improving the human subconscious environment rather than illuminating the environment. Fluorescent tube light has been considered detrimental to productivity if used on its own.

The luminaire provided should be consistent with the level of illumination required, the environment into which it is to be fitted and the cost. Any recommended lux values should be used as minima. Additional illumination should be provided for special cases within a generally illuminated area. The colour of the containment structure will influence the resulting illumination from a particular source. Several small light sources are often better than one large light source. Diffused sources are better than direct ones.

Tungsten filament and tungsten–halogen lamps work direct on line. Fluorescent tubes and high intensity discharge lamps (HID) need additional electrical components to enable the current to 'strike' along the tube.

High intensity discharge lamps take some 20 min to reach their operational lighting intensity. They are thus inconvenient for use where frequent switching is required. Frequent switching of a fluorescent tube can shorten tube life. There should be no effect on tungsten filament through switching. The light intensity diminishes with age.

The various light source efficiencies have a bearing on an overall heat energy balance in a room (*Table 23.7*).

Table 23.7 Luminaire life and efficacy

Nominal life (h)	Lamp type	Efficacy (1 m/W)
7500	High pressure sodium (gold colour light)	70–100
7500	HID mercury vapour (+halide colour renderer)	62–72
7500	Fluorescent tubes ('white' colour)	54–67
7500	HID mercury vapour (+fluorescent coat)	35–50
7500	Fluorescent tubes (colour rendered)	33–40
2000	Tungsten–halogen	16–22
1000	Tungsten–filament	10–18

Technical assistance can be gained from the Area Electricity Board. Reference may be made to CP 324.101 and the Electrical Development Association.

GAS

Natural gas

Installation of natural gas supply is the responsibility of the British Gas Corporation (BGC), or its approved contractors. Compliance with Regulation Parts I to VII is necessary. Also included is guidance for appliances.

The 'supply' is the service up to the outlet of the primary meter, and is the responsibility of the BGC. The gas safety regulations apply throughout an installation.

Corrosion is a major hazard on any installation. Electrical and gas services should not be laid alongside each other. It is usual practice to install the meter at a convenient low point and then radiate service pipework outwards and upwards. This will drain any condensate which may (rarely) occur.

The danger of puncturing non-ferrous pipe below floors and corrosion by cement of copper pipe should be avoided. Sleeves and support for surface pipes away from and through walls should be provided. Bends and angles producing a loss in pressure should be avoided.

Flues should be provided. The combustion product is water and carbon dioxide and, should an unlit appliance be left on, the flue will aid ventilation as well as venting the condensate to the outside.

Normal operating pressure is about 1.5 kPa.

Natural gas burners tend to be noisy due to their air consumption. More than double the air of town gas (volume based) is needed. The energy value is $37 \times 10^3 \, \text{J/m}^3$.

Liquified petroleum gas (LPG)

A product of the oil industry its price rises with oil price rises. The convenience of the heat form, and an energy value some three times that of natural gas, may make its use plausible.

The service may suffer from loss of flame in draught due to low pressure. Propane boils at $-42\,°\text{C}$ and butane at $-0.5\,°\text{C}$. Propane may provide the better service where high demands in winter are needed.

Small bottles, to 50 kg mass, should not be stored in cupboards, put near a heat source or insulated. The valve must be uppermost to allow gas and not liquid into the system.

Pipes can be of smaller diameter than for natural gas. Change over to natural gas should be considered and choice of pipe size made accordingly. Solid drawn copper pipe should be used for preference. Blockage of the small burner jets can occur.

Red lead paint should not be used as a pipe jointing compound. Solder joints should be replaced by brazed joints when near the heat source.

The BGC, the supplier of LPG, or approved agents should install any fixtures, furniture or appliances.

Reference should be made to the Building Regulations and amendment, the Gas Safety Regulations 1972 (SI 1972:1178), the Gas Act 1972, CP 331, CP 338, CP339 and BS 5258.

Air

Compressed air services can be extremely useful in workshops. Safety can be brought more readily to air than electricity-operated hand tools. Compressed air is, however, extremely dangerous. No laxity in maintenance can be tolerated. Insurers pay particular attention to this service. Corrosion from condensate is the main problem in use. This can be dealt with adequately by pipe and fixture design.

Though a very useful source of power, it can be very expensive to run if leakage is allowed. Flow meters may prove useful loss monitors on large systems.

BS 1710 depicts a colour code for pipes that can be used to show contents and the use to which those contents may be put.

HEATING

Comfortable living conditions, acceptable working environment, or an environment conducive to crop or animal production, are provided by heating systems when ambient temperatures fall below an acceptable level.

Heavy work areas require some 13 °C, sedentary work 18 °C and crop and animal units special conditions.

Humidity, ventilation, insulation and fuels supply and storage are integral elements of any system. Fail-safe devices are important.

Heating units should be positioned to provide an equitable environment and no great temperature gradients. Open flues, or heater vents may create draughts, which may be detrimental to animals and plants.

When designing systems macro-decisions as to fuel type and heater type are made from use, building and economic considerations. Safety and micro-consider-ations, such as resultant condensate, venting of hot water pipework or vibration in air ducts, should be taken into account.

The SI unit of heat is the kilocalorie, this being equivalent to 4186 J.

A range of fuels is available (*see Table 23.8*). Many are self stacking. Alternatives to traditional fuels are gaining merit, but generally require courage to install and technical expertise to utilise fully.

Table 23.8 Energy value of various fuels

Fuel	Gross calorific value (MJ/kg)
Anthracite and good coal	35
Peat (14% DM)	19
Sawdust	18
Wheat straw (10% DM)	17.7
Oil—Class C Kerosene	46.4
Class G heavy	42.5
Methane	$37/m^3$
Biogas (60–70% methane)	$22–26/m^3$

Coal fired independent boilers or room heaters with high output back boilers are popular domestic heaters. Gas heaters with a flue to the outside through a wall or conventional flues can provide hot water or direct space heating. Electricity, gas or hot water can be used to heat air for warm air installations. These tend to be cheaper than indirect hot water systems, but may not be practical because of safety problems.

Oil burning is about 75% efficient and is a useful concentrated stored form of heat, less subject to industrial action or national network breakdown. Overhead infrared or under-floor cable electric heaters are useful for partial heating systems.

Wood and other vegetable matter burners can provide useful quantities of heat. Solar gain units for low grade heating can be useful but are expensive to install. Heat may be stored in 'rocks' or used directly as hot water in short-term storage.

Heat pumps offer greater opportunity on farms. Recovered heat from milk is a useful source of low temperature water. Pumping heat from other heat sources such as solar panels, or water can be at a coefficient of performance of 3:1. Conventional heat units have a life expectancy of some ten years. Better than double this may be reasonably expected from a heat pump, but a relatively low concentration of heat is a major problem.

Biological degradation of animal faeces and urine can provide useful heat sources; a ten year 'payback' on digesters would be sought.

A constant 30–35 °C is held in a 3–15% solid content mix. Temperature maintenance consumes some 35% of the gases produced. This may rise to 100% in winter when production is most needed elsewhere. Regular throughput of material is essential with stirring to avoid scum or sediment formation.

Pig, cattle and poultry slurries are held for about 8, 17 and 17 d, and produce about 0.3, 0.2 and 0.28 m^3/kg waste, respectively.

The system is of advantage where by-products for gas production can be utilised. The gas of some 65% methane content has a value of 22 to 26 MJ/m^3. A 50 dairy cow system produces some 3.75 kW. Gas storage is a problem and immediate use is inevitable.

Water power can be useful. The capital investment tends to be high and use of water is as determined by the Area Water Authority. A 60% efficiency for electricity production would be typical from an impulse turbine/alternator set.

FIRE

When involved in a construction exercise, reference must be made to BS 5502:1978 and Building Regula-tions Parts G and L.

Insulation, claddings or structural skins should not be of material that produces toxic fumes, is highly flammable, produces dense smoke or constitutes a major hazard to the building concerned. Easily taken stock exit routes, allowing escape largely unaided by humans in an emergency, should be considered. Smoke vents designed into a structure can only aid livestock and human escape.

Working areas in buildings should have adequate means of escape and preferably should be illuminated.

Alarm systems may be considered unnecessary or expensive. These installations can only be to advan-tage. Their use may double if linked to a systems failure unit on, e.g. a controlled environment house. Systems installed as permanent fixtures other than alarm may prove as useful.

Building layout should provide fire breaks, ease of access for emergency services and have ample water between 6 and 100 m away. Static tanks of 20 m^3 capacity for suction hose use should be provided if adequate pipe supply is not available. Hose reels should comply with BS 5274 and hydrants to BS 3251.

High fire risk areas such as fuel stores should provide 2 h fire resistance by the walls, or have 12 m separation from any other boundary. Boiler rooms, maintenance workshops, plant material drying com-partments and potentially explosive fertiliser stores should provide 1 h fire resistance by walls or 6 m separation. The separation of buildings physically to provide fire breaks can be extended to cover the inside of a particular building. Compartmentalisation is advisable, particularly where livestock are concerned with undivided floor areas normally exceeding 500 m^2.

Pesticides can be particularly dangerous in a fire. Even if they themselves are not involved their storage proximity may cause a hazard to fire defence. A separated, lockable fireproof bunker should be provided. Chemical contents list should be kept elsewhere from the store.

Feed grinding and feed mixing should offer 0.5 h fire resistance or be 3 m from other boundaries.

Water or foam extinguishers should not be used where an electrocution hazard is present from live electrical sources.

Dry powder gains its action from smothering the fire. The powder may settle, impairing performance, particularly if mounted on a vehicle or vibrating structure. BCF should be used with caution inside buildings. Throat irritation will signify the build up to dangerous levels of toxic combustion products (*see Table 23.9*).

Table 23.9 Portable fire fighting appliances

Water	5–9 ℓ capacity, the contents of which (water) are ejected by puncturing a compressed gas container or by nozzle control on the previously air pressurised unit.
Foam	5–9 ℓ capacity of foam making liquid ejected by puncturing a compressed gas container or by nozzle control on the previously air pressurised unit. The nozzle design creates foam from the liquid.
Dry powder	2–12 kg of dry powder expelled, completely or partially if necessary, through control of the nozzle in the larger sizes, by compressed gas.
CO$_2$	1–6 kg of carbon dioxide gas under pressure. Ejection by nozzle control.
BCF	1–5 kg of BCF gas under pressure. Ejection by nozzle control.
Blankets	Asbestos sheets of 1 m^2 or preferably glass fibre sheet.

Old units are coloured red. New units have colours depicting contents. Red—water, Yellow—foam, Red—fire blanket, Blue—dry powder, Black—CO$_2$ gas, Green—BCF. They should be suitably sighted to give ease of vision, access and use.

Suitability

Carbonaceous materials (e.g. wood, paper) where recombustion may occur due to contained heat	Water
Flammable liquids	Foam, dry powder, BCF
Electrical	CO$_2$, dry powder, BCF
Special risks	Obtain local fire brigade assistance

ELECTRICAL SERVICES

Compliance with the detail laid down by the Institution of Electrical Engineers (IEE) should be observed at all times.

Electrical service outlets in excess can be useful, but equally can be hazardous and expensive. General installation recommendations for milking installations are in BS 5502:1978 or CP 3007:1965 (*see also Table 23.10*).

The supply can be 'mains' or 'private'.

Table 23.10 Cable sizes

SI (mm^2)	Rating (amps)	Use
Cable sizes		
1.0	11	Lighting and small heaters
1.5	13	Ring circuits
2.5	18	Heaters to 3 kW
4.0	24	Radial circuits
6.0	31	Cooker circuits
10.0	42	Large heaters
Flexibles		
0.50	3	
0.75	6	
1.00	10	
1.50	15	
2.50	20	

Colour codes Dimension	Old	New	Wire
2 Core flexible	Red	Brown	Live
	Black	Blue	Neutral
3 Core flexible 1 phase	Red	Brown	Live
	Black	Blue	Neutral
	Green	Green and yellow stripe	Earth
3 Core flexible 3 phase	Red	Brown ⎫ Ends coded	Brown ⎧ Live
	Yellow	Brown ⎬	Yellow ⎨ Live
	Blue	Brown ⎭	Blue ⎩ Live

Mains

For practical purposes the mains voltage supply is constant. The current taken by any appliance is thus dependent upon its resistance, or, in the case of alternating current (AC), its impedance. The supply cable size for an installation will influence the voltage. The IEE regulations recommend a maximum drop of 2.5%.

Large demanders of power may receive their supply as 11 kV (11 000 volts) and transform their own to a usable voltage. This has pricing advantages. Reference to the Area Electricity Board for the most advantageous price should be made. A tariff structure for capital cost and for demand is available.

Typically the electrical pressure is 415 V for three phase supply or 240 V if single phase supply. Where hand power tools are used consideration should be given to 110 V supply. Where long distances are involved higher voltage should be used (nationally 700 V/km).

Power for single phase supply,
watts (W) = VIpf

Power for three phase supply,
watts (W) = $\sqrt{3}$ VIpf

The line amperage a system demands

$$= \frac{kW \times 1000}{\text{line voltage} \times 1.732 \times pf}$$

$$= \frac{\text{brake horsepower} \times 746}{\text{line voltage} \times 1.732 \times pf \times \text{efficiency}}$$

where V, volts; A, ampere (amps); pf, power factor are electrical terms.

Voltage 'pushes' current through a resistance (or impedance). Power factor is an electrical inefficiency experienced with AC supplies. Where windings, motors and welders are included in the circuit the pf effect can be very adverse. Electricity bills are adjusted according to the pf if it falls below a certain level. The supply equipment provided for a machine can have its current rating exceeded if the pf is less than 1. Capacitors, or condensers, can be used to rectify this. Their application is specific. Of particular note are electric welding sets which could require expensive correction capacitors.

At first sight 110 V needing greater current loading for a particular power consumption may seem wrong, as it is current that kills. However, 110 V has less chance than 240 V to push that current through the human body to earth and so give a shock.

Conversely 415 V does enable a good power level to be achieved for a given current loading. It is particularly suitable where motors of 3 kW plus capacity are used.

Wiring diagrams and switch labelling associated with the installation are strongly recommended.

Assessment, prior to installation, of the likelihood of damage to wiring, switch boxes or plant and equipment should be made and measures taken to avoid risk. Mechanical, chemical or heat damage should be anticipated and avoided. Trailing cables in corrosive liquids, common in agriculture, is a particular hazard.

All electrical conductions should be of a size suitable for the demand, and nature of installation. Explosive atmospheres created by dust need special attention. Grain storage areas can sometimes be explosive and require flameproof installations.

All main switch gear should be firmly mounted on a board. The main fuses and meters are the property of the relevant electricity board. Reasonable access should be provided. The remainder of the installation is the property owners.

Underground cables should be laid 0.6 m down; 1.0 m is recommended where drainage of land is likely and 0.75 m under roads. Dry ducting is admirable for indoor use. Where entry to a building is not against an outside wall suitable 'through-foundation' ducting should be provided.

Fuses are provided to protect the wiring service. It is usual to fit a single pole and neutral (SPN) 60 A switch fuse near to the incoming main. A 16 mm^2 cable connects to the meter from this fuse. A 100 A fuse is used where night storeheaters are used on domestic premises. Suitable fuses are provided according to anticipated demand from the nature of the work.

To monitor dangerous earth leakage, or live/neutral imbalance of voltage or current, earth leakage circuit breakers (ELCBs) should be used. An ELCB cannot be recommended highly enough. Whilst area electricity boards will test circuits installed for continuity before supply connection, and provide an earth lead, the user is responsible, however, for providing that earth.

Water pipe continuity is nowadays suspect due to the use of plastic pipe, and thus it should not be used as an earth rail but metal water and gas pipes still require bonding.

An earth wire should link all services. Double insulated hand tools may work off two wires, live and neutral. It is increasingly popular to use conducting material mainly for light fittings and a suitable earth is thus advisable.

Outside overhead cabling between buildings is largely unsupported. Regulation B127 of IEE regulations specifically refers to this detail. Underground cable should be suitably water and armour protected. Damp is a particular hazard to electrical services. Where water has not been catered for, i.e. in a normally dry operating motor, which is then left unused in a damp atmosphere, the insulation may break down.

The main consumer switch–fuse unit may contain an ELCB; lighting circuits at 5 A, lighting circuit and small heater/motor outlet at 10 A, single socket outlet, or heater outlet, or three kW demand point at 15 A, and for cooker, or for domestic ring circuit at 30 A.

Electrical motors should be provided with a suitable starter and an isolator which should be within easy reach of the operator. Three phase electric motors or equipment designed to be used from one of various outlets should have all the outlets matched such that the motor does not run in reverse. Should one phase cause the overload circuit breaker to operate it must be remembered that the remaining two phases could be live, even though the motor *does not revolve*. This is particularly so with small motors using simple switching.

Small hand held motors, such as on power tools, are of the brush type. These series universal motors consume current proportional to the torque and should not be used unless they can revolve at near full operating speed. Similar conditions occur with brushless motors. Stall conditions on starting may damage seriously the supply service.

Consideration should be given to the consumer demand on the main. A 25% tolerance above anticipated demand should be allowed on any installed fixture.

Likewise fixed and mobile generators (alternators) used for supplies other than the 'main' must be capable of not only maintaining an adequate voltage under normal loading, but of starting loads on such as motors. The running current loading may be worked out. The starting current of motors can be six times the running current, even with starters such as star–delta switches.

Private

Alternators for private supply can be purchased. Stationary engine driven units or tractor powered units may be satisfactory to enable tasks to be completed singly, i.e. corn milling or milking.

The cost of such units is high but the risk of a break in service from equipment failure or industrial action must match this cost.

Emergency lighting may also be considered. At any escape route 2 lx must be provided within 5 s of failure of the normal service. Permanently on charge accumulators provide the power; 2 h of light are usually provided. There may be a 'discharged accumulator' problem if the supply fails more than 2 h prior to being needed!

The likelihood of lightning strike can be assessed and precautions taken. Tower silos are particularly prone; CP 326:1965 should be consulted.

24

Farm machinery

P. H. Bomford

THE AGRICULTURAL TRACTOR

The tractor provides power for almost all mobile operations on the farm, as well as for many stationary processes. Power is defined as the rate of doing *work*, and work is done when a *force* acts through a *distance*. One Newton metre (Nm) of work (or 1 Joule) is done when a force of one Newton acts through a distance of one metre. The same units define *energy*, which is the potential to do work. For the measurement of *power*, a time factor is included, and the unit of measurement is the joule per second, or watt (W). A rate of work of 1000 Newton metres per second is the kilowatt (kW).

Power for the tractor is produced by its engine, and this power is made available in two main forms; as pull at the drawbar, to operate trailed equipment, or as rotary power at the power take-off (pto) shaft. A small proportion of the engine's power is available through the tractor's hydraulic system.

The 'size' of a tractor is generally described by quoting the power produced by its engine. Since the full power of the engine is not available to do work outside the tractor, power which is available at the pto is used, in some countries, as a more useful indication of the work a tractor may be able to carry out.

The engine

The tractor's engine converts the chemical energy contained in diesel fuel into rotary power at the flywheel. Two-thirds or more of the energy value of the fuel is lost as waste heat via the exhaust and cooling systems. Flywheel power, often called brake power because it is measured by applying a braking load to the engine, is the product of the engine *speed*

(N rev/min) and the *torque* (T Nm), or twisting effect that the engine can maintain at that speed. The relationship between these factors is:

$$\text{power (kW)} = \frac{2\pi N \text{ (rev/min) T (Nm)}}{60\,000}.$$

A typical tractor engine produces little torque below 500 rev/min. Maximum torque is reached at about 1400 rev/min, and torque then decreases with increasing speed to 85–90% of maximum at full speed, which is 2000–2800 rev/min. Brake power, however, increases almost linearly with engine speed, and maximum power is produced only at maximum engine speed.

The conventional engine is the best mobile power source at present available in terms of efficiency, weight, availability of fuel and cost. However, it is by no means perfect. It is noisy, it vibrates, and it produces exhaust gases which pollute the atmosphere. It is made up of many hundreds of individual components and has many points of wear. The life of a well maintained tractor engine, before it needs a major overhaul to renew worn parts, is between 4000 and 7000 h of work. The life of other mobile engines ranges from 200 h for small air-cooled engines, to 1500–2000 h for a car engine, to a maximum of 12 000 h for heavy duty industrial engines.

The faster an engine rotates, the greater the amount of power (and fuel) that is consumed in just keeping the engine turning at that speed, in relation to the power that is available at the flywheel. If it is not necessary to run an engine at full speed, because maximum power is not needed, operating the engine more slowly will save fuel. Slower operation also increases the engine's reliability and prolongs working life.

Many diesel engines are fitted with turbochargers,

to increase power output. A turbochanger is an exhaust-driven rotary compressor which forces more air into the engine. This allows more fuel to be burnt, releasing more energy and producing more power. A power increase of up to 30% may be achieved, but this will subject the engine to greater thermal and mechanical stresses. Mechanical components, and the cooling and lubricating systems, must be upgraded accordingly. Since the increase in power is achieved with no increase in engine speed or size, frictional losses do not increase in proportion, and the increase in power is combined with an increase in engine efficiency.

The transmission

In order to deliver a full range of engine power at a wide range of forward speeds, the engine is connected to the wheels by a transmission offering from 6–30 gear ratios. With so many ratios available it is necessary for the driver to change gear often in order to match power and forward speed to changing conditions. Various devices are provided, to make this task easier. Synchromesh transmissions synchronise the speeds of rotating parts, to allow quiet gearchanging on the move. Semi-automatic transmissions change gear hydraulically by releasing one clutch and engaging another. The commonest application of a semi-automatic transmission is in a 'high-low' change which inserts an extra ratio between each pair of existing gears; complete transmissions can operate this way, providing up to ten forward and two reverse ratios. Semi-automatic transmissions deliver less of the engine's power to the wheels than conventional systems because power is lost due to friction and in operating the hydraulic control system of the transmission itself.

Some tractors are fitted with hydrostatic transmissions which provide an infinitely variable range of ratios with a stepless single-lever change even from forward to reverse. This system is ideal for the operation of trailed pto-driven machines such as balers or forage harvesters. Forward speed can be continuously adjusted to match crop and ground conditions while maintaining a constant engine (and pto) speed. For the same reason, hydrostatic transmissions are fitted to many combines and self-propelled forage harvesters. Where a major proportion of the tractor's power is to be used in traction, the lower efficiency of the hydrostatic transmission (80% as compared to 95% for a simple gear transmission) means that less power is available at the wheels from a given engine power.

Drawbar power

A tractor develops drawbar (DB) pull as a result of the gross tractive effort developed between its drive wheels and the soil. After this tractive effort has overcome the rolling resistance of the tractor's own wheels moving the tractor along, any remaining force is available at the drawbar.

DB pull = gross tractive effort − rolling resistance

As will be seen, maximum drawbar pull is achieved both by maximising the gross tractive effort, and by minimising the 'parasitic' effect of rolling resistance.

Drawbar power is the product of drawbar pull and forward speed, as expressed by the equation:

$$\text{DB power (kW)} = \frac{\text{DB pull (kN)} \times \text{speed (km/h)}}{3.6}$$

Gross tractive effort

When the lugs of a drive wheel or track bite into the soil, the rearwards thrust of the lugs tends to push the soil back. As the soil trapped between the lugs is sheared from the underlying soil, a horizontal force is developed. The further the soil is displaced, the greater is this force.

As the tractor moves forward, exerting a pull on some following attachment, the soil is pushed backwards a little. The percentage the soil is moved back, in relation to the distance the tractor would move forward on a rigid surface with no drawbar load, is called the slip. A wheeled tractor develops its maximum drawbar power at 20–25% slip, but maximum tractive efficiency occurs at only 10–15% slip. Any slip at all means that some of the tractor's power is being lost in pushing soil backwards instead of pushing the tractor forwards, but since no pull can be generated without some slip, this must be accepted.

The shear strength of the soil under the tractor's wheel or track depends to a small degree on the rate of shearing; a tractor operating at a higher speed can generate a slightly greater pull because of this property. The two major factors affecting soil strength are its coefficient of internal friction and its cohesion.

The more weight that is applied to the soil under the driving wheel, the greater will be its frictional strength, and the greater will be the tractive effort generated at a particular level of slip. Coefficients of internal friction range from below 0.2 for a plastic clay, to over 0.8 for a coarse sandy soil. An 'average' figure is 0.6, which means that 60% of the vertical force applied to the soil by the driving wheels would be available as gross tractive effort. Excess water acts as a lubricant between the soil particles, and can reduce the coefficient of friction almost to zero.

Cohesion is the strength with which the soil clings together, even when no weight is applied to it. A typical value for a friable soil is 30 kN/m^2, with a range from zero for very coarse-textured soils to a maximum of 60 kN/m^2 in some clay soils. The greater the area of cohesive soil that is put in shear, the greater will be the tractive force generated. Compacting a loose soil, for example by running a wheel over it, will increase its cohesive strength so that a following wheel can generate a greater tractive effort than it could if running on uncompacted soil. Cohesive strength is high in undisturbed soils.

Frictional tractive effort is increased by increasing the weight on the driving wheels or tracks. The loading may be by means of iron weights, or water ballast in the tyres, or the tractor can be made to carry part of the weight of the implement it is pulling. This last approach has the advantage that when the implement is detached from the tractor, so is the extra weight.

As the tractor pulls a load, the resistance of the load tends to tip the tractor backwards about a point on the ground beneath the rear axle. This has the effect of transferring weight from the front to the rear wheels, and thus increasing the tractive effort. The height of the hitch point must be kept low enough to eliminate any risk of overturning the tractor. Front-end weights may be fitted to ensure that enough weight remains on the front axle to give steering control. Four-wheel drive tractors, and crawlers, gain no benefit from weight transfer, since all the weight of the machine is carried on the driving members. On four-wheel drive conversions of two-wheel drive tractors, weight transfer can remove much of the weight from the front axle under good tractive conditions so that the powered front axle contributes little to traction.

Cohesive tractive effort is increased by increasing the contact area between drive member and soil. The fitting of larger section rear tyres or dual wheels, or the use of four-wheel drive or crawler tracks all have this effect. Additional soil may be put in shear by the use of grousers, strakes or spade lugs, which can penetrate through a slimy surface layer into stronger underlying soil. However, traction in this case is increased at the expense of reduced tractive efficiency since power is lost in digging into the soil. The use of tyres at no more than the recommended inflation pressure for the load carried will ensure the maximum safe contact area between tyre and soil.

Rolling resistance

In order to roll a wheel along a surface, a force must be applied to overcome its rolling resistance. Rolling resistance increases with the load carried by the wheel, with the amount of flexing of the tyre, and with the amount of deformation of the surface, or sinkage. Increasing wheel diameter reduces rolling resistance, and increasing wheel width increases rolling resistance. Minimum rolling resistance is achieved by operating a large-diameter, narrow rigid wheel on a rigid surface.

Tyres

Most tractors rely on rubber tyres to transmit the power of the engine to the soil and to generate tractive effort. So long as lug height is not less than 20 mm, tread pattern has little effect on overall tractive performance. There is negligible tractive advantage in increasing lug height above this value, as taller lugs will deform under load and lose their bite into the soil.

The familiar chevron tread pattern has the advantage that it is self-cleaning under quite sticky conditions so that the tread bars can continue to bite into the soil when other tread patterns would become completely clogged with mud. Because the tread is more rigidly supported, radial tyres have been shown to increase traction by 5% or more, and to have a longer wear life. However, this rigidity reduces the self-cleaning ability of the tyre so that performance in sticky conditions deteriorates more rapidly than that of conventional crossply tyres.

The size of a tractor rear tyre is described by two dimensions (in inches). The first of these gives the maximum width of the tyre section, and the second indicates the diameter of the wheel rim on which the tyre is mounted. A 50 kW tractor can be fitted with 13.6×38, or 18.4×30 rear tyres, which are similar in overall diameter. However, the wider 18.4×30 tyre has a larger ground contact area, and will give a 5–15% increase in tractive performance. It is not possible to use wide section tyres for row-crop work, but for most tillage and haulage operations there is no restriction on tyre width.

Compaction

Soil compaction by heavy machines is a problem not only of rutting the soil surface, but also of compression of the soil itself which reduces pore space, inhibits water movement, and increases the formation of clods. Compaction, under any particular combination of soil conditions, is largely a function of ground pressure; the higher the pressure, the more dense the soil becomes, and the greater the depth to which compaction occurs. Ground pressure is reduced by spreading the weight of the machine over a greater ground contact area, using wider section tyres or dual wheels. Although the degree and depth of compaction is reduced, more soil is compacted by the wider contact surface. Minimum degree, depth and volume of compaction is achieved by a long, narrow contact patch, or by the use of wheels in tandem. Rolling resistance is also minimised by this practice.

Compaction is also increased under a wheel which is operating at high slip, the maximum effect occurring at a slip range of 15–25%. Since this is above the level at which maximum tractive efficiency occurs, it is advisable, for reasons of efficiency as well as reducing compaction, to operate a tractor at loads which only require a slip of 10–15%. Working rate is maintained by pulling these lighter loads at higher speeds.

The most effective method of controlling compaction is to use the lightest machines that will do the job, keep off the soil when it is wet, and reduce the number of passes over the field to the minimum.

Four-wheel drives

Driving all four wheels of a tractor offers a number of benefits:

(1) greater soil contact area produces a greater cohesive pull, which is particularly advantageous under wet conditions when friction is low;
(2) the pull is shared between four drive wheels, reducing each wheel's slip, compaction and sinkage;
(3) the powered front wheels give improved steering control in wet conditions;
(4) the tractor's brakes are effective on all four wheels.

The best tractive performance is produced by systems where all four wheels are the same size. Smaller front wheels, used in many conversions of two-wheel drive tractors, give a tighter steering lock but have smaller contact area and higher rolling resistance than full-sized front wheels. Larger equal-wheel tractors overcome the problem of poor manoeuvrability by using centre-pivot steering. The tractor bends to 45 degrees in either direction to give a small turning radius.

If the front wheels are to contribute fully to drawbar pull, they must carry as much weight as the rear wheels when the tractor is working. Where a nose-heavy weight distribution is not designed into the tractor, much front ballast weight must be added to keep the front axle fully loaded despite the effect of weight transfer.

The power take-off (pto)

Between 80 and 94% of the power of a tractor's engine can be transmitted through the pto shaft, as against only 50–70% through the wheels under average tractive conditions. There are two internationally standardised shaft sizes and speeds, a six-spline shaft turning at 540 rev/min and a 21-spline shaft turning at 1000 rev/min. Rotation is clockwise when seen from the rear of the tractor.

Since rotary power is the product of torque and rev/min, it can be seen that the faster pto speed can transmit almost twice the power at any given torque, or through a shaft of a particular size since the power-carrying capacity of a shaft is limited by the torque it can withstand. Many tractors have dual pto systems to accommodate all types of machines.

A fixed ratio between the engine and the pto sometimes allows the engine to develop its maximum power at the standard pto speed. Where this is not so, it is generally possible to over-speed the pto so that engine power can be maximised.

The pto is connected to the engine by its own clutch. If this can be operated quite separately from the transmission clutch, the system is called an 'independent' pto. A two-stage clutch pedal controlling the transmission at half depression and disconnecting the pto at full depression, gives a 'live' pto.

Hydraulic systems

Approximately 20% of a tractor's power is available through the hydraulic system. This is adequate for light work such as the operation of the three-point linkage, tipping trailers, most front-end loaders and a few light duty excavator attachments. Larger hydraulically operated machines such as high-lift loaders or hedge cutters must have their own hydraulic power units, driven by the tractor's pto.

Hydraulic power (in kilowatts) is a function of fluid flow rate and pressure, and is represented by the expression

$$\frac{\text{flow rate } (\ell/\text{min}) \times \text{pressure (bar)}}{600} = \text{power (kW)}.$$

Tractor hydraulic systems operate at pressures of 140–170 bar, with flow rates of 20–100 ℓ/min.

The tractor's three-point linkage is of standard dimensions and pin sizes (Category I, II or III or combinations according to tractor size). It is able to carry mounted equipment for transport, and to control the working depth of many types of soil engaging implements. Three alternative control systems may be available.

Draught control

This system adjusts the working depth of ploughs or other high-draught implements to maintain a constant draught or tractive load. Draught is sensed by springs in the upper or lower linkage; changes in draught cause the hydraulic system to raise or drop the linkage. A 'response' adjustment controls the rate at which these hydraulic corrections are made; slow response gives the smoothest work, but fast response may be needed if the ground is uneven.

Position control

This system will hold the linkage at a constant height relative to the tractor. This is generally used for transporting equipment in a raised position but may also be used to control the working position of some machines.

Pressure or traction control

A partial pressure is applied to the lifting ram. This takes a controlled amount of weight from a mounted implement (or a trailed implement if a special linkage is used), to increase the traction of the tractor rear wheels. The partial lifting of the implement must not affect its work.

Most tractors have available as options two pairs of external hydraulic couplings, controlled by separate double-acting control valves, and also a simple coupling which may be used to operate a tipping trailer. The double-acting valves may be used to control a front-end loader, rear fork-lift or other accessory, saving the cost of purchasing separate control valves for each attachment. On some tractors, the three-point linkage cannot be used while the external hydraulic system is in operation.

External hydraulic hoses are usually connected to the tractor by way of snap-on couplings, which will pull out and seal themselves if the machine becomes detached from the tractor. All couplings are a source of contamination to the tractor's hydraulic system, and care must be taken to avoid the entry of dirt when attaching and storing hydraulic accessories.

Health and safety for the tractor driver

Until recently, the tractor driver was subject to a number of risks from his occupation. The major ones were being crushed by an overturning tractor, and suffering hearing loss from long exposure to noise, particularly from the exhaust.

Tractor overturns most commonly occur sideways, when operating on steep slopes or driving too close to gulleys, ditches or steep banks. Heavy rollers or other trailed machines can push the tractor down hills. Backward overturns occur less frequently, usually from attempting to pull a heavy load from a high hitch point.

Since the introduction of BS approved safety cabs on all new tractors, the number of deaths from tractor overturns has diminished dramatically. There are almost no recorded instances of drivers being killed *inside* safety cabs. In an overturn accident, it is most important to hang on and stay inside the cab until the tractor has come to rest completely.

Tractors are noisy, typically exposing an unprotected driver to a noise level of 95–105 dB(A). In the short term, such noise levels can cause fatigue, and leave a ringing sensation in the ears when the noise has stopped. The long-term effect of exposure to high levels is to produce permanent hearing loss.

The ears can be protected by acoustic ear plugs or earphone-type protectors. Both are effective and cheap, but may not be comfortable, especially in hot weather. Quiet cabs fitted to new tractors must reduce the tractor's noise to no more than 85 dB(A) inside the cab. Because of the logarithmic scale used, a reduction

of 10 dB(A) represents a halving of the noise level to which the tractor driver is exposed.

Since much of the noise from a tractor is airborne, most of the effectiveness of the cab is lost if it is necessary to leave a door or window open, for access to implement controls or for ventilation in hot weather. Remote controls can eliminate the first problem, while the provision of refrigerated air-conditioning systems can (at a price) do away with the second.

CULTIVATION MACHINES

The operations which are necessary for the conversion of a field carrying the remains of a previous crop into a suitable seedbed to accept the following crop are traditionally divided into two categories. Primary cultivations are those involved in breaking-up the soil initially, followed by secondary cultivations which refine the soil to produce a final smooth, level, firm seedbed.

Subsoilers

On heavy or poorly-structured soils it is occasionally necessary to loosen the soil to a greater depth than that reached by normal cultivations, in order to improve drainage and root penetration. Subsoilers for this purpose can operate at depths from 300–600 mm, and at spacings as close as 1 m.

The subsoiler consists of one or more heavy vertical tines, with a replaceable point or foot. A knife-edged vertical tine is common, but a flat leading edge in front of a tapering tine requires a lower draught force, as does a tine which is angled forward at 45 degrees. Both these latter alternatives have the disadvantage that subsoil will slide up the flat front of the blade and be left on the surface. All tines, but particularly those which are angled forward, lift the soil into a bulge or 'surcharge' ahead of the tine. It is important that no depth wheel or other component is positioned where it will prevent this action from taking place.

Below a particular depth, called the 'critical depth', at 200–400 mm according to soil conditions, the tine will just cut a slit through the soil, rather than bursting it upwards in a wide V. The width of soil loosened by the subsoiler can be considerably increased with only a small increase in draught force, by the addition to the foot of horizontal wings 300 mm wide.

In some situations, a dry, hard surface crust may confine the underlying soil and prevent it from being loosened and burst upwards by deep tines. Short leading tines, positioned well ahead of the subsoiler tines, can break up the surface crust and therefore allow the subsoiler to be fully effective.

Ploughs

For centuries, the plough has been the main implement for primary cultivation. As well as loosening and breaking-up the ground, it inverts the top soil to bury weeds and trash. The soil loosening function can be performed by many other machines, but if soil inversion and burying is required, the plough must be used. In many situations, the value of inverting the top 200 mm of soil is being questioned, and some very successful cultivation systems are designed to keep the upper 100 mm, plus organic matter, on the surface by only cultivating to this depth. Similarly, the adoption of straw burning as a husbandry practice has reduced the need for a machine which can bury straw. The widespread use of herbicides for stubble cleaning and sward destruction has had a comparable effect.

The plough is made up of a number of bodies, each of which turns one furrow, mounted on a rigid frame. Each body covers a width of 300–400 mm; this is normally fixed, but in a few cases furrow width can be adjusted to suit changing field conditions. The total width of all the furrow slices turned by the plough constitutes its effective working width, which can range from 300 mm to over 4 m. Many different bodies are available to suit a full range of field conditions. A 'general purpose' body has a slow curvature which leaves the furrow slice intact, while a 'digger' body has a very abrupt curvature, to shatter the soil more effectively. 'High-speed' bodies reduce draught and leave neater work at high operating speeds.

Most ploughs can be fitted with spring-loaded release mechanisms which allow the whole body to fold back if it strikes an obstruction. This is particularly necessary where large ploughs are operated at high speeds. A small investment in protective devices can reduce the risk of long and costly delays and expensive repairs at a busy time.

The soil-engaging parts of the plough body are the coulter, a vertical knife or freely rotating disc, which makes a cut to divide the furrow slice from the unploughed ground, the share, which undercuts the furrow slice, and the mouldboard, which lifts and turns the furrow slice to leave it in its final inverted position. A small secondary body, or skim, may be used to shave off an upper corner of the furrow slice to ensure that no surface vegetation remains exposed. All soil-engaging parts are subject to wear, and are individually replaceable.

Since the plough body pushes soil to the side (to the right on conventional ploughs), it follows that the soil pushes the plough equally in the opposite direction. This side force is resisted by a flat plate, the landside, which bears against the vertical edge of the unploughed ground. A correctly adjusted plough exerts no side force on the tractor which pulls it.

While the construction of the plough sets the width of most of the furrow slices, the width of the front furrow slice is governed by the distance between the front body and the inside edge of the tractor rear wheel which is running in the previous furrow. The wheel must be set at the correct width from the tractor's centre line, as specified by the plough manufacturer. A further adjustment can be made to the plough itself, to steer it closer or farther from the tractor wheel until the front furrow slice is at the correct width.

Ploughing depth, which should not exceed two-thirds of the width of the furrow slice if the slices are to turn satisfactorily, may be set by means of an adjustable depth-wheel, if one is provided. More commonly, the tractor's hydraulic draught control system is used. The use of a depth wheel gives more accurate control, and is not affected by changes in soil conditions, but the use of draught control without a depth wheel puts more weight on the tractor's rear wheels. This improves traction and also eliminates the extra rolling resistance of the trailed wheel. Long ploughs, of four furrows or more, often have a depth wheel at the rear to supplement the draught control system.

In addition to the side-force difficulty mentioned earlier, the one-sided action of the plough poses problems in working a field, since it cannot simply be drawn up and down like most other machines. The field must be marked out in 'lands', which are then worked separately. At the centre of the land a ridge is formed from the soil turned inwards by one or more passes of the plough in each direction, and then the plough works up and down each side of the ridge, 'gathering' the soil, until the land is ploughed. The finished field shows a ridge at the centre of each land, and a double furrow between adjoining lands. The furrows, in particular, can affect subsequent operations carried out in the field at least to the harvesting of the next crop.

The problems of unproductive time spent marking the field out into lands, time wasted travelling along the headland from one land to the next, and the uneven surface produced by the conventional plough have led to the development and widespread adoption of the reversible plough. In this machine a second set of bodies, which turn the soil to the left, is mounted on the beam in opposition to the right-hand set. The front end of the beam is attached to the tractor's three-point linkage by a headstock which can rotate the plough through almost 180 degrees, when it is raised, and thus transpose the two sets of bodies. By using alternate sets of bodies, the reversible plough can work across a field from one side to the other with no marking-out, no ridges or furrows, and a minimum of idle time on the headland. A higher rate of work can be expected as a result of the better use of time, although the heavier plough will be harder to pull and will cost about 60% more than a conventional plough with the same number of bodies. The extra weight gives good penetration into hard soil, but the extra complexity of the rotating mechanism and headstock can sometimes cause trouble.

Rates of work of (0.8–2.0 ha h)/100 kW rated tractor power can be achieved, depending on soil conditions and working depth. Plough and tractor should be matched so that the power of the engine is fully utilised at a speed of 5–8 km/h. Fully loading the tractor at slow speeds causes high losses due to wheelslip, while operating at very high speeds puts great strain on the plough and leads to high draught forces, as draught increases with the squared power of forward speed.

The chisel plough

This machine is not a plough at all, but was hailed as a replacement for the mouldboard plough when the first units were imported from the USA. A heavy, wheeled, frame carries a number of curved, spring-mounted tines with replaceable points. The tines, which are spaced out on three or four crossbars of the frame to give good clearance for trash, break the soil at intervals of 300–500 mm across the working width of the machine. Working depth is controlled by adjustable wheels, which also can carry the heavier machines when out of work.

Comparable home-produced machines have rigid tines, often protected by a shear-bolt. Most tines are flat-fronted and some are raked forward, as discussed under 'subsoilers'. Rates of work are double those which can be attained with a mouldboard plough, but at least two passes are necessary to achieve the same amount of soil loosening. There is little inversion, or burying of surface material.

The disc plough

The disc plough resembles a mouldboard plough in layout, but the bodies have been replaced by angled, inclined, free turning concave discs. Large stationary scrapers are fitted to prevent soil build-up on the discs and to increase the turning action on the soil.

The soil is loosened and mixed, rather than inverted; where erosion is a problem, partially-buried crop residues can be very effective in binding and stabilising the soil.

The machine is difficult to adjust, and the work produced does not look like conventional ploughing. Where obstacles such as roots or rocks abound, the discs avoid damage by rolling over the obstructions rather than catching under them.

Although not much used in this country, the disc plough and its derivatives are widely accepted in areas where their special properties can be used advantageously.

The rotary digger

Unlike the cultivation machines mentioned so far, which are pulled along, the rotary digger receives its power to cultivate the soil from the tractor pto. This is a very efficient method of transmitting power and is little affected by wet soil conditions which can at times make traction difficult or impossible.

A heavy 2.4 m wide horizontal spindle, rotating at 50–120 rev/min carries L-shaped blades at a diameter of 760 mm. As the machine moves forward, the blades penetrate the ground to a depth of 200 mm and tear out chunks of soil 150–250 mm in length. A rate of work 1.5 times that for conventional ploughing with the same tractor can be achieved, and a tractor of 60 kW can produce an output of 1 ha/h. There is good soil inversion in wet conditions, and very little smearing.

When set to a long bite length, the machine generates a forward push of 1.5–2.0 tonnes force. This may be absorbed by three chisel tines at 300 mm depth or two subsoiler tines at 400 mm depth. Combining two operations in this way saves both labour and energy, while a comparison with separate ploughing and subsoiling shows an energy saving of almost 50% and a labour saving of 70%.

While the rotary digger does not do the work of a plough, one pass can move as much soil as two passes with a chisel plough. When tines are fitted very good soil mixing is achieved.

The rotary cultivator

The rotary cultivator, which is similar in layout to the rotary digger, may be used for primary or secondary cultivation work. The standard rotor carries right- and left-handed L blades and turns at 120–270 rev/min to produce either a coarse or a fine tilth. A rear hood may be raised to allow clods to be thrown out, or lowered to give a further shattering effect as the clods strike the inner surface of the hood. Working depth, to 200 mm, is controlled by a land wheel or a rear crumbler roller. An alternative spiked rotor may be used for seedbed preparation, and a bridge link may be used to operate a drill directly behind the machine to save time and labour.

If the ground is hard or stony, some machines can be fitted with front loosening tines which reduce rotor power requirement and extend blade or spike life. The chopping and mixing action of the rotary cultivator is ideal for the incorporation of crop residues, fertilisers or chemicals into the soil. Repeated use of the rotary cultivator may break the soil down into fine dust, and there is some risk of polishing the underlying uncultivated soil if conditions are hard. The machine has a high power requirement of 15–30 kW/m of width, but where one pass of the machine can replace several passes with alternative machines this will be acceptable.

The spring-tine cultivator

This is a versatile machine for secondary cultivations. A grid frame 1.5–10 m wide carries a number of S-shaped spring tines distributed over the grid so that one tine passes through each 100–200 mm strip of soil. Replaceable points of various widths are available. The machine is normally tractor-mounted, but is carried in work by adjustable depth control wheels. Machines above 3 m in width are made up of a central section with two hinged wings which fold for transport.

The spring tines vibrate as the machine moves forward, which is very effective in shattering clods. Trash and hard clods are brought up to the surface. The machine leaves the soil furrowed at about 600 mm intervals, corresponding with the spacing of the last row of tines. Crumbler rollers or light harrows, available as extras to fit on the rear of the machine, will produce a smooth surface.

The spring-tine cultivator has a good mixing action at speeds above 7 km/h and may be used for the incorporation of soil chemicals. Two passes are recommended, at an angle to each other.

The disc harrow

The tandem disc harrow is made up of four 'gangs' of concave discs, each gang clamped to an axle which is free to turn in sealed bearings. The front pair of gangs is angled to turn soil outwards, while the rear gangs draw soil inwards. Seen from above, the four gangs form a wide X.

An alternative layout, known as the offset disc harrow, has only one front and one rear gang of discs, each of which extends across the full width of the machine. The gangs meet at one side and are separated at the other to achieve the same relationship of angles and disc orientation as one-half of the tandem machine.

The front gangs, which must penetrate firm soil, are often fitted with scalloped discs. The rear gangs use plain discs, which last longer and move more soil.

The action of the disc harrow is to cut downwards into the soil, and to turn and mix the material thus loosened. Clods and subsoil are not brought to the surface. Two fast passes of the machine can effectively incorporate chemicals into the soil.

Penetration may be increased by increasing the angle of the discs to the direction of forward travel, but a more effective method of increasing penetration is to add weight to the frame of the machine.

Tractor mounted disc harrows are available but, since much of the penetrating effect depends on

weight, heavier trailed machines are better able to deal with tough or hard soil conditions. Large units can perform primary cultivations, particularly in lighter soils.

These large machines are available in widths up to 4 m, with discs of 760 mm diameter, and weighing 1 tonne/m of width. A machine 4 m wide, pulled by a tractor of 120 kW, can cultivate at rates exceeding 3 ha/h.

Pto-driven secondary cultivation machines

In addition to the rotary cultivator, described above, other power-driven machines are available for secondary cultivation work. Many use vertical spiked tines rotating in pairs about a vertical axis, or reciprocating across the width of the machine. The speed of the tines is much greater than the forward speed of the tractor and their shattering action is thus more effective than that of rigid tines being pulled through the soil. The action of the tines also levels and compacts the seedbed, without raking up much buried trash or subsoil. Bridge links may be used in conjunction with some of these machines, to combine the operations of seedbed preparation and drilling.

Power requirement is 15–30 kW/m of width. Tine life can be short if soils are abrasive and this can increase the operating cost of the machine. Where one pass of the powered machine can replace several passes with conventional machines, this cost is likely to be justified.

Harrows

A harrow consists of a large number of small tines or spikes carried on rigid or flexible (chain) frames. As they are dragged along, the action of the tines is to sort, level and compact the seedbed, the degree of penetration depending on the weight of the frame and the size of the spikes. Harrows are also used after drilling to ensure seed is covered. Chain harrows may be used on grassland to break up matted swards or to spread dung after grazing.

The power requirement of harrows is very low, so it is uncommon for any tractor to be fully loaded by a set of harrows of normal width. Harrows may be used in combination with other machines—spring tine cultivators, seed drills—to achieve two operations in one pass.

Rollers

Ridged, or cambridge, rollers, made up of a number of ribbed cast iron wheels on a free turning axle, are often used in seedbed preparation to crush clods, compact the soil and leave a smooth finish. The compacting effect of a roller depends on its weight, decreases with increasing roller diameter and decreases with increasing forward speed. Since cambridge rollers weigh 300–400 kg/m of width, and are used at fairly high speeds, their compacting effect is generally small and confined to the top few centimetres of soil only. Light, rigid-tined implements can be just as effective in increasing soil compaction, and do not leave a tight 'skin' at the surface.

Cambridge rollers are available in widths up to 7 m, the larger sizes being made up of a central section with two separate smaller units trailed from either end. The units are hitched in line for transport.

Smooth rollers are commonly used for levelling grassland and pressing-in stones in spring to prepare for hay or silage harvesting later in the season. Rollers up to 3.6 m in width are available. By adding ballast to the hollow cylindrical rollers, weights from 0.5–1.3 tonnes/m width can be applied.

Rollers weighing up to 5 tonnes or more can sometimes take control on steep ground. Care must be exercised in matching the roller to a tractor of adequate weight and in operating safely on hillsides.

Some secondary cultivation machines may be fitted with a 'crumbler' roller. This is an open-cage roller, with a surface of spaced straight steel rods held in position by two or more integral wheels. The whole unit is free to turn in bearings at either end. As the roller moves forward, its weight is concentrated successively on each steel rod, which is quite effective in crushing clods at that point. There is also a good levelling effect.

Combination seedbed-preparation machines

Since many seedbed-finishing operations demand little power, a number of manufacturers have developed machines which combine several operations and can thus utilise a tractor's power more completely. Such items as ridged rollers, rigid or spring tines, crumblers or toothed rollers may be combined in such a way as to lift out and crush clods repeatedly, leaving a fine and level seedbed.

Reduced cultivations

Where a deep seedbed is not necessary (as it is for many root crops, for example), there can be savings in energy and increases in work rate if the soil is only cultivated to a depth of 100 mm. This system has been used successfully, even in heavy soils, for cereal production, without loss of yield. In many situations this practice has also led to long-term improvements in the structure of the upper soil layers.

The soil may be worked to 100 mm depth by means of a heavy spring tined cultivator or a rigid-tined cultivator, with tines spaced at 200 mm overall. Adjustable wheels control working depth. Three or four passes are needed before drilling in heavy soils.

Combination machines, generally using gangs of discs in conjunction with banks of heavy spring-tines, can reduce the number of passes needed to produce a satisfactory seedbed. A recently developed example, 3 m wide, requires a tractor of 75–105 kW, and can cultivate up to 2.5 ha/h. Two or three passes are generally sufficient to produce a seedbed.

FERTILISER APPLICATION MACHINES

Ninety per cent of the nation's fertiliser is applied as dry granular, prilled, crystalline or powdered materials. In some areas liquid fertiliser solutions are available; they have the advantage of being handled easily by pumping, but the disadvantage that they are generally less concentrated than solid fertilisers and thus more material must be handled in order to apply a given weight of nutrients to the land. Also, a storage tank must be installed on the farm to hold at least part of a year's supply of fertiliser.

Liquid fertiliser is applied by spraying machines very similar to those described below under the heading 'Crop sprayers'.

Dry fertiliser may be applied by a combine drill, simultaneously with planting the crop. Then it is not practicable to apply fertiliser at high rates, because the small carrying capacity of most combine drills would necessitate frequent refilling stops.

Broadcasters

Most dry fertiliser is applied by broadcasting machines. A hopper of 250–1000 kg capacity on mounted machines or up to 8 tonnes on trailed machines discharges fertiliser onto a spreading mechanism comprising a single or double spinning disc, or an oscillating spout. If pto speed is correctly set, the fertiliser can be spread to an effective width of 5–14 m, according to the particular machine and the material used. The material is distributed in a pattern which is heaviest directly behind the machine, gradually reducing to zero at a distance of 4–8 m to each side. As long as this pattern is symmetrical and the reduction of rate is constant with increasing distance from the path of the machine, a return bout at the correct spacing will produce an even application rate across the complete area. Light materials are not spread as widely as heavy ones, so that it is extremely important to follow the manufacturers' bout width recommendations for the type of fertiliser being applied.

Application rates of 20–2500 kg/ha are possible with broadcasting machines, at speeds up to 12 km/h. Application rate is controlled by the rate at which fertiliser is metered onto the spreading mechanism, and by forward speed. Some trailed machines have land-wheel driven metering devices which makes the application rate independent of forward speed.

The machine can be calibrated at a particular setting by operating it, stationary, indoors for a

Table 24.1 Time taken to cover 1 ha, according to working width and forward speed (min)

Effective width (m)	*Speed* (km/h)								
	5	6	7	8	9	10	12	14	16
3	40	33.3	28.6	25	22.2	20.0	16.7	14.3	12.5
4	30	25	21.4	18.8	16.7	15.0	12.5	10.7	9.38
5	24	20	17.1	15.0	13.3	12.0	10.0	8.57	7.50
6	20	16.7	14.3	12.5	11.1	10.0	8.33	7.14	6.25
7	17.1	14.3	12.2	10.7	9.52	8.57	7.14	6.12	5.36
8	15	12.5	10.7	9.38	8.33	7.50	6.25	5.36	4.69
10	12	10	8.57	7.5	6.67	6.00	5.00	4.29	3.75
12	10	8.3	7.14	6.25	5.56	5.00	4.17	3.57	3.13

Notes:
(1) Non-productive time for such operations as filling hoppers, adjustment or turning is not included.
(2) Combinations of width and speed not shown in the table may be evaluated by means of the formula
   $t = 600/(w \times s)$ where t is the time in min to cover 1 ha, w is the working width in m and s is the speed, in km/h.

measured time, collecting and weighing the fertiliser delivered. The time needed to cover 1 ha at a known speed and spreading width (*Table 24.1*) can be used to convert the delivery rate/min to that /ha. This test can also show up any difference in the amount of fertiliser thrown to right and to left, although the exact spread pattern cannot be determined.

In addition to the factors of correct pto speed and forward speed, good condition of spreading mechanism, correct height above target and machine set level, the accuracy with which the operator maintains his required width of spread is critical to accuracy and evenness of application. Foam marker nozzles on a boom are very satisfactory, and the same attachment can also be used on a sprayer to spread the cost, which only amounts to a few pence/ha. Far more than this amount can be wasted by incorrect fertiliser application.

Fertiliser, particularly in combination with water from the atmosphere, can be very corrosive to mild steel, iron or aluminium components. Manufacturers use corrosion resistant materials such as plastics or stainless steel for some parts of the broadcaster, but it is important also that the machine be easy to dismantle and clean at the end of the season. Vulnerable components can be brushed or washed clean, and coated with oil to protect them in good condition for the following season.

Full-width spreaders

The broadcaster is an inexpensive machine which can be very accurate if calibrated often and operated correctly. However, the variation of application rate with width, and the difficulty of maintaining the correct bout width have led manufacturers to develop much more sophisticated machines which spread evenly over a known and constant width. Most of these machines take the form of a central hopper (mounted to 1 tonne, trailed to 4 tonnes) with a metering mechanism, carrying a folding boom up to

12 m wide along which the fertiliser is conveyed by an air blast. The fertiliser emerges from a series of nozzles, spaced as close as 600 mm in some cases, to form an even overlapping pattern across the full width of the boom.

Where the metering is by a wheel-driven force-feed mechanism, the machine may be calibrated by rotating the mechanism by hand for a specified number of turns and collecting the fertiliser in the tray provided. These machines can be extremely accurate, at application rates from 1 kg/ha for certain granules or seeds, up to 2 tonnes or more per ha. However, the machine must be in good condition throughout, pto speed must be correct, and spacing must be within 0.5 m of the correct width. Inaccurate driving can leave strips without any fertiliser, or apply double rates. Again, the advantages of using a marker device are very clear.

PLANTING MACHINES

The objective of any planting operation is to produce the desired population of vigorous, healthy plants.

Grain drills

These machines can plant a range of seeds from grass or clover to beans. Seed is discharged in rows, in an even trickle, but not as individual seeds. Combine drills also deliver fertiliser, generally to lie close to the seed in the row.

Seed (and fertiliser) hoppers extend across the full width of the machine, which ranges from 2–6 m. Carrying capacity is 100–120 kg/m width on mounted machines, 220–300 on trailed models. Where a combined grain and fertiliser hopper is provided, it is often possible to reverse a central divider to give ratios of either 1:1 or 1:2 in weight of grain to weight of fertiliser, according to application rate. Removal of the divider allows the full hopper to be used for grain.

Table 24.2 The number of seeds (or plants)/m² according to row width and spacing within the row

Within-row spacing (mm)	*Row width* (mm)										
	110	*120*	*125*	*130*	*135*	*140*	*165*	*170*	*175*	*180*	*190*
10	909	833	800	769	741	714	606	588	571	555	526
15	606	556	533	513	494	476	404	392	381	370	351
20	455	417	400	385	370	357	303	294	286	278	263
25	364	333	320	308	296	286	242	235	229	222	211
30	303	278	267	256	247	238	202	196	190	185	175
35	260	238	229	220	212	204	173	168	163	159	150
40	227	208	200	192	185	179	152	147	143	139	132
Metres of row/m²	9.09	8.33	8.0	7.69	7.41	7.14	6.06	5.88	5.71	5.55	5.26

Seed is dropped in rows 120–135 mm or 175–180 mm apart, the number of rows per machine being from 15 to 50. Narrower rows put seed further apart in the row (*see Table 24.2*), but make the machine more complicated and expensive. Closely spaced coulters are more prone to blockage, and less weight can be applied to each one if penetration is difficult.

The metering mechanism is driven from one of the ground wheels. There is one metering roller per row. As the roller turns in the bottom of the hopper, serrations grip the seeds and carry them out to a position where they are discharged down the seed tube. The seed rate is adjusted by changing the speed of the roller in relation to forward speed, or by changing the proportion of the roller exposed to the seed.

To deal with different sizes of seed, the internal force feed mechanism has alternative openings for small or large seeds. The unwanted opening is closed off by a hinged flap. Some manufacturers can supply alternative sets of metering rollers with small or large serrations, for the same purpose.

Individual rows can be shut off by slides in the bottom of the hopper. Thus a crop such as rape can be planted in wider rows than those used for corn, by shutting off two out of every three rows. Small internal hoppers can concentrate the seed over only those rows that are being planted. Automatic or semi-automatic 'tramlining' attachments can be fitted to close off certain pre-selected rows every two, three or four passes across the field, to leave unseeded strips for further passes of sprayers or fertiliser machines.

Most modern drills are supplied with calibrating trays to collect the seed, and handles to operate the mechanism, for the equivalent of a known area so that the correct setting for a particular batch of seed can be established before drilling begins. Static calibration does not however take into account the effect of forward speed or of wheelslip which can vary from 0.5–15% according to soil and tyre conditions. A final calibration where the drill is run at full planting speed over a measured area (*see Table 24.3*) can ensure that the amount of seed delivered is correct under field conditions.

Seed population can be further checked after drilling, and plant population after emergence, with reference to *Table 24.2*.

Seed is released by the metering mechanism up to 600 mm above the ground, and falls downwards through a telescopic or concertina tube. The tube spreads out the flow of seed from small groups which leave some types of metering device, into an even stream. Fertiliser or seed dressing can build up inside these tubes; the action of flexible tubes tends to dislodge these deposits as the coulters move up and down.

The seed is placed in the ground by the coulter. Ideally, each seed will be planted at the same depth, and surrounded by firm, moist, warm soil. Where the soil has been thoroughly prepared, either single-disc or Suffolk coulters are used.

The single curved-disc coulter uses an angled disc to cut a groove in the soil, and the seed drops into this groove. Some soil falls back onto the seed after the coulter has passed. The disc coulter can cut through loose surface trash, but stones can push it up out of the ground, resulting in uneven planting depth.

Table 24.3 Calibration distances for machines of different widths

Working width (m)	*Distance* (m) *to cover*	
	$\frac{1}{10}$ ha	$\frac{1}{25}$ ha
1	1000	400
1.5	667	267
2	500	200
2.5	400	160
3	333	133
3.5	286	114
4	250	100
5	200	80.0
6	167	66.7
7	143	57.1
8	125	50.0
9	111	44.4
10	100	40.0
11	90.9	36.4
12	83.3	33.3

The Suffolk coulter is a fixed blade, shaped like the front of a boat. It presses a groove in the soil, into which the seed falls. It can work in stony ground, and can give very even planting depth in well-cultivated soil. If loose surface trash is present, the Suffolk coulter will rake it up and frequent blockages will reduce the rate of work.

Both types of coulter are carried on spring-loaded arms; increasing spring tension will increase planting depth. Individual coulters can move vertically to follow uneven ground. In both cases, there is advantage in using a following harrow to improve seed cover; a trailed harrow is ungainly to use, and many manufacturers can supply integral spring-tined harrows for use with their drills. Where additional compaction is required, a very few makes of drill can be fitted with a heavy narrow press wheel behind each coulter.

In order to match-up adjacent passes of the machine, most drills can be fitted with disc markers which leave a small furrow that is followed by the front wheel of the tractor at the next pass across the field.

Direct drills

With the increasing use of direct drilling, several manufacturers now produce machines capable of planting seed into uncultivated land.

Most machines use a standard hopper and metering mechanism, but these are mounted on a very heavy chassis, with provision for carrying additional weight to increase penetration of the coulters if ground conditions are particularly hard. Special coulters open the ground and place the seed.

A popular coulter is the triple-disc type. A single vertical disc is followed by a pair of discs in V-formation. The tip of the V is ahead of the lowest point, so that the soil is pushed apart as the coulter moves along. Seed and fertiliser fall from the seed tube into the groove; there is no covering device.

The coulters are carried on independent spring arms; loading can be increased by lowering the frame of the drill onto the springs, or by applying hydraulic pressure to the top of the springs through an elaborate equalising linkage. There are many alternative types of coulter, using discs, disc and tine combinations, cultivator tines or narrow rotary cultivators to open up the soil. Spray nozzles may be positioned so that a strip of vegetation can be killed along the line of each row.

Because the direct drill is a heavier machine than the conventional version, rates of work are slower and power requirement is higher. However, since many operations can be eliminated by the practice of direct drilling, the overall requirement for labour and fuel/ha is reduced to less than 20% of that for conventional cultivation and planting systems.

Precision seeders

Where crops are grown as individual, separate, spaced plants, it is necessary to plant seeds singly, rather than in a stream. This demands a metering mechanism which is capable of picking out individual seeds.

The most common metering device is the cell. Cells of the correct size to match the seed (which should itself be graded or pelleted to improve accuracy) are carried in the outer rim of a wheel, or as perforations in a flexible belt. The cells pass under the seed as it lies in a small hopper, and the seeds drop into the cells by gravity. After removal from the hopper, the seed falls from the cell to the ground by gravity, sometimes aided by an ejector. The seeds may be thrown rearwards as they leave the seeder, to counteract the machine's forward speed and reduce the tendency of the seeds to roll or bounce along the ground.

The most satisfactory results are achieved at low speeds (3–5 km/h), since this allows a sufficient time for the seeds to find their way into the cells. With increasing speed, more and more cells remain unfilled, and the number of gaps in the row of seeds increases.

An improved design of cell-wheel planter separates the two functions of selecting and feeding the seeds. A slow moving many-celled selector wheel turns in the bottom of the hopper. The seeds are delivered singly into a fast-turning feeder wheel, which ejects the seeds at ground level, with a rearwards velocity equal to the forward speed of the machine. Satisfactory precision planting is possible at speeds of 10–12 km/h.

More positive seed selection is achieved by the vacuum seeder. The vacuum is provided by a pto driven fan. Single seeds are sucked into a ring of small perforated depressions on one side of a thin circular disc. As the disc rotates, the seeds are retained and lifted out of the hopper. An adjustable finger displaces any doubles. When the seed is directly above the coulter, the vacuum is cut off and the seed drops to the ground, again with a rearwards impetus in many cases. Some vacuum machines can plant seeds accurately at 8–10 km/h. One size of disc can meter a considerable range of seed sizes, so the time and cost normally involved in changing from one set of cells to another between crops is much reduced.

The metering mechanism is driven from a ground wheel on each seeder unit (the cheapest system to buy) or from a master landwheel which drives all units. Changing the drive ratio, by means of stepped pulleys or gears, changes the rotation of the mechanism in relation to the ground covered, and thus the seed spacing. On most machines, cell wheels or belts can be changed for ones with space for more or less seeds per turn, which will also change the spacing.

Once the machine has been prepared for work, by matching cells to seed size and by adjusting drive ratios, it is important to check performance in the field, at normal planting speed. A portion of every row should be uncovered so that spacing and evenness may be assessed, and corrected if necessary. *Table 24.4* shows the relationship between spacing, row width and plant population.

Seed population must not be confused with plant population, which will be lower in proportion to the emergence percentage of the crop involved. The figure for sugar beet is 60–65%; extra seeds must be planted, to allow for this loss.

Table 24.4 Plant (or seed) population (10^3/ha) according to row width and spacing

Spacing (mm)	Row width (mm)					
	400	500	600	700	800	900
100	250	200	167	143	125	111
125	200	160	133	114	100	88.8
150	167	133	111	95.2	83.3	74.1
175	143	114	95.2	81.6	71.4	63.5
200	125	100	83.3	71.4	62.5	55.6
250	100	80.0	66.7	57.1	50.0	44.4
300	83.3	66.7	55.5	47.6	41.7	37.0
400	62.5	50.0	41.7	35.7	31.3	27.8
km of row/ha	25.0	20.0	16.7	14.3	12.5	11.1

Precision seeders are made up of a number of single-row units, each with its own seed hopper, attached by flexible links to a tractor-mounted toolbar. By moving the units along the toolbar, row widths down to 200 mm are possible on some machines. Closer spacings are possible by staggering units in two rows. The number of rows of the seeder should be a multiple of the number of rows to be harvested simultaneously, for example a crop to be harvested by a three-row harvester should be planted by a 6, 9, 12, 18 or 24 row planter. This avoids the risk of misalignment which might occur if the harvester had to overlap two passes of the seeder.

The seed is released very close to the ground, so that the spacing is not affected by the presence of a seed tube. This gives the machine very little ground clearance; soil must be smooth, trash free and well cultivated. A typical unit has a front depth wheel, followed by a boat-shaped coulter which presses a groove in the soil. The coulter may be lowered to increase planting depth. Behind the coulter is a covering device, followed by a second depth wheel.

Accessories are available on some machines to push aside loose dry clods, or to level and compact an uneven tilth. Angled discs or blades can draw soil over the seed, and heavy narrow press wheels can improve compaction round the seed. Use of the correct accessories can ensure fast, even germination under a range of soil conditions, but care must be taken not to disturb the seed spacing with any following treatment. Seeder units may also be equipped for the application of liquid or granular chemicals.

Seeder units are very compact and close to the ground, so that it is difficult for the operator to see whether or not they are working satisfactorily. Some machines can be fitted with simple electric monitors to show whether the mechanism of each unit is turning, or whether the seed hoppers are empty. A more sophisticated device causes a light beam to be broken by each seed as it is released. Any interruption in seed flow sets off an alarm, so that the fault can be identified and corrected immediately.

Fluid seeding

This technique, developed at the National Vegetable Research Station, makes it possible to plant seeds which have been pre-germinated under controlled conditions. This can give very even, early emergence of the crop, although it is not precision seeding.

The seeds are pre-germinated in a liquid medium in a temperature-controlled tank. A 'gel' containing the chitted seeds is then planted through a small nozzle, delivered at a controlled rate by a peristaltic pump. The concentration of seeds in the solution, and the output of the pump, may be adjusted to give the desired seeding rate. Because of the number of operations and machines involved, this is a high cost system which must pay for itself by producing better yields of valuable or difficult crops.

Potato planting machines

The potato planter must deal with a large weight of 'seed' material, 2–3.5 tonnes/ha. The material must be handled carefully, especially if chitted seed is used.

Since the crop is grown in a ridge, a deep tilth must be prepared before planting to allow the ridges to be formed. In some cases, ridges are formed before planting but most modern potato planters form ridges from flat ground as part of the planting operation. It is important at all stages of production to avoid any activity which will press the soil into clods, as these are difficult to separate from potatoes at harvest time. If potatoes are to be grown in stony soil, it is desirable to pick up and crush or remove stones from the topsoil, or to gather them into windrows which are buried away from the crop.

Hand-drop planters

These machines, which generally plant two rows at a time into ridged land, are based on the mouldboard type ridger. Seats are provided on the planter for one operator per row, and seed is carried in trays or a hopper. A bell, operated by cams on a ground wheel, sounds at regular intervals of forward travel, and the operators drop a potato into a tube at each ring. The potatoes fall into an open furrow and are then covered with soil as the machine moves slowly forward. A two-row machine can plant 1.25 ha/d.

Semi-automatic planters

The task of the operator is less onerous on this machine as he fills up a series of small trays or cups as they move. These containers subsequently release the tubers at the correct intervals as the machine moves along. One operator per row is still required.

Automatic planters

Operators are not required on automatic planters, as a series of cups or fingers picks up the tubers out of a bulk hopper (375–500 kg capacity for a two-row planter) and releases them close to the ground at the correct spacing. A coulter makes a small furrow to receive the seed, which is then covered by disc or mouldboard ridgers.

The metering mechanism is a single or double row of cups on an endless belt or chain, or a series of spring-loaded fingers on the face of a large disc. Some machines have a make-up device which can add a potato to the metering device if an empty cup is detected. Feeding rates of up to 500 tubers/min are achieved on some cup-feed machines, giving ground speeds to 6.5 km/h, or a rate of work of 1.4 ha/h for a four-row machine. A more typical figure for a four-row machine is 5 ha/d.

Not all automatic planters are gentle enough to deal with chitted seed, but most manufacturers produce special models for this purpose. Potatoes with short chits are the most susceptible to damage.

An alternative method of metering is to feed tubers onto a pair of flat belts at a rate which varies with tuber size to give a constant weight of tubers/ha at a variable spacing. Irregular spacing may reduce yields slightly, but the machine is gentle and can plant chitted seed at up to 11 km/h.

Because of the large weights of seed to be handled, as well as the fertiliser which is often applied by the planter, an efficient transport system is necessary if working rates are to be maximised.

Transplanters

Many vegetable and nursery crops are traditionally planted out as transplants. Bare-rooted transplants are planted by semi-automatic machines. The operator, seated on each single-row unit, selects plants and feeds them roots-up into a gripping device which may consist of a pair of flexible discs or a series of spring-loaded fingers. As this device rotates, it places the plant roots in a furrow which has been opened by a leading coulter, and releases the plants at the correct spacing. A pair of inclined wheels under the operator's seat gathers and firms the soil round the plant roots.

Spacing is adjusted by changing the gearing between the ground-wheel and the planting mechanism, or by changing the number of plant-places on the mechanism. Individual units can be spaced as closely as 600 mm; narrower rows can be planted by two or more banks of units. As with precision seeding, this operation demands a smooth, even soil surface with a good tilth.

Rates of work of up to 1500 plants/h per operator are possible; a five-row machine operated by a team of six, and well serviced with supplies of plants, can plant 1 ha of cabbages in 7 h, or a 1 ha of lettuce in 13 h.

Block transplanters

To avoid the growth check which occurs when quite large plants are uprooted, handled in boxes for a while and then re-planted, many growers are considering transplanting very small plants growing in small blocks of peat compost.

The transplanter presses square studs into the ground, and operators drop individual blocks into the depressions thus produced. Rain or irrigation washes the soil closely round the block, and growth continues without a check. Because the blocks are handled as well as the plants, materials handling becomes an even larger part of the operation. The use of smaller blocks combats the problem to some extent. Automatic systems for seeding and transplanting peat blocks are under development.

CROP SPRAYERS

Most chemicals are applied as liquid solutions or suspensions, generally water-based. The liquid is applied as fine droplets to give an even cover (*Table 24.5*) and because small particles are more likely to be retained on the leaf than large drops, which can roll off (*Table 24.6*).

Table 24.5 Effect of droplet size on closeness of spray cover

Size of droplet (μm)	Number of droplets/cm² at 45 ℓ/ha application rate
400	13
250	55
150	254
50	1750

Table 24.6 Retention of spray droplets on barley leaves

Size of droplets (μm)	% retained on leaf
300	15
200	35
100	80+

However, very fine drops fall slowly through the air, and are therefore subject to drift (*Table 24.7*).

Table 24.7 Effect of droplet size on spray drift

Size of droplet (μm)	Rate of drop in still air (m/s)	Drift when dropped from 0.5 m, in a 1.5 m/s wind (m)
250	1	0.75
100	0.5	2.7
70	0.25	5.4
50	0.125	10.8
10	0.005	270

Very fine droplets can also lose weight by evaporation in dry weather, which can compound the problem of drift.

Apart from the problem of drift, it can be seen that reducing droplet size offers two major advantages:

(1) more close coverage of plant or soil. A more concentrated spray solution can be distributed evenly and a tankful of spray will treat a greater area;
(2) better retention on leaves, so that a reduced application of chemical is necessary to give the required protection. There is less risk of the build-up of residues in the soil.

Reduced application rates are suggested, according to droplet size, as follows: 250 μm droplets, 20 ℓ/ha; 60 μm droplets, 1–2 ℓ/ha; 30 μm droplets, 0.1 ℓ/ha. The chemical would however be applied at a higher concentration than at present.

Most field sprayers are fitted with hydraulic nozzles. When supplied with spray solution under pressure (2–4 bar) these nozzles produce a fan or cone of droplets. As pressure increases so does output, and finer droplets are formed. The typical range of droplet sizes produced simultaneously, is 10–700 μm. The bulk of the nozzle's output is in droplets of around 400 μm, and the chemical concentration is such that one 'dose' of chemical is carried in a droplet this size. Application rates are from 60–350 ℓ/ha. Clearly, if the bulk of the nozzle's output was in the form of smaller droplets, spray concentration could be increased so that each of these smaller droplets could carry a full 'dose', with resulting reductions in the amount of water to be transported and applied. The chemical manufacturers' recommendations must be followed.

Sprayer components

The tank, generally of a plastic material to resist corrosion, is of 200–1000 ℓ capacity on mounted machines, and up to 3500 ℓ on trailed models. The contents can be agitated by a pressure jet or by the more positive mechanical paddle. Agitation is especially important where suspensions are applied. The tank has a filler opening and strainer on top. Since the top of the tank may be quite high, many machines

also have a low level hopper from which the chemical can be washed into the main tank via the pump. The tank is fitted with a fine strainer at its outlet to protect other components from blockage.

The pump is of positive displacement, delivering 30–150 ℓ/min at 540 rev/min, at pressures of 2–4 bar. Roller-vane pumps are cheapest but are subject to wear; diaphragm pumps are used on the majority of machines. Piston pumps are available, and can operate to pressures of 40 bar.

An adjustable pressure regulator maintains the correct spray pressure, and allows excess liquid to return to the tank once a pre-set pressure has been built up. Some pressure regulators contain a forward-speed measuring device which increases or decreases pressure in line with changes in forward speed. This reduces the possibility of errors in application rate because of speed variations.

The boom, which carries the spray nozzles, can be from 6–24 m in width; 12 m is a common size as it fits in with tramline systems. It is generally made in three sections, the outer two of which fold up for transport. Each section of the boom can be shut off independently.

The function of the boom is to carry the nozzles at a set height above the target, usually 0.5 m, with minimal vertical or longitudinal bounce. Boom height can be adjusted to maintain the desired clearance above a growing crop. Vertical bounce is countered by mounting the boom in a damped linkage to isolate it from the rocking of the machine. There may also be considerable longitudinal oscillations at the boom tip. The ground speed of the outer nozzles may vary from backwards at half tractor speed to forwards at 1.5 times tractor speed. This effect can also be reduced by flexible mounting.

The nozzles are mounted along the boom at intervals of 0.5 m, and must be held at the recommended height above target if application is to be even. The nozzle is the only metering device on the machine; application rate is changed by altering system pressure, or by fitting nozzles of a different size. Manufacturers sometimes code different size nozzles in different colours, to make selection easier. The interchangeable nozzle tips may be made of brass, plastics, stainless steel or ceramics; durability of these materials increases in the same order. With wear, the aperture in the nozzle increases in size, letting through more spray. The spray pattern also changes so that more material is applied directly under the nozzle and less between nozzles. Nozzle output/min should be checked regularly, and nozzles which are more than 10% off target should be replaced. All nozzles should be replaced at least annually. The cost of a set of nozzles is no more than the cost of 2 ℓ of some spray chemicals, which is a very small price to pay for accuracy.

Many nozzle holders are equipped with check valves which shut off the flow of spray when pressure drops to 0.2–0.3 bar. This prevents dribble from the nozzles after the machine has been shut off at the end of a run. Corrosion-proof strainers may be fitted to protect the nozzle from blockage. It is important to match the screen size of the strainer to the aperture of the nozzle; if the strainer is too coarse, the nozzle is not protected; too fine, and the screen may clog up with small particles which could otherwise pass freely through the nozzle.

Sprayer calibration

The value of spray chemicals handled by a sprayer in a year is usually far greater than the value of the sprayer itself. Time spent calibrating and checking the machine can yield large dividends in terms of more efficient use of chemicals. The volume of water delivered/min by each nozzle at a given gauge pressure can be measured using a special meter, or by collecting the liquid in cans or buckets for a timed interval and then measuring it in a graduated cylinder.

This allows all individual nozzles to be checked for uniformity, as well as measuring the total output of the machine/min. From *Table 24.1*, the time taken to spray 1 ha at a given forward speed can be determined, and the output/ha of the sprayer can be calculated. If thick emulsion-type sprays are used, the sprayer must be calibrated using the spray itself, with appropriate caution, since the use of water will give a high reading.

Correct forward speed is essential to precision spraying; tractor speedometers, if fitted, must be calibrated to take account of alternative tyre sizes and field conditions. Speed monitors, using independent ground wheels or radar sensors, are available, and some monitors will also display flow rates and application rate/ha.

Recent developments in spraying

As mentioned earlier, hydraulic nozzles produce a wide spread of droplet sizes, from those which are too heavy to stick to a leaf, to those which are light enough to be carried away by a breeze. Rotary atomisers have been shown to produce a very narrow spectrum of droplet sizes as the liquid spins off the edge of a rotating disc.

In the controlled droplet application (CDA) sprayer, droplets of 250–300 µm are produced. These resist drift and give good cover using less spray. Application rates of 20–40 ℓ/ha are recommended, which treats a very large area per tankful.

The ultra low volume (ULV) system uses a droplet size of 70 µm, at very low application rates. These droplets are carried by the wind to give a very thorough cover of the crop with excellent penetration into dense foliage. Herbicides and poisonous materials are not applied, because of the risk of drift, but satisfactory results can be achieved with approved fungicides and insecticides. Special oil-based formulations are used, to reduce the risk of evaporation while the tiny droplets are in the air. Only chemicals formulated for this system of application may be used, and manufacturers' recommendations must be followed.

In order to move small droplets more positively towards the target one system applies an electrostatic charge of up to 25 000 V to the droplets as they leave the nozzle. The voltage gradient from the charged nozzle to the neutral crop or ground propels the droplets toward the target at perhaps 20 times the speed of similar sized non-charged particles, much reducing the risk of drift. The charged droplets are attracted onto the surface of the crop ensuring that a large proportion of the spray actually ends up on the crop. This 'recovery index' is reported to be 2.5 times as high as that for uncharged droplets of the same size. Application rates as low as 0.5 ℓ/ha are recommended.

Servicing sprayers

Although the field sprayer is a wide machine, and high rates of work are possible, a typical work rate for a 12 m machine is only 2.5–3.5 ha/h. A recent ADAS survey showed that over 70% of machines studied spent more than half their 'working' time on such tasks as filling the tank, or travelling to the water supply. By good organisation, some machines were able to spend 70% of their time spraying. Fast refilling was achieved by the provision of a water tanker on the headland, often fitted with a 450 ℓ/min centrifugal pump. This cut down travelling time almost to zero and severely reduced filling-up time in comparison to the time taken to fill the tank via a hose pipe from the mains. Where only a few days are available for the application of a particular chemical, this level of attention to detail can greatly increase the productive capacity of the spraying machine. The survey also pinpointed the importance of driving at the correct bout width, and demonstrated clearly the advantages of a bout-marker system.

HARVESTING FORAGE CROPS
Mowing and swath treatment machines

Whether they are to be conserved as hay or as silage, the majority of forage crops are first cut and then allowed to wilt or dry. The length of this drying period will vary from 24 h to several days, according to the conservation system.

Maximum drying rate of the cut swath depends on:

(1) free movement of air through the swath;
(2) exposure of the wetter, thicker butts to the air and sun;

(3) bruising and scuffing of the outer cuticle to allow water vapour to escape more easily;

(4) flattening of stems to reduce the distance water must move from centre to surface.

This 'conditioning' must be achieved without the penalty of excessive fragmentation and leaf loss which can result from over-vigorous treatment. Losses increase with the length of time that the crop is drying in the field, so a treatment which gives a rapid drying rate for a 24 h wilted silage crop may well be too severe for a 5 d hay crop.

The ability to leave a very short stubble may seem to be advantageous but in fact a longer stubble, of 50 mm or more, has been shown to give faster drying rates and better regrowth than where the crop is shaved to the ground. The risks of contaminating the crop with soil, and also of damaging the machine, are reduced considerably by leaving a longer stubble.

The cutterbar mower

This machine cuts by shear as the reciprocating knife sections move across the stationary ledger plates. Knives must be sharpened often. Work rate is 0.2–0.6 ha/h for a 1.5 m machine. A leaning or lodged crop reduces the work rate. The machine is very economical to use as it only requires 3–5 kW at the pto and it is also the cheapest type of mower, being less than half the price of its nearest rival. The cut swath lies quite flat, with the butts covered; it must be tedded or conditioned immediately after cutting if a fast drying rate is to be achieved.

Impact mowers

All other mowers cut by impact. A cutting edge moving at 50–90 m/s smashes its way through the base of the crop. The crop is severed whether or not the machine is sharp, although blunt cutting edges may absorb 50% more power. Power requirement is much greater than for the cutterbar machine; the highest power is consumed when mature, stemmy crops are cut. There is danger both to the machine and to the user if it strikes a stone or other obstacle. Blades can be broken and fragments can be thrown out. Guards and shields must be kept in place.

Impact machines can offer very fast working speeds with little maintenance or downtime, and are little affected by the condition of the crop. Against these advantages must be set the machine's higher initial cost and greater power requirement.

Drum and disc mowers

Two or more rotors turn on vertical spindles. Small flat knives are attached to the periphery of the rotors, which are from 350 mm to 1 m in diameter. As the rotors turn the knives cut a series of horizontal arcs as the machine moves forward. Tip speeds of 80–90 m/s ensure a clean cut.

The rotors of a disc machine are driven from below, and the cut material passes over them; the alternative is a drum-shaped rotor assembly driven from above. In the latter case the cut material passes back between adjacent pairs of drums. If the cut material does not pass quickly back off the machine there is a risk of double-cutting, where small fragments are severed from the butt end of the crop and are subsequently lost in the stubble.

Power requirement is 35 kW for a 1.5 m machine. Speeds of up to 20 km/h are possible, and work rates of up to (1 ha h)/m of cut can be achieved. Widths of cut from 1.5–3 m are common. Cost per unit width increases sharply above 1.5 m, starting at double the price of a cutterbar mower. The machines are complex; some have many fast-moving parts which may rotate at speeds in excess of 3000 rev/min. A high level of maintenance is vital if a long working life is to be achieved.

The cut swath is slightly fluffed-up, but the butts are covered. After an initial period of rapid drying, the overall rate is little better than that from a reciprocating mower. Tedding or conditioning is essential.

Flail mowers

A heavy horizontal spindle extends across the width of the machine, which is typically 1.5–1.8 m. Shovel-shaped cutting flails are arranged in two or four rows across the spindle. As the spindle rotates at 800–1200 rev/min the flails scoop forward into the crop at 50–60 m/s. This action severs the crop at its base and carries it up over the rotor, from which point it is guided to fall back into a loose swath which is adjustable in width. The action of the flails also bruises, scuffs and mixes the crop, a very effective aid to fast drying. Over-severe 'conditioning' can result in considerable crop loss. Broad flails and slow rotor speed give the gentlest treatment, and *vice versa*.

The power requirement is high, at 50 kW for a 1.8 m machine. Cost is three times that of a cutterbar mower. Rate of work is (0.5–0.7 ha h)/m of cut. Drying rate of the swath produced by this machine is equal or better than that from any other treatment or combination of treatments, but losses can be high in hay crops.

A serious problem with flail mowers is their tendency to shave off or 'scalp' any bumps or hillocks they encounter, because there can be no skids or wheels except at the extreme sides of the machine or behind the cutting rotor, Some makes can be fitted with 'anti-scalping rollers' which minimise but do not eliminate the problem. The lifting action of the flails, valuable in dealing with lodged crops, can also raise loose soil and mix it with the lacerated crop. So long as the soil contamination problem can be overcome, this is a rugged reliable machine, to be recommended for the silage producer.

Mower-conditioners

As mentioned above, the swath produced by cutterbar or drum/disc machines is not 'set up' for rapid drying without further treatment to achieve the conditions described on p. 509. The combination of a mower with a 'conditioner' can perform both these operations in a single pass and leave the swath in an ideal state for fast drying. A mower-conditioner requires 20% more power than the equivalent mower and may cost 80% more. Work rate will be lower than that of the mower alone.

Two types of conditioner may be used. The original machines, introduced from the USA, have a pair of horizontal rollers which crush and bruise the crop before dropping it into a loose swath. These machines are best suited to stemmy crops such as lucerne, but deal less well with heavy, leafy crops.

More recently, flail conditioners have been developed. The cut crop is lifted, mixed, turned and scuffed by a rotor carrying narrow fixed or hinged flails. This machine deals more thoroughly with thick, leafy crops, where the rigidly mounted resilient flails are able to penetrate throughout the material. It is also less expensive than the roller machine, and the severity of treatment can be easily adjusted by moving a baffle which can delay the passage of the crop so that the flails apply more impacts to it. The flail conditioner is being increasingly adopted by mower-conditioner manufacturers.

The use of any steel-tined machine carries with it the risk that a broken tine may occasionally be left in the swath and cause serious damage to following machinery. The introduction of stiff nylon-bristle conditioners, may eliminate this problem.

Swath treatment machines

Once cut, and possibly conditioned, further treatment to assist drying requires fluffing-up (tedding) or spreading-out the crop to maintain air circulation, and this must be followed by windrowing the crop to match the next stage of the harvest operation or to reduce the risk of spoilage if rain is expected. The crop should be tedded at least daily in good weather.

Spreading the crop to increase its exposure to sun and wind has been shown to increase the drying rate of hay under good weather conditions. Drying rates for silage crops, or for hay under poor weather conditions can be reduced by spreading. Crop losses are increased both by the scattering action and because a tractor must run over part of the crop in order to gather it back into windrows.

Most swath treatment machines use spring-steel tines to move the crop. If these are set too close to the ground the risk of breakage is increased. Farmers often modify their machines by adding small clips or ties so that broken tines are retained rather than being allowed to fall into the windrow.

Combination machines

Some machines are able to perform all the functions of tedding, turning and windrowing. They cover a 3–5 m width, and generally consist of two pto-driven tined rotors which rotate on vertical spindles, some with movable swath guides behind. The whole machine is tilted forward to ted or scatter the crop, or levelled to produce a windrow.

Most machines can be made to rotate more slowly if gentle handling is required when the crop becomes dry and brittle or in difficult crops, such as lucerne, where leaf loss is a problem. It is not always easy to produce an even swath, but this is most important if a high harvesting rate is to be achieved by following machines.

'Specialised' machines

An alternative swath treatment system is to use separate machines for the different functions. Spreading machines of 4–7 m width use four or six pto-drive tined rotors. Wide machines are articulated to follow ground contours.

A machine which can ted a mown swath and return it to the ground in a fluffy windrow is the horizontal rotor machine. Forward-acting spring tines lift and mix the crop, which then drops between a pair of adjustable guides to give a wide or narrow windrow. These machines, in one-, two- or three-row form, are not popular at present, and may be obtained cheaply at farm sales.

A strip of spread material 3–5 m in width may be gathered into a windrow by a single-rotor machine on which groups of up to four tines sweep the ground as they pass across the front of the machine, and then fold back to release the crop and leave it in a windrow at the side. A guide, of canvas or of steel rods, assists in the formation of an even swath.

A gentle method of gathering spread material into a windrow, or of combining two or more rows into one, is by means of the finger-wheel side rake. A series of four to eight spring-tined wheels roll along the ground while held at an angle to the direction of travel. The tines sweep the ground and roll the crop from one wheel to the next across the width of the machine. The tines travel slowly in relation to the crop, and the rolling action retains small fragments of the crop that might otherwise be lost in the stubble. This action also tightens the swath and can reduce air circulation and the rate of drying if used at an early stage. The machine is inexpensive to buy and requires little power. Front mounting can allow two rows to be put into one by the tractor which also powers the baler or the forage harvester.

Barn hay drying

When a hay crop has dried to 35% MC in the field, further drying will only take place if the relative humidity of the air is low enough to extract moisture from the crop against the attraction of the plant material (*Table 24.8*).

Table 24.8 The equilibrium relationship between air relative humidity (RH) and hay moisture content (MC)

Air RH (%)	Hay MC (%) *below which moisture will not be removed*
95	35
90	30
80	21.5
77	20
70	16
60	12.5

This final stage can be speeded up, and exposure time in the field reduced by 1 or 2 d, if the hay is removed from the field at moisture contents of 50–35%, and dried with forced air in a barn or stack. The result will be a better quality product with reduced field losses and weather dependence, although it is produced at a higher cost/tonne than field cured hay.

Conventional bales are generally dried in the store which they will occupy until used. Batch systems involve a second move of the crop which increases cost. The air is blown into the stack by a fan, via a false floor or a large duct under the stack, or a vertical duct up the centre of the (square) stack. Mesh floor systems require an airflow of $(0.25 \text{ m}^3\text{s})/\text{m}^2$ of floor area, while the vertical duct system demands $0.005 \text{ m}^3/\text{s}$ per bale, at pressures of 60–85 mm water gauge.

Large rectangular bales can be batch-dried by building four or five arches of 11 bales each, with a tunnel at the centre. A portable fan unit delivers $0.5 \text{ m}^3/\text{s}$ or air per bale, and drying is completed in about two weeks. The bales can then be moved into a permanent store.

Where the air temperature can be raised by 5 °C above ambient, drying can be continuous. Without heat, drying can only continue while the ambient air is dry enough to extract water. This requires constant monitoring of the relative humidity, and the drying process will be extended over a longer period.

Forage harvesters

The job of the forage harvester is to gather up crops which are to be conserved as silage or, occasionally, fed directly to livestock. The crop must be delivered into a trailer or lorry at an appropriate chop length to suit the conservation system. A short chop length requires more power and more expensive machines. Chopped material becomes more compact in load and store. Fine chopping improves fermentation, particularly of high dry-matter forage.

A 10 m^3 trailer can carry 1500 kg of fresh grass chopped to 200–250 mm, 2300 kg if chopped to 100 mm, and 2500 kg if chopped to 25 mm. For satisfactory compaction and fermentation in a clamp silo, a chop length of 200 mm is recommended for material up to 20% DM, 130 mm for material of 20–25% DM, 80 mm for 25–30% DM and 25 mm for material above 30% DM. For a tower silo a chop length of 25 mm or less is required.

Most forage harvesters discharge directly into trailers and cannot work at all unless a trailer is available. To keep the machine working at full capacity, careful organisation of transport is of paramount importance.

Flail forage harvester

This is a simple and rugged machine which can cut a standing crop or pick up a windrow. A horizontal spindle across the width of the machine carries two or four rows of pivoted flails with sharpened cutting edges. These cut or scoop up the crop as the spindle rotates at 1000–1600 rev/min. The crop is thrown up a chute and directed into the trailer by a rotating spout and a movable deflector flap, controlled by the driver.

Cutting widths range from 1100–1500 mm and the machine requires a tractor of 35–50 kW to produce an overall rate of work of 7–10 tonnes/h. Harvested material is lacerated and broken by the flails; the severity of treatment can be adjusted by moving a baffle closer or further from the flails. Chop length averages 150–300 mm, but particle sizes range from 50 mm to original length.

Side-mounted machines allow the trailer to be pulled directly behind the tractor for good manoeuvrability, stability and ease of hitching. For a one-man operation, the forage harvester is quickly detached, allowing the tractor and trailer to travel to the silo. Trailed machines normally run directly behind the tractor with the trailer behind. This makes a less stable combination and changing trailers takes longer.

Double-chop forage harvester

This machine combines a flail head, using edge-cutting flails, with a flywheel cutter and blower. This gives a more positive chopping action although chop length is not precisely controlled. The machine was originally designed for direct-cutting forage as part of a zero-grazing system, but is also suitable for picking up wilted material.

Typical chop length range is 50–150 mm and the machine requires 40–55 kW to achieve a nett output of 8–12 tonnes/h. This is a trailed offset machine which pulls the trailer directly in line with the tractor.

Fine-chop forage harvester

Several manufacturers offer fine-chop machines which give better control of chop length without the complexity of the most sophisticated precision-chop machines. The wilted crop is picked up by a spring-tined reel and fed by a pair of rollers into a cylinder cutter. This is similar to the mechanism of a lawn mower, where up to 12 knives arranged round the periphery of a cylinder shear the crop against a stationary shear bar. Chop length is increased by feeding in the crop faster or by removing some knives from the cylinder. The chopping mechanism, which must be sharpened twice a day and adjusted regularly, also throws the crop up the chute and into a trailer.

Chop lengths of 20–80 mm are possible, at outputs of 15–25 tonnes/h and require a tractor of 50–80 kW. Putting two or more windrows together ahead of the machine can increase output by reducing the forward speed necessary to fully utilise its capacity.

Precision-chop forage harvester

This type of forage harvester was introduced from the USA; at present domestic, European and North American manufacturers supply the market. As well as a tined pickup for windrowed crops, interchangeable headers are available for direct-cutting and for harvesting maize.

All machines use cylinder choppers, but the crop is metered into this mechanism by two pairs of spring-loaded feed rollers. Quick-change gears allow selection of alternative chop lengths (range 5–50 mm), and a second control allows the feed rollers to be stopped, or reversed, to clear a blockage. A refinement available on some makes is the fitting of a magnetic metal-detector in the intake or feed mechanism. If a particle of ferrous metal is detected, the intake and feed mechanism is stopped instantly. After reversing, the particle of metal may be picked out before harvesting continues. This device protects the machine from expensive breakdowns, and also prevents small metal fragments from being incorporated into the silage and later fed to livestock.

Machines are of two types; 'cut and throw', where the cutting cylinder throws chopped material directly into the chute, and 'cut and blow' where a separate blower performs this function. This allows the use of a recutter screen to ensure that material is finely chopped before it can leave the chopping mechanism. Where a separate blower is used, the discharge chute can be positioned in line behind the tractor, rather than directly behind the chopper. This makes it easier to discharge into a trailer towed alongside, which is often necessary with high-output machines. All machines have knife touch-up stones, and many also have power-driven rebevelling grinders to keep the cutting knives in peak condition.

Trailed offset machines come in a range of sizes to match 50–110 kW tractors. Outputs range from 10–30 tonnes/h, according to available power, crop density, chop length and field organisation. Self-propelled harvesters, generally contractors' machines, range in power from 110–220 kW; most are based on the working parts of trailed machines, and so share their adjustments and attachments. Outputs of 30–80 tonnes/h are possible, but this depends very much on the availability of large, even windrows and plenty of empty trailers. To reduce overhead costs, the machine's working season may be extended by the harvesting of a succession of crops throughout the summer and autumn.

Self-loading forage (SLF) wagons

An alternative one-man system of loading and transporting crops for clamp silage, the SLF wagon has been introduced from Europe and gained popularity over the last decade. The wagon, of 3.5–5 tonnes capacity, loads itself by means of a reel pickup under the drawbar and a short vertical conveyor which packs the crop into the enclosed body. Up to 21 stationary knives may be positioned against the conveyor, to slice up the crop as it passes by. Chop lengths of 80–220 mm are achieved when knives are sharp. A chain and slat conveyor on the floor of the wagon can move the load rearwards as the machine fills and can discharge the contents through a rear gate in 3–10 min at the silo. When the trailer is loaded, the pickup is raised hydraulically and the combination is then driven to the silo.

The SLF wagon has a low requirement for power at the pto, but a tractor of 35–50 kW is necessary to control the heavy trailer on hilly ground; three or four loads/h can be achieved over a transport distance of 500 m or less. The chop length is longer than that recommended for wilted silage; the use of a flail mower to cut the crop can give increased fragmentation.

Baled silage

Silage crops may be handled as unit loads if they are baled using large round balers or high-density large rectangular balers. These balers can also be used for packaging straw, but are not suitable for baling hay unless it is drier than 20% MC. The heavy wet crop causes more wear to the baler than a similar amount of dry material.

Bales are handled by tractors with front and rear forks, and may be loaded onto flat trailers for haulage over greater distances. Individual bales may be ensiled in large polythene bags, or a stack may be built and covered with plastic, similar to the conventional clamp but without the cost of the walls. No further compaction is necessary, as the bales are compressed already. A high stack may be built from ground level without any need for a machine to be driven onto the material.

Silo filling

Handling at the silo must keep pace with harvesting equipment. A tractor with rear push-off buckrake can handle up to 30 tonnes/h. Greater outputs are possible with a front mounted buckrake, and more still from four-wheel drive handling vehicles with fork capacities of 1–2 tonnes. In addition to spreading the crop the machine also compresses it. Final densities of 700–850 kg/m^3, after fermentation, are common.

Silage may also be stored in tower silos, which reduce storage losses and permit every stage of silage handling to be mechanised. Tipping trailers deliver precision-chopped forage into a stationary dump box, which feeds it at up to 30 tonnes/h into a forage blower drive by a tractor of 30–50 kW. The forage is blown up a vertical pipe 225 mm in diameter, over a curved section and against a spreading device in the centre of the roof of the silo. Spreading is essential if the silo is to be filled evenly and its potential storage capacity is to be fully utilised. A minimum filling rate of 2 m depth/d ensures satisfactory fermentation.

During the early stages of fermentation, gases such as carbon dioxide and oxides of nitrogen are evolved. If the silo is partly filled, these gases accumulate above the silage and can kill anyone climbing down onto the silage, perhaps to remove plastic sheeting after a weekend. Great care must be taken to clear any such hazards before entering the silo. *Never* enter a silo or other enclosed space (empty slurry tank, grain silo) without a rope attached, and the supervision of companions capable of giving assistance (without entering the silo themselves) should an emergency arise.

Baling and bale handling

A convenient system for handling hay (and straw) is to package it into bales. Thse may be moved by hand or grouped into larger unit loads for machine handling. Large bales must be handled by machine. The process adds nothing to the final usefulness or value of the product, indeed this may be reduced. Handling as bales can offer a least-cost solution to the problem of moving bulky materials on the farm.

The conventional pickup baler

This machine produces bales of section 350 × 460 mm, and variable length. Larger sections such as 400 ×

450 mm are used on some high-output machines. Bale density, which is adjustable, varies from 160–220 kg/m^3, and bale weights range from 15–40 kg.

The offset spring-tined pickup lifts the crop into the machine. Its height is controlled by skids or small wheels at either side.

The crossfeed moves wads of material from behind the pickup and feeds them laterally into the bale chamber each time the ram draws back.

The heavy ram slices off each wad from any trailing material, and forces it down the bale chamber towards the rear of the machine. As the compressed material moves along, driven by the ram working at 60–95 strokes/min, it carries with it two loops of string which will eventually form the securing bands of the bale.

When sufficient material has passed the bale length sensor, the tying cycle is initiated. This is timed to occur when the ram is forward. The needles bring up the ends of the strings to encircle the bale, place them in the knotters, and withdraw, trailing out a new string ready for the next bale. The knotters tie each bale string by forming a loop and pulling the two ends through. The strings are cut at the knotters to release the bale, which continues rearward past the adjustable restrictor that controls bale density, and is finally discharged via the bale chute.

Many balers have potential work rates of 15 tonnes/h or more; typical field performance is 6–8 tonnes/h. Large, even windrows allow the machine to operate steadily at high rates of throughput.

Handling conventional bales

The conventional bale was originally intended to be a suitable unit for man-handling, and can still be handled that way if labour is available. It remains a convenient unit for feeding to livestock.

Many mechanised bale-handling systems are available. Most make larger units by forming the bales into groups of 6, 8, 10, 20, or multiples of these numbers. A collector sledge may be pulled behind the baler, and the bales released at intervals to form windrows. These windrows are formed into the desired groups by hand. Alternatively, the groups can be formed automatically on an 'accumulator', which releases each group when it is complete. These groups are dotted all over the field unless a further collector is drawn behind.

Groups of bales are picked up by specialised lifting attachments on tractor fore-end loaders or other handling machines, and may be built up into larger stacks or loaded directly onto trailers. Stacks of 40–144 bales may be picked up as units by specialised trailers and moved directly to store.

Self-loading wagons can pick up individual bales, as dropped by the baler. This eliminates any baling delays caused by the presence of accumulators behind the baler. Individual bales are built up by the machine into stacks of 88 or 105. The stack can be taken to store and set down as a unit, or bales can be unloaded one by one into an elevator if time allows.

Ideally, a bale handling system will deal with bales at all stages—field stacking, transport from field to store, loading the store, and unloading from store to final use. It is very unusual for a conventional-bale handling system to fulfil all these requirements; many are only used for loading trailers in the field, which is a very small part of their potential.

A recent ADAS survey has indicated the labour requirement for many systems of handling bales from windrow to store. The figures shown below are averages, and considerable reductions were achieved by the best or 'premium' operators.

Bale/collect; hand-load trailers; hand stack:
   83 man min/tonne
Bale/accumulate 8s; front-end loader to trailer:
   hand stack: 49 man min/tonne
Bale/accumulate 10s; field stack 100s:
   specialised trailer to store 26 man min/tonne
Bale/self-loading wagon 88s; place stack direct into
   store: 21 man min/tonne

It must be remembered that labour requirement is only one of a number of criteria on which a bale-handling system is chosen. Other factors may be overall rate of work, capital costs of the equipment, or the suitability of a system for an existing storage building. The final solution, as with other materials handling problems, must deal with bales at least cost when all forms of cost are taken into account.

Big balers

For many years, the development of a larger package for handling hay and straw has been seen as a logical move towards handling by machine rather than by man-power. Since large bales, of 250 kg to over 1 tonne, can never be moved by hand, the bale-handling system must carry out all the functions mentioned above, in moving the bale from field to end-use.

Apart from low density large rectangular bales, little drying can take place after baling, and so most machines are not suitable for making hay unless it is quite dry (below 20% MC) before baling. As mentioned earlier, large balers can be used for packaging silage crops, which spreads their use over a longer season.

Large rectangular low-density baler

This British development produces a bale 1.5 × 1.5 × 2.4 m in size, weighing 250–300 kg in straw, 450–700 kg in dry hay.

The machine requires an even, rectangular swath at least 1 m in width. The tractor (40–60 kW) straddles this swath which is picked up at the front of the baler. It is packed into a large bale-chamber by reciprocating

tines. When the tractor driver judges the bale-chamber to be full he initiates a tying cycle, and the bale is then pushed out through a rear tail gate as the next bale begins to form.

Bales are dropped all over the field, and are handled by a squeeze-loader. They form more stable stacks if turned on one side. The bales can be pulled apart by hand, or may be self-fed as units.

Rates of work of 12.5 tonnes/h can be achieved, and the handling system can be at least as economical of labour as the best conventional-bale handling systems. Because of the rectangular shape and low density, 80–110 kg/m³, of the bales, barn-drying is quite practicable.

Large round baler

These machines, from American and European manufacturers, produce a cylindrical bale, either 1.5 m wide × up to 1.8 m diameter or 1.2 m wide × up to 1.5 m diameter. Bale weights range from 150–800 kg, with densities from 90–140 kg/m³. The bale is produced in a cylindrical chamber in the form of a roll. The bale has good weather resistance, but is less easy to stack than the rectangular bale. To improve the bale's suitability for hay making, a number of manufacturers produce 'soft centre' machines which form the centre of the bale from a mass of loose material and then roll the outer layers more tightly around it. As with other large balers, this machine has no crossfeed, so careful swath building and care in aligning windrow and pickup are essential for the production of good-shaped bales.

When the bale is sufficiently large, the operator stops forward travel and initiates the wrapping of six or more turns of string round the bale as the baler continues to turn. The string is cut, but not knotted, and the bale is discharged through a rear gate. If the windrows are close together it may be advisable to reverse the baler out of the row so that the bale can be dropped in a position where it will not interfere with the next pass of the machine.

Round bales can be carried by a grapple or prong attachment on the front or rear of a tractor, or may be loaded onto flat trailers. The smaller size allows two bales to be loaded across the width of a trailer or lorry, which is not possible with the larger size.

The bales may be self-fed, or can be unrolled on the ground over a distance of 25 m. Mechanical unrollers, shredders and tub grinders may also be used to break up the bale.

Large rectangular high density baler

Packages weighing 1 tonne or more are produced by these very large machines, which require a tractor of at least 75 kW. Bale size is 1.2 × 1.3 × 2.4 m, at a density of 300 kg/m³ or more. This system is suitable for handling very dry hay, straw or silage crops.

The bales are too heavy to be handled by most tractor front-end loaders, but are well within the capabilities of specialised handling vehicles fitted with grapple or squeeze-loader attachments.

GRAIN HARVESTING

The combine harvester

Grain crops (and also pulses, oil seeds and other seed crops) are now universally harvested by the combine harvester. This machine combines four major functions: it gathers the crop, threshes the seed out of the ear, separates the seed from the straw and finally cleans the seed of chaff, weed seeds and dust. After temporary storage in a hopper on the machine the grain is delivered into a bulk trailer or lorry.

Most machines are self-propelled with cutting widths ranging from 2.1–6.7 m; many machines have two or three alternative cutting widths. Engines of 32–170 kW are fitted. Although the 'size' of a combine is often described by quoting its cutting width, a more useful indication of its harvesting capacity might be its power. The factor which actually limits a combine's rate of work is its ability to shake out the grain from the remainder of the 'material other than grain' (MOG) in the separating mechanism. This ability varies from crop to crop according to maturity, moisture content, straw-length and threshing adjustments and so cannot be quoted as a constant figure.

A typical grain harvesting season contains 12–20 dry working days (up to 200 working hours), according to geographical location. While it is common to select a combine size which will provide approximately 1 m of cutting width for every 35 ha of grain (up to 60 ha in a few cases), an alternative approach is to specify a harvesting capacity which can deal with an expected crop within the working time available in the region. As explained earlier, no firm output figures can be given, but the following approximations are satisfactory for planning purposes; machines of 38–55 kW can harvest 4–7 tonnes of grain/h, 60–80 kW, 6–10 tonnes/h; 80–100 kW, 8–12 tonnes/h; over 100 kW, 11–16 tonnes/h.

Inadequate combine capacity leads to an extended harvest period with greater weather risk and rapidly rising levels of crop loss from shedding, both in the field and at the combine intake. A doubling of pre-harvest losses can occur after a delay of only two weeks in some crops.

Most combines operate below full capacity. More than two-thirds of machines are used less than 200 h/year. The availability of excess harvesting capacity is a form of insurance against bad weather at harvest time, which also unfortunately leads to high overhead costs/ha harvested. A further form of insurance against a poor harvest season is the installation of grain drying equipment (*see* p. 294).

Gathering

The crop is gathered into the machine at the 'table' or 'header'. It is drawn in by a rotating reel, which can be moved forward, lowered and speeded up to deal with a lodged crop. Unnecessary speed or reach can cause shedding, particularly in over-ripe crops. Crop lifters can be used to slide the crop gently up onto the table; in one trial the use of crop lifters in lodged barley was found to halve grain losses at this point. Header losses can also be reduced by cutting only against the lean of the crop.

The crop is cut by a reciprocating knife, with serrated blades to grip the hard stems. Control of the height of cut, which is adjusted hydraulically, is the most demanding task facing the combine operator. Many larger machines have automatic table-height controllers, and other machines have flotation devices which allow the table to follow ground contours.

The cut crop falls into the table and is drawn towards its centre by a double flight auger. Loss of grain from the table itself is minimised by the length and profile of the deck, which is shaped to retain loose grain should any be threshed out of the ears by the reel or the auger. The crop leaves the header via a conveyor in a band 0.9–1.6 m wide to match the width of subsequent mechanisms in the machine.

To increase the versatility of the machine, most tables can be detached quickly for transport or storage, and alternative gathering equipment is available for such crops as maize, windrowed crops, sunflowers or edible beans.

Threshing

The crop is threshed by a rasp-bar cylinder, or 'drum', working against an open grate concave which matches its curvature for almost one-third of its circumference. The impacts of the beaters on the crop at up to 30 m/s, and of the moving crop on the stationary concave, break the seeds loose. Severity of treatment is adjusted to match the type and condition of the crop by altering cylinder speed or varying the gap between concave and cylinder, through which the crop is forced. Excessive threshing severity will break grains, while insufficient treatment will leave grains attached to the ears which will be discharged with the straw and lost.

In addition to the threshing action of the cylinder and concave, 70–80% of the threshed grain is separated from the MOG by passing outwards through the spaces in the concave.

Separation

Separation of the remaining threshed grain from the straw is the function of the straw walkers, which offer a separating area of 2.3–5.2 m². To shake out the grain, the threshed mass is agitated as it moves rearwards, finally to be discharged onto the ground.

The thicker the mat of straw on the straw walkers, the more difficult is the task of separation, and the situation is even worse if the straw is damp or broken-up. Beyond a certain MOG throughput, different for every machine, satisfactory separation cannot be completed within the length of the straw walkers and grain is carried over the back of the machine in increasing quantities. Electronic loss monitors may be fitted at the rear of the straw walkers; with their aid (if regularly calibrated), the forward speed (and hence throughput) of the machine can be adjusted to maintain an acceptable level of loss at this point. Automatic control of forward speed is possible on some machines.

Many manufacturers have supplemented the action of their straw walkers with rotary or reciprocating beaters, which are effective in shaking out the mat of straw and allowing more of the trapped grains to escape. A more radical approach is embodied in the axial flow machine, where a very large drum combines an initial threshing section with a rotary separator. The crop is handled fast, in a thin layer, which gives more positive separation.

Cleaning

Impurities are removed from the crop by a combination of size and density separation. An adjustable air blast moves less dense particles to the rear of the machine, while two, or three, sieves of total area 1.95–4.97 m² prevent particles larger than those required from passing through. Adjustable sieves are convenient, but interchangeable fixed-size sieves are more precise in their dimensions, and also permit greater throughputs of smaller-sized seeds. Grain losses over the sieves can be detected by a grain-loss monitor in the same way as straw walker losses.

Grain losses

Losses of grain from the combine can occur at several points, for some of the following reasons:

(1) gathering losses—reel too aggressive; crop over-ripe; knife too high; crop lodged; inaccurate driving;
(2) threshing losses—cylinder too slow; concave clearance too large; broken grain;
(3) separating losses—forward speed too high; straw walkers blocked;
(4) cleaning losses—too much wind; sieves insufficiently opened; sieves blocked.

Table 24.9 shows acceptable loss levels. In some cases, a greater straw walker loss may be acceptable, as this will allow harvesting to progress at a faster rate.

Table 24.9 Acceptable grain losses from a combine harvester in good harvesting conditions

Gathering loss	0.2%	12 kg/ha in a 6 tonne/ha crop
Threshing loss	0.1%	6 kg/ha in a 6 tonne/ha crop
Straw walker loss	1.0%	60 kg/ha in a 6 tonne/ha crop
Sieve loss	0.2%	12 kg/ha in a 6 tonne/ha crop

Grain losses are measured in the field by collecting and quantifying the lost grain from a known area. If the combine is stopped when harvesting an 'average' part of the field, losses may be assessed in the standing crop in front of the combine to give pre-harvest losses. After reversing the combine a short distance, lost grain from beneath the centre of the machine in its initial position is the sum of pre-harvest and gathering losses. Behind the machine lies the straw, with unthreshed grains still attached, and the sum of all types of loss. By subtraction, the extent of each individual type of loss may be determined, and the appropriate corrections can be made.

A convenient unit area is 1 m², which should extend across the full width of cut and can conveniently be marked out with a string and four pegs. All lost grain within this area is then gathered up. *Table 24.10* shows the relationship between the number of grains found and the loss of grain/ha.

Table 24.10 A method of assessing grain losses

Crop	Seeds/m² to equal 1 kg/ha
Barley	1.7
Maize	0.2
Oats	1.6
Rape	20
Wheat	2

Effect of slopes on grain losses

Much grain is grown on sloping land. The effect of sloping land on grain losses is as follows:

(1) up hill—small increase in sieve losses; small decrease in straw walker losses;
(2) down hill—no effect on sieve losses; small increase in straw walker losses;
(3) side slope—small increase in straw walker losses; large increase in sieve losses.

Hillside combines can automatically level the body of the machine on land sloping as steeply as 30 degrees up, 10 degrees down and 40 degrees to the side, a combination also being possible. Side-hill combines can accommodate side slopes only, to 11.5 degrees. Within these limits, the adverse effects of slopes are completely eliminated, and maximum throughput is possible.

Hillside combines cost about 70% more than the equivalent conventional model, while the side-hill type costs 20% more.

Monitors

Many combines are equipped with sealed sound-proof cabs, often air-conditioned. Even without this barrier it is very difficult for the operator to keep in touch with all the processes and components in the very complex machine. In addition to loss monitors, most of the larger machines have as standard, or can be fitted with, a range of 'function monitors' which detect blockages or malfunctions and alert the operator before a major breakdown can occur.

ROOT HARVESTING

Potato harvesters

Lifters

Where labour is available, potatoes, particularly for the valuable early market, may be harvested by hand. The tubers are picked from the ground after the ridge has been broken up by a lifter or 'potato digger'. There are two types, a power driven tined 'spinner', where the ridge is broken up and spread over previously picked ground to expose the tubers, or the elevator-digger on which the ridge is undercut by a flat share and then conveyed up and back by an elevator of round steel crossbars or 'webs'. Loose soil falls between the webs and the tubers are carried over the back to fall on top of the sifted soil.

Complete harvesters

In order to reduce the high labour demand at potato harvest, the elevator-digger has been developed into a complete harvester, which can deliver the potatoes directly into sacks, boxes or bulk trailers. The web elevator has been extended, haulm-removing devices added, and most machines have a divided horizontal conveyor where a gang of four to eight pickers can remove stones and clods, or tubers. The tubers are then directed into the chosen handling system.

Many growers have adopted unmanned machines, typically harvesting two rows at a time, where any clods, stones or trash that cannot be separated by the machine are harvested and dealt with subsequently at the store by hand sorting, or by automatic sorting machines. Rates of work of 0.75–1.25 ha/row daily can be expected.

The mechanisation of the potato harvest has brought with it the two problems of crop losses and crop damage. Most of the 1.25–2.5 tonnes/ha that are left in or on the ground in a typical potato field are lost at the share of the machine. If the share is too shallow, tubers will be missed or sliced. If soil does not pass freely back over the share, perhaps under loose or wet conditions when working downhill, or when the share is trailing haulm, soil and tubers will be 'bulldozed'

ahead of the share and some will be lost by spillage to either side.

Improved harvester designs incorporate fine control of share depth, and may include automatic depth control. To move the soil more positively back over the share, a powered double disc design has been developed. The hydraulically driven discs are inclined forward and inward, and the ridge moves back between them as they rotate. This design can also minimise clod formation and reduce draught.

Tuber damage, typically as high as 20% severe damage, occurs mainly on the web elevator which is also a frequent source of breakdowns. Forward speed of the machine and the amount of web agitation must be carefully matched to web speed to maintain a cushion of soil over the elevator. Designers are reducing the damage potential of machines by minimising the length of web elevators and eliminating drops and changes of direction of the tubers.

A further site for tuber damage is the drop into the bulk trailer. This should be no more than 200 mm, which requires constant adjustment of discharge conveyor height as the trailer fills. Great care must also be taken when discharging the hopper of tanker type machines.

Tubers can also be damaged by contact with stones during harvesting or transport—in stony ground the removal or windrowing of stones can significantly reduce tuber damage, as well as speeding the harvest operation.

Sugar beet harvesters

The entire sugar beet crop is harvested by machine. The harvesting process involves first cutting off the tops, which may be conserved for livestock feeding but are more usually discharged back onto the ground. The roots are then lifted from the soil, most commonly by a pair of inclined, sharp edged, spoked wheels which cut into the soil to either side of the row and lift the root between them while soil can fall away through the spokes. A minority of machines use fixed shares. The beet is then raised by a web elevator which further cleans the crop, before discharging into an integral hopper or into a trailer running alongside.

The most popular machines are single-row trailed tanker models and single row self-propelled harvesters, while three-row trailed machines are favoured by larger growers. The largest producers, cooperatives and contractors may use five or six-row machines, either self-propelled or mounted around a four-wheel drive tractor of 75 kW or more. Rates of work of 0.125–0.2 ha/h per row can be expected.

Crop losses result from incorrect topping, failing to gather whole roots into the machine, and from leaving the lower parts of roots in the ground. Typical losses are 2.5 tonnes/ha, while over 30% of growers leave as much as 5 tonnes/ha. One-third of these losses are visible as loose beet on the surface, the remainder being below ground. One 0.6 kg root lying on the surface in a 20 m length of row represents a loss of 2 tonnes/ha.

The two most common faults in machine operation which lead to crop losses are failing to operate the lifters sufficiently deeply, and failing to position the lifters exactly on the row. The latter fault can be eliminated by the provision of automatic steering on the lifters, or by the use of quite inexpensive guide skids to centre the lifters on the row.

Faulty topping units can damage the beet, and may loosen the roots in the ground so that they are knocked out of the row and are not gathered in by the lifting mechanism. Both the topping knife and the feeler wheels, which adjust the knife to the correct position below the beet crown and steady the root while it is topped, should be sharpened regularly. Lightweight topping units which locate the beet crown by feeler blades or spikes have been introduced. These can respond more rapidly to beet of different heights, and will work accurately at speeds up to 8 km/h. (Traditional toppers are not accurate at speeds beyond 4.5 km/h.)

Harvested beet are often piled at the roadside for considerable periods before being transported to the factory. Losses are minimised if a concrete area is available for storing and loading the crop. The roots are removed from the pile by front-end loaders; industrial loaders can move up to 90 tonnes/h, rough terrain fork lifts 60 tonnes/h, tractor front-end loaders 20–30 tonnes/h; slatted buckets allow some soil to escape, and this process can be continued by passing the beet through a cleaner-loader on its way to the lorry.

FEED PREPARATION MACHINERY

Grinding

Cereals and other ingredients of livestock feeds are ground into fine particles for a number of reasons:

(1) to improve digestion by breaking up the protective outer skin and increasing the surface area of the material;
(2) to increase intake;
(3) to aid mixing of ingredients into homogeneous final diets.

Desirable features of a grinding machine include:

(1) an end product of uniform particle size, with a minimum of dust;
(2) fineness adjustable to suit different classes of livestock;
(3) minimum power requirement in relation to output;
(4) ability to operate reliably without supervision.

Hammer mills

Grain is metered into the mill where it is sheared by sharp-edged beaters rotating at up to 100 m/s. Moving particles from this action are sheared again by the sharp edges of the perforated screen which surrounds the rotor. The process continues until the particles are small enough to pass through the perforations of the screen. A range of screens from 1–9.5 mm caters for all requirements. Most mills are equipped with an integral pneumatic conveying system which aids the flow of grain into the mill and also delivers the meal into a hopper or mixer. Less power is needed to convey the meal by auger, and some manufacturers take advantage of this alternative system. Throughput of a particular mill decreases with fineness and with increasing moisture content of the grain; 14–16% MC is considered ideal. Edges on rotor and screen become blunt with wear, and must be turned or renewed to restore performance. Typical output would be (45–85 kg h)/kW of power available; machines range in size from 2.2–37 kW, those up to 7.5 kW being the most popular on farms.

The high working speed of the rotor makes the mill vulnerable to damage if metal objects enter it, so most machines are protected by magnets, which will retain ferrous metals. A considerable amount of fine dust is produced by this method of grinding, necessitating the use of cyclones and dust socks.

Hammer mills can operate unattended for several hours; many small machines are run at night, drawing corn from an overhead hopper which has been filled to the required level for making up a batch of feed. When the hopper is empty, a switch in its base is released, and this shuts down the mill.

Plate mills

The plate mill grinds by abrasion and shear. Grain up to 18% MC is broken up between the serrated surfaces of a pair of cast iron discs, one of which rotates as it is pressed against the other. Reduced clearance between the discs results in a finer product and a lower throughput. A more uniform meal is produced, with less dust than the hammer mill. Discharge is by gravity only, so a conveying system must be added unless the mill is positioned directly above a receptable for its output. The mill produces (60–110 kg of meal h)/kW of power available.

Roller mills

A roller mill crushes grain between a pair of flat surfaced or serrated cylindrical rollers. The rollers are spring-loaded against one another, and spring tension can be adjusted. Dry grain is cracked to produce a coarse meal, while grain of higher moisture content is flattened and crushed into flakes. Moisture content of 18% or more is recommended and the machine lends itself well to a system of high-moisture grain storage. Dry grain can be moistened to improve its rolling characteristics.

Flaked grain is almost dust-free and hence more palatable to livestock than dusty materials. Break-up of the fibrous portion of the grain can be minimised, making a suitable product for feeding to ruminants.

An output of (90–120 kg h)/kW can be expected, and machines are available in sizes to 7.5 kW with electric drive, as well as larger pto-driven models. Discharge is by gravity, so that further handling or receiving equipment must be installed. Flaked material is fragile; handling and mixing equipment must be selected and operated with this in mind if the product is to be preserved in this desirable form.

Mixing

When all ingredients of a ration are available in ground form they may be mixed into a homogeneous final product. Mixes are generally prepared in batches of 0.5–1 tonne. Ingredients are weighed into the mixer in most cases, but some ingredients may be added by volume into a calibrated container. Since different ingredients are of different densities, the container must be calibrated with care if the final diet is to be as required. The volume occupied by 1 tonne of various ingredient is shown in *Table 24.11* (*see also Table 24.12*).

Table 24.11 The volume of 1 tonne of various livestock feedstuffs

	m^3
Wheat	1.29
Oats	1.96
Barley	1.43
Beans	1.21
Wheat meal	2.13
Oat meal	2.38
Barley meal	1.96
Bean meal	1.85
Grass meal	2.84–3.90
Dry beet pulp	5.67–3.90
Fish meal	1.83–2.08
Soya bean meal	1.48–1.83
Pelleted ration	1.60–1.69
Crumbed ration	1.83

Table 24.12 Typical annual concentrate consumptions of various types of livestock

	Tonnes
Dairy cow	1.25–2.0
Sow or gilt	1.0–1.8
Baconers (from 8 weeks)	0.7–1.0
Porkers (from 8 weeks)	0.5–0.7
1000 laying hens	35–45
1000 broilers	15–20

Larger operators may find it more convenient to produce diets continuously rather than in batches; equipment for this purpose is produced by several manufacturers.

Vertical mixers

The vertical mixer is an upright cylinder, narrowing to a funnel-shaped base. Capacities of 0.5, 1 and 2 tonnes are available. Ingredients may be added through the top, often from an upper floor or granary, or via a small hopper at the base. A central vertical auger circulates and mixes the ingredients. Adherence to the recommended mixing period of 12–20 min is essential; too short a period results in incomplete mixing, while an excessive time may result in separation of some ingredients. Automatic timers are available.

The mix is discharged from an outlet or sacking spout just above the base of the machine.

Horizontal mixers

The horizontal mixer is a rectangular box with a sloping bottom. A loading hopper is positioned at the low end, while discharge spouts are located under the high end of the machine. The flat top carries the drive motor and can accommodate a grinding machine discharging its output directly into the mixer.

The ingredients are mixed by the action of a chain and slat conveyor which moves up the sloping bottom of the mixer and returns across the top. Mixing time is short (2–3 min) and the gentle action is very suitable for rations containing flaked ingredients.

Blenders

A blender prepares the diet as a continuous process. All ingredients are metered out simultaneously in a stream which is mixed as it is carried away by an auger conveyor.

The metering devices are generally short auger conveyors, the speeds of which are adjusted to give the desired flow rate of each ingredient. Since the metering is in fact volumetric, each auger must be carefully and regularly calibrated to ensure that the proportions, by weight, of each ingredient in the mix remain correct.

Mill mixers

Complete mill mixer units are sold by several manufacturers. They normally consist of a small mill, served by a corn hopper and delivering into a mixer. Controls are pre-wired to the motor, so that the unit can be installed on the farm at minimum cost.

An alternative to fixed equipment, which still permits the farmer to include his own corn in his livestock rations, is the use of a contractor operated mobile mill mixer. This is a large, high-output machine mounted on a lorry which can be brought onto the farm at intervals to prepare batches of feedstuffs to the farmer's requirements.

Cubers

Meal may be made more palatable to livestock if it is pressed into 'cubes', 'crumbs', 'pellets', 'nuts', 'pencils' or 'cobs'. This is achieved by forcing the meal into a funnel-like 'die', of a size appropriate to the desired product. The pressure, plus the heat which is generated by friction, forms a durable cube which may be further strengthened by the use of additives such as molasses, water or beet pulp. Freshly made cubes must be cooled before being stored in bulk, and cubed material must be handled gently.

Cubing is a slow, power-consuming process, and die life is relatively short, making the cuber a rather expensive addition to a farm scale provender plant.

MILKING EQUIPMENT
Milking parlours

Most dairy animals are milked in parlours, a system in which the milking equipment and operator(s) are static, and the cows move from a collecting yard to be milked, and then depart, to be replaced by further batches of cows. This system gives economies in labour over the more traditional cowshed milking, where the cows are tethered in a large building and the operators move the milking equipment from cow to cow as milking proceeds.

The most popular form of parlour is the 'herringbone', where a row of standings is constructed on each side of a sunken operator's pit. The cows face slightly outwards, so that their rear ends are in towards the operator. A feeder is provided for each cow to allow individual rationing. Milking units and milk handling pipelines are centrally installed. In many cases, only one line of cows can be milked at a time, while the other batch is being installed, fed and washed. The provision of a second set of milking units allows milking to begin on all cows as soon as udder washing is complete, regardless of the progress of the previous batch, and can increase the throughput of cows. Herringbone parlours are described by quoting the number of milking units and then the number of standings, for example an 8/16, where there are two rows of eight standings and one set of eight milking units, or a 20/20.

Milking performance in a conventional herringbone parlour is 50 cows/man-hour, with up to 75 possible. Further refinements in feeding and milking equipment, and in the shape and layout of the parlour, can reduce the man-time/cow still more. Outputs of 120 cows/man-hour are being achieved, while research

workers quote figures of 150 or more as being attainable in the near future.

Milking machines

Milk is extracted from the udder by applying a vacuum of 50 kPa to the teats. A higher vacuum will extract the milk faster but the risk of teat damage is also increased. Although a constant vacuum is applied to the interior of the flexible teat-cup liner, this vacuum is only applied to the teat for two-thirds or half of the time. The alternation of vacuum with atmospheric pressure in the space between the liner and the teat-cup, controlled by the pulsator, allows the liner to collapse round the teat and shield it from the vacuum for 0.33 s, at 1 s intervals. This allows blood to circulate more freely, and maintains udder health.

The pulsation ratio, the ratio of the time the teats are exposed to vacuum to the time the liner is collapsed, is typically 2:1 as described above, although some equipment operates at 1:1. Pulsation rate is normally 60 cycles/min. Milking rate can be increased to some extent by widening the pulsation ratio or by increasing the pulsation rate.

The four teat-cups are connected to the claw, which provides pulsation and vacuum connections to each teat-cup. This whole assembly is the cluster. Some claws have a built in shut-off valve to prevent air entry when the cluster is removed from the cow or if it falls off during milking. Alternatively, a pinch valve on the long milk tube can perform the first function. A small hole in the claw allows air bubbles to enter the milk, making it easier for the vacuum to raise it up the long milk tube.

Most parlours are provided with graduated glass recorder jars of 23 ℓ capacity, normally at operator's head level. The milk from each cow accumulates in the corresponding jar, so that yield can be recorded or a sample taken.

An individual batch of milk can be rejected, if necessary, without contaminating the entire system. The recorder jar has in some cases been replaced by a milk meter. When installed in the long milk tube, the milk meter indicates milk yield and collects a representative sample of the milk for butterfat analysis. When yields are not being recorded, the meters are not installed and milk passes directly into the pipeline system. A milk flow indicator may be used to let the operator know when a cow has finished milking, or a transparent section of tube or a transparent milk filter may be included in the long milk tube.

A further refinement is to couple the automatic detection of the end of milk flow with the automatic removal of the cluster from the cow (ACR). The cluster is raised by a cord powered by a vacuum operated piston, while the milking vacuum is shut off to release the udder and prevent air entry. If an individual cow has an erratic milk output, ACR can terminate milking before all her milk has been removed.

Milk enters the pipeline either from a transfer valve at the base of the recorder jar, or directly from the milk meter or the long milk tube. An air bleed into the claw or the transfer valve helps the milk to flow towards the source of vacuum until it reaches the releaser vessel (capacity 25 ℓ) which is normally mounted on the milk room wall. An electric milk pump is actuated by a high level of milk in the vessel to withdraw milk against the vacuum and discharge it through a filter into the bulk tank.

Pipeline milking systems and all their components are designed to be cleaned and sterilised in place. Circulation cleaning is a routine by which the system is first rinsed through with hot water, then a hot solution of detergent and sodium hypochlorite is circulated through the system for 5–10 min followed by a final water or hypochlorite rinse. The various liquids are drawn from a trough or tank into the wash line, and enter each cluster via a set of four jetters which fit over the ends of the teat-cups. From here, the liquid follows the route of the milk, described above, until it reaches the releaser pump which either returns it to the trough for recirculation or discharges it from the system.

A reduced consumption of water and of electricity for water heating is achieved by the process of acidified boiling water (ABW) cleaning. Sterilisation is mainly by heat, as the entire system is raised to 77 °C for at least 2 min; 14–18 ℓ/unit of near-boiling water (96 °C), acidified by a small quantity of sulphamic or nitric acid, is drawn once through the system at a rate controlled by an inlet orifice, and discharged by the releaser pump. In addition to a mild sterilising action, the acid helps to keep the glass tubes clear of hard-water salts.

Milk cooling and storage

Milk is normally cooled, and stored prior to collection, in a stainless steel bulk tank of 600–4000 ℓ capacity. When holding 40% of its nominal capacity (from the evening milking) which has been cooled to 4.4 °C overnight, the remaining 60% of capacity must be cooled from 35 °C to the same level within 30 min of the end of filling, at an ambient temperature of 32.2 °C.

The milk is cooled by contact with the chilled inside surface of the tank; an agitator paddle keeps the milk in circulation to assist in this process. The agitator must also be able to homogenise a sample of milk of 4.5% butterfat so that after 2 min operation the milk may be sampled to give an accuracy of ±0.05 units of butterfat percentage. The tank is insulated so that the chilled milk will not rise in temperature by more than 1.7 °C over an 8 h period when the ambient temperature is 32.2 °C.

Most bulk tanks are cooled by an ice bank. A medium sized refrigeration unit, the evaporator coils of which are submerged in water in the space between the inner and outer walls of the tank, operates for many h/d to build up an ice bank. A control will stop the refrigerator when the ice bank is sufficiently large to cope with the expected milk yield and ambient temperature. At milking time, the circulating warm milk inside the tank gives up its heat which is absorbed as latent heat by the melting ice. This system allows a relatively low-powered refrigeration unit to give a rapid rate of milk cooling.

Where very large quantities of milk are handled, it may be more economical to cool the milk in a separate plate-type heat exchanger before it is stored in an insulated tank.

The conventional bulk tank will cool about 45 ℓ of milk for each kW of energy consumed. This figure may be almost doubled if the milk can be pre-cooled in a heat exchanger before it reaches the bulk tank. If a free source of cold water, such as a spring, is available, or if large quantities of purchased water are to be used elsewhere, this water can be used to carry away as much as half the heat from the milk. A heat pump may also be used to recover some of the energy from the milk and use it to heat washing water.

Bulk tanks are always cold, so that heat sterilisation is impossible; chemical sterilisation must be used. Most tanks are equipped with a programmed cleaning unit which, by means of a spray head, rinses, washes, sterilises and finally rinses the tank. This cycle is set in motion by the tanker driver after he has collected the milk.

Parlour feeding

Feeding of the cows in the parlour occupies them during milking and also allows individual rationing of each cow. In a small parlour, the operator can recognise each cow, remember her ration, and operate a volumetric feeder a certain number of times to deliver the correct ration into a trough in front of her. In some cases the cows also learn how to operate the feeders, and can draw additional rations for themselves by banging the unit.

In order to take up less of the operator's time, parlour feeding systems have been developed in a number of ways:

(1) vacuum operation of the feeders from a central control panel, to reduce the distance walked;
(2) a programmable unit which releases the correct ration at the correct position when the cows' identification numbers are manually dialled or punched in;
(3) machine recognition of individual cows at each standing by means of a coded transponder worn around the neck, followed by the release of the correct quantity of feed. The memory unit is

updated regularly, to adjust the ration according to yield and stage of lactation.

Further refinements to the computer systems can include the automatic recording of yield or milk flow rate, and the automatic monitoring of milk conductivity (to detect mastitis) or milk temperature (to detect ill-health or 'heat').

As milking routines speed up, the cow has less and less time to eat her ration. While a typical 6/12 parlour gives her time to eat 5.5 kg, a 12/12 reduces her waiting time, and she only has time to consume 3.5 kg of feed. In order to satisfy the requirements of high-yielding cows, a number of manufacturers offer out-of-parlour feeders, which respond only to cows wearing transponders with a particular coding. A batch of feed is delivered to the animal, and then no further batch can be delivered to that individual for a pre-set period, in order to control her concentrate intake. Some systems also record the amount of food delivered to each cow in this way.

MANURE HANDLING AND SPREADING

Farmyard manure

The traditional method of dealing with manure is to provide straw or other absorbent material as bedding, and to handle the combined product of faeces, urine and bedding as farmyard manure. This increases the volume of material to be handled by 10–25% in comparison with the figure mentioned in *Table 24.13*.

Table 24.13 Daily production of faeces and urine by livestock

Calves 3–6 months	7 ℓ	Pigs 9–12 weeks	3.5 ℓ
Young stock			
6–15 months	14 ℓ	Pigs 16–20 weeks	7.5 ℓ
Heifers	21 ℓ	Sows	11.5 ℓ
Dairy cows	45 ℓ	1000 broiler	
		chickens	80 ℓ
1000 turkey growers	130 ℓ	1000 laying hens	140 ℓ

Where there is sufficient clearance, farmyard manure may be stored in place until it can be spread. Alternatively, a double-handling system is used, involving removal to a temporary storage site followed by spreading onto the land at a convenient time of year.

Using a tractor mounted front-end loader, a rough terrain fork lift with manure bucket, or an industrial loader similarly equipped, farmyard manure can be loaded into trailers or spreaders at a rate of 20–60 tonnes/h.

Spreading on the land is by rear discharge or side discharge, spreader, which have carrying capacities of 2–10 tonnes.

Rear discharge machines resemble trailers. They are equipped with a slatted floor conveyor to move the load rearwards into a beating and shredding mechanism. This distributes a swath of manure behind and to either side of the machine. Some of these spreaders can be adapted to handle other materials.

Side discharge spreaders are cylindrical in shape, with an opening along the top portion for loading and discharge. A shaft through the centre of the cylinder carries numerous chain flails, and a rigid flail at either end. As the shaft rotates, manure is discharged first from the ends of the load, and then progressively towards the centre of the load. Spreading is to one side of the machine only; a wide range of materials from strawy farmyard manure to semi-liquids can be carried and spread.

Liquid manure

In livestock housing systems where little or no bedding is provided, the manure consists only of faeces and urine. This is removed from contact with the animals by allowing it to fall through a slatted floor into a space below, or, as in the case of cubicle housing for dairy cows, dunging areas are scraped daily by a tractor rear-mounted scraper.

It is rarely possible to spread manure daily, so a storage tank must be included in the system to hold, typically, at least three months' production. A longer storage period may permit most of the manure to be applied to stubbles in autumn, or some other seasonal priority.

Size of store

When calculating the size of store necessary for a particular length of storage, it is advisable to add a factor of 40% to the figure given in *Table 24.13*, to allow for spilt drinking water and any washing water which will find its way into the store. If rainwater cannot be excluded, an additional volume V ℓ must be added, where V = (mm of rainfall over the storage period) × (m² of area draining into the store + m² surface area of the store).

Size of tanks

While liquid manure stores can be excavated and lined, or constructed of concrete or masonry to any size, many farmers install prefabricated stores which are bolted together from curved coated steel panels, mounted on a ground level concrete base. Stores are available in diameters of 4.5–25 m, heights of 1.2–6 m, and capacities up to 3000 m³. When planning a manure handling system, provision must always be made for increasing its capacity, even if no immediate increase in livestock numbers is envisaged. An increase in store height to give greater capacity is only feasible if the original store was designed to accept this increase.

The storage tank may be below the level of the yard, in which case it can be filled directly by the scraper. Alternatively, partially or completely above ground tanks may be used. A small below-ground sump, covered by a heavy grid, receives the scraped manure from the yard. This is transferred into the main tank at intervals by a pump.

Successful operation of a liquid manure system demands care and good management at all stages, to include the following points.

(1) Exclusion of any fibrous material: hay, silage, straw. This will otherwise block up pumps, and contribute to crust formation.
(2) Pumping of the manure can be facilitated by the addition of water to the dung and urine, although this also increases the volume of material to be handled. A pump capable of delivering 75 ℓ/s of a manure with 'gravy-like' consistency may only be able to move 10 ℓ/s of a thick, porridge-like slurry.
(3) If the manure is to be handled by pumps and tanks, the contents of the tank must be thoroughly agitated at least fortnightly to prevent the formation of surface crusts in the case of cattle manure, or sludge deposits in the case of pig manure. The pump which is used for handling the manure can also be used for this purpose. Movable jets can be provided at various levels in the tank, and the output of the pump is concentrated through any jet to break up the surface crust, dislodge sludge deposits, or just mix the contents of the tank. Thorough agitation is also essential immediately before the tank is emptied. A recent development is the use of air bubbles to agitate the tank contents. Air is introduced into the bottom of the tank from a pattern of permanent plastic tubes, and can achieve a very thorough mixing for a very small power input (5 kW or less). Very frequent agitation can be programmed into the controls of the electric air pump, without the need for constant supervision or for a tractor to provide power.

Liquid manure is unloaded from the store directly by a centrifugal or positive-displacement pump, or by vacuum into a tanker of 3500–13 500 ℓ capacity. After transport to the field, a built-in spreading mechanism distributes the manure as the machine moves forward. Considerable odour may be produced, particularly if the liquid manure is thrown high into the air. As an alternative, some tankers can be fitted with heavy tine 'injectors' which can place the manure below soil surface. This minimises odour and also reduces the loss of nutrients by run-off.

Where the manure is to be spread within 0.5 km of the store, organic irrigation may be considered. A pump capable of handling semi-liquids distributes the manure through mains of 100 or 125 mm diameter to a raingun which covers 0.1–0.2 ha at a setting. Mobile irrigators may be used, so that a very large area may be covered without re-setting. The addition of at least an equal volume of water to the manure is essential for ease of pumping. Again, odour may be a problem; wind direction should be favourable before operations are commenced.

If the manure store can be entered by a ramp down from ground level, it can be emptied by a self-loading tanker which is backed down the ramp. Alternatively, a four-wheel drive loader with large capacity bucket may be used to unload the manure at rates of 60–80 m^3/h into tankers or tipping trailers.

Slurry separators

By passing liquid manure over a series of sieves, rollers or vacuum devices it is possible to separate most of the liquid, leaving a fairly dry material which can be stored and handled in the same way as farmyard manure. The liquid fraction can be handled at a high rate by means of irrigation equipment or tankers.

25

Farm buildings

J. L. Carpenter

GENERAL

Investment

The level of capital invested in building structures on individual holdings may vary from 10–90% of the business total, reflecting the great variety of purpose, intensity of use, and method of construction to be found.

Buildings can be broadly categorised according to their production function. They may:

(1) assist in productive processes, e.g. potato chitting, bacon fattening and flowering of chrysanthemums;
(2) facilitate marketing operations, e.g. grain cleaning and storage;
(3) provide general farm services, e.g. workshops and fertiliser storage; or
(4) control an environment essential for production, e.g. mushroom houses and veal calf housing.

In practice the majority of buildings have functions under one roof which are combinations of two or more of the above. These multiple activities often make it difficult to establish the true value of proposed systems when assessing the potential relationship between capital input and resultant contribution to farm business output.

Other factors which govern investment levels include type of farming, scale of operation and management inclinations. The result of all these design input parameters is a range of structure types, from those which provide minimal stock protection at one extreme to total environment control for cropping at the other.

Planning and design

Despite this great variation there are some consider-
ations which are common to the planning processes of all farm buildings. These may be summarised as follows:

(1) suitability for purpose and use;
(2) appropriate cost, relative to permanence and maintenance;
(3) flexibility of design to allow for
   (a) expansion,
   (b) production of changes,
   (c) alternative (or alternate) use;
(4) minimal labour and energy use;
(5) minimal risk
   (a) to health of crop, stock or manpower,
   (b) of fire,
   (c) of polluting the surroundings;
(6) utilisation of common/standard/local materials (with account taken of DIY element);
(7) harmony with surrounding environment;
(8) appropriate use of site topography for aspect/drainage/shelter/and gravity movement of materials;
(9) finance available;
(10) impact of legislation and codes of practice;
(11) potential modification or extension of existing buildings and allowance for age and technological differences.

Additionally, specific use buildings will have particular requirements concerned with their productive functions, e.g. precise environment control (refer to later sections relating to individual building types).

Recent trends

There are three areas of farm building design interest which are subject to rapid change at the present time. One is the increasing application of legal and other authoritarian constraints to farm construction in the UK, the second concerns the environmental impact of modern agricultural structures and the third derives from the need to control internal building environments more precisely.

Legislation has affected farm buildings in the past in many ways, both directly, e.g. by the Town and Country Planning Acts and indirectly, e.g. by the Control of Pollution Acts, but hitherto, decision making regarding acceptability was made difficult because of a lack of precise standards for comparison and regulation. Recently, in January 1980, a new British Standard (BS 5502/1980) covering all aspects of agricultural building works was introduced. This establishes a clear and precise set of criteria against which building performance and satisfaction can be measured. It can be used within the existing legal framework to create a greater pressure for high quality workmanship and good building practices on farm sites (*see Appendix 25.1*).

The rural environment and its protection from industrial erosion has become a great public concern; this must inevitably affect attitudes to farm buildings as they probably have a greater immediate impact on the landscape than any other farming factor. Thus great care must be taken with the choice of building layouts, siting, shapes, colours and textures in order to create a harmonious balance between the completed structure, farming technology and the existing country scene.

The need to increase productivity has created pressure for better and more effective use of expensive space, particularly within specific purpose buildings. As the 'shell' of a building represents a large proportion of the total investment and is a relatively 'fixed' cost, regardless of the way the system inside behaves, much design effort in recent years has been put into achieving more precise environments by manipulation of cladding, services and mechanical/electrical control mechanisms. The use of sophisticated electronic computer technology and building materials based on plastic derivatives will greatly assist in this aim of accurate environmental control. The most advanced examples are to be found at present in the glasshouse section of the agricultural industry and in some types of crop storage but much thought is currently being given to animal housing methods, and control systems are improving rapidly for these.

STEPS IN THE DESIGN OF FARM BUILDINGS

Preparation

(*See Appendix 25.2* for advisory organisations which may be of assistance.)

(1) Establish the intention or productive purpose of

the buildings and examine the motives, financial and otherwise of investing in a building for this purpose.
(2) Consider the implications of such buildings inputs and outputs during construction and operation on the rest of the farm environment and business activities.
(3) Calculate in a simple fashion the overall size of the building unit and from this calculate, very approximately, the probable cost (*see Table 25.1*).
(4) Establish the financial constraints and the likelihood of grant aid and tax concessions being available for the proposed work.
(5) Make comparisons with other farm building systems which perform the same functions, and discuss the alternatives with persons having some experience of them (*see Appendix 25.2*).

Table 25.1 Ranges of cost for new agricultural buildings, per production unit, at 1980 prices (£)[1]

	Gross cost (£)
(1) Cow accommodation per adult animal	
Housing and feed system	200–550
Parlour	80–100
Parlour equipment	70–125
Additional automatic parlour equipment	150
Bulk milk tank	50
Slurry collection and containment	150–250
Probable total	500–1200
(2) Beef accommodation per animal—housing only	200–450
(3) Calf accommodation per animal 3–6 months old—housing only	60–120
(4) Bull accommodation/adult animal—pens with handling equipment	7500+
(5) Sows, dry—housing per animal	80–120
(6) Sows, farrowing, housing + equipment per animal	400–1000
(7) Boar accommodation, single pens only	1400+
(8) Weaner pigs, housing per animal	30–60
(9) Fattening pigs/animal 3–5 months old	40–80
(10) Ewes and lambs—housing only, per ewe	60–80
(11) Yearling sheep—housing and feedstore, per animal	30–50
(12) Sheep handling systems/m²	20
(13) Potatoes (bulk store)/tonne	40–60
(14) Grain (bulk store)/tonne	30–50
(15) Grain (bin containment)/tonne	50–90
(16) Silage/tonne—clamps	c. 20
(17) Silage/tonne—towers	c. 80
(18) Hay, bales—store only/tonne	c. 50

[1] (Subject to scale economies, variation in tenders, generosity of specification, etc.)

NB: Low end of ranges indicates simple, cheap system with minimum control sophistication and high DIY element. High end of ranges presumes long life expectancy, complex environment, low labour input for operation, high mechanisation level.

(6) Refer to planning/structure/safety/pollution/ health regulations for building controls (*see Appendix 25.1*).

(7) Examine possible sites for buildings, consider size, access (vehicular and animal) and circulation, soil type, drainage, environmental impact, aspect, availability of services, slope.

(8) Allow for expansion.

Concept confirmation

(*See* sections on specific buildings and structure design.)

(1) Calculate the area of building from the needs of production. Examine alternative internal arrangements, including required servicing areas. Provide for standard size of equipment and preferred sizes of buildings materials and structure. Allow for feeding, circulating, cleaning, handling, emptying space as appropriate.

(2) Consider aspects of materials handling and adjust flow lines of materials, animals, people, machines. Arrange so as to create minimum obstruction or confusion and shortest lines of travel. Avoid changes of direction where possible.

(3) Establish the number and size of fittings and servicing equipment to be installed.

(4) Height of building must be decided by reference to storage needs, size of occupants, manoeuvring space, doorway clearances, air exchange volumes and the external environmental effect the structure will have on its surroundings. The interaction between clear space inside and roof shape and slope should be taken into account, as well as the relationship between the weather and building height and shape.

(5) Decisions can now be made with regard to preferred materials, manufacturers and their product ranges to match the requirements of building size, arrangement and fittings. It is generally desirable at this point to obtain advice from expert bodies unless the building is very simple; and it may be preferable to employ professional help from this point in the design process on a fee paying basis. There are many advisory bodies who exist to disseminate information about farm building design and construction (*Appendix 25.2*) but most of them cater for sectional interests only and are not able to advise over the wide spectrum usually required for complete systems.

Detailing the building

(1) Following the establishment of general requirements, methods of cladding, roofing and roof support can now be chosen with reference to insulation values, rigidity and strength, damp and water proofing, and smoothness for cleaning.

(2) The form of foundation and groundworks for the structure can be ascertained by reference to the load imposed by the structure and its contents, the weather effects on the building and the soil which is to act as the supporting medium. This part of the design should only be carried out by experts fully conversant with both the site and the principles of foundation construction. The size and depth of foundations may, in some areas, be very critical, so 'rule of thumb' techniques can only be used under very restricted conditions (*see* later in section on foundations). These do not allow for the large potential variability deriving from soil, climate and loading conditions. A site survey should be carried out at this stage of the proceedings.

(3) Drainage works should be considered and problems of diversion, linking or modification should be established and accounted for in the proposals. Care should be taken to discharge clean surface water to water courses or soakaways and foul water to appropriate holding or processing systems.

(4) Further consideration of the environmental control features may be made at this stage of design. Required temperatures, humidities, light levels and air flows should be established and calculations of insulation, fan power, duct size and positioning should be carried out where appropriate (*see* later section on control of the environment).

Drawings for the construction (*Figure 25.1*)

(1) Drawings should be prepared to scales of 1:50 or 1:100 showing:

(a) the layout in plan view indicating position of doors, barriers, feeders, circulation space, drainage, water and electricity supplies in relation to frame members, walls and other structural features at or above ground level (some of these might be more clearly shown on separate additional drawings to avoid confusion of lines and shading);

(b) the building in elevation to demonstrate shape, volume, cladding materials and external features likely to be obtrusive;

(c) cross sections of the structure demonstrating material details and thicknesses, positions of service fixings and the nature of space usage in the building.

(2) A scale map of 1:2500 should be used to show the location and general arrangement of the building relative to its surroundings. (Permission may be granted for a copy to be made for personal use from an Ordnance Survey Map of this scale.)

528

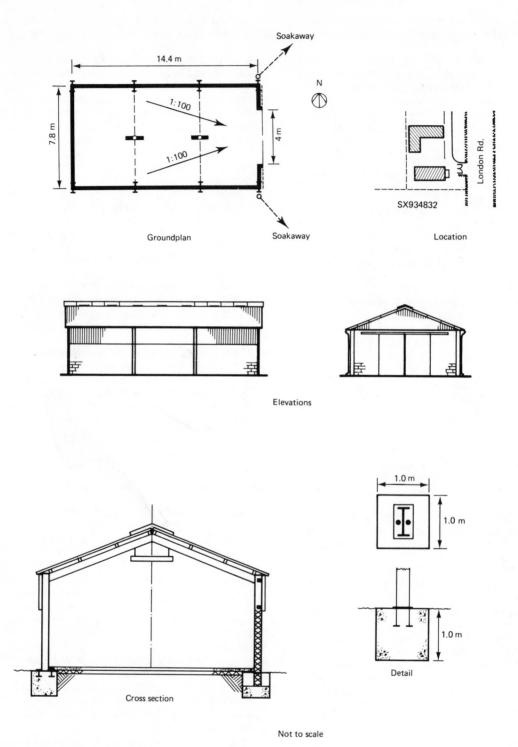

Groundplan · Soakaway · Location

Elevations

Cross section · Detail

Not to scale

Figure 25.1 Building drawings

(3) Further drawings to scales of 1:10, 1:20 and 1:50 may be prepared to show constructional features such as foundations, framing and fixings, drainage schemes, damp and vapour proofing, insulation methods and mechanical services. These are generally the province of professional experts, surveyors and architects or may be provided by commercial building firms as part of a sales service (*see Figure 25.1*).

(4) Planning and grant-aiding authorities may need to see some or all of these drawings to ensure that the law is being complied with in every respect and that the building conforms to the requirements of BS 5502 (*see Appendix 25.1*). (In practice it is advisable to consult the appropriate bodies at an early stage of design thinking. It is not good, or safe, procedure to request retrospective approval.)

Pricing and general arrangements

(1) After plans, elevations and sections have been drawn up, prices or tenders can be obtained. With the current uncertainty about the effects of inflation on costs, builders and estimators are finding it very difficult to provide firm prices for building work. They may prefer to operate on a 'cost plus' (cost of materials and labour plus a reasonable addition for expenses and profit) or a 'price, ex-works, at date of dispatch' basis. It is important to bear in mind the possible extra complications of Value Added Tax in pricing procedures when building costs are obtained piecemeal.

(2) It is very common for agricultural construction work to be carried out in separate stages by different types of commercial organisation. For instance, frame members and part-cladding for the structure might be provided first, by a largescale manufacturer. This would be erected by a sub-contracted site crew, whilst the rest of the work might be carried out afterwards by a local building firm, who might in their turn sub-contract some specialist jobs such as electrical installations. This system of building construction is very flexible and allows for elements of DIY to be introduced, or the use of 'labour-only' contracting, giving relatively low costs. An alternative contract arrangement (more commonly found outside agriculture) in which a 'main contractor' takes responsibility for most of the site works, and possibly for their supervision, generally leads to less management confusion and a more orderly 'attendance' of tradesmen at the site as required. This militates also for a higher quality of workmanship overall.

(3) It should be made clear to all parties what work on site is expected of them before contracts are signed and the work commences. Not all requirements can be predicted before construction starts but the chance of situations arising in which variation orders have to be issued to contractors should be minimised, as these can lead to an insidious rise in costs. A contingency sum should perhaps be set aside, in the original design estimate, to cover this set of circumstances.

(4) Suitable insurance cover should be arranged for the sitework and building during and after construction.

SITE WORKS STAGES

Preparation—setting out a simple rectangular building

(1) The turf and surface soil is stripped from the site, along with trees, shrubs, bracken, etc. complete with roots, if possible. The subsoil thus exposed is 'levelled' over an area exceeding that of the building. If the site slopes then levelling will be needed to reduce the surface to a satisfactory gradient as required by the plans.

(2) Divert existing drains which cross the site (or relay in a straight line, surrounded by 150 mm of 1:2:4 (C20P) concrete (*see* later section on concrete) deeper than proposed works and on suitable falls.

(3) Stretch a line between pegs to represent the position of the longest side of the building but much greater in length such that the pegs are well clear of the proposed corners as at (a) in *Figure 25.2*.

(4) Mark one of the corner positions with another peg (b) and at right angles from the first line stretch another to mark a second building side (c). An accurate right angle may be made by using a builder's square, a surveyor's site square, or the method of Pythagoras, (3:4:5 triangulation giving 6 m along one side, 8 m along a second and 10 m across the measured hypotenuse).

(5) In a similar manner mark the third and fourth sides (d). The dimensions and position of the building are now shown in plan by the lines and corner intersections. It is usual for these to be the outside edges of the building above ground but they may represent centrelines or inside surfaces. It is important to plan according to fixed reference surfaces and levels throughout the construction and to avoid changing them unless it is absolutely necessary.

(6) Check the diagonals (e) of the building outline— they should be the same.

(7) Profile boards (f) are now situated beyond each corner well clear of the building area. These are pieces of timber (e.g. 100 mm × 25 mm) pegged into the ground having permanent markers (nails or sawcuts) indicating the position of some aspect of construction, for example the edges of the foundation trench. Another set of marks on the board might show the position of the wall to be built on the foundation.

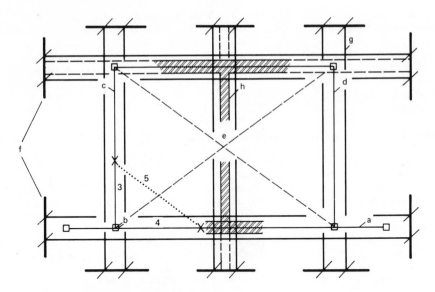

Figure 25.2 Setting out

(8) Lines stretched from marker to marker (g) provide an accurate facsimile of the ground plan full size on the surface to be worked on. Profile boards at right angles will represent corner details and others may show internal features such as walls and doorways (h). The lines are only a guide and are taken away before construction begins, being replaced by silver sand trickled along, chalk marking or paint. The profile boards are retained until such time as the reference marks on them have no value at an advanced stage of construction.

(9) The heights and levels over the site should be set out and checked against a vertical reference stake driven into the ground firmly, clear of all site activity. Marks on the stake should indicate ground level where the stake was driven in, a general reference height, say 1 m above ground level, and other points necessary to the construction such as damp course level.

Construction—groundworks

(*See* sections on foundations and drainage.)

(1) Trenches are dug out to profiled width and correct depth with vertical sides and firm bottom. Any areas showing signs of softness are stabilised with stone ballast or concrete and rammed down. Water and water-softened soil should be removed prior to concrete pouring. In very soft soil conditions it may be necessary to support the trench sides with struts and board temporarily, not least to ensure safety of workers.

(2) To ensure horizontal levelling and the correct thickness of materials used, drive pegs into the soil at the trench bottom so that the tops represent the finished foundation surface. Use a spirit level and a straight board over short distances but a surveyor's level, either automatic or water tube type, (*see* later section on surveying) should be used when the distances to be pegged exceed 10 m in total.

(3) Foundations should always be horizontal; on sloping sites a 'stepped' foundation is required. The depth of each step must be a multiple of whole building units, e.g. block(s) or brick(s). Thus:

$$\text{Distance between steps} = \frac{\text{Depth of step} \times 100}{\text{Slope \%}}.$$

An overlap must be allowed for on each step such that the slab above projects over the slab below substantially more than the step 'depth'.

(4) The concrete mixture used for foundation works will normally be 1:2.5:4 (*see* section on concrete mixing) but may be 1:3:6 in certain situations. The Ministry of Agriculture, Fisheries and Food's 'Standard Costs, Part I: Specification' uses a recent British Standard (BS 5328:1976) to indicate quality requirements for grant aid. The concrete used must be that defined as C7P, which implies precise mixing, batching, laying and curing techniques as well as mix of materials, to give high quality, high strength concrete.

(5) The top surface of the foundation concrete must be level, smooth and horizontal to facilitate the laying of brick, block or framework.

(6) Drainage trenches are dug out at the same time as the foundations and any work which affects both is carried out first. For instance any pipes to be positioned under or through strip foundations should be placed and fixed before the concrete is poured.

(7) Site services are laid onto the site and are led to the main distribution points for connection on a temporary basis during construction and permanently when the building nears completion.

Construction—frame and cladding (*see Figures 25.3 and 25.4*)

(1) Each foundation pad supports and locates a frame member. The location system (socket, bolt assembly, etc.), should be positioned exactly to receive the stanchion of the frame without distortion. Some frame manufacturers may wish to cast pad foundations themselves for this reason. Small frames may rely on their own weight only to hold them in position.

(2) The frame members will arrive on site, sectionalised, for assembly prior to lifting into position by mobile crane. The span sections are lightly bolted and stabilised by use of longitudinal members, e.g. the purlins, until the whole structure is complete at which time the joints are made in a permanent fashion.

(3) Steel, concrete and timber frames have different shapes and material characteristics which require different assembly and fixing techniques (*see Figure 25.4*). The manufacturer of the frame will give specific guidance as to the fixing and foundation requirements of particular structures.

(4) Once the frame is secured roof cladding is fixed, starting from one gable end away from the direction of the prevailing wind. Care is needed to align the corrugated sheets squarely with the verge and eaves of the building, or there may be a danger of too little or too much overhang as the laying of sheets progresses down the length of the building.

(5) After both slopes are covered the ridge capping is fixed in place. The overlap on both sheets and capping should be such as to minimise penetration by the prevailing wind.

(6) The gable ends and side sheeting are positioned, cut as necessary and secured, followed by the guttering, at a suitable slope, and the 'down' pipes to channel the surface water to ground level.

(7) Special features like translucent sheeting, eaves fillers, ventilation ducting, insulation or anticondensation materials may be incorporated as the roof is clad.

Construction—masonry and floor laying

(1) With the frame in place blocks or bricks may be laid on strip or 'trench-fill' foundations between stanchion members up to damp-course level, at 150 mm above outside ground surface.

(2) Doorways and other gaps are left in the wall as appropriate with timber 'liners' or with fixing blocks left in the masonry for subsequent connections.

(3) The walls will be built according to the design specification based on the original requirements, and may incorporate cavity construction, piers and/or other reinforcements, insulation, water proofing or a cleanable surface.

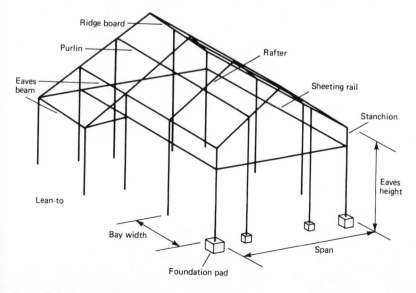

Figure 25.3 Structure elements

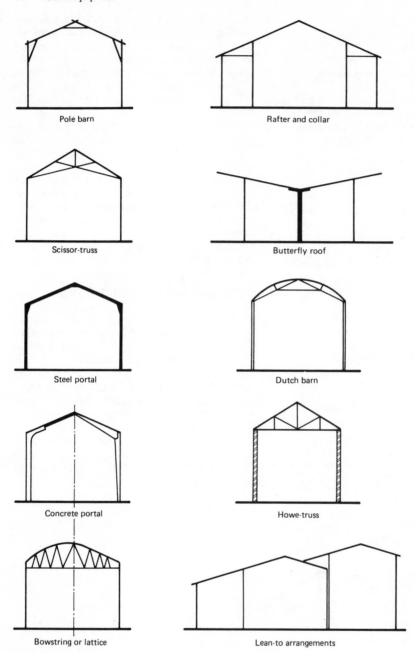

Figure 25.4 Frame systems

(4) When the walls have reached damp proof course (dpc) level the floor can conveniently be laid. One form of construction sometimes used in agriculture for lightweight buildings is the 'ring-beam' foundation system. This is a method in which floor is cast in concrete together with the frame and wall supports, usually reinforced at the edges. The surface should already be level and free of top-soil. Pegs are placed to indicate the finished surface and hardcore is laid and vibration rolled into a hard, dense mass. The top of this is then 'blinded' by sand to fill all surface cracks and provide a smooth plane on which to place the damp proof membrane. Polythene sheet, 125 μm thick, or building paper, is laid, with a miminum overlap of 150 mm at the sheet edges, to cover the

floor and is bonded to the wall dpc by bitumen based paint or mastic.

(5) The floor concrete (*see* section on mass concrete later) should be poured in convenient shuttered bays, not exceeding 10 m × 3 m in area, for manageability and timeliness. It should be to C20P or C25P (BS 5328/1976) specification and of a suitable thickness, usually a minimum of 100 mm. It must be tamped and surfaced by floating or other method, preferably with the use of vibration equipment for consolidation. After which it should be left to 'cure' for at least 7 d and preferably 14 d, before the enclosing shuttering is removed or it is required to bear a load.

(6) An additional surface may be called for in some buildings, for instance in a farm dairy where a granolithic 'screed' is necessary. This is a thin (up to 50 mm) top layer of fine granite aggregate concrete mix, applied to the base mix before curing has finished, and preferably within 72 h of laying. This is a skilled job, usually best left to specialist tradesmen.

Construction—internal finishing and services

(1) The internal fittings such as water supplies, electrical installations, joinery work and masonry finishing (plastering, etc.) are normally the province of highly skilled tradesmen, whose quality can make or mar a building. In this work there is an element of flexibility of design and decision making which is not possible in other areas of construction. 'On-site' decisions are usual and a craftsman would be expected to think for himself in respect of details like pipe and wire runs and fixing methods on all but the largest farm construction. However there are some pitfalls on which advice needs to be given, e.g. door security in piggeries, moisture proofing of electrical equipment in dairy parlours and positioning of drinking bowls.

(2) The complexity of the work to be done and the need for a careful order of procedure to be followed means that the timeliness of the attendance of the specialists is often important to the smooth transition of the stages of building construction to avoid time wasting and mistakes. Overall supervision is crucial to the successful completion of a complicated building and the employment of an experienced professional to do this, is recommended.

BUILDING MATERIALS AND TECHNIQUES

Mass concrete

Concrete is usually a mixture composed of ordinary Portland cement (as opposed to rapid hardening Portland cement and other types), coarse aggregate, fine aggregate and water. Aggregate may consist of gravel or crushed stone, of a light or a heavy nature, and may be round or sharp (angular). The sand or fine aggregate fills in the voids between the coarse particles, the cement coating adhering to both. This produces an almost solid matrix when set. All aggregates must be clean and free from vegetable matter, salt and clay. They must be dry for accurate volume measurements to be made prior to mixing. 'All-in' or 'as-dug' aggregates should not be used. Always use clean water, preferably from a drinking source. Although it is possible to mix concrete by hand in small batches it is not recommended practice as the product will not be of a high enough quality to conform with current standards. The most reliable way of apportioning mix ingredients is by 'weigh batching' that is weighing out enough materials to mix with one bag (50 kg) of cement, or a multiple number of bags, and to load the quantities into a suitable sized mixing machine in one whole batch.

The time of mixing should be long enough to ensure that all the aggregate particles are coated evenly with cement to give the mix a consistent greyish coloration, normally a minimum of 2 min is required for this, after the last ingredient has been placed in the drum. Most mixers work best when the coarse aggregate is placed in the drum first, followed by the water, then half the sand, then the cement and finally the second half of the sand. Some 'sticky' aggregates may require that this order be varied to get the best possible mix.

The quantities of cement, aggregate and water will vary according to the job to be done.

Table 25.2 shows the normal proportions by volume, for the work loads expected in three types of circumstance, on farm sites.

The size of aggregate must be taken into account when selecting an appropriate mix. Thin section or reinforced concrete requires stone which is relatively small in size, graded from 6–20 mm, whereas large mass concrete can satisfactorily enclose aggregate of up to 40 mm size grade. Fine aggregate is 5 mm or less in size, usually, but must not contain too much dust.

Water quantities are critical—almost as important to the ultimate strength of the concrete as the cement itself. Too much weakens the concrete by cracking when it dries out, too little will give ineffective bonding due to lack of chemical reaction in the cementing process. There is therefore an optimum amount, but as workability is also governed by the water content of a mix, very often concrete is mixed and laid much too wet to increase the facility of placement. The moisture content can be checked by use of the 'slump' test. Special testing cones can be purchased or hired but a steel bucket may be used to provide a rough guide; the bucket is filled with concrete and is then inverted onto a flat surface (steel/wood plate), the bucket is drawn gently away from the contained concrete leaving a cone-shaped

Table 25.2 Concrete mixes for various purposes proportional by volume

	C7P Large bulk volumes, e.g. deep foundations		C20P Average use, floors, roadways		C25P Heavy use and thin sections	
Cement	2 vol	1 bag	3 vol	1 bag	3 vol	1 bag
Fine aggregate (damp)	5 vol	190 kg	5 vol	115 kg	4 vol	90 kg
Coarse aggregate	8 vol	270 kg	9 vol	195 kg	7 vol	170 kg
Water	1–1.25 vol	30 ℓ	1.25–1.5 vol	23 ℓ	1 vol	17 ℓ
Approx. yield	7–8 vol	200 ℓ	9–10 vol	170 ℓ	7–8 vol	140 ℓ
Previous designation	1:3:6		1:2:4		1:1.5:3	

heap, the height of which has 'slumped'. A slump of more than 25% of the height of the bucket shows the mix is too wet. Using the correct test equipment 50 mm ± 25 mm should be attained at the correct water content. If such a mix is made it will probably require vibrating equipment to place and consolidate (tamp) it satisfactorily. These can also be hired from the same hire centres as will provide the slump testers.

All concrete should be laid as soon as it is mixed and always before 1 h has elapsed. It should be placed carefully into position from a low height avoiding any movement which might cause desegregation of the particles, for instance rough barrowing. There is a brief period of 'setting' (about ½ to 2½ h) followed by 'hardening', a chemical process which carries on for a long period of time gradually achieving greater strength. During this latter 'curing' period the concrete should be undisturbed, and should be kept from drying out and frost-free. The actual curing time depends on the conditions of mixing and the environment as well as the mix but is never less than 7 d and in most cases it should be 14 d. Under no circumstances should concrete be mixed when the aggregates are frosted and in very cold conditions (less than 4 °C) expert advice should be sought as to methods of using antifreeze (*not* motor car type) enabling concreting to be carried out.

Pre-cast concrete

Many concrete items used in construction are 'precast', that is, they are made away from the point of use, e.g. concrete blocks, drain covers, frame members.

Concrete blocks

These are supplied in various sizes and aggregate material to suit the requirements of the structure. The majority of those used in farm construction work are of 'dense' aggregate, although 'cellular' blocks are used from time to time for insulation purposes. These, latter, have a bath sponge appearance, due to the foaming method of manufacture; although of low heat conductivity they are relatively weak and are porous to water passage. (In fact, all concrete blocks are poor

moisture barriers, relative to bricks.) Block sizes are standardised and have nominal face measurements of either 450 mm × 225 mm or 400 mm × 200 mm by thicknesses of commonly 75 mm, 100 mm, 140 mm or 215 mm. The last two sizes may be 'hollow', i.e. have two central hollow core spaces moulded into them. The actual sizes of the blocks received on a building site may differ from the nominal size by 10 mm, which is a mortar gap allowance, so they may actually measure 440 mm × 215 mm and 390 mm × 190 mm respectively.

Concrete block walling

Most walls constructed in farm buildings are of 140 mm or 215 mm wide units; both of which may for some purposes require strengthening. This is carried out traditionally by either the incorporation of piers or the use of steel rods or netting in the bond pattern of the wall. Piers are part of the wall which are thickened by the insertion of blocks such that a vertical ridge is formed, usually one block wide up the wall. The use of steel in various forms is intended to provide strength in tension situations which the concrete blocks lack. So long as the steel is correctly placed and bonded into the wall such that it all behaves as one material, the steel reinforcement system has greater possibilities. In practice piering and reinforcement are used together to achieve maximum strength, one method supplementing the other.

See Table 25.3 regarding wall stability and further 'rule of thumb' relationships of wall width and strength.

Table 25.3. Block thickness and wall stability unreinforced

	Block size (mm)	100	140	215
Walls lacking top support	Max. length (m)	3.6	4.5	9
	Normal max. related height (m)	1.8	3	3.6
With continuous top support, e.g. rafters, framing	Max. length (m)	4.5	6	7.5
	Normal max. related height (m)	4.5	6	7.5

NB Walls designed to withstand wind and stock pressures; not to retain stored grain, etc. For larger stock use the larger block sizes.

It should be noted that many agricultural buildings carry wall loads which are lateral and quite severe. These must be treated with great care in design and 'rules of thumb' are not recommended for these.

'Cavity' walls are not common in agriculture, as they are complicated to build and can be expensive. Their main purposes are keeping the interior free from rain penetration and adding insulation to the structure. They consist of two 'skins' of 100 mm concrete blocks 'tied' together by means of steel or plastic connectors (ties) across a 50–60 mm gap left between the skins. The ties are placed in the wall, as it is built, about 1 m apart horizontally and 0.5 m vertically. It is important to ensure that no mortar or rubbish is allowed to settle on the ties as it will form a bridge across the cavity and allow water to transmit from the wet outside skin to the drier inside surface. The internal skin may be of other than dense block material, for instance it may be of cellular blocks so increasing the thermal insulation further. The cavity wall is a relatively weak structure to lateral pressure and must not be expected to retain bulk grain or potatoes unless specially designed to do so.

The bonding pattern used, in most concrete block walls is 'stretcher bond' (*see Figure 25.5*), that is, every joint between blocks in each course occurs halfway between joints in the course above and below, thus the distance in between vertical joints is 225 mm or 200 mm in succeeding courses. When corners (quoins) are reached, the bond pattern is retained by the use of one-quarter and three-quarter blocks in alternate courses but in no circumstances should the distance between adjacent course vertical joints be less than one-quarter block width.

Long walls (where the length exceeds the height by one and a half to two times or where there is no natural bond break within 6 m) should have an expansion break to accommodate thermal and moisture movement. This is provided by a slot cut in the outer block surface at least 50 mm deep or possibly completely through the wall. In the latter case the stability is retained by the use of steel dowels set in the mortar on one side of the break and greased before laying in the other side. The slot should be filled with mastic at the surface to provide a weather-seal.

The mortar used in concrete block laying should be 1:2:9 or 1:1:6 by volume proportion with cement, lime and builder's sharp sand, respectively. The lime may be replaced with proprietary ingredients which make the mortar 'fatty', that is, slip easily into place between the blocks. These mixes are weak relative to block strength so that if there is a tendency to crack, any such cracks will appear less unsightly and be of less significance to wall stability when they are confined to the mortar gap.

Concrete block walls may, depending on their purpose, require waterproofing. This is generally done by use, in the first instance, of a render coating of mortar-like material, in two coats, the first, at 10 mm thick, of 1:1:6 mix as before and the second at 6 mm thick, of 1:0.25:3. Alternatively, the block pores can be filled in and the mortar gaps flush pointed to give a flat surface which can be painted with chlorinated rubber, epoxy-resin or acrylic resin bonding paints.

A new approach to block laying combines the simplicity of 'dry-laying' that is no mortar usage, and render-coating for water proofing by the use as a wall fixative of a mix of cement, sand, glass fibre and bonding chemical to produce a hard, durable coat which stabilises and creates the strength of the wall from the outside surfaces.

Concrete frame members

These are almost always precast, although very simple ones may be cast *in situ*, for instance in the case of door lintels. They contain steel reinforcement at a position determined by the need to carry tensile loads (*see Figure 25.6*) for which concrete has only one-tenth of its capacity to carry compression loads. The need

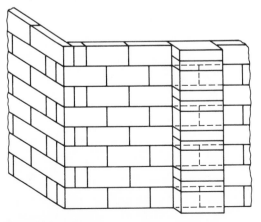
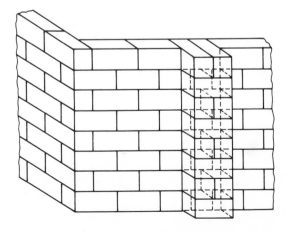

Figure 25.5 Wall bonds

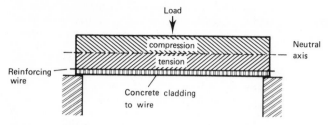

Figure 25.6 Simple reinforcement in a short concrete beam

for a high degree of reliability and thus quality control ensures that most of these are factory made items. They may be of a type known as 'pre-stressed', in which the steel wire in the beams, etc. is pre-tensioned before it has concrete settled round it in the moulds. This technique increases the possible compressive load which can be brought to bear on a member before the tension bearing reaches a limiting value. The net effect of this is to enable economies to be made in all parts of a frame due to the reduction of surplus concrete, which in plain reinforcing systems is not 'working hard'.

Timber

Timber used in construction should be well seasoned, pressure treated with biocides and free from warping, knots, sapwood, bark and shakes (splits). Softwood is used for nearly all but decorative or highly specialised building work due to the high cost and difficult working of most hardwoods. Sawn softwood can vary in standard off-the-shelf size from 75 mm × 16 mm in section to 225 mm × 100 mm. Larger sizes may be less easy to obtain. Lengths are now sold by the metric 'foot' (300 mm) as follows, 1.8 m, 2.1 m, 2.4 m—by 0.3 m–5.1m but for larger quantities, by the cubic metre, in particular sizes. Beyond a length of 5.4 m a premium price may have to be paid. All dimensions are likely to be nominal and account must be taken of sawing and planing allowances which may be as much as 2 mm on each dimension, less than the nominal.

There are three main biotic decay agents of timber in UK climatic conditions, the so-called 'woodworm', (actually a beetle and its larvae, *Anobium punctatum*) 'dry-rot' (a fungus, *Merulius lacrymans*), and 'wet-rot', which may be caused by several fungal agents. Elimination or eradication of these pests is usually a matter of avoiding poor climatic or environmental conditions, for instance by keeping the timber dry when exposed. Trouble can be avoided at the outset by choosing an appropriately good quality of timber and treating it with permanent preservative biocides. Repair of timber which is already infected is difficult and sometimes, as in the case of dry rot, almost impossible without wholesale replacement.

Recently the use of 'stress-graded' timber has become a requirement in some types of structure. This is timber which has been inspected visually for flaws such as knots, size imperfections, shakes and other defects. Marks made on the timber indicate the limitations to its use in structural work. GS (general structural) has been passed as acceptable for general use, e.g. as floor joisting, SS (special structural) implies a more particular usage (*see* BS 4978 on stress grading of timber).

For information concerning screws and nails for timber work *see Tables 25.4* and *25.5*.

There are two general design concerns, other than biotic decay, expressed about timber in structures; one is its apparent susceptibility to fire risk, this is not as important as might be suspected. Timber in large sections tends to char rather than burn through, particularly in the case of 'treated' timber, this is due to its low thermal conductivity: hence if a proper

Table 25.4 Wood screw sizes (slotted countersunk steel)

Gauge	Diameter of unthreaded shank (mm)	Lengths available (mm)
0	1.58	6, 10, 12
1	1.78	5, 6, 8, 10, 12, 16
2	2.08	6, 8, 10, 12, 16, 19, 25
3	2.39	6, 8, 10, 12, 16, 19, 22, 25
4	2.74	6, 8, 10, 12, 16, 19, 22, 25, 32, 38, 45, 50
5	3.10	6, 8, 10, 12, 16, 19, 22, 25, 32, 38, 45, 50
6	3.45	8, 10, 12, 16, 19, 22, 25, 32, 38, 45, 50, 57, 63, 70, 76
7	3.81	10, 12, 16, 19, 22, 25, 32, 38, 45, 50, 57, 63, 76, 89, 101
8	4.17	10, 12, 16, 19, 22, 25, 32, 38, 45, 50, 57, 63, 76
9	4.52	12, 16, 19, 22, 25, 32, 38, 45, 50, 57, 63, 76, 82, 89, 101, 124, 127
10	4.88	12, 16, 19, 22, 25, 32, 38, 45, 50, 57, 63, 76, 82, 89, 101, 124, 127, 152
12	5.59	16, 19, 22, 25, 32, 38, 45, 50, 57, 63, 76, 82, 89, 101, 124, 127, 152
14	6.30	19, 22, 25, 32, 38, 45, 50, 57, 63, 76, 89, 101, 124, 127, 152
16	7.01	25, 32, 38, 45, 50, 57, 63, 70, 76, 89, 101, 127, 152
18	7.72	32, 38, 45, 50, 63, 76, 101, 114, 127
20	8.43	38, 50, 63, 76

From BS 1210: 1963

Table 25.5 Common nail sizes

Length (nominal) (mm)	Round plain head nails, diameter (approx.) (mm)	Approximate number of nails/kg
100	3.7–4.9	22–14
75	2.6–4.1	57–23
50	2.0–3.3	147–58
40	1.8–2.6	236–118
30	1.6–2.3	383–172
25	1.6–2.0	434–294
	Cut floor brads	
75	3.3	20
50	2.6	55
40	2.3	80

From BS 1202:1966

safety factor for structural strength is used a considerable amount of the timber must be burnt away before failure occurs.

The second concern relates to its flexibility. It tends to lack 'stiffness' by comparison with the other two structural materials, so its use is restricted in simple beam or lintel framing to short spans, above which the timber must be used in a lattice or similar pattern, in order to achieve the required strength with rigidity.

Timber boards and composites

The cost of timber has risen rapidly in recent years and will continue to do so in the foreseeable future. As a consequence materials have been developed which utilise more fully the properties of the lesser quality proportion of timber previously thrown to waste, and at the same time make better use of the good quality material. Some composite materials are made partly or even wholly from sources other than wood, e.g. recent innovations in building boards which are combinations of plaster board and glass or asbestos fibres, also plastic based insulation boards in combination with wood particle board. The framework used for mounting these in structures is generally based on the traditional 'studwork' used for timber planking (*see Figure 25.7a*) but they are sometimes supported on rails in a similar manner to asbestos cement corrugated sheeting (*see Figure 25.7b*).

External grade (WBP, weather and boil proof) plywood sheets make excellent cladding for walls and roofs (with a felt covering) though care must be taken in fixing if warping is to be avoided. All edges should be secured with galvanised nails or screws, with cross nailing for hardboard at 600 mm centres. Generally, this means main studs at 1.2 m centres, equal to the sheets, a central intermediate stud and cross rails or nogging at 600 mm vertical centres. With hardboard, an inferior but cheap alternative, edges should be nailed every 200 mm or screwed every 300 mm. Only an oil-tempered hardboard should be used (*see later*).

Timber boarding can be used as wall cladding. For superior work, such as controlled environment houses, the boards are horizontal or vertical with tongued and grooved joints. For normal barn cladding, boards are horizontal with lapped joints (shiplap boarding) or, preferably, they will be tapered with lapped joints (weather boarding). Boards should be 25 mm thickness, though 19 mm is a normal economy on cheaper work. In the latter case they should be galvanised clout-nailed to vertical studs placed at 450 mm centres.

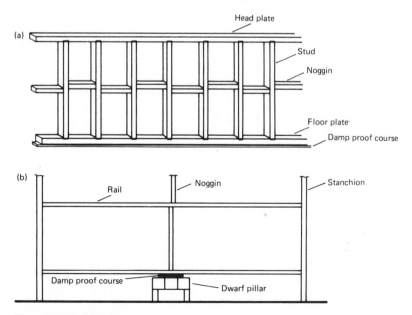

Figure 25.7 Wall framing

For proofing against wind and snow the boards should be lined with building paper to the studwork.

Vertical boards may be used to provide ventilation to cattle yards, when nailed to horizontal rails with about 13 mm gaps between adjacent ones. The boards are 125 or 150 mm × 25 mm and should be above stock height with ridge ventilation to assist air circulation. In some situations, plastic-coated or galvanised netting, sacks or hurdles can be an alternative ventilation cladding. Yorkshire boarding is used to roof cattle yards, being an alternative to slotted asbestos or aluminium sheets. Boards should be 150 mm × 25 mm with two longitudinal drainage grooves, one near each topside edge. They should be nailed over a spacer which lifts the boards 6 mm above the purlins. Gaps of 6–12 mm are left between boards.

Internal lining boards of different materials are available. Flat asbestos-cement in sheets 2.4 × 1.2 m come in several qualities, from 3 mm up to 12.5 mm thick. Asbestos-cement linings to timber frames or panels for gates can be used for controlled environment housing. Similarly, oil-tempered hardboard may be a useful cladding material for this purpose as it resists water absorption. This comes in thicknesses of 3–8 mm and sheet sizes up to 1.2 × 3.6 m. Insulating fibreboards, usually 12–18 mm thick, have low impact or structural strength and absorb moisture, though they can be obtained with a foil-lining or bitumen impregnation. Sheets are up to 1.2 × 3.6 m in area.

Lightweight cladding for frames and simple structures

Framed construction can be clad with lightweight, uninsulated sheets fixed with nails or special bolts to wall rails or roof purlins. Corrugated asbestos-cement is the normal cladding, but has low impact strength and becomes brittle with age. (Note crawling boards are to be used on this type of roof.) Sheet lengths are in increments of 150 mm from about 1 to 3 m. Sheets when laid with a single corrugation lap, have an effective width of 1 m. The minimum end lap should be 150 mm and for roofs of under 15 degrees pitch should be sealed with a bitumen cord. Maximum purlin and rail spacing should be 1.35 and 1.8 m, respectively. Sheets are available with slots in the crown of the corrugation which can assist ventilation of roofs for cattle. Translucent plastics/glass fibre sheets for roof-lights or wall panels follow the same profile and sheet size as asbestos–cement. Clear plastic ('perspex') sheets are available, increasing light transmittance, but are much more expensive. Sheets are fixed with hook bolts over steel rails or purlins or with drive screws into timber, the latter being cheaper. Bolts and screws should be galvanised, protected externally with plastic washers and caps. The holes for fixings should always be drilled in asbestos cement sheeting.

The common sheet profiles in asbestos cement have radial corrugations with a nominal pitch of 75 mm, 150 mm and 300 mm. The smaller is the pitch the thinner and lighter is the sheet and the closer the purlin support must be. It is usual to support each sheet in at least three places. Where four sheets coincide (and overlap), it is necessary to cut two of the sheets on a mitred angle to lay them as flat as possible. There are many acessory sheets and cladding members available for asbestos cement to cover awkward areas and difficult shapes, for example roof edges and curved roof sections.

Corrugated galvanised steel is an alternative sheet material, being lighter and stronger in positions liable to impact. Sheet lengths are in increments of 150 mm from 1 to 3.5 m. Exposed sites should have a double corrugation lap, semi-exposed sites a one and a half corrugation lap. Lengths up to 7.5 m are now manufactured though not in general use, being particularly suitable for low-pitch roofs due to the elimination of end laps. End laps and fixing methods are similar to those for asbestos-cement except that mitring of the sheet corners is unnecessary due to the relative thinness.

Corrugated aluminium can be used, though it is too costly for general use. Profiles and widths are similar to corrugated steel but lengths can be obtained up to 12 m. Purlins can be lighter and wider apart. Apart from reduced weight, aluminium is a good heat reflector, helping to keep buildings warm in winter and cool in summer. Thus, the material is particularly useful for controlled environment buildings.

There has been a recent revival of interest in slotted roofing which originally was restricted to wooden slat systems. This has now been reintroduced in the form of corrugated sheeting laid from ridge to eaves as normal but with no side overlap and a small gap (15–50 mm wide) left between sheet rows. The sheets are laid with the edges of the corrugations upwards at the slots. This system reduces sheet costs in a new roof and additionally saves the cost of a purpose built ventilation system let into the roof.

Frames and simple roof support structures (*see Figures 25.3, 25.4,* and *25.7*)

The term 'portal' frame should specifically be related to support frames which provide clear, unimpeded headroom to the ridge inside the building, but the term is often corrupted to mean any factory made, site assembled, large, clear span system. The common components are the upright supports or stanchions, the arms or rafters running from the eaves to the ridge, purlins, to support the roof sheets, eaves beams and ridge boards. Many types of connector are used in assembling the structural members, including bolts, clamps, nails, staples, welds and even glue, each relating to an appropriate material and structure system.

There are preferred size ranges for the major frame dimensions to which most manufacturers adhere. Bay widths are offered at 3 m, 3.6 m, 4.8 m or 6 m, with 4.8 m being the most common: 4.5 m may still be available from some firms, being the older 15′ dimension, metricated for catalogue purposes. Span widths are more variable than bays ranging from 5.4–24 m commonly, but particular manufacturers will tend to keep to more limited sections of the total range, appropriate to type of frame and material used. As a general rule the wider the span is, the shallower the roof slope has to be and the stronger the framework must be as a result. Thus very wide buildings are often provided by means of central, main spans with 'lean-tos' on one or both sides. These may extend the span by as much as 11 m per 'lean-to'. The roof slope angle may vary from 30 degrees for a small building in timber to 4 degrees for a large concrete structure. The common pitches are 10, 11, 15, 16, 20 and 22 degrees from the horizontal. The purlins are usually of the same material as the rest of the frame for the sake of manufacturing simplicity, with the exception that timber may often be supplied in non-timber frames, where easy sheet fixing by nailing is required. The purlin spacing down the roof will depend more on the length of sheets to be fixed than on strength considerations. Each sheet must be supported in three places, thus they are usually placed not more than 1.35 m apart.

Occasionally, in a narrow building, it is more appropriate and economic to use a roof truss instead of a frame system. This may be of steel, bolted or welded type, or of timber in a number of patterns, from the simple traditional nailed or bolted variety to the sophisticated design of glued or stapled lattice-work. The simplest type of frame of all is the 'beam and post' construction of which there are many versions from the rolled steel joist (RSJ) to the box beam in 'glue-lam' plywood. The choice of these is usually left to structural engineers and should not be attempted by the non-professional, but in very simple cases where timber joisting is required for heavy roofing or suspended floors a traditional 'rule of thumb' may be used:

$$\text{Depth of timber in mm} = \frac{125 \times \text{span (m)}}{3} + 50.$$

This provides the required depth of a joist system of stress graded timber, measuring 50 mm thick spaced at 400 mm centres along a strong support wall and strutted by means of noggin pieces halfway across the span. As timber is expensive, particularly in 'heavy' sections, economics will dictate the limitation to the span capacity of this method but timber's characteristic of 'giving' under load will also limit the maximum distance over which it may be unsupported, usually 6 m.

Foundations

Foundations are intended to sustain building structures against three main loading problems in agriculture:

(1) sinkage—when the building weight (its own structure load and that imposed upon it by contents) exceeds the support capacity of the soil beneath, downward or sideward movement occurs;
(2) water effect:
  (a) volume changes due to soil moisture content variation (sand soils best, clay and peat worst) cause structure stress and movement,
  (b) frost action giving volume and structure changes in soil (minimum depth for foundations related to climatic considerations),
  (c) leaching and erosion undermining the base for support;
(3) cartwheeling—where lateral (side) loads applied by building contents, e.g. stored grain, turn the structure base through 90 degrees.

Foundations must resist other forces such as vibration and load cycling but fortunately these are comparatively rare in agriculture and can be ignored for practical design purposes.

Thus foundations must be

(1) correct in area (load imposed × soil bearing capacity × safety factor) which is sometimes for simplicity to be a standard recommended size;
(2) at a correct depth (about 600 mm to 1.2 m according to soil and climate);
(3) of a correct width to prevent cartwheeling and provide for easy construction practices, e.g. trench and hole sizes are dictated partly by the skill and ease of digging, also masonry units require a margin of measurement accommodation;
(4) of a thickness to be strong enough to resist failure in compression, i.e. to avoid a hole being punched through by the load applied (it may be necessary to reinforce with steel to achieve this);
(5) of sufficient material quality to resist failure by chemical reaction and premature loading during curing.

The actual size and depth is thus inherently related to particular sites and building circumstances, and there can be no safe 'rules of thumb' for every occasion. Advice on foundation design should always be obtained from an expert source, or the building manufacturer, where appropriate, before commitment is made to a type of construction. Agricultural buildings, however, are generally only very light-weight structures, with few complicated imposed loads, so some recommendations can be made for a few specific load conditions as shown in *Table 25.6* for strip foundations (*see Figure 25.8*).

Table 25.6 Foundation dimensions

Strip foundations	Clay	Soft subsoils	Sand, gravel, chalk
Depth (normal climate)	1 m	600 m	500 m
Depth (frost prone)	1.5 m	1.1 m	1 m
Minimum width (100 mm wall)	300 m	500 m	300 m
Width with working space (100 mm wall)	400 mm	500 mm	400 mm
Minimum width (270 mm wall)	600 mm	700 mm	450 mm
Width with working space (270 mm wall)	600 mm	700 mm	600 mm
Minimum thickness	200 mm	300 mm	150 mm (200 mm in soft conditions)

For *'trench-fill' systems* read as strip foundations for depth, always use minimum width. Thickness is depth minus 75 mm. To be safe in uncertain soil conditions use minimum width required in soft soil.

NB This table should only be used where the soil on which the foundation rests is firm, undisturbed, not subject to frost or water action and can support the loads implied by the table conditions.

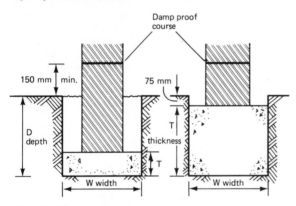

Figure 25.8 Strip and trenchfill foundations

Pad foundations

These provide support for frame uprights and locate them laterally. Manufacturers of frames will provide detailed drawings of the socket, bolt or dowel fixings and their positions in the pads. It is not possible to give guidelines for these as they vary tremendously with size and type of frame and soil properties.

Drainage

Two categories of drainage must be considered and treated separately in practice (*see* section on services for further information):

(1) surface water, that is water from rainfall run-off, which should be 'clean'; in effect this implies from roofs only;

(2) foul water (or soiled water) including sewage and slurry flows from all sources and other 'dirty'

water such as cow yard washings. This category must in no circumstances be discharged into water courses without purification to ensure that it conforms with the various anti-pollution requirements.

Roadways and external works

Within a farm there are usually three types of road: the access road connecting the farmstead with the highway, the roads which are the links between buildings and the internal roads serving the fields. Each serves a different purpose and needs different treatment to be economic relative to the investment required.

The access road is the artery of the farm and needs not only a good layout but a reasonably good surface. The width should not be less than 3 m to prevent concentrated wear and to allow for wide, modern vehicles. Passing bays should be provided at not more than 150 m intervals.

The weight of traffic on a farm road is light, weight in this case being the frequency of traffic not wheel loads. A light four-wheel lorry sometimes has a greater wheel load than a heavy vehicle with six wheels. It is speed plus weight, however, that wears out a road and, therefore, speed should be restricted. Modern bulk tankers for grain, feed or milk may be in excess of 20 tonnes. Steep gradients should have concrete 'grips' to assist wheels. Some farm roads have particular problems such as cloven hoofs and acid attack from silage and slurry.

Roadways around the buildings are an important feature of the planning and vary in width according to the amount of traffic discharging on them. They should be laid clear of the buildings, except where they serve as links. Large vehicles entering a cul-de-sac must be provided with space to turn. All bends and turning circles should be not less than 16 m diameter.

Internal roads usually serve to link fields and buildings together and take mostly tractors, trailers and stock.

Construction

To keep down the cost of making farm roads local material should be used wherever possible: chalk, quarry waste, ashes, hardcore from old buildings, flints, burnt shale, crushed limestone, are all materials which make satisfactory roads. The finished surface may then be treated with a coat of tar or bitumen to keep the water out of the subsoil, but this will not form a hard-wearing road.

Keeping the subsoil dry is the art of successful roadmaking.

It is not usual therefore to excavate into the ground

to make a road, though 'vegetable' soils should be removed; it should instead be built up from the ground surface. Where a road is made down into the subsoil because of poor support conditions the lower 'in-soil' part is made of introduced stone or hardcore and the surface will still be above the surrounding ground.

On sloping sites, a land drain on the rising gradient side of the road should be provided. All work should be carried out in dry weather, preferably during the summer months. Any junction with a highway must be carried out to the satisfaction of the County Council, possibly in pre-coated stone laid on a suitable stone base. It is often possible to have this work done by a highway authority when a roadmaking gang are working in the area. An existing farm road can be given a wearing surface of pre-coated stone laid between 50 and 75 mm thick and then well rolled. The work should be completed whilst the material is warm. Alternatively, uncoated aggregate can be laid first and grouted afterwards with tar or bitumen. If the surface is good it merely needs surfacing with a coating of bitumen or tar applied through a watering can and covered over afterwards. All sharp bends should be constructed in concrete as the greatest amount of wear takes place on corners.

Wheel track roads in concrete cost almost as much as a full width road, owing to the labour required in forming the track and are therefore not to be recommended.

Concrete should be 150 mm thick on at least 150 mm compacted hardcore, with a mesh reinforcement when bulk tankers use the road. Internal roads around the building should be constructed in C7P concrete 100 mm thick, laid on at least 100 mm compacted hardcore after the vegetable material has been removed, and in C20P if there is a chance of use before 7 d curing has been allowed or that crawler tractors might run on it.

Reinforcement is only necessary in soil of low bearing capacity such as peat, fenland silt and the like. Expansion joints are necessary every 5 m length of road and the best way is to lay the concrete in a 5 m length, miss out the next length, but start with the farther one, then return and fill in the one between. If this is properly carried out, expansion strips of wood or fibre boards are unnecessary. When laid, the concrete should be cured properly.

To save expense passing bays can be constructed of hardcore: they should be 2.5 m wide by 12 m long.

Light, internal farm roads are made by laying 100–150 mm of hardcore on top of the surface of the subsoil after it has been sprayed with herbicide and well rolled. Larger stone is used at the edges and smaller stone fills the middle. A drainage furrow may be ploughed alongside the edge which also acts as a support against the spreading of the surface. Any wet spots should be drained before the stone is laid down and consolidated.

Bridges, culverts and cattle grids

Where ditches have to be crossed, a bridge can be formed from precast concrete pipes of 150–750 mm diameter by 75 mm and 750–1800 mm by 150 mm steps. The 'invert' or bottom of the pipe should be laid about 150 mm above the natural bed of the ditch. Placing a pipe below the bed reduces the effective diameter of the pipe if it settles, and again when silting up, as it invariably does, thereby restricting the flow of water. Oil drums, with the ends knocked out, laid end to end and surrounded with concrete, are an alternative to pipes.

Spans of up to 6 m can be effectively and economically bridged in prestressed standard type concrete beams, laid on a mass concrete abutment, or on a suitable concrete base laid on either side of the banks.

Unit beams vary in size and shape according to manufacturer, but 225 mm wide by 150 mm deep is an average: they can be laid with farm labour, and are often cheaper and certainly better than railway sleepers.

A cattle grid soon pays for itself in time saved opening and shutting gates on a busy farm road, but it needs careful thought before siting, owing to the legal complications which can arise from 'rights of way' and safety considerations. For this reason also, DIY grids are not always desirable. Any cattle grid must be stock proof against all types found in the locality, it should be 3 m long and the full road width with permanent fencing both sides the full length of the grid. Alongside there should be a by-pass gate 1.5 m away measuring at least 3.6 m wide.

Fences

Permanent fencing is specified in BS 1722 Fences, and BS 3954 Farm stock fences. To contain livestock, fences should be 1.1–1.25 m high. Fences can have concrete or timber posts with strained wire or woven wire or with timber rails between posts. The type of fence, rail spacing, etc., depends largely on the type of stock to be contained. Angle irons and rabbit wire can be used for light work.

Reinforced concrete posts and wire

Reinforced precast concrete posts should be made to BS 1722 and bought from a reputable firm. It is impossible to make them properly using farm labour. Corner posts and straining posts should be set 900 mm in concrete and strutted with 100 × 75 mm struts. Straining posts are spaced at 20–40 m centres and intermediate posts at 2–3 m centres. If metal spacers are used at under 3.3 m centres, intermediate posts may be at 50 m centres and straining posts at 200 m centres. To contain most types of stock, except lambs and piglets, seven galvanised wires are needed, starting with the bottom one 100 mm above the ground. The others are spaced at 125, 150, 175, 200, 225 and 250 mm centres but six, or even four wires are sufficient for some type of stock, the top wire being barbed. The wires are spaced by means of galvanised eyeletted strainers and winders.

Timber posts

Timber posts and wire fences are constructed in a similar manner to concrete. All softwood timber should be pressure treated with preservative at the sawmill or factory of origin.

Galvanised woven wire stockproof fences in steel or aluminium alloy can be obtained in an infinite number of varieties in every size and are secured to timber posts at 2–3 m centres. A strand of barbed wire is usually run along the top of the posts.

Post and rail fences

Posts should be 125 × 100 mm at 2.7 m centres; with four 75 × 32 mm sawn rails, the first 150 mm above the ground and the remainder at 275, 300 and 300 mm centres. An intermediate 75 × 50 mm post is driven into the ground between the main posts.

Rabbit-proof fencing comprises 32 mm mesh fencing nailed to timber posts, the construction of which is similar to the timber post and wire fence, with the exception of the intermediate posts which may be increased in distance to 3.8 m centres. The top of the netting is clipped to a strand of barbed wire set 900 mm above the ground, and 150 mm above that a tripwire is stretched. The bottom 150 mm of the netting should be bent to lay flat on the ground, towards the rabbit side of the fence, on top of which turves should be laid; a further wire should be set about 450 mm above the ground to support the netting in the centre.

Gates

The standard timber five-barred gate is 3.6 m wide by 1.2 m high. Rails and braces should be bolted together with galvanised bolts. The top rail, or back, and second rail from the bottom should be the only two rails morticed through the head.

There are a variety of fasteners and hinges on the market, suitable for all types of gates and hangings. Most gates are hung 18–25 mm off centre to ensure self-closing. Implements such as combine harvesters, balers and drills may need a 5 m wide gate opening. A single gate cannot be economically made in this width, in either timber or steel, but many proprietary tubular or framed mesh gates are available in pairs. They are usually graded for heavy stockyard or light duties, with openings of 2.5–5 m.

SITE AND FIELD SURVEYING

Ranging out lines

To range a line between two points A and B on fairly level land, sight from A, insert ranging poles at C, D and E, working from B towards A (*Figure 25.9*). The bottom of the ranging poles should be observed.

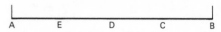

Figure 25.9 Surveying

To range a line between two points A and B with undulating land between so that B cannot be seen from A (*see Figure 25.10*).

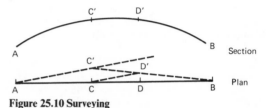

Figure 25.10 Surveying

Two assistants take positions C' and D', so that C' can see B and D' sees A. They then direct each other in turn into line with A and B until points C and D are reached, when no further shift is possible. ACDB is then a straight line.

To range out a line between two points A and B with a steep slope intervening, as in *Figure 25.11*.

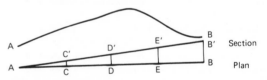

Figure 25.11 Surveying

Range out a trial line AB' in the general direction of B, marking stations C', D', E'. Measure AB', noting the distances from A to C', D' and E'. B'B, the ranging error, is also measured. Consider E'E, D'D and C'C drawn parallel to B'B. Triangles AB'B and AE'E are similar. Hence

$$E'E/AE' = B'B/AB'$$

Therefore

$$E'E = (B'B \times AE')/AB'$$

D'D and C'C can be calculated in the same way. E'E, D'D and C'C are then moved parallel to B'B to give straight line ACDEB.

To continue a chain line obstructed by a building. Range out the chain line to the building. Choose two points A and B on this line, erect perpendiculars AC and BD of equal length to clear the building. From C, range a line through D, past the building and choose

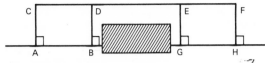

Figure 25.12 Surveying

two points E and F. Drop perpendiculars EG and FH equal in length to AC and DB (*see Figure 25.12*). From G, continue the chain line through H. To give greater accuracy AB and GH should be three times the length of the perpendiculars.

For small areas a simple chain survey may be made, using the metre chain. This chain is 20 or 30 m in length and subdivided every 200 mm, with tallies every 2 m from the ends. Note that 10 000 m² = 1 ha.

The area to be surveyed is divided into any convenient number of triangles.

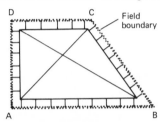

Figure 25.13 Surveying

Stations ABCD (*Figure 25.13*) are chosen in the corners of the fields. Lines DB, AB, BC, CD and DA are ranged out, dividing the greater portion of the area into two triangles. Line AC may also be ranged out and measured as a check line. When measuring AB, BC, CA and DC offset measurements are taken to the boundary, wherever it changes direction, as indicated by the broken lines.

Taking line AB as an example, measurements in links of 200 mm could be entered in the field book as shown in *Figure 25.14*. From the field notes the area may be drawn to a convenient scale, and the area calculated from the plan as shown for *Figure 25.15*.

Curved boundaries are replaced by equalising lines, so that figure EFGH is equal in area for all practical purposes to the original field.

Draw in diagonal HF and erect perpendiculars EJ and KG.

Scale off these lines.

Area = (EJ + KG)/2 × HF

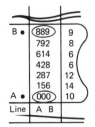

Figure 25.14 Surveying

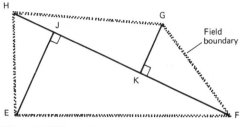

Figure 25.15 Surveying

Ordnance survey

The maps of most importance to agriculturists are the 1:10 000, (which replaces the 6 in), and the 1:2500.

Map 1:10 000 is useful to the estate owner. It covers an area 5 km square but some coastal sheets are larger. All boundaries—country, parliamentary, urban, rural and parish—are shown. Also all enclosures, buildings and the like are included. In the vicinity of towns, streets are widened to accommodate street names. Instrumentally determined contour lines are drawn in at 10 m, 5 m or as a direct metric equivalent to the older imperial measure. Bench marks are given to two places of decimals of a metre, whilst numerous spot levels are shown.

Map 1:2500 (25.344 in to the mile) is termed the parish or farmer's map. Each map covers an area 1000 m², some sheets consisting of two adjacent maps. All features shown on the 1:10 000 map are shown, except contour lines, and enclosures bear a reference number with the area to three places of decimals of a hectare.

Simple levels used on building sites

Three instruments are used commonly by builders for levelling on small area construction sites; they are types using fluids (water usually), automatic levels and 'dumpy' levels.

(1) The fluid level is the simplest in principle, being two transparent tubes, equipped with scales in millimetres and connected by a flexible tube containing coloured solution. As the tubes are raised and lowered the fluid level remains the same in each tube and vertical distances can be referenced to the liquid surface positions and the tube calibration (*see Figure 25.16*). These are limited distance instruments up to 30 m usually, but can measure out of line of sight.

(2) So-called 'automatic' levels are available which require minimal instrument adjustment once they are placed suitably on a tripod. Their optical system is designed to obviate the need for precise horizontal alignment which is necessary with other equipment, and to indicate by the viewfinder when the tripod requires repositioning to give the

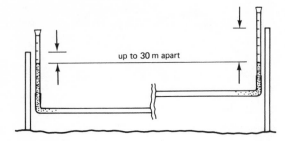

Figure 25.16 Liquid level

correct field of view. They also are short distance instruments (up to 20 m) and can only be used with a clear line of sight, but they are simple to interpret. The second operator holding the sight staff reads the measurements on the staff in this system (*see Figure 25.17*).

(3) Dumpy levels are telescope devices with the addition of siting crosswires built into the optics, which are used to pin-point readings on a target staff (*see Figures 25.18 and 25.19*). The telescope must be set up precisely horizontal. The staff consists of a telescopic scale of three or four sections, which can extend rigidly to a height of up to 6 m from the ground. The markings on the scale are graduated in centimetres, decimetres and metres, the latter two in figures.

To take satisfactory measurements from the staff the 'plane of collimation' must be exactly horizontal at all times, thus great care is needed when the level is set up, to make sure that the spirit bubble on the top of the instrument is used in two planes at right angles to position the telescope. Unlike the automatic level the operator viewing the staff through the sight records the measurements and logs their variation. Care should be taken to choose a 'station' for the level which will scan as much of the site as possible and that the base lines used are interrelated if the tripod must be moved, as shown in *Figure 25.19*.

It is usually convenient for building sites to be 'pegged' as levels are taken, such that the bottom of the staff rests on the top of each peg and records the same distance to the line of collimation seen through the level as in *Figure 25.17*.

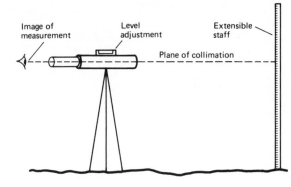

Figure 25.18 Dumpy level

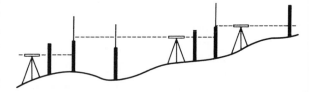

Figure 25.19 Use of levels

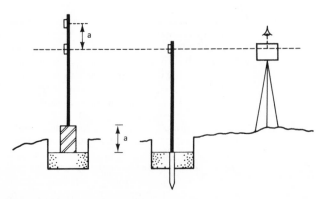

Figure 25.17 Automatic level

ENVIRONMENTAL CONTROL

Two main building characteristics other than lighting are generally involved in the basic design of environmental control systems for farm structures; these are insulation values and air-flow patterns. Both affect heat flows and temperature levels, the first in a passive sense, the second more actively.

Insulation

Insulation values are a measure of a negative effect, that is, they portray the ability of a material to prevent heat passing through; thus in the commonly used 'U' value assessment of the property of thermal transmittance, larger figures indicate poorer performance (*see Table 25.7*). There are three 'mechanisms' involved in thermal transmittance, namely conduction, convection and radiation. Conduction concerns the passage

Table 25.7 Examples of 'U' value on thermal transmittance

	$W/m^2\,°C$
Concrete block, dense, 150 mm, solid	3.58
Brick, solid 115 mm	3.64
Brick wall, 280 mm 2 leaf, cavity, inside plastered	1.70
Brick outer with aerated concrete block inner, 260 mm, cavity inside plastered	1.13
Concrete block wall, 225 mm, hollow	2.5
Concrete block wall, 255 m, cavity outside rendered, inside plastered, aerated block inner leaf	1.19
Concrete block wall, hollow, 230 mm, rendered and plastered, aerated	1.70
Corrugated asbestos cement sheet in steel framing	6.53
Ditto + 25 mm expanded polystyrene	0.97
Ditto + 50 mm wood wool slab	1.19
Ditto + 25 mm glass fibre quilt	0.57
Ditto + 12 mm insulating fibre board	2.04
20 mm timber board on studding with 50 mm glass fibre in-fill and hardboard inner cladding	0.62
Corrugated steel sheet	8.52
Ditto + insulation board	4.82
Concrete on ground or hardcore fill	1.13
Aluminium foil + paper and mineral backing	1.08
Windows single glazed	5.68
Windows double glazed 6 mm air space	3.29
Windows double glazed 20 mm air space	2.84
Roof lights as for windows + 20%	

three-quarters of maximum exposure level

NB Insulation value is dependent on the moisture content (or lack of it) thus a water vapour proofing agent is essential for a continuing satisfactory insulation level from a material.

of heat by particle to particle contact, convection involves heat flow by air mass movement within materials, and radiation passes heat from place to place by waveform, even across vacuum space. To provide effective barriers to heat passage each of these mechanisms must be minimised. This may be done by choice of material with poor particle contact characteristics, many small-size discrete trapped air pockets and reflective surfaces of silver or white coloration. It must also be capable of water rejection or be vapour-proofed against a damp entry, otherwise water will replace the air masses, converting poor conductivity into good.

'U' values are expressed as heat passage (watts per square metre per degree Celsius) $(W/m^2\,°C)$ through a surface area under the 'pressure' of a given temperature gradient. They are used in the equation

$$Q = UA(t_i - t_o)$$

where Q is the quantity of heat passing, U is the structure thermal transmittance, A is the structure area, and t_i and t_o are the temperatures on inside and outside faces. For instance in the case of a hollow block wall, 6 m long and 2.5 m high, where the temperatures are 20 °C and 5 °C respectively, then the quantity of heat transmitted through will be

$$Q = 1.7 \times (6 \times 2.5) \times (20 - 5)W = 382.5\ W\ (J/s)$$

It should be noted that 'U' value is affected by the nature of the external aspect and environment and has no absolute values for a material; an exposed condition can increase it by as much as 10% whereas a very sheltered position may reduce it. Other considerations such as cleanliness, paint coloration, and dampness will also affect it.

'K' values $(W/m°C)$ are another measure of heat transmission through materials. These thermal conductivities are absolute and do not depend on the condition or environment of use. They are quoted per unit of thickness, which must, therefore, be taken account of, before the value can be of use in calculations. Additionally characteristics such as surface transmission are not accounted for and must be added in before a meaningful result is obtained. Thus these are more likely to be of use to the specialist design engineer than the practical builder.

A further value sometimes met with is 'R' value $(m°C/W)$, which stands for resistivity and is the reciprocal of conductivity

$$(R = 1/K)$$

This may be used in preference to 'K', as the values may be added together to produce an overall resistivity which has direct relevance.

For an understanding of the effect which insulation has on temperature gradient *see Figure 25.20*. This stresses the importance of providing adequate vapour-proofing. Without this, vapour pressure will drive water into the materials to the point where the temperature drop reaches the dew point of the water vapour and condensate appears and accumulates. Many agricultural buildings have high moisture content atmospheres together with surrounding high-insulation values, giving potential failure conditions.

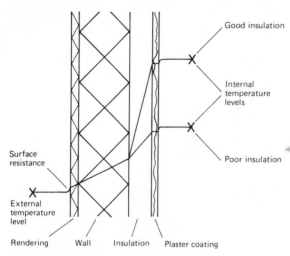

Figure 25.20 Temperature gradients through insulation

Ventilation

This can be provided by natural means such as open front or open ridge; it may be semi-controlled as in simple exhaust-fan systems; or totally controlled with complex duct and pressurising fan equipment.

The usual purpose of air exchange in UK livestock buildings is to lower temperature, so keeping the environment in the thermoneutral zone for the animals concerned. The actual required temperature is very variable depending on the species, age, acclimatisation and hardiness of the strain (*see* specific housing requirements later).

As the external temperature is continually changing a constant series of adjustments may be necessary and in particular there will be a great difference in ventilation rate between winter and summer; perhaps as much as 20 times the amount in warm months compared to cold. In practice ventilation during the deep winter period is only necessary to get rid of stale air and adverse gas accumulations (ammonia for instance), and usually happens incidentally through the opening of doors, leaks via drains, poor fitting windows and eaves cracks. The main problem during winter is to avoid draughty conditions which are likely to give thermal shock to the occupants, particularly where young stock is housed. This means that the velocity of the incoming air must be reduced and the air stream deflected to a level where it does not impinge directly on the livestock before it has had a chance to mix with sufficient warm air to become innocuous.

Ventilation systems are usually sized and installed to counteract summer conditions in which the heat from the stock cannot be lost through the walls and roof in sufficient quantity to stabilise the temperature at a reasonable level. In this situation the air flow usually has a cooling effect. When the outside temperature rises above the optimum required, cooling ceases, and the internal temperature must also rise, causing discomfort at the least and if it continues may lead to acute distress. The only certain solution to this problem is to provide air-conditioning, in which the incoming air is cooled artificially before entering the environment. This is too expensive at the moment to consider for most circumstances, except for fruit and vegetable storage.

The moisture contained in the air is of greater importance, from time to time, than heat due to potential difficulties with condensation or fungal attack in produce stores. Ventilation in such circumstances may be difficult to control as the exchanged air often lowers the temperature of the environment and supplementary heating becomes necessary. (In reverse, water evaporating from a porous surface is used occasionally to lower the temperature of air flowing past, but raises the humidity level in so doing. This technique is only possible in dry, low humidity climates.) The two factors, temperature and air moisture-content are linked by the fact that relative humidity is dependent on temperature level. As the temperature rises the amount of moisture in an air mass required to cause saturation increases. Thus the following equation defining relative humidity (RH) changes with saturation level:

$$\text{Relative humidity }\% = \frac{\text{actual mass of water vapour in air (kg/kg dry air) at T}^\circ}{\text{saturated capacity mass of water vapour in air (kg/kg dry air) at T}^\circ} \times \frac{100}{1}.$$

As temperature decreases so the saturation level reduces and with a given water vapour mass, known as 'specific' or absolute humidity, the RH will increase, until at 100% any further temperature reduction will cause condensation. This is referred to as dew point, and has great importance for insulation design in relation to vapour-proofing.

Ventilation air flows may result from natural causes, induced by wind and 'stack effect' (temperature gradient) or they may be 'forced' by either exhausting or pressurising fan power. Natural flows (*see Figure 25.21*) can be positive and helpful or confusing and negative. They can only be controlled in a passive sense so are not tolerable where a precise environment is required. They generally find their most appropriate use in low temperature, shelter-only, stock housing.

Forced ventilation, relying on fan power, is more controllable but still suffers from the variability of

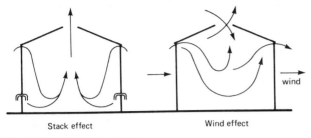

Stack effect   Wind effect

Figure 25.21 Natural ventilation

wind forces and pressure changes in the external environment. In addition adverse temperature gradients and draughty air-leakages can cause particular difficulties for exhaust fan systems. Pressurising methods of ventilating, in which air is blown into the house are more satisfactory and easier to control. However they cost more and require carefully designed ducting arrangements to distribute the air evenly, avoiding high velocities and potential thermal shock conditions.

The number of fans used will depend on the type of system, 'suck' or 'blow', the amount of air to be moved, the appropriateness of the structure and whether ducting can be installed easily. Using few fans will be easy to control and cheaper but may need ducting to achieve a reasonable air-flow distribution; they may also result in a greater noise disturbance and the risk of total failure is greater. The cost of using many fans, as an alternative is likely to be greater and control may become complex, but such an arrangement will minimise the risk of a total ventilation breakdown, especially if they are mounted in the roof and extract (*see Figure 25.22*).

Air inlets and outlets should be designed and sited to match each other so that there will be no restriction placed on the system, as this will cause leakage through other parts of the structure. They should also be proof against direct interference from wind forces, externally.

The most important purpose of a ventilation system is to maintain a heat 'balance' in the building environment. It is the only mechanism which is controllable in a 'minute to minute' sense commonly available in most livestock house or crop stores. (Heat is added by the use of infrared emitters in some pig environments and high value crops may be stored, long term, with the help of air conditioning or refrigerating plants.)

Heat exchanged through ventilation may be shown:

Heat loss or gain (Q) = m × s × (te − ti)

where m is the mass of air exchanged per unit time, s is the specific heat of air (0.36 J/kg effectively for the purpose of environmental control calculations), and te and ti are the temperatures of exhaust and inlet.

Thus a heat balance for the total environment can be made, where Q stands for heat quantity in each part of the equation:

Q animals or stored produced + Q lights, men, motors, etc. + Q condensation + Q heaters
= Q insulation + Q evaporation + Q waste removal + Q ventilation + Q refrigeration.

Table 25.8 shows some heat outputs expected from stock and stored products in protected environments.

Table 25.8 Approximate heat outputs of products in agricultural environments

Stock	kJ/h per 50 kg liveweight
Cattle, adult	250–300
Calves, up to three months	60–100
Sows, dry	600
Sows, farrowing	750
Piglets, weaners	70–150
Baconers, heavy hogs	400–500
Sheep	300
Broiler poultry at three weeks	600
Adult poultry	1000+

Crop	W/tonne
Potatoes, immature at loading	72
Potatoes, mature at loading	18
Potatoes, mature at 5 °C storage	8
Onions, bulb at 2 °C storage	10
Sprouts, at 2 °C storage	60
Cabbage, winter at 2 °C storage	12
Apples, at 3–4 °C storage	15–30 (depends on variety)

548

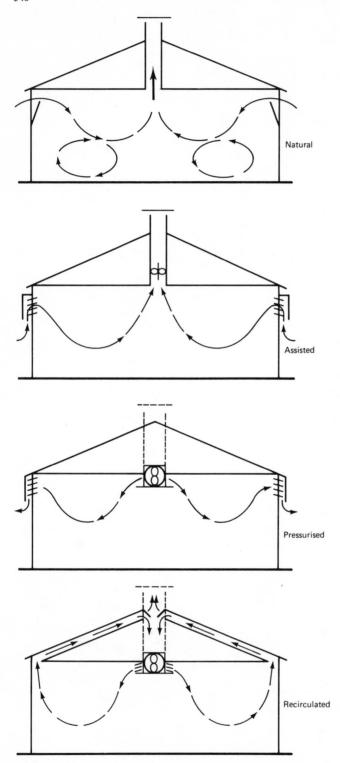

Figure 25.22 Ventilation systems

SPECIFIC PURPOSE BUILDINGS

1 Cattle accommodation

1.1 Dairy cows

Space for loose housing.

4.5–5.5 m²/small animal, of which 3.2–3.5 m² are lying area; and 5.5–8.7 m²/large animal, of which 3.7–6.5 m² are lying area.

Space for restrained housing (cowsheds, double range, for single range *see* 1.4).

7.5–9.2 m total house width (± feed passages) by 2.0–2.15 m length for every double standing.

Cubicle/kennel dimensions plus necessary feeding and circulation space.

1.05–1.2 m wide × 2.0–2.3 m deep (for small and large cows) each cubicle, + 2.5–5.5 m² additional (*see Figures 25.23–25.27*).

Minimum height to eaves.

2.2 m (for cubicle systems)–3 m. Depending on bedding depth allowance, access requirements, slatted floor height, etc. to be added to basic minimum shown.

Temperature and insulation.

Not important down to 5 °C, no control of environment usually. 2 °C is 'lower critical' temperature. Insulation limited to bedding or, occasionally, flooring.

Ventilation.

Natural, provided by open ridge or space boards in roof and upper wall area. Where mechanical used, 7–25 ℓ/s per 50 kg liveweight (winter–summer).

Bedding, water and feed consumption.

Straw 750–1500 kg/cow per winter.
Wood shavings 500–1000 kg/cow per winter.
Silage 20–40 kg/d per cow.
Hay 5–10 kg/d per cow.
Concentrates 5–7 kg/d per cow (for 15–20 ℓ of milk).
Water 35–70 ℓ/d per cow (from trough with allowance of 200 cm²/cow or 1 drinking bowl/10 cows).

Effluent and drainage.

Urine 25–60 ℓ/d per cow   ⎫   depends on feed and
Dung 10–35 ℓ/d per cow   ⎬   housing system.
Drainage slopes on floors from 1:50 to 1:100.

Feed space.

550–700 mm trough length/cow.
100–150 mm self-feed silage face/cow with unrestricted access.
250–300 mm self-feed silage face/cow with restricted access.
600–700 mm silage and hay manger/cow.

Other considerations.

Races and handling equipment should be adjacent to and integrated with the housing.
Slats may be part of floor or whole floor, may have drain channel or tank under, to contain 3–6 months' accumulation, may be raised 2+ m to provide total storage for permanently housed stock. They may be of timber, concrete or steel with general dimensions of 125 mm wide with 38 mm gap, 150 mm deep and 25 mm taper (*see Figure 25.28*).
Floors and steps with non-slip surface required.
Need calving and isolation boxes at 10 m²/animal, separated.

General.

Consider cow management and handling, effluent problems, feed store siting, cow comfort, access for mechnical equipment.

1.2 Calves

Space for controlled environment housing.

0.9 m²–1.5 m² in single pens of approximately 1 m × 1.5 m maximum for small calves (*see Figure 25.29*), up to 14 d old.

0.9 m² increasing to 2.5 m², 14 d old to 3 months old.

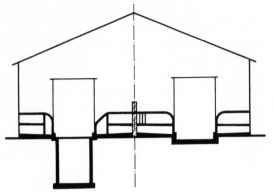

Figure 25.23

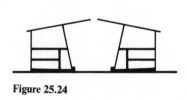

Figure 25.24

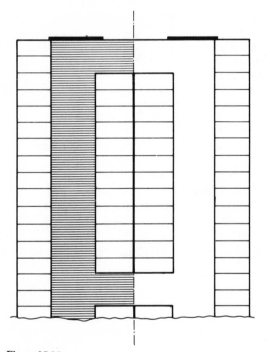

Figure 25.25

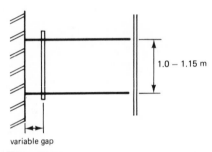

variable gap

Figure 25.26

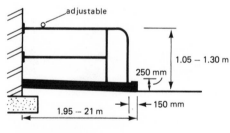

adjustable

1.05 – 1.30 m

250 mm

1.95 – 21 m

150 mm

50 mm galvanised tube or
timber posts 100 x 50 mm and
150 x 38 mm rails

Figure 25.27

1.0 – 1.15 m

Figure 25.28

Space in open yards, 3–6 months old.

2.5–3.0 m² with three or four per pen (*see Figure 25.30*).

Height recommendations.

2.5 m to ceiling minimum. Barrier height 1.0 m minimum, up to 1.4 m for larger calves to provide isolation between pens.

Temperature.

20 °C for young calves, 15 °C at 1 month old and 12 °C at 3 months.

Insulation of small calf environments.

Floor damp-proofed and having 'U' value of 0.5 W/m² °C. Roof, double-skinned or with ceiling, of 'U' value 0.6 W/m² °C maximum. Walls, cavity or other, with 'U' value of 0.9 W/m² °C maximum.

Ventilation.

Adequate, draught-free ventilation essential to good health. Controlled environment should be capable of 25–30 ℓ/s per 50 kg liveweight in winter. Natural ventilation in housing for over 3 months old stock, with hopper windows or space boarding and ridge ventilation.

Feed and water consumption.

Feed from bucket or machine at pen front or trough for older calves, all near passageway. Milk substitute and small quantity of concentrate (0.7 kg/d) then up to 2.5 kg/d up to 3 months. 15–25 ℓ/d per calf from water bowls or bucket near passage.

Effluent output.

5 ℓ/d–15 ℓ/d depending on food, size and age.

Flooring.

Hard, cleanable, durable, resistant to disinfectant. Surfaced with wood float in exposed areas if concrete. Drain to passageway at 1:20, and via step of 100 mm at edge of pen to drain channel in passage.

Other considerations.

Solid or 'see-through' pen divisions. Opinions vary as to value of each because of hygiene and social aspects.
Care with surfaces as calves chew and lick everything.
Drainage must never interconnect pen to pen.
Humidity should be kept down to 70%.

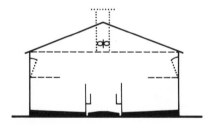

Figure 25.29

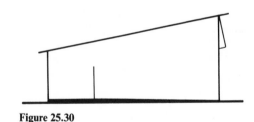

Figure 25.30

1.3 Beef animals

Space for straw bedded animals (*see Figures 25.31* and *25.32*).

1.8–3.7 m² per animal at 12 months, 4.7–6.0 m² per 2 year old. (4.0 m² with self-feed silage systems) 10–20 in group to one court or pen.

Space for beasts on slats.

1.4–1.9 m² per 12 month old.
2.0–2.5 m² per 2 year old.
20–25 animals per court or pen.

Height to the eaves.

3.0 m minimum (allow extra for dung build up of 1.2 m in strawed yards, or slurry depth under slatted floor).

Temperature and insulation.

Not critical within range 2–20 °C.
Insulation not required.

Ventilation.

Natural, except in intensive units, with space boarded upper walls and ventilated ridge. Forced systems must be capable of ranges 15–100 ℓ/s per animal in summer for 1–2 year olds and 5–20 ℓ/s per animal in winter.

Bedding and fodder consumption.

Straw 750–1500 kg/beast for winter.
Silage 30–40 kg/beast daily.
Rolled barley 7–10 kg/beast daily.

Trough space.

500 mm length/9–12 month old beast.
600 mm length/18 month old beast.
700 mm length/2 year old beast.
150 mm/animal for *ad libitum* fed young stock.

Water consumption.

25–45 ℓ/d per head.
Allow 200 cm² trough per head or one drinking bowl per 10 animals.

Other considerations.

Kennels and cubicles may be appropriate (*see* 1.1 Dairy cows) also 'topless cubicles'.
Safety of personnel; (especially with bull beef animals) requirement for sturdy barriers and doors.
Slats may be same as 1.1 Dairy cows description; may be smaller for younger animals—100 mm wide with 30 mm gap.
Flooring to be hard surfaced with concrete open areas. Ramp gradients not to exceed 1:10.

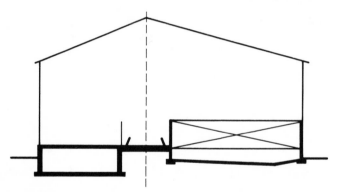

Figure 25.31

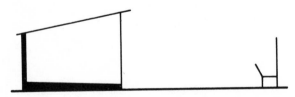

Figure 25.32

1.4 Milk production buildings and bull pens

Cowshed, single range (*see Figure 25.33*, for double range *see* 1.1 Dairy cows).

Overall width 4.45–5.35 m.
Standing length 1.4–1.6 m.
Standing breadth (double) 2.0–2.15 m.
Dung channel width 0.9 m.

Feed passage width 0.9 m.
Back walk, minimum 1.2 m.
Height to eaves 2.5 m.

Abreast parlour (six standings) (*see Figure 25.34*).	Overall depth 4.8–5.8 m. Overall width 6.9 m. Side exit passage 0.3 m. Minimum headroom over standings 2.0 m (plus 0.45 m step to standings: gives total height minimum 2.45 m).
Tandem parlour (six standings) (*see Figure 25.35*).	Overall width 5.1 m. Overall length 9.3 m. Passage widths 1.0 m (one end, across, and two exit passages). Minimum headroom over standings 2.0 m. Depth of operator pit 0.75 m. Width of operator pit 1.8 m.
Herringbone parlour (eight standings) (*see Figure 25.36*).	Overall width 4.8 m. Overall length 6.9 m (passageway where required 1.0 m). Depth of operator pit 0.75 m. Width of operator pit 1.2 m. Minimum headroom over standing 2.0 m. Breast rail clearance from wall 0.4 m. May be open at entry end.
Chute parlour (*see Figure 25.35*).	As tandem, less end and exit side passages and with narrower pit. Six standing measures 7.4 m × 3.1 m overall.
Rotaries, tandem (*see Figure 25.37*).	Space occupied overall from 5.5 m × 5.5 m for six milking points up to 15.5 m × 15.5 m for 18 points
abreast.	Space occupied overall from 8.1 m × 8.1 m for 12 milking points up to 14.0 m × 14.0 m for 30 points.
herringbone.	Space occupied overall from 7.0 m × 7.0 m for 12 milking points up to 14.0 m × 14.0 m for 28 points.
Polygon parlours.	Four ranks (or three in 'trigon') of standings with up to eight cows per rank, in flattened diamond shape arrangement, 16–40 cows at one time in parlour. 16–18 m wide × 18–33 m long, with 7–9 m at pit centre.
Collecting yards and circulation space.	Circular or rectangular, open or covered (in which case ventilation should not be through the parlour) allowance of 0.9–1.4 m² per cow.
Drainage.	Smooth floated channels or vitrified pipe. Overall falls 1:100 or 150 over lengths of buildings. 1:25 for standings and falls to outlets and gullies. Channels to have 50 mm high minimum rise at edges.
Water supplies.	1.0–3.0 ℓ/cow daily udder wash allowance. 20–40 ℓ/milking unit daily circulation cleaning allowance. 0.5–1.5 m³/d for pressure hose facility.
Services.	Pipes and fittings required for vacuum, milk transfer, cleaning systems; storage facilities for hot water, warm water, milk. All accessible electricity supplies and fittings should be low voltage (110, 24 or 12 V). Consider heating installations against freezing of pipes and for operator comfort.
Light levels and sources.	Natural preferred from above for general illumination. Local high intensity artificial light at udder level. Visual contact to be maintained between, parlour, dairy and collecting yard.
Flooring.	Hard, durable, resistant to detergents and disinfectants with non-slip surface (granite chippings and carborundum dust) which is cleanable by pressure hose.

Walls.

Other considerations.

Dairies.

Bull pens.

Rendered with waterproof, cleanable, smooth surface to a height of 1.37 m minimum (1.7 m preferable) painted with epoxy resin or chlorinated rubber finish. No insulation provided.

Steps up and down should be non-slip with 150–200 mm 'rise' and 600 mm 'going'. Heelstones on pit edges should be a minimum of 150 mm above standing level.
Sliding doors or rubber plastic flap doors required with automatic opening and closing as appropriate.

See Figure 25.38.

See Figure 25.39.

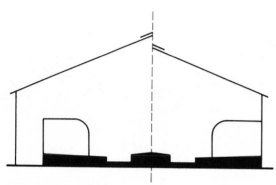

Figure 25.33

Figure 25.34

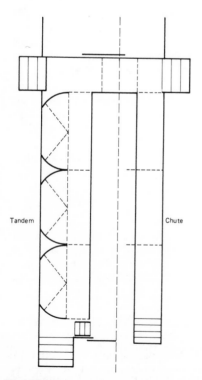

Figure 25.35

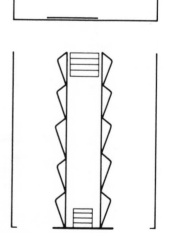

Figure 25.36

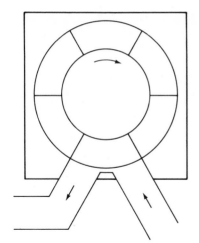

Figure 25.37

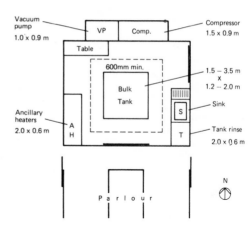

Figure 25.38

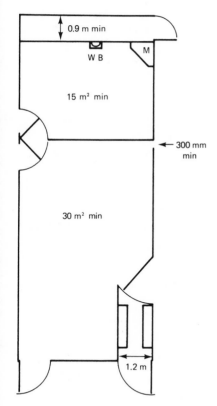

Figure 25.39

2 Pig housing

2.1 Pig fattening accommodation

Space within controlled environment housing.

Porkers, 0.6–0.7 m² of which 0.25 m² is dunging area (or 0.15 m² slatted area).

Baconers, 0.75–0.9 m² of which 0.3 m² is dunging area (or 0.2 m² slatted area).

Heavy hogs, 0.9–1.1 m² of which 0.35 m² is dunging area (or 0.25 m² slatted area).
10–20 pigs per pen usual, some systems up to 30.

Pen dimensions (trough length usually governs length and shape of pen and thus the house) (*see Figures 25.40–25.44*).

Optimum depth of lying area 1.8 m. Trough width 300–400 mm. Trough length per pig 250, 325, 400 mm for porkers, baconers and heavy hogs, respectively. (*Ad libitum* feeding systems only require 225 mm for all classes of fattening pig.) Some systems have no trough allowance—floor feeding.

Height of building.

As low as possible. Passageways for operatives give minima 2.15 m normal, 2.7 m minimum with tractor scraping, 3.0 m with catwalk over pen walls for feeding.
Eaves height 1.7 m over slatted areas or 1.1 m with strawed yards outside and pophole access.

Passage widths.

Dung passages, 1.05–1.35 m slatted outer, 1.15–1.65 m slatted central, 1.15–1.5 m concrete solid, outer.
Feed passages, 1.0–1.2 m.

Temperature and insulation.

Minimum of 15 °C at floor level up to 22 °C mean throughout environment, 5 °C less at floor level with straw bedding. 2 °C less, general, for pigs over four months old. Roof and walls should have 'U' value of 0.5 and 1.0 W/m² °C maximum respectively. Floor in lying area should also be insulated and vapour proofed to high standard.

Ventilation.

2.5 ℓ/s per 50 kg liveweight in winter.
12.5 ℓ/s per 50 kg liveweight in summer.

Feed and water consumption.

2–4 kg/pig daily of meal.
4.5–9 ℓ/pig daily through trough with meal; or bowls at dunging area/lying area junction, or in dunging passage.

Effluent and drainage.

7 ℓ/d from baconer on dry feed. 14 ℓ/d from baconer on *ad libitum* wet feed. Floor falls 1:15 in lying area; 1:50 to 1:100 for dung passages or channels.
Slats, concrete or steel (T bars) with 35 mm–75 mm top face and 15–21 mm gap width (depends on pig size) (*see Figure 25.28*).

Other considerations.

Chew-proof doors and partitions with secure latching and fixings.

2.2 Farrowing accommodation

Space for crate.

2.3–2.45 m × 0.6 m wide × 1.4m high. (Plinth may be provided of 200–300 mm high.)

Space for circulation around crate.

Passages at rear minimum of 1.4 m. Feed passages at head end minimum of 0.8 m. Eaves height 2.4 m from passage floor.

Space for creeps surrounding crate (*see Figure 25.45*).

0.1 m² per piglet allowance. 0.4–0.5 m wide × 1.4 m long (length of crate). Covered with lid and heat source above piglets below lid. Bottom rail or pophole access for piglets. Rail height 250–300 mm or 3–4 popholes.

Traditional farrowing places/pens with weaning space and creep possibly (*see Figure 25.46*).

2.4 m deep × 3.0 m long (pens either side of feed passage). Eaves height 2.4 m. Creep rails on rear and side wall nearest creep, set 250 mm out from walls and 250 mm above floor.
Creep at pen front, 0.9 m × 0.8 m area with access from pen via popholes and creep rail. Lid and heat source over.

Open front pen (*see Figure 25.47*).

1.5 m wide × 4.5–5.5 m deep × 1–1.5 m at rear eaves (mono-pitch with front eaves of 2.1–2.4 m). Creep railing against rear wall. Removable farrowing crate rails linked to creep, arranged along the pen at centre rear.

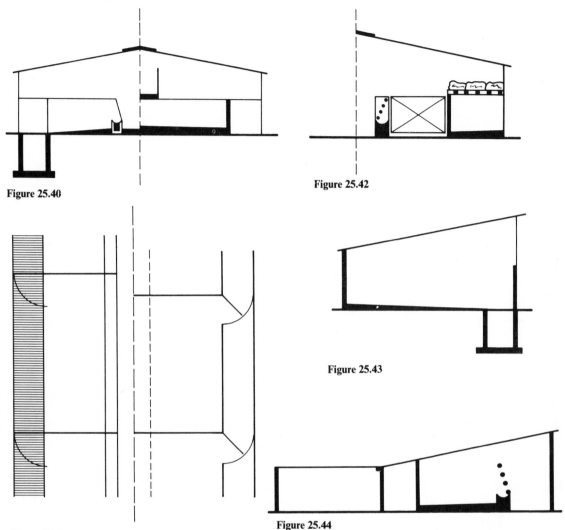

Figure 25.40

Figure 25.42

Figure 25.43

Figure 25.44

Figure 25.41

Temperature and insulation.	General areas 16–20 °C. Creep at 27 °C initially, reducing to 24 °C over three to four weeks from birth. 'U' values of 0.5 and 1.0 W/m² °C for roof and walls respectively. Crate and creep areas to be well-insulated concrete.
Ventilation.	Natural in open systems and in small, pen-type housing (up to five sows). 5 ℓ/s per sow in winter and 15 ℓ/s per sow in summer in controlled environment houses.
Feed and water consumption.	5–8 kg/d per lactating sow. 0.6 m long trough. 18–25 ℓ/d per lactating sow. 4–5 ℓ/d per piglet at four to five weeks old—provided by drip feed or trough away from creep feeder.
Effluent and drainage.	10–15 ℓ/d per sow. Drainage falls in lying and creep areas to be 1:20, 1:40 in general areas. Steps to drain or dung areas not to exceed 50 mm rise. Slats or weldmesh area in rear half of crate. Where used, slats consist of slabs of concrete with 10 mm slots and 75 mm slat tops. Crate

	raised 400–500 mm above surrounding levels at rear to accommodate slurry build up and drainage flows.
Materials of walls and floor.	Easy cleaned and resistant to strong disinfectants. Floor flat but not smooth trowelled—no tamping grooves to impede drainage.
Other considerations.	Robust fittings, including water bowls required. Strong post and frame fixings. May have crates in an 'open' arrangement with or without low partitioning between each, situated on a plinth or on a well-sloped floor. All crates should be removable for easy cleaning of crate sections and floor.

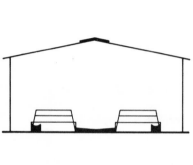

Figure 25.45

Figure 25.46

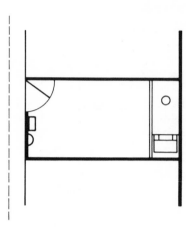

Figure 25.47

2.3 Dry sows and boar accommodation

Space allowances
  in open yards,
  in sow stalls,
 	
  in sow pens,
  in boar pens *(see Figure 25.48).*
	2.7–3.7 m² per sow of which 1.2–1.5 m² is covered lying area. Group size normally up to 20 per pen or yard.
	2.1 m × 0.6 m plus passages of 1.05 m minimum for feeding and dunging. Two types—short stall divisions (of 1 m long) and full length divisions (of 2.2 m). Short type requires restraint collars.
	2.1 m × 1.5 m per individual sow.
	4.5 m × 3.0 m pen of which 4.5 m² is lying area (2.4 m × 1.8 m minimum).

Temperature and insulation.	21 °C maximum, 14 °C minimum. Roof and walls of 0.5 and 1.0 W/m² °C 'U' value respectively.
Ventilation.	Natural usually adequate but forced ventilation in some totally enclosed houses at 50 ℓ/s per pig maximum.
Feed and water consumption.	2–3 kg/d per pregnant sow in individual feeders. 3–5 kg/d per boar from a trough of 0.6 m length. 5–9 ℓ/d per sow or boar.
Effluent and drainage.	20–25 ℓ/d per animal. Floor falls laid to 1:40 generally; 1:20 to discharge into open gullies. House end falls 1:50–100. (Slats may be used *see* 2.2 Farrowing accommodation.)
Other considerations.	Boars kept with visual contact of sows in extensive houses. Sow restraint may reduce bullying, can produce sores.

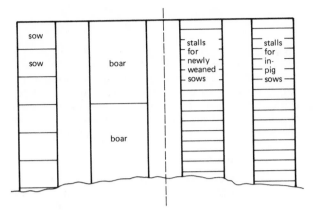

Figure 25.48

2.4 Weaner housing

Space requirements (*see Figures 25.49* and *25.50*).	0.15–0.25 m² per pig in 'flat deck' accommodation. Eaves or ceiling height 2.0 m. Groups of 1, 5, 10, 20 are common. 0.5–0.6m² per pig in strawed yards of which up to 0.3 m² is lying area if open fronted or 'verandah' type. Eaves height 0.9–2 m depending whether kennel or framed type. Groups of 10–20 usual.
Temperature and insulation.	20–22 °C inside kennels and in controlled environment housing. Roofing or kennel lid 'U' value should be 0.3–0.5 W/m² °C. Wall 'U' value should be 0.5 W/m² °C. Heat usually needed in winter—gas heaters common.
Ventilation.	2.5 ℓ/s per 50 kg liveweight in winter. 15.0 ℓ/s per 50 kg liveweight in summer. Care to ventilate avoiding passage of noxious gases from dung through house in total control systems.
Feed and water consumption.	*Ad libitum* feeding from 50 kg capacity hoppers. 1–2 kg/d per pig. Some systems use troughs at 200 mm per pig. Should be spill-proof 1–2 ℓ/d per pig from nipple drinkers over well-drained area. Very young pigs might need temporary ramp to reach drinkers.
Effluent and drainage.	1–2 ℓ/d per pig. Either open dunging area outside kennel or weldmesh/expanded metal part or whole floor. Weldmesh is 76 mm × 13 mm × 8 swg (10 swg for smaller) over slurry pit 1–2 m deep or raised 0.6 m in total control housing.
Materials.	Prefabricated buildings with easy erect, easy clean and control facilities.

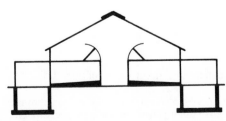

Figure 25.49

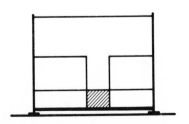

Figure 25.50

3 Sheep accommodation

3.1 Sheep housing

Space requirements (*see Figure 25.51*).

Slatted floor allowance for large ewes 0.9–1.1 m² per ewe; for small ewes, 0.7–1.0 m² per ewe; and for ewe hoggs, 0.5–0.8 m² per hogg. Solid floor allowance for large ewes 1.2–1.4 m² per ewe; for small ewes, 1.0–1.3 m² per ewe; for ewe hoggs, 0.7–1.0 m² per hogg; for Welsh mountain lamb, 0.4 m² per lamb; and for Scotch Blackface lambs, 0.5 m² per lamb. 30–60 ewes per pen usual. Feeding passage 0.9–1.1 m wide. (Floor area allowed may be less for sheared ewes.)

Temperature and ventilation.

Ambient of surroundings, down to 2–5 °C. No insulation necessary.

Ventilation.

Adequate air exchange essential to maintain health of stock. Natural from space boarding walls and ridge ventilation (open 300–600 mm, relative to building span). Must avoid draughts. Building may be open at one side.

Bedding, feed and water consumption.

0.8–1.5 kg/d of straw per ewe in yards.
0.25–2.0 kg/d per ewe of concentrate.
1–1.5 kg/d per ewe of hay.
Trough and manger space 250–300 mm per lamb, rising to 400–500 mm per ewe, accessible from outside pen.
Up to 5 ℓ/d per animal from a water trough 0.3 × 0.6 m in each pen. Running water (to overflow) desirable, particularly with hill breeds.

Effluent and drainage.

4–5 ℓ/d per adult animal. Slats are wood, 38–50 mm top width × 25–30 mm thickness, with 15–20 mm gaps. Leave 0.65 m clearance underneath for one season's holding capacity. Floor falls of concrete to be 1:100 to drainage collection point.

Special areas
  Lambing accommodation

Clean area, isolated from general holding area. Adequate services. Pens of 0.75–1.2 m × 1.2–1.8 m

  Shearing accommodation.

0.5–0.6 m² per animal holding area. 1.5 m² per shearer.

Other considerations.

Provide races and satisfactory entry and exits to dipping and handling areas. Ramps to and from house to be 1:10 maximum. Provide long-term fodder storage within building envelope to provision against poor weather periods. 9–13 m spans preferred.

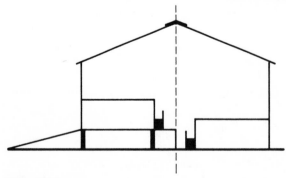

Figure 25.51

3.2 Sheep handling areas

Space allocations.

Gathering and dispersal pens to contain groups of up to 100 sheep at 0.5–0.6 m² per animal.
Catching pens to contain 10–15 sheep at 0.3–0.4 m² per animal, may be circular with centre pivot gate to crowd into race, dip, etc. or rectangular.

Central races 3.0–6.0 m long × 0.4 m wide × 0.8–1.1 m high.

Footbath and race 3 m long × 0.25 m wide at bath (0.15 m deep with 0.1 m fluid depth) and 0.4 m wide at top of boards, about 1.0 m high.

Dip tank (*see Figure 25.52*) capacity from 0.75 m³ for 300 ewe flock, to 2.5 m³ for flocks of 1000 +.

Ancillary holding pens for up to 50 sheep, short term, at 0.4 m² per animal.

Other considerations.

Allow 2.5 ℓ/sheep of dip fluid. Floors should be roughened concrete with slope to drain channels of 1:50–100, or may be hard, dry soil in well drained areas, or may be of introduced stone.

Drain baths, tanks etc. to an approved holding or disposal point.

Pen posts for rails should be away from sheep, either on non-sheep side or with shoulder protector boards.

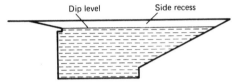

Figure 25.52

4 Poultry housing

4.1 Laying hens

Space requirements (*see Figures 25.53–25.56*).

Deep litter birds, 0.27–0.36 m² per bird, 0.2–0.25 m² per bird when one third floor area is slatted.

Eaves height 1.5–1.8 m.

Cages for battery layers measure 430 mm deep × 450 mm high × 300/400/500 mm for single/double/treble occupants per cage. Arrange in tiers of up to three cages high.

Passages 1.2 m wide along house (at sides and between rows of cages), 1.5 m across end of house.

Eaves height 2.1–2.4 m.

Temperature and insulation.

13–16 °C in traditional deep litter. 21–24 °C in controlled environment housing with 'U' value of 0.5 W/m² °C or less for roof and walls.

Ventilation.

0.25 ℓ/s per hen in winter, 2.8–3.0 ℓ/s per hen in summer. Relative humidity should not exceed 70%, and air delivery speed should not exceed 2 m/s.

Litter, feed and water consumption.

250 mm of wood shavings, chopped straw or newspaper; litter depth maintained throughout production cycle. 0.1–0.3 kg/d per bird from trough the size of which allows 60 mm per bird in deep litter or full width of cage in batteries. 20–30 ℓ/d per 100 birds from constant level troughs or two nipple drinkers per cage in batteries. Drinkers no more than 3 m apart in deep litter houses.

Effluent and drainage.

Approximately 0.1 ℓ/d per bird. Mechanical scraper removal from vertical tiers of cages, dropping pit under staggered tiering. Drain channel for trough overflow let into floor. General floor falls of 1:100. Deep-pits in some slatted floor houses.

Other considerations.

Lighting effects laying performance—refer to special manuals for artificial lighting services.

Egg room floor allowance at 9.0 m² per 1000 birds.

Egg boxes should measure 320 mm × 310 mm × 310 mm.

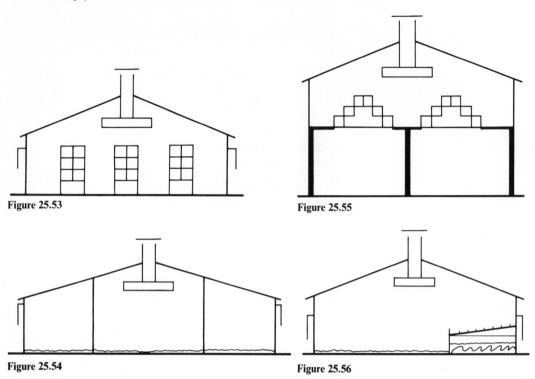

Figure 25.53

Figure 25.55

Figure 25.54

Figure 25.56

4.2 Broiler and capon housing

Space requirements (*see Figure 25.54* for general arrangement).

0.09 m² per bird, up to ten weeks old, kept in batches of 5000–7000. 0.2 m² per bird up to 16 weeks old, kept in batches of up to 2500.
Eaves height 1.5 m minimum.

Temperature and insulation.

16–24 °C with a maximum 'U' value of 0.5 W/m² °C for roof and walls.

Ventilation.

Natural possible for larger birds when controlled by baffling and ducting to avoid draughts. 0.2 m²/100 birds of vent area required. 0.2–2 ℓ/s per bird required in controlled environment houses for winter–summer conditions.

Feed and water consumption.

0.2–0.4 kg/d per bird with 150 mm length of trough allowed. 13–20 ℓ/d per 100 birds from 600 mm minimum of water trough length per 100 birds.

Effluent and drainage.

About 0.5 ℓ/d per 100 birds up to ten weeks old. About 1 ℓ/d per 100 birds up to 16 weeks. General floor falls to drain outlet of 1:100 for washing and disinfecting.

Other considerations.

Hard floors preferable under litter of 100 mm deep. Platform-tiered, so called 'aviary' system, can increase stocking rate. Light intensity and daylength are important for maximum growth rate—*see* specialist texts.

4.3 Turkey housing

Space requirements.

Breeding stock, 0.5–0.9 m² per bird in groups of 25–30. Fattening stock, 0.1–0.5 m² per bird from three weeks to 16 weeks+. Up to 0.3 m² per bird in controlled environment housing.
Eaves height 1.8 m for 16 week olds.

Environment.	Natural ventilation usually, with limited control to provide weather protection and shelter. Insulated roof and walls for winter fattening systems.
	'Verandah' building usually adequate. 18 °C optimum temperature.
Feed and water consumption.	0.4–0.6 kg/d per bird at 16 weeks old. 3.5–7.0 m trough per 100 birds or 50 mm per adult if tubular feeders used. 12–70 ℓ/d per 100 birds from three weeks to 16 weeks old. Spill resistant water troughs required.
Effluent and drainage.	0.3–0.6 ℓ/d per adult bird. Floor falls at 1:100 where concrete. Where slatted allow 0.6 m space below floor. Area of wire or slats must not exceed one-third of total flooring.
Other considerations.	Lighting may be used in 'open' housing systems as a deterrent to foxes, etc. Fox proof netting walls are essential.

5 Crop storage

5.1 Grain storage

		Barley	*Beans*	*Linseed*	*Maize*	*Oats*	*Peas*	*Rye*	*Wheat*
Space requirements (based on 14% MC and 30 degree repose angle).	m³/t	1.45	1.2	1.45	1.35	2.0	1.3	1.45	1.3
	kg/m³	705	833	705	740	513	785	705	785
Capacity of level, 2 m fill, of loose grain, per 4.8 m bay and 6 m span.	t	42	51	42	47	29	44	42	44
Capacity of above 2 m fill when heaped, per 4.8 m bay and 6 m span.	t	57	—	—	64	39	—	—	60
Tonnes of product 2 m high, level fill per m² of floor area.	t	1.34	1.67	1.34	1.48	1.0	1.54	1.34	1.54
Capacity, in tonnes, of 4.5 m diameter silo 9 m high.	t	102	—	—	—	76	220	98	110

Store types.	Circular bins 2–6 m in diameter, up to 6 m high (height not to exceed twice the diameter)—(*see Figures 25.57* and *25.58*). Materials may be hessian and wire mesh, oil-tempered hardboard, galvanised corrugated steel, galvanised slotted steel, plyboard, welded steel plate, glass- or enamel-surfaced steel or concrete stave with steel hoops. Floor may be plain concrete, perforated mesh or sloped steel (45 degrees). Emptied by 'air-sweep', slope or auger conveyor.
	Square or rectangular bins 2.4–4.5 m with heights up to 9 m (6 m commonly for drying purposes). Materials may be corrugated steel sheet (square profile), galvanised steel pressed panels or reinforced concrete block with substantial piering and frame support. Flooring will be plain, ventilated or sloped as with circular type.
	Loose bulk ('on-the-floor') storage uses standard building types and frames; with or without simple retaining walls for containment at the edges or at dividing partitions (*see Figure 25.59*).
	Wet grain storage, in circular sealed silos (or with non-sealed roofing), using glass fused or enamelled steel. (Allow for density of product at 14% MC plus 5–10%.) Danger from CO_2 build up.

Wet grain heaped 'on-the-floor' in clear span framed buildings, preserved by propionic acid application. (Density as above.) Refrigerated grain stored, at 18–22%. MC in bins at a density of 3–5% more than at 14% MC.

Ventilation and drying.

Either high or low volume flow available for most grain storage systems giving either a drying or a conditioning effect. Require main ducts at centre or sides with laterals and diffusers to distribute air evenly throughout the mass. Air systems may also act as grain conveyors. Fans may be noisy.

Floor.

Hard, durable, smooth, dust-free and dry. 150 mm of concrete, reinforced on soft, difficult soils and at edges when supporting bins; steel floated and incorporating a damp proof membrane. May have to accommodate air ducts or conveyors.

Walls.

Thrust resistant (*see Figure 25.60*) and air tight. Must be designed with adequate foundations. Building frame may support thrust if designed to do so.

Other considerations

High power consumption likely in some stores.
Overhead hoppers to fill lorries, with high delivery rate conveying systems to match. Large headroom and strong framing required.
Rankine's formula expresses the relationship between pressure on wall, density, height of loading and angle of repose of contained materials. Thus when $\theta = 30$ degrees

$$p = wd\frac{(1 - \sin\theta)}{(1 + \sin\theta)} \qquad p = wd\frac{(1 - \frac{1}{2})}{(1 + \frac{1}{2})} = \frac{wd}{3},$$

where p is pressure at height d from the top surface, w is density and θ is the angle of repose.
Building should be vermin and bird-proof. Doors should provide clearance for tipped trailers. Provision should be made for heavy vehicles in the vicinity of grain stores. Special consideration should be given to fire hazard in grain drying areas. Deep pits, elevator equipment and temporary storage prior to drying may be required.

5.2 Hay, silage, straw and concentrate fodder accommodation

Space required for bulky low density materials (m³/tonne).

Hay, loose 9.0.
Hay, medium density baled 6.0.
Hay, barn dried 7.5.
Barley, straw, medium density baled 11.5.
Wheat, straw, medium density baled 13.0.
Grass, wilted 2.2.
Grass, silage, consolidated 1.4.
Grass, tower silage, high dry matter 1.25.
Grass, high temperature dried (75 mm long) 8.5.

Volumes and tonnages.

A 4.8 m bay of a 9.0 m span building will contain:
- 35 tonnes of baled hay at a height of 4.8 m.
- 69 tonnes of consolidated silage at a height of 2.4 m.
- 20 tonnes of baled barley straw at a height of 5.4 m.

Store types.

Standard frame with open or closed sides, may have special walling and floor system (*see Figures 25.61* and *25.62*). 9–18 m spans preferred.

Flooring.

Earth or rammed hardcore on well-drained soils, for simple storage of baled hay and straw. Raised floor of weldmesh, etc. to provide plenum chamber for drying of hay in batch or storage driers.

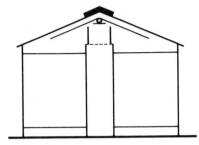

Figure 25.57

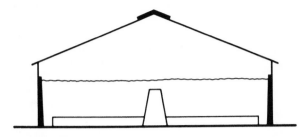

Figure 25.59

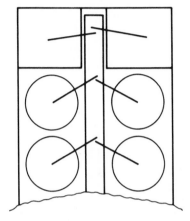

Figure 25.58

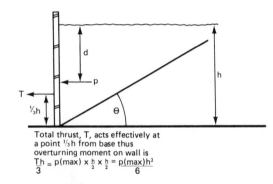

Total thrust, T, acts effectively at
a point $\frac{1}{3}$h from base thus
overturning moment on wall is
$$\frac{Th}{3} = p(\max) \times \frac{h}{2} \times \frac{h}{2} = \frac{p(\max)h^2}{6}$$

Figure 25.60

	Silage requires durable, acid-resistant surface of heavy duty concrete, laid to adequate side and end falls and incorporating drainage gullies, leading to holding or disposal systems which must be approved.
Walls.	Thrust resistant for silage (*see* section 5.1) which are a specialist design problem. Should have rails on top for safety of operatives during silage consolidation. Tower silos are similar to wet grain silos (*see* section 5.1) except the height may exceed 20 m. 15 m minimum wall length required to accommodate loading ramp.

Hay barn walls may be removable in the form of wooden baulks or steel sheet or they may, in the case of loose material be of wire netting. Some types of walling must be air tight in storage drying. Top of wall near eaves should be slatted for up to 1 m to allow air movement during drying.

Roofing.

No roof required for silage clamps. Sealed tops for some tower silos with air pressure release bag incorporated. Hay barn roofing should be ventilated and of non-condensation drip-forming type.

Other considerations.

Fire risk should be minimised. Ease of access to farm roads and livestock buildings. Power requirement may be high in storage driers or tower silo systems. Drying fans tend to be noisy in operation. Danger from CO_2 build up in silage towers.

Space for concentrates.

Meal, loose, 2.0 m³/tonne or 500 kg/m³.
Pellets, loose, 1.7 m³/tonne or 600 kg/m³.
Nuts, loose, 1.4 m³/tonne or 700 kg/m³
(50 kg bags stack two high, occupy 2.0 m³/tonne or 500 kg/m³.

These figures are only approximate and depend on the constituents. For bin store floors an angle of 60 degrees should be used for self-emptying facilities.

Space for equipment used in milling and mixing.

Horizontal mixer, 4–10 m² × 2–3 m high.
Vertical mixer, 3–5 m² × 3–6 m high.
Crushing mill, 2–3 m² × 1–1.5 m high.
Hammermill, 1–6 m² × 1–1.5 m high (5 m with cyclone above).
Allow 1–1.5 m headroom for overhead conveying.

Other considerations.

Noise and dust are potential health hazards. Risk of explosion in a dusty atmosphere. Very high power requirement by most barn machinery.

Mechanical handling requires smooth running, hard flooring. Use of gravity desirable in conveying—overhead hoppers for lorry and trailer filling with 10–20 tonnes capacity for big units. High clearance needed.

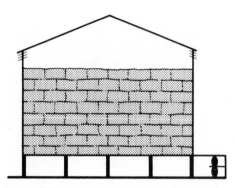

Figure 25.61

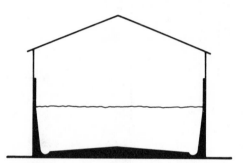

Figure 25.62

5.3 Vegetable storage

Space requirements and environment.	m³/tonne	Mean temperature to store long term	Maximum height
Potatoes, bulk (*see Figures 25.63* and *25.64*).	1.56	Ware 4–5 °C	4.0 m (6 m if airflow can reduce temperature gradient).
Potatoes, in pallet boxes. (1.2 m × 1.4 m × 0.8–1.0 m).	2.4	Ware—as above. Crisping and processing, 6–8 °C	4–6 boxes (4–5 m).
Onions, dry (*see Figures 25.63* and *25.64*).	2.0	2–3 °C	4–5 m
Onions, green topped.	3.0	Ambient temperature +3–6 °C until dry.	2.5 m
Onions, refrigerated.	2.0	0.5–1.5 °C	3.0 m
Onions dry, in boxes (1.1 m × 1.3 m × 1.05 m high).	3.0	Ambient temperature +3–6 °C.	Boxes stacked to eaves.

Store types.

Standard, clear span frames with heavy duty concrete floor. May have air-flow and conveying ducting.

Floor.

As section 5.1 plus smooth running floor for pallet handling equipment.

Walls.

Thrust resistant (*see Figure 25.60* and section 5.1 for Rankine's formula, angle of repose for potatoes is 30–40 degrees) except for pallet box stores where cheaper walling may be used. 'U' value of 1.0 maximum W/m² °C for walls (and roof) is necessary to protect potatoes from frost. Care must be taken to avoid 'cold bridging' across the walls via reinforcing or framing systems, Walls should be smooth and cleanable.

Other considerations.

Avoid materials in structure with a high thermal conductivity. Use straw as insulation layer on top of stack, and bales as temporary wall insulators.

Should be a concrete apron of 2.5 m minimum width in front of loading doors.

Porch desirable to enable store to be filled to end wall, potatoes rest on removable boards which carry stack weight as it is built up.

Chitting houses for potatoes (*see Figure 25.65*).

Potatoes stacked in trays, with access to illumination from movable lighting columns suspended from ceilings. Trays are $750 \times 450 \times 150$ mm on pallets 1.5×0.9 m in area, stacked to depth of 3.5 m maximum. Alleyways 0.5 m minimum between rows. 1 m side alleys. Generous end space in building. Should be frost-proof and force ventilated, with heating equipment available when necessary.

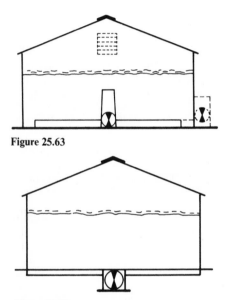

Figure 25.63

Figure 25.64

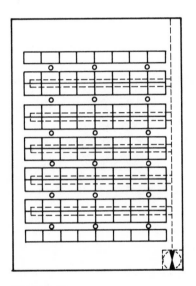

Figure 25.65

6 Servicing and general buildings

6.1 *Fertiliser storage*

Space required.

Loose, bulk, 1.1 m³/tonne average for prilled fertiliser.

50 kg bags, stacked, 1.2 m³/tonne (at 180 mm thickness per bag, with maximum of 10 high).

'Big' bags containing 0.5 tonnes, stack two or three high; are nearly equivalent to loose in space occupied.

Other considerations.

Stored in standard frame building with weather protection and anti-condensation provisions to prevent dripping onto bags. Thrust resistant walls for bulk fertilisers with corrosion resistant surfacing. Ends and top of heap covered with plastic sheeting. Walls and floor clearances of 0.5 m and 0.15 m respectively for bags (on pallets or boards). Provide temporary divisions to isolate different fertiliser types.

6.2 Slurry and farmyard manure storage

Space required.

Straw and dung, farmyard manure, 1.0–1.4 m³/tonne or 700–1000 kg/m³.
Slurry, glutinous, 1.0–1.1 m³/tonne, 900–1000 kg/m³.

Storage systems.

40–50% dry matter, handled straight from bedded area or stored in dungstead (walled enclosure with moisture retaining floor falls). Move by fore-end loader.
25–35% dry matter, glutinous, non-flowing, in thin layers on flat, hard surfaces, moved by scraping into pit for pumping to enclosure as above with solid matter or into above-ground tank.
10–20% dry matter, plus washing water, etc. pumped from sumps, holding tanks and catch pits in drainage and slat systems into above-ground slurry stores, circular (20–1000 m³) or rectangular (100–2000 m³) or reservoirs or lagoons, excavated into the ground and sealed against leakage.
Up to 10% dry matter (excluding straw) held in lagoon or reservoir for active oxidation or passive storage.

6.3 Machinery storage

General.

Height and width are often more of a problem with storage than area or volume occupied, due to doorway restrictions, eaves heights and turning circles. Site building near good hard road, central to farm operations, adjacent to workshop and cleaning facilities. Provide sufficient light, natural and artificial, for inspection and manoeuvring.

Clearance requirements for tractors and implements.

Medium size tractor, 3.0 m high × 1.85 m width (may be extended to 2.5 m wheel track), area 6.75 m², with a turning circle of 7.5 m diameter (exhaust should have 0.5 m minimum height clearance from structure). Medium size tractor plus loader, turning circle, 10 m diameter.
Combine harvester, 3.0–6.0 m wide × 7.5–10 m long × 3.0–4.0 m high.
Trailers 2.0–2.43 m wide × 4.0–5.5 m long × up to 4.5 m high (tipped).
Lorries 2.43 m wide × 4.5–9.1 m long (rigid chassis) or 10.0–12.5 m long (articulated) × up to 4 m high.
Plough, reversible, three furrow, 1.5 m wide × 2.5 m long.
Disc harrow, 2.5–5 m wide × 1.9 m long.
Seed and fertiliser drills, 2.5–4 m wide (trailed) or 2.5–6.5 m wide (mounted) × 2.0–3.0 m long.
Balers, medium density, 2.5–3.0 m wide × 4.5–5.5 m long.
Beet harvesters, 2.5–4 m wide × 5.5–7 m long × 3.5 m high.

Other considerations.

Hard, well-drained, floor (but not necessarily concrete). High eaves at doorway (upstand in roof at this point?). Translucent sheeting in roof to give high level of natural light. Concrete at point of entry to building with generous hard standing outside, adjacent.
Limited number of socket outlets for electricity supply.

6.4 Fuel storage

Diesel store only (petrol storage restricted and licensed, *see Appendix 25.1*).

Tank sizes
  1.0 m × 1.0 m × 1.0 m, 1137 ℓ (250 gallons)
  1.5 m × 0.9 m × 0.9 m, 1137 ℓ (250 gallons)
  2.0 m × 1.25 m × 1.25 m, 2728 ℓ (600 gallons).
Black mild steel to BS 799, sited high enough on support walls

(225 mm) to discharge to highest tank-filler when almost empty. Tank should have fall away from delivery pipe to drain cock. Dip stick reading with ladder access or sight-gauge required.

6.5 Workshop accommodation

Area required and facilities.

One or two bays of standard frame building of 9–15 m span. Pit, 2.0 m × 0.8–1.0 m wide × 1.7 m deep with steps at each end, water proofed, with drain sump; placed centrally in one of the bays with access from main door to workshop. Clear wall length of 5 m for benching with electric power and spot lighting. Concrete area of 50 m^2 minimum outside main workshop door with good drainage falls to sump or trapped catchpit. Floor inside of 150 mm concrete, reinforced to take heavy equipment. Overhead gantry for crane desirable.

6.6 Farm office accommodation

Space required (may be controlled by Offices, Shops and Railway Premises Act 1973).

10 m^2 minimum, 15–20 m^2 preferred, with washroom area and WC of 3 m^2 adjacent, all served by porch or lobby of 3 m^2 minimum. Natural light required through one wall of office, preferably with view of main farm access road. Electric power, heating, water and telephone services essential. Large area of clear wall space desirable. Drainage and other building features may be regulated by Building Regulations and inspected during construction by Local Authority Building Control Officer.

APPENDIX 25.1

The principal regulations concerning the design, erection or alteration of farm buildings (excluding finance)

Town & Country Planning Act (1971)

This Act follows, replaces and/or consolidates those of 1947, 1959, 1962, 1968 and all Orders made under those previous acts.

General principle

This is one of control of development; requires permission to be sought before any building or engineering operation is carried out or any change of use or purpose of land or building is instituted, or new access is made on to classified road. Agricultural construction works (in Class 6) are broadly permitted under the Act through Article 3 of the Town and Country Planning General Development Order of 1973, subject to certain limitations:

(1) the holding is of more than 1 acre,
(2) the sole purpose is agricultural,
(3) the item is not in connection with a dwelling house,
(4) the area does not exceed 465 m^2 in aggregate with any other works engineered within 90 m and within two years of the start of construction,
(5) the height does not exceed 12 m (3 m if within 3 km of an airfield),
(6) no part is within 25 m of the metalled portion of a classified road.

Planning permission is not required for earthworks and excavations so long as they are 25 m at least away from classified roads.

Planning permission *is required* for roadside selling stands and advertisement hoardings (T and CP (Control of Advertisements) Regulations 1969 and (Amendment) 1972).

Where planning permission is required, certificates, ensuring that owners and tenants affected by the development have been notified of the intentions, must accompany the application. Outline permission may be applied for, in which case no development may start until full permission is granted. Five years is the normal length of time for planning permission to be granted. Require re-application after that. Applicant may appeal against refusal within six months of the original decision.

Town & Country Amenities Act 1974

This takes over some of the functions of T & CP Acts in respect of conservation areas; with particular reference to historic buildings and areas of special scenic beauty, e.g. National Parks.

Lists of buildings which may not be redeveloped

without permission are maintained by Local Authorities and the owners are now notified. Planning permission must be sought for any development in an area of special scenic beauty.

All buildings in a designated conservation area are subject to control and demolition, alteration or improvement requires an application for listed building consent.

Tree preservation orders

These may be made for the preservation of trees at any time and also for the planting of trees during development. Preservation includes action against topping, trimming or wilful destruction. Tree must be of a minimum size and preservation is subject to ministerial confirmation. Application can be made to cut down, etc. preserved trees.

The Forestry Act 1967

This is intended to control tree husbandry but has some bearing on development work. May cut down trees not listed up to 23.4 m^3 (or 4.25 m^3 for sale) in any quarter of a year. Further felling requires Forestry Commission agreement unless it is done under planning permission or to obviate danger.

Public Health Acts 1936, 1961

Building Regulations 1976 (as amended).

Wide powers to control building structures to achieve public health and safety. Controls structural strength, drainage and sanitation, area, volume, environment, permanence. Some buildings are partially exempt—including Class 6 (farm buildings): single storey buildings exclusively used for the accommodation or storage of plant, machines, animals, crops and in which only agricultural workers will be employed and which is detached from other buildings. Class 6 buildings require only compliance in respect of submission of plans, giving notices of works and water supply and drainage (also fire resistance, i.e. chimneys and fences). (Not done widely in practice.) Consent is required before farm effluent is discharged in public sewers. Intensive farm enterprises covered by *Public Health (Drainage of trade premises) Act 1937.*

Control of Pollution Act 1974

Consolidates previous legislation. It is an offence to pollute streams (any ditch or watercourse, in practice, even if dry for a proportion of year) or to discharge liquid waste underground except by agreement with the Water Authority—usually apply a BOD level of 25 ppm as requisite.

Health and Safety at Work Act 1974

Sets out safety provisions:

(1) The Agriculture (Ladders) Regulations, 1957:
 (a) enclosed stairs—one handrail,
 (b) handrails on each open side of all stairs which are 1 m or more high and 30 degrees or more from vertical,
 (c) handholds provided at top of any stairs which are less than 30 degrees to vertical, all handrails to be strong, smooth and rigid.
(2) The Agriculture (Safe-guarding of workplaces) Regulations, 1959:
 (a) pits or openings in floors more than 1.53 m deep should have a rail of between 0.92–1.07 m high or a fence not less than 0.92 m high or a cover (including a grid) so long as it gives no less protection than the rails,
 (b) floor edges (other than wall openings) to have guard rail or fence,
 (c) wall openings—guarded by door or fence or guard rail (not openings of less than 1.22 m from top to bottom or those to stairs or those with sills more than 1.61 m from floor). Handhold provided for when doors, etc. are removed for, e.g. movement of materials.
 (Do not apply during periods of construction or alteration.)
(3) The Agriculture (Stationary Machinery) Regulations, 1959:
 (a) prime movers to have cut off, clearly labelled with directions,
 (b) no worker contact with belts, shafts, pulleys, etc. when moving,
 (c) adequate natural or artificial light where machine is used.

Responsibility for safety lies with employer and employed *equally*.

Food and Drugs Act 1955

The Milk and Dairies (General) Regulations 1959

Part V—regulates water supplies and general building construction in food producing area.
Part VI—suitable structural finishes, good drainage with traps, falls, pipes, etc., cooling of milk.

Petroleum (Consolidation) Act 1928

Petroleum Spirit (Motor Vehicles) Regulations 1929—regulating licensing of petrol storage in excess of 9 ℓ.

Other Acts

Other acts having some bearing on building construction in agriculture.

Housing Acts—fitness for habitation of dwellings.

Offices, Shops and Railway Premises Acts—staff working conditions.

Caravan Sites (Licences Application) Order 1960 under T & CP Act 1959 consolidated in T & CP Act 1971—controls through Local Authority licensing the precise management of sites in respect of area, numbers and hygiene. Some sites exempt from control, e.g. workers' caravans.

Water Resources Act 1968—Water Authorities set up and abstraction rigidly controlled.

Land Drainage Act 1930—controls, bridges and roadworks.

Highways Acts—rights of way.

Highway (Provision of Cattle Grids) Act 1950—empowers Local Authority to construct grids on classified roads and rights of way.

National Parks Act 1949 (*see* Town & Country Amenities Act 1974, p 569)—controls pathways, etc. in the countryside.

Agriculture (Miscellaneous Provisions) Act 1968—codes of practice in animal environments to avoid so-called 'factory farming' conditions and in particular fire hazards—flame proofing and escape routes being required.

In addition regulations not given similar power of law which may have the same effect however are:

IEE Regulations—electricity supply systems and fixtures in building structures and elsewhere.

British Standards Codes of Practice in both construction and use of buildings and equipment. Note BS 5502 particularly (*see later*).

Farm Capital Grant Scheme Standards may be very forceful and all-embracing for construction (but are subject to rapid changes caused by variations in economic and political climates).

Private Wayleaves obtained by private Act of Parliament to cross land or boundaries, e.g. electricity pylon systems or gas pipelines.

Bramble Committee Report into housing conditions for livestock.

BS 5502: 1980 is intended to embrace all agricultural structures and impose upon them precise structural and environmental requirements which will result in satisfactory operation for a predicted and assured lifetime and safe working conditions during that time. All buildings conforming to the British Standard will carry a plate showing their classification, manufacturer and date. Buildings not conforming to British Standards will not be acceptable for grant aid directly. Classification relates to occupancy, predicted safe lifespan, type of activity intended for, and the construction materials safe stressing limitations, together with its general suitability for the purpose and use. Part 2 of BS 5502 incorporates animal welfare guidance and lays down environmental standards desirable for healthy livestock in humane production systems.

APPENDIX 25.2

List of advisory organisations

The Agricultural Engineers' Association
6 Buckingham Gate
London, SW1.

Asbestos Cement Manufacturers' Association
16 Woodland Avenue
Lymm, Cheshire.

British Constructional Steelwork Association
1 Vincent Square
London, SW1P 2PJ.

British Glasshouse Manufacturers' Association
Simpsons of Spalding Ltd
Branton's Bridge
Bourne Road
Spalding, Lincs, PE11 3LP.

British Precast Concrete Federation Ltd
60 Charles Street
Leicester.

British Veterinary Association
7 Mansfield Street
London, W1N 0AT.

British Wood Working Federation
82 New Cavendish Street
London, W1M 8AD.

Building Research Establishment
Buckwalls Lane
Garston
Watford
Herts.

Cement & Concrete Association
Wexham Springs
Slough
Bucks, SL3 6PL.

Chipboard Promotion Association
7A Church Street
Esher, Surrey.

Constructional Steel Research and Development Organisation (CONSTRADO)
12 Addiscombe Road
Croydon, Surrey, CR9 3JH.

Council for Small Industries in Rural Areas
141 Castle Street
Salisbury, Wilts, SP1 3TP.

The Country Landowners Association
16 Belgrave Square
London, SW1X 8PQ.

Electricity Council Marketing Department
Farm Electric Centre
National Agricultural Centre
Stoneleigh
Warwickshire, CV8 2LS.

The Farm Buildings Association
Roseleigh
Deddington
Oxford.

The Farm Buildings Information Centre
National Agricultural Centre
Stoneleigh,
Kenilworth
Warwickshire.

Fibre Building Board Development Organisation Ltd
Stafford House
Norfolk Street
London WC2.

Federation of Master Builders
33 John Street
London, WC1N 2BB.

The Finnish Plywood Association
21 Panton Street
London, SW1Y 4DR.

The Fire Protection Association
Aldermary House
Queen Street
London EC4.

The Institution of Civil Engineers
Great George Street
London SW1.

Institute of Structural Engineers
11, Upper Belgrave Street
London, SW1X 8BH.

The Milk Marketing Board
Thames Ditton Surrey
Surrey.

Ministry of Agriculture, Fisheries & Food
Great Westminster House
Horseferry Road
London WC1.
(and local regions)

The National Council of Building Material Producers
26 Store Street
London, WC1E 7BT.

The National Farmers' Union
Agricultural House
Knightsbridge
London, SW1X 7NJ.

National Institute of Agricultural Engineering
Wrest Park
Silsoe
Bedfordshire.

National Institute for Research in Dairying
Shinfield
Reading
Berks.

Plywood Manufacturers Association of British
  Columbia
81, High Holborn
London WC1.

Potato Marketing Board
50 Hans Crescent
Knightsbridge
London SW1X 0NB.

Poultry Research Centre
West Mains Road
Edinburgh 9.

Royal Agricultural Society of England
National Agricultural Centre
Stoneleigh
Warwickshire.

The Royal Institute of British Architects
66 Portland Place
London, W1N 4AD.

The Royal Institution of Chartered Surveyors
12 Great George Street
London, SW1P 3AD.

Scottish Farm Buildings Investigation Unit
Craibstone
Bucksburn
Aberdeen, AB2 9T2.

Steel Cladding Association
European Profiles Ltd
Llandybie
Ammanford
Dyfed.

Timber Research & Development Association
Hughendon Valley
High Wycombe
Bucks, HP14 4ND.

Part 4

Farm management

26

The Common Agricultural Policy of the European Economic Community

P. W. Brassley

BACKGROUND, INSTITUTIONS AND THE LEGISLATIVE PROCESS

The need for a Common Agricultural Policy

When the European Economic Community was created it needed a Common Agricultural Policy (CAP) for two reasons: first, the agricultural industries in each of the member states were subject to government intervention; and second, if this intervention was to continue, it had to be compatible with the other provisions of the EEC.

Left to themselves, the markets for agricultural products in developed economies will produce fluctuating, and, in terms of purchasing power, gradually declining incomes for farmers, as a result of fluctuating and declining real prices of agricultural products. The price of agricultural products, like that of any other product in a free market, depends upon the balance of demand and supply. If demand increases less than supply it will fall. The demand for agricultural products in a developed country does not usually increase very rapidly, unless those products can be sold cheaply on the world market. Supplies can either be produced at home or imported. As a result of technological change, producing increased output per hectare and per unit of labour, supplies tend to increase more quickly than demand. So in the long run prices tend to fall. Demand is also relatively unresponsive to price changes, so short-run supply variations, caused by fluctuations in climate and the incidence of disease, produce short-run fluctuations in prices.

Short-term fluctuations in a free market may result in the demise of businesses which would be viable at average levels of input and output prices, and the possibility of this event creates a disincentive to investment in such farms. Long-run price falls result

in lower incomes for those farms which are unable to reduce costs or increase output, and if such low incomes are unacceptable they may cease trading altogether. It is usually found that farm incomes have to fall to very low levels before farmers leave the agricultural industry, and in any case the land that they no longer farm is often taken over by another farmer. The total output of the industry is therefore maintained, so there is no tendency for prices to rise. While farmers with low incomes may be found in all areas, there may be some regions which are particularly disadvantaged by physiographical or structural factors, and if agriculture is the major industry in such regions the whole region may be affected by income and outmigration problems.

In a free trading economy farm income problems may be further intensified by the availability of agricultural products at prices below the domestic supply price. Imports will not only restrict domestic price rises but will also increase the total import bill, which may be considered a problem in countries with balance of payments difficulties. However, the government of a country which has no difficulty in exporting enough to pay for food imports may still consider it unwise to be totally reliant on imported food supplies: unforeseen emergencies may give rise to difficulties in obtaining supplies, or long-term world price rises, due, for example, to population increases, may in time reduce the price advantages of imports. A capacity for rapid expansion of domestic agricultural production may therefore be thought desirable, even if considerable reliance is placed upon imports. In short, free markets for agricultural products in developed free market economies may produce low farm income, balance of payments and potential supply security problems.

By the middle of the 1950s the agricultural industries in all western European countries faced all these problems to some extent, and governments had evolved a number of policies for dealing with them. In the continental European countries agricultural policies usually involved a system of import controls and price support which kept price levels higher than they would have been in a free market and so maintained agricultural incomes and self-sufficiency levels. When the discussions which led to the formation of the EEC were held in 1955 and 1956 it was agreed that the exclusion of agriculture from the general common market was impossible: without free trade in farm products national price levels could differ, and those countries with the lowest food prices would have the lowest industrial costs, thus undermining the common policies which would be introduced for other industries. Although there was a general similarity between the existing agricultural policies, the differences in detail were so numerous that the simple continuation of existing policies would have been impossible. Not only did Community countries require an agricultural policy, they required a common agricultural policy.

The formation of the Common Agricultural Policy

The Treaty which established the EEC was signed in Rome on 25 March 1957. Articles 38–47 of the Treaty apply directly to agriculture, and of these Articles 39 and 40 are probably the most significant. Article 39.1 lays down the objectives of the Common Agricultural Policy:

(1) to increase agricultural productivity by promoting technical progress and by ensuring the rational development of agricultural production and the optimum utilisation of the factors of production, in particular labour;
(2) thus to ensure a fair standard of living for the agricultural community, in particular by increasing the individual earnings of persons engaged in agriculture;
(3) to stabilise markets;
(4) to ensure the availability of supplies;
(5) to ensure that supplies reach consumers at reasonable prices.

Article 40 lays down the broad guidelines for the various policy instruments by which these objectives are to be achieved, and Article 43 indicates the procedure to be followed in reaching agreement on the detailed provisions of the CAP. Article 43 also provides for discussions to be held in order that member states and EEC institutions can decide upon their requirements, after which the Commission is required to submit proposals to the Council of Ministers. The Commission produced these proposals in 1960, and although they have been altered in detail they form the basis of the CAP described below. It is interesting to note that four principles which are often assumed to lie behind the CAP system (a single market, joint financing, Community preference, and a comparable earned income for efficient farms), appear neither in the Commission proposals nor in the Treaty of Rome, but seem to have appeared early in the working life of the CAP. These developments, and the ability of Community policy makers to produce new policies to cope with new problems, mean that the decision-making process in the Community must be examined before the mechanisms of the CAP can be outlined.

The process of decision making in the EEC

The Treaty of Rome lays down the basis of the CAP but decisions still have to be made about what policy shall be and how it should work in detail. There are therefore a large number of formal and informal bodies involved in decision making, of which the two most important are the Commission and the Council of Ministers.

The *Commission of the European Communities* is presided over by 14 Commissioners. It is divided into 20 Directorates-general (DGs): DG VI is responsible for Agriculture, and is the responsibility of one Commissioner. In some ways the Commission is like a national civil service, but it has extra functions: it handles the day to day administration of Community policies, but also proposes policies and takes infringements to the Court. After consultation with the appropriate bodies, Commission proposals are referred to the Council of Ministers.

The *Council of Ministers* is the major legislative body of the Community, and consists of a minister from each member state, depending upon the subject under discussion: the Minister of Agriculture for agricultural matters, finance ministers for financial matters, and so on. It is normally the final decision-making body for all legislation, although when problems cannot be resolved they are now often passed on to the European Council, a meeting of the heads of government of the member states which has no legal basis in any treaty and which developed from earlier summit meetings. Its decisions have to be passed back to the Council of Ministers to be given legal validity.

Much of the background work for Council of Ministers meetings is handled by the *Committee of Permanent Representatives* (COREPER), which draws up the agenda, agreeing on non-controversial items and leaving only items on which it cannot agree for the Council of Ministers. Since the Council of Agricultural Ministers has a great volume of work it is treated slightly differently, and the *Special Committee for Agriculture* (SCA) does much that would normally be done by COREPER, although COREPER retains responsibility for financial and some other matters.

The *Court of Justice* rules on the application and interpretation of Community law, which takes precedence over national law. There are two categories of Community law: first, all the founding treaties; second, the Regulations, Directives and Decisions proposed by the Commission and agreed by the Council of Ministers. There are important differences between these last three. A Regulation is automatically and directly applicable to all people and governments in all member states. A Directive is binding as to its intentions on governments, which must then pass national legislation to give it effect, so that it is more flexible than a Regulation. A Decision is as binding and immediately applicable as a Regulation, but only on the people or governments to whom it is addressed. The Commission may, in consultation with a Management Committee, also make Regulations, Directives or Decisions necessary to carry out the wishes of the Council of Ministers on a day to day basis.

These are the major decision-making bodies of European Communities, but there are also many other institutions which have an advisory or consultative role in the decision-making process. They fall into three categories: European institutions, national government or Parliamentary bodies, and non-governmental organisations, and are consulted at different stages of the administrative process. The ideas behind a new item of policy may come from any of them, although the formal initiation of any policy proposal must begin with the Commission.

In the day to day administration of the CAP, and in the formulation of policy proposals, the Commission is advised by *Management Committees*, consisting of civil servants from the Ministries of Agriculture in the member states. These have been set up for each of the commodities which are the subject of a CAP market regime. The Standing Committee on agricultural structures and the Committee of the European Agricultural Guidance and Guarantee Fund have similar compositions and roles. Each of the Management Committees has a corresponding *Advisory Agricultural Committee*, half of which consists of farmers' representatives provided by the Committee of Professional Agricultural Organisations (COPA) and the General Committee of Agricultural Cooperation (COCEGA). The other half is comprised of representatives of food manufacturers, trade unions and consumers from organisations such as the European Bureau of Consumers Organisations (BEUC). The meetings of these committees ensure that the Commission is made aware of the views of a wide range of non-governmental bodies on any proposal. In addition, the Commission may also consult directly with COPA, COCEGA, the Consumers Consultative Committee, technical experts, trade and professional bodies, and any other special interest groups affected by specific proposals.

When, after these consultations, a proposal has been approved by the Commission, it is formally submitted to the Council of Ministers. In order to make a decision the ministers first need to know the views of their own national governments and parliaments. In the UK these views are formulated through the preparation of briefs and explanatory memoranda in the Ministry of Agriculture, Fisheries and Food (MAFF), and by the reports produced by the House of Commons Select Committee on European Legislation and the House of Lords Select Committee on the European Communities.

The views of the major European consultative organisations, the Parliament and the Economic and Social Committee, are also sought. The agricultural committee of the *European Parliament* reports on Commission proposals, and this report, when passed by the full Parliament, is passed to the Council of Ministers as a Resolution or Opinion. A similar procedure is adopted by the *Economic and Social Committee* (ESC), a body set up by the Treaty of Rome to advise the Council and Commission. It consists of members who are appointed by member states in their personal capacities, and who may broadly be divided into employers, trade unionists and independent members. Each of these national and European institutions may also solicit advice from, or be lobbied by, special interest groups or individuals affected by policy proposals in the same informal way as the Commission. The final decision on whether or not the proposal shall become law is then made by the Council of Ministers.

CAP PRICE MECHANISMS

General outline

The underlying principle of the CAP price mechanism is that the producer should receive a price for his produce which is determined by market forces. But these market forces are controlled so that the market price fluctuates only within predetermined upper and lower limits. Thus the farmer is protected against excessively low prices, and the consumer against excessively high prices.

The mechanism by which this protection is carried out recognises that farm products may be produced either within the EEC or outside it. The demand for farm products is relatively constant, so if the market was supplied only from within the EEC the major reason for low prices would be an excess supply. The effects of this can be mitigated by artificially increasing demand. Therefore the EEC enters the market to buy up agricultural products for storage when prices are low, a process known as '*intervention*'. If prices subsequently rise this stored produce can be released on to the market again. If they do not it may be sold on the world market with the aid of subsidies known as '*export restitutions*' or 'refunds'.

Many farm products can also be supplied from countries outside the EEC (known as 'third countries').

If this non EEC supply was available at less than the EEC market price it would reduce the EEC market price, and so the entry of such produce is subject to *import levies* to raise its price. When EEC prices are high, perhaps as a result of supply shortages, the entry of produce from third countries serves to reduce prices by increasing supplies.

The national agency which deals with intervention, import levies, export restitutions, and all the other payments made under the Guarantee section of the EAGGF (*see below*) in the UK is the *Intervention Board for Agricultural Produce* (IBAP). It does not physically handle any goods itself, but supervises the work of other bodies such as the Home Grown Cereals Authority for cereals and the Meat and Livestock Commission for beef. Butter sold into intervention is stored in commercial cold stores under IBAP contracts.

The financial resources needed for the activities of the IBAP are obtained from the *European Agricultural Guidance and Guarantee Fund* (EAGGF—often referred to by its French acronym FEOGA). The Guarantee section of the Fund, which is the larger part, finances the measures used to support commodity prices within the community. The Guidance section finances structural policy (*see below* p. 583). The Fund is in turn financed from the EEC budget (of which it forms the major part), which has three main sources of revenue: levies on imports of agricultural products from third countries, common Customs Duties, and up to 1% of the Value Added Tax collected in member states, calculated on a uniform base.

Essentially, therefore, the price received by the EEC farmer will be at least the price at which intervention begins, and at most the price at which imported goods can be sold, assuming that EEC prices are higher than world prices. Between these two extremes market prices will fluctuate according to the balance of supply and demand.

This basic system is used for regulating markets in most of the major products produced by EEC farmers, including cereals, milk products, sugar, beef and veal, pigmeat, poultrymeat and eggs. Different systems are used for sheepmeat, fruit and vegetables and other crops. At the time of writing there is no EEC price mechanism for potatoes or wool. There are significant differences between the detailed regulations for each commodity, and these are described below, together with the terms applied to the management of each market.

Cereals

The price which the EEC wishes the farmer to receive for his cereals is known as the *target price*. When the market price of cereals falls below the target price to the *intervention price*, intervention agencies may buy cereals for storage, thus preventing further price falls. Cereals sold to intervention stores must be of

intervention quality, and in 100 tonne lots, so it is usually merchants and cooperatives rather than farmers which sell into intervention, and the farmer receives the intervention price indirectly. Intervention prices increase in monthly steps throughout the year to encourage orderly marketing. There is a single intervention price for common wheat, barley and maize, another price for durum wheat (used in the manufacture of pasta) and a *reference price* for wheat of breadmaking quality which functions as an intervention price but compensates for the lower yield expected from breadmaking varieties by being higher than the common wheat intervention price.

The price at which cereals imported into the EEC from third countries can be sold is also determined from the target price. It is assumed that the highest price of cereals will be found at Duisburg in the Ruhr district of West Germany, and that they will be imported at Rotterdam. In order to ensure that the price of imported grain will not be less than the target price a minimum import price, known as the *threshold price*, is calculated by subtracting the transport costs from Rotterdam to Duisburg from the target price. If imported grain price quotations are less than the threshold price a *variable levy* is imposed, which is equal to the difference between the threshold price and the minimum import offer price, calculated daily.

The export of cereals from the community is also regulated. When the EEC market price is higher than the world market price (as it has been for most of the period for which the EEC has existed) an *export refund*, of the difference between the EEC market price and the world market price, is paid. If the EEC market price is lower than the world market price an *export levy* is charged. Thus the community producer will always receive the community price, no more and no less.

These systems are summarised in the diagram below:

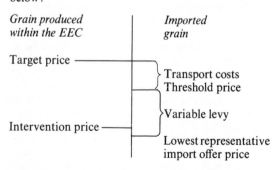

Milk and milk products

Market support for milk and milk products is also based on a system of target and intervention prices for EEC production together with a threshold price for imports. The system is necessarily more complex than support for cereals because a greater range of products is involved, and because a wide variety of additional measures has been introduced in an attempt to overcome the problems of surplus which have existed in the milk market for several years. Nevertheless, the overall aim, that producers should receive the target price for milk of 3.7% fat content delivered to a dairy, remains.

The target price is supported by intervention in the market for butter, skimmed milk powder, and three varieties of Italian cheese. The intervention price set for these products reflects the quantity of milk required for their production, so that the intervention price for butter is normally about two and a half times higher than that of skimmed milk powder. Intervention may take place either by buying in to intervention store or by aids to private storage by processors. In the UK the price received by producers is controlled by the activities of the Milk Marketing Boards. Some of the activities of the MMBs appeared to contradict the provisions of the Treaty of Rome, but the Community permitted their continued existence because they appeared to maintain the consumption of liquid milk at a higher level than that found in most other European countries. Their constitutions were also changed so that they became, in effect, cooperatives. The returns to UK farmers therefore result from the pool price determined by the MMBs from their sales of milk to the liquid and manufacturing markets. The manufacturing price of milk realised by the MMBs depends upon the intervention prices of butter and skimmed milk powder, but the maximum retail and the wholesale prices of liquid milk continue to be determined by the UK Agriculture ministers. These are set at levels which do not interfere with CAP mechanisms.

As with cereals, the intra-EEC price of imported milk products is fixed by the threshold price determined by the Community. Threshold prices are determined for 12 'pilot' dairy products (three types of powder, five types of cheese, two types of condensed milk, butter and lactose) and from the lowest representative offer price for these, import levies for imported milk products are determined. Export refunds are also available to enable milk products to be sold on the world market. The importance of dairy farming to community processors, and in particular to those operating small businesses, has meant that the target and intervention prices set by the Council of Ministers have been high enough to maintain production and restrict demand. A number of additional measures have therefore been introduced at various times in an attempt to increase consumption or reduce supplies. Among those designed to increase consumption are: disposal subsidies for skimmed milk powder used in the production of animal feeds or casein; consumer subsidies for butter; sales of butter at reduced prices from intervention stocks to the armed services, non profit making institutions (e.g. hospitals) and persons receiving social assistance; and sales of liquid milk at reduced prices to schoolchildren.

The *co-responsibility levy* is designed to reduce supplies by reducing the revenue received by milk producers. It is deducted from payments made to producers by the buyers of their milk, and used to finance programmes to increase milk consumption. Under the terms of the original enabling legislation the maximum rate of levy was set at 4% of the target price, but the actual rate of levy is set by the Council of Ministers. In practice, it has always been less than 4%, and lower rates than normal are usually applied to producers farming in less favoured areas. Provision has also been made for a *supplementary co-responsibility levy* to be applied following years in which total milk deliveries increase. Again the rate of levy is set by the Council of Ministers.

Sugar

The sugar regime is basically similar to that for cereals, but production quotas are imposed and special arrangements are made for imports from the sugar producing African, Caribbean and Pacific (ACP) countries which are signatories of the Lomé Convention.

The target price is fixed for white sugar, and from this the intervention price is derived. The minimum price for beet paid by sugar factories for production within the *basic quota* (the A quota) is calculated from the intervention price, taking into account the sugar yield of the beet, together with processing and delivery costs. Production outside the A quota, but within the *maximum quota* (the B quota), is subject to a levy of up to 30% of the intervention price, designed to assist in financing export restitutions. Production in excess of both quotas must be exported to third countries without the aid of export refunds.

The calculation of the threshold price for white sugar imports is different from that of cereals. It is obtained by adding the storage levy and transport costs to the target price, the import levy being the difference between the minimum import offer price and the threshold price. However, the ACP countries are allowed to sell up to 1.2 million tonnes of white sugar (or its equivalent in raw sugar) at or above the guaranteed price.

Beef and veal

Market support for beef and veal is based on intervention, but with some significant differences from the systems described above. The average price which should be realised over the year is known as the *guide price*, and intervention may take the form of aids for private storage or buying in, usually of primal cuts rather than live animals. The producer receives the *buying in price*, which is calculated from the intervention price by using coefficients and the killing out percentage. *Reference prices*, or weighted average prices in representative markets, are calculated each week for each member state and for the Community as a whole. Intervention may occur over the whole Community if the reference price is less than 93% of the guide price. Alternatively, when the Community reference price is less than 98% of the guide price, and the reference price in a member state is less than the intervention price, intervention may also occur.

Imports of beef and veal from third countries are subject to Customs Duties (e.g. 16% on live animals and 20% on fresh, chilled or frozen carcasses) in addition to variable levies. The *basic levy* is the difference between the guide price and the free at frontier offer price (duty paid), but the amount of the basic levy which is payable by importers is dependent upon the relationship between the Community reference price and the guide price, ranging from 114% of the basic levy when the reference price is less than 90% of the guide price to zero when the reference price is more than 106% of the guide price. These arrangements affect less than half of all imports. There are other arrangements which apply to trade under the terms of the General Agreement on Tariffs and Trade or the Rome Convention, which have the effect of easing the entry of imports, particularly of frozen beef.

In the UK only there is a variation on the standard regime. A *target price* is set, which must be no more than 85% of the guide price when averaged over the year, although it may rise to 88% of the guide price in any one week to encourage orderly marketing. If the UK reference price is less than the target price a *variable premium* may be paid to producers selling cattle in that week to make up the difference. Beef may still be sold into intervention, but in this case the variable premium is deducted from the buying in prices.

Sheepmeat

The sheepmeat regime, which was introduced in 1980, revolves around the *basic price*, the level of which depends upon the market situation in sheepmeat and its close substitutes, and costs of production. The market price is supported by aids to private storage which may be paid if the EEC market price falls below 90% of the basic price. If the EEC market price falls below 85% of the basic price sheepmeat may be bought into intervention between mid July and mid December. A *reference price*, based on 1979 market prices, is also set for the first four years of the regime. If the average market price for the whole year is below this, *annual compensatory premia* may be paid to producers to bridge the gap. As is customary, these institutional prices are denominated in European

Currency Units (ECUs), which are converted into national values at representative rates of exchange, but, since no monetary compensatory amounts are payable on trade between member states, currency fluctuations can produce differences in institutional prices between member states (the reasons for this are explained on p. 582 below). EEC producers are protected from competition from non member states by the imposition of quotas on imports from such states and by a 10% customs duty.

The regime provides for the use of a different system in the UK. This replaces private storage and intervention by a *variable premium* system. A *guide price* is set at the same level as the EEC intervention price. If the average market price falls below this a variable premium is paid to bring the price received by the producer up to the guide price. However, this only applies to sheepmeat bought and consumed within the UK, and an export charge, equal to the variable premium, is paid on sheepmeat exported to other Community countries.

Pigmeat

Pigmeat production is cyclical, and therefore each year a *basic price* for standard quality carcasses is set at a level which will produce reasonable stability in the market without producing structural surpluses. The market price may be supported by intervention, when the *reference price* (i.e. the weighted average EEC price) falls below 103% of the basic price, in one of two ways: either by the provision of aids to private storage by traders, or by buying in by intervention agencies. In fact, such buying in, at a *buying in price* of between 78 and 92% of the basic price, happens only rarely. The provision of private storage aids is more common, although these are not mandatory, and the Commission may allow reference prices to remain some way below the basic price, particularly when pig numbers in the Community are tending to increase.

Pigmeat imports from third countries are controlled by *sluicegate prices*, designed to prevent offers of pigmeat to Community traders at prices less than world production costs, and *basic levies*, which take account of the fact that cereal feed costs in third countries may be below those within the EEC. The sluicegate price is therefore an estimate of world production costs, made up of cereal and other feed costs together with overhead costs. The basic levy represents the difference between world and EEC production costs together with an additional 7% of the sluicegate price to protect the domestic industry. Thus the minimum price at which imported pigmeat may be sold within the EEC is the sluicegate price plus the basic levy, since if pigmeat is offered at less than the sluicegate price a *supplementary levy* is applied.

Poultrymeat and eggs

There are no internal support measures for poultrymeat or eggs, and all market support is by regulation of the trade with third countries, which itself is not extensive. Sluicegate prices and basic levies are calculated in the same way as those for pigmeat, and applied in the same way.

Supplementary levies are also applied if the import offer price is less than the sluicegate price.

Fruit and vegetables

Market support for fruit and vegetables is normally operated through the activities of producers' organisations and by import controls. A *basic price* is set which is the average of market prices in the lowest price areas in the Community over the preceding three years, but takes expected cost increases into account. If market prices fall below basic prices producers' organisations may take produce off the market for use as animal feed or for processing. The major element in market support is the use of import controls. A *reference price*, similar to the basic price but taking transport costs into account, is calculated, and if imports are delivered from third countries at less than the reference price for more than two days, countervailing charges are applied.

Other commodities

Although the markets for the major commodities which are produced in the EEC are supported by various systems involving some form of intervention and import control, there are other commodities, less important in terms of total output, which are supported by other means. The crops considered below are those which are grown commercially in the UK, but the CAP also includes regimes for a wide variety of other products, such as rice, tobacco, wine, a number of oilseeds, and olive oil, which do not fall into this category, although several of them are of major significance in the areas where they are grown.

Oilseeds

Oilseeds including oilseed rape and sunflower seeds, are supported by a production subsidy which is paid to the crusher. A target price is set each year and the size of the subsidy is calculated from the difference between the target price and the world market price.

Peas and field beans

These are supported by an aid paid to feedstuffs manufacturers purchasing these products from, and paying a minimum price to, growers under contract. An *activating price* (equivalent to a target price) is fixed each year and a subsidy equal to 45% of the difference between the activating price and the world price of soya meal is paid.

Dried fodder

Dried fodder includes both dehydrated potato products which are unfit for human consumption, and artificially dried legume forage crops. A guide price is fixed each year and a variable aid, based on the relationship between the guide price and the world price, is paid to processors. They also receive a production aid based on a flat rate/kg.

Hemp and flax

Hemp and flax grown mainly for fibre also receive a fixed payment/ha sown and harvested.

Seeds

Seeds, including grass, legume, flax and hemp seeds produced for sowing, receive a flat rate payment/100 kg of seed.

Hops

Hops are supported by a fixed payment/ha of registered area. In addition, launching aids are provided to encourage the formation of hop producer groups.

MONETARY ARRANGEMENTS IN THE CAP

The target, intervention and other institutional prices which have just been described are common to all member states of the Community and must therefore be denominated in a unit which is also common to all. The unit used for this purpose (and for other purposes in the EEC) is the *European Currency Unit* (ECU). However, the ECU differs from other currencies, such as the pound sterling or the French franc, in that it is not used in everyday transactions among people in Community countries, If it is to be used to determine the price received by a farmer when, for example, selling his grain into intervention, it must be converted into the currency of the country in which this transaction occurs. An exchange rate must therefore exist between the ECU and the national currencies of member states, and this exchange rate is calculated by taking a weighted average of the values of the currencies of all member states.

If the exchange rates of member states' currencies were always fixed in relation to each other the conversion of a price denominated in ECUs to an equivalent national currency value would present no difficulties. But in reality currency values are not absolutely fixed, although the relative variation allowed to the currencies of all EEC members states, except the UK, is limited by the provisions of the *European Monetary System* (EMS) to which they belong. Since a variation in the exchange rate of an EMS currency will affect the value of the weighted average of all currencies, the value of the ECU may change, and the relative values of national currencies and ECUs may vary from day to day according to movements in the foreign exchange markets. An institutional price denominated in ECUs would therefore change in terms of, for example, pounds sterling, from day to day. The confusion that this would produce is avoided by having fixed exchange rates, known as *representative rates* (or, colloquially, *green rates*) for agricultural purposes. The *green pound* is thus a pound whose value in terms of ECUs is fixed until it is changed by a decision approved by the Council of Ministers.

In overcoming one problem, however, a new one arises. Since green rates of exchange are fixed, while market rates are variable, at any one time the two may come to differ significantly. This is particularly likely if one currency is conspicuously stronger or weaker than other European currencies. If this did not happen the intervention price in strong currency countries would tend to fall, and *vice versa*. This is demonstrated by the following (fictitious) example:

If the UK green rate is £1 = 1.61641 ECU and a commodity has an intervention price of 1 ECU/kg, then intervention price is 1/1.61641 = £0.618655/kg.

If the UK market rate rises to £1 = 1.85436 ECU and the UK green rate is adjusted to be equal to the market rate, then an intervention price of 1 ECU/kg is equivalent to 1/1.85436 = £0.53927/kg.

In a weak currency country the effect of keeping green rates at the same level as falling market exchange rates would be an increase in intervention prices measured in terms of national currencies. In practice some EEC countries (e.g. Denmark) have tended to keep green and market exchange rates at about the same level, while others (e.g. the UK and West Germany) have allowed considerable discrepancies to develop. The price received by farmers in EEC countries therefore depends not only upon the price determined by negotiation in the Council of Ministers but also upon a decision made by national governments and agreed in the Council of Ministers about the relationship between green and market rates of exchange. The original concept of the CAP as providing a single intervention price throughout all member states need no longer apply if national governments decide otherwise.

It follows from this that, if intervention prices are effectively higher in one country than another, traders may wish to take advantage of the higher prices. This may also be demonstrated by an example:

If the intervention price of butter is 2844.8 ECU/tonne:

In the UK, converting to £ sterling at a green rate of £1 = 1.61641 ECU,

Intervention price of butter is 2844.8/1.61641 = £1759.9495/tonne.

In France, converting to French francs at a green rate of 1 ECU = 5.847 FF,

Intervention price of butter is 2844.8 × 5.847 = 16 633.545 FF/tonne.

But in the foreign exchange markets, £1 sterling = 10.72209 FF.

So 16 633.545 FF
$$= £1551.3342 \left. \frac{(16\,633.545)}{(10.72209)} \right/ \text{tonne of butter.}$$

A trader selling into intervention would therefore lose £208.6153/tonne by deciding to sell butter in France rather than the UK.

In order to prevent the distortions which would be produced by such differences in intervention prices a system of border taxes or subsidies, known as *monetary compensatory amounts* (MCAs), has been introduced. These are applied as subsidies on exports from countries with an undervalued green rate to countries with an overvalued green rate, and *vice versa*. The MCA is calculated by expressing the difference between market rates and green rates as a percentage, as in the example below:

If the market rate of exchange is £1 = 1.85436 ECU, and the green rate of exchange is £1 = 1.61641 ECU, then UK green rate is undervalued by

$$100 - \frac{1.61641}{1.85436} \times 100\% = 12.8\%.$$

Since a 1% franchise is applied to revaluing currencies the monetary percentage used in the MCA calculation is £12.8–1 = 11.8%.

In the case of trade in butter this MCA percentage is applied to the intervention price of butter to calculate the MCA thus:

Intervention price (= 2844.8 ECU) × MCA% (11.8%) = MCA tax/tonne (= 335.6864 ECU).

Converting this at the green rate (£1 = 1.61641 ECU)

MCA tax/tonne = 335.6864/1.61641
            = £207.674.

The export of butter from France to the UK would therefore be subject to an MCA payment of £207.674/tonne, and trade in the opposite direction would attract a subsidy of the same amount. Monetary compensatory amounts are also applied to the import levies charged on trade with third countries.

STRUCTURAL POLICY

In addition to the common price and market measures, financed through the guarantee section of the EAGGF, the CAP also includes a number of structural measures, financed through the guidance section of the EAGGF. In this context the term 'structural' refers to such aspects of the farming industry as farm size, vocational training, cooperation, processing and marketing.

When, in 1960, the Commission first made proposals outlining the form that the CAP should take, they accepted that there were a number of structural problems within the Community. In 1964, when the detailed rules concerning the operation of the EAGGF were formulated, it was decided that the expenditure on the guidance section should not exceed one-third of the expenditure on the guarantee section. (In fact, as guarantee expenditure rapidly expanded, this limit was never reached, and in 1966 it was replaced by a restriction to a fixed sum each year.) Of the assistance available under the terms of the first guidance section regulations (R 17/64), about half was used for projects concerned with the structure of production, such as land improvements or irrigation and drainage. Most of the rest was used for improving market structures.

It was originally expected that assistance from the guidance section would only be available for projects which formed part of a wider Community project, the framework of which was to have been laid down by the Commission. However, the Commission proposals were never finally adopted, and grants were given for individual projects. By the late 1960s it became apparent that much of the available grant aid was being allocated to the wealthier farming regions, and that a structural problem remained which was not being solved. The major structural problem in the EEC arises from the existence of many small farm businesses. This does not automatically imply that the larger business is necessarily more efficient than the small, but simply that the larger business can make lower profits per unit of output and survive because it produces more units of output to cover fixed costs. Other conclusions follow from this: small farm businesses may have difficulty in acquiring capital, and they are forced to concentrate on the more intensive crop and livestock enterprises. The price and marketing policies of the CAP are not designed to solve these problems. Indeed, insofar as they support farm incomes, and prevent the disappearance of small businesses from the farming industry, they perpetuate them. As has been pointed out above (p. 575) there are a number of reasons for doing this, such as the desire to maintain national food self sufficiency

and prevent rural depopulation. But by the late 1960s it was felt that the balance might have tilted too far, and that measures designed to decrease the number of small businesses and increase the viability of those remaining might be required.

Accordingly, the Commission Memorandum on the Reform of EEC Agriculture (usually known as the Mansholt Plan), published in 1968, proposed widespread changes designed to reduce the number of small farmers. Although these proposals were not fully accepted they led, after much discussion, to the formulation of a number of directives designed to ameliorate the structural problems of the Community. The precise means by which the directives should be implemented is left to individual governments.

In the UK the Agriculture and Horticulture Development Scheme (AHDS) has been set up under the terms of Directive 72/159 *on the modernisation of farms*, which provides for assistance to be given to farms which do not provide an income comparable with non-farm incomes in the same region (the 'comparable income criterion'), but which could provide such an income after the completion of an approved development plan. The assistance available is of several types: capital grants are available at varying rates to finance the planned developments, guarantees on loans may be given if the loans would not otherwise be available, and developing farmers have a prior claim on land released under the retirement directive (*see below*). An additional guidance premium is available to farmers intending to specialise in the production of beef or sheepmeat, but grants for pig production are restricted to those capable of providing 35% of the food for the pigs kept on the unit, and grants for dairy units are restricted to those with a maximum of 40 cows. No grants are available for egg and poultrymeat production or for dwelling houses. In order to qualify for assistance under this scheme a farmer or grower must practise farming as his main occupation, possess adequate professional skills, produce an approved plan outlining the developments he expects to make, and keep specified accounts for the period during which he is working to this plan. Farms which, because they cannot satisfy these requirements, or because they already have a comparable income, are not eligible for this assistance, may still receive assistance from nationally financed schemes such as the Agriculture and Horticulture Grant Scheme (AHGS), although the rates of grant available through this scheme are not as high as the AHDS rates.

In the case of farms which could not, even after further investment, provide a comparable income, and whose occupiers are too old to change jobs easily, measures available under Directive 72/160 provide for early retirement and reallocation of the land to viable farms. In the UK this policy is carried out through the Farm Structure (Payments to Outgoers) Scheme 1976, which is available to farmers aged between 55 and 65 years. Their land must be leased or sold to a farmer carrying out an AHDS development plan, and in return they receive an annuity (paid until they reach the age of 65) or a lump sum, the size of which depends upon the area of land released.

Farmers may have difficulty in making comparable incomes not only because their farms are small or under-capitalised, but also because they are unaware of opportunities available to them, or lack necessary skills. In order to alleviate this problem farmers and their families may receive socio-economic guidance and training in occupational skills under the provisions of Directive 72/161. In the UK socio-economic guidance is provided through the advisory services, by advisers covering such areas as rural community problems, labour relationships and, especially, sources of extra income both within and outside agriculture. Training in agricultural skills is mainly provided by the Agriculture, Horticultural and Forestry Industry Training Boards. Part of the cost of providing all of these services is met by the EAGGF.

The widespread application of the three directives discussed above would result, in the long run, in the disappearance of many small farm businesses which were incapable of producing comparable incomes. In areas where the productivity of land is low for reasons of climate, altitude, slope, infertility or other natural difficulties, farms would have to be very large in order to produce comparable incomes. A large proportion of those currently employed in agriculture would have to leave the land, and some of the land might no longer be used for agriculture, if the comparable income criterion were strictly applied. But this could give rise to further problems: those leaving the land might also have to leave such areas in order to find work, and the appearance of the countryside might change significantly. In short, depopulation and conservation problems would arise, and these would affect not only the remaining residents but also the population of the Community as a whole, since these areas are often those which are heavily used by tourists. In order to overcome these problems the Community produced Directive 75/268 *on mountain and hill farming and farming in certain less favoured areas.*

In the UK the areas designated as less favoured under this directive are hill areas in danger of depopulation in Scotland, Wales, Northern Ireland, northern England and south west England. Several types of assistance are available. Farmers with at least 3 ha of land who undertake to remain in agriculture for at least five years may receive an annual compensatory allowance based on the number of breeding beef cows and sheep kept, up to a stocking rate of just over one cow or six sheep/ha. They also receive higher rates of grant under the AHDS and the AHGS, and the regulations for the attainment of a comparable income are amended to make it easier for them to qualify for assistance through the AHDS.

When introducing the first guidance section regulations the Community recognised that agricultural prosperity is not only dependent upon price support and efficient farm structure, but also that both producers and consumers benefit from efficient marketing. The CAP therefore provides for grants of up to 25% for projects to improve the processing and marketing of agricultural products. Grants are also available to assist the formulation and operation of producer groups concerned with the marketing of fruit, vegetables and hops.

Further details of all these schemes may be obtained from MAFF.

Further reading

FENNELL, R. (1979). *The Common Agricultural Policy of the European Community.* St. Albans, Herts: Granada

MARSH, J. S. and SWANNEY P. J. (1980). *Agriculture and the European Community.* London: George Allen & Unwin

27

Farm business management

G. W. Furness

INTRODUCTION

The management role of a farmer or farm manager has two dimensions; policy determination or organisation and day to day running or operation of the activities on the farm to achieve the business objectives. Existing farm businesses already have certain resources of land, buildings, workers, machinery, livestock and working capital. Management makes decisions on the enterprises to be carried and their sizes and on the methods and timing of production. In operating the business, numerous decisions and actions must be taken daily concerning work to be done, buying of requirements and the selling of products. Decisions on major changes of policy and financing are infrequent on most farms, but may be vital to the long-term progress of the business and a clear appreciation of the business objectives is necessary to avoid conflicting actions.

Farm management is much more than the collection and analysis of financial records. Such data do provide information on the past and aid future decisions and actions. The financial success of a business is demonstrated in the annual accounts and balance sheets. Analyses of these and supplementary records provide evidence of achievements to date and highlight the strong and weak aspects of performance. Planning techniques, ranging from simple budgeting to linear programming by computer, seek ways of making better use of available resources and examine possible changes in policy.

Labour management is critical to success in any business and procedures to help management are explained. Decisions on use of capital in the farm business may be of particular importance. Interpretation of balance sheets and the preparation of cash flow budgets assist decisions on borrowing and methods of investment appraisal test the worthwhileness of possible developments.

When the policy for the farm business is decided, management should set production and income targets and control operations during the year. The value of financial and budgetary control procedures to identify departures from the planned returns and expenditures is outlined.

As farms vary in size and type so the needs for and role of business management techniques differ. In the main, this chapter is directed to those who are personally concerned with the assembly and interpretation of management information. Large businesses develop procedures suiting their own needs, frequently with the aid of outside agencies, some using integrated computer routines. Alternatively, the farm office may have an advanced recording, business analysis and forecasting programme possibly with its own microcomputer. However sophisticated the records and treatment of data, the basic principles for assisting management are the same. The value of management information depends on its relevance when problems are met and whether it enables better decisions and more appropriate actions to be taken.

Farm accounts and business records

Farm accounts are required for the assessment of tax payable. However, when suitably prepared, they have much wider value and application for farm management purposes. They can provide information on efficiency of management of the farm business and on the performance of its component production activities. Data may also be extracted as the base for business planning decisions and for the preparation of forward budgets. Detailed farm records, accounts and budgets are particularly valuable when making a case for borrowing funds and they are required for some grant aided farm development schemes.

The system of record keeping adopted by the farmer determines the detail which can be incorporated into the annual accounts and the extent of the analysis of performance. With supplementary physical information on the sizes of enterprises and on the use of inputs and quantities of products, the financial breakdown of past and current achievements should indicate any weaknesses in the existing organisation and operation of the activities on the farm.

Two financial statements are usually prepared at the end of each financial year,

(1) the *trading and profit and loss acount,* and
(2) the *balance sheet.*

The trading and profit and loss account shows the profit or loss obtained during the period under review, usually one year. It includes an opening valuation of livestock, crops and stocks of purchased materials; the farm revenue and expenditure, including depreciation of equipment, for the period and a closing valuation of livestock, crops and stocks of purchased materials on the farm at the end of the financial period.

The balance sheet shows the net capital position of the farm business at a particular date, usually the last day of the farm's financial year. Examples of these two accounts are given in *Tables 27.1* and *27.2*.

The financial year can close at any suitable date. For farm management purposes it is advisable to choose a date when the amounts of live and dead stock are at low levels and when there is relatively little activity on the farm. Two sets of data are required, details of payments and receipts, preferably analysed in an appropriate business records book, and the annual inventory and valuation. Additional information is needed to produce detailed *gross margin accounts*. Hence, the records kept must be related to the form of accounts and method of analysis to be adopted. Where a farm has a number of end products, it is advisable to identify sales items and stock valuations for each of the main enterprises. An

'enterprise' is defined as a sector of the farm business with an actual or potential sale product (or joint products). The amounts and costs of inputs (e.g. concentrate feedstuffs, seeds, fertilisers) which are distinctly utilised by separate enterprises should also be identified where possible. In the following pages it is assumed that the aim of the recording system will be to produce gross margin accounts or, at least, to measure the output of each enterprise and the costs of concentrate feedstuffs used by each livestock enterprise. Where records are insufficiently detailed for these purposes, some assessment of business performance can still be made using a comprehensive or whole-farm approach as outlined in pp. 599–601.

RECORD OF RECEIPTS AND PAYMENTS

Throughout the year details of all monies received and all monies paid out should be entered into a farm business records book (sometimes separate books are used for each enterprise or department) or on specially designed pages provided by an accounting agency. These entries should cover all farm transactions and any other payments out of, or receipts lodged in, the farm bank account together with summaries of cash transactions. The book should be written up regularly, probably at not less than weekly intervals. Normally, separate pages are used for receipts and payments. Every item is entered twice, first in a total (or 'Amount') column and then under the appropriate heading(s) in the columns provided for analysis. For example, receipts may be entered under: cattle, milk, sheep and wool, pigs, barley, potatoes, machinery, sundries, private receipts and value added tax (VAT) refunded. On the payments pages columns may be identified for: purchases of cattle, sheep and pigs and for dairy cow food, pig food, vet and medicines, dairy stores, seeds, fertilisers, sprays, vehicle and machinery repairs, fuel and electricity, wages, new equipment and capital improvements, VAT paid and recoverable, private drawings and tax and so on. It is important to allow sufficient columns to enter the sales from each enterprise and to show the main categories of expenses. Payments should be entered net of VAT where this is reclaimable and also net of discounts. All transactions should be recorded in appropriate detail, showing physical quantities and numbers bought or sold as well as values. Under any main heading, code letters may be used to identify particular items of costs or sales. For example, the various categories of cattle might be shown either in separate columns or within one column by means of code letters for week-old calves, culled cows, breeding

Table 27.1 Trading and profit and loss account for the year ended 31 March 198–

Opening valuation	£	£	Sales	£	£
Dairy cows	32 400		Cattle	10 600	
Other cattle	10 700		Sheep	8200	
Sheep	7000		Pigs	16 100	
Pigs	4400		Milk	46 500	
Crops	4900		Wool	500	
Materials in store	6800		Wheat	6800	
		66 200	Barley	5200	
			Straw	400	
Expenditure			Potatoes	13 000	
Purchased bull	600				107 300
Purchased sheep	2600				
Purchased boar	200		Sundry revenue	300	300
Purchased feed for			Value of produce consumed	560	560
Dairy cows	12 900				
Other cattle	1200		Closing valuation		
Sheep	200		Dairy cows	31 600	
Pigs	4500		Other cattle	12 000	
Seeds	3190		Sheep	7800	
Fertilisers	11 400		Pigs	4500	
Sprays	2400		Crops	6900	
Contract charges	2400		Materials in store	7400	
Sundry crop costs	3100				70 200
Sundry livestock costs	2730				
Vet and medicines	1740				
Wages—casual workers	2300				
Wages and insurances—regular	15 200				
Vehicles and machinery repairs, fuel and insurances[1]	6200				
Electricity and heating[1]	1800				
Rent and rates[1]	4800				
Interest paid	4500				
Building repairs and maintenance	1200				
Miscellaneous expenses[1]	3400				
		88 560			
Depreciation of vehicles and machinery	11 000	11 000			
Farm profit		12 600			
		178 360			178 360

[1] After deductions for private use

Table 27.2 Balance sheet as at 31 March 198–

Liabilities	£	£	Assets	£
Creditors		9600	Cash in hand	300
Tax owed		1500	Debtors	6500
Bank overdraft		8700	Balance in bank	—
Term loan		20 000	Valuations:	
Capital account:			Crops	6900
Net capital at beginning of year	374 000		Stores	7400
Add:			Non breeding livestock	13 300
Farm profit	12 600		Breeding livestock	42 600
Revaluations[1]	29 500		Vehicles and machinery	49 800
	416 100			
Less:				
Non-cash receipts	560			
Private drawings	8540			
Net capital (Owners' equity)		407 000	Land and buildings	320 000
		446 800		446 800

[1] Adjustments to vehicles and machinery values, before calculation of depreciation, and to property valuation

animals, fat or store cattle. Similarly, a variable premium received, or Hill Livestock Compensatory Allowances for breeding cows, should be identified so that they can be attributed to the cattle finishing or rearing enterprises respectively.

Instead of analysing items into a number of columns, an alternative method is to break down cheque payments, or the lodgements of receipts, into the appropriate items and make single entries with sufficient description or a specific coding system to facilitate analysis by a computer programme.

The sources of information for entries into the receipts and payments analyses are paying-in slips and cheque book stubs, backed up by sales advice notes and statements or invoices. As many accounts as possible should be paid by cheque and each time a cheque is drawn, full details should be entered on the counterfoil. If the cheque covers several items, e.g. feedstuffs and fertilisers, the amount should be analysed at the time of payment ready for allocation into the appropriate columns of the payments analysis pages. When transferring figures from the cheque counterfoil to the analysis book, it is useful to enter the cheque number alongside so that every transaction can be traced back to the bank account. When cheques and cash received are paid into the bank, details of what they are for should be entered in a book of paying-in slips. Periodically during the year the figures in the analysis book should be reconciled with those of the bank statement. Payments made directly by the bank, such as standing orders and bank charges, should be entered in the analysis pages, as should any receipts which have been transferred directly into the bank account.

At the year end, the actual bank balance should agree with: opening bank balance plus total receipts minus total payments entered in the bank columns of the analysis book after allowing for any unpresented cheques.

Cash payments

If cash is required to meet farm expenses a round figure of say £200 should be drawn from the bank for this purpose. The amount is recorded as an expense in the payments analysis in the usual way and entered in a column marked *Petty cash*. In addition, the amount is entered as a receipt on a separate page set aside for cash transactions. On this, all cash payments and receipts not lodged in the bank are shown separately. In this way it is possible to keep a record of the amount of petty cash used and how it is used. At the end of the year, when the accounts are drawn up, any relatively major items on the petty cash page can be added to appropriate totals from the analyses of payments and receipts and the remainder included as *Sundries*.

Where cash is drawn from the bank for wages, the total amount need only be entered in the payments analysis records but a *Wages book* should be kept in which full details of the weekly wages are recorded.

When cash for private expenditure is required a round sum should be drawn and entered in the payments analysis as *Personal drawings*. Any household payments made from the farm bank account by cheque should be entered in the same way giving the name of the person to whom the sum has been paid.

Contra accounts

Some transactions are settled by contra in cases where the farmer may owe money to a merchant, or supplier of services, who in turn owes money for goods supplied to him. For example, the farmer may owe a contractor £400 for baling whilst the contractor owes the farmer £250 for 5 tonnes of hay collected from the field. Only the sum of £150 may change hands, but the two distinct transactions must be shown with full details in the payments and receipts analysis pages. Similarly, when a second-hand vehicle or piece of equipment is traded-in for a new one, two transactions showing the full cost of the new item and the value of that traded in must be recorded.

The items in the analysis columns of the payments and receipts pages should be totalled and the sum of the column totals on each page should equal the total of the Amount column plus Contra items if the analyses have been carried out correctly.

Debtors and creditors

At the end of the financial year the farmer may not have been paid for some goods which have been sold. A list headed *Debtors* should be prepared of accounts due to the farm including any sums not yet received for stock or produce which have been sold and any outstanding grants which are expected for work done during the financial year.

Also, at the year end there may be accounts payable by the farm and, subsequently, bills may be received for goods delivered or work done during the farm account year. A list of such items headed *Creditors* should be made, this may be compiled readily if a file is kept of all unpaid accounts.

Annual inventory and valuations

An annual inventory is important. At the date fixed for the financial year end, some of the year's production may still be on the farm as increases in stocks or additional animals, or alternatively, part of sales during the year may have been achieved through running down stocks on hand. It is therefore necessary to make an inventory each year of all livestock, grain, seed, fertilisers, crops in the ground, vehicles and implements on the farm. In continuing farm businesses a similar inventory will have been made at the end of the previous financial year. However, with a change of ownership or tenancy, the first inventory is at *Ingoing* and may be compiled from documents prepared at the time showing the stock taken over. There is no need to make a complete inventory of machinery and plant each year. To the opening valuation, any machinery purchased during the year is added and any machines sold or scrapped and their book values are deleted.

Opening and closing valuations are based on inventories at the beginning and end of the accounting period. The valuation of livestock and crops is a process of estimation and the basis depends on the purpose for which the accounts are required. For farm management guidance this is to reflect the market value of crops and animals produced in or retained through the financial year. The basis should be the same for both opening and closing valuations.

Livestock

In the inventory these should have been classified under appropriate headings according to function and age group. Whether for breeding, production, or sale, each animal or group of animals should be valued at current market value, less cost of marketing; fluctuations in market value which are expected to be temporary should be ignored. (In a period of rising values, increases in unit values contribute to the farm profit; this is appropriate for trading livestock which are likely to be sold within the following year and, for breeding animals, it enables book values in the profit and loss account and balance sheet to be shown at realistic levels. However, for management purposes, particularly the calculation of gross margins, it is convenient in updating the valuations of breeding livestock of similar quality to revise both the opening and closing valuations using the same per unit figures.)

Saleable crops in store

These should be valued at estimated market value less costs still to be incurred, such as marketing expenses.

Saleable crops ready for harvesting but still in ground

These should preferably be valued like saleable crops in store, less estimated harvesting costs.

Growing crops (and cultivations)

These are valued at variable costs incurred to the date of valuation. (Residual manurial values need to be taken into account only on change of tenancy.)

Fodder stocks (home grown)

They should be valued at accrued variable costs where gross margins are to be calculated for the livestock enterprises; minor changes in stocks due to weather conditions or length of winter can be ignored, but where stocks are varied because of policy changes the associated variable costs should be indicated.

Stocks of purchased materials (including fodder)

These are valued normally at cost net of any discounts and subsidies.

Vehicles, machinery and equipment

They may frequently be valued at original or historic cost (net of grants), less accumulated depreciation to date of valuation. Depreciation is usually calculated on the reducing balance method. However, in a period of inflation, it is advisable to adjust values of vehicles and machinery to current levels by applying appropriate indices to historic cost data and to calculate depreciation on the revised figures.

Buildings and fixed equipment

These will be included in the value of an owner-occupied farm which, for inclusion in the balance sheet, should be assessed at conservative market value. Where provided by a tenant, fixed assets which are still in use should preferably be revalued by applying appropriate indices to historic cost data. Depreciation is calculated on the revised figures usually on a straight line basis, i.e. related to the effective life of the assets.

Depreciation allowances for machinery

In determining farm profit an appropriate rate of depreciation to allow for wear and tear is applied to the current value of each class of vehicle and machine. The practice for tax purposes of writing-off the whole of the cost of a new machine in the year of purchase is inappropriate for management accounts. Tables giving rates of depreciation according to type of machine are available and used by accountants, but it is sufficient in most instances to calculate depreciation of cars, tractors, other self-propelled vehicles, harvesters, sprayers and the like at 25% and of other longer lasting equipment at 12.5% of the current replacement value.

PREPARATION OF ACCOUNTS

The profit and loss account

After checking that the totals of the analysis columns agree with total receipts and total payments, it is necessary to adjust the relevant column totals by subtracting opening debtors and creditors (since these refer to actions carried out in the previous financial year) and adding closing debtors and creditors. The various items of revenue and expenditure now relate to the 12 month period of farm operation and are ready to be transferred to the trading and profit and loss account. The sum of the year's depreciation allowances on vehicles and machinery should be included on the expenditure side. In addition, certain non-cash benefits should be estimated and may be entered on the revenue side of the account. Amongst these are:

(1) the value of farm produce consumed in the house;
(2) an allowance for the use of the farmhouse as a private dwelling;
(3) allowances for the use of coal, oil, electricity, telephone and the farm car for private purposes. (Alternatively, items (2) and (3) may be deducted from the relevant items of expenses.)

When the value of these items has been assessed, the details of appropriate opening valuations and of closing valuations should be entered into the account as illustrated in *Table 27.1*. The profit is calculated as the balance between revenue plus closing valuations minus opening valuations and expenses. If revenue and closing valuations do not exceed opening valuations plus expenses the balance represents a loss on the year's operation.

Private expenses and private receipts are not transferred to the farm trading account. Also, capital expenditure on new buildings or improvements, the benefit of which covers a number of years, is not included in the profit and loss account, though a depreciation allowance, or annual share of the net costs, may be included in the management analysis.

The balance sheet

The values of all the farm's assets at the end of the year are usually summarised on the right-hand side of the balance sheet. Included are debts receivable and any cash or bank balances.

Any money owed by the farm is shown on the left-hand side and the difference between the totals of the two sides represents the net capital (net worth or owner's equity) of the farm. This net capital figure may be verified in a small 'capital account' which reconciles the changes from the net capital position at the beginning of the year by adding the trading profit, private receipts and holding gains* and subtracting

---

* Due to any adjustments made for inflationary changes during the year to the opening valuations of breeding livestock, machinery retained or purchased, land and fixed equipment.

private payments and non-cash benefits (*see Table 27.2*).

Where the farm is owner-occupied the assets include the value of the property. As already mentioned, for management purposes, it is important that the assets are realistically valued. If the farm is operated as a partnership, the net capital holdings of each partner are shown. Interpretation of balance sheet data is outlined on p. 612.

ADDITIONAL RECORDS

Apart from showing the profit on the year's activities a profit and loss account of the form shown in *Table 27.1* provides few pointers for management. Rearrangement of the information is necessary to enable the outputs of enterprises and the efficiency of use of inputs to be calculated. A more complete picture of the value of what is produced on a farm can be obtained if there is information on the total quantities of saleable crops grown, how much of each is fed or used as seed and the values of these amounts at market prices. These values can then be included as part of farm output and treated as costs of the enterprises using them.

The additional desirable records to be assembled, where relevant, are summarised below:

(1) crop areas,
(2) total yield and value of each cereal crop,
(3) total yield and value of straw utilised,
(4) total yield of potatoes including seed and stockfeed,
(5) quantities and value of milk fed to calves, used in the farmhouse and by employees.

Each month (or week, if more convenient) record

(1) numbers of births and deaths by class of livestock,
(2) number and value of
    (a) calving heifers transferred to a breeding herd,
    (b) calves weaned from a suckler herd and retained for finishing,
    (c) week-old calves transferred from a dairy herd to heifer rearing or beef rearing activities,
    (d) weaned lambs retained for winter finishing,
    (e) home-reared in-pig gilts transferred to a breeding herd,
    (f) store pigs transferred from a rearing unit to fattening,
(3) numbers of each class of livestock on the farm at each month end.

This information makes it possible to reconcile livestock numbers and calculate the output of each enterprise (*see Tables 27.3 and 27.4*). Information on the weights of store stock transferred from rearing to finishing enterprises and of stock purchased for finishing, as well as the weights of finished animals sold and stock in closing valuations, also makes possible the calculation of liveweight gains and feed conversion ratios.

In addition, records should be kept indicating the quantities and costs of key inputs used by each of the various enterprises. Besides the allocation of purchased feeds, it is most important to cover the use of home grown grain by livestock enterprises. To facilitate this, it is advisable to maintain a barn record of the ingredients used and the allocation of home mixed rations.

Where a purchased item is attributable to one enterprise, e.g. seed potatoes, and this is shown in the payments analysis record no further information is needed. But where quantities of materials are purchased in advance and eventually used by a number of enterprises, as is the case with 'straight' feedstuffs by different classes of stock or fertilisers for various crops and grassland, intermediate records are needed. A reconciliation summary of the following form allows a comprehensive check on usage of fertilisers. A similar layout can be used to reconcile supplies and allocation of feedstuffs.

However, in certain forms of production, distinct batches of animals (or birds) are produced. For these, records enabling the output and variable costs of each batch to be calculated can be valuable especially where there is more than one batch within a year or where the production period spans the financial year ending date. Where a foreman or herdsman is in charge of a particular section of the farm he will usually keep the primary data and make these available to the farm office, the manager or the farmer.

Generally, it is not difficult to combine suitable production and control records with the accounting system, but some farmers join enterprise recording schemes to overcome the need to design their own forms and to prepare management analyses. Such schemes supply forms and a requirement to submit these monthly is a useful discipline. The agency may update previous information and feedback analyses of results, these may be compared with target data or related to the results of other farms with the same enterprise. Computerised systems are available, either off-farm or on-farm at a price, to provide anything from the monitoring of a single enterprise to cash

Summary of fertiliser use

Type of fertiliser

	A		B		C		etc.
	Qty (tonnes)	Cost (£)	Qty (tonnes)	Cost (£)	Qty (tonnes)	Cost (£)	
In stock at beginning of year							
Purchased: (from payments analyses)							
TOTAL AVAILABLE							
Used for: (from field records)							
Wheat							
Winter barley							
Spring barley							
Potatoes							
Fodder roots							
Silage area							
Grazing areas							
TOTAL USED							
In stock at end of year							

Further details of the use of seeds, sprays, baler twine, drying charges and so on can often be taken from field records or diary entries. Similarly, livestock expenses such as haulage, veterinary fees and medicines, and use of homegrown straw may be identified by class of stock.

This list of records has been presented in a comprehensive way to cover most situations. The procedures should be adapted to individual farms. A farm which has only two or three enterprises, all its land in grass and where feedstuffs and other requisites are identifiable from the purchase documents, needs few additional records apart from stock numbers.

flows and farm financial performance to date, gross margin accounts and performance ratios, a balance sheet and comparisons with targets or the results of earlier years.

GROSS MARGIN ACCOUNTS

Where a farm business is made up of a number of enterprises and when there is flexibility in the area of

Table 27.3 Livestock enterprise outputs, variable costs and gross margins

	Dairy cows		Cattle rearing		Sheep		Pig rearing	
	(No.)	(£)	(No.)	(£)	(No.)	(£)	(No.)	(£)
Output								
A.								
Closing valuation	79	31 600	54	11 500	E 150	7500	41	2700
					(+220)			
			B 1	500	R 4	300	122	1800
Sales:								
Cull breeding stock	21	6700			30	700	14	900
Other livestock	53	3500	1	400	193	7500	660	15 200
Produce		46 500				500		
Transfers out:								
Breeding stock			20	10 000				
Other livestock	24	1500						
Produce		400						
Used in house and by workers		300			2	100	2	100
Subsidies								
Deaths	1	—	2	—	11	—	31	—
Sub total (a)	178	90 500	78	22 400	610	16 600	870	20 700
Less								
B.								
Opening valuation	81	32 400	53	10 700	E 136	6800	42	2800
					(+200)			
					R 3	200	110	1600
Purchases:								
Breeding stock			B 1	600	42	2600	1	200
Other livestock								
Transfers in								
Breeding stock	20	10 000						
Other livestock			24	1500				
Births	77				229		717	
Sub total (b)	178	42 400	78	12 800	610	9600	870	4600
ENTERPRISE OUTPUT (a)–(b)		48 100		9600		7000		16 100
Variable costs	(Qty)	(£)	(Qty)	(£)	(Qty)	(£)	(Qty)	(£)
C.								
Purchased concentrates	90	13 000	5	1200	1.5	200	27	4500
Home grown grain	5	500	8.5	890	3	300	53.5	5500
Home grown straw transfer	20	300	35	700			5	100
Stockfeed potatoes		240						
Milk to calves				400				
Vet and medicines		1000		210		230		300
Casual labour and contract charges		1560		300		170		700
Service and recording fees transport and sundries								
Total variable costs (excl. forage)		16 600		3700		900		11 100
Share of grassland var. costs		4700		1600		1700		—
TOTAL VARIABLE COSTS		21 300		5300		2600		11 100
ENTERPRISE GROSS MARGIN		26 800		4300		4400		5000

land devoted to each crop and in the sizes of enterprises, gross margin accounts can be of great assistance to management. Analysis of the accounts can point the way to desirable changes in the relative sizes of the different farm enterprises or to increased profits by simplification and emphasis on those which are most rewarding.

The procedure depends on the identification of the items contributing to the output of each enterprise and on the allocation of the specific variable costs incurred in production. The *gross margin* of an enterprise is its *enterprise output* less its *variable costs*. The objective in the preparation of gross margin accounts is

(1) to calculate the gross margin generated by each enterprise and the Total Farm Gross Margin, as

the sum of the enterprise gross margins plus miscellaneous revenue,

(2) to summarise the main categories of fixed costs and calculate the Total Fixed Costs of the farm business in the accounting period, and

(3) show the Farm Profit (or Loss) as the difference between Total Farm Gross Margin and Total Fixed Costs.

Enterprise outputs

The enterprise output for each crop and livestock enterprise is calculated from the information assembled in the payments and receipts analyses (after adjustment for debtors and creditors) the valuations and the additional records on transfers of livestock or produce between enterprises. For a *livestock enterprise* the enterprise output is arrived at as the difference between

(1) the sum of returns for livestock and produce sold, subsidies, the market value of transfers out of the enterprise (e.g. calves to a rearing or fattening unit, milk to calf rearing, weaning pigs to a fattening unit), the value of produce used in the farmhouse and by workers (but not paid for) and the closing valuation of livestock, and

(2) the sum of the opening valuation of livestock, purchases of livestock and the market value of transfers in from another enterprise (e.g. calving heifers brought into a breeding herd).

In a table such as *Table 27.3*, a column is identified for each enterprise and the above items are listed in sections A and B to facilitate the calculation of enterprise output. Revenue for stock and produce sold and the value of transfers out normally make up most of the output, but where store animals are purchased, only the amount added to their value is output of the finishing farm. The procedure also adjusts for valuation changes; sales would be greater than production within the year if closing valuation is lower than opening valuation, but they would not fully reflect output if there are more or bigger animals in closing valuation than in opening valuation. Entry of the numbers of animals in transactions and at the beginning and end of the year together with numbers of births and deaths provides a check on the accuracy of the records; the sub-totals (a) and (b) should reconcile.

When calculating the enterprise output of a *crop enterprise* for a *financial year*, the opening valuation of produce at the beginning of the year is subtracted from the sum of returns from: sales within the year, subsidies, the value of produce used as feed, seed or bedding and the closing valuation of produce on the farm at the year end. In this procedure the enterprise output is increased or reduced by any difference

between the opening stocks and the proceeds from their disposal whether sold or charged to another enterprise (or the same enterprise in a subsequent year). For most management purposes it is preferable to calculate crop enterprise outputs for a *harvest year*, i.e. to include only the proceeds from or value of the crop produced in the year under consideration. Where there were opening stocks from previous crops, the actual disposal returns or transfer values should be separately identified and any surpluses or deficits realised can be shown in the summary of contributions to total farm gross margin. When accounting for financial year transactions, two columns in a summary table can be used to separate the margins from different seasons' crops.

Variable costs

The variable costs attributable to individual farm enterprises have two characteristics. They vary in approximate direct proportion to changes in the size of the enterprise and they can be readily allocated to the specific enterprise.

For crops they include: the costs of fertilisers, purchased seed, sprays, casual labour and contract charges (these amounts should be identifiable from the payments analysis after deducting opening and including closing creditors plus opening stocks less closing stocks), together with any transfers of seed from previous crops or opening valuation.

For livestock enterprises variable costs include expenditure on purchased feeds and the market value of home-grown cereals fed (with adjustments in each case for any difference between opening and closing stocks of feed) plus veterinary fees, medicines, specific contract and haulage charges, service and recording fees and straw for bedding. Subtraction of the sum of these items from enterprise output gives the enterprise gross margin for non-land using enterprises and the 'gross margin excluding forage costs' for grazing livestock enterprises.

For the latter a further step is needed to apportion grassland and forage variable costs. The total of such costs can be assembled as in the right-hand column of *Table 27.4* and is usually allocated to the cattle and sheep enterprises in proportion to the number of grazing livestock units on the farm during the year. This allocation should be made when it is meaningful to calculate the gross margins of grazing livestock enterprises after deduction of forage variable costs and is most useful for comparisons with the gross margins from crop enterprises on a mixed farm.

To calculate the grazing livestock units which have utilised the grassland and forage produced on the farm, it is necessary to multiply the average number of each class and age group of animal on the farm by an appropriate coefficient (grazing livestock unit) and to arrive at the totals for each enterprise. The

Table 27.4 Crop outputs, variable costs and gross margins

	Wheat (tonnes)	(£)	Barley (tonnes)	(£)	Potatoes (tonnes)	(£)		Grassland and forage (£)
Output								
A								
Closing valuation			30	3200	50	3000		500
Main product:								
Sales of produce	60	6800	50	5200	220	13 000		
Transfers out—feed			70	7190				
Transfers out—seed			1	110	10	600		
Used in house and by workers					1	60		
Secondary product:								
Closing valuation			10	200				
Sales			22	400				
Transfers out	20	300	40	800	16	240		
Subsidies								
Sub total (a)		7100		17 100		16 900		500
Less:								
B								
Main product—opening valuation	10	1100	20	2000	15	1200		500
Secondary product—opening valuation			5	100				
Sub total (b)		1100		2100		1200		500
ENTERPRISE OUTPUT (a)–(b)		6000		15 000		15 700		—
Variable costs								
C	(Qty)	(£)	(Qty)	(£)	(Qty)	(£)	(Qty)	(£)
Seed: Purchased		300	3.5	790	15	1600		400
Home grown			1	110	10	600		
Fertilisers		700		2000		1800		6400
Sprays		300		700		1000		400
Contract charges		600		1800				
Casual labour		100		300		1600		300
Sundries (including levies, twine)		400		700		1500		500
TOTAL VARIABLE COSTS		2400		6400		8100		8000
ENTERPRISE GROSS MARGIN		3600		8600		7600		

Table 27.5 Grazing livestock units

Type of livestock	Livestock unit[1]
Dairy cows	1.0
Beef cows (excluding calf)	0.8
Other cattle over 2 years old	0.8
Other cattle 1 to 2 years old	0.6
Other cattle under 1 year old	0.4
Lowland ewes (including lambs under 6 months old)	0.2
Hill ewes	0.1
Rams and other sheep over 6 months old	0.1

[1] These 'units' are to be applied to the average number of animals on the farm over 12 months, calculated from the sum of monthly numbers ÷ 12.

coefficients in *Table 27.5* are meant to reflect the annual intake of grass and bulk fodder of each class of animal in relation to that of a dairy cow counted as one grazing livestock unit*.

The allocation of variable costs is demonstrated in *Table 27.6*, where the area (80 ha) of grass and forage

* Another set of coefficients is given in Appendix II of *Definitions of Terms used in Agricultural Business Management* (MAFF) based on the total energy requirements of each class of animal from both forage and concentrates.

on the example farm is also allocated in proportion to the livestock units in each grazing enterprise.

If one enterprise makes specific use of a source of grazing (e.g. open hill) or fodder not covered by these variable costs, an appropriate adjustment should be made to the number of grazing livestock units sharing in the jointly used grass and forage. On some farms sales of hay are significant and should be treated as a cash crop with appropriately allocated variable costs. Only the remaining variable costs should be attributed to grazing livestock. Returns from occasional minor sales of hay can be deducted from the variable costs of grassland before allocation to the grazing animals.

The shares of grassland variable costs arrived at in *Table 27.6* are then entered in *Table 27.3* and subtracted from the 'gross margins excluding forage variable costs' to give the gross margins of the grazing livestock enterprises.

At the foot of each column in *Tables 27.3* and *27.4* the enterprise gross margin is shown. These figures are summaries of the contribution of each enterprise to Total Farm Gross Margin. From this total the fixed costs of the farm have to be paid with the balance representing the farm profit. Instead of the columnar layout of these tables, a separate page may be used to

Table 27.6 Allocation of grassland variable costs to grazing livestock

Type of livestock	Average no. for 12 months	Grazing livestock units	Share of grassland Var. costs (£)	(ha)
Dairy cows	80	80	4700	47
Bull	1	0.8 ⎫		
Dairy heifers over 2 years	5	4.0 ⎪		
Dairy heifers 1–2 years	22	13.2 ⎬ 27.2	1600	16
Dairy heifers < 1 year	23	9.2 ⎭		
Lowland ewes	143	28.6 ⎫ 28.9	1700	17
Rams	3	0.3 ⎭		
Totals		136.1	8000	80

prepare a gross margin account for each enterprise. Basically, this has section A on the right (credit side) and sections B and C on the left (debit side). The enterprise gross margin is then calculated as the balancing item. This arrangement provides more room to show details of physical inputs and production and, on the same page, enterprise efficiency ratios may be derived.

Fixed costs

The term *fixed costs* refers to the nature of the inputs: land, regular labour and machinery complement and various business overhead expenses. Characteristically,

(1) the costs incurred would not be influenced by fairly small changes in sizes of the enterprises on the farm (e.g. depreciation and upkeep costs of machinery, rent and rates, farm insurances, telephone, professional fees);

(2) the resources may be used at various times during the accounting period in the production of all or a number of enterprises (regular labour, general machinery); and

(3) the costs are not readily allocated to specific enterprises (e.g. tractor fuel, electricity, lime, bank interest charges).

The above items are associated with the whole farm business and should be summarised under appropriate headings. Referring back to *Table 27.1* (the traditional trading and profit and loss account) it may be noted that almost all the valuations, the returns (income) items and the expenditures on livestock have been brought together in arriving at the enterprise outputs and that the totals of variable cost items have been apportioned to the enterprises which use them in *Tables 27.3* and *27.4*. The remaining items make up the fixed costs. These may be transferred directly from the payments analysis after adjustment for opening and closing creditors and deductions for private use. Depreciation charges for vehicles and machinery should be calculated as indicated earlier and allowances may be included for depreciation of buildings and improvements on owned farms and for tenant's fixtures on rented farms.

Gross margin and fixed costs summary

A summarised gross margin profit and loss account may be drawn up with the gross margins generated by the enterprises plus sundry revenue, such as wayleaves and payments for contract work done on the right-hand (credit) side. The fixed costs may be shown as debit items on the left-hand side and the difference between Total Fixed Costs and Total Farm Gross Margin is the Farm Profit for the year. Alternatively, the same data may be set out as in column 1 of *Table 27.7*. The summary of account data in this form is convenient for the analysis of performance and considerations of how the farm business can be modified to improve profits. This is dealt with on p. 603.

Farm profit

The farm profit shown on the trading and profit and loss account or on the gross margin profit and loss account is derived as the difference between the value of what is produced and the costs incurred in the accounting period. For comparison with results and ratios achieved on other farms, or to arrive at a more standardised margin, various imputed costs may be included and certain expenses omitted. It is not necessary to make these adjustments for internal assessment of the farm performance though interpretation of the profit margin must recognise the inputs still to be 'rewarded'. Where comparisons are to be made with published farm management standards, the farm margin should be calculated according to conventional definitions, otherwise interpretation may be misleading.

Net farm income

This is the reward to the farmer and spouse for their manual work, for management of the farm business and the return on investment in 'tenant-type' assets. To calculate this 'margin' for an individual farm, imputed wages must be included in costs for any unpaid family (or partner's) labour other than the farmer and spouse; also, a rental value (based on rents paid for comparable tenanted farms) is charged for

Table 27.7 Gross margin and fixed costs summary for year ended 31 March 198–

	(£)	Ha used	Gross margin/ha (£)	Average number	Gross margin/head (£)
Enterprise gross margins					
Wheat	3600	10	360		
Barley	8600	30	287		
Potatoes	7600	10	760		
Dairy cows	26 800	47	570	80	335
Cattle rearing	4300	16	269	27.2 livestock units	158
Sheep	4 400	17	259	143 ewes	31
Pigs	5000			40 sows	125
Sundry revenue	300				
TOTAL FARM GROSS MARGIN	60 600	130			

Fixed costs	(£)		Per ha of farm (£)
Regular labour	15 200		117
Vehicle and machinery			
running expenses	6100		47
depreciation	11 000		84
Farm fuel	1800		14
Rent (50 ha) and rates	4800		37
Building repairs and maintenance	1200		9
General overheads	3400		26
Interest paid	4500		35
TOTAL FIXED COSTS	48 000	130	369
Farm profit	12 600		
Add back: Interest	+4500		
Less: Rental value (80 ha)	−8000		
Net farm income	9100	130	70
Less: farmer and spouse labour	5800		
Management and investment income	3300	130	25
Adjusted fixed costs	57 300	130	441

the use of owned land. Interest paid, any wage included in expenses for the farmer's wife and costs of land ownership are omitted.

Management and investment income

This is the reward for management, both paid and unpaid, and for the return on tenant-type capital invested in the farm business, whether borrowed or not. To arrive at this, adjusted fixed costs should include actual or imputed charges for the use of all inputs except management and investment in operating capital.

Accordingly, notional wages are included as expenses to the business for the manual work of the farmer and spouse but paid management costs are omitted.

Adjusted fixed costs

These are arrived at by making the above adjustments to the actual fixed costs of the farm business. It follows that Total Farm Gross Margin minus Adjusted Fixed Costs gives Management and Investment Income.

ENTERPRISE COSTINGS AND NET MARGINS

Various advisory bodies, costing agencies and commercial firms provide a service or encourage farmers to keep account data for specific enterprises on their farms. Where this is a major enterprise, such as pig or beef cattle rearing and finishing, or a dairy herd, the economic efficiency of the enterprise is a major influence on the profit of the whole farm and any pointers to improved performance should be of benefit.

The costing procedures usually embrace the calculation of the enterprise gross margin and require the information discussed earlier in relation to whole farm gross margin accounting. However, some go further and attempt to identify the fixed or overhead costs which may be attributable to the enterprise which is costed. Thus, wages of specialist herdsmen and charges for the time spent by other workers (including the farmer and spouse) on the particular enterprise will be recorded. Specific machinery expenses and machinery operating costs may be calculated including a charge for the use of tractors based on the hours of work associated with the enterprise.

Rents can be readily attributed for land-using enterprises and allowances made for depreciation of specialist equipment and housing. In some cases, interest charges for the 'working capital invested' in the enterprise may be included. Deduction of actual or imputed charges for some or all of the above inputs from the enterprise gross margin results in an enterprise *net margin*.

There is no standard definition of 'margin' or 'net margin'. Interpretation of a net margin requires care; it is influenced by the fixed or overhead cost items deducted in the calculation, by the reality for the recorded farm of any standard charges used, or interest charges included and by the extent of allowances made for 'other overhead expenses' for which there is no reliable basis of allocation to the enterprises making up a farm business. For comparison with similarly prepared data from other farms, the figures are of value to assess enterprise efficiency and resource use. However, as indications of the contribution of the enterprise to the farmer's profit or personal income they can be very misleading. If charges have been included for his own labour, for interest on capital which is not borrowed and a rent for land he owns, it must be recognised that these expenses to the enterprise are really parts of farm profit. Moreover, the net margin does not indicate the effect on the farmer's profit of removing the enterprise or of expanding it. In any enterprise costing scheme the purpose and method of calculating a net margin must be examined before decisions affecting the scale of production are taken.

Analysis of farm business performance

INTERPRETATION OF GROSS MARGIN ACCOUNTS

The information presented in *Table 27.7* helps the farmer or manager to discover which enterprises have contributed most to farm profit and, with reference to the use made of limited inputs, which should be expanded or contracted. Moreover, the analysis should consider the scope for improvements in the efficiency of each enterprise and whether any of the fixed costs are unreasonably high.

Since the farm profit is the difference between total farm gross margin and total fixed costs, means of increasing farm profit must be sought by examining ways of

(1) increasing total farm gross margin with the present fixed costs or a less than proportional rise in fixed costs, or
(2) reducing fixed costs whilst maintaining, or not reducing to the same extent, the total farm gross margin.

Relative contribution of enterprises

To examine the total farm gross margin, note how much of this is contributed by non-land using enterprises such as pigs, poultry or barley beef; apart from any miscellaneous revenue, the remainder has been derived from the land of the farm. Could this be increased, given the other resources available? The answer depends on

(1) the types of land using enterprises (the greater the proportion of land devoted to enterprises producing a high gross margin/ha the higher the total gross margin will be), and
(2) the actual level of gross margin achieved/ha compared with the potential contribution from each enterprise.

Comparison of the proportion of each enterprise's contribution to the total gross margin from land enterprises with the proportion of land used can be instructive; frequently, less than half of the land is generating three-quarters or more of the total gross margin from land.

More direct pointers to (1) and (2) above, are obtained by calculating the gross margin achieved/ha devoted to each enterprise (*see* column 3 of *Table 27.7*). It is clear which have contributed the highest gross margins/ha and which have produced relatively little. If similar differences can be expected in future, these gross margins/ha indicate the possible direction of adjustments to the farm plan. Secondly, the /ha results can be compared with figures of average, or better than average, performance from similar farms. Interpretation will also be assisted by the calculation of gross margin/head for dairy cows, beef cows, or ewes and /livestock unit for cattle rearing. Gross margin/ha for these classes of stock is the product of stocking rate and margin/head and it is important to examine each element.

As most pigs and poultry enterprises make no direct use of land their efficiency has to be studied in terms of gross margin/head, e.g. sow, fattening pig, laying hen or broiler.

The standards for comparison may be obtained from University and College Agricultural Economics Departments, ADAS, or other agencies such as the Milk Marketing Board, Meat and Livestock Commission, consultants and commercial firms conducting farm business or enterprise recording schemes. If gross margins/ha or /head are low the explanation is sought by examining the components of enterprise output and variable costs. Output may be low due to poor yields and prices or variable costs may be high relative to the level of production as indicated for the main enterprises below.

Examination of enterprise efficiency

Sale crops

The profitability of crop production is closely related to financial output/ha. The output/ha of each cash crop may be calculated from the gross margin accounts and compared with local averages. Physical yield/ha will be the main explanation for a high or low output. In some cases poor prices may have been obtained because of poor quality, or condition, or sale at a time of plentiful supplies, e.g. at harvest. The contribution of secondary products such as straw can also be important in achieving a high output. Explanations for high variable costs generally have to be pursued on the individual farm. High use of inputs can be justified by high yields, so the main objective is to avoid wasteful use and to obtain supplies at the most reasonable price possible. Farms which use contractors for harvesting or other operations will incur higher variable costs than those using their own machinery and labour.

Non land-using enterprises

Low output/head causing low gross margin should also be explored by reference to figures normally readily available on yield or rate of production/head. The average price obtained depending on quality or grade of fat pig or eggs, also has an important effect. In addition output is adversely affected by high mortality, high prices for stores or replacement stock. For these enterprises, variable costs are dominated by concentrate feedstuffs. Food conversion efficiency ratios can point to problems though these may be poor due to unsatisfactory output as well as wasteful feeding methods. Great attention must be paid to minimising costs through sensible buying and the use of appropriate least-cost rations.

Grazing livestock enterprises

There are two stages in the examination of gross margin/ha. First, consideration of the stocking rate of grassland and factors affecting it and second, the influences on gross margin/head. The most valuable means of assessing the efficiency of use of grass and forage hectares is to calculate the stocking rate by conversion of the various classes and ages of grazing animals on the farm to livestock units using the factors given in *Table 27.5*. The total of grazing livestock units should be divided by the adjusted hectares of grassland and forage which have supported them for the year. On many farms the hectares used will be the whole of the grass area, plus any kale or root crops grown. Areas of rough grazings are converted to the grassland equivalent and where hay is sold or there has been a marked increase in the amount of fodder on hand at the year end compared with the beginning of the year, an appropriate reduction should be made in arriving at the hectares which have supported stock. Where hay or other bulk fodder is purchased, the equivalent in extra hectares may be added to the farm hectares used.

With good land and skilled management 2.5 or more grazing livestock units may be carried /ha of grass whereas on other farms the grass and forage may keep less than 1.5 grazing livestock units/ha. Interpretation of the figure calculated for an individual farm must take account of the locality, the level of concentrate feeding and availability of by-products from cash crops. Further explanations for poor results may be sought in terms of fertiliser use, management of grazing, conservation practices and yields of kale and root crops.

Gross margin/head is strongly influenced by enterprise output/head, though for these classes of stock, the availability and quality of grass and conserved forage are important in determining the requirements for and hence costs of concentrates fed. Again, simple measures such as milk yield/cow (calculated as total quantity produced divided by average number of cows on the farm for the year), lambing percentage of breeding ewes, average weight of suckled calves and liveweight gain of fattened cattle are relevant in turn. Prices obtained are influenced by quality and particularly season of sale. In fattening enterprises the value of output is depressed by the payment of unduly high prices for stores, and returns from breeding herds or flocks are influenced by the proportions replaced in the year. Care is needed to relate the level and cost of concentrates fed to the production response. Modest output with a relatively high level of concentrate feeding suggests inadequate grassland production, poor conservation or inefficiency in stock management. Frequently, more efficient grazing livestock enterprises incur higher variable costs for the grassland and forage used than farms earning low gross margins/head or /ha.

While examining the performances of individual enterprises it is important to note which weaknesses could be overcome in the short term, enabling future planning to be based on improved unit gross margins. Other factors such as poor quality of land or low genetic potential of breeding stock cannot be overcome quickly and the performance to be expected must be reflected in planning data.

Examination of fixed costs

By definition these costs are not easily changed but it is important to determine whether they are unduly high in relation to total farm gross margin or in comparison with other farms of similar type and size. The total of fixed costs/effective ha of the whole farm and /£100 total farm gross margin may be compared with average figures for groups of similar farms. Where the overall level is high further examination should indicate which of the main categories of fixed costs is high/ha of the farm. Costs somewhat higher than average may be justified by a very intensive farm system and the achievement of a high total farm gross margin. Further insight into the levels of labour and machinery costs may be gained by the calculation and examination of the following measures.

Labour

Calculation of *total farm output | £100 total labour costs* and/or of *total farm gross margin | £100 regular labour costs* may be appropriate. For the former, the total of enterprise output plus sundry revenue is divided by the total costs of regular and casual labour and the notional cost for the manual work of the farmer and spouse. For the second measure, the total farm gross margin is divided by the total of *regular* labour costs plus the value of the manual work of the farmer and spouse. A low figure for either of these measures in relation to the results of similar farms may be due to high labour costs, the reason for which should be sought, or it could result from a farming system which is either of low intensity, or is obtaining unusually low yields or prices for products.

The *man-work index* (or labour efficiency ratio) is an overall measure which relates the calculated requirement of labour in standard man-days* to grow the crops and look after the stock on a farm to the amounts of labour in man-days available. The available labour includes the work done by the farmer and spouse. This measure can only indicate where a problem exists. To improve labour efficiency on a small farm it is usually necessary to try to raise output by increasing stock numbers or acquiring more land. On a large farm the possibilities of reducing the labour force through mechanisation, improved building layout or changes in enterprises need more detailed study.

In EEC farm accounting *labour income/man employed* is commonly calculated for comparisons between types and sizes of farms and as a basis for assessing the ability of farm development plans to provide those employed on a farm with an income comparable to that of workers outside agriculture.

* A list of standard labour requirements is given in Appendix I of *Definitions of Terms used in Agricultural Business Management*, MAFF 1977.

The figure may be calculated from the account data by adding back to net farm income the wages of regular workers, deducting an interest charge on tenants' capital and dividing the resulting labour income by the number of regular workers including the farmer.

Labour and machinery

Greater efficiency in labour use may be achieved through mechanisation and it is important that machinery and power expenses are considered in conjunction with use of labour. Machinery expenses may be high and it might not have proved possible to reduce labour costs accordingly. This would be revealed if *total farm gross margin/£100 total regular labour and machinery expenses* is calculated and compared with the standard for similar farms.

Two other measures for assessing the efficiency of machinery use are

(1) *total farm gross margin/£100 machinery expenses*; and
(2) *machinery and power costs/1000 standard tractor hours†*.

To calculate the latter measure a table of the average hours of tractor work required/ha for various crops grown and/head of each class of stock on a farm is required. Where machinery costs are high relative to a satisfactory output, the individual items making up the total of machinery and power costs should be examined to see if economies can be made.

COMPARATIVE ANALYSIS OF WHOLE FARM ACCOUNT DATA

Where enterprise gross margins cannot be calculated easily and accurately comparative analysis of account data may be of assistance to management. This procedure has been widely used in the past, based on a re-arrangement of figures in a trading and profit and loss account into Gross Output (with breakdown under main headings) and Costs of purchased inputs. As indicated earlier it is preferable, and not usually difficult, to proceed to the calculation of enterprise outputs and the Total Farm Output. When there is insufficient reliable information to allocate variable inputs to the farm enterprises, the totals of these items, based on expenditure plus opening valuation minus closing valuation, together with home produced feedstuffs (other than forage) and home grown seeds may be summarised as set out in *Table 27.8*. Total Farm Output minus Total Variable Costs equals Total Farm Gross Margin. The latter minus Total Adjusted

† A list is given in *Farm Management Pocket Book*, J. Nix, Wye College (University of London). Eleventh edition, p. 87.

Table 27.8 Whole-farm management account summary

	Your farm		Comparable group/adj. ha (£)
	Total (£)	Per adj. ha (£)	
Enterprise outputs			
Cereals	21 000	161	
Potatoes	15 700	121	
Dairy cows	48 100	370	
Cattle rearing	9600	74	
Sheep and wool	7000	54	
Pigs	16 100	124	
Sundry revenue	300	2	
TOTAL FARM OUTPUT	117 800	906	
Variable costs			
Feed: purchased	18 900	145	
home produced	8930	68	
Seeds: purchased	3090	24	
home grown	710	5	
Fertilisers	10 900	84	
Casual labour	2300	17	
Contract charges	2400	18	
Other inputs: crops	5500	42	
livestock	4870	37	
TOTAL VARIABLE COSTS	57 200	440	
TOTAL FARM GROSS MARGIN	60 600	466	
Fixed costs			
Regular labour: paid	15 200	117	
farmer and spouse	5800	45	
Machinery: depreciation	11 000	84	
running expenses	6100	47	
farm fuel, electricity	1800	14	
Rent and rates: paid	4800	37	
imputed	8000	62	
Other fixed costs	4600	35	
TOTAL ADJUSTED FIXED COSTS	57 300	441	
Management and investment income	3300	25	
Add: farmer and spouse labour	5800	45	
Net farm income	9100	70	
Add: Imputed rent	8000	62	
Subtract: Interest paid	4500	35	
FARM PROFIT	12 600	97	
Total farm ouput/£100 total inputs		103	
Total farm gross margin/£100 total regular labour		289	
Total farm gross margin/£100 total regular labour and machinery costs		152	
Management and investment income/£100 average operating capital		2.9	
Rent, management and investment income/£100 total capital		2.7	
Farm profit, less farmer and spouse labour/£100 average owner's equity		1.7	

Fixed Costs (assembled as in gross margin analysis) equals Management and Investment Income.

Each of these totals and their various components, as in *Table 27.8*, can be divided by the effective hectares of the farm providing a column of /ha figures for comparison with the average results of farms of similar type and size, or those of the top quarter or third of recorded farms when ranked by Management and Investment Income /ha. Comparative data derived from the results of farms representing the main types and business sizes in each region are provided by University and College Agricultural Economics Departments.

In this form of comparison the results must be

viewed as a whole, the explanation for a low management and investment income/ha being explored as follows. If total farm output/ha is low in comparison with other farms, the explanation may be too high a proportion of the farm devoted to low output/ha enterprises, or a low stocking rate, or, the physical yields and prices obtained for produce may be relatively low. Where the problem is not one of 'output', the explanation for a poor result must be either high total variable costs, resulting in a low total farm gross margin, or high total adjusted fixed costs. It may be possible to observe which category of variable costs is high if the farm business being examined is comparable in cropping and stock numbers to the group average. However, differences, particularly in pigs and poultry enterprises, complicate such comparisons and a breakdown of the use of feedstuffs to livestock enterprises is desirable for satisfactory identification of inefficient use.

Procedures for examining fixed costs are as outlined on p. 599. The whole farm performance can be further examined by means of a series of ratios relating 'output' to measures of the major inputs. The following are some of those most often calculated.

Total farm output/£100 total inputs where total inputs is total variable costs plus adjusted fixed costs (including the value of farmer and spouse labour); this measure indicates the overall efficiency of use of inputs. An assessment of the productivity of the land on the farm may be derived from *total farm output/ha* if supplementary enterprises are unimportant, but as this can be boosted by high levels of purchased feed, a more reliable measure is *total farm gross margin/ha*. Similar to the latter is *total farm output less feed and seeds costs/ha* (also, referred to as net output/ha). On livestock farms the overall stocking rate, calculated as *grazing livestock units/ha of grass and forage* (adjusted if necessary for rough grazings and purchased forage) as described on p. 598, is a very convenient measure of land use. Overall use of labour and machinery is indicated by *total farm output/£100 total labour cost* and */£100 total labour and machinery costs* or by *total farm gross margin related to total regular labour costs*.

As outlined earlier, *management and investment income/ha* is the reward for managing the business and the return on all operating (tenant type) capital invested. A measure of the *return on capital* is obtained by dividing the management and investment income for the farm × 100 by the average of the opening and closing valuations of livestock, crops on hand, materials in store and vehicles and machinery. Alternative measures of return on capital more suitable for an owner-occupied farm are *management and investment income plus rental value × 100 divided by average total capital invested* (average operating capital plus the value of owned land and buildings) or *farm profit less the value of farmer and spouse manual work × 100 divided by average owner equity*. The latter expresses the return to the farmer (after payment of interest and a charge for his own labour) as a percentage of his own capital investment in the farm.

These comprehensive measures show how a farm business is faring in comparison with other farms or, if calculated over a series of years, the trend within the farm studied. They can provide reassurance when things are going well or indicate broadly where a problem exists if returns are poor. Actual reasons for poor performance require further analyses of yields and prices of products and quantities and costs of inputs related to output. Even for fairly specialised farm systems an analysis of enterprise output and allocation of concentrate feeds to livestock enterprises are preferable to provide management with sound bases for seeking improvements.

Farm business planning

The term 'farm business plan' refers to the combination and sizes of enterprises making up the farm business. Modifications to the 'plan' may require changes in the sizes of existing enterprises and introduce new enterprises using the present basic resources, or re-organisation with changes in the resources—usually regular labour, capital (including buildings and machinery) and possibly land. Any change in the farm plan or in the methods of production must be in accordance with the objectives of the farmer and farm family (or partners or directors).

It is assumed here that improvement of farm profit (or net farm income) is the objective. However, maintenance of current income with a change in resources or after giving up an enterprise which has featured in the farm system to date (e.g. potatoes or dairy cows), could be studied in a similar manner. Business decisions concerned with increasing the capital value of herds or the farm property or modifications to ease the workload and reduce management stress would be treated differently.

The objectives or aims may change through time; as one milestone is reached, or is within sight, another aspiration may be identified or new problems emerge which modify the previous main objective. It is important to identify the business problem or the farmer's reason for considering a change and to be aware of his main medium-term objective. Solutions to immediate difficulties must not conflict with the medium and longer-term aims.

METHODS OF FARM PLANNING

If a particular decision is to be made on the worthwhileness of the substitution of one enterprise for another, or on a possible change of method, preparation of a partial budget may be sufficient to provide an answer. Where a more comprehensive review is needed of the activities on a farm which has a number of enterprises—a systematic, logical approach involving replanning the whole farm business should be adopted. Gross margin planning, programme planning or linear programming may be used depending on the complexity of the problem.

Preparation of partial budgets

Partial budgets are prepared to test the effects of changes in an existing plan of farming, or in the methods used, rather than wholesale replanning. The procedure is one of examining marginal changes. In a particular decision situation many aspects of farm output and costs may be unaffected; a partial budget takes account only of variations in *existing* levels of costs and revenues. The variable costs of the changing activities have to be included and any alterations needed in 'fixed cost' items are also entered. Items which should appear in the budget are less likely to be missed if a set form is employed, the most convenient being to group the entries under the headings *Extra costs incurred* and *Revenue foregone* on the one hand and *Costs saved* and *Revenue gained* on the other.

In the main, the types of changes which may be examined by means of partial budgets are:

(1) expansion of an existing enterprise;

(2) introduction of a new enterprise or the dropping of a present activity;

(3) substitution of enterprises, i.e. expanding one enterprise while contracting one or more others;

(4) changes in methods of production, e.g. use of own machinery instead of a contractor, home milling and mixing feed instead of the purchase of compounds, or the growing of fodder roots to replace hay or silage for feeding beef cattle or sheep.

The method of partial budgeting is best illustrated by an example. The dairy herd on a farm might be increased from 40 to 60 cows partly by improving the stocking rate of grass from 1.67 to 2.0 grazing livestock units/ha and partly by rearing ten fewer steer calves each year for beef. It is assumed that the existing labour can manage the bigger herd after alterations to buildings. All the items of costs and revenues which would be affected by the changes are included in the budget (*see Table 27.9*).

If the prices and quantities used in *Table 27.9* are appropriate for the time of the change in farm plan, the result should be an increase in farm profit of £3000. It should be noted, however, that no charge has been made for interest on the capital invested in additional livestock. The budget shows the differences in the estimated costs and returns in a financial year after the change has been completed. To achieve the herd increase the capital requirement for the extra cows, less funds released from the rearing enterprise, could be appreciable and if this has to be borrowed the interest charges have to be met out of the addition to farm profit. Moreover, the yields and prices used in the budget should be reviewed to examine the stability of the profit margin. Setting out the detailed quantities and prices of inputs and of products enables the farmer to see what he must achieve to obtain the income improvement. Care should be taken not to be too optimistic in assessing the returns expected from new ventures.

Table 27.9 Partial budget for effect of increase in dairy herd

	£		£
Extra costs		*Costs saved*	
Additional fertilisers on grass 11 tonnes	1300	Concentrates for rearing 10 calves	700
Variable costs of making extra silage	300	Concentrates for 10 beef cattle	700
Concentrates for extra cows 30 tonnes	4300	Miscellaneous costs of rearing	200
Vet, AI and miscellaneous costs for 20 cows	800		
4 extra replacement cows	2200		
Capital and interest charges on building extension at £10 000	2000		
Revenue foregone		*Revenue gained*	
10 beef cattle	4600	Milk of 20 cows	13 600
		4 extra cull cows	1300
Balance = Addition to farm profit	3000	29 extra calves	2000
	18 500		18 500

Where alternative investments or modifications of plan are possible separate budgets should be prepared to test the effects on farm profit and to decide on the most appropriate course of action. Where gross margin data are available, these can be used to indicate more directly the most rewarding enterprise changes provided the impacts on fixed costs are also assessed. A budget for marginal changes using unit gross margins is illustrated in the following calculation for the substitution of 8 ha of winter barley for potatoes with savings from a reduction of one full-time worker and the depreciation of potato equipment which can be sold.

	£
Expected gross margin from incoming enterprise or expansion:	3000
Less: Loss of gross margin from enterprise given up or contracted	7000
Change in total farm gross margin (a)	−4000
Additions to fixed costs due to the change	300
Less: Savings in fixed costs (if any)	5300
Change in total fixed costs (b)	−5000
Change in farm profit (a)−(b)	+£1000

When the expected increase in gross margin is greater than the addition to fixed costs (or, as illustrated, the fall in gross margin is less than the anticipated saving in fixed costs) the change should be worthwhile provided interest charges on any extra borrowed capital have been taken into account.

With this summarised approach, it is important that the farmer appreciates the yields expected and the assumptions on input changes underlying the gross margins/ha or /head which are used in the budget.

Gross margin budgeting

Examination of the performance of a farm business by means of the gross margin analysis procedure outlined on pp. 597–599 should have revealed the strengths and weaknesses of the present farm enterprises and the efficiency of use of the fixed inputs. The next step is to decide the improvements which management could make in the short term through greater attention to the essential features of crop and livestock husbandry, greater control on the use of inputs and better marketing.

Achievable improvements can be incorporated into revised unit gross margins to be used in considering adjustments to or major replanning of the farm business. In addition, the gross margins obtained /ha or /head should be 'normalised' for any unusual yields, costs or prices which are unlikely to recur in a normal future season. In the preparation of unit gross margins for planning the evidence from past performance should be adjusted, where appropriate, to take account of expected future yields, future product prices (and subsidies), future usage of variable inputs (e.g. feed and fertilisers), future prices of variable inputs.

Further adjustments to unit gross margins should be made

(1) where they would be influenced by an intended change in methods (e.g. changing from contract harvesting to use of own machinery), and
(2) for improvements in stocking rate. Keeping a greater number of animals/ha can appreciably alter the gross margin obtained/unit of land used by grazing livestock enterprises.

Consideration must also be given to enterprises which, though not part of the farm business to date, could have a role to play in future. These could be substitutes for some of the present activities or additions and might include, new break crops, the introduction of intensive sheep or retention of calves for fattening. Data on the potential gross margins/ha may be assembled from experimental and advisory sources and from farms with experience of the new activity, but such information must be related to the circumstances of the farm. For gross margin budgeting it is convenient to draw up a list of enterprises which the farmer is willing to include in future plans together with their expected gross margins /ha or /head.

A second list should be drawn up identifying the constraints or limitations on the sizes of enterprises. Features of resource availability and limitations may include the following.

Land

Area suitable for cropping (sub-divided, if some fields have different potential gross margins), areas of grassland suitable for grazing and conservation, how much can only be grazed and how much is inaccessible to a dairy herd.

Cropping limits

Due to quotas, contracts, disease risks or weed problems.

Labour and machinery

The availability of regular skilled and casual labour, particularly in seasons of peak requirement, may be critical. If so, information is needed on the rates at which major farm operations are carried out, the normal sizes of gangs for different jobs and the days available to complete operations such as sowing and harvesting.

Buildings

The housing capacity for various classes of stock puts short-term limits on livestock numbers and the availability of capital and labour may affect the longer-term maxima.

Capital

The availability of funds, including credit, may be a limitation on developments, though it may be worthwhile drawing up a long-term plan presuming capital is available. Short-term plans or the phasing of developments may have to be related to funds available. Alternatively, with a computer programme, a multi-stage plan can be devised with annual developments based on funds generated in the previous period.

Personal preferences

Whilst a farmer is most likely to succeed with activities in which he is interested, preferences should be questioned if they are likely to prevent the achievement of the main objective of the farm businesses. However wishes to retain an enterprise, such as rearing heifer replacements, or for minimum sizes of selected activities should be noted.

Drawing up the plan

In drawing up the plan on a farm where land is the most limiting resource, enterprises should be combined in turn into a farming system by selecting in descending order of gross margin/ha and allocating land to the relevant limit. This ensures that the farm land will be devoted to the potentially most rewarding enterprises.

When all the land has been allocated it is necessary to check that the plan is workable at the periods of peak labour requirement and with the available capital. If not, it will be necessary to cut back the selected enterprise which has the lowest gross margin/unit of the most limiting resource and substitute an enterprise with a high gross margin in relation to that resource. Alternatively, on some farms it may be worthwhile re-examining the availability of the critical resource and seeking methods of overcoming the difficulty. Possibilities to be explored are employment of contractors in busy periods, use of casual or seasonal help, a greater degree of mechanisation of a key operation and overtime or shift working. Generally, where labour is scarce, cereals and combinable crops will be preferred to intensive vegetables and potatoes. Similarly, grazing cattle, winter fattening cattle where housing facilities are convenient, and sheep (except for lambing and shearing times) have lower labour requirements relative to land use than dairy cows or intensive calf rearing.

Table 27.10 Example gross margin plans

Enterprise	Normalised gross margin/ha (£)	Present system (No.)	(ha)	Gross margin (£)	Normalised gross margin/ha with higher stocking rate (£)
Potatoes	900		10	900	900
Dairy cows	570	80	47	26 800	650
Dairy followers	275	27.2[1]	16	4400	315
Winter wheat	400		10	4400	400
Winter barley					350
Beef rearing and finishing[2]					340
Breeding ewes	290	143	17	4900	330
Spring barley	300		30	900	300
	per sow				per sow
Breeding sows	100	40		4000	100
TOTAL GROSS MARGIN			130	62 100	
Fixed costs					
Regular labour				17 000	
Machinery running expenses and farm fuel				6800	
Machinery depreciation				12 000	
Rent and rates				5000	
Annual charge on additional buildings					
Building repairs and general overheads				5200	
TOTAL FIXED COSTS[3]				46 000	
FARM PROFIT				16 100	

[1] Livestock units
[2] Only in the absence of dairy cows.
[3] Excluding interest charges.

If capital is scarce, cereal and other crops have fairly low though seasonal working capital requirements while contract rearing of dairy heifers and the keeping of sheep have much lower requirements than most cattle enterprises using grassland.

When the plan which makes the best use of land has been selected the contribution that supplementary enterprises (pigs, poultry or barley beef) can make to the total gross margin should be explored. These may use surplus buildings and labour which is available or they could be means of expanding the business using specialised housing and workers.

Next, the items making up fixed costs should be examined for possible reductions or necessary increases with the new system.

By subtracting expected total fixed costs from the calculated total farm gross margin for the selected plan, the new farm profit is obtained. This should be compared with the profit likely to be earned by continuing the present farm system, assessed by applying the normalised gross margins to the hectares of crops and numbers of livestock and subtracting the probable future fixed costs. The improvement in farm profit to be realised by the new farm plan must be related to the amount of capital needed for implementation. If funds have to be borrowed the increased profit should cover expected interest charges and loan repayments even in a difficult season. Details of cash flow budgeting and investment appraisal are dealt with in the section on financial management.

Before finally adopting a revised farm plan, drawn up as above, the potential profits from alternative systems using different amounts of fixed resources may be estimated in a similar manner. An appreciably different farm plan may be necessary if one less regular worker is employed on a small or medium sized farm or if a major building constraint could be overcome. Moreover, a decision on whether or not to include a particular enterprise such as potato growing or the retention or introduction of a dairy herd would be assisted by the examination of farm plans including and without such activities. The differences in expected farm profits from the systems should then be assessed in relation to risk factors, management burdens, capital requirements, likely interest charges and ability to repay borrowed funds. Careful and systematic examination of possible changes in the farm plan, and especially the impact on fixed costs and interest charges, is essential for sound decisions to be taken. When land is to be purchased, or substantial capital investments are to be made, full appraisal of future income prospects and external commitments is needed. The sensitivity of the profits expected from alternative plans in relation to poorer prices, lower yields or higher cost increases than anticipated and the implications for management should also be examined. Moreover, the most suitable plan determined at one point in time may need to be revised if there are changes in relative prices or when replacing major items of equipment.

A convenient format for the presentation of alternative plans is illustrated in *Table 27.10*. In this

Plan 1			Plan 2			Plan 3		
(No.)	(ha)	Gross margin (£)	(No.)	(ha)	Gross margin (£)	(No.)	(ha)	Gross margin (£)
	15	13 500		—			20	18 000
80	40	26 000	130	65	42 250		—	
20[1]	10	3150	34[1]	17	5350		—	
	15	6000		12	4800		20	8000
	15	5250		24	8400		20	7000
						100	33	11 200
200	20	6600		—		170	17	5600
	15	4500		12	3600		20	6000
40		4000	40		4000	70		7000
	130	69 000		130	68 400		130	62 800
		17 200			12 000			11 400
		7300			6400			7000
		12 000			11 200			12 000
		5000			5000			5000
					4000			1000
		5400			5800			5000
		46 900			44 400			41 400
		22 100			24 000			21 400

example, the present crops and livestock, with the normalised gross margins (*see* p. 603) shown in the first column, would produce a total gross margin of £62 100. After subtracting the expected fixed costs, for a year ahead when an alternative plan could be adopted, the farm profit would be £16 100 before deduction of interest paid on borrowed funds. In the fifth column the gross margins for grazing livestock enterprises have been adjusted for a modest improvement in stocking rate from 1.7 to 2.0 grazing livestock units /ha with provision for additional applications of fertilisers on grassland.

The farm has 30 ha of permanent grass, and at least one-fifth of the croppable area is to be in temporary grass. Because of locality and soil conditions, it is unlikely that winter cereals could be planted after harvesting potatoes. At present buildings limit the dairy herd to 80 cows. As the farmer wishes to continue to rear his own heifer replacements the gross margin from dairy cows plus necessary replacements should be viewed in combination at £583/ha*, though additional heifers could be reared at £315/ha. Beef cattle rearing would only be considered if milk production were given up. The present labour force is the farmer, two sons and an elderly general worker.

Plan 1 represents a tightening up of the existing system. Whilst retaining the present dairy herd, there is reluctance to expand potatoes to more than 15 ha but, beyond this, wheat, winter barley and sheep are expanded with reductions in dairy followers and spring barley. Changes in fixed costs would be limited to small increases in overtime working, machinery running and miscellaneous expenses. Farm profit is calculated to be £6000 more than from continuation of the present system.

In Plan 2, expansion of the dairy herd to 130 cows is examined whilst dropping the potato crop and ewe flock and releasing one farm worker. Extensions of buildings, silage and slurry facilities for the larger dairy herd and additional followers would require a net investment of £40 000. This more specialised milk and cereals farm system should have lower fixed costs and produce a potentially higher farm profit. However, the additional profit as compared with Plan 1 would be insufficient to pay interest charges if the extra capital needed for buildings and livestock had to be borrowed.

A more drastic change of system is illustrated in Plan 3, this would mean giving up milk production and expanding cropping as far as practicable, rearing 100 purchased calves for sale as 20 month old beef cattle in adapted buildings, some expansion of the pig enterprise and a reduced labour force. Less capital would be required though there would be greater seasonal fluctuations and appreciable variations in seasonal labour demands. Fixed costs would be reduced but returns from crops and beef cattle would be more variable than those of the present mixed system and there would be more marketing decisions. Nevertheless, if capital is scarce and if the future returns from crops, beef cattle and sheep were considered better than from milk production, this option might be considered.

It is important to note that the differences in total profit from the three plans are small and would change if interest charges were taken into account. Year to year variations are likely to be much greater than these differences and as five of the enterprises in this example have fairly similar potential gross margins of £300 to £350/ha, the choice between the plans must take account of capital availability, demands on labour and management and prospects in the longer term. However, all show ways of obtaining a higher profit than would be achieved by continuing the present system and there is scope for further improvement in stocking rate and in the performance of most of the enterprises.

In recent years, more and more farm businesses have become specialised, e.g. in cereal growing or dairying or, in hill areas, they are limited to beef cattle and sheep rearing. On such farms there is less scope for re-organising the combination of enterprises to improve farm profits. The efficiency with which the main activities are run and the containing of fixed costs are the vital features for management. However, variations in production practices can be examined. These may include the main season of calving cows or of lambing ewes in relation to potential returns over feed costs, the holding over of store stock for sale in spring rather than autumn, or purchase of different weights or types of cattle for fattening.

Gross margins for such alternative forms of production within enterprises can be helpful, but it is not always necessary or convenient to use 'gross margins after deduction of grassland and forage variable costs'. Most of the costs of grass production on an all grass farm will be the same for a given stocking rate and only the margin of enterprise output over purchased feed and direct variable costs may be needed to examine alternative uses of grassland. Model budgets which enable the systems margins to be compared over ranges of prices for livestock and products and with differing feed prices can be of assistance. While specific alternatives can be fairly quickly compared by desk calculations, much more information may be obtained from computer models.

For more complex whole farm planning problems, where resources other than land are severely limiting, sophisticated planning procedures are needed.

---

* For each hectare devoted to cows (gross margin £650) approx. 0.25 ha are needed for heifer rearing (gross margin £79); combined, 1.25 ha produce £729 gross margin or £583/ha.

Programme planning

This is a more formal procedure intermediate between gross margin planning and linear programming, seeking to maximise farm profit when two or more resources are limiting. Knowledge of the procedure helps in understanding the data requirements for linear programming and the principles of marginal substitution and optimum allocation of limited resources in achieving maximum profit. In complex farm planning problems it is preferable to enlist the help of advisers or consultants with computer facilities.

Table 27.11 Unit resource requirements

Crop	Labour h/ha		
	Spring	Late summer	Autumn
Potatoes	18	6	36
Sugar beet	10	2	30
Winter wheat	1	8	6
Spring barley	6	8	—

Table 27.12 Gross margin/unit of resource

Crop	£/ha of land	£/h of labour		
		Spring	Late summer	Autumn
Potatoes	900	50	150	25
Sugar beet	600	60	300	20
Winter wheat	400	400	50	67
Spring barley	300	50	37	∞

The essentials of the programme planning procedure are:

(1) prepare a table giving the resource requirements of each enterprise;
(2) calculate the gross margins of viable enterprises/unit of each potentially scarce shared resource; and
(3) if, in the selection procedure a second resource is exhausted while some of the first resource is unused, substitute the enterprise not yet in the plan (or not yet to its maximum) which gives the highest return to the second resource for some (or all) of the enterprise already in the plan giving the lowest return. Substitution should continue until no improvement in total gross margin can be achieved.

A simplified example referring to the 180 ha cash crop area of a farm with four regular field workers is presented in *Tables 27.11–27.13*. The labour requirements in three possibly critical seasons for each of four crops are included in *Table 27.11*. The expected gross margins/ha and /man-hour of labour needed in these critical periods are illustrated in *Table 27.12*. In *Table 27.13* the top line shows the available resources and the plan builds up by selecting enterprises giving the best gross margin/ha. Potatoes and sugar beet are limited to 20 ha each and wheat to one-third of the cereal area. At the second selection autumn labour limits the area of sugar beet and prevents the inclusion of wheat at stage 3. Here barley is limited by the available spring labour and 16 ha would remain unused. Examining the returns to autumn labour in *Table 27.12* it is clear that wheat gives a better return

Table 27.13 Planning to 'scarce' resources—land and seasonal labour

Enterprise	No. of units	Gross margin/ unit	Land	Labour			Gross margin
				Spring	Late summer	Autumn	
		(£)	(ha)	(h)	(h)	(h)	(£)
Available resources			180	1300	1600	1290	
(1) Potatoes	20	900	20	360	120	720	18 000
Remainder			160	940	1480	570	
(2) Sugar beet	19	600	19	190	38	570	11 400
Remainder			141	750	1442	0	
(3) Spring barley	125	300	125	750	1000	0	37 500
Remainder			16	0	442	0	(66 900)
(4a) Reduce sugar beet	−4	600	+4	+40	+8	+120	−2400
Remainder			20	40	450	120	
(4b) Wheat	20	400	20	20	160	120	8000
Remainder			0	20	290	0	(72 500)
(5a) Reduce sugar beet	−6	600	+6	60	12	180	−3600
(5b) Reduce barley	−24	300	+24	144	192	0	−7200
Remainder			30	224	494	180	
(5c) Increase wheat	30	400	30	30	240	180	12 000
Remainder			0	194	254	0	
				Total crops gross margin			73 700

than sugar beet and from *Table 27.11* the release of 1 ha of sugar beet would enable 5 ha of wheat to be grown. Thus stage 4 illustrates the effect of reducing the area of sugar beet included in the plan by 4 ha, enabling this land and the unused 16 ha to be devoted to wheat with an increase in gross margin of £5600. After this substitution the land area and autumn labour are fully utilised. Reference to *Table 27.13* shows that sugar beet and wheat would provide a higher return to land than some of the barley included in the plan and that wheat might use some of the autumn labour devoted to sugar beet. Since barley makes no demand on autumn labour wheat can only be increased profitably with reductions in both the sugar beet and barley areas. In stage 5, a cut back of 6 ha devoted to sugar beet would release sufficient autumn labour to grow 30 ha of wheat and if 24 ha are switched from spring barley growing, the combined effect is to increase the wheat area to the maximum permitted, raising the total gross margin by £1200 to the highest level from these crops. This is achieved with a cropping plan of 20 ha potatoes, 9 ha sugar beet, 50 ha wheat, 101 ha spring barley.

To the gross margin from the cropped area, the gross margins from grassland and other enterprises on the farm should be added before subtracting the total fixed costs to arrive at farm profit.

The presentation of the gross margin approach to farm planning as illustrated earlier in this chapter is simplified. Nevertheless, it illustrates the essential steps, which are:

(1) to keep records in such a way that the variable costs of each enterprise can be determined;
(2) to examine critically the gross margin of each enterprise, and to seek out and adopt possible improvements;
(3) to examine the basic or fixed inputs for any possible changes in methods which could lead to a reduction in fixed costs;
(4) finally, after taking into account possible improvements, to increase the sizes of the most profitable enterprises to the maximum permitted by the resources of land, labour and the dictates of good husbandry.

Linear programming

Applications of this mathematical technique in farm management include:

(1) the preparation of optimum farm plans, when resources and constraints are specified; and
(2) the formulation of least-cost feed rations of specified composition from available ingredients.

In farm planning a computer is needed to find the optimum solution when there is a wide choice of enterprises, various methods of operation and a number of potentially limiting factors. Advisers and consultants have access to various computer programmes and normally a farmer with a complex farming problem should seek their aid. Using linear programming much more data can be handled and more comprehensive solutions obtained than by budgeting or programming by hand.

The objective of the exercise must be carefully specified and accurate data provided if meaningful solutions are to be obtained. The aim in farm planning is to find that combination of enterprises and activities which maximises the total of their net revenue within the specified constraints.

The data are assembled in a table or *matrix*, as in *Table 27.14*. Usually set out across the top are the enterprises or activities possible on the farm and the net revenues*/unit of these activities. Besides existing and potential enterprises, other possibilities can be explored, such as buying or contract rearing heifer replacements, or buying hay and straw, feeding cattle with different types of fodder, hiring casual labour or varying the harvesting methods. Down the left-hand side are listed the resources, such as different classes of land and labour in the various seasons, with the available amounts shown in the first column. Also listed are the constraints on activities; these may be rotational, institutional, e.g. quotas, or physical limits due to the capacities of buildings and so on.

In the body of the matrix are entered the resource requirements and output contributions/unit of each activity (reading down each column); or the demand for and supply of a given resource or constraint (reading across each row). The provision of information on seasonal labour requirements of enterprises may be difficult, as also are estimates of labour hours available for field operations in particular months or periods of the year. Some standards are published by University Agricultural Economics Departments, but it is often better to assess labour requirements for the actual farm. Besides the above data necessary to produce an optimum solution, facilities are usually available to *overwrite* some of the figures so that further solutions can be calculated—perhaps with different gross margins, a change in the amount of labour, or the exclusion of a particular activity.

Having supplied this data matrix, a computer program tape directs the computations. These, very briefly, involve the selection of a plan followed by many systematic substitutions, within the constraints, until it is impossible to make any changes which will

---

* An activity is a specified production process, its net revenue is the difference between the gross revenue of the activity (if any) and the activity variable costs. Net revenue for most crop enterprises is their gross margin, for grazing livestock enterprises often gross margin excludes forage variable costs and for a buying activity it is the variable cost.

Table 27.14 Linear programming data matrix

			1	2	3	4	5	6	7	8	9	10	11	12	13	14	15	16	17	18
Activities			Potatoes	S barley	W barley	W wheat	Leys	Make hay	Make silage	Perm-grass grazed	Perm-grass cut	Dairy cows	Dairy replacements	Beef cows	Beef fatteners	Buy hay	Hire casual labour	Sell straw	Grazing transfer	Labour transfer
			1 ha	1 ha	1 ha	1 ha	1 ha	1 ha	1 ha	1 ha	1 ha	1 c	1 dh	1 c	1 fb	1 t	1 h	1 t	1 t	
Net revenues (£)			900	350	400	450	−110	−40	−60	−40	−60	400	250	160	100	−60	−2	15	0	0
Constraints																				
Croppable land	19	150 ha	1	1	1	1	1	0	0	0	0	0	0	0	0	0	0	0	0	0
Potato limit	20	30 ha	1	0	0	0	0	0	0	0	0	0	0	0	0	0	0	0	0	0
Barley limit	21	0	5	−3	−3	5	2	0	0	0	0	0	0	0	0	0	0	0	0	0
Winter barley limit	22	50 ha	0	0	1	0	0	0	0	0	0	0	0	0	0	0	0	0	0	0
Wheat permit	23	0	−1	0	0	1	−0.3	0	0	0	0	0	0	0	0	0	0	0	0	0
Straw use	24	0 tonnes	0	−2.5	−2.5	−2.5	0	0	0	0	0	0	1	0.75	0.75	0	0	1	0	0
Cow housing limit	25	85	0	0	0	0	0	0	0	0	0	1	0	0	0	0	0	0	0	0
Hay purchase limit	26	40 tonnes	0	0	0	0	0	0	0	0	0	0	0	0	0	1	0	0	0	0
Dairy followers	27	0	0	0	0	0	0	0	0	0	0	−1	5	0	0	0	0	0	0	0
Housing space	28	400 m²	0	0	0	0	0	0	0	0	0	0	3.5	5	4.5	0	0	0	0	0
Permanent grass	29	20 ha	0	0	0	0	0	0	0	1	1	0	0	0	0	0	0	0	0	0
Permanent grass cutting	30	12 ha	0	0	0	0	0	0	0	0	1	0	0	0	0	0	0	0	0	0
Forage tie line	31	0	0	0	0	0	−1	1	1	0	−0.75	0	0	0	0	0	0	0	1	0
Grazing requirements	32	0 ha	0	0	0	0	0	−0.4	−0.1	−0.6	0	0.24	0.34	0.28	0	0	0	0	−1	0
Hay requirements	33	0 tonnes	0	0	0	0	0	−5	0	0	0	0.2	0	0	0	−1	0	0	0	0
Silage requirements	34	0 tonnes	0	0	0	0	0	0	−30	0	0	8	6	7	5	0	0	0	0	0
Labour early spring	35	1032 h	3.5	4	0.5	0.5	0.7	0	0	0	0	3	3.2	1.5	1.5	0	0	0	0	0
Labour late spring	36	1568 h	6	1	1	1	1.2	0	7	1.2	1.2	2.5	1.8	1.5	1.0	0	0	0	0	0
Labour early autumn	37	1796 h	2	3	2	3.2	0.7	0	6	0.7	0.7	2.5	3.5	2	1.5	0	0	0.4	0	0
Labour late autumn	38	1400 h	40	1	7	7	0.2	0	0	0	0	3.5	3.2	1.5	1.5	0	0	0	0	1
Casual labour	39	1000 h	0	0	0	0	0	0	0	0	0	0	0	0	0	0	1	0	0	0
Casual labour potatoes	40	0 h	80	0	0	0	0	0	0	0	0	0	0	0	0	0	−1	0	0	−1

Table 27.15 Selected activities and stability of solution

Solution	Total net revenue £83 750			
Activity	Size	Net revenue/unit (£)	Upper limit of net revenue (£)	Lower limit of net revenue (£)
Selected:				
Potatoes	13.9 ha	900	941	538
Spring barley	60.9 ha	350	353	320
Winter wheat	28.0 ha	450	584	433
Dairy cows	85 cows	400	Inf	270
Dairy replacements	17 heifers	250	301	−400
Beef fatteners	76 head	100	935	99.50
Sell straw	149 tonnes	15	15.50	0
Buy hay	3.4 tonnes	−60	0	−64.70
Not selected:			Decrease in total net revenue if introduce 1 unit (£)	
Winter barley	0	400	0	
Beef cows	0	160	105	
Make hay	0	−40	23.44	

increase the total net revenue. The computer print-out gives the optimum solution and other useful information, some of which is illustrated in *Tables 27.15* and *27.16*. These include the levels of activities in the optimum solution and the total net revenue (*Table 27.15*). Such figures should be examined carefully as the selected levels of activities may need adjustment for a practical farm plan. Hectares and stock numbers may be given to two decimal places and should be rounded up or down, and enterprises selected at impractically small levels have to be omitted. If an unreasonable or unsound selection is found, it is probable that an important constraint was overlooked in preparing the data matrix and a re-run may be necessary.

Details given in the print-out show which resources are fully used and the marginal value products of these (*Table 27.16*). These figures show how much extra net revenue could be obtained if one more unit could be found of resources which prove to be restricting in the

Table 27.16 Resources used (row information)

(a) Binding constraints	Marginal value product of 1 unit of resource (£)
Croppable land	383/ha
Permanent grass	243/ha
Permanent grass cutting	51/ha
Cow housing	130/cow place
Cattle housing	0.12/m²
Late autumn labour	4.90/h
Casual labour for potatoes	2.90/h
(b) Constraints unused	
Potato limit	16.1 ha
Barley limit	121.6 ha
Early spring labour	270 h
Late spring labour	726 h
Early autumn labour	772 h
Hay purchase	36.6 tonnes

plan formulation, e.g. the rent which could be paid for an extra hectare of land. Even though it may not be possible to acquire more land, in this example if extra cow housing could be provided for an annual cost of under £130/cow place the total gross margin could be increased. Similarly, up to £4.90/h could be paid for autumn labour as this is severely limiting the area of potatoes grown. Also shown in *Table 27.16b* are the unused resources, these may indicate where there is scope for cost savings.

Indications of the stability of the plan may be gained from information on the amounts the net revenue of each selected activity could rise or fall before more or less of the activity would be included to maximise net revenue (assuming no changes in other coefficients). In *Table 27.15* the inclusion of most activities is highly stable, though the number of cattle fattened would change if the net revenue/head fell slightly and the area of spring barley would increase if its net revenue rose above £353/ha. It is also evident that the net revenue/head from beef cows would have to rise significantly before they could be considered and that a decision to make, instead of purchase, hay for calf feeding would depress the total net revenue by £23/tonne.

Additional solutions and data resulting from over-writing (e.g. allowing the dairy herd to increase to, say, 120 cows or changing the available labour or unit requirements) would add much extra information to be considered before deciding on the course of action adopted.

There are a number of sophisticated developments of the basic linear programming technique illustrated. These have advantages in formulating and providing a selection of plans for farms where manifold choices of activities and methods exist. Preparation of the necessary data is time consuming. It is important to do this thoroughly, since the computer merely proc-esses the figures fed in and solutions reflect the accuracy of the information. If standard programs

and standard input–output coefficients are used, interpretation of the solutions must take account of the relevance of the data to the actual farm. In drawing up the data matrix the comprehensive examination of what is being done and what may be possible on a farm provides most rewarding insights into its organisation and operation.

PLAN IMPLEMENTATION AND BUDGETARY CONTROL OF ENTERPRISES

Once the overall farm plan, or simple changes to the existing system, have been decided the farmer or manager has to act on the decisions. He must ensure that the proposed changes are put into effect. At its simplest, the organisational plan may be seen as the crops to be grown and the stock to be kept in the coming year. Where a number of years are needed to implement a longer-term development, possibly re-quiring investment in buildings or land improvements and expansions of production stock, careful phasing is needed. Revenue should be increased as early as possible and adverse cash flows minimised by delaying expenditure on fixed equipment until it is needed. The use of cash flow budgets is described on p. 616.

Having coordinated development and acquired the necessary inputs, management has to control events and monitor the performance of enterprises and of staff. The farmer/manager runs the business through-out the year and must try to achieve planned production levels while using inputs efficiently. This means not only doing jobs at the right time and avoiding waste and losses but, in particular, using variable inputs such as fertilisers, sprays and feedstuffs when they will obtain economic responses.

Once a production cycle has commenced, little economic monitoring of crops is possible and for fat lamb or suckler calf production, periodic checking on growth and concentrate feeding (if any) should be adequate. Actual performance has to be measured over the season and may be assessed as part of overall farm business control measures. Towards and after a year end, examination of the annual farm results against a budget prepared at the beginning of the year indicates how far expectations have been achieved. However, for continuously producing enterprises it is unwise to delay appraisal and evaluation for so long. Budgetary control within the year is needed for enterprises such as milk and egg production, pig breeding and fattening, and continuous veal calf and cattle raising activities.

Many organisations and firms run schemes to assist in monitoring or controlling production. Some of these prepare monthly and cumulative summaries of the essential output and cost items and provide comparisons with other participating farms. Some of the difficulties which arise through monthly fluctuations in input:output ratios, and variations in the values of stock on hand, are overcome by presenting rolling averages for 6 or 12 month periods.

Though useful such schemes do not amount to budgetary control unless at the start of the year detailed targets have been agreed against which actual performances are examined as the year progresses. These targets need to be worked out in physical and financial terms for production and for the important inputs, usually concentrate feedstuffs. For a dairy herd the expected number of cows, calving pattern, herd yield and intended level of feeding are used to prepare forecasts of monthly milk sales, feed inputs and margins over concentrate feed, or over feed and forage variable costs. Only those items which are amenable to adjustment in the course of production need to be included in short-period calculations.

Where a business is made up of separate farms or departments, top management is better informed of their progress if a system of budgetary control is established and maintained. The targets set must be agreed by those responsible for day to day running of the enterprises and should be the best estimates of what can reasonably be expected to occur. If these are somewhat better than in a previous year, achievement will be a challenge or incentive to herdsmen or section managers. Many farmers have targets they personally wish to reach but it may be unwise to pitch these so high that staff feel they are unattainable.

In budgetary control the reasons why actual results vary from those planned should be analysed so that action can be taken to correct adverse deviations. An example for the operation of a pig rearing enterprise over six months is given in *Table 27.17*. Though a detailed budget and shorter periods are preferable for control, to simplify interpretation, attention here is focused on the proceeds from weaners sold and the use of feedstuffs. At the start there are 100 sows and 20 in-pig gilts, with the intention of increasing the herd to 110 sows within the period. The forecast number of weaners is 1045 selling at £23 each and food consumption 0.87 tonnes/sow, with meal costing £160/tonne. In the event, through heavy culling and losses, only 106 sows produced 954 weaners, but to try to maintain planned revenue weaners were held on to a heavier average weight achieving a better price. In column 3 differences between planned and actual total sales and total feed costs may not appear high but analysis shows the effects of some of the deviations from plan are substantial, particularly when the lower price for meal should have increased the enterprise margin. Failure to achieve the planned number of sows reduced the margin by about £300 but the main weakness was the lower number of pigs reared/sow. Moreover, while some of the higher feed/weaner, shown in the performance data at the foot of the table, results from the sow feed being borne by fewer weaners, the reward for keeping weaners to a heavier weight is uncertain.

Table 27.17 Analysis of differences between actual and budgeted performance

		Target	*Actual*	*Difference*	*Contribution to differences[1] in total sales and feed costs (£)*
Production[2]					
Plan	Sows farrowing	110	106	−4	−882 (a)
Performance level	Weaners/sow	9.5	9.0	−0.5	−1348 (b)
Price	£/weaner	23	24.50	+1.50	+1568 (c)
TOTAL SALES OF WEANERS (£)		24 035	23 373	−662	−662
Input					
Plan	Sows	110	106	−4	−585 (a)
Input/unit	Feed/sow (kg)	870	943	+73	+1245 (b)
Cost/unit	£/tonne	160	155	−5	−479 (c)
TOTAL FEED COST (£)		15 312	15 493	+181	+181
Margin of sales over feed costs (£)		8723	7880	−843	−843
Margin/weaner (£)		8.35	8.26		
Feed cost/£100 pig sales (£)		63.7	66.3		
Feed cost/weaner (£)		14.65	16.24		
Feed/weaner (kg)		92	105		

[1] Method of calculation:
(a) Difference in number × actual performance or input level and actual price.
(b) Planned number × difference in performance or input level and actual price.
(c) Planned number and performance or input level × difference in price.
[2] Weaner pig production for 6 months.

Regular checks of this kind identify problems arising and alert management to try to put matters right. Also, there may be opportunities to exceed the forecast performance; these should be fully exploited. Some changes will occur which are outside of the control of those concerned with the enterprise and adjustments may be made for those which were unforeseen when the budget was prepared.

Emphasis has been placed on the value of recording and analysing relevant physical and financial information to aid ongoing business control. However, trends in these short-term measures and examination of the annual results in relation to the medium and longer-term prospects for the enterprises may require reappraisal of the farm business plan. Along with consideration of technological developments and market opportunities, so far as they affect his farm, the farmer should reaffirm, or if necessary modify, his strategy, re-set targets, draw up new enterprise budgets and an overall forward budget for the next year.

The activities of spotting opportunities and analysing problems, planning, decision making, implementing, recording and controlling and reappraisal are often integrated and some are concurrent. They make up farm business management and are identifiable in the role and work of an effective farmer. Undefined objectives and lack of attention to some of these activities, at the right time, frequently lead to frustration and disappointing results.

Financial management

SOURCES AND ALLOCATION OF FUNDS

The net profit shown on a trading and profit and loss account may give very little indication of the funds available to the farmer for capital expenditure, private spending or repayment of loans. Frequently, in an expanding farm business, much of the year's net income is tied up during the year in physical assets. Also, in a period of inflation where rising values of, for example, young stock and produce have contributed to farm profit, such increases may not be viewed as spendable income.

A simple reorganisation of information in the trading account and balance sheet provides a valuable insight into the disposition of funds during the financial year. In this summary, spending on capital improvements and for private purposes is included. Funds accrue from the trading profits, adjusted for non-cash transactions, from a run-down of stocks, sales of equipment, grants, borrowing and gifts. They are used for capital improvements, increases in live and dead stock on hand, purchases of machinery and possibly land, for private expenditure, to pay tax and repay loans. A breakdown should be made against the headings shown in *Table 27.18*. If there has been no change in cash-in-hand the totals of the two columns should balance.

If future farm profit has been estimated in farm planning or forward budgeting, the format of *Table 27.18* can be used for an *annual* cash flow budget. The sum of expected farm profit, grants, valuation decreases, sales of equipment and so on, may be compared with the sum necessary for capital spending, increases in valuation implicit in the plan and estimated private drawings and tax. If funds available from the business are likely to be too low to meet expected outgoings, additional borrowing will be needed. When surplus funds become available, they can be used to repay loans or to make further improvements. Clearly, the same conclusions can be reached from a budget, which instead of starting from estimated profit, shows details of the year's expected receipts (from sales of livestock and produce, machinery, grants and private sources) minus the sum of planned payments (for production requisites, machinery, capital improvements and private drawings).

BALANCE SHEET INTERPRETATION

Preparation of the annual balance sheet was outlined in 'Farm Accounts' (p. 590). A balance sheet shows the nature and value of assets at the particular date and the amounts and sources of borrowed funds. The asset values should be realistic; if land is shown at original purchase price and machinery at depreciated historic costs, revaluation to market values is desirable for management purposes and to establish the true net worth of the business.

It is preferable to study a series of balance sheets, as these indicate whether the net capital of the business is growing or declining and, similarly, what is happening to indebtedness and to total assets. Total assets may be increased either from net profit or through increased borrowing, but net capital declines if private drawings exceed net profit plus private receipts for the year.

Interpretation is assisted by sub-division of assets into *fixed* and *current* assets. Fixed assets include land and buildings, machinery and breeding livestock, all of which are used to generate production for more than one season. Current assets include physical working assets, which are normally converted into

Table 27.18

Sources	£	Allocation	£
Farm profit		Increase in valuation (of livestock, crops and stores)	
Less:			
Non-cash receipts and allowances		Decrease in creditors	
Add:		Increase in debtors	
Depreciation allowances		Capital improvements	
Imputed costs		Purchases of machinery and equipment	
Decreases in valuation (of livestock, crops and stores)		Purchase of land	
Increase in creditors			
Decrease in debtors		Interest payments (if not included in trading account)	
Sales of machinery			
Sale of land		Private drawings	
Capital grants		Tax paid	
Capital introduced and private receipts			
		Repayments of loans	
Additional loans		Reduction in bank overdraft (or increase in bank balance)	
Increase in bank overdraft (or fall in bank balance)			
Total funds available		Total allocated	

cash within a year or so, e.g. growing livestock and harvested or growing crops, and liquid assets which are cash or 'near cash' such as sundry debtors and the balance in the bank. Liabilities are sub-divided into:

(1) current liabilities which include sundry creditors, tax owed, short-term loans and the bank overdraft;
(2) long-term loans and mortgages which are unlikely to be recalled at short notice; and
(3) net capital or owner's equity.

The latter is the residual or balancing item, it is calculated by subtracting the sum of external liabilities from total assets. The data in a balance sheet may be arranged differently. For example, current liabilities may be subtracted from total assets to show 'net assets' followed by deduction of long-term liabilities to give net capital.

The stability of the business and its further borrowing capacity may be judged from ratios of the main items identified above. The *ratio of net capital to total assets* indicates the owner's stake in the business and is a measure of credit worthiness. If external liabilities are high in relation to assets, the substantial interest charges and repayments can be met satisfactorily only if the farm profit is high and stable. When external liabilities are low, i.e. net capital represents a high proportion of total assets, the farm business should be in a position to borrow further funds for worthwhile improvements or expansion.

The *gearing ratio*, i.e. the ratio of long-term loans to owner equity (net capital) similarly indicates the dependence of the business on long-term loans usually bearing fixed interest charges. A business in which all funds are provided by the owner has no gearing; it is low geared when such loans are small in relation to net capital. There is high gearing when the ratio of fixed interest long-term loans to net capital is high and such a business is vulnerable if profits are low or losses are incurred over a period of years.

In addition, attention must be paid to ability to meet current liabilities. This may be judged from the *liquidity ratio* (liquid assets: current liabilities) or the *current ratio* (current assets: current liabilities). If current liabilities could not be met without cashing an appreciable part of physical working assets there could be strain on the earning capacity of the business, e.g. through the sale of unfinished livestock or of produce at an inappropriate time. Where current liabilities exceed the total of current assets the basic fabric of production could be threatened if creditors pressed for payment. In this situation, breeding livestock or equipment might have to be sold to meet current liabilities. It is generally unwise to finance the purchase of fixed assets by means of short-term credit, particularly when fixed assets form a high proportion of total assets. When substantial building improvements are to be made or breeding herds expanded using borrowed money, long-term loans should be negotiated so that the liquidity problems which may arise with dependence on short-term credit sources are avoided.

The above ratios are inter-related and should be viewed together; it may be misleading to focus attention on one figure. As already mentioned, it is preferable to examine balance sheets and the ratios derived from them for three or more years to detect trends in the capital structure and indebtedness of the business. Even then because of variations in the timing of purchases and sales some ratios are influenced by the time of year when the balance sheet is prepared. Most cropping farms have high stocks of produce on hand or in the ground at the end of September. A farm on which large numbers of cattle are winter fattened is likely to have a high level of borrowing in December before cattle are sold, while the position at 31 March may differ from year to year depending on the proportion of animals sold before or after that date.

Bank managers are not only interested in credit worthiness, they require evidence of sound use of capital to date, this would be indicated by growth in net worth, healthy ratios and a record of having met obligations in the past. When approached for additional credit they may wish to consider the purpose of the loan and the budgeted evidence of the expected returns, showing ability to cover interest and capital repayments and withstand risks. The soundness of an investment proposal should be examined using one of the techniques discussed in the next section.

CAPITAL INVESTMENT

Farmers' decisions on use of capital depend on their business objectives and current levels of income. Much of the investment in farming is made out of current earnings. On farms making appreciable profits the objective may be to retain capital in the business, to improve the value of the property or to improve working conditions. In such cases, the immediate return on capital is not of the same importance as on a farm striving to achieve the highest possible returns, particularly if substantial expansions are being made with borrowed funds bearing a high interest rate.

Where a high return on available capital is the objective, careful appraisal should be made of all investment opportunities. Past information of returns on tenant's capital in the farm business is of no direct value. Also, capital investment in existing buildings and machinery which will not be changed need not be considered.

Information for each possible development is required under two headings:

(1) the effect on annual farm profit of the new development or expansion; and
(2) the additional (net) capital needed.

The expected change in farm income can be calculated using a partial budget, but if gross margins are being used, changes in fixed costs and gross margins forfeited by the removal or reduction of an enterprise must be taken into account. Similarly, the additional capital is that required for new buildings, extra machinery, livestock and working capital less any capital released by removal of an enterprise. For marginal adjustments to enterprises the capital requirement will be the cost of extra animals plus variable costs which would be incurred until receipts cover current expenses.

Expansions of the areas of cash crops can often be made with relatively little additional capital. Also, the capital requirements for sheep to utilise a given area of land are much less than for dairy cows or beef cattle at similar stocking rates. Where cattle would require extra housing and provision of additional silage and slurry facilities, the capital requirement would be even greater. When a development can be made in a short space of time and if the returns are likely to continue at a fairly regular level, simple methods of investment appraisal may be useful.

Pay-back period

This is the number of years needed for additional income to repay the capital outlay. In this approach, additional income is calculated omitting depreciation charges, but after deducting interest; taxation allowances can also be incorporated. The procedure is adequate if speed of repayment is all important and there is uncertainty about the future; however, there must be some expectation of returns after the pay-back period for a project to be worthwhile and the choice between alternatives should not be made only on the basis of the shortest pay-back period.

Rate of return

The annual addition to farm income after deduction of depreciation charges is expressed as a percentage of the additional capital invested. Sometimes capital invested is taken as the initial or peak requirement and sometimes it is based on half the initial capital in machinery, equipment and buildings plus any additional outlay on livestock and working capital, to represent the average investment over the life of the project. In comparing two or more projects either approach may indicate the most rewarding investment prospect. Return on initial capital is preferable for examining a single investment decision, but it is a harsh test and further consideration should be given to projects which just fail to show an acceptable return.

There are two problems with this method: differences in annual cash flows are ignored, and projects with different length of life cannot be satisfactorily compared. Discounting is necessary to handle these difficulties.

Discounted cash flow methods

The net cash flows of projects may differ. In some (e.g. starting a breeding herd from a nucleus of bulling heifers or establishing an orchard) investment is made over period of years and little income may be generated initially. For other projects the initial net returns may be high, possibly due to tax allowances, but then tail-off through time. These time considerations are crucial. The present value of future annual returns decreases with time and the longer the span the greater the uncertainty of results. In general, an investment which generates returns more quickly is to be preferred because the earnings are more certain and can be reinvested. For this reason it is necessary to discount future net cash flows and bring them to present values for comparison of alternative investments.

The first step is to estimate the net cash flows over the economic lives of the investment opportunities. For each of these the net cash flow in any year is the surplus of expected receipts over payments (including any further injections of capital). Adjustments can be made for expected tax allowances following expenditure on machinery and buildings, but depreciation is not deducted. The time period over which returns are considered should represent the anticipated economic life of the investment; ten or sometimes 15 years are most usual for developments on livestock farms. An assessment should be made of the *terminal value* of any stock or equipment which would be realisable at the end of the period. This sum is included with the cash flow of the final year. It is likely to be significant where a development includes the establishment or expansion of a herd of breeding livestock and where land is purchased or reclamation work of a permanent nature is carried out.

Net present value (NPV)

For an investment to be profitable the sum of the discounted expected net returns must exceed the initial capital investment:

$$NPV = \sum_1^n \frac{Ai}{(1+r)^i} - P$$

where P = original capital, Ai = net cash flow in the year i, r = appropriate discount rate of interest.

The appropriate discount rate is not easily determined, but many farmers have views on what is the lowest acceptable return for investment. This may be related to borrowing rate if outside capital is to be used, or the rate which could be earned from investment off the farm if own funds are being reinvested. In either case allowance should be made for risks associated with the type of development and for variations which can occur in external interest rates.

Where two or more projects are being compared, that which gives the highest NPV is normally preferred if a similar capital sum is invested in each case and the earning period is the same.

Tables 27.19 and *27.20* give discount factors for calculating the present value of a cash flow receivable in an individual year, and the present value of constant annual cash flows receivable up to the year selected. From *Table 27.19*, when the discount rate is 16%, the present value of £1000 receivable in year 4 is £1000 × 0.552 = £552, and from *Table 27.20*, if £2000 a year is received for ten years the present value is £2000 × 4.83 = £9660.

Examples of the calculation of the present values (discounted at 20%) of two cash flows are given in *Table 27.21*. If each of these is generated by an investment of £20 000, the net present value for A is

Table 27.19 Discount factors for calculating the present value of a future sum receivable in year n

Years (n)	Discount rate of interest (%)									
	8	10	12	14	16	18	20	25	30	35
1	0.926	0.909	0.893	0.877	0.862	0.847	0.833	0.800	0.769	0.741
2	0.857	0.826	0.797	0.769	0.763	0.718	0.694	0.640	0.592	0.549
3	0.794	0.751	0.712	0.675	0.641	0.609	0.579	0.512	0.455	0.406
4	0.735	0.683	0.636	0.592	0.552	0.516	0.482	0.410	0.350	0.301
5	0.681	0.621	0.567	0.519	0.476	0.437	0.401	0.328	0.269	0.223
6	0.630	0.564	0.507	0.456	0.410	0.370	0.335	0.262	0.207	0.165
7	0.583	0.513	0.452	0.400	0.354	0.314	0.279	0.210	0.159	0.122
8	0.540	0.467	0.404	0.351	0.305	0.266	0.233	0.168	0.123	0.091
9	0.500	0.424	0.361	0.308	0.263	0.225	0.194	0.134	0.094	0.067
10	0.463	0.386	0.322	0.270	0.227	0.191	0.161	0.107	0.073	0.050
11	0.429	0.350	0.287	0.237	0.195	0.162	0.135	0.086	0.056	0.037
12	0.397	0.319	0.257	0.208	0.168	0.137	0.112	0.069	0.043	0.027
13	0.368	0.290	0.229	0.182	0.145	0.116	0.093	0.055	0.033	0.020
14	0.340	0.263	0.205	0.160	0.125	0.099	0.078	0.044	0.025	0.015
15	0.315	0.239	0.183	0.140	0.108	0.084	0.065	0.035	0.020	0.011

Table 27.20 Discount factors for calculating the present value of a constant annual cash flow receivable in years 1 to n

Years (n)	Discount rate of interest (%)									
	8	10	12	14	16	18	20	25	30	35
1	0.93	0.91	0.89	0.88	0.86	0.85	0.83	0.80	0.77	0.74
2	1.78	1.74	1.69	1.65	1.61	1.57	1.53	1.44	1.36	1.29
3	2.58	2.49	2.40	2.32	2.25	2.17	2.11	1.95	1.82	1.70
4	3.31	3.17	3.04	2.91	2.80	2.69	2.59	2.36	2.17	2.00
5	3.99	3.79	3.60	3.43	3.27	3.13	2.99	2.69	2.44	2.22
6	4.62	4.36	4.11	3.89	3.68	3.50	3.33	2.95	2.64	2.39
7	5.21	4.87	4.56	4.29	4.04	3.81	3.60	3.16	2.80	2.51
8	5.75	5.33	4.97	4.64	4.34	4.08	3.84	3.33	2.92	2.60
9	6.25	5.76	5.33	4.95	4.61	4.30	4.03	3.46	3.02	2.67
10	6.71	6.14	5.65	5.22	4.83	4.49	4.19	3.57	3.09	2.72
11	7.14	6.50	5.94	5.45	5.03	4.65	4.33	3.66	3.15	2.75
12	7.54	6.81	6.19	5.66	5.20	4.79	4.44	3.73	3.19	2.78
13	7.90	7.10	6.42	5.84	5.34	4.91	4.53	3.78	3.22	2.80
14	8.24	7.37	6.63	6.00	5.47	5.01	4.61	3.82	3.25	2.81
15	8.56	7.61	6.81	6.14	5.58	5.09	4.68	3.86	3.27	2.83

more than £6000 and for B −£970. Project A is preferred, as it leaves a surplus after applying a 20% interest rate, whereas B with delayed cash flows does not.

Discounted yield (internal rate of return)

This is the rate of discount which when applied to each annual cash flow makes the sum equal to the original capital outlay. The value of 'r' is sought which makes

$$P = \Sigma_1^n \frac{Ai}{(1+r)^i}$$

'r' may be found by trial and error following the NPV type calculation; successive discount rates are applied until one results in a negative NPV. The yield lies between this discount rate and a lower one with a

Table 27.21 Calculation of net present value (discount rate of interest, 20%)

Year	Discount factor (20%)	Project A		Project B	
		Net cash flow (£)	Present value (£)	Net cash flow (£)	Present value (£)
1	0.833	7000	5831	1000	833
2	0.694	9000	6246	3000	2082
3	0.579	7000	4053	4000	2316
4	0.482	6000	2892	5000	2410
5	0.401	5000	2005	6000	2406
6	0.335	4000	1340	6000	2010
7	0.279	4000	1116	6000	1674
8	0.233	4000	932	6000	1398
9	0.194	4000	776	6000	1164
10	0.161	6000	966	17000	2737
Total present value of cash flow			26 157		19 030
Capital outlay, year 0			20 000		20 000
Net present value			£6157		−£970

positive NPV and may be calculated by interpolation. For example with B in *Table 27.21*, the NPV using a discount rate of 18% is £936 and the yield is 19%.

The discounted yield can be compared with the rate of interest on borrowed funds plus an allowance for risk. The discounted yields from different projects indicate that which is likely to produce the highest return on capital. Risk and other factors being equal, this project should be selected but it is important to consider the amount of capital invested in each project. A project giving a high return on a small outlay adds less to farm income than a larger investment providing a somewhat lower (but still acceptable) yield. The NPV method is preferable for ranking a range of alternatives.

The sensitivity of the calculated discounted yield should be explored, particularly when a major development is being examined of which there is little experience and where the outcome is uncertain. This is best carried out by systematically varying the underlying assumptions. For example, what would be the effect on 'internal rate of return' of a 20% increase in the cost of the initial capital investment? This could easily occur through delays in establishing the project. In a similar way, the effects of a 10% reduction in prices, 10% lower yields, or a 20% increase in variable inputs could be analysed. If, under all probable variations, the project appears likely to be worthwhile, it may be pursued with confidence. However, where feasibility would be dependent on the achievement of particular levels of performance or price, careful consideration must be given to the likelihood of obtaining these.

CASH FLOW BUDGETS

The preparation of cash flow budgets is a valuable management discipline. They are used:

Table 27.22 Quarterly cash flow budget for year 198–

Period	October–December		January–March		April–June		July–September		Annual total	
	Physical	Financial (£)	Physical	Financial (£)	Physical	Financial (£)	Physical	Financial (£)	Physical	Financial (£)
Receipts										
Fat cattle	—		80	32 000	100	42 000	38	15 000	218	89 000
Fat pigs	250	14 500	250	15 000	350	21 000	400	24 000	1250	74 500
Sows and boars	5	300	6	400	8	500	8	500	27	1700
Straw			30 t	800			50 t	1000	80 t	1800
Sundries						100				100
Machinery and vehicles sold				1500						1500
Capital grants								2400		2400
TOTAL RECEIPTS (R)		14 800		49 700		63 600		42 900		171 000
Payments										
Store cattle	180	54 000					50	16 000	230	70 000
Cattle sundries		500		400		400		200		1500
Boars			2	300					2	300
Pig meals	40 t	6000	47 t	7000	72 t	10 000	110 t	15 000	269 t	38 000
Pigs sundries		200		300		400		400		1300
Barley seed			6 t	1400					6 t	1400
Fertilisers			20 t	2300					20 t	2300
Sprays and twine						500		800		1300
Grassland fertilisers					25 t	2800	10 t	1100	35 t	3900
Silage contractor						2200				2200
Silage sundries						500				500
Wages		1200		1200		1300		1300		5000
Machinery running costs		1400		1400		1600		1600		6000
Farm fuel		400		500		300		500		1700
Rents[1] and rates				2000				2500		4500
Building repairs and miscellaneous		1000		1000		1200		1200		4400
Machinery and vehicles purchased		1400		9000						10 400
Capital improvements		8000								8000
Private drawings and tax		1500		1500		1500		1500		6000
Interest		5200				5400				10 600
TOTAL PAYMENTS (P)		80 800		28 300		25 900		44 300		179 300
Quarterly balance (R − P)		− 66 000		+ 21 400		+ 37 700		− 1400		− 8300
Initial balance (opening overdraft)		− 28 000		− 94 000		− 72 600		− 34 900		− 28 000
CUMULATIVE BALANCE		− 94 000		− 72 600		− 34 900		− 36 300		− 36 300

[1] Part of farm

(1) to determine the future levels of a bank overdraft, including the peak requirement to implement a modified or new farm plan in a pre-determined way;

(2) to decide the rate at which improvements can be made and a new plan introduced whilst holding borrowing within a specific ceiling;

(3) after decisions have been taken on the course of action, as the base for budgetary control during the financial year;

(4) to examine the worthwhileness of alternative investment opportunities.

For purposes (1), (2) and (3) the budget is drawn up showing all receipts and payments monthly, bi-monthly or quarterly. The choice of period depends on the volume and frequency of sales and purchases and the accuracy with which they can be forecast. When not using a computer program to prepare the budget, bi-monthly or quarterly estimates reduce the amount of calculation and allow for some flexibility in the timing of transactions. However, the intervals chosen should not be so long that they disguise the maximum requirements for funds. Most of the main banks supply forms for and explanatory booklets on the preparation of cash flow budgets and encourage (or require) the completion of these by farmers who are borrowing for substantial farm developments.

All expected receipts, i.e. for produce and livestock, grants, sales of machinery and private income are shown, as are all payments—trading, capital and private expenditure—in the periods they are likely to occur. The example layout in *Table 27.22* is for a quarterly budget, this could be extended to 12 pairs of centre columns if monthly details are likely to be important. The items shown in the rows should be altered to the needs of the farm. Form MA7, Multi-Stage Cash Flow Projection used by the Agricultural Development and Advisory Service is another suitable example, though on this receipts and payments are arranged into trading, capital and private sections.

At the foot of the table the quarterly (or monthly) balances of receipts minus payments are calculated. From these and the opening overdraft position the likely interest charge is estimated and the payments total and quarterly balance are adjusted accordingly. A cumulative balance is built up from the quarterly balances. If the farm has an overdraft at the beginning of the financial year this may be regarded as an opening negative balance which is reduced by a positive quarterly balance or increased if payments exceed receipts in a quarter. It is useful to record the physical quantities and the numbers on which are based the estimated receipts and payments in the budget. These may be written into columns provided for the purpose.

In the preparation of the budget, items which have a seasonal pattern such as purchases of seeds and fertilisers and sales of fat or store lambs, sugar beet and vegetable crops are entered accordingly. Sales of milk, eggs and pigs normally occur throughout the year but the pattern of production may be influenced by, e.g. the main season of calving of dairy cows or of flock replacement for laying hens. Moreover, changes in the size of herd at a particular time affect the subsequent levels of production, and hence receipts, and the monthly requirements for concentrate feed-stuffs. Where few changes in the programme of production are contemplated, the pattern of receipts and payments in the past year provides a basis for budgeting ahead. More detailed or sophisticated procedures can be used to calculate the expected monthly levels of production; e.g. using lactation curves with the number of cows and heifers expected to calve each month and the average herd yield, the volumes of milk sales can be predicted. Such estimates when combined with anticipated monthly prices for milk indicate the expected income from milk sales and, in combination with the feeding system for the herd, the expected monthly concentrate feed requirements.

Expenditure on wages, machinery running expenses and other overheads can be based on previous patterns (suitably adjusted for changing unit costs) or spread fairly evenly over the periods into which the year is divided. However, care is needed to identify when large items of non-recurrent payments are likely so that the cumulative balances can be interpreted with confidence. For a complex farm business a computer financial budgeting programme is very convenient but it is not necessary to produce valid guides for many types of farm. The discipline of thinking through the consequences for expenditure and receipts following planned changes is a particularly important aspect of management and enables difficulties to be anticipated or avoided.

A cash flow budget should be extended for a second or third year until the plan is fully implemented and it is calculated the overdraft will have been repaid or will be reducing satisfactorily. A soundly prepared budget of this type indicates to a bank manager the total requirements for a farm business development and when he can expect a loan to be repaid. If the farm has a borrowing limit and the budget indicates that this is likely to be exceeded in certain periods, action must be taken to delay some of the planned spending or to bring sales forward in order to stay within the limit.

The use of annual cash flow data in investment appraisal is dealt with on p. 616.

Financial control

A cash flow budget for the agreed plan is valuable for control purposes. A similarly laid out page can be used to enter the actual receipts and payments as each period is completed. Alternatively, the form can be drawn up with blank columns alongside each budget column in which to enter the actual transactions as each period ends. By comparing actual receipts and expenses periodically with the budgeted items it is possible to tell whether the outcome for the year is likely to be on target or differ sufficiently to call for changes in planned transactions.

When actual results during the year differ from the plan the reasons for variation should be sought. On the output side this may be because of:

(1) failure to implement the plan, e.g. fewer cows than intended may have been introduced into the herd;
(2) variation in yield (or physical production/animal); or
(3) a difference in price received/unit of production.

Inputs are similarly studied to see whether the main cost items are off target. This would be understandable if due to variations from planned numbers, but reasons for differences in use (e.g. of feed)/unit of production should be followed up. Costs may also have risen because of unanticipated price rises. Some of the causes may be outside the farmer's control but those which are not can be pinpointed and action should be taken to rectify adverse deviations. Fixed cost items have often been taken for granted, but in periods when product prices are not rising as fast as costs, it is most important to seek ways of economising on items which do not adversely affect output.

After about nine months of a financial year it is usually possible to estimate the farm's sales and expenses to the year end, the probable closing valuations and depreciation allowances. A forecast can then be made of the year's net profit. This may be useful for decisions relating to bonuses and the timing of machinery purchases in relation to tax allowances.

Financial controls are normally backed up by physical control measures and records and usually these contribute to the accounting system. Reference has been made earlier (p. 611) to individual enterprise control procedures. A farmer who has one main enterprise need concentrate only on the production aspects of that main activity and make periodic checks that fixed costs, capital and private expenditure to date are not out of line with targets for the year. The aim must be to avoid unnecessary complexity by selecting aids directly useful to management in achieving the objectives of the farm business.

28

Farm staff management

E. J. Sobey

THE NEED

The farmer, as a business manager, needs to apply professional management techniques to his small, but vital labour force. The abilities of each employee should be known and utilised in the most effective and economic way. Thus the employer is involved periodically in exercising skill and knowledge of man management. The application of staff skills may be related to the seasonal requirements of the farm through the use of such techniques as labour profiles and gang-work schedules. The skills may be related to day to day control and motivation of the labour force, to the administration of personal needs—pay, training safety and equipment or recruitment to meet vacancy or expansion in accordance with a manpower-plan.

Relevant techniques will be discussed in the following order:

(1) farm requirement: labour profiles, gang-work schedules;
(2) manpower planning: the manpower plan, job descriptions, personal specification;
(3) obtaining the right labour: recruitment and selection, induction training, continuation training;
(4) motivation: motivation theory, pay policy, including job evaluation, incentives;
(5) working conditions: need for procedures, facilities.

FARM REQUIREMENTS

Most farms operate with regular labour acquired over a period of years, but additionally have periods when extra labour may be required, e.g. silage making, fruit or potato picking. If the farm needs are not carefully

calculated timeliness of operations is likely to suffer or the manager may pay for extra labour or worse creating extra enterprise complications to 'use up' spare labour capacity.

Labour profiles

Seasonal labour profiles can be calculated so that exact requirements are known and may be plotted as a bar chart (*Figure 28.1*). These charts are also extremely helpful when considering a change in the size of an enterprise or the introduction of a new one. A labour profile can also show casual labour requirements.

Planning for skilled men and machines

Gang-work day (GWD) charts (*Figure 28.2* and *Table 28.1*) aim to ensure that enough men and machines are available to carry out operations at the correct time and within the time available for the operation.

Construction of gang-work day charts

It is seldom necessary to chart the whole year, only those periods which are peak periods usually need be charted.

Stage 1

Collect data:
(1) cropping and stocking,
(2) regular labour employed,
(3) machinery and implements available,
(4) gang size and machines required for each operation,
(5) rate of work of each operation,

Man hours/month

Crop	ha	Jan	Feb	Mar	Apr	May	Jun	Jly	Aug	Sept	Oct	Nov	Dec
Winter wheat	15			15	15				68	119	93	38	3
Spring barley	48.5		48	195	122	35			195	342	68		
Maincrop potatoes	11			82	168	22	43	43	10	296	540		
Sugar beet	6	7		43	26	75	81	9		26	94	109	16
Grassland	38.5			25	25		225	125	25	25			
Kale	10.5						105	26	5				
Total		7	48	360	356	132	454	203	303	808	795	147	19

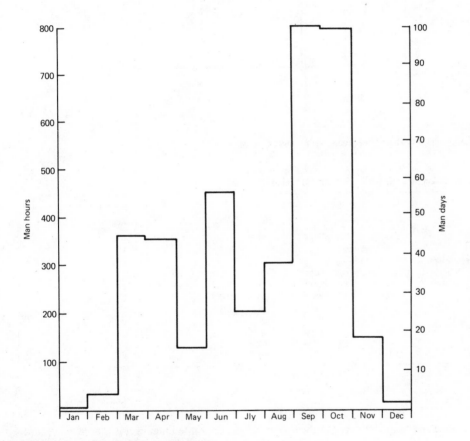

Figure 28.1 Day worker requirements (hours per month measured in the example: average and premium performances may be obtained from the *Farm Management Pocket Book* by John Nix)

(6) earliest start and latest finish date for each operation,

(7) number of days on which field work is possible in each month.

Stage 2

Tabulate the information for each operation, under the headings—area, time period, working days available, work rate and gang size. Calculate the gang-work days required for each operation.

Stage 3

Draw chart. Operations using large gangs are the least flexible and should be put first.

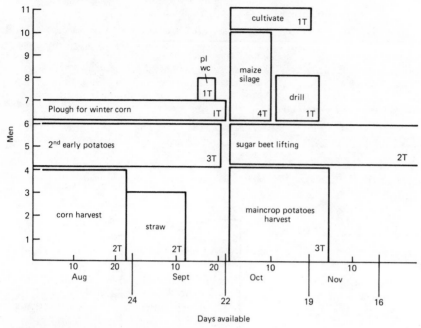

Figure 28.2 Gang-work day chart (T = number of tractors required)

Table 28.1 Information for gang-work day chart (315 ha arable land)

Activity	(ha)	Period	(d)	Work rate (ha/GWD)	GWD reqd	Gang size Regular	Casual	Tractors
Corn harvest	221	Aug – Sept 15	34	10.1	22	4	—	2
Straw cart	60.7	Sept	22	4	15	3	1	2
Second early potato harvest	18.2	Aug – Sept	46	0.4	45	2	4	3
Main potato harvest	12.1	Oct – Nov 7	23	0.5	24	4	6	3
Sugar beet	30.5	Oct – Dec	50	0.7	44	2	—	2
Maize silage	33	Oct 1 – 20	14	3.3	10	4	—	4
Winter corn								
Ploughing	80.1	Aug – Sept	46	1.6	50	1	—	1
Cultivate	80.1	Oct	19	4	20	1	—	1
Drill	80.1	Oct	19	8.1	10	2	—	1

Performance figures obtained by work study measurement (after Niemeyer).

Initially assume operations can start at the earliest start date.

Charting is a matter of trial and error, like a jigsaw.

A chart with no spaces has no flexibility for poorer than average seasons or other delays.

Casual labour may be written in or 'ballooned' on top.

MANPOWER PLANNING

One problem for farmers, often not recognised in time, is that they and their workforce grow old together. The position then facing the heir or equally an appointed manager is that he may shortly have to replace all the staff fairly quickly with the consequent loss of knowledge.

Age is, of course, not the only factor; changes in farm size, type of enterprise, technology, or national agricultural policy can all affect farm staffing.

As has been demonstrated already by the use of labour demand charts the day to day requirement should be known. Planning the work force for the future requires different techniques:

(1) examination of the external resources;
(2) examination of internal constraints and resources.

External resources

People

National and local statistical returns of school leavers. National and local information on (apprentice, college, university) trained leavers.

Information on local (and national) availability of skilled, semi-skilled and general labour (Department of Employment and trade journals).

Neighbours and competitors' pay rates and employment conditions—pensions, perquisites, etc.

Availability of 'contract' labour—short or medium term.

Availability of casual, part-time and pensioner staff.

Vacation period labour, students of all disciplines and nationalities.

Technology

Development work currently taking place in ARC, institutes and commercial companies which may reduce or increase labour requirement, remove the need for old 'skills' or introduce new skills.

Major changes in farming fashions and markets.

Government policy—national and international, in particular legislation affecting conditions under which the work is done, e.g. safety legislation.

Internal constraints and resources

People

Age distribution of current labour force.

Present skills of labour force.

Health and mobility of present labour force.

Ability of management to train, or to obtain services of appropriate commercial, training board or college training for staff.

Willingness of present staff to undertake new, expanded or diminished work loads.

Potential, i.e. internal promotional possibilities.

By careful examination of all factors, the family farm or the large estate can formulate a plan for the future so that retirement, resignation and sickness are handled with the minimum of 'disturbance' to the farm business. Training for present and future needs can then be given at the appropriate time and if necessary recruitment, or at the other extreme, redundancy, can be given the maximum time and attention. The plan should demonstrate:

(1) the required number of staff,
(2) the required skills for every major task,
(3) that staff effort is concentrated on the business objectives,
(4) that staff are deployed in appropriate tasks.

Internal succession

Formalised succession charts are a management tool in larger farm businesses and allied industries. A simple example (*Figure 28.3*) based on an arable farm employing one foreman, one head tractor driver and six other tractor drivers (TD) illustrates the principles.

In this situation the foreman is due to retire within three years, there appears to be a ready made successor in the head tractor driver, who is gaining experience of man management with two other tractor drivers responsible to him. Should the head tractor driver decide to leave, TDs 1 and 2 are probably too old to promote, TD 6 too young. TDs 3, 4 and 5, should be given careful consideration; if their supervisory potential is considered unsuitable it would reduce risk to replace TD 2, who retires within the year, with a man of foreman potential. The replacement of TD 2 on retirement is a key appointment.

Job analysis—job description

This is the basis of several management techniques and is a statement of:

(1) all the parts of the work to be done,
(2) the nature and extent of responsibilities for people, materials, machines and/or events,

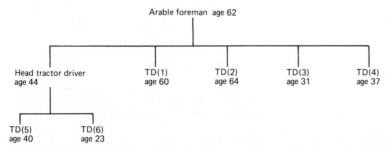

Figure 28.3 Succession plan

(3) the physical effort and endurance required, and
(4) the working conditions under which the job is performed.

The statement can then be used in three primary ways:

(1) *Job evaluation* for the purpose of relating pay to job level. There is an intermediary stage in this, which is *job assessment*. At this point there is some determination of the factors in skills, responsibility, physical effort, mental effort and working conditions which are involved in meeting the demands of the *job description*.
(2) In the development of training programmes for groups of workers or individuals by testing actual performance levels against optimum standards and applying planned training to cover deficiencies and improve the speed of learning.
(3) In the basic analysis of a job prior to recruitment under any set of circumstances, vacancy or expansion to form a platform upon which to build the selection process (*Figure 28.4*).

General description of job and title

Key tasks   Date

1
2
3
4   (preferably not more than 8 - 10)
5
6
7
8

Accountable to . . .
Responsibilities . . .
Supervisory:
Tools:
Materials:
Reports:
Records:
Working conditions
Hazards
Safety:
Health:

Figure 28.4 Job analysis form

Personal specification

This stage is specific to the recruitment process in determining what kind of man/woman the employer wants to fill the job (*Figure 28.5* on pages 626–627).

It will be noted that sections 3, 4 and 5 have two levels: Essential = the minimum necessary standard to perform the job; Preferred = the ideal level of education, training or experience to be able to meet all the requirements of the job at present and to be able to expand if required beyond the present work situation.

Sections 6–10 have two vertical divisions to assist the interviewer to consider the importance of any particular factor, e.g. 6A vision—if employing a hill shepherd it is essential that the shepherd should have good eyesight. Means of checking—take the applicant to a viewpoint and ask him to identify a distant object or objects.

Section 8—motor requirements refers to the coordination of body, hands and feet.

OBTAINING THE RIGHT LABOUR

Recruitment and selection

Objectives of selection

To select a new employee who is likely to be able to perform the tasks to the employer's satisfaction after training: in other words—that he can do the job, or can learn the job, and that he *will* do the job.

Therefore the selector's task is to predict each applicant's probable ability in the job.

Preparation and process

The process of recruitment and selection thus needs care, particularly if this is a relatively infrequent operation, and the opportunities for practice are limited. Proceed as follows:

(1) prepare a job description—on paper;
(2) prepare a personal specification—on paper.

The combination of these two tasks is *job analysis* and should take into consideration the main parts of the work to be done, the difficulty of learning or teaching any part, the areas of high risk (damage to stock, crops) and then the special attributes that a man will require in order to carry out the work.

(3) (a) Draft advertisement,
    (b) put out contact enquiries,
    (c) look at situations wanted;
(4) decide where and when to place advertisement;
(5) decide where and when to hold interviews.

Timing is of great importance at this planning stage in order to give possible applicants the opportunity to arrange time off for interview, etc.

(6) Examine and *answer* all replies;
(7) decide who you want to interview and when;

(8) prepare to hold interviews, including tests if applicable.

It should not be necessary to remind a potential employer to answer all letters and telephone enquiries.

(9) Hold interviews (and tests);
(10) *decide*
    (a) *Yes*, there is a suitable candidate, *or*
    (b) *No*, re-examine job/personal specification and advertisement;
(11) make offer—take up references—reply to interviewees;
(12) prepare for arrival of new employee and induct on arrival.

Practice

The interview is generally used as the main selection technique.

Whilst it is unimportant as to which particular form of interview is adopted, the selector has certain factors to establish in order to make his prediction on suitability to perform the task.

(1) Check facts stated on letter or application form.
(2) Ensure that there are no unexplained career 'gaps'.
(3) If the selector is himself knowledgeable in the technical areas relevant to the job—assess the candidate's level of acquired skills.
(4) Assess the consistency of the candidate and his realism in terms of the goals he has set himself, i.e. motivation.
(5) Decide what his/her likely impact would be in the work situation.

The question of impact is often given too much prominence in selection; however if you are unable to decide which of two men, who are technically and otherwise acceptable, to choose try asking the following questions of yourself:

(1) Which employee would you prefer to invite into your own home?
(2) Which family would be most acceptable in your community?
(3) How do you think each person would react if he did not get his own way?

Preparing for interview

Administrative arrangements should include:

(1) timetable of interviews;
(2) notification to candidates, and to others taking part in the selection;
(3) adequate waiting and toilet facilities;
(4) a light cool room in which the conversation cannot be overheard;
(5) complete freedom from any form of interruption.

The interviewer should then prepare by re-reading the job description, personal specification, and the personal history sheet filled in by the applicant. He can then decide the topics to be covered during the interview, and any areas which need special clarification or confirmation. Prepare a check list of these points.

Conducting an interview

Observing a set of rules will not automatically bring about a good result. A knowledge of some basic principles and good preparation will however help the interviewer.

Every interview should have a beginning—setting the candidate at ease and showing an interest in the candidate, a middle which is a thorough and logical exploration of experience, background, interests, medical history, attitudes and plans. The conclusion should also be planned to enable the candidate to raise points which may not have been covered in the middle stage. At this point the candidate should be told what is the next stage of the process and how he/she will be informed, also travel or other claims should be dealt with and the applicant looked after until leaving the farm.

The social skills of the interviewer lie in encouraging the candidate to talk freely so that open ended questions, which gives the candidate the opportunity to expand are preferred to closed questions which require only yes/no answers or leading questions which suggest to the candidate the expected answer.

Induction training

Introduction

Recruitment and selection is not complete, when an offer of employment having been made is accepted. The new employee now has to be introduced to the business, an event which may be only days after interview or could be months in the case of management or senior technical staff. Unless the introduction is thoroughly prepared and controlled the whole exercise of recruitment could be wasted, that is, the newly appointed employee may leave the business thoroughly disheartened by attitudes formed in the first few days or weeks.

Industrial psychologists have examined this situation and have produced the following simple analysis. The new employee is tied by his fear of an unknown situation so he is afraid to ask questions when he should do so. This in turn limits his usefulness, prevents him from thinking clearly and taking in information and finally gives a wrong impression of him to his fellow workers and supervisors.

If this is kept in mind by the manager, supervisor or trainer the planning of what is called Induction training becomes straightforward. The new employee

1.	Personal data
	Preferred age: Home circumstances:
2.	Physique
A. B. C. D. E.	Health and stamina Disabilities (not acceptable) Speech Manner Appearance
3.	Basic education
	Essential: Preferred:
4.	Technical training
	Essential: Preferred:
5.	Work experience
	Essential: Preferred:

	Job requirement	Example from job content	Means of checking
6.	Sensory discrimination A. Vision B. Hearing C. Touch D. Taste and smell		
7.	Mental requirements A. General aptitudes 1. Comprehension of terms 2. Reasoning and learning 3. Memory for names, faces and detail 4. Initiative & planning 5. Care, attention and accuracy. B. Special aptitudes 1. Verbal fluency 2. Verbal expression 3. Numerical aptitude 4. Spatial aptitude 5. Mechanical comprehension		

Figure 28.5 Personal specification

Job requirement	Example from job content	Means of checking
8. Motor requirements		
9. Character traits required A. Stability B. Industry C. Perseverence D. Loyalty E. Self-reliance F. Get along with Others G. Leadership		
10. Other requirements 1. Military commitments 2. Union membership 3. Licences		

Figure 28.5 (cont.)

has the opportunity and guidance to learn about the people with whom and for whom he works, the rules and conditions under which he works, the products/crops of the business and the different parts of the business with which he may not usually come into contact. To this end each manager/supervisor responsible for new staff should have a written plan for each type of employee under him, which must be kept up to date. The detail will be different in each business and in different parts of each business but should always have a simple objective. The sooner a new member of staff can begin to earn his wages the better for him and the company he joins.

Aims and objectives

The general aim of an induction programme is therefore to assist the newcomer to become familiar with work place conditions and be effective in his/her job as soon as possible.

The detailed objectives and content varies greatly depending on the individual and the job. Some examples are shown in *Table 28.2*.

Organisation

Induction training must be planned. The objectives above indicate that a great deal of information is required and various people will be involved during this period of learning. A programme should therefore be drawn up to ensure that this learning takes place at the right time, in the right place and at an acceptable rate. The main factors to consider are shown in *Table 28.3*.

Table 28.2

On completion of induction the trainee will be acquainted with:

(1) The general objectives of the business.
(2) The main products of the business.
(3) The general lay-out of the holding and the identity of the various units.
(4) Other employees and their authority.
(5) Conditions of employment.
(6) The main duties of his/her job.
(7) Future training to be received.
(8) General rules and conditions required by the employer and matters concerning trainees' welfare, e.g. housing.
(9) Safety regulations and their application to the holding.

Table 28.3

(1) Who will organise and conduct induction training?
(2) When will it start?
(3) How long will the programme need to be?
(4) Where will the various parts of the programme take place?
(5) If equipment for demonstration is required, is it available?
(6) Is transport available if required?
(7) Are the various people involved in the induction programme free when required?
(8) Will all new employees have the same programme?
(9) Can visual aids be used in presenting information, i.e. farm maps, recording systems?

Summary

Induction training should be carefully tailored to suit the individual concerned. Objectives should have been considered carefully and a programme of activity set out to achieve these objectives. It is essential that the programme is carried out smoothly to ensure effective learning and to gain the goodwill of the new employee. The programme must therefore be carefully organised by a person of authority.

Continuation training

The establishment of training needs is derived from three main sources:

(1) a continuous review of employee performance throughout the year;
(2) an assessment of future supervisory requirement;
(3) an assessment of future skills requirement.

The first of these is not an opportunity to apportion blame nor an assessment of personality but an ongoing and participative understanding of the task performance and involvement. The second, is normally evident from a manpower plan, particularly in a farm expansion phase. The last is a constant requirement to keep up with technological change or a major change of farm policy, e.g. the introduction of a new enterprise.

Improving present performance

The best place for most training in relation to present performance is on the farm. There are many agents for the farmer to use as well as his own skill. Manufacturers, particularly manufacturers of agricultural machinery find it is in their own interest, as well as that of their customer farmers, to ensure that the farm employees who use the equipment can use it properly, safely and effectively. Similarly it is possible, with the help of the Agricultural Training Board, to obtain the services of a trained instructor for most other farm tasks.

Alternatively there are many 'off the farm' short and day release courses available. Local Education Authority courses in County Agricultural Colleges or in the Local Technical College, ATB courses locally or at Stoneleigh and commercially sponsored short courses, together with demonstrations and discussion groups, etc. organised by ADAS, NIAB, research stations and so on.

For these and all other training courses a word of caution. It is necessary to examine carefully the stated aims or objectives of any course and the level of trainee for whom it is designed. Much money and time is wasted sending people on unsuitable courses, often to their own frustration and especially damaging if at too high a level for the employee away from the home environment possibly for the first time for many years.

Training for future needs or potential

The dangers in training for future needs lie mainly in the timing of courses, too far in advance leads to frustration of the employee and possible loss, too late, to anxiety that he will be unprepared for the next step when it occurs. It involves an examination not only of present skills but also of attitudes and almost certainly will involve the employee in training away from his home farm. Certainly such courses as 'Effective Supervision' mounted by the ATB have as one objective a number of changes of attitude in potential and newly appointed supervisors regarding increased responsibility as well as knowledge and skill in man management techniques.

Training programmes

After consideration of the factors above an annual training programme for all staff can be drawn up as a programme to fit farm need and course availability as well as individual leave arrangements. It would normally include:

(1) the individual training targets,
(2) the training to take place on the farm,
(3) the training to take place off the farm, e.g. courses,
(4) long-term training, e.g. full time or block release courses at College of Agriculture, University, etc. where appropriate.

A comprehensive and job related training programme is a helpful motivating force if used correctly.

Validation and evaluation of training

Training of itself is of little use without the opportunity to follow initial training with practice and experience. Validation of training therefore is an exercise which only asks the question 'Did the training given achieve its stated aim?'. Evaluation asks the farmer to consider the benefits which can be listed as:

New staff becoming more effective more quickly.
A more effective and reliable work force.
Improved staff relations.
Better work organisation.
Greater safety.
Lower operating costs.
Better supervision.
Increased job satisfaction and pride in work.
Greater confidence to undertake present and new work.
Overall more effective farming.

MOTIVATION THEORY

In the early twentieth century the developers of management science looked upon man the worker as a flexible but inefficient machine which would work reasonably well if the tasks were planned, orders given and work was supervised. Later it was discovered that pay for the work done, even when financial incentives were applied, did not increase the quality or quantity of work performed for more than a short period of time and, with the development of social security in the western world—work or starve is not a single option. More recent management theorists have examined the traditional approach to man management and suggested some practical alternatives.

Abraham Maslow

He evolved the theory of the 'Hierarchy of needs'. This suggests that as a lower order of need is satisfied a new higher need is revealed. Thus a satisfied need ceases to motivate and is often illustrated in the form of a pyramid (*Figure 28.6*).

Figure 28.6

Douglas McGregor

He developed a theory to explain the traditional approach to direction and control of staff which he called *Theory X*. He compared this with the idea that people at work prefer involvement as *Theory Y*.

Theory X

(1) The average human being has an inherent dislike of work and will avoid it if he can,
(2) Because of this characteristic, most people must be coerced, controlled, directed, threatened with punishment to get them to put forth adequate effort toward the achievement of organisational objectives.

(3) The average human being prefers to be directed, wishes to avoid responsibility, has relatively little ambition and wants security above all,

(Often known as the 'carrot and big stick' method of management as if people were donkeys.)

Theory Y

(1) The expenditure of physical and mental effort in work is as natural as play or rest.
(2) External control and the threat of punishment are not the only means for bringing about effort toward organisational objectives. Man will exercise self-direction and self-control in the service of objectives to which he is committed.
(3) Commitment to objectives is related to the rewards associated with their achievement.
(4) The average human being learns, under proper conditions, not only to accept but to seek responsibility.
(5) The capacity to exercise a relatively high degree of imagination, ingenuity and creativity in the solution of organisational objectives as widely, now narrowly distributed in the population.
(6) Under the conditions of modern industrial life, the intellectual potentialities of the average human being are only partially utilised.

Frederick Herzberg

He was the originator of the Motivation/Hygiene theory. He discovered that a well motivated employee derived satisfaction through five factors all related to the job itself:

(1) 'doing the job';
(2) 'liking the job';
(3) 'achieving success in doing the job';
(4) 'receiving recognition for doing the job'; and
(5) 'thinking that doing the job will lead to better prospects'.

The fact that he might be paid more for continued success at the job came sixth in this table of motivation. These factors are called 'satisfiers'.

'Hygiene factors'

At the other end of the motivational scale there are 'dissatisfiers'. These all relate to the context in which the job is done. Bad policy and administration, poor supervision, waste, duplication of effort, unfair personnel policy and constant criticism are examples that create dissatisfaction. Remuneration again appears in the scale—insufficient salary or wage is a 'dissatisfier'.

Herzberg calls the dissatisfiers the 'hygiene factors' likening them to the role of public health in the control of disease. Good hygiene eliminating disease factors does not cure diseases but it is important in preventing their occurrence. Similarly, the absence of dissatisfiers in a work situation may prevent an employee being unhappy in his employment, but their absence will not motivate an employee.

It is important to realise that the causes of satisfaction are not the same factors as the causes of dissatisfaction.

If a factor giving satisfaction is absent, the result is not dissatisfaction but just no satisfaction and similarly if a cause of dissatisfaction is absent the result is not satisfaction but no dissatisfaction.

Equity theory (ascribed to several authors)

It is often said that Herzberg's theory, so relevant to the farm situation does not take sufficient account of money as a motivator. This is not the case as has been shown above but he does disregard money as a comparison. Equity theory suggests that people at work can be demotivated by comparing their earnings with the earnings of others and the work performed by each.

My pay for *my efforts* should be comparable to *his pay* for *his efforts*. This type of motivator equation takes us into the world of pay differentials and the human rationalisation of '*I* would not do that job: *He* (the comparison) deserves how (little/much) he gets'. The difficulty for the man manager arises when he is unable or unwilling to adjust the pay and the operator reduces his efforts to make the equation more equal again. This last theory is very much the problem of the late 1970s and 1980s.

Management needs to search for the factors within the farm business which will motivate their work force. Removing demotivating factors usually involves direct expenditure of money (improvements in accommodation, etc.) and supervisory training. Motivating the staff needs identification of goals, stimulating the interest in achieving personal and organisational targets and making sure that these goals, once identified are realisable in some measure.

Payment policy and incentives

General

The payment of a wage or salary for work performed is a factor in motivating an employee. It is at the same time a major tool to use in achieving the personal objectives of a business. Therefore, the objectives of any payment policy may be clearly expressed as:

(1) to attract and retain sufficient staff of sufficient quality to meet the objectives of the business;
(2) to provide, with other forms of encouragement,

sufficient incentive to staff to give more than minimum effort;
(3) to achieve the first two objectives at minimum long-term cost.

In practice this means that the policy must provide competitive wages with other businesses, considered to be 'fair' internally between different grades and skills, not out of proportion to the overall cost/returns of the business and yet offering at the same time the necessary incentive mentioned before.

Internal constraints

The policy should be logical, tidy and internally consistent. This requires careful planning and at times careful adjustment to bring into line those accidents of payment, such as the high wage a single specialist can expect until there are more equally skilled available. This happened over most of Europe ten or so years ago with computer programmers—very few available, so if a company needed such an employee they had to pay a wage for some years far above the real value of the service provided, and out of line with the promotion/reward payment policy. As programmers became more available their wages tended not to be increased at the same rate as other technologists and managers, so that today the wage is generally in line with their contribution to the business and other staff.

The exception to this is still found in those companies where there was no policy, or it was not adequately controlled.

The policy should be easily understood and operated; should have flexibility to deal with the planning and cost control systems on the business.

Payment strategy

There are two groups of factors, both of which change with time, which have to be examined before a policy can be established. These are 'external factors' and 'internal factors'. Examples of these are given in *Figure 28.7*.

It is possible to see almost any combination of these factors leading to a payment policy and adding the question 'does the business want to be a market leader or follower on pay'? to arrive at a pay structure which meets all the requirements of the three clauses of the general statement.

Pay

Agricultural wage minima are determined by national negotiations. The Agricultural Wages Board defines categories of workers and lays down the minimum rate which must be paid to full time and part time workers with differential rates, according to age up to 20, against agreed hours. It also sets minimum rates

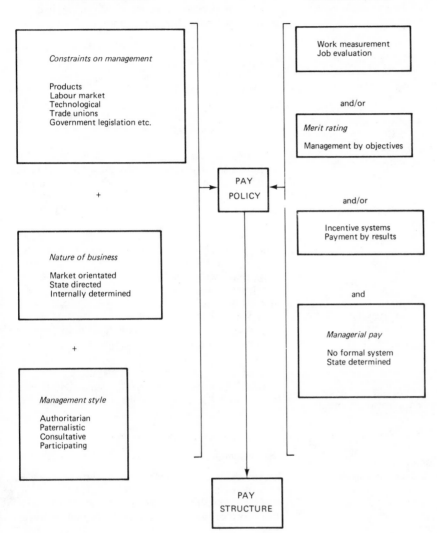

EXTERNAL FACTORS

INTERNAL FACTORS

Constraints on management

Products
Labour market
Technological
Trade unions
Government legislation etc.

+

Nature of business

Market orientated
State directed
Internally determined

+

Management style

Authoritarian
Paternalistic
Consultative
Participating

Work measurement
Job evaluation

and/or

Merit rating

Management by objectives

and/or

Incentive systems
Payment by results

and

Managerial pay

No formal system
State determined

PAY
POLICY

PAY
STRUCTURE

Figure 28.7

for classes of workers, for overtime and night work and the calculation of appropriate stoppages for board and lodging, even a rate for the shepherd's first and subsequent dogs. These rates are normally negotiated annually and are due for payment on publication of the Wages Board order or from the set date specified therein.

The wages board in no way prevents an employer from paying a higher rate if he wishes or the job market compels him.

Wage payment

Wages, as opposed to salaries, are normally calculated on an hourly basis and paid in cash weekly. Employers may ask hourly paid staff to accept payment by bank or giro transfer to reduce the risk in having large sums of money in the farm safe, but employees have the right to insist on cash payment. Employers may not ask any employee to accept payment in kind.

Incentives

Types of incentive

Incentives can be classified under three broad headings:

(1) non-financial;
(2) semi-financial;
(3) financial.

Non-financial

Security of employment
Working conditions, hours and holidays.
Opportunities for promotion.
Opportunities for training and education.
Job satisfaction.
Good management.
Good communications.

These are difficult to measure but are more vital to the success of a payment structure than any specially constructed incentive schemes. The best incentive, in all circumstances, is the evidence of good management continuously in action.

Semi-financial

Pension, insurance and sick pay.
Subsidised meals and social facilities.
Farm perquisites—free milk, fuel, etc.
Free or low-rent housing.
Company car.
Prize award schemes.

This is the area of fringe benefits. As distinct from the benefits in the section above these rewards can be given a firm financial value.

Prize award schemes

This type of scheme is the odd man out in the list of fringe benefits because it depends upon the attainment of a specific target.

Financial

Wages and salaries	—job evaluation
Ad hoc	—annual bonuses
	—profit sharing
	—suggestion schemes
Performance based	—merit rating
	—share of production
	—rate fixing
	—related to specific targets
Work study based	—directly proportional to output
	—geared schemes
	—high stable earnings

Job evaluation Job evaluation is a formal process used to compare one job with another in terms of skill, responsibility, physical requirements in order to arrive at a consistent and equitable pay structure in an organisation.

A number of grading systems are in use. The most widely used is the 'points rating system' where each job is analysed under a number of headings and a points value put upon each factor. Certain key jobs are analysed in depth to provide bench-marks as a reference point for the remainder.

Merit rating Merit rating or performance appraisal aims at a periodic assessment of an individual's performance based on such factors as:

(1) quality of work,
(2) consistency of output,
(3) reliability,
(4) willingness to cooperate.

Individuals assessed in this way are paid weekly fixed sums until re-examination takes place.

Share of production These are usually company wide with incentive schemes based on added value of reduction in costs. A standard labour cost is formulated and the actual labour cost in any one accounting period is related to this to determine the bonus distribution.

Rate fixing Prices or times for jobs are estimated from experience, comparison with other jobs, or from rough and ready timings. They are usually agreed after some bargaining between operatives and management.

In industry this method is often used in heavy engineering or maintenance work and in farming most piece-work rates have no firmer base than this.

Related to specified targets This describes those financial incentives related to output or a performance factor and not time or the work content of a job, e.g. quantity of milk.

Implementation

It is essential to ensure that an incentive scheme is both necessary and will make a contribution to profitability either directly by increased output or lower costs or indirectly by improved quality or lower waste.

In general a bonus scheme should meet the following conditions:

(1) The results against which the incentives are to be paid must be influenced by the participant's skill, judgement or effort.
(2) The rewards offered should be related to the effectiveness of the effort applied.
(3) The scheme should be easily understood and the terms and conditions set out and agreed.
(4) The bonus earnings should be simple to calculate and to verify.
(5) Payments should be made as frequently and soon as practicable after the performance.
(6) The bonus earning opportunities should offer a significant addition to the basic wage. (A target of 25% of the basic wage is suggested as a normal base.)

(7) Quality standards must be rigorously maintained.

Other desirable factors are:

(1) On the whole individual incentive schemes are more effective than group schemes.
(2) Bonuses should be the same for all taking part in a particular scheme.
(3) There should be little or no mixing of day work and incentive work.

WORKING CONDITIONS

Need for procedures

Unfortunately, in any employment situation, there may be a need to dismiss an employee. It may be that the job has simply ceased to exist or possibly there may be severe misconduct, Much legislation since the Contract of Employment Act 1963 has provided justifiable protection for employees and has laid down guides for employers. Most particularly minimum standards are set for the form and period of notice to be given under differing circumstances. As this legislation is liable to be changed in detail by a party in government and to meet changing political and economic conditions, current minimum requirements are not stated here. The details and supporting employers' and employees' guides are fully available from the offices of the Department of Employment or from HMSO.

It must be made quite clear that Government lays down minimum standards, there is nothing to prevent an employer offering better terms either in length of notice or for severance money. There is one definite onus on the employer that he should record in detail every disciplinary action and that he should have clear procedures, steps to follow, as should the employees to deal with grievances or appeals on a disciplinary decision. Again guidelines are available from the Department of Employment, from the NFU and the NUAAW.

Facilities

The farmer is one of the small number of employers who may have an interest not only in employees at work but also after work; the worker's home frequently being on or part of the estate.

As has been mentioned under motivation a sound and comfortable house will not necessarily make an employee happy at work, but bad housing will certainly make him unhappy. Similarly provision of washing and changing facilities may be essential under safety provisions; it requires more imagination than money to make facilities acceptable and pleasant for staff.

Retirement

Approaching retirement age, staff should have the subject discussed with them by the employer. It may be that the retiring employee has everything planned or he/she may be resentful. It is a subject requiring much tact on the employer's part but it can be greatly eased by having a check list.

Has the employee a house to live in after retirement? (The farm-provided house may be required for a successor.)

Will the employee want part time or occasional employment for a few more years and can the farm offer the opportunity?

Will assistance be required in moving?

Has the farm in the past given rights (e.g. rabbit shooting) which may prove an embarrassment if continued or terminated?

Has the employee sufficient pension (this should be discussed as far before retirement as possible if the farm does not have a pension scheme for employees)?

Are any of the perquisites to be continued and for how long if offered?

Whilst the labour force on a farm may be small in numbers by comparison with industry, and relationships between employer and employee less distant, there is no less requirement for the employer to have a sound policy for the most flexible resource—people.

References

(1) HERZBERG, F. (1966). *Work and the Nature of Man.* Cleveland: World Publishing Co.
(2) MCGREGOR, D. (1960). *The Human Side of Enterprise.* London: McGraw Hill
(3) MASLOW, A. (1954). *Motivation and Personality.* New York: Harper Bros

29

Agricultural law

G. Spring

INTRODUCTION

This section is designed to provide an introduction to law as it affects the agricultural industry. It should, however, be recognised at the outset that most people concerned in it will need to take professional advice on legal matters sooner or later and it is important for those involved in management to establish relations of mutual confidence with a firm of solicitors well versed in rural business. Solicitors may be regarded as the general practitioners of the legal profession but through them it is possible to obtain the services and advice of specialists in advocacy and in specific areas of law; such specialists practise as barristers. Nevertheless some knowledge of the relevant law will help agriculturalists to assess when professional advice is necessary and also to alert them to the heavy duties laid upon them by the law.

These duties are such that the farmer will need not only the assistance of lawyers but of insurers and it is essential that the skills of a competent insurance broker are utilised in covering the risks involved. In this connection also the work of the industry's own organisations should be recognised, for example, the NFU, the CLA and the NUAAW. Not only do they advise and support their members in handling their own legal problems, but they also act as pressure groups that can lobby for alterations in the law and monitor changes proposed by government and by institutions of the EEC. Law is not static, if a sufficient number of people want it changed it can and will be changed.

The chapter contains an outline of the law as it relates to ownership, possession and occupation of agricultural land and to the liabilities of those who live by it as employers, employees and as self-employed persons. It does not seek to provide a comprehensive guide to all the rules and regulations that affect the industry but to show the general pattern of law as it affects agriculture and to point to the sources of fuller information and new decisions and enactments.

THE ENGLISH LEGAL SYSTEM

English law shares with our agriculture a long history, many of its features retain traces of medieval origin, but in recent times large accretions and alterations have shown it to be as lively and innovative as modern farming. Our law is very different in substance and procedure from continental European practice—although this is not to say that ideas of justice differ—and it should be noted that in the rulings of the Court of Justice of the European Communities there now exists a unifying factor of particular importance in relation to agricultural law. But agricultural law is not a separate part of English law; the same general principles apply in agricultural as in other areas. Thus an appreciation of the setting in which it occurs is necessary for its understanding.

Law consists of those rules of conduct that the courts will enforce and in this connection we include in the term 'court' all those bodies recognised by the judges as having an obligation to act judicially. So in addition to the civil and criminal courts we recognise various tribunals and other bodies set up by act of Parliament. Civil courts decide disputes between fellow citizens where one, the plaintiff, alleges that he has suffered injury or loss by the unlawful act or omission of another, the defendant. If the defendant is adjudged legally responsible for the injury he will be required to compensate the plaintiff, normally by a money payment called damages. When a state agency causes such an injury an action brought against it will also be a matter of civil law. The criminal courts decide cases where the state is involved as prosecutor against a citizen accused of committing a criminal offence. Accused persons, if found guilty, are punished by fine or imprisonment or both. The two sets of courts are kept separate; Magistrates Courts and the Crown Court hold criminal trials while County Courts and the High Court hear ordinary civil cases. Certain matters are however reserved for specialised tribunals of which the Agricultural Land Tribunal, the Lands Tribunal and Industrial Tribunals are particularly relevant to the agricultural industry. At the top of the hierarchy of courts the Court of Appeal and 'the House of Lords sitting as a court' provide an appeal structure with the possibility of reference to the European Court on questions of EEC law.

Most hearings in court are concerned with matters of fact, in criminal trials the prosecution will have to prove 'beyond all reasonable doubts' that the defendant has committed the crime of which he is accused; in civil trials the court will decide 'on the balance of probabilities' whether the events concerned took place as alleged by the plaintiff or not. But some cases also turn partly or wholly on questions of law—for example the question whether buildings used for the production and rearing of trout should be exempt from rating because they are agricultural buildings and as such de-rated. The courts recently decided that trout are not 'livestock', because this term covers only mammals and birds, and therefore such buildings must be rated, Cresswell v BOC Ltd (1980).

There is a subsequent history to this particular matter (*see below*) but it is to be hoped that when an agriculturalist wishes to know the law on a certain topic a source will already exist in clear terms and be readily available without the need for litigation. There are two principal sources of law—decided cases and statute. Case law has been built up into a system termed the common law because when a superior court makes a decision turning on a point of law that decision becomes a precedent binding on judges dealing with subsequent cases involving the principle. Statute consists of Acts of Parliament and Statutory Instruments, i.e. orders and regulations made under the authority of acts of Parliament by duly authorised Ministers. EEC Treaty provisions are incorporated into English law by virtue of the European Communities Act 1972 and that act also gives statutory force to Regulations passed by the EEC Council. EEC Directives normally require legislative action by our own Parliament before implementation but they are capable in certain circumstances of having direct effect. Decisions of the European Court also create precedents which our courts must follow.

A statutory provision overrules any common law precedents that directly conflict with it—thus the Local Government, Planning and Land (No. 2) Act 1980 has de-rated commercial fish farms, effectively setting aside the decision in Cresswell v BOC. Nowadays the definition of criminal offences and the powers of courts to penalise offenders depend almost exclusively upon statute and a vast range of legislation covers fiscal and commercial matters together with the social activities of government, e.g. housing, employment, public health, and social security. Even in the field of property law, contract and tort (*see below*), formerly the domain of common law, parliament has codified or amended rules developed in the courts.

Thus to find the law relevant to a particular topic it is necessary to know if statutory rules apply—for instance in respect of security of tenure for full-time farm workers occupying service cottages the position is governed by the Rent (Agriculture) Act 1976, and questions concerning the rights of the farmer and worker when employment comes to an end can only be resolved by reference to that act. When the topic is one that is covered by common law the decisions of relevant cases can be found in the Law Reports where decisions of significance are recorded. In practice, a legal practitioner will depend upon books of reference and texts dealing with specific areas of the law to enable him to find the statutes and cases that are relevant. The agriculturalist must know about changes in the law relating to his business. Details of many new statutory rules are publicised by government agencies, e.g. on employment law by the Department of Employment, on safety regulations by the Health and Safety Commission, but to keep up to date a farmer should read the professional journals. Criminal offenders are liable to punishment, those committing civil wrongs to pay damages or to have their activities stopped by an order termed an injunction made by a court. These are the means by which the law is enforced. By the standards of other legal systems enforcement of civil judgements in England is reasonably effective. Debtors may have their property sold up, bankrupt persons are subject to severe business disabilities. Flouters of injunctions may be imprisoned for contempt. Although legal procedures can be protracted and nothing is to be gained by suing an impecunious defendant the farmer should be prepared, when it is businesslike to do so, to assert his legal rights just as much as he should be careful to fulfil his legal duties.

LEGAL ASPECTS OF THE OWNERSHIP, POSSESSION AND OCCUPATION OF AGRICULTURAL LAND

Most agricultural enterprises are run by owner–occupiers or tenants of agricultural holdings. In many cases these will be single individuals or partnerships and the persons involved will be legally responsible for any obligations that arise. Where the enterprise is run by a limited company the company is recognised as a legal person with rights and duties separate from those of its members. The directors of an enterprise organised in this way must however recognise their obligations under the Companies Acts which contain provisions designed to prevent them from using the advantages of corporate personality to defraud creditors, members and employees of the company.

English law recognises only two ways by which land may be held: freehold and leasehold. If it is desired to tie up freehold land in the ownership of succeeding generations within a family this can only be done by the creation of a trust set up in accordance with certain rules—known as Equity—developed in the Court of Chancery, now a part of the High Court. This definitely requires professional expertise.

A freeholder has the largest possible freedom to decide how to use his land that the law recognises. It has never been possible for a freeholder to do whatever he pleases to the detriment of neighbours but in recent times to the limits already imposed by the law of nuisance have been added the constraints of town and country planning and the compulsory acquisition of land. Public and private rights of way may also diminish a freeholder's privacy and land can be lost by adverse possession. A freeholder's rights may be reduced by restrictive covenants and his obligations increased by a mortgage. Coal and oil found under the soil do not belong to the freeholder but to the state.

The freedom of landlords and tenants to negotiate the terms on which property is let has been greatly circumscribed by statute in recent years—in no area more so than in agriculture—and this is true not only as between landowner and tenant farmer but between farmer and farm employee. An account of the present position is given below but it should be emphasised that important regulations are made from time to time adjusting the details. It is therefore unwise to rely on the text of an Act without checking for recent amendments.

Both the owner–occupier and the tenant farmer have duties as occupiers towards neighbours, visitors and trespassers. These duties are in general the result of principles developed from case law but they also arise as a result of Public Health Acts and from Health and Safety legislation.

Farm tenancies

The law relating to the letting of agricultural property differs from the general law on leaseholds. It is therefore appropriate to consider it in more detail. In the first half of the twentieth century the normal agricultural tenancy developed under the common law was a tenancy from year to year terminable at six months' notice by landlord or tenant. Statutory provisions have now changed this law radically.

Security of tenure

Security for tenant farmers was first introduced as a permanent measure in the Agricultural Holdings Act 1948 by placing restrictions on the landlord's right to give notice to quit to his tenant. These provisions as subsequently amended have been re-enacted in the Agricultural Holdings (Notice to Quit) Act 1977. If the landlord serves a 12 months' notice to quit and the tenant does nothing, the notice will be legally effective, but the tenant may, within one month, serve a counter-notice if he does not wish to go. In this case the landlord's notice will not have effect unless the Agricultural Land Tribunal consents to its operation.

The tenant will be deprived of security of tenure only on a limited number of grounds, i.e. that the purpose for which the landlord requires possession is:

(1) in the interests of good husbandry;
(2) in the interests of sound estate management;
(3) in the interests of agricultural research or education, or for the provision of small-holdings and allotments; or
(4) that greater hardship will be caused by the withholding than by the granting of consent.

If the Tribunal finds one or more of these grounds proved, it has nevertheless to refuse consent 'if in all the circumstances it appears to them that a fair and reasonable landlord would not insist on possession'.

The tenant's right to serve a counter-notice is excluded by a strictly limited list of cases covering consent by the Tribunal for the reasons given above, failure to remedy a breach of the contract of tenancy or bankruptcy of the tenant, or bad husbandry, or failure to pay rent due by the tenant. Under the 1977 Act there are special rules regarding notices to remedy breaches of the tenant's obligations to keep fixed equipment including hedges, ditches, roads and ponds in good order and the tenant may go to arbitration if he disputes the landlord's claim. An arbitrator can be chosen by agreement of the parties or one of them can ask the Ministry of Agriculture, Fisheries and Food to appoint one from his panel.

Until the Agriculture (Miscellaneous Provisions) Act 1976 the death of a tenant enabled the landlord to serve an incontestable notice to quit. Under Part II of that Act, however, security of tenure can now be claimed for up to three generations including the first occupier. Eligible persons can apply to the Tribunal for a tenancy of a holding whose tenant has died. To be eligible a person must be the widow, widower, brother, sister or child, natural or adopted, of the deceased, have derived his or her livelihood from agricultural work on the holding for five years out of the previous seven (up to three years spent at college or university will count) and not be the occupier of another viable commercial unit. An eligible applicant must also be found by the Tribunal to be suitable to take over the holding; suitability is judged by the agricultural experience, age, health and financial standing of the applicant.

If a landlord serves a notice to quit within three months of the death of a tenant and if no application is made by a suitable successor this will end the tenancy. If an application is made the landlord can dispute the application on grounds of unsuitability or ineligibility of the applicant or on grounds of good enate management or hardship but the 'fair and reasonable landlord test' still applies.

Many applications have been made since 1977 and a large body of decisions has already been built up including two judgements of the House of Lords, Jackson v Hall (1980) 1 All ER 177, which decided that an applicant could not avoid ineligibility by disposing of a commercial unit after the tenant's death and Williamson v Thomson and Carpenter (1980) 1 All ER 177, which decided that an applicant who occupied a commercial unit as a partner could not get a succession tenancy, as 'occupier' in the Act does not mean 'sole occupier'.

Obligations of landlord and tenant

The landlord and tenant of an agricultural holding will be bound by the terms of their agreement but there has been considerable intervention by statute so as to modify and extend their contractual obligations. Either party can insist on a written agreement and the terms may be fixed by arbitration. The parties can fix the rent at the start of the agreement and change it at any time by agreement. The 1948 Act provides for the rent reviews at three year intervals and either party can demand an arbitration to fix it. Although the parties may agree their respective maintenance and insurance obligations 'model repair clauses' are set out in the Agriculture (Maintenance, Repair and Insurance of Fixed Equipment) Regulations published in 1948 and 1973. Improvements undertaken by the landlord with the tenant's agreement may lead to an increase in rent. If the tenant carries out improvements he will be entitled under certain circumstances to compensation at the end of the tenancy in accordance with the 1948 Act and consequent regulations. The Act also deals with the tenant's right to remove fixtures and the landlord's right to purchase them. When a tenancy comes to an end there will normally be a settlement of claims as between landlord and tenant, by arbitration if necessary, for disturbance and delapidations.

In general it may be said that a tenant farmer has freedom to crop a holding as he sees fit despite contrary indications in the tenancy agreement. There are, however, limits on a tenant's freedom in the last year of a tenancy when items of manurial value may not be sold or removed from the holding and the tenant may not establish a cropping scheme at any time that is not in accordance with the practice of good husbandry.

Occupation of agricultural land without security of tenure

If prior approval is given by the Minister of Agriculture, which will only be done for special reasons, an agreement for a letting for a specified period will be terminable when the period expires. By a quirk of the 1948 Act a tenancy for more than one year but less than two does not have statutory protection, nor does a short grazing agreement. Thus an agreement for grazing or mowing provided it is for a period of less than a year gives no protection to the taker but it must not permit ploughing or re-seeding or use of buildings unless they are simply shelters for grazing stock. Permission given to an employee to use agricultural land by a landlord employer during the period of employment does not create an agricultural lease. Partnerships between farmers and owners of agricultural land for agricultural business purposes do not create tenancies in favour of farmers. It must, however, be remembered that a partner is liable for his fellow partner's debts while a lease creates no such liability as between landlord and tenant.

Service occupation agreements

When houses are provided for employees by an owner–occupier or a tenant farmer the arrangement between them will normally be a service agreement acceptable to the parties. In accordance with the Protection from Eviction Act 1977 an employee other than a full-time farm worker will have six months' security of tenure when his contract of employment ends, subject to recourse to the County Court by the employer if the latter has urgent need of the accommodation. A full-time farm worker however may have rent deducted from his wages only as laid down by the current Agricultural Wages Board Order and has security of tenure of his house under the Rent (Agriculture) Act 1976. Provision is made in the Act, which otherwise gives the farm worker the same protection as that enjoyed by statutory tenants under the Rent Act 1977, for the County Court to grant an order for the employer to recover possession where suitable alternative accommodation is provided by a housing authority. Local housing authorities are obliged under the 1976 Act to rehouse the outgoing worker when the employer needs the house for another farm worker on grounds of agricultural efficiency and on this question the employee or the local authority can obtain the advice of an Agricultural Dwelling House Advisory Committee.

Occupier's liability

Any person recognised by the law as the occupier of land owes duties towards persons who enter upon it and towards those who are in its vicinity. These duties are imposed by the law of tort which covers wrongs caused by a person's failure to carry out duties imposed by law as contrasted with duties imposed by a contract. The common law has evolved a number of such duties which have been recognised in judicial decisions as falling within the categories of trespass, nuisance, negligence and strict liability to which further rules have been added by statute.

A farmer is more likely to suffer from trespassers against whom an action for damages or an injunction may be brought, than to be a trespasser himself. However, if his animals trespass onto neighbouring land he will be liable for the damage they cause unless it was the fault of his neighbour or another person and could not have been reasonably anticipated. In this connection it should be remembered that a farmer is responsible for keeping his own stock in; he cannot complain if his animals escape through his neighbour's fence. Much of the law on liability for animals is covered by the Animals Act 1971 which also changed the rules about animals straying onto or off the highway. A person who negligently allows this to happen is now liable for damage done. But there is no duty to fence in animals grazed on common land by those who have the right to do so if it is customarily unfenced. The Act also permits a farmer to shoot dogs worrying livestock. However, he must notify the police within 48 h in order to have a defence if he is sued by the dog's owner.

Trespassers enter upon other people's land at their peril; it might be thought, therefore, that a farmer would be under no duty to take care to prevent them suffering harm while on his land. But as a result of recent decisions, particularly British Railways Board v Herrington (1972) AC 877 and Pannett V.P.G. Guinness & Co. Ltd. (1972) 2 QB 599, a farmer would be expected to take care to prevent child trespassers from encountering hazards, especially when the hazards are attractive to children and children may be expected to approach them. So far as adult trespassers are concerned a notice warning of known hazards is probably sufficient.

Persons entering land lawfully, as guests or for payment or because they have a statutory right of entry (e.g. Health and Safety Inspectors) are owed a duty by the occupier defined in the Occupiers Liability Act 1957 as 'a duty to take such care as in all the circumstances of the case is reasonable to see that the visitor shall be reasonably safe in using the premises for the purposes for which he is invited or permitted by the occupier to be there'. The Act permits the occupier to 'restrict, modify or exclude' the common duty of care by contract or adequate notice but since the passing of the Unfair Contract Terms Act 1977 it is normally impossible for an occupier to exclude liability for personal injury caused by his own negligence.

If a farmer's methods create a nuisance such as excessive smells, noise or pollution then a neighbouring owner or occupier can bring an action for damage,

and for an injunction to prevent repetition. Country dwellers are expected by the common law to put up with a reasonable degree of inconvenience arising from agriculture but intensive livestock practices have resulted in several successful cases against farmers in recent years. When a nuisance interferes with the rights or convenience of the public or a section of it then it is a public nuisance. An individual who suffers more than most can sue but the perpetrator is open to prosecution and local authorities have powers under the Public Health Acts 1936 and 1961 and the Control of Pollution Act 1974 to order abatement through the Magistrates' Courts. Decayed buildings, contaminated water sources, cesspools, rubbish dumps and waste land all fall under the rules in the Public Health Acts and there are specific provisions in the Refuse Disposal (Amenity) Act 1978 that give local authorities power to deal with abandoned vehicles and other rubbish. There is no liability at common law for the normal spread of weeds or wild pests, but the Ministry of Agriculture can take proceedings under the Weeds Act 1959 and the Pests Act 1954 against the occupier of land harbouring them.

As a result of the leading case of Rylands v Fletcher (1868) LR 3 HL 330 an occupier is strictly liable for the escape of any potentially dangerous thing kept on his land if it escapes and injures neighbouring property or people. Defences to such a claim are very limited and the occupier is liable even if the escape was caused by a contractor. The liability is, however, limited to non-natural uses of land and this probably means that things commonly brought onto agricultural land for agricultural purposes give rise to liability only if released negligently.

A farmer will cover his liabilities in tort as occupier by insurance but it should be remembered that insurance covers only for legally enforceable claims. Agriculture as an industry has a bad record so far as accidents are concerned and the utmost vigilance is required for the law does not attempt to compensate for the incompensatable such as the death of a child drowned in a slurry pit and damages cannot make good injury caused to an active adult crushed under a runaway machine.

Employer's liabilities

A farm enterprise has the same legal obligations towards its employees as any other business. The law on contracts of employment, job security and safety has recently become increasingly technical and although its main features are indicated below, farm management will require more detailed information—which must be kept up to date—than is here outlined.

The obligations which a farmer undertakes as an employer—whether towards an employee or other people—will arise only when there is a contract of employment between him and the employee. When work is done by a contractor the employer will not normally be responsible for wrongs done by the contractor unless on express instructions from the employer. If a third party, such as a road user run down by a negligent tractor driver, suffers injury because of the act of an employee done in the course of employment then the employer will be vicariously liable for his employee's wrongful act. The employee will be liable personally but this will usually not help the employer greatly as the injured person will almost certainly proceed against him as he, the employer, is likely to be insured against claims.

In fact, while it is not compulsory for an employer to be insured against claims by non-employees, he is obliged by the Employers Liability (Compulsory Insurance) Act 1969, to insure against liability for injury or disease sustained by his employees in the course of employment in Great Britain. Such injury will, more often than not, come about because of the negligence of another employee.

A contract of employment, known as a contract of service, is often hard to distinguish from a contract for services such as those rendered by a contractor but most of the modern employment Acts only apply to the former. A farmer has a duty to take reasonable care in choosing a contractor of repute but there it ends. With 'labour only' contracts, on a relief milking scheme for example, the distinction between employee and contractor may be a fine one; the law however looks at the reality of the situation and not at the words used by the parties to describe themselves.

There are statutory restrictions on the recruitment of employees and contract workers, notably the Sex Discrimination Act 1975 which makes it unlawful for employers and managers to discriminate against women or men or married persons when advertising for, or engaging employees, and the Race Relations Act 1976 which makes it unlawful to discriminate on grounds of colour, race, ethnic or national origins. The first Act does not apply to firms where five or fewer persons are employed but the second applies universally. This legislation has to be taken into account also when promotion or redundancy is considered.

Under the Protection of Employment (Consolidation) Act 1978 an employee is entitled to be given a written statement containing specified particulars of the terms and conditions of employment. It must contain details of pay, hours of work, holidays, pensions, sick pay, job description and grievance procedures. Any changes made must be notified. Personal correspondence is not necessary, for example the current Agricultural Wages Orders can be displayed to inform staff of changes in rates of pay. Other statutory obligations of an employer include the giving of time off for certain purposes, e.g. pregnancy, trade union activities and public duties and, so far as agricultural workers are concerned, the payment of wages at or above the rates laid down under the Agricultural Wages Act 1948.

When a contract of employment ends, in the absence of prior arrangements, the common law provides only for summary dismissal of an employee for misconduct or the giving of reasonable notice. Because of the one-sidedness of this position legislation has been passed to give employees the right to minimum periods of notice, redundancy payments and to compensation or reinstatement if unfairly dismissed. First introduced by the Contracts of Employment Act 1963, Redundancy Payments Act 1965 and Industrial Relations Act 1971 the rules have been constantly revised so that the current legislation is to be found in the Employment Protection (Consolidation) Act 1978, as implemented by subsequent Orders and amended by the Employment Act 1980. In effect an employee under retirement age is entitled to notice according to length of service, to redundancy payments in accordance with age and length of service and to remedies for unfair dismissal when the employer dismisses for an inadequate reason or in an unfair manner, e.g. gives the employee little warning or opportunity to justify his actions. When a farmer takes on with a farm its previous owner's or tenant's workers he will also take on responsibility for their previous service on that farm so far as notice and redundancy payments are concerned, Lloyd v Brassey (1969) 1 All ER 382. A farmer who has given recognition to a trade union must consult in advance about redundancies. The unfair dismissal provisions apply to employees of over one year's standing except for new employees in firms with 20 employees or less who only qualify after two years. All disputes arising under this statute are heard by Industrial Tribunals and, despite some folklore to the contrary, tribunals take account of the practicalities of agricultural employment and employers who take care to act with circumspection and follow their own disciplinary rules are seldom penalised.

All employers and employees owe each other mutual duties during the course of employment; the employee to give faithful service and the employer to take reasonable care for the employees' safety. This common law duty has been reinforced by statute, notably the Health and Safety at Work Act 1974 which imposes general obligations on employers to ensure as far as is reasonably practicable that persons in their employment are not exposed to risks to their health and safety. The Act imposes a similar duty on employees and the self-employed with regard to themselves, fellow workers and other persons including children. Under this and earlier acts many Regulations have been made covering inter alia Stationary and Field Machinery, Workplaces, Tractor-cabs, Power Take-offs, Circular Saws and Poisonous Substances. The act empowers Health and Safety Inspectors to enter farms and issue Improvement or Prohibition Orders in respect of dangerous equipment.

Contravention of the Act may lead to prosecution, fine and imprisonment of both owners, and managers of defaulting firms; moreover an employee injured on account of an employer's failure to provide safe working conditions or equipment will be able to sue for damages. Under the Employers Liability (Defective Equipment) Act 1969 it will be no defence for an employer to plead that he purchased faulty equipment from a reputable manufacturer if the manufacturer was negligent. Every business is required to have a Safety Policy and in general the law not only expects employers to provide employees with healthy and safe conditions of work but to take reasonable steps to see that safety equipment such as protective clothing is used and safe procedures followed.

CONCLUSION

Many more aspects of law than those mentioned so far have effects on agriculture and agricultural businesses. Farmers, as business men, are closely affected by the law of contract, sale of goods, consumer protection and fair trading and, in their particular profession, by rules concerning agricultural cooperatives, animal health and poisonous wastes, to name some of the more obvious. In addition, there are the regulations on competition, monopolies, free movement of goods and the Common Agricultural Policy under the Treaty of Rome.

It is difficult to point to any area of agricultural law today and declare that it is likely to stand still. Certainly as economic, ethical and recreational pressures grow the significance of law in its impact on agriculture will increase and ignorance of it will be no excuse.

Further reading

As a general introduction to the English legal system and the general principles of English law:

Introduction to English Law (1980), 10th edition, P. S. James, London: Butterworths.

For an up-to-date coverage of commercial law:
Commercial Law (1980), 5th edition, G. Borrie. London: Butterworths.

On labour law:
Selwyn's Law of Employment (1980), 3rd edition, N. M. Selwyn. London: Butterworths.

On most aspects of farm tenancies, land ownership and occupation:
Essential Law for Landowners and Farmers (1980), 1st edition, M. Gregory and M. Parrish. London: Granada Publishing.

For basic knowledge of EEC law:
Law and Institutions of the European Communities (1976), 2nd edition, D. Lasok and J. W. Brodge. London: Butterworths.

and for further detail:
The Substantive Law of the EEC (1980), 1st edition, D. Wyatt and A. Dashwood. London: Sweet and Maxwell.

30

Health and safety in agriculture

J. Weeks

INTRODUCTION

It is beyond dispute that work in agriculture is more hazardous than in the days when the horse was not only the source of power but controlled the speed of farm operations. The urgent need for home produced food in the Second World War began what now can be seen as an industrial revolution bringing a massive increase in mechanisation, the ready availability of electricity on every farm and an unprecedented growth in the use of pesticides. Unfortunately while these mechanical and chemical aids have taken much of the hard slog out of farm jobs they have brought increased risks to those who work in the industry.

Apart from the Threshing Machines Act 1878 and the Chaff Cutting Machines (Accidents) Act 1897 – both now repealed – the first significant legislation for the health and safety of workers in agriculture was the Agriculture (Poisonous Substances) Act 1952 followed by the Agriculture (Safety, Health and Welfare Provisions) Act 1956. The former enabled regulations to be made to protect workers from the risk of poisoning when using certain specified substances. The second Act had a much wider application relative to accidents and health hazards on the farm including the provision of sanitary and washing facilities for workers and safeguards for young persons when lifting weights. Regulations subsequently were made to protect workers against bodily injury or injury to health arising out of the use of machinery, for providing a safe place of work and for the avoidance of accidents to children.

In 1972 the Committee on Safety and Health at Work under the chairmanship of Lord Robens published its Report containing a number of recommendations for improving the health and safety of persons at work. From the Robens Report stemmed the Health and Safety at Work, etc. Act 1974 which came into force on 31 July 1974. The Act covers all persons at work (except domestic servants in private households) and includes employers and the self employed. It is estimated that something like 6 million additional workers—including many in agriculture—are now covered by health and safety legislation for the first time. Furthermore the Act protects the health and safety of the general public who may be affected by work activities.

THE HEALTH AND SAFETY AT WORK, ETC. ACT 1974

The Act comprises four parts:

Part 1 dealing with health and safety and welfare in relation to work;
Part 2 relating to the Employment Medical Advisory Service;
Part 3 amends the law relating to building regulations and is the responsibility of the Secretary of State for the Environment;
Part 4 contains a number of miscellaneous and general provisions.

Its purpose is to provide the legislative framework to promote, stimulate and encourage high standards of health and safety at work. The Act is an enabling measure and is superimposed over existing health and safety legislation including the 1952 and 1956 Acts and their regulations. All earlier legislation remains in force until replaced by revised and updated provisions in the form of regulations and approved Codes of Practice prepared in consultation with industry so as to create an integrated body of requirements enforced on a common basis. An early example of revision was the replacement of the

641

Poisonous Substances Regulations made under the 1952 Act by the Health and Safety (Agriculture) (Poisonous Substances) Regulations 1975 imposing obligations on employers, employees and for the first time the self employed.

Approved Codes of Practice may either supplement or be an alternative to regulations where appropriate. They have a special legal status in that their requirements are not themselves statutory but may be used in criminal proceedings as evidence that statutory requirements have been contravened.

Fire prevention and precautions are referred to in Part 4 Miscellaneous and General Provisions of the 1974 Act which amended the Fire Precautions Act 1971 to provide for the general fire precautions at the majority of places of work to be dealt with by the fire authorities and the Home Departments. The Health and Safety Commission and its Executive will, however, remain responsible for control of 'process' risks, i.e. risks of outbreak of fire associated with particular processes or particular substances.

The Act provides for a Health and Safety Commission consisting of a full-time Chairman and up to nine part-time members representing the Confederation of British Industry, the Trades Union Congress and local authorities. The Commission is appointed by the Secretary of State for Employment and has taken over from Government Departments including the Ministry of Agriculture, Fisheries and Food and the Department of Agriculture for Scotland the responsibility for developing policies in the health and safety field. It submits to Government proposals for new or revised regulations and approved Codes of Practice and keeps under review the adequacy of legal requirements generally for health and safety. The Commission has general oversight of the work of the Health and Safety Executive and has power to delegate to that Executive any of its functions.

The Health and Safety Executive is the operative arm of the Commission and is responsible for advising on and enforcing the relevant statutory provisions including earlier and existing legislation. The Executive is headed by a Director General who is assisted by a Deputy Director and a third member. The safety inspectorates for the different industries, including agriculture, form part of the Executive instead of being scattered and working independently within several Government Departments as previously. They have readily available to them the support of the Employment Medical Advisory Service and of groups of specialists dealing across the board in such subjects as dust, noise, chemicals and asbestos.

The Employment Medical Advisory Service (EMAS) is the medical arm of the Health and Safety Commission and the main channel of medical advice to the Commission and to the inspectorates in the Health and Safety Executive. The Service gives advice over the whole range of the Commission's activities including assisting the Agricultural Inspectorate in their investigations into incidents at work involving pesticides. It is involved also with the inspectorates in occupations for which there was no previous statutory responsibility, e.g. the health and safety of persons employed in laboratories and research establishments who are now covered by the 1974 Act. The EMAS continue to give medical advice to a number of training, employment and rehabilitation agencies.

HM Agricultural Inspectorate is headed by the Chief Agricultural Inspector and covers England, Wales and Scotland. It has a total strength of approximately 200 inspectors working full time on health and safety of which a small number are at headquarters and the remainder work in the field. All have a wide experience of farming and farm mechanisation. Inspectors are mainly accommodated in the same premises as other inspectors of the Health and Safety Executive but with some additional offices in rural localities where an Executive office would not otherwise be justified. The Agricultural Inspectorate maintains a close liaison with the Ministry of Agriculture, Fisheries and Food and the Department of Agriculture for Scotland including the Agricultural Development and Advisory Service at Headquarters, regional and divisional offices and the Scottish Colleges of Agriculture.

The Agriculture Industry Advisory Committee (AIAC) is one of a number of committees set up by the Health and Safety Commission to provide a direct and permanent source of advice from the various industries. The Chairman of the AIAC is HM Chief Agricultural Inspector and its 12 members are drawn from both sides of the agricultural industry. Other people with a wide knowledge and experience of the industry are co-opted and MAFF and DAFS are represented by assessors. The terms of reference of the Committee are to consider and advise the Commission on:

(1) the protection of people at work from hazards to health and safety arising from their occupation within the agricultural industry and the protection of the public from related hazards arising from such activities; and
(2) other associated matters referred to them by the Commission or the Executive.

SOME OF THE STATUTORY REQUIREMENTS OF THE 1974 ACT

In addition to the need to comply with earlier and continuing legislation the 1974 Act imposes new responsibilities—general duties—on employers and others.

Employers have a general duty under the Act to ensure, so far as is reasonably practicable, the health, safety and welfare at work of all their employees by:

(1) maintaining safe systems of work;
(2) ensuring the safe use, handling, storage and transport of articles and substances;
(3) providing adequate instruction, training and supervision;
(4) maintaining safe premises and other places of work;
(5) providing a safe working environment and adequate welfare arrangements.

Employers are required also to prepare a written statement of their general policy, organisation and arrangements for health and safety at work, to keep it up to date by revision and to bring the content to the notice of employees.

Employers should not charge employees for anything provided to comply with specific health and safety requirements designed for their protection, e.g. the provision of protective clothing when using specified poisonous substances.

Recognised trade unions may appoint safety representatives to represent the employees.

The intention is that employers should look at the conduct of their undertaking as a whole to ensure so far as is reasonably practicable the health and safety of their employees, and that other people, such as contractors and visitors to the farm as well as the general public (including children) are not affected adversely by their activities.

Self-employed persons (including employers) need to ensure, so far as is practicable, that their activities do not endanger others and they are required to provide information to the public about a potential hazard to health and safety. Like employees, the self-employed also have a duty in respect of their own health and safety. The intention here is that the self-employed should protect themselves as well as others.

Employees are required to take reasonable care to avoid injury to themselves or to others by their work activities and to cooperate with employers and others in meeting statutory requirements. The aim is that employees—like the self-employed (including individual employers)—must observe the requirements of the law and positively exercise reasonable care to ensure the health and safety of themselves and others. The Act also requires employees not to interfere with or misuse anything provided to protect their health and safety or welfare in compliance with the Act.

Persons concerned with premises have a duty, in relation to persons who are not their employees but use non-domestic premises as a place of work or use plant or substances provided for their use, to ensure, so far as is reasonably practicable, that the premises including access and egress and any plant or substance provided is safe and without risks to health.

Designers, manufacturers, importers, suppliers and installers of machinery, equipment or substances for use at work, must ensure that, so far as is reasonably practicable, they are safe when properly used. They are required to test articles for safety in use, or arrange for this to be done by a competent authority. They must also supply information about the use for which an article or substance is designed, and include any conditions of use regarding its safety. Similarly anyone who installs or erects any article for use at work must ensure that, so far as is reasonably practicable, it does not constitute a risk to health and is safe for use. This section of the Act places the responsibility on the manufacturer to build safety into a product during manufacture, on the dealer to provide information of the use for which the machine or substance was designed and the installer to ensure that, so far as is reasonably practicable, the machine when installed is safe and without risks to health when properly used.

Enforcement

Should an inspector discover a contravention of one of the provisions of the existing Acts or regulations or a contravention of the 1974 Act and its regulations he can:

(1) issue a *prohibition notice* if there is a risk of serious personal injury, to stop the activity giving rise to this risk, until the remedial action as supplied in the notice has been taken. The notice can be served on the person undertaking the activity or on the person in control of it at the time;
(2) issue an *improvement notice* if there is a legal contravention of any of the relevant statutory provisions to remedy the fault within a specified time. This notice should be served on the person who is considered to be contravening the legal provisions, or it can be served on any person on whom responsibilities are placed, whether it is an employer, or an employed person, or a supplier of equipment or materials;
(3) *prosecute* any person contravening a relevant statutory provision instead of or in addition to serving a notice.

Contravention of some requirements can lead to prosecution summarily in a Magistrate's Court and for other either summarily or on indictment in the Crown Court in England and Wales or the Sheriff Court in Scotland. The maximum fine, on summary conviction for most offences, is £1000. There is no limit to the fine on conviction on indictment, and imprisonment up to two years can be imposed for certain offences. If a person on whom an Improvement or Prohibition Notice is served fails to comply with it he is liable to prosecution, and failure to comply with a Prohibition Notice could lead to imprisonment. A person on whom a Notice is served may appeal against the Notice, or on any terms of it, to an Industrial Tribunal.

REGULATIONS MADE UNDER THE 1974 ACT RELEVANT TO AGRICULTURE

The Health and Safety (Agriculture) (Poisonous Substances) Regulations 1975

These regulations made under the 1974 Act prescribe the precautions to be taken including the wearing of protective clothing when poisonous substances specified in the regulations are used in agriculture. They apply to employers and employees and to self-employed persons.

Although not made under the 1974 Act the following legislation is important to the safe use of pesticides:

The Farm and Garden Chemicals Regulations 1971

These regulations, made under the Farm and Garden Chemicals Act 1967, require the name of a substance to be clearly written on the label and apply to retail sales of products sold for use in Great Britain as weed killers, pesticides or growth controllers on farms or in gardens.

Deposit of Poisonous Waste Act 1972

The Act makes it an offence to deposit on land poisonous, noxious or polluting waste such as to give rise to an environmental hazard.

The Packaging and Labelling of Dangerous Substances Regulations 1978

The regulations implement for Great Britain the provisions of an EEC Directive for the classification, packaging and labelling of dangerous substances.

The Notification of Accidents and Dangerous Occurrences Regulations 1980

Impose responsibilities on employers to report accidents resulting in the death of or major injury to persons at work. Dangerous occurrences which are defined in the regulations are required to be similarly reported.

The Safety Signs Regulations 1980

The regulations provide that safety signs for persons at work and colours in strips identifying places where there is danger to their health or safety shall comply to BS 5378: Part 1: 1980.

THE AGRICULTURE (SAFETY, HEALTH AND WELFARE PROVISIONS) ACT 1956 AND REGULATIONS

This Act—much of which has now been repealed—places obligations on an employer to provide suitable and sufficient washing facilities and sanitary conveniences. Employers are also required to ensure that a worker under 18 years of age does not lift, carry or move a load so heavy as to be likely to cause him injury.

The Agriculture (First Aid) Regulations 1957

These prescribe the prescriptions and quantities of first aid requisites which must be provided by an employer.

The Agriculture (Power take-off) Regulations 1957

Prescribe the obligations of employers of workers for the guarding and safe use of the power take-off and power take-off shaft on agricultural tractors and machines.

The Agriculture (Avoidance of Accidents to Children) Regulations 1958

The regulations make it illegal to allow children under the age of 13 to drive or ride on agricultural tractors and self-propelled machines and to ride on machines and implements.

The Agriculture (Circular Saws) Regulations 1959

Set out the requirements for the guarding, maintenance and operation of circular saws. Workers under the age of 16 may not operate or assist at a circular saw and workers between 16 and 18 may do so only when supervised by an experienced person over 18 years of age.

The Agriculture (Safeguarding of Workplaces) Regulations 1959

Prescribes safety requirements in places where agricultural workers are employed and cover the construction and maintenance of floors and stairways, the provision of handrails and the guarding of apertures in floors, walls and the edges of floors.

The Agriculture (Stationary Machinery) Regulations 1959

Lay down guarding and safety requirements for stationary machinery.

The Agriculture (Lifting of Heavy Weights) Regulations 1959

Prescribe that the maximum weight of a sack or bag and its contents which a worker may lift or carry unaided is 180 lb. The regulations do not affect the provision of the 1956 Act that a young person under the age of 18 years may not be allowed to lift, carry or move a load so heavy as likely to cause an injury.

The Agriculture (Threshers and Balers) Regulations 1960

Require the guarding of stationary threshers, hullers, balers and trussers.

The Agriculture (Field Machinery) Regulations 1962

These regulations provide for the guarding and the provision of safety devices on farm machines which are not stationary machines and include trailers and power driven hand tools.

The Agriculture (Tractor Cabs) Regulations 1974

Regulations were introduced in 1967 for the fitting of an approved safety cab or frame to all new wheeled tractors weighing 11 cwt or more sold after 31 August 1970 for use in agriculture. The 1974 regulations required from 1 June of that year that all such new tractors be fitted with a protective cab in which the noise level at the driver's ear does not exceed 90 dB(A). From 1 September 1977 all tractors are required to be fitted with an approved protective cab or frame when driven by a worker.

CONCLUSION

The prevention of ill health and accidents at work is a complex problem. The word accident is often used to imply that an accident was unavoidable because all the circumstances could not be foreseen and were beyond a person's control. Accidents do not happen; they are caused. They can be prevented providing the causes are analysed and positive steps taken to eliminate those causes. It is not just a question of providing the safeguards required by legislation but setting down systems of work for the various operations and getting others to carry them out. It was the Robens Report which pointed out that a necessary ingredient for success is the development of a greater health and safety awareness in management who create risks and by the employees who have to work with them.

The Agricultural Inspectorate enforce and advise on the requirements of the Health and Safety at Work, etc. Act 1974 and the various regulations specific to the industry. Their objective is to reduce the number of accidents to an acceptable minimum and to improve the farm working environment. Although some progress has been made much remains to be done. The fitting of safety cabs on tractors, improved machinery design including better operator comfort, increased automation and the use of less toxic pesticides have all helped toward the desired objective. These must continue but at the same time there is a need for greater awareness of health and safety by those who work on farms. No farm programme can be complete unless full consideration is given to the elimination of risks to the health and safety of those who have to carry out the different operations and of the public, including children, who may be affected by them. This is a management function. The benefits to be gained extend far beyond the prevention of health risks and of accidents. They include fewer stoppages, reduced absence from work, a happier and healthier workforce and greater overall efficiency.

This brief description of health and safety in agriculture in Great Britain is by way of guidance only and is not intended to be an authoritative interpretation of the Acts and regulations, copies of which can be obtained from HMSO or through any bookseller. In addition a wide range of leaflets and guidance notes explaining and giving advice on the regulations and accident prevention generally are available from HM Agricultural Inspectors in their local offices.

References

Safety and Health at Work—Report of the Committee 1970–72 Cmnd 5034

The Health and Safety at Work, etc. Act 1974 Chapter 37

The Health and Safety (Agriculture) (Poisonous Substances) Regulations 1975 SI 1975 No 282

The Farm and Garden Chemicals Regulations 1971 SI 1971 No 729

The Deposit of Poisonous Waste Act 1972 Chapter 21

The Employers Health and Safety Policy Statements (Exception) Regulations 1975 SI 1975 No 1584

Safety Representatives & Safety Committees Regulations 1978 SI 1977 No 500

The Notification of Accidents and Dangerous Occurrences Regulations 1980 SI 1980 No 804

The Safety Signs Regulations 1980 SI 1980 No 1471

The Agriculture (Safety, Health and Welfare Provisions) Act 1956 Chapter 49

The Agriculture (First Aid) Regulations 1957 SI 1957 No 940

The Agriculture (Ladders) Regulations 1957 SI 1957 No 1385

The Agriculture (Power take-off) Regulations 1957 SI 1957 No 1386

The Agriculture (Avoidance of Accidents to Children) Regulations 1958 SI 1958 No 366

The Agriculture (Circular Saws) Regulations 1959 SI 1959 No 427

The Agriculture (Safeguarding of Workplaces) Regulations 1959 SI 1959 No 428

The Agriculture (Stationary Machinery) Regulations 1959 SI 1959 No 1216

The Agriculture (Lifting of Heavy Weights) Regulations 1959 SI 1959 No 2120

The Agriculture (Threshers and Balers) Regulations 1960 SI 1960 No 1199

The Agriculture (Field Machinery) Regulations 1962 SI 1962 No 1472

The Agriculture (Tractor Cabs) Regulations 1974 SI 1974 No 2034

31

Agricultural computers

C. L. Pugh
with introduction by J. F. Birtles

Introduction

J. F. Birtles

FOREWORD

Much of this chapter is by definition associated with 'in-house' equipment. It is however generally accepted that only a relatively small proportion of farms and estates can alone justify or indeed are likely to instal their own computer within the next few years. A short section has therefore been included to cover in outline the likely future activities of computing bureau companies providing services to the agricultural sector.

THE ROLE OF THE COMPUTING SERVICES INDUSTRY

A difficulty is currently being encountered by computing services when discussing their requirements with farmers. There exists at the present time an artificially high level of expectation by the agricultural community on the practical and economic computing alternatives that are immediately available to them.

It is to be hoped that this problem, which is particularly associated with farms whose size, scale of operations and other practical constraints effectively excludes the operation of an in-house installation, is only a temporary one. Progress to extend computing facilities widely within agriculture will be adversely affected until a more practical and down to earth approach to this aspect of computing reasserts itself.

There are a number of ways in which the computing services industry is likely to be involved. Broadly speaking (and one would be ill advised to speak otherwise) the key lies in reducing relative cost and in the ease of communications between a 'user' working from his own office and the computer installation which may be operating from a location some considerable distance away. This being operated, maintained, serviced and secured by trained operators and supporting staff. Little expertise will be required by the 'user' who will be able to treat the complexities of computing as his predecessors have done with the typewriter and the accounting machine. Concurrent with this development could be the extended use of small, cheap, easily operated equipment dedicated to the control and management of single farm enterprises.

The variety of services made available by centralised installations is likely to increase as 'communications' become easier and cheaper.

Whatever 'central', 'host', 'parent', 'parish' computing solution (or combination of alternatives) is included in the services made available there is likely to be a move towards a communicating network of compatible facilities. Much will depend on the value and desirability of 'instant' information and the cost effectiveness associated with it. Compatibility of equipment and software could in due course allow the establishment of the much talked about, occasionally maligned and arguably sinister 'data base' system.

At the end of the day perhaps the most important decision to be made in the farm or estate office is not the choice of how various computing facilities are to be obtained as the choice of the supplier whether it be of equipment or services. That choice is critical as it is clearly essential that the chosen supplier will still be in operation when those who have found the going too competitive have left the field leaving their casualties behind them.

Farm computers

C. L. Pugh

EARLY DEVELOPMENTS AND APPLICATIONS

In the universities

The gradual appearance of electronic data processing equipment in the late 1950s and early 1960s led the agricultural economists of the time to examine the contribution that computers could make to farm management problems. With computing at that time a highly expensive facility, attention was centred on problems of high complexity, where an investment in computing seemed to have the best chance of a valuable payoff. Such problems tended to be those of particular interest to academics, whose concern with formal applications of economic theory was at least as important as the value of the exercise to the farming profession. Thus the early literature is dominated by work on farm planning applications based on variants of linear programming, which can be shown to be nicely coincident with the economic theories of production.

However, criticism of linear and other mathematical programming techniques, often based on totally theoretical argument, led to the favouring of non-optimising approaches to the farm planning problem. Thus simulation, based not on the search for optimal enterprise levels, but on a more realistic examination of several or indeed many trial plans, began to find favour during the late 1960s. However, although prodigious quantities of computer time were expended, little attention was paid to the usefulness (or lack of it) to the farming industry.

The advisory services

As a result of those early, mostly university based exercises, the Advisory Services had begun to look for computerised approaches which could be put into field use as management advisory techniques. Considerable effort was expended on putting linear programming, Monte Carlo programming and others into a format which would be useful to advisers in the field. Further efforts were expended on training advisers in the use and interpretation of these methods. Similar attempts were made by some commercial firms. However, there were several problems with the whole approach which, it can be seen with hindsight, precluded the widespread adoption of all these farm planning techniques. It is worthwhile considering them in a little more detail, as they illustrate the directions in which farm computing is likely to progress in the 1980s.

Problems with farm planning techniques

The first problem was that of operational inconvenience. The problem had to be examined at farm level by a trained adviser, put into computer format, then sent for processing at a remote, central computer unit, and finally interpreted at farm level again by the adviser. This procedure in itself runs counter to the way in which farmers plan, advisers advise, and information is used at farm level.

The second problem is that while the farm planning problem contains the largest quota of academic interest, it is by no means the dominant concern of managers and farmers. Quite simply, in the conditions of uncertainty that are predominant in farm situations, an 'optimal' enterprise combination is not a meaningful concept. The combination of enterprises is more likely to be determined by non- or semi-economic criteria and structural organisation or reorganisation is simply not the major problem or decision which farmers and farm managers face. The development of elaborate techniques which addressed themselves to this problem was therefore unlikely to succeed, and indeed it did not. By the early 1970s, computers had had virtually no direct effect on the management of farms at all in spite of two decades of research and development.

Book-keeping and accounting

In fact, the most significant indicators of the directions for farm computing were occurring during the late 1960s and 70s within the many large organisations which service the farmer—the banks, merchants, marketing boards, electricity boards—indeed in any large organisation which computerised not its corporate planning but its routine book-keeping, paperwork and data storage. For it is in these areas that computers have the most to offer to farm management; it is in these areas that a farm manager spends his time; and in these areas that in many cases he is least effective, and is therefore able to be assisted. However, the way in which computer facilities developed and came to the agricultural industry in the UK completely prevented these potential applications being realised.

Another interesting feature of UK agriculture which further held back the development of useful farm computer facilities was the absence of any central agency involved in account keeping and data collection. In many countries (Denmark, France and Canada for example) there existed central organisations which combined the role of data collection with that of producing annual accounts data to meet taxation or management requirements. The availability of computer power to such central agencies made this function an obvious candidate for computerisation. The existing systems of data capture could be linked to a centrally sited computer for organisation and storage of the large volumes of data, and for printing out reports for the farmer. Thus computer power could be applied to the task most suited to it—volume data processing—offering particular efficiencies since the dual role of data collection and providing farm management information could be incorporated in a single operation.

In the UK, there is no scope for such centralisation. The collection of accounts data for government usage has been performed for many years by a network of university departments which largely operate independently, thus keeping the clerical workload relatively small in each case. The individual centres have gradually computerised their own operations, but the obvious aggregate case for computerisation which would have arisen had the whole function been centralised was not present. Furthermore, there would certainly not have been sufficient economies from centralising the data processing load to justify disturbing an efficient existing data collection network.

A similar situation exists on the account keeping side. There has never been any central agency for producing annual account data for farmers. Farmer's needs for management and taxation accounts have been traditionally met by accountants, and there have been relatively few of the statutory requirements which caused the initiation of centralised systems in other countries.

The UK agricultural sector, therefore, in the absence of 'umbrella' organisations set up on behalf of farm businesses by NFU, Government, or private enterprise, was relatively unaffected by the computer until the late 1970s and the birth of the microcomputer. However, an important further development was taking place during the period 1970–77.

Farm accounting bureaux

This period saw the emergence of the farm accounting bureau. The basic concept of such a service is that simple farm records are kept and recorded on the farm in a suitably determined format. At monthly or quarterly intervals, these records are dispatched to a central site where they are processed into suitable management accounts which are then returned to the farm. The use of a computer is by no means essential to such a service, and indeed in the early days, accounting machines or even simply manual clerical

procedures were used. However, as computer power became available at less and less cost, these bureaux incorporated computer power into their service as much as to enable an expanded service to be provided as to offer any particular advantages in terms of facilities.

These bureaux attracted considerable success during this period, although by no means overwhelming the industry. However, their significance is that they were indicating the way in which farm computing was moving, in that the emphasis was on regular record and account keeping—the production of standard management information to assist in the control of the business. It is this area where heavy demand for data manipulation and storage is created if the job is to be done effectively, and for this reason it lends itself admirably to computerisation.

Other mail-in services

As well as the emergence of accounting bureaux as described above, there was also during the seventies considerable expansion of the computerised services offered to the farmer by the agricultural supply sector. These services include the dairy and pig costing services offered by the feed and fertiliser firms, feed planning and ration formulation, cash flow and other forecasting services and many others. ADAS also developed a range of similar services. All these bridge the gap between, on the one hand, a straight bureau service designed to make computerised services available to farm businesses, and, on the other hand, an existing manual service which is computerised as a matter of operational efficiency and convenience, but which might as well be manually operated. In fact, it illustrates the degree of penetration of the computer into everyday use in the course of the last ten years that by the end of the decade it is probably more unusual for a service not to be computerised than for the reverse to be true. The emphasis nowadays is on what the service has to offer rather than on the fact that it uses a computer.

These mail-in services—both the straight account keeping bureaux, the computer based services offered by the feed and fertiliser firms, and the incidentally computerised services (such as National Milk Records)—represent the bulk of computer use by farms in the late 1970s. Unfortunately it is not possible to quantify the extent of this use. At one extreme, all farmers are now exposed to computerised accounts systems by their suppliers. Likewise all users of National Milk Records now receive computerised reports and action lists. On the other hand many fewer positively seek out and use a computerised service and very few in relative terms actually use fee paying as opposed to free services. As a guide, though it is impossible to establish accurately, probably no more than 500 farms use fee paying accounts bureaux, while perhaps 15% of dairy farmers utilise fee paying costing schemes. Many more use the various free services provided, but many of these are so simple that they cannot be classified as real computer use.

In conclusion, therefore, the development of farm computing has been firmly based on mail-in services, but with only modest penetration for any but the simplest and cheapest level of service.

CURRENT DEVELOPMENTS

The microcomputer

During the 1970s, dramatic technological advances were being made in the methods of constructing computers. Based on the development of silicon chip technology, the cost of manufacture of the basic components of computers has been reduced by many orders of magnitude. As a rough guide, it is reckoned that the cost of computing has halved every two or three years since about 1960. In real terms this has meant that a complete computer system which would not have disgraced a university less than 20 years ago can now be purchased (its modern microcomputer equivalent) for only a few thousand pounds. Central Processing Units (CPUs), which are the heart of any computer, now cost only a few pounds, as opposed to many thousands of pounds only a few years ago. A complete 'home computer', containing all the basic elements required to write and operate computer programs, can now be had for less than £100.

This sharp reduction in the price of computing power has been accompanied by a comparable reduction in size, so that rather than requiring a large room, air conditioning and specialist staff to keep it running, the modern microcomputer now occupies only desktop space, and needs no special operating conditions other than a power supply. Needless to say, these truly revolutionary advances have had and will continue to have all manner of effects on the way computing is carried out in all fields.

The general effect of the dramatic reduction in the price of computing power has been that the size of problem which can now justify the cost of applying computer power is enormously reduced. If a complete computer sytem can now be operated for (say) £2000 per annum, as opposed to £20 000 per annum, then a whole range of problems which are excluded by simple economics from computer application by the higher cost now become eligible for consideration at the lower cost. It so happens that this particular reduction brings into the range of economic consideration many situations found on farms, from simple account keeping, through monitoring of livestock enterprises, to controlling environments.

Possible applications for small computers within farming

There are three distinct, though overlapping areas where computers find application in the farming industry which can be summarised under the headings *automation, communication* and *data processing*. It should be emphasised that virtually all of these applications were prohibited until the arrival, in the last two or three years of the 1970s, of the low cost microcomputer. All of the applications described in this present section, except those which have some historical precedent (mail-in accounting and monitoring schemes) were quite simply unfeasible until the last few years, both for economic and technical reasons.

Automation

In this section are grouped those applications where a computer or microprocessor is used to control the operation of some process or system, reducing or eliminating the need for manual operator intervention. The ability of the computer to control logical processes by responding to particular information stimuli with appropriate controlling decisions makes it able to be built into virtually automatic control systems in many farming situations. Some of these are relatively complicated, such as the automatic control of glasshouse or livestock environment, graindriers or the feeding of livestock in response to automatic yield, lactation and bodyweight measurements. Others are much simpler, and are closer to the controlled process itself. For instance, electronic control of motor ignition, instrumentation on vehicles, the collection of data from weather, geophysical or biological monitoring stations around the farm are all possible and applications are under development. Indeed, the range of applications for the microprocessor/microcomputer in direct and indirect automation is enormous, and has only just begun to be perceived. Over the next decade, as development of these possibilities into marketable systems takes place, it can be confidently predicted that at least as much on farm use of computer electronics will come from this type of application as from the more conventional data processing applications which are currently much more in the public eye.

The process of development described is time consuming. It takes longer to develop a foolproof and robust piece of electronic automation equipment than it does to develop a computer system for conventional data processing. For this reason there has been, and will continue to be a lag between achievement of technological and economic feasibility and the emergence of reliable products. There is considerable potential, however, for development in this area.

Communication

The developments in computer technology over the last few years has formed the basis of a substantial revolution in the dissemination of information. This has hitherto been the province of the printed word for the most part, with some contribution from broadcast information. However, only the most frequently changing information has been available via the broadcasting medium (e.g. news, weather, market reports), while printed information suffers from the high cost of repeated and frequent updating. Two new types of communication medium have been launched recently which overcome both these problems using the by now ubiquitous domestic television as a receiver (and possibly sender) of information.

The simplest of these uses spare capacity in the television airwaves to broadcast a range of selectable information which a conventional television, suitably modified, can access when required. This system, which operates somewhat like an 'information channel' can allow only a limited number of alternative 'frames', since all the 'frames' must be broadcast simultaneously. However, there are no extra facilities required other than a fairly simple decoding device for the receiver. Clearly this medium is most suitable for information which has a wide potential coverage, and which needs updating at regular intervals. Thus it is particularly useful for weather forecasts, stock exchange data and the like.

The second system, which has been pioneered by the British postal service, puts the user and his modified television set in contact with a very large central computer via the telephone system. For the cost of a local telephone call the user can call up any 'page' of information from a very large range which is stored in the central computer. He can also, if required, send responses back to the computer (giving orders, reservations, etc.). The information bank is effectively unlimited and is available for any organisation which wishes to put information onto the system for general or restricted access. This system (called Prestel) offers a very significant new medium of communication with advantages both to the information user and to the information provider. It seems certain to become widely used for many sorts of information, a significant proportion of which is currently disseminated on paper. Furthermore it will probably serve for many as the first stage in using electronic technology and will therefore play a significant part in the adoption and uptake of computer technology at the individual level. It also offers considerable scope for further developments in communications involving simple transfer of material from organisation to organisation. Thus the network could find itself as a medium of transfer for almost any sort of data between firms or individuals, offering significant advantages over existing postal, telephone or telegraphic services.

Conventional data processing

The third sector of application for computer technology in farming is the use of the computer in its typical role of data processing, organisation and storage. This type of application is directed to the provision of management information by means of the computer's ability to receive, store, access, calculate and output data in large volumes and at high speeds. The applications are invariably based on activities which are already to some extent carried out in the farm office, but which are likely to be done by computer more efficiently, more accurately or more quickly than by hand, or indeed can be done rather than not done by virtue of limited man-hours available for manual calculations and manipulations. The applications are found in management activities where there are inherently large volumes of information created, and these centre on two main areas of the farm business—account keeping and livestock enterprise control. There are of course others, such as payroll administration, but these are of lesser importance in agriculture, though they will be discussed briefly later in this section.

Accounting and record keeping

This activity is central to farm management, regardless of the type of farm. Most farmers employ a simple analysed cash-book system which is quite sufficient for most purposes. The major purpose of such bookkeeping is to provide tax and VAT records. Such typical farms would not normally have any chance of justifying the owning or operating of a computer, though perhaps it might make the account keeping simpler, or possibly more accurate.

However, on larger farms and estates, where many enterprises, and possibly several farms need to be accounted separately, and where it is not possible for the farmer or manager to keep up to date using simple, historical accounts a more sophisticated accounting procedure is necessary, which can soon become very time consuming if performed by hand. Particular complexities arise when

(1) large stock records are required,
(2) large numbers of transactions are found,
(3) many enterprises and cost centres are involved,
(4) monthly budgetary control is carried out.

When several of these are found, the availability of a computer system may well ease the load on clerical staff, lead to better accounting, and permit more detailed analysis than could be achieved by hand.

Livestock monitoring

The main livestock enterprises where sophisticated recording and analysis are useful are pig and dairy production. The basic need in dairy production monitoring is to keep a regular, updated record of individual cow lactations, and then to aggregate to provide herd or group analysis as required. To achieve this, details of individual milk yields must be recorded weekly or monthly, along with feed consumption, and other important events such as services, calvings, and veterinary treatments. While these can readily be kept by hand for a small or medium sized herd, for the larger herd, the task becomes cumbersome, particularly if considerable further analysis is to be carried out. With the aid of a well designed computer system, the same basic records can be used to provide valuable detailed analysis for herd management. In particular details of the physical and financial performance of groups of cows (e.g. heifers, calving in a particular month, etc.) can be produced when required, as well as herd performance details such as calving index, rolling margins, milk yields and many more. In addition, 'action lists' can be produced which provide details of expected events, particular cows for attention, exception reports and others for use by herdsmen. Another very useful feature, though by no means common is an annual forecast of milk production and feed use to predict future margins which might be expected.

In pig production, many similar records need to be kept, except that outputs are not measured in weekly milk produced, but in progeny weaned. Details are kept of individual sows, breeding groups or bloodlines, and performance is measured for them in details such as progeny reared, mortality, weight gain, farrowing index and feed consumption. Since the breeding aspect is of comparatively greater importance in sows than in cows, there is a greater emphasis on breeding records in the former case, with performance being closely monitored in relation to breeding. This, coupled with the larger numbers in sow herds, makes sow record keeping and analysis extremely complex. The use of a computer is therefore likely to find considerable justification in terms of potential improved performance.

Feed planning

In all livestock production, feed inputs represent a major cost. Ration formulation to produce balanced rations at economic cost is also a highly complex operation, and the potential of the low cost feed mix was established in the earliest years of computing. The arrival of the microcomputer has simply widened the area of application from specialist feed compounders to ordinary producers. It is also possible to calculate feed requirements more accurately using a computer particularly in the case of ruminants and when using the metabolisable energy system with its relatively complex calculations. Feed calculations therefore represent an important part of the potential

use of computers on livestock farms, particularly in the case of the larger unit where the possibility of on-farm mixing can be considered.

Other applications in farming

Though accounting, dairy and pig monitoring represent the larger portion of farm data processing activity, there are a number of other areas where profitable use may be made of available computer power.

Payroll

Though payroll calculations are not particularly arduous on any but the most untypical farm, it is a job tailor-made for computerisation. It could in fact justify owning a computer on its own in the case of large payrolls (at least 50 workers paid weekly).

Arable monitoring and record keeping

Though it could be argued that keeping field records is easily enough done on a map or in a notebook, systems are now being presented for keeping arable records and providing analysis for use in subsequent crop planning. These have particular applications where the cropping system is particularly complex, including a range of seed rates, fertilisers and spray applications depending on soil type, previous cropping or other factors. However, it is unlikely that genuine economic justification could be found in this area alone at the present time.

Budgeting and forward planning

As discussed earlier in this chapter, virtually all the early computer work in agriculture was directed to this area (with conspicuous lack of uptake). A computer—particularly sited close at hand in the farm office—can undoubtedly be useful in preparing budgets and forecasts. However, the typical farmer calculates budgets at only infrequent intervals, perhaps once, twice or four times per year; this is simply not sufficient justification on its own for owning a computer system. In fact there is a demand for planning/forecasting facilities from land agents, planning consultants and similar agencies whose job involves regularly producing sets of budgets and forecasts.

In conclusion, therefore, an important general point can be made, namely that in farm situations, there are likely to be several areas of use for a computer, few or even none of which will justify the cost of installing a computer individually. However since there is virtually certain to be spare capacity in any farm based machine, it is possible to build up an aggregate justification based on use in several different areas.

Current hardware developments

It must be stressed that the real development in farm computers has only become possible since the late 1970s when the cost of microprocessors fell sharply at the same time as the capability of equipment reached sufficient levels to enable reasonably priced but effective systems to be made available. The last two years (until the end of 1980) have seen the development of comparably priced software and at the time of writing it can be said that the farming industry is now well supplied with reliable computer systems. It is equally true that uptake has been extremely limited in that probably only a very few hundred bona fide systems have been sold. The stage of mass adoption remains definitely in the future, though how far into the future is by no means clear.

A typical hardware configuration might consist of a 'central processor unit' with 48 or 64 thousand 'bytes'* of 'memory'; together with this will be several 'disc units' which are used to store programs and data when the machine is not in use. There will also be a 'printer' for producing written reports, and a 'visual display unit' or screen to display transient information, and a 'keyboard' to allow the user to control operations. Such a system, with a range of suitable programs ('software') might cost in the order of £5000, but a wide range of prices can be found, depending on the complexity of the system, the hardware used and the pricing policy of the firm.

The trend has been firmly towards single intermediary firms supplying both hardware and software, in marked contrast to the situation a few years ago, when it was not unusual to purchase hardware and software from different sources. There has also been the virtually complete replacement of tape based storage systems by disc systems. These in turn have rapidly increased both in speed of access and density of storage, so that systems often have half a 'megabyte' (1 megabyte = 1 000 000 bytes) of 'online' disc storage and in many cases more. More recently, 'hard' or 'fixed' discs with much greater capacity and even faster retrieval times, have become available. In due course the 'bubble memory' seems likely to further revolutionise the storage and retrieval of data. Printing devices have also become more sophisticated, more reliable and less expensive in the last few years. Simple matrix printers can now be purchased for a few hundred pounds and for more sophisticated uses such as 'word processing', the 'daisy wheel' printer provides high quality printing and costs from £1000 upwards.

---

* The computer industry is full of its own jargon. While every attempt has been made to avoid such terminology, in this section some technical terms have had to be used. These have been enclosed in single quotes. A 'byte' is a 'word' or value made up of 8 'bits' or binary digits.

Software

As recently as 1978, it was well nigh impossible to find a good source of proven software for a farm computer system. The few pioneers who had installed systems by this time were either busy trying to make them work, or formulating plans to market their experience or expertise, in order to justify what might have been upwards of £20 000 invested.

However, the next two years saw the development of several specialist farm computer packages, combining the supply of hardware with particular software, and backing this up with training and support. Although sales have proved disappointingly small, some of these firms have now reached a state of development of their software whereby it very adequately meets the needs of the bulk of farm businesses.

In nearly every case, the central theme is account keeping, with additional packages to include livestock monitoring, arable systems, feed planning and several other areas where a market is perceived to exist.

The problem area for many potential users remains the desire to be non-standard in some small way. Clearly, if the systems are to be marketed inexpensively, there can be little or no customising to meet clients' particular requirements. However many farm businesses have small parts of their management system which are not compatible with standard packages, and though probably not critical to the operation, it can be an obstacle to adoption of a computerised approach. A few firms have specialised in such customised operations, but in general it is only much larger installations which can justify the extra costs involved.

In general, it may be concluded that the first generation of farm computer software has been adequately developed and marketed. Considerable investment has been made in these developments in the expectation of a large market for farm systems. In the present small market, rather oversupplied with products, the systems are in terms of unit costs, modestly priced, and offer rather good value for money.

Costs of farm systems

Great care is necessary in quoting prices at a particular time when prices in general are rising fast, and computer costs in general are falling. They must inevitably soon be out of date. However, if only as a historical record, a brief indication of prices as at the end of 1980 might be useful.

A typical complete farm computer system, capable of dealing with most farm needs (accounts, dairy, pigs, payroll, etc.) might be expected to cost around £5000. Of this the hardware would cost some £3500 if purchased separately, with £1500 providing for software, support and other services supplied.

There are advertised complete systems ranging from £3500 to well over £10 000, with 'customised', specially written installations for the very large farm or estate costing towards £20 000. At the lower end of the market, there are one or two systems (at the time of writing) supplied for specific purposes (e.g. dairy management) only which are advertised at £1500 upwards. At the simplest end of the market the enthusiast may buy his own hardware and attempt to write his own programs (if he has both time and inclination in copious quantities) for only a few thousand pounds, even less if simple equipment is accepted in the first instance.

The price trend is interestingly divergent at the present time. The cost of software development, being highly labour intensive, is increasing all the time, thus pushing up the cost of both 'one off' custom-built systems and of new developments in generalised systems. At the same time, with hardware costs still tending to fall, and with the possibility of larger volume markets to spread development costs, it is reasonable to expect prices of generalised systems to fall in real terms, though this is very dependent on firms achieving substantial penetration of the farming industry, and thus reaching economic levels of volume sales.

Justification on the farm

At the present time, a farmer purchasing a computer can expect to spend at least £5000 and quite possibly more on his basic system. Amortised over a small number of years, and allowing for some additional operating costs (maintenance, for example), an annual charge of approximately £2000 has to be justified. This justification has to come from two main areas. Firstly, savings can arise directly from reductions in existing costs of running the office and management system. These might come from reduced fees for existing book-keeping services, livestock costings or auditing fees. However, it is only in a small minority of cases that significant savings will be made from direct cost reductions of this sort. Experience has shown that when a small office staff has to absorb and operate a computer—a complex piece of equipment— it is likely to require more man-hours rather than less in total, while producing a much enhanced level of information. If there are very large economies to be made, furthermore, it is quite likely that they could be made without recourse to a computer.

The second area of justification comes from the value of the extra information provided. The installation of a computer in a farm office provides the potential to produce much more information, in a much more succinct form and in a more timely fashion, than can be produced by hand. It is the value of this extra information to the management of the business that justifies, or fails to justify, a computer system on the farm.

There are a substantial number of farms in the UK which could probably justify the capital outlay in terms of either cost savings, better information or both. It has been estimated that some 5% of farms could make economic use of computers at the present time. However, this figure, though no doubt increasing, does leave a large percentage which in current conditions cannot make the justification. Furthermore, only a small part of this already small proportion have installed computers to date, and the rate of uptake continues to be slow.

FUTURE PROSPECTS

Reasons for slow adoption

Although adoption of farm computers has been slower than expected, there are several reasons why this should be the case.

The first of these is that the farmer is being asked to invest capital in a completely strange and new technology. Most equipment adopted by farmers is at least comprehensible enough in function for its basic mode of operation to be understood. The operation of a computer is, by contrast, outside the knowledge of the businessman. At the same time its benefits, as outlined above, are indirect, rather than direct. There is no clear cut cost/benefit advantage to be seen. Adoption is in this case, therefore, almost an act of faith, and it is easy to sympathise with farmers in this respect. In such circumstances it would be reasonable to expect farmers to prefer to hire a complete (computerised) service, rather than attempt to 'go it alone' with their own equipment. However, such services are hard to find—probably because those who might offer such services (farm secretaries, accountants, consultants, land agents) are themselves somewhat overawed by the high technology of computers, but this has provided opportunities for specialists who have and are developing in this area.

The other main reason why farmers have been slow to adopt computer technology is the fear of obsolescence. It is clear even to the layman that computer technology is developing at a dramatic pace. Whenever the plunge is taken, there is the fear that in a short time either prices will fall or that the same money will buy a better package, or both. This is almost certainly true. To judge the moment when 'to wait longer will be more expensive than to go ahead' is not easy, and it is particularly not so when much of this benefit will be of an indirect nature, not in the form of measurable cost savings.

It could be argued that, however underpriced from a supplier's point of view, the farm computer in the majority of cases will be difficult to justify until prices fall considerably in real terms. Not until they can be priced, and viewed by purchasers, as 'expendable' rather than luxury capital which competes with other significant capital expenditure, will mass adoption take place. On the other hand there already exists a significant, and growing minority who can readily justify the capital outlay and who are in fact exhibiting an excessive, if understandable caution.

Alternatives to the on-farm computer

If it is accepted that in due course the bulk of progressive farmers will make use of computers, it is helpful to look at ways in which adoption can progress which fall short of the farmer purchasing and operating his own computer.

Sharing facilities

Since the total capacity of even a small computer system is likely to be far in excess of an average farmer's needs, there is an obvious case for sharing computer facilities. This can take many forms. At the simplest, a farmer who owns a machine can simply make it available to other farmers for a charge, thus offsetting his own investment. The question of how much additional service will be provided is of considerable importance, as there is a great difference between simply allowing another farmer access to the system and actually processing his data for him.

An interesting variant of this theme is the 'parish computer' concept, in which cooperating farmers (or other businessmen) share the cost of a full size computer installation, including operating 'staff'. They thus create for themselves a full 'bureau' service, removing the need for any of them to shoulder the responsibility of owning and operating their own individual system. This most appealing concept could well be extended to provide other group services such as telex, viewdata or other communications equipment, or even simply secretarial services. There are several firms developing this concept, though at the time of writing, there are none known to the author to have been operating for any length of time. It must be added, furthermore, that the British farming industry has proved generally unresponsive to the benefits of formal cooperation, tending only to recognise its virtues when the cooperative venture is large enough to resemble a conventionally constituted business. Nevertheless, the neighbourhood computer certainly offers solutions to many of the problems facing individual farmers seeking computer facilities.

Terminals/networks

A development from the sharing of computer facilities is the use of formal 'timesharing' computer systems. In this case a 'host' computer (of some considerable size) is accessed by users via a 'terminal'. The terminal is a relatively simple input/output device—usually keyboard, display and possibly printer—which is linked via the telephone network to the host computer. Once the link is established the terminal operator is more or less in control of all the features of the host machine, and can carry on all his computing requirements while he is 'online'. A 'host' computer may support any number of terminals according to its design and power, from just one to a whole network. Software is provided at the host station, so the user only needs the relatively simple terminal in order to use the system, wherein lies its appeal. A possible limitation to this approach is the high potential cost of telephone accounts.

Such terminal networks are not widely available to farms at present, though it is known that certain existing 'mail in' bureau systems are actively planning to make access by terminal possible. However, there are several ways in which terminal systems can be usefully deployed in agriculture. The first is, as described, where the mail-in services allow terminal access to speed up data collection and conversion. It can be envisaged that a subsequent number of existing mail-in services will seek to augment their services by providing such a facility. It can also be predicted that terminals suitable for this sort of work will become quite common items of equipment on farms, able to link to a variety of services—data processing and communications—and costing only a few hundred pounds.

The second area of potential for terminal networks is in the same style as the 'parish computer' described above. It is a simple extension to equip the local computer as a 'host' and allow users to submit data by telephone linked terminal. This has the advantage that the cost of the telephone link is kept down to local rate. (Long distance telephone rates as presently charged permit only relatively brief periods of link-up before the communication charges begin to dominate the total costs involved.) It is also possible that such local host computers could be sited in the offices of such professionals as land agents or accountants. Such professional offices already have a need for computer facilities 'in house', and the possibility of using surplus capacity to service a network of (probably existing) clients would certainly be appealing.

Problems with terminal systems

The largest problem with network systems is in the initiation stage. There is little doubt that many farmers would make use of it, particularly the 'parish computer' if one existed. However, there will be no 'parish computer' until some enterprising group or individual shoulders the responsibility for setting it up, and commits funds to it. This in turn will not happen until a ready market is identified. The problem lies in identifying a market where currently one is not perceived to exist. This problem applies even more so to the large scale network where a heavy investment in hardware, software and support is needed before the service can be launched, and again the market will not evolve until the service is launched. For this reason probably the best hope for terminal systems lies in providing terminal access to existing services.

The other main problem lies in the cost of telephone charges. At present these are charged at normal voice line rates as per a normal telephone call. This can soon become very expensive indeed. However, it is possible that as such applications develop, data communications charges will fall sharply, since a data link employs only a small fraction of the total capacity of a normal telephone voice line. When this occurs, terminal networks will certainly have an increasing appeal.

Other future developments

As has been noted above, most of the farm computer developments remain in the future. Adoption is only in its infancy at the present time. By the time the farming industry is making extensive use of computers, the style of application will no doubt have evolved. There follows an assessment of some areas in which this evolution might take place.

Data capture

At present, virtually all data must be entered into the computer by manual keyboard, a time consuming task, which adds significantly to the cost. It seems likely that much of these raw data will find their way into the computer whether automatically or by voice input. Measuring equipment (e.g. for milk yield) will be linked to feed data directly to the computer as it is produced. Voice recognition technology, currently in its infancy, will in due course allow data to be spoken straight into computer memory, by-passing the keyboard.

The information processing module

The farm computer may well in time develop into a central, multi-purpose information processor and controller. Linked to a range of other equipment both on the farm and at remote locations, it will be the hub of the information flow for the whole farm. Thus information will be received from monitoring stations around the farm, some processes and systems may be automatically controlled, with the computer sifting and organising the data needed for immediate decision, storing data away for archives, accessing information from central data bases via a communications link, and in due course preparing routine reports for management information. In a slight variation, there may develop ranges of completely independent self-contained process controllers which are so relatively inexpensive that there is no economy to be had in centrally siting them.

CONCLUSION

The 1980s mark the beginning of a new era of information processing in which great changes can be expected in all branches of data processing, communication and automation. Farming is likely to see its fair share of these changes, though given the small size of business and the wide geographical spread, it is not the prime candidate for computerisation. These developments will alter completely the way in which information is collected, processed, disseminated, analysed and used. However disturbing some aspects of this revolution might seem, it is not a trend that can be resisted. Farmers, furthermore, have a record of enthusiastic adoption of new technology, and in time will undoubtedly absorb information technology into their working practices. It will be a salutory lesson to look back in 1990 or 1995 with the benefit of hindsight to see how dramatic have been the developments during the intervening period.

Glossary of units

METRIC UNITS AND 'IMPERIAL' CONVERSION FACTORS

The established metric unitary system used in the UK is the 'Systeme Internationale d'Unites (SI)' which replaces 'Imperial Measure'.

The system consists of basic units from which all measurements are specified by use of multiplying factors. The most important of these factors in common usage are:

Multiplying factor		Prefix	Symbolised
0.000 000 000 001	(10^{-12})	pico	p
0.000 000 001	(10^{-9})	nano	n
0.000 001	(10^{-6})	micro	μ
0.001	(10^{-3})	milli	m
0.01	(10^{-2})	centi	c
1000	(10^{3})	kilo	k
1 000 000	(10^{6})	mega	M
1 000 000 000	(10^{9})	giga	G

Examples of the use of these factors are

millilitre (ml), i.e. 0.001×1 litre and kilogram (kg), i.e. 1000×1 gram.

In general, measurements should be specified in terms of units such that the 'whole number' part of the measurement is between 1 and 10 000, e.g. not 100 000 g, and 20 mm not 0.020 m.

The six primary units used in SI are:

	Unit	Symbol	Conversion factor SI to Imperial	
Length	metre	m	to ft	3.281
Weight	kilogram	kg	to lb	2.204
	tonne	t	to ton	0.984
Time	second	s		
Temperature	degree kelvin or Celsius	K or °C	to °F	$1.8\,(+32)$
Luminous intensity	candela per square metre	cd/m²	to cd/ft²	0.0929
Electric current	ampere	A		1.000

The above are the basis of many derived practical units which include:

	Unit	Symbol	Conversion factor SI to Imperial	
Force	newton	$N \equiv kg\ m/s^2$	to lbf	0.2248
Work energy	joule	$J \equiv N\ m$	to ft lbf	0.7376
			to Btu	0.9479×10^{-3}
			to Therms	9.479×10^{-9}
			to Wh	2.7×10^{-3}
Power	watt	$W \equiv N\ m/s$	to HP	1.341×10^{-3}
Pressure	pascal	Pa or $N/m^2 \equiv n/m$	to in H_2O	4.015×10^{-3}
			to in Hg	2.953×10^{-4}

Pressure is also expressed in the 'non-preferred' unit of the millibar (one mbar is equivalent to 100 N/m^2)

to lbf/in^2	1.45×10^{-4}
to ton f/ft^2	9.324×10^{-6}
to tonf/in^2	6.475×10^{-8}

Other units commonly used in agriculture are:

	Unit	Symbol and conversion factor	
Calorific value	kilojoules per cubic metre	kJ/m^3 to Btu/ft^3	2.684×10^{-2}
U-value	watts per square metre degree Celsius	W/m^2 °C to Btu/ft^{2}h °F	0.176
Specific heat capacity	kilojoules per kilogram degree Celsius	kJ/kg °C to Btu/lb °F	0.239
Moisture capacity	grams per cubic metre	g/m^3 to grains/100 ft^3	43.6
Illumination level	lux	lx to 1m/ft^2	0.0929
Land area	hectare	ha to acres	2.47
Fuel consumption	kilometres per litre	km/ℓ to miles/gal	2.825
Rate of usage	grams per square metre	g/m^2 to oz/yd^2	0.0295
	kilograms per hectare	kg/ha to lb/acre	0.892
	litres per hectare	ℓ/ha to gal/acre	0.089
Crop yields	tonnes per hectare	t/ha to ton/acre	0.398
		to cwt/acre	7.97
Density	tonnes per cubic metre	t/m^3 to ton/yd^3	0.753
	kilograms per cubic metre	kg/m^3 to lb/ft^3	0.0624
Speed	kilometres per hour	km/h to miles/h	0.621
	metres per second	m/s to ft/min	196.9
Flow rate	cubic metres per second	m^3/s to ft^3/min	2.118×10^3
	litres per second	ℓ/s to gal/min	13.2

  With considerable amounts of imperial standard equipment still in use it will be necessary to convert from metric to imperial measure for many years to come. Accurate conversion will always be the best solution, though quick approximations of equivalence are of value to the farmer.
  The following figures are useful to remember and do not incur great error:

Length

$$25\ mm \simeq 1\ in$$
$$300\ mm \simeq 1\ ft$$
$$1\ m \simeq 1.1\ yd$$
$$8\ km \simeq 5\ mile$$

Area

$$1000\ mm^2 \simeq 1.5\ in^2$$
$$1\ m^2 \simeq 10\ ft^2$$
$$4\ ha \simeq 10\ acre$$

Volume

$$100\ cm^3 \simeq 6\ in^3$$
$$3\ m^3 \simeq 100\ ft^3$$
$$3\ m^3 \simeq 4\ yd^3$$
$$4\ m^3 \simeq 100\ bushel$$
$$9\ ℓ \simeq 2\ gal$$
$$1\ m^3 \simeq 220\ gal$$

Mass

$$30 \text{ g} \simeq 1 \text{ oz}$$
$$1 \text{ kg} \simeq 2 \text{ lb}$$
$$50 \text{ kg} \simeq 1 \text{ cwt}$$
$$1 \text{ t or } 1000 \text{ kg} \simeq 1 \text{ ton}$$

Force and pressure

$$4.5 \text{ N} \simeq 1 \text{ lbf}$$
$$10 \text{ kN} \simeq 1 \text{ tonf}$$
$$7 \text{ kN/m}^2 \simeq 1 \text{ lbf/in}^2$$

Heat energy

$$1 \text{ kJ} \simeq 1 \text{ Btu}$$
$$3 \text{ W or J/s} \simeq 10 \text{ Btu/h}$$
$$100 \text{ MJ} \simeq 1 \text{ therm}$$

Rate of use and yield

$$4.5 \text{ kg/ha} \simeq 4 \text{ lb/acre}$$
$$125 \text{ kg/ha} \simeq 1 \text{ cwt/acre}$$
$$5 \text{ t/ha} \simeq 2 \text{ ton/acre}$$
$$11 \text{ } \ell/\text{ha} \simeq 1 \text{ gal/acre}$$
$$100 \text{ m}^3 \simeq 1 \text{ acre-in} \equiv 250 \text{ m}^3/\text{ha}$$

Rate of flow

$$1 \text{ m}^3/\text{s} \simeq 2000 \text{ ft}^3/\text{min}$$
$$1 \text{ } \ell/\text{s} \simeq 13 \text{ gal/min}$$

Common metric equivalences

$$1 \ell \equiv 1000 \text{ cm}^3$$
$$1 \text{ ha} \equiv 10\,000 \text{ m}^2$$
$$1 \text{ m}^3 \equiv 1000 \text{ } \ell$$
$$1 \text{ km}^2 \equiv 100 \text{ ha}$$
$$1 \text{ MN/m}^2 \equiv 1 \text{N/mm}^2 \ (\equiv 1 \text{ MPa})$$

Index